Advanced Wireless Networks

Advanced Wireless Networks

4G Technologies

Savo G. Glisic
University of Oulu, Finland

John Wiley & Sons, Ltd

Copyright © 2006 John Wiley & Sons Ltd, The Atrium, Southern Gate, Chichester,
West Sussex PO19 8SQ, England

Telephone (+44) 1243 779777

Email (for orders and customer service enquiries): cs-books@wiley.co.uk
Visit our Home Page on www.wiley.com

All Rights Reserved. No part of this publication may be reproduced, stored in a retrieval system or transmitted in any form or by any means, electronic, mechanical, photocopying, recording, scanning or otherwise, except under the terms of the Copyright, Designs and Patents Act 1988 or under the terms of a licence issued by the Copyright Licensing Agency Ltd, 90 Tottenham Court Road, London W1T 4LP, UK, without the permission in writing of the Publisher. Requests to the Publisher should be addressed to the Permissions Department, John Wiley & Sons Ltd, The Atrium, Southern Gate, Chichester, West Sussex PO19 8SQ, England, or emailed to permreq@wiley.co.uk, or faxed to (+44) 1243 770620.

Designations used by companies to distinguish their products are often claimed as trademarks. All brand names and product names used in this book are trade names, service marks, trademarks or registered trademarks of their respective owners. The Publisher is not associated with any product or vendor mentioned in this book.

This publication is designed to provide accurate and authoritative information in regard to the subject matter covered. It is sold on the understanding that the Publisher is not engaged in rendering professional services. If professional advice or other expert assistance is required, the services of a competent professional should be sought.

Other Wiley Editorial Offices

John Wiley & Sons Inc., 111 River Street, Hoboken, NJ 07030, USA

Jossey-Bass, 989 Market Street, San Francisco, CA 94103-1741, USA

Wiley-VCH Verlag GmbH, Boschstr. 12, D-69469 Weinheim, Germany

John Wiley & Sons Australia Ltd, 42 McDougall Street, Milton, Queensland 4064, Australia

John Wiley & Sons (Asia) Pte Ltd, 2 Clementi Loop #02-01, Jin Xing Distripark, Singapore 129809

John Wiley & Sons Canada Ltd, 22 Worcester Road, Etobicoke, Ontario, Canada M9W 1L1

Wiley also publishes its books in a variety of electronic formats. Some content that appears in print may not be available in electronic books.

British Library Cataloguing in Publication Data

A catalogue record for this book is available from the British Library

ISBN-13 978-0-470-01593-3 (HB)
ISBN-10 0-470-01593-4 (HB)

Typeset in 10/12pt Times by TechBooks, New Delhi, India.
Printed and bound in Great Britain by Antony Rowe Ltd, Chippenham, Wiltshire.
This book is printed on acid-free paper responsibly manufactured from sustainable forestry in which at least two trees are planted for each one used for paper production.

To my family

Contents

Preface

The major expectation from the fourth generation (4G) of wireless communication networks is to be able to handle much higher data rates, which will be in the range of 1Gb in the WLAN environment and 100 Mb in cellular networks. A user, with a large range of mobility, will access the network and will be able to seamlessly reconnect to different networks, even within the same session. The spectra allocation is expected to be more flexible, and even flexible spectra sharing among the different subnetworks is anticipated. In such a 'composite radio environment' (CRE), there will be a need for more adaptive and reconfigurable solutions on all layers in the network. For this reason the first part of the book deals with adaptive link, MAC, network and TCP layers including a chapter on crosslayer optimization. This is followed by chapters on mobility management and adaptive radio resource management. The composite radio environment will include presence of WLAN, cellular mobile networks, digital video broadcasting, satellite, mobile *ad hoc* and sensor networks.

Two additional chapters on *ad hoc* and sensor networks should help the reader understand the main problems and available solutions in these fields. The above chapters are followed by a chapter on security, which is a very important segment of wireless networks.

Within the more advanced solutions, the chapter on active networks covers topics like programmable networks, reference models, evolution to 4G wireless networks, 4G mobile network architecture, cognitive packet networks, the random neural networks based algorithms, game theory models in cognitive radio networks, cognitive radio networks as a game and biologically inspired networks, including bionet architecture.

Among other topics, the chapter on networks management includes self-organization in 4G networks, mobile agent-based network management, mobile agent platform, mobile agents in multioperator networks, integration of routing algorithm and mobile agents and *ad hoc* network management.

Network information theory has become an important segment of the research, and the chapter covering this topic includes effective capacity of advanced cellular network, capacity of *ad hoc* networks, information theory and network architectures, cooperative transmission in wireless multihop *ad hoc* networks, network coding, capacity of wireless networks using

MIMO technology and capacity of sensor networks with many-to-one transmissions. Two additional chapters, energy efficient wireless networks and QoS management, are also included in the book.

As an extra resource a significant amount of material is available on the book's companion website at *www.wiley.com/go/glisic* in the form of three comprehensive appendices: Appendix A provides a review of the protocol stacks for the most important existing wireless networks, Appendix B presents a comprehensive review of results for the MAC layer and Appendix C provides an introduction to queueing theory.

The material included in this book is a result of the collective effort of researchers across the globe. Whenever appropriate, the reference to the original work, measurement results or diagrams is made. The lists of references includes approximately 2000 titles.

Discussions and cooperation with Professor P. R. Kumar, of the Coordinated Science Laboratory, University of Illinois at Urbana-Champaign, had a significant impact, especially on the network information theory material presented in the book. Professor Imrich Chlamtac, of University of Texas at Dallas helped a great deal with the material regarding bioinspired nets. Professor Carlos Pomalaza-Raes, of Indiana-Purdue University, USA, inspired the presentation on *ad hoc* and sensor networks. Professor Kaveh Pahlavan of Worchester Polytechnic Institute, Massachusetts, inspired the presentations of the WLAN technology. Dr. Moe Win of Massachusetts Institute of Technology provided a set of original diagrams on Ultra Wide Band Channel measurements.

The author would also like to thank Professor P. Leppanen, J.P. Mäkelä, P. Nissinaho and Z. Nikolic, for their help with the graphics.

Savo G. Glisic
Oulu

1

Fundamentals

1.1 4G NETWORKS AND COMPOSITE RADIO ENVIRONMENT

In the wireless communications community we are witnessing more and more the existence of the *composite radio environment (CRE)* and as a consequence the need for *reconfigurability* concepts. The CRE assumes that different radio networks can be cooperating components in a heterogeneous wireless access infrastructure, through which network providers can more efficiently achieve the required capacity and quality of service (QoS) levels. Reconfigurability enables terminals and network elements to dynamically select and adapt to the most appropriate radio access technologies for handling conditions encountered in specific service area regions and time zones of the day. Both concepts pose new requirements on the management of wireless systems. Nowadays, a multiplicity of radio access technology (RAT) standards are used in wireless communications. As shown in Figure 1.1, these technologies can be roughly categorized into four sets:

- Cellular networks that include second-generation (2G) mobile systems, such as Global System for Mobile Communications (GSM) [1] , and their evolutions, often called 2.5G systems, such as enhanced digital GSM evolution (EDGE), General Packet Radio Service (GPRS) [2] and IS 136 in the USA. These systems are based on TDMA technology. Third-generation (3G) mobile networks, known as Universal Mobile Telecommunications Systems (UMTS; WCDMA and cdma2000) [3] are based on CDMA technology that provides up to 2 Mbit/s. In these networks 4G solutions are expected to provide up to 100 Mbit/s. The solutions will be based on a combination of multicarrier and space–time signal formats. The network architectures include macro- micro- and picocellular networks and home (HAN) and personal area networks (PAN).
- Broadband radio access networks (BRANs) [4], or wireless local area networks (WLANs) [5], which are expected to provide up to 1 Gb/s in 4G. These technologies are based on orthogonal frequency division multiple access (OFDMA) and space–time coding.

Advanced Wireless Networks: 4G Technologies Savo G. Glisic
© 2006 John Wiley & Sons, Ltd.

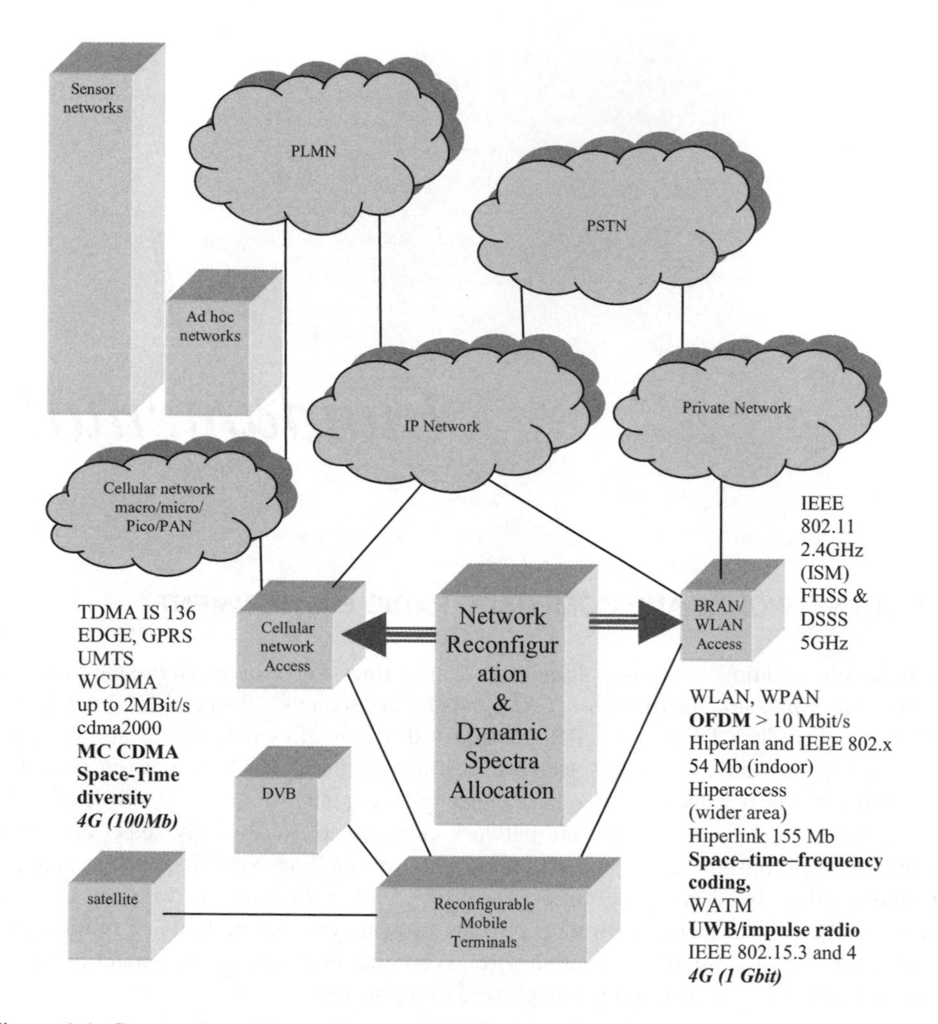

Figure 1.1 Composite radio environment in 4G networks.

- Digital video broadcasting (DVB) [6] and satellite communications.
- *Ad hoc* and sensor networks with emerging applications.

Although 4G is open for new multiple access schemes, the CRE concept remains attractive for increasing the service provision efficiency and the exploitation possibilities of the available RATs. The main assumption is that the different radio networks , GPRS, UMTS, BRAN/WLAN, DVB, and so on, can be components of a heterogeneous wireless access infrastructure. A network provider (NP) can own several components of the CR infrastructure (in other words, can own licenses for deploying and operating different RATs), and can also cooperate with affiliated NPs. In any case, an NP can rely on several alternative radio networks and technologies to achieve the required capacity and QoS levels, in a cost-efficient manner. Users are directed to the most appropriate radio networks and technologies, at different service area regions and time zones of the day, based on profile requirements and network performance criteria. The various RATs are thus used in a complementary manner rather than competing with each other. Even nowadays a mobile handset can make a handoff between different RATs. The deployment of CRE systems can be facilitated by the *reconfigurability* concept, which is an evolution of software-defined radio [7, 8]. CRE requires terminals that are able to work with different RATs and the existence of multiple radio networks, offering alternative wireless access capabilities to service area regions. Reconfigurability supports the CRE concept by providing essential technologies that enable terminals and network elements to dynamically (transparently and securely) select and adapt to the set of RATs, that is most appropriate for the conditions encountered in specific service area regions and time zones of the day. According to the reconfigurability concept, RAT selection is not restricted to those technologies pre-installed in the network element. In fact, the required software components can be dynamically downloaded, installed and validated. This makes it different from the static paradigm regarding the capabilities of terminals and network elements.

The networks provide wireless access to IP-based applications, and service continuity in light of intrasystem mobility. Integration of the network segments in the CR infrastructure is achieved through the management system for CRE (MSCRE) component attached to each network. The management system in each network manages a specific radio technology; however, the platforms can cooperate. The fixed (core and backbone) network will consist of public and private segments based on IPv4 and IPv6-based infrastructure. Mobile IP (MIP) will enable the maintenance of IP-level connectivity regardless of the likely changes in the underlying radio technologies used that will be imposed by the CRE concept. Figures 1.2 and 1.3 depict the architecture of a terminal that is capable of operating in a CRE context. The terminals include software and hardware components (layer 1 and 2 functionalities) for operating with different systems. The higher protocol layers, in accordance with their peer entities in the network, support continuous access to IP-based applications. Different protocol busters can further enhance the efficiency of the protocol stack. There is a need to provide the best possible IP performance over wireless links, including legacy systems.

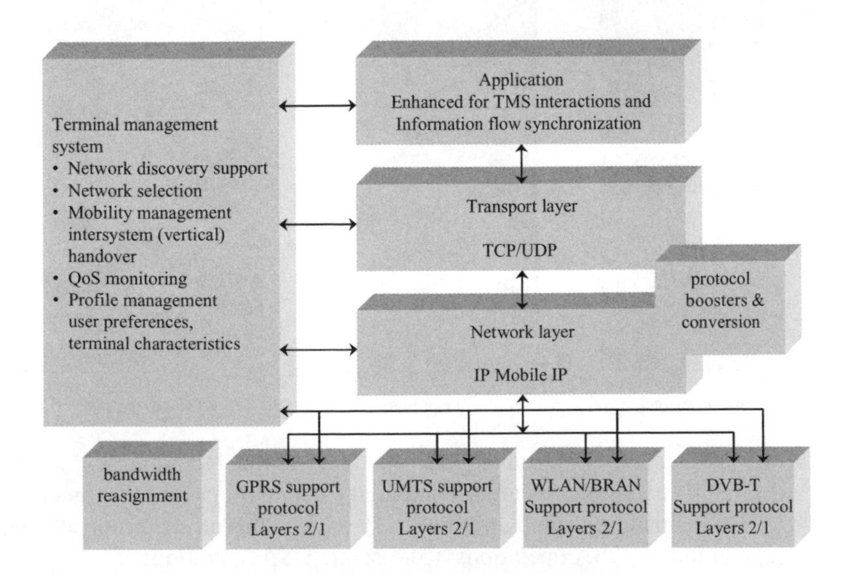

Figure 1.2 Architecture of a terminal that operates in a composite radio environment.

(a)

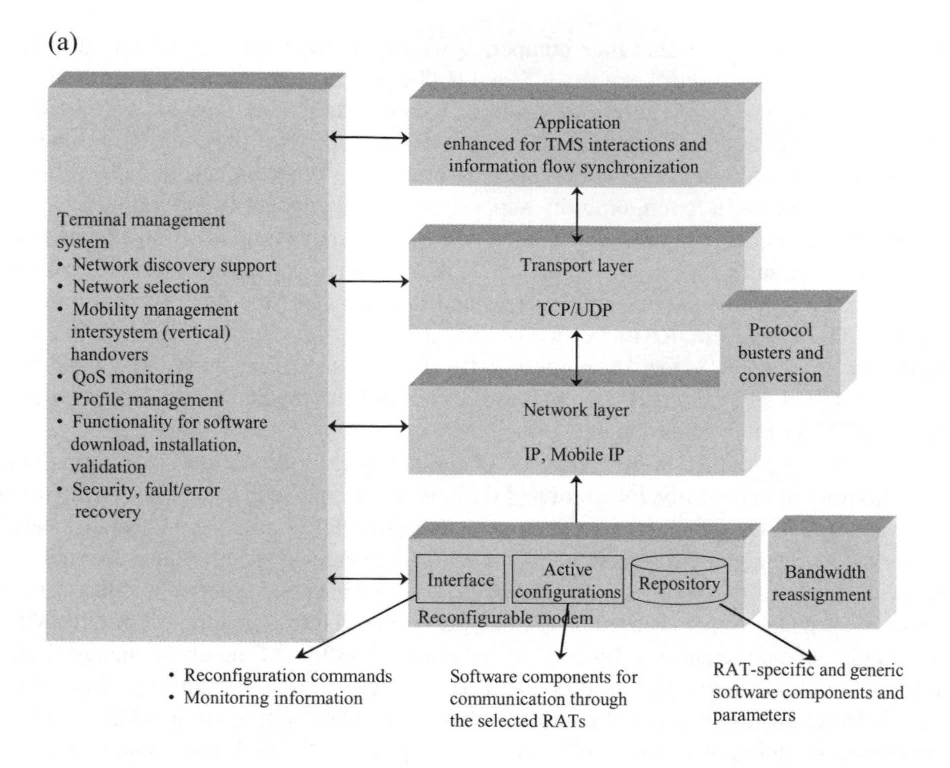

(b)

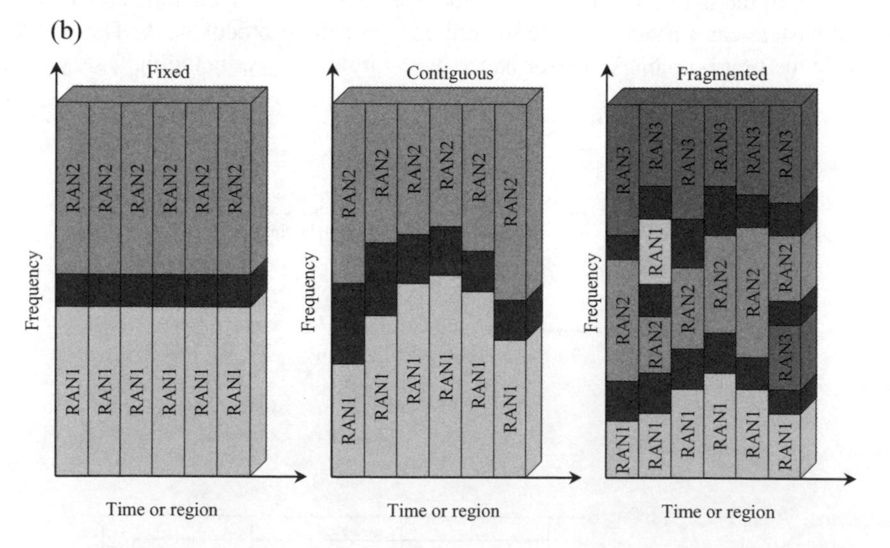

Figure 1.3 (a) Architecture of terminal that operates in the reconfigurability context. (b) Fixed spectrum allocation compared to contiguous and fragmented DSA. (c) DSA operation configurations: (1) static (current spectrum allocations); (2) continuous DSA operations; (3) discrete DSA operations.

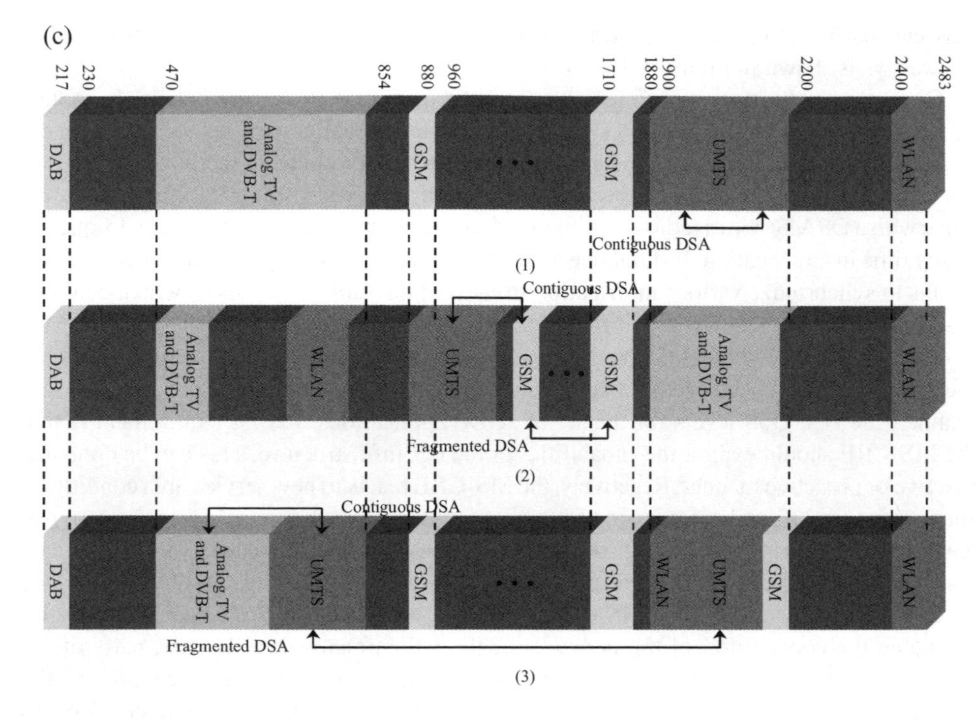

Figure 1.3 (*Continued*)

Within the performance implications of link characteristics (PILC) IETF group, the concept of a performance-enhancing proxy (PEP) [9–12] has been chosen to refer to a set of methods used to improve the performance of Internet protocols on network paths where native TCP/IP performance is degraded due to the characteristics of a link. Different types of PEPs, depending on their basic functioning, are also distinguished. Some of them try to compensate for the poor performance by modifying the protocols themselves. In contrast, a symmetric/asymmetric boosting approach, transparent to the upper layers, is often both more efficient and flexible. A common framework to house a number of different protocol boosters provides high flexibility, as it may adapt to both the characteristics of the traffic being delivered and the particular conditions of the links. In this sense, a control plane for easing the required information sharing (cross-layer communication and configurability) is needed. Furthermore, another requirement comes from the appearance of multihop communications as PEPs have traditionally been used over the last hop, so they should be adapted to the multihop scenario. Most communications networks are subject to time and regional variations in traffic demands, which lead to variations in the degree to which the spectrum is utilized. Therefore, a service's radio spectrum can be underused at certain times or geographical areas, while another service may experience a shortage at the same time/place. Given the high economic value placed on the radio spectrum and the importance of spectrum efficiency, it is clear that wastage of radio spectrum must be avoided. These issues provide the motivation for a scheme called dynamic spectrum allocation (DSA), which aims to manage the spectrum utilized by a converged radio system and share it between participating radio networks over space and time to increase overall spectrum efficiency, as shown in Figure 1.3(b, c).

Composite radio systems and reconfigurability, discussed above, are potential enablers of DSA systems. Composite radio systems allow seamless delivery of services through the most appropriate access network, and close network cooperation can facilitate the sharing not only of services, but also of spectrum. Reconfigurability is also a very important issue, since with a DSA system a radio access network could potentially be allocated any frequency at any time in any location. It should be noted that the application layer is enhanced with the means to synchronize various information streams of the same application, which could be transported simultaneously over different RATs. The terminal management system (TMS) is essential for providing functionality that exploits the CRE. On the user/terminal side, the main focus is on the determination of the networks that provide, in a cost-efficient manner, the best QoS levels for the set of active applications. A first requirement is that the MS-CRE should exploit the capabilities of the CR infrastructure. This can be done in a reactive or proactive manner. Reactively, the MS-CRE reacts to new service area conditions, such as the unexpected emergence of hot spots. Proactively, the management system can anticipate changes in the demand pattern. Such situations can be alleviated by using alternate components of the CR infrastructure to achieve the required capacity and QoS levels. The second requirement is that the MS-CRE should provide resource brokerage functionality to enable the cooperation of the networks of the CR infrastructure. Finally, parts of the MS-CRE should be capable of directing users to the most appropriate networks of the CR infrastructure, where they will obtain services efficiently in terms of cost and QoS. To achieved the above requirements an MS architecture such as that shown in Figure 1.4 is required.

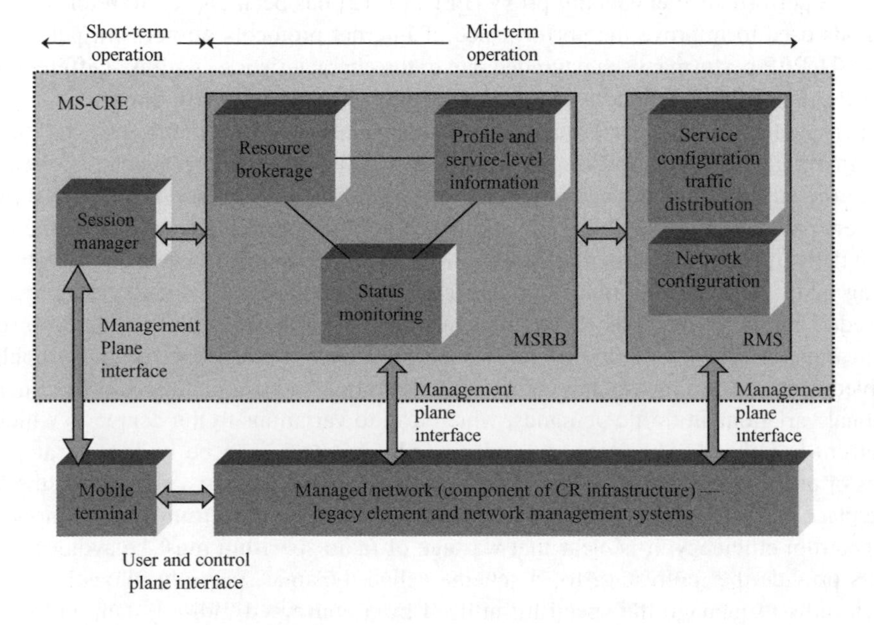

Figure 1.4 Architecture of the MS-CRE.

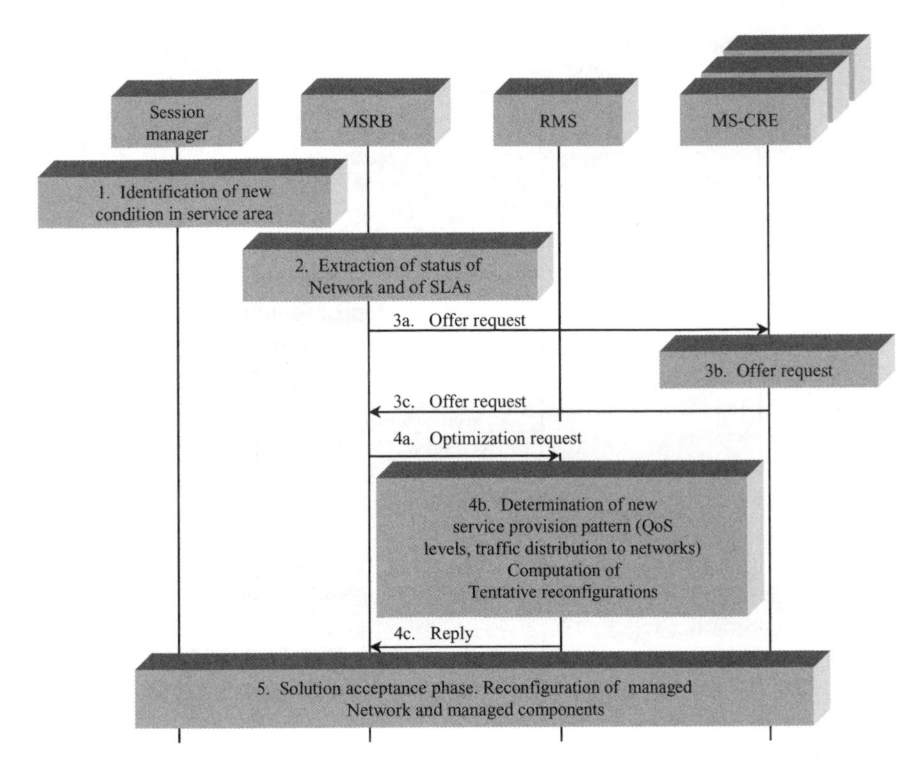

Figure 1.5 MS-CRE operation scenario.

The architecture consists of three main logical entities:

- monitoring, service-level information and resource brokerage (MSRB);
- resource management strategies (RMS);
- session managers (SMs).

The MSRB entity identifies the triggers (events) that should be handled by the MS-CRE and provides corresponding auxiliary (supporting) functionality. The RMS entity provides the necessary optimization functionality. The SM entity is in charge of interacting with the active subscribed users/terminals. The operation steps and cooperation of the RMS components are shown in Figures 1.5 and 1.6, respectively. In order to get an insight into the scope and range of possible reconfigurations, we review in Appendix A (please go to www.wiley.com/go/glisic) the network and protocol stack architectures [1–58] of the basic CRE components as indicated in Figure 1.1.

1.2 PROTOCOL BOOSTERS

As pointed out in Figure 1.2, an element of the reconfiguration in 4G networks is protocol booster. A protocol booster is a software or hardware module that transparently improves protocol performance. The booster can reside anywhere in the network or end systems, and may operate independently (one-element booster), or in cooperation with other protocol

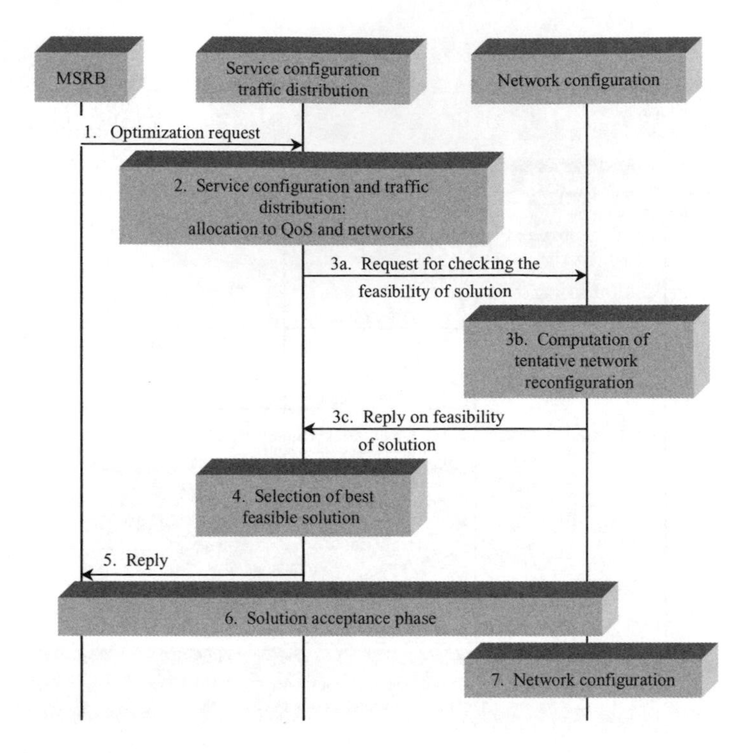

Figure 1.6 Cooperation of the RMS components.

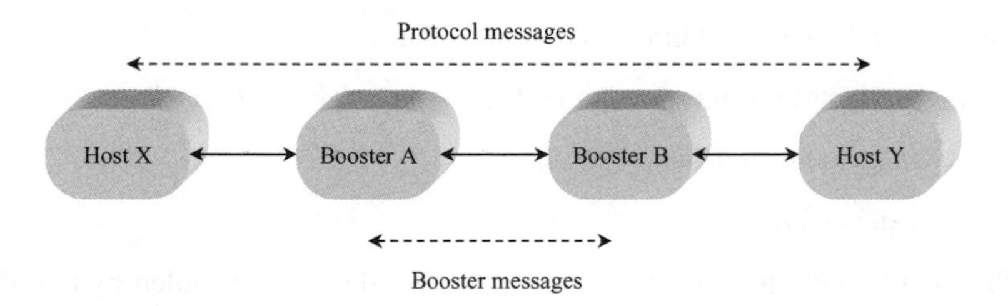

Figure 1.7 Two-element booster.

boosters (multielement booster). Protocol boosters provide an architectural alternative to existing protocol adaptation techniques, such as protocol conversion. A protocol booster is a supporting agent that by itself is not a protocol. It may add, delete or delay protocol messages, but never originates, terminates or converts that protocol. A multielement protocol booster may define new protocol messages to exchange among themselves, but these protocols are originated and terminated by protocol booster elements, and are not visible or meaningful external to the booster. Figure 1.7 shows the information flow in a generic two-element booster. A protocol booster is transparent to the protocol being boosted. Thus, the elimination of a protocol booster will not prevent end-to-end communication, as would, for example, the removal of one end of a conversion [e.g. transport control protocol/Internet protocol (TCP/IP) header compression unit [13]]. In what follows we will present examples of protocol busters.

1.2.1 One-element error detection booster for UDP

UDP has an optional 16-bit checksum field in the header. If it contains the value zero, it means that the checksum was not computed by the source. Computing this checksum may be wasteful on a reliable LAN. On the other hand, if errors are possible, the checksum greatly improves data integrity. A transmitter sending data does not compute a checksum for either local or remote destinations. For reliable local communication, this saves the checksum computation (at the source and destination). For wide-area communication, the single-element error detection booster computes the checksum and puts it into the UDP header. The booster could be located either in the source host (below the level of UDP) or in a gateway machine.

1.2.2 One-element ACK compression booster for TCP

On a system with asymmetric channel speeds, such as broadcast satellite, the forward (data) channel may be considerably faster than the return (acknowledgment, ACK) channel. On such a system, many TCP ACKs may build up in a queue, increasing round-trip time, and thus reducing the transmission rate for a given TCP window size. The nature of TCP's cumulative ACKs means that any ACK acknowledges at least as many bytes of data as any earlier ACK. Consequently, if several ACKs are in a queue, it is necessary to keep only the ACK that has arrived most recently. A simple ACK compression booster could insure that only a single ACK exists in the queue for each TCP connection. (A more sophisticated ACK compression booster allows some duplicate ACKs to pass, allowing the TCP transmitter to get a better picture of network congestion.) The booster increases the protocol performance because it reduces the ACK latency, and allows faster transmission for a given window size.

1.2.3 One-element congestion control booster for TCP

Congestion control reduces buffer overflow loss by reducing the transmission rate at the source when the network is congested. A TCP transmitter deduces information about network congestion by examining ACKs sent by the TCP receiver. If the transmitter sees several ACKs with the same sequence number, then it assumes that network congestion has caused a loss of data messages. If congestion is noted in a subnet, then a congestion control booster could artificially produce duplicate ACKs. The TCP receiver would think that data messages had been lost because of congestion, and would reduce its window size, thus reducing the amount of data it injected into the network.

1.2.4 One-element ARQ booster for TCP

TCP uses ARQ to retransmit data unacknowledged by the receiver when a packet loss is suspected, such as after a retransmission time-out expires. If we assume the network of Figure 1.7 (except that booster B does not exist), then an ARQ booster for TCP: (1) will cache packets from host Y; (2) if it sees a duplicate acknowledgment arrive from host X and it has the next packet in the cache, then deletes the acknowledgment and retransmits the next packet (because a packet must have been lost between the booster and host X); and (3) will delete packets retransmitted from host Y that have been acknowledged by host X. The ARQ booster improves performance by shortening the retransmission path. A typical application would be if host X were on a wireless network and the booster were on the interface between the wireless and wireline networks.

1.2.5 A forward erasure correction booster for IP or TCP

For many real-time and multicast applications, forward error correction coding is desirable. The two-element forward error correcting (FEC) booster uses a packet forward error correction code and erasure decoding. The FEC booster at the transmitter side of the network adds parity packets. The FEC booster at the receiver side removes the parity packets and regenerates missing data packets. The FEC booster can be applied between any two points in a network (including the end systems). If applied to IP, then a sequence number booster adds sequence number information to the data packets before the first FEC booster. If applied to TCP (or any protocol with sequence number information), then the FEC booster can be more efficient because: (1) it does not need to add sequence numbers; and (2) it could add new parity information on TCP retransmissions (rather than repeating the same parities). At the receiver side, the FEC booster could combine information from multiple TCP retransmissions for FEC decoding.

1.2.6 Two-element jitter control booster for IP

For real-time communication, we may be interested in bounding the amount of jitter that occurs in the network. A jitter control booster can be used to reduce jitter at the expense of increased latency. At the first booster element, timestamps are generated for each data message that passes. These timestamps are transmitted to the second booster element, which delays messages and attempts to reproduce the intermessage interval that was measured by the first booster element.

1.2.7 Two-element selective ARQ booster for IP or TCP

For links with significant error rates, using a selective automatic repeat request (ARQ) protocol (with selective acknowledgment and selective retransmission) can significantly improve the efficiency compared with using TCP's ARQ (with cumulative acknowledgment and possibly go-back-N retransmission). The two-element ARQ booster uses a selective ARQ booster to supplement TCP by: (1) caching packets in the upstream booster; (2) sending negative acknowledgments when gaps are detected in the downstream booster; and (3) selectively retransmitting the packets requested in the negative acknowledgments (if they are in the cache).

1.3 HYBRID 4G WIRELESS NETWORK PROTOCOLS

As indicated in Appendix A (please go to www.wiley.com/go/glisic), there are two basic types of structure for WLAN:

(1) *Infrastructure WLAN – BS-oriented network.* Single-hop (or cellular) networks that require fixed base stations (BS) interconnected by a wired backbone.

(2) *Noninfrastructure WLAN – ad hoc WLAN.* Unlike the BS-oriented network, which has BSs providing coverage for mobile hosts (MHs), *ad hoc* networks do not have any centralized administration or standard support services regularly available on the network to which the hosts may normally be connected. MHs depend on each other for communication.

The BS-oriented network is more reliable and has better performance. However, the *ad hoc* network topology is more desirable because of its low cost, plug-and-play property, flexibility, minimal human interaction requirements, and especially battery power efficiency. It is suitable for communication in a closed area – for example, on a campus or in a building.

To combine their strength, possible 4G concepts might prefer to add BSs to an *ad hoc* network. To save access bandwidth and battery power and have fast connection, the MHs could use an *ad hoc* wireless network when communicating with each other in a small area. When the MHs move out of the transmitting range, the BS could participate at this time and serve as an intermediate node. The proposed method also solves some problems, such as a BS failure or weak connection under *ad hoc* networks. The MHs can communicate with one another in a flexible way and freely move anywhere with seamless handoff.

There have been many techniques or concepts proposed for supporting a WLAN with and without infrastructure, such as IEEE802.11 [14], HIPERLAN [15], and *ad hoc* WATM LAN [16]. The standardization activities in IEEE802.11 and HIPERLAN have recognized the usefulness of the *ad hoc* networking mode. IEEE 802.11 enhances the *ad hoc* function to the MH. HIPERLAN combines the functions of two infrastructures into the MH. Contrary to IEEE802.11 and HIPERLAN, the *ad hoc* WATM LAN concept is based on the same centralized wireless control framework as the BS-oriented system, but insures that MH designed for the BS-oriented system can also participate in *ad hoc* networking. Both the BS oriented and *ad hoc* networks have some drawbacks. In the BS-oriented networks, BS manages all the MHs within the cell area and controls handoff procedures. It plays a very important role for WLAN. If it does not work, the communication of MHs in this area will be disrupted. Under this situation, some MHs could still transmit messages to each other without BS. Therefore, to increase the reliability and efficiency of the BS-oriented network, MH-to-MH direct transmission capability can be added. However, this is restricted to at most two hops such that this new enhancement will not increase the protocol complexity too much.

In the *ad hoc* networks, it is not easy to rebuild or maintain a connection. When the connection is built, it will be disrupted any time one MH moves out of the connection range. So, as a compromise, the MHs could communicate with each other over the wireless media, without any support from the infrastructure network components within the signal transmission range. Yet when the transmission range is less than the distance between the two MHs, the MHs could change back to the BS-oriented systems. MH would be able to operate in both *ad hoc* and BS-oriented WLAN environments.

Two different methods – one-hop and two-hop direct transmission within the BS-oriented concept – will be considered. The first method is simple and controlled by the signal strength. The second method should include the data forwarding and implementation of routing tables.

1.3.1 Control messages and state transition diagrams

To integrate the BS-oriented method and the direct transmission method, we define some control messages:

(1) ACK/ACCEPT/REJECT – used to indicate the acknowledgment, acceptance, or denial of connection or handoff request;

(2) CHANGE – used by MH to inform the sender to initiate the handoff procedure;

(3) DIRECT – used by MH to inform BS that the transmission is in direct transmission mode;

(4) SEARCH – used to find the destination; each MH receiving this message must check the destination address for a match;

(5) SETUP – used to establish a new connection;

(6) TEARDOWN – used for switching from BS-oriented handoff to direct transmission; it will let BS release the channel and buffer;

(7) AGENT – used by the MH whose BS fails to accept another MH acting as a surrogate for transmission;

(8) BELONG – used by a surrogate MH to accept another MH's WHOSE-BS-ALIVE request;

(9) WHOSE-BS-ALIVE – used by the MH whose BS failed to find a surrogate MH.

Since a mobile host may be in BS-oriented, one-hop direct-transmission mode or two-hop direct-transmission mode, it is important that we understand the timing for mode transition.

Figure 1.8 shows the state transition diagram. The meaning and timing of each transition are explained below:

(1) The receiver can receive the sender's signal directly.

(2) The receiver is a neighbor of a neighbor of the sender.

(3) Neither case 1 nor 2.

(4) The receiver can no longer hear the sender's signal; however, a neighbor of the sender can communicate with the receiver directly.

(5) The receiver discovers that it can hear the sender's signal directly.

(6) The receiver can no longer hear the sender's signal, and none of the sender's neighbors can communicate directly with the receiver.

(7) The receiver discovers that it can hear the sender's signal directly.

(8) No neighbors of the sender can communicate with the receiver directly.

(9) The sender's original relay neighbor fails. However, the sender can find another neighbor that can communicate with the receiver directly.

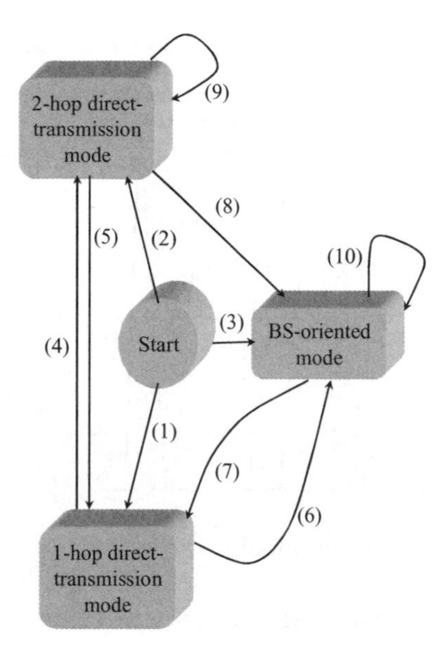

Figure 1.8 Transition diagram for transmission mode.

(10) The handoff from one BS to another occurs. In Figure 1.8, we note the following two points: (a) when a mobile host starts communication, it could be in any mode depending on the position of the receiving mobile host; and (b) the transition from the BS-oriented to two-hop direct-transmission mode is not possible because the communicating party cannot know that a third mobile host exists and is within range.

1.3.2 Direct transmission

Direct transmission defines the situation where two mobile hosts communicate directly or use a third mobile host as a relay without the help of base stations. This section considers the location management and handoff procedures when the MH moves around. These functions are almost the same as the traditional ones. However, the system must decide whether one-hop direct transmission, two-hop direct transmission or BS-oriented transmission method should be used. When the sender broadcasts the connection request message, both the BS and the MH within the sender's signal covering area receive this message. Each MH receiving the message checks the destination ID. If the destination ID matches itself, the transmission uses the one-hop direct-transmission method. (If we allow two-hop direct transmission, each receiving mobile host must check its neighbor database to see if the destination is currently a neighbor of itself.) Otherwise, the BS is used for connection. When the destination moves out of the covering area, the BS has to take over. On the other hand, when the MH moves into the covering range, the receiver has the option to stop going through the BS and changes to one-hop direct transmission.

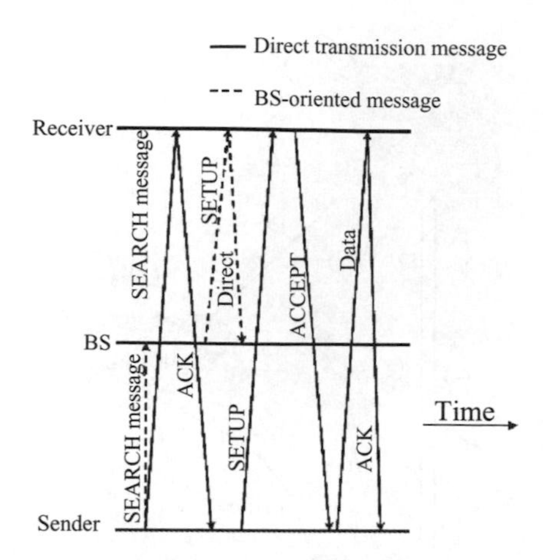

Figure 1.9 Using one-hop direct-transmission mode.

1.3.3 The protocol for one-hop direct transmission

For one-hop direct-transmission mode, each case of protocol operations is described in more detail as follows:

1.3.3.1 One-hop direct-transmission mode

(1) The sender broadcasts a SEARCH message. Every node in the signal covering range (including the BS) receives the message, as shown in Figure 1.9.

(2) If the receiver is within the range, it receives the message and finds out the destination is itself. It responds with the message ACK back to the sender.

(3) At the same time, the BS also receives the SEARCH message. It locates the MH and sends the SETUP message to the destination. For direct transmission, the destination receives the SETUP message and sends the DIRECT message to BS. Otherwise, it sends an ACCEPT message to the BS, and the communication is in BS-oriented mode.

(4) The sender continues transmitting directly until the MH moves out of the covering area.

1.3.3.2 BS-oriented Mode

(1) The sender broadcasts a SEARCH message. If the receiver is out of the covering range, it will not receive the message, as in Figure 1.10. (It is possible that two-hop direct-transmission mode can be used. This will be explained later.)

(2) However, the BS of the sender always receives the SEARCH message. It queries the receiver's position and sends the SETUP message to the destination.

(3) When the destination receives the SETUP message, it sends an ACCEPT message to the BS.

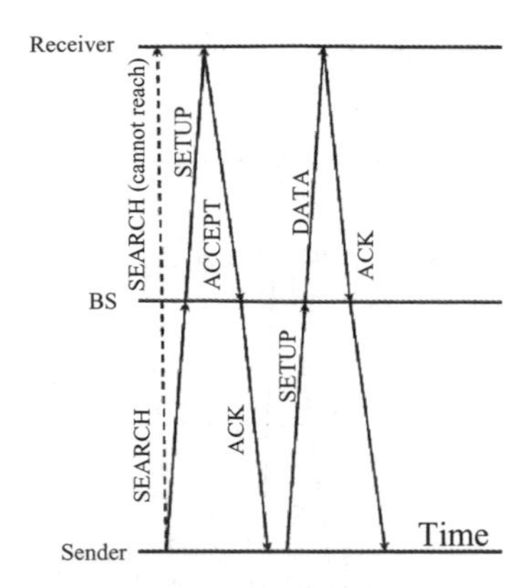

Figure 1.10 Using the BS-oriented mode.

(4) The communication continues by BS-oriented mode until the distance between the two MHs is close enough and the receiver wants to change to direct transmission mode.

1.3.3.3 Handoff – out of direct transmission range

In direct transmission mode, when the destination detects that the strength of the signal is less than an acceptable value, handoff should be executed. The procedures are described as follows:

(1) The destination sends the CHANGE message to its sender.

(2) As the sender receives the CHANGE request, it will send out the SEARCH message again. Then a BS-oriented mode will be used or a two-hop direct-transmission mode, as explained later.

1.3.3.4 Handoff-BS to one-hop direct transmission

If the receiver detects that it is within the covering region of the sender's signal and the signal is strong enough, it has the option of switching from the BS-oriented mode to one-hop direct-transmission mode. Each step is described below and presented in Figure 1.11.

(1) The sender sends the SEARCH message out. Then the one-hop direct-transmission will be established.

(2) After the sender receives the ACCEPT message from the receiver, it sends a TEARDOWN message to the BS and breaks the connection along the path.

1.3.4 Protocols for two-hop direct-transmission mode

Two-hop direct-transmission mode will cover a wider area than one-hop direct-transmission mode. It allows two mobile hosts to communicate through a third mobile host acting as a relay. Therefore, each mobile host must implement a neighbor database to record its current

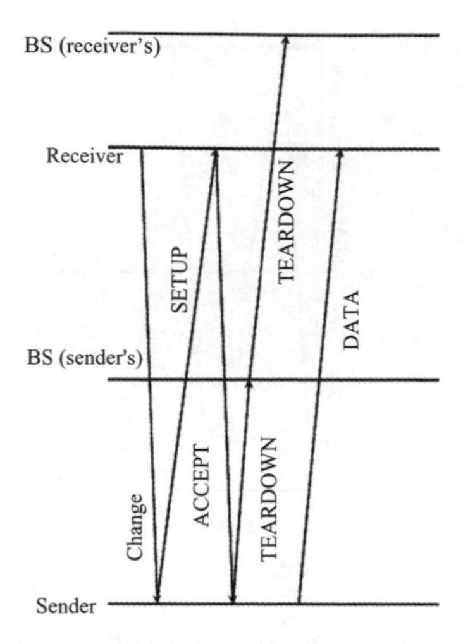

Figure 1.11 BS-oriented handoff to one-hop direct transmission.

neighbors. (A neighbor of a mobile host is another mobile host that can be connected directly with radio waves and without the help of base stations.) Furthermore, we must handle the case when the two-hop direct-transmission connection is disrupted because of mobility, for example, the relay MH moves out of range or the destination moves out of range. With one-hop or two-hop direct transmission, the system reliability is increased since some mobile hosts can still communicate with others even if their base stations fail. However, we still limit the number of hops in direct transmission to two for the following reasons:

(1) The routing will become complicated in three or more hops direct transmission. In multihop direct communications, handling many routing paths wastes the bandwidth in exchanging routing information, time stamps, avoiding routing update loop, and so on.

(2) Problems of *ad hoc* networks, such as routing and connections maintenance, are more manageable.

(3) If we allow three-hop direct transmission, the number of links in the air (at least three) not involving routing exchange will be larger than the BS-oriented mode (always two). In the long run, the battery power consumption will be more than the BS-oriented mode.

1.3.4.1 The neighbor database

In two-hop direct-transmission mode, each MH maintains a simple database (see Table 1.1) to store the information of neighboring MHs within its radio covering area. Each mobile host must broadcast periodically to inform the neighboring MHs of its related information. For example, the neighbor database of MH1 in Figure 1.12 is shown in Table 1.1. In the table, the BS-down field indicates whether or not the neighboring mobile host can detect a nearby base station. A value *True* means that the mobile host cannot connect to a base station.

Table 1.1 The neighbor database for MH1 in Figure 1.13

Neighbor ID	BS-down
MH2	False
MH4	False
MH5	False

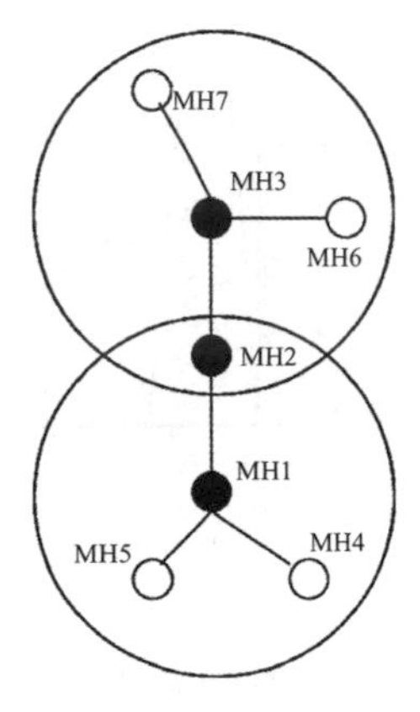

Figure 1.12 Two-hop direct-transmission zone.

1.3.4.2 Two-hop direct-transmission mode

This situation applies when the sender and destination are both within an intermediate's coverage area. The sender transmits the data to the destination through an intermediate MH. The connection setup procedures are as follows (see Figure 1.13):

(1) If an MH is within the transmission range, it receives the sender's message. There are two cases: (a) when the receiver finds the destination is itself, it sends the message ACCEPT back to the sender; the connection will be the one-hop direct transmission; and (b) if the destination is not itself, the mobile host checks the neighbor database. If the destination is in the database, it forwards the SETUP message to make connection to the destination. After the destination accepts the connection setup, it sends ACCEPT back to the sender.

(2) If the destination receives many copies of SETUP, it only accepts the first one. The redundant messages are discarded. The other candidate intermediate nodes will give up after timeout.

(3) At the same time, the BS receives the SEARCH message. It queries the MH and sends the SETUP message to the destination. For direct transmission, the destination receives the SETUP message and sends the DIRECT message to its BS. Otherwise, the destination accepts the connection from the base station.

(4) The communication continues transmitting directly until the transmission path is broken.

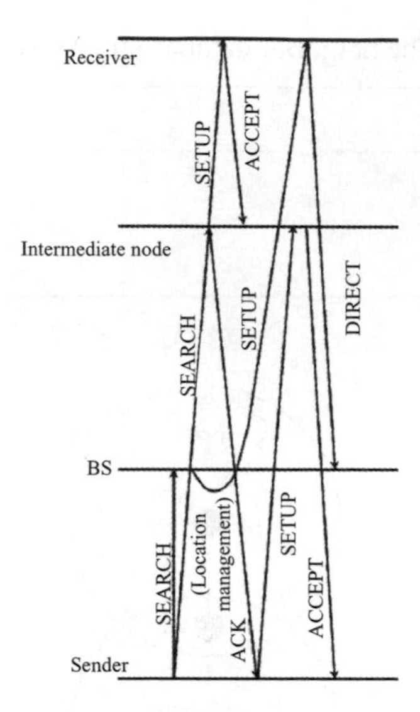

Figure 1.13 Two-hop direct-transmission mode.

1.3.4.3 Handoff – two-hop direct-transmission mode to two-hop direct-transmission mode, one-hop direction-transmission mode or BS-oriented mode

If the destination or the intermediate node finds that the strength of the signal is less than a critical value, the handoff procedure is requested and executed at this time. The system will try to find another direct transmission path. If a direct transmission path is not found, the BS-oriented mode will take over. The handoff procedures are as follows:

(1) The destination or intermediate node sends the CHANGE message to the sender for changing connection.

(2) As the sender received the CHANGE request, it reinitiates the connection by sending out SEARCH. The next several steps are the same as in the initial connection setup.

1.3.4.4 How to solve the problem of BS failure

The method is also robust against BS failure in the middle of a connection. In Figure 1.14, when MH3 finds that its BS (BS2) failed, it performs the following steps until its BS is alive again (see Figure 1.15 for message flow).

(1) MH3 broadcasts the WHOSE-BS-ALIVE message to the neighbors.

(2) If one MH, say, MH2, receives the message and its BS is still alive, it records the sender ID and sends the BELONG message back to the sender.

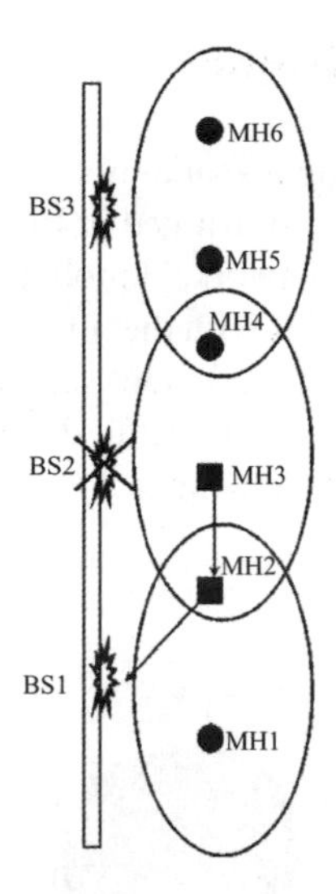

Figure 1.14 The BS failure problem.

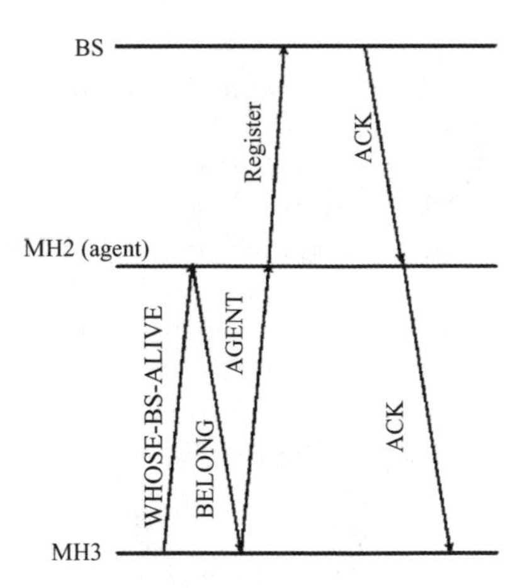

Figure 1.15 MH whose BS failed uses neighbor as an agent.

(3) MH3 receives the BELONG message and records MH2's ID. Then it sends the AGENT message to MH2.

(4) After the agent (MH2) receives the AGENT message, it represents MH3 to register its location to its BS (BS1).

(5) MH2 will now relay the information to and from MH3. When MH2 or MH3 is leaving the other's covering area, MH3 gives up the current agent and repeats steps 1–4 to find another agent. MH2 then removes the registration of MH3 from BS1.

1.4 GREEN WIRELESS NETWORKS

4G wireless networks might be using a spatial notching (angle α) to completely suppress antenna radiation towards the user, as illustrated in Figures. 1.16 and 1.17. These solutions will be referred to as 'green wireless networks' for obvious reasons. In a mobile environment in the periods when the notch coincides with the direction of the base station (access point) the multihop protocol, as discussed in the previous section, can be used. In addition, to reduce the overall transmit power, a cooperative transmit diversity, discussed in Section 19.4, and adaptive MAC protocol, discussed in Appendix B (please go to www.wiley.com/go/glisic), can be used.

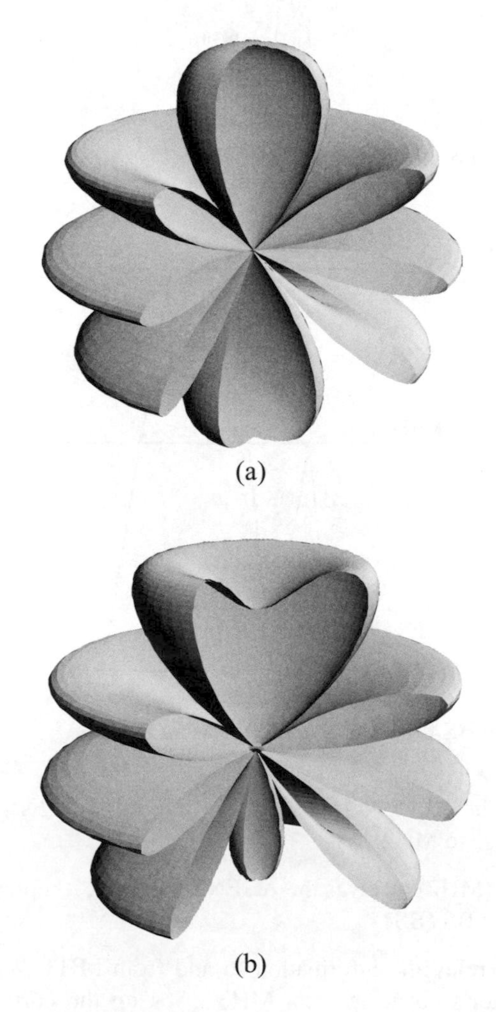

Figure 1.16 Three-dimensional amplitude patterns of a two-element uniform amplitude array for $d = 2\lambda$, directed towards (a) $\theta_0 = 0°$, (b) $\theta_0 = 60°$.

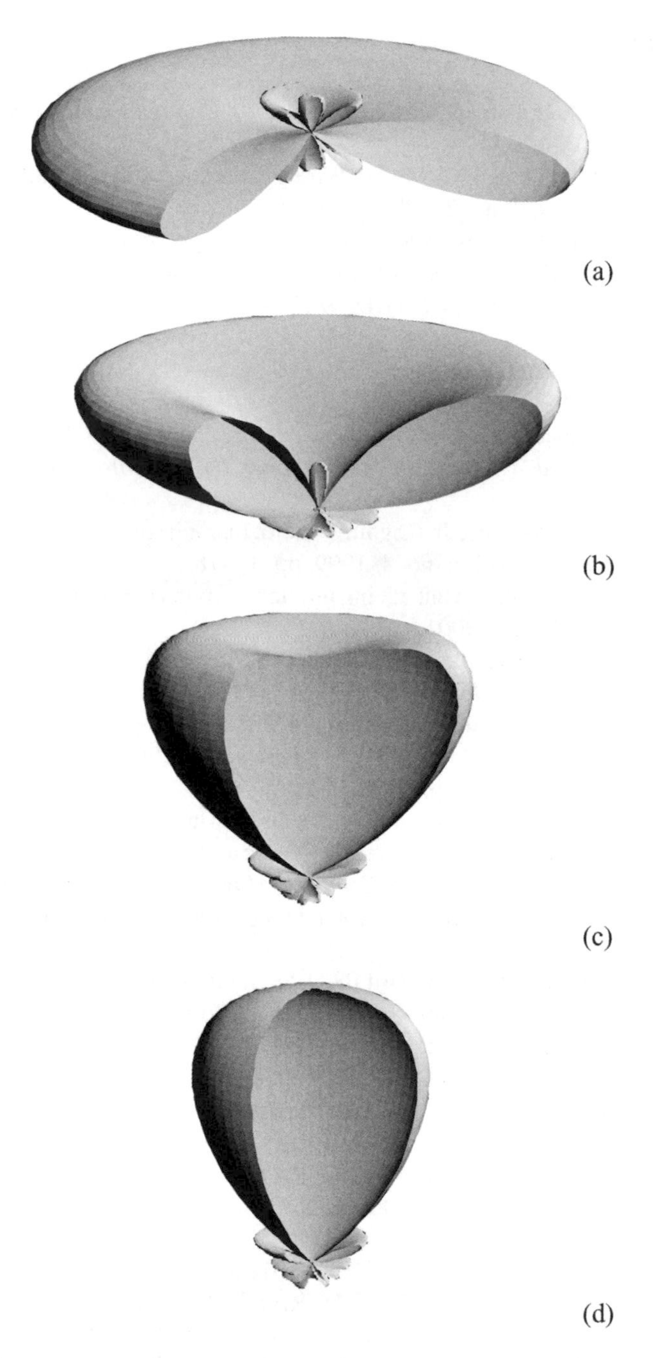

Figure 1.17 Three-dimensional amplitude patterns of a 10-element uniform amplitude array for $d = \lambda/4$, directed towards (a) $\theta_0 = 0°$, (b) $\theta_0 = 30°$, (c) $\theta_0 = 60°$, (d) $\theta_0 = 90°$.

REFERENCES

[1] M. Mouly and M.-B. Pautet. *The GSM System for Mobile Communications*. Palaiseau: France, 1992.

[2] R. Kalden, I. Meirick and M. Meyer. Wireless internet access based on GPRS, *IEEE Pers. Commun.*, vol. 7, no. 2, 2000, pp. 8–18.

[3] 3rd Generation Partnership Project (3GPP), www.3gpp.org

[4] J. Khun-Jush, P. Schramm, G. Malmgren and J. Torsner. HiperLAN2: broadband wireless communications at 5 GHz. *IEEE Commun. Mag.*, vol. 40, no. 6, 2002, pp. 130–137.

[5] U. Varshney. The status and future of 802.11-based WLANs, *IEEE Comput.*, vol. 36, no. 6, 2003, pp. 102–105.

[6] Digital Video Broadcasting (DVB), www.dvb.org, January 2002.

[7] S. Glisic. *Advanced Wireless Communications: 4G Technology*. John Wiley & Sons: Chichester, 2004.

[8] J. Mitola III and G. Maguire Jr. Cognitive radio: making software radios more personal, *IEEE Pers. Commun.*, vol. 6, no. 4, 1999, pp. 13–18.

[9] J. Border *et al.* Performance enhancing proxies intended to mitigate link-related degradations. RFC 3135, June 2001.

[10] D.C. Feldmeier, A,J. McAuley, J.M. Smith, D.S. Bakin, W.S. Marcus and T.M. Raleigh, Protocol boosters, *IEEE JSAC*, vol. 16, no. 3, 1998, pp. 437–444.

[11] M. García *et al.* An experimental study of Snoop TCP performance over the IEEE 802.11b WLAN. *5th Int. Symp. Wireless Personal Multimedia Commun.*, Honolulu, HI, Vol. III, October 2002, pp. 1068–1072.

[12] L. Muñoz, M. Garcia, J. Choque, R. Aguero and P. Mahonen, Optimizing internet flows over IEEE 802.11b wireless local area networks: a performance enhancing proxy based on forward error correction, *IEEE Commun. Mag.*, vol. 39, no. 12, 2001, pp. 60–67.

[13] V. Jacobson. TCP/IP compression for low-speed serial links, RFC 1144, February 1990.

[14] *Wireless LAN*. IEEE Draft Standard P802.11, January 1996.

[15] Radio equipment and systems (RES); High performance radio local area network (HIPERLAN); Functional specification. ETSI, France, Draft prETS 300 652, 1995.

[16] D. Evans, Y. Du, C. Herrmann, S.N. Hulyalkar and P. May. Wireless ATM LAN with and without infrastructure. In *2nd IEEE Int. Workshop Broadband Switching Systems*, Taipei, Taiwan, 1997, pp. 120–128.

[17] 3GPP Technical Specification 25.401 UTRAN Overall Description.

[18] 3GPP Technical Specification 25.410 UTRAN In Interface: General Aspects and Principles.

[19] 3GPP Technical Specification 25.411 UTRAN Iu Interface: Layer 1.

[20] 3GPP Technical Specification 25.412 UTRAN Iu Interface: Signalling Transport.

[21] 3GPP Technical Specification 25.413 UTRAN Iu Interface: RANAP Signalling.

[22] 3GPP Technical Specification 25.414 UTRAN Iu Interface: Data transport and Transport Signalling.

[23] 3GPP Technical Specification 25.415 UTRAN Iu Interface: CN-RAN User Plane Protocol.

[24] 3GPP Technical Specification 25.420 UTRAN Iur Interface: General Aspects and Principles.

[25] 3GPP Technical Specification 25.421 UTRAN Iur Interface: Layer 1.

[26] 3GPP Technical Specification 25.422 UTRAN Iur Interface: Signalling Transport.
[27] 3GPP Technical Specification 25.423 UTRAN Iur Interface: RNSAP Signalling.
[28] 3GPP Technical Specification 25.424 UTRAN Iur Interface: Data Transport and Transport Signalling for CCH Data Streams.
[29] 3GPP Technical Specification 25.425 UTRAN Iur Interface: User Plane Protocols for CCH Data Streams.
[30] 3GPP Technical Specification 25.426 UTRAN Iur and Iub Interface Data Transport and Transport Signalling for DCH Data Streams.
[31] 3GPP Technical Specification 25.427 UTRAN Iur and Iub Interface User Plane Protocols for DCI-1 Data Streams.
[32] 3GPP Technical Specification 25.430 UTRAN Iub Interface: General Aspects and Principles.
[33] 3GPP Technical Specification 25.431 UTRAN Iub Interface: Layer 1.
[34] 3GPP Technical Specification 25.432 UTRAN Iub Interface: Signalling Transport.
[35] 3GPP Technical Specification 25.433 UTRAN Iub Interface: NBAP Signalling.
[36] 3GPP Technical Specification 25.434 UTRAN Iub Interface: Data Transport and Transport Signalling for CCH Data Streams.
[37] 3GPP Technical Specification 25.435 UTRAN Iub Interface: User Plane Protocols for CCH Data Streams.
[38] 3G TS 25.301 Radio Interface Protocol Architecture.
[39] 3G TS 25.302 Services Provided by the Physical Layer.
[40] 3G TS 25.303 UE Functions and Interlayer Procedures in Connected Mode.
[41] 3G TS 25.304 UE Procedures in Idle Mode.
[42] 3G TS 25.321 MAC Protocol Specification.
[43] 3G TS 25.322 RLC Protocol Specification.
[44] 3G TS 25.323 PDCP Protocol Specification.
[45] 3G TS 25.324 Broadcast/Multicast Control Protocol (BMC) Specification.
[46] 3G TS 25.331 RRC Protocol Specification.
[47] 3G TS 24.008 Mobile Radio Interface Layer 3 Specification, Core Network Protocols – Stage 3.
[48] 3G TS 33.102 3G Security; Security Architecture.
[49] GSM 04.18 Digital Cellural Telecommunications System (Phase 2+); Mobile Radio Interface Layer 3 Specification, Radio Resource Control Protocol.
[50] IETF RFC 2507 IP Header Compression.
[51] 3G TS 25.305 Stage 2 Functional Specification of Location Services in UTRAN.
[52] 3G TS 33.105 3G Security; Cryptographic Algorithm Requirements.
[53] G. Armitage and K. Adams. Packet reassembly during cell loss, *IEEE Network*, vol. 7, no. 5, 1995, pp. 26–34.
[54] U. Black. *ATM Volume I: Foundation for Broadband Networks*. Prentice-Hall: Upper Saddle River, NJ, 1992.
[55] M. Garrett. A service architecture for ATM: from applications to scheduling, *IEEE Network*, vol. 10, no. 3, 1996, pp. 6–14.
[56] D. McDysan and D. Spohn. *ATM: Theory and Application*. McGraw-Hill: New York, 1999.
[57] K. Sato, S. Ohta and I. Tokizawa. Broad-band ATM network architecture based on virtual paths, *IEEE Trans. Commun.*, vol. 38, no. 8, 1990, pp. 1212–1222.
[58] T. Suzuki. ATM adaptation layer protocol, *IEEE Commun. Maga.*, vol. 32, no. 4, 1994, 80–83.

2

Physical Layer and Multiple Access

In this chapter we will briefly summarize the signal formats used in the existing wireless systems and point out possible ways of evolution towards the 4G system. The focus will be on ATDMA, WCDMA, OFDMA, MC CDMA and UWB signals [1–54].

2.1 ADVANCED TIME DIVISION MULTIPLE ACCESS-ATDMA

In a TDMA system each user is using a dedicated time slot within a TDMA frame as shown in Figure 2.1 for GSM or in Figure 2.2 for ADC (american digital cellular system). Additional data about the signal format and system capacity are given in Tables 2.1 and 2.2. The evolution of the ADC system resulted in the TIA (Telecommunications Industry Association) universal wireless communications (UWC) standard 136. The basic system parameters are summarized in Table 2.3. The evolution of GSM resulted in a system known as EDGE with parameters also summarized in Table 2.3.

If TDMA is chosen for 4G, the signal formats are further enhanced by using multidimensional trellis (space–time–frequency) coding and advanced signal processing [54]. This is also combined with Orthogonal frequency division multiplex (OFDM) and Multicarrier code division multiple access(MC CDMA) signal formats described below.

2.2 CODE DIVISION MULTIPLE ACCESS

Code division multiple access (CDMA) technique is based on spreading the spectra of the relatively narrow information signal S_n by a code c, generated by much higher clock (chip) rate. Different users are separated using different uncorrelared codes. As an example, the

© 2006 John Wiley & Sons, Ltd.

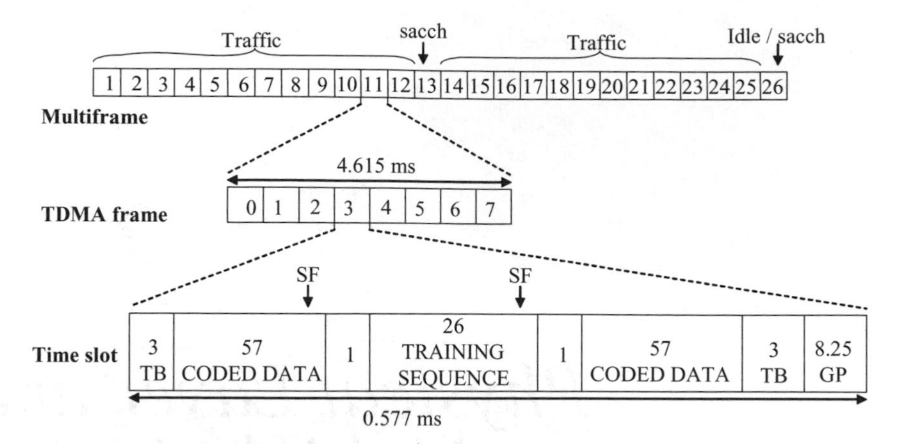

Figure 2.1 Digital cellular TDMA systems: GSM slot and frame structure showing 130.25 bits/time slot (0.577 ms), eight time slots/TDMA frame (full rate) and 13 TDMA frames/multiframe (TB = tail bits, GP = guard period, SF = stealing flag).

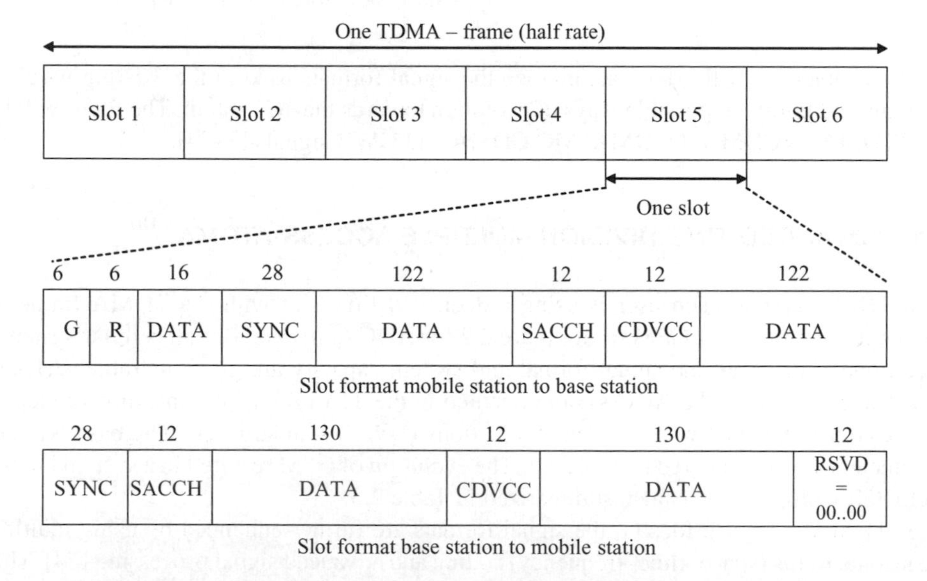

Figure 2.2 ADC slot and frame structure for down- and uplink with 324 bits/time slot (6.67 ms) and 3(6) time slots/TDMA frame for full-rate (half-rate) (G = guard time, R = ramp-up time, RSVD = reserved bits).

narrowband signal in this case can be a PSK signal of the form

$$S_n = b(t, T_m) \cos \omega t \tag{2.1}$$

where $1/T_m$ is the bit rate and $b = \pm 1$ is the information. The baseband equivalent of Equation (1.1) is

$$S_n^{\mathrm{b}} = b(t, T_m) \tag{2.1a}$$

Table 2.1 TDMA system parameters

	Europe (ETSI)	North America (TIA)	Japan (MPT)
Access method	TDMA	TDMA	TDMA
Carrier spacing	200 kHz	30 kHz	25 kHz
Users per carrier	8 (16)	3 (6)	3 (tbd)
Modulation	GMSK	$\pi/4$ DPSK	$\pi/4$ DPSK
Voice codec	RPE 13 kb/s	VSELP 8kb/s	tbd
Voice frame	20 ms	20 ms	20 ms
Channel code	Convolutional	Convolutional	Convolutional
Codec bit rate	22.8 kb/s	13 kb/s	11.2 kb/s
TDMA frame duration	4.6 ms	20 ms	20 ms
Interleaving	~40ms	27 ms	27 ms
Associated control channel	Extra slot	In slot	In slot
Handoff method	MAHO	MAHO	MAHO

ETSI, European Telecommunications Standards Institute; MPT, Mobile portable terminal; TDMA, time division multiple access.

Table 2.2 Approximate capacity in Erlang per km^2 assuming a cell radius of 1 km (site distance of 3 km) in all cases and three sectors per site. The Lee–Merit is number of channels per site assuming an optimal reuse plan

	Analog FM	GSM pessimistic	GSM optimistic	ADC pessimistic	ADC optimistic	JDC pessimistic	JDC optimistic
Bandwidth	25 MHz	25 MHz		25 MHz		25 MHz	
Number of voice channels	833	1000		2500		3000	
Reuse plan	7	4	3	7	4	7	4
Channels/site	119	250	333	357	625	429	750
Erlang/km^2	11.9	27.7	40.0	41.0	74.8	50.0	90.8
Capacity gain	1.0	2.3	3.4	3.5	6.3	4.2	7.6
(Lee–Merit gain)	(1.0)	(2.7)	(3.4)	(3.8)	(6.0)	(4.0)	(7.2)

The spreading operation, presented symbolically by operator $\varepsilon(\,)$, is obtained if we multiply narrowband signal by a pseudo noise (PN) sequence (code) $c(t, Tc) = \pm 1$. The bits of the sequence are called chips and the chip rate $1/T_c \gg 1/T_m$. The wideband signal can be represented as

$$S_w = \varepsilon(S_n) = cS_n = c(t, T_c)b(t, T_m)\cos\omega t \tag{2.2}$$

The baseband equivalent of Equation (2.2) is

$$S_w^b = c(t, T_c)b(t, T_m) \tag{2.2a}$$

Table 2.3 Parameters of UWC-136 and EDGE signal

Key characteristic	TIA UWC-136	GSM radio interface (for reference only)
Multiple access	TDMA	TDMA
Band width	30/200/1600 kHz	200 kHz
Bit rate	48.6 kb/s 72.9 kb/s 270.8 kb/s 361.1 kb/s 722.2 kb/s 2.6 Mb/s 5.2 Mb/s	270.8kb/s for EDGE 812.5 kb/s
Duplexing	FDD/TDD	FDD
Carrier spacing	30/200/1600 kHz	200 kHz
Inter BS timing	Asynchronous (synchronization possible)	Asynchronous (synchronization possible)
Inter-cell synchronization	Not required	Not required
Base station synchronization	Not required	Not required
Cell search scheme	L1 power-based, L2 parameter-based, L3 service/network/ operator-based	L1 power-based, L2 parameter-based, L3 service/network/ operator-based
Frame length	40/40/4.6/4.6 ms	4.6 ms
HO	HHO	HHO
DL Data modulation	$\pi/4$ DPSK $\pi/4$ coherent QPSK 8 PSK GMSK Q-O-QAM B-O-QAM	GMSK 8 PSK
DL Power control	Per slot and per carrier	Per slot
DL Variable rate accommodation	Slot aggregation	Slot aggregation
UL Data modulation	$\pi/4$ DPSK $\pi/4$ coherent QPSK 8 PSK GMSK Q-O-QAM B-O-QAM	GMSK 8 PSK
UL Power control	Per slot and per carrier	BS-directed MS power control

(*Continued*)

Table 2.3 (*Continued*)

Key characteristic	TIA UWC-136	GSM radio interface (for reference only)
UL Variable rate accommodation	Slot aggregation	Slot aggregation
Channel coding	Punctured convolutional code (R = 1/2, 2/3, 3/4, 1/1) Soft or hard decision coding	Convolutional coding Rate dependent on service
Interleaving periods	0/20/40/140/240 ms	Dependent on service
Rate detection	Via L3 signaling	Via stealing flags
Other features	Space and frequency diversity; MRC/ 'MRC-like' Support for hierarchical structures	MRC
Random access mechanism	Random access with shared control feedback (SCF), also reserved access	Random
Power control steps	4 dB	2 dB
Super frame length	720/640 ms (hyperframe is 1280 ms)	720 ms
Slots/frame	6 per 30 kHz carrier 8 per 200 kHz carrier 16–64 per 1.6 MHz carrier	8
Focus of backward compatibility	AMPS/IS54/136/GSM	GSM

HHO, hard handoff; DL, downlink; UL, uplink.

Despreading, represented by operator $D(\ )$, is performed if we use $\varepsilon(\ )$ once again and bandpass filtering, with the bandwidth proportional to $2/T_m$ represented by operator $BPF(\ )$, resulting in

$$\mathrm{D(Sw)} = \mathrm{BPF}[\varepsilon(\mathrm{Sw})] = \mathrm{BPF}(cc\,b\cos\omega t) = \mathrm{BPF}(c^2\,b\cos\omega t) = b\cos\omega t \tag{2.3}$$

The baseband equivalent of Equation (2.3) is

$$D\left(S_{\mathrm{w}}^{\mathrm{b}}\right) = LPF\left[\varepsilon\left(S_{\mathrm{w}}^{\mathrm{b}}\right)\right] = LPF[c(t, T_{\mathrm{c}})c(t, T_{\mathrm{c}})b(t, T_m)] = LPF[b(t, T_m)] = b(t, T_m) \tag{2.3a}$$

where $LPF(\ )$ stands for low pass filtering. This approximates the operation of correlating the input signal with the locally generated replica of the code $\mathrm{Cor}(c, S_{\mathrm{w}})$. Nonsynchronized despreading would result in

$$D_\tau(\ ); \mathrm{Cor}(c_\tau, S_{\mathrm{w}}) = \mathrm{BPF}[\varepsilon_\tau(S_{\mathrm{w}})] = \mathrm{BPF}(c_\tau cb\cos\omega t) = \rho(\tau)b\cos\omega t \tag{2.4}$$

In Equation (2.4) BPF would average out the signal envelope $c_\tau c$, resulting in $E(c_\tau c) = \rho(\tau)$. The baseband equivalent of Equation (2.4) is

$$D_\tau(\,);\quad \mathrm{Cor}\left(c_\tau, S_{\mathrm{w}}^{\mathrm{b}}\right) = \int_0^{T_m} c_\tau S_{\mathrm{w}}^{\mathrm{b}}\,\mathrm{d}t = b(t, T_m)\int_0^{T_m} c_\tau c\,\mathrm{d}t = b\rho(\tau) \tag{2.4a}$$

This operation extracts the useful signal b as long as $\tau \cong 0$, otherwise the signal will be suppressed because $\rho(\tau) \cong 0$ for $\tau \geq T_{\mathrm{c}}$. Separation of multipath components in a RAKE receiver is based on this effect. In other words, if the received signal consists of two delayed replicas of the form

$$r = S_{\mathrm{w}}^{\mathrm{b}}(t) + S_{\mathrm{w}}^{\mathrm{b}}(t-\tau)$$

the despreading process defined by Equation (2.4a) would result in

$$D_\tau(\,);\quad \mathrm{Cor}(c, r) = \int_0^{T_m} cr\,\mathrm{d}t = b(t, T_m)\int_0^{T_m} c(c + c_\tau)\,\mathrm{d}t = b\rho(0) + b\rho(\tau)$$

Now, if $\rho(\tau) \cong 0$ for $\tau \geq T_{\mathrm{c}}$, all multipath components reaching the receiver with a delay larger then the chip interval will be suppressed. If the signal transmitted by user y is despread in receiver x the result is

$$Dxy(\,);\ BPF[\varepsilon_{xy}(S_w)] = BPF(c_x c_y\, b_y \cos\omega t) = \rho_{xy}(t) b_y \cos\omega t \tag{2.5}$$

So in order to suppress the signals belonging to other users (multiple access interference, MAI), the cross-correlation functions should be low. In other words if the received signal consists of the useful signal plus the interfering signal from the other user:

$$r = S_{\mathrm{w}x}^{\mathrm{b}}(t) + S_{\mathrm{w}y}^{\mathrm{b}}(t) = b_x c_x + b_y c_y \tag{2.6}$$

the despreading process at receiver of user x would produce

$$D_{xy}(\,);\quad \mathrm{Cor}(c_x, r) = \int_0^{T_m} c_x r\,\mathrm{d}t = b_x\int_0^{T_m} c_x c_x\,\mathrm{d}t + b_y\int_0^{T_m} c_x c_y\,\mathrm{d}t = b_x\rho_x(0) + b_y\rho_{xy}(0) \tag{2.7}$$

When the system is properly synchronized $\rho_x(0) \cong 1$, and if $\rho_{xy}(0) \cong 0$ the second component representing MAI will be suppressed. This simple principle is elaborated in WCDMA standard, resulting in a collection of transport and control channels. The system is based on 3.84 Mcips rate and up to 2 Mb/s data rate. In a special downlink high data rate shared channel the data rate and signal format are adaptive. There will be mandatory support for QPSK and 16 QAM and optional support for 64 QAM based on UE capability, which will proportionally increase the data rate. For details see www.3gpp.com. CDMA is discussed in detail in Glisic [54, 55].

2.3 ORTHOGONAL FREQUENCY DIVISION MULTIPLEXING

In wireless communications, the channel imposes the limit on data rates in the system. One way to increase the overall data rate is to split the data stream into a number of

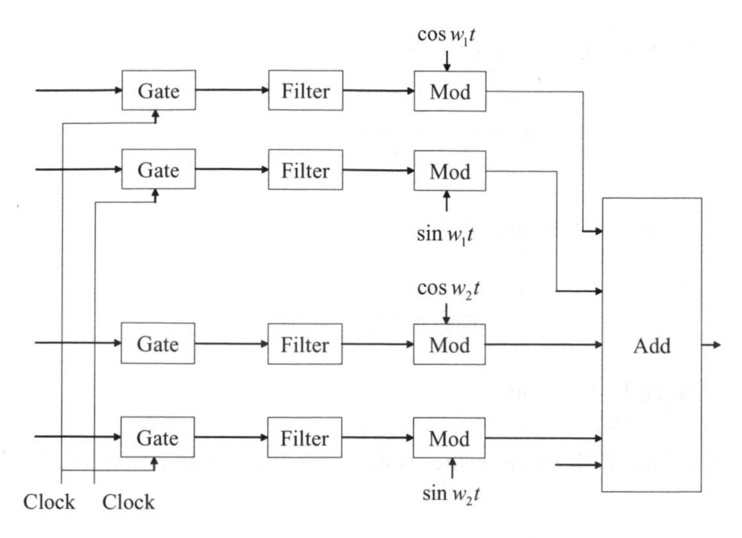

Figure 2.3 An early version of OFDM.

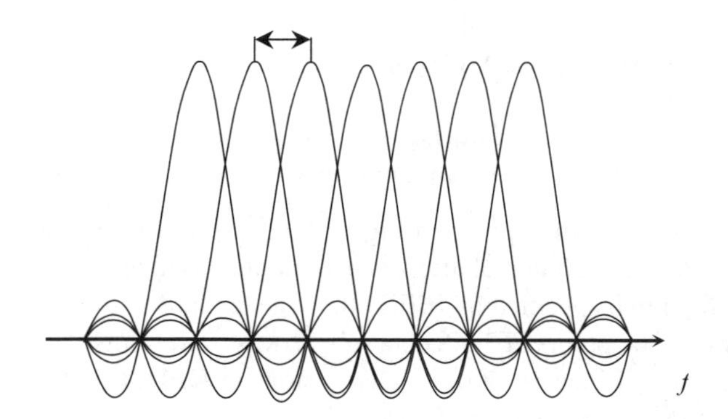

Figure 2.4 Spectrum overlap in OFDM.

parallel channels and use different subcarriers for each channel. The concept is presented in Figure 2.3 and represents the basic idea of the orthogonal frequency division multiplexing (OFDM) system. The overall signal can be represented as

$$x(t) = \sum_{n=0}^{N-1} \left\{ D_n e^{j2\pi(n/N)f_s t} \right\}; \quad -\frac{k_1}{f_s} < t < \frac{N + k_2}{f_s} \tag{2.8}$$

In other words complex data symbols $[D_0, D_1, \ldots, D_{N-1}]$ are mapped in OFDM symbols $[d_0, d_1, \ldots, d_{N-1}]$ such that

$$d_k = \sum_{n=0}^{N-1} D_n e^{j2\pi(kn/N)} \tag{2.9}$$

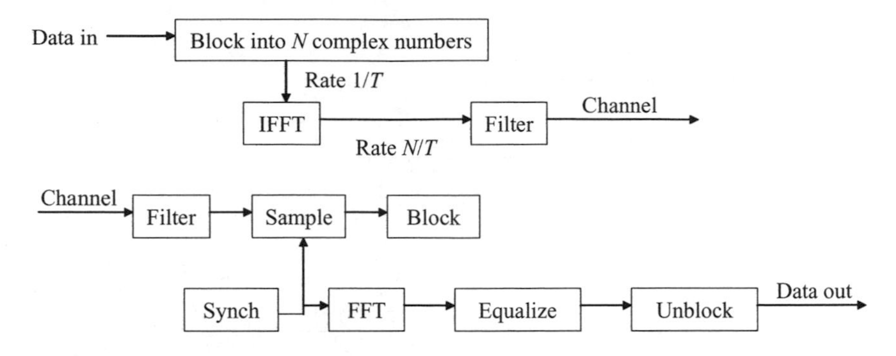

Figure 2.5 Basic OFDM system.

The output of the FFT block at the receiver produces data per channel. This can be represented as

$$\begin{aligned}
\tilde{D}_m &= \frac{1}{N}\sum_{k=0}^{N-1} r_k \mathrm{e}^{-j2\pi m(k/2N)} \\
r_k &= \sum_{n=0}^{N-1} H_n D_n \mathrm{e}^{j2\pi(n/2N)k} + n\,(k) \\
\tilde{D}_m &= \begin{cases} H_n D_n + N(n), n = m \\ N(n)\,, n \neq m \end{cases}
\end{aligned} \tag{2.10}$$

The system block diagram is given in Figure 2.6. In order to eliminate residual intersymbol interference, a guard interval after each symbol is used as shown in Figure 2.7. An example of OFDM signal specified by IEEE 802.11a standard is shown in Figure 2.8. The signal parameters are: 64 points FFT, 48 data subcarriers, four pilots, 12 virtual subcarriers, DC component 0, guard interval 800 ns. Discussion on OFDM and an extensive list of references on the topic are included in Glisic [54].

2.4 MULTICARRIER CDMA

Good performance and flexibility to accommodate multimedia traffic are incorporated in multicarrier (MC) CDMA, which are obtained by combining CDMA and OFDM signal formats. Figure 2.9 shows the DS-CDMA transmitter of the jth user for binary phase shift keying/coherent detection (CBPSK) scheme and the power spectrum of the transmitted signal, respectively, where $G_{\mathrm{DS}} = T_m/T_{\mathrm{c}}$ denotes the processing gain and $C^j(t) = [C_1^j\ C_2^j \cdots C_{G_{\mathrm{DS}}}^j]$ the spreading code of the jth user. Figure 2.10 shows the MC-CDMA transmitter of the jth user for CBPSK scheme and the power spectrum of the transmitted signal, respectively, where G_{MC} denotes the processing gain, N_{C} the number of subcarriers, and $C^j(t) = [C_1^j\ C_2^j \cdots C_{G_{\mathrm{MC}}}^j]$ the spreading code of the jth user. The MC-CDMA scheme is discussed assuming that the number of subcarriers and the processing gain are the same.

However, we do not have to choose $N_{\mathrm{C}} = G_{\mathrm{MC}}$, and actually, if the original symbol rate is high enough to become subject to frequency selective fading, the signal needs to be first S/P-converted before spreading over the frequency domain. This is because it is

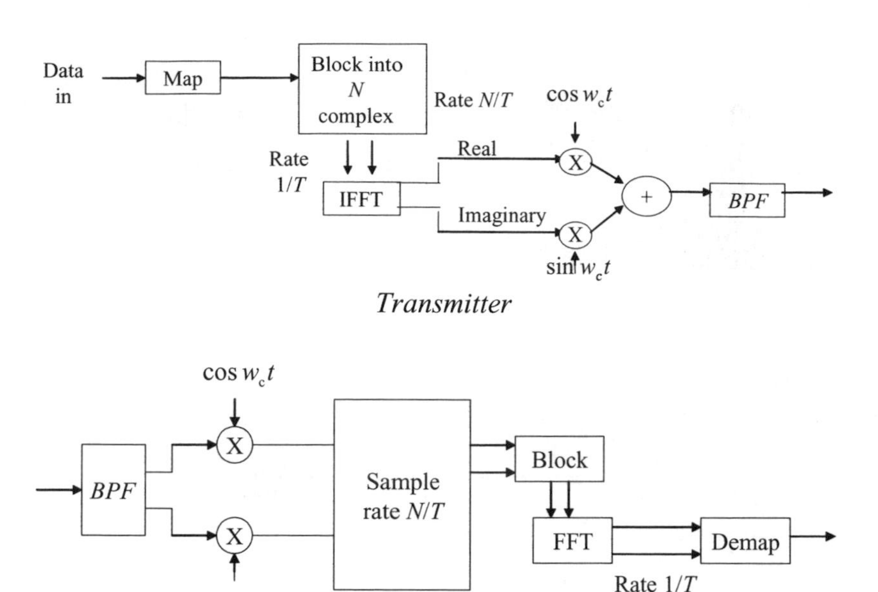

Figure 2.6 System with complex transmission.

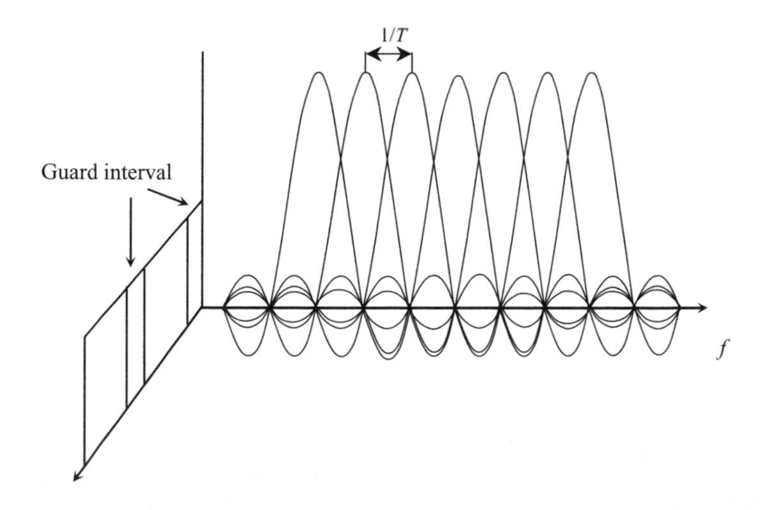

Figure 2.7 OFDM time and frequency span.

crucial for the multicarrier transmission to have frequency nonselective fading over each subcarrier.

Figure 2.11 shows the modification to ensure frequency nonselective fading, where T_S denotes the original symbol duration, and the original data sequence is first converted into P parallel sequences, and then each sequence is mapped onto G_{MC} subcarriers ($N_C = P \times G_{MC}$).

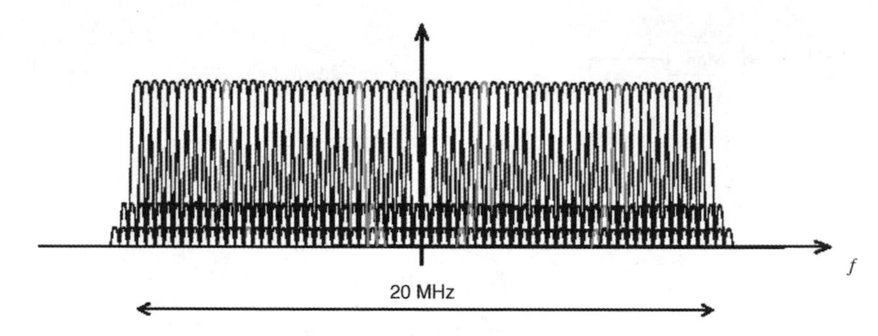

Figure 2.8 802.11a/HIPERLAN OFDM.

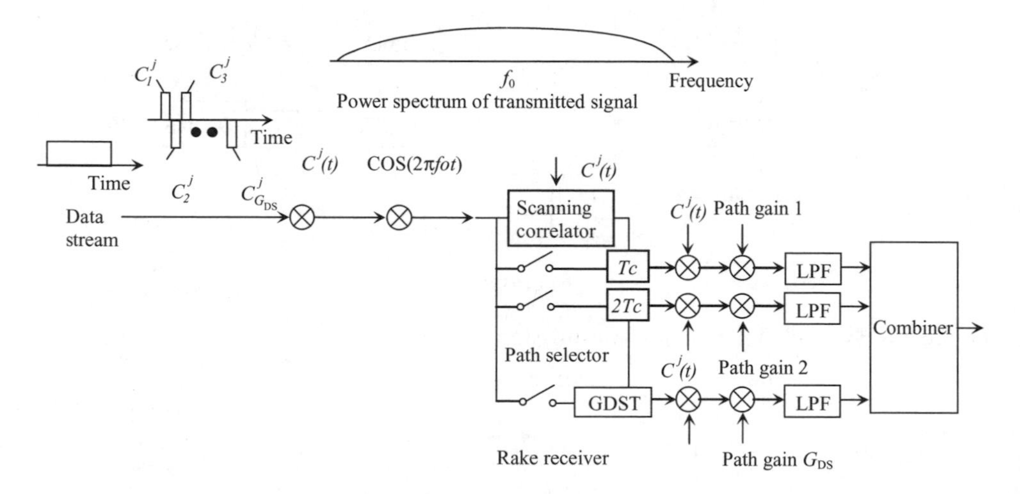

Figure 2.9 DS-CDMA scheme.

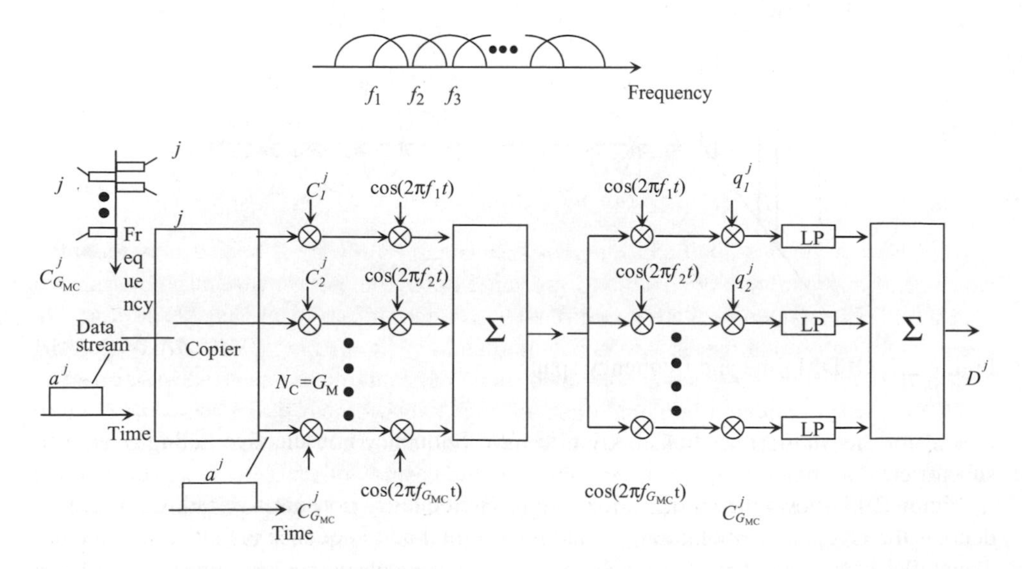

Figure 2.10 MC-CDMA scheme.

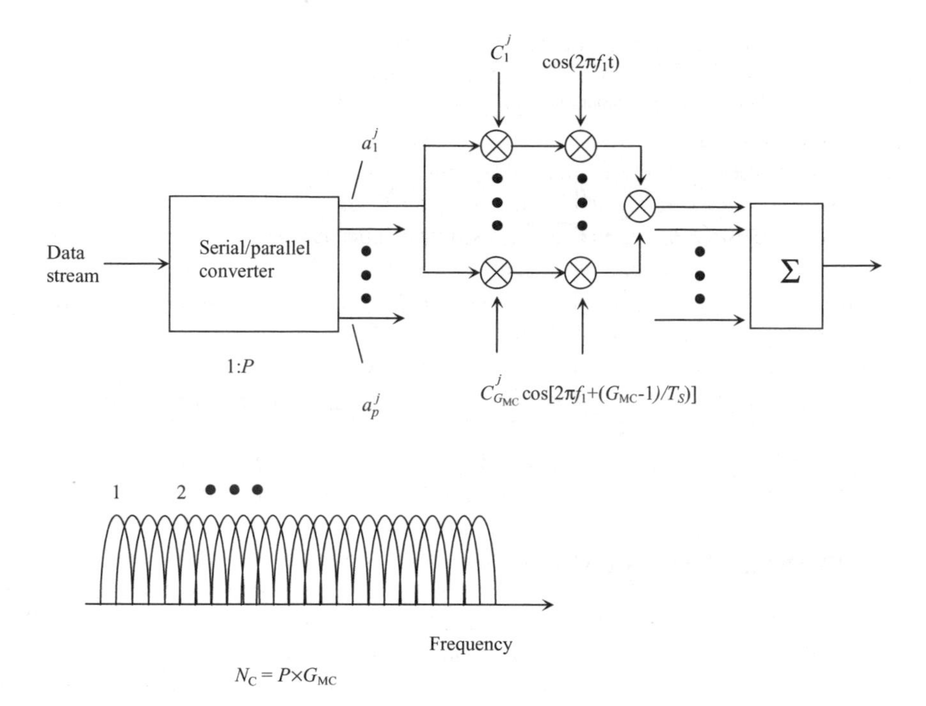

Figure 2.11 Modification of MC-CDMA scheme: spectrum of its transmitted signal.

The multicarrier DS-CDMA transmitter spreads the S/P-converted data streams using a given spreading code in the time domain so that the resulting spectrum of each subcarrier can satisfy the orthogonality condition with the minimum frequency separation. This scheme was originally proposed for an uplink communication channel, because the introduction of OFDM signaling into the DS-CDMA scheme is effective for the establishment of a quasi-synchronous channel.

Figure 2.12 shows the multicarrier DS-CDMA transmitter of the jth user and the power spectrum of the transmitted signal, respectively, where G_{MD} denotes the processing gain, N_{C} the number of subcarriers, and $C^j(t) = [C_1^j \; C_2^j \cdots C_{G_{\mathrm{MD}}}^j]$ the spreading code of the jth user.

The multitone MT-CDMA transmitter spreads the S/P-converted data streams using a given spreading code in the time domain so that the spectrum of each subcarrier prior to spreading operation can satisfy the orthogonality condition with the minimum frequency separation. Therefore, the resulting spectrum of each subcarrier no longer satisfies the orthogonality condition. The MT-CDMA scheme uses longer spreading codes in proportion to the number of subcarriers, as compared with a normal (single carrier) DS-CDMA scheme; therefore, the system can accommodate more users than the DS-CDMA scheme.

Figure 2.13 shows the MT-CDMA transmitter of the jth user for the CBPSK scheme and the power spectrum of the transmitted signal, respectively, where G_{MT} denotes the processing gain, N_{C} the number of subcarriers, and $C^j(t) = [C_1^j C_2^j \cdots C_{G_{\mathrm{MT}}}^j]$ the spreading code of the jth user. All these schemes will be discussed in details in Glisic [54].

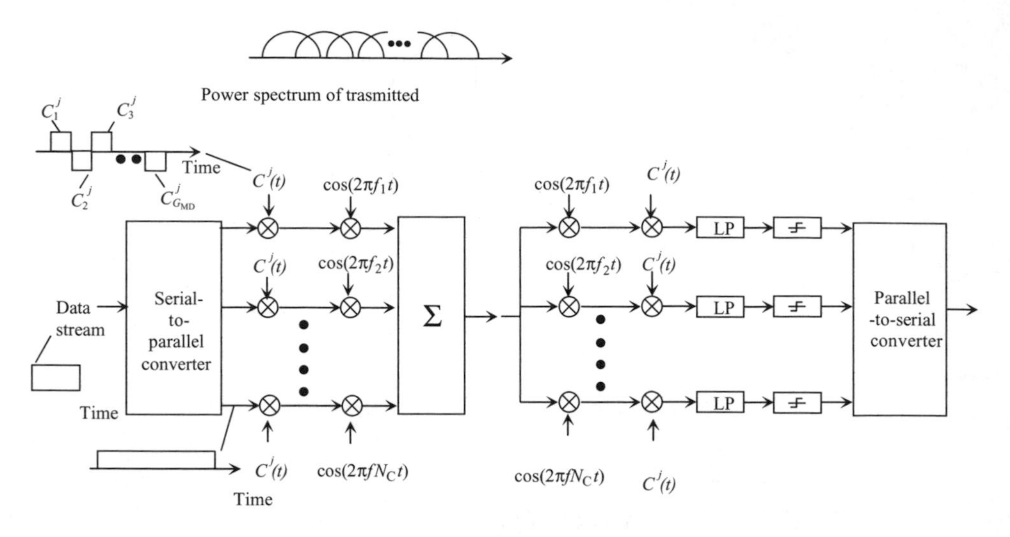

Figure 2.12 Multicarrier DS-CDMA scheme.

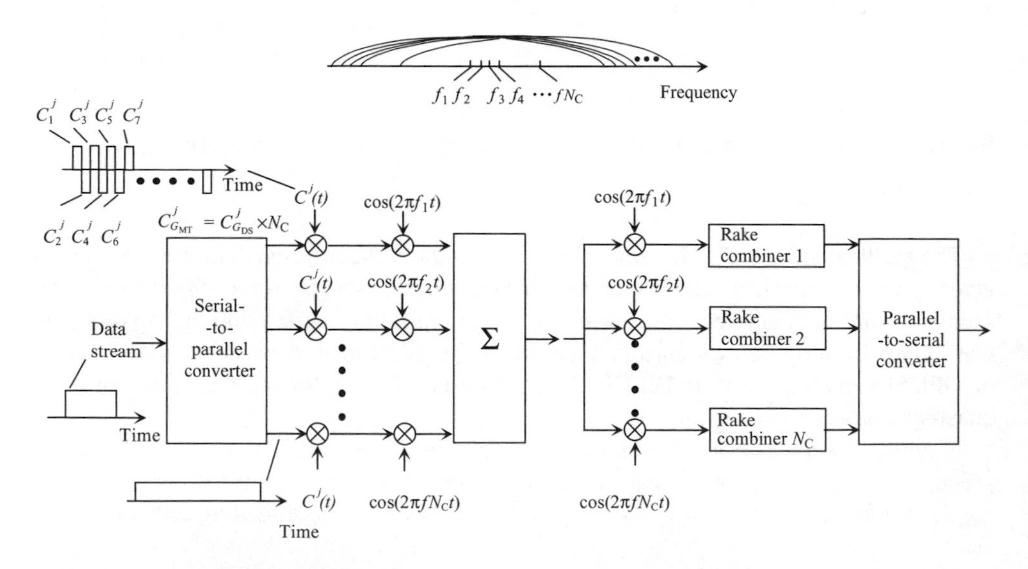

Figure 2.13 MT-CDMA scheme.

2.5 ULTRAWIDE BAND SIGNAL

For the multipath resolution in indoor environments a chip interval of the order of few nanoseconds is needed. This results in a spread spectrum signal with the bandwidth in the order of few GHz. Such a signal can also be used with no carrier, resulting in what is called impulse radio (IR) or ultrawide band (UWB) radio. A Typical form of the signal used in this case is shown in Figure 2.14. A collection of pulses received on different locations within the indoor environment is shown in Figure 2.16. UWB radio is discussed in detail in Glisic [54]. In this section we will define only a possible signal format.

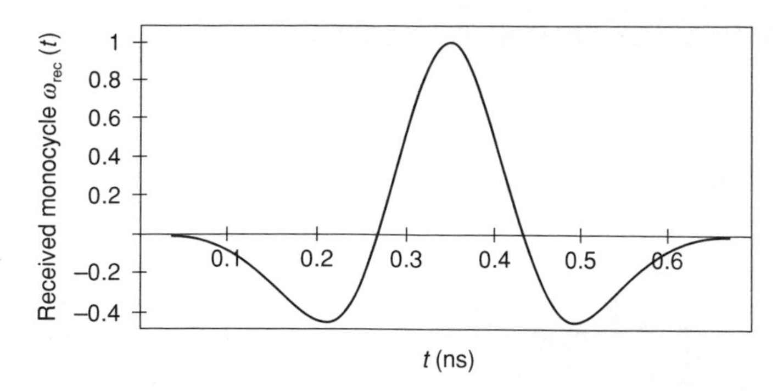

Figure 2.14 A typical ideal received monocycle $\omega_{\text{rec}}(t)$ at the output of the antenna subsystem as a function of time in nanoseconds.

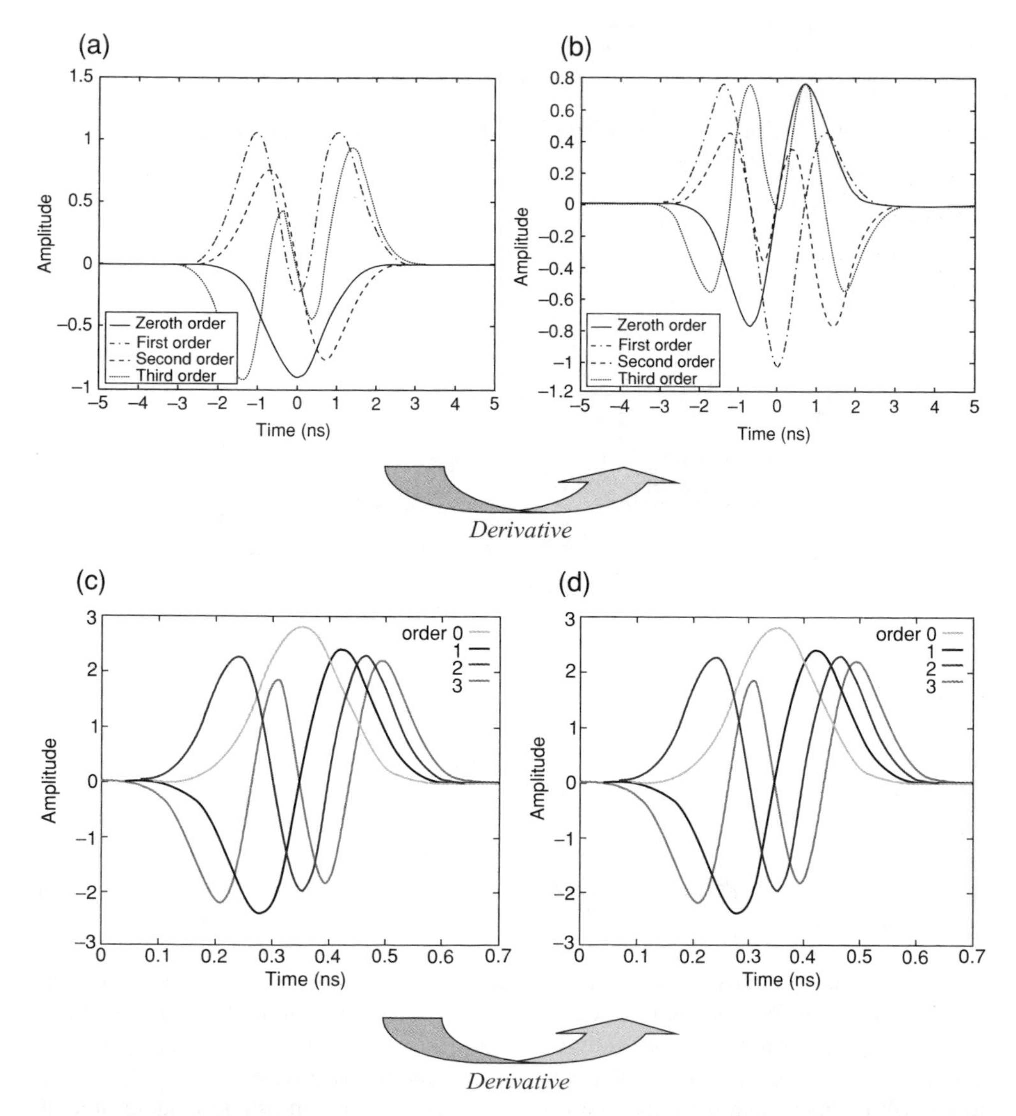

Figure 2.15 Modified Hermite pulse with Gram-Schimidt orthogonalization. (a) Generated pulses; (b) transmit pulses.

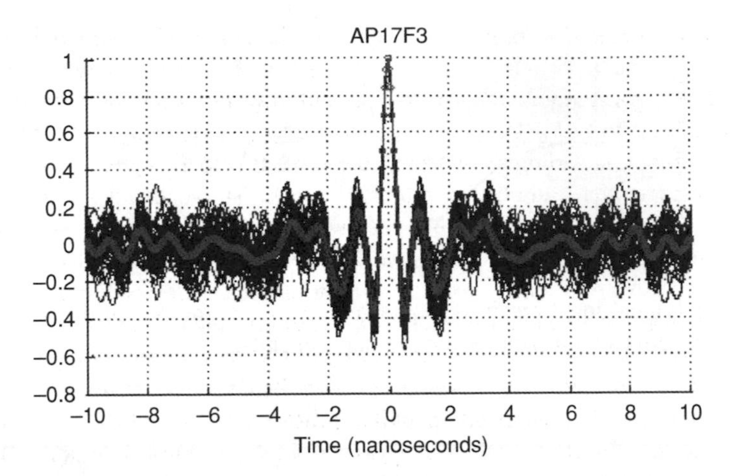

Figure 2.16 A collection of received pulses in different locations. (Reproduced by permission of IEEE [53].)

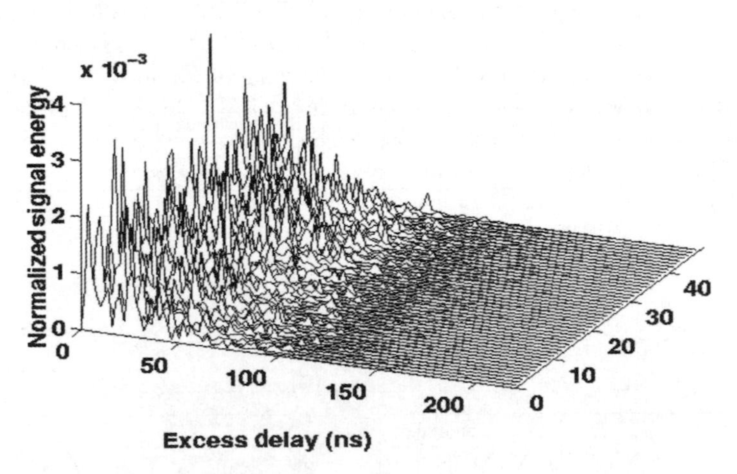

Figure 2.17 A collection of channel delay profiles. (Reproduced by permission of IEEE [52].)

A typical time-hopping format used in this case can be represented as

$$s_{\mathrm{tr}}^{(k)}\left(t^{(k)}\right) = \sum_{j=-\infty}^{\infty} \omega_{\mathrm{tr}}\left(t^{(k)} - jT_{\mathrm{f}} - c_j^{(k)}T_{\mathrm{c}} - \delta d_{\lfloor j/N_{\mathrm{s}}\rfloor}^{(k)}\right) \tag{2.11}$$

where $t^{(k)}$ is the kth transmitter's clock time and T_{f} is the *pulse repetition time*. The transmitted pulse waveform ω_{tr} is referred to as a *monocycle*. To eliminate collisions due to multiple access, each user (indexed by k) is assigned a distinctive time-shift pattern $\{c_j^{(k)}\}$ called a *time-hopping sequence*. This provides an additional time shift of $c_j^{(k)}T_{\mathrm{c}}$ seconds to the jth monocycle in the pulse train, where T_{c} is the duration of addressable time delay bins. For a fixed T_{f} the *symbol rate* R_{s} determines the number N_{s} of monocycles that are modulated by a given binary symbol as $R_s = (1/N_s T_f)\, s^{-1}$. The modulation index δ is chosen to optimize performance. For performance prediction purposes, most of the time the data sequence $\{d_j^{(k)}\}_{j=-\infty}^{\infty}$ is modeled as a wide-sense stationary random process composed of equally likely symbols. For data a pulse position data modulation is used.

When K users are active in the multiple-access system, the composite received signal at the output of the receiver's antenna is modeled as

$$r(t) = \sum_{k=1}^{K} A_k s_{\rm rec}^{(k)}(t - \tau_k) + n(t) \tag{2.12}$$

The antenna/propagation system modifies the shape of the transmitted monocycle $\omega_{\rm tr}(t)$ to $\omega_{\rm rec}(t)$ on its output. An idealized received monocycle shape $\omega_{\rm rec}(t)$ for a free-space channel model with no fading is shown in Figure 2.14.

The optimum receiver for a single bit of a binary modulated impulse radio signal in additive white Gaussian noise (AWGN) is a correlation receiver

$$\text{'decide } d_0^{(1)} = 0\text{' } if$$

$$\underbrace{\sum_{j=0}^{N_s-1} \overbrace{\int_{\tau_1 + jT_f}^{\tau_1+(j+1)T_f} r(u,t)\, \upsilon\left(t - \tau_1 - jT_f - c_j^{(1)} T_c\right) \mathrm{d}t}^{Pulse\ correlator\ output\ \triangleq\ \alpha_j(u)} > 0}_{Test\ statistic\ \triangleq\ \alpha(u)} \tag{2.13}$$

where $\upsilon(t) \stackrel{\Delta}{=} \omega_{\rm rec}(t) - \omega_{\rm rec}(t - \delta)$.

The spectra of a signal using TH are shown in Figure 2.18. If instead of TH a DS signal is used, the signal spectra is shown in Figure 2.19(a) for pseudorandom code and Figure 2.19(b) for a random code. The FCC (Frequency Control Committee) mask for indoor communications is shown in Figure 2.18. Possible options for UWB signal spectra are given in Figures 2.21 and 2.22 for single band and Figure 2.23 for multiband signal format. For more details see www.uwb.org and www.uwbmultiband.org.

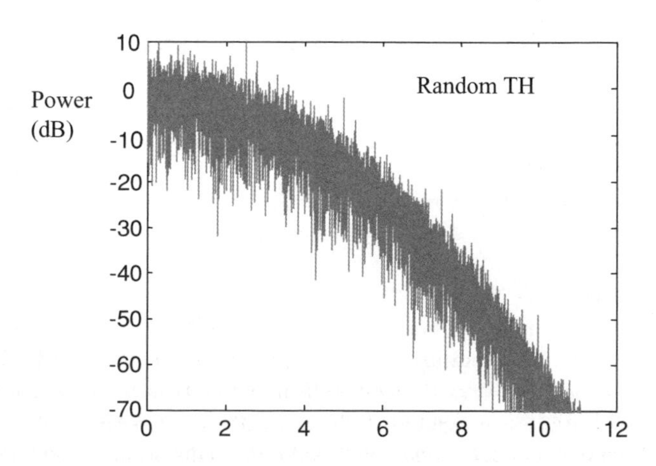

Figure 2.18 Spectra of a TH signal.

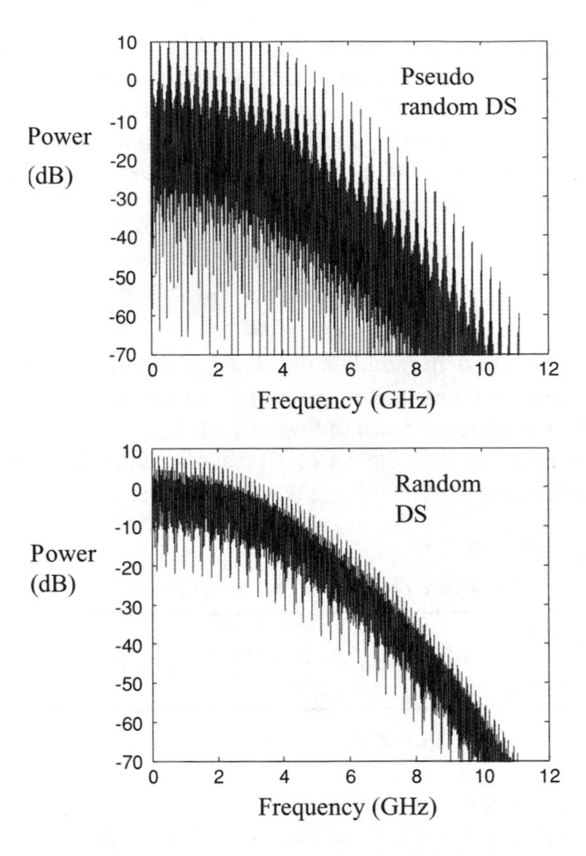

Figure 2.19 Spectra of pseudorandom DS and random DS signal.

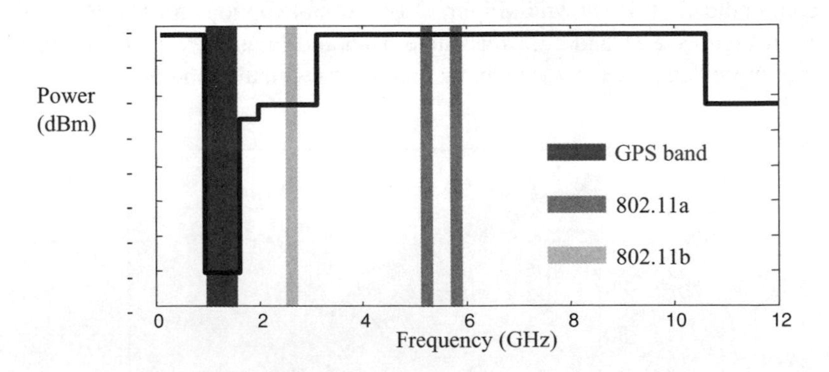

Figure 2.20 FCC frequency mask.

The optimal detection in a multiuser environment, with knowledge of all time-hopping sequences, leads to complex parallel receiver designs [2]. However, if the number of users is large and no such multiuser detector is feasible, then it is reasonable to approximate the combined effect of the other users' dehopped interfering signals as a Gaussian random process. All details regarding the system performance can be found in Glisic [54].

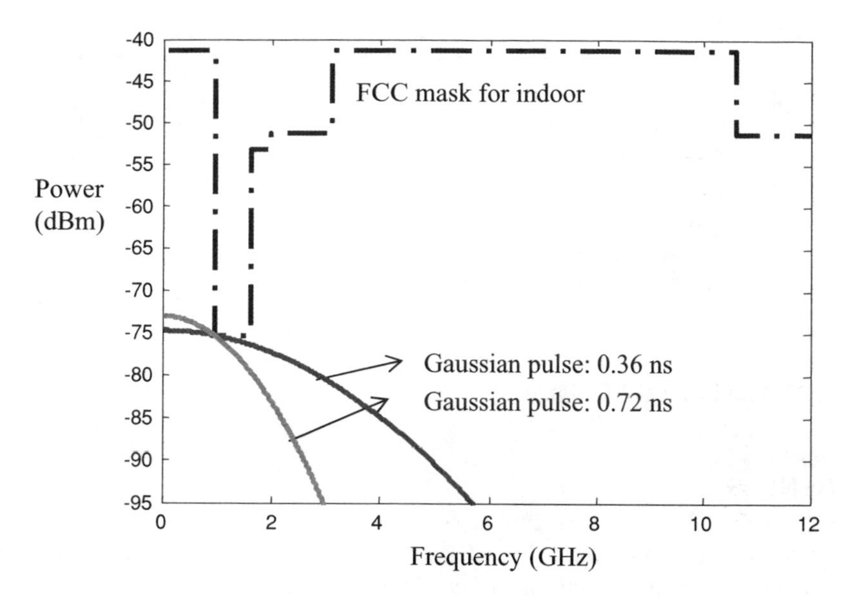

Figure 2.21 FCC mask and possible UWB signal spectra.

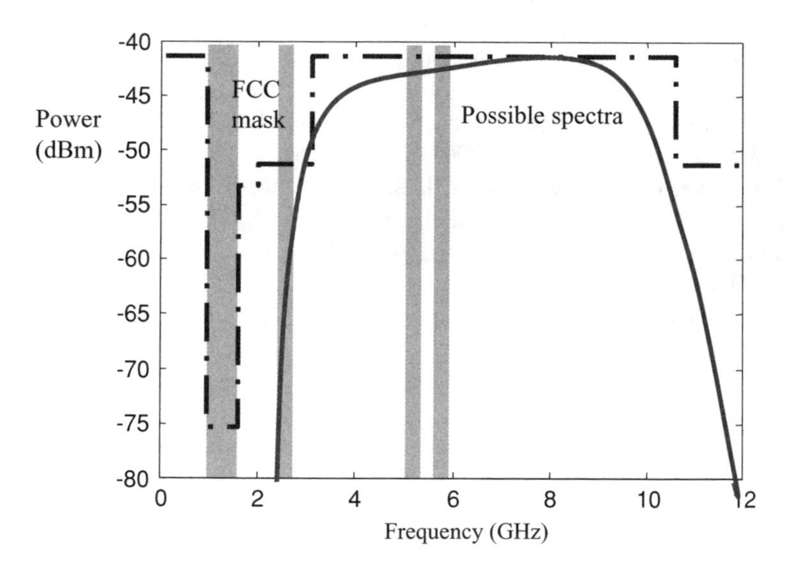

Figure 2.22 Single-band UWB signal.

2.6 MIMO CHANNELS AND SPACE TIME CODING

In order to increase the capacity, 4G networks use the previously described signal formats in MIMO (multiple input multiple output) channels combined with space-time coding. More details on these technologies can be found in Glisic [54].

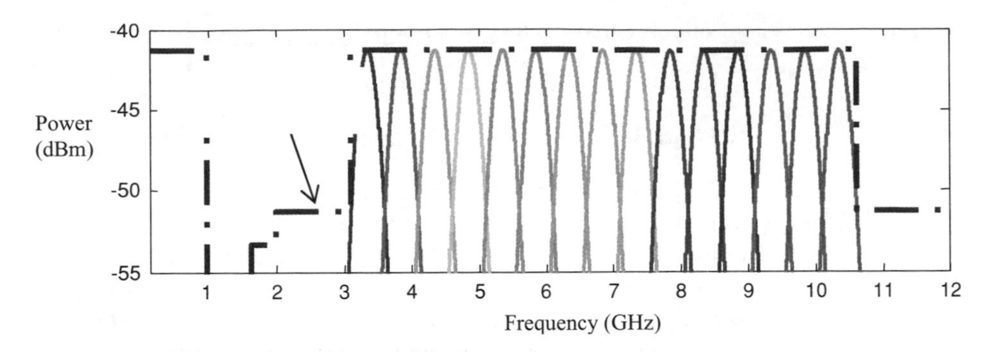

Figure 2.23 Multiband UWB signal.

REFERENCES

[1] F. Adachi, Evolution towards broadband wireless systems, in *5th Int. Symp. Wireless Personal Multimedia Communications*, Honolulu, HI, vol. 1, 27–30 October 2002, pp. 19–26.

[2] Jun-Zhao Sun, J. Sauvola and D. Howie, Features in future: 4G visions from a technical perspective, *IEEE Global Telecommunications Conference, GLOBECOM '01*, vol. 6, 25–29 November 2001, pp. 3533–3537.

[3] D.I. Axiotis, F.I. Lazarakis, C. Vlahodimitropoulos and A. Chatzikonstantinou, 4G system level simulation parameters for evaluating the interoperability of MTMR in UMTS and HIPERLAN/2, in *4th Int. Workshop on Mobile and Wireless Communications Network*, 9–11 September 2002, pp. 559–563.

[4] A. Mihovska, C. Wijting, R. Prasad, S. Ponnekanti, Y. Awad, and M. Nakamura, A novel flexible technology for intelligent base station architecture support for 4G systems, in *5th Int. Symp. Wireless Personal Multimedia Communications*, vol. 2, 27–30 October 2002, pp. 601–605.

[5] D. Kitazawa, L. Chen, H. Kayama and N. Umeda, Downlink packet-scheduling considering transmission power and QoS in CDMA packet cellular systems, in *4th Int. Workshop on Mobile and Wireless Communications Network*, 9–11 September, 2002, pp. 183–187.

[6] L. Dell'Uomo and E. Scarrone, An all-IP solution for QoS mobility management and AAA in the 4G mobile networks, in *5th Int. Symp. Wireless Personal Multimedia Communications*, vol. 2, 27–30 October 2002, pp. 591–595.

[7] E.R. Wallenius, End-to-end in-band protocol based service quality and transport QoS control framework for wireless 3/4G services, in *5th Int. Symp. Wireless Personal Multimedia Communications*, vol. 2, 27–30 October 2002, pp. 531–533.

[8] M. Benzaid, P. Minet and K. Al Agha, Integrating fast mobility in the OLSR routing protocol, in *4th Int. Workshop on Mobile and Wireless Communications Network*, 9–11 September 2002, pp. 217–221.

[9] G. Kambourakis, A. Rouskas and S. Gritzalis, Using SSL/TLS in authentication and key agreement procedures of future mobile networks, in *4th Int. Workshop on Mobile and Wireless Communications Network*, 9–11 September 2002, pp. 152–156.

[10] M. van der Schaar and J. Meehan, Robust transmission of MPEG-4 scalable video over 4G wireless networks, in *Proc. Int. Conf. Image Processing*, 24–28 June 2002, pp. 757–760.

[11] Qing-Hui Zeng, Jian-Ping Wu, Yi-Lin Zeng, Ji-Long Wang and Rong-Hua Qin, Research on controlling congestion in wireless mobile Internet via satellite based on multi-information and fuzzy identification technologies, in *Proc. 2002 Int. Conf. Machine Learning and Cybernetics*, vol. 4, 4–5 November 2002, pp. 1697–1701.

[12] *Proc. 5th Int. Symp. Wireless Personal Multimedia Communications.* (catalog no. 02EX568), vol. 1, 27–30 October 2002.

[13] T. Sukuvaara, P. Mahonen and T. Saarinen, Wireless Internet and multimedia services support through two-layer LMDS system, in *IEEE Int. Workshop on Mobile Multimedia Communications (MoMuC '99)*, 15–17 November 1999, pp. 202–207.

[14] C.C. Martin, J.H. Winters and N.R. Sollenberger, Multiple-input multiple-output (MIMO) radio channel measurements, in *Proc. 2000 IEEE Sensor Array and Multichannel Signal Processing Workshop*, 16–17 March 2000, pp. 45–46.

[15] J.M. Pereira, Fourth generation: now, it is personal!, in *11th IEEE Int. Symp. Personal, Indoor and Mobile Radio Communications, PIMRC 2000*, vol. 2, 18–21 September 2000, pp. 1009–1016.

[16] T. Otsu, N. Umeda and Y. Yamao, System architecture for mobile communications systems beyond IMT-2000, in *IEEE Global Telecommunications Conference, GLOBECOM '01*, vol. 1, 25–29 November 2001, pp. 538–542.

[17] Yi Han Zhang, D. Makrakis, S. Primak and Yun Bo Huang, Dynamic support of service differentiation in wireless networks 2002, in *IEEE CCECE 2002. Canadian Conf. on Electrical and Computer Engineering*, vol. 3, 12–15 May 2002, pp. 1325–1330.

[18] Jun-Zhao Sun and J. Sauvola, Mobility and mobility management: a conceptual framework, in *10th IEEE Int. Conf. Networks, ICON 2002*, 27–30 August 2002, pp. 205–210.

[19] V. Vassiliou, H.L. Owen, D.A. Barlow, J. Grimminger, H.-P. Huth and J. Sokol, A radio access network for next generation wireless networks based on multi-protocol label switching and hierarchical mobile IP, in *Proc. IEEE 56th Vehicular Technology Conference, VTC 2002*, vol. 2, 24–28 September 2002, pp. 782–786.

[20] P. Nicopolitidis, G.I. Papadimitriou, M.S. Obaidat and A.S. Pomportsis, 3G wireless systems and beyond: a review, in *9th Int. Conf. Electronics, Circuits and Systems,* vol. 3, 15–18 September 2002, pp. 1047–1050.

[21] *Proc. IEEE Wireless Communications and Networking Conference, WCNC 2002* (catalog, no. 02TH8609), vol. 1, 17–21 March 2002.

[22] J. Borras-Chia, Video services over 4G wireless networks: not necessarily streaming, in *IEEE Wireless Communications and Networking Conf., WCNC2002*, vol. 1, 17–21 March 2002, pp. 18–22.

[23] B.G. Evans and K. Baughan, Visions of 4G, *Electron. Commun. Eng. J.*, vol. 12, no. 6, 2000, pp. 293–303.

[24] J. Kim and A. Jamalipour, Traffic management and QoS provisioning in future wireless IP networks, *IEEE Person. Commun.* (see also *IEEE Wireless Commun.*), vol. 8, no. 5, 2001, pp. 46–55.

[25] A.H. Aghvami, T.H. Le and N. Olaziregi, Mode switching and QoS issues in software radio, *IEEE Person. Commun.* (see also *IEEE Wireless Commun.*), vol. 8, no. 5, 2001, pp. 38–44.

[26] T. Kanter, An open service architecture for adaptive personal mobile communication, *IEEE Person. Commun.* (see also *IEEE Wireless Commun.*), vol. 8, no. 6, 2001, pp. 8–17.
[27] H. Sampath, S. Talwar, J. Tellado, V. Erceg and A. Paulraj, A fourth-generation MIMO-OFDM broadband wireless system: design, performance, and field trial results, *IEEE Commun. Mag.*, vol. 40, no. 9, 2002, pp. 143–149.
[28] W. Kellerer and H.-J. Vogel, A communication gateway for infrastructure-independent 4G wireless access, *IEEE Commun. Mag.*, vol. 40, no. 3, 2002, pp. 126–131.
[29] V. Huang and Weihua Zhuang, QoS-oriented access control for 4G mobile multimedia CDMA communications, *IEEE Commun. Mag.*, vol. 40, no. 3, 2002, pp. 118–125.
[30] P. Smulders, Exploiting the 60 GHz band for local wireless multimedia access: prospects and future directions, *IEEE Commun. Mag.*, vol. 40, no. 1, 2002, pp. 140–147.
[31] Y. Raivio, 4G-hype or reality, in *Second Int. Conf. 3G Mobile Communication Technologies.* (Conference Publication no. 477), 26–28 March 2001, pp. 346–350.
[32] L. Becchetti, F. Delli Priscoli, T. Inzerilli, P. Mahonen and L. Munoz, Enhancing IP service provision over heterogeneous wireless networks: apath toward 4G, *IEEE Commun. Mag.*, vol. 39, no. 8, 2001, pp. 74–81.
[33] T. Abe, H. Fujii and S. Tomisato, A hybrid MIMO system using spatial correlation, in *5th Int. Symp. Wireless Personal Multimedia Communications*, vol. 3, 27–30 October 2002, pp. 1346–1350.
[34] S. Lincke-Salecker and C.S. Hood, A supernet: engineering traffic across network boundaries, in *36th Annual Simulation Symp.*, 30 March–2 April 2003, pp. 117–124.
[35] Y. Yamao, H. Suda, N. Umeda and N. Nakajima, Radio access network design concept for the fourth generation mobile communication system, in *Proc. IEEE 51st Vehicular Technology Conf.*, VTC 2000, Tokyo, vol. 3, 15–18 May 2000, pp. 2285–2289.
[36] E. Ozturk and G.E. Atkin, Multi-scale DS-CDMA for 4G wireless systems, *IEEE Global Telecommunications Conf., GLOBECOM '01*, vol. 6, 25–29 November 2001, pp. 3353–3357.
[37] L. Dell'Uomo and E. Scarrone, The mobility management and authentication/authorization mechanisms in mobile networks beyond 3G, in *12th IEEE Int. Symp. Personal, Indoor and Mobile Radio Communications*, vol. 1, 30 September–3 October 2001, pp. C-44–C-48.
[38] K.J. Kumar, B.S. Manoj and C.S.R. Murthy, On the use of multiple hops in next generation cellular architectures, in *10th IEEE Int. Conf. Networks, ICON 2002*, 27–30 August 2002, pp. 283–288.
[39] S.S. Wang, M. Green and M. Malkawi, Mobile positioning and location services, in *IEEE Radio and Wireless Conf. RAWCON*, 11–14 August 2002, pp. 9–12.
[40] M. Motegi, H. Kayama and N. Umeda, Adaptive battery conservation management using packet QoS classifications for multimedia mobile packet communications, in *Proc. 56th IEEE Vehicular Technology Conf. VTC 2002*, vol. 2, 24–28 September 2002, pp. 834–838.
[41] Ying Li, Shibua Zhu, Pinyi Ren and Gang Hu, Path toward next generation wireless internet-cellular mobile 4G, WLAN/WPAN and IPv6 backbone, in *Proc. 2002 IEEE Region 10 Conf. Computers, Communications, Control and Power Engineering TENCOM '02*, vol. 2, 28–31 October 2002, pp. 1146–1149.

[42] R.C. Qiu, Wenwu Zhu and Ya-Qin Zhang, Third-generation and beyond (3.5G) wireless networks and its applications, in *IEEE Int. Symp. Circuits and Systems, ISCAS 2002*, vol. 1, 26–29 May 2002, pp. I-41–I-44.

[43] C. Bornholdt, B. Sartorius, J. Slovak, M. Mohrle, R. Eggemann, D. Rohde and G. Grosskopf, 60 GHz millimeter-wave broadband wireless access demonstrator for the next-generation mobile internet, in *Optical Fiber Communication Conf. and Exhibit, OFC*, 17–22 March 2002, pp. 148–149.

[44] Jianhua He, Zongkai Yang, Daiqin Yang, Zuoyin Tang and Chun Tung Chou, Investigation of JPEG2000 image transmission over next generation wireless networks, in *5th IEEE Int. Conf. High Speed Networks and Multimedia Communications*, 3–5 July 2002, pp. 71–77.

[45] E. Baccarelli and M. Biagi, Error resistant space-time coding for emerging 4G-WLANs, in *IEEE Wireless Communications and Networking, WCNC 2003*, vol. 1, 16–20 March 2003, pp. 72–77.

[46] W. Mohr, WWRF – the Wireless World Research Forum, *Electron. Commun. Eng. J.*, vol. 14, no. 6, 2002, pp. 283–291.

[47] T. Otsu, I. Okajima, N. Umeda and Y. Yamao. Network architecture for mobile communications systems beyond IMT-2000, *IEEE Person. Commun.* (see also *IEEE Wireless Commun.*), vol. 8, no 5, October 2001, pp. 31–37.

[48] A. Bria, F. Gessler, O. Queseth, R. Stridh, M. Unbehaun, Jiang Wu, J. Zander and M. Flament, 4th-Generation wireless infrastructures: scenarios and research challenges, *IEEE Person. Commun. (see also IEEE Wireless Commun.*), vol. 8, no 6, 2001, pp. 25–31.

[49] F. Fitzek, A. Kopsel, A. Wolisz, M. Krishnam and M. Reisslein, Providing application-level QoS in 3G/4G wireless systems: a comprehensive framework based on multirate CDMA, *IEEE Wireless Commun.* (see also *IEEE Person. Commun.*), vol. 9, no 2, April 2002, pp. 42–47.

[50] B. Classon, K. Blankenship and V. Desai, Channel coding for 4G systems with adaptive modulation and coding, *IEEE Wireless Commun.* (see also *IEEE Person. Commun.*), vol. 9, no. 2, 2002, pp. 8–13.

[51] Yile Guo and H. Chaskar, Class-based quality of service over air interfaces in 4G mobile networks, *IEEE Commun. Mag.*, vol. 40, on 3, 2002, pp. 132–137.

[52] D. Cassioli, M.Z. Win and A.F. Molisch, The ultra-wide bandwidth indoor channel: from statistical model to simulations, *IEEE J. Selected Areas in Commun.*, vol. 20, no. 6, 2002, pp. 1247–1257.

[53] M.Z. Win and R.A. Scholtz, Characterization of ultra-wide bandwidth wireless indoor channels: a communication-theoretic view, *IEEE J. Selected Areas in Communi.*, vol. 20, no. 9, 2002, pp. 1613–1627.

[54] S. Glisic, *Advanced Wireless Communications, 4G Technology*. John Wiley & Sons Ltd: Chichester, 2004.

[55] S. Glisic, *Adaptive WCDMA, Theory and Practice*. John Wiley & Sons Ltd Chichester, 2004.

3

Channel Modeling for 4G

3.1 MACROCELLULAR ENVIRONMENTS (1.8 GHz)

In this section we briefly present statistical properties of azimuth and delay spread in macrocellular environments. For more details see Glisic [1]. The analysis is based on data reported from a measurement campaign in typical urban, bad urban and suburban (SU) [3] areas. In the experiment a base station (BS) equipped with an eight-element uniform linear antenna array and a mobile station (MS) with an omnidirectional dipole antenna are used. The MS is equipped with a differential global positioning system (GPS) and an accurate position encoder so its location is accurately known by combining the information from these two devices. MS displacements of less than 1cm can, therefore, be detected. The system is designed for transmission from the MS to the BS. Simultaneous channel sounding is performed on all eight branches, which makes it possible to estimate the azimuth of the impinging waves at the BS. The sounding signal is a maximum length linear shift register sequence of length 127 chips, clocked at a chip-rate of 4.096 Mbs. This chip rate was initially used in WCDMA proposals in Europe.The testbed operates at a carrier frequency of 1.8 GHz. Additional information regarding the stand-alone testbed can be found in Pedersen *et al.* [2], Frederiksen *et al.* [3] and Algans *et al.* [4]. A summary of macrocellular measurement environments is given in Table 3.1.

Channel azimuth-delay spread function at the BS is modeled as

$$h(\phi, \tau) = \sum_{l=1}^{L} \alpha_l \delta(\phi - \phi_l, \tau - \tau_l) \tag{3.1}$$

where the parameters α_l, τ_l and ϕ_l are the complex amplitude, delay and incidence azimuth of the lth impinging wave at the BS. In general, $h(\phi, \tau)$ is considered to be a time-variant function, since the constellation of the impinging waves is likely to change as the MS moves

Advanced Wireless Networks: 4G Technologies Savo G. Glisic
© 2006 John Wiley & Sons, Ltd.

Table 3.1 Summary of macro cellular measurement environments

Class	BS antenna height	Description of environment
TU, typical urban	10 and 32 m	The city of Aarhus, Denmark. Uniform density of buildings ranging from four to six floors. Irregular street layout. Measurements were carried out along six different routes with an average length of 2 km. No line-of-sight between MS and BS. MS–BS distance varies from 0.2 to 1.1 km
TU	21 m	Stockholm city, Sweden (area 1). Heavily built-up area with a uniform density of buildings, ranging from four to six floors. Ground is slightly rolling. No line-of-sight between MS and BS. MS–BS distance varies from 0.2 to 1.1 km
BU, bad urban	21 m	Stockholm city, Sweden (area 2). Mixture of open flat areas (river) and densely built up zones. Ground is slightly rolling. No line-of-sight between MS and BS. MS–BS distance varies from 0.9 to 1.6 km
SU suburban	12 m	The city of Gistrup, Denmark. Medium-sized village with family houses of one to two floors and small gardens with trees and bushes. Typical Danish residential area. The terrain around the village is rolling with some minor hills. No line-of-sight between MS and BS. MS–BS distance varies from 0.3 to 2.0 km

along a certain route. The local average power azimuth-delay spectrum is given as

$$P(\phi, \tau) = E\left\{\sum_{l=1}^{L} |\alpha_l|^2 \,\delta(\phi - \phi_l, \tau - \tau_l)\right\} \tag{3.2}$$

From Equation (3.2), local power azimuth spectrum (PAS) and the local power delay spectrum (PDS) are given as

$$P_{\mathrm{A}}(\phi) = \int P(\phi, \tau)\, \mathrm{d}\tau \tag{3.3}$$

$$P_{\mathrm{D}}(\tau) = \int P(\phi, \tau)\, \mathrm{d}\phi \tag{3.4}$$

The radio channels' local azimuth spread (AS) σ_{A} and the local delay spread (DS)σ_{D} are defined as the root second central moments of the corresponding variables. The values of the local AS and DS are likely to vary as the MS moves within a certain environment. Hence, we can characterize σ_{A} and σ_{D} as being random variables, with the joint pdf $f(\sigma_{\mathrm{A}}, \sigma_{\mathrm{D}})$.

Their individual PDFs are

$$f_{\mathrm{A}}(\sigma_{\mathrm{A}}) = \int f(\sigma_{\mathrm{A}}, \sigma_{\mathrm{D}})\, \mathrm{d}\sigma_{\mathrm{D}} \tag{3.5}$$

$$f_{\mathrm{D}}(\sigma_{\mathrm{D}}) = \int f(\sigma_{\mathrm{A}}, \sigma_{\mathrm{D}})\, \mathrm{d}\sigma_{A} \tag{3.6}$$

The function $f(\sigma_{\mathrm{A}}, \sigma_{\mathrm{D}})$ can be interpreted as the global joint PDF of the local AS and DS. If the expectation in Equation (14.2) is computed over the radio channel's fast fading component, we can furthermore apply the approximation

$$\int\int P(\phi, \tau)\, \mathrm{d}\phi\, \mathrm{d}\tau = h_{\mathrm{channel}} \cong h_{\mathrm{loss}}(d)\, h_{\mathrm{s}} \tag{3.7}$$

where h_{channel} is the radio channel's integral path loss, $h_{\mathrm{loss}}(d)$ is the deterministic long-term distance-dependent path loss, while h_{S} is the channel's shadow fading component, which is typically modeled by a log–normal distributed random variable [5.6]. The global PDF of h_{S} is denoted $f_{\mathrm{S}}(h_{\mathrm{S}})$. The global degree of shadow fading is described by the root second central moment of the random shadow fading component expressed in decibel, i.e.

$$\sigma_{\mathrm{S}} = \mathrm{Std}\left\{10 \log_{10}(h_{\mathrm{S}})\right\} \tag{3.8}$$

where Std {} denotes the standard deviation. Empirical results for cumulative distribution functions (CDF) for σ_{A} and σ_{D} are given in Figures 3.1 and 3.2, respectively. The log–normal fit for σ_{A} results is given as

$$\sigma_{\mathrm{A}} = 10^{\varepsilon_{\mathrm{A}} X + \mu_{\mathrm{A}}} \tag{3.9}$$

where X is a zero-mean Gaussian distributed random variable with unit variance, $\mu_{\mathrm{A}} = E\left\{\log_{10}(\sigma_{\mathrm{A}})\right\}$ is the global logarithmic mean of the local AS, and $\varepsilon_{\mathrm{A}} = \mathrm{Std}\left\{\log_{10}(\sigma_{\mathrm{A}})\right\}$ is the logarithmic standard deviation of the AS.

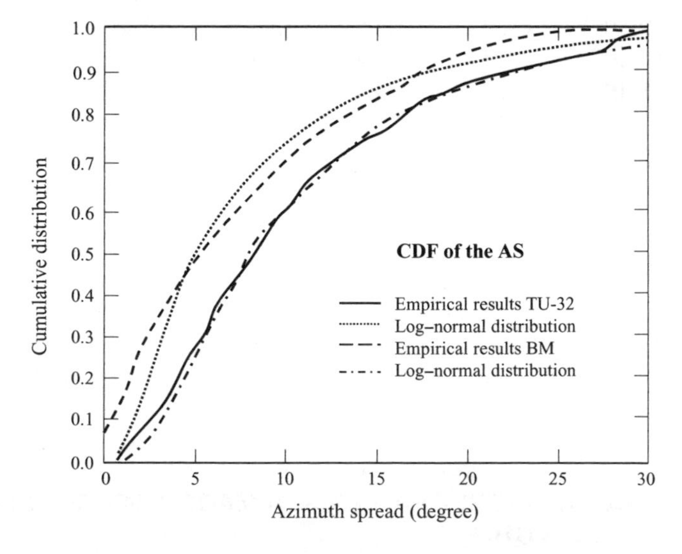

Figure 3.1 Examples of empirical CDF of AS obtained in different environments. The CDF of a log–normal distribution is fitted to the empirical results for comparison. (Reproduced by permission of IEEE [4].)

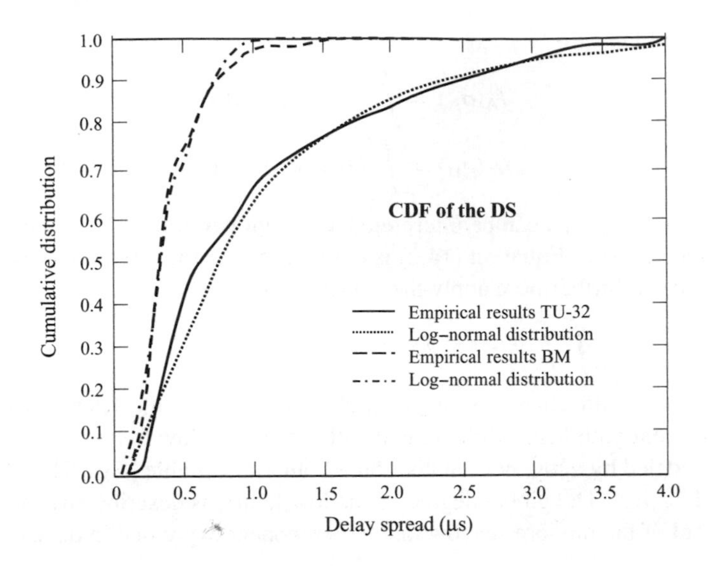

Figure 3.2 Examples of empirical CDFs of the DS in different environments. The CDF of a log–normal distribution is fitted to the empirical results for comparison. (Reproduced by permission of IEEE [4].)

Table 3.2 Summary of the first and second central moments of the AS, DS, and shadow fading in the different environments (reproduced by permission of IEEE [4])

Class	σ_S	$E\{\sigma_A\}$	μ_A	ε_A	$E\{\sigma_D\}$	μ_D	ε_D
TU-32	7.3 dB	8^0	0.74	0.47	0.8 μs	−6.20	0.31
TU-21	8.5 dB	8^0	0.77	0.37	0.9 μs	−6.13	0.28
TU-21	7.9 dB	13^0	0.95	0.44	1.2 μs	−6.08	0.35
BU	10.0 dB	7^0	0.54	0.60	1.7 μs	−5.99	0.46
SU	6.1 dB	8^0	0.84	0.31	0.5 μs	−6.40	0.22

Similarly,

$$\sigma_D = 10^{\varepsilon_D Y + \mu_D} \tag{3.10}$$

where Y is a zero-mean Gaussian distributed random variable with unit variance, $\mu_D = E\left\{\log_{10}(\sigma_D)\right\}$ is the global logarithmic mean of the local DS, and $\varepsilon_D = \text{Std}\left\{\log_{10}(\sigma_D)\right\}$ is the logarithmic standard deviation of the DS. A summary of the results for these parameters is given in Table 3.2. For characterization of shedowing fading see References [1–30].

3.2 URBAN SPATIAL RADIO CHANNELS IN MACRO/MICROCELL ENVIRONMENT (2.154 GHz)

The discussion in this section is based on the experimental results collected with a wideband channel sounder using a planar antenna array [31]. The signal center frequency was 2154 MHz and the measurement bandwidth was 100 MHz. A periodic PN-sequence, 255 chips

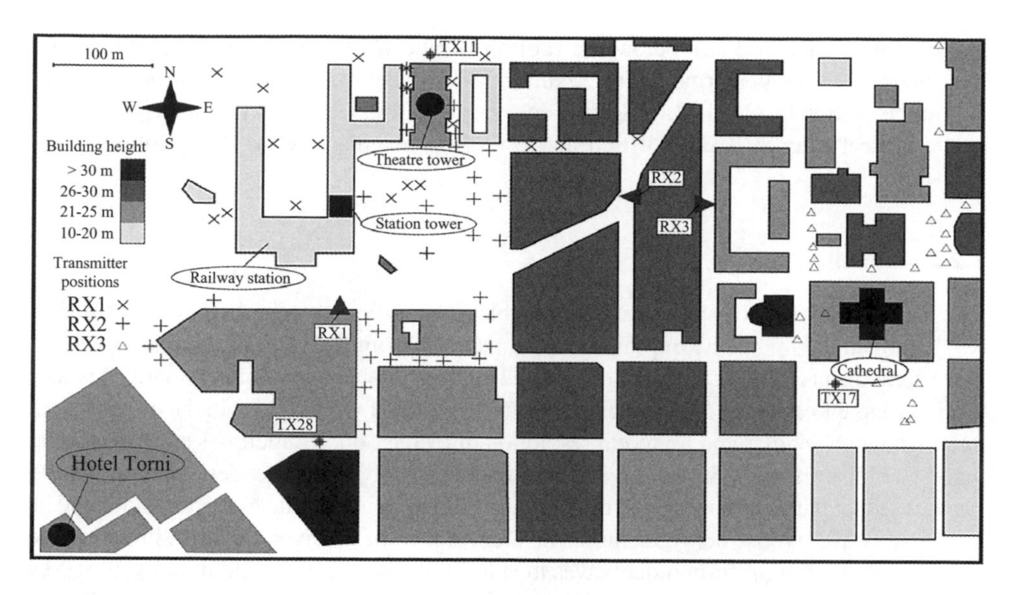

Figure 3.3 The measurement area with all three RX sites; TX-positions of the sample plots are marked. (Reproduced by permission of IEEE [31].)

long, was used. The chip rate was 30 MHz, the sampling rate 120 MHz, giving a oversampling factor 4. The correlation technique was used for the determination of the impulse response. Hence, the delay range was 255/30 MHz = 8.5 μs, with a resolution of 1/30 MHz = 33 ns. The transmit antenna at the MS was a vertically polarized omnidirectional discone antenna. The vertical 3 dB beamwidth was 87° and the transmit power 40 dBm. Approximately 80 different transmitter positions were investigated.

The receiving BS was located at one of three different sites below, at and above the rooftop level (RX1–RX3, see Figure 3.3). A 16-element physical array with dual polarized $\lambda/2$-spaced patch antennas was combined with a synthetic aperture technique to build a virtual two-dimensional (2-D) antenna structure. The patches were linearly polarized at 0° (horizontal direction) and 90° (vertical direction). With these 16 × 62 elements the direction of arrival (DOA) of incoming waves both in azimuth (horizontal angle) and elevation (vertical angle) could be resolved using the super-resolution Unitary ESPRIT algorithm [32–34]. Note that the number of antenna elements limits the number of identifiable waves, but not the angular resolution of the method. Together with a delay resolution of 33 ns, the radio channel can be characterized in all three dimensions separately for the two polarizations. Array signal processing, including estimation of the DOAs and a comparison of ESPRIT with other algorithms, can be found in Glisic [1].

One prerequisite for the applicability of the synthetic aperture technique is that the radio channel is static during the whole data collection period. To avoid problems, the whole procedure was done at night with minimum traffic conditions.

3.2.1 Description of environment

A typical urban environment is shown in Figure 3.3 [31] with three receiver locations (RX1–RX3) marked by triangles pointing in the broadside direction of the array. Figure 3.3 also shows all the corresponding TX positions. The location RX 1 (height $h_{\mathrm{RX}} = 10$ m) was a typical microcell site below the rooftop height of the surrounding buildings, and measurements were performed with 20 different TX positions. RX 2 (height $h_{RX}-27$ m) was at the rooftop level, and 32 TX positions were investigated. RX 3 (height $h_{RX} = 21$ m) was a typical macrocell BS position above rooftop heights, and 27 TX positions were measured.

3.2.2 Results

The measurement results show that it is possible to identify many single (particular, different) multipath components, impinging at the receiver from different directions. However, these components are not randomly distributed in the spatial and temporal domain; they naturally group into clusters. These clusters can be associated with objects in the environment due to the high angular and temporal resolution of evaluation. (Sometimes even individual waves, within a cluster, can be associated with scattering objects.) The identification of such clusters is facilitated by inspection of the maps of the environment. A cluster is defined as a group of waves whose delay, azimuth and elevation at the receiver are very similar, while being notably different from other waves in at least one dimension. Additionally all waves inside a cluster must stem from the same propagation mechanism. The definition of clusters always involves a certain amount of arbitrariness. Even for mathematically 'exact' definitions, arbitrary parameters (e.g. thresholds or number of components) must be defined. Clustering by human inspection, supported by maps of the environment, seems to give the best results. The received power is calculated within each cluster (cluster power) by means of unitary ESPRIT and a following beam-forming algorithm. The results are plotted in the azimuth-elevation-, azimuth-delay- and elevation-delay-planes

According to the obvious propagation mechanism, each cluster is assigned to one of three different classes:

- *Class 1, street-guided propagation* – waves arrive at the receiver from the street level after traveling through street canyons.
- *Class 2, direct propagation over the rooftop* – the waves arrive at the BS from the rooftop level by diffraction at the edges of roofs, either directly or after reflection from buildings surrounding the MS. The azimuth mostly points to the direction of the transmitter with some spread in azimuth and delay.
- *Class 3, reflection from high-rise objects over the rooftop* – the elevation angles are near the horizon, pointing at or above the rooftop. The waves undergo a reflection at an object rising above the average building height before reaching the BS. The azimuth shows the direction of the reflecting building; the delay is typically larger than for class 1 or class 2. The sum of the powers of all clusters belonging to the same class is called class power. In some cases the propagation history is a mixture of different classes, e.g. street guidance followed by diffraction at rooftops. Such clusters are allocated to the class of the final path to the BS.

For the evaluation of delays we define the vector **P** containing the powers of the clusters and vector $\boldsymbol{\tau}$ of corresponding mean delays. A particular cluster i has mean delay τ_i, and power P_i. The relation between the delays $\boldsymbol{\tau}$ and the powers **P** is modeled as exponential:

$$P_n \propto P(\tau_n) = a\mathrm{e}^{-\tau_n/b} \tag{3.11}$$

Table 3.3 The model parameters a and b for both received polarizations (VP and HP) averaged over all available clusters. The transmitter was VP

	VP-VP	VP-HP
a	-3.9 dB	-3.6 dB
b	8.9 dB/μs	11.8 dB/μs

Table 3.4 Average delay of the strongest cluster of RX1, RX2 and RX3

RX	Average delay VP-VP μs	Average delay VP-HP μs
1	0.068	0.071
2	0.38	0.28
3	0.11	0.048

Experimental data are fit into the model (3.11) using the least square (LS) estimation. The logarithmic estimation error $\mathbf{v}$ is defined as

$$\mathbf{v} = 10 \log \mathbf{P} - 10 \log \mathbf{s}(\theta) \tag{3.12}$$

and its standard deviation σ_v as

$$\sigma_v = \sqrt{\text{var}\{\mathbf{v}\}} \tag{3.13}$$

The summary of the results for parameters a and b is shown in Tables 3.3 and 3.4.

$$P_i = a e^{-\tau_i/b}$$

In equation (14.21) σ_v was defined as the standard deviation of the logarithmic estimation error in dB. This estimation error was found to be log–normally distributed and, up to a delay of about 1 μs, σ_v is independent of the delay τ. The value of σ_v is 9.0 and 10.0 dB (co-and cross-polarization), respectively, averaged over the first microsecond. The average powers for different classes of clusters are shown in Table 3.5. Additional data on the topic can be found in References [31–41].

3.3 MIMO CHANNELS IN MICRO- AND PICOCELL ENVIRONMENT (1.71/2.05 GHz)

The model presented in this section is based upon data collected in both picocell and microcell environments [43]. The stochastic model has also been used to investigate the capacity of MIMO radio channels, considering two different power allocation strategies, water filling and uniform and two different antenna topologies, 4×4 and 2×4. It will be demonstrated that the space diversity used at both ends of the MIMO radio link is an efficient technique in picocell environments, achieving capacities within 14 and 16 b/s/Hz in 80 % of the cases for a 4×4 antenna configuration implementing water filling at a signal-to-noise ratio (SNR) of 20 dB.

Table 3.5 Averaged class powers of RX1, RX2 and RX3

RX	Class	Horizontal power, percentage of total power	Vertical power, percentage of total power
1	1	96.5 %	95.7 %
	2	2.4 %	3.8 %
	3	1.1 %	0.4 %
2	1	93.5 %	97.2 %
	2	4.0 %	2.7 %
	3	2.5 %	0.1 %
3	1	46.7 %	78.0 %
	2	37.2 %	12.8 %
	3	16.0 %	9.2 %

The basic parameters of the measurements set-up are shown in Figure 3.4. The following notaion is used in the figure: $d_{\mathrm{MS-DS}}$ stands for distance between MS and BS; $h_{\mathrm{BS}}$for the height of BS above ground floor, and AS for azimuth spread [43].

The vector of received signals at BS can be represented as $\mathbf{y}(t) = [y_1(t), y_2(t), \ldots, y_M(t)]^{\mathrm{T}}$, where $y_m(t)$ is the signal at the mth antenna port and $[\cdot]^{\mathrm{T}}$ denotes transposition. Similarly, the signals at the MS are $\mathbf{s}(t) = [s_1(t), s_2(t), \ldots s_N(t)]^{\mathrm{T}}$. The NB MIMO radio channel $\mathbf{H} \in X^{M\times N}$, which describes the connection between the MS and the BS, can be

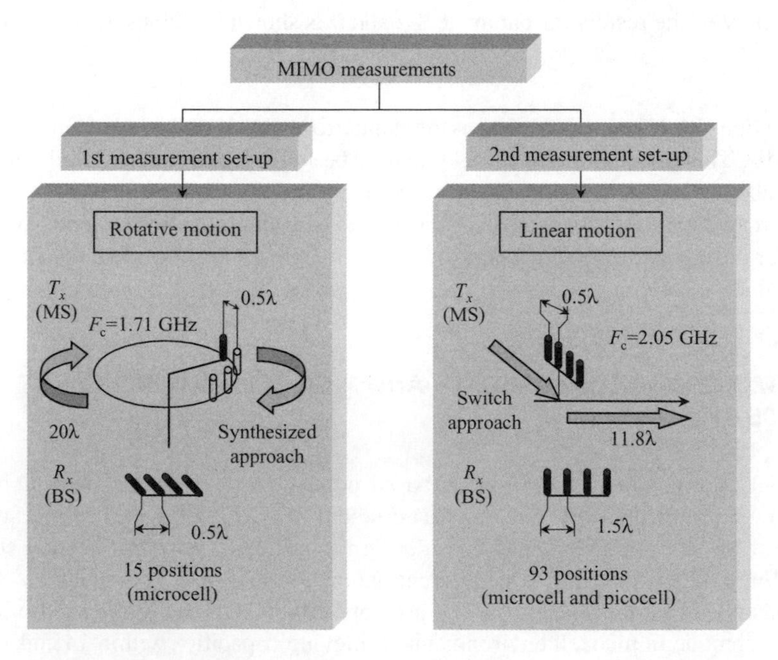

Figure 3.4 Functional sketch of the MIMO model. (Reproduced by permission of IEEE [43].)

expressed as [1]:

$$\mathbf{H} = \begin{bmatrix} \alpha_{11} & \alpha_{12} & \cdots & \alpha_{1N} \\ \alpha_{21} & \alpha_{22} & \cdots & \alpha_{2N} \\ \vdots & \vdots & \ddots & \vdots \\ \alpha_{M1} & \alpha_{M2} & \cdots & \alpha_{MN} \end{bmatrix} \tag{3.14}$$

where α_{mn} is the complex transmission coefficient from antenna at the MS to antenna at the BS. For simplicity, it is assumed that α_{mn} is complex Gaussian distributed with identical average power. However, this latest assumption can be easily relaxed. Thus, the relation between the vectors $\mathbf{y}(t)$ and $\mathbf{s}(t)$ can be expressed as

$$\mathbf{y}(t) = \mathbf{H}(t)\mathbf{s}(t) \tag{3.15}$$

In the sequel we will use the following correlations:

$$\rho^{\mathrm{BS}}_{m_1 m_2} = \langle \alpha_{m_1 n}, \alpha_{m_2 n} \rangle \tag{3.16}$$

$$\rho^{\mathrm{MS}}_{n_1 n_2} = \langle \alpha_{m n_1}, \alpha_{m n_2} \rangle \tag{3.17}$$

$$\mathbf{R}_{\mathrm{BS}} = \begin{bmatrix} \rho^{\mathrm{BS}}_{11} & \rho^{\mathrm{BS}}_{12} & \cdots & \rho^{\mathrm{BS}}_{1M} \\ \rho^{\mathrm{BS}}_{21} & \rho^{\mathrm{BS}}_{22} & \cdots & \rho^{\mathrm{BS}}_{2M} \\ \vdots & \vdots & \ddots & \vdots \\ \rho^{\mathrm{BS}}_{M1} & \rho^{\mathrm{BS}}_{M2} & \cdots & \rho^{\mathrm{BS}}_{MM} \end{bmatrix}_{M \times M} \tag{3.18}$$

$$\mathbf{R}_{\mathrm{MS}} = \begin{bmatrix} \rho^{\mathrm{MS}}_{11} & \rho^{\mathrm{MS}}_{12} & \cdots & \rho^{\mathrm{MS}}_{1N} \\ \rho^{\mathrm{MS}}_{21} & \rho^{\mathrm{MS}}_{22} & \cdots & \rho^{\mathrm{MS}}_{2N} \\ \vdots & \vdots & \ddots & \vdots \\ \rho^{\mathrm{MS}}_{N1} & \rho^{\mathrm{MS}}_{N2} & \cdots & \rho^{\mathrm{MS}}_{NN} \end{bmatrix}_{N \times N} \tag{3.19}$$

The correlation coefficient between two arbitrary transmission coefficients connecting two different sets of antennas is expressed as

$$\rho^{n_1 m_1}_{n_2 m_2} = \langle \alpha_{m_1 n_1}, \alpha_{m_2 n_2} \rangle \tag{3.20a}$$

which is equivalent to

$$\rho^{n_1 m_1}_{n_2 m_2} = \rho^{MS}_{n_1 n_2} \rho^{BS}_{m_1 m_2} \tag{3.20b}$$

provided that Equations (14.24a) and (14.24b) are independent of n and m, respectively. In other words, this means that the spatial correlation matrix of the MIMO radio channel is the Kronecker product of the spatial correlation matrix at the MS and the BS and is given by

$$\mathbf{R}_{\mathrm{MIMO}} = \mathbf{R}_{\mathrm{MS}} \otimes \mathbf{R}_{\mathrm{BS}} \tag{3.21}$$

where $\otimes$ represents the Kronecker product. This has also been confirmed in yu *et al.* [44].

3.3.1 Measurement set-ups

The TX is at the MS and the stationary RX is located at the BS. The two set-ups from Figure 3.4 provide measurement results with different correlation properties of the MIMO channel for small antenna spacings of the order of 0.5λ or 1.5λ . The BS consists of four parallel RX channels. The sounding signal is a MSK-modulated linear shift register sequence of a length of 127 chips, clocked at a chip rate of 4.096 Mcs. At the RX, the channel sounding is performed within a window of 14.6 μs, with a sampling resolution of 122 ns (half-chip period) to obtain an estimate of the complex IR. The narowband (NB) information is subsequently extracted by averaging the complex delayed signal components. A more thorough description of the stand-alone testbed (i.e. RX and TX) is documented in References [23, 45]. The description of the measurement environments is summarized in Table 3.6.

A total of 107 paths were investigated within these seven environments. The first measurement set-up was used to investigate 15 paths in a microcell environment, i.e. environment A in Table 3.6. The MS was positioned in different locations inside a building while the BS was mounted on a crane and elevated above roof-top level (i.e. 9 m) to provide direct line-of-sight to the building. The antenna was located 300 m away from the building. The second set-up was used to investigate 92 paths for both microcell and picocell environments,

Table 3.6 Summary and description of the different measured environments (reproduced by permission of IEEE [43])

Cell type	Environment	MS locations	Measurement set-up	Description
Microcell	A	15	1st	The indoor environment consists of small offices with windows metallically shielded –300 m between MS and BS
	B	13	2nd	The indoor environment consists of small offices –31–36 m between MS and BS
Picocell	C	21	2nd	The indoor environment is the same as in A
	D	12	2nd	Reception hall – large open area
	E	18	2nd	Modern open office with windows metallicaly shielded
	F	16	2nd	The indoor environment is the same as in B
	G	12	2nd	Airport – very large indoor open area

i.e. environment B and C–G, respectively, as shown in Table 3.6. The distance between the BS and the MS was 31–36 m for microcell B, with the BS located outside.

3.3.2 The eigenanalysis method

The eigenvalue decomposition (EVD) of the instantaneous correlation matrix $\mathbf{R} = \mathbf{H}\mathbf{H}^{\mathrm{H}}$(not to be confused with $\mathbf{R}_{\mathrm{MIMO}}$), where $[\cdot]^{\mathrm{H}}$ represents Hermitian transposition, can serve as a benchmark of the validation process. The channel matrix $\mathbf{H}$ may offer K parallel subchannels with different mean gains, with $K = \mathrm{Rank}\,(\mathbf{R}) \leq \min\,(M, N)$ where the functions Rank $(\cdot)$ and min $(\cdot)$ return the rank of the matrix and the minimum value of the arguments, respectively [27]. The kth eigenvalue can be interpreted as the power gain of the kth subchannel [27]. In the following, λ_k represents the eigenvalues.

3.3.3 Definition of the power allocation schemes

In the situation where the channel is known at both TX and RX and is used to compute the optimum weight, the power gain in the kth subchannel is given by the kth eigenvalue, i.e. the SNR for the kth subchannel equals

$$\gamma_k = \lambda_k \frac{P_k}{\sigma_{\mathrm{N}}^2} \tag{3.22}$$

where P_k is the power assigned to the kth subchannel, λ_k is the kth eigenvalue and σ_{N}^2 is the noise power. For simplicity, it is assumed that $\sigma_{\mathrm{N}}^2 = 1$. According to Shannon, the maximum capacity normalized with respect to the bandwidth (given in terms of b/s/Hz-spectral efficiency) of parallel subchannels equals [46]

$$C = \sum_{k=1}^{K} \log_2 (1 + \gamma_k) \tag{3.23}$$

$$= \sum_{k=1}^{K} \log_2 \left(1 + \lambda_k \frac{P_k}{\sigma_N^2}\right) \tag{3.24}$$

where the mean SNR is defined as

$$\mathrm{SNR} = \frac{E\,[P_{\mathrm{RX}}]}{\sigma_{\mathrm{N}}^2} = \frac{E\,[P_{\mathrm{TX}}]}{\sigma_{\mathrm{N}}^2} \tag{3.25}$$

Given the set of eigenvalues $\{\lambda_k\}$, the power P_k allocated to each subchannel k was determined to maximize the capacity using Gallager's water filling theorem [27] such that each subchannel was filled up to a common level D, i.e.

$$\frac{1}{\lambda_1} + P_1 = \cdots = \frac{1}{\lambda_K} + P_k = \cdots D \tag{3.26}$$

with a constraint on the total TX power such that

$$\sum_{k=1}^{K} P_k = P_{\mathrm{TX}} \tag{3.27}$$

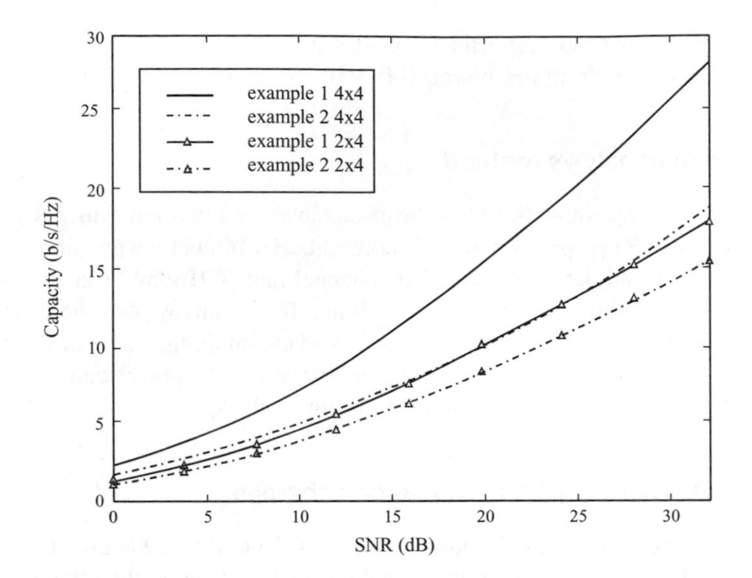

Figure 3.5 Capacity (10 % level) vs SNR for example 1 (picocell decorrelated) and example 2 (microcell-correlated). The water filling power allocation scheme.

where P_{TX} is the total transmitted power. This means that the subchannel with the highest gain was allocated the largest amount of power. In the case where $1/\lambda_k > D$ then $P_k = 0$. When the uniform power allocation scheme is employed, the power P_k is adjusted according to$P_1 = \cdots = P_K$. Thus, in the situation where the channel is unknown, the uniform distribution of the power is applicable over the antennas [27] so that the power should be equally distributed between the N elements of the array at the TX, i.e.

$$P_n = \frac{P_{TX}}{N}, \quad \forall n = 1 \ldots N \tag{3.28}$$

Some results are given in Figure 3.5.

3.4 OUTDOOR MOBILE CHANNEL (5.3 GHz)

In this section we discuss the mobile channel at 5.3 GHz. The discussion is based on measurement results collected at six different sites [47]. *Site A* is an example of a dense urban environment; the transmitting antenna was about 45 m above ground level, representing a case with the BS antenna over rooftops. *Site B* is a dense urban residential environment. Here the transmitting antenna was placed at a mast with a height of 4 m, which is a typical case with the BS antenna lower than rooftops. The measurement routes for this site are shown in Figure 3.6(a). The receiving antenna mobile station was at a height of 2.5 m on top of a car for both of the sites mentioned above. *Site C* is located in a typical city center. The goal was to place the transmitter at some elevation relative to ground, but still keep it below the rooftops. The transmitting antenna was placed at a height of 12 m and the receiving antenna was on top of a trolly with a height of 2 m above ground level. In site C, the rotation was measured using a directive horn antenna. The 3 dB beamwidth of the

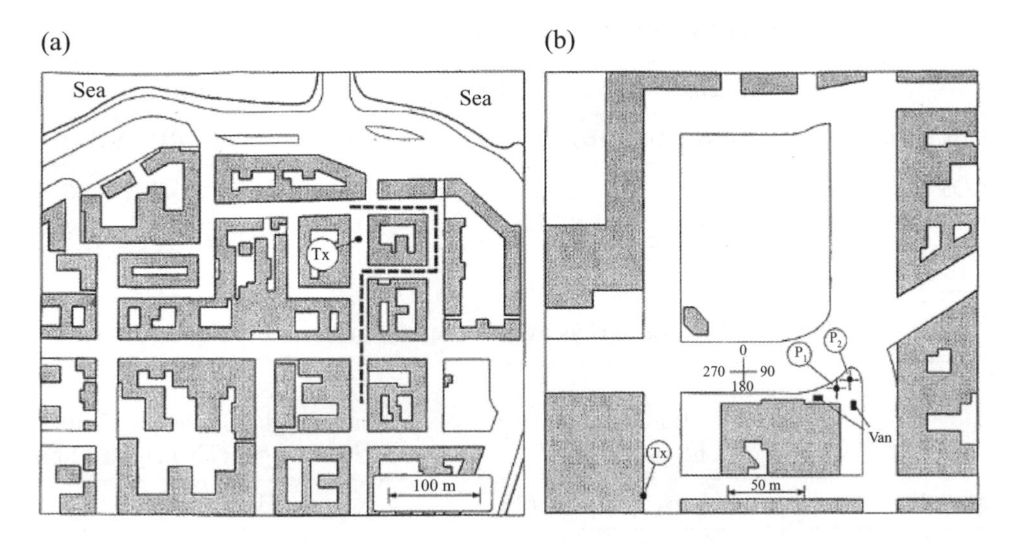

Figure 3.6 (a) Measurements routes for site B with TX height of 4 m. (b) Rotation measurements in an urban environment. (Reproduced by permission of IEEE [47].)

horn antenna was 30° in the H-plane and 37° in the E-plane and the peak sidelobe level was 26 dB. The specific environment for rotation measurements is shown in Figure 3.6(b). *Site D* represents semiurban/semirural residential area. The three-story buildings are the tallest ones around, and the transmitting antenna was placed over rooftops at the height of 12 m from ground level. *Site E* was selected to represent the rural case. The transmitting antenna was placed on top of a 5 m mast at the hilltop so that the antenna was about 55 m above the surrounding area. *Site F* represents a typical semiurban/urban case. The transmitter was placed on top of a 5 m mast. The receiving antenna was always on top of a car at a height of 2.5 m. The routes were measured using the wideband channel sounder. System parameters are summarized in Table 3.7

Table 3.7 System configuration for mobile measurements in urban (U), suburban (S) and rural (R) areas

Receiver	Direct sampling/5.3 GHz
Transmitter power	30 dBm
Chip frequency	30 MHz
Delay range	4.233 μs
Doppler range	124 Hz (U), 62 Hz (S, R)
Measurement rate	248 sets/s (U), 124 sets/s (S, R)
Sampling frequency	120 Ms/s
IR/wavelength	5 (U), 4.2 (S,R)
Receiver velocity	2.80 m/s (U), 1.67 m/s (S, R)
Antennas and polarization	Omni-directional antenna with 1 dBi gain; vertical polarization

Table 3.8 Path loss models for urban environments

	TX height: 4 m			TX height: 12m			TX height: 45m		
Urban models	n	b (dB)	Std (dB)	n	b (dB)	Std (dB)	n	b (dB)	Std (dB)
LOS	1.4	58.6	3.7	2.5	35.8	2.9	3.5	16.7	4.6
NLOS	2.8	50.6	4.4	4.5	20.0	1.7	5.8	−16.9	2.8

Table 3.9 Path loss models for suburban and rural environments

	Rural TX height: 55 m			Suburban Line-of-sight (TX height: 5 m); no Line-of-sight (TX height: 12 m)		
Models	n	b (dB)	Std (dB)	n	b (dB)	std (dB)
LOS	3.3	21.8	3.7	2.5	38.0	4.9
NLOS	5.9	−27.8	1.9	3.4	25.6	2.8

3.4.1 Path loss models

The following model is used for path loss

$$PL\,(\text{dB}) = b + 10n\log_{10} d \tag{3.29}$$

where $d_0 = 1$ m, n is attenuation exponent, b is the intercept point in the semilog coordinate, and $d(m)$ is the distance from receiver to transmitter. The measurement distances are about 30–300 m in this case. Some results are given in Tables 3.8 and 3.9. The measurements results for delays are summarized in Table 3.10

3.4.2 Path number distribution

The multipath number distribution was regarded as Poisson's and modified Poisson's distributions in Suzuki [48] and modified Poisson's distribution has been shown to have good agreement with the experimental results in some cases. However, the modified Poisson's distribution does not have an explicit expression, but just a process. Therefore, it is not convenient for practical use. In Zhao *et al.* [47], another simple and useful path number distribution was suggested by considering that the path number variation of radio waves in land mobile communications is a Markov process at finite state space, and it was shown to have good agreement with the experimental results. The path number distributions given by Poisson and Gao can be expressed as

$$P\,(N) = \frac{\eta^{N_\mathrm{T}-N}}{(N_\mathrm{T}-N)!}\mathrm{e}^{-\eta} \tag{3.30}$$

$$P\,(N) = C_{N_\mathrm{T}}^{N}\frac{\eta^{N_\mathrm{T}-N}}{(1+\eta)^{N_\mathrm{T}}} \tag{3.31}$$

Table 3.10 Measured values for mean excess delay and rms delay spread

( ) Tx height in meters			Urban		Suburban		Rural	
Mean excess delay (ns)			38	(4)				
	LOS		42	(12)	36	(5)	29	(55)
			102	(45)				
	NLOS		70	(4)	68	(12)		
rms delay spread (ns)			44	(4)				
	Mean	LOS	43	(12)	25	(5)	22	(55)
			88	(45)				
		NLOS	44	(4)	66	(12)		
			25	(4)				
	Median	LOS	31	(12)	13	(5)	15	(55)
			86	(45)				
		NLOS	37	(4)	63	(12)		
	CDF		93	(4)				
	<90 %	LOS	64	(12)	57	(5)	44	(55)
			120	(45)				
		NLOS	63	(4)	105	(12)		

where N is variable and C means combination. N_T is the maximum number of paths that the mobile can receive. The parameters η and N_T can be fitted by the experimental data. For Poisson's probability density function (PDF), the mean path number is $\langle N \rangle = \eta$. For Gao's PDF, the mean value is $\langle N \rangle = N_\mathrm{T}/(1+\eta)$. The empirical path number distributions for the outdoor measurements are fitted using Equations (3.30) and (3.31), respectively. The path numbers are obtained from measured data by counting the peaks of the power delay profiles. The best fit is obtained by minimizing the following standard deviation:

$$\mathrm{Std} = \sqrt{\frac{1}{N_\mathrm{T}} \sum_{i=1}^{N_\mathrm{T}} \left(p^i - p_\mathrm{e}^i\right)^2} \tag{3.32}$$

where p_e^i is the experimental probability corresponding to path number i and p^i is the fitted probability using Equations (3.30) and (3.31). The fitted parameters are available in Table 3.11. If the dynamic range is cut at different levels, for example, -25, -20 and -15 dB, the fitting parameters in Equations (3.30) and (3.31) will be changed, but the path number distributions still follow Poisson's and Gao's distributions.

3.4.3 Rotation measurements in an urban environment

The rotation measurements at points P_1 and P_2 were performed at site C shown in Figure 3.6(b). The transmitter height was 12 m and the receiver was on a rotating stand at the height of 1.6 m and close to the receiver is a large open square. In the experiments, large

Table 3.11 Path number distributions for outdoor environments

		Urban				Suburban		Rural
		TX 4 m		TX 12 m	TX 45 m	TX 12 m		TX 55 m
		LOS	NLOS	LOS	LOS	LOS	NLOS	LOS
η	Poisson	2.8	4.2	3.3	6.0	1.2	4.5	1.8
	Gao	4.7	4.5	3.5	2.7	9.0	3.3	4.0
	N_T	16	21	14	22	13	20	9
$\langle N \rangle$	Poisson	2.8	4.2	3.3	6.0	1.2	4.5	1.8
	Gao	2.8	3.8	3.2	6.0	1.3	4.7	1.8
	Experiment	3.4	4.2	3.5	6.2	2.4	5.0	1.7

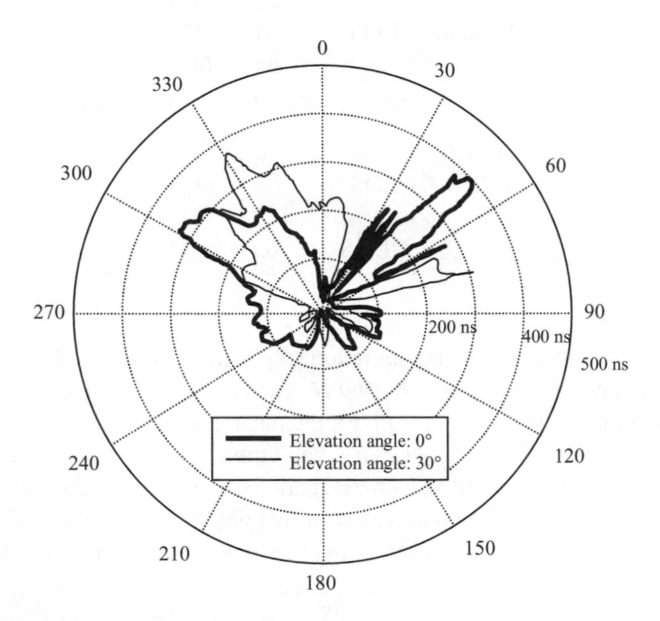

Figure 3.7 RMS delay spread with different rotation angles in the azimuth plane.

excess delays up to 1.2 μs and rms delay spread, shown in Figure 3.7, of about 0.42 μs are found.

The power angular profiles (PAPs) $P_r(\phi)$ of the measurements were calculated using the maximal ratio combining algorithm in the delay domain [50]:

$$P_r(\phi) = \alpha_{\text{cal}} \int_{\tau_{\min}}^{\tau_{\max}} |h(\tau, \phi)|^2 \, d\tau \tag{3.33}$$

where α_{cal} is a factor which is obtained from the calibration measurement with a cable and an attenuator, $\tau_{\min}$ and $\tau_{\max}$ are the delays of the first and last detectable IR components, and ϕ is the angle of arrival of the waves in the azimuth plane. In the rotation measurements, the dynamic range is cut at -26 dB relative to the strongest path. Some results for angular profiles of relative received power are shown in Figure 3.8. The number of paths as a function of asimuth angle is shown in Figure 3.9.

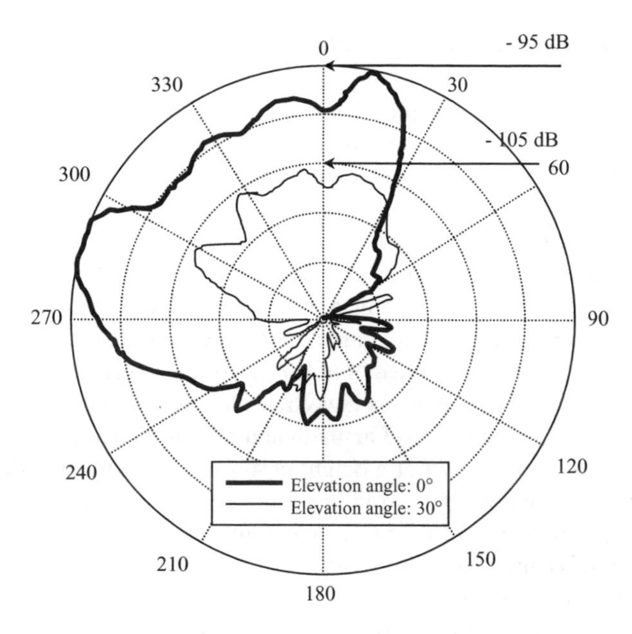

Figure 3.8 Angular profiles of relative received power. (Reproduced by permission of IEEE [47].)

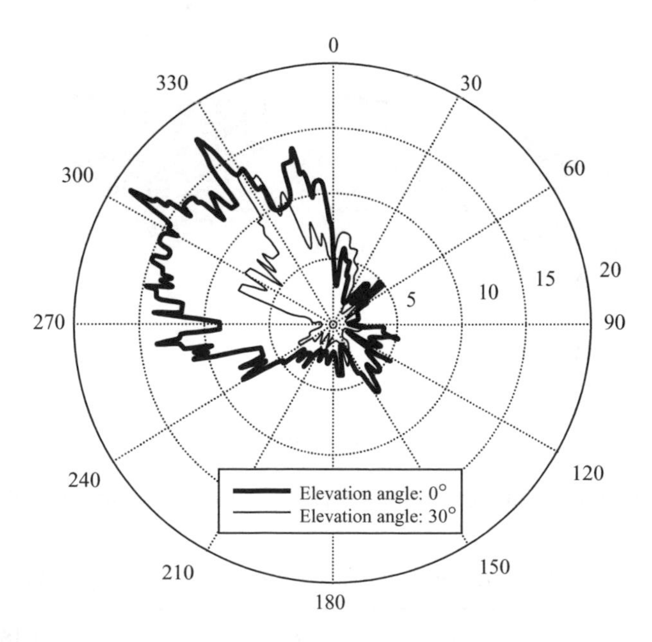

Figure 3.9 Numbers of paths with different rotation angles in the azimuth plane. (Reproduced by permission of IEEE [47].)

3.5 MICROCELL CHANNEL (8.45 GHz)

In this section spatio-temporal channel characterization in a suburban non-line-of-sight microcellular environment is discussed. Figure 3.10 shows a map of the environment under consideration [49]. This is a residential area with predominantly wooden houses of 8 m average height, considered to be a typical suburban microcellular environment of size $600 \times 600\,\text{m}^2$. The traffic was very light and the environment was considered to be static throughout the experiments. The non-line-of-sight (NLOS) transmitter and receiver shown in Figure 3.10 are considered. The distance between the transmitter and the receiver was 219 m. As a transmitter antenna, simulating the MS, a vertically polarized omnidirectional half-wave sleeve dipole was set at a height of 2.7 m. At the receiving point, corresponding to the BS, the azimuth-delay profile was measured. This was done by rotating a vertically polarized parabolic antenna which had azimuth and elevation beamwidths of about 4°, at an azimuth step of 4°, and was set at a height of 4.4 m. At each azimuth step, the delay profile was measured using a delay profile measurement system described in References [49, 50]. At a center frequency of 8.45 GHz and with a chip rate of 50 Mcs, a seven-stage M sequence with a dynamic range of 42 dB was transmitted. A correlation receiver was used to produce the delay profile.

The electric parameters used in the ray-tracing simulation are presented in Table 3.12. The wooden houses are modeled by the concrete, as the surfaces of the houses were covered with concrete-like paint. To compare with the experimental results, the directivity of the antenna and the autocorrelation of the pseudo-random noise (PN) sequence were convolved with the result of the ray-tracing simulation. Therefore, the path gain includes the antenna gain.

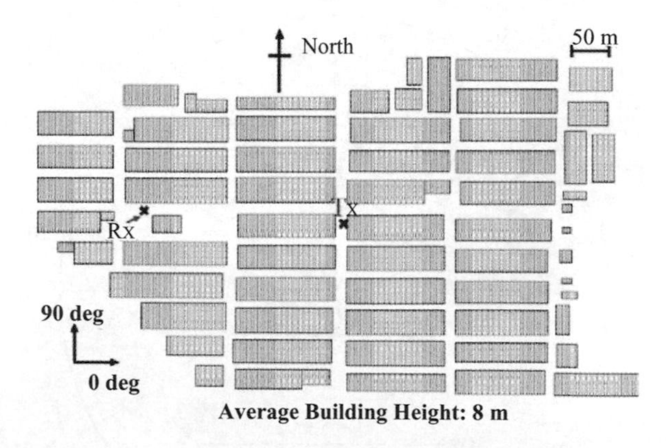

Figure 3.10 The microcellular environment. (Reproduced by permission of IEEE [49].)

Table 3.12 Electrical parameters used in the ray-tracing simulation

	ε_r	σ(s/m)
Concrete [16]	5.5	0.023
Foliage [15]	1.2	0.0003
Ground [18]	15.0	1.3
Metal	—	∞

3.5.1 Azimuth profile

Azimuth profiles are obtained by summing up the power of azimuth-delay profiles with respect to the delay time. The forward arrival waves within the range from −40 to 44° were observed in both experimental and simulation data sets. The experimental profile had a floor level of about 30 dB below the peak, but this level was low enough so that its effect on the transmission property was negligibly small.

3.5.2 Delay profile for the forward arrival waves

For the forward arrival waves (−40 to 44°), the delay profiles are obtained by summing up the azimuth-delay profiles with respect to the azimuth. The experimental result exhibits an exponential decay. The results of least squares fitting can be expressed as

$$P(\tau) = -0.038\tau - 28.6 \tag{3.34}$$

where P is the path gain in dB, and τ is the delay time in nanoseconds.

3.5.3 Short-term azimuth spread for forward arrival waves

The short-term AS for the forward arrival waves, σ_φ, is defined as

$$\sigma_\varphi(\tau) = \sqrt{\langle\varphi^2(\tau)\rangle - \langle\varphi(\tau)\rangle^2} \tag{3.35}$$

where $\{\cdot\}$ is the average of the φ weighted by the power for a fixed delay time, τ. At each delay time, the threshold level is set at 30 dB below the peak in the profile in order to calculate the short-term AS in Equation (3.35). The CDF of the short-term AS is shown in Figure 3.11. The solid line indicates the experimental result and the dotted line indicates the simulation result. It is noted that the simulation result is obtained within a range of

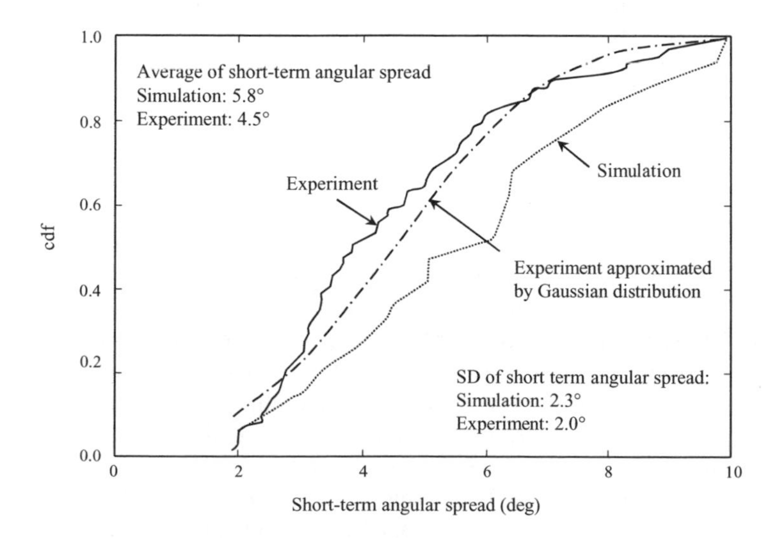

Figure 3.11 The CDF of the short-term AS for forward arrival waves. Solid line: experiment; dotted line: simulation. Reproduced by permission of IEEE [49].)

delay times from 700 to 880 ns. Although the distribution functions of the experiment and the simulation look slightly different, their average and their standard deviation are both in agreement. A Gaussian distribution has been used as an approximation in the figure.

3.6 WIRELESS MIMO LAN ENVIRONMENTS (5.2 GHz)

The presentation in this section is based on results of a measurement campaign [52] in two courtyards in the 5.2 GHz band assigned for wireless LANs [e.g. HIPERLAN (see www.etsi.org) or IEEE 802.11a]. These standards specify wireless communication between computers, which is a compelling application for MIMO systems. For measurement a channel sounder with a bandwidth of 120 MHz, connected via a fast RF switch to a uniform linear receiver antenna array, was used. This array consisted of $N_R = 8$ antenna elements ($\pm 60^\circ$ element-beamwidth), plus two dummy elements at each end of the array. All these components together constituted a single-directional channel sounder. A virtual array at the transmitter consisted of a monopole antenna mounted on an *X*–*Y*-positioning device with stepping motors.

The experiment started by positioning a transmit antenna at a certain position. At the receiver, the RF switch was connected to the first antenna element of the array, so that the transfer function (measured at 192 frequency samples) from the first transmit to the first receive element of the array was measured. Then, the switch was connected to the next receive antenna element, and the next transfer function was measured. The measurement of all those transfer functions was repeated 256 times, in order to assess the time variance of the channel (see below). Then, the transmit antenna was moved to the next position, and the procedure was repeated. $N_T = 16$ transmit antenna positions were used situated on a cross (i.e. eight positions on each axis of the cross) and bursts of complex channel transfer functions were recorded. Any virtual array requires that the channel remains static during the measurement period. One complete measurement run (2×8 antenna positions at TX $\times$ 8 spatial samples at RX $\times$ 192 frequency samples and 256 temporal samples gives $16 \times 8 \times 192 \times 256 = 6\,291\,456$ complex samples) took about 5 min.

3.6.1 Data evaluation

Starting from the four-dimensional transfer function (time, frequency, position of RX antenna, position of TX antenna), the Doppler-variant transfer function by Fourier transforming the 256 temporal samples (with Hanning windowing) is computed first. Next, all components that do not exhibit zero Doppler shift (Doppler filtering) are eliminated. Those components correspond, for example, to MPCs scattered by leaves moving in the wind. The eliminated components carry on the order of 1 % of the total energy. The three-dimensional (static) transfer function obtained in that way is then evaluated by Unitary ESPRIT [34] to estimate the delays, τ_i. Unitary ESPRIT is an improved version of the classical ESPRIT algorithm discussed in Glisic [1]. They both estimate the signal subspace for extraction of the parameters of (spatial or frequency) harmonics in additive noise. One important step in ESPRIT is the estimation of the model order. Different methods have been proposed in the literature for that task. The relative power decreases between neighboring eigenvalues with additional correction by visual inspection of the *Scree Graph*, showing that the eigenvalue is an option used for generating the results presented in the sequel.

After estimation of the parameters τ_i, we can determine the corresponding 'steering' matrix, $\mathbf{A}_\tau$. Subsequent beamforming with its Moore–Penrose pseudoinverse [34,53–56] $\mathbf{A}_\tau^+$ gives the vector of delay-weights for all $\mathbf{x}_\mathrm{R}$,$\mathbf{x}_\mathrm{T}$

$$\mathbf{h}_\tau\left(\mathbf{x}_\mathrm{T},\mathbf{x}_\mathrm{R}\right)=\mathbf{A}_\tau^+\mathbf{T}_\mathrm{f}\left(\mathbf{x}_\mathrm{T},\mathbf{x}_\mathrm{R}\right) \tag{3.36}$$

where $\mathbf{T}_f$ is the vector of transfer coeffcients at the 192 frequency sub-bands sounded. This gives us the transfer coeffcients from all positions $\mathbf{x}_\mathrm{T}$ to all positions $\mathbf{x}_\mathrm{R}$ separately for each delay τ_i. Thus, one dimension, namely the frequency, has been replaced by the parameterized version of its dual, the delays.

For the estimation of the direction of arrivals (DOA) in each of the two-dimensional transfer functions, ESPRIT estimation and beamforming by the pseudo-inverse are used

$$\mathbf{h}_{\varphi\mathrm{R}}\left(\tau_i,\mathbf{x}_\mathrm{T}\right)=\mathbf{A}_{\varphi\mathrm{R}}^+\mathbf{h}_{x\mathrm{R}}\left(\tau_i,\mathbf{x}_\mathrm{T}\right) \tag{3.37}$$

Finally for the direction of departure (DOD) we have

$$\mathbf{h}_{\varphi\mathrm{T}}\left(\tau_i,\varphi_{\mathrm{R},i,j}\right)=\mathbf{A}_{\varphi\mathrm{T}}^+\mathbf{h}_{x\mathrm{T}}\left(\tau_i,\varphi_{\mathrm{R},i,j}\right) \tag{3.38}$$

Figure 3.12 illustrates these steps.

The procedure gives us the number and parameters of the MPCs, i.e. the number and values of delays, which DOA can be observed at these delays and which DOD corresponds to each DOA at a specific delay. Furthermore, we also obtain the powers of the multipath channels (MPCs). One important point in the application of the sequential estimation procedure is the sequence in which the evaluation is performed. Roughly speaking, the number of MPCs that can be estimated is the number of samples we have at our disposal.

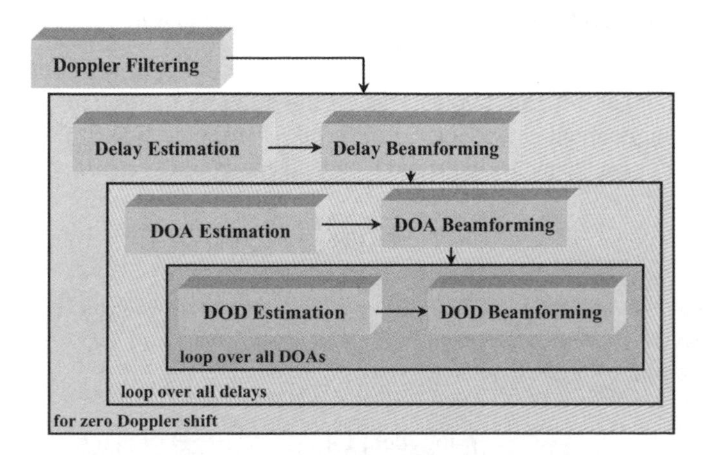

Figure 3.12 Sequential estimation of the parametric channel response in the different domains: alternating estimation and beamforming. (Reproduced by permission of IEEE [52].)

3.6.2 Capacity computation

In a fading channel, the capacity is a random variable, depending on the local (or instantaneous) channel realization. In order to determine the cdf of the capacity, and thus the outage capacity, we would have to perform a large number of measurements either with slightly displaced arrays, or with temporally varying scatterer arrangement. Since each single measurement requires a huge effort, such a procedure is highly undesirable.

To improve this situation, an evaluation technique that requires only a single measurement of the channel is used. This technique relies on the fact that we can generate different realizations of the transfer function by changing the phases of the multipath components. It is a well-established fact in mobile radio that the phases are uniformly distributed random variables, whose different realizations occur as transmitter, receiver or scatterers move [27]. We can thus generate different realizations of the transfer function from the mth transmit to the kth receive antenna as

$$\begin{aligned} h_{k,m}(f) = \sum_i a_i \exp\left\{-j\tfrac{2\pi}{\lambda} d\left[k \sin\left(\phi_{\mathrm{R},i}\right) + m \sin\left(\phi_{\mathrm{T},i}\right)\right]\right\} \\ \times \exp\left(-j2\pi f \tau_i\right) \exp\left(j\alpha_i\right) \end{aligned} \tag{3.39}$$

where α_i is a uniformly distributed random phase, which can take on different values for the different MPCs numbered i. Note, however, that α_i stays unchanged as we consider different antenna elements k and m. To simplify discussion, we for now consider only the flat-fading case, i.e. $\tau_i = 0$. We can thus generate different realizations of the channel matrix $\mathbf{H}$

$$\mathbf{H} = \begin{pmatrix} h_{11} & h_{12} & \cdots & h_{1N_T} \\ h_{21} & h_{22} & \cdots & h_{2N_T} \\ \cdots & \cdots & \cdots & \cdots \\ h_{N_{\mathrm{R}}1} & h_{N_{\mathrm{R}}2} & \cdots & h_{N_{\mathrm{R}}N_{\mathrm{T}}} \end{pmatrix} \tag{3.40}$$

by the following two steps:

(1) From a single measurement, i.e. a single snapshot of the channel matrix, determine the DOAs, and DODs of the MPCs as described earlier in the section.

(2) Compute synthetically the impulse responses at the positions of the antenna elements, and at different frequencies. Create different realizations of one ensemble by adding random phase factors (uniformly distributed between 0 and 2π) to each MPC. For each channel realization, we can compute the capacity from [97]

$$C = \log_2 \det\left(\mathbf{I} + \frac{\rho}{N_T}\mathbf{H}^{\mathrm{H}}\mathbf{H}\right) \tag{3.41}$$

where ρ denotes the SNR. $\mathbf{I}$ is the identity matrix and superscript H means Hermitian transposition. For the frequency-selective case, we have to evaluate the capacity by integrating over all frequencies

$$C = \int \log_2 \det\left[\mathbf{I} + \frac{\rho}{N_{\mathrm{T}}}\mathbf{H}^{\mathrm{H}}(f)\mathbf{H}(f)\right] \mathrm{d}f. \tag{3.42}$$

Here, $\mathbf{H}(f)$ is the frequency-dependent transfer matrix. The integration range is the bandwidth of interest.

3.6.3 Measurement environments

As an example the following scenarios are evaluated with the procedure described above [52]:

- *Scenario I – a courtyard with dimensions* 26 × 27*m, open on one side*. The RX-array broadside points into the center of the yard; the transmitter is located on the positioning device 8 m away in LOS.
- *Scenario II – closed backyard of size* 34 × 40 *m with inclined rectangular extension*. The RX-array is situated in one rectangular corner with the array broadside of the linear array pointing under 45° inclination directly to the middle of the yard. The LOS connection between TX and RX measures 28 m. Many metallic objects are distributed irregularly along the building walls (power transformers, air-condition fans, etc.). This environment looks very much like the backyard of a factory (Figure 3.13).
- *Scenario III – same closed backyard as in Scenario II but with artificially obstructed LOS path*. It is expected that the metallic objects generate serious multipath and higher-order scattering that can only be observed within the dynamic range of the device if the LOS path is obstructed.
- *Scenario IV* – same as scenario III but with different TX position and LOS obstructed. The TX is situated nearer to the walls. More details about the senarios can be found in Steinbauer *et al.* [57].

Some of the measurements results for these scenarios are presented in Figure 3.14

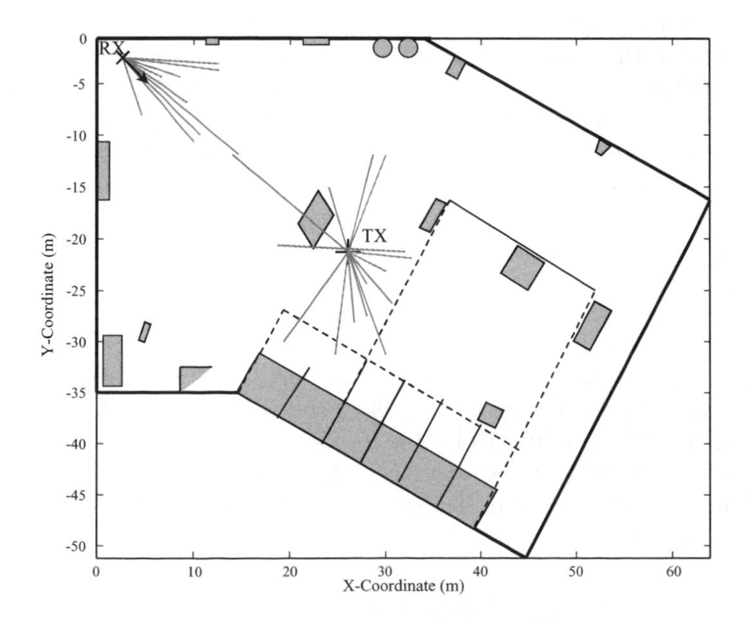

Figure 3.13 Geometry of the environment of scenarios II–IV (backyard) in top view. Superimposed are the extracted DOAs and DODs for scenario III. (Reproduced by permission of IEEE [52].)

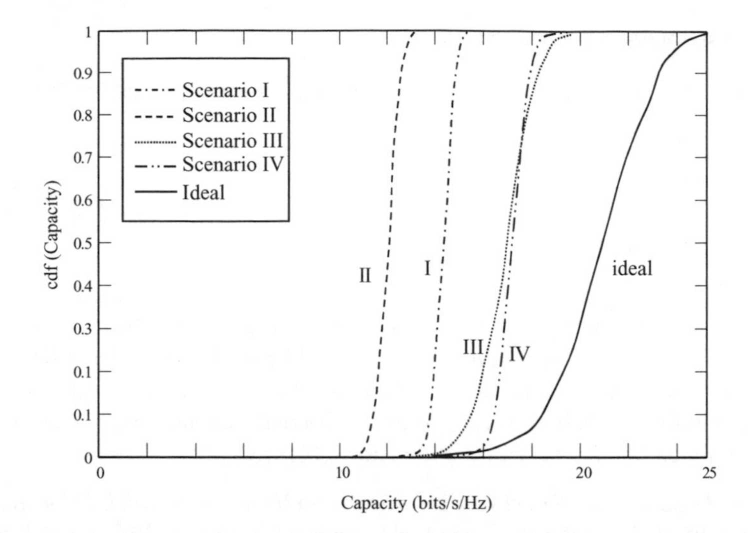

Figure 3.14 The CDFs of the MIMO channel capacity encountered in scenarios I–IV, and the cdf for an ideal channel. The SNR is 20 dB, and 4 × 4 antenna elements were used.

3.7 INDOOR WLAN CHANNEL (17 GHz)

In this section we discuss the indoor radio propagation channel at 17 GHz. The presentation is based on results reported in Rubio *et al.* [58]. Wideband parameters, such as coherence bandwidth or rms delay spread, and coverage are analyzed for the design of an OFDM-based broadband WLAN. The method used to obtain the channel parameters is based on a simulator described in Rubio *et al.* [58]. This simulator is a site-specific propagation model based on three-dimensional (3-D) ray-tracing techniques, which has been specifically developed for simulating radio coverage and channel performance in enclosed spaces such as buildings, and for urban microcell and picocell calculations. The simulator requires the input of the geometric structure and the electromagnetic properties of the propagation environment, and is based on a full 3-D implementation of geometric optics and the uniform theory of diffraction (GO/UTD). Examples of the measurement environments are given in Figure 3.15. The results for coherence bandwidth $B_c = 1/\alpha\tau_{rms}$ are given in Table 3.13 and Figure 3.16.

A further requirement related to the correct and efficient channel estimation process by the receiver is the selection of a number of subcarriers in OFDM satisfying the condition of being separated between approximately $B_c/5$ and $B_c/10$. Results for delay spread are shown in Figure 3.17 and Tables 3.14 – 3.17.

The results for the path loss exponent and k factor are given in Figure 3.18 and Table 3.18 and Table 3.19. For channel modeling purposes, the mean power of the received signal will be represented as

$$P_{RX}|_{dB} = P_{TX}|_{dB} + G_{TX}|_{dB} + G_{RX}|_{dB} - L_{fs}|_{dB} + 10 \cdot \log\left[\int_0^{\infty} PDP(t)\ dt\right] \qquad (3.43)$$

where T_{TX} is the mean power at the transmitting antenna input, G_{TX} is the transmitting antenna gain while G_{RX} is the receiving antenna gain. L_{fs} is free space propagation losses,

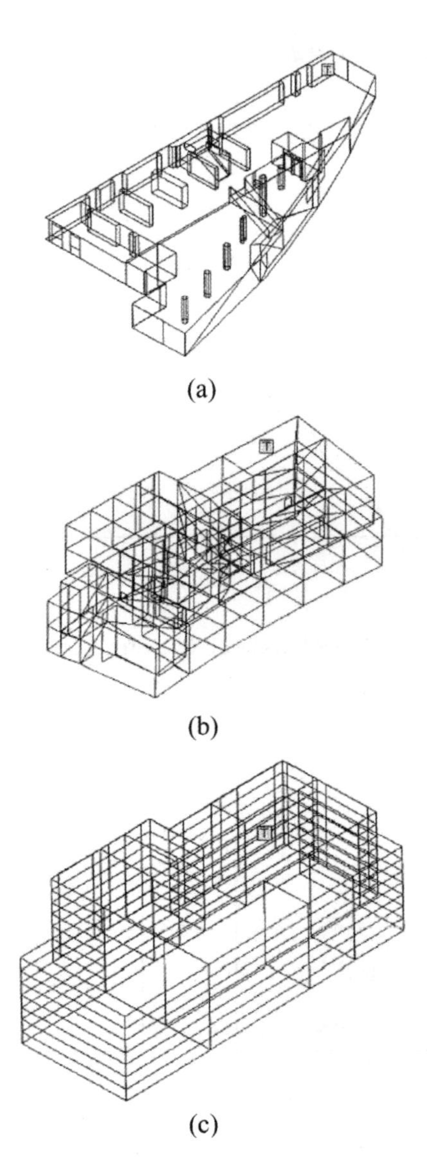

Figure 3.15 (a) ETSIIT hall ($49 \times 26\,m$); (b) DICOM, floors 2 and 3 ($34 \times 20\,m$); (c) office building ($72 \times 38\,m$); 3-D representations 63. (Reproduced by permission of IEEE [58].)

Table 3.13 B_c at 17 GHz

Place	Coherence bandwidth (MHz)	
	Mean	Standard deviation
Hall	24.85	12.35
Floors	14.44	9.85
Building	22.86	10.24
Total	20.72	11.56

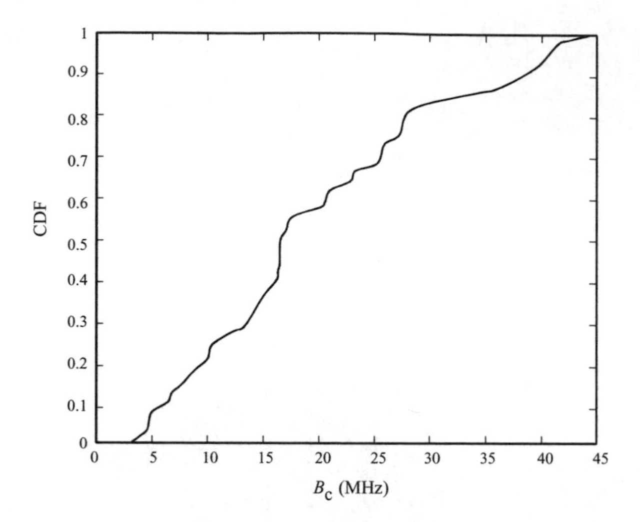

Figure 3.16 B_c CDF at 17 GHz. (Reproduced by permission of IEEE [58].)

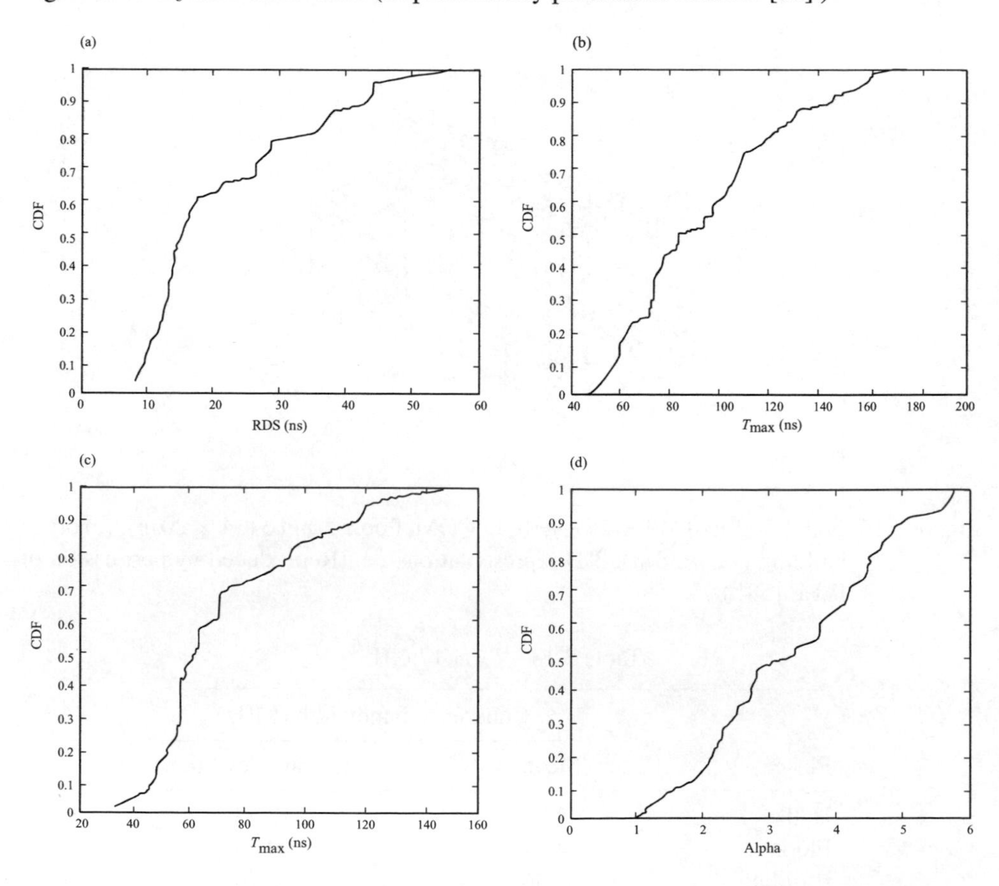

Figure 3.17 (a) The RMS delay spread CDF ($B_c = 1/\alpha\tau_{rms}$). (b) Maximum delay CDF, 30 dB criterion. (c) Maximum delay CDF, 20 dB criterion. (d) Alpha CDF.

Table 3.14 The RMS delay spread CDF. (Reproduced by permission of IEEE [58])

CDF value	RDS value
0.2	12.1 ns
0.4	14.3 ns
0.6	17.5 ns
0.8	34.3 ns
1	58.3 ns

RDS, root delay spread.

Table 3.15 Maximum delay CDF, 30 dB criterion. (Reproduced by permission of IEEE [58])

CDF value	T_{max} value
0.2	62 ns
0.4	76 ns
0.6	101 ns
0.8	122 ns
1	197 ns

Table 3.16 Maximum delay CDF, 20 dB criterion. (Reproduced by permission of IEEE [58])

CDF value	T_{max} value
0.2	51 ns
0.4	56 ns
0.6	69 ns
0.8	94 ns
1	156 ns

Table 3.17 Alpha CDF, $B_c = 1/\alpha\tau_{rms}$

CDF value	Alpha value
0.2	2.17
0.4	2.67
0.6	3.75
0.8	4.44
1	5.78

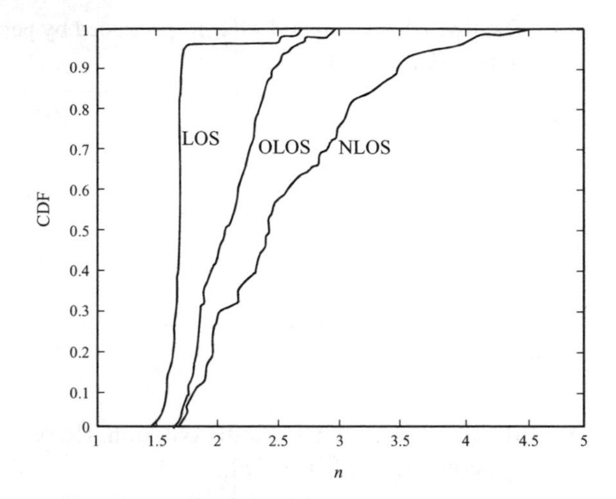

Figure 3.18 CDF of path loss exponent n.

Table 3.18 Mean values of n

Type of path	LOS	OLOS	NLOS
n Mean value	1.68	2.14	2.61

Table 3.19 Fading statistic over distance, LOS case

Radius (m)	K factor
4	17
5	10
6	9
7	8
8	6
9	5
10	1

given by

$$L_{\mathrm{fs}}|_{\mathrm{dB}} = 32.45\ \mathrm{dB} + 20 \cdot \log_{10}(d_{\mathrm{km}} + f_{\mathrm{MHz}})$$

and PDP(t) the modeled power delay profile. Once the PDF is modeled, to obtain the discrete channel impulse response, h_i, we only have to add a random phase to the square root of each delay bin amplitude, as follows:

$$h_i = \sqrt{p_i}\mathrm{e}^{j\phi_i} \quad \phi_i \quad r.v. \quad unif[0, 2\pi] \tag{3.44}$$

where h_i is the ith bin of the modeled channel impulse response and p_i, the module of the ith bin of the modeled power delay profile.

It can be assumed that phases of different components of the same channel impulse response are uncorrelated at the frequency of interest (17 GHz), because their relative range is higher than a wavelength, even for high-resolution models [59]. As the total bandwidth assigned to the communication is 50 MHz, a selection of 10 ns for the bin size must be made. Using 99 % of the total power criterion for the maximum duration of the PDF, the former bin size selection leads to a total of nine taps for the LOS case and 17 for the NLOS case.

The statistical variability of the bin amplitudes has been modeled following different probability density functions. Taking into account the fact that the area of service of future applications (SOHO – small office, home office) has small ranges, the variability has been analyzed considering a medium-scale, that is, the environment is divided in to the LOS area and the NLOS one. In the LOS case, a Frechet PDF [60] is chosen for the first bin and exponential PDFs for the rest. A continuous random variable X has a Frechet distribution if its PDF has the form

$$f(x;\sigma;\lambda) = \frac{\lambda}{\sigma}\left(\frac{\sigma}{x}\right)^{\lambda+1} \exp\left\{-\left(\frac{\sigma}{x}\right)^{\lambda}\right\};x \geq 0; \quad \sigma, \lambda > 0 \tag{3.45}$$

A Frechet variable X has the CDF

$$F(x;\sigma;\lambda) = \exp\left\{-\left(\frac{\sigma}{x}\right)^{\lambda}\right\} \tag{3.46}$$

This model has a scale structure, with σ a scale parameter and λ a shape parameter.A continuous random variable X has an exponential distribution if its PDF has the form

$$f(x;\mu) = \frac{1}{\sigma}\exp\left\{-\left(\frac{x-\mu}{\sigma}\right)\right\};x \geq 0; \quad \mu, \sigma > 0 \tag{3.47}$$

This PDF has location-scale structure, with a location parameter, μ, and a scale one, σ. The CDF of the exponential variable X is

$$F(x;\mu) = 1 - \exp\left\{-\left(\frac{x-\mu}{\sigma}\right)\right\} \tag{3.48}$$

These PDFs were considered the most suitable after a fitting process. The NLOS case needs a combination of exponential and Weibull PDFs for the first bin and exponential PDFs for the others. A continuous random variable X has a Weibull distribution if its PDF has the form

$$f(x;\sigma;\lambda) = \frac{\lambda}{\sigma}\left(\frac{x}{\sigma}\right)^{\lambda-1} \exp\left\{-\left(\frac{x}{\sigma}\right)^{\lambda}\right\};x \geq 0; \quad \sigma, \lambda > 0 \tag{3.49}$$

While the CDF is

$$F(x;\sigma;\lambda) = 1 - \exp\left\{-\left(\frac{x}{\sigma}\right)^{\lambda}\right\} \tag{3.50}$$

This model has a scale structure, that is, σ is a scale parameter, while λ is a shape parameter. Tables 3.20 and 3.21 show the probability density functions employed for LOS and NLOS channel models [58].

For both tables, the units of σ parameters are Hz (s^{-1}), while λ has no units. These units have no physical correlation but make the last term of Equation (3.43) nondimensional, as it represents a factor scale between the free space behavior and the real one. The mean

Table 3.20 Wind-flex channel model PDFs, LOS case. (Reproduced by permission of IEEE [58])

Bin 1	Frechet ($\sigma = 2.66 \times 10^8, \lambda = 7$)	Bin 4	exp ($\sigma = 1.45 \times 10^7$)	Bin 7	exp ($\sigma = 0.41 \times 10^7$)
Bin 2	exp ($\sigma = 5.44 \times 10^7$)	Bin 5	exp ($\sigma = 1.03 \times 10^7$)	Bin 8	exp ($\sigma = 0.27 \times 10^7$)
Bin 3	exp ($\sigma = 2.51 \times 10^7$)	Bin 6	exp ($\sigma = 0.79 \times 10^7$)	Bin 9	exp ($\sigma = 0.71 \times 10^7$)

Table 3.21 Wind-flex channel model PDFs, NLOS case. (Reproduced by permission of IEEE [58])

Bin 1	0.5*[exp ($\sigma = 4.378 \times 10^6$)+ Weibull($\sigma = 4.207 \times 10^7, \lambda = 5$)	Bin 7	exp ($\sigma = 1.88 \times 10^5$)	Bin 13	exp ($\sigma = 9.21 \times 10^4$)
Bin 2	exp ($\sigma = 3.04 \times 10^6$)	Bin 8	exp ($\sigma = 2.51 \times 10^5$)	Bin 14	exp ($\sigma = 1.27 \times 10^5$)
Bin 3	exp ($\sigma = 2.47 \times 10^6$)	Bin 9	exp ($\sigma = 5.69 \times 10^5$)	Bin 15	exp ($\sigma = 2.76 \times 10^4$)
Bin 4	exp ($\sigma = 2.14 \times 10^6$)	Bin 10	exp ($\sigma = 1.53 \times 10^5$)	Bin 16	exp ($\sigma = 6.71 \times 10^4$)
Bin 5	exp ($\sigma = 1.1 \times 10^6$)	Bin 11	exp ($\sigma = 3.29 \times 10^5$)	Bin 17	exp ($\sigma = 6.42 \times 10^4$)
Bin 6	exp ($\sigma = 3.71 \times 10^5$)	Bin 12	exp ($\sigma = 2.67 \times 10^5$)		

value of the probability density functions is so high due to the ulterior integral over the time (in seconds) required, and the PDF duration (tens of nanoseconds). As expected, the mean value of the first bin is the highest, since it includes the direct ray (LOS case). Additional details on the topic can be found in References [59–71].

3.8 INDOOR WLAN CHANNEL (60 GHz)

Based on the results reported in Hao *et al.* [72], in this section we present spatial and temporal characteristics of 60 GHz indoor channels. In the experiment, a mechanically steered directional antenna is used to resolve multipath components. An automated system is used to precisely position the receiver antenna along a linear track and then rotate the antenna in the azimuthal direction, as illustrated in Figure 3.19. The precisions of the track and spin positions are less than 1 mm and 1°, respectively. When a highly directional antenna is used, the system provides high spatial resolution to resolve multipath components with different angles of arrival (AOAs). The sliding correlator technique was used to further resolve multipath components with the same AOA by their times of arrival (TOAs). The spread spectrum signal has a RF bandwidth of 200 MHz, which provids a time resolution of approximately 10 ns.

For this measurement campaign, an open-ended waveguide with 6.7 dB gain is used as the transmitter antenna and a horn antenna with 29 dB gain is used as the receiver antenna. These antennas are chosen to emulate typical antenna systems that have been proposed for millimeter-wave indoor applications. In these applications a sector antenna is used at the transmitter and a highly directional antenna is used at the receiver. Both antennas are vertically polarized and mounted on adjustable tripods about 1.6 m above the ground. The theoretical half-power beamwidths (HPBW) are 90° in azimuth and 125° in elevation for the open-ended waveguide and 7° in azimuth and 5.6° in elevation for the horn antenna.

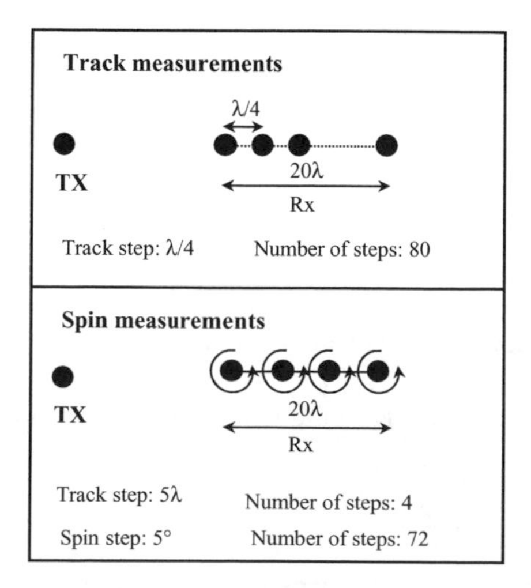

Figure 3.19 Track and spin measurement procedure.(Reproduced by permission of IEEE [72].)

3.8.1 Definition of the statistical parameters

3.8.1.1 Path loss and received signal power

The free-space *path loss* at a reference distance of d_0 is given by

$$\overline{PL}_{\text{fs}}(d_0) = 20\log\left(\frac{4\pi d_0}{\lambda}\right) \tag{3.51}$$

where λ is the wavelength. Path loss over distance d can be described by the path loss exponent model as follows:

$$\overline{PL}(d)[\text{dB}] = PL_{\text{fs}}(d_0)[\text{dB}] + 10n\log_{10}(d/d_0) \tag{3.52}$$

where $\overline{PL}(d)$ is the average path loss value at a transmitter – receiver (TR) separation of d and n is the *path loss exponent* that characterizes how fast the path loss increases with the increase in TR separation. The path loss values represent the signal power loss from the transmitter antenna to the receiver antenna. These path loss values do not depend on the antenna gains or the transmitted power levels. For any given transmitted power, the received signal power can be calculated as

$$P_{\text{r}}[\text{dBm}] = P_{\text{t}}[\text{dBm}] + G_{\text{t}}[\text{dB}] + G_{\text{r}}[\text{dB}] - \overline{PL}(d)[\text{dB}] \tag{3.53}$$

where G_{t} and G_{r} are transmitter and receiver gains, respectively. In this measurement campaign, the transmitted power level was 25 dBm, the transmitter antenna gain was 6.7 dB, and the receiver antenna gain was 29 dB.

3.8.1.2 TOA parameters

TOA parameters characterize the time dispersion of a multipath channel. The calculated TOA parameters include mean excess delay ($\bar{\tau}$), rms delay spread (σ_τ), and also timing jitter [$\delta(x)$] and standard deviation [$\Delta(x)$], in a small local area. Parameters of $\bar{\tau}$ and σ_τ are given as [72]:

$$\bar{\tau} = \frac{\sum_{i=1}^{N} P_i\tau_i}{\sum_{i=1}^{N} P_i}, \sigma_\tau = \sqrt{\overline{\tau^2} - (\bar{\tau})^2}, \overline{\tau^2} = \frac{\sum_{i=1}^{N} P_i\tau_i^2}{\sum_{i=1}^{N} P_i} \tag{3.54}$$

where P_i and τ_i are the power and delay of the ith multipath component of a PDF, respectively, and $N$is the total number of multipath components. Timing jitter is calculated as the difference between the maximum and minimum measured values in a local area. Timing jitter $\delta(x)$ and standard deviation $\Delta(x)$ are defined as

$$\delta(x) = \max_{i=1}^{M}\{x_i\} - \min_{i=1}^{M}\{x_i\}, \Delta(x) = \sqrt{\overline{x^2} - (\bar{x})^2}$$

$$\bar{x} = \frac{1}{M}\sum_{i=1}^{M} x_i, \overline{x^2} = \frac{1}{M}\sum_{i=1}^{M} x_i^2 \tag{3.55}$$

where x_i is the measured value for parameter $x(\bar{\tau}\ or\ \sigma_\tau)$ in the ith measurement position of the spatial sampling and M is the total number of spatial samples in the local area. For example, for the track measurements, M was chosen to be 80.

Mean excess delay and rms delay spread are the statistical measures of the time dispersion of the channel. Timing jitter and standard deviation of $\bar{\tau}$ and σ_τ show the variation of these parameters over the small local area. These TOA parameters directly affect the performance of high-speed wireless systems. For instance, the mean excess delay can be used to estimate the search range of rake receivers and the rms delay spread can be used to determine the maximum transmission data rate in the channel without equalization. The timing jitter and standard deviation parameters can be used to determine the update rate for a rake receiver or an equalizer.

3.8.1.3 AOA parameters

AOA parameters characterize the directional distribution of multipath power. The recorded AOA parameters include angular spread Λ, angular constriction γ, maximum fading angle θ_{max} and maximum AOA direction. Angular parameters Λ, γ and θ_{max} are defined based on the Fourier transform of the angular distribution of multipath power, $p(\theta)$ [74]:

$$\Lambda = \sqrt{1 - \frac{||F_1||^2}{||F_2||^2}},\ \gamma = \frac{||F_0 F_2 - F_1^2||}{||F_0||^2 - ||F_1||^2},\ \theta_{max} = \frac{1}{2}\text{Phase}\left\{F_0 F_2 - F_1^2\right\} \tag{3.56}$$

where

$$F_n = \int_0^{2\pi} p(\theta)\exp(jn\theta)\,\mathrm{d}\theta \tag{3.57}$$

F_n is the nth Fourier transform of $p(\theta)$. As shown in Durgin and Rappaport [74], angular spread, angular constriction and maximum fading angle are three key parameters to characterize the small-scale fading behavior of the channel. These new parameters can be used for diversity techniques, fading rate estimation, and other space–time techniques. Maximum AOA provides the direction of the multipath component with the maximum power. It can be used in system installation to minimize the path loss. The results of measurements for the parameters defined by Equations (3.51)–(3.57) are given in Table 3.22–3.24 and Figure 3.20. More details on the topic can be found in References [74–85].

3.9 UWB CHANNEL MODEL

UWB channel parameters will be discussed initially based on measurements results in Cassioli *et al.* [86]. The measurements environment is presented in Figure 3.21 and the signal format used in these experiments in Figure 3.22. The repetition rate of the pulses is 2×10^6 pulses/s, implying that multipath spreads up to 500 ns could have been observed unambiguously. Multipath profiles with a duration of 300 ns were measured. Multipath profiles were measured at various locations in 14 rooms and hallways on one floor of the building presented in Figure 3.21. Each of the rooms is labeled alpha-numerically. Walls around offices are framed with metal studs and covered with plaster board. The wall around the laboratory is made from acoustically silenced heavy cement block. There are steel core support pillars throughout the building, notably along the outside wall and two within the laboratory itself. The shield room's walls and door are metallic. The transmitter is kept stationary in the central location of the building near a computer server in a laboratory

Table 3.22 Spin measurements: transmitter–receiver separations in meters, time dispersion parameters ($\bar{\tau}$ and σ_τ) in nanoseconds, angular dispersion parameters (Λ and γ) are dimensionless, maximum fading angle ($\theta_{\max}$) and AOA of maximum multipath (max AOA) in degrees, ratio of maximum multipath power to average power (peak/avg) in decibels and maximum multipath power ($P_{\max}$) in dBm [72]

Site	No.	TR	$\bar{\tau}$	σ_τ	Λ	γ	$\theta_{\max}$	Max *AOA*	Peak/avg	$P_{\max}$	Comments
LOS, hallway Durham Hall	1.1	5	80.0	14.7	0.46	0.83	−80.7	−4.0	12.3	−14.9	
	1.2	10	52.0	18.8	0.44	0.74	−86.6	4.0	12.0	−18.2	
	1.3	20	85.9	40.1	0.56	0.28	−61.9	8.0	14.5	−28.8	
	1.4	30	116.6	38.7	0.42	0.22	−66.4	5.0	14.7	−28.3	Open area
	1.5	40	84.9	60.0	0.69	0.25	4.3	5.0	13.9	−38.2	
	1.6	50	52.1	26.1	0.66	0.26	8.2	10.0	13.3	−38.2	
	1.7	60	53.2	30.3	0.78	0.36	4.0	2.0	13.2	−40.8	
LOS, hallway Whittemore	2.1	5	51.0	20.7	0.48	0.88	−73.5	5.0	12.5	−13	
	2.2	10	62.1	29.4	0.66	0.79	−72.3	21.0	11.4	−21.7	Intersection
	2.3	20	90.7	14.6	0.36	0.43	−73.8	4.0	12.9	−29.8	
	2.4	30	41.2	12.3	0.41	0.15	−64.8	10.0	13.8	−31.7	
	2.5	40	83.7	53.8	0.72	0.19	5.0	1.0	13.2	−36.0	
LOS, room Durham Hall	3.1	4.2	42.6	16.2	0.86	0.64	−79.2	0.0	12.5	−11.8	Corner
	3.2	3.3	47.7	17.5	0.81	0.70	−79.1	5.0	13.1	−12.1	Center
LOS, room Whittemore	4.1	7.1	46.6	13.0	0.84	0.55	−88.0	−60.0	12.3	−26.8	Corner
	4.2	3.8	64.3	13.3	0.62	0.74	−89.6	−1.0	13.1	−25.6	Center
	4.3	5.2	66.3	17.7	0.73	0.84	−35.2	49.0	14.0	−30.4	Corner, ⊥ to TX
	4.4	4.2	77.8	13.3	0.78	0.72	−38.2	−49.0	14.2	−28.6	Corner, ⊥ to TX
Hallway toroom	5.1	2.4	49.1	21.4	0.81	0.13	−76.3	0.0	12.0	−6.0	LOS
	5.2	2.4	41.6	18.1	0.74	0.44	−89.6	5.0	10.3	−14.1	Through wall
	5.3	2.4	95.8	14.6	0.63	0.40	−88.1	0.0	12.1	−5.6	LOS
	5.4	2.4	80.3	16.0	0.68	0.27	72.3	5.0	11.9	−8.9	Through glass
Room to room	6.1	3	42.7	16.6	0.80	0.40	−25.3	52.0	11.5	−36.4	Through wall
LOS, outdoor parking lot	7.1	1.9	41.3	17.4	0.12	0.97	−81.2	2.0	13.9	−15.0	TX pattern
	7.2	1.9	56.6	16.1	0.49	0.94	−66.7	20.0	8.5	−29.9	RX pattern
LOS, outdoor	8.1	2	24.4	7.7	0.26	0.76	−66.3	3.0	13.9	−10.1	Near Durham Hall

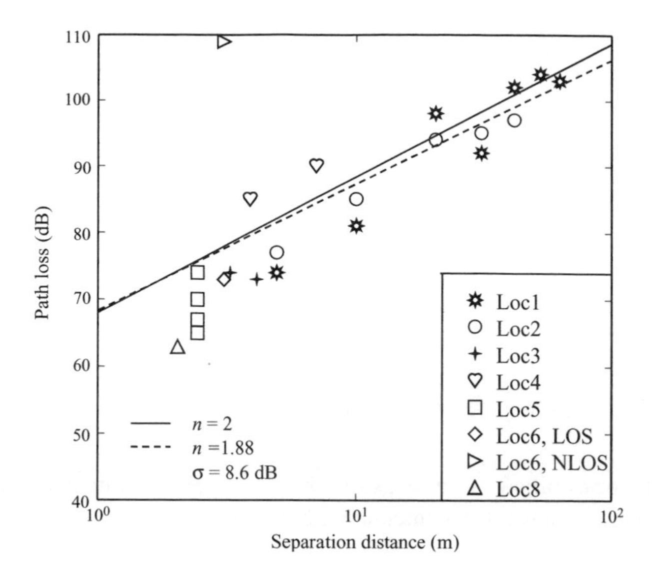

Figure 3.20 Scatter plot of the measured path loss values.

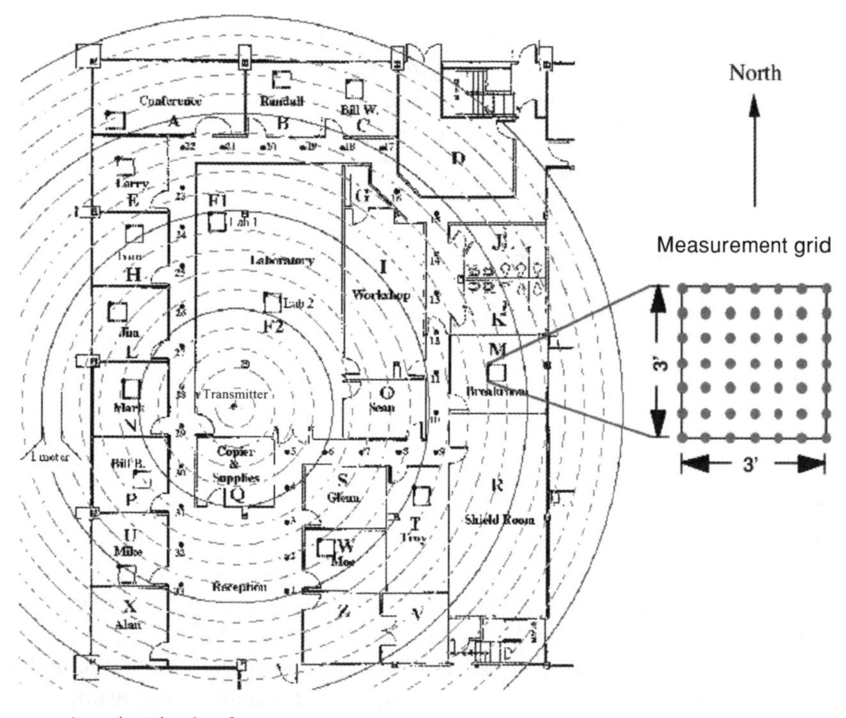

Figure 3.21 The floor plan of a typical modern office building where the propagation measurement experiment was performed. The concentric circles are centered on the transmit antenna and are spaced at 1 m intervals. (Reproduced by permission of IEEE [87].)

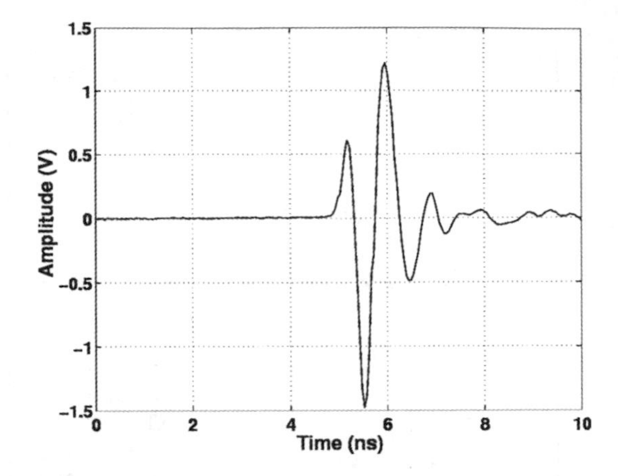

Figure 3.22 The transmitted pulse measured by the receiving antenna located 1 m away from the transmitting antenna with the same height.

denoted by F. The transmit antenna is located 165 cm from the floor and 105 cm from the ceiling.

In each receiver location, impulse response measurements were made at 49 measurement points, arranged in a fixed-height, 7×7 square grid with 15 cm spacing, covering 90×90 cm. A total of 741 different impulse responses were recorded. One side of the grid is always parallel to north wall of the room. The receiving antenna is located 120 cm from the floor and 150 cm from the ceiling.

3.9.1 The large-scale statistics

Experimental results show that all small-scale averaging SSA-PDPs exhibit an exponential decay as a function of the excess delay. Since we perform a delay axis translation, the direct path always falls in the first bin in all the PDPs. It also turns out that the direct path is always the strongest path in the 14 SSA-PDPs even if the LOS is obstructed. The energy of the subsequent MPCs decays exponentially with delay, starting from the second bin. Let $\overline{G_k} \triangleq A_{\text{Spa}}\{G_k\}$ be the locally averaged energy gain, where $A_{\text{Spa}}\{\cdot\}$ denotes the spatial average over the 49 locations of the measurement grid. The average energy of the second MPC may be expressed as fraction $r$of the average energy of the direct path, i.e. $r = \overline{G}_2/\overline{G}_1$. We refer to r as the *power ratio*. As we will show later, the SSA-PDP is completcly characterized by $\overline{G}_1$, the power ratio r, and the decay constant ε (or equivalently, by the total average received energy $\overline{G}_{\text{tot}}$, r and ε).

The power ratio $r$and the decay constant ε vary from location to location, and should be treated as stochastic variables. As only 14 values for ε and r were available, it was not possible to extract the *shape* of their distribution from the measurement data. Instead, a model was assumed *a priori* and the *parameters* of this distribution were fitted. It was found that the log–normal distribution, denoted by $\varepsilon \sim \mathcal{LN}(\mu_{\varepsilon_{\text{dB}}}; \sigma_{\varepsilon_{\text{dB}}})$, with $\mu_{\varepsilon_{\text{dB}}} = 16.1$ and $\sigma_{\varepsilon_{\text{dB}}} = 1.27$, gives the best agreement with the empirical distribution. Applying the same procedure to characterize the power ratios rs, it was found that they are also log–normally distributed, i.e. $r \sim \mathcal{LN}(\mu_{r_{\text{dB}}}; \sigma_{r_{\text{dB}}})$, with $\mu_{r_{\text{dB}}} = -4$ and $\sigma_{r_{\text{dB}}} = 3$, respectively.

Table 3.23 Track measurement results: TR separations in meters, time dispersion parameters ($\bar{\tau}$ and σ_τ) in nanoseconds, variations of time dispersion parameters ($\delta\bar{\tau}$, $\Delta\bar{\tau}$, $\delta\sigma_\tau$ and $\Delta\sigma_\tau$) in nanoseconds and average received power (P_{rx}) in dBm [72]

Site	LOC no.	TR	$\bar{\tau}$	σ_τ	$\delta\bar{\tau}$	$\Delta\bar{\tau}$	$\delta\sigma_\tau$	$\Delta\sigma_\tau$	P_r	Comments
LOS, hallway Durham Hall	1.1	5	1.20	6.95	6.33	1.91	1.20	0.29	−13.7	
	1.2	10	6.16	5.88	5.06	1.20	6.16	1.73	−20.3	
	1.3	20	32.61	47.25	32.89	8.43	32.61	9.02	−36.6	
	1.4	30	15.50	31.15	10.16	3.43	15.50	5.69	−31.2	Open area
	1.5	40	27.60	37.04	25.89	8.81	27.60	9.76	−40.5	
	1.6	50	46.42	28.17	36.70	8.10	46.42	10.73	−42.8	
	1.7	60	6.38	22.57	5.99	1.82	6.38	1.57	−41.5	
LOS, hallway Whittemore	2.1	5	2.22	6.24	7.52	2.38	2.22	0.73	−16.7	
	2.2	10	2.78	6.48	8.24	2.61	2.78	0.82	−24.4	Intersection
	2.3	20	2.3	4.56	7.81	2.55	2.30	0.55	−32.86	
	2.4	30	22.02	33.87	13.17	4.60	22.02	6.30	−34.7	
	2.5	40	77.3	45.07	105.04	34.41	77.30	25.86	−36.3	
LOS, room Durham Hall	3.1	4.2	0.74	4.85	6.20	1.88	0.74	0.20	−12.1	Corner
	3.2	3.3	0.92	4.95	5.97	1.87	0.92	0.23	−12.9	Center
LOS, room Whittemore	4.1	7.1	2.74	4.72	11.16	3.08	2.47	0.36	−29.7	Corner
	4.2	3.8	2.4	4.98	11.11	3.17	2.40	0.47	−24.2	Center
	4.3	5.2	12.88	31.10	26.36	6.86	12.88	2.95	−56.2	Corner, ⊥ to TX
	4.4	4.2	21.3	33.94	31.5	7.4	21.3	5.43	−57.9	Corner, ⊥to TX
Hallway to room	5.1	2.4	0.83	5.50	2.41	0.69	0.83	0.32	−5.5	LOS
	5.2	2.4	2.46	7.41	2.61	0.84	2.46	0.94	−14.3	Through wall
	5.3	2.4	0.71	5.36	1.30	0.41	0.71	0.25	−6.7	LOS
	5.4	2.4	1.16	5.19	1.85	0.61	1.16	0.36	−9.1	Through glass
Room to room	6.1	3	10.67	14.72	23.07	6.62	10.67	1.30	−12.8	LOS
	6.2	3	14.82	21.78	34.30	8.57	14.82	3.37	−48.3	Through wall
LOS, outdoor	8.1	2	7.63	24.59	10.24	2.66	7.63	1.75	−2.4	Near Durham Hall

Table 3.24 Measured penetration losses and results from literature

Material	Penetration loss	Reference
Composite wall with studs not in the path	8.8	[72]
Composite wall with studs in the path	35.5 dB	[72]
Glass door	2.5 dB	[72]
Concrete wall 1 week after concreting	73.6 dB	[75]
Concrete wall 2 weeks after concreting	68.4 dB	[75]
Concrete wall 5 weeks after concreting	46.5 dB	[75]
Concrete wall 14 months after concreting	28.1 dB	[75]
Plasterboard wall	5.4–8.1 dB	[76]
Partition of glass wool with plywood surfaces	9.2–10.1 dB	[76]
Partition of cloth-covered plywood	3.9–8.7 dB	[76]
Granite with width of 3 cm	>30 dB	[77]
Glass	1.7–4.5 dB	[77]
Metalized glass	>30 dB	[77]
Wooden panels	6.2–8.6 dB	[77]
Brick with width of 11 cm	17 dB	[77]
Limestone with width of 3 cm	>30 dB	[77]
Concrete	>30 dB	[77]

By integrating the SS A-PDP of each room over all delay bins, the total average energy $\overline{G}_{\text{tot}}$ within each room is obtained. Then its dependence on the TR separation is analyzed. It was found that a breakpoint model, commonly referred to as *dual slope* model, can be adopted for path loss PL as a function of the distance, as

$$PL = \begin{cases} 20.4\log_{10}(d/d_0), & d \leq 11 \text{ m} \\ -56 + 74\log_{10}(d/d_0), & d > 11 \text{ m} \end{cases} \tag{3.58}$$

where PL is expressed in decibels, $d_0 = 1$ m is the reference distance, and d is the TR separation distance in meters. Because of the shadowing phenomenon, the $\overline{G}_{\text{tot}}$ varies statistically around the value given by Equation (3.59). A common model for shadowing is log–normal distribution [87, 88]. By assuming such a model, it was found that $\overline{G}_{\text{tot}}$ is log–normally distributed about Equation (3.58), with a standard deviation of the associated normal random variable equal to 4.3.

3.9.2 The small-scale statistics

The differences between the PDPs at the different points of the measurement grid are caused by small-scale fading. In 'narrowband' models, it is usually assumed that the magnitude of the first (quasi-LOS) multipath component follows Rician or Nakagami statistics and the later components are assumed to have Rayleigh statistics [89]. However, in UWB propagation, each resolved MPC is due to a small number of scatterers, and the amplitude distribution in *each* delay bin differs markedly from the Rayleigh distribution. In fact, the presented analysis showed that the best-fit distribution of the small-scale magnitude statistics is the Nakagami distribution [90], corresponding to a gamma distribution of the

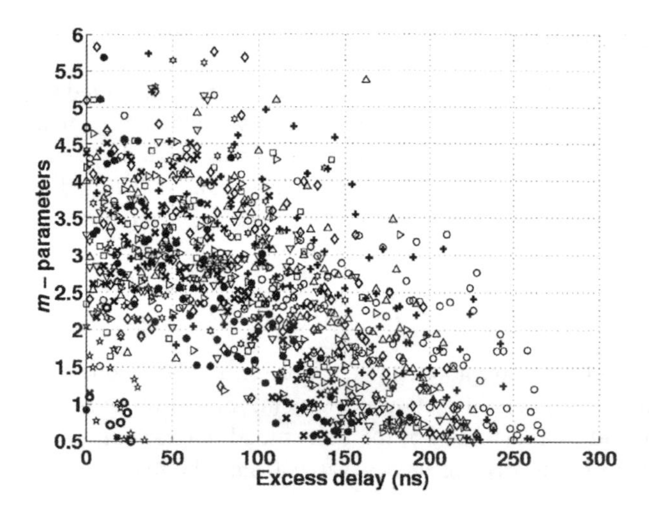

Figure 3.23 Scatter plot of the m-Nakagami of the best fit distribution vs excess delay for all the bins except the LOS components. Different markers correspond to measurements in different rooms. (Reproduced by permission of IEEE [86].)

energy gains. This distribution has been used to model the magnitude statistics in mobile radio when the conditions of the central limit theorem are not fulfilled [91]. The parameters of the gamma distribution vary from bin to bin: $\Gamma(\Omega;\, m)$ denotes the gamma distribution that fits the energy gains of the local PDPs in the kth bin within each room. The Ω_k are given as $\Omega_k = \overline{G}_k$, i.e. the magnitude of the SSA-PDP in the kth bin. The m_k are related to the variance of the energy gain of the kth bin. Figure 3.23 shows the scatter plot of the m_k, as a function of excess delay for all the bins (except the LOS components). It can be seen from Figure 3.23 that the m_k, values range between 1 and 6 (rarely 0.5), decreasing with the increasing excess delay. This implies that MPCs arriving with large excess delays are more diffused than the first arriving components, which agrees with intuition.

The m_k parameters of the gamma distributions themselves are random variables distributed according to a truncated Gaussian distribution, denoted by $m \sim \mathcal{T}_{\mathcal{N}}(\mu_m;\, \sigma_m^2)$, i.e. their distribution looks like a Gaussian for $m \geq 0.5$ and zero elsewhere

$$f_m(x) = \begin{cases} K_m \mathrm{e}^{-\left[(x-\mu_m)^2/2\sigma_m^2\right]}, & \text{if} \quad x \geq 0.5 \\ 0, & \text{otherwise} \end{cases} \tag{3.59}$$

where the normalization constant K_m is chosen so that the integral over the $f_m(x)$ is unity. The mean and variance of such Gaussian distributions that fit the m_k as a function of the excess delay, are given by [17]

$$\mu_m(\tau_k) = 3.5 - \frac{\tau_k}{73} \tag{3.59a}$$

$$\sigma_m^2(\tau_k) = 1.84 - \frac{\tau_k}{160} \tag{3.59b}$$

where the unit of τ_k is nanoseconds.

3.9.3 The statistical model

The received signal is a sum of the replicas (echoes) of the transmitted signal, being related to the reflecting, scattering and/or deflecting objects via which the signal propagates. Each of the echoes is related to a single such object. In a narrowband system, the echoes at the receiver are only attenuated, phase-shifted and delayed, but undistorted, so that the received signal may be modeled as a linear combination of N_{path}-delayed basic waveforms $w(t)$

$$r(t) = \sum_{i=1}^{N_{\text{path}}} c_i w(t - \tau_i) + n(t) \tag{3.60}$$

where $n(t)$ is the observation noise. In UWB systems, the frequency selectivity of the reflection, scattering and/or diffraction coefficients of the objects via which the signal propagates can lead to a distortion of the transmitted pulses. Furthermore, the distortion and, thus, the shape of the arriving echoes, varies from echo to echo. The received signal is thus given as

$$r(t) = \sum_{i=1}^{N_{\text{path}}} c_i \tilde{w}_i(t - \tau_i) + n(t) \tag{3.61}$$

If the pulse distortion was greater than the width of the delay bins (2 ns), one would observe a significant correlation between adjacent bins. The fact that the correlation coefficient remains very low for all analyzed sets of the data implies that the distortion of a pulse due to a single echo is not significant, so that in the following, Equation (3.60) can be used.The SSA-PDP of the channel may be expressed as

$$\bar{g}(\tau) = \sum_{k=1}^{N_{\text{bins}}} \overline{G}_k \delta(\tau - t_k) \tag{3.62}$$

where the function $\bar{g}(\tau)$ can be interpreted as the average energy received at a certain receiver position and a delay τ, normalized to the total energy received at one meter distance, and N_{bins} is the total number of bins in the observation window. Assuming an exponential decay starting from the second bin, we have

$$\bar{g}(\tau) = \overline{G}_1 \delta(\tau - \tau_1) + \sum_{k=2}^{N_{\text{bins}}} \overline{G}_2 \exp[-(\tau_k - \tau_2)/\varepsilon] \delta(\tau - t_k) \tag{3.63}$$

where ε is the decay constant of the SSA-PDP. The total average energy received over the observation interval T is

$$\overline{G}_{\text{tot}} = \int_0^T \bar{g}(\tau)\, d\tau = \overline{G}_1 + \sum_{k=2}^{N_{\text{bins}}} \overline{G}_2 \exp[-(\tau_k - \tau_2)/\varepsilon] \tag{3.64}$$

Summing the geometric series gives

$$\overline{G}_{\text{tot}} = \overline{G}_1 [1 + rF(\varepsilon)] \tag{3.65}$$

where $r = \overline{G}_2 / \overline{G}_1$ is the power ratio, and

$$F(\varepsilon) = \frac{1 - \exp[-(N_{\text{bins}} - 1)\Delta\tau/\varepsilon]}{1 - \exp(-\Delta\tau/\varepsilon)} \approx \frac{1}{1 - \exp(-\Delta\tau/\varepsilon)} \tag{3.66}$$

The total normalized average energy is log–normally distributed, due to the shadowing, around the mean value given from the path loss model (3.58)

$$\overline{G}_{\text{tot}} \sim \mathcal{L}_{\mathcal{N}}(-PL;\ 4.3) \tag{3.67}$$

From Equation (3.10), we have for the average energy gains

$$\overline{G}_k = \begin{cases} \dfrac{\overline{G}_{\text{tot}}}{1+rF(\varepsilon)}, & \text{for } k=1 \\ \dfrac{\overline{G}_{\text{tot}}}{1+rF(\varepsilon)} r\mathrm{e}^{-(\tau_k-\tau_2)/\varepsilon}, & \text{for } k=2,\ldots,N_{\text{bins}} \end{cases} \tag{3.68}$$

and Equation (3.62) becomes

$$\bar{g}(\tau) = \frac{\overline{G}_{\text{tot}}}{1+rF(\varepsilon)} \left\{ \delta(\tau-\tau_1) + \sum_{k=2}^{N_{\text{bins}}} \left[r\mathrm{e}^{-[(\tau_k-\tau_2)/\varepsilon]} \right] \delta(\tau-t_k) \right\} \tag{3.69}$$

3.9.4 Simulation steps

In the model, the local PDF is fully characterized by the pairs $\{G_k,\ \tau_k\}$, where $\tau_k = (k-1)\Delta\tau$ with $\Delta\tau = 2\,\text{ns}$. The G_k are generated by a superposition of large and small-scale statistics. The process starts by generating the total mean energy $\overline{G_{\text{tot}}}$ at a certain distance according to Equation (3.67). Next, the decay constant ε and the power ratio $r$are generated as lognormal distributed random numbers

$$\varepsilon \sim \mathcal{L}_{\mathcal{N}}(16.1;\ 1.27) \tag{3.70}$$

$$r \sim \mathcal{L}_{\mathcal{N}}(-4;\ 3) \tag{3.71}$$

The width of the observation window is set at $T = 5\varepsilon$. Thus, the SSA-PDP is completely specified according to Equation (3.69). Finally, the local PDPs are generated by computing the normalized energy gains $G_k^{(i)}$ of every bin k and every location i as gamma-distributed independent variables. The gamma distributions have the average given by Equation (3.68), and the m_ks are generated as independent truncated Gaussian random variables

$$m_k \sim \mathcal{T}_{\mathcal{N}}\left(\mu_m(\tau_k);\ \sigma_m^2(\tau_k)\right) \tag{3.72}$$

with $\mu_m(\tau_k)$ and $\sigma_m^2(\tau_k)$ given by Equation (3.59). These steps are summarized Table 3.25. Some results are shown in Figure 3.24.

3.9.5 Clustering models for the indoor multipath propagation channel

A number of models for the indoor multipath propagation channel [92–96] have reported a clustering of multipath components, in both time and angle. In the model presented in Spencer *et al.* [95], the received signal amplitude β_{kl} is a Rayleigh-distributed random variable with a mean-square value that obeys a double exponential decay law, according to

$$\overline{\beta_{kl}^2} = \overline{\beta^2(0,0)}\mathrm{e}^{-T_l/\Gamma}\mathrm{e}^{-\tau_{kl}/\gamma} \tag{3.73}$$

where $\overline{\beta^2(0,0)}$ describes the average power of the first arrival of the first cluster, T_l represents the arrival time of the lth cluster, and τ_{kl} is the arrival time of the kth arrival within the

Table 3.25 Statistical models and parameters

	Global parameters $\Rightarrow \overline{G}_{\text{tot}}$ and $\overline{G}_k$
Path loss	$PL = \begin{cases} 20.4\log_{10}(d/d_0), & d \leq 11\text{ m} \\ -56 + 74\log_{10}(d/d_0), & d > 11\text{ m} \end{cases}$
Shadowing	$\overline{G}_{\text{tot}} \sim \mathcal{L}_{\mathcal{N}}(-PL;\ 4.3)$
Decay constant	$\varepsilon \sim \mathcal{L}_{\mathcal{N}}(16.1;\ 1.27)$
Power ratio	$r \sim \mathcal{L}_{\mathcal{N}}(-4;\ 3)$
	Local parameters $\Rightarrow G_k$
Energy gains	$G_k \sim \Gamma(\overline{G}_k;\ m_k)$
	$m_k \sim \mathcal{T}_{\mathcal{N}}\left(\mu_m(\tau_k);\ \sigma_m^2(\tau_k)\right)$
m Values	$\mu_m(\tau_k) = 3.5 - \dfrac{\tau_k}{73}$
	$\sigma_m^2(\tau_k) = 1.84 - \dfrac{\tau_k}{160}$

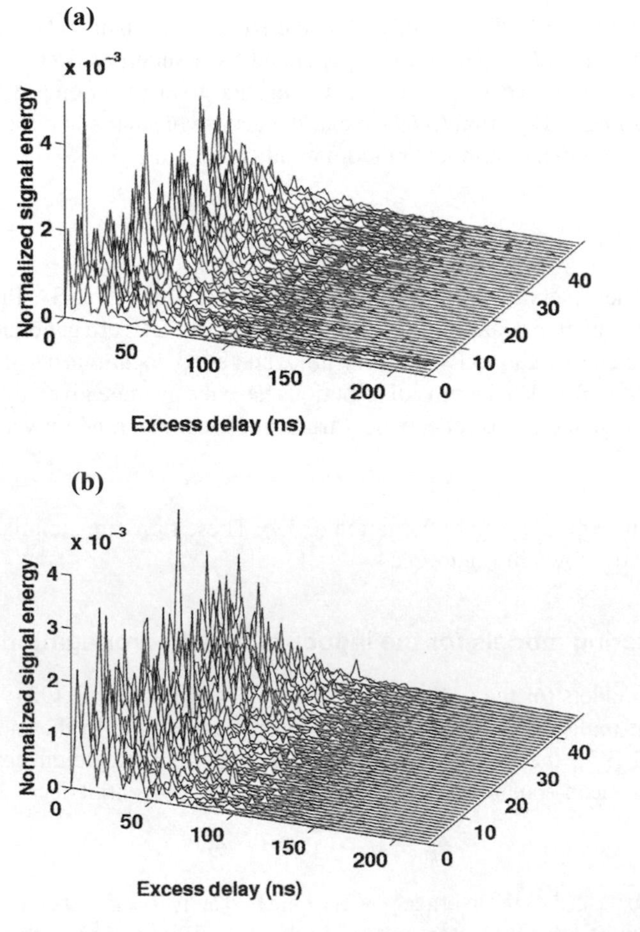

Figure 3.24 (a) The measured 49 local PDPs for an example room. (b) Simulated 49 local PDPs for an example room. (Reproduced by permission of IEEE [86].)

lth cluster, relative to T_l. The parameters Γ and γ determine the intercluster signal level rate of decay and the intracluster rate of decay, respectively. The parameter Γ is generally determined by the architecture of the building, while γ is determined by objects close to the receiving antenna, such as furniture. The results presented in Spencer *et al.* [95] make the assumption that the channel impulse response as a function of time and azimuth angle is a separable function, or

$$h(t,\theta) = h(t)\,h(\theta) \tag{3.74}$$

from which independent descriptions of the multipath time-of-arrival and angle-of-arrival are developed. This is justified by observing that the angular deviation of the signal arrivals within a cluster from the cluster mean does not increase as a function of time.

The cluster decay rate Γ and the ray decay rate γ can be interpreted for the environment in which the measurements were made. For the results, presented later in this section, at least one wall separates the transmitter and the receiver. Each cluster can be viewed as a path that exists between the transmitter and the receiver along which signals propagate. This cluster path is generally a function of the architecture of the building itself. The component arrivals within a cluster vary because of secondary effects, e.g. reflections from the furniture or other objects. The primary source of degradation in the propagation through the features of the building is captured in the decay exponent Γ. Relative effects between paths in the same cluster do not always involve the penetration of additional obstructions or additional reflections, and therefore tend to contribute less to the decay of the component signals. Results for $p(\theta)$ generated from the data in Cramer *et al.* [97] are shown in Figure 3.25. *Interarrival times* are hypothesized [95] to follow exponential rate laws, given by

$$\begin{aligned} p\left(T_l \mid T_{l-1}\right) &= \Lambda e^{-\Lambda(T_l - T_{l-1})} \\ p\left(T_{kl} \mid T_{k-1,l}\right) &= \lambda e^{-\lambda(T_l - T_{l-1})} \end{aligned} \tag{3.75}$$

where Λ is the cluster arrival rate and λ is the ray arrival rate. Channel parameters are summarized in Table 3.26.

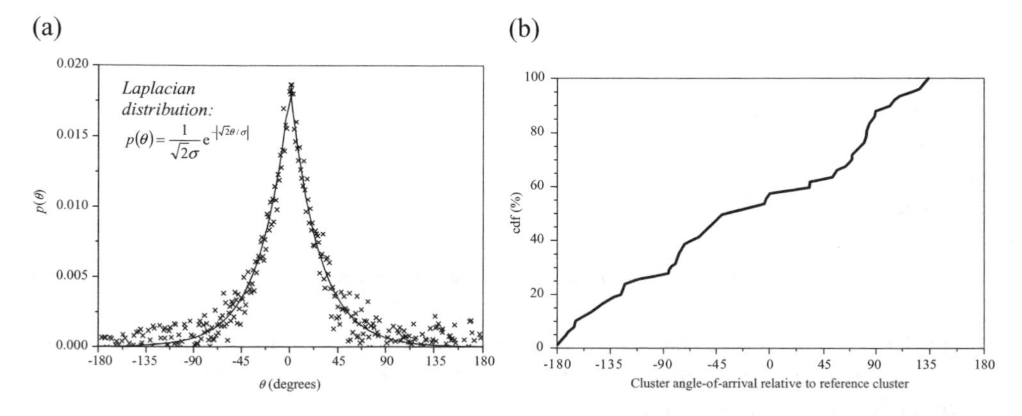

Figure 3.25 (a) Ray arrival angles at 1° of resolution and a best fit Laplacian density with $\sigma = 38°$. (b) Distribution of the cluster azimuth angle-of-arrival, relative to the reference cluster. (Reproduced by permission of IEEE [97].)

Table 3.26 Channel parameters

Parameter	UWB [97]	Spencer *et al.* [95]	Spencer *et al.* [95]	Saleh and Valenzuela [94]
Γ	27.9 ns	33.6 ns	78.0 ns	60 ns
γ	84.1 ns	28.6 ns	82.2 ns	20 ns
$1/\Lambda$	45.5 ns	16.8 ns	17.3 ns	300 ns
$1/\lambda$	2.3 us	5.1 ns	6.6 ns	5 ns
σ	37°	25.5°	21.5°	—

3.9.6 Path loss modeling

In this section we are interested in a transceiver operating at approximately 2 GHz center frequency with a bandwidth in excess of 1.5 GHz, which translates to sab-nanosecond time resolution in the CIRs.

3.9.6.1 Measurement procedure

The measurement campaign is described in Yano [98] and was conducted in a single-floor, hard-partition office building (fully furnished). The walls were constructed of drywall with vertical metal studs; there was a suspended ceiling 10 feet in height with carpeted concrete floor. Measurements were conducted with a stationary receiver and mobile transmitter; both transmit and receive antennas were 5 feet above the floor. For each measurement, a 300 ns time-domain scan was recorded and the LOS distance from transmitter to receiver was recorded. A total of 906 profiles were included in the dataset with seven different receiver locations recorded over the course of several days. Except for a reference measurement made for each receiver location, all successive measurements were NLOS links chosen randomly throughout the office layout that penetrated anywhere from one to five walls. The remainder of datapoints were taken in similar fashion.

3.9.6.2 Path loss modeling

The average pathless for an arbitrary TR separation is expressed using the power law as a function of distance. *The indoor* environment measurements show that, at any given d, shadowing leads to signals with a path loss that is log–normally distributed about the mean [99]. That is:

$$PL(d) = PL_0(d_0) + 10N \log\left(\frac{d}{d_0}\right) + X_\sigma \tag{3.76}$$

where $N$is the pathloss exponent, X_σ is a zero mean log–normally distributed random variable with standard deviation σ (dB), PL_0 is the free space path loss at reference distance, d_0. Some results are shown in Figure 3.27.

Assuming a simple RAKE with four correlators where each component is weighted equally, we can calculate the path loss vs distance using the peak channel impulse response (CIR) power plus RAKE gain, $PL_{\text{PEAK+RAKE}}$, for each CIR, as shown in Figure 3.26(c). The

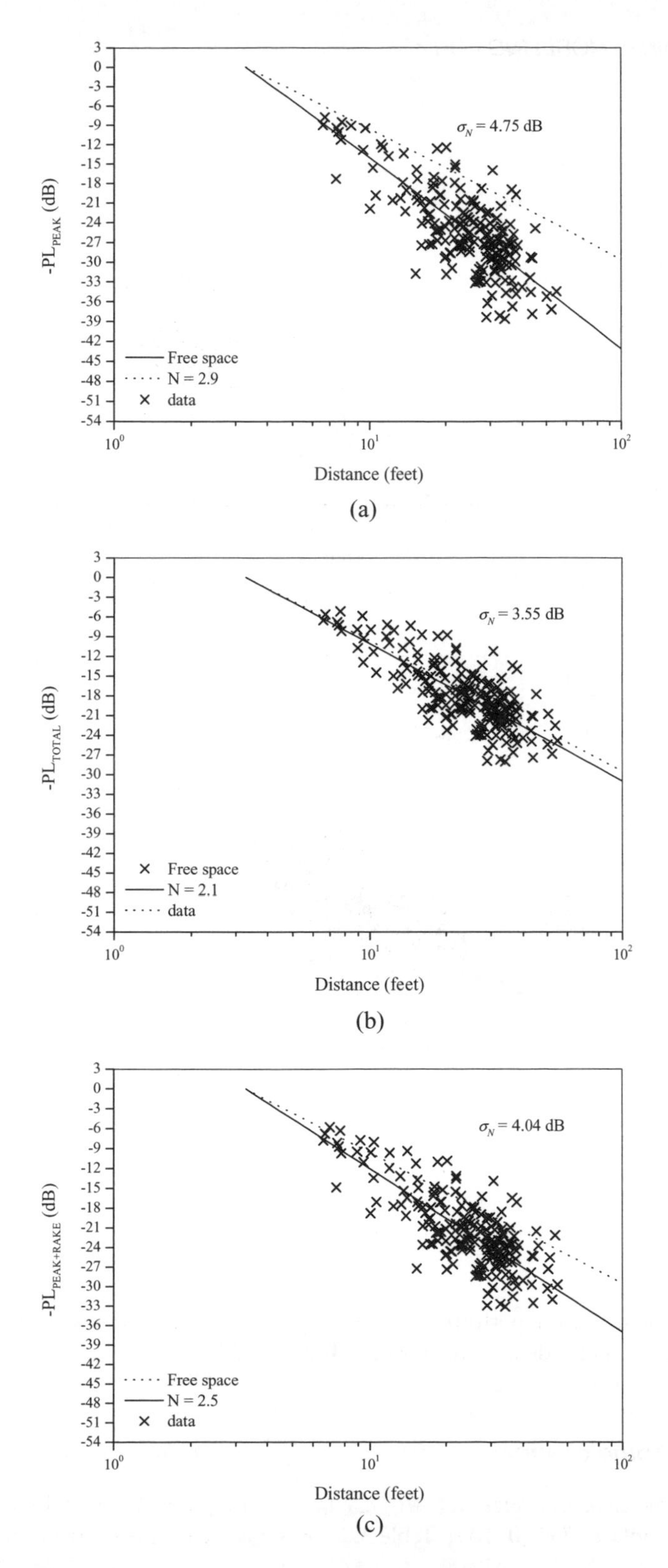

Figure 3.26 (a) Peak PL vs distance; (b) total PL vs distance; (c) peak PL + rake gain vs distance. (Reproduced by permission of IEEE [98].)

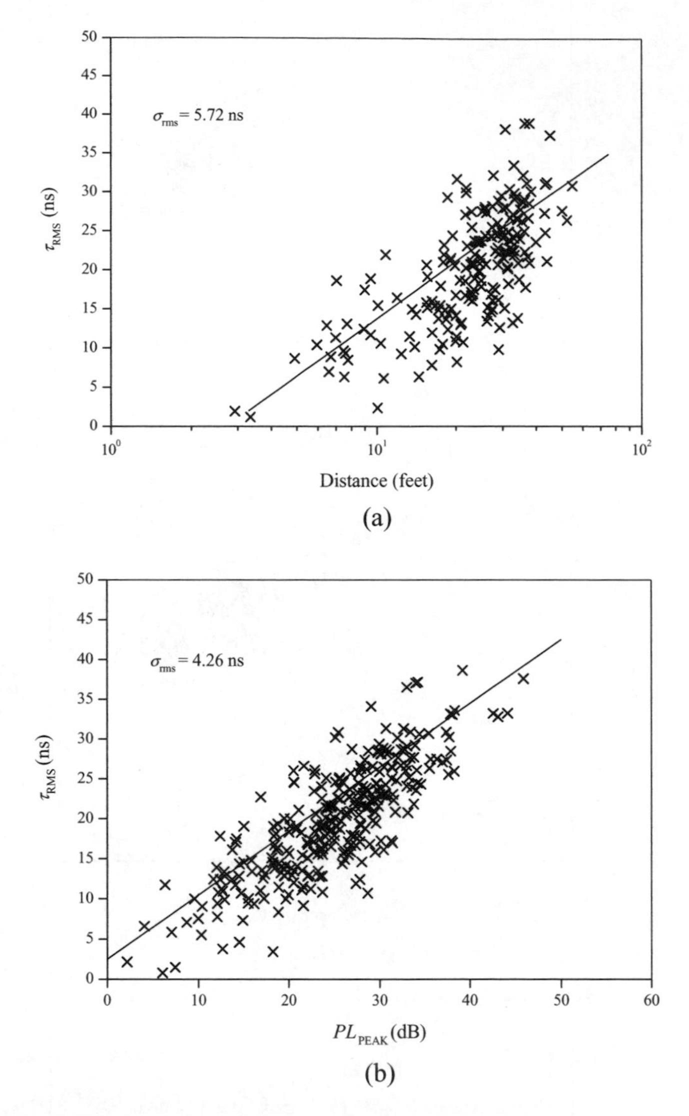

Figure 3.27 (a) The RMS delay spread vs distance; (b) RMS delay spread vs path loss. (Reproduced by permission of IEEE [98].)

exponent N obtained from performing a least-squares fit is 2.5 with a standard deviation of 4.04 dB. The results for delays are shown in Figure 3.27.

3.9.6.3 In-home channel

For the in-home channel Equation (3.76) can be also used to model path loss. Some results are shown in Table 3.27 [100–103]. Table 3.28 presents the results for delay spread in home channel [100].

Table 3.27 Statistical values of the path loss parameters

	LOS		NLOS	
	Mean	SD	Mean	SD
PL_0 (dB)	47		51	
N	1.7	0.3	3.5	0.97
σ (dB)	1.6	0.5	2.7	0.98

Table 3.28 Percentage of power contained in profile, number of paths, mean excess delay and RMS delay spread for 5, 10, 15, 20 and 30 dB threshold levels

	50 % NLOS				90 % NLOS			
Threshold	Power %	L	τ_m(ns)	τ_{RMS}(ns)	Power %	L	τ_m(ns)	τ_{RMS}(ns)
5 dB	46.8	7	1.95	1.52	46.9	8	2.2	1.65
10 dB	89.2	27	7.1	5.77	86.5	31	8.1	6.7
15 dB	97.3	39	8.6	7.48	96	48	10.3	9.3
20 dB	99.4	48	9.87	8.14	99.5	69	12.2	11
30 dB	99.97	60	10.83	8.43	99.96	82	12.4	11.5

REFERENCES

[1] S. Glisic, *Advanced Wireless Communications, 4G Technology*. Wiley: Chichester 2004.

[2] K.I. Pedersen, P.E. Mogensen and B.H. Fleury, A stochastic model of the temporal and azimuthal dispersion seen at the base station in outdoor propagation environments, *IEEE Trans. Vehicle Technol.*, vol. 49, 2000, pp. 437–447.

[3] F. Frederiksen, P. Mogensen, K.I. Pedersen and P. Leth-Espensen, A software testbed for performance evaluation of adaptive antennas in FH GSM and wideband-CDMA, in *Proc. 3rd ACTS Mobile Communication Summit,* vol. 2, Rhodes, June 1998, pp. 430–435.

[4] A. Algans, K.I. Pedersen and P.E. Mogensen, Experimental analysis of the joint statistical properties of azimuth spread, delay spread, and shadow fading, *IEEE J. Select. Areas Commun.*, vol. 20, 2002, pp. 523–531.

[5] D.C Cox, R. Murray and A. Norris, 800 MHz attenuation measured in and around sub-urban houses, *AT&T Bell Lab. Tech. J.*, vol. 673, 1994.

[6] R.C. Bernhardt, Macroscopic diversity in frequency reuse systems, *IEEE J. Select. Areas Commun.*, vol. 5, 1987, pp. 862–878.

[7] H. L. Bertoni, *Radio Propagation for Modern Wireless System.* Prentice-Hall: Englewood Cliffs, NJ, 2000.

[8] L. Correia, *Wireless Flexible Personalised Communications-Cost 259 Final Report.* Wiley: New York, 2001.

[9] J. Winters, Optimum combining in digital mobile radio with cochannel interference *IEEE Trans. Vehicle Technol.*, vol. VT-33, 1984.

[10] K.I. Pedersen, P.E. Mogensen and B.H. Fleury, Spatial channel characteristics in outdoor environments and their impact on BS antenna system performance, in *Proc. IEEE Vehicular Technology Conf. (VTC'98)*, Ottawa, May 1998, pp. 719–724.
[11] K.I. Pedersen and P.E. Mogensen, Evaluation of vector-RAKE receivers using different antenna array configurations and combining schemes, *Int. J. Wireless Inform. Networks*, vol. 6. 1999, pp. 181–194.
[12] V. Veen and K.M. Buckley, Bcamforming: a versatile approach to spatial filtering, *lEEE Acoust., Speech Signal Processing (ASSP) Mag.*, vol, 5, 1988, pp. 4–24.
[13] J. Liberti and T. Rappaport, A geometrically based model for line-of-sight multipath radio channels, in *Proc. IEEE Vehicular Technology Conf. (VTC 96)*, May 1996, pp. 844–848.
[14] M. Lu, T. Lo and J. Litva, A physical spatio-temporal model of multipath propagation channels, in *Proc. IEEE Vehicular Technology Conf. (VTC 97)*, May 1997, pp. 810–814.
[15] O. Nørklit and J.B. Andersen, Diffuse channel model and experimential results for antenna arrays in mobile environments, *IEEE Trans. Antennas Propagat.*, vol. 96, 1998, pp. 834–840.
[16] C. Cheon, H.L. Bertoni and G. Liang, Monte Carlo simulation of delay and angle spread in different building environments, in *Proc. IEEE Vehicular Technology Conf. (VTC'00)*, Boston, MA, vol.1, 20–28 September 2000, pp. 49–56.
[17] U. Martin, A directional radio channel model for densely built-up urban areas, in *Proc. European Personal Mobile Communications Conf. (EPMCC)*, Bonn, October 1997, pp. 237–244.
[18] M. Gudmundson, Correlation model for shadow fading in mobile radio systems, *IEEE Electron. Lett.*, vol. 27, 1992, pp. 2126–2145.
[19] A. Mawira, Models fur the spatial correlation functions of the log-normal component of the variability of VHF/UHF field strength in urban environment, in *Proc. Personal, Indoor and Mobile Radio Communications (PIMRC'92)*, London, 1992, pp. 436–440.
[20] A. Gehring, M. Steinbauer, I. Gaspard and M. Grigat, Empirical channel stationarity in urban environments, in *Proc. European Personal Mobile Communications Conf. (EPMCC)*, Vienna, February, 2001.
[21] T.B. Sørensen, Correlation model for shadow fading in a small urban macro cell, in *Proc. Personal Indoor and Mobile Radio Communications (PIMRC'98)*, Boston, MA, 1998.
[22] Algorithms and antenna array recommendations, Public deliverable from European ACTS, TSUNAMI II Project, Deliverable code: AC020/AUC/A1.2/DR/P/005/bl, May 1997.
[23] Commission of the European Communities, Information technologies and sciences – digital land mobile radio communications, COST 207, 1998.
[24] P.A. Bello, Characterization of randomly time-variant linear channels, *IEEE Trans. Commun. Syst.*, vol. CS-11, 1963, pp. 360–393.
[25] J.A. Fessler and A. Hero, Space-alternating generalised expectation-maximization algorithm, *IEEE Trans. Signal Processing*, vol. 42, 1994, pp. 2664–2677.
[26] B.H. Fleury, M. Tschudin, R. Heddergott, D. Dahlhaus and K. I. Pedersen, Channel parameter estimation in mobile radio environments using the SAGE algorithm, *IEEE J. Select. Areas Commun.*, vol. 17, 1999, pp. 434–450.

[27] K.I. Pedersen, B.H. Fleury and P.E. Mogensen, High resolution ot electromagnetic waves in time-varying radio channels, in *Proc. Int. Symp. Personal, Indoor and Mobile Radio Communications (PIMRC'97)*, Helsinki, September 1997, pp. 650–654.

[28] K.I. Pedersen, P.E. Mogensen , B.H. Fleury, F. Frederiksen and K. Olesen, Analysis of time, azimuth and doppler dispersion in outdoor radio channels, in *Proc. ACTS Mobile Communication Summit '97,* Aalborg, October 1997, pp. 308–313.

[29] M. Hata, Empirical formula for propagation loss in land mobile radio service, *IEEE Trans. Vehicle Technol.*, vol. VT-29, 1980, pp. 317–325.

[30] L.J. Greenstein, V. Erceg, Y.S. Yeh and M.V. Clark, A new path-gain/delay-spread propagation model for digital cellular channels, *IEEE Trans. Vehicle Technol.*, vol. 46, 1997, pp. 477–485.

[31] M. Toeltsch, J. Laurila, K. Kalliola, A.F. Molisch, P. Vainikainen and E. Bonek, Statistical characterization of urban spatial radio channels, *IEEE J. Select. Areas Commun.*, vol. 20, no. 3, 2002, pp. 539–549.

[32] R. Roy, A. Paulraj and T. Kailath, ESPRIT – a subspace rotation approach to estimation of parameters of cisoids in noise, *IEEE Trans. Acoust., Speech, Signal Process.*, vol. 32, 1986, pp. 1340–1342.

[33] M. Zoltowski, M. Haardt and C. Mathews, Closed-form 2-D angle estimation with rectangular arrays in element space or beamspace via unitary ESPRIT, *IEEE Trans. Signal Process.*, vol. 44, 1994, pp. 316–328.

[34] M. Haardt and J. Nossek, Unitary ESPRIT: how to obtain an increased estimation accuracy with a reduced computational burden, *IEEE Trans. Signal Process.*, vol. 43, 1995, pp. 1232–1242.

[35] S.C. Swales, M. Beach, D. Edwards and J.P. McGeehan, The performance enhancement of multibeam adaptive base-station antennas for cellular land mobile ratio systems, *IEEE Trans. Vehicle Technol.*, vol. 39, 1990, pp. 56–67.

[36] R.B. Ertel, P. Cardieri, K.W. Sowerby, T.S. Rappaport and J.H. Reed, Overview of spatial channel models for antenna array communications systems, *IEEE Personal Commun.*, vol. 5, no. 1, 1998, pp. 10–22.

[37] U. Martin, J. Fuhl, I. Gaspard, M. Haardt, A. Kuchar, C. Math, A.F. Molisch and R. Thomä, Model scenarios for direction-selective adaptive antennas in cellular mobile communication systems – scanning the literature, *Wireless Personal Commun. Mag. (Special Issue on Space Division Multiple Access)*, vol. 11, no. 1, 1999, pp. 109–129.

[38] K. Pedersen, P. Mogensen and B. Fleury, A stochastic model of the temporal and azimuthal dispersion seen at the base station in outdoor propagation environments, *IEEE Trans. Vehicle Technol.*, vol. 49, no. 2, 2000, pp. 437–447.

[39] J. Fuhl, J.-P. Rossi and E. Bonek, High resolution 3-D direction-of-arrival determination for urban mobile radio, *IEEE Trans. Antennas Propagat.*, vol. 4, 1997, pp. 672–682.

[40] A. Kuchar, J.-P. Rossi and E. Bonek, Directional macro-cell channel characterization from urban measurements, *IEEE Trans. Antennas Propagat.*, vol. 48, 2000, 137–146.

[41] K. Kalliola, H. Laitinen, L. Vaskelainen and P. Vainikainen, Real-time 3D spatial-temporal dual-polarized measurement of wideband radio channel at mobile station, *IEEE Trans. Instrum. Meas.*, vol. 49, 2000, pp. 439–448.

[42] J. Kivinen, T. Korhonen, P. Aikio, R. Gruber, P. Vainikainen and S.-G. Häggman, Wideband radio channel measurement system at 2 GHz, *IEEE Trans. Instrum. Meas.*, vol. 48, 1999, pp. 39–44.
[43] J.P. Kermoal, L. Schumacher, K.I. Pedersen, P.E. Mogensen and F. Frederiksen, A stochastic MIMO radio channel model with experimental validation, *IEEE J. Selected Areas Commun.*, vol. 20, 2002, pp. 1211–1226.
[44] K. Yu, M. Bengtsson, B. Ottersten, D. McNamara, P. Karlsson and M. Beach, Second order statistics of NLOS indoor MIMO channels based on 5.2 GHz measurements, in *Proc. GLOBECOM'01*, San Antonio, TX, 2001, pp. 156–160.
[45] F. Frederiksen, P. Mogensen, K.I. Pedersen and P. Leth-Espensen, A 'Software' testbed for performance evaluation of adaptive antennas in FH GSM and wideband-CDMA, in *Conf. Proc. 3rd ACTS Mobile Communication Summit*, vol. 2, Rhodes, June 1998, pp. 430–435.
[46] J.B. Andersen, Array gain and capacity for known random channels with multiple element arrays at both ends, *IEEE J. Select. Areas Commun.*, vol. 18, 2000, pp. 2172–2178.
[47] X. Zhao, J. Kivinen, P. Vainikainen and K. Skog, Propagation characteristics for wideband outdoor mobile communications at 5.3 GHz, *IEEE J. Selected Areas Commun.*, vol. 20, 2002, pp. 507–514.
[48] H. Suzuki, A statistical model for urban radio propagation, *IEEE Trans. Cammun.*, vol. 25, 1977, pp. 673–680.
[49] J. Takada, Fu Jiye, Zhu Houtao T. Kobayashi, Spatio-temporal channel characterization in a suburban non line-of-sight microcellular environment, *IEEE J. Selected Areas Commun.*, vol. 20, no. 3, 2002, pp. 532–538.
[50] J. Kivinen, X. Zhao and P. Vainikainen, Empirical characterization of wideband indoor radio channel at 5.3 GHz, *IEEE Trans. Antennas Propagat.*, vol. 49, 2001, pp. 1192–1203.
[51] H. Masui, K. Takahashi, S. Takahashi, K. Kage and T. Kobayashi, Delay profile measurement system for microwave broadband transmission and analysis of delay characteristics in an urban environment, *IEICE Trans. Electron.*, vol. E82-C, 1999, pp. 1287–1292.
[52] A.F. Molisch, M. Steinbauer, M. Toeltsch, E. Bonek and R.S. Thomä, Capacity of MIMO systems based on measured wireless channels, *IEEE J. Select. Areas Commun.*, vol. 20, 2002, pp. 561–569.
[53] J. Parks-Gornet and I.N. Imam, Using rank factorization in calculating the Moore-Penrose generalized inverse, in *IEEE Southeastcon '89. 'Energy and Information Technologies in the Southeast'*, vol. 2, 9–12 April 1989, pp. 427–431.
[54] J. Tokarzewski, System zeros analysis via the Moore–Penrose pseudoinverse and SVD of the first nonzero Markov parameter, *IEEE Trans. Autom. Control*, vol. 43, no. 9, 1998, pp. 1285–1291.
[55] L.P. Withers, Jr, A parallel algorithm for generalized inverses of matrices, with applications to optimum beamforming, in *IEEE Int. Conf. Acoustics, Speech, and Signal Processing, ICASSP-93*, vol. 1, 27–30 April 1993, pp. 369–372.
[56] Shu Wang and Xilang Zhou, Extending ESPRIT algorithm by using virtual array and Moore-Penrose general inverse techniques, in *IEEE Southeastcon '99*, 25–28 March 1999, pp. 315–318.
[57] M. Steinbauer, A.F. Molisch and E. Bonek, The double-directional radio channel, *IEEE Antennas Propagat. Mag.*, 2001, pp. 51–63.

[58] M.L. Rubio, A. Garcia-Armada, R.P. Torres and J.L. Garcia, Channel modeling and characterization at 17 GHz for indoor broadband WLAN, *IEEE J. Select. Areas Commun.*, vol. 20, 2002, pp. 593–601.
[59] H. Hashemi, The indoor radio propagation channel, *Proc. IEEE*, vol. 81, no. 7, 1993, pp. 943–968.
[60] K. Bury, *Statistical Distributions in Engineering*. Cambridge University Press: Cambridge, 1999.
[61] R.P. Torres, L. Valle, M. Domingo, S. Loredo and M. C Diez, CIN-DOOR: an engineering tool for planning and design of wireless systems in enclosed spaces, *IEEE Anennas Propagat. Mag.*, vol. 41, 1999, pp. 11–22.
[62] S. Loredo. R.P. Torres, M. Domingo, L. Valle and J.R. Pérez, Measurements and predictions of the local mean power and small-scale fading statistics in indoor wireless environments, *Microvave Opt. Technol. Lett.*, vol. 24, 2000, pp. 329–331.
[63] R.R. Torres, S. Loredo, L. Valle and M. Domingo, An accurate and efficient method based on ray-tracing for the prediction of local flat-fading statistic in picocell radio channels, *lEEE J. Select. Areas Commun.*, vol. 18, 2001, pp. 170–178.
[64] S. Loredo, L. Valle and R.P. Torres, Accuracy analysis of GO/UTD radio channel modeling in indoor scenarios at 1.8 and 2.5 GHz, *IEEE Anennas Propagat. Mag.*, vol. 43, 2001, pp. 37–51.
[65] A. Bohdanowicz, Wide band indoor and outdoor radio channel measurements at 17 GHz, Ubicom. Technical Report/2000/2, Febuary 2000.
[66] A. Bohdanowicz, G. J. M. Janssen and S. Pietrzyk, Wide band indoor and outdoor multipath channel measurements at 17 GHz, in *Proc. IEEE Vehicular Technology Conf.*, Amsterdam, vol. 4, 1999, pp. 1998–2003.
[67] L. Talbi and G. Y. Delisle, Experimental characterization of EHF multipath indoor radio channels, *IEEE J. Select. Areas Commun.*, vol. 14, 1996, pp. 431–439.
[68] T.S. Rappaport and C.D. McGillem, UHF fading in factories, *IEEE J. Select. Areas Commun.*, vol. 7, 1989, pp. 40–48.
[69] T.S. Rappaport, Indoor radio communications for factories of the future, *IEEE Commun. Mag.*, 1989, pp. 15–24.
[70] A.F. Abou-Raddy and S. M. Elnoubi, Propagation measurements and channel modeling for indoor mobile radio at 10 GHz, in *Proc. IEEE Vehicular Technology Conf.*, vol. 3, 1997, pp. 1395–1399.
[71] A.A.M. Saleh and R.A. Valenzuela, A statistical model for indoor multipath propagation, *IEEE J. Select. Areas Commun.*, vol. SAC-5, 1987, pp. 128–137.
[72] Hao Xu, V. Kukshya and T.S. Rappaport, Spatial and temporal characteristics of 60-GHz indoor channels, *IEEE J. Selected Areas Commun.*, vol. 20, no. 3, 2002, pp. 620–630.
[73] T.S. Rappaport, *Wireless Communications: Principles and Practice*. Prentice-Hall: Englewood Cliffs, NJ, 1996.
[74] G. Durgin and T.S. Rappaport, Theory of multipath shape factors for small-scale fading wireless channels, *IEEE Trans. Antennas Propagat.*, vol. 48, 2000, pp. 682–693.
[75] T. Manabe, Y. Miura and T. Ihara, Effects of antenna directivity and polarization on indoor multipath propagation characteristics at 60 GHz, *IEEE J. Select. Areas Commun.*, vol. 14, 1996, pp. 441–448.
[76] K. Sato, T. Manabe, T. Ihara, H. Saito, S. Ito, T. Tanaka, K. Sugai, N. Ohmi, Y. Murakami, M. Shibayama, Y. Konishi and T. Kimura, “Measurements of reflection

and transmission characteristics of interior structures of office building in the 60-GHz band," *IEEE Trans. Antennas Propagat.*, vol. 45, 1997, pp. 1783–1792.

[77] B. Langen, G. Lober and W. Herzig, Reflection and transmission behavior of building materials at 60 GHz, in *Proc. IEEE PIMRC'94*, The Hague, September 1994, pp. 505–509.

[78] J.P. Rossi, J.P. Barbot and A.J. Levy, Theory and measurement of the angle of arrival and time delay of UHF radiowave using a ring array, *IEEE Trans. Antennas Propagat.*, vol. 45, 1997, pp. 876–884.

[79] Y.L.C. de Jong and M.H.A.J. Herben, High-resolution angle-of-arrival measurement of the mobile radio channel, *IEEE Trans. Antennas Propagat.*, vol. 47, 1999, pp. 1677–1687.

[80] H. Droste and G. Kadel, Measurement and analysis of wideband indoor propagation characteristics at 17 GHz and 60 GHz, in *Proc. IEE Antennas Propagation, Conf. (publication no. 407)*, April 1995, pp. 288–291.

[81] C.L. Holloway, P.L. Perini, R.R. Delyzer and K.C. Allen, Analysis of composite walls and their effects on short-path propagation modeling, *IEEE Trans. Vehicle Technol.*, vol. 46, 1997, pp. 730–738.

[82] W. Honcharenko and H. Bertoni, Transmission and reflection characteristics at concrete block walls in the UHF bands proposed for future PCS, *IEEE Trans. Antennas Propagat.*, vol. 42, 1994, pp. 232–239.

[83] P. Smulders and L. Correia, Characterization of propagation in 60 GHz radio channels, *Electron. Commun. Eng. J.*, 1997, pp. 73–80.

[84] H. Xu, T. S. Rappaport, R. J. Boyle and J. Schaffner, Measurements and modeling for 38-GHz point-to-multipoint radiowave propagation, *IEEE J. Select. Areas Commun.*, vol. 18, 2000, pp. 310–321.

[85] H. Xu, V. Kukshya and T. Rappaport, Spatial and temporal characterization of 60 GHz channels, in *Proc. IEEE VTC'2000*, Boston, MA, vol. 1, 24–28 September, 2000, pp. 6–13.

[86] D. Cassioli, M.Z. Win and A.F. Molisch, The ultra-wide bandwidth indoor channel: from statistical model to simulations, *IEEE J. Selected Areas Commun.*, vol. 20, no. 6, 2002, pp.1247–1257.

[87] L.J. Greenstein, V. Erceg, Y. S. Yeh and M. V. Clark, A new path-gain/delay-spread propagation model for digital cellular channels, *IEEE Trans. Vehicle Technol.*, vol. 46, 1997, pp. 477–485.

[88] V. Erceg, L.J. Greenstein, S.Y. Tjandra, S.R. Parkoff, A. Gupta, B. Kulic, A. A. Julius and R. Bianchi, An empirically based path loss model for wireless channels in suburban environments, *IEEE J. Select. Areas Commun.*, vol. 17, 1999, pp. 1205–1211.

[89] E. Failli, ed., *Digital Land Mobile Radio. Final Report of COST 207*. Commission of the European Union: Luxemburg 1989.

[90] M. Nakagami, The m-distribution – a general formula of intensity distribution of rapid fading, in *Statistical Methods in Radio Wave Propagation*, W. C. Hoffman, ed. Pergamon: Oxford, 1960, pp. 3–36.

[91] W.R. Braun and U. Dersch, A physical mobile radio channel model, *IEEE Trans. Vehicle. Technol.*, vol. 40, 1991, pp. 472–482.

[92] H. Hashemi, The indoor radio propagation channel, *Proc. IEEE*, vol. 81, 1993, pp. 943–968.

[93] J.C. Liberti and T.S. Rappaport, *Smart Antennas for Wireless Communications: IS-95 and Third Generation CDMA Applications.* Prentice-Hall: Englewood Cliffs, NJ, 1999.
[94] A.A.M. Saleh and R.A. Valenzuela, A statistical model for indoor multipath propagation, *IEEE J. Select. Areas Commun.*, vol. 5, 1987, pp. 128–137.
[95] Q. Spencer, M. Rice, B. Jeffs and M. Jensen, A statistical model for the angle-of-arrival in indoor multipath propagation, in *Proc. IEEE Vehicle Technol. Conf.*, 1997, pp. 1415–1419.
[96] Q. Spencer, B. Jeffs, M. Jensen and A. Swindlehurst, Modeling the statistical time and angle of arrival characteristics of an indoor multipath channel, *IEEE J. Select. Areas Commun.*, vol. 18, 2000, pp. 347–360.
[97] R.J.-M. Cramer, R.A. Scholtz and M.Z. Win, Evaluation of an ultra-wide-band propagation channel, *IEEE Trans. Antennas Propagat.*, vol. 50, no. 5, 2002, pp. 561–570.
[98] S.M. Yano, Investigating the ultra-wideband indoor wireless channel, *IEEE 55th Vehicular Technology Conf.*, VTC, vol. 3, 6–9 May 2002, pp. 1200–1204.
[99] D. Cassioli, M.Z. Win and M.F. Molisch, A statistical model for the UWB indoor channel, *IEEE VTC*, vol. 2, 2001, pp. 1159–1163.
[100] S.S. Ghassemzadeh, R. Jana, C.W. Rice, W. Turin and V. Tarokh, A statistical path loss model for in-home UWB channels, *2002 IEEE Conf. Ultra Wideband Systems and Technologies, Digest of Papers*, 21–23 May 2002, pp. 59–64.
[101] W. Turin, R. Jana, S.S. Ghassemzadeh, C.W. Rice and T. Tarokh, Autoregressive modelling of an indoor UWB channel, *IEEE Conference on Ultra Wideband Systems and Technologies, Digest of Papers*, 21–23 May 2002, pp. 71–74.
[102] S.S. Ghassemzadeh and V. Tarokh, UWB path loss characterization in residential environments, *IEEE Radio Frequency Integrated Circuits (RFIC) Symp.*, 8–10 June 2003, pp. 501–504.
[103] S.S. Ghassemzadeh and V. Tarokh UWB path loss characterization in residential environments, *IEEE MTT-S International Microwave Symposium Digest*, vol. 1, 8–13 June 2003, pp. 365–368.
[104] G.H. Golub and C.F. Van Loan, *Matrix Computations*, 3rd edn. The Johns Hopkins University Press: Baltimore, MD, 1996.
[105] P. Petrus, J.H. Reed and T.S. Rappaport, Effects of directional antennas at the base station on the Doppler spectrum, *IEEE Commun. Lett.*, vol. 1, 1997, pp. 40–42.
[106] W.C.Y. Lee, Estimate of local average power of a mobile radio signal, *IEEE Trans. Vehicle Technol.*, vol. 34, 1985, pp. 22–27.
[107] E. Green, Radio link design for micro-cellular systems, *Br. Telecom. Technol. J.*, vol. 8, 1990, pp. 85–96.
[108] U. Dersch and E. Zollinger, Physical characteristics of urban micro-cellular propagation, *IEEE Trans. Antennas Propagat.*, vol. 42, 1994, pp. 1528–1539.
[109] Y. Karasawa and H. Ivai, Formulation of spatial correlation statistic in Nakagami-Rice fading environments, *IEEE Trans. Antennas Propagat.*, vol. 48, 2000, pp. 12–18.

4

Adaptive and Reconfigurable Link Layer

4.1 LINK LAYER CAPACITY OF ADAPTIVE AIR INTERFACES

In wireless systems, the channel reliability is affected by several phenomena, such as the propagation properties of the environment and mobility of terminals. To compensate for these impairments, various techniques are used at the physical layer, including adaptive schemes, which dynamically modify the transceiver structure. As a consequence of varying channel conditions and dynamically changing transceivers structures, the available information data rate at the link layer is in general time-varying. The channel seen from above the physical and link control layer, which will be referred to as a medium access control (MAC) channel, must be characterized with sufficient accuracy but still by a simple model that can be used in the analysis of the higher network layers. In the resulting MAC channel model used in this chapter, physical layer characteristics, as well as the physical channel and some implementation losses, are taken into account. The efficiency of the model is improved, avoiding bit level or signal level and detailed channel statistics calculations. The result is a model that can be easily used for analytical purposes, as well as for efficient modeling of the MAC channel behavior in network simulators. Owing to its modular structure, the analysis can be extended to more general and possibly complicated systems. In general, the physical layer, or layer 1 (L1), provides a virtual link of unreliable bits. For the sake of simplicity, in this chapter we will use the term *physical layer* (PHY) when referring to the protocol stack portions in which no distinction is made regarding the information carried by the bits. The portions in which such a distinction is made will be referred to as *upper layers*. This corresponds to L2–L7 of the International Standardization Organization–Open System of Interconnections (ISO-OSI) model.

The model discussed in this chapter includes a physical channel and the physical layer (Figure 4.1) [1]. The same portion of the protocol stack is covered in a link layer model,

Advanced Wireless Networks: 4G Technologies Savo G. Glisic
© 2006 John Wiley & Sons, Ltd.

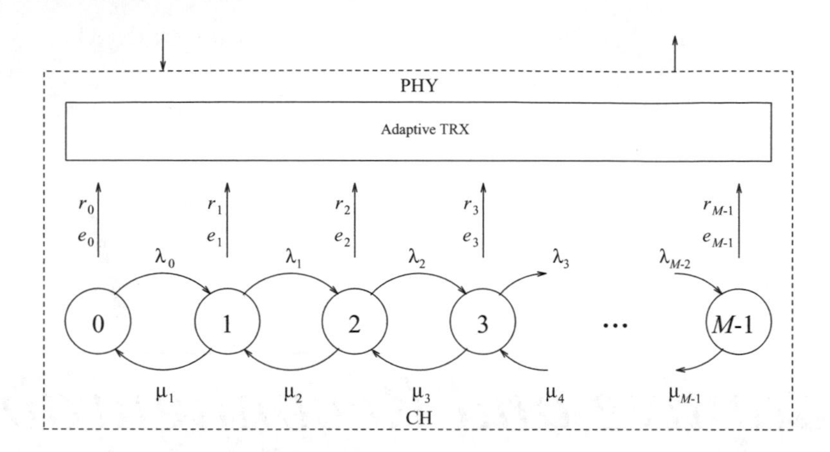

Figure 4.1 The MAC channel model that includes the overall behavior of the physical channel and the PHY. According to the channel state, the PHY mode is chosen. As a consequence, in each state, data rate r_i and error rate e_i are defined. The model includes imperfections in the implementation of the adaptation method.

called the effective capacity link model [2], which models directly some *link layer parameters* used in queuing analysis without including *imperfections* of the physical layer or *adaptation* in the link layer. A similar definition of MAC channel is given in Liebl *et al.* [3], where a model for *packet losses* is included, taking into account physical channel, modulation and channel coding, and some other functions of the data link layer, but only for transceivers with *fixed* structure.

In order to extend Gilbert [4], Gilbert–Elliot [5], Fritchman [6] or bipartite models [7], as hidden Markovian models [8], to a finer granularity of error rates or larger dynamic range with adaptation techniques at the physical layer, finite-state Markov chains with larger numbers of states have been introduced, e.g. in Wang and Moayeri [9]. The basic approach is to quantize the signal envelope or the range of the signal to interference-plus-noise ratio (SINR), and associate each state of the Markov chain with a given range, to model the packet error process in the presence of a fading channel. In Steffan [10], the error process is modeled as the 'arrival' of errors, with arrival rate varying according to a Markov-modulated process (MMP). In the existing literature, the Markov model is often considered for Ricean or Rayleigh fading channels. It has also been applied to other statistical models, such as Nakagami fading channels [11], or built from error traces obtained from measurements [12]. The validity of the first-order Markov model is addressed in References [13–18]. The accuracy of the model can be assessed by using information theoretic [15] or probabilistic approaches [18]. In Babich and Lombardi [18], it is reported that the first-order Markovian model, in which the states represent discrete, non-overlapping intervals of the signal envelope's amplitude, is not suitable to model very slow fading channels unless the analysis is carried out over a short time duration. However, it is also pointed out that this applies to bit-level systems, whereas the model can be valid for block-level systems. The performance of adaptive radio links has been studied in presence of an AWGN channel [19], or fading channels but *without* channel coding [20], or for coded systems analyzed under *specific* channel coding schemes and decoding methods [21, 22].

The goal of this chapter is not to capture effects of higher frequency fading, but rather to represent the behavior of the signal quality level in terms of presence in a region, and to link it to the behavior of the service offered by the PHY in terms of data rate and error rate, with a granularity given by the number of PHY modes of the system. When the number of regions is small, e.g. less than eight, the quality level exhibits lower frequencies owing to practical constraints discussed in the following sections. In this chapter, a continuous, rather than discrete, time channel model is used. The benefit of this approach is 3-fold. First, the model may be integrated better in some broader analytical models, and the simulation model that can be derived from it can be efficiently integrated in event-driven simulators, which are usually more efficient. Second, the model presented in this chapter integrates both channel model and link adaptation in a flexible way open to generalizations. Third, some imperfections such as channel estimation error, estimation delay and feedback error, as well as implementation implications, like switching hysteresis, are included in the model.

4.1.1 The MAC channel model

We are interested in modeling the properties of the service capacity offered by the PHY to the upper layers, rather than in the rigorous characterization of the physical channel. The service capacity is basically given by the gross transmission bit rate and the errors at the receiver. In absence of link adaptation, i.e. with one fixed PHY mode, the gross bit rate is constant across channel states and the error rate absorbs the variability of the channel. Conversely, with link adaptation, the error rate is kept bounded whereas the bit rate changes with the state of the channel. Our model captures the dynamics of those metrics by modeling the channel as seen from above the PHY (Figure 4.1). The model therefore includes the physical channel and the PHY characteristics. In this chapter this model is referred to as the MAC channel. The MAC channel is modeled as a finite Markov chain (Figure 4.1), in which each state corresponds to a PHY mode and is hence associated with a specific transceiver configuration. In an adaptive modulation and coding (AMC) system, a mode corresponds to a modulation and channel coding schemes pair, and hence to the transmit bit rate and the error rate. The link service capacity characterized in this chapter, denoted $R_c(t)$ and defined later, is actually a stochastic process $R_c(t;\zeta)$, which depends on the PHY mode in use, which is selected depending on the signal quality level estimate $\hat{\gamma}(t;\zeta)$. In the notation to follow, unless otherwise specified, the dependence on time t and realization ζ will be omitted.

4.1.2 The Markovian model

Packets or symbol channel errors may be observed at sampling instants using a model time unit equal to packet or symbol interval. This approach is often used for error models at both symbol and packet level error analysis [9, 23]. In this case, if a Markovian model is adopted, the resulting model is discrete time.

Channel conditions may be represented by the value of some metric of the link quality γ, whose domain is divided in non-overlapping regions. In the model considered in this chapter, changes in the state of the system coincide with boundary crossings of this metric. We model the true quality level as a continuous-time Markov chain with the state of the model representing the signal quality region. Typical fading channels exhibit correlation between successive *values* and therefore the system cannot be considered as memory-less if samples of those values are analyzed. However, in our system the process state is represented by the γ*-region* which corresponds to a PHY mode. This correspondence has been outlined above and will be described in detail in the following section. The number of PHY modes is typically small. For example, the high-speed downlink packet access (HSDPA) of the UMTS terrestrial radio access (UTRA) envisages seven modes, whereas IEEE 802.11a and the ETSI HIPERLAN/2 have seven and eight modes, respectively.

In continuous time model, the system does not sample the specific value of the quality metric, but rather the region in which the quality metric falls. It is clear that for a small number of states the higher frequency correlation in the physical channel is smoothed out when considering the jumps from one state to another. The assessment of the validity of the Markov property assumption may be feasible for simple statistical physical channel models, but may be impossible with more complicated channel models or for models derived from measurements. The framework presented in this chapter is intended to be applicable to generic channel processes that are continuous, based on known statistics or on heuristics, and that can be represented by a small number of states. The true channel quality level is modeled as a continuous time Markov chain (CTMC).

Let us consider a continuous-time finite Markov chain obtained by a stochastic process $\gamma(t;\zeta)$ defined over a time interval $T \subseteq (-\infty, +\infty)$ which assumes values on a continuous set $\Gamma \subseteq \mathbb{R}$ split in a finite number of contiguous, nonoverlapping intervals. If the process is continuous, then the CTMC is a birth–death process, since the new value assumed by a continuous function after an infinitesimal time interval Δt can only belong to the same or a neighboring interval. For our purpose, it is sufficient to assume that the process is continuous. So, looking at the time instants in which γ crosses the target levels, the process can jump only to neighboring states.

We now define the elements of the *state transition rate matrix*, Q, also known as infinitesimal generator or intensity matrix. With state S_i associated with $\gamma \in [\gamma_i, \gamma_{i+1})$, see Equation (4.3), the generic elements of $\mathbf{Q}$ can be written as

$$\begin{aligned} q_{k,k+1} &= \frac{p(k, k+1; t, t+\Delta t)}{\Delta t} = \frac{\Pr\{\text{crossing upwards in } \Delta t | S_k\}}{\Delta t} \\ &= \frac{N^+_{k+1}\Delta t}{\pi_k}\frac{1}{\Delta t} = \frac{N^+_{k+1}}{\pi_k} \end{aligned} \tag{4.1}$$

where N^+_k is the expected number of times level k is crossed upwards in a second (the level crossing rate, LCR), and π_k is the probability of state S_k. Going from S_k to S_{k+1} is defined by crossing the threshold γ_{k+1}. Similarly,

$$q_{k,k-1} = \frac{p(k, k-1; t, t+\Delta t)}{\Delta t} = \frac{N^-_k}{\pi_k} \tag{4.2}$$

For practical reasons described later, a certain hysteresis at switching points is introduced. In this case, the definition of LCRs N^+_{k+1} and N^-_k is modified by considering the levels $\gamma_{k+1} + \varphi^+_{k+1}$ and $\gamma_k - \varphi^-_k$, respectively. The interpretation of the term π_k with hysteresis is discussed later. Note that, by rewriting Equation (4.1) as $p(k, k+1; t, t+\Delta t) = q_{k,k+1}\Delta t + o(\Delta t)$, replacing $\Delta t = T_u$, where T_u is the time unit of a discrete time model, we obtain the expression of the transition probability of the discrete time Markov chain representing symbol or packet errors, used (without the concept of hysteresis) in Brümmer [8] and related papers [12].

The quantities in Equations (4.1) and (4.2), namely the level crossing rate and state probabilities, are found in literature for some common processes like Rice and Nakagami. For more complicated channel models or for measurements traces when the terms in the right-hand sides of Equations (4.1) and (4.2) are not known, those quantities can be easily evaluated from the observation of the measurement traces or from link-level simulations of the possibly complicated models, by observing that the LCRs and state probabilities are $N_k^{\pm} \approx n_k^{\pm}/t$ and $\pi_k \approx t_k/t$. In these expressions, n_k^+ and n_k^- are the number of crossing levels γ_k upwards and downwards, respectively, t_k is the time spent in region S_k, and t is the total observation time. By replacing these expressions in Equations (4.1) and (4.2), we obtain $q_{k,k+1} = n_{k+1}^+/t_k$ and $q_{k,k-1} = n_k^-/t_k$.

4.1.3 Goodput and link adaptation

The received signal quality can be measured, for example with the SINR, but in the following we will denote with γ a generic metric for the quality level. The signal quality is a function of a number of contributions, related to both the transceiver structure and settings and the communication environment. Changes in the time-varying signal quality are due to various factors which change with different speed. Changes occurring in periods comparable with bit rate duration, $O(T_{\text{bit}})$, or shorter, are dealt with typically with diversity gain. Slower changes, occurring in periods comparable with packet interval, $O(T_{\text{pkt}})$, are handled by changing PHY schemes and possibly by proper scheduling. Changes of $O(T_{\text{frame}})$ and greater are handled typically with scheduling or renegotiations and reconfigurations at higher layers. In addition to diversity gain and multiplexing gain [24], link adaptation techniques are used to enhance performance. As a response to the received signal quality γ, link adaptation algorithms initiate changes in the transceiver structure K_r, keeping a performance metric e between certain boundaries: $e(K_r, \gamma) \in \{\text{acceptable values}\}$. Link adaptation strategies include changing both modulation and channel coding schemes (referred to in the literature as adaptive modulation and coding), and transmission power, or the combinations of the two [25]. A larger number of transceiver configuration parameters can be adaptively adjusted according to some strategy. This leads to the concept of PHY mode, extension of the AMC and in line with the idea of software-defined radio (SDR).

The definition of the reference and control signals that are fed into the adaptation algorithm is the basis of adaptive systems. These are the target quality and the actual signal quality, respectively. Basically, there are two possible methods for the definition of the control signal. The first is to let the transmitter know what the measured quality at the receiver is (Figure 4.2). Being a closed loop solution, it has the drawbacks that the signal that is used by the algorithm does not reflect the channel conditions at the transmission time, and that the control message itself is subject to errors in the feedback channel. The other possibility is that the transmitter autonomously estimates the channel, based on the quality measured at its own side (Figure 4.3). This open loop scheme assumes reciprocity of the direct and feedback channels. UMTS TDD is an example of the system where such approach is possible. In both cases, a related problem is how transmitter and receiver agree their configuration. With closed loop, the channel status information is sent from the receiver to the transmitter, e.g. by piggybacking the information on outgoing packets. In this

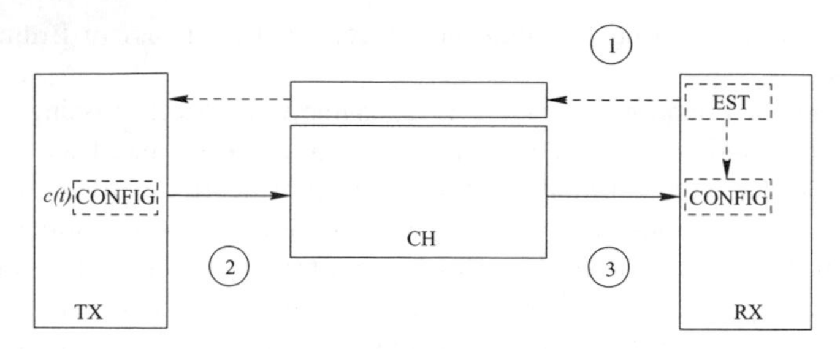

Figure 4.2 Adaptive radio with receiver-controlled link adaptation, or closed loop control of mode switching. The receiver does the channel estimation and communicates the channel quality or directly commands the transmitter the mode to be used. Numbers indicate the sequence order of operations. Solid lines are data transmission, whereas dashed lines are control signaling. The metric used as a control signal, $c(t) \approx \hat{\gamma}(t - \tau_e)$, is a delayed estimate of the true channel quality value, possibly affected by errors in the feedback channel.

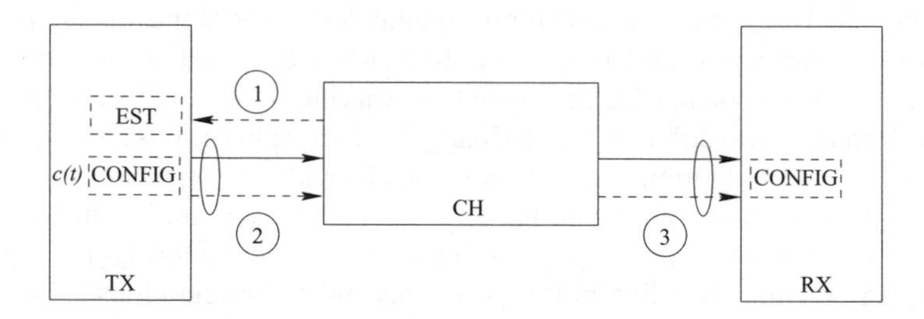

Figure 4.3 Adaptive radio with transmitter-controlled link adaptation, or open loop control of mode switching. The transmitter first estimates the state of the channel based on the received signal, and then transmits data. Some coded information about the chosen mode must be sent, or complexity at the receiver must be added to estimate the mode. Numbers indicate the sequence order of operations. Solid lines are data transmission, whereas dashed lines are control signaling. The metric used as control signal, $c(t) = \tilde{\gamma}(t)$, is an approximated estimate of the true channel quality value. The value is also affected by the lack of channel reciprocity in two directions.

receiver-controlled scheme the PHY mode will be obviously known at the receiver side. Conversely, in the open loop case, the mode can be either sent by the transmitter, causing overhead, or obtained blindly from the receiver, leading possibly to an increase in complexity and to PHY mode acquisition errors. It must be emphasized that, with both open and closed loop schemes, it is practically impossible for the transmitter to know what the actual channel state will be at the transmission instant at the receiver. Imperfections in the adaptation chain have an impact on the effectiveness of the algorithm. Indeed, the closed loop scheme is affected by delay, which is not less than the round trip propagation delay, and possibly by channel state estimation error. The open loop is affected by a much smaller estimation delay, but the estimate may not match the state at reception site and time. In addition to that, the information signaled from the receiver to the transmitter is prone to errors in the communication channel. All these aspects are taken into account in the model for imperfections described in the next section.

The error rate is a fundamental performance measure for QoS for both delay-insensitive and delay-sensitive services. According to QoS requirements, certain target error rates can be identified, and from them, the domain of the signal quality metric can be divided into M, nonoverlapping intervals. Each region is associated with a state of the CTMC $\mathcal{S} = \{S_0, S_1, \ldots, S_{M-1}\}$. According to the adaptation algorithm, one of the M PHY modes (a modulation and channel coding scheme pair) available in the system is associated with each region of the signal quality metric: $\mathcal{M} = \{M_0, M_1, \ldots, M_{M-1}\}$. Each state is therefore characterized by the bit rate and the error rate: $S_i \Rightarrow M_i \Rightarrow r_i,\ e_i$. The set of the threshold levels, the mode switching points, $\{\gamma_i\}$, $i = \{0, \ldots, M\}$, is defined so that:

$$\begin{aligned} &\gamma \in [\gamma_i, \gamma_{i+1}) \Rightarrow S_i \Rightarrow M_i, \quad 0 < i \leq M-1 \\ &\gamma \in (\gamma_0, \gamma_1) \Rightarrow S_0 \Rightarrow M_0 \end{aligned} \tag{4.3}$$

where $\gamma_0 = -\infty$ and $\gamma_M = +\infty$. For the PHY mode M_i the selection must comply with the requirement $e_i \leq e_{\max}$, $0 \leq i < M$, where $e_{\max}$ is the maximum tolerable error rate, i.e. the target error rate. By noting that $e(\gamma)$ is a monotonic decreasing function of γ, we have $\gamma_i = e^{-1}(e_{\max})$. If the statistics of the channel process are known, the nominal value of the signal quality metric in each region can be defined as follows:

$$\bar{\gamma}_i = \int_{\gamma_i}^{\gamma_{i+1}} \gamma p_\gamma(\gamma)\,\mathrm{d}\gamma \Bigg/ \int_{\gamma_i}^{\gamma_{i+1}} p_\gamma(\gamma)\,\mathrm{d}\gamma \tag{4.4}$$

If not, it can be simply represented by the center value in the range.

Switching points can be determined also using methods other than the one described above, including optimization [26]. The model described in this chapter is independent of the method used.

4.1.4 Switching hysteresis

As with all real control systems, if the same threshold value was chosen for both process rising and falling thresholds, the system could exhibit too frequent mode switching. A solution to this known problem is to introduce a hysteresis, fixing two distinct values for falling and rising thresholds. If γ^- and γ^+ are falling and rising threshold values, respectively, the actual threshold values will be defined as:

$$\gamma_i^\pm = \gamma_i \pm \varphi_i^\pm \tag{4.5}$$

With $\varphi_i^+ = \varphi_i^- = \varphi_i = \varphi$, 2φ is the width of the hysteresis region. Another choice is to select the switching thresholds $\forall k$ accordingly to the following equation:

$$\int_{\gamma_k-\varphi_k^-}^{\gamma_k} p_\gamma(\gamma)\,\mathrm{d}\gamma = \int_{\gamma_k}^{\gamma_k+\varphi_k^+} p_\gamma(\gamma)\,\mathrm{d}\gamma \tag{4.6}$$

This definition can be applied independently of the method used to define the ideal threshold levels. For example, an effective definition of switching points could be obtained by imposing an upper limit on the probability of spurious switching in a given time.

4.1.4.1 Net bit rate

The error rate used for expressing QoS requirements is usually the packet error rate (PER) since it is closer to the user's perception of quality. At the link layer, a packet is considered valid if no error is present or all errors have been corrected. In our model, the correction capability of the code is included in its coding gain, therefore, for the packet to be valid, no residual error is allowed. The packet error rate p_{E} is therefore given by:

$$p_{\mathrm{E}} = 1 - \{1 - p_{\mathrm{e}}[\gamma G_{\mathrm{c}}(\gamma)]\}^{L_{\mathrm{p}}} \tag{4.7}$$

where $p_{\mathrm{e}}(\gamma)$ is the bit error rate of the uncoded system, $G_{\mathrm{c}}(\gamma)$ is the coding gain, γ is the SINR per bit, and L_{p} is the packet length in bits. Any channel coding scheme can be included in this model provided that its performance is summarized in the coding gain curve, $G_{\mathrm{c}}(\gamma)$. In Equation (4.7), errors in the header and in the payload are assumed equiprobable. In reality, errors in the header are more critical because they may cause erroneous addressing and insertion errors. If distinct channel coding schemes are adopted for the two parts, with a stronger code for the header, the probability of a novalid packet is $p'_{\mathrm{E}} = 1 - (1 - p_{\mathrm{E,h}})(1 - p_{\mathrm{E},i})$ where the probabilities of erroneous header and payload are computed separately as $p_{\mathrm{E,h}} = 1 - \{1 - p_{\mathrm{e}}[\gamma G_{\mathrm{c}}(\gamma)]\}^{L_{\mathrm{h}}}$ and $p_{\mathrm{E},i} = 1 - \{1 - p_{\mathrm{e}}[\gamma G_{\mathrm{c}}(\gamma)]\}^{L_i}$, respectively.

Besides of its advantages in achieving reliable transfer of information, channel coding protection implies a loss in capacity due to the introduced redundancy. Moreover, each protocol layer adds its own header to the packet as it is proceeding downwards through the protocol stack. We assume in the following that the same channel coding scheme is applied to both header and payload. The generalization when distinct schemes are adopted is straightforward. The header efficiency is defined as the ratio between the payload size and the total packet size without redundancy (i.e. header and payload together, $L_{\mathrm{d}} = L_i + L_{\mathrm{h}}$): $\eta_{\mathrm{h}} = L_i/L_{\mathrm{d}}$. The channel coding efficiency depends on the added redundancy and is represented by its code rate $\eta_{\mathrm{c}} = L_{\mathrm{d}}/(L_{\mathrm{d}} + L_{\mathrm{o}})$. Since the packet size after channel coding and packetization can be written as $L_{\mathrm{p}} = L_i + L_{\mathrm{h}} + L_{\mathrm{o}} = L_i/(\eta_{\mathrm{h}}\eta_{\mathrm{c}})$, the bit rate for a given PHY mode (still including residual errors) can be written as

$$r_i = \eta_{\mathrm{h}}\eta_{\mathrm{c},i}k_i/T_{\mathrm{s}} \tag{4.8}$$

where k_i is the number of bits per symbol of the ith mode modulation scheme, and T_{s} is the symbol duration.

4.1.5 Link service rate with exact mode selection

The link service capacity is the rate at which error-free information units are transmitted through the channel and is referred to as 'goodput'. It is a function of the gross bit rate or throughput, which includes the corrupted information units, r, and the residual error rate after correction, e. By using a similar expression for the link service rate as given in Kim and Li [27] for ideal systems with fixed PHY, and extending it to time-varying systems, the goodput at a given instant t can be expressed as

$$R'_{\mathrm{c}}(t) = [1 - e(t)]r(t) \tag{4.9}$$

In Equation (4.9), $r(t)$ is the gross channel transmission rate, or throughput, which depends on modulation constellation size, code rate, overheads, etc., and $e(t)$ is the residual error rate, which depends essentially on modulation scheme, channel coding gain, diversity combining methods, and physical channel conditions. The residual error rate after reassembling e has the same value of the packet error rate defined in Equation (4.7) i.e. $e = p_{\mathrm{E}}$.

We extend further Equation (4.9) to adaptive systems. In adaptive systems the transceiver configuration settings change in time depending on the channel state at time t, denoted by $S(t) = i$ or S_i. The PHY mode ideally associated with each region and used at time t is denoted with $M(t) = i$ or M_i. To simplify notation, whenever possible we drop the dependence on time t. In presence of ideal link adaptation, the service capacity is obtained averaging across all the modes, resulting in

$$\bar{R}_{\mathrm{c}}' = \sum_{i=0}^{M-1} [1 - e_i] r_i \pi_i \tag{4.10}$$

where the subscript i denotes the mode, M is the number of implemented modes, and π_i is the probability of the channel being in state i and using the ith mode.

4.1.5.1 Model of impairments

In Equation (4.10), the service capacities for all modes can be represented by a vector because the effective and chosen states are assumed to coincide. In case of erroneous mode selection, the gross bit rate depends on the *used* mode whereas the error rate depends on both the used mode and the *effective* state. Hence, in the presence of imperfections the two contributions must be separated. In the sequel, we will use the following notation.

The *throughput vector* $\mathbf{r} = \{r_i\} \in \mathbb{R}^M$ includes the transmit rates for each PHY mode and is defined as $\mathbf{r}^{\mathrm{T}} \doteq [\,r_0 \quad \cdots \quad r_{M-1}\,]$ with r_i given by Equation (4.8). The *normalized goodput matrix* $\mathbf{Y} = \{y_{ij}\} \in \mathbb{R}^{M \times M}$ includes the effects of errors due to corruption in the channel for each PHY mode and each state, and expresses the useful bit rate normalized to the transmit rate in each mode. The generic element of matrix $\mathbf{Y}$ is defined as $y_{ij} \doteq 1 - e_{ij}$, where e_{ij} denotes the error rate when the mode M_i is used and the channel is in effective state S_j. The elements of π, $\pi^{\mathrm{T}} \doteq [\,\pi_0 \quad \cdots \quad \pi_{M-1}\,]$, are the state probabilities of the CTMC. The mode selection is made by the adaptation algorithm based on an estimated channel conditions. If the adaptive system rapidly follows without errors the channel state, the average goodput, Equation (4.10), is expressed with the new notation by:

$$\bar{R}_{\mathrm{c}}\big|_{\mathrm{ideal}} = \sum_{i=0}^{M-1} r_i \,(1 - e_{ii})\, \pi_i \tag{4.11}$$

or, in compact form:

$$\bar{R}_{\mathrm{c}}\big|_{\mathrm{ideal}} = \mathbf{r}^{\mathrm{T}} \operatorname{diag}(\mathbf{Y}\Pi) \tag{4.12}$$

where $\Pi = \operatorname{diag}(\pi)$ is the diagonal matrix having as diagonal elements the elements of vector π, and $\operatorname{diag}(\mathbf{A}) \doteq [a_{ii}]$ is the operator that extracts the vector of diagonal elements of matrix $\mathbf{A}$: $\operatorname{diag}(\mathbf{A}) \doteq (\mathbf{A} \odot \mathbf{I})\mathbf{u}$, where $\mathbf{I}$ is the identity matrix, $\mathbf{u}$ is the unity vector having all elements 1, and $\odot$ is the Hadamard–Schur product.

4.1.6 Imperfections in the adaptation chain

A real adaptive communication system is prone to erroneous setup of the transceiver configuration. Estimate errors, estimation delay and feedback errors have so far been neglected. The actual properties of the related errors depend on the particular system configuration. In this chapter, we are not interested in studying a specific system, but rather we want to model common sources of errors in a flexible way open to generalizations.

In reality, the true channel condition is hidden. A delayed or erroneous version of it is visible instead and, based on that, the PHY mode is chosen. The parameter used for the mode switching can be expressed as $\hat{\gamma}(t) = \gamma(t - \tau_e) + \varepsilon_e + \varepsilon_f$, where γ and $\hat{\gamma}$ are the true and the estimated value, respectively, ε_e is the estimate error, ε_f is the feedback error, and τ_e is the estimation delay or estimation lag. Among these three terms, the first one has more important effect in the open loop case, whereas, in the closed loop case, the other two terms are dominant.

The model of imperfections is depicted in Figure 4.4. At estimation time t the *true* channel quality metric falls in the kth region, say $S(t) = k$, and the metric is estimated. Based on this *estimate* $\hat{S}(t) = h$, the mode is selected, where indices k and h may differ because of errors in the estimation process. Since there is equivalence between the cardinality of the PHY modes set $\mathcal{M}$ and the signal quality metric regions set $\mathcal{S}$, the code can unambiguously

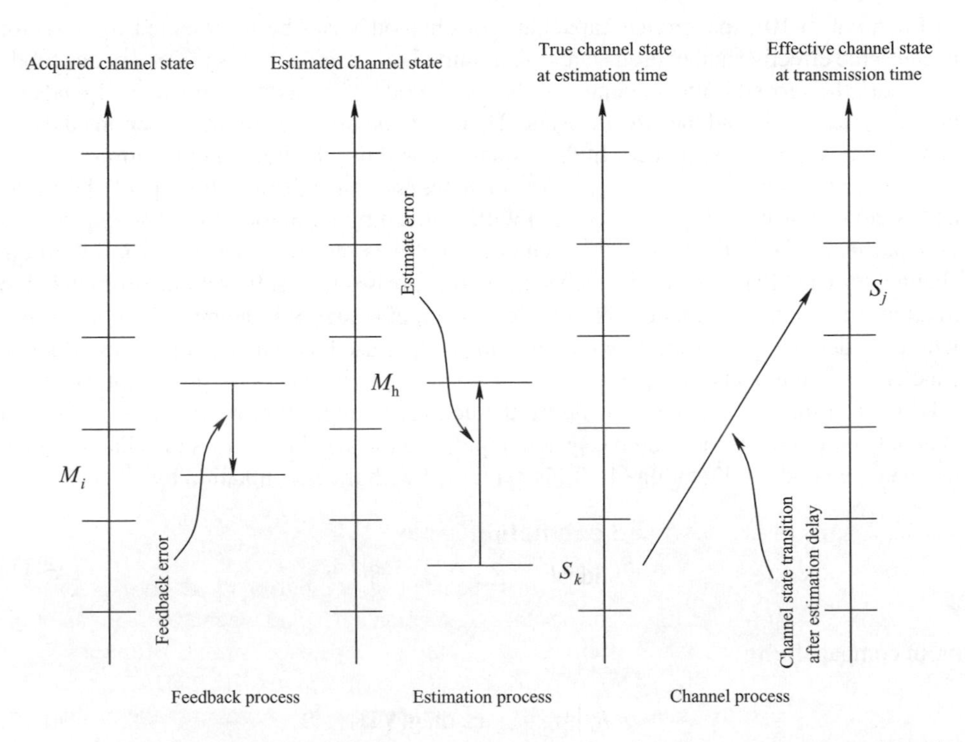

Figure 4.4 The model of imperfections of the channel estimation process. Based on the true channel state at estimation time, an estimate subject to errors is obtained. Its identification code is sent as a message through the error-prone feedback channel and is acquired as the selected state. During the estimation delay, the channel state may change due to its dynamics.

represent either the quality metric region or the mode itself. After the estimation delay, the mode, subject to errors in the feedback channel, is transmitted and $M(t+\tau_e)=i$ is *acquired* at the transmitter (or at the receiver, in the open loop case). This mode is used for transmission (Figure 4.2; or reception, Figure 4.3). In the mean time, the *effective* channel quality metric falls in general in a new region, $S(t+\tau_e)=j$, due to possible channel state transitions during the estimation delay τ_e. The model described here is unaffected whether estimator and mode selector are implemented at transmitter or receiver side, provided that proper values are given to related quantities. For example, in the open loop case, errors on the control message sent to the receiver through a separate channel are modeled with the feedback error. The final effect of the entire process is that in general mode M_i is used at transmission time, when the channel is in effective state S_j. According to the notation above, the probability of this event can be written as $P(M_i, M_h, S_k, S_j)$. Because the estimation process and the channel process are clearly independent, and assuming that the feedback channel is independent of the direct channel, we can write

$$\begin{aligned} P(M_h, S_k, M_i, S_j) &= P(M_h, S_k)P(M_i, S_j\,|M_h, S_k) \\ &= P(M_h, S_k)P(M_i\,|M_h, S_k)P(S_j\,|M_i, M_h, S_k) \\ &= P(M_h, S_k)P(M_i\,|M_h)P(S_j\,|S_k) \\ &= P(S_k)P(M_h\,|S_k)P(M_i\,|M_h)P(S_j\,|S_k). \end{aligned} \tag{4.13}$$

The three last terms in Equation (4.13) are treated separately in the sequel and their effect on the goodput is then combined by averaging independently the appropriate expressions.

4.1.7 Estimation process and estimate error

In the absence of estimation delay and feedback errors, a possibly erroneous estimate of the metric at time t is used to choose the mode at time t, i.e. $\hat{\gamma}(t) \to M(t)$. The *mode estimation probability matrix* $\mathbf{H}^{(e)} = \{h_{hk}^{(e)}\} \in \mathbb{R}^{M\times M}$ models the effects of the imperfections in the estimation process due to noise. Its generic element $h_{hk}^{(e)}$ represents the probability of selecting mode M_h when the true channel is in state S_k at estimation time:

$$\mathbf{H}^{(e)} = \{h_{hk}^{(e)}\ B\ \Pr\{M_h \mid S_k\}\} \tag{4.14}$$

Equation (4.14) represents the second term in Equation (4.13).

4.1.8 Channel process and estimation delay

Now we study the probability that the true state has changed during the delay τ_e. Consider the presence of estimation delay and absence of estimate error and feedback errors. In other words, the mode at time $t+\tau_e$ is chosen depending on the exact estimate of the metric at time t, i.e. $\gamma(t) \to M(t+\tau_e)$. The *delayed channel transition matrix* $\mathbf{H}^{(d)} = \{h_{kj}^{(d)}\} \in \mathbb{R}^{M\times M}$ models the effects of estimation delay. Its generic element $h_{kj}^{(d)}$ represents the probability that given the true channel was in state k at estimation time, at transmission time the channel is in state j:

$$\mathbf{H}^{(d)} = \{h_{kj}^{(d)} \doteq \Pr\{S(t+\tau_e)=j\,|S(t)=k\}\} \tag{4.15}$$

Equation (4.15) represents the last term in Equation (4.13).

4.1.9 Feedback process and mode command reception

The code representing the PHY mode or, equivalently, the related signal quality region, may be corrupted during its transmission through the feedback channel. The *mode acquisition probability matrix* $\mathbf{H}^{(f)} = \{h_{ih}^{(f)}\} \in \mathbb{R}^{M\times M}$ models the probability of mode error due to errors in the feedback channel and has a generic element given by:

$$\mathbf{H}^{(f)} = \{h_{ih}^{(f)} \doteq \Pr\{M_i|M_h\}\} \tag{4.16}$$

Equation (4.16) represents the third term in Equation (4.13).

4.1.10 Link service rate with imperfections

Combining Equations (4.11), (4.12) and (4.13), and remembering the model in Figure 4.4, we get the expression of the *average goodput*, i.e. the link service rate in presence of imperfections:

$$\bar{R}_c = \sum_{i=0}^{M-1} r_i \sum_{j=0}^{M-1} (1-e_{ij}) \sum_{h=0}^{M-1} h_{ih}^{(f)} \sum_{k=0}^{M-1} h_{hk}^{(e)} \pi_k h_{kj}^{(d)} \tag{4.17}$$

It is easy to see that the average goodput in Equation (4.17) can be written in compact form as:

$$\bar{R}_c = \mathbf{r}^T \mathrm{diag}(\mathbf{Y}\boldsymbol{\Theta}^T) \tag{4.18}$$

where $\boldsymbol{\Theta} = \{\vartheta_{nj}\} \in \mathbb{R}^{M\times M}$ is the *mode-channel probability matrix*:

$$\Theta = \mathbf{H}^{(f)}\mathbf{H}^{(e)}\Pi\mathbf{H}^{(d)} = \Pi\mathbf{H}^{(d)}\mathbf{H}^{(f)}\mathbf{H}^{(e)} = \boldsymbol{\Pi}\Omega \tag{4.19}$$

having generic element

$$\vartheta_{nj} = \sum_{h=0}^{M-1} h_{nh}^{(f)} \sum_{k=0}^{M-1} h_{hk}^{(e)} \sum_{m=0}^{M-1} \pi_{km} h_{mj}^{(d)} = \sum_{h=0}^{M-1} h_{nh}^{(f)} \sum_{k=0}^{M-1} h_{hk}^{(e)} \pi_{kk} h_{kj}^{(d)} \tag{4.20}$$

Equation (4.20) models the probability of using mode M_n with the channel in the effective state S_j. Ω is the *equivocation matrix*, which includes the effects of link layer adaptation process. The innermost summation in Equation (4.17), $\mathbf{H}^{(e)}\boldsymbol{\Pi}\mathbf{H}^{(d)} = \sum_{k=0}^{M-1} h_{hk}^{(e)}\pi_k h_{kj}^{(d)}$, is the probability of estimating the mode M_h when the channel is in effective state S_j (at transmission time), averaged over all possible true channel states S_k (at estimation time). The term $\boldsymbol{\Theta} = \mathbf{H}^{(f)}\mathbf{H}^{(e)}\boldsymbol{\Pi}\mathbf{H}^{(d)} = \sum_{h=0}^{M-1} h_{ih}^{(f)} \sum_{k=0}^{M-1} h_{hk}^{(e)}\pi_k h_{kj}^{(d)}$ is the probability of using mode M_i when the channel is in state S_j at transmission time, averaged over all possible estimated modes M_h. The term $\mathrm{diag}(\mathbf{Y}\boldsymbol{\Theta}^T) = \sum_{j=0}^{M-1}(1-e_{ij})\sum_{h=0}^{M-1} h_{ih}^{(f)} \sum_{k=0}^{M-1} \pi_k h_{hk}^{(e)} h_{kj}^{(d)}$ is the *normalized average goodput vector* representing the goodput when using mode M_i, averaged over all possible channel states at transmission time S_j. Finally, the average goodput is given by averaging the quantity above over all possible used modes, weighted by the transmit rate, thus removing the previous normalization by the product $\mathbf{r}^T$ $\mathrm{diag}(\mathbf{Y}\boldsymbol{\Theta}^T)$, leading to Equation (4.17). The *scalar normalized average goodput* is obtained by summing the elements of the normalized average goodput vector:

$$\bar{R}_r = \left|\mathrm{diag}(\mathbf{Y}\boldsymbol{\Theta}^T)\right| \tag{4.21}$$

$\bar{R}_r$ corresponds to the average goodput of a system adopting an hypothetical set of PHY modes all having unitary transmission rate. The use of this quantity will be elaborated later. In the ideal case of perfect link adaptation, the channel quality metric is estimated without errors, tracked without delay, and communicated to the transmitter without errors, i.e. $\mathbf{H}^{(e)} = \mathbf{H}^{(d)} = \mathbf{H}^{(f)} = \mathbf{I}$, $\boldsymbol{\Omega}$ reduces to the identity matrix, and we have $\boldsymbol{\Theta} = \boldsymbol{\Pi}$. As a result, under the assumptions defined above, the definition of average goodput given in Equations (4.17) and (4.18) reduces to the one of the ideal case, Equations (4.11) and (4.12). Single PHY mode systems are considered in this model as a special case with errorless feedback channel, $\mathbf{H}^{(f)} = \mathbf{I}$, instantaneous estimation, $\mathbf{H}^{(d)} = \mathbf{I}$, and static mode selection leading to a matrix $\mathbf{H}^{(e)}$ having as nonzero elements all 1 in the ith row, if mode M_i is implemented in the system: $h_{hk}^{(e)} = \delta_{hi}\delta_{kk}$, where δ_{ij} is the Kronecker delta.

4.1.10.1 System performance

For illustration purposes, we apply the above analysis tool to the study of a sample system and investigate its sensitivity to some imperfection or implementation constraints.

4.1.10.2 System parameters

The PHY modes include a no-transmission mode for values of the signal quality that do not comply with minimum service requirements. The modulation schemes used in the numerical analysis are BPSK, QPSK, 8QAM and 16QAM. The channel coding scheme used is the half rate convolutional code having generator polynomials $g_0 = 133_8$ and $g_1 = 171_8$, and constraint length $K = 7$. Coding gain curves for the half rate and punctured versions of that mother code, having $\eta_c = 2/3$ and $\eta_c = 3/4$, are obtained from standard sources [28, 29]. The minimum signal to interference-plus-noise ratio for the *coded* system, $\gamma_{\min}$, is obtained from $\text{BER}_{\max} = \text{BER}\left(\gamma'_{\min}\right)$, knowing the coding gain for the given channel coding scheme: $\gamma_{\min} = \gamma'_{\min}/G_c$. Inverses of the BER of all schemes can be efficiently obtained from the approximated parametric function $\text{BER}\left(\gamma'\right)$, valid for all schemes [25], or by numerical inversion of exact expressions [30]. In the system used for numerical analysis, switching thresholds are obtained from the required PER for a packet length of 2048 b.

We assume Ricean fading process [31]. The expression for the LCR for the case of a line-of-sight component with zero Doppler frequency, uncorrelated Gaussian noise components, and Jakes-shaped Doppler power spectral density or isotropic scattering can be found in [32, 33]. Under the assumptions above, and using the corresponding LCR, expressions (4.1) and (4.2) can be written as

$$q_{k,k+1} = f_{\max}\sigma_0\sqrt{\pi}\, p_\xi(\xi_{k+1}; K_f, \sigma_0)/\Delta P_\xi(\xi_k, \xi_{k+1}; K_f, \sigma_0)$$

and

$$q_{k,k-1} = f_{\max}\sigma_0\sqrt{\pi}\, p_\xi(\xi_k; K_f, \sigma_0)/\Delta P_\xi(\xi_k, \xi_{k+1}; K_f, \sigma_0)$$

where $f_{\max}$ is the maximum Doppler frequency, σ_0 is the standard deviation of the quadrature Gaussian components, $p_\xi(\xi_k; K_f, \sigma_0)$ is the value of the PDF of the Ricean distribution having Rice factor K_f at signal level ξ_k, and $\Delta P_\xi(\xi_k, \xi_{k+1}; K_f, \sigma_0)$ is the difference of the Rice cumulative distribution function (CDF) at points ξ_{k+1} and ξ_k.

Table 4.1 Channel model parameters

i	π_i	N_i^-	N_{i+1}^+	λ_k	μ_k
0	5.12×10^{-2}	—	1.98×10^{-1}	3.86	—
1	1.71×10^{-1}	1.92×10^{-1}	5.44×10^{-1}	3.18	1.12
2	4.42×10^{-1}	5.41×10^{-1}	6.57×10^{-1}	1.48	1.22
3	3.05×10^{-1}	6.58×10^{-1}	1.20×10^{-1}	0.394	2.16
4	3.05×10^{-2}	1.27×10^{-1}	—	—	4.17

State probabilities, level crossing rates downwards and upwards, and birth and death rates for the CTMC with five states used in the examples.

4.1.10.3 Channel model statistics

Parameters of the channel model for the example system are given in Table 4.1. Match of the statistics of the channel model is verified by comparing [34] the histogram obtained by simulation of the CTMC with the theoretical solution of the CTMC and with the theoretical PDF of the underlying fading process.

Figure 4.5 shows a snapshot of the time series obtained by simulations with above CTMC model with $f_{\max} = 300$ Hz. It can be seen that very large changes in the metric (about one order of magnitude) occur in less than 1 ms around 3.5 ms, very rapid changes occur around 7 ms, and finally relatively long periods with steady values are seen on the right-hand side of the diagram. The system with $f_{\max} = 1$ Hz simulated over 50 000 values, corresponding to a simulated time of $t_{\text{sim}} \approx 4.6$ h, exhibits $\tau_{\text{it,min}} \approx 6.2$ μs and $\tau_{\text{it,max}} \approx 5.5$ s as minimum and maximum of the inter-transition time, which depends on both the Doppler frequency of the fading process and the size of the signal quality regions.

4.1.11 Sensitivity of state probabilities to hysteresis region width

Consider the probability of staying in region S_k, π_k. The kth state is entered by crossing levels $\gamma_k + \varphi_k^+$ or $\gamma_{k-1} - \varphi_{k-1}^-$, when entering from below and above, respectively. This implies that in this case the actual width of the region is narrowed. Similarly, the state is left by crossing levels $\gamma_k - \varphi_k^-$ or $\gamma_{k-1} + \varphi_{k-1}^+$, when going below and above, respectively, and the actual width is now broadened. Although the ingress rate roughly equals the egress rate, as it can be seen in Table 4.1, the impact of hysteresis also depends on the state probabilities. In fact, the state probabilities after introducing the hysteresis can be written as:

$$\pi_i^{(h)} = \pi_i + \Delta_{\text{CDF}}^{(i,+)}\pi_i + \Delta_{\text{CDF}}^{(i,-)}\pi_i - \Delta_{\text{CDF}}^{(i+1,-)}\pi_{i+1} - \Delta_{\text{CDF}}^{(i-1,+)}\pi_{i-1} + o(\Delta_{\text{CDF}}) \tag{4.22}$$

where $\pi_i^{(h)}$ and π_i are the state probabilities with and without hysteresis, respectively, and $\Delta_{\text{CDF}}^{(i,\pm)}$ is the half-width of the hysteresis region i at upper (+) and lower (−) side, respectively. In case of equal and symmetric hysteresis regions and neglecting higher order infinitesimals, we have

$$\pi_i^{(h)} \approx \pi_i + \Delta_{\text{CDF}}\,(2\pi_i - \pi_{i+1} - \pi_{i-1}) \tag{4.23}$$

Figure 4.6 shows the sensitivity of state probabilities to Δ_{CDF} in this latter case, with regions defined as in Equation (4.6). For larger values of Δ_{CDF}, states with larger probability, the

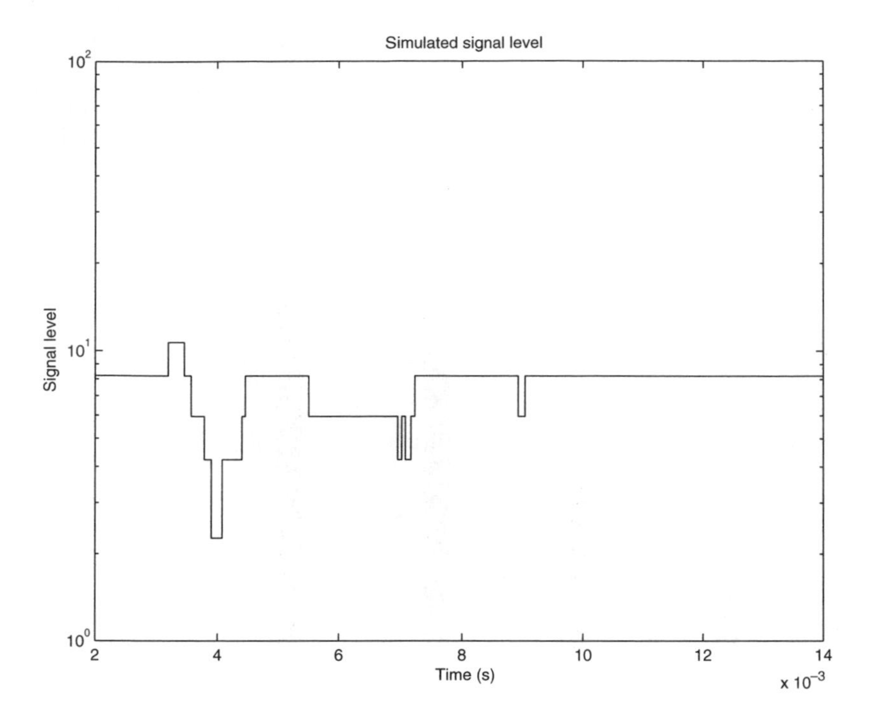

Figure 4.5 Converted snapshot of a signal level, as a function of time, obtained from the time series generated by simulations with CTMC model for $f_{\max} = 300$ Hz. It can be seen that very large changes in the metric (about one order of magnitude) occur in less than 1 ms around 3.5 ms, very rapid changes occur around 7 ms, and finally relatively long periods with steady values are seen on the right-hand side of the diagram.

central ones in our case, have their probability further increased by the introduction of hysteresis. It can be observed that up to about $\Delta_{\mathrm{CDF}} = 10^{-3}$, state probabilities are almost unaffected.

4.1.12 Estimation process and estimate error

Under the assumption of constant noise power, the problem of studying the SINR is reduced to the study of the signal envelope. The envelope is estimated with an estimation error. The distribution of the estimation error depends on the specific adopted estimation technique and analytical and/or empirical distributions for estimation error are generally unknown [20]. For these reasons, and because we want to illustrate the model rather than to provide performance analysis for a specific scenario, no assumption is done here on the structure of the estimator. For the example below, we assume the error to be Gaussian distributed. Under these assumptions, it is easy to see that the generic element of matrix $\mathbf{H}^{(\mathrm{e})}$, the probability of selecting mode M_h when the true state is S_k, can be written as:

$$h_{hk}^{(\mathrm{e})} = 0.5\left[\mathrm{erfc}\left(\frac{\inf(\xi_h) - \bar{\xi}_k}{\sigma_{\mathrm{e}}}\right) - \mathrm{erfc}\left(\frac{\sup(\xi_h) - \bar{\xi}_k}{\sigma_{\mathrm{e}}}\right)\right] \tag{4.24}$$

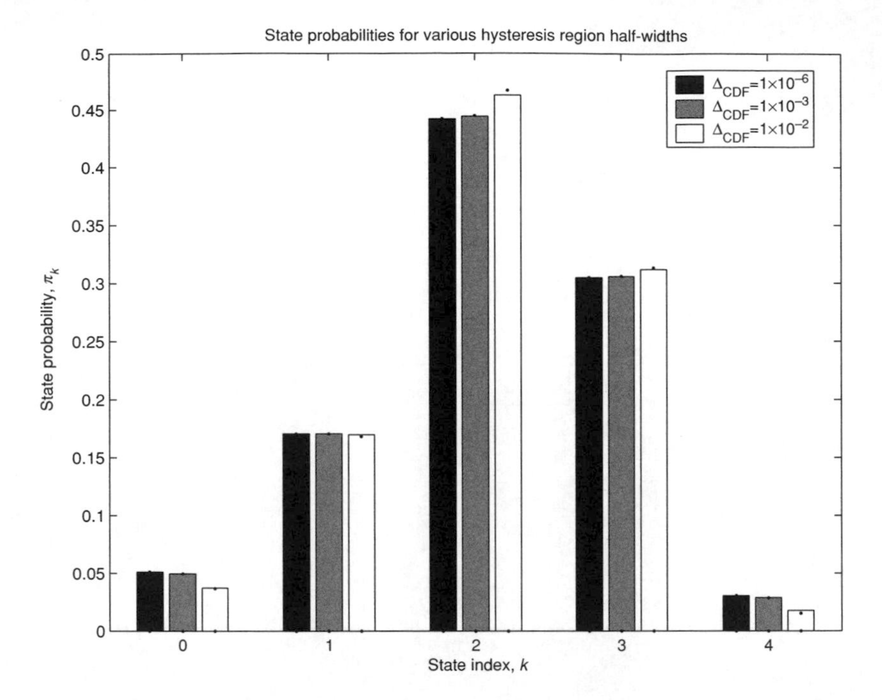

Figure 4.6 Sensitivity of state probabilities to Δ_{CDF}. Up to about $\Delta_{CDF} = 10^{-3}$, state probabilities are unaffected. For larger values of Δ_{CDF}, states with large probability, the central ones in our case, have their probability further increased by the introduction of hysteresis. Bars are simulated values, whereas points are the theoretical values given by Equation (4.23).

In Equation (4.24) erfc(g) is the error function complement, σ_e is the standard deviation of the estimation error, $\bar{\xi}_k$ is the nominal value of the metric in the kth interval S_k, and $\inf(\xi_h)$ and $\sup(\xi_h)$ are the lower and upper boundaries, respectively, of interval S_h. Figure 4.7 shows the comparison of simulation and theoretical results for the sensitivity of the average goodput $\bar{R}_c$ on estimate errors, under the assumptions given above in the non-coherent case, assuming absence of estimation delay and error-free feedback channel.

4.1.12.1 Channel process and estimation delay

The variability in time of the quality metric depends on both the fading characteristics (coherence time) and the width of the state regions and is expressed by the inter-transition time, i.e. the time between state changes. In order to simplify the analysis, we assume that the estimation delay is negligible compared with the minimum inter-transition time. For our analysis it is sufficient to assume that $\forall i \in [0, M-1] \quad \lambda_i \tau_e = q_{i,i+1}\tau_e = 1$ and $\mu_i \tau_e = q_{i,i-1}\tau_e = 1$, where M is the cardinality of sets of PHY modes and quality regions. For the Poisson process, the reciprocal of the transition rate is the mean inter-transition time. The previous constraint states basically that the estimation delay should be much smaller than the mean inter-transition time.

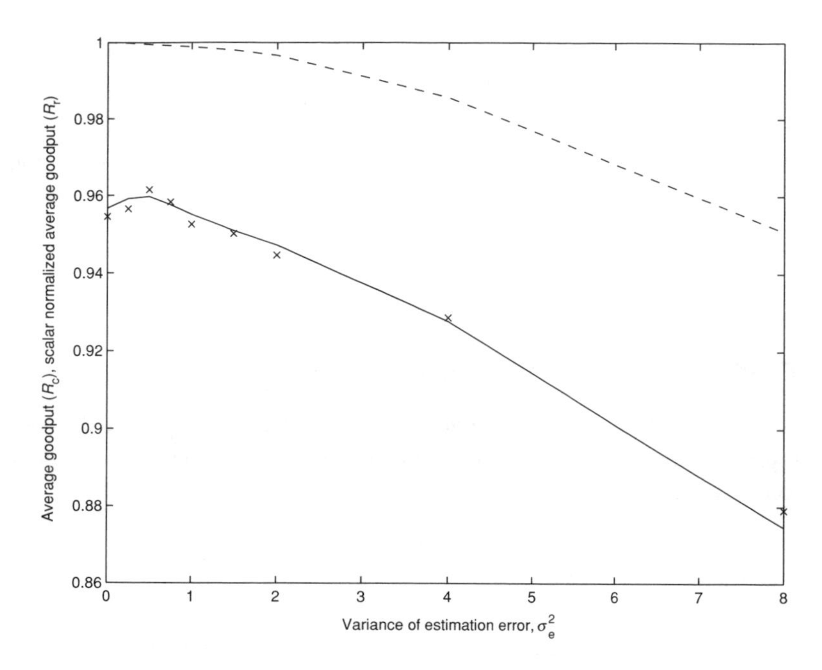

Figure 4.7 Sensitivity of average goodput to variance of channel estimation error. Average goodput $\bar{R}_c$ (solid line), scalar normalized average goodput $\bar{R}_r$ (dashed line), and simulations (crosses).

Under the previous assumptions, at most one state change can occur during τ_e, and the probability of channel state transition can be written as:

$$h_{ij}^{(d)} = \begin{cases} \lambda_i \tau_e & j = i + 1 \\ \mu_i \tau_e & j = i - 1 \\ 0 & |i - j| > 1 \\ 1 - \lambda_i \tau_e - \mu_i \tau_e & j = i \end{cases} \tag{4.25}$$

It is straightforward to see that

$$\mathbf{H}^{(d)} = \mathbf{I} + \mathbf{Q}\tau_e \tag{4.26}$$

where $\mathbf{Q}$ is the rate matrix of the CTMC and $\mathbf{I}$ is the identity matrix. In particular, in case of instantaneous channel estimation, matrix $\mathbf{H}^{(d)}$ reduces to the identity matrix. The probability of no transition from the ith state is $1 - (\lambda_i + \mu_i)\tau_e$. This term is null for $\tau_e = 1/(\lambda_i + \mu_i)$. Therefore, the quantity

$$T_c = \min_{i=\{0,\dots,M-1\}} \{1/(\lambda_i + \mu_i)\}$$

has the meaning of coherence time in our model. Considering the case of perfect estimation and errorless feedback channel, and under the previous assumptions for the estimation delay model, the average goodput is expressed by:

$$\begin{aligned} \bar{R}_c(\tau_e) &= \mathbf{r}^T \operatorname{diag}(\mathbf{Y}\boldsymbol{\Theta}^T) = \mathbf{r}^T \operatorname{diag}[\mathbf{Y}\left(\mathbf{I} + \mathbf{Q}^T\tau_e\right)\boldsymbol{\Pi}] \\ &= \mathbf{r}^T \operatorname{diag}(\mathbf{Y}\boldsymbol{\Pi}) + \tau_e \mathbf{r}^T \operatorname{diag}(\mathbf{Y}\mathbf{Q}^T\boldsymbol{\Pi}) = \bar{R}_c(0) + \tau_e \Psi_d \end{aligned} \tag{4.27}$$

where

$$\Psi_{\mathrm{d}} = \frac{\partial \bar{R}_{\mathrm{c}}}{\partial \tau_{\mathrm{e}}} = \mathbf{r}^{\mathrm{T}} \operatorname{diag}(\mathbf{Y}\mathbf{Q}^{\mathrm{T}}\mathbf{\Pi}) \tag{4.28}$$

is the drift of the average goodput from the ideal conditions due to the estimation delay. Replacing $\mathbf{Q}$ in Equation (4.28) it is easy to see that the explicit expression of Ψ_{d} is:

$$\Psi_{\mathrm{d}} = \frac{\partial \bar{R}_{\mathrm{c}}}{\partial \tau_{\mathrm{e}}} = \sum_{i=1}^{M} r_i \pi_i \left[\mu_i(e_{i,i} - e_{i,i-1}) + \lambda_i(e_{i,i} - e_{i,i+1})\right] \tag{4.29}$$

4.1.13 Feedback process and acquisition errors

To identify a PHY mode, a control message with $\lceil \log_2 M \rceil$ bits is used. The distance in bits among all pairs of codewords is given by the elements of a symmetric matrix having null elements on the diagonal, and defined by

$$\mathbf{D} = \{d_{ij}\} B \sum_{n=1}^{\lceil \log_2 M \rceil} w_i^{(n)} \otimes w_j^{(n)} \tag{4.30}$$

where $w_i^{(n)}$ is the nth bit of the ith codeword and $\otimes$ denotes the modulo 2 bit-wise product of the codewords. The element of matrix $\mathbf{H}^{(\mathrm{f})}$ is then given by:

$$h_{ih}^{(\mathrm{f})} = p_{\mathrm{e,r,f}}^{d_{ih}} (1 - p_{\mathrm{e,r,f}})^{\lceil \log_2 M \rceil - d_{ih}} \tag{4.31}$$

where $p_{\mathrm{e,r,f}}$ is the residual bit error probability in the feedback channel. Parameter $p_{\mathrm{e,r,f}}$ is a function of channel coding gain and SINR through the bit error curve of the uncoded system for the given modulation, and is given by $p_{\mathrm{e,r,f}} = p_{\mathrm{e}}^{(\mathrm{f})}[G_{\mathrm{c}}(\gamma)\gamma]$.

One way of coding mode identifiers is adopting the Gray code. Assume that the message is transmitted always using the strongest mode, BPSK-1/2, in our case, but the error probability depends on the actual channel state. The error rate does depend on the state of the feedback channel. Under the assumption of symmetric channel, the channel state at control message transmission time is assumed to be the state of the direct channel τ_{e} before transmission time, i.e. the state at estimation time. Under this assumption, the delay in the adaptation chain is concentrated in the estimation delay model. With these assumptions, it is seen that the effect of feedback errors is negligible. Therefore, less relaxed assumptions on this feedback error model might not lead to very different results. In fact, as it is shown in Figure 4.8, in which the BER in the feedback channel $p_e^{(\mathrm{f})}$ is assumed fixed and independent of the state of the direct channel, the impact is no longer negligible only for values of $p_e^{(\mathrm{f})}$ so large as to be out of range in adaptive systems and often in communications systems in general.

4.2 ADAPTIVE TRANSMISSION IN *AD HOC* NETWORKS

The adaptive-transmission protocol described in this chapter is intended for mobile packet radio networks (PRNs), in which half-duplex radios employ direct-sequence spread spectrum (DS-SS). PRNs are examples of the *ad hoc* networks described in detail in Chapter 13. One of the key features of such networks is the lack of a fixed infrastructure or central controller to determine the power levels, code rates and other transmission parameters that

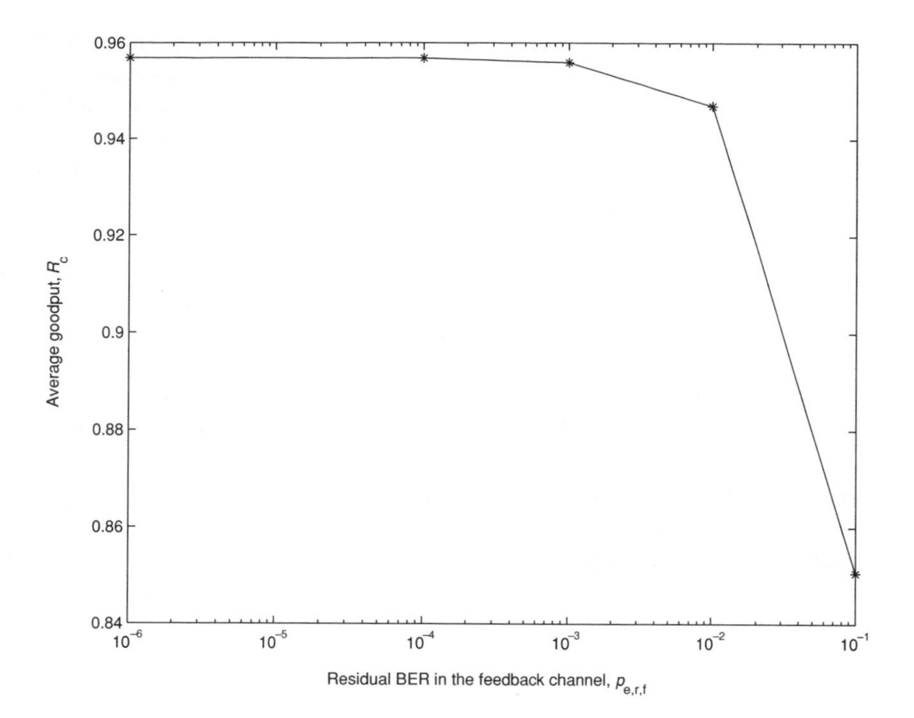

Figure 4.8 Sensitivity of average goodput to errors in the feedback channel. In this figure, the residual BER in the feedback channel is assumed fixed and independent of the state of the direct channel. The impact is no longer negligible only for values of $p_{\mathrm{e,r,f}}$ so large to be out of range in adaptive systems and often in communications systems in general.

should be used by the terminals in the network. The adaptive-transmission protocol is a fully distributed protocol that determines the transmission parameters for a communication link from statistics obtained in the receiver on the link.

The receiver includes an automatic gain control (AGC) subsystem, one or more programmable matched filters (with different processing gain), a soft-decision decoder, and a subsystem that generates a postdetection signal quality (PDSQ) statistic. Three statistics, which can be derived in each receiver, are used to decide on the receiver configuration.

The PDSQ statistic is based on the output of the matched filter. It represents the desired signal level. In general this statistics can have high $s(h)$ or low $s(l)$ value. The symbol-error rate (SER) statistic is determined from an error count that is derived in the decoding subsystem. The error count can be obtained by re-encoding the information bits at the output of the decoder and comparing the encoded symbols with hard decisions that are based on the corresponding outputs of the matched filter. SER can have high $e(h)$ or low $e(l)$ value. The AGC statistic is an estimate of the total received power that indicates the level of interference and can have high $i(h)$ or low $i(l)$ value. Using a combination of the PDSQ, SER, and AGC statistics, the receiver selects the transmission parameters for the next packet, and it sends this information to the transmitter in an acknowledgment packet or as part of the exchange that occurs if a reservation protocol is used for channel access (e.g. in a clear-to-send message). The PDSQ, SER, and AGC statistics derived in the receiver provide the only CSI that is available to the adaptive-transmission system. The transmission parameters that are adapted are the power level, spreading factor and code rate. As part of the design process for the adaptive transmission protocol, a set of thresholds, referred to as the *adaptation thresholds*, must be chosen. More details on statistics measurements and the threshold selection can be found in Block and Pursley [35]. The adaptive-transmission protocol uses the thresholds in conjunction with the AGC, PDSQ and SER statistics from past receptions to determine the transmission parameters for the next packet transmission.

A combination of measured values *s*( ) *e*( ) and *i*( ) is used in the adaptation algorithm to increase $I(\ )$, decrease $D(\ )$ or not to change $N(\ )$, the code rate R, power level S or processing gain G. *NI*( ), for example, represents the action 'not change' or 'increase' the parameter where 'not change' is the first choice that will be used in the first following packet. If the performance with the first choice is not satisfactory, the second choice will be used in the next transmission. Only one parameter is changed in the next transmission interval.

In general, the goal is to keep S low (save energy and minimize multiple access interference) and R high (minimize redundancy and maximize the throughput). High G would improve interference suppression but at the same time reduce the throughput for the fixed chip rate. Some examples of the control action given the measured statistics are:

(1) $s(l),\ e(l),\ i(l) \Rightarrow I(R),\ D(S),\ D(G)$

(2) $s(l),\ e(h),\ i(l) \Rightarrow ND(R),\ NI(S),\ N(G)$

(3) $s(l),\ e(h),\ i(h) \Rightarrow ND(R),\ NI(S),\ I(G)$

Extension of the control rules to other combination of the measured statistics is straightforward.

The results shown in this section are for a communication system with binary differential phase-shift key (PSK) data modulation and binary DS-SS. Error correction is achieved with binary convolutional coding and soft-decision Viterbi decoding. Convolutional codes of rates 1/2 and 3/4 are available to the adaptive-transmission protocol. The code of rate 1/2 has constraint length 7 and minimum free distance 10.

For performance results, the spreading sequence is a maximal-length linear-feedback shift-register sequence of period 1023. An interfering signal, if present, uses a different m-sequence of length 1023. The spreading factors available to the adaptive-transmission protocol are 32, 64 and 128, and the number of information bits per packet is fixed at 1000. For the results, each simulation concludes after 50 000 packets are demodulated and decoded correctly. The sampling interval for the PDSQ statistic is $t_0 = T_c$ (chip interval).

The channel model incorporates range attenuation, shadow loss, MAI, and multipath propagation. Each of these four characteristics may be fixed or time-varying. For our purposes, a *Markov channel* is a communication channel for which at least one characteristic is time-varying, and each time-varying characteristic is modeled as a discrete-time Markov process. The Markov model for range attenuation is illustrated in Figure 4.9 [35], and the Markov model for the loss due to shadowing is illustrated in Figure 4.10. The model for the interference caused by other transmissions in the network consists of the two Markov chains, illustrated in Figure 4.11. Finally, each multipath component is modeled by the Markov chain in Figure 4.12. Each Markov chain may change states from one packet transmission to the next, but the state is assumed to be constant over the duration of the packet transmission.

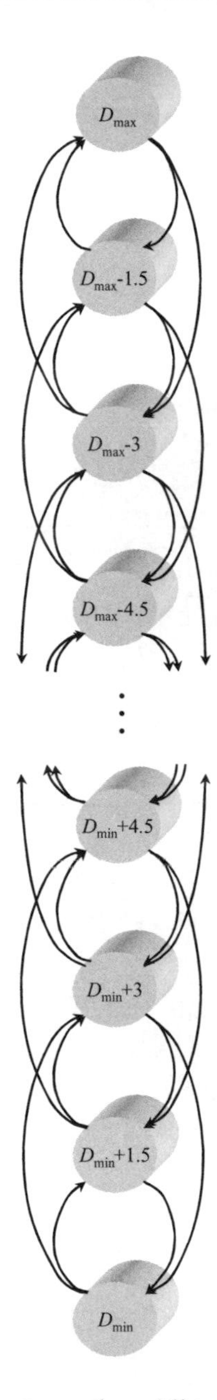

Figure 4.9 Markov chain for range attenuation. All transition probabilities are $p = 0.05$.

Because a Markov channel is represented by one or more finite-state Markov chains, each of which has multiple states, it is actually a collection of a number of different channels. Such a collection is referred to as a *channel class*. The channel class denoted by C_1 consists of the collection of all channels for which there is no shadow loss, multipath or MAI. For class C_1, the only channel characteristic that has multiple states is the range loss, which takes values from D_{min} to D_{max} in steps of 1.5 dB, as illustrated in Figure 4.9. The values for

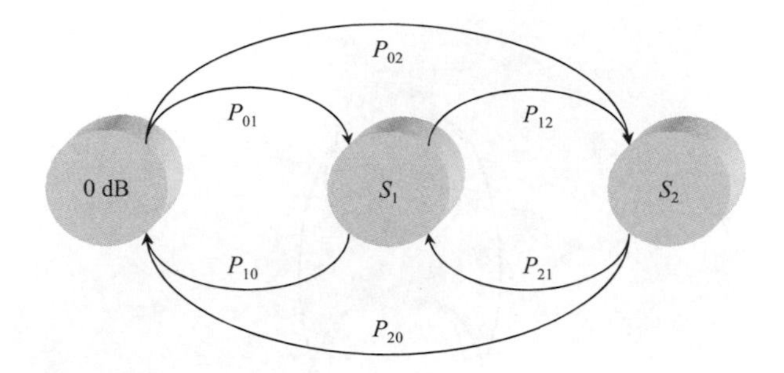

Figure 4.10 Markov chain for shadowing.

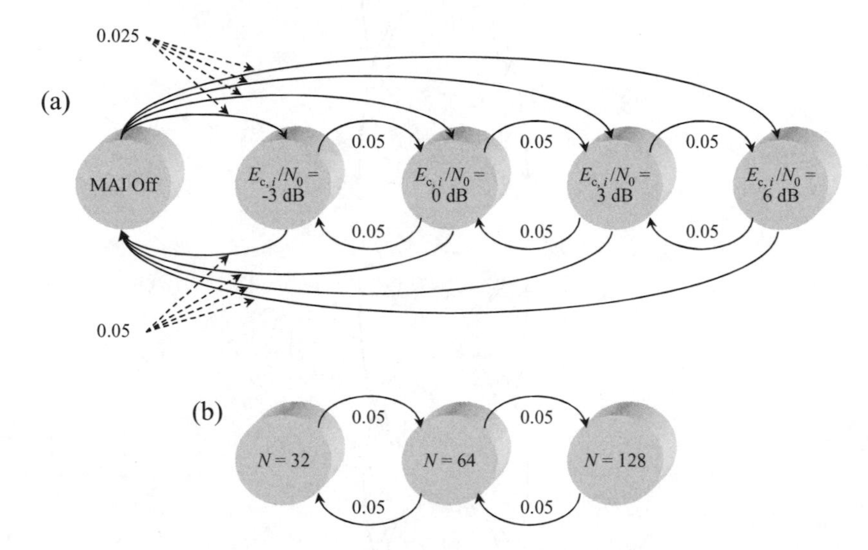

Figure 4.11 Markov chains for the interfering signal. The Markov chain in (a) determines the received power of the interferer, and the chain in (b) determines the spreading factor of the interferer.

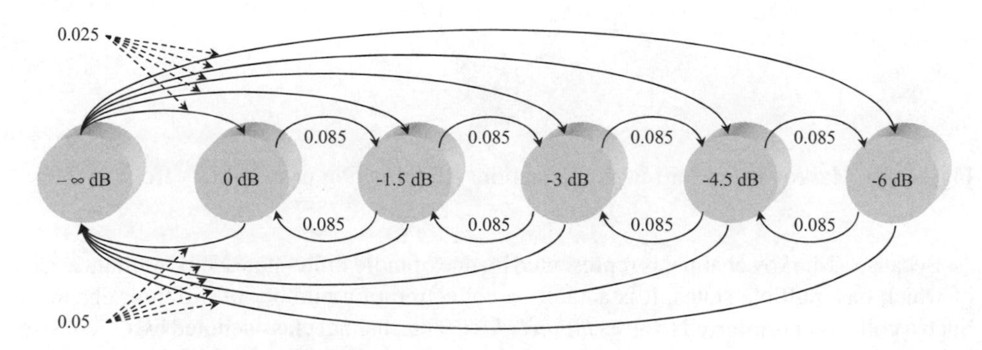

Figure 4.12 Markov chain for a single multipath component.

Table 4.2 Channel class parameters

Channel class	C_1	C_2	C_3	C_4	C_5	C_6
D_{min} (dB)	89	90	90	90	89	90
D_{max} (dB)	137	117	129	114	137	129
S_1 (dB)	0	10	0	10	0	0
S_2 (dB)	0	20	0	15	0	0
P_{01}, P_{02}	N/A	0.015	N/A	0.015	N/A	N/A
P_{12}, P_{21}	N/A	0.015	N/A	0.015	N/A	N/A
P_{10}, P_{20}	N/A	0.1	N/A	0.1	N/A	N/A
Interference	No	No	Yes	Yes	No	
K_{max}	1	1	1	1	3	3

D_{min} and D_{max} for class C_1 are 89 and 137 dB, respectively. Each of the other five channel classes also has multiple states for the range loss, but the values for D_{min} and D_{max} are different (see Table 4.2). The transmitter has six power levels arranged in steps of 6 dB.

In the model, the minimum power level provides an SNR of $E_c/N_0 = 0$ dB at the receiver if the propagation loss is 90 dB. The parameter E_c is the received energy per chip for the desired signal. The value 0 dB is referred to as the reference value for E_c/N_0, and 90 dB is the reference value for the propagation loss. If the chip rate is 16 Mchips/s, $E_c/N_0 = 0$ dB gives $S/N_0 \approx 72$ dB, where S is the received power for the desired signal. If the transmitter uses its maximum power level and the propagation loss is 137 dB, the maximum value for class, then the SNR at the receiver is $E_c/N_0 = -17$ dB. The maximum power level is 30 dB above the minimum power level, and the maximum propagation loss is 47 dB greater than its reference level, so E_c/N_0 is 17 dB less than its reference value. Class C_2 has variable shadow loss in addition to variable range attenuation. Each state is identified by its corresponding shadow loss, S_i, $0 \leq i \leq 2$, as illustrated in Figure 4.10. The value of S_0 is 0 dB, and the values of S_1 and S_2 for class C_2 are 10 and 20 dB, respectively, as shown in Table 4.2. Shadow losses of 20 dB or more have been observed in urban environments (see Chapter 3). The transition probability from S_i state to S_j is denoted by P_{ij}. The values of the transition probabilities are also given in Table 4.2.

Notice that D_{max}, the maximum range loss, is reduced from 137 to 117 dB in going from class C_1 to class C_2. The reason for this is to have the same maximum value for the total propagation loss for the two classes. In this way, the same set of transmitter power levels is appropriate for both classes. If we were to keep the maximum range attenuation at 137 dB and allow a maximum shadow loss of 20 dB, then the maximum transmitter power would have to be increased by 20 dB to provide the same SNR for the worst channel in the class. Because the minimum power level should remain the same in order to provide the same SNR for the best channel in the class, either the step size or the number of steps would have to be increased. It is more meaningful to keep the transmitter capability the same for different channel classes.

Channel class C_3 has intermittent MAI. The strength and spreading factor for the interference are governed by the Markov chains illustrated in Figure 4.11, where $E_{c,i}$ denotes the energy per chip for the interference signal. The model for the interfering signal includes the use of a rate-$\frac{1}{2}$ code, so the spreading factor determines the duration of the interference.

The maximum range attenuation is adjusted so that the maximum transmitter power can provide an adequate packet-error probability for the worst channel in the class, which corresponds to the maximum MAI and a range attenuation of 129 dB, the value of D_{max} shown in Table 4.2.

Multipath is present in channel class C_4 in addition to the primary signal. The strengths of the different multipath components relative to the primary component are controlled by independent Markov chains, each having the form shown in Figure 4.12.

The delays of the multipath components relative to the primary component are chosen randomly and without replacement from the set $\{nT_c;\ 1 \leq n \leq 150\}$ (i.e. no two components have the same delay). The adaptive protocol is compared with a protocol that has fixed transmission parameters and a protocol with *perfect CSI* (channel state information). As the name implies, the latter protocol is always told the state of the channel for previous packet receptions, and so it is a model for an adaptive-transmission protocol that is perfect in its channel measurements and estimates of the channel state. Note that perfect knowledge of the past states does not imply perfect knowledge of the channel state for the next transmission, thus, the results for the protocol with perfect CSI do not provide an upper bound on the performance of adaptive-transmission protocol. The protocol with perfect CSI chooses the transmission parameters according to the channel state that was in effect during the most recent reception.

Each of the three protocols attempts to maximize the throughput efficiency, subject to constraints on the packet-error probability. The *throughput efficiency*, a measure of the throughput per unit energy, is defined as the ratio of the number of packets successfully received to the total energy expended. The constraint on average packet-error probability is $P_E \leq 0.05$. In some circumstances, the protocols may be able to improve the throughput efficiency by tolerating large numbers of errors in very poor channel conditions, while still meeting the constraint on P_E. However, this is unfair, in the sense that the protocol would not provide an average packet-error probability of 0.05 over a time interval in which the channel state was very poor. To avoid this, it is required that each protocol adapt the parameters in order to provide a packet-error probability of 0.05 or less for each channel in the class. Note that meeting the latter constraint does not ensure that the former constraint is also met. For example, after a change in channel state, the error rate may be high until the protocol has adapted all parameters to the new state. The performance comparisons are for two scenarios. In the first scenario, each of the three transmission protocols knows the channel class but not the current channel state. The protocol with perfect CSI is told the state of the channel for the previous transmission, but the other two protocols are not given any information about the previous channel state. Each protocol chooses its transmission parameters in the best way for the given channel class. For example, the protocol with fixed transmission parameters uses the minimum power level, minimum spreading factor, and maximum code rate that provide the required packet-error probability for the worst channel in the given class. Similarly, the adaptive-transmission protocol uses the adaptation thresholds that are best suited for the given class.

In Figure 4.13, the throughput efficiency of each protocol is shown for each class. As expected, the protocol with fixed transmission parameters performs substantially worse than the protocols that adapt the transmission parameters. The protocol with perfect CSI has slightly better performance than adaptive-transmission protocol for classes C_3, C_4, and C_5. For the remaining classes, the two protocols give nearly the same performance. Even though classes C_1 and C_2 have the same worst-case propagation loss, the results in Figure 4.13

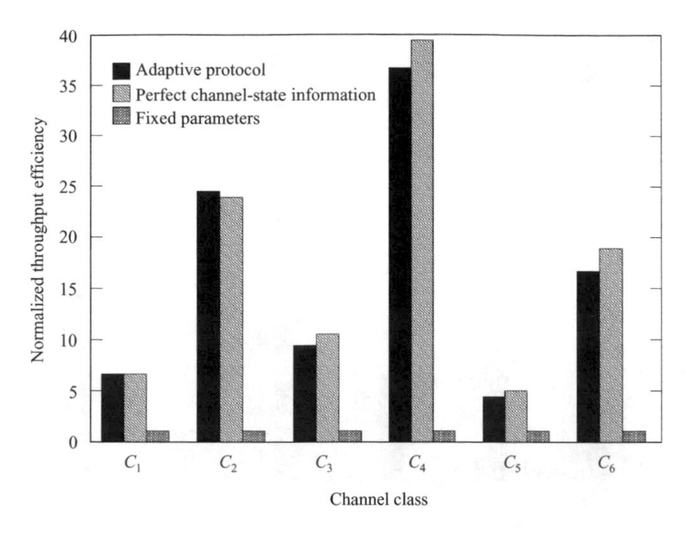

Figure 4.13 Throughput efficiency for protocols with a known channel class. (Reproduced by permission of IEEE [35].)

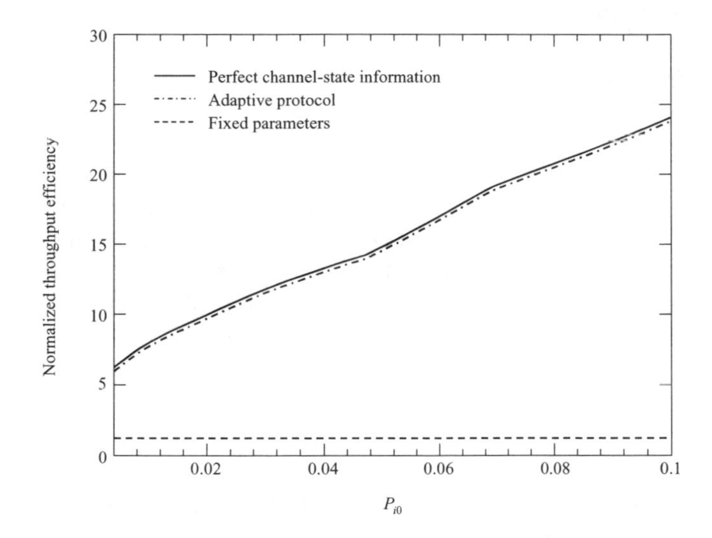

Figure 4.14 Throughput efficiency for channel class C as parameters $P_{10} = P_{20}$ of the Markov chain of Figure 4.10 are varied. Adaptation thresholds are constant. (Reproduced by permission of IEEE [35].)

show that the throughput efficiency is much higher for class C_2, in spite of the possibility of shadowing that is included in this class. This difference in throughput efficiency is due only to the choice of channel class parameters. For the parameters given in Table 4.2, class C_1 is much more likely than class C_2 to exhibit maximum or near-maximum propagation loss. In Figure 4.14, the throughput efficiencies are shown for a range of values of P_{10} and P_{20} from the Markov chain of Figure 4.10. As these probabilities are increased from 0.01 to 0.1, the probability of having no loss due to shadowing increases, so the throughput efficiency

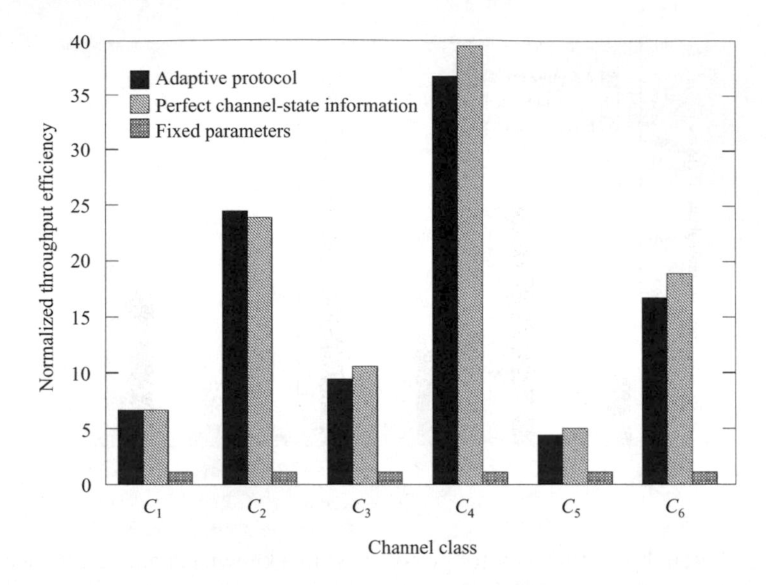

Figure 4.15 Throughput efficiency for protocols with an unknown channel class. (Reproduced by permission of IEEE [35].)

also increases for each of the adaptive-transmission protocols. Increases in P_{10} and P_{20} are offset by corresponding decreases in P_{11} and P_{22}, which are the probabilities of staying in states 1 and 2, respectively.

In the second scenario, the transmission protocols do not even know the channel class. The design of the protocols must account for the full range of channel conditions that can be experienced for all six classes. The performance of a protocol in this scenario is a good measure of its robustness. In Figure 4.15, the throughput efficiency for each protocol is shown for the six channel classes. If an adaptive protocol is used, the average energy needed to meet the requirement on the error probability is substantially less than for fixed transmissions. The protocol with perfect CSI gives the best performance, but adaptive-transmission protocol does nearly as well.

4.3 ADAPTIVE HYBRID ARQ SCHEMES FOR WIRELESS LINKS

This section considers the problem of using error-control coding in the data link layer to achieve reliable communication over a wireless link. Broadly speaking, there are two types of error control: forward error correction (FEC), considered so far in this chapter, and ARQ. ARQ is efficient when the channel condition is good or moderately good, but as the channel condition deteriorates, ARQ throughput performance becomes unacceptably poor. In this chapter we consider: (1) an adaptive hybrid of FEC and ARQ using the Reed–Solomon (RS) code; and (2) adaptive frame-length control. RS codes are known to provide excellent error-correction capability, especially in terms of bursty error suppression. The RS code rate and frame length are chosen adaptively based on the estimated channel condition to maximize the throughput performance. We also consider error control on the MAC header since: (1) without proper error protection of the header, it would be of no help to apply any error control on the user data; and (2) the MAC header is also part of the data link layer. More discussion on adaptive frame length and coding can be found in References [36–39]. Three error-control schemes (referred to as FEC1, FEC2 and FEC3) are considered depending on: (1) how many RS code segments are used for each packet; and (2) how a packet with uncorrectable errors is retransmitted. The computation complexity of an error-control scheme is closely related to battery power consumption. Encoding/decoding processes with RS codes are known to consume substantial battery power. In this chapter we use the central processing unit (CPU) time for encoding/decoding of an RS codec implementation as the measure of computational complexity. While the actual complexity of the RS code will depend on a particular implementation, it is believed that the CPU time measurement can be a good reference. The three schemes are compared in terms of throughput performance and computation complexity.

4.3.1 RS codes

For the error control of user data, (N, K, q) RS codes over $GF(q)$ are used, in which the codeword size $N \leq q - 1$ and the number of information symbols $K < N$. A q-ary symbol is mapped to b bits, so $q = 2^b$. RS codes are known to have the maximum error-correction capability for given redundancy, i.e. a maximum distance separable (MDS) code. For an (N, K, q) MDS code, the minimum distance $d_{\min}$ is determined as $d_{\min} = N - K + 1$, where the error correction capability $t = (d_{\min} - 1)/2 = (N - K)/2$, i.e. any combination of t symbol errors within N symbols can be corrected. The code rate r_c is defined as $r_c = K/N$. One can easily see that the more parity symbols (i.e. larger $N - K$), the better error-correction capability. RS codes are also known to be efficient for handling bursty errors. For example, with $(N, K, 2^b)$ RS code with the error correction capability t, as many as bt bit errors can be corrected in the best case when all of b bits in each of bt-bit symbols are erroneous (i.e. bursty errors). However, only t bit errors can be corrected in the worst case when only one bit in each of tb bit symbols is erroneous (i.e. nonbursty errors). Originally, the codeword size of (N, K, q) RS code is determined to be $q - 1$. However, a shorter codeword can be obtained via *code shortening*. For example, given an (N, K, q) code, $K - s$ information symbols are appended by s zero symbols. These K symbols are then encoded to make an N symbol-long codeword. By deleting all s zero symbols from the codeword, we can obtain $(N - s, K - s)$ code. For decoding this shortened code, the original (N, K) decoder can still be used by appending zero symbols between $K - s$ information symbols and $N - K$ parity symbols. Shortened RS codes are also MDS codes. Code shortening is especially useful for transmitting information with less than K symbols.

4.3.2 PHY and MAC frame structures

In this segment parameters of the popular WaveLAN modem are used. PHY and MAC overheads of the WaveLAN are shown in Figure 4.16 [40]. In WaveLAN, no error-correction coding is implemented; only the CRC code for error detection is used. In this section the MAC frame structure is modified as shown in Figure 4.17 for error-control schemes. First, the MAC header size is increased by one byte. This additional byte is used to give the

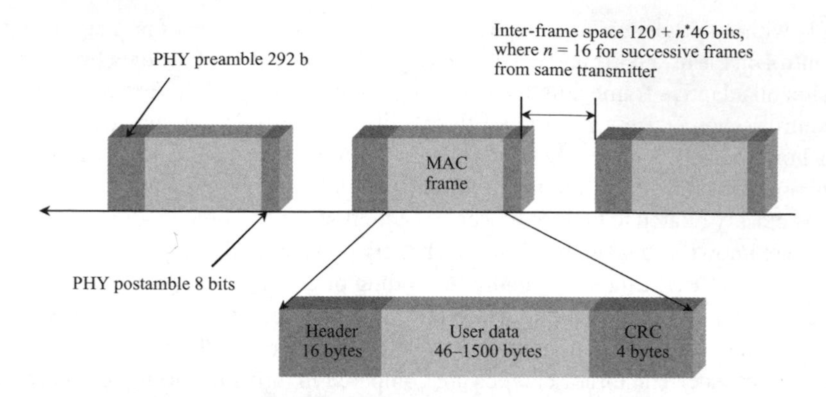

Figure 4.16 PHY and MAC overheads for WaveLAN.

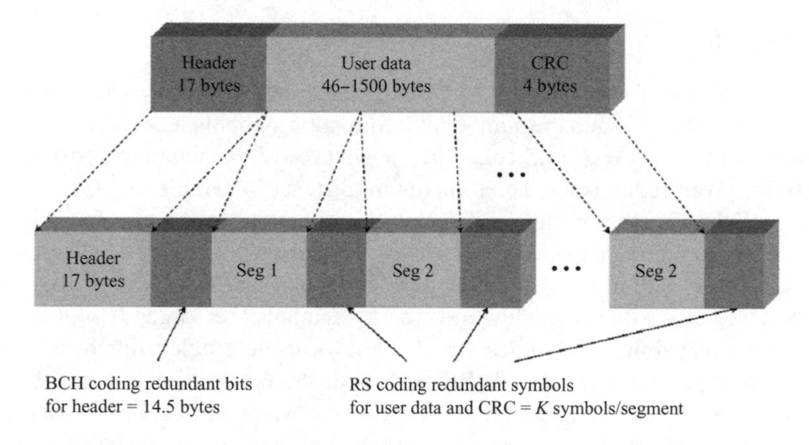

Figure 4.17 The newly modified MAC frame structure.

error-control scheme information like the RS-code rate used for the subsequent user data, which can be adaptively changed. For the MAC frame header, the binary Bose–Chaudhuri–Hocquenghen (BCH) code is adopted, so parity bits (equivalently, 14.5 bytes) are appended to the header. Using this code, up to $t = 15$ bit errors can be corrected. Note that this BCH coding is fixed for every MAC frame so that the receiver can receive/decode it without any extra information to check if the received MAC frame is destined for itself.

The user data plus CRC is divided into M segments, and each segment is encoded by an RS code. The segmentation and encoding procedures are detailed in the next section. The user data is assumed to be an IP packet with a 20-byte-long header and, throughout this section, the terms 'user data' and 'packet' are used interchangeably. In order to examine the performance of error-control schemes, we specify various overheads of interest as follows:

- H – the length of MAC header overhead including BCH redundancy $= 31.5(= 17 + 14.5)$ bytes $= 252$ b;
- O – the length of PHY overhead $= 420(= 120 + 292 + 8)$ b (assuming $n = 0$);
- C – the length of CRC and IP header overheads $= 24(= 4 + 20)$ bytes $= 192$ b.

Table 4.3 Nine RS codes considered for adaptation

N		255			511			1023	
K	229	153	77	459	307	153	921	613	307
q		256			512			1024	
r_c	0.9	0.6	0.3	0.9	0.6	0.3	0.9	0.6	0.3
t	13	51	89	26	102	179	51	205	358

4.3.3 Error-control schemes

Nine RS codes as shown in Table 4.3 are considered. The reason for limiting the number of codes is that a different encoder–decoder pair is needed for each (N, K, q) RS code. Described below are error-control schemes for the link layer.

4.3.3.1 A hybrid of FEC and ARQ

A hybrid of FEC and ARQ is used. The receiver attempts to correct errors first and, if the errors cannot be corrected, retransmission of the packet is requested. When errors are successfully corrected, an acknowledgment is transmitted to the sender and, when errors are detected but not correctable, a *negative acknowledgment* (NAK) is sent. In this section the *selective-repeat* (SR) ARQ is assumed. The sender keeps transmitting packets without waiting for ACK/NAK of those packets already transmitted. If NAK of a transmitted packet is received, or if neither ACK nor NAK of a packet is received within a timeout interval, the packet is retransmitted. The timeout interval is determined based on the roundtrip time. As will be clear below, the schemes considered here do not depend on ARQ, and hence they can be used in conjunction with any other ARQ schemes, such as stop-and-wait (SW) and go-back-*N* (GBN).

Throughput performance and complexity may depend on the underlying ARQ scheme. The SR-ARQ scheme is known to achieve better throughput performance than SW and GBN-ARQ at the expense of higher complexity.

An ACK/NAK packet consists of four bytes, in which the first two bytes are for the frame number of the packet associated with ACK/NAK, the third byte is to inform (1) whether it is an ACK or NAK, and (2) the adapted code rate. The last byte is a checksum. As discussed in Section 4.1, the code rate is mainly adapted by the receiver depending on the estimated channel state/condition; the code-rate information is fed back to the sender within each ACK/NAK packet. These four bytes are encoded by (148, 32) BCH code, which is a shortened code from (255, 139) BCH code adopted for the MAC header error protection.

4.3.3.2 Sender side

An IP packet is fragmented into a number of segments, referred to as *user data*, depending on the maximum user data size that can be accommodated in a MAC frame, or the maximum transmission unit (MTU). The MTU of the WaveLAN is originally 1.5 kbytes. In this segment, MTU is adapted dynamically as required by the underlying error-control scheme. The CRC is calculated for both the user data and MAC header. The combined user data and

CRC are then divided into M segments, where each of the first $M - 1$ segments is Kb bits long and the length of the last segment is $\leq Kb$ bits. Note that MTU can be determined as $MKb - 32$ bits, where -32 represents the CRC overhead. Each segment is encoded with $(N, K, 2^b)$ RS code. For the last segment, code shortening might be needed. The MAC header is then encoded by (255, 139) BCH code. Note that one byte of the header is used to represent the information of the error-control code such as N, K, and M. For the schemes discussed in this segment, only a set of codes is used, so one byte suffices to specify all the necessary information. The MAC frame forms the structure, as shown in Figure 4.17.

4.3.3.3 Receiver side

The error control at the receiver works as follows:

(1) The header of the received frame is decoded at the MAC sublayer. If the header is successfully decoded and, if the frame is found to be destined for itself, the user data with M RS-coded segments as well as the RS code information from the header are sent upward to the link layer.

(2) At the link layer, each of M RS-coded segments is first decoded by the RS decoder, then the entire frame (including the header and user data) is checked by the CRC decoder.

As described in the next section, the RS decoder at the receiver first attempts to correct errors. If the errors are uncorrectable, only their presence will be detected. Note that errors can be detected in three different places in the receiver: the MAC sublayer by the BCH decoder, the link layer by the RS decoder, and finally by the CRC decoder. This three-level error detection detects virtually all uncorrected errors.

4.3.3.4 Three error-control schemes

The following three possible error-control schemes using the above error-control facilities are considered.

- *FEC1* – $M = 1$, when a long RS code with a large N is used for each packet.
- *FEC2* – $M \neq 1$ and the entire packet is retransmitted if any RS code segment has uncorrectable errors.
- *FEC3* – $M \neq 1$ and each code segment with uncorrectable errors is requested to be retransmitted.

Then, a set of those code segments received in error is retransmitted by the sender in one MAC frame. Note that, for *FEC3*, a simple form of ACK and NAK is not enough; more on this issue will be discussed later. A packet decoded without any RS decoding failure can have undetected errors as explained in the next section. These errors are most likely to be detected by the CRC. In this case, the sender is requested to retransmit the entire packet for all three schemes.

4.3.3.5 Adaptation rule

Analysis in Choi and Shin [36] has shown that FEC2 provides the best compromise between the performance and complexity. Here we consider how to adapt the error-control code based on adaptive FEC2. First, we define the set of codes for adaptive use as c_0, c_1, c_2, c_3, where

$$\begin{aligned} c_0 &\Rightarrow \textit{no coding} \\ c_1 &\Rightarrow (255, 229, 256) \\ c_2 &\Rightarrow (255, 153, 256) \\ c_3 &\Rightarrow (255, 77, 256) \end{aligned}$$

According to Choi and Shin [36], adaptation points (or symbol error probabilities at which the code rate would be adapted) are determined by $P_\mathrm{s} = 7.5 \times 10^{-5}, 0.034, 0.17$. We use a modified set of the adaptation points as

$$\begin{aligned} P_\mathrm{s}^{0,1} &= 6.5 \times 10^{-5} \\ P_\mathrm{s}^{1,2} &= 0.025 \\ P_\mathrm{s}^{2,3} &= 0, 15 \end{aligned}$$

The reasons for choosing values lower than the original ones include: (1) adaptation at a higher P_s than the original adaptation point can result in a very low throughput while adaptation at a lower P_s would not; and (2) an adaptation rule based on the original probabilities might result in too frequent adaptations (for the first reason).

To develop an adaptation rule, we need to divide time-varying error patterns into two categories. One is the long-term variation in which error characteristics vary over several seconds to 10 seconds. A time-varying distance between the sender and receiver and shadowing can cause this type of variation. The other is the short-term variation in which error characteristics vary over several milliseconds or less. The multipath fading environment resulting from multipath propagation of the transmitted signal and the mobile's relative movement can cause this type of variation.

Similar to Section 4.1, error-control code adaptation is based on the channel state estimation, which is done using the decoding results at the receiver. That is, the receiver determines the desired code rate based on the decoding result of a received packet, then informs it to the sender through the ACK/NAK of each received packet. Because our adaptation is not based on any prediction but on the observation and feedback, it might not be able to handle the short-term variation of the channel condition well depending on the actual channel behaviour. However, it is effective for long-term variations. The code currently set at the receiver is denoted by c.

4.3.3.6 Increase of code rate

When a packet is successfully received (i.e. without any decoding failure), the receiver considers if it is time to change the code rate. Based on the currently set code at the receiver, the code rate is increased adaptively as follows:

- When $c = c_i$ for $i = 2, 3$, the code is adapted to c_{i-1} if the number e_M of symbol errors out of $N_{M,L}$ RS symbols in the last frames is smaller than $P^{(i-1),i} N_{M,L}$ for the smallest L such that $P^{(i-1),i} N_{M,L} > 1$.

- When $c = c_1$, the code is adapted to c_0 if there was only a single or no symbol error during the last L frames received for the smallest L such that, $P^{0,1} N_{M,L} > 1$ where $N_{M,L}$ is the total number of RS symbols in the last L frames. Note that the number of symbol errors in an RS code segment is readily available from the decoder unless the decoding fails.

4.3.3.7 Decrease in code rate

When a packet needs to be retransmitted due to a CRC error for $c \neq c_0$, the code rate is not adapted in order to defer the adaptation decision until the next packet is received since a CRC error is very rare and will not usually happen over consecutive packets. On the other hand, when $c = c_i$ for $i = 1, 2$ and a packet needs to be retransmitted due to RS decoding failures, the code rate is decreased by one step, i.e. $c_i \Rightarrow c_{i+1}$ for $i = 1, 2$, if any of the following conditions occurs:

- when all RS segments of a packet result in decoding failures;
- when more than a half of all RS segments in the last two packets result in decoding failures;
- when three of the last 10 received packets result in decoding failures.

The above adaptation rule for code-rate decrease may appear very *ad hoc* but, with the third condition, we try to limit the retransmission probability to under 0.3 since a similar or better throughput efficiency can otherwise be achieved with the next lower rate code. The first and second conditions, on the other hand, render prompt adaptation for an abrupt change of the channel condition.

Now, when $c = c_0$, the code is adapted to c_1 if there were more than one CRC error detections in the last 10 received packets in order to keep the retransmission probability under 0.2. When the code rate is adapted to the decrease, and it is known to the sender, the sender encodes packets with the adapted code rate. The packets which need to be retransmitted are also re-encoded. Note that, for example, 229 symbols encoded by (255, 229, 256) RS code cannot be reencoded by (255, 153, 256) RS code. For a packet encoded by c_i code to be re-encoded by code c_{i+1}, the user data is divided first into two equal-sized packets (i.e. IP fragmentation) if the user data cannot be accommodated in c_{i+1}, then each encoded by c_{i+1}.

Finally, the code rate can also be decreased by the sender's decision. That is, upon expiration of the retransmission time out, the code is set to c_3. The rationale behind this adaptation is that a time-out happens due to the corruption of the MAC header or the loss of ACK/NAK while both the header and ACK/NAK are protected by very strong codes, thus implying that the channel is very bad.

4.3.4 Performance of adaptive FEC2

4.3.4.1 Link model

A very simple wireless network environment is used to evaluate the proposed adaptive error-control scheme. There is only one sender and one receiver in the network, where the sender is assumed to have an infinite amount of information to send. Therefore, all the packets are transmitted in a full-length MAC frame composed of six RS code segments. Immediately

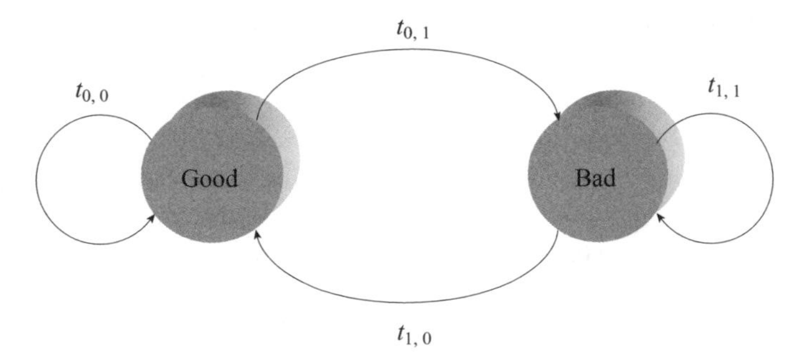

Figure 4.18 Two-state Markov chain model for the channel.

after receiving/decoding a packet, the receiver sends an ACK/NAK for the packet. Upon receiving the ACK/NAK, the sender determines whether to transmit a new packet or retransmit the previously sent one. When the retransmission time-out happens consecutively, the retransmissions are backed off exponentially. That is, for 2, 3, 4, ... consecutive time-outs, the retransmission is deferred for the time of 1, 2, 4, ... full-length packet transmissions, because consecutive time-out expirations implies that the channel is too bad for a packet to go through successfully.

4.3.4.2 Channel model

The channel is modelled by a two-state Markov chain, which is widely accepted to model the multipath fading channel, as shown in Figure 4.18. The channel state is either *good* or *bad* and can change on bit boundaries, that is, the channel condition stays in a state during one bit duration. Following the widely accepted Rayleigh fading model, which corresponds to the case of no line-of-sight path between the sender and receiver, we can derive the transition probabilities $t_{0,1}$ and $t_{1,0}$ as follows. First, Rappaport [41] provides the *level-crossing rate*, defined as the expected rate at which the Rayleigh fading envelope, normalized to the local root mean square (rms) signal level R_{rms}, crosses a threshold level in a positive-going direction:

$$N_{\mathrm{R}} = \sqrt{2\pi}\, f_{\mathrm{m}} \rho \mathrm{e}^{-\rho^2} \tag{4.32}$$

where the maximum Doppler frequency is given by $f_{\mathrm{m}} = \nu/\lambda$ for the mobile speed v and the wavelength λ of the carrier and the normalized threshold fading envelope is given by $\rho = R/R_{\mathrm{rms}}$. Next, the *average fade duration*, defined as the average period of time for which the received signal is below the threshold level, is given by [41]

$$T_{\mathrm{f}} = \frac{\mathrm{e}^{\rho^2} - 1}{\rho f_{\mathrm{m}} \sqrt{2\pi}} \tag{4.33}$$

Using the above formulas and assuming steady-state conditions, the probabilities μ_0 and μ_1 that the channel is in good and bad states, respectively, are given by [38]

$$\mu_0 = \frac{1/N_{\mathrm{R}} - T_{\mathrm{f}}}{1/N_{\mathrm{R}}} \quad \text{and} \quad \mu_0 = \frac{T_{\mathrm{f}}}{1/N_{\mathrm{R}}} \tag{4.34}$$

Finally, the state transition probabilities can be approximated by [42]

$$t_{0,1} = \frac{N_{\mathrm{R}}}{R_{\mathrm{t}}^0} \quad \text{and} \quad t_{1,0} = \frac{N_{\mathrm{R}}}{R_{\mathrm{t}}^1} \tag{4.35}$$

where $R_{\mathrm{t}}^k = R_{\mathrm{t}}\mu_{\mathrm{k}}$, and R_{t} is the symbol transmission rate. With the transmission rate $R_{\mathrm{t}} =$ 1 Mbps, the mobile speed $v = 2$ km/h (pedestrian speed), the carrier frequency $f = 900$ MHz ($\lambda = c/f = 1/3m$), and the normalized threshold fading envelope $\rho = 0.3$, we obtain

$$\mu_0 = 0.914, \quad \mu_1 = 0.0861$$

and

$$t_{0,1} = 1.253 \times 10^{-6}, \quad t_{0,0} = 1 - t_{0,1} \qquad t_{1,0} = 1.331 \times 10^{-5}, \quad t_{1,1} = 1 - t_{1,0}$$

Assuming the binary phase shift keying (BPSK) modulation, the bit error rates (BERs) $P_{\mathrm{b},0}$ and $P_{\mathrm{b},1}$ when the channel is in *good* and *bad* states, respectively, can be calculated as [43]

$$P_{\mathrm{b},i} = \int_{\gamma_i}^{\gamma_i-1} P_{\mathrm{b}/\gamma} f_i(\gamma)\,\mathrm{d}\gamma \tag{4.36}$$

where the BER for a given SNR γ is

$$P_{\mathrm{b}/\gamma} = Q(\sqrt{2\gamma}) \tag{4.37}$$

the conditional distribution of the instantaneous SNR γ in a given state with the mean SNR $\bar{\gamma}$ is

$$f_i(\gamma) = \frac{\frac{1}{\bar{\gamma}}\mathrm{e}^{-\gamma/\bar{\gamma}}}{\mathrm{e}^{-\gamma_i/\bar{\gamma}} - \mathrm{e}^{-\gamma_{i-1}/\bar{\gamma}}} \tag{4.38}$$

for $\gamma_i < \gamma < \gamma_{i-1}$ and the set $\{\gamma_{-1}, \gamma_0, \gamma_1\} = \{\infty, \rho^2\bar{\gamma}, 0\}$. Note that the mean SNR $\bar{\gamma}$ depends on the transmitted power, signal attenuation over the channel, and others.

4.3.5 Simulation results

Figure 4.19 shows the throughput efficiencies as the mean SNR increases for five different cases of adaptive FEC2: (1) two marked with 'fixed w/K' are the nonadaptive versions of FEC2 with (255, K, 256) codes; (2) one marked with 'w/exp backoff' is the adaptive FEC2 we are considering; (3) one marked with 'w/o exp backoff' is the adaptive FEC2 without exponential back-off; and (4) one marked with 'ideal' is an ideal version of FEC2, in which the code rate is determined after seeing errors in the received packet; that is, if these errors are uncorrectable even with $r_{\mathrm{c}} = 0.3$ code, the packet is assumed not to be transmitted at all (i.e. the transmission is deferred), while the code which can correct all these errors with the maximum code rate is selected otherwise. Basically, there is no retransmission in the ideal version. This ideal and unrealistic version can be used as a reference (or an upper bound) of the throughput efficiency. The figure demonstrates in which region and how much the adaptive scheme offers better performance.

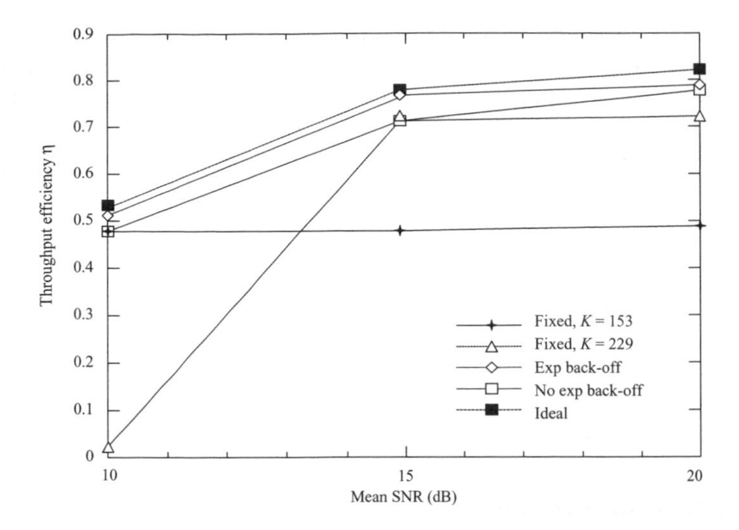

Figure 4.19 Throughput efficiency vs mean SNR.

4.4 STOCHASTIC LEARNING LINK LAYER PROTOCOL

Link layer ARQ protocols such as SW ARQ and SR ARQ, considered so far, recover from errors by retransmitting the erroneously transmitted packets regardless of the channel state. Such persistent retransmission schemes may lead to wasted energy consumption due to unnecessary packet losses and therefore additional retransmissions [44]. Discussions on this problem can be found in References [44–47]. In this section, we discuss a link layer algorithm that can produce soft predictions of the next network state given the past states, i.e. probabilities of the next state being any of the possible states are estimated.

The goal is to predict the wireless network state without *a priori* knowledge of the state transition probabilities and accordingly change the link layer transmission control for energy efficiency. Towards this objective a stochastic iterative learning control technique is used based on a stochastic learning automaton [48, 49]. The state of the wireless network is learnt on the fly by the transmitter via simple adaptive stochastic learning based on feedback from the receiver regarding the *quality* of the wireless link. Then a link layer control algorithm optimizes the transmission policy accordingly. That is, in a time slot the link layer control protocol either chooses to transmit a packet or not based on the channel state prediction produced by the stochastic learning algorithm. Energy efficiency is traded off for delay during this process.

4.4.1 Stochastic learning control

A stochastic learning control algorithm based on variable structure learning automaton (VSLA) [48, 49] in general is characterized by a 5-tuple,$\{\alpha, \beta, p, \tau, C\}$ where $\alpha = \{\alpha_1, \alpha_2, \ldots, \alpha_r\}$ is the set of r actions from which the automaton can choose any action at time n, denoted by $\alpha_i(n)$, $\beta = \{0, 1\}$ is the set of binary response (feedback) generated by an unknown random environment for the chosen action. $\beta = 0$ corresponds to a reward to the automaton for the chosen action while $\beta = 1$ corresponds to a penalty. $p = \{p_1, p_2, \ldots, p_n\}$ contains the set of probabilities with which the corresponding actions are chosen. τ is the stochastic iterative learning algorithm according to which the elements of the set p are updated over time. $C = \{c_1, c_2, \ldots, c_r\}$ is the set of penalty probabilities conditioned on the chosen action, i.e. $c_i = \Pr[\beta = 1 \mid \alpha_i]$ that characterizes the unknown environment. It is reasonable to assume that arg min C is unique for a given random environment. The VSLA attempts to learn the optimal action (policy) that results in the minimum value in C. The VSLA can be characterized by the following equation:

$$p(n+1) = \tau\{\alpha(n), \beta(n), p(n)\} \tag{4.39}$$

where the ith element of the set $p(n)$ at time n is

$$p_i(n) = \Pr(\alpha(n) = \alpha_i),\ i = 1, 2, \ldots, r$$
$$\sum_{i=1}^{r} p_i(n) = 1,\ \forall n \text{ and } p_i(1) = 1/r,\ \forall i \tag{4.40}$$

4.4.2 Adaptive link layer protocol

The two-state Markov chain model (Gilbert–Elliot model) from Figure 4.18 is also used in this section. The transmission energy (E_t), mean number of retransmissions ($\bar{R}$) and the packet error rate (P_{ep}) for an ARQ scheme are related as [45]

$$\begin{aligned} E_t &= (\bar{R}+1)P/C \\ \bar{R} &= (1-P_{ep})\sum_{i=1}^{R-1} iP_{ep}^{i} + RP_{ep}^{R} \\ P_{ep} &= P_G\left[1-(1-P_{eG})^L\right] + P_B\left[1-(1-P_{eB})^L\right] \end{aligned} \tag{4.41}$$

In Equation (4.41) E_t is the total transmission energy, P is the transmit power, C is the bit rate of the channel, P_{eG} and P_{eB} are the bit error rates in states G (good) and B (bad), respectively, L is the packet length in bits, R is the maximum number of retransmissions allowed, P_{ep} is the packet error probability and, P_G and P_B are the steady state probabilities of G and B states. As seen from Equation (4.41), the total energy consumed is directly proportional to the mean number of retransmissions, indicating that considerable energy can be saved if the number of retransmissions is reduced significantly. Therefore, smart link layer protocols that recognize the wireless channel state and adaptively control the transmission/retransmission policy could result in significant energy savings.

In this section an adaptive link layer protocol is presented that adjusts itself over time to arrive at the optimal policy for a fixed but unknown finite state wireless channel model. The link layer protocol employs a VSLA with two actions in its action set: α_1, *transmit a packet*; and α_2, *do not transmit a packet*.

(1) If the transmitter sends a packet, it waits for a fixed time period to receive an acknowledgement. If an ACK is received within this period then the action α_1 is rewarded ($\beta = 0$); if not a time-out occurs and α_1 is penalized ($\beta = 1$).

(2) Then, a reward–penalty learning algorithm, as discussed later in this section [see Equation (4.42)], is used to update the action probabilities of the set $p(n) = \{p_1(n), p_2(n)\}$.

(3) The next transmission decision is chosen according to the probability distribution $p(n)$.

This process continues until the probability vector $p(n)$ converges when the state of the channel is predicted accurately one time unit ahead. We assume that the reverse channel or the feedback channel is error free. The effect of an erroneous feedback channel and feedback delays are similar to those discussed in Section 4.1. It is also assumed that an ACK is generated for every packet transmitted. Depending on the needs of the situation, the feedback and updating can be also done after every few packets with some minor changes to the above algorithm.

4.4.2.1 Reward-penalty algorithm

When α_1 is chosen then $p_1(n)$ and $p_2(n)$ are updated linearly depending on $\beta(n)$; however, when α_2 is chosen both $p_1(n)$ and $p_2(n)$ are left unchanged. This is because, when the link layer chooses not to transmit a packet, there will be no feedback from the receiver. This stochastic learning control algorithm that predicts the channel state and controls the retransmission policy defined as $p\,(n+1) = \tau\,\{\alpha(n), \beta(n), p(n)\}$ can be described as

$$\begin{aligned}
&\text{if} \quad \alpha(n) = \alpha_1 \quad \text{and} \quad \beta(n) = 0\\
&p_1(n+1) = p_1(n) + a(1 - p_1(n) - A)\\
&p_2(n+1) = p_2(n) - a(p_2(n) - A)\\
\\
&\text{if} \quad \alpha(n) = \alpha_1 \quad \text{and} \quad \beta(n) = 1\\
&p_1(n+1) = p_1(n) - b(p_1(n) - B)\\
&p_2(n+1) = p_2(n) + b(1 - p_2(n) - B)\\
\\
&\text{if} \quad \alpha(n) = \alpha_2\\
&p_1(n+1) = p_1(n)\\
&p_2(n+1) = p_2(n)
\end{aligned} \tag{4.42}$$

In Equation (4.42), a and b are reward and penalty parameters that control the speed and accuracy with which the VSLA predicts and tracks the time-varying wireless channel state. Parameters A and B control the probability with which the link layer can explore both actions even while deviating from the currently most probable action. They also determine the amount of exploration vs exploitation during learning.

4.4.2.2 Performance results

The simulation scenarios consists of two nodes, with one node acting as a transmitter while the other is a receiver. The channel between the two nodes is simulated as a two-state Markov model (Figure 4.18). With probability 1 packets are dropped in the B state. Various values for the packet loss rate in the G state were chosen to observe the performance. A transmission rate of 11 Mbps was assumed as it is typical in a present-day wireless network such as a wireless LAN. The packet size L was fixed at a constant 1000 bytes. No upper limit on the number of retransmissions was set. In each simulation the number of packets transmitted by the sending node to the receiving node was 5000 and the results were averaged. The parameters $A = B = 0.02$ was fixed throughout the simulations and $a = b = 0.3$ unless otherwise stated. If the link layer transmitted a packet then a timer was

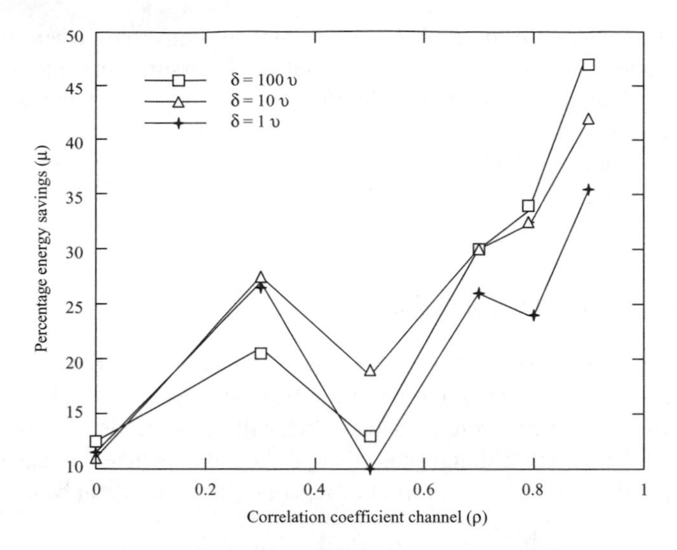

Figure 4.20 Percentage energy savings vs channel burstiness when packet error rate in G state is 0 %. Performance for different choices for the delay period δ.

set equal to 2.5 times the time taken by a packet to reach the destination. If an ACK was not received before the timer expires (time-out) then the packet was considered lost and the action probabilities updated. If the chosen action was not to transmit, then a back-off timer with a time period δ was used. After the expiry of the time period δ the transmitting node again decided the next action to be generated depending on the action probabilities in $\{p(n)\}$.

Figure 4.20 shows the percentage savings in transmission energy (μ) obtained using the adaptive link layer protocol in comparison with an SW ARQ protocol for 0 % packet loss rate in the G state. Recall that the packet loss rate in the B state was assumed to be 100 %. With respect to notation from Figure 4.18, the correlation coefficient of the channel ρ for $t_{0,0} = t_{1,1}$ is $\rho = t_{0,0} - t_{0,1} = t_{1,1} - t_{1,0}$ is a measure of the burstiness of the channel. In these simulations the parameter δ was varied in multiples of v, where $1/v$ is the transmission rate. As seen from the figure the energy savings are substantial (nearly 50 %) when the correlation coefficient of the channel (ρ) is high. This is because when the correlation coefficient is high the learning algorithm is able to predict the state of the channel more accurately as it utilizes the higher 'memory' in the channel. When the correlation coefficient is small, the energy savings reduce (to nearly 10 %) as the behavior of the channel becomes less predictable. Nevertheless, for both high and low values of ρ the adaptive link layer protocol conserves energy when compared with the traditional ARQ scheme.

The tradeoff in higher energy savings is an increased mean delay per packet transmission Δ, as seen in Figure 4.21. We observe that the difference in additional mean packet delay (w.r.t. the SW ARQ) between $\delta = v$ and $\delta = 10v$ is reasonably small. We reiterate that δ is a user-defined parameter which can be tuned to a specific application's mean delay requirements. Finally, we note that similar performances can be observed when the proposed method is also compared with other traditional ARQ protocols.

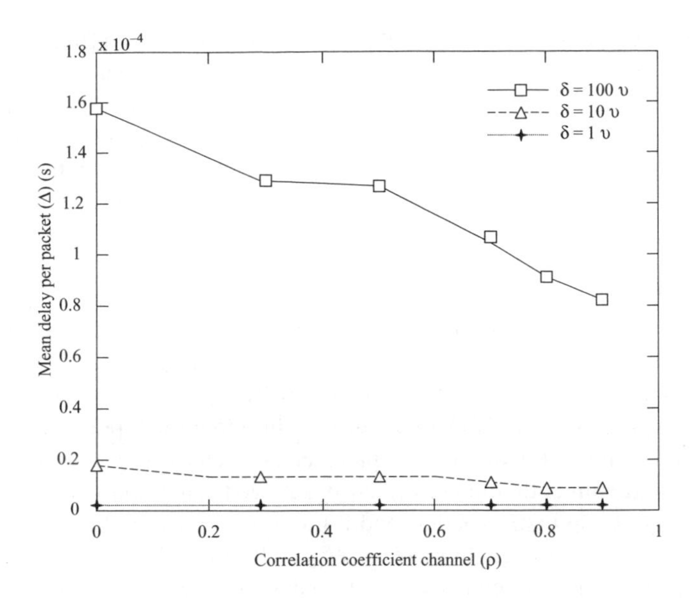

Figure 4.21 Mean delay per packet vs channel burstiness when G state packet loss rate is equal to 0 %.

4.5 INFRARED LINK ACCESS PROTOCOL

Recent growth in laptop computers and on portable devices, such as personal digital assistants (PDAs) and digital cameras, has led to an increasing demand for information transfer from or between portable devices [52]. Digital representation of information is expanding to new devices such as video and photocameras. New devices have 'computer-like' capabilities for storing and retrieving information such as mobile phones and portable information gathering appliances. Computer manufacturers have adopted the Infrared Data Association (IrDA) standard [53] and almost every portable computer and all Windows CE device on the market today contains an infrared (IR) port according to standards developed by IrDA. Laptop computers, PDAs, digital cameras, mobile phones and printers are examples of devises with IrDA links. More than 40 000 000 devices are shipped each year with IrDA ports [54] capable of using the unregulated IR spectrum for their cable-less communication needs. The IrDA standard addresses low-cost, indoor, short-range, half duplex, point-to-point links [55]. The IrDA physical layer specification (Ir-PHY) [56] supports optical links from 0 to at least 1 m, an angle of $\pm 15°$ at a BER of less that 10^{-8}. Ir-PHY version 1.0 serial IR (SIR) specification [57] supported data rates up to 115.2 kb/s using standard serial hardware, Ir-PHY version 1.1 fast IR (FIR) [58] extended the data rate to 4 Mb/s, and finally Ir-PHY version 1.3 very fast IR (VFIR) [59] specification added the 16 Mb/s link rate. The IrDA hardware is controlled by a link-layer protocol, the IrDA link access protocol (IrLAP) [60, 61]. IrLAP is based on the widely used high-level data link control (HDLC) protocol operating in NRM (see Chapter 1). The performance of IrDA optical wireless links may be measured by the throughput which can be drawn at the IrLAP layer. In Vitsas and Boucouvalas [62, 63] performance evaluation of the IrLAP protocol and deriving optimum values for link layer parameters for maximizing throughput is presented. In the literature, a mathematical model for the IrLAP throughput using the concept of a frame's 'virtual transmission time' is presented [52, 53] based on the HDLC analysis model presented in Williams [54]. However, this model does not lead to a simple formula for the IrLAP throughput. In Vitsas and Boucouvalas [57, 58], a new mathematical model using the average window transmission time (WTT) is developed. By taking advantage of IrLAP half-duplex operation, this model leads to a simple closed form formula for IrLAP throughput. The formula relates throughput to physical layer parameters, such as link BER, link data rate and minimum turnaround time, and with link layer parameters such as frame size window size and frame overhead. As this equation gives an intuitive understanding of the performance of IrDA links, it is very valuable for designers and implementers of such links. By setting the first derivative equal to zero, the optimum values for window size and frame length that maximize throughput are derived. Formulas for all IrLAP time-consuming tasks are also presented, allowing evaluation of link parameter values to throughput performance. IrLAP performance is examined for various link parameters, such as BER, data rate and window size and compared with optimum performance achieved using optimum window size and frame length values. Optimum window and frame-size values can be easily implemented and result in significant throughput increase, especially for links experiencing high BERs. However, implementing optimum frame-size values on retransmissions requires buffer reorganization at a low level.

4.5.1 The IrLAP layer

IrLAP is the IrDA data link layer. It is designed based on the pre-existing HDLC and synchronous data link control protocols (see Chapter 1, also Reference [60]). IrLAP stations operate in two modes: the normal disconnect mode during the contention period and the NRM during the connection period. In the contention period, a station advertises its existence to the neighboring stations along with the link parameters it supports and wishes to employ during the connection period. One of the participating stations becomes the primary station. Any station may claim to become the primary station but, at the end of the contention period, only one station is granted the primary role and all other stations are assigned the secondary role. All data traffic during the connection period is sent to or from the primary station. A secondary station wishing to communicate to another secondary station does so through the primary station. The parameters negotiated and agreed on during the contention period are given below:

(1) *Data rate* (C) – this parameter specifies the station's transmission rate.

(2) *Maximum turnaround time* (T_{max}) – this parameter specifies the maximum time interval a station can hold transmission control. For data rates less than 115.2 kb/s, the maximum turnaround time must be 500 ms. A smaller value may be agreed between the two stations for 115.2 kb/s or higher data rates.

(3) *Data size* (l) – this is the maximum length allowed for the data field in any received information frame (I frame). This parameter has an upper limit of 2048 bytes (16 kb).

(4) *Window size* (W_{max}) – this is the maximum number of unacknowledged frames I a station can receive before it has to acknowledge the number of frames received

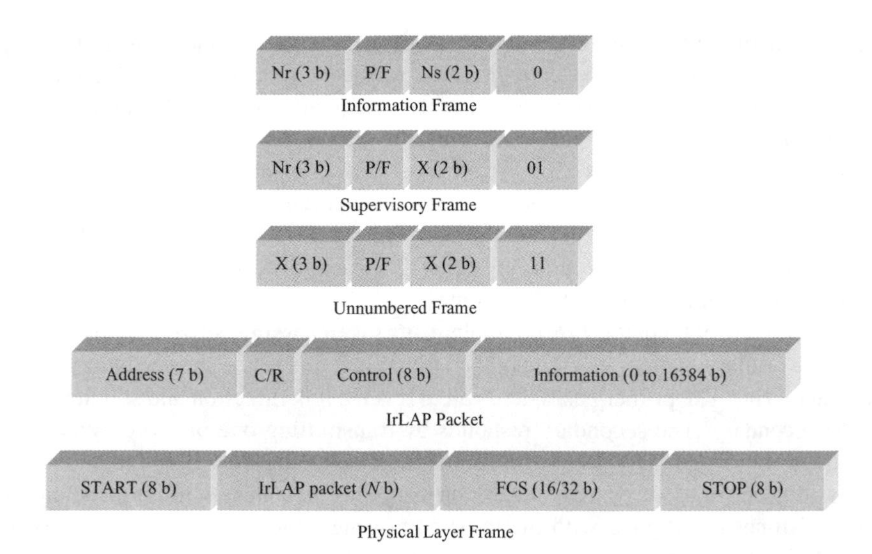

Figure 4.22 IrDA SIR and FIR frame structure.

correctly. An acknowledgement may be requested by the transmitting station before the window size is reached. This parameter has an upper limit of 7 for data rates up to 4 Mb/s and 127 for 4 and 16 Mb/s [60, 61].

(5) *Minimum turnaround time* (t_{ta}) – this is the time required by the station's receive circuit to recover after the end of a transmission initiated from the same station (turnaround latency). Each station must wait a *minimum turnaround time* delay when moving from receive mode to transmit mode to ensure that the receive circuit of the station that was transmitting is given enough time to recover. This is the time required to change link direction.

Both stations must use the same data rate. However, parameters (2)–(5) are negotiated and agreed independently for each station. The IrDA frame structure is shown in Figure 4.22. The FCS contains a 16 b CRC for data rates up to 4 Mb/s and a 32 b CRC for 4 Mb/s and higher rates. IrLAP employs the following frame types:

(1) *Unnumbered frames* (*U frames*) are used for link management. '*U* frame' functions include discovering and initializing secondary stations, reporting procedural errors not recoverable by retransmissions, etc.

(2) *I frames* carry information data across the link during the connection period. The *I*-frame control field contains send and receive frame counts to ensure ordered frame reception.

(3) *Supervisory frames* (*S frames*) assist in information data transfer, although *S* frames never carry information data themselves. They are used to acknowledge correctly received frames, request an acknowledgement from the communicating station, convey station conditions, etc.

The control field contains an identifier, which determines the frame type. Depending on frame type, the control field may contain a send sequence number N_s, used to number the transmitted frames. It may also contain a receive sequence number N_r, used to indicate the expected sequence number of the next I frame. SIR and FIR specifications employ an 8 b-long control field (Figure 4.22). N_s and N_r occupy 3 b each in the control field, thus, N_s and N_r cycle through values from 0 to 7 and maximum window size is 7. Very fast infrared (VFIR) specification extended the control field to 16 b for the 4 and 16 Mb/s data rate IrDA links. In this case, N_s and N_r occupy 7 b each, cycling through values 0 to 127 and a maximum window size of 127 is allowed.

Within the control field, the P/F bit implements token passing between stations. When it is set by the primary station, it is the poll (P) bit. When it is set by the secondary station, it is the final (F) bit. The primary uses the P bit to reverse link direction and solicit a response from the secondary. The secondary responds by transmitting one or more frames and by setting the F bit of the last frame it transmits, thus, reversing link direction and returning transmission control to the primary. IrLAP primary and secondary stations also employ the P timer. P timer is assigned with the maximum turnaround time (T_{max}) agreed between stations during the contention period and represents the maximum time a station can hold transmission control. Each station starts the P timer upon reception of a frame with the P/F bit set and stops the P timer when it transmits a frame with the P/F bit set. If the P timer expires, meaning that the station holds transmission control longer than allowed, the station immediately sends a receive ready (RR) S frame with the P/F bit set to pass transmission control. The primary station also employs an F timer to limit the time a secondary station can hold transmission control. The primary starts the F timer upon transmission of a frame with the P bit set and stops the F timer upon reception of a frame with the F bit set. F-timer expiration means that the secondary failed to return transmission control within the agreed time period. Since the secondary's P -timer operation guarantees that this never happens, F-timer expiration can only be explained by the loss of either the frame containing the P bit or the frame containing the F bit. The primary resolves this situation by transmitting an RR frame with the P bit set when the$F$ timer expires.

4.5.2 IrLAP functional model description

In this section, the transmission of a large amount of information data from the primary to the secondary station is considered. The saturation case is assumed, where the primary station always has information data ready for transmission.

The parameters used in the model are shown in Table 4.4. In the contention period, the primary station determines the window size N it will employ. N represents the maximum number of I frames the primary can transmit before soliciting an acknowledgement. The maximum window-size parameter W_{max} is negotiated and agreed between the two stations during the contention period. However, the maximum time a station can hold transmission control T_{max} must always be obeyed and, according to IrLAP specification [60], T_{max} combined with frame length and link data rate may limit the window size applied. In other words, if the time needed for transmitting W_{max} frames carrying 'frame length' information bytes at the link data rate exceeds T_{max}, then a smaller window size must be employed. Thus, N is given by

$$N = \min\{W_{max}, \text{floor}(T_{max}/t_I)\} \tag{4.43}$$

Table 4.4 System parameters

Parameter	Description	Unit
C	Link data bit rate	b/s
p_b	Link bit error rate	—
p	Frame error probability	—
l	*I*-frame length data length	b
l'	*S*-frame length/*I*-frame overhead	b
t_I	Transmission time of an *I*-frame	s
$t_{I\max}$	Transmission time of an *I*-frame with 16 k bits user data	s
t_S	Transmission time of an *S*-frame	s
t_{ta}	Minimum turn-around time	s
t_{ack}	Acknowledgement time	s
T_{max}	Maximum turn-around time	s
t_{Fout}	*F*-timer time-out period	s
W_{max}	Maximum window size	Frames
N	Window size	Frames
D_f	Frame throughput	Frame/s
D_b	Data throughput	bits/s

where *min* is 'the lesser of' and *floor* is 'the largest integer not exceeding'. In this section, T_{max} is always fixed to 500 ms. The information transfer procedure used in the model is presented in Figure 4.23. Each node holds three variables, V_s for counting I frames transmitted, V_r for counting I frames received, and w indicating the number of the remaining I frames the station can transmit before reversing link direction. The primary also employs an F timer for limiting the secondary's transmission period. When the primary station sends an I frame, the N_s subfield of the frame's control field is assigned the current V_s value and is increased by one (modulo 8 or 128 depending of the control field size employed). The primary also makes a buffer copy of the frame for possible retransmissions. Since the primary always has information ready for transmission, it immediately checks the w value. If w is not equal to 1, the primary reduces w by 1, transmits the I frame with the P bit not set and the actions previously described are repeated. When w reaches 1, indicating that the next frame should be the last frame in the window transmission, the primary sets the bit P to poll the secondary and transmits the I frame. The primary also assigns N to w for the next N window frame transmission and starts the F timer.

When the secondary station receives an I frame, it compares the received frame sequence N_s value with the expected V_r value. If N_s equals V_r (the received frame is in sequence), V_r is increased by 1 (modulo 8 or 128) and information data is extracted and passed to the upper layer. If the received frame is not in sequence (one of the previous I frames in current window transmission was lost due to a CRC error), the frame is discarded and V_r remains unchanged. The secondary station also checks the P bit. If the P bit is set and, as the current model assumes that the secondary station never has information for transmission, it awaits a *minimum turnaround time* t_{ta} to allow for the hardware recovery latency and transmits an S frame with the F bit set. The S frame's N_r field contains V_r, a value informing the primary

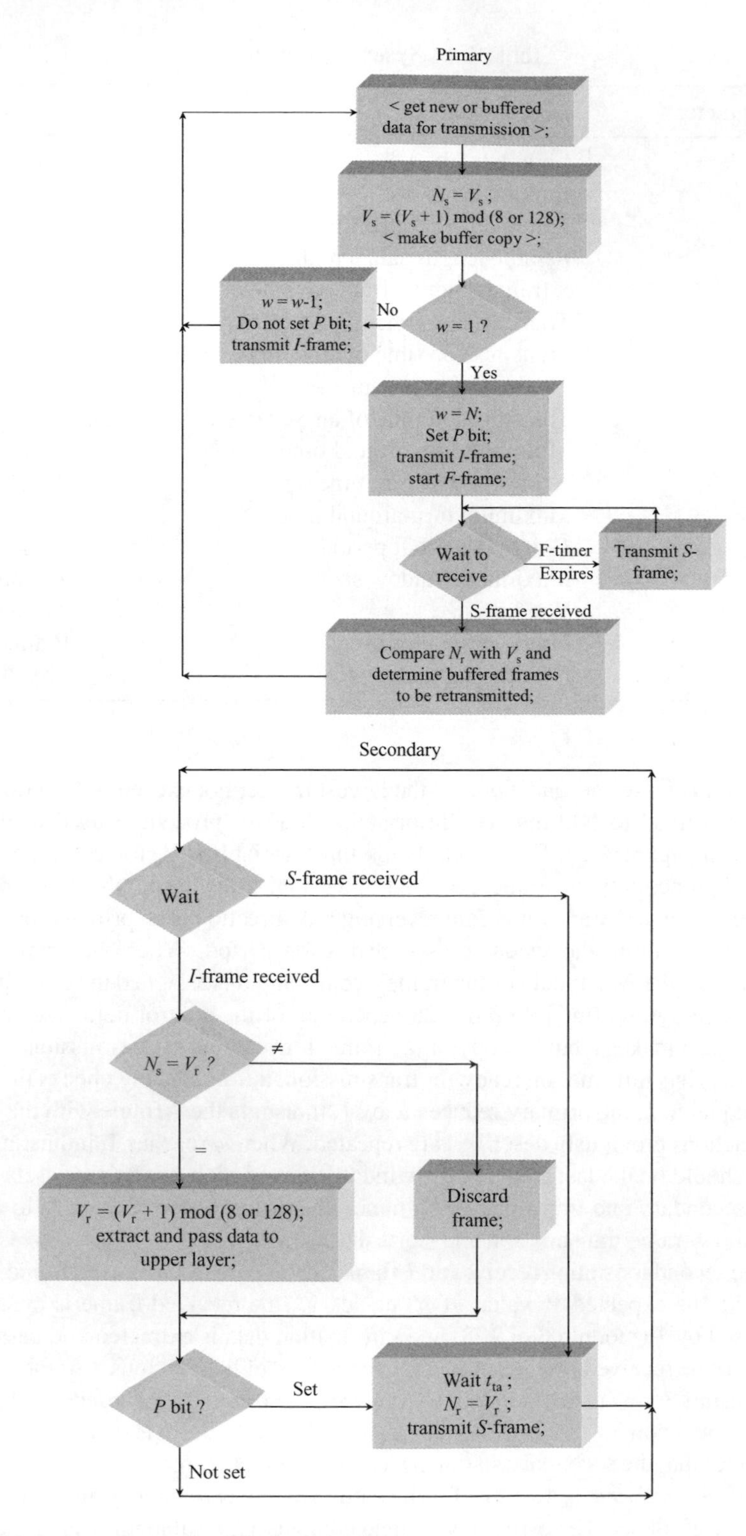

Figure 4.23 Information transfer procedure.

of the number of I frames received correctly and in sequence in the previous window transmission. When the primary receives the frame, it resumes I-frame transmission as transmission control was returned to the primary by means of the F bit. The primary first compares the received S frame's N_r with its current V_s value. If N_r equals V_s (all frames in the previous window transmission were received correctly by the secondary), the primary transmits I frames containing new information data to the secondary. If N_r is not equal to V_s, one or more I frames in the previous window transmission are lost. The primary retransmits buffered frames starting from the indicated N_r position before new data can be transmitted.

If the last I frame that contains the P bit is lost, the secondary station fails to respond as it does not realize that it has transmission control. The situation is resolved by the primary's F-timer expiration. The primary realizes that the secondary failed to respond within the agreed time period and transmits an S frame forcing the secondary to respond. In the current model, S frames are considered small enough to always be received error-free.

The saturation case model considered in this section can be summarized as follows. The transmitting station always has information ready for transmission. As a result, it transmits a window of N consecutive I frames and reverses the link direction by setting the P bit in the last I frame. The receiver awaits a minimum turnaround time and responds with an RR S frame indicating the expected sequence number of the next frame. RR frames always have the F bit set. The transmitter determines the number of frames correctly received before any error(s) occurred and repeats the erroneous frame and the frames following it, in the next window, followed by new frames to form a complete N frame transmission. If the last frame in a window transmission is lost, the receiver fails to respond as the P bit is lost. When F timer expires, the primary station sends an RR S frame with the P bit set forcing the secondary station to acknowledge correctly received frames. Additional discussion on infrared link access protocol can be found in References [62–65].

REFERENCES

[1] U. Chelentano and S. Glisic, Effective link layer capacity, in *PIMRC05*, Berlin, 12–14 September 2005, paper I06-2.

[2] D. Wu and R. Negi, Effective capacity: a wireless link model for support of quality of service, *IEEE Trans. Wireless Commun.*, vol. 2, no. 4, 2003, pp. 630–643.

[3] G. Liebl, T. Stockhammer and F. Burkert, Modelling and simulation of wireless packet erasure channels, in *Proc. Tech/MRPG Symp. Wireless Personal Commun.*, Blacksburg, VA, May 2000, pp. 203–214.

[4] T. Rancurel, D. Roviras, J. Conan and F. Castanie, Expression of the capacity for the Gilbert channel in presence of interleaving, *Proc. Int. Conf. Acoustics Speech and Signal Processing (ICASSP)*, vol. 4, 2000, pp. 2533–2536.

[5] T. Rancurel, D. Roviras, J. Conan and F. Castanie, Effect of interleaving on a Markov channel, *Proc. Int. Conf. Acoustics Speech and Signal Processing (ICASSP)*, vol. 5, 2000, pp. 2661–2664.

[6] C. Pimentel and I.F. Blake, Modeling burst channels using partitioned Fritchman's Markov models, *IEEE Trans. Vehicle Technol.*, vol. 47, no. 3, 1998, pp. 885–899.

[7] A. Willig, A new class of packet- and bit-level models for wireless channels, in *Proc. IEEE Symp. Personal Indoor Mobile Radio Communications,* Lisboa, September 2002.

[8] J.N.L. Brümmer, Characterization of digital channels using hidden Markov models, *Proc. Symp. Communication Signal Processes*, 11 September 1992, pp. 183–188.
[9] H.S. Wang and N. Moayeri, Finite-state Markov channel – a useful model for radio communication channels, *IEEE Trans. Vehicle Technol.*, vol. 44, no. 1, 1995, pp. 163–171.
[10] H. Steffan, Adaptive generative radio channel models, in *Proc. IEEE Symp. Personal Indoor Mobile Radio Communications*, vol. 1, 1994, pp. 268–273.
[11] C.-D. Iskander and P.T. Mathiopoulos, Analytical level crossing rates and average fade durations for diversity techniques in Nakagami fading channels, *IEEE Trans. Commun.*, vol. 50, no. 8, 2002, pp. 1301–1309.
[12] F. Babich and G. Lombardi, A Markov model for the mobile propagation channel, *IEEE Trans. Vehicle Technol.*, vol. 49, no. 1, 2000, pp. 63–73.
[13] M. Zorzi and R.R. Rao and L.B. Milstein, ARQ error control for fading mobile radio channels, *IEEE Trans. Vehicle Technol.*, vol. 46, no. 2, 1997, pp. 445–455.
[14] H.S. Wang, On verifying the first-order Markovian assumption for a Rayleigh fading channel model, in *Proc. IEEE Int. Conf. Universal Personal Commun.*, 1994, pp. 160–164.
[15] H.S. Wang and P.-C. Chang, On verifying the first-order Markovian assumption for a Rayleigh fading channel model, *IEEE Trans. Vehicle Technol.*, vol. 45, no. 2, 1996, pp. 353–357.
[16] A. Abdi, Comments on verifying the first-order Markovian assumption for a Rayleigh fading channel model, *IEEE Trans. Vehicle Technol.*, vol. 48, no. 5, 1999, p. 1739.
[17] C.C. Tan and N.C. Beaulieu, On first-order Markov modeling for the Rayleigh fading channel, *IEEE Trans. Commun.*, vol. 48, no. 12, 2000, pp. 2032–2040.
[18] F. Babich and G. Lombardi, On verifying a first-order Markovian model for the multi-threshold success/failure process for Rayleigh channel, *Proc. IEEE Symp. Personal Indoor Mobile Radio Communications*, vol. 1, 1997, pp. 12–16.
[19] D. Qiao and S. Choi, Goodput enhancement in IEEE802.11a wireless LAN via link adaptation, in *Proc. IEEE Int. Conf. Commun.*, Helsinki, June 2001.
[20] N.C. Ericsson, Adaptive modulation and scheduling of IP traffic over fading channels, in *Proc. IEEE Vehicle Technology Conf. (VTC'99)*, Amsterdam, vol. 2, 19–22 September 1999, pp. 849–853.
[21] A.J. Goldsmith and S.-G. Chua, Variable-rate variable-power MQAM for fading channels, *IEEE Trans. Commun.*, vol. 45, no. 10, 1997, pp. 1218–1230.
[22] S. Vishwanath and A. Goldsmith, Adaptive turbo-coded modulation for flat-fading channels, *IEEE Trans. Commun.*, vol. 51, no. 6, 2003, pp. 964–972.
[23] V. Tralli and M. Zorzi, Markov models for the physical layer block error process in a WCDMA cellular system, *Proc. IEEE Global Telecommun. Conf.*, vol. 2, 17–22 November 2002, pp. 1925–1929.
[24] A.J. Paulraj, D.A. Gore, R.U. Nabar and H. Bölcskei, An Overview of MIMO communications – A key to gigabit wireless, *IEEE Proc.*, vol. 92, no. 2, 2004, pp. 198–218.
[25] S.-T. Chung and A.J. Goldsmith, Degrees of freedom in adaptive modulation: a unified view, *IEEE Trans. Commun.*, vol. 49, no. 9, 2001, pp. 1561–1571.
[26] L. Hanzo, C.H. Wong and M.S. Yee, *Adaptive Wireless Transceivers*. Wiley: Chichester, 2002, pp. 191–255.
[27] Y.Y. Kim and S.-Q. Li, Capturing important statistics of a fading/shadowing channel for network performance analysis, *IEEE J. Select. Areas Commun.*, vol. 17, no. 5, 1999, pp. 888–901.

[28] D. Haccoun and G. Begin, High-rate punctured convolutional codes for Viterbi and sequential decoding, *IEEE Trans. Commun.*, vol. 37, no. 11, 1989, pp. 1113–1125.
[29] S. Wei and D.L. Goeckel, Error statistics for average power measurements in wireless communication systems, *IEEE Trans. Commun.*, vol. 50, no. 9, 2002, pp. 1535–1546.
[30] J.G. Proakis, *Digital Communications*, 2nd edn. McGraw-Hill: Singapore, 1989.
[31] N. Wax (ed.), *Selected Papers on Noise and Stochastic Processes*. Dover: New York, 1954, pp. 133–294.
[32] M. Pätzold and F. Laue, Level-crossing rate and average duration of fades of deterministic simulation models for rice fading channels, *IEEE Trans. Vehicle Technol.*, vol. 48, no. 4, 1999, pp. 1121–1129.
[33] N.C. Beaulieu and X. Dong, Level crossing rate and average fade duration of MRC and EGC diversity in Ricean fading, *IEEE Trans. Commun.*, vol. 51, no. 5, 2003, pp. 722–726.
[34] A.M. Law and W.D. Kelton, *Simulation Modeling and Analysis*. Industrial Engineering Series, 2nd edn. McGraw-Hill: Singapore, 1991, pp. 372–416.
[35] F.J. Block and M.B. Pursley, A protocol for adaptive transmission in direct-sequence spread-spectrum packet radio networks, *IEEE Trans. Commun.*, vol. 52, no. 8, 2004, pp. 1388–1397.
[36] P. Lettieri and M.B. Srivastava, Adaptive frame length control for improving wireless link throughput, range, and energy efficiency, in *Proc. IEEE INFOCOM'98*, 29 March to 2 April 1998, pp. 564–571.
[37] I. Joe, An adaptive hybrid ARQ scheme with concatenated FEC codes for wireless ATM, in *Proc. ACM/IEEE MobiCom'97*, 1997, pp. 131–138.
[38] D.A. Eckhardt and P. Steenkiste, Improving wireless LAN performance via adaptive local error control, in *Proc. IEEE ICNP'98*, 1998, pp. 327–338.
[39] M. Elaoud and P. Ramanathan, Adaptive use of error-correcting codes for real-time communication in wireless networks, *Proc. IEEE INFOCOM'98*, 1998, pp. 548–555.
[40] S. Choi and K.G. Shin, A class of adaptive hybrid ARQ schemes for wireless links, *IEEE Trans. Vehicular Technology*, vol. 50, no. 3, May 2001, pp. 777–790.
[41] T.S. Rappaport, *Wireless Communications: Principle and Practice*. Prentice-Hall: Englewood Cliffs, NJ, 1996.
[42] P. Lettieri, C. Fragouli and M.B. Srivastava, Low power error control for wireless links, in *Proc. ACM/IEEE MobiCom'97*, 1997, pp. 139–150.
[43] Y.H. Lee and S.W. Kim, Adaptive data transmission scheme for DS/SSMA system in bandlimited Rayleigh fading channel, in *Proc. IEEE ICUPC'95*, 1995, pp. 251–255.
[44] M. Zorzi and R.R. Rao, Error control and energy consumption in communications for nomadic computing, *IEEE Trans. Comput. (Special Issue On Mobile Computing)*, vol. 46, 1997, pp. 279–289.
[45] P. Lettieri, C. Schurgers and M.B. Srivastava, Adaptive link layer strategies for energy efficient wireless networking, *Wireless Networks*, vol. 5, 1999, pp. 339–355.
[46] M. Zorzi and R.R. Rao, Energy constrained error control for wireless channels, *IEEE Person. Commun.*, vol. 4, 1997, pp. 27–33.
[47] C. Chien, M. Srivastava, R. Jain, P. Lettieri, V. Aggarwal and R. Sternowski, Adaptive radio for multimedia wireless links, *IEEE JSAC*, vol. 17, no. 5, 1999, pp. 793–813.
[48] K. Narendra and M. Thathachar, *Learning Automata: an Introduction*. Prentice Hall: Englewood Cliffs, NJ, 1989.
[49] M. Norman, Some convergence theorems for stochastic learning models with distance diminishing operators, *J. Math. Psychol.*, vol. 5, 1968, pp. 61–101.

[50] V. Vitsas and A.C. Boucouvalas, Optimization of IrDA IrLAP link access protocol, *IEEE Trans. Wireless Commun.*, vol. 2, no. 5, 2003, pp. 926–938.

[51] V. Vitsas and A.C. Boucouvalas, Throughput analysis of the IrDA IrLAP optical wireless link access protocol, in *Proc. 3rd Conf. Telecommunications*, Figueira da Foz, 23–24 April 2001, pp. 225–229.

[52] D.J.T. Heatly, D.R. Wisely, I. Neild and P. Cochrane, Optical wireless: the story so far, *IEEE Commun. Mag.*, vol. 36, 1998, pp. 72–82.

[53] A.C. Boucouvalas and Z. Ghassemlooy, Editorial, special issue on optical wireless communications, *Proc. Inst. Elect. Eng. J. Optoelectron.*, vol. 147, 2000, p. 279.

[54] S.Williams, IrDA: past, present and future, *IEEE Person. Commun.*, vol. 7, 2000, pp. 11–19.

[55] I. Millar, M. Beale, B.J. Donoghue, K.W. Lindstrom and S. Williams, The IrDA standard for high-speed infrared communications, *Hewlett-Packard J.*, vol. 49, no. 1, 1998, pp. 10–26.

[56] *IrDA, Serial Infrared Physical Layer Specification – Version 1.4*. Infrared Data Association, 2001.

[57] *IrDA, Serial Infrared Physical Layer Specification – Version 1.0*. Infrared Data Association, 1994.

[58] *IrDA, Serial Infrared Physical Layer Specification – Version 1.1*. Infrared Data Association, 1995.

[59] *IrDA, Serial Infrared Physical Layer Specification for 16 Mb/s Addition (VFIR) – Errata to version 1.3*. Infrared Data Association, 1999.

[60] *IrDA: Serial Infrared Link Access Protocol (IrLAP) – Version 1.1*. Infrared Data Association, 1996.

[61] *IrDA: Serial Infrared Link Access Protocol Specification for 16 Mb/s Addition (VFIR) – Errata to Version 1.1*. Infrared Data Association, 1999.

[62] P. Barker, A.C. Boucouvalas and V. Vitsas, Performance modeling of the IrDA infrared wireless communications protocol, *Int. J. Commun. Syst.*, vol. 13, 2000, pp. 589–604.

[63] P. Barker and A.C. Boucouvalas, Performance analysis of the IrDA protocol in wireless communications, in *Proc. 1st Int. Symp. Communication Systems Digital Signal Processing*, Sheffield, 1998, April, 6–8, pp. 6–9.

[64] W. Bux and K. Kummerle, Balanced HDLC procedures: a performance analysis, *IEEE Trans. Commun.*, vol. 28, 1980, pp. 1889–1898.

[65] S. Williams and I. Millar, The IrDA platform, in *Proc. 2nd Int. Workshop Mobile Multimedia Communications*, Bristol, 11–14 April 1995.

5

Adaptive Medium Access Control

Introductory material on MAC is presented in Appendix B (please go to www.wiley.com/go/glisic). Within this chapter we focus only on few specific solutions in WLAN, *ad hoc* and sensor networks.

5.1 WLAN ENHANCED DISTRIBUTED COORDINATION FUNCTION

The last few years have witnessed an explosive growth in 802.11 WLAN [1]. Unfortunately, the current 802.11 MAC does not possess any effective service differentiation capability, because it treats all the upper-layer traffic in the same fashion. Hence, a special working group, IEEE 802.11e [2–12], was established to enhance the 802.11 MAC to meet QoS requirements for a wide variety of applications. The 802.11e EDCF (extended distributed coordination function) is an extension of the basic DCF mechanism of current 802.11 (Chapter 1). Unlike DCF, EDCF is not a separate coordination function, but a part of a single coordination function of 802.11e called the hybrid coordination function (HCF). The HCF combines both DCF and PCF (point coordination function) from the current 802.11 specification with new QoS specific enhancements. It uses EDCF and a polling mechanism for contention-based and contention-free channel access, respectively. In EDCF, each station can have multiple queues that buffer packets of different priorities. Each frame from the upper layers bears a priority value which is passed down to the MAC layer. Up to eight priorities are supported in a 802.11e station and they are mapped into four different access categories (AC) at the MAC layer [3]. A set of EDCF parameters, namely the arbitration interframe space (AIFS[AC]), minimum contention window size (CWMin[AC]) and maximum contention window size (CWMax[AC]), is associated with each access category to differentiate the channel access. *AIFS*[*AC*] is the number of time slots a packet of a given

© 2006 John Wiley & Sons, Ltd.

AC has to wait after the end of a time interval equal to a short interframe spacing (SIFS) duration before it can start the backoff process or transmit. After $i (i \geq 0)$ collisions, the *backoff counter* in 802.11e is selected uniformly from $[1, 2^i \times CWMin[AC]]$, until it reaches the *backoff stage i* such that $2^i \times CWMin[AC] = CWMax[AC]$. At that point, the packet will still be retransmitted, if a collision occurs, until the total number of retransmissions equals the maximum number of allowable retransmissions (*RetryLimit*[*AC*]) specified in IST WSI [13], with the backoff counters always chosen from the range [1, *CWMax*[*AC*]]. Since multiple priorities exist within a single station, it is likely that they will collide with each other when their backoff counters decrement to zero simultaneously. This phenomenon is called an *internal collision* in 802.11e and is resolved by letting the highest priority involved in the collision win the contention. Of course, it is still possible for this winning priority to collide with packets from other station(s).

Performance example – the basic parameters used in simulation are:

Packet payload size 8184 bits at 11 Mbps
MAC header 272 bits at 11 Mbps
PHY header 192 bits at 1 Mbps
ACK 112 bits + PHY header
Propagation delay 1 μs
Slot time 20 μs
SIFS 10 μs

The QoS parameters, i.e. *CWMin*[*AC*], *CWMax*[*AC*] and *AIFS*[*AC*], used in the following discussion are similar to the values specified by IEEE 802.11e Working Group for voice and video traffic [12].

Simulation with the same parameters is also presented for example, in Tao and Panwar [14]. In Figure 5.1, each station contains two priorities, which are only differentiated by the internal collision resolution algorithm discussed before. As expected, internal collision resolution by itself can provide some differentiation for channel access between different priorities.

The two priorities in each station are further differentiated by *AIFS* and *CWMin/CWMax* in Figures 5.2 and 5.3, respectively. It can be seen that *AIFS* may have a more marked effect on service differentiation than *CWMin/CWMax* alone. When all QoS mechanisms in 802.11e EDCF are enabled, the resulting throughput is shown in Figure 5.4. Comparing Figure 5.4 with Figure 5.2, we find that the QoS differentiation effect of *AIFS* is almost identical to the aggregate impact of *AIFS* plus *CWMin/CWMax*.

All the figures reveal that, as the number of stations in the network grows, the throughput for each priority as well as the total throughput drop fairly fast, especially when the QoS-specific parameters are small. Under heavy load assumption, the throughput for low priority often decreases to almost zero before the number of stations reaches 10. For this region, the high priority traffic dominates.

5.2 ADAPTIVE MAC FOR WLAN WITH ADAPTIVE ANTENNAS

Smart antennas (or adaptive array antennas) have some unique properties that enable us to achieve high throughputs in *ad hoc* network scenarios. A transmitter equipped with a

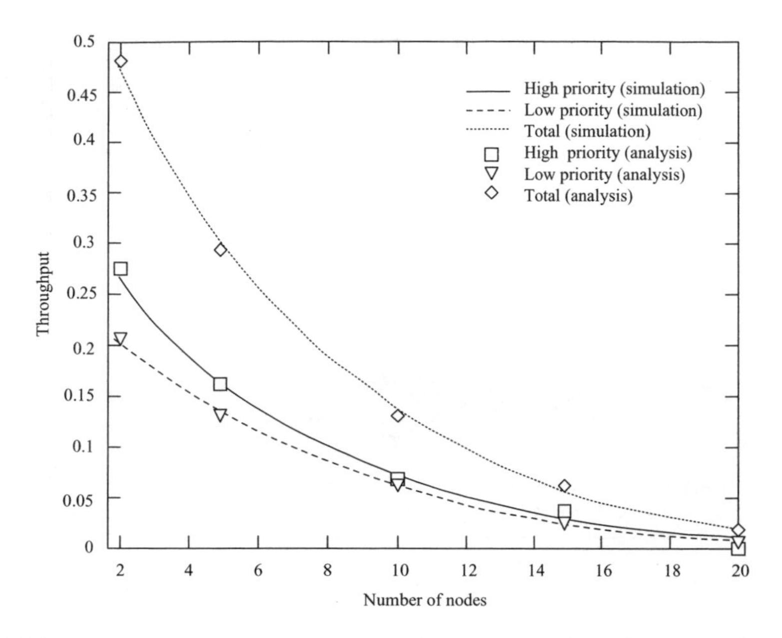

Figure 5.1 Two priorities differentiated only by internal collision resolution: CWMin/Max[0] = 8/16, CWMin/Max[1] = 8/16, AIFS[0] = 2, AIFS[1] = 2.

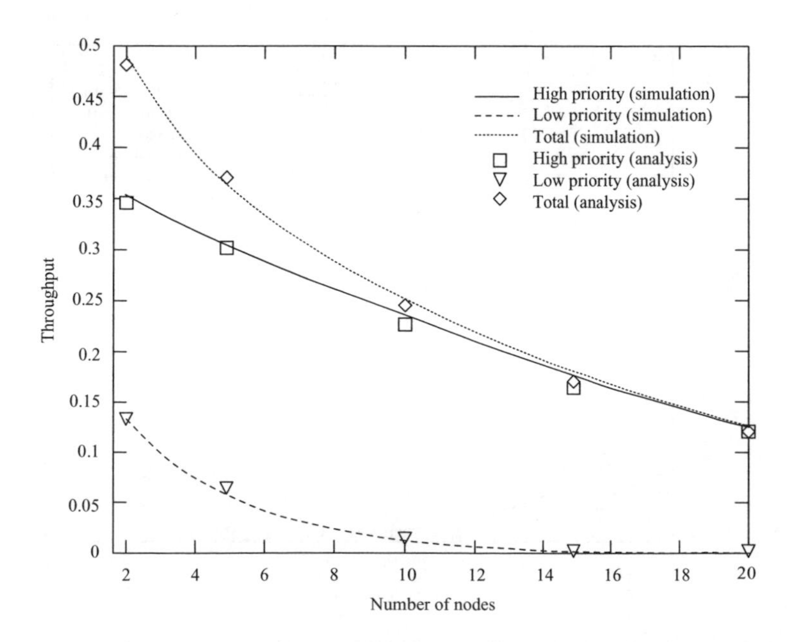

Figure 5.2 Two priorities with different AIFS values: CWMin/Max[0] = 8/16, CWMin/Max[1] = 8/16, AIFS[0] = 2, AIFS[1] = 3.

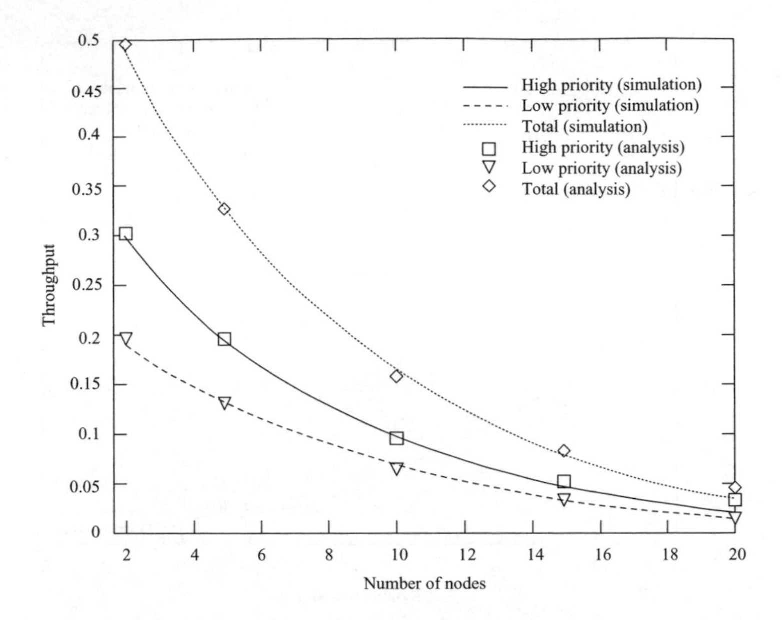

Figure 5.3 Two priorities with different CWMin and CWMax. AIFS[0] = 2, AIFS[1] = 2, CWMin/Max[0] = 8/16, CWMin/Max[1] = 10/20.

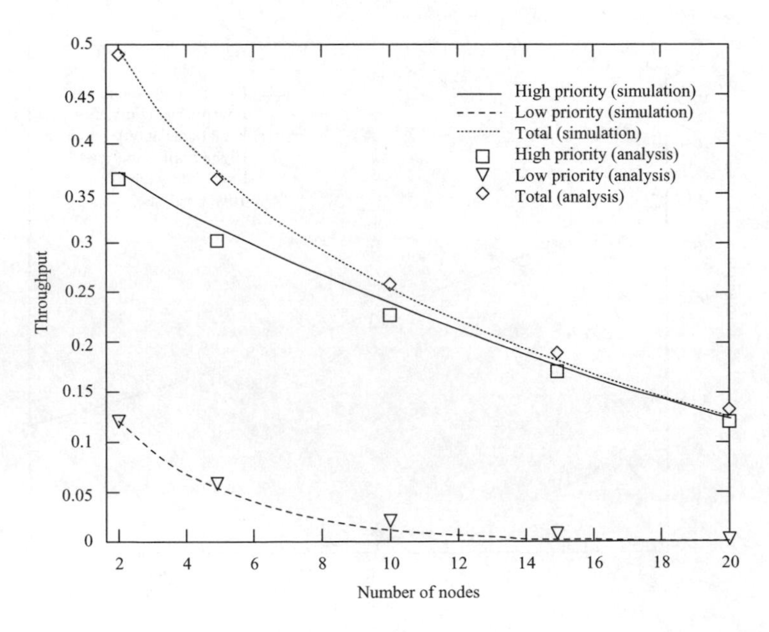

Figure 5.4 Two priorities with different CWMin, CWMax and AIFS: AIFS[0] = 2, AIFS[1] = 3, CWMin/Max[0] = 8/16, CWMin/Max[1] = 10/20.

smart antenna can form a directed beam towards its receiver and a receiver can similarly form a directed beam towards the sender, resulting in very high gain. A receiver can also identify the direction of multiple simultaneous transmitters by running DOA algorithms and use this information to determine the directions in which it should place the *nulls*. Placing nulls effectively cancels out the impact of interfering transmitters. In this paper we present a simple 802.11b-based MAC protocol called Smart-802.11b that explicitly uses these three properties of smart antennas (beamforming, DOA and nulling) to achieve high throughputs. The two protocols are called *Smart-Aloha* [15, 16] and *Smart-802.11b* and, as the name implies, these two protocols are modifications to the well-known Aloha and 802.11b protocols. In both cases, functionality at the MAC layer is added to allow it to directly control the antenna: *the MAC layer controls the direction of the beam and the direction of the nulls*. In addition, the antenna provides the MAC layer with DOA information for all transmissions it can hear along with signal strength information. The main results are that these protocols show a very high throughput while maintaining fairness. Table 5.1 summarizes the main throughput results of the MAC protocols designed for directional antenna equipped nodes [15–26].

5.2.1 Description of the protocols

Consider the case when a node a needs to transmit a packet to node b which is its one-hop neighbor. It is assumed that a knows the angular direction of b and it can therefore form a beam in the direction of b. However, to maximize SINR, b should also form a beam towards a and form nulls in the direction of all other transmitters. In order to do this, b needs to know two things – first, that a is attempting to transmit to it, and second, the angular direction of all the other transmitters that interfere at b. The two protocols discussed in this section answer these two questions somewhat differently.

Smart-Aloha is a slightly modified version of the standard *Slotted-Aloha* protocol. To transmit a packet, a transmitter forms a beam towards its receiver and begins transmission. However, it prefaces its packet transmission with the transmission of a short (8 byte) *pure tone* (this is a simple sinusoid). Idle nodes remain in an omnidirectional mode and receive a complex sum of all such tones (note that the tones are identical for all nodes and thus we cannot identify the nodes based on the tone) and run a DOA algorithm to identify the direction and strength of the various signals. An idle node then beamforms in the direction of the maximum received signal strength and forms nulls in other directions, and receives the transmitted packet. If the receiver node was the intended destination for the packet, it immediately sends an ACK using the already formed directed beam. On the other hand, if the packet was intended for some other node, then the receiver discards it.

A sender waits for an ACK immediately after transmission of the packet and if it does not receive the ACK, it enters backoff in the standard way. Thus, the Smart-Aloha protocol follows a *Tone/Packet/Ack* sequence. The intuition behind the receiver beamforming in the direction of the maximum signal is that, because of the directivity of the antenna, there is a high probability that it is the intended recipient for the packet. However, in some cases, as in Figure 5.5, the receiver d incorrectly beamforms towards a because a's signal is stronger than b's. While this is not a serious problem in most cases, we can envision scenarios where the $b \rightarrow d$ transmission gets starved due to a large volume of $a \rightarrow c$ traffic. A possible

Table 5.1 Performance of MAC protocols using adaptive antennas. (Reproduced by permission of IEEE [27].)

Prior work	Characteristics of simulation experiments	Maximum throughput					
[17]	Switched beam antenna 45° beamwidth, 10 dB gain, 250 m range for omni, 900 m directional 4CBR sources, 75 kbps to 2 Mbps each	Random topology			Mesh topology		
		($N = 25.4$ hops)					
		MMAC	DMAC	802.11	MMAC	DMAC	802.11
		1000 kbps	400	200	800	300	200
		(5×)	(2×)	(1×)	(4×)	(1.5×)	(1×)
[18]	Multi-beam antenna (1, 2, 4, beams each) 30° beamwidth, 2 Mbps channel slotted (8 ms slot), 16 kbit packet (throughput converted to bps from pkts/slot/net)	Fully connected (20 nodes)			Multi-hop (100 nodes, 5 hops)		
		1 beam	2	4	1	2	4
		12 Mbps	30	60	60	150	300
		(Max over ROMA, UxDMA)					
[19]	Adaptive antenna: 4 × 4, 8 × 8 planar arrays, TDMA-802.11, 1-hop	4 × 4			8×8		
		(55 nodes)					
		8 pkts/packet time			9 pkts/packet time		
[20]	Switched beam, 60° beamwidth	Proposed		DRTS/DCTS		CSMA/CA	
		(50 nodes)					
		3.5 Mbps		2.5		2	

[21]	Circular adaptive antenna array, beamwidth 64°, 8 dB gain (improvement over 802.11)	25 nodes (grid)			225 nodes (grid)		
		No PC	Global PC	Local PC	No PC	Global PC	Local PC
				(PC, power control)			
		1.3×	1.7×	2.1×	2.6×	4.75×	5.25×
[22]	Ideal adaptive antenna 20 nodes, no nulling (improvement over omni case) Packet transmission is directional at sender/receiver	*Protocol*			*Beamwidth*		
		O, omnidirectional			(20 nodes, degree = 7.5)		
		D, directional		90°	60°	30°	10°
		ORTS/DCTS		35 %	57 %	100 %	142 %
		DRTS/DCTS		64 %	107 %	143 %	186 %
		DRTS/OCTS		28 %	43 %	n/a	57 %
		ORTS/OCTS		29 %	50 %	86 %	121 %
		STDMA		n/a	400 %	n/a	400 %
[23]	Six-element circular antenna array (10 fixed patterns, no adaptation) 45° beamwidth, 100 nodes, 1500 m^2 2-ray propagation model, no nulling				No mobility		
		Omni		RX directional		DVCS	DVCS-Ideal
				TX omnidirectional		TX, RX directional	
		400 kbps		800 kbps		1.4 Mbps	2.2 Mbps

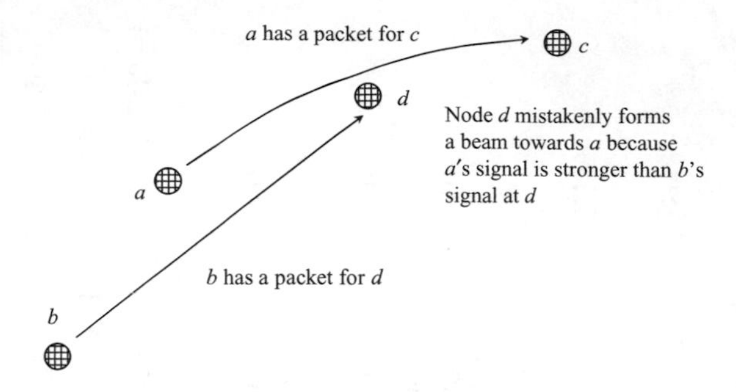
a has a packet for c
c
d
Node d mistakenly forms
a beam towards a because
a's signal is stronger than b's
signal at d
a
b has a packet for d
b

Figure 5.5 False beamforming. (Reproduced by permission of IEEE [27].)

optimization is a *single-entry cache* scheme which works as follows:

- If a node beamforms incorrectly in a given timeslot, it remembers that *direction* in a single-entry cache.
- In the next slot, if the maximum signal strength is again in the direction recorded in the single-entry cache, then the node ignores that direction and beamforms towards the second strongest signal. If the node receives a packet correctly (i.e. it was the intended recipient), it does not change the cache. If it receives a packet incorrectly, it updates the cache with this new direction.
- If there is no packet in a slot from the direction recorded in the cache, the cache is reset.

The *Smart-802.11b* protocol is based on the 802.11b standard. As in the case of the Smart-Aloha protocol, transmitters beamform towards their receivers and transmit a short *sender-tone* to initiate communication. However, unlike Smart-Aloha, the transmitter does not immediately follow the tone with a packet. Instead, it waits for a *receiver-tone* and only then transmits its packet. After transmission of a packet, it waits for the receipt of an ACK. If there is no ACK, it enters backoff as in 802.11b. Figure 5.6 presents a state diagram of tone-based protocol. The behavior of the protocol in various states can be summarized as follows.

5.2.1.1 Idle

In case a node has no packet to send, it will remain in the idle state and set its antenna to operate in the omnidirectional mode. If it receives a sender-tone from some other node, it will move into the data receive wait state. On the other hand, if it wishes to send data, it will beamform in the direction of the receiver. It chooses a random number [0–CW] and sets the CW (contention window) timer 1. When the CW timer expires, it sends a sender-tone in the direction of the receiver and moves to the ACK wait state. If, before the CW timer expires, the node receives a sender-tone from another node, it will freeze its CW timer and move to data receive wait state.

5.2.1.2 Data receive wait

A node will move to this state in the event it receives a sender-tone. The node will beamform towards the sender and then randomly defer transmitting the receiver-tone by choosing a

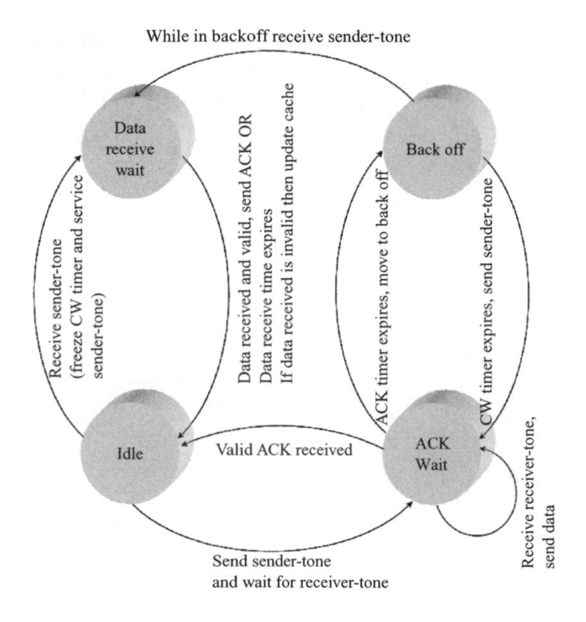

Figure 5.6 State diagram of the Smart-802.11b protocol.

random waiting period of [0–32] $\times$ 20 μs. The reason for deferring the reply is to minimize the chance of several receiver-tones colliding at sender 2. After transmitting a receiver-tone, the node remains in this state for 2τ (twice the maximum propagation delay + tone transmission time). If it does not hear a transmission, it returns to the idle state. If it hears the start of a transmission, it remains in this state and receives the packet. It then discards the packet if the packet was meant for some other node If, however, the packet was meant for it, then it sends an ACK.

5.2.1.3 Ack wait

If the sender node receives a receiver-tone before the tone RTT timer goes off (which is twice the tone transmission time plus propagation delay), it will transmit the data packet. Reception of a valid ACK will move the node to the idle state, and if packets are there in the queue then it will schedule the one at the head of the queue. The node will move to the backoff state under two conditions: (1) a receiver-tone did not arrive; (2) an ACK was not received following transmission of the data packet.

5.2.1.4 Backoff

The node computes a random backoff interval (as in 802.11) and remains in backoff for this time period (it also resets its antenna to omnidirectional mode). If, however, a sender-tone is received, it freezes the backoff timer and enters the data receive wait state. If the node is in backoff, upon expiration of the timer, it retransmits the sender-tone, increments the retransmit counter and enters the ACK wait state. A packet is discarded after the retransmit counter exceeds Max Retransmit = 7, as in the IEEE 802.11 standard.

The reception of a data packet by a node may be interfered with by transmissions of sender-tones, receiver-tones or other data packets (since the protocol does not take care of hidden terminals). A node engaged in receiving a data packet can dynamically form nulls towards new interferers, but this process takes some time (*we model this time as the length of a sender-tone*). Thus, the data packet will have errors due to this interference. This error is mitigated by relying on FEC codes as used in IEEE 802.11e, where (224, 208) shortened RS codes are used. In 802.11e, an MAC packet is split into blocks of 208 octets and each block is separately coded using an RS encoder. A (48, 32) RS code, which is also a shortened RS code, is used for the MAC header, and CRC-32 is used for the FCS.

Performance example – the simulation parameters are:

Background noise + ambient noise = 143 dB
Propagation model free space
Bandwidth 1000 kHz
Min frequency 2402 MHz
Data rate 2000 kbps
Carrier sensing threshold + 3 dB
Minimum SINR 9 dB
Bit error based on BPSK modulation curve
Maximum radio range 250 m
Packet size 16 kb
Simulation time 200 s
Single hop: number of nodes 20, area 100 × 100 m
Multihop: number of nodes 100, area 1500 × 1500

The existing 802.11b implementation in OPNET is modified to create Smart-802.11b. The modifications included adding the two tones (sender and receiver) as well as changing the FEC to the 802.11e specification.

The performance of the protocol is presented for a single-hop case with 20 nodes and a five-hop case with 100 nodes using of 16 KB packets. The 16 antenna elements (for an effective beamwidth of 400) were used. Figure 5.7 presents the aggregate one-hop throughput as a function of arrival rate for the one-hop case. One can see that 802.11b achieves a maximum throughput of 1 Mbps while Smart-802.11b achieves a high of 8.5 Mbps and Smart-Aloha achieves a high of approximately 10.5 Mbps. In fact, the throughput of Smart-802.11b and Smart-Aloha increases with arrival rate because of good spatial reuse of the channel. Figure 5.8 plots the aggregate throughput of the protocol for the 100-node five-hop case; 802.11b reaches a maximum throughput of well below 0.5 Mbs while Smart- 802.11b reaches a maximum of 50 Mbs and Smart-Aloha reaches a maximum throughput of 60 Mbs. Again, the better spatial reuse of the channel given the directivity of the antenna is the reason for this performance improvement.

5.3 MAC FOR WIRELESS SENSOR NETWORKS

This section discusses an MAC protocol designed for wireless sensor networks (S-MAC). As will be discussed in Chapter 14, wireless sensor networks use battery-operated computing and sensing devices. A network of these devices will collaborate for a common application such as environmental monitoring. Sensor networks are expected to be deployed in an *ad hoc* fashion, with nodes remaining largely inactive for long time, but becoming suddenly active when something is detected. These characteristics of sensor networks and applications

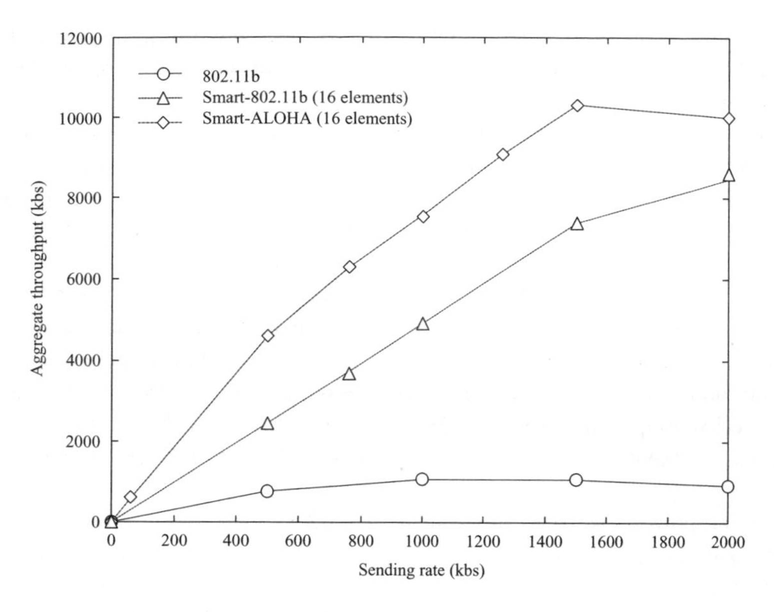

Figure 5.7 Single-hop case with 20 nodes. (Reproduced by permission of IEEE [27].)

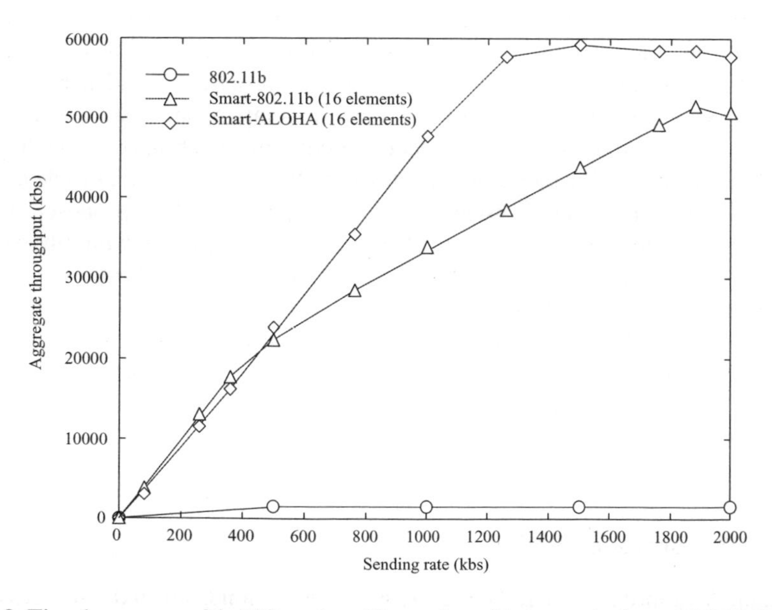

Figure 5.8 Five-hop case with 100 nodes. (Reproduced by permission of IEEE [27].)

motivate an MAC that is different from traditional wireless MACs such as IEEE 802.11, described in previous sections, in several ways. Energy conservation and self-configuration are primary goals, while per-node fairness and latency are less important. S-MAC uses a few novel techniques to reduce energy consumption and support self-configuration. It enables low-duty-cycle operation in a multihop network. Nodes form *virtual clusters* based on common sleep schedules to reduce control overhead and enable traffic-adaptive wake-up. S-MAC uses in-channel signaling to avoid overhearing unnecessary traffic. Finally, S-MAC applies *message passing* to reduce contention latency for applications that require in-network data processing.

Woo and Culler [28] examined different configurations of carrier sense multiple access (CSMA) and proposed an adaptive rate control mechanism, whose main goal is to achieve fair bandwidth allocation to all nodes in a multihop network. There is also some work on the low-duty-cycle operation of nodes, which are closely related to S-MAC. The first example is Piconet [29], which is an architecture designed for low-power *ad hoc* wireless networks. Piconet also puts nodes into periodic sleep for energy conservation. However, there is no coordination and synchronization among nodes about their sleep and listen time. The scheme to enable the communications among neighboring nodes is to let a node broadcast its address when it wakes up from sleeping. If a sender wants to talk to a neighbor, it must keep listening until it receives the neighbor's broadcast. In contrast, S-MAC tries to coordinate and synchronize neighbors' sleep schedules to reduce latency and control overhead.

Perhaps the power-save (PS) mode in IEEE 802.11 DCF is the most related work to the low-duty-cycle operation in S-MAC. Nodes in PS mode periodically listen and sleep, just like that in S-MAC. The sleep schedules of all nodes in the network are synchronized together. The main difference from S-MAC is that the PS mode in 802.11 is designed for a single-hop network, where all nodes can hear each other, simplifying the synchronization.

As observed by Woo and Culler [28], in multihop operation, the 802.11 PS mode may have problems in clock synchronization, neighbor discovery and network partitioning. In fact, the 802.11 MAC in general is designed for a single-hop network, and there are questions about its performance in multihop networks [30]. In comparison, S-MAC is designed for multihop networks, and does not assume that all nodes are synchronized together. Finally, although 802.11 defines PS mode, it provides very limited policy about *when* to sleep, whereas in S-MAC, a complete system is defined. Tseng *et al.* [31] proposed three sleep schemes to improve the PS mode in the IEEE 802.11 for its operation in multihop networks. Among them the one named periodically fully awake interval is the closest to the scheme of periodic listen and sleep in S-MAC. However, their scheme does not synchronize the sleep schedules of any neighboring nodes. The control overhead and latency can be large. For example, to send a broadcast packet, the sender has to explicitly wake up each individual neighbor before it sends out the actual packet. Without synchronization, each node has to send beacons more frequently to prevent long-term clock drift.

5.3.1 S-MAC protocol design

S-MAC includes approaches to reducing energy consumption from all the sources of energy waste such as: (a) idle listening; (b) collision; and (c) overhearing and control overhead. Before describing the components in S-MAC, we first summarize assumptions about the wireless sensor network and its applications.

Sensor networks will consist of large numbers of nodes to take advantage of short-range, multihop communications to conserve energy (see Chapter 14). Most communications will occur between nodes as peers, rather than to a single base station. In-network processing is critical to network lifetime, and implies that data will be processed as whole messages in a store-and-forward fashion. Packet or fragment-level interleaving from multiple sources only

Figure 5.9 Neighboring nodes A and B have different schedules. They synchronize with nodes C and D respectively.

increases overall latency. Finally, we expect that applications will have long idle periods and can tolerate latency on the order of network messaging time.

5.3.2 Periodic listen and sleep

As stated above, in many sensor network applications, nodes are idle for a long time if no sensing event happens. Given the fact that the data rate is very low during this period, it is not necessary to keep nodes listening all the time. S-MAC reduces the listen time by putting nodes into periodic sleep state. Each node sleeps for some time, and then wakes up and listens to see if any other node wants to talk to it. During sleeping, the node turns off its radio, and sets a timer (alarm clock) to wake itself later.

A complete cycle of listen and sleep is called a *frame*. The listen interval is normally fixed according to physical-layer and MAC-layer parameters, such as the radio bandwidth and the contention window size. The *duty cycle* is defined as the ratio of the listen interval to the frame length. The sleep interval can be changed according to different application requirements, which actually changes the duty cycle. For simplicity, these values are the same for all nodes. All nodes are free to choose their own listen/sleep schedules. However, to reduce control overhead, we prefer neighboring nodes to synchronize together. That is, they listen at the same time and go to sleep at the same time. It should be noticed that not all neighboring nodes can synchronize together in a multihop network. Two neighboring nodes A and B may have different schedules if they must synchronize with different nodes, C, and D, respectively, as shown in Figure 5.9.

Nodes exchange their schedules by periodically broadcasting a SYNC packet to their immediate neighbors. A node talks to its neighbors at their scheduled listen time, thus ensuring that all neighboring nodes can communicate even if they have different schedules. In Figure 5.9, for example, if node A wants to talk to node B, it waits until B is listening. The period for a node to send a SYNC packet is called the *synchronization period*. One characteristic of S-MAC is that it forms nodes into a flat, peer-to-peer topology. Unlike clustering protocols, S-MAC does not require coordination through cluster heads. Instead, nodes form virtual clusters around common schedules, but communicate directly with peers. One advantage of this loose coordination is that it can be more robust to topology change than cluster-based approaches. The downside of the scheme is the increased latency due to the periodic sleeping. Furthermore, the delay can accumulate on each hop. Later on, a technique that is able to significantly reduce such latency will be presented.

5.3.3 Collision avoidance

If multiple neighbors want to talk to a node at the same time, they will try to send when the node starts listening. In this case, they need to contend for the medium. Among contention protocols, the 802.11 does a very good job on collision avoidance. S-MAC follows similar procedures, including virtual and physical carrier sense, and the RTS/CTS (request to send/clear to send) exchange for the hidden terminal problem [32]. There is a duration field in each transmitted packet that indicates how long the remaining transmission will be. If a node receives a packet destined to another node, it knows how long to keep silent from this field. The node records this value in a variable called the network allocation vector (NAV) [33] and sets a timer for it. Every time the timer fires, the node decrements its NAV until it reaches zero. Before initiating a transmission, a node first looks at its NAV. If its value is not zero, the node determines that the medium is busy. This is called 'virtual carrier sense'. Physical carrier sense is performed at the physical layer by listening to the channel for possible transmissions. Carrier senses time is randomized within a contention window to avoid collisions and starvations. The medium is determined as free if both virtual and physical carrier senses indicates that it is free.

All senders perform carrier sense before initiating a transmission. If a node fails to get the medium, it goes to sleep and wakes up when the receiver is free and listening again. Broadcast packets are sent without using RTS/CTS. Unicast packets follow the sequence of RTS/CTS/DATA/ACK between the sender and the receiver. After the successful exchange of RTS and CTS, the two nodes will use their normal sleep time for data packet transmission. They do not follow their sleep schedules until they finish the transmission. With the low-duty-cycle operation and the contention mechanism during each listen interval, S-MAC effectively addresses the energy waste due to idle listening and collisions. In the next section, details of the periodic sleep coordinated among neighboring nodes will be presented. Two techniques will be presented that further reduce the energy waste due to overhearing and control overhead.

5.3.4 Coordinated sleeping

Periodic sleeping effectively reduces energy waste on idle listening. In S-MAC, nodes coordinate their sleep schedules rather than randomly sleep on their own. This section details the procedures that all nodes follow to set-up and maintain their schedules. It also presents a technique to reduce latency due to the periodic sleep on each node.

5.3.5 Choosing and maintaining schedules

Before each node starts its periodic listen and sleep, it needs to choose a schedule and exchange it with its neighbors. Each node maintains a *schedule table* that stores the schedules of all its known neighbors. It follows the steps below to choose its schedule and establish its schedule table.

(1) A node first listens for a fixed amount of time, which is at least the synchronization period. If it does not hear a schedule from another node, it immediately chooses its own schedule and starts to follow it. Meanwhile, the node tries to announce the schedule by broadcasting a SYNC packet. Broadcasting a SYNC packet follows the normal contention procedure. The randomized carrier sense time reduces the chance of collisions on SYNC packets.

(2) If the node receives a schedule from a neighbor before choosing or announcing its own schedule, it follows that schedule by setting its schedule to be the same. Then the node will try to announce its schedule at its next scheduled listen time.

(3) There are two cases where a node receives a different schedule after it chooses and announces its own schedule. If the node has no other neighbors, it will discard its current schedule and follow the new one. If the node already follows a schedule with one or more neighbors, it adopts both schedules by waking up at the listen intervals of the two schedules.

To illustrate this algorithm, consider a network where all nodes can hear each other. The node that starts first will pick up a schedule first, and its broadcast will synchronize all its peers on its schedule. If two or more nodes start first at the same time, they will finish initial listening at the same time, and will choose the same schedule independently. No matter which node sends out its SYNC packet first (wins the contention), it will synchronize the rest of the nodes.

However, two nodes may independently assign schedules if they cannot hear each other in a multihop network. In this case, those nodes on the border of two schedules will adopt both. For example, nodes A and B in Figure 5.9 will wake up at the listen time of both schedules. In this way, when a border node sends a broadcast packet, it only needs to send it once. The disadvantage is that these border nodes have less time to sleep and consume more energy than others.

Another option is to let a border node adopt only one schedule – the one it receives first. Since it knows that some other neighbors follow another schedule, it can still talk to them. However, for broadcasting, it needs to send twice to the two different schedules. The advantage is that the border nodes have the same simple pattern of periodic listen and sleep as other nodes.

It is expected that nodes only rarely see multiple schedules, since each node tries to follow an existing schedule before choosing an independent one. However, a new node may still fail to discover an existing neighbor for several reasons. The SYNC packet from the neighbor could be corrupted by collisions or interference. The neighbor may have delayed sending a SYNC packet due to the busy medium. If the new node is on the border of two schedules, it may only discover the first one if the two schedules do not overlap.

To prevent the case that two neighbors miss each other forever when they follow completely different schedules, S-MAC introduces periodic neighbor discovery, i.e. each node periodically listens for the whole synchronization period. The frequency with which a node performs neighbor discovery depends on the number of neighbors it has. If a node does not have any neighbors, it performs neighbor discovery more aggressively than in the case where it has many neighbors. Since the energy cost is high during the neighbor discovery, it should not be performed too often. In a typical implementation, the synchronization period is 10 s, and a node performs neighbor discovery every 2 min if it has at least one neighbor.

5.3.6 Maintaining synchronization

Since neighboring nodes coordinate their sleep schedules, the clock drift on each node can cause synchronization errors. Two techniques can be used to make it robust to such errors: (1) all exchanged timestamps are relative rather than absolute; and (2) the listen period is significantly longer than clock drift rates. For example, the listen time of 0.5 s is more than 10 times longer than typical clock drift rates. Compared with TDMA schemes with very short time slots, S-MAC requires much looser time synchronization. Although the long listen time can tolerate fairly large clock drift, neighboring nodes still need to periodically update each other with their schedules to prevent long-term clock drift. The synchronization

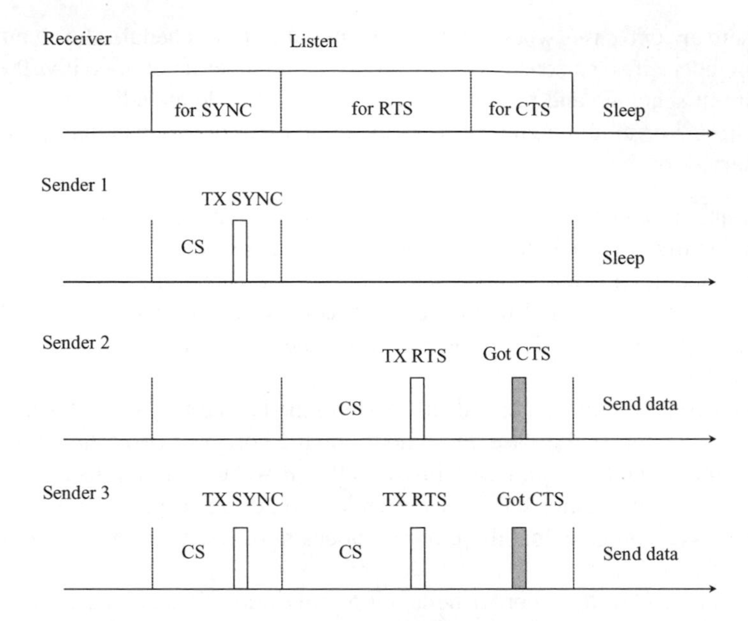

Figure 5.10 Timing relationship between a receiver and different senders. CS stands for carrier sense.

period can be quite long. The measurements show that the clock drift between two nodes does not exceed 0.2 ms/s.

As mentioned earlier, schedule updating is accomplished by sending a SYNC packet. The SYNC packet is very short, and includes the address of the sender and the time of its next sleep. The next sleep time is relative to the moment that the sender starts transmitting the SYNC packet. When a receiver gets the time from the SYNC packet, it subtracts the packet transmission time and uses the new value to adjust its timer. In order for a node to receive both SYNC packets and data packets, its listen interval is divided into two parts. The first one is for SYNC packets, and the second one is for data packets, as shown in Figure 5.10. Each part has a contention window with many time slots for senders to perform carrier sense. For example, if a sender wants to send a SYNC packet, it starts carrier sense when the receiver begins listening. It randomly selects a time slot to finish its carrier sense. If it has not detected any transmission by the end of that time slot, it wins the contention and starts sending its SYNC packet. The same procedure is followed when sending data packets.

Figure 5.10 shows the timing relationship of three possible situations that a sender transmits to a receiver. Sender 1 only sends a SYNC packet. Sender 2 only sends a unicast data packet. Sender 3 sends both a SYNC and a data packet.

5.3.7 Adaptive listening

The scheme of periodic listen and sleep is able to significantly reduce the time spent on idle listening when traffic load is light. However, when a sensing event indeed happens, it is desirable that the sensing data can be passed through the network without too much delay. When each node strictly follows its sleep schedule, there is a potential delay on each hop, whose average value is proportional to the length of the frame. For this reason, a mechanism is introduced to switch the nodes from the low-duty-cycle mode to a more active mode in this case.

S-MAC uses an important technique, called *adaptive listen*, to improve the latency caused by the periodic sleep of each node in a multihop network. The basic idea is to let the node that overhears its neighbor's transmissions [ideally only request to send (RTS) or clear to send (CTS)] wake up for a short period of time at the end of the transmission. In this way, if the node is the next-hop node, its neighbor is able to immediately pass the data to it instead of waiting for its scheduled listen time. If the node does not receive anything during the adaptive listening, it will go back to sleep until its next scheduled listen time.

Let us look at the timing diagram in Figure 5.10 again. If the next-hop node is a neighbor of the sender, it will receive the RTS packet. If it is only a neighbor of the receiver, it will receive the CTS packet from the receiver. Thus, both the neighbors of the sender and receiver will learn about how long the transmission is from the duration field in the RTS and CTS packets. So they are able to adaptively wake up when the transmission is over.

The interval of the adaptive listening does not include the time for the SYNC packet as in the normal listen interval (see Figure 5.10). SYNC packets are only sent at scheduled listen time to ensure all neighbors can receive it. To give the priority to the SYNC packet, adaptive listen and transmission are not performed if the duration from the time the previous transmission is finished to the normally scheduled listen time is shorter than the adaptive listen interval.

One should note that not all next-hop nodes can overhear a packet from the previous transmission, especially when the previous transmission starts adaptively, i.e. not at the scheduled listen time. Therefore, if a sender starts a transmission by sending out an RTS packet during the adaptive listening, it might not get a CTS reply. In this case, it just goes back to sleep and will try again at the next normal listen time.

5.3.8 Overhearing avoidance and message passing

Collision avoidance is a basic task of MAC protocols. S-MAC adopts a contention-based scheme. It is common that any packet transmitted by a node is received by all its neighbors even though only one of them is the intended receiver. Overhearing makes contention-based protocols less efficient in energy than TDMA protocols.

5.3.9 Overhearing avoidance

In 802.11 each node keeps listening to all transmissions from its neighbors in order to perform effective virtual carrier sense. As a result, each node overhears many packets that are not directed to itself. It is a significant waste of energy, especially when node density is high and traffic load is heavy. S-MAC tries to avoid overhearing by letting interfering nodes go to sleep after they hear an RTS or CTS packet. Since DATA packets are normally much longer than control packets, the approach prevents neighboring nodes from overhearing long DATA packets and following ACKs. The question is which nodes should sleep when there is an active transmission in progress.

In Figure 5.11, nodes A, B, C, D, E and F form a multihop network where each node can only hear the transmissions from its immediate neighbors. Suppose node A is currently

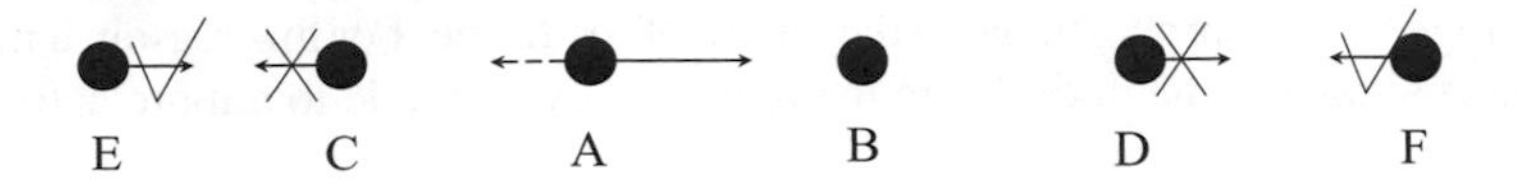

Figure 5.11 Which nodes should sleep when node A is transmitting to B?

transmitting a data packet to B. Which of the remaining nodes should go to sleep during this transmission? Remember that collision happens at the receiver.

It is clear that node D should sleep since its transmission interferes with B's reception. Nodes E and F do not produce interference, so they do not need to sleep. Should node C go to sleep? C is two hops away from B, and its transmission does not interfere with B's reception, so it is free to transmit to its other neighbors, like E. However, C is unable to get any reply from E, e.g. CTS or data, because E's transmission collides with A's transmission at node C. So C's transmission is simply a waste of energy. Moreover, after A sends to B, it may wait for an ACK from B, and C's transmission may corrupt the ACK packet. In summary, *all immediate neighbors of both the sender and receiver should sleep after they hear the RTS or CTS until the current transmission is over, as indicated in Figure 5.11*. Each node maintains the NAV to indicate the activity in its neighborhood. When a node receives a packet destined to other nodes, it updates its NAV using the duration field in the packet.

A nonzero NAV value indicates that there is an active transmission in its neighborhood. The NAV value decrements every time when the NAV timer fires. Thus, a node should sleep to avoid overhearing if its NAV is not zero. It can wake up when its NAV becomes zero. We also note that in some cases overhearing is indeed desirable. Some algorithms may rely on overhearing to gather neighborhood information for network monitoring, reliable routing or distributed queries. If desired, S-MAC can be configured to allow application-specific overhearing to occur. However, it is suggested that algorithms without requiring overhearing may be a better match to energy-limited networks. For example, S-MAC uses explicit data acknowledgments rather than implicit ones [28].

5.3.10 Message passing

A *message* is the collection of meaningful, interrelated units of data. The receiver usually needs to obtain all the data units before it can perform in-network data processing or aggregation. The disadvantages of transmitting a long message as a single packet is the high cost of re-transmitting the long packet if only a few bits have been corrupted in the first transmission. However, if we fragment the long message into many independent small packets, we have to pay the penalty of large control overhead and longer delay. This is so because the RTS and CTS packets are used in contention for each independent packet. A possibility is to fragment the long message into many small fragments, and transmit them in a burst. Only one RTS and one CTS are used. They reserve the medium for transmitting all the fragments. Every time a data fragment is transmitted, the sender waits for an ACK from the receiver. If it fails to receive the ACK, it will extend the reserved transmission time for one more fragment, and re-transmit the current fragment immediately. As before, all packets have the duration field, which is now the time needed for transmitting all the remaining data fragments and ACK packets. If a neighboring node hears an RTS or CTS packet, it will go to sleep for the time that is needed to transmit all the fragments. Each data fragment or ACK also has the duration field. In this way, if a node wakes up or a new node joins in the middle of a transmission, it can properly go to sleep whether it is the neighbor of the sender or the receiver. If the sender extends the transmission time due to fragment losses or errors, the sleeping neighbors will not be aware of the extension immediately. However, they will learn it from the extended fragments or ACKs when they wake up.

The purpose of using ACK after each data fragment is to prevent the hidden terminal problem in the case that a neighboring node wakes up or a new node joins in the middle. If the node is only the neighbor of the receiver but not the sender, it will not hear the data fragments being sent by the sender. If the receiver does not send ACKs frequently, the new node may mistakenly infer from its carrier sense that the medium is clear. If it starts transmitting, the current transmission will be corrupted at the receiver.

It is worth noting that IEEE 802.11 also has fragmentation support. In 802.11 the RTS and CTS only reserve the medium for the first data fragment and the first ACK. The first fragment and ACK then reserve the medium for the second fragment and ACK, and so forth. For each neighboring node, after it receives a fragment or an ACK, it knows that there is one more fragment to be sent. So it has to keep listening until all the fragments are sent. Again, for energy-constrained nodes, overhearing by all neighbors wastes a lot of energy.

The 802.11 protocol is designed to promote fairness. If the sender fails to get an ACK for any fragment, it must give up the transmission and re-contend for the medium so that other nodes have a chance to transmit. This approach can cause a long delay if the receiver really needs the entire message to start processing. In contrast, message passing extends the transmission time and re-transmits the current fragment. It has less contention and a small latency. S-MAC sets a limit on how many extensions can be made for each message where the receiver is really dead or the connection lost during the transmission. However, for sensor networks, application-level performance is the goal as opposed to per-node fairness.

5.3.10.1 Performance examples

In Ye *et al.* [34], The simulation results are obtained for the system with the following set of parameters:

Radio bandwidth	20 kbs
Channel coding	Manchester
Control packet length	10 bytes
Data packet length	up to 250 bytes
MAC header length	8 bytes
Duty cycle	1–99 %
Duration of listen interval	115 ms
Contention window for SYNC	15 slots
Contention window for data	31 slots

The modulation scheme is the amplitude shift keying (ASK). The power consumptions of the radio in receiving, transmitting and sleep modes are 14.4 mW, 36 mW and 15 W, respectively. The topology is a two-hop network with two sources and two sinks, as shown in Figure 5.12. Packets from source A flow through node C and end at sink D, while those from B also pass through C but end at E. The traffic load is changed by varying the inter-arrival period of messages. If the message inter-arrival period is 5 s, a message is generated

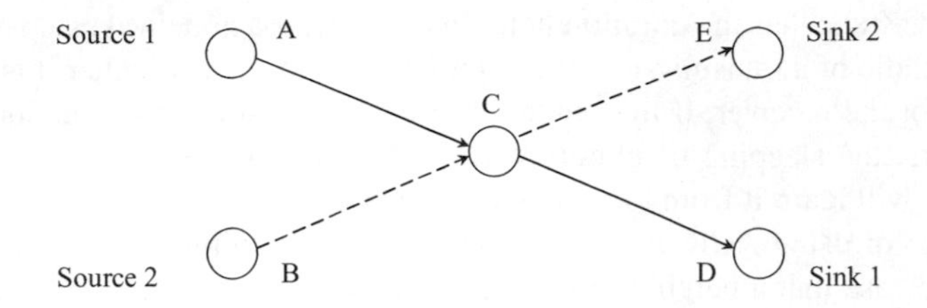

Figure 5.12 Two-hop network with two sources and two sinks.

every 5 s by each source node. In this experiment, the message inter-arrival period varies from 1 to 10 s.

For the highest rate with a 1 s inter-arrival time, the wireless channel is nearly fully utilized due to its low bandwidth. For each traffic pattern, 10 independent tests are done when using different MAC protocols. In each test, each source periodically generates 10 messages, which in turn is fragmented into 10 small data packets (40 bytes each). Thus, in each experiment, there are 200 data packets to be passed from their sources to their sinks. The energy consumption of the radio on each node to pass the fixed amount of data is measured . The actual time to finish the transmission is different for each MAC module. In the 802.11-like MAC, the fragments of a message are sent in a burst, i.e. RTS and CTS are only used for the first fragment.

The 802.11-like MAC without fragmentation, which treats each fragment as an independent packet and uses RTS/CTS for each of them, is not measured, since it is obvious that this MAC consumes much more energy than the one with fragmentation. In S-MAC message passing is used, and fragments of a message are always transmitted in a burst. In the S-MAC module with periodic sleep, each node is configured to operate in the 50 % duty cycle.

Figure 5.13 shows the average energy consumption on the source nodes A and B. The traffic is heavy when the message inter-arrival time is less than 4 s. In this case, 802.11 MAC uses more than twice the energy used by S-MAC. Since idle listening rarely happens, energy savings from periodic sleeping is very limited. S-MAC achieves energy savings mainly by avoiding overhearing and efficiently transmitting long messages. When the message inter-arrival period is larger than 4 s, traffic load becomes light. In this case, the complete S-MAC protocol has the best energy performance, and far outperforms 802.11 MAC. Message passing with overhearing avoidance also performs better than 802.11 MAC. However, as shown in the figure, when idle listening dominates the total energy consumption, the periodic sleep plays a key role in energy savings.

Compared with 802.11, message passing with overhearing avoidance saves almost the same amount of energy under all traffic conditions. This result is due to overhearing avoidance among neighboring nodes A, B and C. The number of packets sent by each of them is the same in all traffic conditions.

5.4 MAC FOR *AD HOC* NETWORKS

A key component in the development of single channel *ad hoc* wireless networks is the MAC protocol with which nodes share a common radio channel. Of necessity, such a protocol

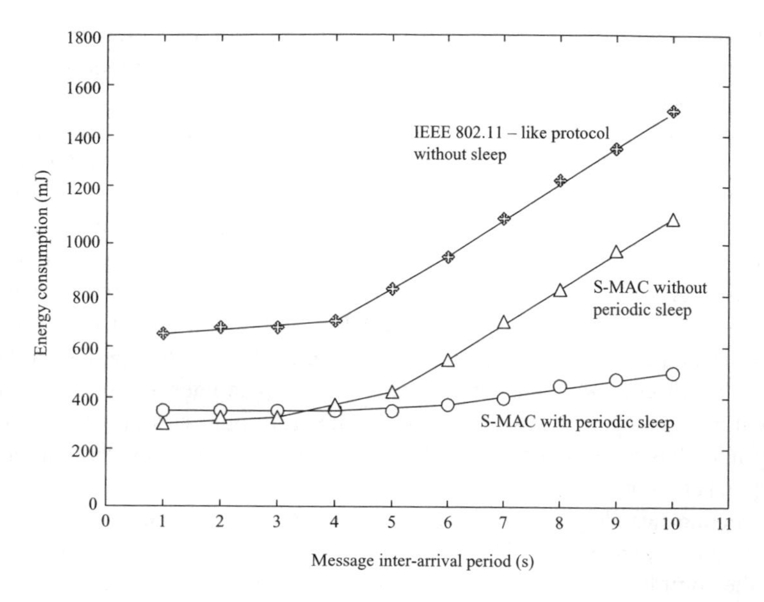

Figure 5.13 Mean energy consumption on radios in each source node. (Reproduced by permission of IEEE [34].)

has to be distributed. It should provide an efficient use of the available bandwidth while satisfying the QoS requirements of both data and real-time applications. CSMA is one of the most pervasive MAC schemes in *ad hoc* wireless networks. CSMA is a simple distributed protocol whereby nodes regulate their packet transmission attempts based only on their local perception of the state, idle or busy, of the common radio channel.

Packet collisions are intrinsic to CSMA. They occur because each node has only a delayed perception of the other nodes' activity. They also happen due to hidden nodes: two transmitting nodes outside the sensing range of each other may interfere at a common receiver. Many types of CSMA exist, but invariably the nodes that participate in a collision schedule the retransmission of their packets to a random time in the future, in the hope of avoiding another collision. This strategy, however, does not provide QoS guarantees for real-time traffic support.

MAC schemes for *ad hoc* wireless networks have been proposed, aimed either at improving the throughput over that of CSMA or at providing QoS guarantees for real-time traffic support. Among the first group of schemes is the multiple access collision avoidance protocol (MACA) [35], which forms the basis of several other schemes. With MACA, a source with a packet ready for transmission first sends a request-to-send (RTS) minipacket, which if successful elicits a clear-to-send (CTS) minipacket from the destination. Upon reception of the CTS minipacket, the source sends its data packet. In environments without hidden nodes, MACA may improve the throughput of the network over that attained with CSMA because collisions involve only short RTS minipackets rather than normal data packets as in CSMA. MACA also alleviates the hidden nodes problem because the CTS sent by the destination serves to inhibit the nodes in its neighborhood, i.e. exactly those nodes that may interfere with the ensuing packet transmission from source to destination. The floor acquisition multiple access (FAMA) class of protocols [36] includes several variants of MACA, one of which is immune to hidden nodes [37]. These protocols, however, have not been designed for QoS: control minipackets are subject to collisions, and their retransmissions are randomly scheduled.

The group allocation multiple access (GAMA) [38, 39] is an attempt to provide QoS guarantees to real-time traffic in a distributed wireless environment. In GAMA, there is a contention period where nodes use an RTS–CTS dialog to explicitly reserve bandwidth in the ensuing contention-free period. A packet transmitted in the contention-free period may maintain the reservation for the next cycle. The scheme is asynchronous and developed for wireless networks where all nodes can sense, and indeed receive, the communications from their peers. MACA/packet reservation (MACA/PR) [40] is a protocol similar to GAMA, but an acknowledgment follows every packet sent in contention-free periods to inform the nodes in the neighborhood of the receiver whether or not another packet is expected in the next contention-free cycle. These schemes deviate from pure carrier sensing methods in that every node has to construct channel-state information based on reservation requests carried in packets sent onto the channel.

In this section, we elaborate on the black-burst (BB) contention mechanism presented in Sobrinho and Krishnakumar [41]. With this mechanism, real-time nodes contend for access to the common radio channel with pulses of energy, BBs, the lengths of which are proportional to the time that the nodes have been waiting for the channel to become idle. The scheme is distributed and is based only on carrier sensing. It gives priority access to real-time traffic and ensures collision-free transmission of real-time packets. When operated in an *ad hoc* wireless LAN, it further guarantees bounded real-time delays. In addition, the BB contention scheme can be overlaid on current CSMA implementations, notably that of IEEE 802.11 standard for wireless LANs, with only minor modifications required to the real-time transceivers: the random retransmission scheme is turned off, and in substitution, the possibility of sending BBs is provided.

5.4.1 Carrier sense wireless networks

Carrier sense wireless networks are designed in such a way that the distance from which a node can sense the carrier from a given transmitter is different and typically larger than the distance from which receivers are willing to accept a packet from that same transmitter. In addition, the carrier from a transmitter can usually be sensed at a range beyond the range at which the transmitter may cause interference. To account for these differences, a wireless network is modeled as a set of nodes N, interconnected by links of three different types. Node i has a communication link with node j, if and only if in the course of time, it has packets to send to node j. Node i has an interfering link with node j if and only if any packet transmission with destination j that overlaps in time at j with a transmission from i is lost.

The lost packets are said to have collided with the transmission from i. Finally, node i has a sensing link with node j, if and only if a transmission by node i prevents node j from starting a new transmission, i.e. node i inhibits node j. The communication, interference and sensing graphs are denoted by $G_C = (N, L_C)$, $G_I = (N, L_I)$, and $G_S = (N, L_S)$, respectively, where, L_C, L_I and L_S are the edge sets (the links). The communication graph is a directed graph, whereas the interfering and sensing graphs are undirected. We assume that if node i has a communication link with node j, then i and j also have an interfering

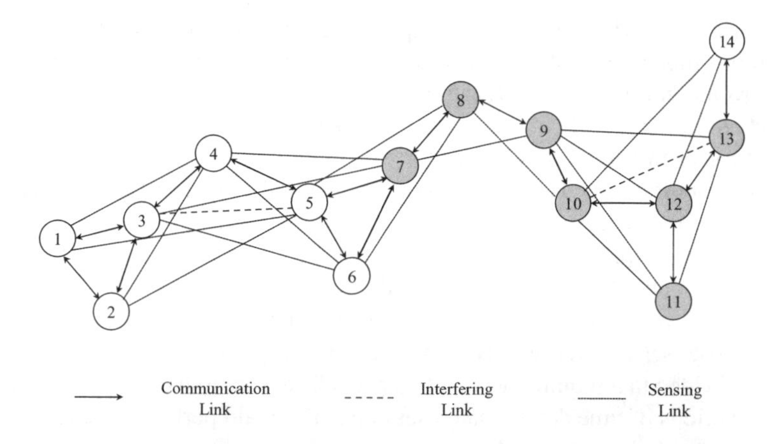

Figure 5.14 A wireless network without hidden nodes. The shaded nodes form the set N_S (9).

link between them. Similarly, an interfering link is also a sensing link, but not conversely. That is, $L_I \subset L_S$: G_I is a spanning subgraph of G_S. Any node has an interfering and sensing link with itself, since whenever a node transmits, it cannot simultaneously receive or start another transmission. As an example in the wireless network of Figure 5.14, node 9 has a communication link with node 10, and thus these nodes have both an interfering and a sensing link between them. Nodes 10 and 13 have an interfering link, and thus they also have a sensing link between them. Finally, nodes 9 and 13 have only a sensing link between them. The links from a node to itself are not explicitly represented.

A path delay is associated with each sensing link to account for the propagation delay separating the nodes, the turn-around (round trip) time of the wireless transceivers, and the sensing delay. The path delay of link *ij* is denoted by τ_{ij}. Since the sensing graph is undirected, $\tau_{ij} = \tau_{ji}$. The path delays further satisfy the two conditions $\tau_{ij} > 0$ and $\tau_{ik} + \tau_{kj} > \tau_{ij}$, for $ik, kj, ij \in L_S$.

Let τ @ $\max(\tau_{ij})$. The sets $N_I(i)$ and $N_S(i)$ represent the nodes that are neighbors of i, i included, in the interfering and sensing graphs, respectively. In Figure 5.14, $N_I(10) = \{9, 10, 11, 12, 13\}$ and $N_S(9) = \{7, 8, 9, 10, 11, 12, 13\}$. For communication link *ij*, the set of nodes which are interfering neighbors of j but are not sensing neighbors of i, i.e. the set, $N_I(j) \cap [N - N_S(i)]$, is the set of nodes hidden from *ij*. A node in this set will not sense an ongoing packet transmission from i to j and may initiate its own packet transmission that will collide at j. In a wireless network without hidden nodes, we have $N_I(j) \subset N_S(i)$ for every $ij \in L_C$. The network of Figure 5.14 does not have hidden nodes. Nevertheless, the common radio channel can be reused in space. For example, a packet transmission from node 9 to node 8 can coexist in time without collisions with a packet transmission from node 5 to node 7. We use the term 'wireless LAN' for wireless networks in which $G_I = G_S$ forms a complete graph. In a wireless LAN, all nodes can sense each other's transmissions. The CSMA/CA protocol of the IEEE 802.11 standard defines three interframe spacings, $t_{short}, t_{med}, t_{med} \geq 2\tau + t_{short}$ and $t_{long}, t_{long} \geq 2\tau + t_{med}$. If a node with a packet that is ready for transmission has perceived that the channel is idle during a long interframe spacing of length t_{long}, the node immediately starts the transmission of the packet. Otherwise, it waits until that condition is satisfied and enters into backoff.

Likewise, a node whose packet has experienced c consecutive collisions enters into backoff. In this mode, the node chooses a random number of slots s uniformly distributed between zero and min $\{32 \times 2^c - 1, 255\}$ and sets a timer with an initial value $s \times t_{slot}$ units of time, where t_{slot}, $t_{slot} \geq 2\tau$, is the length of a slot. The timer counts down only while the channel has been perceived idle for more than t_{long} units of time – it is frozen during a medium busy condition – and the packet is (re)transmitted as soon as the timer reaches zero. A node learns of the success or failure of its transmission through a positive acknowledgment scheme; the recipient of a correctly received packet sends back an acknowledgment minipacket within an interval of time of length t_{short}.

BB contention is a MAC mechanism developed to provide QoS guarantees to real-time traffic over carrier sense wireless networks. The real-time applications considered are those like voice and video that require more or less periodic access to the common radio channel during long periods of time denominated sessions. The main performance requirement for these applications is bounded end-to-end delay, which implies a bounded packet delay at the MAC layer. This is the goal of BB contention. Real-time nodes contend for access to the channel after a medium interframe spacing of length t_{med}, rather than after the long interframe spacing of length t_{long} used by data nodes. Thus, real-time nodes as a group have priority over data nodes.

Instead of sending their packets when the channel becomes idle for t_{med}, real-time nodes first sort their access rights by jamming the channel with pulses of energy, denominated BBs. The length of a BB transmitted by a real-time node is an increasing function of the contention delay experienced by the node, measured from the instant when an attempt to access the channel has been scheduled until the channel becomes idle for t_{med}, i.e. until the node starts the transmission of its BB. To account for the path delays in the network, BBs are formed by an integral number of black slots, each of length t_{bslot}, with t_{bslot} not smaller than the maximum round-trip path delay 2τ. Now, we would like the BBs sent by distinct real-time nodes when the channel becomes idle for t_{med} to differ by at least one black slot. To this end, we assume that every real-time packet transmission lasts at least a certain time t_{pkt} and that real-time nodes only schedule their next transmission attempts – to a time t_{sch} in the future – when they start a packet transmission. If a node starts a packet transmission at time u and that transmission is successful, that means that no other real-time node started a packet transmission during an interval of length $2t_{pkt}$ around time u. Therefore, the next scheduled attempt made by the node in question is also staggered in time by t_{pkt} from the scheduled access attempts made by the other nodes. Counting the number of black slots to be sent in a BB in units of t_{pkt}, we obtain the desired property that distinct nodes contend with BBs comprising different numbers of black slots. Following each BB transmission, a node senses the channel for an observation interval of length t_{obs} to determine without ambiguity whether its BB was the longest of the contending BBs. The winning node will transmit its real-time packet successfully and schedule the next transmission attempt. On the other hand, the nodes that lost the BB contention wait for the channel to once again become idle for t_{med}, at which time they send new longer BBs. In conclusion, once the first real-time packet of a session is successfully transmitted, the mechanism ensures that succeeding real-time packets are also transmitted without collisions. In the end, real-time nodes appear to access a dynamic time division multiplexing (TDM) transmission structure without explicit slot assignments or slot synchronization. In the sequel a detailed description of the access rules followed by every real-time node is presented. Every real-time packet lasts for at least a certain amount of time t_{pkt} $t_{pkt} \geq 2\tau$, when transmitted on the channel. At the beginning of a session, a real-time node uses conventional CSMA/CA rules, possibly with a more expedited retransmission algorithm, to convey its first packet until it is successful. Subsequent packets are transmitted according to the mechanisms, described below, until the session is dropped.

Whenever a real-time node transmits a packet, it further schedules its next transmission attempt to a time $t_{\rm sch}$ in the future, where $t_{\rm sch}$ is the same for all nodes. Suppose, then, that a real-time node has scheduled an access attempt for the present time. If the channel has been idle during the past medium interframe interval of length $t_{\rm med}$, the node starts the transmission of a BB. Otherwise, it waits until the channel becomes idle for $t_{\rm med}$ and only then starts the transmission of its BB. The length b of the BB sent by the node is a direct function of the contention delay it incurred, $d_{\rm cont}$:

$$b(d_{\rm cont}) = \left(1 + \left\lfloor \frac{d_{\rm cont}}{t_{\rm unit}} \right\rfloor\right) t_{\rm bslot}$$

where $t_{\rm bslot}$ is the length of a black slot, the parameter $t_{\rm unit}$ is the unit of time used to convert contention delays into an integral number of black slots, and $\lfloor x \rfloor$ is the floor of x, i.e. the largest integer not larger than x. Correct operation of the scheme requires that $t_{\rm unit} \leq t_{\rm pkt}$. After exhausting its BB transmission, the node waits for an observation interval $t_{\rm obs}$, the length of which has to satisfy $t_{\rm obs} \leq t_{\rm bslot}$ and $t_{\rm obs} \leq t_{\rm med}$, to see if any other node transmitted a longer BB, implying that it would have been waiting longer for access to the channel. If the channel is perceived idle after $t_{\rm obs}$, then the node (successfully) transmits its packet. On the other hand, if the channel is busy during the observation interval, the node waits again for the channel to be idle for $t_{\rm med}$ and repeats the algorithm.

The start of packet transmissions from different nodes is shifted in time by at least $t_{\rm pkt}$. Since it is only when a node initiates the transmission of a packet that it schedules its next transmission attempt to a time $t_{\rm sch}$ in the future, the contention delays of different nodes will likewise differ by at least $t_{\rm pkt}$. Therefore, taking $t_{\rm unit} \leq t_{\rm pkt}$, the BBs of different nodes differ by at least one black slot, and thus every BB contention period produces a unique winner. That winner is the node that has been waiting the longest for access to the channel. The observation interval $t_{\rm obs}$ cannot last longer than the black slot time, i.e. $t_{\rm obs} \leq t_{\rm bslot}$, so that a node always recognizes when its BB is shorter than that of another contending node. It also has to be shorter than the medium interframe spacing, i.e. $t_{\rm obs} \leq t_{\rm med}$, to prevent real-time nodes from sending BBs by the time that a real-time packet transmission is expected. Overall, the BB contention scheme gives priority to real-time traffic, enforces a round-robin discipline among real-time nodes, and results in bounded access delays to real-time packets.

BB contention can also be used to support real-time sessions with different bandwidth requirements, which might be useful for multimedia traffic. On the one hand, distinct real-time sessions may have the corresponding nodes send packets of different sizes when they acquire access rights to the channel. On the other hand, the BB mechanism can be enhanced to accommodate real-time sessions with different scheduling intervals as long as the set of values allowed for the scheduling interval $t_{\rm sch}$ is finite and small. In the latter case, BB contention proceeds in two phases. Real-time nodes first sort their access rights based on contention delays as before. However, it is now possible for two nodes with different scheduling intervals to compute BBs with the same number of black slots. Hence, after this first phase, a real-time node contends again with a new BB, the length of which univocally identifies the scheduling interval being used by the node.

5.4.2 Interaction with upper layers

5.4.2.1 Operation with feedback

If a real-time node were alone in the network, two consecutive real-time packet transmissions belonging to the same session would be separated in time by exactly t_{acc}, $t_{acc} @ t_{sch} + t_{bslot} + t_{obs}$. The access delays measure the deviation from this ideal situation. Specifically, an access delay is the time that elapses from the moment an access attempt occurs until the node is able to transmit the corresponding real-time packet, corrected for $t_{bslot} + t_{obs}$. For $n \geq 2$, the nth access delay associated with a session is denoted by $d^{(n)}$ and is given by $d^{(n)} = (u^{(n)} - u^{(n-1)} - t_{acc})$, where $u^{(n)}$ is the instant of time when the node started the transmission of its nth packet. Given the maximum length of data packets, the rate of real-time sessions and number of real-time nodes, the BB mechanism guarantees that the access delays are bounded and usually by a very small value d_{max}.

When a node is the source node of a session, the contents of its real-time packets can reflect the access delays incurred in contending for access to the channel. Typically, a real-time application generates blocks of information bits at regular intervals of time, of length much smaller than t_{acc}. The block delay is the time interval that elapses from the moment an information block is made available by the application until it is successfully transmitted at the MAC layer (corrected for $t_{bslot} + t_{obs}$ and neglecting processing delays). The relation between access and block delays depends on how the application blocks of information are packetized for transmission at the MAC layer. One possibility is to have the MAC layer convey in a packet all the information blocks generated up to the instant when the node is about to start a packet transmission. The length of a real-time packet would thus grow with the access delay incurred by the node. The block delay of the oldest block conveyed in the packet would consist of t_{acc}, plus the corresponding access delay: the block delay would never exceed $t_{acc} + d_{max}$. In general, however, it is not feasible to assemble a packet at the time that its transmission should start, and further, the MAC layer usually contains a single buffer that we must ensure is filled with a packet by the time access to the channel is granted.

For a realistic alternative within the spirit of this section, consider a simplified communication architecture in which a real-time application puts its generated blocks of information into an application buffer. Whenever the node successfully transmits a packet it signals the application, which will assemble the next packet with all the blocks of information currently queued at the application buffer, plus the blocks that will be generated during the next interaccess interval of length t_{acc}. At this later time, the packet is delivered to the MAC layer for transmission. With this procedure, the MAC layer always has a packet ready for transmission by the time it acquires undisputed access to the channel. When a node transmits its nth packet at time $u^{(n)}$, it leaves in the application buffer the blocks of information generated during the previous $d^{(n)}$ units of time; they will be part of the contents of the $(n+1)$th packet. The latter packet further incurs an access delay of $d^{(n+1)}$ at the MAC layer. Therefore, the block delay of the oldest block conveyed in the $(n+1)$th packet is not greater than $(d_n + t_{acc} + d_{n+1})$: the block delay during a session never exceeds $(t_{acc} + 2d_{max})$.

5.4.2.2 Operation without feedback

In the previous section, the contents of a real-time packet depended on the access delays incurred by a node. There is a direct coupling between the MAC layer and the real-time application. A simpler communication architecture may be desired in which already assembled packets are passed onto the MAC layer for transmission one by one. This is also the situation encountered when a node is simply relaying real-time packets arriving from a distant source.

Suppose that real-time packets are presented to the MAC layer periodically, one every t_{rdy} units of time. The packet delay is the time that elapses from the moment a packet is available for transmission until it is successfully transmitted at the MAC layer (corrected for $t_{\mathrm{bslot}} + t_{\mathrm{obs}}$). The packet delay of the nth packet $\omega^{(n)}$ is given by $\omega^{(n)} = (u^{(n)} - t^{(n)} - t_{\mathrm{bslot}} - t_{\mathrm{obs}})$, where $t^{(n)}$ is the instant of time when the nth packet becomes ready for transmission, $t^{(n)} = t^{(1)} + (n-1)t_{\mathrm{rdy}}$.

Clearly, we should not choose $t_{\mathrm{sch}} + t_{\mathrm{bslot}} + t_{\mathrm{obs}} = t_{\mathrm{rdy}}$. If that choice was made, the instants when the node accesses the channel would start drifting in relation to the arrival times of new packets, and the node would not keep up with the packet arrival rate. Indeed, the packet delay of the nth packet would be

$$\omega^{(n)} = \omega^{(1)} + \sum_{i=2}^{n} d^{(i)}$$

which would grow monotonically with the number of packets already transmitted.

Consider instead a preventive approach whereby a real-time node schedules its next transmission attempt short of the inter-arrival time for packets t_{rdy}. Specifically, when a real-time node transmits a packet, it schedules the next transmission attempt to time t_{sch} in the future, now with $t_{\mathrm{sch}} = t_{\mathrm{rdy}} - t_{\mathrm{bslot}} - t_{\mathrm{obs}} - \delta$, where δ, $\delta > 0$, is called the slack time. At a scheduled access attempt, a real-time node will only start contending for access to the channel if a real-time packet is available for transmission. Otherwise, it waits for a ready packet and only then starts to contend for access to the channel. The correctness of the BB contention mechanism is preserved as long as the contention delays used to compute the lengths of BBs are always counted from the scheduled access attempts up to the time when the channel becomes idle for t_{med}.

REFERENCES

[1] *IEEE Std 802.11–1999, Part 11: Wireless LAN Medium Access Control (MAC) and Physical Layer (PHY) Specifications, Reference number ISO/IEC 8802-11:1999 (E)*, IEEE Std 802.11, 1999.

[2] *IEEE 802.11 Wireless Local Area Networks*; http://grouper.ieee.org/groups/802/11/.

[3] *IEEE 802.11e/D4.0, Draft Supplement to Part 11: Wireless Medium Access Control (MAC) and Physical Layer (PHY) Specifications: Medium Access Control (MAC) Enhancements for Quality of Service (QoS)*, November 2002.

[4] S. Choi, J. Prado, S. Shankar and S. Mangold, IEEE 802.11e Contention-Based Channel Access (EDCF) Performance Evaluation, in *Proc. IEEE ICC 2003*, Anchorage, AK, May 2003.

[5] P. Garg, R. Doshi, R. Greene, M. Baker, M. Malek and X. Cheng, Using IEEE 802.11e MAC for QoS over wireless, in *Proc. IEEE Int. Performance Computing and Communications Conf.*, Phoenix, AZ, April 2003.

[6] A. Banchs, X. Perez-Costa, and D. Qiao, Providing throughput guarantees in IEEE 802.11e wireless LANs, in *Proc. 18th Int. Teletraffic Congr. (ITC-18)*, Berlin, September 2003.

[7] Y. Xiao, Enhanced DCF of IEEE 802.11e to Support QoS, in *Proc. IEEE WCNC 2003*, New Orleans, LA, March 2003.

[8] S. Mangold, G. Hiertz and B. Walke, IEEE 802.11e wireless LAN – resource sharing with contention based medium access, in *IEEE PIMRC 2003*, Beijing, September 2003.

[9] G. Bianchi, Performance analysis of the IEEE 802.11 distributed coordination function, *IEEE JSAC*, vol. 18, no. 3, 2000, pp. 535–547.

[10] *IEEE Std 802.11b-1999, Supplement to Part 11: Wireless Medium Access Control (MAC) and Physical Layer (PHY) Specifications: Higher-Speed Physical Layer Extension in the 2.4 GHz Band*, September 1999.

[11] Z. Hadzi-Velkov and B. Spasenovski, Saturation throughput-delay analysis of IEEE 802.11 DCF in fading channel, in *Proc. IEEE ICC 2003*, Anchorage, AK, May 2003.

[12] S. Mangold, S. Choi, G. Hiertz, O. Klein and B. Walke, Analysis of IEEE 802.11e for QoS support in wireless LANs, *IEEE Wireless Commun.*, vol. 10, no. 6, 2003, pp. 40–50.

[13] IST WSI, *The Book of Visions 2000: Visions of the Wireless World, Version 1.0*, 2000.

[14] Z. Tao and S. Panwar, An analytical model for the IEEE 802.11e enhanced distributed coordination function, *IEEE Int. Conf. Communications*, vol. 7, 20–24 June 2004, pp. 4111–4117.

[15] H. Singh and S. Singh, Doa-aloha: slotted aloha for *ad hoc* networking using smart antennas, in *IEEE VTC Fall'03*, 6–9 October 2003.

[16] H. Singh and S. Singh, A MAC protocol based on adaptive beamforming for *ad hoc* networks, in *IEEE Pimrc'03*, 7–10 September 2003.

[17] R.R. Choudhury, X. Yang, R. Ramanathan and N.H. Vaidya, Using directional antennas for medium access control in *ad hoc* networks, in *ACM/SIGMOBILE MobiCom 2002*, 23–28 September 2002.

[18] L. Bao and J.J. Garcia-Luna-Aceves, Transmission scheduling in *ad hoc* networks with directional antennas, in *ACM/SIGMOBILE MobiCom 2002*, 23–28 September 2002.

[19] S. Bellofiore, J. Foutz, R. Govindarajula, I. Bahceci, C.A. Balanis, A.S. Spanias, J.M. Capone and T.M. Duman, Smart antenna system analysis, integration, and performance for mobile ad-hoc networks (manets), *IEEE Trans. Antennas and Propagation*, vol. 50, no. 5, 2002, pp. 571–581.

[20] T. ElBatt and B. Ryu, On the channel reservation schemes for *ad hoc* networks utilizing directional antennas, in *WPMC'02*, 2002.

[21] N. Fahmy, T.D. Todd and V. Kezys, *Ad hoc* networks with smart antennas using 802.11-based protocols, in *IEEE ICC'02*, 2002.

[22] G.M. Sanchez, Multiple access protocols with smart antennas n multihop *ad hoc* rural-area networks, M.S. thesis, Royal Institute of Technology, Sweeden, Radio Communication Systems Laboratory, Department of Signals, Sensors and Sytems, June 2002.

[23] R. Bagrodia, M. Takai, J. Martin and A. Ren, Directional virtual carrier sensing for directional antennas in mobile *ad hoc* networks, in *ACM/SIGMOBILE MobiHoc 2002*, October 2002.

[24] T. Nandagopal, T.-E. Kim, X. Gao and V. Bharghavan, Achieving MAC layer fairness in wireless packet networks, in *ACM MOBICOM'00*, August 2000, pp. 87–98.

[25] B. Bensaou, Y. Wang and C.C. Ko, Fair medium access in 802.11 based wireless *ad-hoc* networks, in *ACM MOBIHOC'00*, August 2000, pp. 99–106.

[26] V. Bharghavan, A. Demers, S. Shenker and L. Zhang, Macaw: a media access protocol for wireless lans, in *ACM SIGCOMM'94*, August 1994, pp. 212–225.
[27] H. Singh and S. Singh, Smart-802.11b MAC protocol for use with smart antennas, in *IEEE Int. Conf. Communications*, vol. 6, 20–24 June 2004, pp. 3684–3688.
[28] A. Woo and D. Culler, A transmission control scheme for media access in sensor networks, in *Proc. ACM/IEEE Int. Conf. Mobile Computing and Networking*, Rome, July 2001, pp. 221–235.
[29] F. Bennett, D. Clarke, J.B. Evans, A. Hopper, A. Jones and D. Leask, Piconet: embedded mobile networking, *IEEE Person Commun. Mag.*, vol. 4, 1997, pp. 8–15.
[30] S. Xu and T. Saadawi, Does the IEEE 802.11 MAC protocol work well in multihop wireless *ad hoc* networks?, *IEEE Commun. Mag.*, vol. 36, 2001, pp. 130–137.
[31] Y.-C. Tseng, C.-S. Hsu and T.-Y. Hsieh, Power-saving protocols for IEEE 802.11-based multi-hop ad hoc networks, in *Proc. IEEE INFOCOM*, New York, June 2002, pp. 200–209.
[32] V. Bharghavan, A. Demers, S. Shenker and L. Zhang, MACAW: a media access protocol for wireless lans, in *Proc. ACM SIGCOMM*, London, Septembar 1994, pp. 212–225.
[33] *Wireless LAN Medium Access Control (MAC) and Physical Layer (PHY) Specification*, IEEE Std 802.11, 1999.
[34] W. Ye, J. Heidemann and D. Estrin, Medium access control with coordinated adaptive sleeping for wireless sensor networks, *IEEE/ACM Trans. Networking*, vol. 12, no. 3, June 2004, pp. 493–506.
[35] P. Karn, MACA – A new channel access method for packet radio, in *Proc. ARRL/CRRL Amateur Radio Ninth Computer Networking Conf.*, ARRL, 1990, pp. 134–140.
[36] C. Fullmer and J. Garcia-Luna-Aceves, Floor acquisition multiple access (FAMA) for packet-radio networks, in *Proc. SIGCOMM'95*, Cambridge, MA, 1995, pp. 262–273.
[37] C. Fullmer and J. Garcia-Luna-Aceves, Solutions to hidden terminal problems in wireless networks, in *Proc. SIGCOMM'97*, vol. 2, Cannes, 1997, pp. 39–49.
[38] A. Muir and J. Garcia-Luna-Aceves, Supporting real-time multimedia traffic in a wireless LAN, in *Proc. SPIE Multimedia Computing and Networking 1997*, San Jose, CA, 1997, pp. 41–54.
[39] R. Garces and J. Garcia-Luna-Aceves, Collision avoidance and resolution multiple access with transmission groups, in *Proc. IEEE INFOCOM'97*, Kobe, 1997, pp. 134–142.
[40] C. Lin and M. Gerla, Asynchronous multimedia multihop wireless networks, in *Proc. IEEE INFOCOM'97*, Kobe, 1997, pp. 118–125.
[41] J.L. Sobrinho and A.S. Krishnakumar, Real-time traffic over the IEEE 802.11 medium access control layer, *Bell Labs Tech. J.*, vol. 1, 1996, pp. 172–187.

6

Teletraffic Modeling and Analysis

Traditional traffic models have been developed for wireline networks. These models predict the aggregate traffic going through telephone switches. Queueing theory is the tool which has been traditionally used in the analysis of such systems. A summary of the main results from the queueing theory is included in Appendix C (please go to www.wiley.com/go/glisic). These traditional models do not include subscriber mobility or callee distributions and therefore need modifications to be applicable for modeling the traffic in wireless networks.

6.1 CHANNEL HOLDING TIME IN PCS NETWORKS

Channel holding (occupancy) time is an important quantity in teletraffic analysis of PCS networks. It corresponds to service time in conventional queueing theory. This quantity is needed to derive key network design parameters such as the new call blocking probability and the handoff call blocking probability [1]. The cell residence time is a nonnegative random variable, so a good distribution model for the random variable will be sufficient for characterizing the users' mobility. In this section we use, the *hyper-Erlang distribution model* [2] for such purposes.

The *hyper-Erlang* distribution has the following probability density function and Laplace transform:

$$f_{\rm he}(t) = \sum_{i=1}^{M} \alpha_i \frac{(m_i \eta_i)^{m_i} t^{m_i-1}}{(m_i-1)!} {\rm e}^{-m_i \eta_i t}, \qquad t \geq 0$$

$$f_{\rm he}^{*}(s) = \sum_{i=1}^{M} \alpha_i \left(\frac{m_i \eta_i}{s + m_i \eta_i} \right)^{m_i} \tag{6.1}$$

Advanced Wireless Networks: 4G Technologies Savo G. Glisic
© 2006 John Wiley & Sons, Ltd.

where $\alpha_i \geq 0$, and $\sum_{i=1}^{M} \alpha_i = 1$. $M, m_1, m_2, \ldots,$ m_M are nonnegative integers and $\eta_1, \eta_2, \ldots, \eta_M$ are positive numbers. These distribution functions provide sufficiently general models, i.e. hyper-Erlang distributions are universal approximations.

It can be shown [3] that for a given cumulative distribution function $G(t)$ of a nonnegative random variable we can choose a sequence of distribution functions $G_m(t)$, each of which corresponds to a mixture of Erlang distributions, so that $\lim_{m\to\infty} G_m(t) = G(t)$ for all t at which $G(t)$ is continuous. $G_m(t)$ can be chosen as

$$G_m(t) = \sum_{k=1}^{\infty} \left[G\left(\frac{k}{m}\right) - G\left(\frac{k-1}{m}\right) \right] G_m^k(t), \qquad t \geq 0 \tag{6.2}$$

where $G_m^k(t)$ is the distribution function of an Erlang distribution with mean k/m and variance k/m^2 (the distribution of the sum of k exponential random variables each with mean $1/m$). If $g_m(t)$ and $\overset{*}{g}_m(s)$ are the density function and Laplace transform of $G_m(t)$, and $g_m^k(t)$ is the density function of $G_m^k(t)$ then we have

$$\begin{aligned} g_m(t) &= \sum_{k=1}^{\infty} \left[G\left(\frac{k}{m}\right) - G\left(\frac{k-1}{m}\right) \right] g_m^k(t) \\ \overset{*}{g}_m(s) &= \sum_{k=1}^{\infty} \left[G\left(\frac{k}{m}\right) - G\left(\frac{k-1}{m}\right) \right] \left(\frac{k/m}{s+k/m} \right)^k \end{aligned} \tag{6.3}$$

where the asterisk is used to denote the Laplace transformation.

The resulting distribution is called the 'mixed Erlang distribution'. Their coefficients can be determined from the experimental data. If a finite number of terms is used to approximate the distribution function, the resulting distribution approximates the hyper-Erlang distribution.

To illustate why the distribution $G_m(t)$ provides the universal approximation to general distribution models we show Erlang distribution

$$f_e(t) = \frac{(m\eta)^{mt^{m-1}}}{(m-1)!} e^{-m\eta t}, \qquad t \geq 0 \tag{6.4}$$

in Figure 6.1 by varying the shape parameter m (Appendix C; please go to www.wiley .com/go/glisic). We observe that, as the shape parameter m becomes sufficiently large, the density function approaches the Dirac δ function. Hence, $f_e(t)$ approaches the δ function as m is sufficiently large. From signal processing theory [4], we know that the δ function can be used to sample a function and reconstruct the function from the sampled data (the sampling theorem). We can replace the δ function by the Erlang density function with sufficiently large m, and the resulting approximation is exactly in the form of the hyper-Erlang distribution.

If the cell residence time t is modeled by the hyper-Erlang distribution as in Equation (6.1), its kth moment is given as

$$E\left[t^k\right] = (-1)^k f_{\text{he}}^{*(k)}(0) = \sum_{i=1}^{M} \alpha_i \frac{(m_i+k-1)!}{(m_i-1)!} (m_i\eta_i)^{-k} \tag{6.5}$$

The parameters α_i, m_i and η_i $(i = 1, 2, \ldots, M)$ can be found by fitting a number of moments from field data. Moreover, if the number of moments exceeds the number of variables,

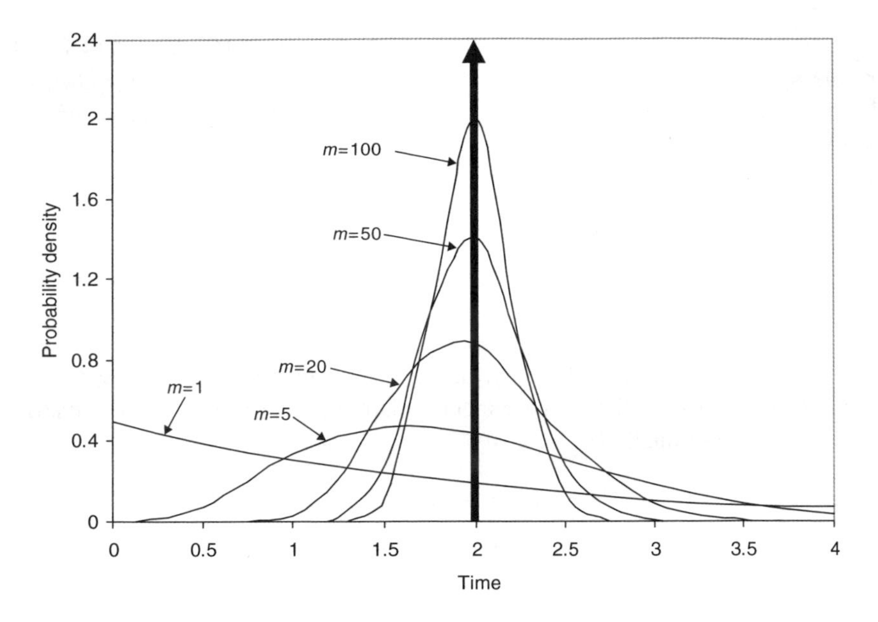

Figure 6.1 Probability density function for Erlang distribution.

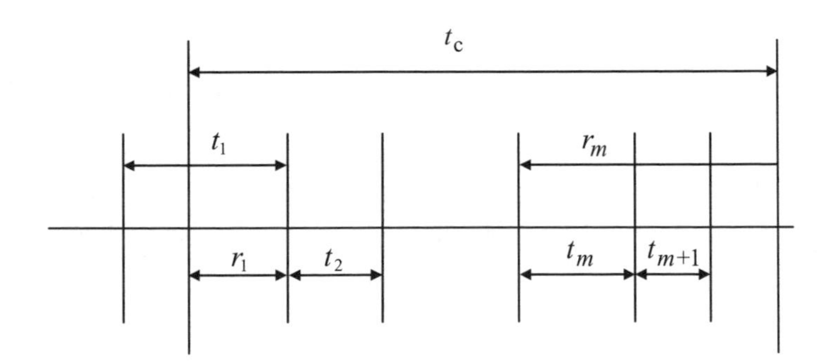

Figure 6.2 The call holding and cell residence times.

then the least-square method can be used to find the best fit to minimize the least-square error.

The channel holding time distribution depends on the mobility of users, which can be characterized by the cell residence time [1–30, 5–31]. In the sequel we use the following notation: t_c, call holding time (exponentially distributed with parameter μ); t_m, cell residence time; r_1, time between the instant a new call is initiated and the instant the new call moves out of the cell if the new call is not completed; $r_m (m > 1)$, residual life time distribution of call holding time when the call finishes mth handoff successfully; and t_{nh} and t_{hh}, the channel holding times for a new call and a handoff call, respectively. Then, from Figure 6.2, the channel holding time for a new call will be

$$t_{nh} = \min\{t_c, r_1\} \tag{6.6}$$

and the channel holding time for a handoff call is

$$t_{hh} = \min\{r_m, t_m\} \tag{6.7}$$

Let t_c, t_m, r_1, t_{hh} and t_{nh} have density functions $f_c(t)$, $f(t)$, $f_r(t)$, $f_{hh}(t)$ and $f_{nh}(t)$ with their corresponding Laplace transforms $f_c^*(s)$, $f^*(s)$, $f_r^*(s)$, $f_{hh}^*(s)$ and $f_{nh}^*(s)$, respectively, and with cumulative distribution functions, $f_c(t)$, $F(t)$, $F_r(t)$, $F_{hh}(t)$ and $F_{nh}(t)$ respectively. From Equation (6.7) we obtain the probability

$$\begin{aligned}\Pr(t_{hh} \le t) &= \Pr(r_m \le t \text{ or } t_m \le t) \\ &= \Pr(r_m \le t) + \Pr(t_m \le t) - \Pr(r_m \le t, t_m \le t) \\ &= \Pr(r_m \le t) + \Pr(t_m \le t) - \Pr(r_m \le t)\Pr(t_m \le t) \\ &= \Pr(t_c \le t) + \Pr(t_m \le t) - \Pr(t_c \le t)\Pr(t_m \le t)\end{aligned} \tag{6.8}$$

which is based on the independency of r_m and t_m, and the memoryless property of the exponential distribution from which we have that the distribution of r_m has the same distribution as t_c. Differentiating Equation (6.8), gives

$$\begin{aligned}f_{hh}(t) &= f_c(t) + f(t) - f_c(t)\Pr(t_m \le t) - \Pr(t_c \le t) f(t) \\ &= f_c(t)\int_t^\infty f(\tau)\,d\tau + f(t)\int_t^\infty f_c(\tau)\,d\tau\end{aligned} \tag{6.9}$$

and

$$f_{hh}^*(s) = f^*(s) + f_c^*(s) - \int_0^\infty e^{-st} f(t)\int_o^t f_c(\tau)\,d\tau\,dt - \int_0^\infty e^{-st} f_c(t)\int_o^t f(\tau)\,d\tau\,dt$$

$$\begin{aligned}f_{hh}^*(s) &= f^*(s) + f_c^*(s) - \mu\int_0^\infty e^{-(s+\mu)t}\int_o^t f(\tau)\,d\tau\,dt - \int_0^\infty e^{-st}(1-e^{-\mu t}) f(t)\,dt \\ &= \frac{\mu}{s+\mu} + \frac{s}{s+\mu} f^*(s+\mu)\end{aligned} \tag{6.10}$$

From the above equations, we obtain the expected handoff call channel holding time as

$$E[t_{hh}] = -f_{hh}^{*\prime}(0) = \frac{1}{\mu}[1 - f^*(\mu)] \tag{6.11}$$

where $f_{hh}^{*\prime}$ is the first derivative of the function. Starting from Equation (6.6) and using a similar argument, we obtain

$$\begin{aligned}f_{nh}(t) &= f_c(t)\int_t^\infty f_r(\tau)\,d\tau + f_r(t)\int_t^\infty f_c(\tau)\,d\tau \\ f_{nh}^*(s) &= \frac{\mu}{s+\mu} + \frac{s}{s+\mu} f_r^*(s+\mu) \\ E[t_{nh}] &= -f_{nh}^{*(1)}(0) = \frac{1}{\mu}\left[1 - f_r^*(\mu)\right].\end{aligned} \tag{6.12}$$

We need to consider the channel holding time distribution for any call (either new call or handoff call), i.e. the channel holding time for the merged traffic of new calls and handoff calls. We will simply call this the channel holding time, using no modifiers such as new call or handoff call. If t_h is the channel holding time and λ_h the handoff call arrival rate, and λ is the new call arrival rate then, $t_h = t_{nh}$ with probability $\lambda(\lambda + \lambda_h)$ and $t_h = t_{hh}$ with probability $\lambda_h(\lambda + \lambda_h)$. Let $f_h(t)$ and $f_h^*(s)$ be its density function and the corresponding

Laplace transform. It is easy to obtain

$$\begin{aligned} f_{\rm h}(t) &= \frac{\lambda}{\lambda+\lambda_{\rm h}} f_{\rm nh}(t) + \frac{\lambda_{\rm h}}{\lambda+\lambda_{\rm h}} f_{\rm hh}(t) \\ f_{\rm h}^{*}(s) &= \frac{\lambda}{\lambda+\lambda_{\rm h}} f_{\rm nh}^{*}(s) + \frac{\lambda_{\rm h}}{\lambda+\lambda_{\rm h}} f_{\rm hh}^{*}(s) \\ E\left[t_{\rm h}\right] &= \frac{\lambda}{\mu(\lambda+\lambda_{\rm h})}\left[1-f_{\rm r}^{*}(\mu)\right] + \frac{\lambda_{\rm h}}{\mu\lambda+\lambda_{\rm h})}\left[1-f^{*}(\mu)\right] \end{aligned} \tag{6.13}$$

When the residual lifetime r_1 of t_1 is exponentially distributed with parameter μ_r, then its Laplace transform $f_{\rm r}^{*}(s)$ is $\mu_{\rm r}/(s+\mu_{\rm r})$. Using this in Equation (6.12), results in

$$f_{\rm nh}^{*}(s) = \frac{\mu}{s+\mu} + \frac{\mu_{\rm r} s}{(s+\mu)(s+\mu+\mu_{\rm r})} = \frac{\mu+\mu_{\rm r}}{s+\mu+\mu_{\rm r}} \tag{6.14}$$

which implies that the new call channel holding time is exponentially distributed with parameter $\mu+\mu_{\rm r}$. Similarly, if the cell residence time t_i is exponentially distributed with parameter η, then the handoff call channel holding time is also exponentially distributed with parameter $\mu+\eta$. In this case, the channel holding time is hyperexponentially distributed. If $\mu_{\rm r}=\eta$, then the channel holding time [see Equation (6.12)] is exponentially distributed with parameter $\mu+\eta$. In fact, since r_1 is the residual life of t_1, from the Residual Life Theorem [21], we have

$$f_{\rm r}^{*}(s) = \frac{\eta\left[1-f^{*}(s)\right]}{s} = \frac{\eta}{s+\eta} = f^{*}(s) \tag{6.15}$$

Hence, the channel holding time is exponentially distributed with parameter $\mu+\eta$ when the cell residence time is exponentially distributed.

Simple results for the conditional distribution for channel holding time when the cell residence time is generally distributed are presented next. Let $f_{\rm cnh}(t)$, $f_{\rm chh}(t)$ and $f_{\rm ch}(t)$ denote the conditional density functions for new call channel holding time, the handoff call channel holding time and the channel holding time, respectively, with Laplace transforms $f_{\rm cnh}^{*}(s)$, $f_{\rm chh}^{*}(s)$ and $f_{\rm ch}^{*}(s)$, and with cumulative distribution functions and $F_{\rm cnh}(t)$, $F_{\rm chh}(t)$ and $F_{\rm ch}(t)$. Let us start with the *conditional distribution* for the handoff call channel holding time. We have

$$\begin{aligned} F_{\rm chh}(h) &= \Pr(t_{\rm hh} \le h \,|\, r_m \le t_m) = \frac{\Pr(r_m \le h, r_m \le t_m)}{\Pr(r_m \le t_m)} = \frac{\int_0^h f_{\rm c}(t) \int_t^\infty f(\tau)\,{\rm d}\tau\,{\rm d}t}{\Pr(r_m \le t_m)} \\ &= \frac{\int_0^h f_{\rm c}(t)\left[1-F(t)\right]{\rm d}t}{\Pr(r_m \le t_m)} \end{aligned} \tag{6.16}$$

Differentiation of both sides gives the conditional density function

$$f_{\rm chh}(h) = \frac{f_{\rm c}(h)\left[1-F(h)\right]}{\Pr(r_m \le t_m)} \tag{6.17}$$

with

$$\begin{aligned} \Pr(r_m \le t_m) &= \int_0^\infty \int_0^t f(t) f_{\rm c}(\tau)\,{\rm d}\tau\,{\rm d}t = \int_0^\infty f(t)\left[1-{\rm e}^{\mu t}\right]{\rm d}t \\ &= 1 - \int_0^\infty f(t){\rm e}^{-\mu t}\,{\rm d}t = 1 - f^{*}(\mu) \end{aligned} \tag{6.18}$$

Using this in Equation (6.16) results in

$$f_{\mathrm{chh}}(h) = \frac{[1 - F(h)]\,\mu \mathrm{e}^{-\mu h}}{1 - f^*(\mu)} \tag{6.19}$$

and

$$f_{\mathrm{chh}}^*(s) = \frac{\mu \int_0^\infty \mathrm{e}^{-(s+\mu)h}\,[1 - F(h)]\,\mathrm{d}h}{1 - f^*(\mu)} = \frac{\mu}{s + \mu}.\frac{1 - f^*(s + \mu)}{1 - f^*(\mu)} \tag{6.20}$$

Similarly, we obtain

$$\begin{aligned} f_{\mathrm{cnh}}(h) &= \frac{[1 - F_{\mathrm{r}}(h)]\,\mu \mathrm{e}^{-\mu h}}{1 - f_{\mathrm{r}}^*(\mu)} \\ f_{\mathrm{cnh}}^*(s) &= \frac{\mu}{s + \mu}.\frac{1 - f_{\mathrm{r}}^*(s + \mu)}{1 - f_{\mathrm{r}}^*(\mu)} \end{aligned} \tag{6.21}$$

The conditional *channel holding time* distribution $f_{\mathrm{ch}}(t)$, $f_{\mathrm{ch}}^*(s)$ is the average of the conditional *new call channel holding time* distribution and *handoff call channel holding time* distribution. In summary, we therefore have

$$\begin{aligned} f_{\mathrm{cnh}}^*(s) &= \frac{\mu}{s + \mu}.\frac{1 - f_{\mathrm{r}}^*(s + \mu)}{1 - f_{\mathrm{r}}^*(\mu)} \\ f_{\mathrm{chh}}^*(s) &= \frac{\mu}{s + \mu}.\frac{1 - f^*(s + \mu)}{1 - f^*(\mu)} \\ f_{\mathrm{ch}}^*(s) &= \frac{\mu}{s + \mu}\left[\frac{\lambda}{\lambda + \lambda_{\mathrm{h}}}.\frac{1 - f_{\mathrm{r}}^*(s + \mu)}{1 - f_{\mathrm{r}}^*(\mu)} + \frac{\lambda_{\mathrm{h}}}{\lambda + \lambda_{\mathrm{h}}}.\frac{1 - f^*(s + \mu)}{1 - f^*(\mu)}\right] \end{aligned} \tag{6.22}$$

If T_{cnh}, T_{chh} and T_{ch} are the expected conditional new call channel holding time, the expected conditional handoff call channel holding time, and the expected conditional channel holding time, respectively, then we have

$$\begin{aligned} T_{\mathrm{cnh}} &= \frac{1}{\mu} + \frac{f_{\mathrm{r}}^{*\prime}(\mu)}{1 - f_{\mathrm{r}}^*(\mu)} \\ T_{\mathrm{chh}} &= \frac{1}{\mu} + \frac{f^{*\prime}(\mu)}{1 - f^*(\mu)} \\ T_{\mathrm{ch}} &= \frac{1}{\mu} + \frac{\lambda}{\lambda + \lambda_{\mathrm{h}}}.\frac{f_{\mathrm{r}}^{*\prime}(\mu)}{1 - f_{\mathrm{r}}^*(\mu)} + \frac{\lambda_{\mathrm{h}}}{\lambda + \lambda_{\mathrm{h}}}\frac{f^{*\prime}(\mu)}{1 - f^*(\mu)} \end{aligned} \tag{6.23}$$

In order to be able to use these results, we need to find the handoff call arrival rate λ_{h}. This parameter depends on the new call arrival rate, the new call blocking probability and the handoff call blocking probability. If p_{bn} and p_{bh} are the new call and handoff call blocking probabilities, respectively, and H is the number of handoffs for a call (its expectation $E[H]$ is also called handoff rate), then using a procedure similar to the one in Fang et al. [12]

gives

$$E[H] = -(1 - p_{bn}) \sum_{P \in \sigma_c} \operatorname{Res}_{s=p} \frac{f_r^*(s)}{s\left[1 - (1 - p_{bh}) f^*(s)\right]} f_c^*(-s) \tag{6.24}$$

where σ_c denotes the set of poles of $f_c^*(-s)$ in the right half of the complex plane and $\text{Res}_{s=p}$ denotes the residue at the pole $s = p$. Since t_c is exponentially distributed with parameter μ, hence $f_c^*(s) = \mu/(s + \mu)$, from the above we obtain

$$E[H] = \frac{(1 - p_{bn}) f_r^*(\mu)}{1 - (1 - p_{bh}) f^*(\mu)} \tag{6.25}$$

Since each unblocked call initiates $E[H]$ handoff calls on average, the handoff call arrival rate can be obtained:

$$\lambda_h = \lambda E[H] = \frac{(1 - p_{bn}) \lambda f_r^*(\mu)}{1 - (1 - p_{bh}) f^*(\mu)} \tag{6.26}$$

As long as $f_r^*(s)$ and $f^*(s)$ are proper rational functions, then the Laplace transforms of distribution functions of all channel holding times (either conditional or unconditional) are all rational functions. To find the corresponding density functions, we only need to find the inverse Laplace transforms. This can be accomplished by using the partial fractional expansion [32].

As an illustration, suppose that $g(s)$ is a proper rational function with poles $p_1, p_2, \ldots, p_k$ with multiplicities $n_1, n_2, \ldots, n_k$. Then $g(s)$ can be expanded as

$$g(s) = \sum_{i=1}^{k} \sum_{j=0}^{n_i} A_{ij} \frac{s^j}{(s + p_i)^{n_i}}$$

where the constants can be found easily by the formula

$$A_{ij} = \frac{d^j}{ds^j} \left[(s + p_i)^{n_i} g(s)\right] \big|_{s=0}, \; j = 0, 1, \ldots, i, \quad i = 1, 2, \ldots, k$$

The inverse Laplace transform L^{-1} gives

$$L^{-1}\left[s^j f(s)\right] = \frac{d^j}{dt^j} \left\{L^{-1}[f(s)]\right\}, L^{-1}\left[1/(s + \beta)^h\right] = \frac{t^h}{h!} e^{-\beta t}$$

and

$$L^{-1}[g(s)] = \sum_{i=1}^{k} \sum_{j=0}^{m_i} A_{ij} \frac{d^j}{dt^j} \left(\frac{t^j}{j!} e^{-p_i t}\right)$$

At this stage it is useful to remember that the inverse Laplace transform of a rational function is in fact the impulse response of a linear system in which the rational function is the system transfer function of the resulting linear system [32] and the cumulative distribution function is the step response of the linear system. In Matlab, the commands *impulse* and *step* can be used to find the density function and the distribution function. When applying the hyper-Erlang distribution model for cell residence time, we can in fact reduce the computation further. Substituting $f^*(s)$ in Equation (6.10) with $f_{he}^*(s)$ given by Equation (6.1), gives

$$f_{\text{hh}}^*(s) = \sum_{i=1}^{M} \alpha_i \left[\frac{\mu}{s+\mu} + \frac{s}{s+\mu} \left(\frac{m_i \eta_i}{s + m_i \eta_i} \right)^{m_i} \right] = \sum_{i=1}^{M} \alpha_i f_{\text{e}}^*(s; m_i, \eta_i) \tag{6.27}$$

where $f_{\text{e}}^*(s; m_i, \eta_i)$ corresponds to the handoff call channel holding time when the cell residence time is Erlang distributed with parameters (m_i, η_i). Thus, the problem reduces to finding the algorithm for computing the channel holding time for the case when the cell residence time is Erlang distributed.

Performance examples: if the cell residence time is Erlang distributed we have

$$f(t) = \beta^m t^{m-1} \text{e}^{-\beta t}/(m-1)!, \quad f^*(s) = [\beta/(s+\beta)]^m$$

where $\beta = m\eta$ is the scale parameter and m is the shape parameter. The mean of this Erlang distribution is η and its variance is $1/(m\eta^2)$. When the mean η is fixed, varying the value m is equivalent to varying the variance and larger m means smaller variance and lesser spread of the cell residence time.

The handoff call channel holding time probability density functions with different variance of cell residence time distributed according to Erlang distribution with the same mean are shown in Figure 6.3. It can be seen that, when the cell residence time becomes less

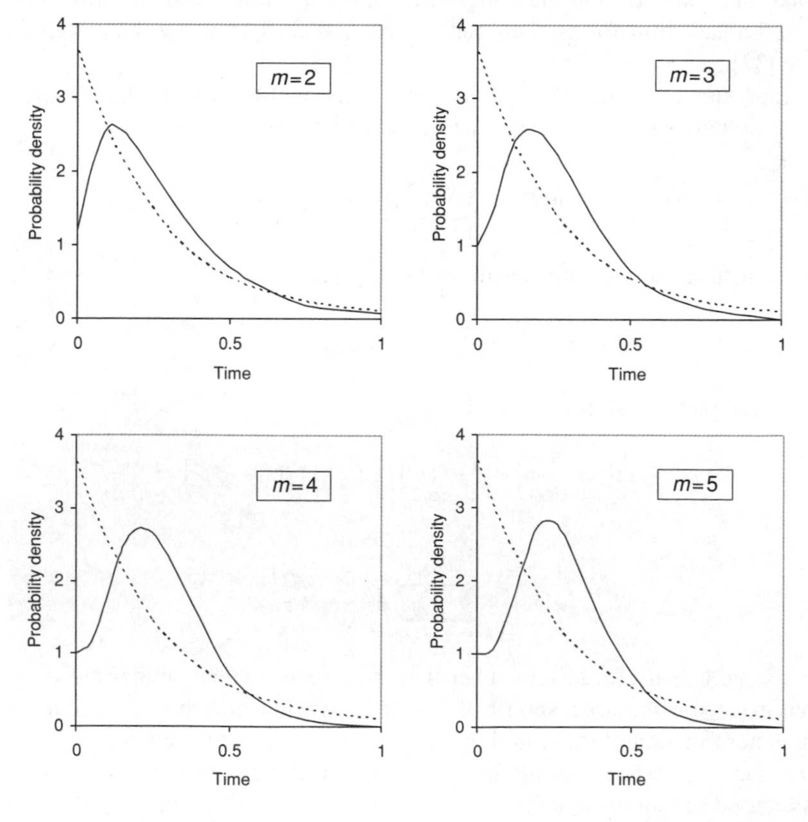

Figure 6.3 Probability density function of handoff call channel holding time (solid line) and its exponential fitting (dashed line) when cell residence time is Erlang distributed with parameter (m, η).

spread, the handoff call channel holding time shows severe mismatch to the exponential distribution.

In the simple case when the cell residence time is hyper-Erlang distributed with two terms,

$$f^*(s) = \alpha_1\left(\frac{m_1\eta}{s+m_1\eta}\right)^{m_1} + \alpha_2\left(\frac{m_2\eta}{s+m_2\eta}\right)^{m_2}$$

$$f^*_{\text{hh}}(s) = \alpha_1\left\{\frac{\mu}{s+\mu} + \frac{s}{s+\mu}\cdot\left(\frac{m_1\eta}{s+m_1\eta}\right)^{m_1}\right\} + \alpha_2\left\{\frac{\mu}{s+\mu} + \frac{s}{s+\mu}\cdot\left(\frac{m_2\eta}{s+m_2\eta}\right)^{m_2}\right\}$$

the results are shown in Figure 6.4. When m_1 and m_2 have different values, the variances of cell residence time are different and the handoff call channel holding time is no longer exponentially distributed.

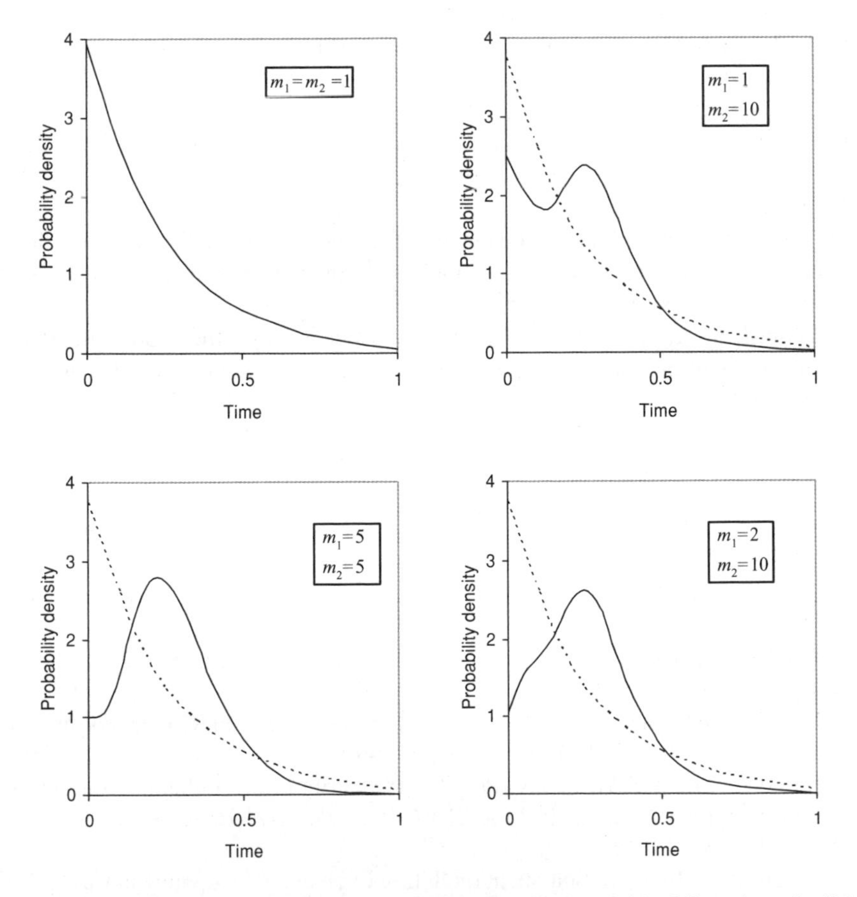

Figure 6.4 Probability density function of handoff call channel holding time (solid line) and its exponential fitting (dashed line) when cell residence time is hyper-Erlang distributed with parameter (m_1, m_2, η).

REFERENCES

[1] D. Hong and S.S. Rappaport, Traffic model and performance analysis for cellular mobile radio telephone systems with prioritized and nonprioritized handoff procedures, *IEEE Trans. Vehicular Technol.*, vol. 35, no. 3, 1986, pp. 77–92.

[2] Y. Fang and I. Chlamtac, Teletraffic analysis and mobility modeling of PCS networks, *IEEE Trans. Commun.*, vol. 47, no. 7, 1999, pp. 1062–1072.

[3] F.P. Kelly, *Reversibility and Stochastic Networks*. Wiley: New York, 1979.

[4] J.G. Proakis, *Digital Communications*, 3rd edn. Prentice-Hall: Englewood Cliffs, NJ, 1995.

[5] V.A. Bolotin, Modeling call holding time distributions for CCS network design and performance analysis, *IEEE J. Select. Areas Commun.*, vol. 12, no. 3, 1994, pp. 433–438.

[6] F. Barcelo and J. Jordan, Channel holding time distribution in cellular telephony, in *Proc. 9th Int. Conf. Wireless Commun. (Wireless'97)*, Alta, Canada, July 9–11, 1997, vol. 1, pp. 125–134.

[7] E. Chlebus and W. Ludwin, Is handoff traffic really Poissonian?, *IEEE ICUPC'95*, Tokyo, 6–10, November, 1995, pp. 348–353.

[8] D.C. Cox, Wireless personal communications: What is it?, *IEEE Personal Commun. Mag.*, pp. 20–35, 1995.

[9] D.C. Cox, *Renewal Theory*. Wiley: New York, 1962.

[10] E. Del Re, R. Fantacci and G. Giambene, Handover and dynamic channel allocation techniques in mobile cellular networks, *IEEE Trans. Vehicular Technol.*, vol. 44, no. 2, 1995, pp. 229–237.

[11] E. Del Re, R. Fantacci and G. Giambene, Efficient dynamic channel allocation techniques with handover queueing for mobile satellite networks, *IEEE J. Selected Areas Commun.*, vol. 13, no. 2, 1995, pp. 397–405.

[12] Y. Fang, I. Chlamtac and Y.B. Lin, Channel occupancy times and handoff rate for mobile computing and PCS networks, *IEEE Trans. Comput.*, vol. 47, no. 6, 1998, pp. 679–692.

[13] Y. Fang, I. Chlamtac and Y.B. Lin, Modeling PCS networks under general call holding times and cell residence time distributions, *IEEE Trans. Networking*, vol. 5, 1997, pp. 893–906.

[14] Y. Fang, I. Chlamtac and Y.B. Lin, Call performance for a PCS network, *IEEE J. Select. Areas Commun.*, vol. 15, no. 7, 1997, pp. 1568–1581.

[15] E. Gelenbe and G. Pujolle, *Introduction to Queueing Networks*. Wiley: New York, 1987.

[16] R.A. Guerin, Channel occupancy time distribution in a cellular radio system, *IEEE Trans. Vehicular Tech.*, vol. 35, no. 3, 1987, pp. 89–99.

[17] B. Jabbari, Teletraffic aspects of evolving and next-generation wireless communication networks, *IEEE Commun. Mag.*, 1996, pp. 4–9.

[18] C. Jedrzycki and V.C.M. Leung, Probability distributions of channel holding time in cellular telephony systems, in *Proc. IEEE Vehicular Technology Conf.*, Atlanta, GA, May 1996, pp. 247–251.

[19] J. Jordan and F. Barcelo, Statistical modeling of channel occupancy in trunked PAMR systems, in *Proc. 15th Int. Teletraffic Conf. (ITC'15)*, V. Ramaswami and P.E. Wirth, Eds. Elsevier Science: Amsterdam, 1997, pp. 1169–1178.

[20] F.P. Kelly, Loss networks, *The Annals of Applied Probability*, vol. 1, no. 3, 1991, pp. 319–378.
[21] L. Kleinrock, *Queueing Systems: Theory*, vol. 1. Wiley: New York, 1975.
[22] W.R. LePage, *Complex Variables and the Laplace Transform for Engineers*. Dover: New York, 1980.
[23] Y.B. Lin, S. Mohan and A. Noerpel, Queueing priority channel assignment strategies for handoff and initial access for a PCS network, *IEEE Trans. Vehicular Technol.*, vol. 43, no. 3, 1994, pp. 704–712.
[24] S. Nanda, Teletraffic models for urban and suburban microcells: cell sizes and handoff rates, *IEEE Trans. Vehicular Technol.*, vol. 42, no. 4, 1993, pp. 673–682.
[25] A.R. Noerpel, Y.B. Lin, and H. Sherry, PACS: personal access communications system-A tutorial, *IEEE Personal Commun.*, vol. 3, no. 3, 1996, pp. 32–43.
[26] P. Orlik and S.S. Rappaport, A model for teletraffic performance and channel holding time characterization in wireless cellular communication with general session and dwell time distributions, *IEEE J. Select. Areas Commun.*, vol. 16, no. 5, 1998, pp. 788–803.
[27] P. Orlik and S.S. Rappaport, A model for teletraffic performance and channel holding time characterization in wireless cellular communication, in *Proc. Int. Conf. Universal Personal Commun. (ICUPC'97)*, San Diego, CA, October 1997, pp. 671–675.
[28] S. Tekinay and B. Jabbari, A measurement-based prioritization scheme for handovers in mobile cellular networks, *IEEE J. Select. Areas Commun.*, vol. 10, no. 8, 1992, pp. 1343–1350.
[29] C.H. Yoon and C.K. Un, Performance of personal portable radio telephone systems with and without guard channels, *IEEE J. Select. Areas Commun.*, vol. 11, no. 6, 1993, pp. 911–917.
[30] T.S. Yum and K.L. Yeung, Blocking and handoff performance analysis of directed retry in cellular mobile systems, *IEEE Trans. Vehicular Technol.*, vol. 44, no. 3, 1995, pp. 645–650.
[31] M.M. Zonoozi and P. Dassanayake, User mobility modeling and characterization of mobility patterns, *IEEE J. Select. Areas Commun.*, vol. 15, no. 7, 1997, pp. 1239–1252.
[32] T. Kailath, *Linear Systems*. Prentice-Hall: Englewood Cliffs, NJ, 1980.

7

Adaptive Network Layer

7.1 GRAPHS AND ROUTING PROTOCOLS

The most important function of the network layer is routing. A tool used in the design and analysis of routing protocols is graph theory. Networks can be represented by graphs where mobile nodes are vertices and communication links are edges. Routing protocols often use shortest path algorithms. In this section we provide a simple review of the most important principles in the field which provides a background to study the routing algorithms.

7.1.1 Elementary concepts

A graph $G(V, E)$ is two sets of objects, vertices (or nodes), set V, and edges, set E. A graph is represented by dots or circles (vertices) interconnected by lines (edges). The magnitude of graph G is characterized by number of vertices $|V|$ (called the order of G) and number of edges $|E|$, size G. The running times of algorithms are measured in terms of the order and size.

7.1.2 Directed graph

An edge $e \in E$ of a directed graph is represented as an ordered pair (u, v), where $u, v \in V$. Here u is the initial vertex and v is the terminal vertex. Also assume here that $u \neq v$. An example with

$$V = \{1, 2, 3, 4, 5, 6\}, |V| = 6$$
$$E = \{(1, 2), (2, 3), (2, 4), (4, 1), (4, 2), (4, 5), (4, 6)\}, |E| = 7$$

is shown in Figure 7.1.

Advanced Wireless Networks: 4G Technologies Savo G. Glisic
© 2006 John Wiley & Sons, Ltd.

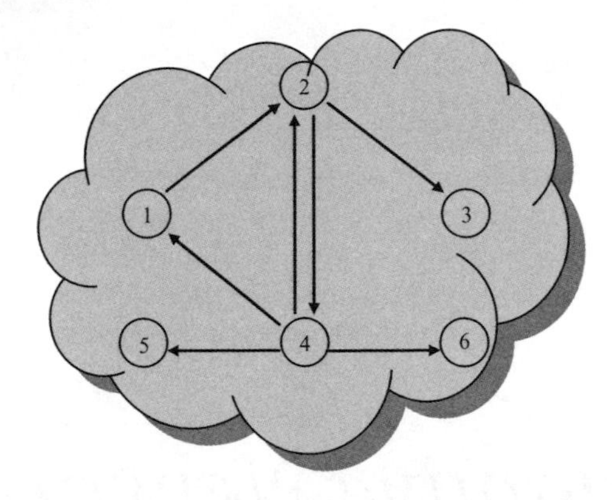

Figure 7.1 Directed graph.

7.1.3 Undirected graph

An edge $e \in E$ of an undirected graph is represented as an unordered pair $(u, v) = (v, u)$, where $u, v \in V$. Also assume that $u \neq v$. An example with

$$V = \{1, 2, 3, 4, 5, 6\}, |V| = 6$$
$$E = \{(1, 2), (2, 3), (2, 4), (4, 1)\}, (4, 5)(4, 6)|E| = 6$$

is shown in Figure 7.2.

7.1.4 Degree of a vertex

Degree of a vertex in an undirected graph is the number of edges incident on it. In a directed graph, the *out degree* of a vertex is the number of edges leaving it and the *in degree* is the number of edges entering it. In Figure 7.2 the *degree* of vertex 2 is 3. In Figure 7.1 the *in degree* of vertex 2 is 2 and the *in degree* of vertex 4 is 1.

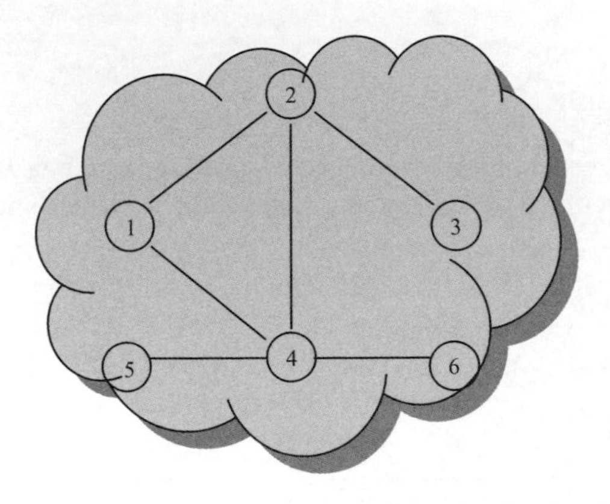

Figure 7.2 Undirected graph.

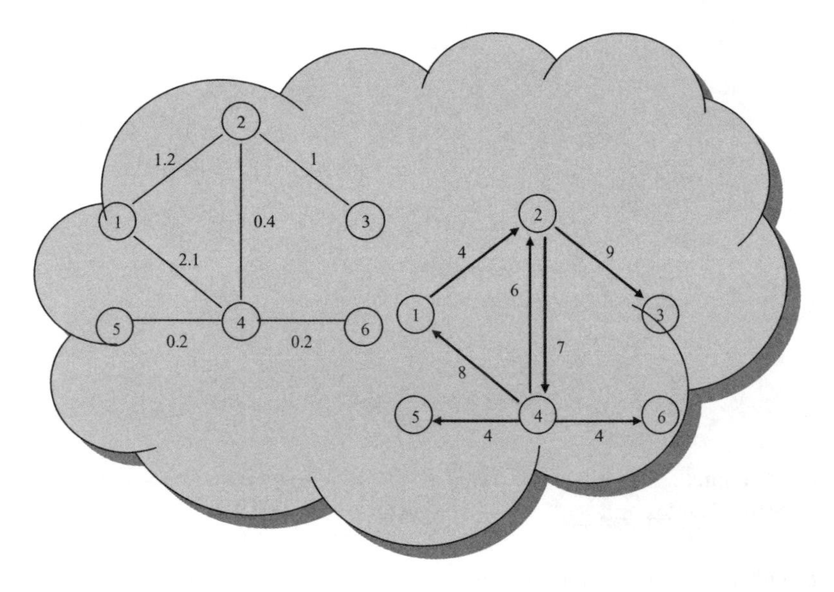

Figure 7.3 Weighted graphs.

7.1.5 Weighted graph

In a weighted graph each edge has an associated weight, usually given by a weight function $w : E \rightarrow R$. Weighted graphs from Figures 7.1 and 7.2 are shown in Figure 7.3. In the analysis of the routing problems, these weights represent the cost of using the link. Most of the time this cost would be delay that a packet would experience if using that link.

7.1.6 Walks and paths

A walk is a sequence of nodes $(v1, v2, \ldots, vL)$ such that $\{(v1, v2), (v2, v3), \ldots, (vL - 1, vL)\} \subseteq E$, e.g. $(V2, V3, V6, V5, V3)$ in Figure 7.4. A *simple path* is a walk with no repeated nodes, e.g. $(V1, V4, V5, V6, V3)$. A *cycle* is a walk $(v1, v2, \ldots, vL)$ where

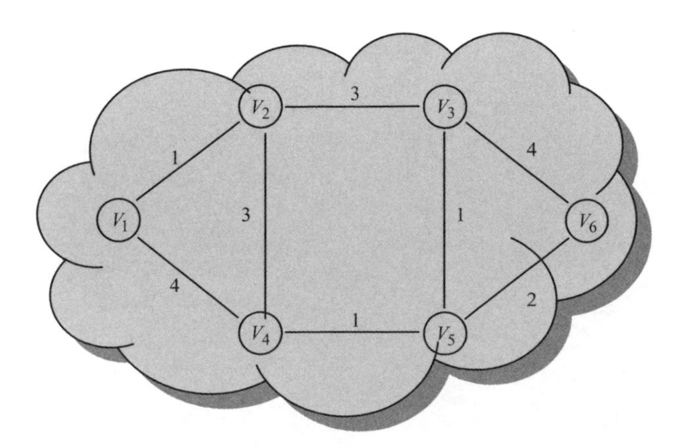

Figure 7.4 Illustration of a walk.

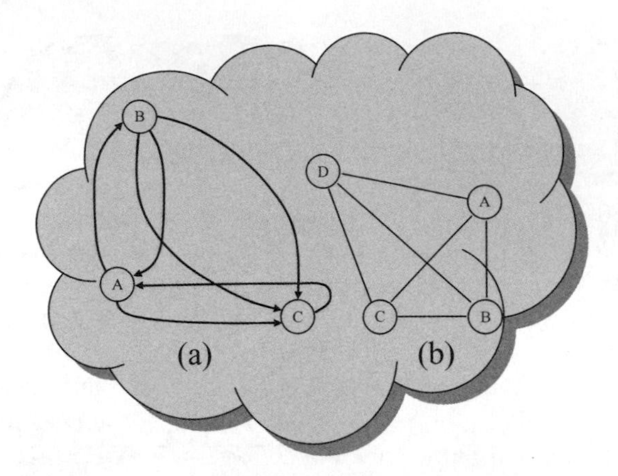

Figure 7.5 Complete graphs: (a) [V nodes and $V(V-1)$ edges] 3 nodes and 3×2 edges; (b) [V nodes and $V(V-1)/2$ edges] 4 nodes and $4 \times 3/2$ edges.

$v1 = vL$ with no other nodes repeated and $L > 3$, e.g. $(V1, V2, V3, V5, V4, V1)$. A graph is called *cyclic* if it contains a cycle; otherwise it is called *acyclic*. A *complete graph* is an undirected/directed graph in which every pair of vertices is *adjacent*. An example is given in Figure 7.5. If (u, v) is an edge in a graph G, we say that vertex v is *adjacent* to vertex u.

7.1.7 Connected graphs

An undirected graph is connected if you can get from any node to any other by following a sequence of edges or any two nodes are connected by a path, as shown in Figure 7.6. A directed graph is *strongly connected* if there is a directed path from any node to any other node. A graph is *sparse* if $|E| \approx |V|$. A graph is *dense* if $|E| \approx |V|^2$.

A *bipartite graph* is an undirected graph $G = (V, E)$ in which V can be partitioned into two sets, $V1$ and $V2$, such that $(u, v) \in E$ implies either $u \in V1$ and $v \in V2$ or $v \in V1$ and $u \in V2$, see Figure 7.7.

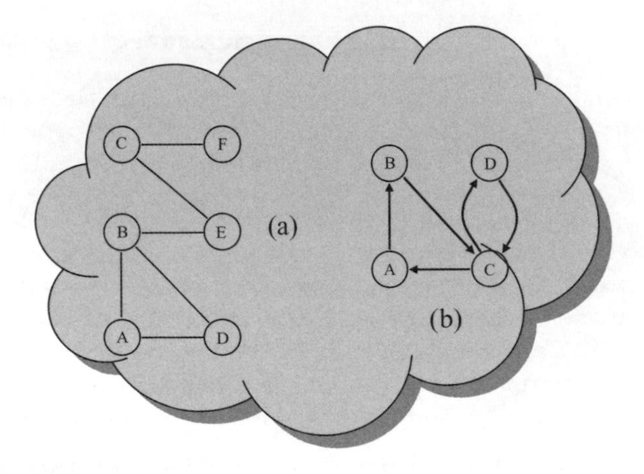

Figure 7.6 Connected graphs.

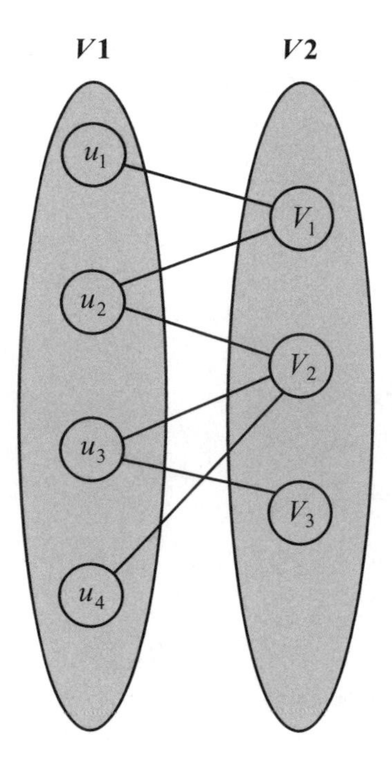

Figure 7.7 Bipartite graph.

7.1.8 Trees

Let $G = (V, E)$ *be an undirected graph. The following statements are equivalent:*

(1) G is a tree;

(2) any two vertices in G are connected by unique simple path;

(3) G is connected, but if any edge is removed from E, the resulting graph is disconnected;

(4) G is connected, and $|E| = |V| - 1$;

(5) G is acyclic, and $|E| = |V| - 1$;

(6) G is acyclic, but if any edge is added to E, the resulting graph contains a cycle.

For an illustration see Figure 7.8.

7.1.9 Spanning tree

A tree (T) is said to span $G = (V, E)$ if $T = (V, E')$ and $E' \subseteq E$. For the graph shown Figure 7.4, two possible spanning trees are shown in Figure 7.9. Given connected graph G with real-valued edge weights ce, a minimum spanning tree (MST) is a spanning tree of G whose sum of edge weights is minimized.

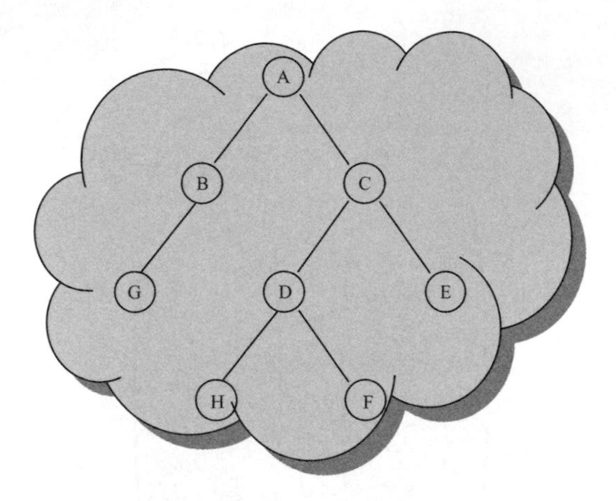

Figure 7.8 Tree.

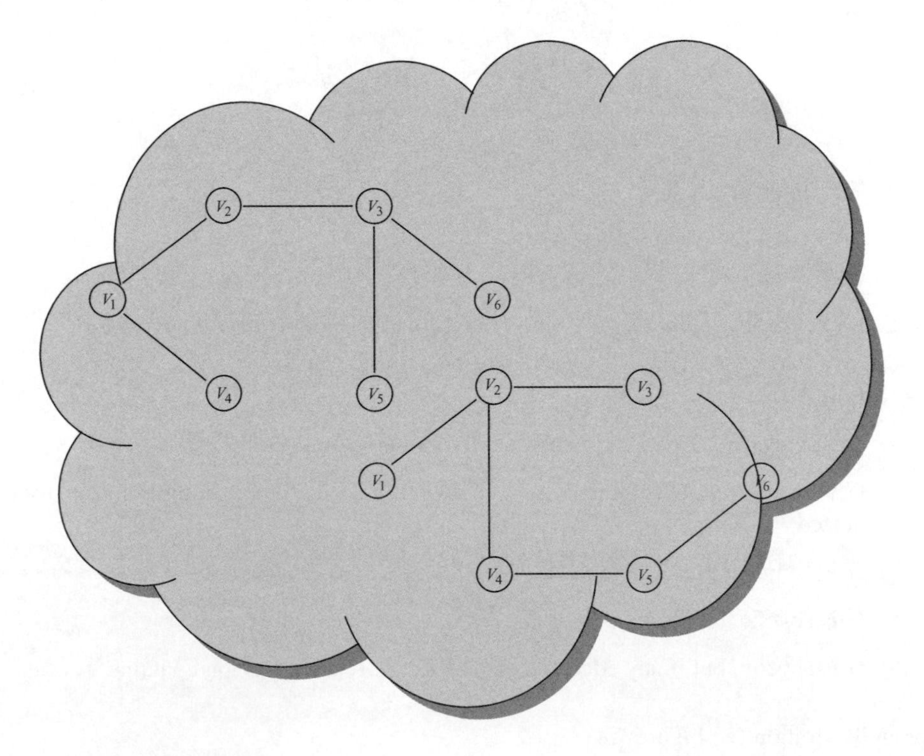

Figure 7.9 Spanning trees.

7.1.10 MST computation

7.1.10.1 Prim's algorithm

Select an arbitrary node as the initial tree (T). Augment T in an iterative fashion by adding the outgoing edge (u, v), (i.e. $u \in T$ and $v \in G - T$) with minimum cost (i.e. weight).

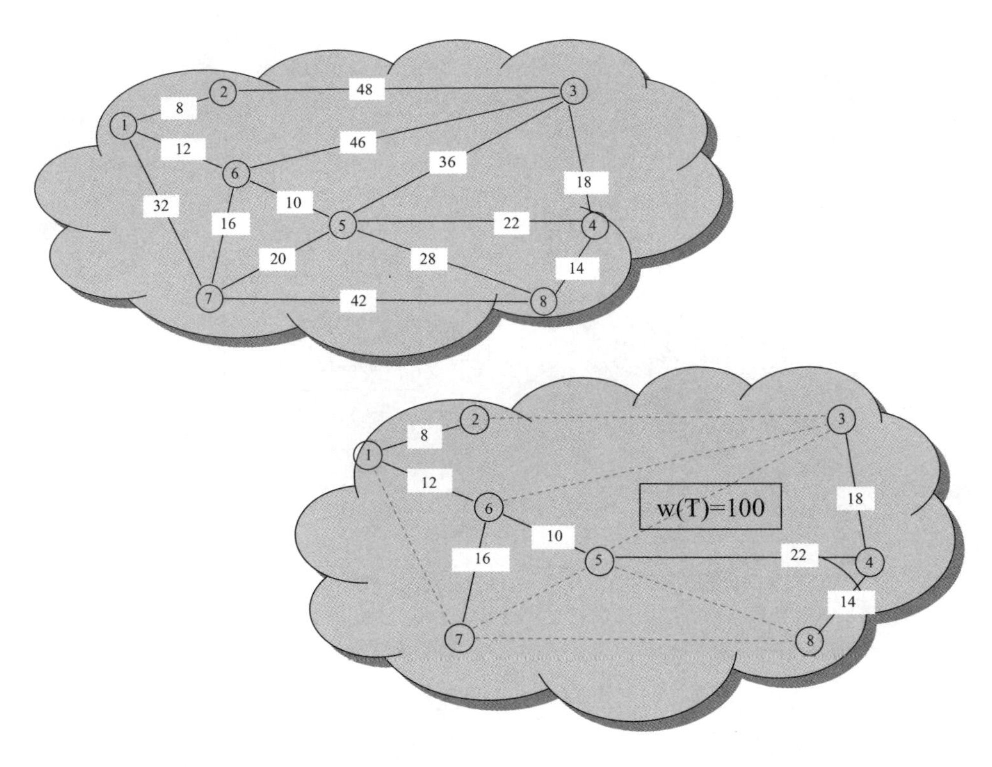

Figure 7.10 Minimum spanning tree.

The algorithm stops after $|V| - 1$ iterations. Computational complexity $= O(|V|2)$. An illustration of the algorithm is given in Figure 7.11.

7.1.10.2 Kruskal's algorithm

Select the edge $e \in E$ of minimum weight $\rightarrow E' = \{e\}$. Continue to add the edge $e \in E - E'$ of minimum weight that, when added to E', does not form a cycle. Computational complexity $= O(|E|x\log|E|)$. An illustration of the algorithm is given in Figure 7.12.

7.1.10.3 Distributed algorithms

For these algorithms each node does not need complete knowledge of the topology. The MST is created in a distributed manner. The algorithm starts with one or more fragments consisting of single nodes. Each fragment selects its minimum weight outgoing edge and, using control messaging fragments, coordinates to merge with a neighboring fragment over its minimum weight outgoing edge.The algorithm can produce an MST in $O(|V|x|V|)$ time provided that the edge weights are unique. If these weights are not unique, the algorithm still works by using the nodes IDs to break ties between edges with equal weight.The algorithm requires $O(|V|x\log|V|) + |E|)$ message overhead. An illustration of the distributed algorithm is given in Figure 7.13.

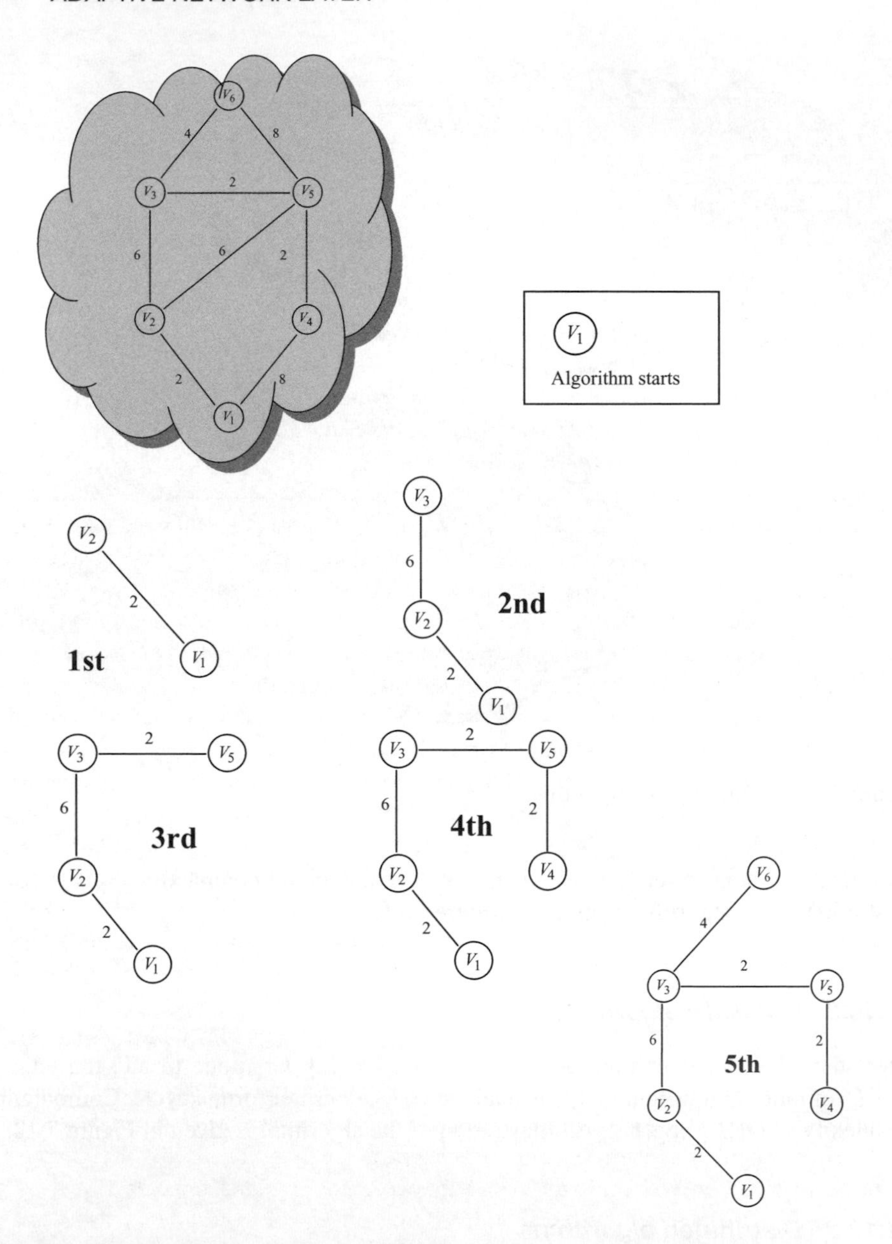

Figure 7.11 MST solution via Prim's algorithm.

7.1.11 Shortest path spanning tree

The shortest path spanning tree (SPST), T, is a spanning tree rooted at a particular node such that the $|V| - 1$ minimum weight paths from that node to each of the other network nodes are contained in T. An example of the shortest path spanning tree is shown in Figure 7.14. Note that the SPST is not the same as the MST.

SPST trees are used for unicast (one to one) and multicast (one to several) routing.

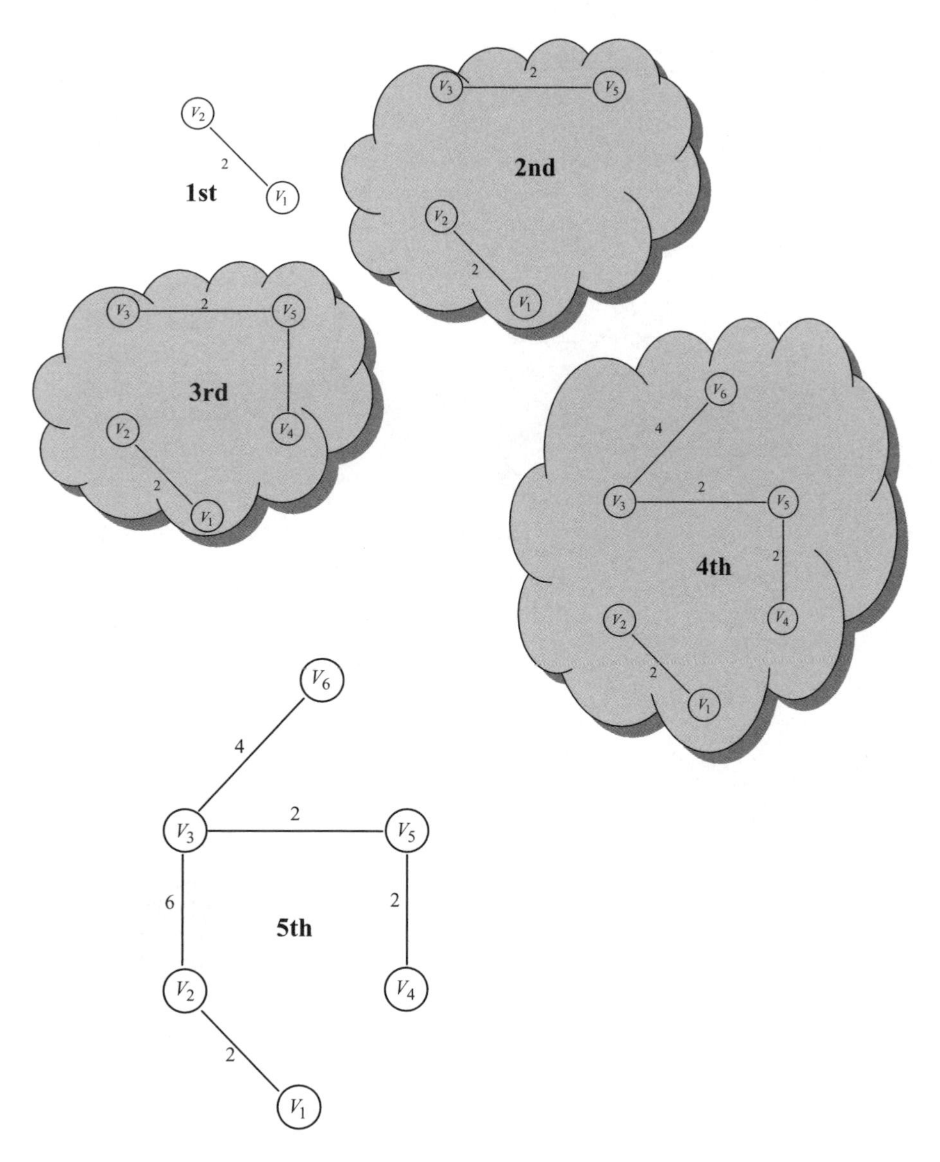

Figure 7.12 MST solution via Kruskal's algorithm.

7.1.11.1 Shortest path algorithms

Let us assume nonnegative edge weights. Given a weighted graph (G, W) and a node s (*source*), a shortest path tree rooted at s is a tree T such that, for any other node $v \in G$, the path between s and v in T is a shortest path between the nodes. Examples of the algorithms that compute these shortest path trees are Dijkstra and Bellman–Ford algorithms as well as algorithms that find the shortest path between all pairs of nodes, e.g. Floyd–Marshall.

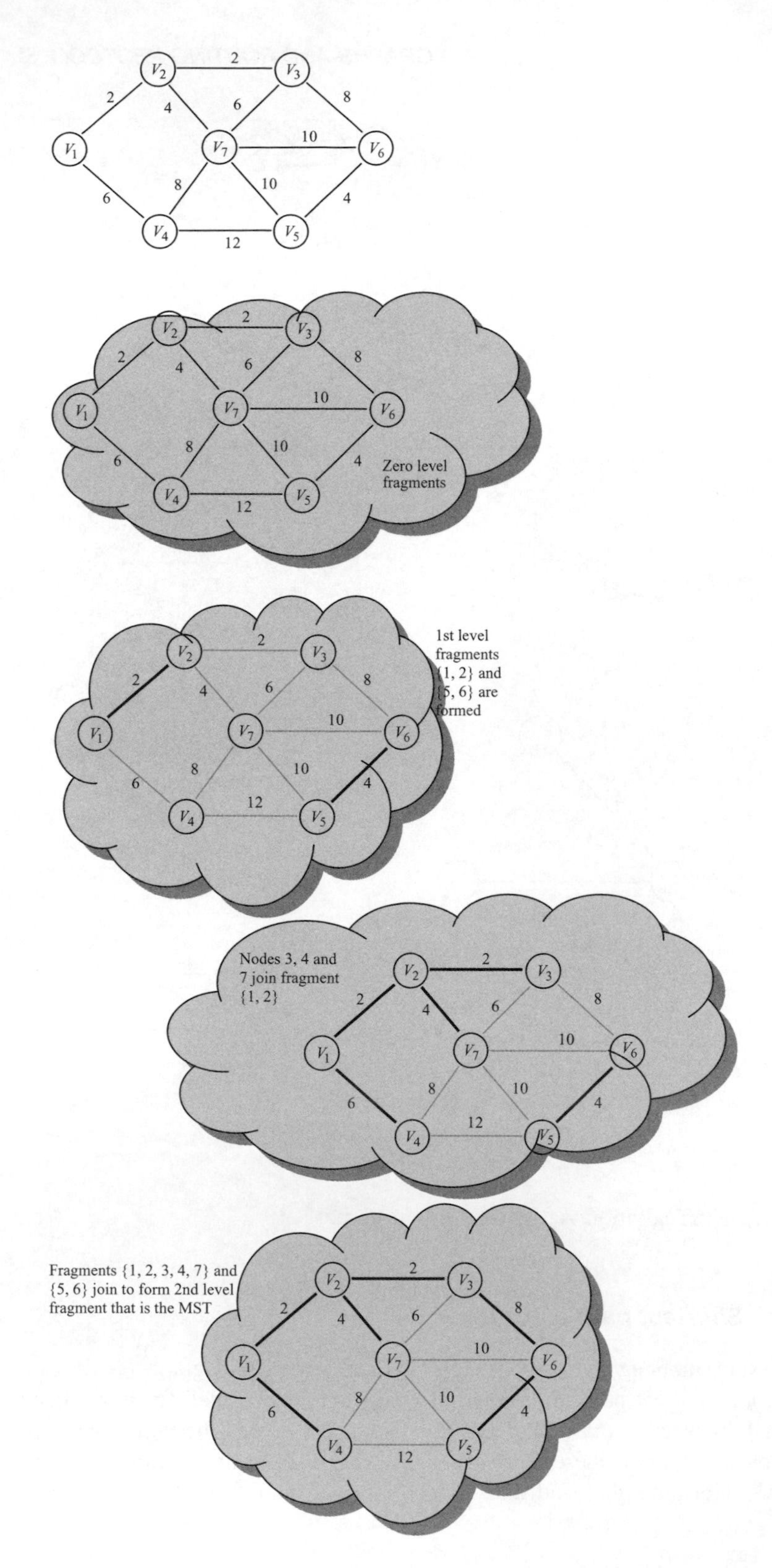

Figure 7.13 Example of the distributed algorithm.

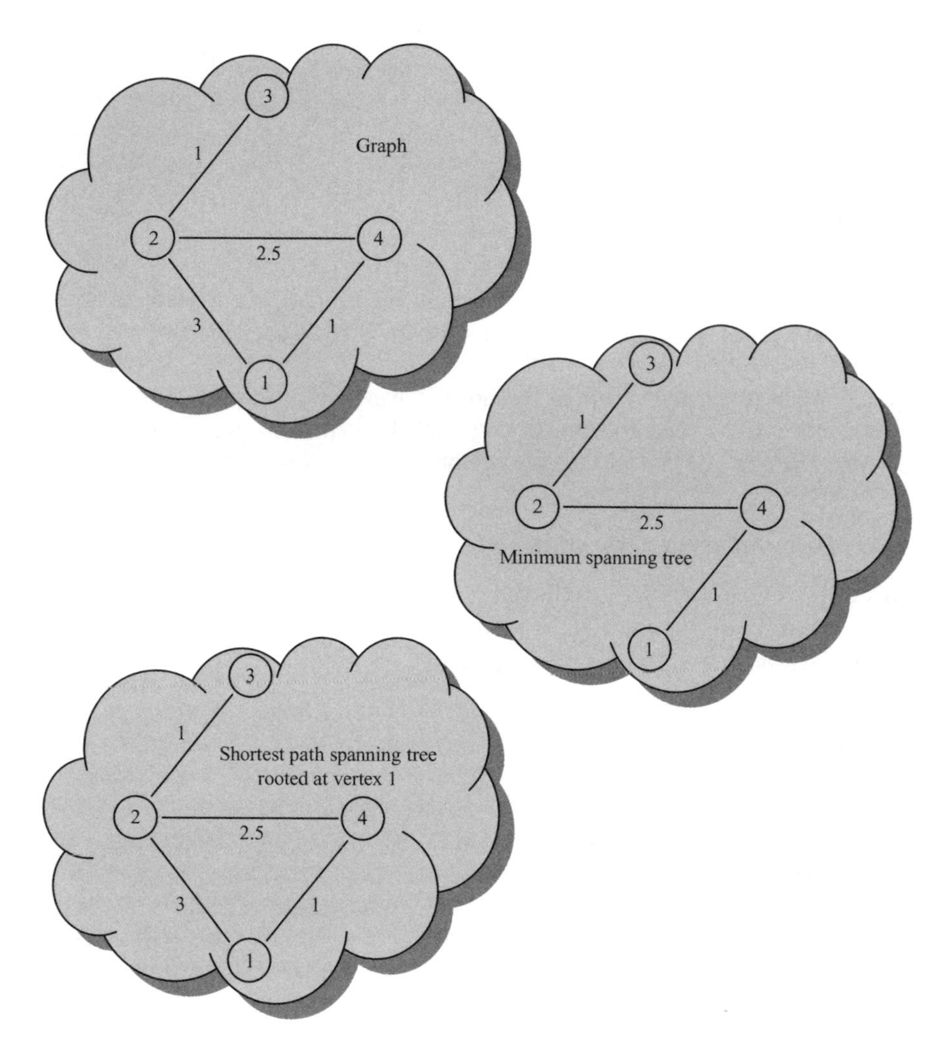

Figure 7.14 Examples of minimum spanning tree and shortest path spanning tree.

7.1.11.2 Dijkstra algorithm

For the source node s, the algorithm is described with the following steps:

$V' = \{s\}; U = V - \{s\};$
$E' = \phi;$
For $v \in U$ do
  $D_v = w(s, v);$
  $P_v = s;$
EndFor
While $U \neq \phi$ do
  Find $v \in U$ such that D_v is minimal;
  $V' = V' \cup \{v\}; U = U - \{v\};$
  $E' = E' \cup (P_v, v);$

For $x \in U$ do
  If $D_v + w(v, x) < D_x$ then
    $D_x = D_v + w(v, x)$;
    $P_x = v$;
EndIf
EndFor
EndWhile

An example of the Dijkstra algorithm is given in Figure 7.15. It is assumed that $V1$ is s and D_v is the distance from node s to node v. If there is no edge connecting two nodes, x and $y \rightarrow w(x, y) = \infty$.

The algorithm terminates when all the nodes have been processed and their shortest distance to node 1 has been computed. Note that the tree computed is not a minimum weight spanning tree. A MST for the given graph is given in Figure 7.16.

7.1.11.3 Bellman–Ford algorithm

Find the shortest walk from a source node s to an arbitrary destination node v subject to the constraints that the walk consists of at most h hops and goes through node v only once. The algorithm is described by the following steps:

$D_v^{-1} = \infty \forall v \in V$;
$D_s^0 = 0$ and $D_v^0 = \infty \forall\, v \neq s, v \in V$;
$h = 0$;
Until ($D_v^h = D_v^{h-1} \forall v \in V$) or ($h = |V|$) do
  $h = h + 1$;
  For $v \in V$ do
    $D_v^{h+1} = \min\{D_u^h + w(u, v)\}$u $\in V$;
  EndFor
  EndUntil

An illustration for *Bellman–Ford Algorithm* is given in Figure 7.17(a).

7.1.11.4 Floyd–Warshall algorithm

The Floyd–Warshall algorithm finds the shortest path between all ordered pairs of nodes (s, v), {s, v} v $\in$ V. Each iteration yields the path with the shortest weight between all pair of nodes under the constraint that only nodes $\{1, 2, \ldots n\}$, $n \in |V|$ can be used as intermediary nodes on the computed paths. The algorithm is defined by the following steps:

$D = W$; (W is the matrix representation of the edge weights)

For $u = 1$ to $|V|$do
For $s = 1$ to$|V|$do
For $v = 1$ to$|V|$do
$D_{s,v} = \min\{D_{s,v}, D_{s,u} + W_{u,v}\}$
EndFor
EndFor
EndFor

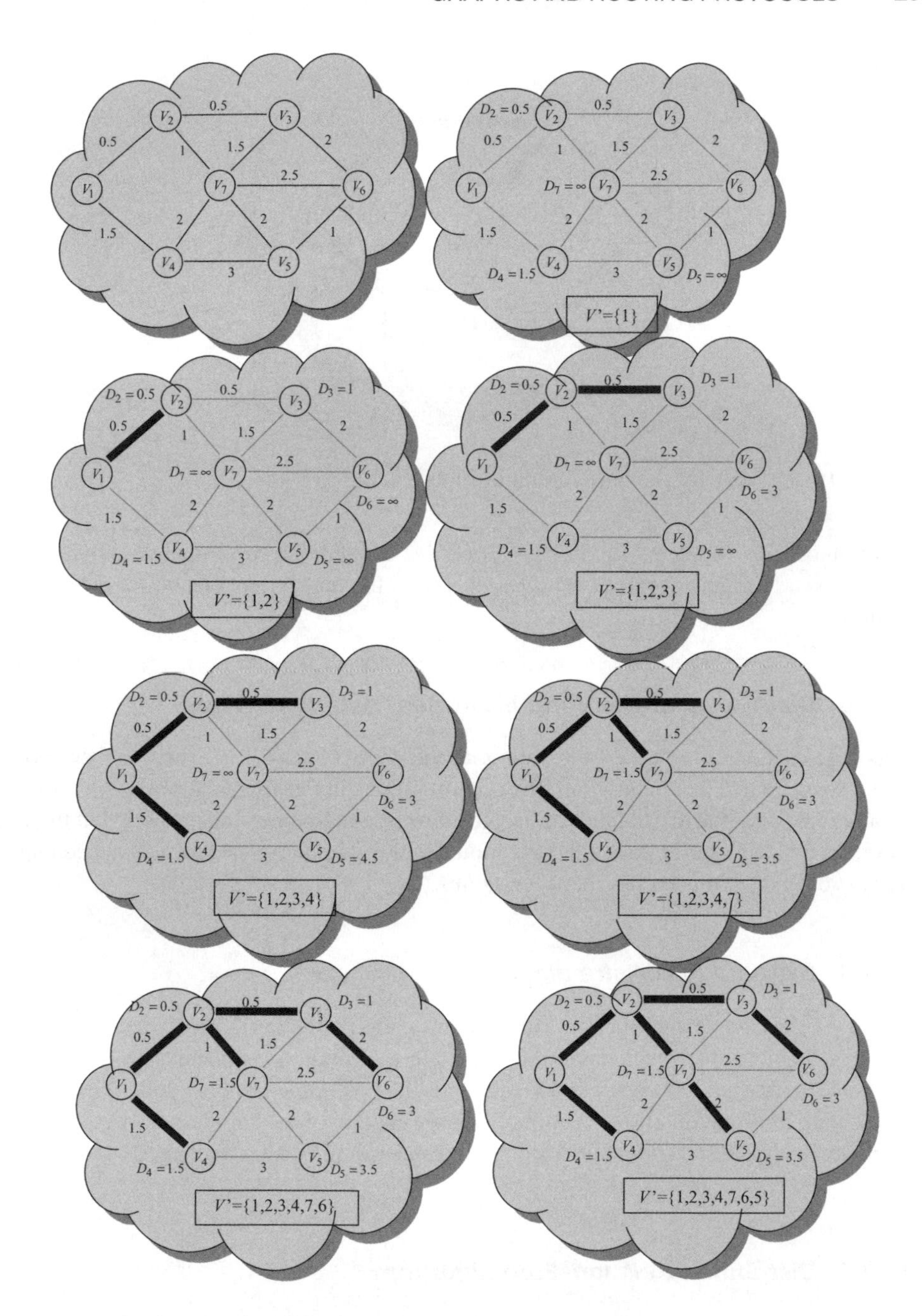

Figure 7.15 Example of Dijkstra algorithm.

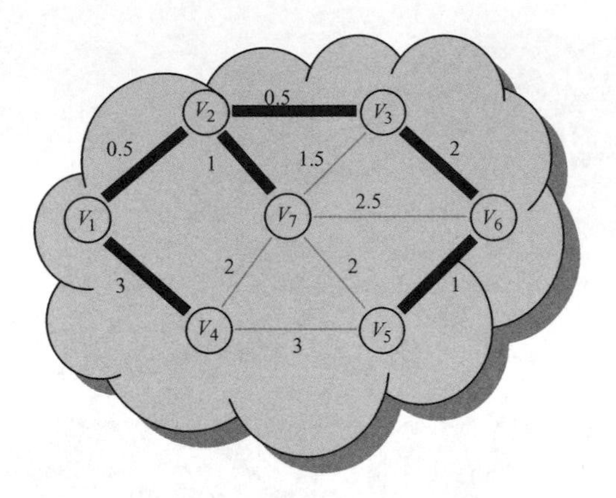

Figure 7.16 An MST for the basic graph in Figure 7.14.

The algorithm completes in O(|V|3) time. An example of the Floyd–Warshall algorithm is given in Figure 7.17(b) with $D = W$ (W is the matrix representation of the edge weights).

7.1.11.5 Distributed asynchronous shortest path algorithms

In this case each node computes the path with the shortest weight to every network node. There is no centralized computation. The control messaging is also required for distributed computation, as for the distributed MST algorithm. Asynchronous means here that there is no requirement for inter-node synchronization for the computation performed at each node or for the exchange of messages between nodes.

7.1.11.6 Distributed Dijkstra algorithm

There is no need to change the algorithm. Each node floods periodically a control message throughout the network containing link state information. Transmission overhead is $O(|V|x|E|)$. The entire topology knowledge must be maintained at each node. Flooding of the link state information allows for timely dissemination of the topology as perceived by each node. Each node has typically accurate information to be able to compute the shortest paths.

7.1.11.7 Distributed Bellman–Ford algorithm

Assume G contains only cycles of *nonnegative* weight. If $(u, v) \in E$ then so does (v, u). The updated equation is

$$D_{s,v} = \min_{u \in N(s)} \{w(s, u) + D_{u,v}\}, \forall v \in V - \{s\}$$

where $N(s) =$ neighbors of $s \rightarrow \forall u \in N(s)$, $(s, u) \in E$. Each node only needs to know the weights of the edges that are incident to it, the identity of all the network nodes and

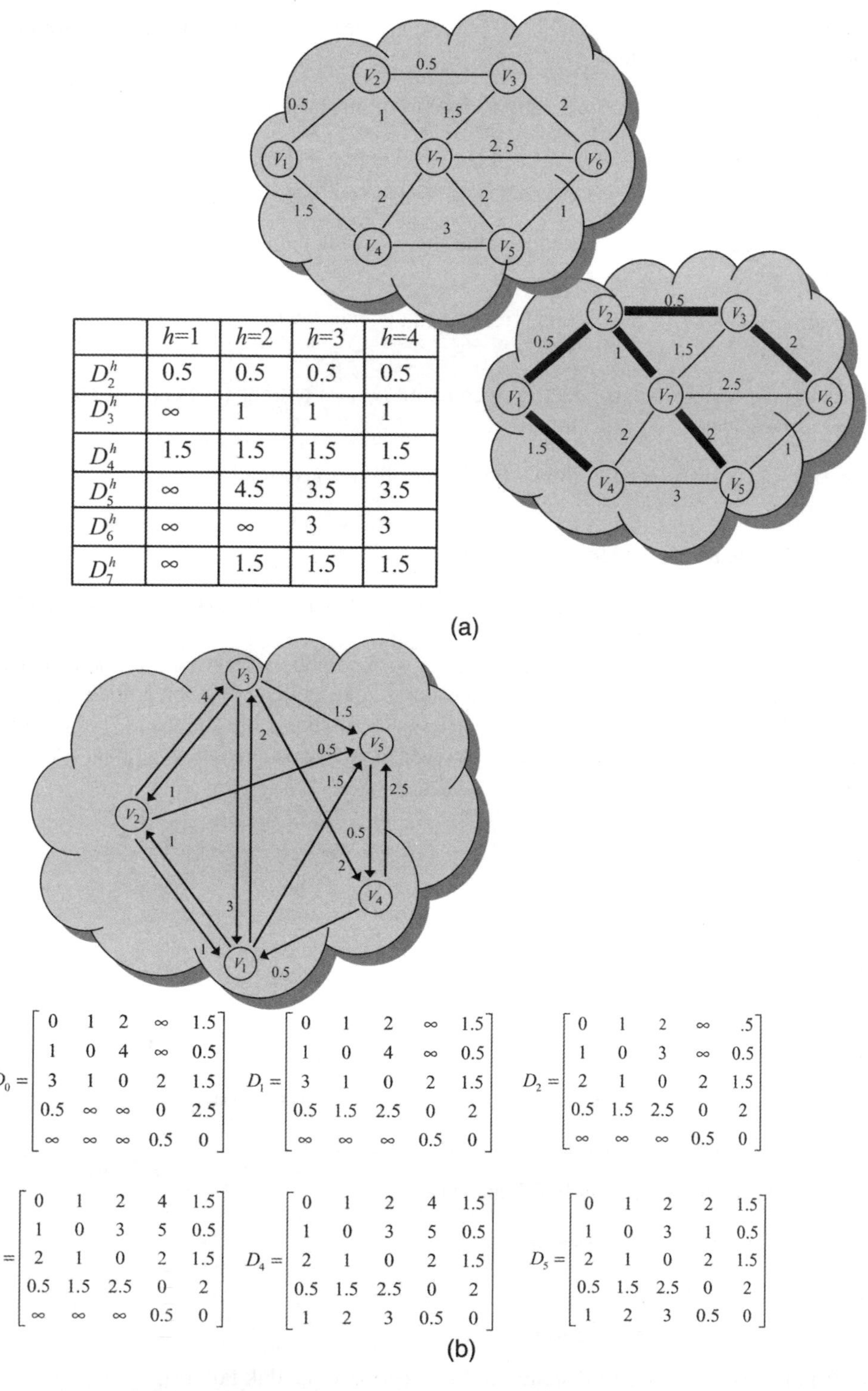

	h=1	h=2	h=3	h=4
D_2^h	0.5	0.5	0.5	0.5
D_3^h	∞	1	1	1
D_4^h	1.5	1.5	1.5	1.5
D_5^h	∞	4.5	3.5	3.5
D_6^h	∞	∞	3	3
D_7^h	∞	1.5	1.5	1.5

$$D_0 = \begin{bmatrix} 0 & 1 & 2 & \infty & 1.5 \\ 1 & 0 & 4 & \infty & 0.5 \\ 3 & 1 & 0 & 2 & 1.5 \\ 0.5 & \infty & \infty & 0 & 2.5 \\ \infty & \infty & \infty & 0.5 & 0 \end{bmatrix} \quad D_1 = \begin{bmatrix} 0 & 1 & 2 & \infty & 1.5 \\ 1 & 0 & 4 & \infty & 0.5 \\ 3 & 1 & 0 & 2 & 1.5 \\ 0.5 & 1.5 & 2.5 & 0 & 2 \\ \infty & \infty & \infty & 0.5 & 0 \end{bmatrix} \quad D_2 = \begin{bmatrix} 0 & 1 & 2 & \infty & .5 \\ 1 & 0 & 3 & \infty & 0.5 \\ 2 & 1 & 0 & 2 & 1.5 \\ 0.5 & 1.5 & 2.5 & 0 & 2 \\ \infty & \infty & \infty & 0.5 & 0 \end{bmatrix}$$

$$D_3 = \begin{bmatrix} 0 & 1 & 2 & 4 & 1.5 \\ 1 & 0 & 3 & 5 & 0.5 \\ 2 & 1 & 0 & 2 & 1.5 \\ 0.5 & 1.5 & 2.5 & 0 & 2 \\ \infty & \infty & \infty & 0.5 & 0 \end{bmatrix} \quad D_4 = \begin{bmatrix} 0 & 1 & 2 & 4 & 1.5 \\ 1 & 0 & 3 & 5 & 0.5 \\ 2 & 1 & 0 & 2 & 1.5 \\ 0.5 & 1.5 & 2.5 & 0 & 2 \\ 1 & 2 & 3 & 0.5 & 0 \end{bmatrix} \quad D_5 = \begin{bmatrix} 0 & 1 & 2 & 2 & 1.5 \\ 1 & 0 & 3 & 1 & 0.5 \\ 2 & 1 & 0 & 2 & 1.5 \\ 0.5 & 1.5 & 2.5 & 0 & 2 \\ 1 & 2 & 3 & 0.5 & 0 \end{bmatrix}$$

Figure 7.17(a) An illustration of the Bellman–Ford algorithm. An example of the Floyd–Warshall algorithm.

estimates (received from its neighbors) of the distances to all network nodes. The algorithm includes the following steps:

- each node s transmits to its neighbors its current distance vector $D_{s,V}$;
- likewise, each neighbor node $u \in N(s)$ transmits to s its distance vector $D_{u,V}$;
- node s updates $D_{s,v}$, $\forall\, v \in V - \{s\}$ in accordance with Equation (7.1);
- if any update changes a distance value then s sends the current version of $D_{s,v}$ to its neighbors;
- node s updates $D_{s,v}$ every time it receives a distance vector information from any of its neighbors;
- a periodic timer prompts node s to recompute $D_{s,V}$ or to transmit a copy of $D_{s,V}$ to each of its neighbors.

An example of distributed Bellman–Ford Algorithm is given in Figure 7.18.

7.1.11.8 Distance vector protocols

With this protocol each node maintains a routing table with entries{Destination, Next Hop, Distance (cost)}.

Nodes exchange routing table information with neighbors (a) whenever table changes; and (b) periodically. Upon reception of a routing table from a neighbor, a node updates its routing table if it finds a 'better' route. Entries in the routing table are deleted if they are too old, i.e. they are not 'refreshed' within a certain time interval by the reception of a routing table.

7.1.11.9 Link failure

A simple rerouting case is shown in Figure 7.19.

- F detects that link to G has failed;
- F sets a distance of ∞ to G and sends update to A;
- A sets a distance of ∞ to G since it uses F to reach G;
- A receives periodic update from C with 2-hop path to G (via D);
- A sets distance to G to 3 and sends update to F;
- F decides it can reach G in four hops via A.

A routing loop case is shown in Figure 7.20.

- The link from A to E fails;
- A advertises distance of ∞ to E;
- B and C had advertised a distance of 2 to E (prior to the link failure);

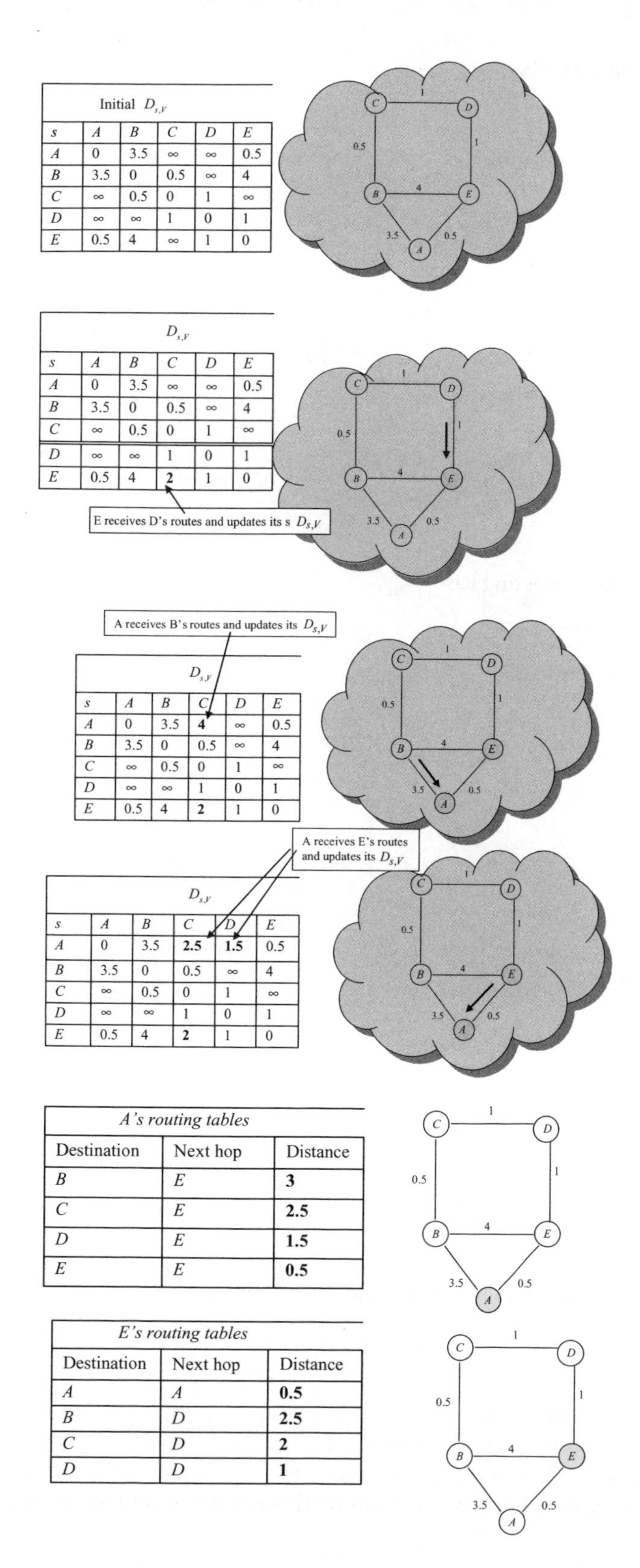
E receives D's routes and updates its s $D_{s,V}$
A receives B's routes and updates its $D_{s,V}$
A receives E's routes and updates its $D_{s,V}$

Initial $D_{s,V}$

s	A	B	C	D	E
A	0	3.5	∞	∞	0.5
B	3.5	0	0.5	∞	4
C	∞	0.5	0	1	∞
D	∞	∞	1	0	1
E	0.5	4	∞	1	0

$D_{s,V}$

s	A	B	C	D	E
A	0	3.5	∞	∞	0.5
B	3.5	0	0.5	∞	4
C	∞	0.5	0	1	∞
D	∞	∞	1	0	1
E	0.5	4	**2**	1	0

$D_{s,V}$

s	A	B	C	D	E
A	0	3.5	**4**	∞	0.5
B	3.5	0	0.5	∞	4
C	∞	0.5	0	1	∞
D	∞	∞	1	0	1
E	0.5	4	**2**	1	0

$D_{s,V}$

s	A	B	C	D	E
A	0	3.5	**2.5**	**1.5**	0.5
B	3.5	0	0.5	∞	4
C	∞	0.5	0	1	∞
D	∞	∞	1	0	1
E	0.5	4	**2**	1	0

A's routing tables

Destination	Next hop	Distance
B	E	**3**
C	E	**2.5**
D	E	**1.5**
E	E	**0.5**

E's routing tables

Destination	Next hop	Distance
A	A	**0.5**
B	D	**2.5**
C	D	**2**
D	D	**1**

Figure 7.18 Distributed Bellman–Ford algorithm example.

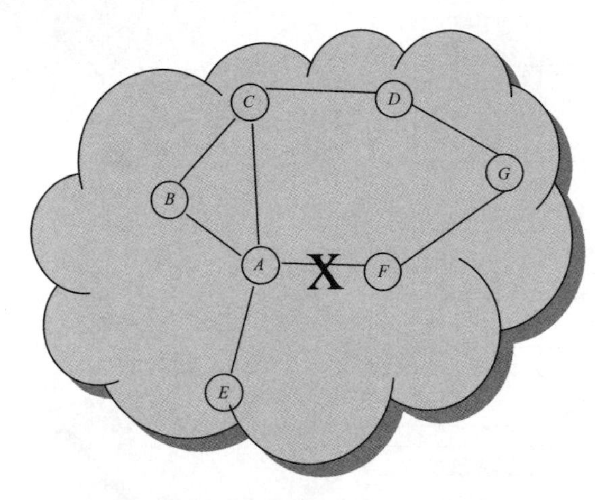

Figure 7.19 Simple rerouting case.

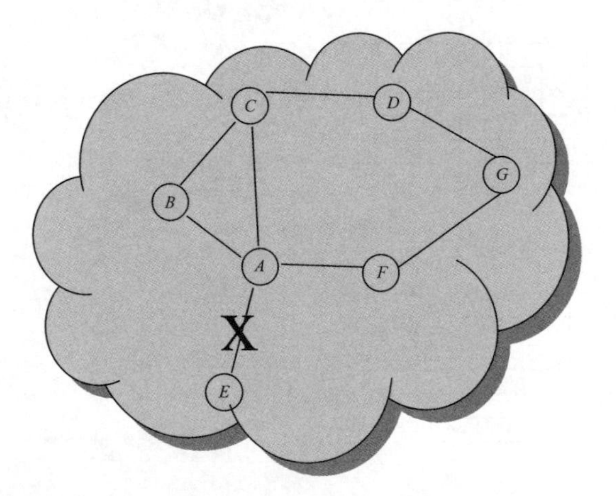

Figure 7.20 Routing loop case.

- upon reception of A's routing update, B decides it can reach E in three hops, and advertises this to A;
- A decides it can read E in four hops, and advertises this to C;
- C decides that it can reach E in five hops . . .

This behavior is called *count-to-infinity.* This problem is further elaborated in Figure 7.21. In the figure routing updates with distance to A are shown. When link from A to B fails, B can no longer reach A directly, but C advertises a distance of 2 to A and thus B now believes it can reach A via C and advertises it. This continues until the distance to A reaches infinity.

Example: routers working in stable state

(A,1) (A, 2) (A, 3) (A, 4)

B C D E

(A, 1) (A, 2) (A, 3)

Example: link from *A* to *B* fails

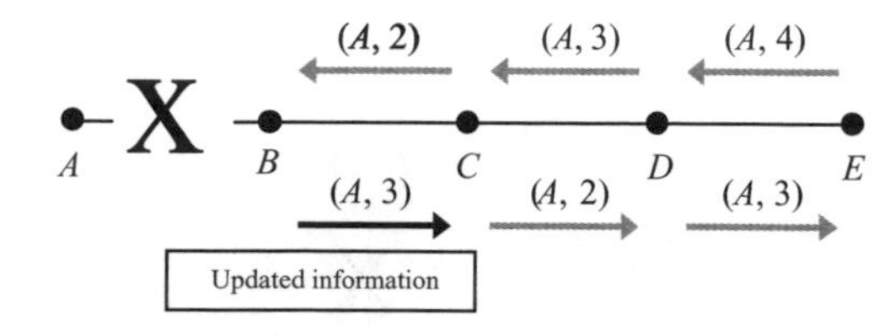

After two exchanges of updates

(A, 4) (A, 3) (A, 4)

X B C D E

(A, 3) (A, 4) (A, 3)

After three exchanges of updates

(A, 4) (A, 5) (A, 4)

A X B C D E

(A, 5) (A, 4) (A, 5)

After four exchanges of updates

(A, 6) (A, 5) (A, 6)

A X B C D E

(A, 5) (A, 6) (A, 5)

After five exchanges of updates

(A, 6) (A, 7) (A, 6)

A X B C D E

(A, 7) (A, 6) (A, 7)

After six exchanges of updates

(A, 8) (A, 7) (A, 8)

A X B C D E

(A, 7) (A, 8) (A, 7)

Figure 7.21 Count-to-infinity problem.

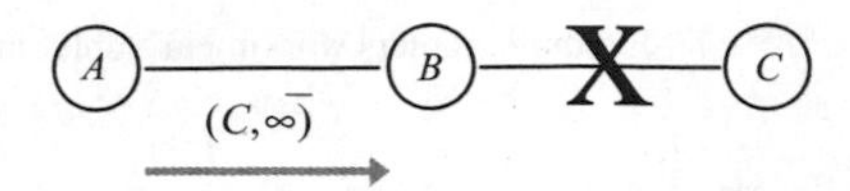

Figure 7.22 Split horizon algorithm.

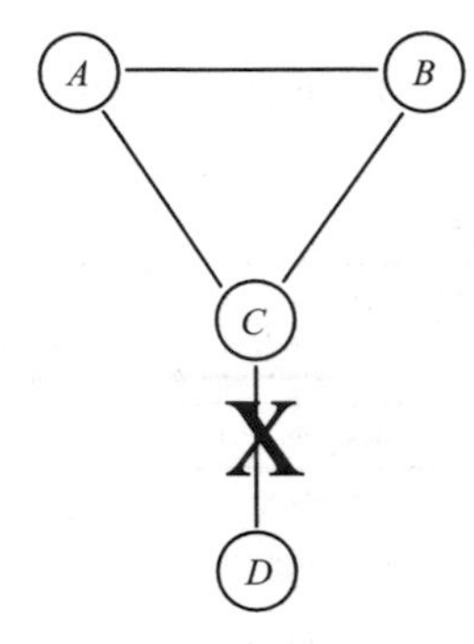

Figure 7.23 Example where split horizon fails.

7.1.11.10 Split horizon algorithm

Used to avoid (not always) the count-to-infinity problem. If A in Figure 7.22 routes to C via B, then A tells B that its distance to C is ∞. As result, B will not route to C via A if the link B to C fails. This works for two node loops, but not for loops with more than two nodes. An example where the split horizon fails is shown in Figure 7.23.

When link C to D breaks, C marks D as unreachable and reports that to A and B. Suppose A learns it first. A now thinks best path to D is through B. A reports D unreachable to B and a route of cost 3 to C. C thinks D is reachable through A at cost 4 and reports that to B. B reports a cost 5 to A who reports new cost to C, etc.

7.1.11.11 Routing information protocol

Routing information protocol (RIP) was originally distributed with BSD Unix and is widely used on the Internet (internal gateway protocol). RIP updates are exchanged in ordinary IP datagrams and RIP sets infinity to 16 hops (cost $\in$ [0–15]). RIP updates neighbors every 30 s, or when routing tables change.

7.2 GRAPH THEORY

The previous section summarized the basic relations and definitions in graph theory. A number of references that cover the graph theory in depth are available. The books [1–10] all discuss the various aspects of graph theory in general, not merely from a communication network point of view. Harary [7] has become a standard in this field. Buckley and Harary [3] provide a lot of material related to the hop distance and graph invariants defined in terms of distance properties, much of which may have applications in communications networks.

Harary *et al.* [8] provide a comprehensive discussion of connectivity in directed networks. Capobianco and Molluzzo [4] present a number of illuminative examples of relations holding among common graph invariants, such as node and edge connectivity, including counter examples which indicate the extent to which these results are best-possible.

Graph-theoretic algorithms are given good coverage in Tarjan [11] and in Evan [12, 13], but [14] is probably the best introduction to this subject. Additional references are [11–28]. Colbourn's book [18] is the only general text available at this time which covers network reliability. McHugh [22] includes a chapter on the implementation of graph algorithms for concurrent and parallel computation. Tenenbaum *et al.* [27, 28] discuss the implementation of some of these algorithms in the Pascal and C languages. See also References [19, 20] for more discussion of using data structures and graph algorithms in these computer languages. Christofides [17] discusses some important graph algorithms in much more detail, including the traveling salesman and Hamiltonian tour problems. A number of these works also treat some topics involved with flows in networks, a subject which we have not discussed so far but one which does have important applications to network connectivity, as discussed in Chapter 13 of this book (see also the article by Estahanian and Hakimi [21]).

References [29–40, 125, 126] discuss the efficiency and complexity of computer algorithms in general, not merely the graph-theoretic algorithms. Here the two books by Aho *et al.* [29, 30] and Knut's three volume series [36–38] are the best general references to the theory and practice of computer algorithms. Garey and Johnson [32] provide the best overall guide to NP-completeness and a compendium of many of those problems that were known to be NP-complete as of 1979. A problem is NP-hard if an algorithm for solving it can be translated into one for solving any other NP (nondeterministic polynomial time) problem. NP-hard therefore means 'at least as hard as any NP problem', although it might, in fact, be harder. There are now more than 1000 known NP-complete problems, many of them in graph theory, and dozens more are discovered every year, so this catalog has rapidly become out of date. As a result there is now an ongoing column on NP-complete problems by Johnson [35] which appears several times a year in the journal *Algorithms*. A number of these columns have discussed NP completeness for problems in communication networks and reliability. Hare [33] gives a very good and exceptionally readable overall account of the current state of the art in algorithmics and a good account of the problems that arise in designing and verifying algorithms for parallel processing. The book by Sedgewick [39] also comes in two other editions which give more details of the implementations of these algorithms in either Pascal or in C.

Problems concerning the statistical dependence of the network component failures are treated in References [41–48]. One should note that the assumption of independent failures can lead to either an overly pessimistic or an overly optimistic estimate of the true network reliability. The paper by Egeland and Huseby [41] gives some results as to how one might determine which of these is the case.

Most of the probabilistic measures of network connectivity lead to computability problems that are NP-hard, so there has been considerable effort in searching for restricted classes of networks for which there are reliability algorithms with a smaller order of complexity [49–59]. For example the papers by Boesch [52] and Pullen [58] consider only constant probability of edge failures. This may reduce the problem to one in graph enumeration, but this problem still has non-polynomial complexity. Similarly, the article by Bienstock [50] considers only planar networks, and he proves the existence of an algorithm whose complexity grows exponentially in the square root of p, rather than p itself. This is still very far from having polynomial growth, and to obtain that complexity even more drastic restrictions are necessary, as shown in the articles by Agrawal and Satayanarana [49] and by Politof and Satyanarayana [57]. Even very regular grid networks in the xy-plane yield NP-hard problems, as shown by the Clark and Colbourn article [54].

Owing to the computational intractability of the network reliability calculations for general probabilistic graphs, there are many papers devoted to the problem of obtaining bounds and approximations of the reliability [60–70]. For the same reason Monte Carlo simulations are also used in this field [71–77]. The NCF (node connectivity factor) and the LCF (link connectivity factor) were introduced in References [78–84] as possible alternatives to the usual reliability measure. They are indicators of how close the network is to being totally disconnected. Unfortunately, the NCF at least is computationally difficult to compute and does not seem to be amenable to simplifying techniques such as factorization or edge reduction used by other methods. Thus it is not yet clear how useful a concept this will prove to be, although if these connectivity factors are available they can be used to identify the most vulnerable components of the network and to adapt the network so as to equalize the vulnerability over its components. References [85–98] are concerned with some aspects of graph connectivity other than the usual path-oriented one, primarily with those deriving from the notion of the diameter of a graph (i.e. the maximum node-to-node hop distance across the graph) or the average node-to-node hop distance. Of special interest here is the notion of leverage, as described in the papers of Bagga *et al.* [85], which is a general method of quantifying changes in graph invariants due to the loss of some network components.

A number of references [99–111] are concerned with a number of other graph invariants that have an obvious connection with the notions of vulnerability and survivability of communications networks. The main concepts here are those of dominance, independence and covering of a graph with respect to either a set of nodes or a set of edges of the underlying graph. These quantities have already been applied to problems involving networks used in scheduling and service facilities, although their applications and usefulness to communications networks remains to be determined. Also, the calculation of some of these quantities can be NP-hard (some in the deterministic sense, others from the probabilistic point of view). This is also an area of very active research.

7.3 ROUTING WITH TOPOLOGY AGGREGATION

The goal of QoS routing is to find a network path from a source node to a destination node which has sufficient resources to support the QoS requirements of a connection request. The execution time and the space requirement of a routing algorithm increase with the size of the network, which leads to the scalability problem. For very large networks, it is impractical to broadcast the whole topology to every node for the purpose of routing. In order to achieve scalable routing, large networks are structured hierarchically by grouping nodes into different domains [113, 114]. The internal topology of each domain is then aggregated to show only the cost of routing across the domain, that is, the cost of going from one *border node* (a node that connects to another domain) to another border node. This process is called *topology aggregation*. One typical way of storing the aggregated topology is for every node to keep detailed information about the domain that it belongs to, and aggregated information about the other domains.

Since the network after aggregation is represented by a simpler topology, most aggregation algorithms suffer from distortion, that is, the cost of going through the aggregated network deviates from the original value [115]. Nevertheless, Hao and Zegura [116] showed that topology aggregation reduces the routing overhead by orders of magnitude and does not always have a negative impact on routing performance. Some aggregation approaches have been proposed. Lee [117] presented algorithms that find a minimum distortion-free representation for an *undirected* network with either a single additive or a single bottleneck parameter. Examples of additive metrics are delay and cost, while an example of a bottleneck parameter is bandwidth. For an additive constraint, it may require $O(|B|^2)$ links to represent a domain in the distortion-free aggregation, where $|B|$ is the number of border nodes in the domain. Awerbuch and Shavitt [118] proposed an algorithm that aggregates *directed* networks with a single additive parameter using $O(|B|)$ links. The algorithm achieves bounded distortion with a worst-case distortion factor of $O(\sqrt{\rho}\log|B|)$, where ρ is the *network asymmetry constant*, defined as the maximum ratio between the QoS parameters of a pair of opposite directed links.

In this section, we discuss networks with two QoS parameters, *delay* and *bandwidth*. Some related work can be found in References [119]–[121]. Lee [117] presented an aggregation method that aggregates an undirected delay-bandwidth sensitive domain into a spanning tree among border nodes. Therefore, there is a unique path between each pair of border nodes after aggregation and the space complexity is $O(|B|)$. The paper showed that a spanning tree can provide a distortion-free aggregation for bandwidth, but not for delay. Iwata *et al.* [120] studied the problem of topology aggregation in networks of six different QoS parameters. The aggregated topology follows the ATM PNNI standard [113]. The authors proposed minimizing the distortion by using a linear programming approach. Both References [117] and [118] assumed certain precedence order among the parameters, so that, among several paths that go between the same pair of border nodes, one path can be selected as the 'best'. The state of a path in a delay-bandwidth sensitive network can be represented as a delay-bandwidth pair [121]. If there are several paths going across a domain, a single pair of values, which is a point on the delay-bandwidth plane, is not sufficient to capture the QoS parameters of all those paths [122].

Korkmaz and Krunz [121] was the first to use a curve on the delay-bandwidth plane to approximate the properties of multiple physical paths between two border nodes, without assuming any precedence among the parameters. A curve is defined by three values: the minimum delay, the maximum bandwidth, and the smallest stretch factor among all paths between two border nodes. The stretch factor of a path measures how much the delay and the bandwidth of the path deviate from the best delay and the best bandwidth of all paths. The curve provides better approximation than a single point, but this approach has several shortcomings. First, the paper did not provide a routing algorithm with polynomial complexity to *find* a feasible path based on the aggregated topology. Instead, it provided an algorithm to *check* if a given path is likely to be feasible. Essentially, the algorithm determined whether the point, defined by the delay/bandwidth requirement, is within the curve defined by the delay, bandwidth and stretch factor of the path. Second, although the paper provided an aggressive heuristic to find the stretch factor of an interdomain path, there are cases where only one QoS metric will contribute to the value, and the information about the other metric is lost. In this section, we discuss a way of representing the aggregated state in delay-bandwidth sensitive networks using line segments. The approach solves some problems in Korkmaz and Krunz [121] and other traditional approaches by introducing a specific QoS parameter representation, a specific aggregation algorithm and the corresponding routing protocol. The algorithm outperforms others due to smaller distortion.

7.4 NETWORK AND AGGREGATION MODELS

A large network consists of a set of domains and links that connect the domains. It is modeled as a directed graph, where link state can be asymmetric in two opposite directions. Figures 7.24 and 7.25 are examples of a network with four domains. There are two kinds of nodes in a domain. A node is called a *border node* if it connects to a node of another domain. A node is an *internal node* if it is not a border node. A domain is modeled as a tuple (V, B, E), where V is the set of nodes in the domain, $B \subseteq V$ is the set of border nodes, and E is the set of directed links among the nodes in V. The entire network is modeled as (G, L), where $G = \{g_i \mid g_i = (V_i, B_i, E_i), 1 \le i \le \eta\}$ is the set of domains, L is a set of links that connect border nodes of different domains, and η is the number of domains in G.

There are several aggregation models for large networks. In this section, we use the topology aggregation model proposed by the private network–network interface (PNNI) [113, 123]. One of the representative topologies in PNNI is the *star* topology. Other popular ones are *simple-node* and *mesh*. In a simple-node topology, a domain is collapsed into one virtual node. This offers the greatest reduction of information as the space complexity after aggregation is $O(1)$, but the distortion is large. The mesh topology is a complete graph among the border nodes. The complexity of this topology is $O(|B|^2)$ and its distortion is much smaller. The star topology is a compromise between the above two. It has a space

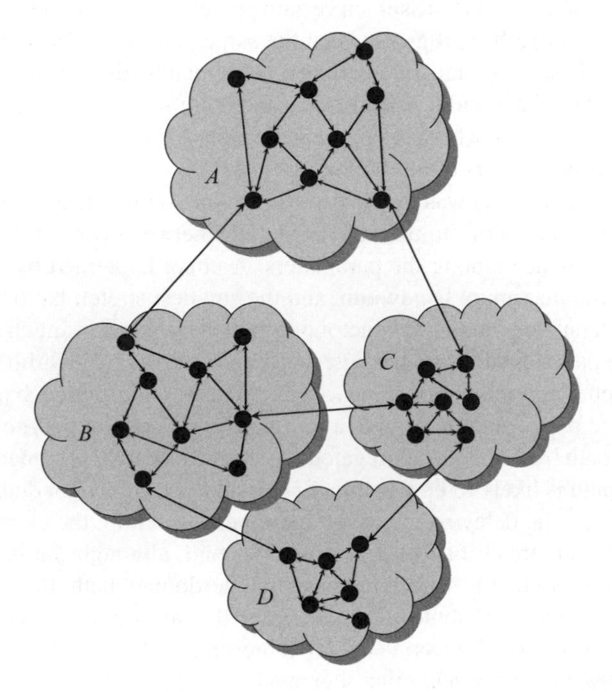

Figure 7.24 Network example.

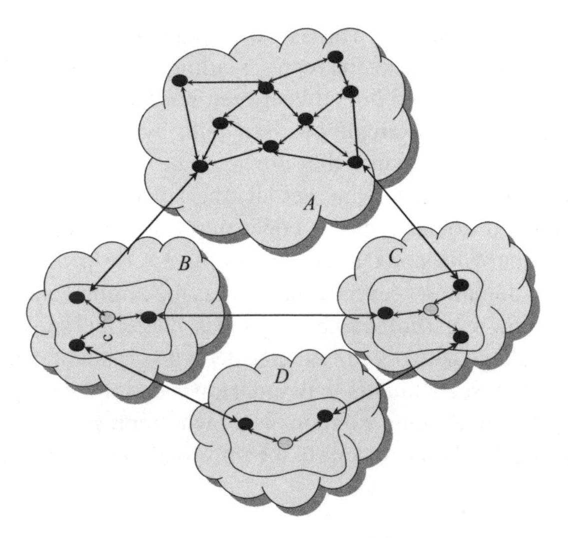

Figure 7.25 Aggregated network from Figure 7.24 with a complete view of domain A.

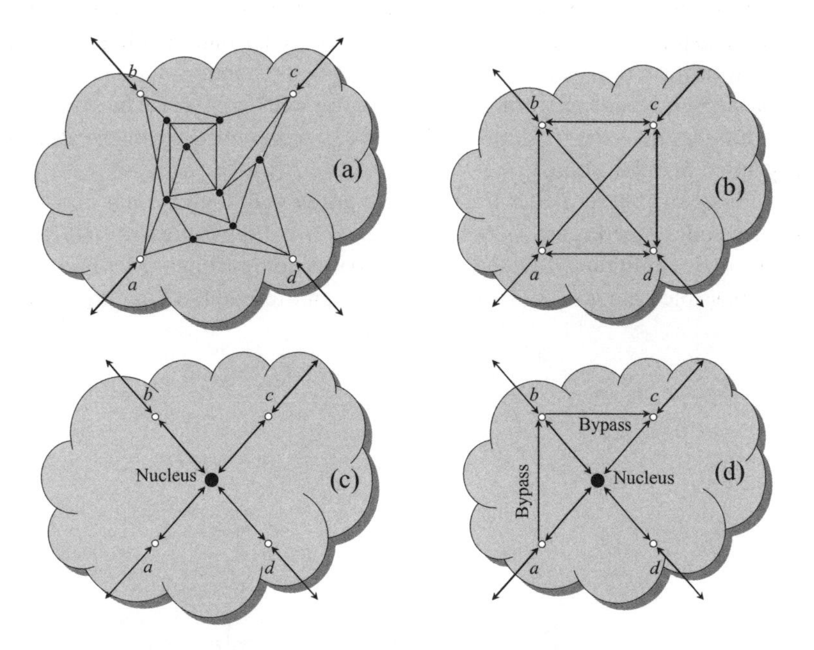

Figure 7.26 Topology aggregation. (a) Domain F. (b) Mesh of the borders. (c) Star representation. (d) Star representation with bypasses.

complexity of $O(|B|)$ and the distortion is between those of a simple node and a mesh. Guo and Matta [124] compare the performances of the above three aggregation methods.

It shows that the star topology outperforms the simple node and performs slightly worse than the mesh in a uniform network. Let us consider the domain in Figure 7.26(a), where nodes a, b, c and d are the border nodes. The mesh aggregation is shown in Figure 7.26(b), and the star aggregation is shown in Figure 7.26(c). In a star topology, the border nodes connect via links to a virtual *nucleus*. These links are called *spokes*. Each link is associated with some QoS parameters. To make the representation more flexible, PNNI also allows a limited number of links connected directly between border nodes. These links are called *bypasses*. Figure 7.26(d) is an example of a star with bypasses. We call the links in an aggregated topology *logical links* since they are not real.

After aggregation, a node in a domain sees all other nodes in the same domain, but only aggregated topologies of the other domains. For example, for the network in Figure 7.24, the aggregated view of the network stored at a node in domain A is shown in Figure 7.25. In such a view, the topology of domain A is exactly the same as the original one but the topologies of the other domains are now represented by border nodes, nuclei and spokes (without bypasses in this example). For a large network, this aggregated view is significantly smaller than the original topology and thus scalability is achieved. However, for the purpose of QoS routing, it is extremely important to develop solutions on how to represent the state information in this aggregated topology and how to control the information loss due to aggregation.

7.4.1 Line segment representation

In this section, we discuss a line-segment representation for the aggregated state information. Given the original topology and a (D, W) representing (delay, bandwidth) pair for each link, we shall first transform every domain to a mesh among the border nodes as an intermediate computation step, and then transform the mesh to a star with bypasses. The state information of a logical link in either the mesh or the final star is represented by line segments, which will be discussed in depth shortly.

A mesh among the border nodes is a complete graph with logical links connecting each pair of border nodes [see Figure 7.26(b)]. A logical link may represent multiple physical paths. For example, in Figure 7.27, there are five paths going from border node a to border node b. One possible path is $a \rightarrow 1 \rightarrow 3 \rightarrow b$. The end-to-end QoS parameter of this path

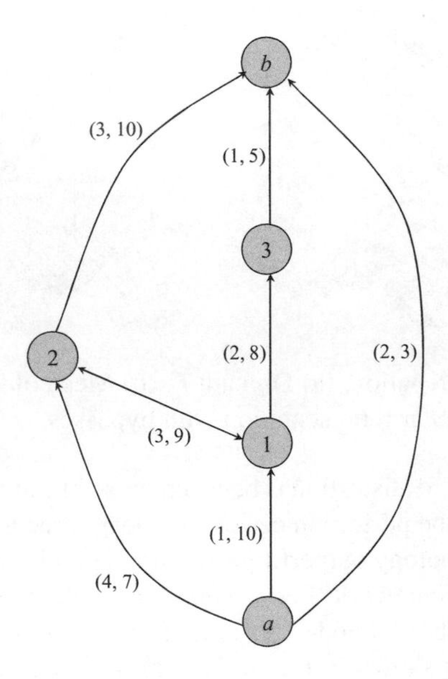

Figure 7.27 Multiple paths.

is (4, 5).We can find the parameters of all other paths and they are (7, 9), (10, 5), (2, 3), and (7, 7). *The equivalent delay is obtained as a sum of the delays on the path and equivalent bandwidth as a minimal value of the bandwidth among the individual segments.* When a logical link is used to represent all these paths, the QoS parameter of the link should be the 'best' parameter among the paths. However, the 'best' QoS parameter may not be defined since there does not exist an absolute order among those pairs. For example, parameter (2, 3) is better than parameter (7, 7) in terms of delay, but not in terms of bandwidth. Fortunately, a partial order can be developed.

By definition a point (x, y) is more representative than a point (x', y') if

- they are not the same, i.e. either $x \neq x'$ or $y \neq y'$; and
- $x \leq x'$ and $y \geq y'$.

Since the QoS parameter is a pair of values that represents a point on the delay-bandwidth plane, we often use *parameter*, *pair* and *point* interchangeably. In the previous example, (7, 9) is more representative than (10, 5) since $7 \leq 10$ and $9 \geq 5$.

Also, by definition, for a given set (S) of points on the delay-bandwidth plane, $(x, y) \in S$ is a representative of S if there does not exist any other point $(x', y') \in S$ which is more representative than (x, y). This means that $\forall(x', y') \in S, x < x'$ or $y > y'$. As an example let S be the set of the delay-bandwidth QoS pairs of the physical paths from a to b in Figure 7.27 [112]:

$$S = \{(4, 5), (7, 9), (10, 5), (2, 3), (7, 7)\}$$

Where (2, 3) is a representative of S, since its delay is less than all other points in S. The other representatives are (4, 5) and (7, 9). All QoS points in S on a delay-bandwidth plane are plotted in Figure 7.28. The shaded area defines the region of *supported services*, that is, any QoS request that falls in that region can be supported by a physical path. The dotted line is a staircase rising from left to right. The representatives are points on the convex corners of the steps. The corresponding paths of the representatives are called nondominated paths

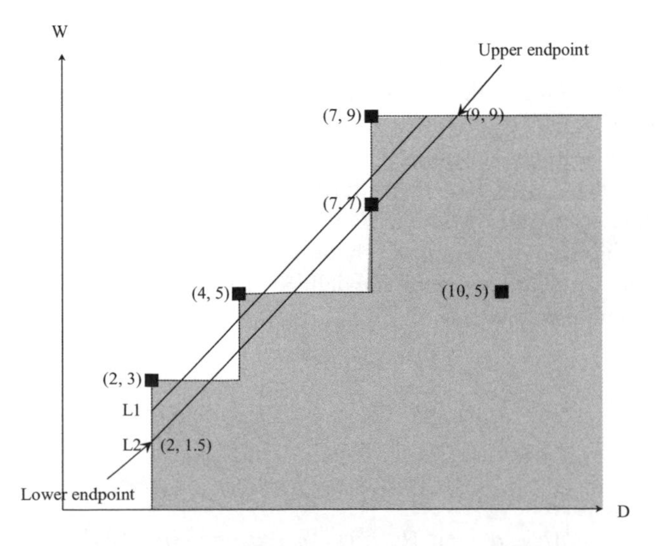

Figure 7.28 Representatives from example. (Reproduced by permission of IEEE [112].)

or Pareto optimal paths in the literature. The size of S depends on how many physical paths there are between the pair of border nodes, which can be exponential. The number of representative points is $|E|$ in the worst case for a domain (V, B, E). That is because there are at most $|E|$ possible bandwidth values for the paths in the domain and thus the staircase can have at most $|E|$ convex corners.

For the purpose of scalability, it is desirable to reduce the memory space for storing the QoS parameter of a logical link to $O(1)$. Hence, we shall neither store all QoS points in S nor store all representative points. The traditional way to solve this problem is to keep only one QoS point per logical link. However, no matter which point we pick, much information is lost.

A possible solution for this problem is to use a line segment that approximates the staircase, e.g. L_1 or $i L_2 n$ (Figure 7.28). Since every line segment can be defined unambiguously by two endpoints, the space needed for a logical link is $O(1)$. The line segment representation strikes a tradeoff between the accuracy in approximating the staircase and the overhead incurred.

After a line segment is chosen, all connection requests that fall under the line segment are accepted. However, it should be pointed out that not all requests under the line segment can actually be supported. For example, in Figure 7.28, if L_1 is selected to approximate the staircase, then the unshaded areas below L_1 represent connection requests that are accepted but not supported by any physical path. When a request is accepted but not supported, the routing process will detect it, and the request will eventually be rejected if a feasible path cannot be found (see the next section). On the other hand, when we reject the connection requests that are above the line segment, we may reject the supported QoS. For example, if a connection request is in the shaded region above L_1 in Figure 7.28, it is rejected although it can be served. Therefore, the choice of line segment depends on the strictness of the desired quality of the service. For instance, in Figure 7.28, both L_1 and L_2 are possible line segments. Using L_2 would probably reject more supported connection requests than L_1, while using L_1 would accept more unsupported requests than L_2. The least square method is used to find a line segment, which takes linear time with respect to the number of representative points. This line segment minimizes the least-square error, i.e. the summation of the squares of the distances from the points to the line.

It was mentioned earlier that any line segment can be defined by two endpoints. Owing to the nature of the staircase, the line always has a nonnegative slope. The endpoint with smaller bandwidth is called the *lower endpoint*, while the other one is called the *upper endpoint*. So a line segment is denoted as [*lower endpoint*, *upper endpoint*]. For example, L_2 in Figure 7.28 is [(2, 1.5), (9, 9)]. We further denote the lower endpoint and the upper endpoint of a line segment l as $l.lp$ and $l.up$, respectively. The delay of a point p is $p.d$ and the bandwidth of a point p is $p.w$. Therefore, is $L_2.lp.d$ is 2 and $L_2.up.w$ is 9. Two operations for line segment parameters are defined. The first operation is the *joint* '+' and *disjoint* '−' operation, defined as

$$\begin{aligned}&[(a, b), (c, d)] + \left[\left(a', b'\right), \left(c', d'\right)\right] = \\ &\left[\left(a + a', \min\left(b, b'\right)\right), \left(c + c', \min\left(d, d'\right)\right)\right]\end{aligned} \tag{7.1}$$

$$\begin{aligned}&[(a, b), (c, d)] - \left[\left(a', b'\right), \left(c', d'\right)\right] = \\ &\left[\left(a - a', \min\left(b, b'\right)\right), \left(c - c', \min\left(d, d'\right)\right)\right]\end{aligned} \tag{7.2}$$

In other words, summing two segments results in a segment whose delay is the sum of the delays of the individuals segments and the bandwidth equals the minimum of the two bandwidths. If $l_1,\ l_2$ and l_3 are three line segments and they satisfy the following conditions:

$$\begin{gathered} l_1.lp.d \geq l_2.lp.d \quad \text{and} \quad l_1.up.d \geq l_2.up.d \\ l_1.lp.w \leq l_2.lp.w \quad \text{and} \quad l_1.lp.w \leq l_2.lp.w \\ l_1 - l_2 = l_3 \end{gathered} \tag{7.3}$$

then

$$l_1 = l_2 + l_3 \tag{7.4}$$

As an example let $l_1 = [(10,\ 4),\ (13,\ 7)]\ and\ l_2 = [(6,\ 5),(11,\ 7)]$. One can see that for this example $l_1.lp.d \geq l_2.lp.d,\ l_1.up.d \geq l_2.up.d,\ l_1.lp.w \leq l_2.lp.w$ and $l_1.lp.w \leq l_2.lp.w$. and we have

$$\begin{aligned} l_1 - \ l_2 &= [(10,\ 4),(13,\ 7)]\ - [(6,\ 5),(11,\ 7)] \\ &= [10 - 6,\ \min(4,\ 5),(13 - 11,\ \min(7,\ 7))] \\ &= [(4,\ 4),(2,\ 7)] \end{aligned} \tag{7.5}$$

$$\begin{aligned} l_2 + l_3 &= l_2 + [(4,\ 4),(2,\ 7)] = [(6,\ 5),(11,\ 7)] + [(4,\ 4),(2,\ 7)] \\ &= [6 + 4,\ \min(5,\ 4),\ (11 + 2,\ \min(7,\ 7))] \\ &= [(10,\ 4),\ (13,\ 7)]\ = l_1 \end{aligned} \tag{7.6}$$

7.4.2 QoS-aware topology aggregation

The topology aggregation algorithm consists of two phases: (1) find a line segment for each logical link in the mesh (complete graph) of the border nodes; and (2) construct a star topology with bypasses from the mesh.

7.4.3 Mesh formation

In this phase, representatives of the paths between each pair of border nodes are found. The representatives for each pair of border nodes can be obtained by running the Dijkstra's algorithm for every link bandwidth value. The links in the domain are first sorted out by descending bandwidths. Let the largest bandwidth be $w_{\max}$. The algorithm starts by finding the smallest delay $d_{\max\_w}$ in a graph that consists of only links of bandwidth $w_{\max}$. The pair $(d_{\max\_w},\ w_{\max})$ is a representative that has the largest bandwidth. The algorithm then inserts the links of the next largest bandwidth value (w') and finds the smallest delay d'. If $d < d_{\max\_w}$, $(d',\ w')$ is another representative. The process ends after the smallest delays for all bandwidth values are identified. The detailed algorithm of finding the representatives can be found in References [125] and [126]. After the representatives are found, linear regression is used to find the line segment for each pair of border nodes.

7.4.4 Star formation

The next step is to aggregate the mesh into a star topology with bypasses. According to the recommendation of PNNI, a default parameter value to spokes can be set. There can be one or more spokes of the default parameter value but at most $3\,|B| - 1$ links can be of other values. As it is possible that all links in our aggregation are of different values, we shall put at most $3\,|B|$ links in the domain topology after aggregation.

Let i and j be two border nodes and n be the nucleus in the star representation. If there is no bypass between i and j, the only path i from to j is $i \to n \to j$ in the star. The goal in this phase is to find the QoS parameters of links $i \to n$ and $n \to j$, such that the line segment of $i \to n \to j$ in the star is the same as the line segment of i$i \to j$n the mesh. Basically, a single link in the mesh has to be 'split' into two links $i \to n$ and $n \to j$ in the star. There are three steps in this phase: (1) find the spokes from the border nodes to nucleus; (2) find the spokes from the nucleus to the border nodes; and (3) find the bypasses between border nodes.

7.4.4.1 Spokes incoming to the nucleus

In order to distinguish the line segments in the mesh from those in the star, superscript 'm' is used for mesh and 's' for star. For instance, let us consider a line segment from i to j. If it is in the mesh, it is denoted l^{m}_{ij}; if it is in the star, it is denoted l^{s}_{ij}.

In order to find spokes, we have to 'break' the line segments in the mesh. From the definition of the joint operation, we have a general idea what the 'broken' line segments look like. The endpoint delays of the spokes l^{s}_{in} and l^{s}_{nj} should be smaller than those of l^{m}_{ij}, while the endpoint bandwidths of the spokes should not be smaller than those of l^{m}_{ij}. The algorithm of finding spokes from border node to nucleus is based on these observations.

Recall that the lower endpoint and the upper endpoint of a line segment l are denoted $l.lp$ and $l.up$, respectively. Given a point p, its delay and bandwidth are denoted as $p.d$ and $p.w$, respectively. The algorithm that finds *spokes incoming to the nucleus* is defined as

$$l^{\mathrm{s}}_{in} = [(\min\_ld,\ \max\_lw), (\min\_ud,\ \max\_uw)] \tag{7.7}$$

where

$$\begin{aligned} \min\_ld &= \min_{j\in B, j\neq i} \{l^{\mathrm{m}}_{ij}.lp.d\} \\ \min\_ud &= \min_{j\in B, j\neq i} \{l^{\mathrm{m}}_{ij}.up.d\} \\ \max\_lw &= \max_{j\in B, j\neq i} \{l^{\mathrm{m}}_{ij}.lp.w\} \\ \max\_uw &= \max_{j\in B, j\neq i} \{l^{\mathrm{m}}_{ij}.up.w\} \end{aligned} \tag{7.8}$$

An example of the program calculating Equation (7.7) can be found in Lui *et al.* [112].

7.4.4.2 Spokes outgoing from the nucleus

Up to this point, we know the mesh, and we also know the set of spokes from borders to the nucleus, which is denoted $S_{b\to n}$. More specifically, we know l^{m}_{ij} as well as l^{s}_{in}, and we want to find l^{s}_{nj}, such that the result of joining l^{s}_{in} and l^{s}_{nj} is l^{m}_{ij}. Since $l^{\mathrm{m}}_{ij}.lp.d \geq l^{\mathrm{s}}_{in}.lp.d$,

$l^{\mathrm{m}}_{ij}.up.d \geq l^{\mathrm{s}}_{in}.up.d, l^{\mathrm{m}}_{ij}.lp.w \leq l^{\mathrm{s}}_{in}.lp.w$, and $l^{\mathrm{m}}_{ij}.up.w \leq l^{\mathrm{s}}_{in}.up.w$, we can obtain l^{s}_{nj} by evaluating $l^{\mathrm{m}}_{ij} - l^{\mathrm{s}}_{in}$, according to Equation (7.3). However, for the same j, $l^{\mathrm{m}}_{ij} - l^{\mathrm{s}}_{in}$ may be different for different i. Since we can have at most one l^{s}_{nj}, we solve this problem by assigning l^{s}_{nj} the average of $l^{\mathrm{m}}_{ij} - l^{\mathrm{s}}_{in}$, for all $i \in B, i \neq j$.

7.4.4.3 Finding bypasses

Owing to the aggregation, $l^{\mathrm{s}}_{ij} = l^{\mathrm{s}}_{in} + l^{\mathrm{s}}_{nj}$ may no longer be the same as l^{m}_{ij} in the mesh. Some may deviate only a little, while others may be quite different. In order to make the aggregation more precise, bypasses are introduced, which are direct links between border nodes . When $l^{\mathrm{s}}_{in} + l^{\mathrm{s}}_{nj}$ deviates a lot from l^{m}_{ij}, a bypass between border nodes i and j is used, and the QoS parameters of the paths from i to j are defined by the bypass instead of $l^{\mathrm{s}}_{in} + l^{\mathrm{s}}_{nj}$. Since there are $2\,|B|$ spokes ($|B|$ outgoing from and $|B|$ incoming to the nucleus), $|B|$ bypasses can be used in the network according to the recommendation of the PNNI standard. Only bypasses between those $|B|$ pairs of border nodes that have the largest deviations are used.

7.4.5 Line-segment routing algorithm

In this segment the line-segment routing algorithm (LSRA) is presented. It is a QoS-based source routing algorithm, which integrates the line segment representation with Dijkstra's algorithm (DA) and the centralized bandwidth-delay routing algorithm (CBDRA) [125, 127]. DA is the fastest known algorithm for finding the least-delay paths. CBDRA first prunes all links that do not satisfy the bandwidth requirement and then applies DA to find the least-delay path. LSRA extends the idea to capitalize the additional information provided by the line segment representation.

Being a centralized routing protocol, LSRA requires that each node keeps the topology of its own domain and the *star-with-bypasses* aggregation of the other domains. As broadcasting takes a lot of time and bandwidth, it is desirable to keep the amount of broadcasted information small. This justifies why the aggregation is a star with bypasses of $O(|B|)$ space instead of a mesh of $O(|B|^2)$ space. Routing is performed at two levels: *interdomain routing* and *intradomain routing*. An interdomain routing path specifies the border nodes of the transit domains, and the intradomain routing finds the subpath within each transit domain. Accordingly, LSRA has two routing phases.

7.4.5.1 Interdomain routing

After obtaining the star-with-bypasses aggregation from the external domains, each node can see all nodes in its own domain and all border nodes of the other domains. There are five steps in the LSRA interdomain routing:

(1) *transform stars with bypasses to meshes* – since nuclei of stars are virtual, the actual routing paths should not include any nucleus;

(2) *prune logical links* – this step prunes the logical links that do not satisfy the bandwidth requirement;

(3) *determine the delays of logical links* – the delay value, supported by a line segment, is a function of the bandwidth requirement; this step determines the delay values of all logical links under the bandwidth requirement;

(4) *prune physical links* – this step prunes the physical links that do not satisfy the bandwidth requirement;

(5) *apply DA on the network* – this step uses DA to find a shortest-delay path to the destination domain.

7.4.5.2 Intradomain routing

After the interdomain routing, an interdomain path is determined. The source node knows how to traverse the nodes in its own domain to a border node, and how to traverse the border nodes of other domains to get to the destination domain. Because the source node does not have the detailed topology of any external domain, it does not know how to fill in the intradomain path segments across the external domains. Therefore, LSRA does the intradomain routing in a distributed fashion. A routing message is sent from the source to travel along the interdomain path. When a border node t of domain g receives the message and the next hop on the interdomain path is another border node t' of g, t locally computes the intradomain path going from itself to t' using CBDRA, since t has the complete knowledge about its own domain g. Node t inserts the intradomain path into the interdomain path that is carried by the message. Then the message is sent to t' along the intradomain path. The message also keeps track of the accumulated delay of the path that has been traversed so far, including the intradomain path segments. This accumulated delay is calculated from the actual link delays as the message travels. If the accumulated delay exceeds *req(d)* the message is forwarded back to the previous node to find an alternate path. This is called *crankback* [113]. If the message successfully reaches the destination node, a feasible path is found.

One should be aware that the expected path delay may be different from the actual path delay due to the network dynamics or the inaccuracy introduced by aggregation. If no intradomain path can be found without violating *req(d)*, the message is sent back to the source to reject the request.

7.4.5.3 Best point algorithm

In this case the best delay and the best bandwidth are used to represent a mesh link. For example, if the delay-bandwidth parameters between a border pair are (2, 3), (4, 4), and (5, 6), then (2, 6) is used to represent the QoS of the mesh link. This optimistic approach is aggressive by choosing the best delay and the best bandwidth that may come from different paths. It suffers from large crankback ratios defined in the sequel. A best point (BP) mesh can be aggregated to a BP star by taking the maximum bandwidth among the mesh links as the spoke bandwidth and by taking half of the average delay among the mesh links as the spoke delay.

7.4.5.4 Worst point

The worst delay and the worst bandwidth are used to represent a mesh link. That is, if the parameters are (2, 3), (4, 4) and (5, 6), then (5, 3) is used. As a result, this algorithm has low success ratios. The minimum bandwidth and average of half of the average delay are used for aggregating a worst point (WP) mesh to a WP star.

7.4.5.5 Modified Korkmaz–Krunz

This is a modified version of the algorithm presented in Korkmaz and Krunz [121], the MKK algorithm. Let us consider a set of delay-bandwidth parameters, which corresponds to a set of paths between two border nodes. The *stretch factor* of a parameter (d, w) is defined as $d/d_{\text{best}} + w_{\text{best}}/w$, where d_{best} and w_{best} are the best delay and the best bandwidth among all parameters, respectively. Refer to the examples used previously for BP and WP, $d_{\text{best}} = 2$ and $w_{\text{best}} = 6$.

The stretch factor of (2, 3) is $2/2 + 6/3 = 3$, and the stretch factors of (4, 4) and (5, 6) are both 3.5. The smallest stretch factor is denoted as *s_factor*, which is 3 in the above case. The mesh link between the two border nodes is represented by a tuple $(d_{\text{best}}, w_{\text{best}}, s\_factor)$. The mesh link is likely to support a connection request [*req*(*d*),*req*(*w*)], if, $d_{\text{best}} \le req(d)$, $w_{\text{best}} \ge req(w)$ and $s\_factor \le req(d)/d_{\text{best}} + w_{\text{best}}/req(w)$ [121].

The transformation from a mesh to a star, in the original KK algorithm, is defined as

$$\begin{aligned} s_{in}.d &= \frac{1}{b-1} \sum_{i=1, i \neq j}^{b} m_{ij}.d \\ s_{nj}.d &= \frac{1}{b-1} \sum_{i=1, i \neq j}^{b} m_{ij}.d \end{aligned} \tag{7.9}$$

where $m_{ij}.d$ is the delay of a mesh link from border node i to border node j, b is the number of border nodes, s_{in} is the delay of a spoke link from i to the nucleus, and s_{nj} is the delay of a spoke link from the nucleus to j.

The problem of this averaging approach is that the delay after transformation deviates a lot from the original delay, because the delay of $s_{in}.d + s_{nj}.d$ may be twice the delay of the mesh link $(m_{ij}.d)$. Therefore, the following modification is used (MKK):

$$\begin{aligned} s_{in}.d &= \frac{1}{2(b-1)} \sum_{i=1, i \neq j}^{b} m_{ij}.d \\ s_{nj}.d &= \frac{1}{2(b-1)} \sum_{i=1, i \neq j}^{b} m_{ij}.d \end{aligned} \tag{7.10}$$

7.4.6 Performance measure

In this kind of analysis the most often used performance measures are: *delay deviation* (DD), *success ratio* (SR) and *crankback ratio* (CBR).

7.4.6.1 Delay deviation

Owing to the distortion of aggregation, the delay between a source node and a destination domain, obtained using the method described in the previous sections, is not accurate. Delay deviation measures the difference between the real delay and the estimated delay, obtained by the aggregation. It is defined as

$$\text{DD} = \left| d_{\text{esp}} - d_{\text{asp}} \right| \tag{7.11}$$

where d_{esp} is the estimated delay of the shortest path and d_{asp} is the actual delay of the shortest path.

7.4.6.2 Success ratio

A feasible request may be rejected due to the imperfect approximation of the delay and bandwidth information during aggregation. The success ratio is used to measure quantitatively how well an algorithm finds feasible paths and it is defined as

$$\text{SR} = \frac{\text{total number of acceptable feasible requests}}{\text{total number of feasible requests}} \tag{7.12}$$

The dividend represents all connection requests that are accepted by both interdomain routing and intradomain routing. The divider is the total number of *feasible* requests (not the total number of requests). Therefore, in simulation results, the success ratio measures the relative performance with respect to the optimal performance (accepting all feasible requests).

7.4.6.3 Crankback ratio

When an algorithm finds an interdomain path from the source node to the destination domain, it may overestimate the bandwidth or underestimate the delay and lead to crankbacks during intradomain routing. Crankback ratio measures how often that happens and is defined as

$$\text{CBR} = \frac{\text{total number requests crankbacked}}{\text{total number of requests accepted by interdomain routing}} \tag{7.13}$$

A good algorithm should have small crankback ratios, high success ratios and small delay deviations for different bandwidth and delay requirements.

7.4.7 Performance example

The *simulation environment* consists of the interdomain topology based on the power-law model [128, 129], and the intradomain topology based on the Waxman model [130]. The degree of a domain is defined as the total number of interdomain links adjacent to the border nodes of the domain. The simulation testbed and parameter configuration are set as follows: 10% of the domains have a degree of one, and the degrees of the other domains follow the power law, i.e. the frequency f_{d} of a domain degree is proportional to the degree $d\ (\geq 2)$ raised to the power of a constant $O = -2.2$. In other words $f_{\text{d}} \propto d^{O}$. After each domain is assigned a degree according to the power law, a spanning tree is formed among the domains to ensure a connected graph. Additional links are inserted to fulfill the remaining degrees of every domain with the neighbors selected according to probabilities proportional to their respective unfulfilled degrees. The Waxman topology for the internal nodes of each domain is formed as follows: the nodes are randomly placed in a one-by-one square, and the probability of creating a link between node u and node v is $p(u, v) \propto \exp[-d(u, v)/\beta L]$.

In this expression $d(u, v)$ is the distance between u and v, $\beta = 0.6$, and L is the maximum distance between any two nodes. The average node degree is 4. The other simulation parameters are as follows: the number of domains in each topology is 200, the number of nodes in each domain is randomly selected between 10 and 40, the delay of a physical link is a random number between 2 and 10 units, and the bandwidth of a physical link is a random number between 1 and 5 units. For delay, a unit can be replaced by a certain number of milliseconds; for bandwidth, a unit can be replaced by a certain number of kilobytes (per second). Each data point is the average of 1000 randomly generated requests. More specifically, given a bandwidth requirement and/or a delay requirement, five topologies are randomly generated. On each topology, 200 requests are generated with the source and the destination randomly selected from the topology. LS, BP, WP and KK are run over these requests, respectively, and the average results give a set of data points in the figures. Simulations with the same set of parameters are presented for example in Liu *et al.* [112].

The simulation results, using the described simulation environment, are shown in Figure 7.29. The horizontal axis is the bandwidth requirement. The results show that the modified algorithm (MKK) has better success ratios than the original algorithm. For CBR, the MKK algorithm is worse than the original KK algorithm, because the spoke delays in the modified KK are smaller than the spoke delays in the original KK. More optimistic estimation leaves more room for overestimation, which may lead to more crankbacks. However, as the crankback ratios of the modified KK are less than 5 % for all bandwidth values (Figure 7.31), it is believed that the performance improvement in success ratios outweighs the degradation of crankback ratios. As a result, the MKK algorithm is used in the rest of the section.

Figure 7.30 shows the delay deviations of LS, BP, WP, and KK. The horizontal axis presents the bandwidth requirement. For fairness, we count the paths of a request only

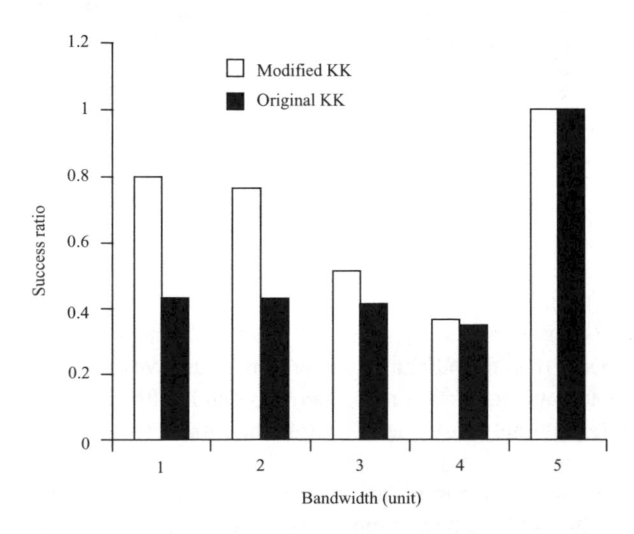

Figure 7.29 Success ratios of two aggregation schemes of the Korkmaz–Krunz method. (Reproduced by permission of IEEE [112].)

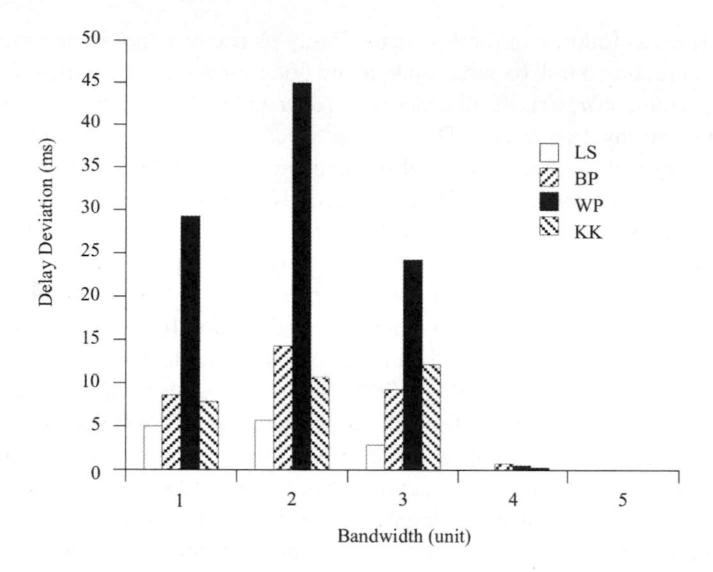

Figure 7.30 Delay deviation of different aggregation schemes. (Reproduced by permission of IEEE [112].)

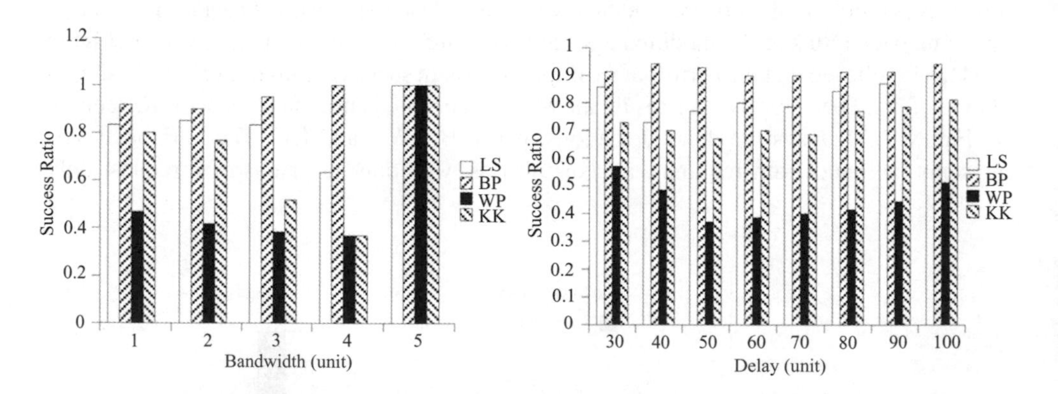

Figure 7.31 Success ratio vs bandwidth and delay. (Reproduced by permission of IEEE [112].)

when all four algorithms accept that request. Figure 7.30 shows that LS has smaller delay deviations than the other algorithms. That is because a line segment approximates the parameter staircase better than the other approaches, and thus gives more accurate information about the delay. The deviations are small when the bandwidth requirements are large (4), because the feasible paths are mostly short (within the same domain or between neighboring domains).

The success ratios are shown in Figure 7.31. It may be surprising that the success ratios are close to 1 when the bandwidth requirement is five units (maximum link capacity). Recall that our success ratio is defined relative to the number of feasible requests, not to the number of all requests. When the bandwidth request is large, there are few feasible requests. These feasible requests are mostly between nodes that are close to each other (in the same domain

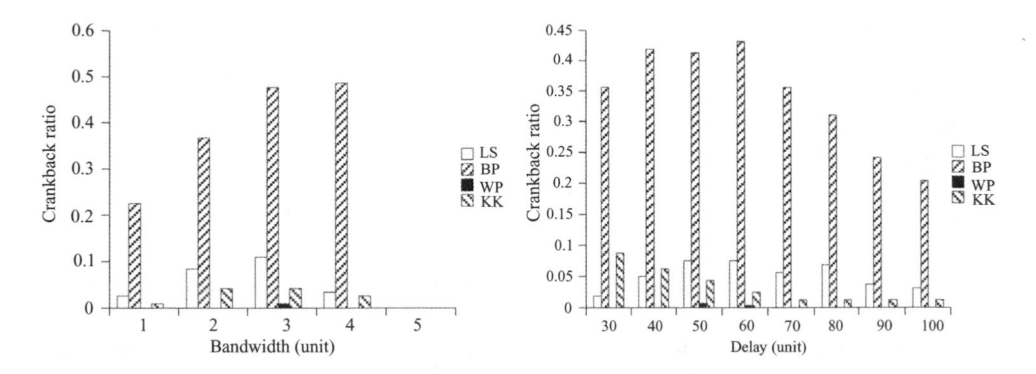

Figure 7.32 Crankback ratio vs bandwidth and delay. (Reproduced by permission of IEEE [112].)

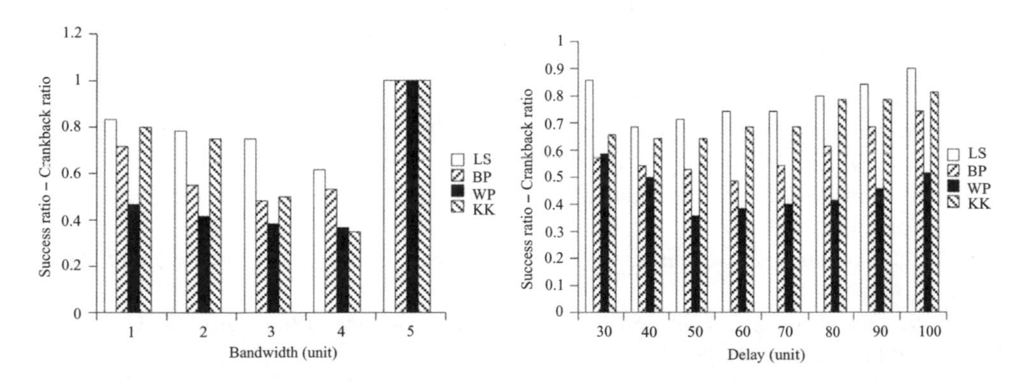

Figure 7.33 Success ratio–crankback ratio. (Reproduced by permission of IEEE [112].)

or neighboring domains), which makes the delay deviation small, as shown in Figure 7.30. That means it is less likely to reject a feasible request due to distortion. Therefore, the success ratio is high.

Figure 7.32 shows the crankback ratios of BP, WP, KK and LS. It is clear that BP has very high crankback ratios compared with other methods. On the contrary, as an aggregated WP topology tends to overestimate the real delay and underestimate the real bandwidth of the paths, WP has very small crankback ratios. Figure 7.33 compares *success ratio–crankback ratio*. BP does not perform well in this comparison due to its high crankback ratios. In contrast, LS has higher success ratios than WP and KK, and has reasonable crankback ratios. It outperforms all other schemes for different bandwidth and delay requirements by making a better tradeoff between success ratio and crankback ratio.

REFERENCES

[1] C. Berge, *Graphs*. North-Holland: New York, 1985.
[2] B. Bollobas, *Graph Theory: an Introductory Course*. Springer: New York, 1979.
[3] F. Buckley and F. Harary, *Distance in Graphs*. Addison-Wesley: Reading, MA, 1990.

[4] M. Capobianco and J.C. Molluzzo, *Examples and Counterexamples in Graph Theory*. North-Holland: New York, 1978.
[5] W.-K. Chen, *Theory of Nets. Flows in Networks*. Wiley: New York, 1990.
[6] N. Deo, *Graph Theory with Applications to Engineering and Computer Science*. Prentice Hall: Englewood Cliffs, NJ, 1974.
[7] F. Harary, *Graph Theory*. Addison-Wesley: Reading, MA, 1969.
[8] F. Harary, R.Z. Norman and D. Cartwright, *Structural Models: an Introduction to the Theory of Directed Graphs*. Wiley: New York, 1965.
[9] F. Harary and E.M. Palmer, *Graphical Enumeration*. Academic Press: New York, 1973.
[10] D.F. Robinson and L.R. Foulds, *Digraphs: Theory and Techniques*. Gordon and Breach: New York, 1980.
[11] R.E. Tarjan, *Data Structures and Network Algorithms*. SIAM: Philadelphia, PA, 1983.
[12] S. Even, *Graph Algorithms*. Computer Science Press: Rockville, MD, 1979.
[13] S. Even, *Algorithmic Combinatorics*. Macmillan: New York, 1973.
[14] A. Gibbons, *Algorithmic Graph Theory*. Cambridge University Press: Cambridge, 1985.
[15] L. Ammeraal, *Programs and Data Structures in C*. Wiley: New York, 1987.
[16] W. Arnsbury, *Data Structures: From Arrays to Priority Queues*. Wadsworth: Belmont, CA, 1985.
[17] N. Christofides, *Graph Theory: an Algorithmic Approach*. Academic Press: New York, 1975.
[18] C.J. Colbourn, *The Combinatorics of Network Reliability*. Oxford University Press: London, 1987.
[19] N. Dale and S.C. Lilly, *Pascal Plus Data Structures* (2nd edn). D. C. Heath: Lexington, MA, 1988.
[20] J. Esakov and T. Weiss, *Data Structures: an Advanced Approach Using C*. Prentice-Hall: Englewood Cliffs, NJ, 1989.
[21] A.H. Estahanian and S.L. Hakimi, On computing the connectivity of graphs and digraphs, *Networks*, vol. 14, 1984, pp. 355–366.
[22] J.A. McHugh, *Algorithmic Graph Theory*. Prentice Hall: Englcwood Cliffs, NJ, 1990.
[23] U. Manber, *Introduction to Algorithms*. Addison-Wesley: Reading, MA, 1989.
[24] E. Minieka, *Optimization Algorithms for Networks and Graphs*. Marcel Dekker: New York, 1978.
[25] C.H. Papadimitriou and K. Steiglitz, *Combinatorial Optimization: Algorithms and Complexity*. Prentice-Hall: Englewood Cliffs, NJ, 1982.
[26] M.N.S. Swarny and K. Thulasiraman. *Graphs, Networks, and Algorithms*. Wiley: New York, 1981.
[27] A.M. Tenenbaum and M.J. Augenstein, *Data Structures Using Pascal* (2nd edn). Prentice-Hall: Englewood Cliffs, NJ, 1986.
[28] A.M. Tenenbaum, Y. Langsarn, and M.J. Augenstein, *Data Structures Using C*. Prentice-Hall: Englewood Cliffs, NJ, 1990.
[29] A.V. Aho, J.E. Hopcroft, and J.D. Ullman, *The Design and Analysis of Computer Programs*. Addison-Wesley: Reading, MA, 1974.
[30] A.V. Aho, J.E. Hopcroft, and J.D. Ullrnan, *Data Structures and Algorithms*. Addison-Wesley: Reading, MA, 1983.

[31] T.H. Cormen, C.E. Leiserson, and R.L. Rivest, *Introduction to Algorithms*. MIT Press: Cambridge, MA, 1990.
[32] M.R. Garey and D.S. Johnson, *Computers and Intractibility. A Guide to the Theory of NP-Completeness*. Freeman: San Francisco, CA, 1979.
[33] D. Hare, *Algorithmics. The Spirit of Computing*. Addison-Wesley: Reading, MA, 1987.
[34] E. Horowitz and S. Sahni, *Fundamentals of Computer Algorithms*. Computer Science Press: Rockville, MD, 1978.
[35] D.S. Johnson, The NP-completeness column: an ongoing guide, *Algorithms* (this column appears several times per year).
[36] D.E. Knuth, *The Art of Computing*: *Vol. 1: Fundamental Algorithms* (2nd edn). Addison-Wesley: Reading, MA, 1973.
[37] D.E. Knuth, *The Art of Computing*: *Vol. 2: Scm inunierical Algorithms* (2nd edn). Addison-Wesley: Reading, MA, 1981.
[38] D.E. Knuth, *The Art of Computing*: *Vol. 3: Sorting and Searching*, Addison-Wesley: Reading, MA, 1973.
[39] R. Sedgewick, *Algorithms* (2nd edn). Addison-Wesley: Reading, MA, 1988.
[40] H.S. Wilf, *Algorithms and Complexity*. Prentice-Hall: Englewood Cliffs, NJ, 1986.
[41] T. Egeland and A.B. Huseby, On Dependence and Reliability Computation, *Networks*, vol. 21, 1991, pp. 521–546.
[42] H. Heffes and A. Kumar, Incorporating dependent node damage in deterministic connectivity analysis and synthesis of networks, *Networks*, vol. l6, l986, pp. Sl–65.
[43] Y.F. Lam and V. Li, On Network Reliability Calculations with Dependent Failures, in *Proc. IEEE 1983 Global Telecommunications Conf.* (GTC '83), San Diego, CA, November, 1977, pp. 1499–1503.
[44] Y.F. Lam and V. Li, Reliability modeling and analysis of communication networks with dependent failures, in Proc. *IEEE INFOCOM*, 3–7 June 1985, pp. 196–199.
[45] Y.F. Lam and V. Li, Reliability modeling and analysis of communication networks with dependent failures, *IEEE Trans. Commun.*, vol 34, 1986, pp. 82–84.
[46] K.V. Lee and V.O. K. Li, A path-based approach for analyzing reliability of systems with dependent failures and multinode components, *Proc. IEEE INFOCOM*, 1990, pp. 495–503.
[47] L.B. Page and J.E. Perry, A model for system reliability with common-cause failures, *IEEE Trans. Reliab.*, vol. 38, 1989, pp. 406–410.
[48] E. Zemel, Polynomial algorithms for estimation of network reliability, *Networks*, vol. 12, 1982, pp. 439–452.
[49] A. Agrawal and A. Satayanarana, An OIE time algorithm for computing the relaibility of a class of directed networks, *Op. Res.*, vol. 32, 1984, pp. 493–515.
[50] D. Bienstock, An algorithm for reliability analysis of planar graphs, *Networks*, vol. 16, 1986, pp. 411–422.
[51] F. Beichelt and P. Tittman, A generalized reduction method for the connectedness probability of stochastic networks, *IEEE Trans. Reliab.*, vol. 40, (1991), pp. 198–204.
[52] F.T. Boesch, On unreliability polynomials and graph connectivity in reliable network synthesis, *J. of Graph Theory*, vol. 10, 1988, pp. 339–352.
[53] A. Bobbio and A. Premoli, Fast algorithm for unavailibility and sensitivity analysis of series–parallel systems, *IEEE Trans. Reliab.*, vol. 31, 1982, pp. 359–361.

[54] B.N. Clark and C.L. Colbourn, Unit disk graphs, *Discrete Math.*, vol. 86, 1990, pp. 165–177.

[55] C.L. Colbourn, Network resiliance, *SIAM J. Algebra Discrete Math.*, vol. 8, 1987, pp. 404–409.

[56] W.H. Debany, P.K. Varshney and C.R.P. Hartman, Network reliability evaluation using probability expressions, *IEEE Trans. Reliab.*, vol. 35, 1986, pp. 161–166.

[57] T. Politof and A. Satyanarayana, A linear-time algorithm to compute the reliability of planar cube-free networks, *IEEE Trans. Reliab.*, vol. 39, 1990, pp. 557–563.

[58] K.W. Pullen, A random network model of message transmission, *Networks*, vol. 16, 1986, pp. 397–409.

[59] O.W.W. Yang, Terminal pair reliability of tree-type computer communication networks, in *Proc. IEEE 1991 Military Communications Conf. (MILCOM '91)*, vol. 3, November 1991, pp. 905–909.

[60] H.M. AboElFotoh and C.J. Colbourn, Computing 2-terminal reliability for radio-broadcast networks, *IEEE Trans. Reliab.*, vol. 38, 1989, pp. 538–555.

[61] M.O. Ball and J.S. Provan, Calculating bounds on reachability and connectedness in stochastic networks, *Networks*, vol. 13, 1983, pp. 253–278.

[62] M.O. Ball and J.S. Provan, Bounds on the reliabilty polynomial for shellable independence systems, *SIAM J. Algebra Discrete Math.*, vol. 3, 1981, pp. 166–181.

[63] T.B. Brecht and C.J. Colbourn, Improving reliabilty bounds in computer networks, *Networks*, vol. 16, 1986, pp. 369–380.

[64] T.B. Brecht and C.J. Colbourn, Multplicative improvements in network reliabilty bounds, *Networks*, vol. 19, 1989, pp. 521–529.

[65] T.B. Brecht and C.L. Colbourn, Lower bounds on two-terminal network reliability, *Discrete Appl. Math.*, vol. 21, 1988, pp. 185–198.

[66] C.J. Colbourn and A. Ramanathan, Bounds for all-terminal reliability by arc packing, *Ars Combinatorica*, vol. 23A, 1987, pp. 229–236.

[67] C.J. Colboum, Edge-packings of graphs and network reliability, *Discrete Math.*, vol. 72, 1988, pp. 49–61.

[68] C.J. Colbourn and D.D. Harms, Bounding all-terminal reliability in computer networks, *networks*, vol. 18, 1988, pp. 1–12.

[69] D. Torrieri, Algorithms for finding an optimal set of short disjoint paths in a communication network, in *Proc. IEEE 1991 Military Communications Cont. (MILCOM '91)*, vol. 1, November, 1991, pp. 11–15.

[70] D.K. Wagner, Disjoint (s, t)-cuts in a network, *Networks*, vol. 20, 1990, pp. 361–371.

[71] G.S. Fishman, A comparison of four Monte Carlo methods for estimating the probability of st-connectedness, *IEEE Trans. Reliab.*, vol. 35, 1986, pp. 145–154.

[72] G.S. Fishman, A Monte Carlo sampling plan for estimating network reliability, *Opns. Res.*, vol. 34, 1986, pp. 581–594.

[73] G.S. Fishman, Estimating the st-reliability function using importance and stratified sampling, *Opns. Res.*, vol. 37, 1989, pp. 462–473.

[74] G.S. Fishman, A Monte Carlo sampling plan for estimating reliability parameters and related functions, *Networks*, vol. 17, 1987, pp. 169–186.

[75] R.M. Karp and M.G. Luby, Monte-Carlo algorithms for enumeration and reliability problems, in *Proc. IEEE 24th Annual Symp. on Foundations of Computer Science*, 7–9 November 1983 Tuscon, AZ, pp. 56–64.

[76] P. Kubat, Estimation of reliability for communication/computer networks – simulation/analytic approach, *IEEE Trans. Commun.*, vol. 37, 1989, pp. 927–933.

[77] L.D. Nd and C.J. Colbourn, Combining Monte Carlo estimates and bounds for network reliability, *Networks*, vol. 20, 1990, pp. 277–298.

[78] K.T. Newport and M.A. Schroeder, Network survivability through connectivity optimization', in *Proc. 1987 IEEE Int. Conf. Communications*, vol. 1, 1987, pp. 471–477.

[79] K.T. Newport and P. Varshney, Design of communications networks under performance constraints, *IEEE Trans. Reliab.*, vol. 40, 1991, pp. 443–439.

[80] K.T. Newport and M.A. Schroeder, Techniques for evaluating the nodal survivability of large networks, in *Proc. IEEE 1990 Military Communications Conf. (MILCOM '90)*, Monterey, CA, 1990, pp. 1108–1113.

[81] K.T. Newport, M.A. Schroeder and G.M. Whittaker, A knowledge based approach to the computation of network nodal survivability, in *Proc. IEEE 1990 Military Communications Conf. (MILCOM '90)*, Monterey, CA, 1990, pp. 1114–1119.

[82] M.A. Schroeder and K.T. Newport, Tactical network survivability through connectivity optimization, in *Proc. 1987 Military Communications Conf. (MILCOM '87)*, Monterey, CA, vol. 2, 1987, pp. 590–597.

[83] G.M. Whittaker, A knowledge-based design aid for survivable tactical networks, in *Proc. the IEEE 1990 Military Communications Conf. (MILCOM '90)*, Monterey, CA, 1990, Section 53.5.

[84] M.A. Schroeder and K.T. Newport, Enhanced network survivability through balanced resource criticality, in *Proc. IEEE 1989 Military Communications Conf. (MILCOM '89)*, Boston, MA, 1989, Section 38.4.

[85] K.S. Bagga, L.W. Beineke, M.J. Lipman and R.E. Pippert, The concept of leverage in network vulnerability, in *Conf. Graph Theory*, Kalamazoo, MI, 1988, pp. 29–39.

[86] K.S. Bagga, L.W. Beineke, M.J. Lipman and R.E. Pippert, Explorations into graph vulnerability, in *Conf. Graph Theory*, Kalamazoo, MI, 1988, pp. 143–158.

[87] D. Bienstock and E. Gyori, Average distance in graphs with removed elements, *J. Graph Theory*, vol. 12, 1988, pp. 375–390.

[88] F.T. Boesch and I.T. Frisch, On the smallest disconnecting set in a graph, *IEEE Trans. Circuit Theory*, vol. 15, 1986, pp. 286–288.

[89] F. Buckley and M. Lewinter, A note on graphs with diameter preserving spanning trees, *J. Graph Theory*, vol. 12, 1988, pp. 525–528.

[90] F.R.K. Chung, The average distance and the independence number, *J. Graph Theory*, vol. 12, 1988, pp. 229–235.

[91] G. Exoo, On a measure of communication network vulnerability, *Networks*, vol. 12, 1982, pp. 405–409.

[92] O. Favaron, M. Kouider and M. Makeo, Edge-vulnerability and mean distance, *Networks*, vol. 19, 1989, pp. 493–509.

[93] F. Harary, F.T. Boesch and J.A. Kabell, Graphs as models of communication network vulnerability: connectivity and persistence, *Networks*, vol. 11, 1981, pp. 57–63.

[94] F. Haraty, Conditional connectivity, *Networks*, vol. 13, 1983, pp. 347–357.

[95] S.M. Lee, Design of e-invariant Networks, *Congressus Nurner*, vol. 65, 1988, pp. 105–102.

[96] O.R. Oellermann, Conditional graph connectivity relative to hereditary properties, *Networks*, vol. 21, 1991, pp. 245–255.

[97] J. Plesnik, On the sum of all distances in a graph or digraph, *J. Graph Theory*, vol. 8, 1984, pp. 1–21.

[98] A.A. Schoone, H.L. Bodlaender and J. van Leeuwer, Diameter increase caused by edge deletion, *J. Graph Theory*, vol. 11, 1987, pp. 409–427.

[99] M.O. Ball, J.S. Provan and D.R. Shier, Reliability covering problems, *Networks*, vol. 21, 1991, pp. 345–357.

[100] L. Caccetta, Vulnerability in communication networks, *Networks*, vol. 14, 1984, pp. 141–146.

[101] L.L. Doty, Extremal connectivity and vulnerability in graphs, *Networks*, vol. 19, 1989, pp. 73–78.

[102] T.J. Ferguson, J.H. Cozzens and C. Cho, SDI network connectivity optimization, in *Proc. the IEEE 1990 Military Communications Conf. (MILCOM '90)*, Monterey, CA, 1990, Sec. 53.1.

[103] J.F. Fink, M.S. Jacobson, L.F. Kinch and J. Roberts, The bondage number of a graph, *Discrete Math*, vol. 86, 1990, pp. 47–57.

[104] G. Gunther, Neighbor-connectedness in regular graphs, *Discrete Appl. Math.*, vol. 11, 1985, pp. 233–242.

[105] G. Gunther, B.L. Hartnell and R. Nowakowski, Neighbor-connected graphs and projective planes, *Networks*, vol. 17, 1987, pp. 241–247.

[106] P.L. Hammer, Cut-threshold graphs, *Discrete Appl. Math.*, vol. 30, 1991, pp. 163–179.

[107] A.M. Hobbs, Computing edge-toughness and fractional arboricity, *Contemp. Math., Am. Math. Soc.*, 1989, pp. 89–106.

[108] T.Z. Jiang, A new definition on survivability of communication networks, in *Proc. IEEE 1991 Military Communications Conf. (MJLCOM '91)*, vol. 3, November 1991, pp. 901–904.

[109] L.M. Lesniak and R.E. Pippert, On the edge-connectivity vector of a graph, *Networks*, vol. 19, 1989, pp. 667–671.

[110] Z. Miller and D. Pritikin, On the separation number of a graph, *Networks*, vol. 19, 1989, pp. 651–666.

[111] L. Wu and P.K. Varshney, On survivability measures for military networks, in *Proc. IEEE 1990 Military Communications Conf. (MILCOM '90)*, Monterey, CA, pp. 1120–1124.

[112] K.-S. Lui, K. Nahrstedt and S. Chen, Routing with topology aggregation in delay-bandwidth sensitive networks, *IEEE/ACM Trans. Networking*, vol. 12, No. 1, 2004 , pp. 17–29.

[113] *Private Network-Network Interface Specification Version 1.0*, March 1996.

[114] Y. Rekhter and T. Li, A border gateway protocol 4 (BGP-4), Network Working Group, RFC 1771, 26–30 March 1995.

[115] R. Guerin and A. Orda, QoS-based routing in networks with inaccurate information: theory and algorithms, *IEEE/ACM Trans. Networking*, vol. 7, 1999, pp. 350–364.

[116] F. Hao and E.W. Zegura, On scalable QoS routing: performance evaluation of topology aggregation, in *Proc. IEEE INFOCOM*, 2000, pp. 147–156.

[117] W. Lee, Minimum equivalent subspanner algorithms for topology aggregation in ATM networks, in *Proc. 2nd Int. Conf. ATM (ICATM)*, 21–23 June 1999, pp. 351–359.

[118] B. Awerbuch and Y. Shavitt, Topology aggregation for directed graphs, *IEEE/ACM Trans. Networking*, vol. 9, 2001, pp. 82–90.

[119] W. Lee, Spanning tree method for link state aggregation in large communication networks, in *Proc. IEEE INFOCOM*, 2–6 April 1995, pp. 297–302.

[120] A. Iwata, H. Suzuki, R. Izmailow and B. Sengupta, QoS aggregation algorithms in hierarchical ATM networks, in *IEEE Int. Conf. Communications Conf. Rec.*, 7–11 June 1998, pp. 243–248.

[121] T. Korkmaz and M. Krunz, Source-oriented topology aggregation with multiple QoS parameters in hierarchical networks, *ACM Trans. Modeling Comput. Simulation*, vol. 10, no. 4, 2000, pp. 295–325.

[122] S. Chen and K. Nahrstedt, An overview of quality-of-service routing for the next generation high-speed networks: problems and solutions, *IEEE Network*, vol. 12, 1998, pp. 64–79.

[123] W. Lee, Topology aggregation for hierarchical routing in ATM networks, in *ACM-SIGCOMM Comput. Commun. Rev.*, vol. 25, 1995, pp. 82–92.

[124] L. Guo and I. Matta, On state aggregation for scalable QoS routing, in *IEEE Proc. ATM Workshop*, May 1998, pp. 306–314.

[125] Z. Wang and J. Crowcroft, Bandwidth-delay based routing algorithms, in *Proc. IEEE GLOBECOM*, vol. 3, 26–29 May 1995, pp. 2129–2133.

[126] D. Bauer, J.N. Daigle, I. Iliadis and P. Scotton, Efficient frontier formulation for additive and restrictive metrics in hierarchical routing, in *IEEE Int. Conf. Communications Conf. Rec.*, vol. 3, June 2000, p. 1353.

[127] Z. Wang and J. Crowcroft, Quality-of-service routing for supporting multimedia applications, *IEEE J. Select. Areas Commun.*, vol. 14, 1996, pp. 1228–1234.

[128] M. Faloutsos, P. Faloutsos, and C. Faloutsos, On power-law relationships of the internet topology, in *Proc. ACM SIGCOMM*, 1999, pp. 251–262.

[129] C. Jin, Q. Chen, and S. Jamin, Inet-3.0: Internet Topology Generator, University of Michigan, Technical Report CSE-TR-456-02, 2002.

[130] B.M. Waxman, Routing of multipoint connections, *IEEE J. Select. Areas Commun.*, vol. 6, 1988, pp. 1617–1622.

[131] T.H. Cormen, C.E. Leiserson, and R.L. Rivest, *Introduction to Algorithms*. MIT Press: Cambridge, MA, 1990.

[132] M. Garey and D. Johnson, *Computers and Intractability: A Guide to the Theory of NP-Completeness*. Freeman: New York, 1979.

8

Effective Capacity

8.1 EFFECTIVE TRAFFIC SOURCE PARAMETERS

For efficient resource allocation for connections requiring QoS, reliable and accurate characterizations of the offered traffic and available effective capacity in the network are needed [1–5, 7–21]. The ATM forum [1] has adopted an open-loop flow control mechanism for ATM networks known as usage parameter control (UPC). A UPC algorithm accepts a cell stream as input and determines whether or not each cell conforms to the parameters of the algorithm. Depending on the network policy, nonconforming cells are either dropped prior to entering the network or tagged as low priority. The UPC function acts as a *traffic policer* at the network access point. In contrast, a *traffic shaper* smoothes out the bitrate variations in a stream by delaying rather than dropping or tagging cells. A user can avoid possible cell loss at the network access point due to UPC by shaping its cell stream to conform to the negotiated UPC. All those actions will result in an *effective traffic source*. The UPC mechanisms proposed by the ATM Forum are variants of the leaky bucket algorithm. The leaky bucket is a simple means of shaping or policing a traffic stream and has desirable attributes in terms of network control [2–4].

A leaky bucket is parameterized by a leak rate μ and a bucket size B. It is convenient to describe the operation of the leaky bucket in terms of a fictitious $x/D/1/B$ queue for a deterministic input (D) served at constant rate μ with capacity B [2]. In this model, each cell arrival to the leaky bucket generates a corresponding customer arrival to the fictitious queue. At the time of a cell arrival, if the number of customers in the fictitious queue is less than B, the cell is said to be *conforming*, otherwise the cell is *nonconforming*. Nonconforming cells are either dropped or tagged as low priority cells.

A leaky bucket shaper parameterized by (μ, B) operates in a manner similar to the leaky bucket policer, except that nonconforming cells are delayed, rather than dropped or tagged. The operation of the leaky bucket shaper can be understood in terms of an $x/D/1$ queue. Upon arrival of a cell, if the number of customers in the fictitious queue is B or more, the

Advanced Wireless Networks: 4G Technologies Savo G. Glisic
© 2006 John Wiley & Sons, Ltd.

cell is placed in an infinite capacity first-in, first-out (FIFO) buffer. Whenever a customer leaves the fictitious queue, a cell (if any) is removed and transmitted from the FIFO buffer.

The *dual leaky bucket* consists of two leaky buckets: a *peak rate* leaky bucket with parameters (λ_p, B_p) and a *sustainable rate* leaky bucket with parameters (λ_s, B_s), where $\lambda_p > \lambda_s$. A cell is conforming with respect to the dual leaky bucket if and only if it is conforming with respect to both the peak and sustainable rate buckets. The sustainable rate λ_s represents the maximum long-run average rate for conforming cell streams. The peak rate λ_p specifies a minimum intercell spacing of $1/\lambda_p$ with burst tolerance B_p. In this section, we shall assume that $B_p = 1$, so that the intercell spacing of conforming streams is constrained to be at most $1/\lambda_p$. Values of $B_p > 1$ allow for cell delay variation (CDV) between the customer premises and the network access point.

The set $(\lambda_p, \lambda_s, B_s)$ constitutes a UPC descriptor that the user negotiates with the network. If the connection is accepted, the user applies a dual leaky bucket shaper to shape its cell stream according to the negotiated UPC. The dual leaky bucket limits the peak rate to λ_p, the long-term average rate to λ_s, and the maximum burst length (transmitted at peak rate) to $B_c = B_s\lambda_p/(\lambda_p - \lambda_s)$.

Deterministic traffic characterization is based on the queueing model of the leaky bucket [5]. Suppose that the cell stream is of finite length and that the peak rate λ_p is known. The cell stream is offered as input to a queue served at constant service rate μ. Let $B_{max}(\mu)$ be the maximum number of cells observed in the system over the duration of the cell stream. The stream then can be characterized by the peak rate and the values $B_{max}(\mu)$ for $0 < \mu < \lambda_p$ as

$$C_D = \left[\lambda_p, B_{max}(\mu), 0 < \mu < \lambda_p\right] \tag{8.1}$$

Equation (8.1) can be interpreted as follows. For each μ, $0 < \mu < \lambda_p$, the cell stream passes through a leaky bucket parameterized by $[\mu, B_{max}(\mu)]$ without being shaped, i.e. the cell stream conforms to the leaky bucket parameters $[\mu, B_{max}(\mu)]$. The notation $B(C_D, \mu) = B_{max}(\mu)$ is introduced to show that the choice of bucket size, for a given leak rate μ, depends on C_D. The characterization is *tight* in the following sense. If the cell stream conforms to a given set of leaky bucket parameters (μ, B'), then necessarily $B' \geq B(C_D, \mu)$. For a given leak rate μ^*, the minimum bucket size for conformance is $B^* \geq B(C_D, \mu^*)$.

Ideal statistical traffic characterization is based on assumption that the peak rate of the cell stream λ_p and the mean rate λ_m are known. Suppose that the cell stream is offered to a leaky bucket traffic shaper with leak rate μ, where $\lambda_m < \mu < \lambda_p$, and bucket size B.

The shaping probability is defined as the probability that an arbitrary arriving cell has to wait in the data buffer of the leaky bucket. In the $x/D/1$ queueing model of the leaky bucket, this event corresponds to a customer arrival to the fictitious queue when the number in the system exceeds B. If $B(\mu, \varepsilon)$ is defined as the minimum bucket size necessary to ensure that the shaping probability does not exceed ε, then

$$C_S(\varepsilon) = \left[\lambda_p, \lambda_m, (\mu, B(\mu, \varepsilon)), \lambda_m < \mu < \lambda_p\right] \tag{8.2}$$

represents statistical traffic characterization.

Approximate statistical traffic characterization is based on the assumption that the cell stream is offered to a queue with constant service rate μ, where $\lambda_m < \mu < \lambda_p$. If W is the steady-state waiting time, in the queue, for an arbitrary cell arriving in the queue, then the complementary waiting time distribution, can be approximated as (see Appendix C; please go to www.wiley.com/go/glisic):

$$P(W > t) \approx a(\mu)e^{-b(\mu)t}, \quad t \geq 0 \tag{8.3}$$

Two asymptotic approximations are of interest:

$$\begin{aligned} e^{bt}P(W > t) &\to a \; ; \; t \to \infty \\ t^{-1}\log P(W > t) &\to -b \; ; \; t \to \infty \end{aligned} \tag{8.4}$$

Also the following approximation is of interest, as suggested in Chang [6]: $a \approx bE(W)$.

In many cases, the approximation in Equation (8.3) with $a = 1$ tends to be conservative, i.e. the bound, $P(W > t) \leq e^{-bt}$, holds under certain conditions [6, 7]. The one-parameter approximation

$$P(W > t) \approx e^{-bt}, \quad t \geq 0 \tag{8.5}$$

is the basis for the theory of *effective bandwidths* that will be discussed below.

More extensive discussions of this class of approximations can be found in References [7, 8]. In this section Equation (8.3) is used as a reasonable approximation for an ATM cell stream modeled as a general Markovian source. Over the relatively short observation windows of interest (on the order of seconds), these approximations are fairly robust for a large class of real traffic sources, in particular, MPEG video sequences. Therefore, parameters $[a(\mu), b(\mu)]$ provide a statistical characterization of the cell stream when offered to a queue with a constant service rate μ. A more complete characterization of the cell stream records the values $[a(\mu), b(\mu)]$ for all μ in the range (λ_m, λ_p) as follows:

$$\hat{C}_S = \left\{\lambda_p, \lambda_m, [a(\mu), b(\mu)] : \lambda_m < \mu < \lambda_p\right\} \tag{8.6}$$

Values of $[a(\mu), b(\mu)]$ are chosen such that Equation (8.3) holds *approximately* for values of t in range of interest. Assuming that Equation (8.3) holds at equality, we have (see Chapter 6)

$$a(\mu)/b(\mu) = E[W] = \tau_r(\mu)E[S_a] + E[Q_a]/\mu \tag{8.7}$$

In Equation (8.7) S_a denotes the number of customers in service, Q_a the number of customers in queue seen by an arbitrary customer arrival, and $\tau_r(\mu)$ the average remaining service time for the customer in service as seen by an arriving customer conditional on there being a customer in service. Bearing in mind that

$$E[S_a] = P(W > 0) = a(\mu) \tag{8.8a}$$

and defining $q(\mu) = E[Q_a]$, Equation (8.7) gives

$$b(\mu) = \frac{\mu a(\mu)}{q(\mu) + \mu a(\mu)\tau_r(\mu)} \tag{8.8b}$$

The estimation of the above parameters is discussed later in this section. Once these parameters are obtained they are used in Equation (8.6) for statistical characterization of a cell stream. The next step is to map the characterization to a suitable UPC descriptor.

8.1.1 Effective traffic source

To avoid UPC violation at the network access point, typically a shaper is applied at the source to force the cell stream to conform to the negotiated UPC, resulting in an *effective traffic source*. If the user can tolerate a larger shaping delay, a UPC descriptor with smaller network cost can be chosen for the connection. So, the problem of selecting the UPC values for a cell stream can be formulated as:

$$\min_{\lambda_s, B_s} c\left(\lambda_p, \lambda_s, B_s\right) \text{ and } P(shaping) \leq \varepsilon \text{ and } \lambda_m \leq \lambda_s \leq \lambda_p \tag{8.9}$$

where parameters (λ_m, λ_p) are assumed to be known *a priori* or obtained through measurements. The *cost function c( )* is an increasing function of each argument and represents the cost to the network of accepting a call characterized by the UPC values $(\lambda_p, \lambda_s, B_s)$. The shaping probability *P(shaping)*, which depends on the cell stream characteristics, represents the probability that an arbitrary arriving cell will be delayed by the shaper.

In the case of an idealized traffic characterization $C_S(\varepsilon)$, defined by Equation (8.2), the values of $B(\mu, \varepsilon) for\, \lambda_m < \mu < \lambda_p$ are recorded, where $B(\mu, \varepsilon)$ is the minimum bucket size that ensures a shaping probability less than ε for a leaky bucket with rate μ.

Alternatively, from the approximate statistical characterization, we can obtain an estimate $B(\hat{C}_s, \mu, \varepsilon)$ for this minimum bucket size. Substituting $B(\hat{C}_s, \mu, \varepsilon)$ in Equation (8.9), we have

$$\min_{\lambda_s} c[\lambda_p, \lambda_s, (B(\hat{C}_S, \lambda_s, \varepsilon)] \text{ and } \lambda_m \leq \lambda_s \leq \lambda_p \tag{8.9a}$$

8.1.2 Shaping probability

In a $x/D/1$ model of the (μ, B) leaky bucket shaper, an arriving cell is delayed *(shaped)* if and only if the number of customers seen in the fictitious queue is B or greater. Otherwise, the cell is transmitted immediately without delay. Suppose that the user wants to choose the UPC descriptor $(\lambda_p, \lambda_s, B_s)$ such that the probability that an arbitrary cell will be shaped is less than ε. For a fixed leak rate μ, we would like to determine the minimum bucket size $B(\mu, \varepsilon)$ required to ensure a shaping probability less than ε. On should remember that the shaping probability is an upper bound for the *violation probability*, i.e., the probability that an arriving cell is nonconforming, in the corresponding UPC policer. Let n_a be a random variable representing the number in the fictitious queue observed by an arbitrary cell arrival. Suppose that the number of customers in the fictitious queue upon cell arrival is given by $n = N$, where $N \geq 1$ is an integer. This means that the new customer sees $N - 1$ customers in the queue and one customer currently in service.

Therefore, the waiting time in the queue W' for this customer lies in the range $[(N - 1)/\mu, N/\mu]$. Also we can say, if W' for this customer lies in this range, then $n = N$ at the arrival epoch. In other words $n = N \Leftrightarrow (N - 1)/\mu < W' < N/\mu$ so $n \geq N \Leftrightarrow (N - 1)/\mu < W'$.

Based on this we can write $P(n_a \geq B) = P[W > (B - 1)/\mu]$. So by using Equation (8.3), the shaping probability can be expressed as

$$P(shaping) = P(n_a \geq B) \approx a(\mu)e^{-b(\mu)(B-1)/\mu}$$

From this relation, for $P(shaping) = \varepsilon$ we get

$$B(\hat{C}_s, \mu, \varepsilon) = [\mu/b(\mu)]\log[a(\mu)/\varepsilon] + 1 \tag{8.10}$$

8.1.3 Shaping delay

We can also specify the shaping constraint in terms of delay. Consider a cell stream passing through a leaky bucket traffic shaper with parameters (μ, B). Recall that the state counter n may be viewed as the number of customers in a fictitious $x/D/1$ queue. If W represents the waiting time in queue experienced by a given customer arriving in the fictitious queue served at rate μ, the *actual* delay experienced by the cell corresponding to the given fictitious customer can be expressed as

$$D = \max[0, W - (B - 1)/\mu)] \tag{8.11}$$

Using Equation (8.3) and W from Equation (8.11) we have

$$\begin{aligned} P(D > D_{\max}) &\approx a(\mu)e^{-b(\mu)D_{\max}}e^{-b(\mu)(B-1)/\mu} \\ E[D] &\approx [a(\mu)/b(\mu)]e^{-b(\mu)(B-1)/\mu} \end{aligned} \tag{8.12}$$

Setting the expected delay equal to the target mean delay $\bar{D}$ gives B as

$$B(C_s, \mu, \bar{D}) = [\mu/b(\mu)]\log[a(\mu)/b(\mu)\bar{D}] + 1 \tag{8.13}$$

which will be referred to as *mean delay constraint curve.*

The *cost function* $c(\lambda_p, \lambda_s, B_s)$ represents the amount of resources that should be allocated to a cell characterized by the UPC parameters $(\lambda_p, \lambda_s, B_s)$. For a source with deterministic periodic on–off sample paths and a random offset, the lengths of the on and off periods are given by

$$T_{on} = B_s/(\lambda_p - \lambda_s) \text{ and } T_{off} = B_s/\lambda_s \tag{8.14}$$

Consider an on–off source model in which the source alternately transmits at rate r during an on-period and is silent during an off-period. The mean on-period and off period lengths are given by β^{-1} and α^{-1}, respectively. For mathematical tractability, Erlang-k distribution for on and off periods is assumed.

The *effective bandwidth* of a source is defined as the minimum capacity required to serve the traffic source so as to achieve a specified steady-state cell loss probability ε_{mux}, where the multiplexer or switch buffer size is denoted by B_{mux}. Using the result from Kobayashi and Ren [9], the effective bandwidth for this source model can be obtained as

$$c(eff) = r\alpha/(\alpha + \beta) + r^2\alpha\beta K/(\alpha + \beta)^3 \tag{8.15}$$

where $K = -\log \varepsilon_{mux}/(2kB_{mux})$. If we set the mean on and off periods as $\beta^{-1} = T_{on}$ and $\alpha^{-1} = T_{off}$, then by using Equation (8.14) in Equation (8.15), we have

$$c(eff) = \lambda_s + \lambda_s(\lambda_p - \lambda_s)KB_s/\lambda_p \tag{8.16}$$

Now, the following cost function for a UPC controlled source can be defined:

$$c(\lambda_p, \lambda_s, B_s) = \min[\lambda_p, c(eff)] = \min[\lambda_p, \lambda_s + \lambda_s\left(\lambda_p - \lambda_s\right)KB_s/\lambda_p] \tag{8.17}$$

The value for parameter K should be chosen such that $c(\lambda_p, \lambda_s, B_s)$ approximates the bandwidth allocation of the network CAC policy. The limiting behavior of the cost function with respect to K is

$$c(\lambda_p, \lambda_s, B_s) \to \lambda_s \text{ as } K \to 0 \quad \text{and} \quad c(\lambda_p, \lambda_s, B_s) \to \lambda_p \text{ as } K \to \infty$$

Real-time UPC estimation is based on observations of the cell stream over a time period T. The procedure consists of the following steps: (1) estimate the mean and peak cell rates; (2) choose a set of candidate sustainable rates; (3) obtain the characterization of the stream; (4) for each candidate sustainable rate, compute the minimum bucket size required to meet a user constraint on shaping delay; and (5) choose the candidate rate that minimizes the cost function.

The *peak and mean rates* can be estimated from observations of the cell stream over a window of length T. The interval is subdivided into smaller subintervals of equal length $T_s = T/M$. For example, in experiments with MPEG video sequences described in Mark and Ramamurthy [5], T_s was set equal to the frame period of 33 ms. If n is the total number of cell arrivals during T and n_i the number of cell arrivals in the ith subinterval, then the mean rate can be estimated as $\hat{\lambda}_m = n/T$. For some applications, the peak rate may be known *a priori*. Otherwise, the peak rate can be estimated by $\hat{\lambda}_p = \max\{n_i/T_s; 1 \le i \le M\}$.

Candidate rates $\mu_1, \ldots, \mu_N$ are chosen between the mean and peak rates. The simplest approach is to choose N and space the candidate rates at uniform spacing between the mean and peak rates. However, with a more judicious method of assignment, the set of candidate rates can be kept small, while still including rates close to the near-optimal sustainable rates. Suppose that a prior estimate of the operating sustainable rate, $\hat{\lambda}_s$, is available. In the absence of prior knowledge, one could assign $\hat{\lambda}_s = (\hat{\lambda}_m + \hat{\lambda}_p)/2$. The candidate rate μ_N is assigned equal to $\hat{\lambda}_s$. The remaining rates are grouped into a set of Nc *coarsely spaced rates* and N_f *finely spaced rates*, with $N = N_c + N_f + 1$ and with N_f even. The coarse rates $\mu_1, \ldots, \mu_{N_c}$ are assigned to be spaced uniformly over the interval $(\hat{\lambda}_s, \hat{\lambda}_p)$ as $\mu_i = \hat{\lambda}_m + i\Delta_c, i = 1, \ldots, N_c$, where Δ_c is defined as $\Delta_c = (\hat{\lambda}_p - \hat{\lambda}_m)/(N_c + 1)$. The remaining N_f fine rates are assigned in the neighborhood of $\hat{\lambda}_s$ as follows: $\mu_{j+N_c} = \hat{\lambda}_s + j\Delta_f,\ j = 1, \ldots, N_f/2$, and $\mu_{j+N_c} = \hat{\lambda}_s - (j - N_f/2)\Delta_f,\ j = N_f/2, \ldots, N_f$, where $\Delta_f = \Delta_c/(N_c + 1)$. Figure 8.1 shows an example of the choice of candidate rates example with $N = 8$, $N_c = 3$ and $N_f = 4$.

Measurement of traffic parameters is based on the bank of queues. The ith fictitious queue is implemented using a counter n_i that is decremented at the constant rate μ_i and incremented by 1 whenever a cell arrives [5]. For each rate μ_i, estimates of the queue parameters $\hat{a}(\mu_i)$ and $\hat{b}(\mu_i)$ are obtained. Each queue is sampled at the arrival epoch of an 'arbitrary' cell. This sampling is carried out by skipping a random number of N_j cells between the $(j\text{-}1)_{st}$ and jth sampled arrivals. Since all of the fictitious queues are sampled at the same arrival epochs, only one random number needs to be generated for all queues at a sampling epoch. Over an interval of length T, a number of samples, say M, are taken from the queues. At the jth sampling epoch, the following quantities are recorded for each queue: S_j, the number of customers in service (so $S_j \in \{0, 1\}$); Q_j, the number of customers in queue; and T_j, the remaining service time of the customer in service (if one is in service). After that the following sample means are computed:

$$\hat{a} = \sum_{j=1}^{M} S_j/M; \; \hat{q} = \sum_{j=1}^{M} Q_j/M; \; \hat{\tau}_r = \sum_{j=1}^{M} T_j/\hat{a}M$$

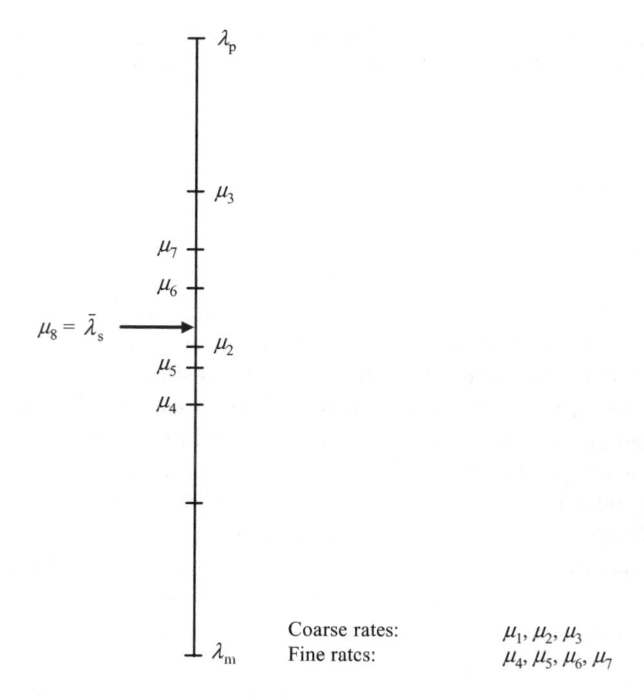

Figure 8.1 Example of candidate rate assignment.

Parameter $\hat{b}$ is then computed by using Equation (8.8) as

$$\hat{b} = \frac{\mu\hat{a}}{\hat{q} + \mu\hat{a}\hat{\tau}_r} \approx \frac{\mu\hat{a}}{\hat{q} + \hat{a}/2}$$

where $\hat{\tau}_r$ is approximated by $\frac{1}{2}\mu$. The approximate statistical characterization is given by

$$\hat{C}_S = \left\{\hat{\lambda}_p, \hat{\lambda}_m \left[\hat{a}(\mu_i),\ \hat{b}(\mu_i)\right] : i = 1, \ldots, N\right\}$$

Mapping to a UPC Descriptor is focused on the case where the user specifies a maximum mean delay $\bar{D}$. The operating rate index is chosen to minimize the cost function defined in Equation (8.17):

$$i^* = \arg \min_{1 \le i \le N} c\left[\hat{\lambda}_p, \mu_i, B_s\left(\hat{C}_S, \mu_i, \bar{D}\right)\right]$$

where $B_s\left(\hat{C}_S, \mu_i, \bar{D}\right)$ is defined in Equation (8.13) and represents the bucket size required to achieve a mean delay of $\bar{D}$ at a leak rate μ. Then the operating UPC descriptor is assigned as follows:

$$\lambda_p = \hat{\lambda}_p, \quad \lambda_s = \mu_{i*}, \quad B_s = B_s\left(\hat{C}_S, \mu_i, \bar{D}\right)$$

If the estimated mean and peak rates are close in value, the stream can be treated as a CBR stream. In this case, the UPC descriptor can simply be assigned according to the peak rate, i.e.

$$\lambda_p = \hat{\lambda}_p, \quad \lambda_s = \mu_{i*}, \quad B_s = 1$$

Some applications, such as video delivery, interactive multimedia sessions, etc., cannot be adequately characterized by a static set of parameters that are expected to hold for the entire session. The ability to renegotiate parameters during a session may be a viable way of efficiently supporting variable bit rate (VBR) traffic [5, 10] with real-time QoS constraints.

8.1.4 Performance example

In the MPEG standard for video compression, video is transmitted as a sequence of frames. For transmission over an ATM network, each MPEG frame must be segmented into ATM cells. In Mark and Ramamurthy [5], two modes of cell transmission over a frame interval are considered. In *nonsmooth mode*, the cells are transmitted at a constant rate λ_p starting at the beginning of the frame interval until the frame is fully transmitted. Then the source is silent until the beginning of the next frame interval. This results in an on–off type stream with correlated on and off periods. In *smooth mode*, the cells for the ith frame are transmitted at a constant rate $f(i)/\tau$ over the frame interval, where $f(i)$ is the total number of cells in the $i-$th frame and τ is the interframe period. Thus, buffering is done to smooth out the transmission of cells of each frame over the corresponding frame interval.

Figure 8.2 shows the characterization $\hat{C}_S$ obtained empirically for an MPEG video stream [5]. For this stream, the empirical mean and peak rates were determined to be 20 and 40.5 cells/ms, respectively. The two streams were offered to 19 queues running in parallel at the rates $\mu = 21, 22, \ldots, 39$ (cells/ms) and the corresponding values for $[a(\mu), b(\mu)]$ were obtained according to the procedure described above. In the nonsmooth mode, $a\,(\mu) \approx 1$ for all values of μ while for smooth mode, $a(\mu)$ monotonically decreases as μ increases from λ_m to λ_p. On the other hand the $b\,(\mu)$ curves for the nonsmooth and smooth modes are close for μ near to the extremes λ_m and λ_p. For values of μ closer to the center of the interval (λ_m, λ_p), the curve for smooth mode lies above the curve for nonsmooth mode. Thus, the characterization $\hat{C}_S$ captures the decrease in traffic burstiness due to intraframe smoothing.

Figures 8.3 and 8.4 demonstrate that, for example, the shaping probability of 0.00001 at cell rate of 28 can be achieved with $B = 1100$ in the case of *smooth mode* while nonsmooth mode would require $B = 2000$ for the same shaping probability. More interesting results on this issue can be found in Mark and Ramamurthy [5].

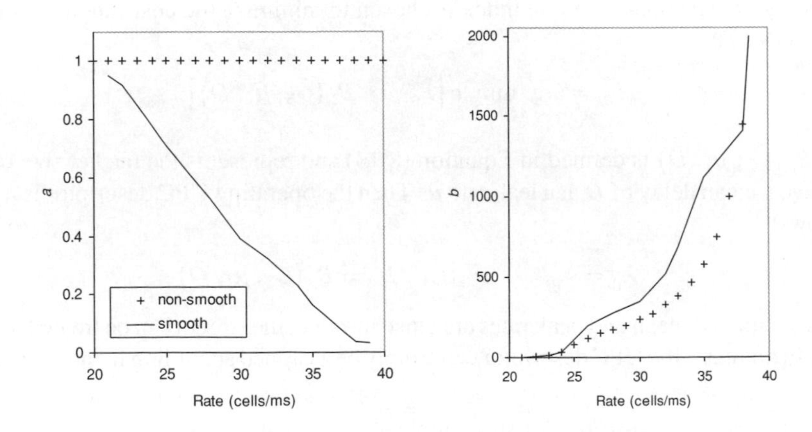

Figure 8.2 (a) a and (b) b parameters for MPEG.

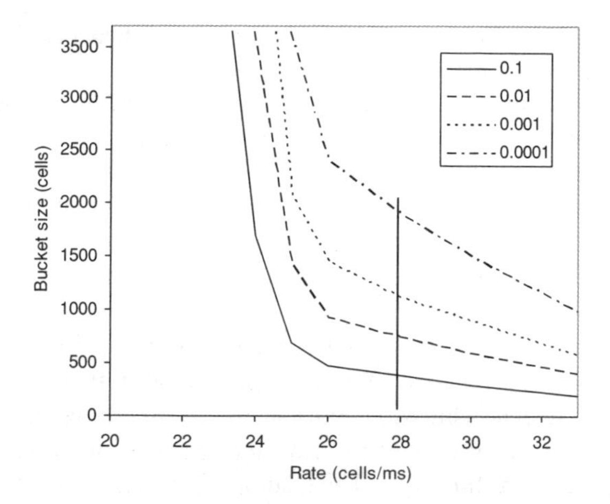

Figure 8.3 Shaping probability curves for nonsmooth mode.

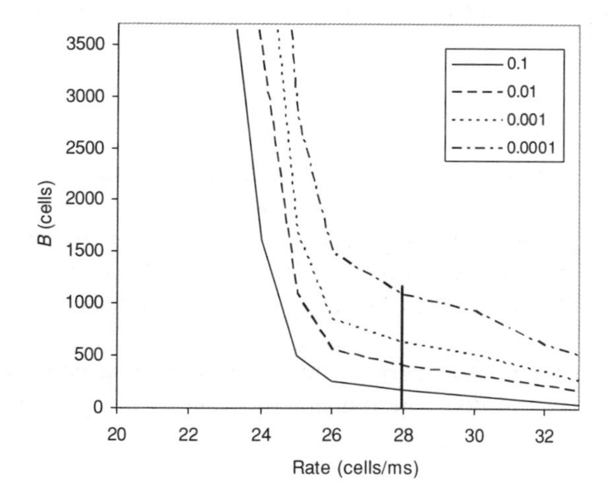

Figure 8.4 Shaping probability curves for smooth mode.

8.2 EFFECTIVE LINK LAYER CAPACITY

In the previous section we discussed the problem of effective characterization of the traffic source in terms of the parameters which are used to negotiate with the network a certain level of QoS. Similarly, on the network side, to enable the efficient support of quality of service (QoS) in 4G wireless networks, it is essential to model a wireless channel in terms of connection-level QoS metrics such as data rate, delay and delay-violation probability. The traditional wireless channel models, i.e. physical-layer channel models, do not explicitly characterize a wireless channel in terms of these QoS metrics. In this section, we discuss *a link-layer channel model* referred to as *effective capacity* (EC) [11]. In this approach, a wireless link is modeled by two EC functions, the probability of nonempty buffer and the QoS exponent of a connection. Then, a simple and efficient algorithm to estimate these EC functions is discussed. The advantage of the EC link-layer modeling and estimation is ease of translation into QoS guarantees, such as delay bounds and hence, efficiency in admission control and resource reservation.

Conventional channel models, discussed in Chapter 3, directly characterize the fluctuations in the amplitude of a radio signal. These models will be referred to as *physical-layer channel* models, to distinguish them from the *link-layer channel* model discussed in this section. Physical-layer channel models provide a quick estimate of the physical-layer performance of wireless communications systems (e.g. symbol error rate vs SNR). However, physical-layer channel models cannot be easily translated into complex link-layer QoS guarantees for a connection, such as bounds on delay. The reason is that these complex QoS requirements need an analysis of the queueing behavior of the connection, which cannot be extracted from physical-layer models. Thus, it is hard to use physical-layer models in QoS support mechanisms, such as admission control and resource reservation.

For these reasons, it was proposed to move the channel model up the protocol stack from the physical layer to the link layer. The resulting model is referred to as EC link model because it captures a generalized link-level capacity notion of the fading channel. Figure 8.5 illustrates the difference between the conventional physical-layer and link-layer model. For simplicity, the 'physical-layer channel' will be called the 'physical channel' and 'link-layer channel' will be referred to as the 'link'.

8.2.1 Link-layer channel model

4G wireless systems will need to handle increasingly diverse multimedia traffic, which are expected to be primarily packet-switched. The key difference between circuit switching and packet switching, from a link-layer design viewpoint, is that packet switching requires *queueing* analysis of the link. Thus, it becomes important to characterize the effect of the data traffic pattern, as well as the channel behavior, on the performance of the communication system.

QoS guarantees have been heavily researched in the *wired* networks [e.g. ATM and Internet protocol (IP) networks]. These guarantees rely on the queueing model shown in

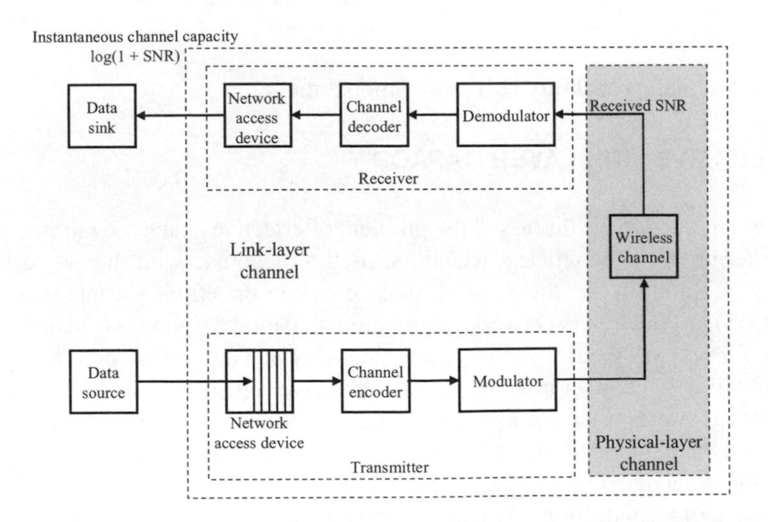

Figure 8.5 Packet-based wireless communication system.

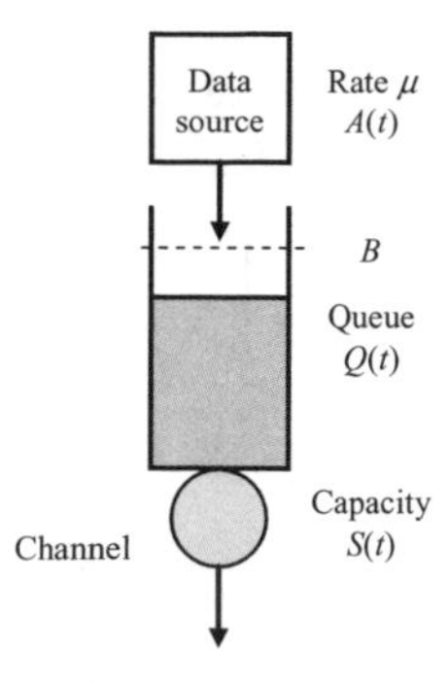

Figure 8.6 Queueing system model.

Figure 8.6 and studied in Chapter 6. This figure shows that the source traffic and the network service are matched using a first-in first-out buffer (queue). Thus, the queue prevents loss of packets that could occur when the source rate is more than the service rate, at the expense of increasing the delay. Queueing analysis, which is needed to design appropriate admission control and resource reservation algorithms, requires source *traffic characterization* and *service characterization*. The most widely used approach for traffic characterization is to require that the amount of data (i.e. bits as a function of time t) produced by a source conform to an upper bound, called the *traffic envelope*, $\Gamma(t)$. The service characterization for guaranteed service is a guarantee of a minimum service (i.e. bits communicated as a function of time) level, specified by a service curve *SC* $\Psi(t)$[12]. Functions $\Gamma(t)$ and $\Psi(t)$ are specified in terms of certain traffic and service parameters, respectively. Examples include the UPC parameters, discussed in the previous section, used in ATM for traffic characterization, and the traffic specification (T-SPEC) and the service specification (R-SPEC) fields used with the resource reservation protocol [12] in IP networks.

A traffic envelope $\Gamma(t)$ characterizes the source behavior in the following manner: over any window of size t, the amount of actual source traffic $A(t)$ does not exceed $\Gamma(t)$ (see Figure 8.7). For example, the UPC parameters, discussed in Section 8.1, specifies $\Gamma(t)$ by

$$\Gamma(t) = \min\left\{\lambda_{\mathrm{p}}^{(s)}t,\ \lambda_{\mathrm{s}}^{(s)}t + \sigma^{(s)}\right\} \tag{8.18}$$

where $\lambda_{\mathrm{p}}^{(s)}$ is the peak data rate, $\lambda_{\mathrm{s}}^{(s)}$ the sustainable rate, and $\sigma^{(s)} = B_{\mathrm{s}}$ the leaky-bucket size [12]. As shown in Figure 8.7, the curve $\Gamma(t)$ consists of two segments: the first segment has a slope equal to the peak source data rate $\lambda_{\mathrm{p}}^{(s)}$, while the second has a slope equal to the sustainable rate $\lambda_{\mathrm{s}}^{(s)}$, with $\lambda_{\mathrm{s}}^{(s)} < \lambda_{\mathrm{p}}^{(s)}$. $\sigma^{(s)}$ is the y-axis intercept of the second segment. $\Gamma(t)$ has the property that $A(t) \leq \Gamma(t)$ for any time t. Just as $\Gamma(t)$ upper bounds the source traffic, a network SC $\Psi(t)$ lower bounds the actual service $S(t)$ that a source will receive. $\Psi(t)$ has the property that $\Psi(t) \leq S(t)$ for any time t. Both $\Gamma(t)$ and $\Psi(t)$ are negotiated during the admission control and resource reservation phase. An example of a network SC is the R-SPEC curve used for guaranteed service in IP networks

$$\Psi(t) = \max\left[\lambda_{\mathrm{s}}^{(c)}(t - \sigma^{(c)}), 0\right] = \left[\lambda_{\mathrm{s}}^{(c)}(t - \sigma^{(c)})\right]^{+} \tag{8.19}$$

where $\lambda_{\mathrm{s}}^{(c)}$ is the constant service rate, and $\sigma^{(c)}$ the delay (due to propagation delay, link sharing, and so on). This curve is illustrated in Figure 8.7. $\Psi(t)$ consists of two segments;

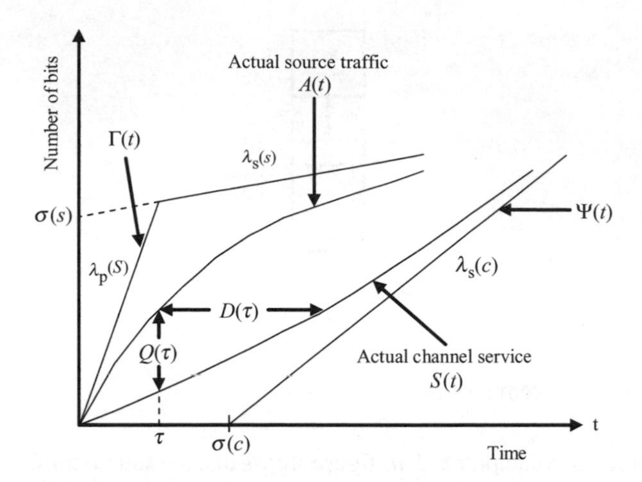

Figure 8.7 Traffic and service characterization. (Reproduced by permission of IEEE [11].)

the horizontal segment indicates that no packet is being serviced due to propagation delay, etc., for a time interval equal to the delay $\sigma^{(c)}$, while the second segment has a slope equal to the service rate $\lambda_s^{(c)}$. In the figure, the horizontal difference between $A(t)$ and $S(t)$, denoted $D(\tau)$, is the delay experienced by a packet arriving at time τ, and the vertical difference between the two curves, denoted by $Q(\tau)$, is the queue length built up at time τ, due to packets that have not been served yet.

As discussed in Chapter 4, providing QoS guarantees over wireless channels requires accurate models of their *time-varying capacity*, and effective utilization of these models for QoS support. The simplicity of the SCs discussed earlier motivates us to define the time-varying capacity of a wireless channel as in Equation (8.19). Specifically, we hope to lower bound the channel service using two parameters: the channel sustainable rate $\lambda_s^{(c)}$, and the maximum fade duration $\sigma^{(c)}$.

Parameters $\lambda_s^{(c)}$ and $\sigma^{(c)}$ are meant to be in a statistical sense. The maximum fade duration $\sigma^{(c)}$ is a parameter that relates the delay constraint to the channel service. It determines the probability $\sup_t \Pr\{S(t) < \Psi(t)\}$. We will see later that $\sigma^{(c)}$ is specified by the source with $\sigma^{(c)} = D_{\max}$, where D is the delay bound required by the source.

However, physical-layer wireless channel models do not explicitly characterize the channel in terms of such link-layer QoS metrics as data rate, delay and delay-violation probability. For this reason, we are forced to look for alternative channel models.

A problem that surfaces is that a wireless channel has a capacity that varies *randomly* with time. Thus, an attempt to provide a strict lower bound [i.e. the deterministic SC $\Psi(t)$, used in IP networks] will most likely result in extremely conservative guarantees. For example, in a Rayleigh or Ricean fading channel, the only lower bound that can be *deterministically* guaranteed is a capacity of zero. The capacity here is meant to be delay-limited capacity, which is the maximum rate achievable with a prescribed delay bound (see Hanly and Tse [13] for details). This conservative guarantee is clearly useless in practice. Therefore, the concept of deterministic SC $\Psi(t)$ is extended to a *statistical* version, specified as the pair $\{\Psi(t),\ \varepsilon\}$. The statistical SC $\{\Psi(t),\ \varepsilon\}$ specifies that the service provided by the channel, denoted $\tilde{S}(t)$, will always satisfy the property that $\sup_t \Pr\left\{\tilde{S}(t) < \Psi(t)\right\} \le \varepsilon$. In other words, ε is the probability that the wireless channel will not be able to support the pledged SC

$\Psi(t)$, referred to as the outage probability. For most practical values of ε, a *nonzero* SC $\Psi(t)$ can be guaranteed.

8.2.2 Effective capacity model of wireless channels

From the previous discussion we can see that for the QoS control we need to calculate an SC $\Psi(t)$ such that, for a given $\varepsilon > 0$, the following probability bound on the channel service $\tilde{S}(t)$ is satisfied:

$$\sup_t \Pr\left\{\tilde{S}(t) < \Psi(t)\right\} \leq \varepsilon \text{ and } \Psi(t) = \left[\lambda_s^{(c)}(t - \sigma^{(c)})\right]^+ \tag{8.20}$$

The statistical SC specification requires that we relate its parameters $\{\lambda_s^{(c)}, \sigma^{(c)}, \varepsilon\}$ to the fading wireless channel, which at the first sight seems to be a hard problem. At this point, the idea that the SC $\Psi(t)$ is a *dual* of the traffic envelope $\Gamma(t)$ is used [11]. A number of papers exist on the so-called *theory of effective bandwidth* [14], which models the statistical behavior of *traffic*. In particular, the theory shows that the relation $\sup_t \Pr\{Q(t) \geq B\} \leq \varepsilon$ is satisfied for large B, by choosing two parameters (which are functions of the channel rate r) that depend on the actual data traffic, namely, the probability of nonempty buffer, and the effective bandwidth of the source. *Thus, a source model defined by these two functions fully characterizes the source from a QoS viewpoint.* The duality between Equation (8.20) and $\sup_t \Pr\{Q(t) \geq B\} \leq \varepsilon$ indicates that it may be possible to adapt the theory of effective bandwidth to SC characterization. This adaptation will point to a new channel model, which will be referred to as the *effective capacity (EC) link model*. Thus, the EC link model can be thought of as the dual of the effective bandwidth source model, which is commonly used in networking.

8.2.2.1 Effective bandwidth

The stochastic behavior of a source traffic can be modeled asymptotically by its effective bandwidth. Consider an arrival process $\{A(t), t \geq 0\}$, where $A(t)$ represents the amount of source data (in bits) over the time interval [0, t). Assume that the asymptotic *log-moment generating function* of $A(t)$, defined as

$$\Lambda(u) = \lim_{t\to\infty} \frac{1}{t} \log E\left[e^{uA(t)}\right] \tag{8.21}$$

exists for all $u \geq 0$. Then, the *effective bandwidth function* of $A(t)$ is defined as [11, 14]

$$\alpha(u) = \frac{\Lambda(u)}{u}, \qquad \forall u \geq 0. \tag{8.22}$$

This becomes more evident if we assume the constant traffic $A(t) = A$ so that Equation (8.21) gives

$$\Lambda(u) = \lim_{t\to\infty} \frac{1}{t} \log E\left[e^{uA(t)}\right] = \lim_{t\to\infty} \frac{1}{t} \log E\left[e^{uA}\right] = \lim_{t\to\infty} \frac{uA}{t} = u \lim_{t\to\infty} \frac{A}{t} = u\alpha(u) \tag{8.21a}$$

Consider now a queue of infinite buffer size served by a channel of *constant service rate* r, such as an AWGN channel. Owing to the possible mismatch between $A(t)$ and $S(t)$, the queue length $Q(t)$ could be nonzero. As already discussed in Section 8.1 [analogous to Equation (8.5)] or by using the theory of large deviations [14] , it can be shown that the probability of $Q(t)$ exceeding a threshold B satisfies

$$\sup_t \Pr\{Q(t) \geq B\} \sim e^{-\theta_B(r)B} \quad \text{as } B \to \infty \tag{8.23}$$

where $f(x) \approx g(x)$ means that $\lim_{x\to\infty} f(x)/g(x) = 1$. For smaller values of B, the following approximation, analogous to Equation (8.3), is more accurate [15]:

$$\sup_t \Pr\{Q(t) \geq B\} \approx \gamma(r)e^{-\theta_B(r)B} \tag{8.24}$$

where both $\gamma(r)$ and $\theta_B(r)$ are functions of channel capacity r. According to the theory, $\gamma(r) = \Pr\{Q(t) \geq 0\}$ is the *probability that the buffer is nonempty* for randomly chosen time t, while the *QoS exponent* θ_B is the solution of $\alpha(\theta_B) = r$. Thus, the pair of functions $\{\gamma(r), \theta_B(r)\}$ model the source. Note that $\theta_B(r)$ is simply the inverse function corresponding to the effective bandwidth function $\alpha(u)$.

If the quantity of interest is the delay $D(t)$ experienced by a source packet arriving at time t, then with same reasoning the probability of $D(t)$ exceeding a delay bound $D_{\max}$ satisfies

$$\sup_t \Pr\{D(t) \geq D_{\max}\} \approx \gamma(r)e^{-\theta(r)D_{\max}} \tag{8.25}$$

where $\theta(r) = \theta_B(r) \times r$ [16]. Thus, the key point is that for a source modeled by the pair $\{\gamma(r),\ \theta(r)\}$, which has a communication delay bound of $D_{\max}$, and can tolerate a delay-bound violation probability of at most ε, the effective bandwidth concept shows that the constant channel capacity should be at least r, where r is the solution to $\varepsilon = \gamma(r)e^{-\theta(r)D_{\max}}$. In terms of the traffic envelope $\Gamma(t)$ (Figure 8.7), the slope $\lambda_s^{(s)} = r$ and $\sigma^{(s)} = rD_{\max}$.

In Section 8.1 a simple and efficient algorithm to estimate the source model functions $\gamma(r)$ and $\theta(r)$ was discussed. In the following section, we use the duality between traffic modeling $\{\gamma(r),\ \theta(r)\}$, and channel modeling to present an EC link model, specified by a pair of functions $\{\gamma^{(c)}(\mu),\ \theta^{(c)}(\mu)\}$. The intention is to use $\{\gamma^{(c)}(\mu),\ \theta^{(c)}(\mu)\}$ as the channel duals of the source functions $\{\gamma(r),\ \theta(r)\}$. Just as the constant *channel rate* r is used in source traffic modeling, we use the constant *source traffic rate* μ in modeling the channel. Furthermore, we adapt the source estimation algorithm from Section 8.1 to estimate the link model parameters $\{\gamma^{(c)}(\mu),\ \theta^{(c)}(\mu)\}$.

8.2.2.2 Effective capacity link model

Let $r(t)$ be the instantaneous channel capacity at time t. Define $\tilde{S}(t) = \int_0^t r(\tau)\,d\tau$, which is the service provided by the channel. Note that the channel service $\tilde{S}(t)$ is different from the actual service $S(t)$ received by the source; $\tilde{S}(t)$ only depends on the instantaneous channel capacity and thus is independent of the arrival $A(t)$. Paralleling the development of Equation (8.21) and (8.22) we assume that

$$\Lambda^{(c)}(-u) = \lim_{t\to\infty} \frac{1}{t} \log E\left[e^{-u\tilde{S}(t)}\right] \tag{8.26}$$

exists for all $u \geq 0$. This assumption is valid, for example, for a stationary Markov-fading process $r(t)$. Then, the *EC function* of $r(t)$ is defined as

$$\alpha^{(c)}(u) = \frac{-\Lambda^{(c)}(-u)}{u}, \qquad \forall u \geq 0 \tag{8.27}$$

Consider a queue of infinite buffer size supplied by a data source of *constant* data rate μ (see Figure 8.6). The theory of effective bandwidth can be easily adapted to this case. The difference is that, whereas in the previous case the source rate was variable while the channel capacity was constant, now the source rate is constant while the channel capacity is variable. Similar to Equation (8.25), it can be shown that the probability of $D(t)$ exceeding a delay bound $D_{\max}$ satisfies

$$\sup_t \Pr\{D(t) \geq D_{\max}\} \approx \gamma^{(c)}(\mu)\mathrm{e}^{-\theta^{(c)}(\mu)D_{\max}} \tag{8.28}$$

where $\{\gamma^{(c)}(\mu),\ \theta^{(c)}(\mu)\}$ are functions of source rate μ. This approximation is accurate for large $D_{\max}$, but we will see later in the simulation results, that this approximation is also accurate even for smaller values of $D_{\max}$.

For a given source rate μ, $\gamma^{(c)}(\mu) = \Pr\{Q(t) \geq 0\}$ is again the *probability that the buffer is nonempty* at a randomly chosen time t, while the *QoS exponent* $\theta^{(c)}(\mu)$ is defined as $\theta(\mu) = \mu\alpha^{-1}(\mu)$, where $\alpha^{-1}(\cdot)$ is the inverse function of $\alpha^{(c)}(u)$. Thus, the pair of functions $\{\gamma^{(c)}(\mu),\ \theta^{(c)}(\mu)\}$ model the link.

So, if a link that is modeled by the pair $\{\gamma^{(c)}(\mu),\ \theta^{(c)}(\mu)\}$ is used, a source that requires a communication delay bound of $D_{\max}$, and can tolerate a delay-bound violation probability of at most ε, needs to limit its data rate to a maximum of μ, where μ is the solution to $\varepsilon = \gamma^{(c)}(\mu)\mathrm{e}^{-\theta^{(c)}(\mu)D_{\max}}$. In terms of the SC $\Psi(t)$ shown in Figure 8.7, the channel sustainable rate $\lambda_s^{(c)} = \mu$ and $\sigma^{(c)} = D_{\max}$.

If the channel-fading process $r(t)$ is stationary and ergodic, then a simple algorithm to estimate the functions $\{\gamma^{(c)}(\mu),\ \theta^{(c)}(\mu)\}$ is similar to the one described in Section 8.1. Paralleling Equation (8.7) we have

$$\frac{\gamma^{(c)}(\mu)}{\theta^{(c)}(\mu)} = E\left[D(t)\right] = \tau_s(\mu) + \frac{E\left[Q(t)\right]}{\mu} \tag{8.29}$$

$$\gamma^{(c)}(\mu) = \Pr\{D(t) > 0\} \tag{8.30}$$

where $\tau_s(\mu)$ is the average remaining service time of a packet being served. Note that $\tau_s(\mu)$ is zero for a fluid model (assuming infinitesimal packet size). Now, the delay $D(t)$ is the sum of the delay incurred due to the packet already in service, and the delay in waiting for the queue $Q(t)$ to clear which results in Equation (8.29), using Little's theorem. Substituting $D_{\max} = 0$ in Equation (2.28) results in Equation (3.30). As in Section 8.1, solving Equation (8.29) for $\theta^{(c)}(\mu)$ gives similarly to Equation (8.8a)

$$\theta^{(c)}(\mu) = \frac{\gamma^{(c)}(\mu) \times \mu}{\mu \times \tau_s(\mu) + E\left[Q(t)\right]} \tag{8.31}$$

According to Equation (8.30) and (8.31) , as in Section 8.1, the functions γ and θ can be estimated by estimating $\Pr\{D(t) > 0\}$, $\tau_s(\mu)$, and $E\left[Q(t)\right]$. The latter can be estimated by taking a number of samples, say N, over an interval of length T, and recording the following quantities at the nth sampling epoch: S_n the indicator of whether a packets is in service ($S_n \in \{0,\ 1\}$), Q_n the number of bits in the queue (excluding the packet in service), and T_n the remaining service time of the packet in service (if there is one in service). Based on the same measurements, as in Section 8.1,

$$\hat{\gamma} = \sum_{n=1}^{N} S_n/N$$

$$\hat{q} = \sum_{n=1}^{N} Q_n/N$$

$$\hat{\tau}_s = \sum_{n=1}^{N} T_n/N$$

are computed and then, from Equation (8.31), we have

$$\hat{\theta} = \frac{\hat{\gamma} \times \mu}{\mu \times \hat{\tau}_s + \hat{q}} \tag{8.32}$$

These parameters are used to predict the QoS by approximating Equation (8.28) with

$$\sup_t \Pr\{D(t) \geq D_{\max}\} \approx \hat{\gamma} e^{-\hat{\theta} D_{\max}} \tag{8.33}$$

If the ultimate objective of EC link modeling is to compute an appropriate SC $\Psi(t)$, then, given the delay-bound $D_{\max}$ and the target delay-bound violation probability ε of a connection, we can find $\Psi(t) = \{\sigma^{(c)}, \lambda_s^{(c)}\}$ by setting $\sigma^{(c)} = D_{\max}$, solving Equation (8.33) for μ and setting $\lambda_s^{(c)} = \mu$.

8.2.3 Physical layer vs link-layer channel model

In Jack's model of a Rayleigh flat-fading channel, the Doppler spectrum $S(f)$ is given as

$$S(f) = \frac{1.5}{\pi f_m \sqrt{1 - (F/f_m)^2}} \tag{8.34}$$

where f_m is the maximum Doppler frequency, f_c is the carrier frequency, and $F = f - f_c$. Below we show how to calculate the EC for this channel [11].

Denote a sequence of N measurements of the channel gain, spaced at a time interval δ apart, by $\mathbf{x} = [x_0, x_1, \cdots, x_{N-1}]$, where $\{x_n, n \in [0, N-1]\}$ are the complex-valued Gaussian distributed channel gains ($|x_n|$ are, therefore, Rayleigh distributed). For simplicity, the constant noise variance will be included in the definition of x_n. The measurement x_n is a realization of a random variable sequence denoted by X_n, which can be written as the vector $\mathbf{X} = [X_0, X_1, \cdots, X_{N-1}]$. The pdf of a random vector $\mathbf{X}$ for the Rayleigh-fading channel is

$$f_{\mathbf{X}}(\mathbf{X}) = \frac{1}{\pi^N \det(\mathbf{R})} e^{-\mathbf{x}\mathbf{R}^{-1}\mathbf{x}^H} \tag{8.35}$$

where $\mathbf{R}$ is the covariance matrix of the random vector $\mathbf{X}$, det($\mathbf{R}$) the determinant of matrix $\mathbf{R}$, and $\mathbf{x}^H$ the conjugate transpose (Hermitan) of $\mathbf{x}$. To calculate the EC, we start with

$$E[e^{-u\tilde{S}(t)}] = E\left[\exp\left[-u\int_0^t r(\tau)\,d\tau\right]\right] \overset{(a)}{\approx} \int \exp\left\{-u\left[\sum_{n=0}^{N-1} \delta \times r(\tau_n)\right]\right\} f_{\mathbf{X}}(\mathbf{x})\,d\mathbf{x}$$

$$\overset{(b)}{\approx} \int \exp\left\{-u\left[\sum_{n=0}^{N-1} \delta \log(1+|x_n|^2)\right]\right\} f_{\mathbf{x}}(\mathbf{x})\, d\mathbf{x} \tag{8.36}$$
$$\overset{(c)}{\approx} \int \exp\left\{-u\left[\sum_{n=0}^{N-1} \delta \log(1+|x_n|^2)\right]\right\}.\frac{1}{\pi^N \det(\mathbf{R})} e^{-\mathbf{x}\mathbf{R}^{-1}\mathbf{x}^{H}}\, dx$$

where (a) approximates the integral by a sum, (b) is the Shannon result for channel capacity (i.e. $\gamma(\tau_n) = \log(1+|x_n|^2)$, and (c) is from Equation (8.35). This gives the EC, Equation (8.27), as

$$\alpha^{(c)}(u) = \frac{-1}{u} \lim_{t\to\infty} \log \int \exp\left\{-u\left[\sum_{n=0}^{N-1} \delta \log(1+|x_n|^2)\right]\right\}.\frac{1}{\pi^N \det(\mathbf{R})} e^{-\mathbf{x}\mathbf{R}^{-1}\mathbf{x}^{H}}\, d\mathbf{x} \tag{8.37}$$

Using the approximation $(a) \log(1+|x_n|^2) \approx |x_n|^2$ for low SNR [11], Equation (8.37) can be further simplified as

$$E[e^{-u\tilde{S}(t)}] \overset{(a)}{\approx} \int \exp\left[-u\delta\left(\sum_{n=0}^{N-1} |x_n|^2\right)\right].\frac{1}{\pi^N \det(\mathbf{R})} e^{-\mathbf{x}\mathbf{R}^{-1}\mathbf{x}^{H}}\, d\mathbf{x}$$
$$\overset{(b)}{=} \int e^{-u\delta\|\mathbf{x}\|^2} \frac{1}{\pi^N \det(\mathbf{R})} e^{-\mathbf{x}\mathbf{R}^{-1}\mathbf{x}^{H}}\, d\mathbf{x} \overset{(b)}{=} \frac{1}{\pi^N \det(\mathbf{R})} \int e^{-\mathbf{x}(\mathbf{R}^{-1}+u\delta\mathbf{I})\mathbf{x}^{H}}\, d\mathbf{x} \tag{8.38}$$
$$= \frac{1}{\pi^N \det(\mathbf{R})} \times \pi^N \det\left[(\mathbf{R}^{-1}+u\delta\mathbf{I})^{-1}\right] = \frac{1}{\det(u\delta\mathbf{R}+\mathbf{I})}$$

where approximation (b) is due to the definition of the norm of the vector $\mathbf{x}$, and (c) the relation $\|x\|^2 = \mathbf{x}\mathbf{x}^{H}$ ($\mathbf{I}$ is identity matrix). Reference [11] considers three cases of interest for Equation (8.38).

8.2.3.1 High mobility scenario

In the extreme high mobility (HM) case there is no correlation between the channel samples and we have $\mathbf{R} = r\mathbf{I}$, where $r = E\,|x_n|^2$ is the average channel capacity. From Equation (8.38), we have

$$E[e^{-u\tilde{S}(t)}] \approx \frac{1}{\det(u\delta\mathbf{R}+\mathbf{I})} = \frac{1}{(ur\delta+1)^N} = \frac{1}{(urt/N+1)^N} \tag{8.39}$$

where $\delta\ t/N$. As the number of samples $N \to \infty$, we have

$$\lim_{N\to\infty} E[e^{-u\tilde{S}(t)}] \approx \lim_{N\to\infty} (urt/N+1)^{-N} = e^{-urt} \tag{8.40}$$

Thus, in the limiting case, the Rayleigh-fading channel reduces to an AWGN channel. Note that this result would not apply at high SNRs because of the concavity of the $\log(\cdot)$ function.

8.2.3.2 Stationary scenario

In this case all the samples are fully correlated, $\mathbf{R} = [R_{ij}] = [r]$. In other words all elements of $\mathbf{R}$ are the same and Equation (8.38) now gives

$$E[\mathrm{e}^{-u\tilde{S}(t)}] \approx \frac{1}{\det(u\delta\mathbf{R} + \mathbf{I})} = \frac{1}{ur\delta N + 1} = \frac{1}{ur \times \dfrac{t}{N} \times N + 1} = \frac{1}{1 + urt} \tag{8.41}$$

8.2.3.3 General case

Denote the eigenvalues of matrix $\mathbf{R}$ by $\{\lambda_n, n \in [0, N-1]\}$. Since $\mathbf{R}$ is symmetric, we have $\mathbf{R} = \mathbf{UDU}^{\mathrm{H}}$, where $\mathbf{U}$ is a unitary matrix, $\mathbf{U}^{\mathrm{H}}$ is its Hermitian, and the diagonal matrix $\mathbf{D} = \mathrm{diag}\,(\lambda_0, \lambda_1, \cdots, \lambda_{N-1})$. From Equation (8.38), we have

$$\begin{aligned} E[\mathrm{e}^{-u\tilde{S}(t)}] &\approx \frac{1}{\det(u\delta\mathbf{R} + \mathbf{I})} = \frac{1}{\det(u\delta\mathbf{UDU}^{\mathrm{H}} + \mathbf{UU}^{\mathrm{H}})} \\ &= \frac{1}{\det[\mathbf{U}\,\mathrm{diag}\,(u\delta\lambda_0 + 1, u\delta\lambda_1 + 1, \ldots, u\delta\lambda_{N-1} + 1)\mathbf{U}^{\mathrm{H}}]} \\ &= \frac{1}{\prod_n (u\delta\lambda_n + 1)} = \exp\left[-\sum_n \log(u\delta\lambda_n + 1)\right] \end{aligned} \tag{8.42}$$

We now use the calculated $E[\mathrm{e}^{-u\tilde{S}(t)}]$ to get

$$\begin{aligned} \Lambda^{(c)}(-u) &= \lim_{t\to\infty} \frac{1}{t} \log E[\mathrm{e}^{-u\tilde{S}(t)}] \overset{(a)}{\approx} \lim_{t\to\infty} \frac{1}{t} \log\exp\left[-\sum_n \log(u\delta\lambda_n + 1)\right] \\ &\overset{(b)}{=} \lim_{\Delta f\to 0} -\Delta f \sum_n \log\left(u\frac{\lambda_n}{B_w} + 1\right) \overset{(c)}{=} -\int \log(uS(f) + 1)\,\mathrm{d}f \end{aligned} \tag{8.43}$$

where (a) follows from Equation (8.42), (b) follows from the fact that the frequency interval $\Delta f = 1/t$ and the bandwidth $B_{\mathrm{w}} = 1/\delta$, and (c) from the fact that the power spectral density $S(f) = \lambda_n / B_{\mathrm{w}}$ and that the limit of a sum becomes an integral. This gives the EC, Equation (8.27), as

$$\alpha^{(c)}(u) = \frac{\int \log(uS(f) + 1)\mathrm{d}f}{u} \tag{8.44}$$

Thus, the Doppler spectrum allows us to calculate $\alpha^{(c)}(u)$. The EC function, Equation (8.44), can be used to guarantee QoS using Equation (8.28).

One should keep in mind that the EC function, Equation (8.44), is valid only for a Rayleigh flat-fading channel, *at low SNR*. At high SNR, the EC for a Rayleigh-fading channel is specified by the complicated integral in Equation (8.37). To the best of our knowledge, a closed-form solution to Equation (8.37) does not exist. It is clear that a numerical calculation of EC is also very difficult, because the integral has a high dimension. Thus, it is difficult to extract QoS metrics from a physical-layer channel model, even for a Rayleigh flat-fading channel. The extraction may not even be possible for more general fading channels. In contrast, the EC link model that was described in this section can be easily translated into QoS metrics for a connection, and we have shown a simple estimation algorithm to estimate the EC model functions.

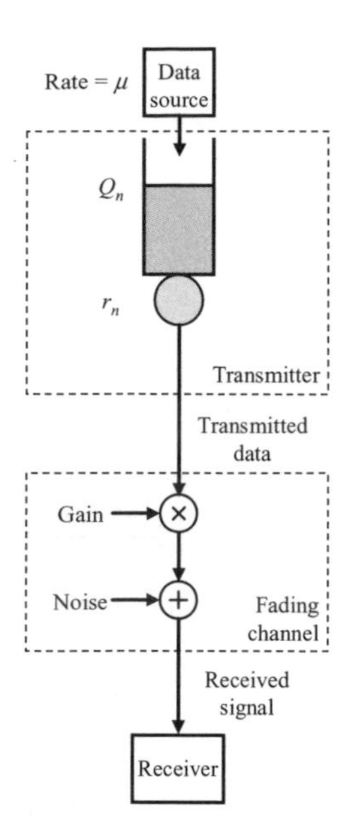

Figure 8.8 Queueing model used for simulations.

8.2.4 Performance examples

The discrete-time system depicted in Figure 8.8 is simulated. The data source generates packets at a *constant* rate μ which are first sent to the (infinite) buffer at the transmitter, whose queue length is Q_n, where n refers to the nth sample interval. The head-of-line packet in the queue is transmitted over the fading channel at data rate r_n. The fading channel has a random channel gain x_n (the noise variance is absorbed into x_n). We use a fluid model that is the size of a packet is infinitesimal. A perfect knowledge of the channel gain x_n (the SNR, really) at the transmitter side is assumed. Therefore, as described in Chapter 4, it can use rate-adaptive transmissions and strong channel coding to transmit packets without errors. Thus, the transmission rate r_n is equal to the instantaneous (time-varying) capacity of the fading channel, defined by the Shannon law, $r_n = B_c \log_2(1 + |x_n|^2)$ where B_c is the channel bandwidth.

The average SNR is fixed in each simulation run. We define r_g as the capacity of an equivalent AWGN channel, which has the same average SNR, i.e. $rg = B_c \log_2(1 + \text{SNR}_{\text{avg}})$ where SNR_{avg} is the average SNR, i.e. $E\,|x_n|^2$. Then, r_n/r_g relation

$$r_n = \frac{r_{\text{awgn}} \log_2(1 + |x_n|^2)}{\log_2(1 + \text{SNR}_{\text{avg}})} \tag{8.45}$$

Simulation parameters as in Wu and Negi [11] were used. Channel samples x_n are generated by the following AR(1) (autoregressive) model: $x_n = kx_{n-1} + v_n$ where the modeling error v_n is zero-mean complex Gaussian with unit variance per dimension and is statistically independent of x_{n-1}. The coefficient k can be determined by the following procedure: (1) compute the coherence time T_c by $T_c \approx 9/16\pi f_m$, where the coherence time is defined as the time over which the time autocorrelation function of the fading process is above 0.5; (2)

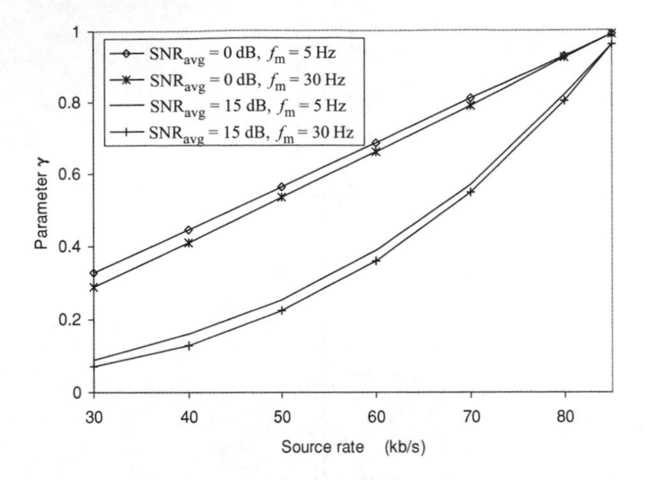

Figure 8.9 Estimated function $\hat{\gamma}(\mu)$ vs source rate μ.

compute the coefficient k by $k = 0.5^{T_s/T_c}$ where T_s is the sampling interval. The other parameters are: $f_m = 5$–30 Hz, $r_g = 100$ kb/s, average SNR $= 0/15$ dB, $T_s = 1$ ms, bit rate $\mu = 30$–85 kb/s.

Figures 8.9 and 8.10 show the estimated EC functions $\hat{\gamma}(\mu)$ and $\hat{\theta}(\mu)$. As the source rate μ increases from 30 to 85 kb/s, $\hat{\gamma}(\mu)$ increases, indicating a higher buffer occupancy, while $\hat{\theta}(\mu)$ decreases, indicating a slower decay of the delay-violation probability. Thus, the delay-violation probability is expected to increase with increasing source rate μ. From Figure 8.10, we also observe that SNR has a substantial impact on $\hat{\gamma}(\mu)$. This is because higher SNR results in larger channel capacity, which leads to smaller probability that a packet will be buffered, i.e. smaller $\hat{\gamma}(\mu)$. In contrast, Figure 8.9 shows that f_m has little effect on $\hat{\gamma}(\mu)$.

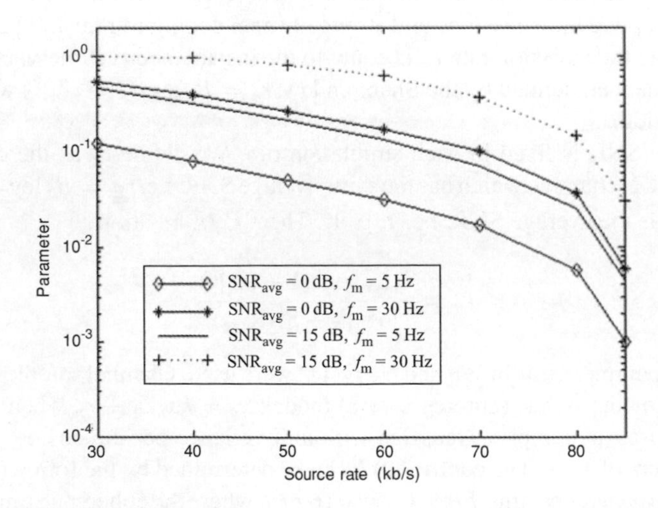

Figure 8.10 Estimated function $\hat{\theta}(\mu)$ vs source rate μ.

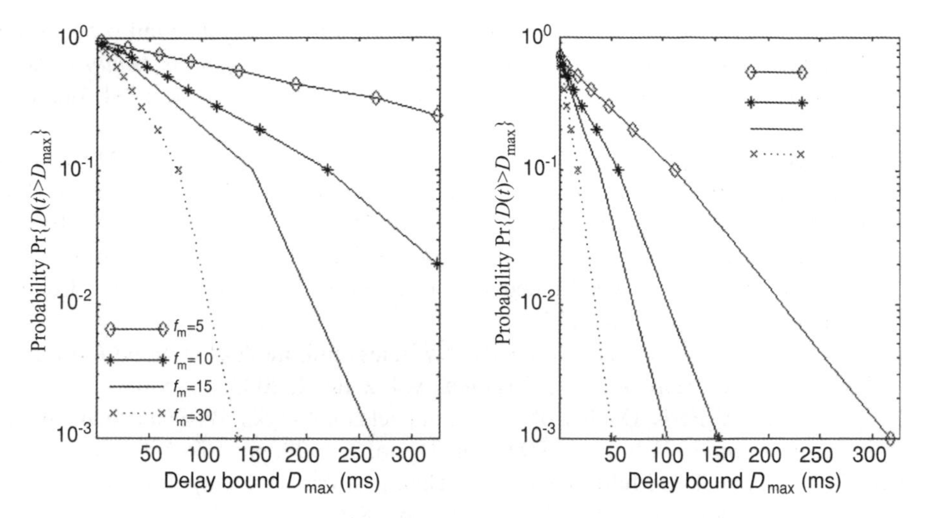

Figure 8.11 Actual delay-violation probability vs $D_{\max}$ for various Doppler rates. (a) Rayleigh fading and (b) Ricean fading $K = 3$.

Figure 8.11 shows the actual delay-violation probability $\sup_t \Pr\{D(t) > D_{\max}\}$ vs the delay bound $D_{\max}$, for various Doppler rates. It can be seen that the actual delay-violation probability decreases exponentially with the delay bound $D_{\max}$, for all the cases. This justifies the use of an exponential bound, Equation (8.33), in predicting QoS, thereby justifying the link model $\{\hat{\gamma}, \hat{\theta}\}$. The figure shows that delay-violation probability reduces with the Doppler rate. This is reasonable since the increase of the Doppler rate leads to the increase of time diversity, resulting in a larger decay rate $\hat{\theta}(\mu)$ of the delay-violation probability. More details on the topic can be found in References [11–17].

REFERENCES

[1] ATM Forum Technical Committee, *Traffic Management Specification, Version 4.0.* ATM Forum, 1996.

[2] A.I. Elwalid and D. Mitra, Analysis and design of rate-based congestion control of high speed networks – I: stochastic fluid models, access regulation, *Queueing Syst.*, vol. 9, 1991, pp. 29–63.

[3] J.S. Turner, New directions in communications (or which way to the information age?), *IEEE Commun. Mag.*, 1986, pp. 8–15.

[4] R.L. Cruz, A calculus for network delay, part I: network elements in isolation, *IEEE Trans. Inform. Theory,* vol. 37, 1991, pp. 114–131.

[5] B.L. Mark, and G. Ramamurthy, Real-time estimation and dynamic renegotiation of UPC parameters for arbitrary traffic sources in ATM networks, *IEEE/ACM Trans. Networking*, vol. 6, no. 6, 1998, pp. 811–828.

[6] C. Chang, Stability, queue length and delay of deterministic and stochastic queueing networks, *IEEE Trans. Automat. Contr.,* vol. 39, 1994, pp. 913–931.

[7] G.L. Choudhury, D.M. Lucantoni and W. Whitt, Squeezing the most out of ATM, *IEEE Trans. Commun.*, vol. 44, 1996, pp. 203–217.

[8] P.W. Glynn and W. Whitt, Logarithmic asymptotics for steadystate tail probabilities in a single-server queue, in *Studies in Applied Probability, Papers in Honor of Lajos Takacs*, J. Galambos and J. Gani, eds, Applied Probability Trust, 1994, pp. 131–156.

[9] H. Kobayashi and Q. Ren, A diffusion approximation analysis of an ATM statistical multiplexer with multiple types of traffic, part I: equilibrium state solutions, in *Proc. 1993 IEEE Int. Conf. Communications*, Geneva, May 1993, vol. 2, pp. 1047–1053.

[10] B.L. Mark and G. Ramamurthy, Joint source-channel control for realtime VBR over ATM via dynamic UPC renegotiation, in *Proc. IEEE Globecom'96*, London, November, 1996, pp. 1726–1731.

[11] D. Wu, and R. Negi, Effective capacity: a wireless link model for support of quality of service, *IEEE Trans. Wireless Commun.*, vol. 2, no. 4, 2003, pp. 630–643.

[12] R. Guerin and V. Peris, Quality-of-service in packet networks: Basic mechanisms and directions, *Comput. Networks, ISDN*, vol. 31, no. 3, 1999, pp. 169–179.

[13] S. Hanly and D. Tse, Multiaccess fading channels: Part II: Delay-limited capacities, *IEEE Trans. Inform. Theory*, vol. 44, 1998, pp. 2816–2831.

[14] C.-S. Chang and J.A. Thomas, Effective bandwidth in high-speed digital networks, *IEEE J. Select. Areas Commun.*, vol. 13, 1995, pp. 1091–1100.

[15] G.L. Choudhury, D.M. Lucantoni and W. Whitt, Squeezing the most out of ATM, *IEEE Trans. Commun.*, vol. 44, 1996, pp. 203–217.

[16] Z.-L. Zhang, End-to-end support for statistical quality-of-service guarantees in multimedia networks, Ph.D. dissertation, Department of Computer Science, University of Massachusetts, 1997.

[17] B. Jabbari, Teletraffic aspects of evolving and next-generation wireless communication networks, *IEEE Pers. Commun.*, vol. 3, 1996, pp. 4–9.

[18] S. Chong and S. Li, (σ; ρ)-characterization based connection control for guaranteed services in high speed networks, in *Proc. IEEE INFOCOM'95*, Boston, MA, April 1995, pp. 835–844.

[19] O. Yaron and M. Sidi, Performance and stability of communication networks via robust exponential bounds, *IEEE/ACM Trans. Networking*, vol. 1, 1993, pp. 372–385.

[20] T. Tedijanto and L. Gun, Effectiveness of dynamic bandwidth management in ATM networks, in *Proc. INFOCOM'93*, San Francisco, CA, March 1993, pp. 358–367.

[21] M. Grossglauser, S. Keshav, and D. Tse, RCBR: a simple and efficient service for multiple time-scale traffic, in *Proc. ACM SigCom'95*, Boston, MA, August 1995, pp. 219–230.

[22] D. Reininger, G. Ramamurthy and D. Raychaudhuri, VBR MPEG video coding with dynamic bandwidth renegotiation, in *Proc. ICC'95*, Seattle, WA, June 1995, pp. 1773–1777.

[23] J. Abate, G.L. Choudhury, and W. Whitt, Asymptotics for steady-state tail probabilities in structured Markov queueing models, *Stochastic Models*, vol. 10, 1994, pp. 99–143.

[24] D.P. Heyman and T.V. Lakshman, What are the implications of long-range dependence for VBR-video traffic engineering? *IEEE/ACM Trans. Networking*, vol. 4, 1996, pp. 301–317.

[25] W. Whitt, Tail probabilities with statistical multiplexing and effective bandwidths in multi-class queues, *Telecommun. Syst.*, vol. 2, 1993, pp. 71–107.

[26] D.E. Knuth, *The Art of Computer Programming, Volume 2: Seminumerical Algorithms, 2nd edn.* Addison-Wesley: Reading, MA, 1981.

[27] A.I. Elwalid and D. Mitra, Effective bandwidth of general Markovian traffic sources and admission control of high speed networks, *IEEE/ACM Trans. Networking*, vol. 1, 1993, pp. 323–329.
[28] A.K. Parekh and R.G. Gallager, A generalized processor sharing approach to flow control in integrated services networks: The singlenode case, *IEEE/ACM Trans. Networking*, vol. 1, 1993, pp. 344–357.
[29] B.L. Mark and G. Ramamurthy, UPC-based traffic descriptors for ATM: How to determine, interpret and use them, *Telecommun. Syst.*, vol. 5, 1996, pp. 109–122.

9

Adaptive TCP Layer

9.1 INTRODUCTION

In this section we first discuss TCP performance independent of the type of network by considering the different possible characteristics of the connection path. We present the problems and the different possible solutions. This study permits us to understand the limitations of the actual solutions and the required modifications to let TCP cope with a heterogeneous Internet on an end-to-end basis. Then, in the rest of the chapter we focus on the specifics of TCP operation in wireless networks.

The TCP provides a reliable connection-oriented in-order service to many of today's Internet applications. Given the simple best-effort service provided by IP, TCP must cope with the different transmission media crossed by Internet traffic. This mission of TCP is becoming difficult with the increasing heterogeneity of the Internet. Highspeed links (optic fibers), long and variable delay paths (satellite links), lossy links (wireless networks) and asymmetric paths (hybrid satellite networks) are becoming widely embedded in the Internet. Many works have studied by experimentation [1], analytical modeling [2] and simulation [3, 4, 12, 17, 18] the performance of TCP in this new environment.

Most of these works have focused on a particular environment (satellite networks, mobile networks, etc.). They have revealed some problems in the operation of TCP. Long propagation delay and losses on a satellite link, handover and fading in a wireless network, bandwidth asymmetry in some media, and other phenomena have been shown to seriously affect the throughput of a TCP connection. A large number of solutions have been proposed. Some solutions suggest modifications to TCP to help it to cope with these new paths. Other solutions keep the protocol unchanged and hide the problem from TCP. In this section we consider the different characteristics of a path crossed by TCP traffic, focusing on bandwidth-delay product (BDP), RTT, noncongestion losses and bandwidth asymmetry. TCP is a reliable window-based ACK-clocked flow control protocol. It uses an additive-increase multiplicative-decrease strategy for changing its window as a function of network

Advanced Wireless Networks: 4G Technologies Savo G. Glisic
© 2006 John Wiley & Sons, Ltd.

conditions. Starting from one packet, or a larger value as we will see later, the window is increased by one packet for every nonduplicate ACK until the source estimate of *network propagation time* (npt) is reached. By the propagation time of the network, sometimes called the *pipe size*, we mean the maximum number of packets that can be fit on the path, which is also referred to as *network capacity*. This is the *slow start* (SS) phase, and the *npt* estimate is called the *SS threshold* (sst). SS aims to alleviate the burstiness of TCP while quickly filling the pipe. Once *sst* is reached, the source switches to a slower increase in the window by one packet for every window's worth of ACKs. This phase, called *congestion avoidance* (CA), aims to slowly probe the network for any extra bandwidth. The window increase is interrupted when a loss is detected. Two mechanisms are available for the detection of losses: the expiration of a retransmission timer (timeout) or the receipt of three duplicate ACKs (fast retransmit, FRXT). The source supposes that the network is in congestion and sets its estimate of the *sst* to half the current window.

Tahoe, the first version of TCP to implement congestion control, at this point sets the window to one packet and uses SS to reach the new *sst*. Slow starting after every loss detection deteriorates the performance given the low bandwidth utilization during SS. When the loss is detected via timeout, SS is unavoidable since the ACK clock has stopped and SS is required to smoothly fill the pipe. However, in the FRXT case, ACKs still arrive at the source and losses can be recovered without SS. This is the objective of the new versions of TCP (Reno, New Reno, SACK, etc.), discussed in this chapter, that call a fast recovery (FRCV) algorithm to retransmit the losses while maintaining enough packets in the network to preserve the ACK clock. Once losses are recovered, this algorithm ends and normal CA is called. If FRCV fails, the ACK stream stops, a timeout occurs, and the source resorts to SS as with Tahoe.

9.1.1 A large bandwidth-delay product

The increase in link speed, like in optic fibers, has led to paths of large BDP. The TCP window must be able to reach large values in order to efficiently use the available bandwidth. Large windows, up to 2^{30} bytes, are now possible. However, at large windows, congestion may lead to the loss of many packets from the same connection. Efficient FRCV is then required to correct many losses from the same window. Also, at large BDP, network buffers have an important impact on performance. These buffers must be well dimensioned and scale with the BDP. *Fast recovery* uses the information carried by ACKs to estimate the number of packets in flight while recovering from losses. New packets are sent if this number falls below the network capacity estimate. The objective is to preserve the ACK clock in order to avoid the timeout. The difference between the different versions of TCP is in the estimation of the number of packets in flight during FRCV. All these versions will be discussed later in more detail. *Reno* considers every duplicate ACK a signal that a packet has left the network. The problem of Reno is that it leaves FRCV when an ACK for the first loss in a window is received. This prohibits the source from detecting the other losses with FRXT. A long timeout is required to detect the other losses. *New Reno* overcomes this problem. The idea is to stay in FRCV until all the losses in the same window are recovered. Partial ACKs are used to detect multiple losses in the same window. This avoids timeout but cannot result in a recovery faster than one loss per RTT. The source needs to wait for the ACK of the retransmission to discover the next loss. Another problem of Reno and New Reno is that they rely on ACKs to estimate the number of packets in flight. ACKs can be lost on the return path, which results in an underestimation of the number of packets that have left the network, and thus an underutilization of the bandwidth during FRCV and, in the case of Reno, a possible failure of FRCV. More information is needed at the source to recover faster than one loss per RTT and to estimate more precisely the number of packets in the pipe. This information is provided by selective ACK (SACK), a TCP option containing the three blocks of contiguous data most recently received at the destination. Many algorithms have been proposed to use this information during FRCV. TCP-SACK may use ACKs to estimate the number of packets in the pipe and SACKs to retransmit more than one loss per RTT. This leads to an important improvement in performance when bursts of losses appear in the same window, but the recovery is always sensitive to the loss of ACKs. As a solution, forward ACK (FACK) may be used, which relies on SACK in estimating the number of packets in the pipe. The number and identity of packets to transmit during FRCV is decoupled from the ACK clock, in contrast to TCP-SACK, where the identity is only decoupled.

9.1.2 Buffer size

SS results in bursts of packets sent at a rate exceeding the bottleneck bandwidth. When the receiver acknowledges every data packet, the rate of these bursts is equal to twice the bottleneck bandwidth. If network buffers are not well dimensioned, they will overflow early during SS before reaching the network capacity. This will result in an underestimation of the available bandwidth and a deterioration in TCP performance. Early losses during SS were first analyzed in Lakshman and Madhow [2]. The network is modeled with a single bottleneck node of bandwidth μ, buffer B, and two-way propagation delay T, as shown in Figure 9.1.

A long TCP-Tahoe connection is considered where the aim of *SS* is to reach quickly without losses *sst*,which is equal to half the pipe size $[(B + \mu T)/2]$. In the case of a receiver that acknowledges every data packet, they found that a buffer $B > BDP/3 = \mu T/3$ is required. Their analysis can be extended to an SS phase with a different threshold, mainly to that at the beginning of the connection, where *sst* is set to a default value. As an example, the threshold can be set at the beginning of the connection to the *BDP* in order to switch to *CA* before the occurrence of losses. This will not work if the buffer is smaller than half the *BDP* (half $\mu T/2$). In Barakat and Altman [5] the problem of early buffer overflow during SS for multiple routers was studied. It was shown that, due to the high rate at which

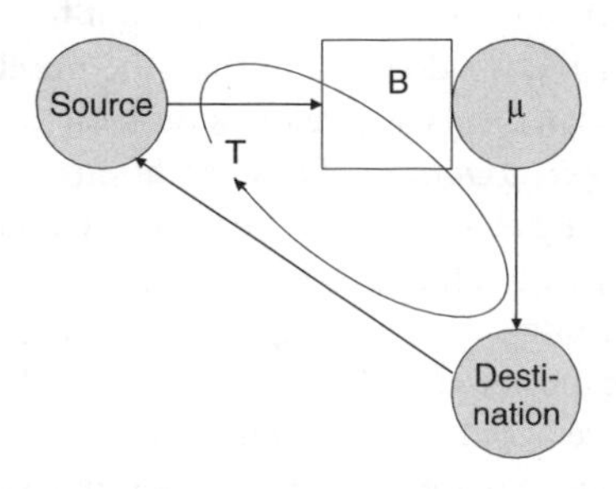

Figure 9.1 One-hop *TCP operation.*

packets are sent during SS, queues can build up in routers preceding the bottleneck as well. Buffers in these routers must also be well dimensioned, otherwise they overflow during SS and limit the performance even though they are faster than the bottleneck. With small buffers, losses during SS are not a signal of network congestion, but rather of transient congestion due to the bursty nature of SS traffic. Now in CA, packets are transmitted at approximately the bottleneck bandwidth. A loss occurs when the window reaches the pipe size. The source divides its window by two and starts a new cycle. To always get a throughput approximately equal to the bottleneck bandwidth, the window after reduction must be larger than the BDP. This requires a buffer B larger than the BDP. Note that we are talking about drop tail buffers, which start to drop incoming packets when the buffer is full. Active buffers such as *random early detection (*RED*)* [3] start to drop packets when the average queue length exceeds some threshold. When an RED buffer is crossed by a single connection, the threshold should be larger than the BDP to get good utilization. This contrasts one of the aims of RED: limiting the size of queues in network nodes in order to reduce end-to-end delay. For multiple connections, a lower threshold is sufficient given that a small number of connections reduce their windows upon congestion, in contrast to drop tail buffers, where often all the connections reduce their windows simultaneously .

9.1.3 Round-trip time

Long RTTs are becoming an issue with the introduction of satellite links into the Internet. A long RTT reduces the rate at which the window increases, which is a function of the number of ACKs received and does not account for the RTT. This poses many problems to the TCP. First, it increases the duration of SS, which is a transitory phase designed to quickly but smoothly fill the pipe. Given the low bandwidth utilization during SS, this deteriorates the performance of TCP transfers, particularly short ones (e.g. Web transfers). Second, it causes unfairness in the allocation of the bottleneck bandwidth. Many works have shown the bias of TCP against connections with long RTTs [2]. Small RTT connections increase their rates more quickly and grab most of the available bandwidth. The average throughput of a connection has been shown to vary as the inverse of T^{α}, where α is a factor between 1 and 2 [2].

Many solutions have been proposed to reduce the time taken by SS on long delay paths. These solutions can be divided into three categories: (1) change the window increase algorithm of TCP; (2) solve the problem at the application level; or (3) solve it inside the network. *On the TCP level* the first proposition was to use a larger window than one packet at the beginning of SS. An initial window of maximum four packets has been proposed. Another proposition, called *byte counting*, was to account for the number of bytes covered by an ACK while increasing the window rather than the number of ACKs. To avoid long bursts in the case of large gaps in the ACK stream, a limit on the maximum window increase has been proposed (*limited byte counting*). These solutions try to solve the problem while preserving the ACK clock. They result in an increase in TCP burstiness and an overload on network buffers. Another type of solution tries to solve the problem by introducing some kind of packet spacing (e.g. rate-based spacing). The source transmits directly at a large window without overloading the network. Once the large window is reached, the ACK clock takes over. This lets the source avoid a considerable part of SS. The problem can be solved at the application level without changing the TCP. A possible solution (e.g. XFTP), consists of establishing many parallel TCP connections for the same transfer. This accelerates the

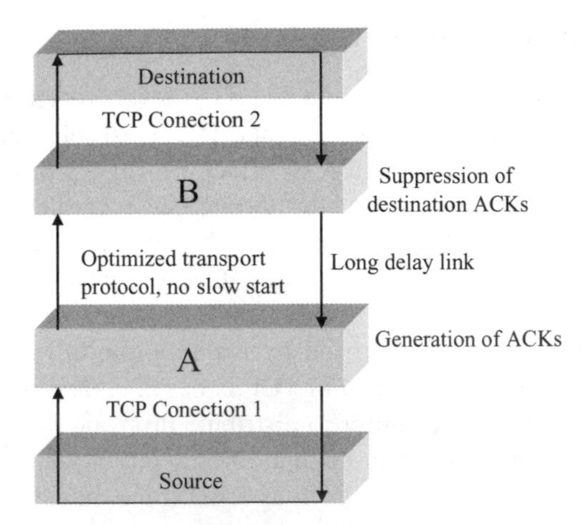

Figure 9.2 Spoofing: elimination of the long delay link from the feedback loop.

growth of the resultant window, but increases the aggressiveness of the transfer and hence the losses in the network. An adaptive mechanism has been proposed for XFTP to change the number of connections as a function of network congestion. Another solution has been proposed to accelerate the transfer of web pages. Instead of using an independent TCP connection to fetch every object in a page, the client establishes a persistent connection and asks the server to send all the objects on it (hypertext transfer protocol, HTTP). Only the first object suffers from the long SS phase; the remaining objects are transferred at a high rate. The low throughput during SS is compensated for by the long time remaining in CA.

The problem can be also solved inside the network rather than at hosts, which is worthwhile when a long delay link is located on the path. In order to decrease the RTT, the long delay link is eliminated from the feedback loop by acknowledging packets at the input of this link (A in Figure. 9.2). Packets are then transmitted on the long delay link using an optimized transport protocol (e.g. STP, described in Henderson and Katz [1]).

This transport protocol is tuned to quickly increase its transmission rate without the need for a long SS. Once arriving at the output (B), another TCP connection is used to transmit the packets to the destination. In a satellite environment, the long delay link may lead directly to the destination, so another TCP connection is not required. Because packets have already been acknowledged, any loss between the input of the link (A) and the destination must be locally retransmitted on behalf the source. Also, ACKs from the receiver must be discarded silently (at B) so as not to confuse the source. This approach is called *TCP spoofing*. The main gain in performance comes from not using SS on the long delay link. The window increases quickly, which improves performance, but spoofing still has many drawbacks. First, it breaks the end-to-end semantics of TCP; a packet is acknowledged before reaching its destination. Also, it does not work when encryption is accomplished at the IP layer, and it introduces a heavy overload on network routers. Further, the transfer is vulnerable to path changes, and symmetric paths are required to be able to discard the ACKs before they reach the source. Spoofing can be seen as a particular solution to some long delay links. It is interesting when the long delay link is the last hop to the destination. This solution is often used in networks providing high-speed access to the Internet via geostationary earth orbit (GEO) satellite links.

9.1.4 Unfairness problem at the TCP layer

One way to solve the problem is action at the TCP level by accelerating the window growth for long RTT connections. An example of a TCP-level solution is, the *constant rate algorithm*. The window is increased in CA by a factor inversely proportional to $(RTT)^2$. The result is a constant increase rate of the throughput regardless of RTT, thus better fairness. The first problem in this proposition is the choice of the increase rate. Also, accelerating window growth while preserving the ACK clock results in large bursts for long RTT connections.

Inside the network, fairness is improved by isolating the different connections from each other. Given that congestion control in TCP is based on losses, isolation means that a congested node must manage its buffer to distribute drops on the different connections in such a way that they get the same throughput. Many buffer management policies have been proposed. Some of these policies, such as RED (random early detection) [3], drop incoming packets with a certain probability when the queue length or its average exceeds a certain threshold. This distributes losses on the different connections proportionally to their throughput without requiring any per-connection state. However, dropping packets in proportion to the throughput does not always lead to fairness, especially if the bottleneck is crossed by unresponsive traffic. With a first-in first-out (FIFO) scheduler, the connection share of the bandwidth is proportional to its share of the buffer. Better fairness requires control of the buffer occupancy of each connection. Another set of policies, known as Flow RED, try to improve fairness by sharing the buffer space fairly between active connections. This ensures that each connection has at least a certain number of places in the queue, which isolates connections sending at small rates from aggressive ones. This improves fairness, but at the same time increases buffer management overhead over a general drop policy such as RED.

The problem of fairness has additional dimensions. Solving the problem at the TCP level has the advantage of keeping routers simple, but it is not enough given the prevalence of non-TCP-friendly traffic. Some mechanisms in network nodes are required to protect conservative TCP flows from aggressive ones. Network mechanisms are also required to ensure fairness at a level below or above TCP, say at the user or application level. A user (e.g. running XFTP) may be unfairly aggressive and establish many TCP connections in order to increase its share of the bandwidth. The packets generated by this user must be considered by the network as a single flow. This requires an aggregation in flows of TCP connections. The level of aggregation determines the level of fairness we want. Again this approach requires an additional effort.

9.1.5 Noncongestion losses

TCP considers the loss of packets as a result of network congestion and reduces its window consequently. This results in severe throughput deterioration when packets are lost for reasons other than congestion. In wireless communications noncongestion losses are mostly caused by transmission errors. A packet may be corrupted while crossing a poor-quality radio link. The solutions proposed to this problem can be divided into two main categories. The first consists in hiding the lossy parts of the Internet so that only congestion losses are detected at the source. The second type of solution consists of enhancing TCP with some mechanisms to help it to distinguish between different types of losses. When hiding noncongestion losses these losses are recovered locally without the intervention of the source. This can be accomplished at the link or TCP level. Two well as known mechanisms are used *as link-level solutions* to improve the link quality: ARQ and FEC. These mechanisms are discussed in Chapters 2 and 4.

TCP-level solutions try to improve link quality by retransmitting packets at the TCP level rather than at the link level. A TCP agent in the router at the input of the lossy link keeps a copy of every data packet. It discards this copy when it sees the ACK of the packet, and it retransmits the packet on behalf of the source when it detects a loss. This technique has been proposed for terrestrial wireless networks where the delay is not so important as to require the use of FEC. The TCP agent is placed in the base station at the entry of the wireless network. Two possible implementations of this agent exist.

The first implementation, referred to as indirect TCP, consists of terminating the originating TCP connection at the entry of the lossy link. The agent acknowledges the packets and takes care of handing them to the destination. A TCP connection well tuned to a lossy environment (e.g. TCP-SACK) can be established across the lossy network. A different transport protocol can also be used. This solution breaks the end-to-end semantics of the Internet. Also, it causes difficulties during handover since a large state must be transferred between base stations. The second implementation (Snoop protocol) respects the end-to-end semantics. The intermediate agent does not terminate the TCP connection; it just keeps copies of data packets and does not generate any artificial ACK. Nonduplicate ACKs sent by the destination are forwarded to the source. Duplicate ACKs are stopped. A packet is retransmitted locally when three duplicate ACKs are received or a local timeout expires. This local timeout is set, of course, to a value less than that of the source. As in the link-level case, interference may happen between the source and agent mechanisms. In fact, this solution is no other than link-level recovery implemented at the TCP level. Again, because it hides all losses, congestion losses must not occur between the Snoop agent and the destination.

9.1.6 End-to-end solutions

The addition of some end-to-end mechanisms to improve TCP reaction to noncongestion losses should further improve performance. Two approaches exist in the literature. The first consists of explicitly informing the source of the occurrence of a noncongestion loss via an explicit loss notification (ELN) signal. The source reacts by retransmitting the lost packet without reducing its window. An identical signal has been proposed to halt congestion control at the source when a disconnection appears due to handover in a cellular network. The difficulty with such a solution is that a packet corrupted at the link level is discarded before reaching TCP, and then it is difficult to get this information. The second approach is to improve the congestion control provided by TCP rather than recovery from noncongestion losses. We mention it here because it consists a step toward a solution to the problem of losses on an end-to-end basis.

The proposed solutions aim to decouple congestion detection from losses. With some additional mechanisms in the network or at the source, the congestion is detected and the throughput reduced before the overflow of network buffers. These examples, which will be discussed later in more detail, are the Vegas version of TCP [4] and the explicit congestion notification proposal. In Vegas, the RTT of the connection and the window size are used to compute the number of packets in network buffers. The window is decreased when this number exceeds a certain threshold. With ECN, an explicit signal is sent by the routers to indicate congestion to TCP sources rather than dropping packets. If all the sources, receivers and routers are compliant (according to Vegas or ECN), congestion losses will considerably decrease. The remaining losses could be considered to be caused mostly by problems other than congestion. Given that noncongestion losses require only retransmission without window reduction, the disappearance of congestion losses may lead to the definition at the source of a new congestion control algorithm which reacts less severely to losses. This ideal behaviour does ont exist in today's networks. In the absence of any feedback from the network as with Vegas, the congestion detection mechanism at the source may fail; here, congestion losses are unavoidable. If the source bases its congestion control on explicit information from the network as with ECN, some noncompliant routers will not provide the source with the required information, dropping packets instead. A reduction of the window is necessary in this case. For these reasons, these solutions still consider losses as congestion signals and reduce their windows consequently.

9.1.7 Bandwidth asymmetry

From the previous discussion we could see that TCP uses the ACK clock to predict what is happening inside the network. It assumes implicitly that the reverse channel has enough bandwidth to convey ACKs without being disturbed. This is almost true with the so-called symmetric networks where the forward and the reverse directions have the same bandwidth. However, some of today's networks (e.g. direct broadcast satellite, 4G cellular networks and asymmetric digital subscriber loop networks) tend to increase capacity in the forward direction, whereas a low-speed channel is used to carry ACKs back to the source. Even if ACKs are smaller in size than data packets, the reverse channel is unable to carry the high rate of ACKs. The result is congestion and losses on the ACK channel. This congestion increases the RTT of the connection and causes loss of ACKs. The increase in RTT reduces throughput and increases end-to-end delay. Also, it slows window growth, which further impairs performance when operating on a long delay path or in a lossy environment. The loss of ACKs disturbs one of the main functionalities of the ACK clock: smoothing the transmission. The window slides quickly upon receipt of an ACK covering multiple lost ACKs, and a burst of packets is sent, which may overwhelm the network buffers in the forward direction. Also, the loss of ACKs slows down the growth of the congestion window, which results in poor performance for long delay paths and lossy links. The proposed solutions to this problem can be divided into receiver-side solutions, which try to solve the problem by reducing the congestion on the return path, and source-side solutions, which try to reduce TCP burstiness. The first receiver-side solution is to compress the headers of TCP/IP packets on a slow channel to increase its capacity in terms of ACKs per unit of time (e.g. SLIP header compression). It profits from the fact that most of the information in a TCP/IP header does not change during the connection lifetime. The other solutions propose reducing the rate of ACKs to avoid congestion. The first proposition is to delay ACKs at the destination. An ACK is sent every d packets, and an adaptive mechanism has been proposed to change d as a function of the congestion on the ACK path. Another option is to keep the destination unchanged and filters ACKs at the input of slow channel. When an ACK arrives, the buffer is scanned to see if another ACK (or a certain number of ACKs) of the same connection is buffered. If so, the new ACK is substituted for the old one. ACKs are filtered to match their rates to the rate of the reverse channel. Normally, in the absence of artificial filtering, ACKs are filtered sometime later when the buffer gets full. The advantage of this solution is that the filtering is accomplished before the increase in RTT. Solutions at the sender side which reduce the burstiness of TCP are also possible. Note that this problem is caused by the reliance of TCP on the ACK clock, and it cannot be completely solved without any kind of packet spacing.

First, a limit on the size of bursts sent by TCP is a possible solution. However, with systematic loss of ACKs, limiting the size of bursts limits the throughput of the connection. Second, it is possible to reconstruct the ACK clock at the output of the slow channel. When an ACK arrives at that point, all the missing ACKs are generated, spaced by a time interval derived from the average rate at which ACKs leave the slow channel. This reconstruction may contain a solution to this particular problem. However, the general problem of TCP burstiness upon loss of ACKs will still remain.

9.2 TCP OPERATION AND PERFORMANCE

The *TCP protocol* model will only include the data transfer part of the TCP. Details of the TCP protocol can be found in the various Internet requests for comments (RFCs; see also Stevens [6]). The versions of the TCP protocol that we model and analyze in this section all assume the same receiver process. The *TCP receiver* accepts packets out of sequence number order, buffers them in a TCP buffer, and delivers them to its TCP user in sequence. Since the receiver has a finite resequencing buffer, it advertises a maximum window size $W_{\max}$ at connection setup time, and the transmitter ensures that there is never more than this amount of unacknowledged data outstanding. We assume that the user application at the TCP receiver can accept packets as soon as the receiver can offer them in sequence and, hence, the receiver buffer constraint is always just $W_{\max}$. The receiver returns an ACK for every good packet that it receives. An ACK packet that acknowledges the first receipt of an error-free in-sequence packet will be called a *first* ACK. The ACKs are *cumulative*, i.e. an ACK carrying the sequence number n acknowledges all data up to, and including, the sequence number $n - 1$. If there is data in the resequencing buffer, the ACKs from the receiver will carry the *next expected* packet number, which is the first among the packets required to complete the sequence of packets in the sequencing buffer. Thus, if a packet is lost (after a long sequence of good packets), then the transmitter keeps getting ACKs with the sequence number of the first packet lost, if some packets transmitted after the lost packet do succeed in reaching the receiver. These are called *duplicate ACKs.*

9.2.1 The TCP transmitter

At all times t, the transmitter maintains the following variables for each connection:

(1) $A(t)$ the *lower window edge*. All data numbered up to and including $A(t) - 1$ has been transmitted and ACKed. $A(t)$ is nondecreasing; the receipt of an ACK with sequence number $n > A(t)$ causes $A(t)$ to jump to n.

(2) $W(t)$ the *congestion window*. The transmitter can send packets with the sequence numbers n, $A(t) \le n < A(t) + W(t)$ where $W(t) \le W_{\max}(t)$ and $W(t)$ increases or decreases as described below.

(3) $W_{\text{th}}(t)$– the *slow-start threshold* controls the increments in $W(t)$ as described below.

9.2.2 Retransmission timeout

The transmitter measures the RTTs of *some* of the packets for which it has transmitted and received ACKs. These measurements are used to obtain a running estimate of the packet *RTT* on the connection. Each time a new packet is transmitted, the transmitter starts a timeout timer and *resets* the already running timeout timer, if any; i.e. there is a timeout only for the last transmitted packet. The timer is set for a *retransmission timeout* (RTO) value that is derived from the RTT estimation procedure. The TCP transmitter process measures time and sets timeouts only in multiples of a *timer granularity*. Further, there is a minimum timeout duration in most implementations. We will see in the analysis that *coarse timers* have a significant impact on TCP performance. For details on RTT estimation and the setting of RTO values, see Reference [6] or [8].

9.2.3 Window adaptation

The basic algorithm is common to all TCP versions [11]. The normal evolution of the processes $A(t)$, $W(t)$ and $W_{\text{th}}(t)$ is triggered by first ACKs (see definition above) and timeouts as follows.

(1) *Slow start:* if $W(t) < W_{\text{th}}(t)$, each first ACK increments $W(t)$ by one.

(2) *Congestion avoidance:* if $W(t) \geq W_{\text{th}}(t)$, each first ACK increments $W(t)$ by $1/W(t)$.

(3) *Timeout* at epoch t sets $W(t^+)$ to one, $W_{\text{th}}(t^+)$ to $\lceil (W(t)/2) \rceil$ and retransmissions begins from $A(t)$.

9.2.4 Packet loss recovery

If a packet is lost, $A(t)$ and $W(t)$ will continue to be incremented until the first ACK for the packet just before the lost packet is received. For a particular loss instance, let their final values be denoted by A and M, respectively; we will call M a *loss window*. Then the transmitter will continue to send packets up to the sequence number $A + M - 1$. If some of the packets sent after the lost packet get through, they will result in duplicate ACKs, all carrying the sequence number A. The last packet transmitted (i.e. $A + \mathrm{M} - 1$) will have an RTO associated with it. The TCP versions differ in the way they recover from loss. We provide some details here; later we will provide more detail on modeling these recovery procedures.

9.2.5 TCP-OldTahoe (timeout recovery)

The transmitter continues sending until packet number $A + M - 1$ and then waits for a coarse timeout.

9.2.6 TCP-Tahoe (fast retransmit [9])

A transmitter parameter K is used, a small positive integer; typically $K = 3$. If the transmitter receives the Kth duplicate ACK at time t (before the timer expires), then the transmitter behaves *as if* a timeout has occurred and begins retransmission, with $W(t^+)$ and $W_{th}(t^+)$ as given by the basic algorithm.

9.2.7 TCP-Reno fast retransmit, fast (but conservative) recovery [6]

Fast-retransmit is implemented, as in TCP-Tahoe, but the subsequent recovery phase is different. Suppose the Kth duplicate ACK is received at the epoch t_0. Loss recovery then starts. Bearing in mind the definitions of A and M above, the transmitter sets

$$W(t_0^+) = \lceil M/2 \rceil + K \quad W_{th}(t_0^+) = \lceil M/2 \rceil \tag{9.1}$$

The addition of K takes care of the fact that more packets have successfully left the network. The Reno transmitter then retransmits *only* the packet with sequence number A. If only one packet is lost then this retransmission will produce an ACK for all of the other packets, whereas if more packets are lost we had better be sure that we are not really experiencing congestion loss. For the ith duplicate ACK received, at say t_{ack}, until recovery completes,

$$W(t_{ack}^+ i) = W(t_{ack} i) + 1 \tag{9.2}$$

Further details of the Reno recovery will be explained by using an example from Kumar [13], where $K = 3$. Suppose $A = 15$, $M = 16$ and packet 15 is lost. The transmitter continues sending packets 16, 17, 18, ..., 30; suppose packet 20 is also lost. The receiver returns ACKs (all asking for packet 15) for packets 16, 17, 18, 19, 21, 29 and 30. Note that the ACK for packet14 would have been the *first* ACK asking for packet 15. When the ACK for packet 18 is received (i.e. the third duplicate ACK), the transmitter sets

$$W(t_0^+) = \lceil M/2 \rceil + K = \lceil 16/2 \rceil + 3 = 11$$
$$W_{th}(t_0^+) = \lceil M/2 \rceil = \lceil 16/2 \rceil = 8$$

A is still 15; thus, packet 15 is retransmitted. Meanwhile, ACKs for packets 19, 21, ..., 30 are also received and based on Equation (9.2) W grows to $11 + 11 = 22$. Since $A = 15$, with $W = 22$, the transmitter is allowed to send packets 15–36; hence, retransmission of packet 15 is followed by transmission of packets 31–36. Receipt of packet 15 results in a first ACK asking for packet 20 (thus first-ACKing packets 15–19) and A jumps to 20. This is called a *partial ACK* since all of the packets transmitted in the original loss window were not ACKed.

If packets 31–36 succeed in getting through, then three duplicate ACKs asking for packet 20 are also obtained, and 20 is retransmitted. This results in a first ACK that covers all of the packets up to packet number 36. At this time t_1, the congestion window is reset as follows and a new transmission cycle starts:

$$W(t_1^+) = \lceil M/2 \rceil \quad W_{th}(t_1^+) = \lceil M/2 \rceil \tag{9.3}$$

Thus, Reno slowly recovers the lost packets and there is a chance that, owing to insufficient duplicate ACKs, the recovery stalls and a timeout has to be waited for. After a timeout, the basic timeout recovery algorithm is applied.

9.2.8 TCP-NewReno (fast retransmit, fast recovery [9])

When duplicate ACKs are received, the first lost packet is resent, but, unlike Reno, upon receipt of a partial ACK after the first retransmission, the next lost packet (as indicated by the partial ACK number) is retransmitted. Thus, after waiting for the first K duplicate ACKs, the remaining lost packets are recovered in as many RTTs. If less than K duplicate ACKs are received, then a timeout is inevitable. Consider the following example [13] to see the difference with Reno. Suppose that after a loss, $A = 7$ and $M = 8$, packets 7, 8, ..., 14 are sent, and packets 7, 8 and 11 are lost. The transmitter receives three duplicate ACKs for packets 9, 10 and 12 (asking for packet 7). A fast retransmit is done (the same as in Reno), i.e. $W = 4 + 3 = 7$, $W_{th} = 4$, and packet 7 is sent. The two ACK's for packets 13 and 14 cause W to become 9 [see Equation (9.2)]. Assuming that packet 7 now succeeds, its ACK (the first ACK asking for 8) would make A equal 8; the transmitter can now send packets 15 and 16 also. *NewReno* would now resend packet 8, whereas *Reno* would wait for three duplicate ACKs; these cannot come since only two more packets have been sent after the retransmission of packet 7. Thus, in case of multiple losses, *Reno* has a higher probability of resorting to a coarse timeout.

9.2.9 Spurious retransmissions

Consider *TCP-OldTahoe*. The retransmission of the first lost packet may result in an ACK that acknowledges all of the packets until the next packet lost in the loss window. This would advance the lower window edge $A(t)$; the congestion window would be increased to two and the next lost packet and its successor would be transmitted, *even if this successor packet had gotten through in its first transmission*. Thus, some of the good packets in the loss window may be retransmitted when retransmission starts; this phenomenon can be seen in the sample path fragments shown in Fall and Floyd [9].

9.2.10 Modeling of TCP operation

The model from Figure.9.3, motivated by many experimental studies of TCP performance over wireless mobile links [7, 10, 13], is used the most often. In this section the additional issue of mobility is not considered [7]; hence, we refer to the wireless link as simply a 'lossy' link. We model only one direction of flow of packets from the LAN host (or base station in cellular system) to the mobile terminal. In this case propagation delays can be neglected. The transmission time of a TCP packet from the LAN host (respectively, the lossy link) is assumed to be exponentially distributed with mean λ^{-1} (respectively, μ^{-1}). By taking $\mu = 1$, time is normalized to the mean packet transmission time on the lossy link. During a bulk transfer over a TCP connection, we would expect that the packets would predominantly be of a fixed length. However, a MAC layer (or radio resource manager) will operate in these systems too. Thus, the randomness in packet transmission times in the model can be taken in to account for the variability in the time taken to transmit a head-of-the-line packet at the transmitter queues. The exponential assumption yields Markov processes and, hence, justifies the use of tools from Chapter 6. Performance analysis of TCP protocols under the above assumptions can be found in Kumar [13]. A sample result is shown in Figure 9.4.

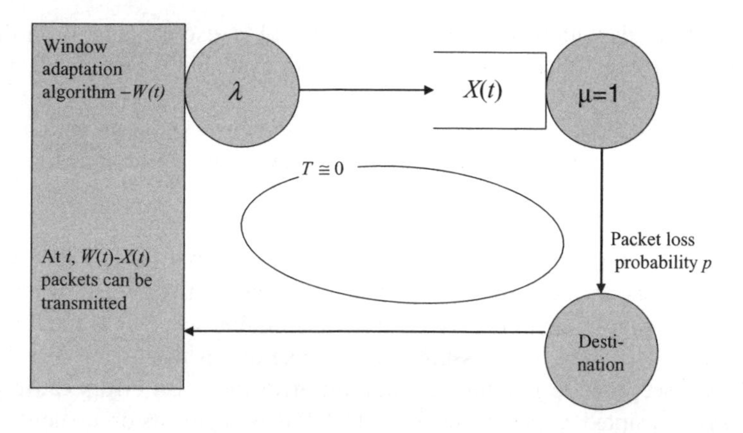

Figure 9.3 Model for the TCP connection. $X(t)$ is the number of TCP packets queued at the intermediate system (IS) at time t.

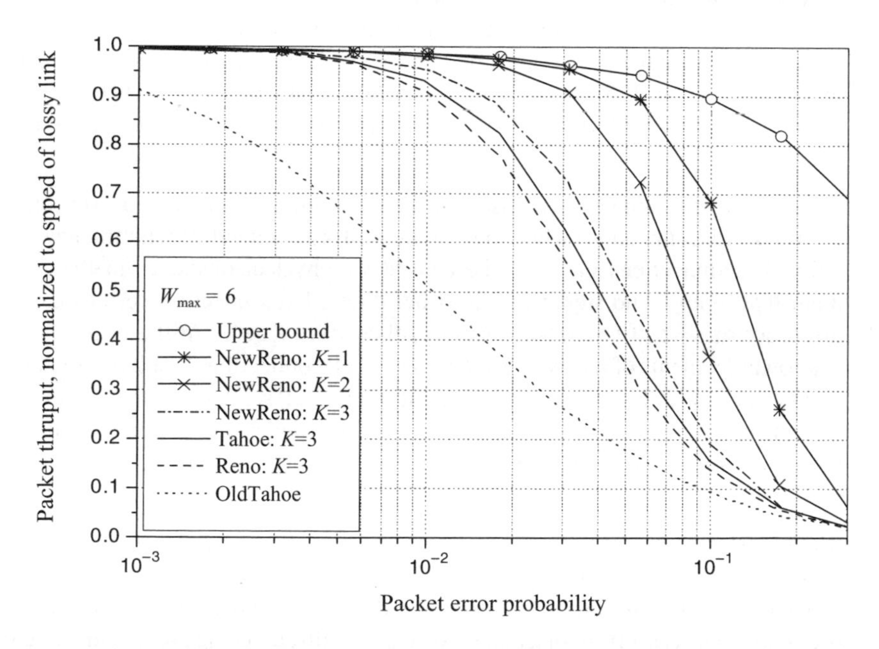

Figure 9.4 Throughput of versions of TCP vs packet-loss probability; $\lambda = 5\mu$; K is the fast-retransmit threshold. (Reproduced by permission of IEEE [13].)

9.3 TCP FOR MOBILE CELLULAR NETWORKS

Many papers have been written proposing methods for improving TCP performance over a wireless link [14–24]. Most of these papers have, however, concentrated on only one problem associated with wireless links, a perceived high BER over the wireless link. While a high BER has significant implications for protocol performance, other limitations of the wireless environment are equally or even more important than high BER. In this section we discuss the effects of periodic disconnections on TCP performance and present a protocol (M-TCP) that successfully deals with this problem. The protocol is also capable of on-the-fly data compression in the event that the wireless bandwidth available is very small. Finally, the protocol also maintains end-to-end TCP semantics.

If we use TCP without any modification in mobile networks, we experience a serious drop in the throughput of the connection. There are several reasons for such a drastic drop in TCP throughput.

(1) The wireless link might suffer from a high bit error rate. While this may be true in some cases, the bit error can be reduced by one or two orders of magnitude by the use of appropriate FEC codes and retransmission schemes at the link layer, see Chapter 4. However, let us assume for the sake of this discussion that the wireless link is susceptible to inordinately high bit error rates. Bit errors cause packets to become corrupted, which results in lost TCP data segments or acknowledgements. When acknowledgements do not arrive at the TCP sender within a short amount of time (the retransmit timeout or RTO which is a multiple of half a second), the sender retransmits the segment, exponentially backs off its retransmit timer for the next retransmission, and closes its congestion window to one segment. Repeated errors will ensure that the congestion window at the sender remains small, resulting in low throughput [15]. It is important to note that FEC may be used to combat high BER, but it will waste valuable wireless bandwidth when correction is not necessary.

(2) In a mobile environment, as a user moves between cells, there is a brief blackout period (or disconnection) while the mobile performs a handoff with the new access point. Disconnections may also be caused by physical obstacles in the environment that block radio signals, such as buildings. Disconnection periods can be of the order of several seconds, causing packet loss or delay in the transmission of acknowledgements of received packets. These disconnections result in lost data segments and lost ACKs which, in turn, result in the TCP sender timing out and closing its congestion window, thus greatly reducing the efficiency of the connection. Since these disconnections tend to be fairly lengthy, forward error correction schemes are ineffective.

(3) It is likely that, in order to provide high-bandwidth wireless connections, cell sizes in 4G systems will have to be reduced. Small cell sizes unfortunately result in small cell latencies that, in turn, cause frequent disconnections as a user roams. All the problems that result from disconnections, as we discussed above, occur more often here.

Another problem caused by small cell latencies and frequent disconnections is that of serial timeouts at the TCP sender. A serial timeout is a condition wherein multiple consecutive retransmissions of the same segment are transmitted to the mobile while it is disconnected. All these retransmissions are thus lost. Since the retransmission timer at the sender is doubled with each unsuccessful retransmission attempt (until it reaches 64 s), several consecutive failures can lead to inactivity lasting several minutes. Thus, even when the mobile is reconnected, no data is successfully transmitted for as long as 1 min. The serial timeouts at the TCP sender can prove to be even more harmful to overall throughput than losses due to bit errors or small congestion windows.

9.3.1 Improving TCP in mobile environments

In order to ensure that the TCP connection to a mobile is efficient, it is necessary to prevent the sender from shrinking its congestion window when packets are lost either due to bit-error or to disconnection. Furthermore, it is important to ensure that, when the mobile is reconnected, it begins receiving data immediately (rather than having to wait for the sender to timeout and retransmit). As we saw earlier, the sender may shrink its congestion window in response to a timeout (caused by lost packets either due to a high BER or disconnections) or in response to duplicate ACKs. As we already discussed in Section 9.1 some solutions attempt to keep the sender's congestion window open by introducing a host in the fixed network who 'spoofs' the sender into thinking everything is fine on the wireless link. Unfortunately, however, these solutions do not work under all scenarios of mobility. Specifically, they all perform poorly when faced with frequent or lengthy disconnections. Furthermore, some solutions fail to maintain end-to-end TCP semantics.

One proposed solution for losses caused by high BER is the Berkeley Snoop Module [16]. The snoop module resides at an intermediate host near the mobile user (typically the base station). It inspects the TCP header of TCP data packets and acknowledgements which pass through and buffers copies of the data packets. Using the information from the headers, the snoop module detects lost packets (a packet is assumed to be lost when duplicate acknowledgements are received) and performs local retransmissions to the mobile. The module also implements its own retransmission timer, similar to the TCP retransmission timeout, and performs retransmissions when an acknowledgement is not received within this interval. An improved version of the snoop module adds selective retransmissions from the intermediate node to the mobile. Another solution to the problem caused by high BER is the I-TCP [14] protocol (indirect-TCP). In the I-TCP protocol a TCP connection between a fixed host and a Mobile Host (MH) is split in two at the Mobile Support Station (MSS) or base station. Data sent to the MH is received and ACKed by the MSS before being delivered to the MH. Note that on the connection between the MSS and MH it is not necessary to use TCP, rather some protocol optimized for the wireless link could be used.

A solution that addresses the problem caused by short disconnections (where one or two segments only are lost) is presented in [21]. It, however, does not split the TCP connection. The solution is based on the following observation: during a handoff, since the MH cannot receive packets, unmodified TCP at the sender will think a congestion has occurred and will begin congestion control (reduce window size and retransmit) after a timeout. The timeout period is long and even though the MH may have completed the handoff it will have to wait for the full timeout period before it begins receiving packets from the sender. The fast retransmit idea presented in Caceres and Iftode [21] forces the MH to retransmit, in triplicate, the last old ACK as soon as it finishes a handoff. This forces the sender to reduce the congestion window to a half and retransmit one segment immediately.

9.3.2 Mobile TCP design

The implementation of M-TCP is influenced, to some degree, by the mobile network architecture. In this section we assume a three-level hierarchy. At the lowest level are the mobile hosts that communicate with MSS nodes in each cell. Several MSSs are controlled by a supervisor host (SH). The SH is connected to the wired network and it handles most of the routing and other protocol details for the mobile users. In addition it maintains connections for mobile users, handles flow-control and is responsible for maintaining the negotiated quality of service. These SHs thus serve the function of gateways.

The design of transport layer is influenced by a number of constraints unique to the mobile environment:

- Available bandwidth within a cell may change dynamically. This leads to difficulties in guaranteeing QoS parameters such as delay bounds and bandwidth guarantees.
- Mobile hosts frequently encounter extended periods of disconnection (due to handoff or physical interference with the signal), resulting in significant loss of data, causing poor TCP and UDP throughput.
- Mobile devices are battery powered and hence power is a scarce resource. Protocols designed for use on mobile platforms must therefore be tailored to be power-efficient.

All of these factors point towards using transport protocols which are optimized specifically for the wireless link. It is not reasonable, however, to assume that the entire installed network base will (or should) upgrade to these mobile-friendly protocols as they may never have occasion to communicate with a mobile user. Hence, the obvious solution is to split transport connections in two. The existing protocols may continue to be used on the fixed network, and the protocols optimized for the wireless environment may be used in the mobile subnetworks. In the following, all connections set up by an MH, where the other endpoint is either another MH or is a fixed host, are split at the SH. Thus, the service-provider sets up a connection with the SH assuming the SH is the other end-point of the connection. The SH sets up another connection to the MH. This allows us to address the problems listed above as follows:

- A bandwidth management module at the SH [19] assigns a fixed amount of bandwidth to each connection (this is recomputed periodically based on the changing needs of other MHs) and ensures that data is transmitted to all MHs at their assigned rate. This mechanism allows implementation of some limited form of QoS guarantees.
- In this architecture, the SH performs local error recovery to combat the efficiency problems resulting from loss over the wireless link.
- The SH tracks mobiles as they roam and uses this information to ensure that the number of duplicate packets transmitted to the mobile are kept small. Since every packet received by the MH has to be processed, consuming battery power, keeping the number of duplicates small reduces power consumption at the MH.

The goal in developing the M-TCP protocol is to provide a general solution to the problem of improving TCP's efficiency for mobile computing applications. Specifically, the following characteristics are desired:

(1) improve TCP performance for mobile clients; (2) maintain end-to-end TCP semantics; (3) be able to deal with the problems caused by lengthy disconnections or by frequent disconnections; (4) adapt to dynamically changing bandwidth over the already starved wireless link; (5) ensure that handoffs (as a mobile roams) are efficient

To this end we chose to use the split connection approach for implementing M-TCP because it fits in well with our general design philosophy articulated in the previous section and because it allows us to modify TCP on the mobile network to make it respond better to disconnections and low (or varying) wireless bandwidth.

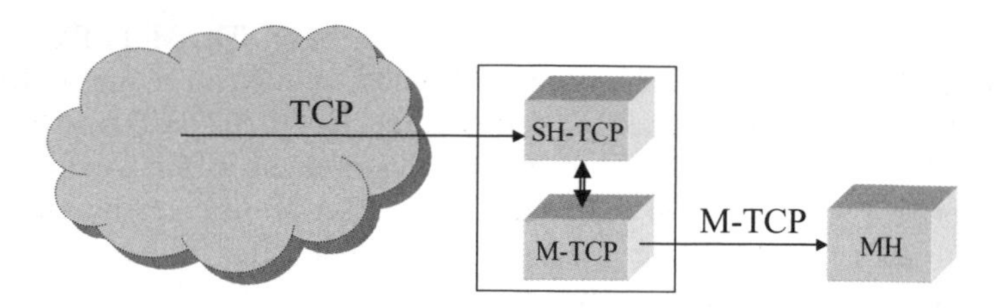

Figure 9.5 M-TCP connection.

So, every TCP connection is split in two at the SH. The TCP sender on the fixed network uses unmodified TCP to send data to the SH while the SH uses M- version of TCP for delivering data to the MH. Figure 9.5 illustrates the way in which a TCP connection is split. The TCP client at the SH (called SH-TCP) receives segments transmitted by the sender and passes these segments to the M-TCP client for delivery to the MH. ACKs received by M-TCP at the SH are forwarded to SH-TCP for delivery to the TCP sender. In the remainder of this section we discuss the behavior of SH-TCP and M-TCP in detail. Before doing so, however, we briefly discuss our assumptions regarding the mobile networking environment.

Wireless bandwidth will always be a precious resource and we therefore believe that it ought to be allocated to users based on their need. In this architecture, we use a bandwidth management module [19], that resides at the SH, to perform this task. This module determines the amount of bandwidth to be allocated to each connection within each cell (a discussion of the possible implementation of this module is given in Chapter 12). Thus, when a mobile opens a TCP connection, it is allocated a fixed amount of bandwidth (this amount may change periodically depending on the needs of other mobiles and the current status of this mobile's connection). What implication does this have for the design of TCP? Clearly, since the bandwidth allocated to the mobile is fixed, there is little reason to implement TCP's sophisticated congestion control mechanisms between the SH and MH. A second assumption we make in this development is that the BER visible at the transport layer will be small. This assumption is based on the fact that the physical layer will keep the BER over wireless channels, in a range close to 10^{-5} . Thus, it is necessary to design a TCP protocol that handles unremedied problems, disconnections and low bandwidth rather than a protocol that works well only in high BER environments.

9.3.3 The SH-TCP client

When SH-TCP receives a segment from the TCP sender, it passes the segment on to the M-TCP client. However, unlike I-TCP, it does not ACK this data until the MH does. SH-TCP is notified of MH ACKs by the M-TCP client running at the SH. It is easy to see that this behavior ensures that end-to-end TCP semantics are maintained. However, how do we ensure that the sender does not go into congestion control when ACKs do not arrive (because the MH was temporarily disconnected and did not receive the data, or could not send the ACKs)? The approach is to *choke* the TCP sender when the MH is disconnected, and allow the sender to transmit at full speed when the MH reconnects by manipulating the TCP sender's window. Specifically: let Wdenote the currently advertised receive window at SH-TCP. Say the window contains $w \leq W$bytes. Assume that the MH has ACKed bytes up to $w' \leq w$. SH-TCP sends an ACK (or ACK) for bytes up to $w' - 1$ in the normal way. As and when the MH ACKs more data, more ACKs are generated but one last byte is always left unacknowledged.

Say the MH disconnects after having ACKed bytes up to w'. The M-TCP client assumes that the MH has been temporarily disconnected because it stops receiving ACKs for bytes transmitted after w'. M-TCP sends an indication of this fact to SH-TCP who then sends an ACK for the w'th byte to the sender. This ACK will also contain a TCP window size update that sets the sender's window size to zero. When the TCP sender receives the window update, it is forced into persist mode. While in this state, it will not suffer from retransmit timeouts and will not exponentially back off its retransmission timer, nor will it close its congestion window. RFC 1122 states that TCP clients go into persist mode when the window is shrunk to zero. However, in order to shrink the window to zero, the receiver must send a new ACK. To grow a window, however, an old ACK can be retransmitted. Note that as long as the SH-TCP sends ACKs for the persist packets sent by the TCP source, the state of the sender does not change no matter how long the disconnection period lasts.

If the MH has not disconnected but is in a cell with very little available bandwidth, SH-TCP still sends an ACK for byte w' with a window size set to 0. SH-TCP estimates the RTT to the TCP sender and estimates the RTO interval. It uses this information to preemptively shrink the sender's window before the sender goes into exponential backoff (to implement this scheme, a timer at the SR is maintained that is initialized to this estimated RTO value).

When an MH regains its connection, it sends a greeting packet to the SH. M-TCP is notified of this event and it passes on this information to SH-TCP which, in turn, sends an ACK to the sender and reopens its receive window (and hence the sender's transmit window). The window update allows the sender to leave persist mode and begin sending data again. Since the sender never timed out, it never performed congestion control or slow-start. Thus the sender can resume transmitting at full-speed! The sender will begin sending from byte $w + 1$ here, although possibly duplicating previous sends, so we do not want to shrink the window more often than necessary.

A potential problem with this protocol is the following: say the sender only sends data to the MH occasionally. Specifically, if the sender only needed to send bytes 0–50 to the MH, according to the above protocol, SH-TCP will ACK bytes up to 49 but will not ACK byte 50. This will cause the sender to timeout and retransmit this byte repeatedly. Eventually, after 12 retransmissions, will the TCP sender give up and quit? This is clearly an undesirable situation. This problem is handled by not allowing SH-TCP to shrink the sender's window if it believes that there will be no more new ACKs from the MH that will allow it to open up the sender's window again. Thus, when SH-TCP thinks that the sender will timeout (and if the MH is in a cell with plenty of available bandwidth), it simply ACKs the last byte. At this point, there will be no saved byte at SH-TCP to allow it to shrink the sender's window. Observe that this is not a problem because the sender did not really have any new segments to send to the MH. As and when the sender does transmit new data to the MH, SH-TCP reverts to its previously described behavior.

9.3.4 The M-TCP protocol

The goal in designing SH-TCP was to keep the TCP sender's congestion window open in the face of disconnection events at the MH. On the wireless side, on the other hand, the goal is to be able to recover quickly from losses due to disconnections, and to eliminate serial timeouts.

In designing M-TCP, a scheme similar to SH-TCP is used except that there is more design freedom since the protocol at both ends of M-TCP can be modified. M-TCP at the mobile receiver is very similar to standard TCP except that it responds to notifications of wireless link connectivity. When M-TCP at the MH is notified (by the communications hardware) that the connection to its MSS has been lost, it freezes all M-TCP timers. This essentially ensures that disconnections do not cause the MH's M-TCP to invoke congestion control. Observe that neither acknowledgements nor data are lost during a disconnection. When the connection is regained, M-TCP at the MH sends a specially marked ACK to M-TCP at the SH which contains the sequence number of the highest byte received thus far. It also unfreezes M-TCP timers to allow normal operation to resume.

At the SH we need to ensure that M-TCP monitors link connectivity and takes appropriate action either when a disconnection occurs or when a reconnection occurs. How does M-TCP determine that the MH has been disconnected? In the design M-TCP monitors the flow of ACKs from the MH in response to segments transmitted on the downlink. If it does not receive an ACK for some time, it automatically assumes that the MH has been disconnected. It then informs SH-TCP of this fact and SH-TCP reacts by putting the TCP sender in persist mode. M-TCP at the SH knows when the MH reconnects because M-TCP at the MH retransmits the last ACK, and marks it as a reconnection ACK, on being reconnected. M-TCP responds by informing SH-TCP who reopens the sender's transmit window. This behavior is based on the assumption that the bandwidth management module at the SH assigns bandwidth to each connection and regulates its usage so there is no need to invoke congestion control at the SH when ACKs are not received from the MH. M-TCP at the SH works as follows.

When a retransmit timeout occurs, rather than retransmit the segment and shrink the congestion window, we force M-TCP into persist mode. It is assumed that the absence of an ACK means that the mobile receiver has temporarily disconnected, hence retransmissions are futile until the MH notifies the SH that it has reconnected.

At first glance the implementation may appear to have a problem when operating in high BER environments. This is because a significant proportion of packet loss in such environments will be due to the high BER and not due to disconnections. In response to this observation the following arguments are used:

- It is believed that in most mobile environments link layer solutions will ensure that the bit error seen at the transport layer is small.
- Even if the mobile environment does have high BER, the M-TCP sender will come out of persist mode quickly and resume regular data transmissions. This is because, when in persist mode, M-TCP will send persist packets to the MH who will be forced to respond when it receives these packets. Typically, the persist packets are generated at increasing intervals starting at 5 s and doubling until the interval becomes 60 s. The interval could be changed to be equal to the RTT between the SH and the MH to ensure earlier recovery.

When the SH receives a specially marked reconnection ACK, it moves back out of persist mode and retransmits all packets greater than the ACKed byte as these must have been lost during the disconnection. If the SH misinterpreted a lengthy delay as a disconnection, the MH will eventually ACK the transmitted packets, which will also move the SH M-TCP out of persist but will not cause all packets to be retransmitted.

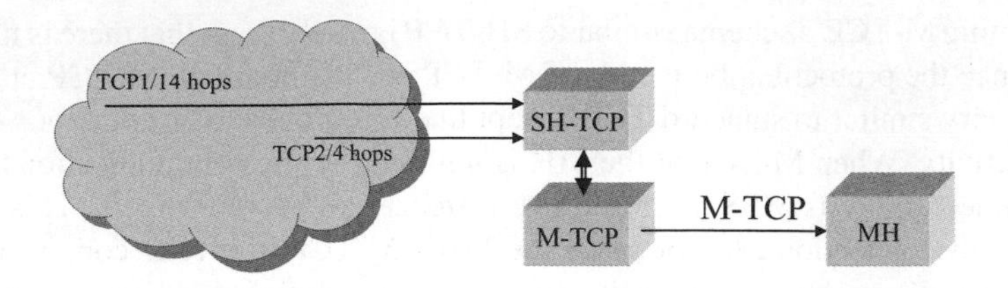

Figure 9.6 Experimental M-TCP connection setup.

9.3.5 Performance examples

Experimental/simulation set up is shown in Figure 9.6. The system parameters are used as in [20]. The wireless link has capacity of 32 kbit. File transfer times for 500 K and 1 MB files (respectively) vs disconnection length is given in Figures 9.7 and 9.8. Latency and disconnection length were normally distributed random variables and the TCP sender was

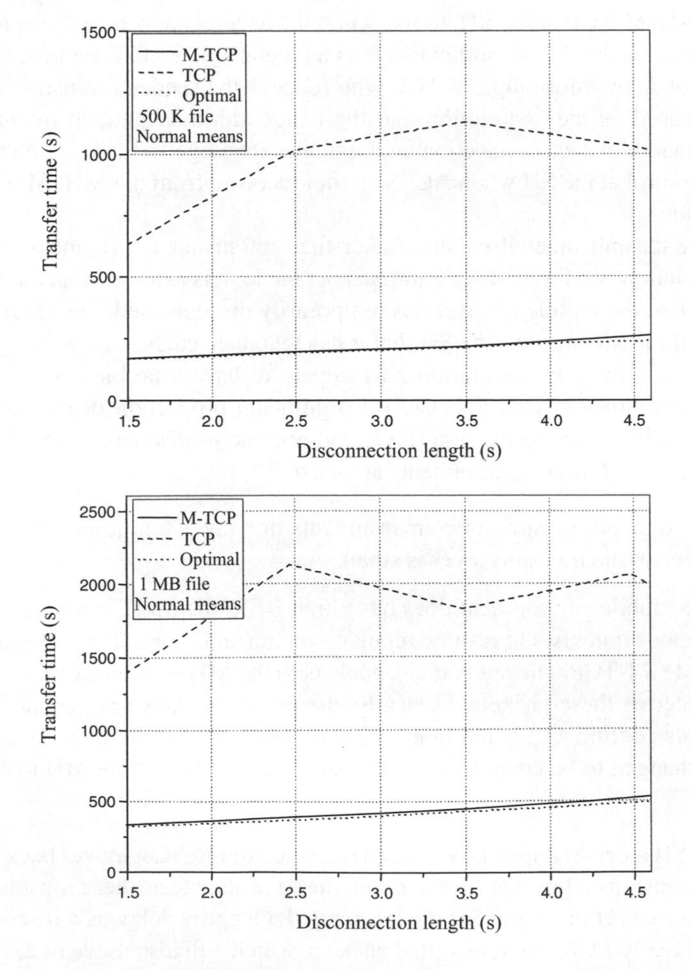

Figure 9.7 M-TCP/TCP vs disconnection length performance – 5 hops TCP connection.

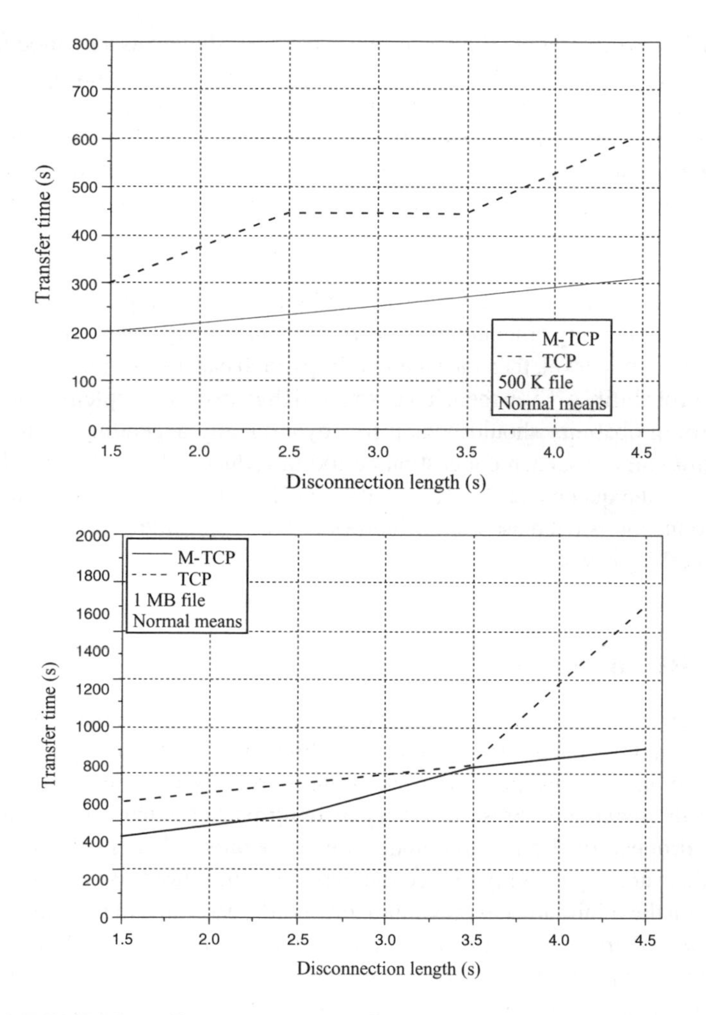

Figure 9.8 M-TCP/TCP vs disconnection length performance – 15 hops TCP connection.

five hops (Figure 9.7) and 15 hops (Figure 9.8) away. We can see that M-TCP offers better performance.

9.4 RANDOM EARLY DETECTION GATEWAYS FOR CONGESTION AVOIDANCE

In this section we discuss *random early detection* gateways for congestion avoidance in packet-switched networks [25]. RED gateways are designed to accompany a transport-layer congestion control protocol. The gateway detects incipient congestion by computing the average queue size. The gateway could notify connections of congestion either by dropping packets arriving at the gateway or by setting a bit in packet headers. When the average queue size exceeds a preset threshold, the gateway drops or marks each arriving packet with a certain probability, where the exact probability is a function of the average queue size.

Prior to RED, *early random drop* gateways have been studied as a method for providing congestion avoidance at the gateway. The initial works proposed gateways to monitor the average queue size to detect incipient congestion and to randomly drop packets when congestion is detected. RED gateways differ from the earlier early random drop gateways in several respects: (1) the literate queue size is measured; (2) the gateway is not limited to dropping packets; and (3) the packet-marking probability is a function of the average queue size.

With an even earlier version called *random drop* gateways, when a packet arrives at the gateway and the queue is full, the gateway randomly chooses a packet from the gateway queue to drop. In the implementation of *early random drop* gateways, if the queue length exceeds a certain drop level, then the gateway drops each packet arriving at the gateway with a fixed drop probability p. Even then, it was stressed that, in future implementations, the drop level and drop probability should be adjusted dynamically depending on network traffic. With *drop tail* gateways, each congestion period introduces global synchronization in the network. When the queue overflows, packets are often dropped from several connections, and these connections decrease their windows at the same time. This results in a loss of throughput at the gateway.

9.4.1 The RED algorithm

The RED gateway calculates the average queue size W_q. The average queue size is compared with a minimum (W_q^-) and a maximum (W_q^+) threshold. When the average queue size is less than the minimum threshold, no packets are marked. When the average queue size is greater than the maximum threshold, every arriving packet is marked. If marked packets are, in fact, dropped or if all source nodes are cooperative, this ensures that the average queue size does not significantly exceed the maximum threshold. When the average queue size is between the minimum and maximum thresholds, each arriving packet is marked with probability p^0, where p^0 is a function of the average queue size Wq. Each time a packet is marked, the probability that a packet is marked from a particular connection is roughly proportional to that connection's share of the bandwidth at the gateway.

Thus, the RED gateway has two separate algorithms. The algorithm for computing the average queue size determines the degree of burstiness that will be allowed in the gateway queue. The algorithm for calculating the packet-marking probability determines how frequently the gateway marks packets, given the current level of congestion. The goal is for the gateway to mark packets at fairly evenly spaced intervals, in order to avoid biases and avoid global synchronization, and to mark packets sufficiently frequently to control the average queue size.

The gateway's calculations of the average queue size take into account the period when the queue is empty (the idle period) by estimating the number m of small packets that could have been transmitted by the gateway during the idle period. After the idle period, the gateway computes the average queue size as if m packets had arrived to an empty queue during that period. As W_q varies from W_q^- to W_q^+, the packet-marking probability p_b varies linearly from 0 to its maximum value P: $p_b \leftarrow P(W_q - W_q^-)/(W_q^+ - W_q^-)$. The final packet-marking probability p^a increases slowly as the count c increases since the last marked packet: $p_b \leftarrow p_b/(1 - cp_b)$. This ensures that the gateway does not wait too long before marking a packet. The gateway marks each packet that arrives at the gateway when $Wq > W_q^+$.

9.4.2 Performance example

In this section we pay attention on the impact of traffic bursteness on the system performance. Bursty traffic at the gateway can result from an FTP connection with a long delay–bandwidth product but a small window; a window of traffic will be sent, and then there will be a delay until the ACK packets return and another window of data can be sent. Variable-bit rate video traffic and some forms of interactive traffic are other examples of bursty traffic seen by the gateway. This section shows that, unlike drop tail or random drop gateways. RED gateways do not have a bias against bursty traffic. For simulation FTP connections with infinite data, small windows and small RTT are used to model the less bursty traffic and FTP connections with smaller windows and longer RTT to model the more bursty traffic, as shown in Figure 9.9. Connections 1–4 have a maximum window of 12 packets, while connection 5 has a maximum window of eight packets. Because node 5 has a large RTT and a small window, node 5 packets often arrive at the gateway in a loose cluster. By this, we mean that, considering only node 5 packets, there is one long inter-arrival time and many smaller inter-arrival times. The same simulation scenario is used for example in Floyd and Jacobson [25].

Figure 9.10 shows the simulation results of the network in Figure 9.7 with drop tail, random drop and RED gateways, respectively. The simulations in Figure 9.10 (a) and (b) were run with the buffer size ranging from eight to 22 packets. The simulations in Figure 9.10 (c) were run many times with a minimum threshold ranging from three to 14 packets and a buffer size ranging from 12 to 56 packets.

For Figure 9.10 (a) and (b), the x-axis shows the buffer size, and the y-axis shows node 5's throughput as a percentage of the total throughput through the gateway. In these simulations, the concern is to examine the gateway's bias against bursty traffic.

For the *drop tail gateway*, Figure 9.11(a) shows the average queue size (in packets) seen by arriving packets at the bottleneck gateway, and Figure 9.11(b) shows the average link utilization on the congested link. Because RED gateways are quite different from drop tail or random drop gateways, the gateways cannot be compared simply by comparing

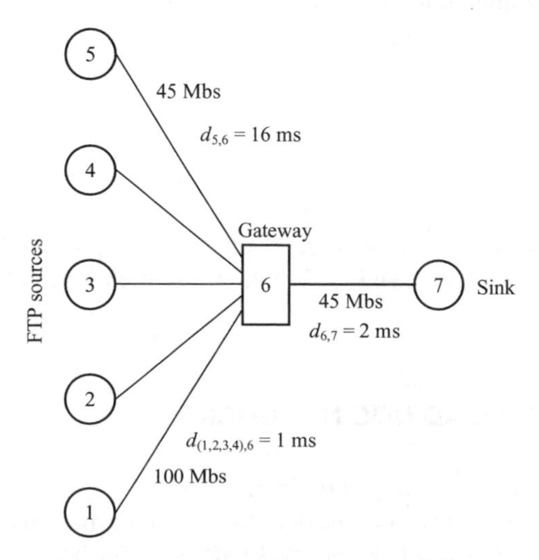

Figure 9.9 A simulation network with five FTP connections.

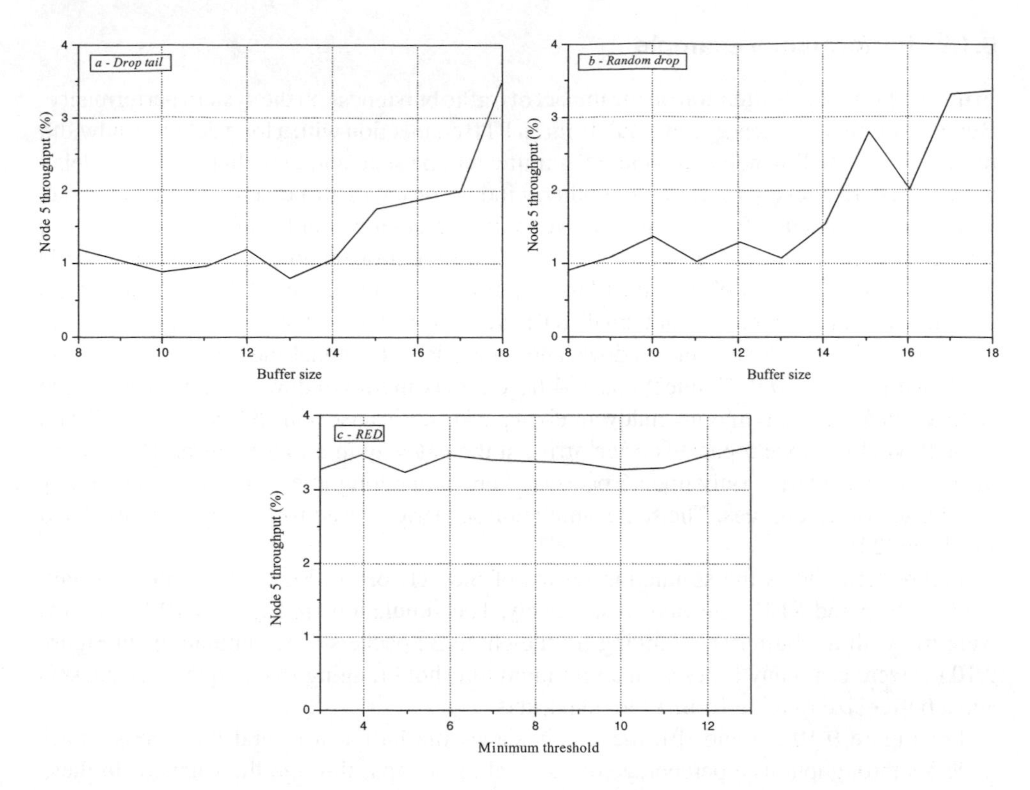

Figure 9.10 Performance of drop tail, random drop and RED gateways.

the maximum queue size. A fair comparison is between a drop rail gateway and an RED gateway that maintains the same average queue size. With drop tail or random drop gateways, the queue is more likely to overflow when the queue contains some packets from node 5. In this case, with either random drop or drop tail gateways, node 5 packets have a disproportionate probability of being dropped; the queue contents when the queue overflows are not representative of the average queue contents. Figure 9.10 (c) shows the result of simulations with RED gateways. The x-axis shows W_{q}^{-}, and the y-axis shows node 5's throughput. The throughput for node 5 is close to the maximum possible throughput given node 5's round trip time and maximum window. The parameters for the RED gateway are as follows: $W_{\mathrm{q}} \leftarrow \left(1 - \omega_{\mathrm{q}}\right) W_{\mathrm{q}} + \omega_{\mathrm{q}} W$ where ω_{q} represents the time constant of the low pass averaging filter and W is the current value of the queue, $\omega_{\mathrm{q}} = 0.002$ and $P = 1/50$ were chosen. In addition $W_{\mathrm{q}}^{+} = 2W_{\mathrm{q}}^{-}$, and the buffer size (which ranges from 12 to 56 packets) is $4W_{\mathrm{q}}^{-}$.

9.5 TCP FOR MOBILE *AD HOC* NETWORKS

So far we have discussed in this chapter problems of TCP design when dealing with congestion, errors in the transmissions, and disconnections due to mobility and fading. Transport connections set up in wireless *ad hoc* networks are even further complicated by problems such as high bit error rates, frequent route changes, and partitions. If we run TCP over such

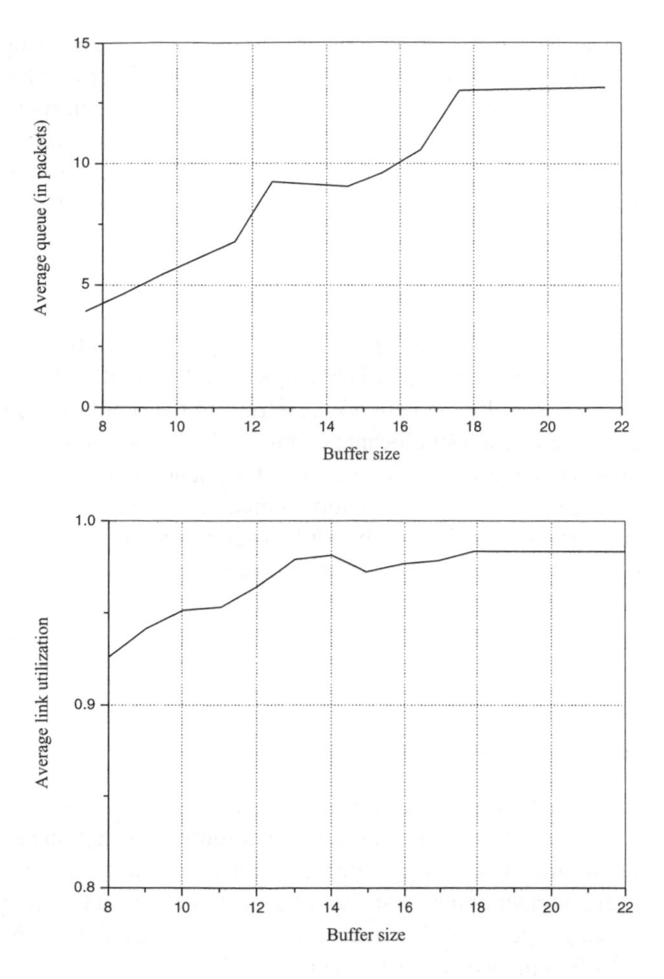

Figure 9.11 Drop tail gateway average queue length and link utilization.

connections, the throughput of the connection is observed to be extremely poor because TCP treats lost or delayed acknowledgments as congestion. In this section, we present an approach where a thin layer between Internet protocol and standard TCP is inserted that corrects these problems and maintains high end-to-end TCP throughput. This sublayer will be referred to as the *ad-hoc* TCP sublayer or A-TCP [26].

9.5.1 Effect of route recomputations

As we will see in Chapter 13, when an old route is no longer available, the network layer at the sender attempts to find a new route to the destination. In dynamic source routing (DSR) this is done via route discovery messages while in destination-sequenced distance-vectoring (DSDV) table exchanges are triggered that eventually result in a new route being found. It is possible that discovering a new route may take significantly longer than the RTO at the sender. As a result, the TCP sender times out, retransmits a packet, and invokes congestion control. Thus, when a new route is discovered, the throughput will continue to be small for some time because TCP at the sender grows its congestion window using the slow start and congestion avoidance algorithm. This is again undesirable behavior because the TCP connection will be very inefficient. If we imagine a network in which route computations are done frequently (due to high node mobility), the TCP connection will never get an opportunity to transmit at the maximum negotiated rate (i.e. the congestion window will always be significantly smaller than the advertised window size from the receiver).

9.5.2 Effect of network partitions

It is likely that the *ad hoc* network may periodically get partitioned for several seconds at a time. If the sender and the receiver of a TCP connection lie in different partitions, all the sender's packets get dropped by the network, resulting in the sender invoking congestion control. If the partition lasts for a significant amount of time (say, several times longer than the RTO), the situation gets even worse because of a phenomeon called *serial timeouts*. A serial timeout is a condition wherein multiple consecutive retransmissions of the same segment are transmitted to the receiver while it is disconnected from the sender. All these retransmissions are, thus, lost. Since the retransmission timer at the sender is doubled with each unsuccessful retransmission attempt (until it reaches 64 s), several consecutive failures can lead to inactivity lasting 1 or 2 *minutes*, even when the sender and receiver get reconnected.

9.5.3 Effect of multipath routing

As we will see in Chapter 13, some routing protocols, such as the temporally ordered routing algorithm (TORA) , maintain multiple routes between source destination pairs, the purpose of which is to minimize the frequency of route recomputation. Unfortunately, this sometimes results in a significant number of out-of-sequence packets arriving at the receiver. The effect of this is that the receiver generates duplicate ACKs which cause the sender (on receipt of three duplicate ACKs) to invoke congestion control.

The congestion window in TCP imposes an acceptable data rate for a particular connection based on congestion information that is derived from timeout events as well as from duplicate ACKs. In an *ad hoc* network, since routes change during the lifetime of a connection, we lose the relationship between the congestion window (CWND) size and the tolerable data rate for the route. In other words, the CWND as computed for one route may be too large for a newer route, resulting in network congestion when the sender transmits at the full rate allowed by the old CWND.

9.5.4 ATCP sublayer

The ATCP sublayer utilizes network layer feedback (from intermediate hops) to put the TCP sender into either a persist state, congestion control state or retransmit state. So, when the network is partitioned, the TCP sender is put into persist mode so that it does not needlessly transmit and retransmit packets. On the other hand, when packets are lost due to error (as opposed to congestion), the TCP sender retransmits packets without invoking congestion control. Finally, when the network is truly congested, the TCP sender invokes congestion control normally.

We do not have to modify standard TCP itself when we want to maintain compatibility with the standard TCP/IP suite. Therefore, to implement this solution, we can insert a thin layer called ATCP (*ad hoc* TCP) between IP and TCP that listens to the network state information provided by ECN [27] messages and by ICMP 'destination unreachable' messages and then puts TCP at the sender into the appropriate state. Thus, on receipt of a 'destination unreachable' message, the TCP state at the sender is frozen (the sender enters the *persist* state) until a new route is found, ensuring that the sender does not invoke congestion control. Furthermore, the sender does not send packets into the network during the period when no route exists between the source and destination.

The ECN, which will be discussed later in more detail, is used as a mechanism by which the sender is notified of impending network congestion along the route followed by the TCP connection. On receipt of an ECN, the sender invokes congestion control without waiting for a timeout event (which may be caused, more often than not, due to lost packets). Thus, the benefits of A-TCP sublayer are:

- Standard TCP/IP is unmodified.
- ATCP is invisible to TCP and, therefore, nodes with and without ATCP can interoperate. The only drawback is that nodes without ATCP will see all the performance problems associated with running TCP over *ad hoc* networks.
- ATCP does not interfere with TCPs functioning in cases where the TCP connection is between a node in the wireline network and another in the wireless *ad hoc* network.

There are in the literature papers dealing with different aspects of TCP problem in *ad hoc* networks. Holland and Vaidya [28], for example, investigate the impact of link breakage on TCP performance in such networks. They use dynamic source routing (DSR), discussed in Chapter 13 (see also Johnson and Maltz [29]), as the underlying routing protocol (simulated in NS2). DSR is an on-demand routing protocol where a sender finds a route to the destination by flooding *route request* packets. DSR's performance is optimized by allowing intermediate nodes to respond to route request packets using cached routes. Unfortunately, if the cached information maintained at a intermediate node is stale, the time it takes to find a new route can be very long (several seconds). Thus, TCP running on top of DSR sees very poor throughput. The paper proposes the use of explicit link failure notification (ELFN) to improve TCP performance. Here, the TCP sender is notified that a link has failed, and it disables its retransmission timer and enters a stand-by mode. In stand-by mode, the TCP sender periodically sends a packet in its congestion window to the destination. When an ACK is received, TCP leaves the stand-by mode, restores its retransmission timers and resumes transmission as normal.

Chandran *et al.* [30] discusses a similar scheme for improving TCP performance in *ad hoc* networks in the presence of failures. Here, the router detecting a failed route generates a route failure notification (RFN) packet toward the source. The TCP source that receives this packet enters a *snooze* state which is very similar to TCP's persist state. When the route is re-established, a route re-establishment notification (RRN) is sent to the source by any router on the previous route that detected the new route. This packet removes the source from the snooze state. In this method, the source continues using the old congestion window size for the new route. This is a problem because the congestion window size is route-specific (since it seeks to approximate the available bandwidth). Chandran *et al.* [30] also does not consider the effects of congestion, out-of-order packets and bit error.

9.5.5 ATCP protocol design

In this section we discuss the design of an ATCP that attempts to provide an integral solution to the problems of running TCP over multihop wireless networks. Such a protocol should have the following characteristics [26]:

(1) It should Improve TCP performance for connections set up in *ad hoc* wireless networks. As already discussed, TCP performance is affected by the problems of high BER and disconnections due to route recomputation or partition. In each of these cases, the TCP sender mistakenly invokes congestion control. A good protocol should provide only retransmissions for packets lost due to high BER, without shrinking the congestion window. Also, the sender should stop transmitting and resume when a new route has been found.the As above, in the case of transient partition, the sender should stop transmitting until it is reconnected to the receiver. In the case of multipath routing when TCP at the sender receives duplicate ACKs, it should not invoke congestion control because multipath routing shuffles the order in which packets are received.

(2) It should maintain TCP's congestion control. If losses are caused due to network congestion, we want TCP to shrink its congestion window in response to losses and invoke slow start.

(3) It should have appropriate CWND. When there is a change in the route (e.g. a reconnection after a brief partition), the congestion window should be recomputed.

(4) It should maintain end-to-end TCP semantics in order to ensure that applications do not crash.

(5) It should be compatible with standard TCP because we cannot assume that all machines deployed in an *ad hoc* network will have ATCP installed.

The above functionalities are implemented within a separate sublayer, the ATCP, placed between the network and TCP layer in the protocol stack (TCP/ATCP/IP stack). This layer monitors the TCP state and the state of the network (based on ECN and ICMP messages) and takes appropriate action. Figure 9.12 illustrates ATCP's four possible states: *normal*, *congested*, *loss* and *disconnected*. When the TCP connection is initially established, ATCP at the sender is in the *normal* state.

In this state, ATCP does nothing and is invisible. Let us now examine ATCP's behavior under four circumstances.

9.5.5.1 Lossy channel

When the connection from the sender to the receiver is lossy, it is likely that some segments will not arrive at the receiver or may arrive *out-of-order*. Thus, the receiver may generate *duplicate* acknowledgment (ACKs) in response to out-of-sequence segments. When TCP receives three consecutive duplicate ACKs, it retransmits the appropriate segment and shrinks the congestion window. It is also possible that, due to lost ACKs, the TCP sender's RTO may expire, causing it to retransmit one segment and invoke congestion control.

ATCP in its *normal state* counts the number of duplicate ACKs received for any segment. When it sees that three duplicate ACKs have been received, it *does not forward* the third

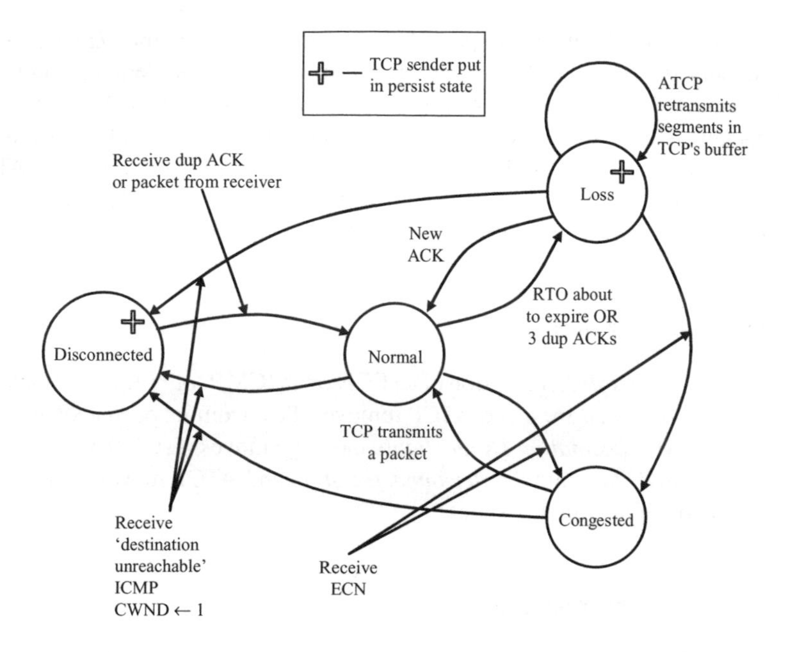

Figure 9.12 State transition diagram for ATCP at the sender.

duplicate ACK but puts TCP in *persist mode*. Similarly, when ATCP sees that TCP's RTO is about to expire, it again puts TCP in *persist mode*. By doing this, we ensure that the TCP sender does not invoke congestion control because that is the wrong thing to do under these circumstances. After ATCP puts TCP in persist mode, ATCP enters the *loss state*. In the *loss state*, ATCP transmits the unacknowledged segments from TCP's send buffer. It maintains its own separate timers to retransmit these segments in the event that ACKs are not forthcoming. Eventually, when a *new* ACK arrives (i.e. an ACK for a previously unacknowledged segment), ATCP forwards that ACK to TCP, which also removes TCP from persist mode. ATCP then returns to its *normal state*.

9.5.5.2 Congested

When the network detects congestion, the ECN flag is set in ACK and data packets. Let us assume that ATCP receives this message when in its *normal state*. ATCP moves into its *congested state* and does nothing. It ignores any duplicate ACKs that arrive and it also ignores imminent RTO expiration events. In other words, ATCP does *not interfere* with TCP's normal congestion behavior. After TCP transmits a new segment, ATCP returns to its *normal state*.

9.5.5.3 Disconnected

Node mobility in *ad hoc* networks causes route recomputation or even temporary network partition. When this happens, we assume that the network generates an ICMP *destination unreachable* message in response to a packet transmission. When ATCP receives this message, it puts the TCP sender into persist mode and itself enters the *disconnected state*. TCP periodically generates *probe packets* while in persist mode. When, eventually, the receiver is connected to the sender, it responds to these probe packets with a duplicate ACK (or a data packet). This removes TCP from persist mode and moves ATCP back into *normal state*. In order to ensure that TCP does not continue using the old CWND value, ATCP sets TCP's *CWND to one segment* at the time it puts TCP in persist state. The reason for doing this is to force TCP to probe the correct value of CWND to use for the new route.

9.5.5.4 Other transitions

When ATCP is in the *loss state*, reception of an ECN or an ICMP *source quench* message will move ATCP into *congested state* and ATCP removes TCP from its persist state. Similarly, reception of an ICMP *destination Unreachable* message moves ATCP from either the *loss state* or the *congested state* into the *disconnected state* and ATCP moves TCP into persist mode (if it was not already in that state).

9.5.5.5 Effect of lost messages

Note that due to the lossy environment, it is possible that an ECN may not arrive at the sender or, similarly, a *destination unreachable* message may be lost. If an ECN message is lost, the TCP sender will continue transmitting packets. However, every subsequent ACK will contain the ECN, thus ensuring that the sender will eventually receive the ECN, causing it to enter the congestion control state as it is supposed to. Likewise, if there is no route to the destination, the sender will eventually receive a retransmission of the *destination unreachable* message, causing TCP to be put into the persist state by ATCP. Thus, in all cases of lost messages, ATCP performs correctly. Further details on protocol implementation can be found in Liu and Singh [26].

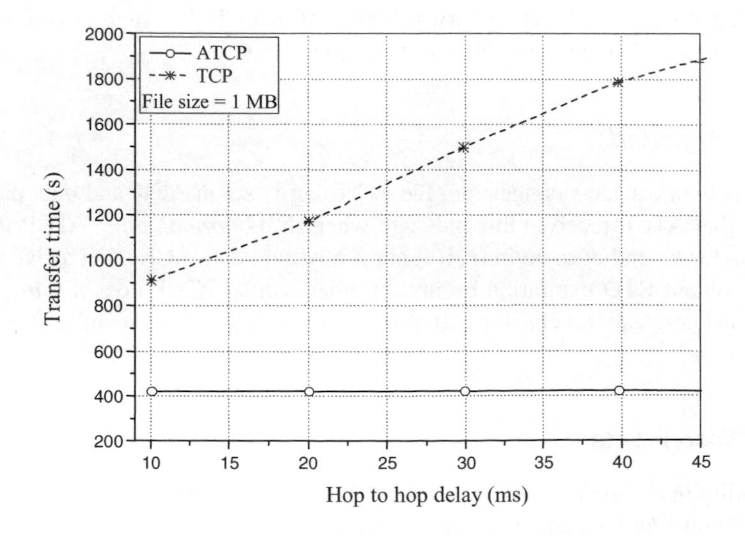

Figure 9.13 ATCP and TCP performance in the presence of bit error only.

9.5.6 Performance examples

An experimental testbed, for the evaluation of ATCP protocol, consisting of five Pentium PCs, each of which with two ethernet cards is described in Liu and Singh [26]. This gives a four-hop network where the traffic in each hop is isolated from the other hops. To model the lossy and low-bandwidth nature of the wireless links, in IP, a 32 kb/s channel is emulated over each hop. Performance examples are shown in Figures 9.13–16(a). A uniformly better performance of ATCP over standard TCP protocols is evident from all these examples.

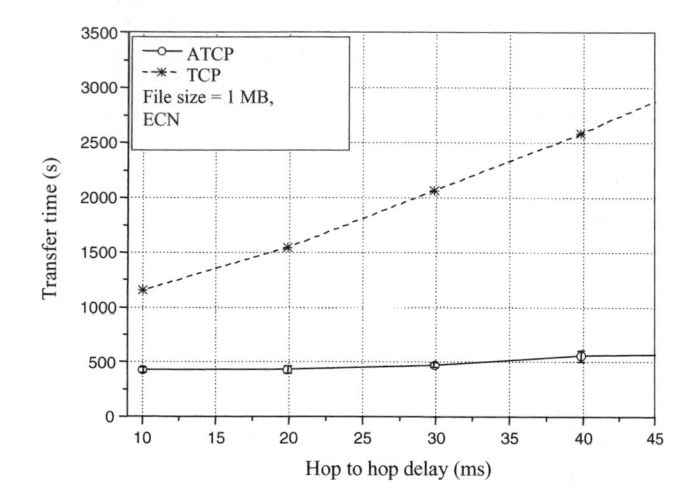

Figure 9.14 ATCP and TCP performance in the presence of bit error and congestion.

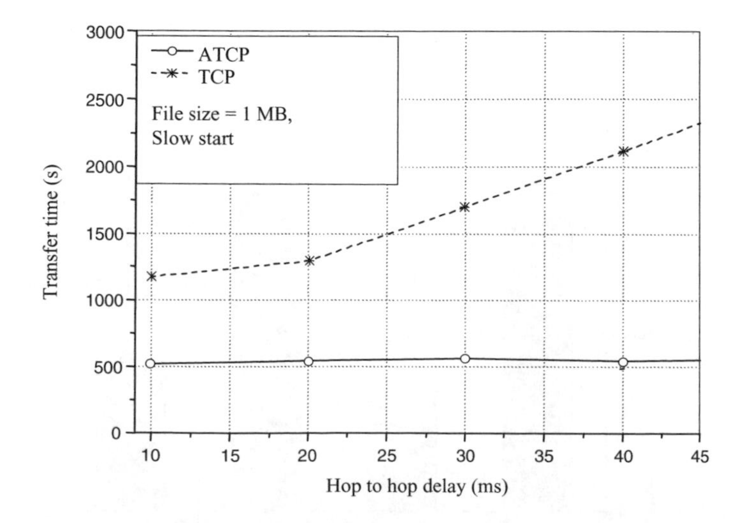

Figure 9.15 ATCP and TCP performance in the presence of bit error and partition.

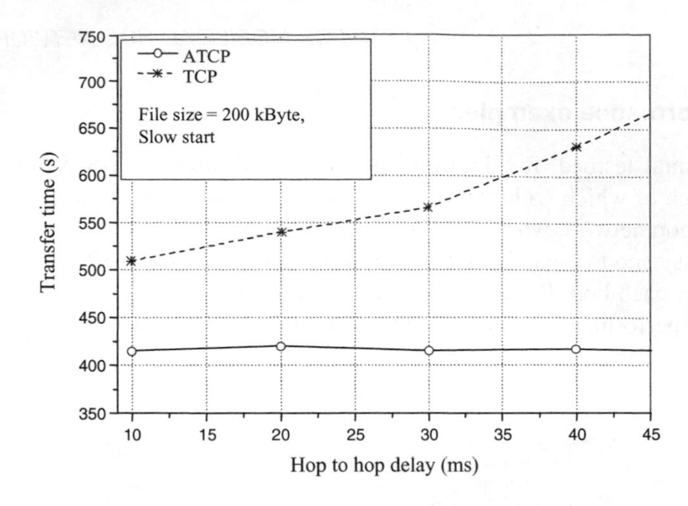

Figure 9.16 ATCP and TCP performance in the presence of bit error and larger partition.

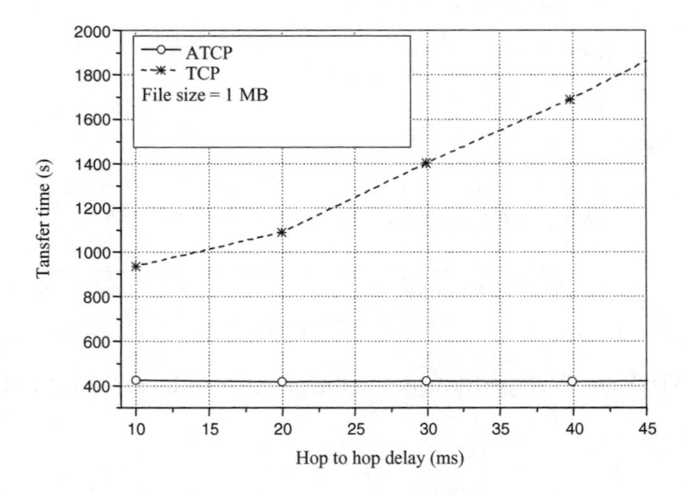

Figure 9.17 ATCP and TCP performance in the presence of bit error and packet reordering.

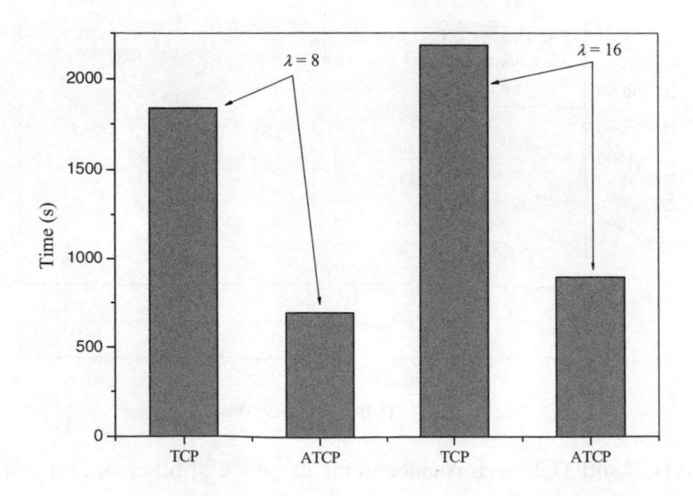

Figure 9.18 TCP and ATCP transfer time for 1 MB data in the general case.

REFERENCES

[1] T. Henderson and R.H. Katz, Transport protocols for Internet-compatible satellite networks, *IEEE JSAC*, vol. 17, no. 2, 1999, pp. 326–344.

[2] T.V. Lakshman and U. Madhow, The performance of TCP/IP for networks with high bandwidth–delay products and random loss, *IEEE/ACM Trans. Networking*, 1997, vol. 5, no. 3, pp. 336–350.

[3] S. Floyd and V. Jacobson, Random early detection gateways for congestion avoidance, *IEEE/ACM Trans. Networking*, vol. 1, no. 4, 1993, pp. 397–413.

[4] L. Brakmo and L. Peterson, TCP Vegas: end to end congestion avoidance on a global internet, *IEEE JSAC*, vol. 13, no. 8, 1995, pp. 1465–1480.

[5] C. Barakat and E. Altman, Analysis of TCP with several bottleneck nodes, *IEEE GLOBECOM*, 1999, pp. 1709–1713.

[6] W.R. Stevens, *TCP/IP Illustrated*, vol. 1. Reading, MA: Addison-Wesley, 1994.

[7] R. Caceres and L. Iftode, Improving the performance of reliable transport protocols in mobile computing environments, *IEEE J. Select. Areas Commun.*, vol. 13, 1995 pp. 850–857.

[8] A. Desimone, M.C. Chuah and O.C. Yue, Throughput performance of transport layer protocols over wireless LAN's, in *Proc. IEEE GLOBECOM'93*, December 1993, pp. 542–549.

[9] K. Fall and S. Floyd, *Comparisons of Tahoe, Reno, and Sack TCP* [Online], March 1996. Available FTP: ftp://ftp.ee.lbl.gov

[10] V.K. Joysula, R.B. Bunt and J.J. Harms, Measurements of TCP performance over wireless connections, in *Proc. WIRELESS'96, 8th Int. Conf. Wireless Communications*, Calgary, 8–10 July 1996, pp. 289–299.

[11] T.V. Lakshman and U. Madhow, The performance of TCP/IP for networks with high bandwidth–delay products and random loss, *IEEE Trans. Networking*, vol. 3, 1997, pp. 336–350.

[12] P.P. Mishra, D. Sanghi and S.K. Tripathi, TCP flow control in lossy networks: Analysis and enhancements, in *Computer Networks, Architecture and Applications*, IFIP Transactions C-13, S.V. Raghavan, G.V. Bochmann and G. Pujolle (eds). Amsterdam: Elsevier North-Holland, 1993, pp. 181–193.

[13] A. Kumar, Comparative performance analysis of versions of TCP in a local network with a lossy link, *IEEE/ACM Trans. Networking*, vol. 6, no. 4, 1998, pp. 485–498

[14] A. Bakre and B.R. Badrinath, I-TCP: indirect TCP for mobile hosts, in *Proc. 15th Int. Con. on Distributed Computing Systems*, Vancouver, June 1995, pp. 136–143.

[15] A. Bakre and B.R. Badrinath, Indirect transport layer protocols for mobile wireless environment, in *Mobile Computing*, T. Imielinski and H.F. Korth (eds). Kluwer Academic: Norwell, MA 1996, pp. 229–252.

[16] H. Balakrishnan, S. Seshan and Randy Katz, Improving reliable transport and handoff performance in cellular wireless networks, *Wireless Networks*, vol. 1, no. 4, 1995.

[17] H. Balakrishnan, V.N. Padmanabhan, S. Seshan and R. Katz, A comparison of mechanisms for improving TCP performance over wireless links, in *ACM SIGCOMM'*96, Palo Alto, CA, August 1996, pp. 256–269.

[18] B.R. Badrinath and T. Imielinski, Location management for networks with mobile users, in *Mobile Computing*, T. Imielinski and H.F. Korth (eds). Kluwer Academic: Norwell, MA 1996, pp. 129–152.

[19] K. Brown and S. Singh, A network architecture for mobile computing, in *Proc. IEEE INFOCOMM'96*, March 1996, pp. 1388–1396.
[20] K. Brown and S. Singh, M-UDP: UDP for mobile networks, *ACM Comput. Commun. Rev.*, vol. 26, no. 5, 1996, pp. 60–78.
[21] R. Caceres and L. Iftode, Improving the performance of reliable transport protocols in mobile computing environments, *IEEE Selected Areas in Commun.*, vol. 13, no. 5, 1994, pp. 850–857.
[22] R. Ghai and S. Singh, An architecture and communication protocol for picocellular networks, *IEEE Personal Commun. Mag.*, no. 3, 1994, pp. 36–46.
[23] K. Seal and S. Singh, Loss profiles: a quality of service measure in mobile computing, *J. Wireless Networks*, vol. 2, 1996, pp. 45–61.
[24] S. Singh, Quality of service guarantees in mobile computing, *J. Comput. Commun.*, vol. 19, no. 4, 1996, pp. 359–371.
[25] S. Floyd and V. Jacobson, Random early detection gateways for congestion avoidance, *IEEE/ACM Trans. Networking*, vol. 1, no. 4, 1993, pp. 397–413.
[26] J. Liu and S. Singh, ATCP: TCP for mobile *ad hoc* networks, *IEEE J. Selected Areas Commun.*, vol. 19, no. 7, 2001, pp. 1300–1315.
[27] http://www-nrg.ee.lbl.gov/ecn-arch/
[28] G. Holland and N. Vaidya, Analysis of TCP performance over mobile *ad hoc* networks, in *Proc. ACMMobile Communications Conf.*, Seattle, WA, 15–20 August 1999, pp. 219–230.
[29] D.B. Johnson and D.A. Maltz, Dynamic source routing in *ad hoc* wireless networks, in *Mobile Computing*, T. Imielinski and H. F. Korth (eds). Norwell, MA: Kluwer Academic, 1996, pp. 153–191.
[30] K. Chandran, S. Raghunathan, S. Venkatesan and R. Prakash, A feedback-based scheme for improving TCP performance in *ad hoc* wireless networks, in *Proc. 18th Int. Conf. Distributed Computing Systems*, Amsterdam, 26–29 May 1998, pp. 474–479.

10

Crosslayer Optimization

10.1 INTRODUCTION

In many algorithms throughout the book the information from physical layer was used to adjust the parameters of the higher layer protocols in the network. In this section we will additionally explicitly discuss the gains that can be achieved by means of a cross-layer approach, where a certain layer of information is passed to the higher layers in order to optimize the system performance [1–49]. In Chapter 9 we discussed the operation of TCP and indicated the importance of being able to distinguish packet losses due to network congestion from those due to degradation of physical channel parameters. The reaction of TCP on the losses due to congestion will be to change the congestion window size and reduce the transmission packet rate, whereas for physical channel degradation simple retransmission will suffice.

Routers in the network indicate congestion by dropping packets, which in turn causes the source to adaptively decrease its sending rate. The ECN mechanism used to notify the receiver whenever congestion occurs in the network was also discussed in the previous chapter. This mechanism works in the following manner: included in a TCP packet's header is the *ECN bit* which is set to zero by the source. If the router detects congestion, it will *set the ECN bit to one*, and the packet is said to be marked. The marked packet eventually reaches the destination, which in turn informs the source about the value of the mark (i.e. the ECN bit value). The source adapts its transmission rate depending on the value of the mark.

The current deployment of the TCP protocol interprets all losses as being congestion-related. Whenever losses occur over a wireless channel, the TCP source reacts to this as though it was due to congestion and thus decreases the packet transmission rate, causing loss in network throughput. A solution that has been proposed to mitigate this problem is to 'smooth' the channel by suitable coding and link-layer ARQ at a faster timescale than that of the TCP control loop so that *the wireless link ideally is perceived as a constant channel,*

Advanced Wireless Networks: 4G Technologies Savo G. Glisic
© 2006 John Wiley & Sons, Ltd.

but with lower capacity. However, in practice, there is still the problem that the TCP sender may not be fully shielded from wireless link losses. This can lead to the TCP congestion control mechanism reacting to packet losses, thus resulting in redundant retransmissions and loss of throughput. However, once ECN-enabled TCP is deployed, where the ECN bit can be used to mark packets to indicate congestion, there is a means of differentiating between *congestion-related loss* and *wireless channel-related loss*.

Thus, the channel need not be smoothed because the ECN mechanism provides a means of explicitly indicating congestion. In Chapter 9 it was demonstrated that, in a single user environment, if packets are marked based solely on congestion information, there is no significant degradation of TCP performance due to the time-varying nature of the wireless channel as compared with wired networks. Such an approach is an example of where a crosslayer view of physical layer information (channel conditions) is used at the network layer to significantly improve network-layer throughput performance.

As a second example consider a conventional cellular system with a fixed base station and a number of mobile users. Data flows (packets) arrive from the core network to the cell base station and are destined for the mobile users, with the packets for each user being queued temporarily at the base station (a separate queue is maintained for each user). The objective of the base station is to schedule these packets to various mobiles in a timely manner. Owing to the asymmetric nature of the traffic we can restrict ourselves to just considering the forward link problem, where the direction of the data flow is from the base station to the mobile user. An efficient way to use the cellular spectrum is to dedicate some bandwidth for data transport and allocate this bandwidth in a dynamic manner among various data users. For instance, time could be divided into fixed-size time slots and users allocated time slots dynamically. The simplest scheduling algorithm is the round-robin mechanism, where users are periodically allocated slots irrespective of whether there is data to be sent to a particular mobile user. Note that this simple TDM scheme still leads to some wasted bandwidth during inactivity. Furthermore, it suffers from a more subtle problem: this approach is independent of the channel state (a channel-state-independent scheduling mechanism).

A channel-state-dependent scheduling algorithm can provide significant gains. As an illustration consider a wireless system consisting of three users as in Figure 10.1. For this example, we consider access time to be slotted, and the channels to be constant over a time slot. For demonstration, let us assume that the channels are either ON or OFF, equally likely, and the channels are independent of each user. Thus, in this system, there are eight possible (instantaneous) channel states for the three independent users, ranging from (ON, ON, ON) to (OFF, OFF, OFF). When a user's channel is ON, one packet can be transmitted successfully to the mobile user during the time slot. The system is assumed to be a TDM system; thus, the base station can transmit to only one user on each time slot. The associated scheduling problem is to decide which user is allowed access to the channel during each particular time slot. A naive scheduling rule would be to employ the well-known round-robin mechanism. In such a scheme, the users are periodically given access to the channel, with each user getting one-third of the slots over time. As the channel of each user is equally likely to be ON or OFF in each time slot, it follows that, over time, on average, each user will get a data rate of $(1/2)(1/3) = 1/6$ packets/slot.

On the other hand, suppose the base station uses knowledge of the instantaneous channel state. Then a simple policy would be to schedule and transmit to a user whose channel is in the ON state. If more than one user's channel is in the ON state, the scheduler could pick a user randomly (equally likely) among those users whose channels are ON, and send

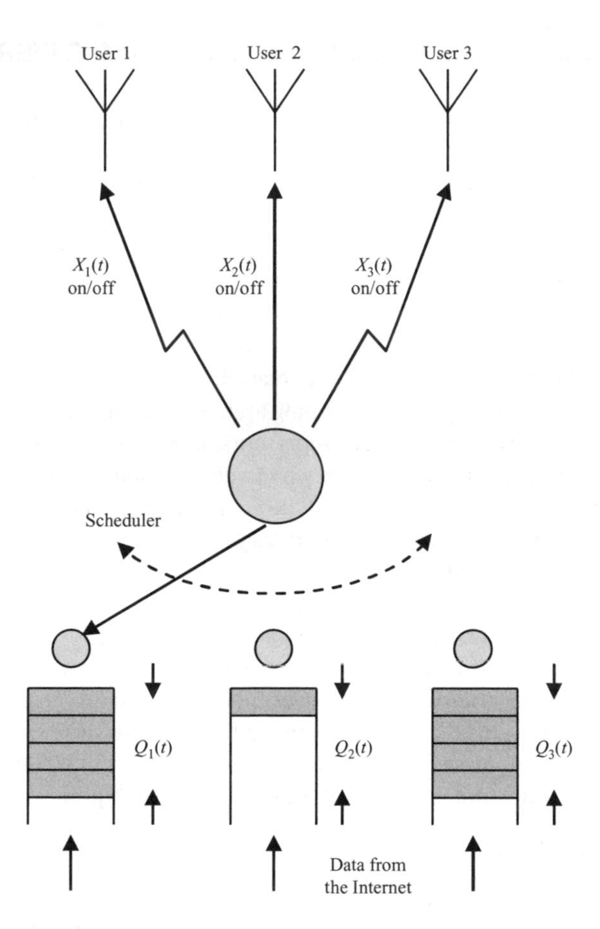

Figure 10.1 Channel state-dependent scheduling.

data to the selected user. This simple example assumes that all users have identical traffic demands. For the case where some users have greater needs than others, high-demand users would be assigned channel access with greater likelihood. In this case *data is not sent by the base station if and only if all users' channels are OFF* (which occurs on average only one-eighth of the time). Thus, the *total data rate* achieved by this state-dependent rule is $1 - \frac{1}{8} = \frac{7}{8}$ packets/slot. As this rule is symmetric across users, it follows that on average, the *data rate per user* is $(1/3)(7/8) = 7/24$ packets/slot, which is almost twice the throughput as the round-robin scheme that provided $\frac{1}{6}$ packets/slot. In addition, the base station does not radiate power during bad channel conditions, thereby decreasing interference levels in the wireless network.

This gain achieved due to channel-state-dependent scheduling is called the *multi-user diversity* gain. This example illustrates the significance of multi-user diversity gain in network scheduling. However, a central question is how to design online algorithms that achieve this gain, while also supporting diverse QoS requirements for various users for realistic channel scenarios. While this question is not yet completely answered, there has been extensive research on various aspects of the problem [2].

10.2 A CROSS-LAYER ARCHITECTURE FOR VIDEO DELIVERY

With time-varying wireless link quality, providing QoS for video applications in the form of *absolute* guarantee [51, 52] may not be feasible. Thus, it is more reasonable to provide QoS in the form of *soft* (or 'elastic') guarantee, which allows QoS parameters in the priority transmission system to be adjusted along with changing channel conditions. The relative QoS differentiation discussed in References [52–54] is one of the possible solutions for 4G adaptive QoS system. Similarly, on the application layer, it is desirable to have a video bit-stream that is adaptive to changing channel conditions. Among several possible approaches for video quality adaptation [55, 56], a scalable video [50] seems to be attractive due to its low complexity and high flexibility in rate adaptation.

Figure 10.2 shows a cross-layer QoS mapping architecture for video delivery over single-hop wireless networks. This architecture represents an end-to-end delivery system for a video source from the sender to the receiver, which includes source video encoding module, cross-layer QoS mapping and adaptation module, link-layer packet transmission module, wireless channel (time-varying and non-stationary), adaptive wireless channel modeling module, and video decoder/output at the receiver. In this context, the wireless channel is modeled at the link layer (instead of the physical layer) since the link layer modeling is more amenable for analysis and simulations of the QoS provisioning system (e.g. delay bound or packet loss rate). For these details see Chapter 8 and the discussion on effective channel capacity. In the link-layer transmission control module in the architecture, a class-based buffering and scheduling mechanism is employed to achieve differentiated services. In particular, K QoS priority classes are maintained with each class of traffic being maintained in separate buffers. A strict (nonpre-emptive) priority scheduling policy is employed to serve

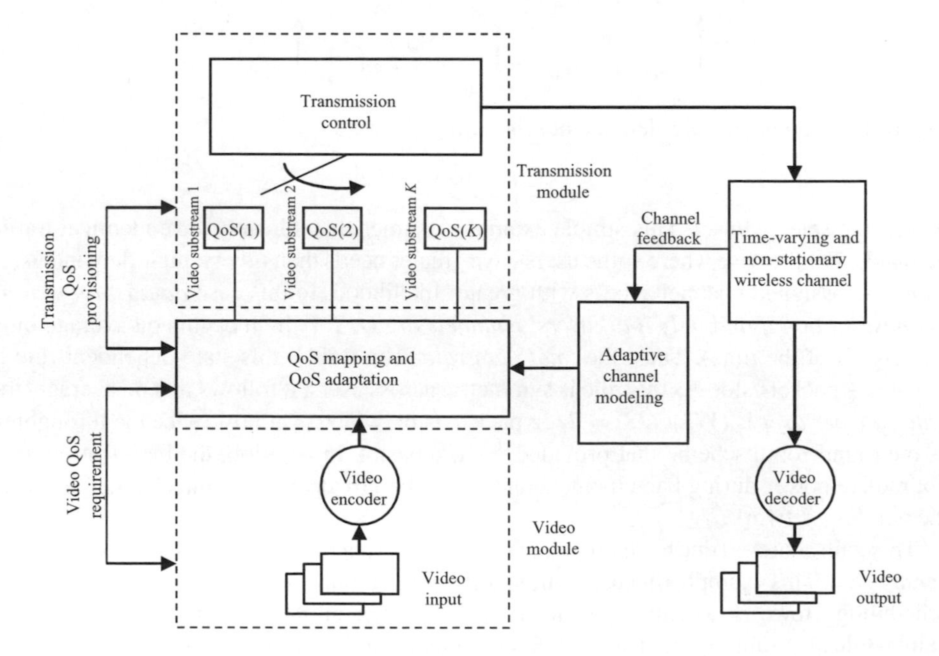

Figure 10.2 A block diagram of a cross-layer QoS management architecture for video delivery over a wireless channel.

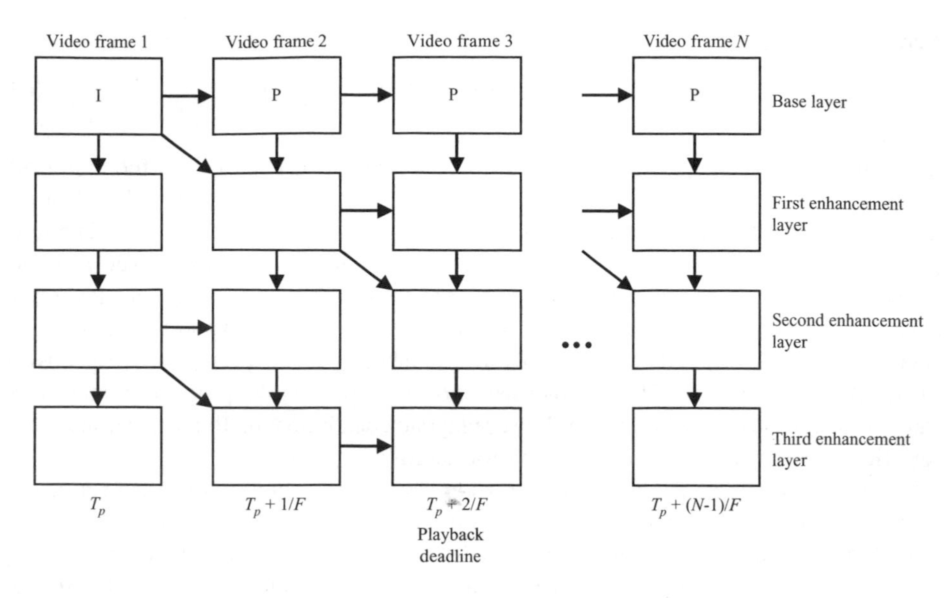

Figure 10.3 GOP structure of MPEG-4 PFGS (progressive fine granularity scalable) scalable video. (Reproduced by permission of IEEE [50].)

packets among the classes. At the video application layer, each video packet is characterized based on its loss and delay properties, which contribute to the end-to-end video quality and service. Then, these video packets are classified and optimally mapped to the classes of link transmission module under the rate constraint [57, 58]. The video application layer QoS and link-layer QoS are allowed to interact with each other and adapt along with the wireless channel condition. The objective of these interaction and adaptation is to find a satisfactory QoS tradeoff so that each end user's video service can be supported with available transmission resources.

Figure 10.3 shows a group-of-picture (GOP) structure of MPEG-4 PFGS (*progressive fine granularity scalable*) [56]. Suppose that there are video frames and there are video layers in one GOP. Therefore, the loss of any video portion will affect the end-to-end video quality due to the interdependency within the encoding structure. Each video layer is packetized into several fixed-size packets before transmission. Each video packet cannot contain video data across video layers or video frames. Packets from the same video layer are put onto the same priority class. Furthermore, suppose that the video playback frame rate at the end user is fixed at F frames/s. If the mobile terminal starts to play back the first video frame of a GOP at time Tp, video frame n in the same GOP should be received and be ready to be displayed before time $T_d(n) = T_p + [(n-1)/F]$ for uninterrupted playback.

Let $p = (p_1, p_2, \ldots, p_M)$ be the mapping policy from M video layers to K priority classes, where $p_j \in \{0, 1, \ldots, K\}$ is the priority class that video layer j is transmitted. $pj = 0$ represents the fact that video layer j is abstained from transmission. The overall expected distortion from the mapping scheme p can be derived using the dependent structure of scalable video [50, 57], as shown in Figure 10.3, which can be expressed as follows:

$$D(p) = D_0 - \sum_{j=0}^{M} \Delta D_{j' \leq j}^{p_j} \tag{10.1}$$

where

$$\Delta D_{j' \leq j}^{p_j} = \Delta D_j \prod_{j' \leq j} \left[1 - \beta_{j'}\left(\theta_{p_{j'}}, \phi_{p_{j'}}\right)\right] \tag{10.2}$$

$D(p)$is the total expected distortion from mapping N scalable frames to K different priority classes with an allocation policy p, D_0 is the expected distortion if no video data are received, ΔD_j is the distortion reduction if video layer j is correctly received. The term $\prod_{j' \leq j} [1 - \beta_{j'}(\theta_{p_{j'}}, \phi_{p_{j'}})]$ in Equation (10.2) is the probability that video layer j and all video layers j', on which video layer depends, are correctly received while video layer j' is transmitted over the priority class $p_{j'}$. The priority class $p_{j'}$ has QoS exponents $\theta_{p_{j'}}$ and $\phi_{p_{j'}}$, which correspond to its guaranteed buffer overflow and delay bound probability, respectively (for details see Chapter 8). On the other hand, $\beta_{j'}(\theta_{p_{j'}}, \phi_{p_{j'}})$ is the probability that video layer j' is lost due to either buffer overflow or playback deadline violation when transmitted over priority class $p_{j'}$ (see again Chapter 8 for details).

The *optimal mapping problem* can be formally stated as follows. Given the set of rate constraints under the priority transmission system in described above, and the expected channel service rate r, which can be considered stationary in a time period t, corresponding to one GOP, what is the optimal mapping policy p from one GOP with N scalable frames (coded in M video layers) to K priority classes such that $D(p)$ is minimized? This can be expressed as

$$\begin{aligned} &\min D(p) \\ &\text{s.t.} \sum_{\forall j, p_j = i} b_j^{p_j} \leq \mu_i(k_i) \cdot t, \; i = 1, \ldots, K \\ &\sum_{i=1}^{K} \mu_i(k_i) < r \end{aligned} \tag{10.3}$$

where $\mu_i(k_i)$is the rate constraint of priority class i, and $b_j^{p_j}$ is the size of video layer j, which will be conveyed by priority class pj. The solution to the optimization problem follows a constrained-based search that exploits the dependency among the layers [57]. This is a source coding problem and will not be discussed within this chapter.

Video adaptation is based on using a set of QoS bounds to characterize the range of video quality requirements and transmission service capabilities. Within this set of bounds, QoS parameters of video and transmission service can be adjusted to cope with the time-varying and nonstationary wireless link quality. Owing to the time-varying characteristics of video content and time-varying wireless channel, the set of bounds is also time-varying. Specifically, the QoS bound for video application at time t can be defined as the video distortion of GOP as

$$\wp(t) = \left\{\varepsilon_{i,L}(\theta_{i,L}), \varepsilon_{i,U}(\theta_{i,L})\right\} \tag{10.4}$$

where $\varepsilon_{i,L}(\theta_{i,L})$ and $\varepsilon_{i,U}(\theta_{i,U})$ are the respective lower and upper bounds of the guaranteed buffer overflow probability by priority corresponding to QoS exponent $\theta_{i,L}$ and $\theta_{i,U}$. Similarly, the rate constraint corresponding to the statistical QoS guarantee can be expressed as

$$\Theta_i = \left\{\mu_i(\theta_{i,L}), \mu_i(\theta_{i,L})\right\} \tag{10.5}$$

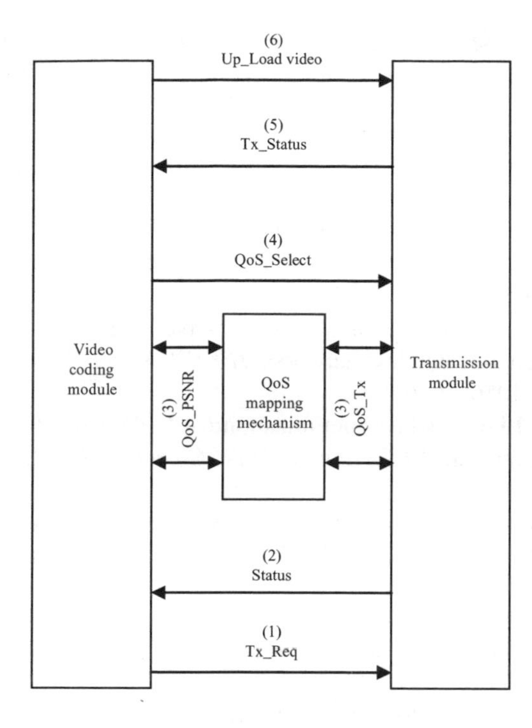

Figure 10.4 Interaction between video applications and priority networks for QoS adaptation.

where $\mu_i(\theta_{i,L})$ and $\mu_i(\theta_{i,U})$ are the respective rate constraints corresponding to $\varepsilon_{i,L}(\theta_{i,L})$ and $\varepsilon_{i,U}(\theta_{i,U})$. Note that the range for guaranteed packet delay can also be obtained from the guaranteed buffer overflow probability (see Chapter 8).

Cross-layer interaction in video QoS adaptation process consists of optimally adjusting the video encoding behavior based on the QoS bounds. Basically the link layer estimates the available effective channel capacity r by using methods described in Chapter 8. Based on this estimate, the encoder optimally maps the frames into the priority classes as defined by Equation (10.3). The process updating rate depends on channel variation rate. The process is illustrated in Figure 10.4. PSNR improvement of the order of 3 dB is demonstrated in such a scheme [50].

REFERENCES

[1] S. Shakkottai, T.S. Rappaport and P.C. Karlsson, Cross-layer design for wireless networks, *IEEE Commun. Mag.*, October 2003, pp. 74–81.

[2] L. Tassiulas and A. Ephremides, Dynamic server allocation to parallel queues with randomly varying connectivity, *IEEE Trans. Info. Theory*, vol. 39, 1993, pp. 466–478.

[3] V. Kawadia and P.R. Kumar, A cautionary perspective on cross-layer design, *IEEE Wireless Commun.* [see also *IEEE Person. Commun.*], vol. 12, no. 1, 2005, pp. 3–11.

[4] Y. Wu, P.A. Chou, Qian Zhang, K. Jain, Wenwu Zhu and Sun-Yuan Kung, Network planning in wireless *ad hoc* networks: a cross-Layer approach, *IEEE J. Selected Areas in Commun.* vol. 23, no. 1, 2005, pp. 136–150.
[5] G. Song and Y. Li, Cross-layer optimization for OFDM wireless networks-part I: theoretical framework, *IEEE Trans. Wireless Commun.*, vol. 4, no. 2, 2005, pp. 614–624.
[6] Q. Liu, S. Zhou and G.B. Giannakis, Queuing with adaptive modulation and coding over wireless links: cross-layer analysis and design, *IEEE Trans. Wireless Commun.*, vol. 4, no. 3, 2005, pp. 1142–1153.
[7] Q. Liu, S. Zhou and G.B. Giannakis, Cross-layer scheduling with prescribed QoS guarantees in adaptive wireless networks, *IEEE J. Selected Areas Commun.*, vol. 23, no. 5, 2005, pp. 1056–1066.
[8] G. Song and Y. Li, Cross-layer optimization for OFDM wireless networks – part II: algorithm development, *IEEE Trans. Wireless Commun.*, vol. 4, no. 2, 2005, pp. 625–634.
[9] P.P. Pham, S. Perreau and A. Jayasuriya, New cross-layer design approach to *ad hoc* networks under Rayleigh fading, *IEEE J. Selected Areas Commun.*, vol. 23, no. 1, 2005, pp. 28–39.
[10] K.B. Johnsson and D.C. Cox, An adaptive cross-layer scheduler for improved QoS support of multiclass data services Wireless systems, *IEEE J. Selected Areas Commun.*, vol. 23, no. 2, 2005, pp. 334–343.
[11] M. Madueno and J. Vidal, Joint physical – MAC layer design of the broadcast protocol in *ad hoc* networks, *IEEE J. Selected Areas Commun.*, vol. 23, no. 1, 2005 pp. 65–75.
[12] S.-H. Lee, E. Choi and D.-H. Cho, Timer-based broadcasting for power-aware routing in power-controlled wireless *ad hoc* networks, *IEEE Commun. Lett.*, vol. 9, no. 3, 2005, pp. 222–224.
[13] L.-C. Wang and C.-H. Lee, A TCP-physical cross-layer congestion control mechanism for the multirate WCDMA system using explicit rate change notification, in *19th Int. Conf. on Advanced Information Networking and Applications*, *AINA 2005*, vol. 2, 25–30 March 2005, pp. 449–452.
[14] Q. Zhang, W. Zhu and Y. -Q. Zhang, End-to-end QoS for video delivery over wireless Internet, *Proc. IEEE*, vol. 93, no. 1, 2005, pp. 123–134.
[15] R. Annavajjala, P.C. Cosman and L.B. Milstein, On source coding, channel coding and spreading tradeoffs in a DS-CDMA system operating over frequency selective fading channels with narrowband interference, *IEEE J. Selected Areas Commun.*, vol. 23, no. 5, 2005, pp. 1034–1044.
[16] K. Weniger, PACMAN: passive autoconfiguration for mobile *ad hoc* networks, *IEEE J. Selected Areas Commun.*, vol. 23, no. 3, 2005, pp. 507–519.
[17] M.R. Souryal, B.R. Vojcic and R.L. Pickholtz, Information efficiency of multihop packet radio networks with channel-adaptive routing, *IEEE J. Selected Areas Commun.*, vol. 23, no. 1, 2005, pp. 40–50.
[18] T.K. Chiew, P. Ferre, D. Agrafiotis, A. Molina, A.R. Nix and D.R. Bull, Cross-layer WLAN measurement and link analysis for low latency error resilient wireless video transmission, in *Int. Conf. Consumer Electronics, ICCE 2005 Digest of Technical Papers*, 8–12 January 2005, pp. 177–178.

[19] H. Gossain, T. Joshi, C. Cordeiro and D.P. Agrawal, A cross-layer approach for designing directional routing protocol in MANETs, in *IEEE Wireless Communications and Networking Conf.*, vol. 4, 13–17 March 2005, pp. 1976–1981.

[20] Hojoong Kwon, Tae Hyun Kim, Sunghyun Choi and Byeong Gi Lee, Lifetime maximization under reliability constraint via cross-layer strategy in wireless sensor networks, in *IEEE Wireless Communications and Networking Conf.*, vol. 3, 13–17 March 2005, pp. 1891–1896.

[21] Xiaodong Wang, Jun Yin, Qi Zhang and D.P. Agrawal, A cross-layer approach for efficient flooding in wireless sensor networks, in *2005 IEEE Wireless Communications and Networking Conf.*, vol. 3, 13–17 March 2005, pp. 1812–1817.

[22] S. Mohapatra, Cornea, H. Oh, K. Lee, M. Kim, N. Dutt, R. Gupta, A. Nicolau, S. Shukla and N. Venkatasubramania, A cross-layer approach for power-performance optimization in distributed mobile systems proceedings, in *19th IEEE Int. Parallel and Distributed Processing Symp.*, 4–8 April 2005, p. 218a.

[23] A. Scaglione and M. van der Schaar, Cross-layer resource allocation for delay-constrained wireless video transmission, in *IEEE Int. Conf. Acoustics, Speech, and Signal Processing (ICASSP '05)*, vol. 5, 18–23 March 2005, pp. 909–912.

[24] I. Martinez and J. Altuna, A cross-layer design for *ad hoc* wireless networks with smart antennas and QoS support, in *15th IEEE Int. Symp. Personal, Indoor and Mobile Radio Communications, PIMRC 2004*, vol. 1, 5–8 September 2004, pp. 589–593.

[25] Qiong Li and M. van der Schaar, Providing adaptive QoS to layered video over wireless local area networks through real-time retry limit adaptation, *IEEE Trans. Multimedia*, vol. 6, no. 2, 2004, pp. 278–290.

[26] Jie Chen, T. Lv and Haitao Zheng, Joint cross-layer design for wireless QoS content delivery, *2004 IEEE Int. Conf. Communications*, vol. 7, 20–24 June 2004, pp. 4243–4247.

[27] Li-Chun Wang, Shi-Yen Huang and A. Chen, On the throughput performance of CSMA-based wireless local area network with directional antennas and capture effect: a cross-layer analytical approach, in *IEEE Wireless Communications and Networking Conf., WCNC*, vol. 3, 21–25 March 2004, pp. 1879–1884.

[28] Jie Chen, Tiejun Lv and Haitao Zheng, Cross-layer design for QoS wireless Communications, in *Proc. 2004 Int. Symp. Circuits and Systems, ISCAS '04*, vol. 2, 23–26 May 2004, pp. 217–220.

[29] A. Maaref and S. Aissa, Combined adaptive modulation and truncated ARQ for packet data transmission in MIMO systems, in *IEEE Global TeleCommunications Conf., GLOBECOM '04*, vol. 6, 29 November–3 December 2004, pp. 3818–3822.

[30] Li-Chun Wang, Ya-Wen Lin and Wei-Cheng Liu, Cross-layer goodput analysis for rate adaptive IEEE 802.11a WLAN in the generalized Nakagami fading channel, in *IEEE Int. Conf. Communications*, vol. 4, 20–24 June 2004, pp. 2312–2316.

[31] Wei Yu, K.J.R. Liu and Z. Safar, Scalable cross-layer rate allocation for image transmission over heterogeneous wireless networks, in *IEEE Int. Conf. Acoustics, Speech, and Signal Processing*, ICASSP '04, vol. 4, 17–21 May 2004, pp. 593–596.

[32] Qingwen Liu, Shengli Zhou and G.B. Giannakis, Cross-layer combining of adaptive modulation and coding with truncated ARQ over wireless links, *IEEE Trans. Wireless Commun.*, vol. 3, no. 5, 2004, pp. 1746–1755.

[33] Li-Chun Wang and Chung-Wei Wang, A cross-layer design of clustering architecture for wireless sensor networks, in 2004 *IEEE Int. Conf. Networking, Sensing and Control*, vol. 1, 21–23 March 2004, pp. 547–552.

[34] A. Maaref and S. Aissa, A cross-layer design for MIMO Rayleigh fading channels, in *Canadian Conf. Electrical and Computer Engineering*, vol. 4, 2–5 May 2004, pp. 2247–2250.

[35] Jianxin Yao, Tung Chong Wong and Yong Huat Chew, Cross-layer design on the reverse and forward links capacities balancing in cellular CDMA systems, in *IEEE Wireless Communications and Networking Conf., WCNC*, vol. 4, 21–25 March 2004, pp. 2004–2009.

[36] H. Boche and M. Wiczanowski, Stability region of arrival rates and optimal scheduling for MIMO-MAC – a cross-layer approach, in *Int. Zurich Seminar on Communications*, 2004, pp. 18–21.

[37] V. Friderikos, L. Wang and A.H. Aghvami, TCP-aware power and rate adaptation in DS/CDMA networks, *IEEE Proc. Communications*, vol. 151, no. 6, 2004, pp. 581–588.

[38] Qingwen Liu, Shengli Zhou and G.B. Giannakis, Cross-layer modeling of adaptive wireless links for QoS support in multimedia networks, in *First Int. Conf. Quality of Service in Heterogeneous Wired/Wireless Networks, QSHINE 2004*, 18–20 October 2004, pp. 68–75.

[39] D. Kliazovich and F. Graneill, A cross-layer scheme for TCP performance improvement in wireless LANs, in *IEEE Global TeleCommunications Conf., GLOBECOM '04*, vol. 2, 29 November–3 December 2004, pp. 840–844.

[40] V. Bhuvaneshwar, M. Krunz and A. Muqattash, CONSET: a cross-layer power aware protocol for mobile *ad hoc* networks, in *IEEE Int. Conf. Communications*, vol. 7, 20–24 June 2004, pp. 4067–4071.

[41] R.A. Berry and E.M. Yeh, Cross-layer wireless resource allocation, *IEEE Signal Process. Mag.*, vol. 21, no. 5, 2004, pp. 59–68.

[42] R. Laroia, S. Uppala and Li Junyi, Designing a mobile broadband wireless access network, *IEEE Signal Process. Mag.*, vol. 21, no. 5, 2004, pp. 20–28.

[43] L. Alonso and R. Agusti, Automatic rate adaptation and energy-saving mechanisms based on cross-layer information for packet-switched data networks, *IEEE Commun. Mag.*, vol. 42, no. 3, 2004, pp. S15–S20.

[44] Fei Yu, V. Krishnamurthy and V.C.M. Leung, Cross-layer optimal connection admission control for variable bit rate multimedia traffic in packet wireless CDMA networks, *IEEE Global TeleCommunications Conf., GLOBECOM '04*, vol. 5, 29 November–3 December 2004, pp. 3347–3351.

[45] U.C. Kozat, I. Koutsopoulos and L. Tassiulas, A framework for cross-layer design of energy-efficient communication with QoS provisioning in multi-hop wireless networks, in *Twenty-third Annual Joint Conf. IEEE Computer and Communications Societies INFOCOM 2004*, vol. 2, 7–11 March 2004, pp. 1446–1456.

[46] L. Lazos and R. Poovendran, Cross-layer design for energy-efficient secure multicast communications in *ad hoc* networks, in *2004 IEEE Int. Conf. Communications*, vol. 6, 20–24 June 2004, pp. 3633–3639.

[47] L. van Hoesel, T. Nieberg, Jian Wu and P.J.M. Havinga, Prolonging the lifetime of wireless sensor networks by cross-layer interaction, *IEEE Wireless Commun.* [*also IEEE Person. Commun.*], vol. 11, no. 6, 2004, pp. 78–86.

[48] M. Bourouha, S. Ci, G.B. Brahim and M. Guizani, A cross-layer design for QoS support in the 3GPP2 wireless systems, in *IEEE Global TeleCommunications Conf. Workshops, GlobeCom Workshops*, 29 November–3 December 2004, pp. 56–61.
[49] Leping Huang, Hongyuan Chen, T.V.L.N. Sivakumar and K. Sezaki, Cross-layer optimized routing for Bluetooth personal area network, in *13th Int. Conf. Computer Communications and Networks, ICCCN 2004*, 11–13 October 2004, pp. 155–160.
[50] W. Kumwilaisak, Y.T. Hou, Q. Zhang, W. Zhu, C.-C. Jay Kuo and Y.-Q. Zhang, A cross-layer quality-of-service mapping architecture for video delivery in wireless networks, *IEEE J. Selected Areas Commun.*, vol. 21, no. 10, 2003, pp. 1685–1698.
[51] I. Stoica, S. Shenkar and H. Zhang, Core-stateless fair queueing: achieving approximately fair bandwidth allocations in high speed networks, *IEEE/ACM Trans. Networking*, vol. 11, 2003, pp. 33–46.
[52] B. Vandalore, R. Jain, S. Fahmy and S. Dixit, AQuaFWiN: Adaptive QoS framework for multimedia in wireless networks and its comparison with other QoS frameworks, in *Proc. IEEE Local Computer Networks*, Boston, MA, 1999, pp. 88–97.
[53] B. Arroyo-Fernandez, J. Dasilva, J. Fernandes and R. Prasad, Life after third-generation mobile Commun., *IEEE Commun. Mag.*, vol. 39, no. 8, 2001, pp. 41–42.
[54] C. Dovrolis, D. Stiliadis and P. Ramanathan, Proportional differentiated services: Delay differentiation and packet scheduling, *IEEE/ACM Trans. Networking*, vol. 10, 2002, pp. 12–26.
[55] D.L. Reyes, A.R. Reibman, S.-F. Chang and J. I.-I. Chuang, Error-resilient transcoding for video over wireless channels, *IEEE J. Select. Areas Commun.*, vol. 18, 2000, pp. 1063–1074.
[56] F. Wu, S. Li and Y.-Q. Zhang, A framework for efficient progressive fine granularity scalable video coding, *IEEE Trans. Circuits Syst. Video Technol.*, vol. 11, 2001, pp. 332–344.
[57] A. Ortega and K. Ramchandran, Rate distortions for image and video compression, *IEEE Signal Process. Mag.*, vol. 15, 2001, pp. 23–50.
[58] J. Shin, J. Kim and C.-C. Jay Kuo, Quality-of-service mapping mechanism for packet video in differentiated services network, *IEEE Trans. Multimedia*, vol. 3, June 2001, pp. 219–231.

11

Mobility Management

11.1 INTRODUCTION

As already indicated in Chapter 1 (see Figure 1.1), 4G wireless networks will integrate services of different segments such as cellular networks, WLAN, WPAN and even LEO satellites. Several alternative backbone networks will be used, like the public land mobile networks (PLMN), mobile Internet protocol (mobile IP) networks, wireless asynchronous transfer mode (WATM) networks, and low Earth orbit (LEO) satellite networks. Regardless of the network, one of the most important and challenging problems for wireless communication and computing is mobility management [1–62]. Mobility management enables telecommunication networks to locate roaming terminals for call delivery and to maintain connections as the terminal is moving into a new service area, process known as *handoff*. The handoff may be executed between different segments (cells) of the same or different systems. *The handoff* event is caused by radio link degradation or initiated by the system that rearranges radio channels in order to avoid congestion. Our focus in this section is on the first kind of handoff, where the cause of handoff is poor radio quality due to a change in the environment or the movement of the wireless terminal. For example, the mobile user might cross cell boundaries and move to an adjacent cell while the call is in process. In this case, the call must be handed off to the neighboring cell in order to provide uninterrupted service to the mobile subscriber. If adjacent cells do not have enough channels to support the handoff, the call is forced to be blocked. In systems where the cell size is relatively small (microcellular systems), the handoff procedure has an important effect on the performance of the system. Here, an important issue is to limit the probability of forced call termination, because from the point of view of a mobile user, forced termination of an ongoing call is less desirable than blocking a new call. Therefore, the system must reduce the chances of unsuccessful handoffs by reserving some channels explicitly for handoff calls. For example, handoff prioritizing schemes are channel assignment strategies that allocate channels to handoff requests more readily than new calls.

Advanced Wireless Networks: 4G Technologies Savo G. Glisic
© 2006 John Wiley & Sons, Ltd.

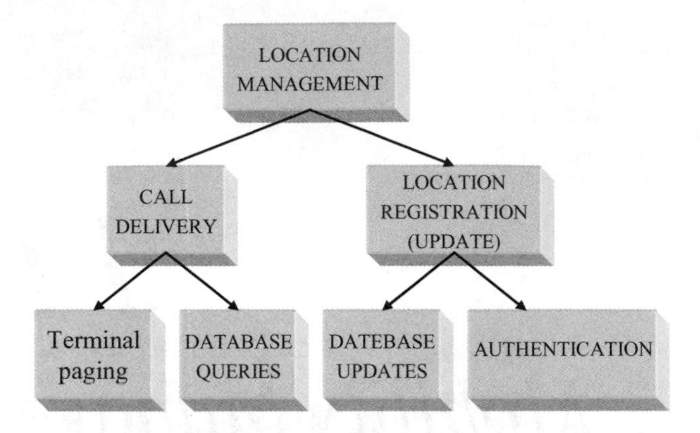

Figure 11.1 Components of location management process.

Thus, mobility management supports mobile terminals, allowing users to roam while simultaneously offering them incoming calls and supporting calls in progress. Mobility management consists of location management and handoff management.

Location management is a process that enables the network to discover the current attachment point of the mobile user for call delivery. The main components of the process are shown in Figure 11.1. The first segment is location registration (or location update). In this stage, the mobile terminal periodically notifies the network of its new access point, allowing the network to authenticate the user and revise the user's location profile. The second segment is call delivery. Here the network is queried for the user location profile and the current position of the mobile host is found. The main issues in location management involve database architecture design, design of messaging procedures and the transmission of signaling messages between various components of a signaling network. Other issues include: security, dynamic database updates, querying delays, terminal paging methods and paging delays.

Handoff (or handover) management enables the network to maintain a user's connection as the mobile terminal continues to move and change its access point to the network. The three-stage process for handoff first involves *initiation*, where the user, a network agent or changing network conditions identify the need for handoff. The second stage is new *connection generation*, where the network must find new resources for the handoff connection and perform any additional routing operations. Under network-controlled handoff (NCHO) or mobile-assisted handoff (MAHO), the network generates a new connection, by finding new resources for the handoff and performing any additional routing operations. For mobile-controlled handoff (MCHO), the mobile terminal finds the new resources and the network approves. The final stage is *data-flow control*, where the delivery of the data from the old connection path to the new connection path is maintained according to agreed-upon QoS. The segments of handoff management are presented in Figure 11.2.

Handoff management includes two conditions: intracell handoff and intercell handoff. Intracell handoff occurs when the user moves within a service area (or cell) and experiences signal strength deterioration below a certain threshold that results in the transfer of the user's calls to new radio channels of appropriate strength at the same (BS). Intercell handoff occurs when the user moves into an adjacent cell and all of the terminal's connections must be transferred to a new BS. While performing handoff, the terminal may connect to multiple BSs simultaneously and use some form of signaling diversity to combine the multiple

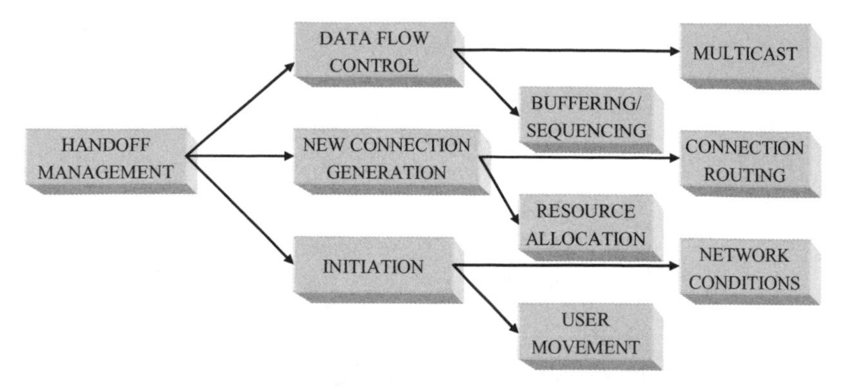

Figure 11.2 Components of handoff management.

signals. This is called *soft handoff*. On the other hand, if the terminal stays connected to only one BS at a time, clearing the connection with the former BS immediately before or after establishing a connection with the target BS, then the process is referred to as *hard handoff*. Handoff management issues are: efficient and expedient packet processing, minimizing the signaling load on the network, optimizing the route for each connection, efficient bandwidth reassignment and refining quality of service for wireless connections. Below we will discuss the handoff management in some of the component networks of 4G integrated wireless network concept as suggested by Figure 1.1.

11.1.1 Mobility management in cellular networks

Mobile terminals (MTs) are free to travel and thus the network access point of an MT changes as it moves around the network coverage area. As a result, the ID of an MT does not implicitly provide the location information of the MT and the call delivery process becomes more complex. The current systems for PLMN location management strategies require each MT to register its location with the network periodically. In order to perform the registration, update and call delivery operations described above, the network stores the location information of each MT in the location databases. Then the information can be retrieved for call delivery.

Current schemes for PLMN location management are based on a two-level data hierarchy such that two types of network location database, the home location register (HLR) and the visitor location register (VLR), are involved in tracking an MT. In general, there is an HLR for each network and a user is permanently associated with an HLR in his/her subscribed network. Information about each user, such as the types of services subscribed and location information, are stored in a user profile located at the HLR. The number of VLRs and their placements vary among networks. Each VLR stores the information of the MTs (downloaded from the HLR) visiting its associated area.

Network management functions, such as call processing and location registration, are achieved by the exchange of signaling messages through a signaling network. Signaling system 7 (SS7), described in Chapter 1 [34, 38, 63], is the protocol used for signaling exchange, and the signaling network is referred to as the SS7 network.

The type of cell site switch (CSS) currently implemented for the PLMN is known as a mobile switching center (MSC). Figure 11.3 shows the SS7 signaling network which

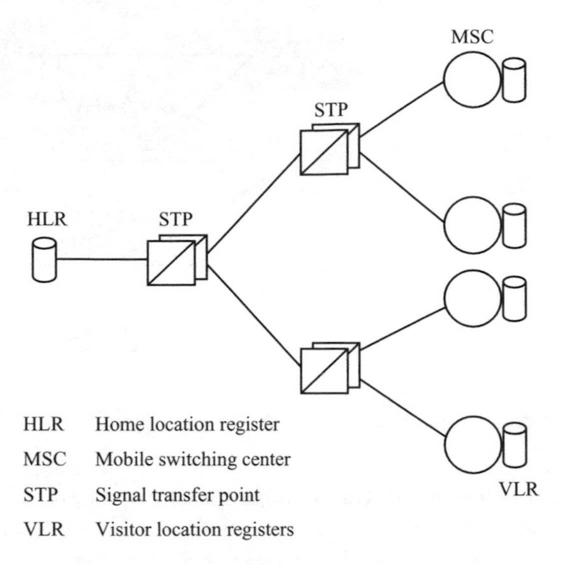

Figure 11.3 Location management SS7 signaling network.

connects the HLR, the VLRs, and the MSCs in a PLMN-based network. The signal transfer points (STPs), as shown in Figure 11.3, are responsible for routing signaling messages.

As mentioned previously, location management includes two major tasks: location registration (or update) and call delivery. In order to correctly deliver calls, the PLMN must keep track of the location of each MT. As described previously, location information is stored in two types of databases, VLR and HLR. As the MTs move around the network coverage area, the data stored in these databases may no longer be accurate. To ensure that calls can be delivered successfully, the databases are periodically updated through the process called location registration.

Location registration is initiated by an MT when it reports its current location to the network. This reporting process is referred to as location update. Current systems adopt an approach such that the MT performs a location update whenever it enters a new location area (LA). Each LA consists of a number of cells and, in general, all base station transmission systems (BTSs) belonging to the same LA are connected to the same MSC.

When an MT enters an LA, if the new LA belongs to the same VLR as the old LA, the record at the VLR is updated to record the ID of the new LA. Otherwise, if the new LA belongs to a different VLR, a number of extra steps are required to: (1) register the MT at the new serving VLR; (2) update the HLR to record the ID of the new serving VLR; and (3) deregister the MT at the old serving VLR. Figure 11.4 shows the location registration procedure when an MT moves to a new LA. The following is the ordered list of tasks that are performed during location registration.

(1) The MT enters a new LA and transmits a location update message to the new BS.

(2) The BS forwards the location update message to the MSC which launches a registration query to its associated VLR.

(3) The VLR updates its record on the location of the MT. If the new LA belongs to a different VLR, the new VLR determines the address of the HLR of the MT from

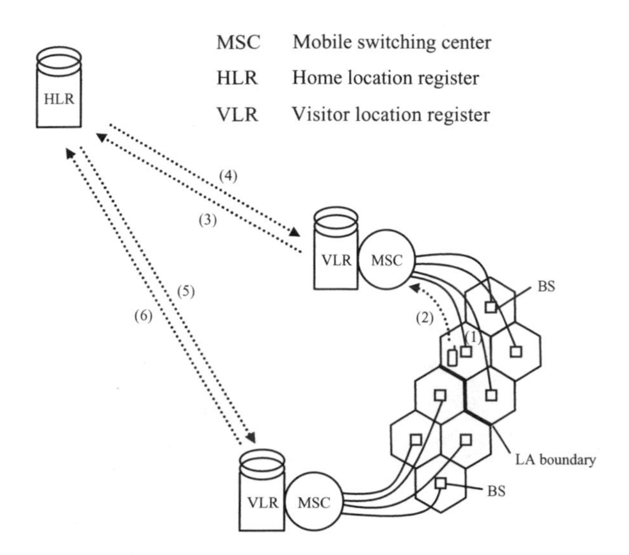

Figure 11.4 Location registration procedures.

its mobile identification number (MIN). This is achieved by a table lookup procedure called global title translation. The new VLR then sends a location registration message to the HLR. Otherwise, location registration is complete.

(4) The HLR performs the required procedures to authenticate the MT and records the ID of the newserving VLR of the MT. The HLR then sends a registration acknowledgment message to the new VLR.

(5) The HLR sends a registration cancellation message to the old VLR.

(6) The old VLR removes the record of the MT and returns a cancellation acknowledgment message to the HLR.

Call delivery consists of two major steps (1) determining the serving VLR of the called MT and (2) locating the visiting cell of the called MT. Locating the serving VLR of the MT involves the following procedure, shown in Figure 11.5:

(1) The calling MT sends a call initiation signal to the serving MSC of the MT through a nearby BS.

(2) The MSC determines the address of the HLR of the called MT by global title translation and sends a location request message to the HLR.

(3) The HLR determines the serving VLR of the called MT and sends a route request message to the VLR. This VLR then forwards the message to the MSC serving the MT.

(4) The MSC allocates a temporary identifier called temporary local directory number (TLDN) to the MT and sends a reply to the HLR together with the TLDN.

(5) The HLR forward this information to the MSC of the calling MT.

(6) The calling MSC requests a call setup to the called MSC through the SS7 network.

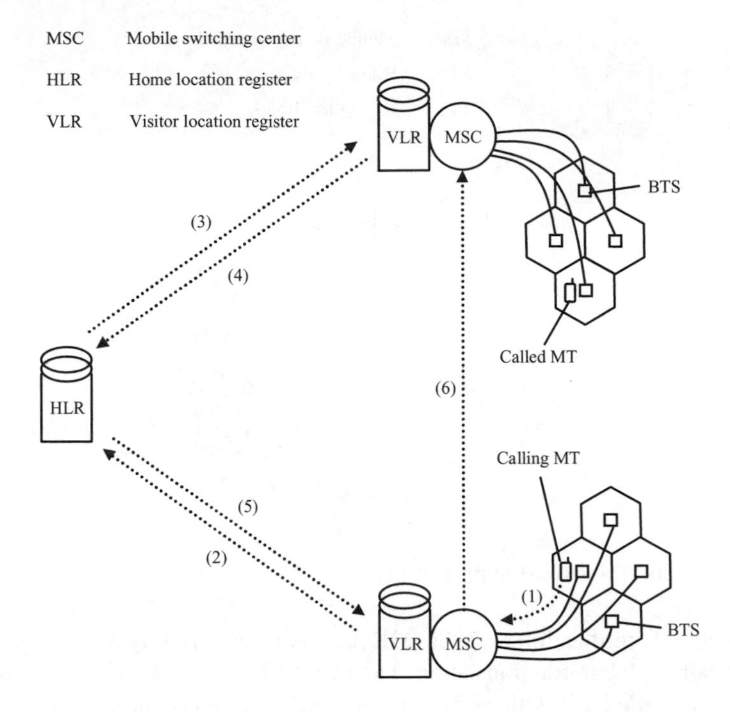

Figure 11.5 Call delivery procedures.

The procedure described above allows the network to set up a connection from the calling MT to the serving MSC of the called MT. Since each MSC is associated with an LA and there is more than one cell in each LA, a mechanism is therefore necessary to determine the cell location of the called MT. In current PLMN networks, this is achieved by a paging (or alerting) procedure, such that polling signals are broadcast to all cells within the residing LA of the called MT. On receiving the polling signal, the MT sends a reply which allows the MSC to determine its current residing cell. As the number of MTs increases, sending polling signals to all cells in an LA whenever a call arrives may consume excessive wireless bandwidth. We describe a number of proposed paging mechanisms for reducing the paging cost later.

11.1.2 Location registration and call delivery in 4G

Location registration involves the updating of location databases when current location information is available. On the other hand, call delivery involves the querying of location databases to determine the current location of a called MT. These can be costly processes, especially when the MT is located far away from its assigned HLR. For example, if the MT is currently roaming USA and its HLR is in Finland, a location registration message is transmitted from USA to Finland whenever the MT moves to a new LA that belongs to a different VLR. Under the same scenario, when a call for the MT is originated from a nearby MT in USA, the MSC of the calling MT must first query the HLR in Finland before it finds out that the called MT is located in the same area as the caller. As the number of mobile subscribers keeps increasing, the volume of signaling traffic generated by location management is extremely high. Methods for reducing the signaling traffic in 4G networks are therefore needed.

Research in this area generally falls into two categories. In the first category, extensions to the existing location management strategy are developed which aim at improving the IS-41 (or GSM/UMTS) scheme while keeping the basic database network architecture unchanged. This type of solution has the advantage of easy adaptation to the current PLMN networks without major modification. These schemes are based on centralized database architectures inherited from the existing standards. Another category of research results lies in completely new database architectures that require a new set of schemes for location registration and call delivery. Most of these schemes are based on distributed database architectures. Some additional research efforts involve: the reverse virtual call setup – a new scheme for delivering mobile-terminated calls [23], an optimal routing scheme based on the ratio of source messaging to location update rates [61], and a single registration strategy for multitier PCS systems [33]. In what follows, we discuss centralized vs distributed database architectures.

11.1.2.1 Centralized database architectures

This solution consists of the two-tier database structure with additional optimizations that aim to reduce the location management cost. The extension to include inter-technology roaming is also expected.

Dynamic hierarchical database architecture

The first centralized database architecture is the dynamic hierarchical database architecture presented in Ho and Akyildtz [18]. The proposed architecture is based on that of the IS-41 standard with the addition of a new level of databases called directory registers (DRs). Each DR covers the service area of a number of MSCs. The primary function of the DRs is to compute periodically and store the location pointer configuration for the MTs in its service area. Each MT has its unique pointer configuration and three types of location pointers are available at the DR:

(1) A local pointer is stored at an MT serving DR, which indicates the current serving MSC of the MT.

(2) A direct remote pointer is stored at a remote DR, which indicates the current serving MSC of the MT.

(3) An indirect remote pointer is stored at a remote DR, which indicates the current serving DR of the MT.

In addition, the HLR of the MT may be configured to store a pointer to either the serving DR or the serving MSC of the MT. In some cases, it may be more cost-effective not to set up any pointers, and the original IS-41 scheme will be used.

As an example, *if the intertechnology roaming is supported*, suppose that the HLR of a given MT is located in Finland and it is currently roaming in Chicago. If a significant number of the incoming calls for the MT are originating from Los Angeles, a direct or indirect remote pointer can be set up for the MT in the DR at the Los Angeles area. When the next call is initiated for this MT from Los Angeles, the calling MSC first queries the

DR and the call can be immediately forwarded to Chicago without requiring a query at the HLR, which is located in Finland. This reduces the signaling overhead for call delivery. On the other hand, the HLR can be set up to record the ID of the serving DR (instead of the serving MSC) of the MT. When the MT moves to another MSC within the same LA in Illinois area, only the local pointer at the serving DR of the MT has to be updated. Again, it is not necessary to access the HLR in Finland. This reduces the signaling overhead for location registration. The advantage of this scheme is that it can reduce the overhead for both location registration and call delivery.

11.1.2.2 Multiple-copy location information strategies

A number of different strategies have been considered to facilitate the search for the user location. The basic idea of the *per-user location caching* strategy [25] is that the volume of signaling and database access traffic for locating an MT can be reduced by maintaining a cache of location information at a nearby STP. Whenever the MT is accessed through the STP, an entry is added to the cache which contains a mapping from the ID of the MT to that of its serving VLR. When another call is initiated for an MT, the STP first checks if a cache entry exists for the MT. If no cache entry for the MT exists, the call delivery scheme as described earlier is used to locate the MT. If a cache entry exists, the STP will query the VLR as specified by the cache. If the MT is still residing under the same VLR, a hit occurs and the MT is found. If the MT has already moved to another location which is not associated with the same VLR, a miss occurs and the call delivery scheme is used to locate the MT.

In order to reduce the number of misses, it is suggested in Lin [35] that cache entries should be invalidated after a certain time interval. Based on the mobility and call arrival parameters, the author [35] introduces a threshold scheme which determines the time when a particular cached location information should be cleared such that the cost for call delivery can be reduced.

User profiles can be *replicated* at selected local databases. When a call is initiated for a remote MT, the network first determines if a replication of the called MTs user profile is available locally. If the user profile is found, no HLR query is necessary and the network can locate the called MT based on the location information available at the local database. Otherwise, the network locates the called MT following the standard procedures. When the MT moves to another location, the network updates all replications of the MT's user profile. This results in higher signaling overhead for location registration. Depending on the mobility rate of the MT and the call arrival rate from each location, this method may significantly reduce the signaling and database access overhead for local management.

11.1.2.3 Local extentions

Pointer forwarding and local anchoring are the strategies where only the far end segment of the rout is modified. The basic idea of the *pointer forwarding* strategy [24] is that, instead of reporting a location change to the HLR every time the MT moves to an area belonging to a different VLR, the reporting can be eliminated by simply setting up a forwarding pointer from the old VLR to the new VLR. When a call for the MT is initiated, the network locates the MT by first determining the VLR at the beginning of the pointer chain and then

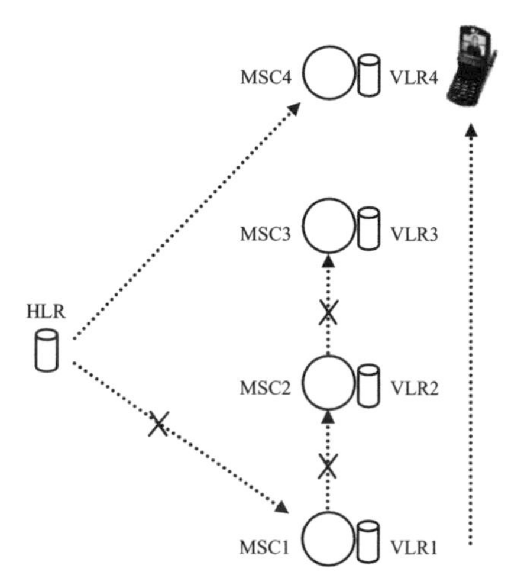

Figure 11.6 Pointer forwarding strategy.

following the pointers to the current serving VLR of the MT. To minimize the delay in locating an MT, the length of the pointer chain is limited to a predefined maximum value K. Figure 11.6 demonstrates the operation of pointer forwarding. A slight modification of the above scheme is the local anchoring [19], where a VLR close to the MT is selected as its local anchor. Instead of transmitting registration messages to the HLR, location changes are reported to the local anchor. Since the local anchor is close to the MT, the signaling cost incurred in location registration is reduced. The HLR keeps a pointer to the local anchor. When an incoming call arrives, the HLR queries the local anchor of the called MT which, in turn, queries the serving VLR to obtain a routable address to the called MT. Figure 11.7 demonstrates the local anchoring scheme.

11.1.2.4 Distributed database architectures

This type of solution is a further extension of the *multiple copy concept* and consists of multiple databases distributed throughout the network coverage area. In a *fully distributed registration scheme* [18] the two-level HLR/VLR database architecture is replaced by a large number of location databases. These location databases are organized as a tree with the root at the top and the leaves at the bottom. The MTs are associated with the leaf (lowest level) location databases and each location database contains location information of the MTs that are residing in its subtree.

Database hierarchy, introduced in Anantharam *et al.* [10] is similar to the fully distributed registration scheme [58]. Here, MTs may be located at any node of the tree hierarchy (not limited to the leaf nodes). The root of the tree contains a database but it is not necessary for other nodes to have databases installed. These databases store pointers for MTs. If an MT is residing at the subtree of a database, a pointer is set up in this database pointing to the next database along the path to the MT. If there is no more database along this path, the pointer points to the residing node of the MT. When a call for an MT is initiated at a node on the tree, the called MT can be located by following the pointers of the MT.

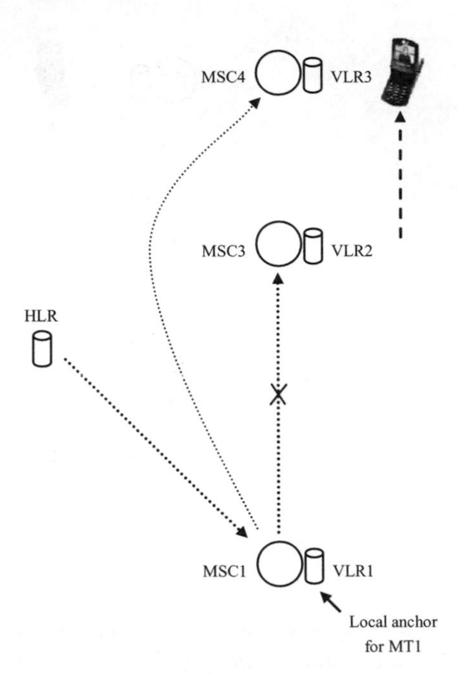

Figure 11.7 Local anchoring scheme.

Another form of distributed concept is *partitioning*. Since the mobility pattern of the MTs varies among locations, partitions can be generated by grouping location servers among which the MT moves frequently. Location registration is performed only when the MT enters a partition.

11.1.2.5 Location update and terminal paging options for 4G

Current PCS networks partition their coverage areas into a number of LAs. Each LA consists of a group of cells and each MT performs a location update when it enters an LA. When an incoming call arrives, the network locates the MT by simultaneously paging all cells within the LA. The main drawbacks of this location update and paging scheme are:

(1) Requiring the network to poll all cells within the LA each time a call arrives may result in excessive volume of wireless broadcast traffic.

(2) The mobility and call arrival patterns of MTs vary, and it is generally difficult to select an LA size that is optimal for all users. An ideal location update and paging mechanism should be able to adjust on a per-user basis.

(3) Finally, excessive location updates may be performed by MTs that are located around LA boundaries and are making frequent movements back and forth between two LAs.

In addition, the LA-based location update and paging scheme is a static scheme as it cannot be adjusted based on the parameters of an MT from time to time. Recent research efforts for 4G have attempted to reduce the effects of these inefficiencies. Excessive location updates are discussed in Rose [51] and Bar-Noy *et al.* [12]. A timer-based strategy that uses a universal timeout parameter is presented in Rose [51], while a tracking strategy for mobile users in PCS networks based on cell topology is explored and compared with the time-based strategy in Bar-Noy *et al.* [12]. For excessive polling, a one-way paging network architecture and the interfaces among paging network elements are examined in Lin [30]. Additional schemes attempt to reduce the cost of finding a user when the MT moves during the paging process [48, 62]. Many recent efforts have focused primarily on dynamic location update mechanisms which perform location update based on the mobility of the MTs and the frequency of incoming calls. In the following a number of dynamic location update and paging schemes is presented.

Location updating

The standard LA-based location update method does not allow adaptation to the mobility characteristics of the MTs. The 4G solutions should allow dynamic selection of location update parameters, resulting in lower cost. *Dynamic LA management* introduces a method for calculating the optimal LA size given the respective costs for location update and cell polling. A mesh cell configuration with square-shaped cells is considered. Each LA consists of $k \times k$ cells arranged in a square, and the value of k is selected on a per-user basis according to the mobility and call arrival patterns and the cost parameters. This mechanism performs better than the static scheme in which LA size is fixed. However, it is generally not easy to use different LA sizes for different MTs as the MTs must be able to identify the boundaries of LAs, which are continuously changing. The implementation of this scheme is complicated when cells are hexagonal shaped, or in the worst case, when irregular cells are used.

Dynamic update schemes are also examined in Reference [13]. In a *time based* scheme an MT performs location updates periodically at a constant time interval ΔT. Figure 11.8 shows the path of an MT. If a location update occurred at location A at time 0, subsequent location updates will occur at locations B, C and D if the MT moves to these locations at times ΔT, $2\Delta T$ and $3\Delta T$, respectively. In a *movement-based* scheme, an MT performs a location update whenever it completes a predefined number of movements across cell boundaries (this number is referred to as the movement threshold). Assuming a movement threshold of three is used, the MT performs location updates at locations B and C, as shown in Figure 11.9. In a *distance-based* scheme an MT performs a location update when its distance from the cell where it performed the last location update exceeds a predefined

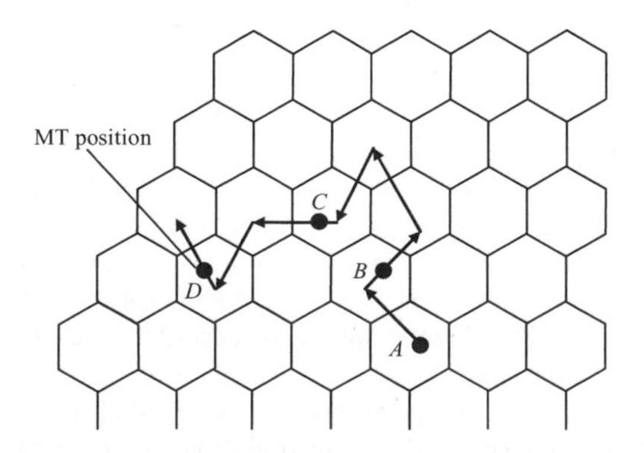

Figure 11.8 Time-based location update scheme.

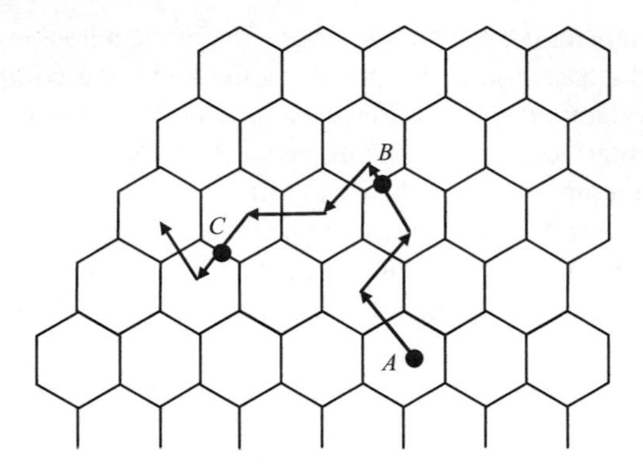

Figure 11.9 Movement-based location update scheme.

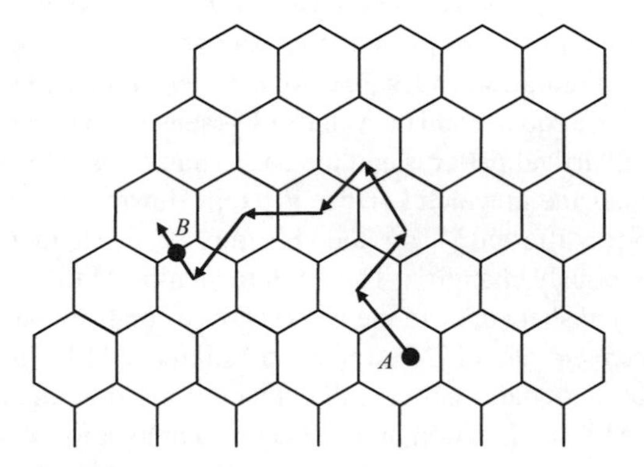

Figure 11.10 Distance-based location update scheme.

value (this distance value is referred to as the distance threshold). Figure 11.10 shows the same path as Figure 11.8. A location update is performed at location B where the distance of the MT from location A exceeds the threshold distance (solid line).

An *iterative distance-based location* update scheme introduces an iterative algorithm that can generate the optimal threshold distance that results in the minimum cost. When an incoming call arrives, cells are paged in a shortest-distance-first order such that cells closest to the cell where the last location update occurred are polled first. The delay in locating an MT is, therefore, proportional to the distance traveled since the last location update. Results demonstrate that, depending on the mobility and call arrival parameters, the optimal movement threshold varies widely. This demonstrates that location update schemes should be per-user-based and should be dynamically adjusted according to the current mobility and call arrival pattern of the user. However, the number of iterations required for this algorithm to converge varies depending on the mobility and call arrival parameters considered. Determining the optimal threshold distance may require significant computation resources at the MT.

A dynamic time-based location update scheme is introduced in Akyildiz and Ho [8]. The location update time interval is determined after each movement based on the probability distribution of the call interarrival time. This scheme does not make any specific assumptions on the mobility pattern of the MT's, and the shortest-distance-first paging scheme as described above is used. It is demonstrated that the results obtained are close to the optimal results given in Ho and Akyildiz [20]. Computation required by this scheme is low and they are, therefore, feasible for application in MTs that have limited computing power. The time required to locate an MT is directly proportional to the distance traveled since the last location update.

Terminal paging optimization provides a trade-off between paging cost and paging delay. Paging subject to delay constraints is considered in Rose and Yates [52]. The authors assume that the network coverage area is divided into LAs, and the probability that an MT is residing in a LA is given. It is demonstrated that, when delay is unconstrained, the polling cost is minimized by sequentially searching the LAs in decreasing order of probability of containing the MT. For constrained delay, the authors obtain the optimal polling sequence that results in the minimum polling cost. However, the authors assume that the probability distribution of user location is provided. This probability distribution may be user-dependent. A location update and paging scheme that facilitates derivation of this probability distribution is needed in order to apply this paging scheme. The tradeoff between the costs of location update and paging is not considered in Rose and Yates [52].

Location update and paging subject to delay constraints is considered in Akyildiz and Ho [20]. Again, the authors consider the distance-based location update scheme. However, paging delay is constrained such that the time required to locate an MT is smaller than or equal to a predefined maximum value. When an incoming call arrives, the residing area of the MT is partitioned into a number of subareas. These subareas are then polled sequentially to locate the MT. By limiting the number of polling areas to a given value such as N, the time required to locate a mobile is smaller than or equal to the time required for N polling operations. Given the mobility and call arrival parameters, the threshold distance and the maximum delay, an analytical model is introduced that generates the expected cost of the proposed scheme. An iterative algorithm is then used to locate the optimal threshold distance that results in the lowest cost. It is demonstrated that the cost is the lowest when the maximum delay is unconstrained. However, by slightly increasing the maximum delay from its minimum value of 1, the cost is significantly lowered. Another scheme using the movement-based location update is reported in Akyildiz and Ho [7]. Similar to Ho and Akyildiz [20], paging delay is confined to a maximum value. Movement-based location update schemes have the advantage that implementation is simple. The MTs do not have to know the cell configuration of the network. The scheme introduced in Akyildiz and Ho [7] is feasible for use in current PLMN networks.

Thus as a summary, remaining open problems for the 4G PLMN-based backbone are: (1) research work should consider the development of dynamic schemes that limit or enhance the distribution of location information on a per-user basis; (2) ongoing research efforts should attempt to reach some middle ground between centralized database architectures and distributed database architectures; and (3) future research should focus on the design of dynamic location update and paging schemes that are simple to implement.

11.1.2.6 Mobility management for mobile IP

Existing standards for terminal mobility over the Internet are described in Perkins [43, 44]. The mobility-enabling protocol for the Internet, *mobile IP*, enables terminals to move from one subnetwork to another as packets are being sent, without interrupting this process. An

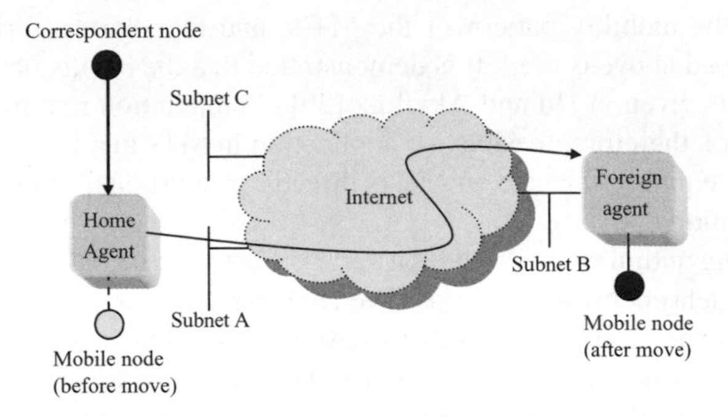

Figure 11.11 Mobile IP architecture.

MN is a host or router that changes its attachment point from one subnet to another without changing its IP address. The MN accesses the Internet via a home agent (HA) or a foreign agent (FA). The HA is an Internet router on the MNs home network, while the FA is a router on the visited network. The node at the other end of the connection is called the correspondent node (CN). A simple mobile IP architecture is illustrated in Figure 11.11. In this example, the CN sends packets to the MN via the MN's HA and the FA. [Note that the term mobile node (MN) is used instead of mobile terminal (MT) in order to follow mobile IP conventions.] As mentioned previously, network organization introduces some differences in the way mobility management is handled over the Internet. For example, mobile IP allows MNs to communicate their current reachability information to their home agent without the use of databases [42]. As a result, mobile IP defines new operations for location and handoff management:

(1) Discovery – how an MN finds a new Internet attachment point when it moves from one place to another.

(2) Registration – how an MN registers with its HA, an Internet router on the MNs home network.

(3) Routing and tunneling – how an MN receives datagrams when it is away from home [43].

Registration operations include mobile agent discovery, movement detection, forming care-of-addresses and binding updates, whereas handoff operations include routing and tunneling. Figure 11.12 illustrates the analogous relationships between the location management operations for mobile IP and those previously described in Figure 11.1 for PLMN.

Location registration

When visiting any network away from home, each MN must have an HA. The MN registers with its home agent in order to track the MN's current IP address. There are two IP addresses associated with each MN, one for locating and the other one for identification. In the standard terminology, the new IP address associated with an MN while it visits a foreign link is called its *care of address* (CoA). The association between the current CoA and the MN's home address is maintained by a *mobility binding*, so that packets destined for the MN may be routed using the current CoA regardless of the MN's current point of attachment to the Internet. Each binding has an associated lifetime period, negotiated during the MN's

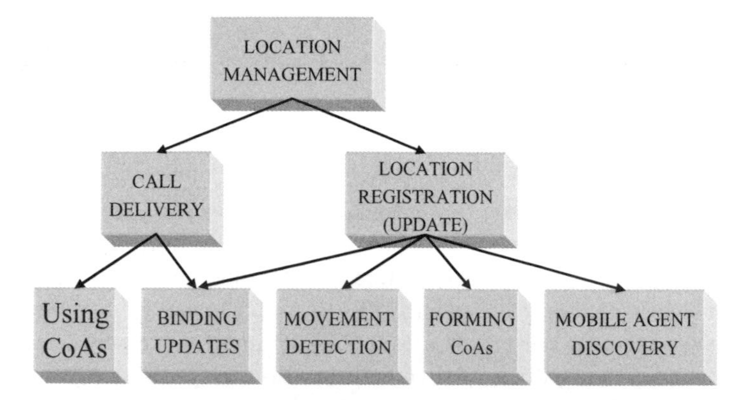

Figure 11.12 Mobile IP location management.

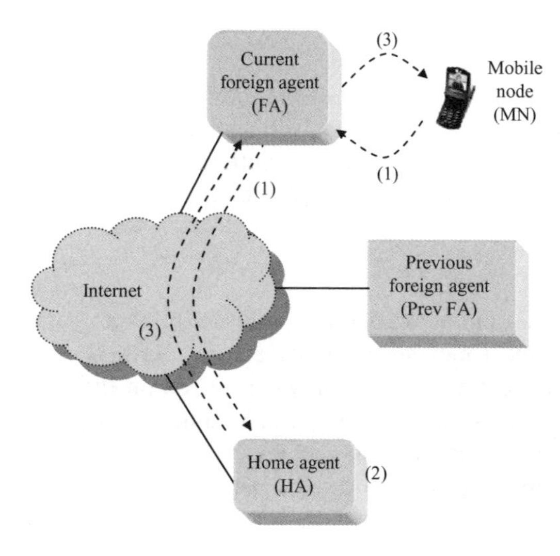

Figure 11.13 Mobile IP location registration.

registration, and after which time the registration is deleted. The MN must reregister within this period in order to continue service with this CoA.

Depending upon its method of attachment, the MN sends location registration messages directly to its HA, or through an FA that forwards the registration to the HA. In either case, the MN exchanges registration request and registration reply messages based on IPv4, as described below and shown in Figure 11.13.

(1) By using a registration request message (the request may be relayed to the HA by the current FA), the MN registers with its HA.

(2) The HA creates or modifies a mobility binding for that MN with a new lifetime.

(3) The appropriate mobile agent (HA or FA) returns a registration reply message. The reply message contains the necessary codes to inform the mobile node about the status of its request and to provide the lifetime granted by the HA [43].

Modifications in IPv6
In IPv6, the FAs in Figure 11.13 no longer exist. The entities formerly serving as FAs are now access points (APs).

Movement detection
For the other backbone networks, the movement of the user is determined by updates performed when the user moves into a new LA. Since mobile IP does not use LAs to periodically update the network, a new feature to determine whether the MN has moved to a new subnet after changing its network APs is used. Mobile agents make themselves known by sending agent advertisement messages. The primary movement detection method for mobile IPv6 uses the facilities of IPv6 neighbor discovery. Two mechanisms used by the MN to detect movement from one subnet to another are *the advertisement lifetime* and the *network prefix.*

Advertisement lifetime
The lifetime field within the main body of the Internet control message protocol (ICMP) router advertisement portion of the agent advertisement is used. A mobile node records the lifetime received in any agent advertisements, until that lifetime expires. If the MN has not maintained contact with its FA, the MN must attempt to solicit a new agent [42].

Network prefix
The second method uses the network prefix, a bit string that consists of some initial bits of an IP address, to detect movement. In some cases, an MN can determine whether or not a newly received agent advertisement was received on the same subnet as the MN's current CoA. If the prefixes differ, the MN can assume that it has moved. This method is not available if the MN is currently using an FA's CoA.

After discovering that MN is on a foreign network, it can obtain a new CoA for this new network from the prefix advertised by the new router and perform location update procedures. For the PLMN, registration is implemented using database storage and retrieval. In Mobile IP, the MN's registration message creates or modifies a mobility binding at the home agent, associating the MN's home address with its new CoA for the specified binding lifetime. The procedure is outlined below and shown in Figure 11.14.

(1) By sending a binding update, the MN registers a new CoA with its HA.

(2) The MN notifies its CN of the MN's current binding information.

(3) The CN and the HA send a binding request to the MN to get the MN's current binding information, if the binding update is allowed to expire.

The MN responds to the binding request with its new binding update. After receiving the new CoA, the CN and HA send a binding acknowledgment to the MN. Once the registration process is complete, call delivery consists of reaching the MN via the new CoAs. A wireless network interface may allow an MN to be reachable on more than one link at a time (i.e. within the wireless transmitter range of routers on more than one separate link). This establishment of coexisting wireless networks can be very helpful for smooth handoff.

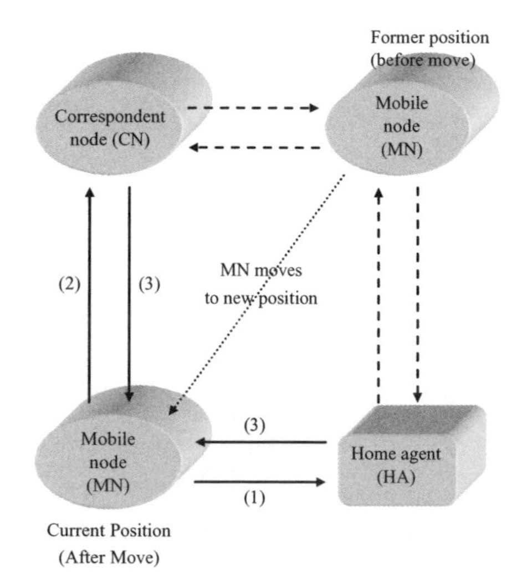

Figure 11.14 Mobile IP location management operations.

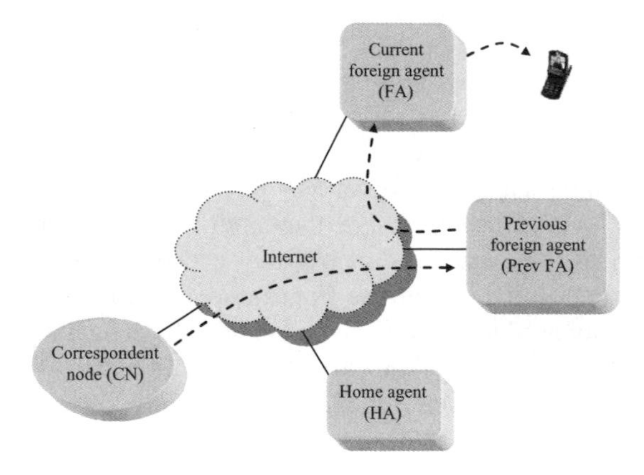

Figure 11.15 Mobile IP smooth handoff with fresh binding at previous FA.

Handoff management

Current routing optimization schemes in IPv4 allow the previous foreign agent (or agents) to maintain a binding for their former mobile visitors, showing a current CoA for each. Then, as packets are sent to the old CoA, the corresponding previous foreign agents can forward the packets to the current CoA of the MN, as demonstrated in Figure 11.15. As a result, an MN is able to accept packets at its old CoA while it updates its home agent and correspondent nodes with a new CoA on a new link. If the previous FA does not have a fresh binding (the binding lifetime has expired), the previous FA forwards the packets to the home agent of the MN, which sends the packets to the CoA from the MN's last location registration update, as shown in Figure 11.16. This can potentially create unnecessary traffic if the HA's binding still refers to the previous FA. Alternatively, the previous FA can invoke the use of special tunnels which forward the packets, but also indicate the need for special handling at the HA.

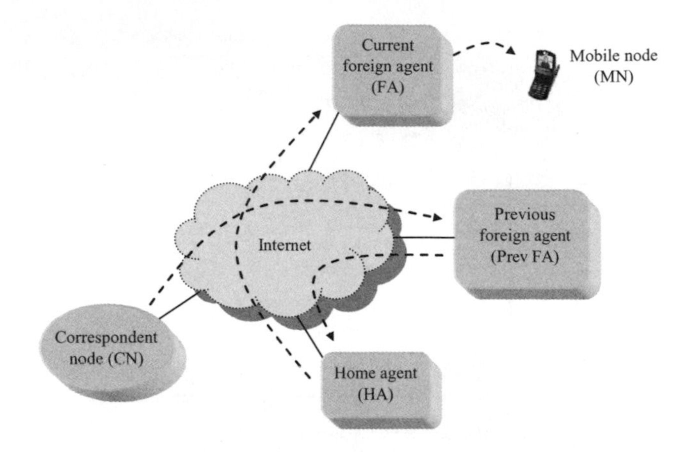

Figure 11.16 Mobile IP smooth handoff without fresh binding at previous FA.

When special tunnels are used, the packets that are sent to the HA are encapsulated with the FA's CoA address as the source IP address. Upon reception of the newly encapsulated packets, the HA compares the source IP address with the MN's most recent CoA. Thus, if the two addresses match, the packets will not be circled back to the FA. However, if the addresses do not match, the HA can decapsulate the packets and forward them to the MN's current CoA, as shown in Figure 11.16 [43]. In IPv6, the smooth handoff procedure is based on routers (IPv6 nodes) instead of FAs.

The process of routing datagrams for an MN through its HA often results in the utilization of paths that are significantly longer than optimal. Route optimization techniques for mobile IP employ the use of tunnels, such as the special tunnels mentioned for smooth handoff, to minimize the inefficient path use. For example, when the HA tunnels a datagram to the CoA, the MN's home address is effectively shielded from intervening routers between its home network and its current location. Once the datagram reaches the agent, the original datagram is recovered and is delivered to the MN. Currently, there are two protocols for routing optimization and tunnel establishment: *route optimization in mobile IP* [64] and *the tunnel establishment protocol* [65].

The basic idea of route optimization is to define extensions to basic mobile IP protocols that allow for better routing, so that datagrams can travel from a correspondent node to a mobile node without going to the home agent first [64]. These extensions provide a means for nodes to cache the binding of an MN and then tunnel datagrams directly to the CoA indicated in that binding, bypassing the MN's home agent. In addition, extensions allow for direct forwarding to the MN's new CoA for cases such as datagrams that are in flight when an MN moves and datagrams that are sent based on an out-of-data cached binding.

In tunnel establishment protocol, Mobile IP is modified in order to perform between arbitrary nodes [65]. Upon establishing a tunnel, the encapsulating agent (HA) transmits PDUs to the tunnel endpoint (FA) according to a set of parameters. The process of creating or updating tunnel parameters is called tunnel establishment. Generally the establishment parameters will include a network address for the MN. In order to use tunnel establishment to transmit PDUs, the home agent must determine the appropriate tunnel endpoint (FA) for the MN. This is done by consulting a table that is indexed by the MN's IP address. Each table entry contains the address of the appropriate tunnel endpoint, as well as any other necessary tunnel parameters. After receiving the packets, the foreign agent may then make use of any of a variety of methods to transmit the decapsulated PDUs so that it can be received by the MN. If the MN resides at this particular FA, no further network operations are required.

4G mobile IP networks will have to additionally address several issues. *Security* is one of the most important. As mentioned for the PLMN, the authentication of the mobile becomes more complex as the MN's address loses its tie to a permanent access point. This allows for a greater opportunity for impersonating an MN in order to receive services. Thus security measures for the registration and update procedures, specifically protecting the CoAs and HAs, must be implemented in order to police terminal use [43]. Some authentication schemes for the MN, the HA, and the FA can be found in Troxel and Sanchez [66]. See also Chapter 15 of this book.

The next issue is *simultaneous binding.* Since an MN can maintain several CoAs at one time, the HA must be prepared to tunnel packets to several endpoints. Thus, the HA is instructed to send duplicate encapsulated datagrams to each CoA. After the MN receives the packets from the CoAs, it can invoke some process to remove the duplicates. If necessary, the duplicate packets may be preserved in order to aid signal reconstruction.

Options for *regionalized registration* should also be considered, the extreme case being the BIONET concept discussed in Chapter 1. Currently, three major concepts have been identificd as potential methods for limiting location update and registration cost. First, there is a need for schemes that manage the local connectivity available to the MN and also to manage the buffering of datagrams to be delivered. Through this, the network can benefit from smooth handoffs without implementing route optimization procedures. Second, a multicast group of foreign agents is needed in order to allow the MN to use a multicast IP address as its CoA. Third, a hierarchy of foreign agents can be used in agent advertisement in order to localize the registrations to the lowest common FA of the CoA at the two points of attachments. To enable this method, the MN has to determine the tree-height required for its new registration message, and then arrange for the transmission of this message to reach each level of the hierarchy between itself and the lowest common ancestor of its new and previous CoA [44].

As already discussed in Chapter 1, 4G is all about integrating different wireless networks. So intertechnology roaming is the central issue in 4G wireless networks. The first steps in that direction have already been undertaken. Recently, there has been some discussion regarding Mobile IP with respect to the third-generation IMT 2000 system. A high-level IP mobility architecture is described in Becker *et al.* [69], in which the diverse nature of today's wireless and wireline packet data networks is explored. To support the seamless roaming among the heterogeneous networks, the mobility management based on mobile IP concepts is extended to the current third generation IMT2000 wireless architecture. Another concept, referred to as simple mobile IP (SMIP) [68], searches for a more simplistic approach to support the mobility of users, compared with the asymmetric triangular approach proposed in IPv6. SMIP employs a more symmetric and distributed solution for location management based on MN connections to fixed network routers that have added mobility functions.

Mobility management in wireless ATM deals with transitioning from ATM cell transport based upon widely available resources over wireline to cell transport based upon the limited and relatively unreliable resources over the wireless channel. For details of networks architectures see Chapter 1. Thus, the issues such as latency, message delivery, connection routing and QoS [67] will be in the focus of discussion. The ATM Forum (WATM Working

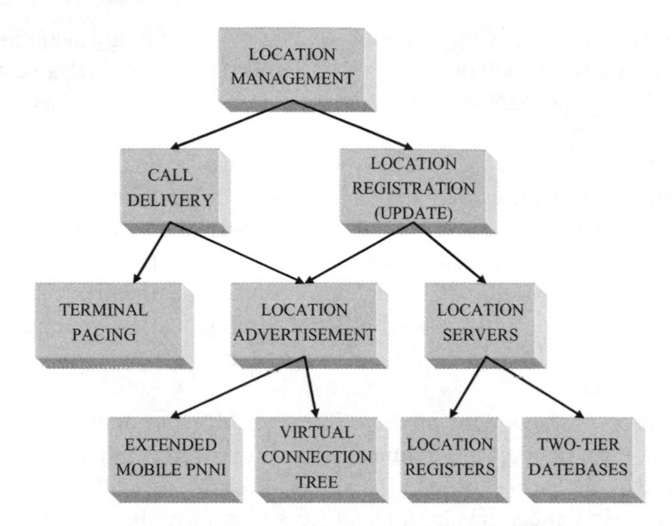

Figure 11.17 ATM location management techniques.

Group), is developing basic mechanisms and protocol extensions for location and handoff management that address these issues. The Forum has specified that new procedures must be compatible with the current ATM standards in order to be implemented with relative ease and efficiency [45]. As a result, many of the procedures are also compatible with PCS, satellite and to a lesser degree mobile IP concepts. In this section, we discuss selected proposed solutions for location management, terminal paging, and handoff.

Options considered for 4G are summarized in Figure 11.17. Proposed protocols for WATM implement location management using three techniques: location servers, location advertisement and terminal paging.

Location servers refer to the use of databases to maintain records of the attachment point of MTs within the network. As discussed earlier, the storage and retrieval process can generate excessive signaling and querying operations. *Location advertisement* avoids the use of databases by passing location information throughout the network on broadcast messages. *Terminal paging* is employed to locate MTs within the service area of its attachment point, as discussed previously.

Location server techniques are based on location servers (the databases) used to store and retrieve a record of the current position of the mobile. They require querying operations, as well as signaling protocols for storage and retrieval. WATM server protocols employ the IS-41/GSM based techniques that were discussed for the PLMN backbone earlier in this section. The first method makes familiar use of the HLR/VLR database structure. The second algorithm, location registers (LRs), uses a hierarchy of databases.

The *two-tier database* uses bilevel databases that are distributed to zones throughout the network, as shown in Figure 11.18. The zones, analogous to the LAs, are maintained by a zone manager. The zone manager, analogous to the mobility service control point (MSCP) of the 4G wireless architecture, controls the zone's location update procedures. The home tier (HLR) of the zone's database stores location information regarding MTs that are permanently registered within that zone, while the second tier (VLR) of the zone's database stores location information on MTs that are visiting the zone. Each MT has a home zone, i.e. a zone in which it is permanently registered.

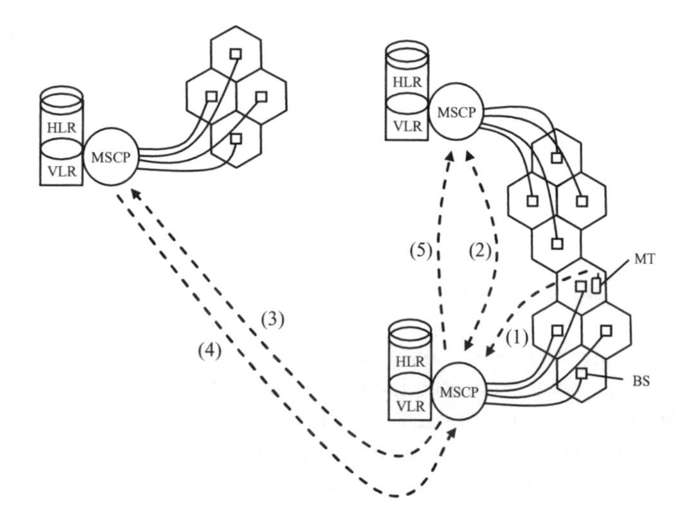

Figure 11.18 Two-tier database scheme.

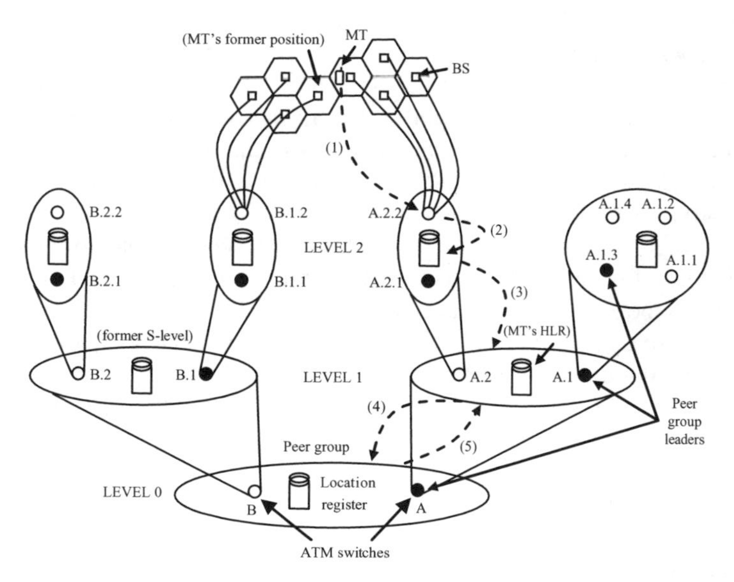

Figure 11.19 LR Hierarchy of WATM location registers scheme.

Upon entering a new zone, the MT detects the new zone identity broadcast from the BSs. The steps for registration, shown in Figure 11.19, are:

(1) The MT transmits a registration request message to the new MSCP that contains its user identification number (UID), authentication data, and the identity of the previous zone.

(2) The current MSCP determines the home zone of the MT from the previous zone.

(3) The current and home MSCPs authenticate the user and update the home user profile with the new location information.

(4) The home zone sends a copy of the profile to the current zone, which stores the profile in the VLR tier of its database.

(5) The current MSCP sends a purge message to the previous zone so that the user's profile is deleted from the previous zone's VLR tier.

Call delivery is achieved by routing the call to the last known zone first. If the MT has moved and has been purged, the call is immediately forwarded to the home zone. The home zone's HLR is queried for the current location of the MT, which is forwarded back to the calling switch. The calling switch can then set up a connection to the current serving switch of the MT.

The advantage of the two-tier scheme is that it keeps the number of queries low, requiring at most two database lookups for each incoming call to find the MT. However, the use of a centralized HLR may cause increased signaling traffic and unnecessary connection set-up delays if the MT makes several localized moves for an extended period of time. A more localized approach may reduce the need for long-distance queries and thereby reduce connection set-up delays.

The *LR hierarchy*-based scheme, described in Veeraraghavan and Dommetry [56] and shown in Figure 11.19, distributes location servers throughout a hierarchical private network-to-network interface (PNNI) architecture. The PNNI procedure is based on a hierarchy of peer groups, each consisting of collections of ATM switches. Each switch can connect to other switches within its peer group. Special switches, *designated peer group leader*, can also connect to a higher ranking leader in the 'parent' peer group. Each peer group also has its own database, or LR, used to store location information on each of the MTs being serviced by the peer group.

The PNNI organization allows the network to route connections to the MT without requiring the parent nodes to have exact location information. Only the lowest referenced peer must record the exact location, and the number of LR updates then corresponds to the level of mobility of the MT. As an illustration, a connection being set up to a MT located at switch $A.2.2$ in Figure 11.19 is first routed according to the highest boundary peer group and switch A. Peer A can then route the connection to its 'child' peer group, level $A.x$ to switch $A.2$. Finally, the connection is routed by $A.2$ to the lowest peer group level to switch $A.2.2$, which resolves the connection to the MT.

Thus, for movement within the $A.2$ peer group, the location update procedure can be localized to only the LR of that peer group. However, a movement from peer group $B.1$ to peer group $A.2$ requires location registration of a larger scope, and the maintenance of a home LR to store a pointer to the current parent peer position of the MT. To limit signaling for the larger scale moves to the minimum necessary level,Veeraraghavan and Dommetry [56] use two scope-limiting parameters, S and L. The S parameter indicates a higher level peer group boundary for LR queries, while the L parameter designates the lowest group. In Figure 11.19, the current S level is level one, while the L level is level two.

When the MT performs a location update by sending a registration notification message to the new BS, this message is relayed to the serving switch, which then stores the MT's location information in the peer group's LR. When the MT powers on or off, this message is relayed up the hierarchy until it reaches the preset boundary S. The S-level register records the entry and then relays the message to the MT's home LR. As an illustration, for movement from position $B.1.2$ to position $A.2.2$, the registration procedure, shown in Figure 11.19, is as follows:

(1) The MT sends a registration notification message to the new BS/switch.

(2) The new switch stores the MT in the peer group's LR.

(3) The peer group relays the new location info to the higher level LR's for routing, stopping at the first common ancestor of the former and current peer groups.

(4) In this case, the former S level is not a common ancestor, so a new S level is designated and the location info stops propagating at the new S level, level 0.

(5) The MT's home LR (located at group A.x) is notified of the new S-level location for the MT.

After the updates are complete, the new switch sends a purge message to the previous switch so that the former location can be removed from the LRs.

Call delivery
An incoming call request can be routed to the last known peer group or switch via the *S*-level LR. If the mobile has moved, the last known switch propagates a location request, querying the upstream LRs until the mobile endpoint's address is recognized by an LR that has a pointer to the mobile's current position. Then the request is sent to the *L*-level LR for that peer group, which resolves the query and sends the location information back to the calling switch.

Finally, if the call request reaches the *S* level before being recognized by an LR, the *S*-level LR forwards the location request directly to the home switch. Since the home LR keeps track of *S*-level changes for its mobile, the home switch can forward the request directly to the correct *S*-level switch, whose LR points to the current peer group position of the MT.

Location advertisement techniques
Although the *LR hierarchy* provides the advantages of simplicity, decreased computation costs and flexibility, the method can still require a substantial signaling and database querying load. This load can be reduced by using location advertisement. For WATM, advertisement refers to the notification of appropriate network nodes of the current location of the MT. The first method, *mobile PNNI*, uses the PNNI architecture described above by removing the LRs and by taking advantage of an internal broadcast mechanism [56]. The second method, *destination-rooted virtual connection trees*, advertises location information via provisioned virtual paths [57]. The third method, *integrated location resolution*, extends the signaling framework of ATM with location information elements that incorporate location resolution into the connection set-up process [4]. For more details see the references.

The *terminal paging problem* has not yet been explored enough for WATM applications, and additional work will be needed in this area.

11.1.2.7 WATM handoff management in 4G wireless networks

Handoff management controls the process of maintaining each connection at a certain level of QoS as the MT moves into different service areas [29]. As illustrated in Figure 11.20,

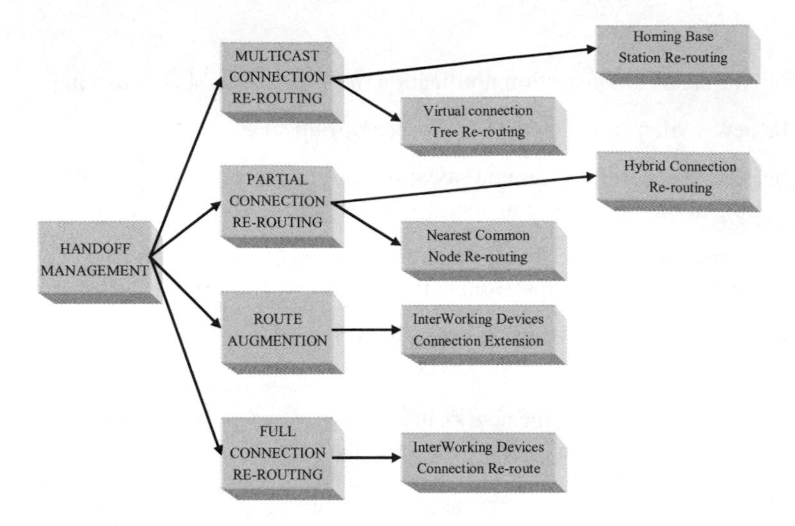

Figure 11.20 WATM handoff management techniques.

potential protocols to be used in 4G, can be grouped into four categories:

(1) full connection rerouting;

(2) route augmentation;

(3) partial connection rerouting; and

(4) multicast connection rerouting.

Full connection rerouting maintains the connection by establishing a completely new route for each handoff, as if it were a brand new call [46]. Route augmentation simply extends the original connection via a hop to the MT's next location [46]. Partial connection rerouting re-establishes certain segments of the original connection, while preserving the remainder [9]. Finally, multicast connection rerouting combines the former three techniques but includes the maintenance of potential handoff connection routes to support the original connection, reducing the time spent in finding a new route for handoff [9]. More details can be found in the above references.

11.1.2.8 Mobility management for satellite networks

In 4G integrated wireless networks the LEO satellites would cover regions where building terrestrial wireless systems are economically infeasible due to rough terrain or insufficient user population. A satellite system could also interact with terrestrial wireless network to absorb the instantaneous traffic overload of the terrestrial wireless network.

LEO satellites are usually those with altitudes between 500 and 1500 km above the Earth's surface [70–72]. This low altitude provides small end-to-end delays and low power requirements for both the satellites and the handheld ground terminals. In addition, inter-satellite links (ISL) make it possible to route a connection through the satellite network without using any terrestrial resources. These advantages come along with a challenge; in contrast to geostationary (GEO) satellites, LEO satellites change their position with reference to a fixed point on the Earth. Owing to this mobility, the coverage region of an LEO satellite is not stationary. A global coverage at any time is still possible if a certain number of orbits and satellites are used. The coverage area of a single satellite consists of small-sized cells, which are referred to as spotbeams. Different frequencies or codes are used in different spotbeams to achieve frequency reuse in the satellite coverage area.

Location management in the LEO satellite network environment represents more challenging problem because of the movement of satellite footprints. As a consequence, an LA cannot be associated with the coverage area of a satellite because of very fast movement of a LEO satellite. Thus, 4G will need the development of new LA definitions for satellite networks as well as the signaling issues mentioned for all of the location management protocols. In del Re [47], LAs are defined using (gateway, spotbeam) pairs. However, the very fast movement of the spotbeams results in excessive signaling for location updates. In Ananasso and Priscoli [73], LAs are defined using only gateways. However, the paging problem has not been addressed in the same reference.

Handoff management ensures that ongoing calls are not disrupted as a result of satellite movement, but rather transferred or handed off to new spotbeams or satellites when necessary. If a handoff is between two spotbeams served by the same satellite, handoff is *intrasatellite*. The small size of spotbeams causes frequent intrasatellite handoffs, which are also referred to as *spotbeam handoffs* [74]. If the handoff is between two satellites, it is referred to as *intersatellite* handoff. Another form of handoff occurs as a result of the change in the connectivity pattern of the network. Satellites near to polar regions turn off their links to other satellites in the neighboring orbits. Ongoing calls passing through these links need to be rerouted. This type of handoff is referred to as *link handoff* [59, 60]. Frequent link handoffs result in a high volume of signaling traffic. Moreover, some of the ongoing calls would be blocked during connection rerouting caused by link handoffs.

11.2 CELLULAR SYSTEMS WITH PRIORITIZED HANDOFF

The handoff attempt rate in a cellular system depends on cell radius and mobile speed as well as other system parameters. As a result of limited resources, some fraction of handoff attempts will be unsuccessful. Some calls will be forced to terminate before message completion. In this section we discuss analytical models to investigate these effects and to examine the relationships between performance characteristics and system parameters. For these purposes some assumptions about the traffic nature are needed. We assume that the new call origination rate is uniformly distributed over the mobile service area with the average number of new call originations per second per unit area Λ_a. A very large population of mobiles is assumed, thus the average call origination rate is for practical purposes independent of the number of calls in progress. A hexagonal cell shape is assumed. The cell radius R for a hexagonal cell is defined as the maximum distance from the center of a cell to the cell boundary. With the cell radius R, the average new call origination rate per cell Λ_R is $\Lambda_R = 3\sqrt{3}R^2\Lambda_a/2$. Average handoff attempt rate per cell is Λ_{Rh}. The ratio γ_0 of handoff attempt rate to new call origination rate (per cell) is $\gamma_0 \triangleq \Lambda_{Rh}/\Lambda_R$. If a fraction P_B of new call origination is blocked and cleared from the system, the average rate at which new calls are carried is $\Lambda_{Rc} = \Lambda_R(1 - P_B)$. Similarly, if a fraction P_{fh} of handoff attempts fails, the average rate at which handoff calls are carried is $\Lambda_{Rhc} = \Lambda_{Rh}(1 - P_{fh})$. The ratio

γ_c of the average carried handoff attempt rate to the average carried new call origination rate is defined as $\gamma_c \triangleq \Lambda_{Rhc}/\Lambda_{Rc} = \gamma_0(1 - P_{fh})/(1 - P_B)$.

The channel holding time T_H in a cell is defined as the time duration between the instant that a channel is occupied by a call and the instant it is released by either completion of the call or a cell boundary crossing by the mobile. This is a function of system parameters such as cell size, speed and direction of mobiles, etc. To investigate the distribution of T_H we let the random variable T_M denote the message duration, that is, the time an assigned channel would be held if no handoff is required. The random variable T_M is assumed to be exponentially distributed $f_{T_M}(t) = \mu_M e^{-\mu_M t}$ with the mean value $\bar{T}_M (\triangleq 1/\mu_M)$. The speed in a cell is assumed to be uniformly distributed on the interval $[0, V_{max}]$.

When a mobile crosses a cell boundary, the model assumes that vehicular speed and direction change. The direction of travel is also assumed to be uniformly distributed and independent of speed. More sophisticated models would assume that the higher the speed the fewer changes in direction are possible.

The random variable T_n is the time a mobile resides in the cell to which the call is originated. The time that a mobile resides in the cell in which the call is handed off is denoted T_h. The pdfs $f_{T_n}(t)$ and $f_{Th}(t)$ will be discussed in Section 11.3.

When a call is originated in a cell and gets a channel, the call holds the channel until the call is completed in the cell or the mobile moves out of the cell. Therefore, the channel holding time T_{Hn} is either the message duration T_M or the time T_n for which the mobile resides in the cell, whichever is less. For a call that has been handed off successfully, the channel is held until the call is completed in the cell or the mobile moves out of the cell again before call completion.

Because of the memoryless property of the exponential distributions, the remaining message duration of a call after handoff has the same distribution as the message duration. In this case the channel holding time T_{Hh} is either the remaining message duration T_M or mobile residing time T_h in the cell; whichever is less. The random variables T_{Hn} and T_{Hh} are therefore given by

$$T_{Hn} = \min(T_M, T_n) \text{ and } T_{Hh} = \min(T_M, T_h) \tag{11.1}$$

The cumulative distribution functions (CDF) of T_{Hn} and T_{Hh} can be expressed as

$$\begin{aligned} F_{T_{Hn}}(t) &= F_{T_M}(t) + F_{T_n}(t)[1 - F_{T_M}(t)] \\ F_{T_{Hh}}(t) &= F_{T_M}(t) + F_{T_h}(t)[1 - F_{T_M}(t)] \end{aligned} \tag{11.2}$$

The distribution of channel holding time can be written as

$$\begin{aligned} F_{T_n}(t) &= \frac{\Lambda_{Rc}}{\Lambda_{Rc} + \Lambda_{Rhc}} F_{T_{Hn}}(t) + \frac{\Lambda_{Rhc}}{\Lambda_{Rc} + \Lambda_{Rhc}} F_{T_{Hn}}(t) \\ &= \frac{1}{1 + \gamma_c} F_{T_{Hn}}(t) + \frac{\gamma_c}{1 + \gamma_c} F_{T_{Hh}}(t) \\ &= F_{T_M}(t) + \frac{1}{1 + \gamma_c} \left[1 - F_{T_M}(t)\right] \left[F_{T_n}(t) + \gamma_c F_{T_n}(t)\right] \end{aligned} \tag{11.3}$$

From the initial definitions,

$$F_{T_H}(t) = \begin{cases} 1 - e^{-\mu_M t} + \frac{e^{-\mu_M t}}{1+\gamma_c} \left[F_{T_n}(t) + \gamma_c F_{T_h}(t)\right], & \text{for } t \geq 0 \\ 0, & \text{elsewhere} \end{cases} \tag{11.4}$$

The complementary distribution function $F^C T_H(t)$ is

$$F^C T_H(t) = 1 - F_{T_H}(t) = F_{T_H}(t)$$
$$= \begin{cases} 1 - e^{-\mu_M t} - \dfrac{e^{-\mu_M t}}{1+\gamma_c}\left[F_{T_n}(t) + \gamma_c F_{T_h}(t)\right], & \text{for} \quad t \geq 0 \\ 0, & \text{elsewhere} \end{cases} \tag{11.5}$$

By differentiating Equation (11.4) we get the probability function (PDF) of T_H as

$$f_{T_H}(t) = \mu_M e^{-\mu_M t} + \frac{e^{-\mu_M t}}{1+\gamma_c}\left[f_{T_n}(t) + \gamma_c f_{T_h}(t)\right] - \frac{\mu_M e^{-\mu_M t}}{1+\gamma_c}\left[F_{T_n}(t) + \gamma_c F_{T_h}(t)\right] \tag{11.6}$$

To simplify the analysis the distribution of T_H is approximated in References [75, 76] by a negative exponential distribution with mean $\bar{T}_H$ ($\triangleq 1/\mu_H$). From the family of negative exponential distribution functions, a function which best fits the distribution of T_H, by comparing $F^C_{T_H}(t)$ and $e^{-\mu_H t}$ is chosen which is defined as

$$\mu_H \Rightarrow \min_{\mu_H} \int_0^\infty \left[F^C_{T_H}(t) - e^{-\mu_H t}\right] dt \tag{11.7}$$

Because a negative exponential distribution function is determined by its mean value, we choose $\bar{T}_H(\triangleq 1/\mu_H)$, which satisfies the above condition. The 'goodness of fit' for this approximation is measured by

$$G = \frac{\displaystyle\int_0^\infty \left|F^C_{T_H}(t) - e^{-\mu_H t}\right| dt}{\displaystyle 2\int_0^\infty F^C_{T_H}(t)\, dt} \tag{11.8}$$

In the sequel the following definitions will be used:

(1) The probability that a new call does not enter service because of unavailability of channels is called the blocking probability, P_B.

(2) The probability that a call is ultimately forced into termination (though not blocked) is P_F. This represents the average fraction of new calls which are not blocked but which are eventually uncompleted.

(3) P_{fh} is the probability that a given handoff attempt fails. It represents the average fraction of handoff attempts that are unsuccessful.

(4) The probability P_N that a new call that is not blocked will require at least one handoff before completion because of the mobile crossing the cell boundary is

$$P_N = \Pr\{T_M > T_n\} = \int_0^\infty \left[1 - F_{T_M}(t)\right] f_{T_n}(t)\, dt = \int_0^\infty e^{-\mu_M t} f_{T_n}(t)\, dt \tag{11.9}$$

(5) The probability P_H that a call that has already been handed off successfully will require another handoff before completion is

$$P_H = \Pr\{T_M > T_h\} = \int_0^\infty [1 - F_{T_M}(t)]\, f_{T_h}(t)\, dt = \int_0^\infty e^{-\mu_M t} f_{T_h}(t)\, dt \qquad (11.10)$$

Let the integer random variable K be the number of times that a nonblocked call is successfully handed off during its lifetime. The event that a mobile moves out of the mobile service area during the call will be ignored since the whole service area is much larger than the cell size. A nonblocked call will have exactly K successful handoffs if all of the following events occur:

(1) It is not completed in the cell in which it was first originated.

(2) It succeeds in the first handoff attempt.

(3) It requires and succeeds in $k - 1$ additional handoffs.

(4) It is either completed before needing the next handoff or it is not completed but fails on the $(k + 1)$st handoff attempt.

The probability function for K is therefore given by

$$\begin{aligned} &\Pr\{K = 0\} = (1 - P_N) + P_N P_{fh} \\ &\Pr\{K = k\} = P_N(1 - P_{fh})(1 - P_H + P_H P_{fh})\{P_H(1 - P_{fh})\}^{k-1}, \quad k = 1, 2, \ldots \end{aligned} \qquad (11.11)$$

and the mean value of K is

$$\bar{K} = \sum_{k=0}^{\infty} k \Pr\{K = k\} = \frac{P_N(1 - P_{fh})}{1 - P_H(1 - P_{fh})} \qquad (11.12)$$

If the entire service area has M cells, the total average new call attempt rate which is not blocked is $M\Lambda_{Rc}$, and the total average handoff call attempt rate is $\bar{K}M\Lambda_{Rc}$. If these traffic components are equally distributed among cells, we have $\gamma_c = (\bar{K}M\Lambda_{Rc})/(M\Lambda_{Rc}) \equiv \bar{K}$.

11.2.1 Channel assignment priority schemes

The probability of forced termination can be decreased by giving priority (for channels) to handoff attempts (over new call attempts). In this section, two priority schemes are described, and the expressions for P_B and P_{fh} are derived. A subset of the channels allocated to a cell is to be exclusively used for handoff calls in both priority schemes. In the first priority scheme, a handoff call is terminated if no channel is immediately available in the target cell (*channel reservation – CR handoffs*). In the second priority scheme, the handoff call attempt is held in a queue until either a channel becomes available for it, or the received signal power level becomes lower than the receiver threshold level (*channel reservation with queueing – CRQ handoffs*).

11.2.2 Channel reservation – CR handoffs

Priority is given to handoff attempts by assigning C_h channels exclusively for handoff calls among the C channels in a cell. The remaining $C - C_h$ channels are shared by both new

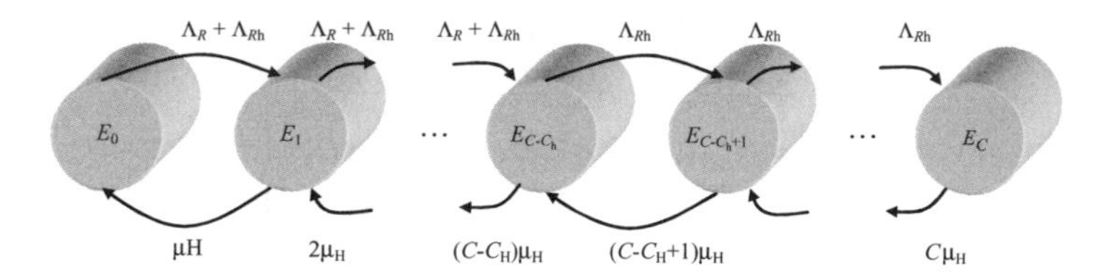

Figure 11.21 State-transition diagram for channel reservation – CR handoffs.

calls and handoff calls. A new call is blocked if the number of available channels in the cell is less than or equal to C_h when the call is originated. A handoff attempt is unsuccessful if no channel is available in the target cell. We assume that both new and handoff call attempts are generated according to a Poisson point process with mean rates per cell of Λ_R and $\Lambda_{R\mathrm{h}}$, respectively. As discussed previously, the channel holding time T_H in a cell is approximated to have an exponential distribution with mean $\bar{T}_\mathrm{H}(\triangleq 1/\mu_\mathrm{H})$. We define the state E_j of a cell such that a total of j calls is in the progress for the base station of that cell. Let P_j represent the steady-state probability that the base station is in state E_j; the probabilities can be determined in the usual way for birth-death processes discussed in Chapter 6. The pertinent state-transition diagram is shown in Figure 11.21.

The state equations are

$$P_j = \begin{cases} \dfrac{\Lambda_R + \Lambda_{R\mathrm{h}}}{j\mu_\mathrm{H}} P_{j-1}, & \text{for} \quad j = 1, 2, \ldots, C - C_\mathrm{h} \\ \dfrac{\Lambda_{R\mathrm{h}}}{j\mu_\mathrm{H}} P_{j-1}, & \text{for} \quad j = C - C_\mathrm{h} + 1, \ldots, C \end{cases} \tag{11.13}$$

As in Chapter 6, by using Equation (11.13) recursively, along with the normalization condition $\sum_{j=0}^{\infty} P_j = 1$, the probability distribution $\{P_j\}$ is

$$P_0 = \left[\sum_{k=0}^{C-C_\mathrm{h}} \frac{(\Lambda_R + \Lambda_{R\mathrm{h}})^k}{k!\mu_\mathrm{H}^k} + \sum_{k=C-C_\mathrm{h}+1}^{C} \frac{(\Lambda_R + \Lambda_{R\mathrm{h}})^{C-C_\mathrm{h}} \Lambda_{R\mathrm{h}}^{k-(C-C_\mathrm{h})}}{k!\mu_\mathrm{H}{}^k} \right]^{-1}$$

$$P_j = \begin{cases} \dfrac{(\Lambda_R + \Lambda_{R\mathrm{h}})^j}{j!\mu_\mathrm{H}^j} P_0, & \text{for} \quad j = 1, 2, \ldots, C - C_\mathrm{h} \\ \dfrac{(\Lambda_R + \Lambda_{R\mathrm{h}})^{C-C_\mathrm{h}} \Lambda_{R\mathrm{h}}^{j-(C-C_\mathrm{h})}}{j!\mu_\mathrm{H}^j} P_0, & \text{for} \quad j = C - C_\mathrm{h} + 1, \ldots, C \end{cases} \tag{11.14}$$

The probability of blocking a new call is $P_\mathrm{B} = \sum_{j=C-C_\mathrm{h}}^{C} P_j$ and the probability of handoff attempt failure P_fh is the probability that the state number of the base station is equal to C. Thus $P_\mathrm{fh} = P_\mathrm{c}$.

11.2.3 Channel reservation with queueing – CRQ handoffs

When a mobile moves away from the base station, the received power generally decreases. When the received power gets lower than a handoff threshold level, the handoff procedure

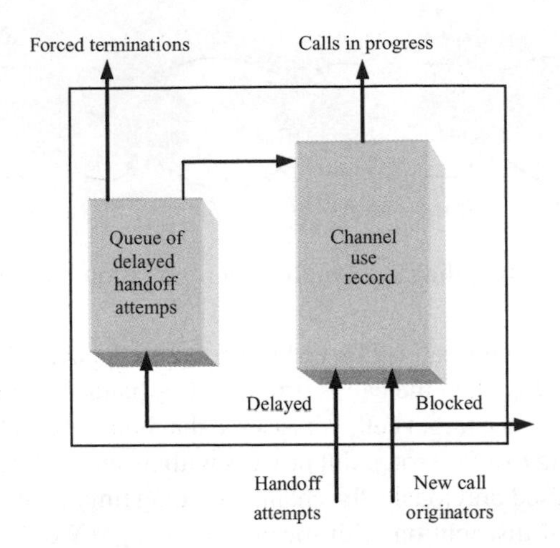

Figure 11.22 Call flow diagram for channel reservation with queueing-CRQ handoffs.

is initiated. The *handoff area* is defined as the area in which the average received power level from the base station of a mobile receiver is between the handoff threshold level (*upper bound*) and the receiver threshold level (*lower bound*). If the handoff attempt finds all channels in the target cell occupied, we consider that it can be queued. If any channel is released while the mobile is in the handoff area, the next queued handoff attempt is accomplished successfully. If the received power level from the source cell's base station falls below the receiver threshold level prior to the mobile being assigned a channel in the target cell, the call is forced into termination. When a channel is released in the cell, it is assigned to the next handoff call attempt waiting in the queue (if any). If more than one handoff call attempt is in the queue, the first-come-first-served queuing discipline is used. The prioritized queueing is also possible where the fast moving (fast signal level losing) users may have higher priority. We assume that the queue size at the base station is unlimited. Figure 11.22 shows a schematic representation of the flow of call attempts through a base station.

The time for which a mobile is in the handoff area depends on system parameters such as the speed and direction of mobile travel and the cell size. We call it the dwell time of a mobile in the handoff area T_Q. For simplicity of analysis, we assume that this dwell time is exponentially distributed with mean $\bar{T}_Q(\triangleq 1/\mu_{\mathrm{H}})$. We define E_j as the state of the base station when *j is the sum of the number of channels being used in the cell and the number of handoff call attempts in the queue*. For those states whose state number j is less than equal to C, the state transition relation is the same as for the CR scheme. Let X be the elapsed time from the instant a handoff attempt joins the queue to the first instant that a channel is released in the fully occupied target cell. For state numbers less than C, X is equal to zero. Otherwise, *X is the minimum remaining holding time* of those calls in progress in the fully occupied target cell. When a handoff attempt joins the queue for a given target cell, other handoff attempts may already be in the queue (each is associated with a particular mobile). When any of these first joined the queue, the *time* that it could remain on the queue *without succeeding* is denoted by T_Q (according to our previous definition). Let T_i be the remaining dwell time for that attempt which is in the ith queue position when another handoff attempt

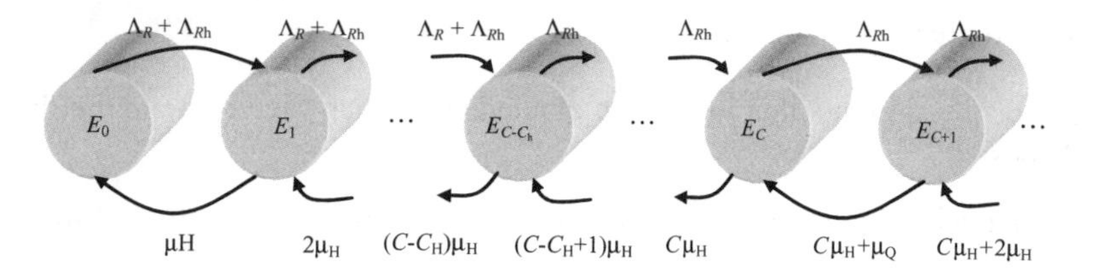

Figure 11.23 State-transition diagram for CRQ priority scheme.

joins the queue. Under the memoryless assumptions here, the distributions of all T_i and T_Q are identical. Let $N(t)$ be the state number of the system at time t. From the description of this scheme and the properties of the exponential distribution it follows that

$$\begin{aligned} P_r\{N(t+h) = C+k-1 | N(t) = C+k\} \\ &= P_r\{X \le h \quad \text{or} \quad T_1 \le h \quad \text{or} \ldots T_k \le h\} \\ &= 1 - P_r\{X > h \quad \text{and} \quad T_1 > h \quad \text{or} \ldots T_k > h\} \\ &= 1 - P_r\{X > h\} P_r\{T_1 > h\} \ldots P_r\{T_k > h\} \\ &= 1 - \mathrm{e}^{-(C_{\mu_\mathrm{H}} + k\mu_Q)h} \end{aligned} \tag{11.15}$$

since the random variables $X, T_1, T_2, \ldots, T_k$ are independent. From Equation (11.15) we see that it follows the birth-and-death process and the resulting state transition diagram is as shown in Figure 11.23.

As before, the probability distribution $\{P_j\}$ is easily found to be

$$P_0 = \left[\sum_{k=0}^{C-C_\mathrm{h}} \frac{(\Lambda_R + \Lambda_{R\mathrm{h}})^k}{k! {\mu_\mathrm{H}}^k} + \sum_{k=C-C_\mathrm{h}+1}^{C} \frac{(\Lambda_R + \Lambda_{R\mathrm{h}})^{C-C_h} \Lambda_{R\mathrm{h}}^{k-(C-C_h)}}{k! \mu_\mathrm{H}^k} \right.$$

$$\left. + \sum_{k=C+1}^{\infty} \frac{(\Lambda_R + \Lambda_{R\mathrm{h}})^{C-C_h} \Lambda_{R\mathrm{h}}^{k-(C-C_h)}}{C! \mu_\mathrm{H}^C \prod_{i=1}^{k-C} (C\mu_\mathrm{H} + i\mu_Q)} \right]^{-1}$$

$$P_j = \begin{cases} \dfrac{(\Lambda_R + \Lambda_{R\mathrm{h}})^j}{j! \mu_\mathrm{H}^j} P_0, & \text{for} \quad 1 \le j \le C - C_\mathrm{h} \\ \dfrac{(\Lambda_R + \Lambda_{R\mathrm{h}})^{C-C_h} \Lambda_{R\mathrm{h}}^{j-(C-C_h)}}{j! \mu_\mathrm{H}^j} P_0, & \text{for} \quad C - C_\mathrm{h} + 1 \le j \le C \\ \dfrac{(\Lambda_R + \Lambda_{R\mathrm{h}})^{(C-C_h)} \Lambda_{R\mathrm{h}}^{j-(C-C_h)}}{C! \mu_\mathrm{H}^C \prod_{i=1}^{j-C} (C\mu_\mathrm{H} + i\mu_Q)} P_0, & \text{for} \quad j \ge C + 1 \end{cases} \tag{11.16}$$

The probability of blocking is $P_\mathrm{B} = \sum_{j=C-C_\mathrm{h}}^{\infty} P_j$. A given handoff attempt that joins the queue will be successful if both of the following events occur before the mobile moves out of the handoff area:

(1) All of the attempts that joined the queue earlier than the given attempt have been disposed.

(2) A channel becomes available when the given attempt is at the front of the queue.

Thus the probability of a handoff attempt failure can be calculated as the average fraction of handoff attempts whose mobiles leave the handoff area prior to their coming into the queue front position and getting a channel. Noting that arrivals that find k attempts in queue enter position $k+1$, this can be expressed as

$$P_{\text{fh}} \triangleq \sum_{k=0}^{\infty} P_{C+k} P_{\text{fh}|k} \tag{11.17}$$

where $P_{\text{fh}|k} = P_r$ {attempt fails given it enters the queue in position $k-1$}.

Since handoff success for those attempts which enter the queue in position $k+1$ requires coming to the head of the queue and getting a channel, under the memoryless conditions assumed in this development, we have

$$\left(1 - P_{\text{fh}|k}\right) = \left[\prod_{i=1}^{k} P\,(i \mid i+1)\right] P_r \left\{\text{get channel in first position}\right\} \tag{11.18}$$

where $P\,(i \mid i+1)$ is the probability that an attempt in position $i+1$ moves to position i before its mobile leaves the handoff area.

There are two possible outcomes for an attempt in position $i+1$. It will either be cleared from the system or will advance in queue to the next (lower) position. It will advance if the remaining dwell time of its mobile exceeds either:

(1) at least one of the remaining dwell times T_j, $j = 1, 2, \ldots, i$, for any attempt ahead of it in the queue; or

(2) the minimum remaining holding time X of those calls in progress in the target cell.

Thus

$$1 - P(i \mid i+1) = P_r\{T_{i+1} \leq X, T_{i+1} \leq T_j, \quad j = 1, 2, \ldots, i\} \quad i = 1, 2, \ldots. \tag{11.19}$$

$$\begin{aligned} 1 - P(i \mid i+1) &= P_r\{T_{i+1} \leq X, T_{i+1} \leq T_1, \ldots, T_{i+1} \leq T_i\} \\ &= P_r\{T_{i+1} \leq \min(X, T_1, T_2, \ldots, T_i)\} \\ &= P_r\{T_{i+1} \leq Y_i\} \quad i = 1, 2, \ldots \end{aligned} \tag{11.19a}$$

where $Y_i \equiv \min(X, T_1, T_2, \ldots, T_i)$. Since the mobiles move independently of each other and of the channel holding times, the random variables, X, T_j, $(j = 1, 2, \ldots, i)$ are statistically independent. Therefore, the cumulative distribution of Y_i in Equation (11.19) can be written as

$$F_{Y_i}(\tau) = 1 - \{1 - F_X(\tau)\}\{1 - F_{T_1}(\tau)\} \ldots \{1 - F_{T_i}(\tau)\}$$

Because of the exponentially distributed variables, this gives

$$F_{Y_i}(\tau) = 1 - \mathrm{e}^{-C\mu_\mathrm{H}\tau}\mathrm{e}^{-\mu_Q\tau}\ldots\mathrm{e}^{-\mu_Q\tau} = 1 - \mathrm{e}^{-(C\mu_\mathrm{H}+i\mu_Q)\tau}$$

and Equation (11.19) becomes

$$\begin{aligned} 1 - P(i \mid i+1) &= P_r\{T_{i+1} \le Y_i\} = \int_0^\infty \{1 - F_{Y_i}(\tau)\} f_{T_{i+1}}(\tau)\,\mathrm{d}\tau \\ &= \int_0^\infty \mathrm{e}^{-(C\mu_\mathrm{H}+i\mu_Q)\tau}\mu_Q\mathrm{e}^{-\mu_Q\tau}\mathrm{d}\tau = \frac{\mu_Q}{C\mu_\mathrm{H}+(i+1)\mu_Q}, \quad i = 1, 2, \ldots \end{aligned} \tag{11.20}$$

The handoff attempt at the head of the queue will get a channel (succeed) if its remaining dwell time T_1 exceeds X. Thus

$$\begin{aligned} P_r\{\text{get channel in front position}\} &= P_r\{T_1 > X\} \text{ and} \\ P_r\,\{\text{does not get channel in front position}\} &= P_r\,\{T_1 \le X\} \\ &= \int_0^\infty \mathrm{e}^{-C_{\mu\mathrm{H}\tau}}\mu_Q\mathrm{e}^{-\mu_Q\tau}\,\mathrm{d}\tau = \frac{\mu_Q}{C\mu_\mathrm{H}+\mu_Q} \end{aligned} \tag{11.21}$$

The probability Equation (11.21) corresponds to letting $i = 0$ in Equation (11.20) Then from Equation (11.18) we have

$$\begin{aligned} 1 - P_{\mathrm{fh}|k} &= \left[\prod_{i=1}^{k} P\,(i|\,i+1)\right] P_r\,\{\text{get channel in first position}\} \\ &= \frac{C\mu_\mathrm{H}+\mu_Q}{C\mu_\mathrm{H}+2\mu_Q}\,\frac{C\mu_\mathrm{H}+2\mu_Q}{C\mu_\mathrm{H}+3\mu_Q}\cdots\frac{C\mu_\mathrm{H}+k\mu_Q}{C\mu_\mathrm{H}+(k+1)\,\mu_Q}\,\frac{C\mu_\mathrm{H}}{C\mu_\mathrm{H}+\mu_Q} \\ &= \frac{C\mu_\mathrm{H}}{C\mu_\mathrm{H}+(k+1)\,\mu_Q} \end{aligned} \tag{11.22}$$

and

$$P_{\mathrm{fh}|k} = \frac{(k+1)\,\mu_Q}{C\mu_\mathrm{H}+(k+1)\,\mu_Q} \tag{11.23}$$

The above equations form a set of simultaneous nonlinear equations which can be solved for system variables when parameters are given. Beginning with an initial guess for the unknowns, the equations are solved numerically using the method of successive substitutions.

A call which is not blocked will be eventually forced into termination if it succeeds in each of the first $(l-1)$ handoff attempts which it requires but fails on the lth. Therefore,

$$P_\mathrm{F} = \sum_{l=1}^{\infty} P_\mathrm{fh}\left[P_n\,(1-P_\mathrm{fh})^{l-1}\,P_\mathrm{H}^{l-1}\right] = \frac{P_\mathrm{fh}P_\mathrm{N}}{1-P_\mathrm{H}\,(1-P_\mathrm{fh})} \tag{11.24}$$

where P_N and P_H are the probabilities of handoff demand of new and handoff calls, as defined previously. Let P_nc denote the *fraction of new call attempts* that will *not* be *completed* because of either blocking or unsuccessful handoff. This is also an important system performance measure. This probability P_{nc} can be expressed as

$$P_{nc} = P_B + P_F(1 - P_B) = P_B + \frac{P_{fh}P_N(1 - P_B)}{1 - P_H(1 - P_{fh})} \tag{11.25}$$

where the first and second terms represent the effects of blocking and handoff attempt failure, respectively. In Equation (11.25) we can guess roughly that, when cell size is large, the probabilities of cell crossing P_N and P_H will be small and the second term of Equation (11.25) (i.e. the effect of cell crossing) will be much smaller than the first term (i.e. effect of blocking). However, when the cell size is decreased, P_N and P_H will increase. The noncompleted call probability P_{nc} can be considered as a unified measure of both blocking and forced termination effects.

Another interesting measure of system performance is the weighted sum of P_B and P_F

$$CF = (1 - \alpha)P_B + \alpha P_F \tag{11.26}$$

where α is in the interval [(0, 1)] and indicates the relative importance of the blocking and forced termination effects. For some applications P_F may be more important than P_B from the user's point of view, and the relative cost α can be assigned using the system designer's judgment.

11.2.4 Performance examples

For the calculations, the average message duration was taken as $\bar{T}_M = 120$s and the maximum speed of a mobile of $V_{max} = 60$ miles/h was used. The probabilities P_B and P_F as functions of (new) call origination rate per unit area Λ_a can be seen in Figure 11.24, with cell radius R being a parameter. A total of 20 channels per cell ($C = 20$) and one channel per cell for handoff priority ($C_h = 1$) was assumed. The CRQ scheme was used for this figure, and the mean dwell time for a handoff attempt $\bar{T}_Q$ was assumed to be $\bar{T}_H/10$. As can be seen, P_F is much smaller than P_B and the difference between them decreases as cell size decreases. As expected, for larger R the effect of handoff attempts and forced terminations on system performance is smaller.

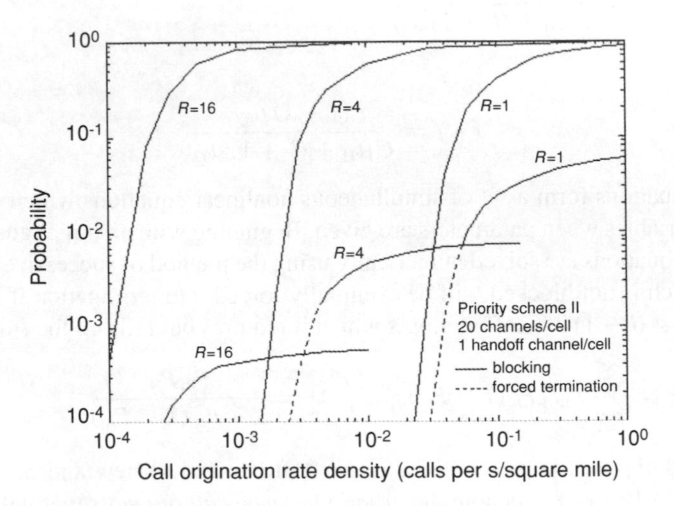

Figure 11.24 Blocking and forced termination probabilities for CRQ priority scheme.

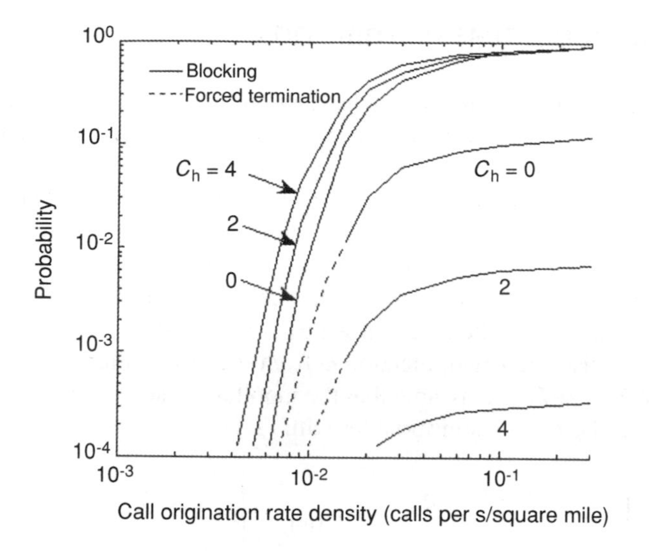

Figure 11.25 Blocking and forced termination probabilities for CRQ systems with 20 channels/cell, $R = 2$ miles.

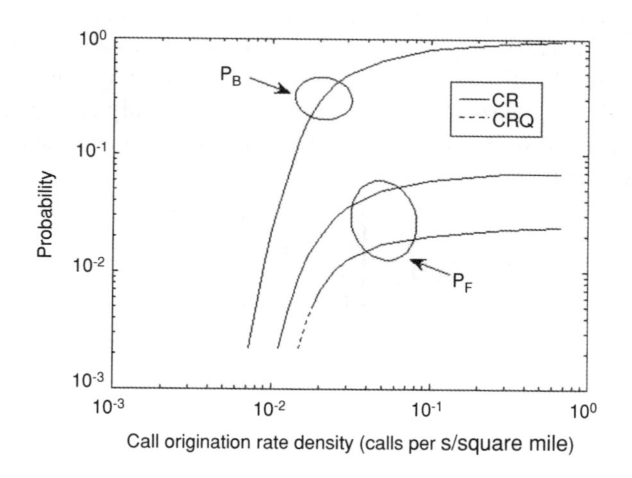

Figure 11.26 Blocking and forced terminations for priority CR and CRQ schemes (20 channels/cell, one handoff channel/cell, $R = 2$ miles).

Figure 11.25 shows P_B and P_F as functions of Λ_a. As the effects of increasing priority given to handoff calls over new calls by increasing C_h, P_F decreases by orders of magnitude with only small to moderate increase in P_B this exchange is important because (as was mentioned previously) forced terminations are usually considered much less desirable than blocked calls.

Blocking and forced termination probabilities for the two priority schemes are shown in Figure 11.26 as functions of call origination rate density Λ_a. The forced termination probability P_F is smaller for th CRQ scheme, but almost no difference exists in blocking probability P_B. We get this superiority of the CRQ priority scheme by queuing the delayed handoff attempts for the dwell time of the mobile in the handoff area.

11.3 CELL RESIDING TIME DISTRIBUTION

In this section we discuss the probability distributions of the residing times T_n and T_{h}. The random variable T_n is defined as the time (duration) that a mobile resides in the cell *in which its call originated*. Also T_{h} is defined as the time a mobile resides in a cell to *which its call is handed off*. To simplify analysis we approximate the hexagonal cell shape as a circle. For a hexagonal cell having radius R, the approximating circle with the same area has a radius, R_{eq}, which is given by $R_{\mathrm{eq}} = \sqrt{(3\sqrt{3}/2\pi)}\,R \approx 0.91R$ and illustrated in Figure 11.27. The base station is assumed to be at the center of a cell and is indicated by a letter B in the figure. The location of a mobile in a cell, which is indicated by a letter A in the figure, is represented by its distance r and direction ϕ from the base station as shown. To find the distributions of T_n and T_{h}, we assume that the mobiles are spread evenly over the area of the cell. Then r and ϕ are random variables with PDFs

$$f_r(r) = \begin{cases} \dfrac{2r}{R_{\mathrm{eq}}^2}, & 0 \le r \le R_{\mathrm{eq}} \\ 0, & \text{elsewhere} \end{cases}, \quad f_\phi(\phi) = \begin{cases} \dfrac{1}{2\pi}, & 0 \le \phi \le 2\pi \\ 0, & \text{elsewhere} \end{cases} \tag{11.27}$$

Next it is assumed that a mobile travels in any direction with equal probability and its direction remains constant during its travel in the cell. If we define the direction of mobile travel by the angle θ (with respect to a vector from the base station to the mobile), as shown in the figure, the distance Z from the mobile to the boundary of approximating circle is $Z = \sqrt{[R_{\mathrm{eq}}^2 - (r\sin\theta)^2]} - r\cos\theta$. Because ϕ is evenly distributed in a circle, Z is independent of ϕ and from the symmetry we can consider the random variable θ is in interval $[0, \pi]$ with PDF

$$f_\theta(\theta) = \begin{cases} \dfrac{1}{\pi}, & 0 \le \theta \le \pi \\ 0, & \text{elsewhere} \end{cases} \tag{11.28}$$

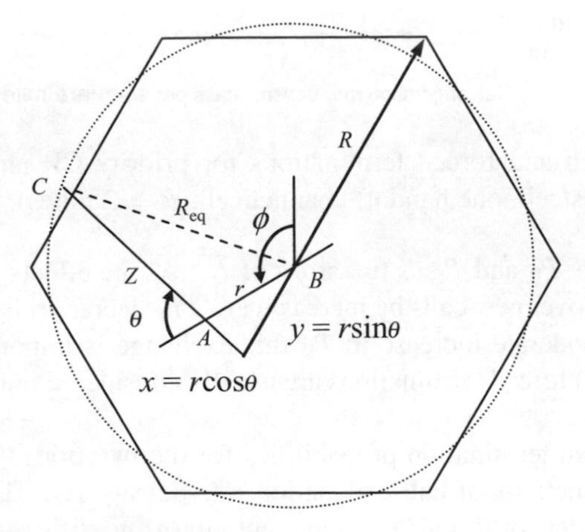

Figure 11.27 Illustration of distance from point A in cell (where call is originated), to point C on cell boundary (where mobile exits from cell).

If we define new random variables x, y as $x = r\cos\theta$, $y = r\sin\theta$, then $Z = (R_{eq}^2 - y^2) - x$ and $W = x$. Since the mobile is assumed to be equally likely to be located anywhere in the approximating circle

$$f_{XY}(x, y) = \begin{cases} \dfrac{2}{\pi R_{eq}^2}, & -R_{eq} \le x \le R_{eq}, \quad 0 \le x^2 + y^2 \le R_{eq}^2, \quad 0 \le y \le R_{eq} \\ 0, & \text{elsewhere} \end{cases}$$

From Equations (11.27)–(11.28), the joint density function of Z and W can be found by standard methods

$$\begin{aligned} f_{ZW}(z, w) &= \frac{|z+w|}{\sqrt{R_{eq}{}^2 - (z+w)^2}} f_{XY}(x, y) \\ &= \frac{2}{\pi R_{eq}^2} \frac{|z+w|}{\sqrt{R_{eq}^2 - (z+w)^2}}, \quad 0 \le z \le 2R_{eq}, -\frac{1}{2}z \le w \le -z + R_{eq} \end{aligned}$$

The PDF of the distance Z is then becomes

$$\begin{aligned} f_Z(z) &= \int_{-z/2}^{R_{eq}-z} \frac{2}{\pi R_{eq}^2} \frac{(z+w)}{\sqrt{R_{eq}^2 - (z+w)^2}} \mathrm{d}w, \quad 0 \le z \le 2R_{eq} \\ &= \begin{cases} \dfrac{2}{\pi R_{eq}^2} \sqrt{R_{eq}^2 - \left(\dfrac{z}{2}\right)^2}, & 0 \le z \le 2R_{eq} \\ 0, & \text{elsewhere} \end{cases} \end{aligned} \tag{11.29}$$

If the speed V of a mobile is constant during its travel in the cell and random variable which is uniformly distributed on the interval $[0, V_{max}]$ with PDF

$$f_V(v) = \begin{cases} \dfrac{1}{V_{max}}, & 0 \le v \le V_{max} \\ 0, & \text{elsewhere} \end{cases}$$

then the time T_n is expressed by $T_n = Z/V$ with PDF

$$\begin{aligned} f_{T_n}(t) &= \int_{-\infty}^{\infty} |w| f_Z(tw) f_V(w) \mathrm{d}w \\ &= \begin{cases} \dfrac{2}{V_{max}\pi R_{eq}^2} \displaystyle\int_0^{V_{max}} w\sqrt{R_{eq}^2 - \left(\frac{tw}{2}\right)^2} \mathrm{d}w, & 0 \le t \le \dfrac{2R_{eq}}{V_{max}} \\ \dfrac{2}{V_{max}\pi R_{eq}^2} \displaystyle\int_0^{2R_{eq}/t} w\sqrt{R_{eq}^2 - \left(\frac{tw}{2}\right)^2} \mathrm{d}w, & t \ge \dfrac{2R_{eq}}{V_{max}} \end{cases} \end{aligned} \tag{11.30}$$

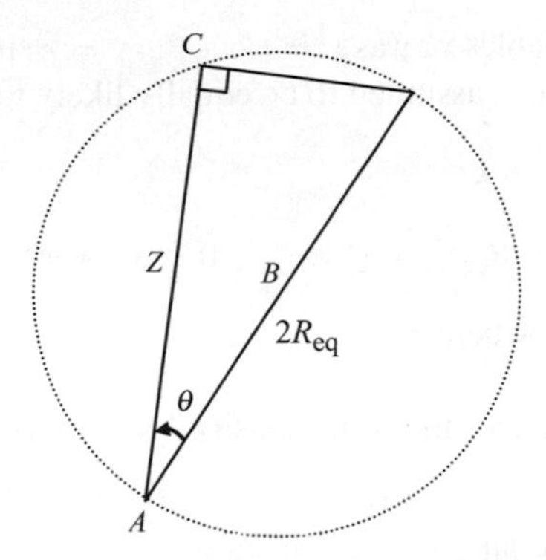

Figure 11.28 Illustration of distance from cell entering point (A on cell boundary), to cell exiting point (C on cell boundary).

$$= \begin{cases} \dfrac{8R_{eq}}{3V_{max}\pi t^2}\left[1 - \sqrt{\left\{1 - \left(\dfrac{tV_{max}}{2R_{eq}}\right)^2\right\}^3}\right], & 0 \le t \le \dfrac{2R_{eq}}{V_{max}} \\ \dfrac{8R_{eq}}{3V_{max}\pi t^2}, \quad t \ge \dfrac{2R_{eq}}{V_{max}} \end{cases}$$

and the cdf of T_n is

$$F_{T_N}(t) = \int_{-\infty}^{t} f_{T_n}(x)\mathrm{d}x\, F_{T_n}$$

$$= \begin{cases} \dfrac{2}{\pi}\arcsin\left(\dfrac{V_{max}t}{2R_{eq}}\right) - \dfrac{4}{3\pi}\tan\left[\dfrac{1}{2}\arcsin\left(\dfrac{V_{max}t}{2R_{eq}}\right)\right] \\ +\dfrac{1}{3\pi}\sin\left[2\arcsin\left(\dfrac{V_{max}t}{2R_{eq}}\right)\right], \quad 0 \le t \le \dfrac{2R_{eq}}{V_{max}} \\ 1 - \dfrac{8R_{eq}}{3\pi V_{max}}\dfrac{1}{t}, \quad t \ge \dfrac{2R_{eq}}{V_{max}} \end{cases} \tag{11.31}$$

To find the distribution of T_h, in the next step we note that, when a handoff call is attempted, it is always generated at the cell boundary, which is taken as the boundary of the approximating circle. Therefore, to find T_h one must recognize that the mobile will move from one point on the boundary to another. The direction of a mobile when it crosses the boundary is indicated by the angle θ between the direction of the mobile and the direction from the mobile to the center of a cell as shown in Figure 11.28 [75, 76].

If the mobile moves in any direction with equal probability, the random variable θ has PDF given by

$$f_\theta(\theta) = \begin{cases} \dfrac{1}{\pi}, & -\dfrac{\pi}{2} \le \theta \le \dfrac{\pi}{2} \\ 0, & \text{elsewhere} \end{cases}$$

The distance Z as shown in Figure 11.28 is $Z = 2R_{\text{eq}}\cos\theta$, which has a CDF given by

$$F_Z(z) = P_r\{Z \leq z\} = \begin{cases} 0, \; z < 0 \\ 1 - \dfrac{2}{\pi}\arccos\left(\dfrac{z}{2R_{\text{eq}}}\right), \; 0 \leq z \leq 2R_{\text{eq}} \\ 1, \; z > 2R_{\text{eq}} \end{cases} \tag{11.32}$$

The PDF of Z is

$$f_Z(z) = \frac{\mathrm{d}}{\mathrm{d}z}F_Z(z) = \begin{cases} \dfrac{1}{\pi}\dfrac{1}{\sqrt{R_{\text{eq}}^2 - \left(\frac{z}{2}\right)^2}}, \; 0 \leq z \leq 2R_{\text{eq}} \\ 0, \; \text{elsewhere} \end{cases} \tag{11.33}$$

The time in the cell T_{h} is the time that a mobile travels the distance Z with speed V, then $T_{\text{h}} = Z/V$. With the same assumption about V, the PDF of T_{h} is

$$f_{T_{\text{h}}}(t) = \int_0^{\infty} |w|\, f_Z(tw) f_V(w)\mathrm{d}w = \begin{cases} \dfrac{1}{\pi V_{\max}} \displaystyle\int_0^{V_{\max}} \dfrac{w}{\sqrt{R_{\text{eq}}^2 - \left(\dfrac{tw}{2}\right)^2}}\mathrm{d}w, \; 0 \leq t \leq \dfrac{2R_{\text{eq}}}{V_{\max}} \\ \dfrac{1}{\pi V_{\max}} \displaystyle\int_0^{2R_{\text{eq}}/t} \dfrac{w}{\sqrt{R_{\text{eq}}^2 - \left(\dfrac{tw}{2}\right)^2}}\mathrm{d}w, \; t \geq \dfrac{2R_{\text{eq}}}{V_{\max}} \end{cases}$$

$$= \begin{cases} \dfrac{4R_{\text{eq}}}{\pi V_{\max}}\dfrac{1}{t^2}\left[1 - \sqrt{1 - \left(\dfrac{V_{\max}t}{2R_{\text{eq}}}\right)^2}\right], \; 0 \leq t \leq \dfrac{2R_{\text{eq}}}{V_{\max}} \\ \dfrac{4R_{\text{eq}}}{\pi V_{\max}}\dfrac{1}{t^2}, \; t \geq \dfrac{2R_{\text{eq}}}{V_{\max}} \end{cases} \tag{11.34}$$

and the CDF of T_{h} is

$$F_{T_{\text{h}}}(t) = \int_{-\infty}^{t} f_{T_{\text{h}}}(x)\,\mathrm{d}x$$

$$= \begin{cases} 0, \; t < 0 \\ \dfrac{2}{\pi}\arcsin\left(\dfrac{V_{\max}t}{2R_{\text{eq}}}\right) - \dfrac{2}{\pi}\tan\left[\dfrac{1}{2}\arcsin\left(\dfrac{V_{\max}t}{2R_{\text{eq}}}\right)\right], \; 0 \leq t \leq \dfrac{2R_{\text{eq}}}{V_{\max}} \\ 1 - \dfrac{4R_{\text{eq}}}{\pi V_{\max}}\dfrac{1}{t}, \; t > \dfrac{2R_{\text{eq}}}{V_{\max}} \end{cases} \tag{11.35}$$

Figure 11.29 shows the mean channel holding time in a cell $\overline{T_{\text{H}}}$. Notice that $\overline{T_{\text{H}}}$ becomes smaller with smaller cell size, but sensitivity to change in cell size is smaller for larger cells. Finally, earlier in Section 11.2 we approximated the cumulative distribution function of the channel holding time in a cell as suggested in References [75, 76]. The goodness-of-fit

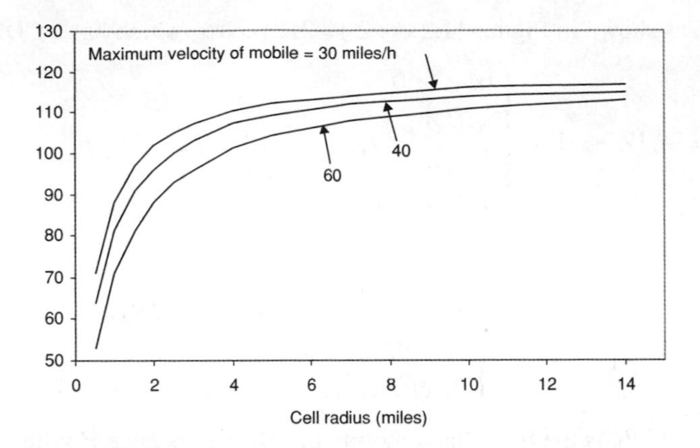

Figure 11.29 Mean channel holding time (s) in cell vs R (average call duration = 120 s).

Table 11.1 The goodness-of-fit G of the approximation

Cell radius, R	G
1.0	0.020220
2.0	0.000120
4.0	0.000003
6.0	0.000094
8.0	0.000121
10.0	0.000107
12.0	0.000086
14.0	0.000066
16.0	0.000053

G of this approximation, defined in Equation (11.8), is shown in Table 11.1 for various cell sizes. We see that G is very small for a wide ranges of cell radius R. These values support the use of the approximation in the calculations.

11.4 MOBILITY PREDICTION IN PICO- AND MICROCELLULAR NETWORKS

It should be expected that 4G networks will further reduce the size of the cells. In a micro- and picocellular network, resources availability varies frequently as users move from one access point to another. In order to deterministically guarantee QoS support for a mobile unit, the network must have prior exact knowledge of the mobile's path, along with arrival and departure times to every cell along the path. With this knowledge, the network can verify the feasibility of supporting the call during its lifetime, as the mobile moves across the network. It is impractical, however, to require the users to inform the network of their exact movement, since they may not know this information *a priori*. Even if the path is known, the exact arrival and departure times to the cells along the path are still hard to determine in advance. Therefore, it becomes crucial to have an accurate mechanism to predict the trajectory of the mobile user.

As an example, the virtual connection tree is designed to support QoS guarantees for mobile units [77]. In this scheme, a number of adjacent cells are grouped into a cell cluster in a static fashion. Upon the admission of a call, the scheme pre-establishes a connection between a root switch and each base station in the cell cluster. The scheme does not take user mobility into consideration to predict the set of base stations which may potentially be visited by the mobile unit. This may result in an unnecessary resource overloading that may underutilize the network resources and cause severe congestion.

The *shadow cluster* (SC) scheme [78] provides a distributed call-admission control (CAC) based on the estimated bandwidth requirements in the SC. An SC is a collection of base stations to which a mobile unit is likely to attach in the future. The admission decision is made in a distributed fashion by all the base stations within the SC. The scheme partitions the time into predefined intervals and verifies the feasibility of supporting calls over those intervals. This requires the communication of large number of messages between base stations during every time interval. Moreover, since bandwidth estimates are calculated at the beginning of each time interval and the admission decisions are made at the end of each time interval, admission of new calls is delayed for at least a time equal to the length of these predefined time intervals.

Both of the above two schemes lack the mechanism to predict the mobile's trajectory and determine the future cells to which the mobile may hand off. Several techniques have been proposed in the literature to address this issue. In Bharghavan and Mysore [79], a profile-based algorithm is proposed to predict the next cell that the mobile unit will hand off, using a user profile and a cell profile, which are simply the aggregate values of the handoff's history. In Liu and Maguire [80], a mobile motion prediction (MMP) algorithm is proposed to predict the future locations of the mobile unit. This algorithm is based on a pattern matching technique that exploits the regularity of the users' movement patterns. The MMP algorithm was further expanded to a two-tier hierarchical location prediction (HLP) algorithm [81]. In the latter case, the two-tiered prediction scheme involves both an intercell and an intracell tracking and prediction component. The first tier uses an approximate pattern matching technique to predict the global intercell direction and the second tier uses an extended self-learning Kalman filter to predict the trajectory within the cell using the measurements received by the mobile unit.

In order to support QoS guarantees of multiple classes of services, the scheme must integrate call and admission control with the mobility profile of the mobile user. The integration of these two components makes it possible to use mobility prediction to verify the feasibility of admitting a new call and make sure that the required QoS can be supported during the lifetime of the call. In other words we should be able to predict location (space) and time when a certain resources well be needed in the network. This concept will be referred to as *space time predictive QoS* or STP QoS. The mobility prediction algorithm must be easy to implement and maintain, since it will be invoked on a per-user basis. Furthermore, the admission control procedure must be invoked only when needed with minimum overhead and in a distributed fashion, where each network cell, potentially involved in supporting the QoS requirements of the call, participates in the decision process [82, 83].

In this section, we present such a framework, which efficiently integrates mobility prediction and CAC, to provide support for PST-QoS guarantees, where each call is guaranteed its QoS requirements for the time interval that the mobile unit is expected to spend within each cell it is likely to visit during the lifetime of the call.

In this framework, efficient support of PST-QoS guarantees is achieved based on an accurate estimate of mobile's trajectory as well as the arrival and departure times for each cell along the path. Using these estimates, the network can determine if enough resources are available in each cell along the mobile's path to support the QoS requirements of the call. The framework is designed to easily accommodate dynamic variations in network resources. The basic components of this framework are:

(1) a predictive service model to support timed-QoS guarantees;

(2) a mobility model to determine the mobile's most likely cluster (MLC); the MLC represents a set of cells that are most likely to be visited by a mobile unit during its itinerary; and

(3) a CAC model to verify the feasibility of supporting a call within the MLC.

The service model accommodates different types of applications by supporting *integral* and *fractional* predictive QoS guarantees over a predefined time-guarantee period. The MLC model is used to actively predict the set of cells that are most likely to be visited by the mobile unit. For each MLC cell, the mobile's earliest arrival time latest arrival time and latest departure time are estimated. These estimates are then used by the CAC to determine the feasibility of admitting a call by verifying that enough resources are available in each of the MLC cells during the time interval between the mobile's earliest arrival time and its latest departure time. If available, resources are then reserved for the time interval between the mobile's earliest arrival time and latest departure time and leased for the time interval between the mobile's earliest and latest arrival times. If the mobile unit does not arrive before the lease expires, the reservation is canceled and the resources are returned to the pool of available resources. The unique feature of the this frame-work is the ability to combine the mobility model with the CAC model to determine the level of PST-QoS guarantees that the network can provide to a call and dynamically adjust these guarantees as the mobile unit moves across the network.

11.4.1 PST-QoS guarantees framework

The first approach, to achieve a high level of QoS support guarantees in mobile environments, is to allocate resources for the duration of the call in all future cells that the mobile unit will visit. This means that the resources within each cell that is to be visited will be held, possibly for the duration of the call, even if the mobile never moves into the cell. This approach is similar to the one proposed in Talukdar *et al.* [85] and will be referred to as a *predictive space* or PS QoS model. Clearly, such an approach will result in underutilization of the network resources as resources are being held, but not used by any call.

The second approach is to only reserve resources in all future cells that the mobile unit may visit for the time interval during which the mobile will reside in each cell. If t_i and t_{i+1} represent the expected arrival and departure times of the mobile unit to cell i along the path, respectively, resources in cell i will only be reserved for the time interval $[t_i, t_{i+1}]$ [84]. Unlike the first approach, this approach is likely to increase resource utilization, since resources in every cell remain available to other calls outside the reservation intervals. This approach, however, is only feasible if exact knowledge of the mobile path and arrival and departure times to every cell along the path is available. Obtaining exact knowledge of mobile mobility is not possible in most cases, due to the uncertainty of the mobile environments and the difficulty in specifying the mobility profiles of mobile units. An acceptable level of service guarantees, however, can be achieved if the path of the mobile can be predicted accurately. This approach is discussed in this section and will be referred to as *predictive space and time* or the PST QoS model. The model attempts to achieve a balance between an acceptable level of service guarantees and a high level of network resource utilization. Based on this model, the required QoS support level is guaranteed by reserving resources in advance in each cell that is *most likely* to be visited by the mobile unit. These reservations only extend for a time duration equal to the time interval the mobile unit is expected to spend within a cell, starting from the time of its arrival time to the cell until its departure time from the cell. In order to characterize the set of 'most likely' cells and capture the level of QoS guarantees requested by the application, the service model uses the following parameters:

(1) the time guarantee period T_G;

(2) a cluster-reservation threshold τ, and a

(3) bandwidth-reservation threshold, γ.

All of these parameters are application-dependent. The parameter T_G specifies the time duration for which the required QoS level is guaranteed; τ defines the minimum percentage of the most likely cells to be visited by the mobile unit that must support the required QoS level for the guarantee period T_G. The parameter γ represents the minimum percentage of the required bandwidth that must be reserved in every cell that is most likely to be visited.

To accommodate different types of applications, the service model provides two types of predictive service guarantees, namely, *integral guaranteed* service and *fractional guaranteed* service. The integral guaranteed service ensures that all cells, which are most likely to be visited by the mobile unit, can support the requested bandwidth requirements for the lifetime of the call. In this case, T_G must be equal to the call duration and τ and γ are both equal to 100 %. The fractional guaranteed service, on the other hand, guarantees that at least τ % of these cells can support at least the γ % of the requested bandwidth requirements for the next T_G interval. A special case arises when either the value of τ or γ is zero. In this case, the service is referred to as *best effort*.

11.4.2 Most likely cluster model

The MLC model considers that the 'most likely to be visited' property of a cell is directly related to the position of the cell with respect to the estimated direction of the mobile unit. This likelihood is referred to as directional probability. Based on this metric, cells that are situated along the mobile unit's direction have higher directional probabilities and are more likely to be visited than those that are situated outside of this direction.

Based on the above, the MLC at any point in time during the lifetime of a call is defined as a collection of contiguous cells, each of which is characterized by a directional probability that exceeds a certain threshold. For each MLC cell, the expected arrival and departure times of the mobile are estimated. Using these estimates, the feasibility of supporting the requested level of timed-QoS guarantees during the mobile's residence time within each cell along path is verified. In the following, we present the method used to predict the direction of a mobile unit and the scheme used to construct its MLC. We then describe the algorithm used to estimate the expected times of arrival and departure of the mobile unit to a given cell within the MLC [84].

The *direction-prediction method* used by MLC to predict the mobile user's direction is based on the history of its movement. It is clear, however, that the prediction method used should not be greatly affected by small deviations in the mobile direction. Furthermore, the method should converge rapidly to the new direction of the mobile unit. To take the above properties into consideration, a first-order autoregressive filter, with a smoothing factor α, is used. More specifically, let D_0 be the current direction of the mobile unit when the call is made. Notice that, when the mobile is stationary within a cell, it is assumed that the current cell is the only member of the MLC, so reservations are done only within the current cell. If D_t represents the observed direction of the mobile unit at time t and $\tilde{D}_t$ represents the estimated direction at time t, the predicted direction $\tilde{D}_{t+1}$ at $t+1$ is obtained as $\tilde{D}_{t+1} = (1-\alpha)\,\tilde{D}_t + \alpha D_t$. In order to track the actual direction of the mobile unit more accurately, the smoothing factor α is computed as $\alpha = cE_s^2/\sigma_{s+1}$ where $0 < c < 1$, $E_s = D_s - \tilde{D}_s$ is the prediction error, and σ_s is the average of the past square prediction errors at time s. σ_s can be expressed as $\sigma_{s+1} = cE_s^2 + (1-c)\,\sigma_s$.

The *directional probability*, at any point in time t, of any cell being visited next by a mobile unit, can be derived based on the current cell, where the mobile resides, and the estimated direction $\tilde{D}_t$ of the mobile unit at time t. The basic property of this probability distribution is that for a given direction, the cell that lies on the estimated direction from the current cell has the highest probability of being visited in the future [83]. Consider a mobile unit currently residing at cell i coming from cell m and let $j = 1, 2, \ldots$, represent a set of adjacent cells to cell i. Each cell j is situated at an angle ω_{ij} from the x-axis passing by the center of cell i, as presented in Figure 11.30. If we define the *directional path* from i to j as the direct path from the center of cell i to the center of cell j, the *directionality* D_{ij} for a given cell j can be expressed as

$$D_{ij} = \begin{cases} \dfrac{\theta_{ij}}{\phi_{ij}}, & \phi_{ij} > 0 \\ \theta_{ij}, & \phi_{ij} = 0 \end{cases} \tag{11.36}$$

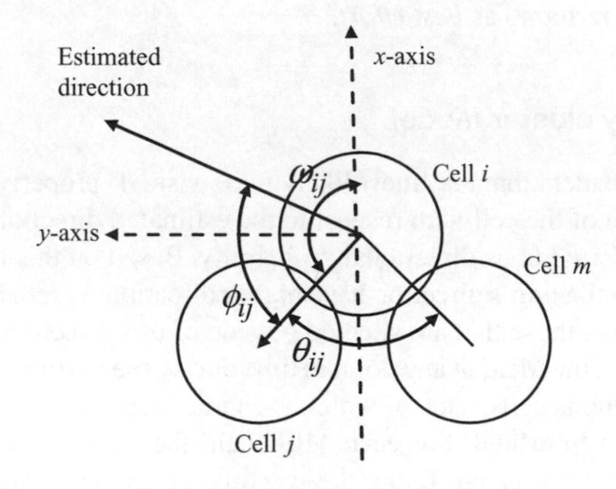

Figure 11.30 Parameters used to calculate the directional probability.

where ϕ_{ij} is an integer representing the deviation angle between the straight path to destination and the directional path from i to j, while θ_{ij} represents the angle between the directional path from m to i and the directional path from i to j.

Based on its directionality D_{ij}, the directional probability $P_{i \to j}$ of cell j being visited next by a mobile unit currently at cell i can be expressed as $P_{i \to j} = D_{ij} / \sum_k D_{ik}$ where k is a cell at the same ring as j with respect to i. A cell k is said to be at ring L with respect to cell i if it is located at a ring L cells away from i. For a given cell i, the directional probabilities $P_{i \to j}$ provide the basis upon which MLCs are formed as the mobile units moves across the network.

11.4.2.1 Forming the most likely cluster

Starting from the cell where the call originated, a mobile unit is expected to progress toward its destination. The mobile unit, however, can temporarily deviate from its long-term direction to the destination, but is expected to converge back at some point in time toward its destination. This mobility behavior can be used to determine the cells that are likely to be visited by a mobile unit.

Let us define the forward span as the set of cells situated within an angle with respect to the estimated direction $\tilde{D}_t$ of the mobile unit as illustrated in Figure 11.31. Based on the directional probabilities and the definition of a forward span, the MLC of a given mobile unit u currently located at cell i, denoted as $C_i^{\mathrm{MLC}}(u)$, can be expressed as $C_i^{\mathrm{MLC}}(u) = \{\text{cells } j \mid \phi_{ij} \le \delta_i, j = 1, 2, \ldots\}$ where ϕ_{ij} is the deviation angle between the straight path to destination and the directional path from i to j. The angle δ_i is defined such that $P_{i \to j} \ge \mu$, where μ represents a system defined threshold on the likelihood that cell is to be visited. More specifically, δ_i can be expressed as $\delta_i = \max |\phi_{ij}|$ such that $P_{i \to j} \ge \mu$.

The next step in the process of forming the MLC is to decide on the size of the MLC window, W_{MLC}, which represents the number of adjacent rings of cells to be included in the MLC. Define $\mathrm{Ring}_{i,j}$ to be the ring at which cell j is located with respect to cell i. Therefore, a cell k is included in $C_i^{\mathrm{MLC}}(u)$, if $\mathrm{Ring}_{i,k} \le W_{\mathrm{LMC}}$ which gives $C_i^{\mathrm{MLC}}(u) = \{\text{cells } j \mid \Phi_{ij} \le \delta_i, \text{ and } \mathrm{Ring}_{i,j} \le W_{\mathrm{MLC}} \quad j = 1, 2, \ldots\}$. The size of the MLC window has a strong impact

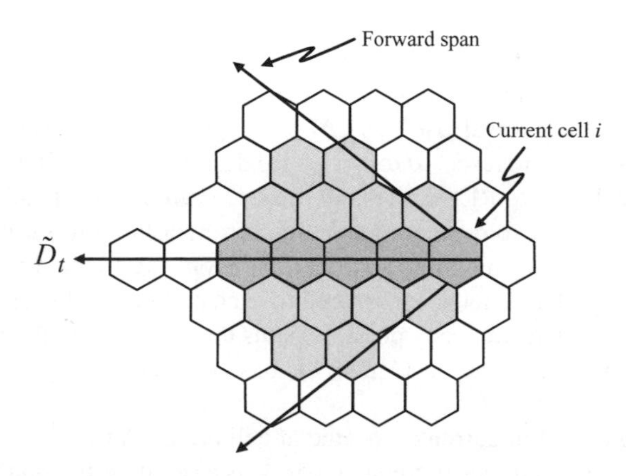

Figure 11.31 Definition of the MLC.

on the performance of the scheme. Increasing the MLC window size, by including more rings, increases the likelihood of supporting the required QoS if the mobile moves along the predicted direction $\tilde{D}(t)$. On the other hand, if the mobile deviates from the predicted direction, increasing the MLC window size may not ensure the continued support of the call, as the mobile unit may move out from the MLC. A possible approach is to reward users who move within the predicted direction by increasing their MLC window size up to a maximum $R_{\max}$. The value of $R_{\max}$ depends on the value of the guarantee period T_G. Higher values of T_G result in larger values of $R_{\max}$.

When the user deviates from the estimated direction, the MLC window size is decreased by an amount proportional to the degree of deviation. As a result, support of the predictable users' QoS requirements can be achieved with high probability, whereas users with unpredictable behavior do not unnecessarily overcommit the network resources. The algorithm dynamically updates the size of the MLC window based on the observed movement patterns of the mobile users. If Δ_t is the measure of the mobile's deviation with respect to the estimated direction at time t, defined as $\Delta_{t+1} = \beta\Delta_t + (1-\beta)|\tilde{D}_t - D_t|$ with $0 < \beta < 1$ and Δ_0 equal to zero, the MLC window size W_{MLC} at time t can be defined as follows:

$$W_{\mathrm{LMC}} = \min\left(R_{\max}, \left\lfloor \left(1 - \frac{\Delta_t}{\pi}\right)^2 \right\rfloor R_{\max}\right) \tag{11.37}$$

The MLC window size is recalculated at every handoff; therefore, the window size shrinks and grows depending on the mobile's behavior.

The method can be easily extended to cellular network with cells of different sizes. When a cellular network has cells of different sizes, the definition of rings is different. The rings are imaginary circles centered at the current cell. The radius of the first ring R_1 is equal to the distance from the center of the current cell to the center of the neighboring cell whose center is farthest away. Consequently, the radius of a ring i, where $i = 1, 2, \ldots,$ is equal $i \times R_1$. Any cell that has its center within the boundaries of a ring is considered in that ring.

The *time of arrival* and *residence time* of the mobile can be estimated for each MLC cell. Based on these estimates, the feasibility of supporting the requested level of timed-QoS guarantees within the residence time can then be verified. The cell residence time within cell j for a mobile unit currently in cell i is characterized by three parameters, namely, expected earliest arrival time $[T_{\mathrm{EA}}(i, j)]$, expected latest arrival time $[T_{\mathrm{LA}}(i, j)]$, and expected latest departure time $[T_{\mathrm{LD}}(i, j)]$. Consequently, $[T_{\mathrm{EA}}(i, j), T_{\mathrm{LD}}(i, j)]$ is the expected residence time of the mobile unit within cell j. This interval is referred to as the resource reservation interval (RRI), while the interval $[T_{\mathrm{EA}}(i, j), T_{\mathrm{LA}}(i, j)]$ is referred to as the resource leasing interval (RLI). Resources are reserved for the entire duration of RRI. However, if the mobile does not arrive to cell before RLI expires, all resources are released and the reservation is canceled. This is necessary to prevent mobile units from holding resources unnecessarily.

In order to derive these time intervals, one can adopt the method used in the SC and consider all possible paths from the current cell to each cell in the cluster [78]. This method can be complex, since there are many possible paths that a mobile unit may follow to reach a cell. The approach taken in the MLC model is based on the concept of most likely paths [84].

Consider a mobile unit u, currently located at cell m, and let $C_m^{\mathrm{MLC}}(u)$ denote its MLC. Define $G = (V, E)$ to be a directed graph, where V is a set of vertices and E a set of edges. A vertex $v_i \in V$ represents MLC cell i. For each cell i and j in $C_m^{\mathrm{MLC}}(u)$, an edge (v_i, v_j)

is in E if and only if j is a *reachable direct neighbor* of i. Each directed edge is (v_i, v_j) in G is assigned a cost $1/P_{i\to j}$.

A path Π between MLC cells i and k is defined as a sequence of edges (v_i, v_{i+1}), $(v_{i+1}, v_{i+2}), \ldots, (v_{k-1}, v_k)$. The cost of a path between MLC cells i and k is derived from the cost of its edges so that the least costly path represents the most likely path to be followed by the mobile. A *k-shortest paths* algorithm [86] is then used to obtain the set K of *k-most likely paths* to be followed by the mobile unit.

For each path $\Pi \in K$ between MLC cell i and j, we define the *path residence time* as the sum of the residence time of each cell in the path. Let Π_s and Π_1 in K, represent the paths with the shortest and longest path residence time, respectively. Π_s is used to derive the expected earliest arrival time , while Π_1 is used to derive expected latest arrival $T_{\mathrm{LA}}(i, j)$. So, $T_{\mathrm{EA}}(i, j)$ and $T_{\mathrm{LA}}(i, j)$ can be expressed, respectively, as

$$T_{\mathrm{EA}}(i, j) = \sum_{k\in\Pi_1} \frac{d(m, k, n)}{\bar{S}_{\max}(k)}, \quad T_{LA}(i, j) = \sum_{k\in\Pi_1} \frac{d(m, k, n)}{\bar{S}_{\min}(k)} \tag{11.38}$$

where $\bar{S}_{\max}(k)$ and $\bar{S}_{\min}(k)$ represent the average maximum and minimum speed for cell k, respectively. $\bar{S}_{\max}(k)$ and $\bar{S}_{\min}(k)$ are provided by the network support based on the observed mobile units' speeds. $d(m, k, n)$ is the main distance within cell k given that cells m, k, and n are three consecutive cells in the path. The value of $d(m, k, n)$ depends on whether cell k is the cell where the call originated, an intermediate cell or the last cell in the path, i.e. cell j

$$d(m, k, n) = \begin{cases} d_{\mathrm{O}}(k, n), & \text{if} \quad k \text{ is the originating cell} \\ d_{\mathrm{I}}(m, k, n), & \text{if} \quad k \text{ is the intermediate cell} \\ d_{\mathrm{LI'}}(k, n), & \text{if} \quad m = n \\ d_{\mathrm{L}}(m, k), & \text{if} \quad k \text{ is the last cell, k} = \text{j} \end{cases} \tag{11.39}$$

When k is the originating cell, the pdf $f_Y(y)$ of the distance Y, within cell k as shown in Figure 11.32, is derived, assuming that the mobile units are evenly spread over a cell area of

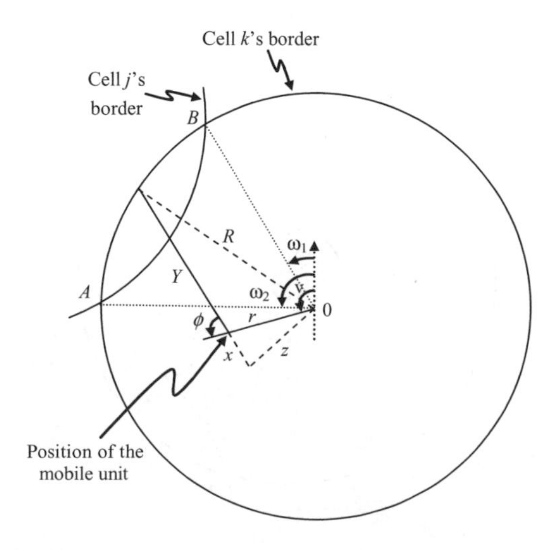

Figure 11.32 Distance Y in originating cell k.

radius R travel along a constant direction within the cell and can exit from the cell from any point along the border with cell n. Therefore, the position of the mobile unit is determined by the angle ν and the distance r from the center of the cell. ν is uniformly distributed between 0 and 2π; r is uniformly distributed between 0 and R. Since ν is uniformly distributed, ϕ is also uniformly distributed between 0 and π. Therefore, $d(k, n)$ is equal to the mean distance $E[Y]$ of the PDF $f_Y(y)$. Based on these assumptions the PDF $f_Y(y)$in a cell where the call originates can be obtained using the standard methods as described in [75]:

$$f_Y(y) = \begin{cases} \dfrac{2}{\pi R^2}\sqrt{\left\{R^2 - \left(\dfrac{y}{2}\right)^2\right\}}, & \text{for} \quad 0 \le y \le 2R \\ 0, & \text{otherwise} \end{cases}$$

which gives

$$d_o(k, n) = E[Y] = \int_0^{2R} y \cdot f_Y(y)\,dy = \frac{8R}{3\pi} \tag{11.40}$$

When k is an intermediate cell, the PDF $f_Y(y)$ of the distance Y, within cell k, as shown in Figure 11.33, is derived assuming that the mobile units enter cell k from cell m at any point along the arc AB of cell k. This arc is defined by the angles β_1 and β_2: The mobile travels along a constant direction within the cell and can exit from cell k to n from any point along the arc CD of cell k, which is defined by the angles ω_1 and ω_2. The direction of the mobile is indicated by the angle ϕ, which is uniformly distributed; $d(m, k, n)$ is equal to the mean distance $E[Y]$ of the PDF $f_Y(y)$, which is derived in Aljadhai and Znati [84] (see also the

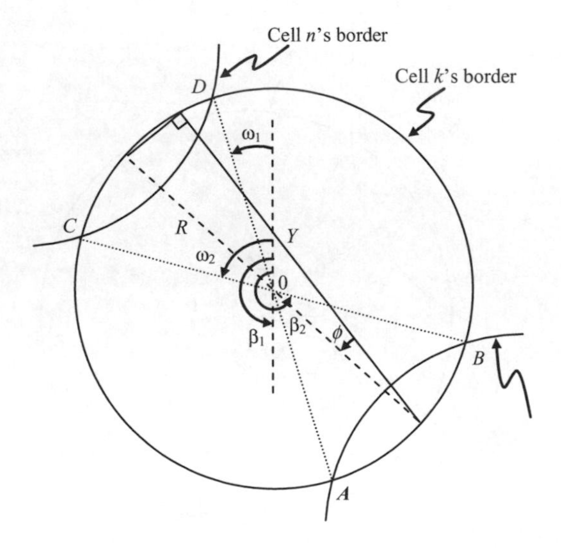

Figure 11.33 Distance Y in an intermediate cell k.

Appendix for details) as

$$d_I(m,\ k,\ n) = E[Y] = \begin{cases} \dfrac{8R}{(\omega_2-\omega_1)(\beta_2-\beta_1)}\left[\sin\left(\dfrac{\beta_2-\omega_2}{2}\right) - \sin\left(\dfrac{\beta_1-\omega_2}{2}\right)\right. \\ \left. \quad -\sin\left(\dfrac{\beta_2-\omega_1}{2}\right) + \sin\left(\dfrac{\beta_1-\omega_1}{2}\right)\right] \quad \text{for} \quad \beta_1 \geq \omega_2 \\ \dfrac{8R}{(\omega_2-\omega_1)\ (\beta_2-\beta_1)}\left[\sin\left(\dfrac{\omega_1-\beta_2}{2}\right) - \sin\left(\dfrac{\omega_1-\beta_1}{2}\right)\right. \\ \left. \quad -\sin\left(\dfrac{\omega_2-\beta_1}{2}\right) + \sin\left(\dfrac{\omega_2-\beta_1}{2}\right)\right] \quad \text{for} \quad \beta_2 \leq \omega_1 \end{cases} \tag{11.41}$$

The mean distance in the last cell in the path is derived as follows:

$$d_{\rm L}(m,\ k) = \max d\,(m, k, q) \quad \forall q \text{ adjacent to } k, q \neq m \tag{11.42}$$

The mean distance in the cell k when the path makes a loop within cell k is derived as follows:

$$d_{\rm LP}(m, k, n) = 2d_o\,(k, n) \tag{11.43}$$

Similarly, the expected latest departure time $T_{\rm LD}(i,\ j)$ from cell j can be computed as:

$$T_{\rm LD}(i,\ j) = T_{\rm LA}(i,\ j) + d(m, k)/\bar{S}_{\rm min}(k) \tag{11.44}$$

The estimates of $T_{\rm EA}(i,\ j)$, $T_{\rm LD}(i,\ j)$ and $T_{\rm LD}(i,\ j)$ for a mobile u currently located at cell i are used to compute RLI and RRI for each cell $j \in C_i^{\rm MLC}(u)$. The CAC uses these values to verify the feasibility of supporting u's call in each cell $j \in C_i^{\rm MLC}(u)$.

A good agreement between the results of the analytical model of the distance, based on Equations (11.40) and (11.41), and the simulation results of mobile units traveling along the same path, is demonstrated in Aljadhai and Znati: [84].

11.4.2.2 Performance example

The MLC CAC scheme is compared with the SC scheme based on the following assumptions [78, 84].

(1) Each cell covers 1 km along a highway. The highway is covered by 10 cells. Mobile units can appear anywhere along a cell with equal probability.

(2) Call holding time is exponentially distributed with mean $T_{\rm H} = 130$ and 180 s.

(3) Total bandwidth of each cell is 40 bandwidth units (BUs). Three types of calls are used: voice, audio, and video, requiring $B_{\rm voice} = 1$ BU, $B_{\rm audio} = 5$ BUs and $B_{\rm video} = 10$ BUs, respectively. The probabilities of each type are, $P_{\rm voice} = 0.7$, $P_{\rm audio} = 0.2$, and $B_{\rm video} = 0.1$.

(4) Mobile units may have one of three different speeds: 70, 90 or 105 km/h. The probability of each speed is 1/3.

(5) In the SC scheme, the time is quantized in time interval of length 10 s.

(6) A reference scheme, referred to as clairvoyant scheme (CS), is introduced. In this scheme, the exact behavior of every mobile unit is assumed to be known at the admission time. CS reserves bandwidth in exactly the cells that the mobile unit will visit and for the exact residence time interval in every cell. Therefore, CS produces the maximum utilization and minimum blocking ratio for a specific dropping ratio, which is zero in this case.

Since mobile units can appear anywhere along a cell, the residence time within the initial cell (the cell in which the call originates) is selected to be uniformly distributed between zero, and a duration equal to cell length/speed. The initial direction probability is assumed to be 0.5 for both possible directions, i.e. left and right directions. After the first handoff, the direction and position of the call become known and, therefore, it is possible to determine the arrival and departure time in other cells.

Figure 11.34 shows the blocking ratio and Figure 11.35 utilization of the three schemes as functions of the call arrival rate. As expected, CS produces the maximum utilization and minimum blocking ratio assuming a zero dropping ratio condition. The utilization in the MLC CAC is better than SC as the call arrival rate exceeds 0.06 for all mean call holding times. Moreover, the call blocking in MLC CAC is much less than that of the SC scheme. This behavior shows that, by simply reserving bandwidth between the earliest arrival time and latest departure time at a cell, the MLC scheme accepts more calls and increases utilization. Moreover, the increase in the utilization in the SC scheme is very slow when the call arrival rate is greater than 0.06. The reason for this behavior is that the SC bases its estimates on the exponential holding time PDF, which decreases as time increases. Therefore, the bandwidth estimates decreases as the distance to a future cell increases. As a result, the chance of dropping the call in subsequent cells is increased unless the minimum survivability estimate is increased. In Figures 11.34 and 11.35, the MLC CAC always outperforms the SC regardless of the mean holding time.

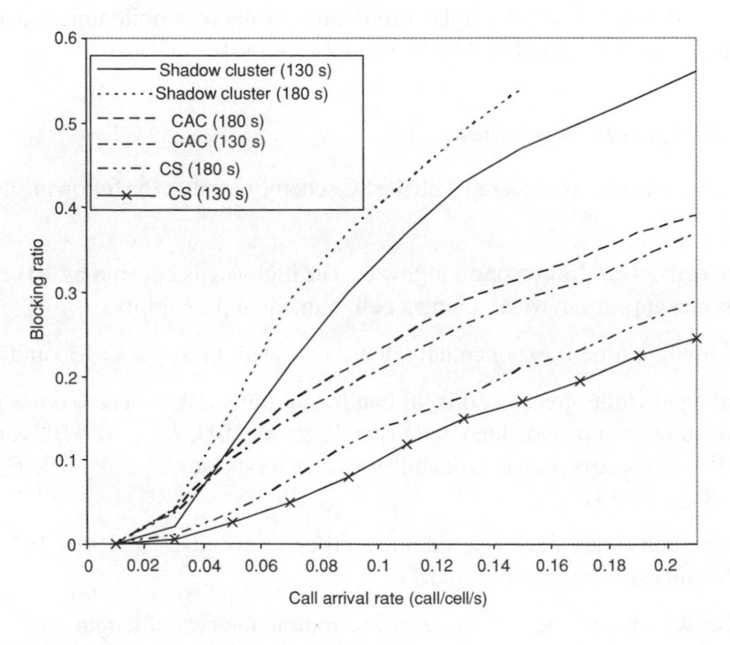

Figure 11.34 Blocking ratio in three systems.

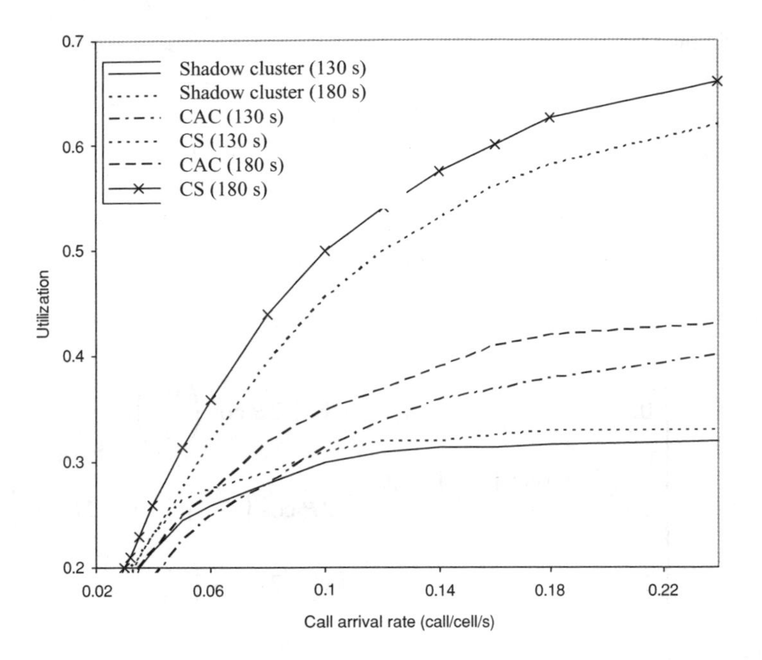

Figure 11.35 Utilization in three systems.

APPENDIX: DISTANCE CALCULATION IN AN INTERMEDIATE CELL

Given an intermediate cell on the path of the mobile unit, the PDF of the distance can be derived based on the angles β and ω, as shown in Figure 11.33. The entry point to the cell is assumed to be the point E, as shown in Figure 11.33. The mobile unit move in a direction evenly distributed leading to the next cell (Figure 11.33), where 2ψ is the range of the direction angle ϕ. The angle β is uniformly distributed between β_1 and β_2. Therefore, $f_\beta(\beta)$ is

$$f_\beta(\beta) = \begin{cases} \dfrac{1}{\beta_2 - \beta_1}, & \beta_1 \le \beta \le \beta_2 \\ 0, & \text{elsewhere} \end{cases} \tag{11.45}$$

Since ϕ is evenly distributed, we have

$$f_\phi(\phi \mid \beta) = \begin{cases} \dfrac{2}{\psi_2 - \psi_1}, & \dfrac{\psi_1}{2} \le \phi \le \dfrac{\psi_2}{2} \\ 0, & \text{elsewhere} \end{cases} \tag{11.46}$$

The CDF of the above function can be represented as

$$F_\phi(\phi \mid \beta) = \begin{cases} 0, & \phi < \dfrac{\psi_1}{2} \\ \dfrac{2\phi - \psi_1}{\psi_2 - \psi_1}, & \dfrac{\psi_1}{2} \le \phi \le \dfrac{\psi_2}{2} \\ 1, & \phi > \dfrac{\psi_2}{2} \end{cases} \tag{11.47}$$

where ψ is defined as

$$\psi_i = \begin{cases} \pi - (\beta - \omega_i), & \beta \geq \omega_2 \\ \pi - (\omega_j - \beta), & \beta < \omega_1 \quad \text{and} \quad i \neq j. \end{cases} \tag{11.48}$$

If Y is the distance traveled from E to X, as in Figure 11.33 then, Y becomes $Y = 2R\cos\phi$ and gives in Equation (11.47) the following four cases.

Case 1: $\psi_2/2 > \psi_1/2 \geq 0$

$$F_Y(y \mid \beta) = \begin{cases} 0, & y < 2R\cos\left(\frac{\psi_2}{2}\right) \\ 1 - \dfrac{2\operatorname{arco}\cos\left(\frac{y}{2R}\right) - \psi_1}{\psi_2 - \psi_1}, & 2R\cos\left(\frac{\psi_2}{2}\right) \leq y \leq 2R\cos\left(\frac{\psi_1}{2}\right) \\ 1, & y > 2R\cos\left(\frac{\psi_1}{2}\right). \end{cases} \tag{11.49}$$

The PDF of y is

$$f_Y(y \mid \beta) = \begin{cases} \dfrac{1}{(\psi_2 - \psi_1)\sqrt{R^2 - \left(\frac{y}{2}\right)^2}}, & 2R\cos\left(\frac{\psi_2}{2}\right) \leq y \leq 2R\cos\left(\frac{\psi_1}{2}\right) \\ 0, & \text{elsewhere} \end{cases} \tag{11.50}$$

The mean distance $E[Y \mid \beta]$ is

$$E[Y \mid \beta] = \left\{ \int_{2R\cos(\psi_2/2)}^{2R\cos(\psi_1/2)} y \cdot \frac{1}{(\psi_2 - \psi_1)\sqrt{R^2 - \left(\frac{y}{2}\right)^2}} \, \mathrm{d}y \right. \tag{11.51}$$

$$= \left\{ \frac{4R}{\psi_2 - \psi_1}\left[\sin\left(\frac{\psi_2}{2}\right) - \sin\left(\frac{\psi_1}{2}\right)\right] \right. \tag{11.52}$$

The mean distance $E[Y]$ for cell $id(m, i, j)$ for a mobile path entering cell i from cell m and exiting cell i to cell j is

$$\begin{aligned} d(m, i, j) = E[Y] &= \int_{\beta_1}^{\beta_2} E[Y \mid \beta] f_\beta(\beta) \, \mathrm{d}\beta \\ &= \left\{ \int_{\beta_1}^{\beta_2} \frac{1}{\beta_2 - \beta_1} \cdot \frac{4R}{\psi_2 - \psi_1}\left[\sin\left(\frac{\psi_2}{2}\right) - \sin\left(\frac{\psi_1}{2}\right)\right] \right. \end{aligned} \tag{11.53}$$

$$= \begin{cases} \dfrac{8R}{(\omega_2-\omega_1)(\beta_2-\beta_1)}\left[\sin\left(\dfrac{\beta_2-\omega_2}{2}\right)-\sin\left(\dfrac{\beta_1-\omega_1}{2}\right)\right. \\ \qquad \left. -\sin\left(\dfrac{\beta_2-\omega_1}{2}\right)+\sin\left(\dfrac{\beta_1-\omega_2}{2}\right)\right] & \beta_1 \geq \omega_2 \\ \dfrac{8R}{(\omega_2-\omega_1)(\beta_2-\beta_1)}\left[\sin\left(\dfrac{\omega_1-\beta_2}{2}\right)-\sin\left(\dfrac{\omega_1-\beta_1}{2}\right)\right. \\ \qquad \left. -\sin\left(\dfrac{\omega_2-\beta_1}{2}\right)+\sin\left(\dfrac{\omega_2-\beta_1}{2}\right)\right] & \beta_2 \leq \omega_1 \end{cases} \tag{11.54}$$

Case 2: $\psi_1/2 < \psi_2/2 \leq 0$

$$F_Y(y \mid \beta) = \begin{cases} 0, & y < 2R\cos\left(\dfrac{\psi_2}{2}\right) \\ 1 - \dfrac{2\arccos\left(\dfrac{y}{2R}\right)-\psi_1}{\psi_2-\psi_1}, & 2R\cos\left(\dfrac{\psi_1}{2}\right) \leq y \leq 2R\cos\left(\dfrac{\psi_1}{2}\right) \\ 1, & y > 2R\cos\left(\dfrac{\psi_2}{2}\right). \end{cases} \tag{11.55}$$

The PDF of y is

$$f_Y(y \mid \beta) = \begin{cases} \dfrac{-1}{(\psi_2-\psi_1)\sqrt{R^2-\left(\dfrac{y}{2}\right)^2}}, & 2R\cos\left(\dfrac{\psi_2}{2}\right) \leq y \leq 2R\cos\left(\dfrac{\psi_1}{2}\right) \\ 0, & \text{elsewhere} \end{cases} \tag{11.56}$$

The mean distance $E[Y \mid \beta]$ is

$$E[Y \mid \beta] = \left\{ \int_{2R\cos(\psi_1/2)}^{2R\cos(\psi_2/2)} y \cdot \frac{1}{(\psi_1-\psi_2)\sqrt{R^2-\left(\dfrac{y}{2}\right)^2}} \mathrm{d}y \right. \tag{11.57}$$

$$= \left\{ \frac{4R}{\psi_2-\psi_1}\left[\sin\left(\frac{\psi_2}{2}\right)-\sin\left(\frac{\psi_1}{2}\right)\right] \right. \tag{11.58}$$

The mean distance $E[Y]$ for cell $id(m, i, j)$ for a mobile path entering cell i from cell m and exiting cell i to cell j is

$$\begin{aligned} d(m,i,j) = E[Y] &= \int_{\beta_1}^{\beta_2} E[Y \mid \beta] f_\beta(\beta)\mathrm{d}\beta \\ &= \left\{ \int_{\beta_1}^{\beta_2} \frac{1}{\beta_2-\beta_1} \cdot \frac{4R}{\psi_2-\psi_1}\left[\sin\left(\frac{\psi_2}{2}\right)-\sin\left(\frac{\psi_1}{2}\right)\right] \right. \end{aligned} \tag{11.59}$$

$$
= \begin{cases} \dfrac{8R}{(\omega_2 - \omega_1)(\beta_2 - \beta_1)} \left[\sin\left(\dfrac{\beta_2 - \omega_2}{2}\right) - \sin\left(\dfrac{\beta_1 - \omega_1}{2}\right) \right. \\ \qquad \left. - \sin\left(\dfrac{\beta_2 - \omega_1}{2}\right) + \sin\left(\dfrac{\beta_1 - \omega_2}{2}\right) \right] & \beta_1 \geq \omega_2 \\ \dfrac{8R}{(\omega_2 - \omega_1)(\beta_2 - \beta_1)} \left[\sin\left(\dfrac{\omega_1 - \beta_2}{2}\right) - \sin\left(\dfrac{\omega_1 - \beta_1}{2}\right) \right. \\ \qquad \left. - \sin\left(\dfrac{\omega_2 - \beta_1}{2}\right) + \sin\left(\dfrac{\omega_2 - \beta_1}{2}\right) \right] & \beta_2 \leq \omega_1 \end{cases} \tag{11.60}
$$

Case 3: $|\psi_1/2| < \psi_2/2$

$$
f_Y(y \mid \beta) = f'_Y(y \mid \beta) + f''_Y(y \mid \beta) \tag{11.61}
$$

$$
F'_Y(y \mid \beta) = \begin{cases} 0, & y < 2R\cos\left(\dfrac{\psi_2}{2}\right) \\ 1 - \dfrac{2\arccos\left(\dfrac{y}{2R}\right) - \psi_1}{\psi_2 - \psi_1}, & 2R\cos\left(\dfrac{\psi_2}{2}\right) \leq y \leq 2R \\ 1, & y > 2R \end{cases} \tag{11.62}
$$

$$
F''_Y(y \mid \beta) = \begin{cases} 0, & y < 2R\cos\left(\dfrac{\psi_2}{2}\right) \\ 1 - \dfrac{2\arccos\left(\dfrac{y}{2R}\right) - \psi_1}{\psi_2 - \psi_1}, & 2R\cos\left(\dfrac{\psi_2}{2}\right) \leq y \leq 2R \\ 1, & y > 2R \end{cases} \tag{11.63}
$$

The PDF of y is

$$
f_Y(y \mid \beta) = f'_Y(y \mid \beta) + f''_Y(y \mid \beta) \tag{11.64}
$$

$$
f'_Y(y \mid \beta) = \begin{cases} \dfrac{1}{(\psi_2 - \psi_1)\sqrt{R^2 - \left(\dfrac{y}{2}\right)^2}}, & 2R\cos\left(\dfrac{\psi_2}{2}\right) \leq y \leq 2R \\ 0, & \text{elsewhere} \end{cases} \tag{11.65}
$$

$$
f''_Y(y \mid \beta) = \begin{cases} \dfrac{-1}{(\psi_2 - \psi_1)\sqrt{R^2 - \left(\dfrac{y}{2}\right)^2}}, & 2R\cos\left(\dfrac{\psi_1}{2}\right) \leq y \leq 2R \\ 0, & \text{elsewhere} \end{cases} \tag{11.66}
$$

The mean distance $E[Y \mid \beta]$ is

$$E[Y \mid \beta] = \begin{cases} \displaystyle\int_{2R\cos(\psi_2/2)}^{2R} y \cdot \frac{1}{(\psi_2 - \psi_1)\sqrt{R^2 - \left(\frac{y}{2}\right)^2}} \mathrm{d}y \\ \displaystyle + \int_{2R\cos(\psi_1/2)}^{2R} y \cdot \frac{-1}{(\psi_2 - \psi_1)\sqrt{R^2 - \left(\frac{y}{2}\right)^2}} \mathrm{d}y \end{cases} \tag{11.67}$$

$$= \left\{ \frac{4R}{\psi_2 - \psi_1} \left[\sin\left(\frac{\psi_2}{2}\right) - \sin\left(\frac{\psi_1}{2}\right) \right] \right. \tag{11.68}$$

The mean distance $E[Y]$ for cell $id(m, i, j)$ for a mobile path entering cell i from cell m and exiting cell i to cell j is

$$d(m, i, j) = E[Y] = \int_{\beta_1}^{\beta_2} E[Y \mid \beta] f_\beta(\beta) \mathrm{d}\beta$$
$$= \left\{ \int_{\beta_1}^{\beta_2} \frac{1}{\beta_2 - \beta_1} \cdot \frac{4R}{\psi_2 - \psi_1} \left[\sin\left(\frac{\psi_2}{2}\right) - \sin\left(\frac{\psi_1}{2}\right) \right] \right. \tag{11.69}$$

$$= \begin{cases} \dfrac{8R}{(\omega_2 - \omega_1)(\beta_2 - \beta_1)} \left[\sin\left(\dfrac{\beta_2 - \omega_2}{2}\right) - \sin\left(\dfrac{\beta_1 - \omega_1}{2}\right) \right. \\ \quad \left. - \sin\left(\dfrac{\beta_2 - \omega_1}{2}\right) + \sin\left(\dfrac{\beta_1 - \omega_2}{2}\right) \right] & \beta_1 \geq \omega_2 \\ \dfrac{8R}{(\omega_2 - \omega_1)(\beta_2 - \beta_1)} \left[\sin\left(\dfrac{\omega_1 - \beta_2}{2}\right) - \sin\left(\dfrac{\omega_1 - \beta_1}{2}\right) \right. \\ \quad \left. - \sin\left(\dfrac{\omega_2 - \beta_1}{2}\right) + \sin\left(\dfrac{\omega_2 - \beta_1}{2}\right) \right] & \beta_2 \leq \omega_1 \end{cases} \tag{11.70}$$

Case 4: $|\psi_1/2| < \psi_2/2$

$$F_Y(y \mid \beta) = F'_Y(y \mid \beta) + F''_Y(y \mid \beta) \tag{11.71}$$

$$F'_Y(y \mid \beta) = \begin{cases} 0, & y < 2R\cos\left(\dfrac{\psi_1}{2}\right) \\ 1 - \dfrac{2\arccos\left(\dfrac{y}{2R}\right) - \psi_1}{\psi_2 - \psi_1}, & 2R\cos\left(\dfrac{\psi_2}{2}\right) \leq y \leq 2R \\ 1, & y > 2R \end{cases} \tag{11.72}$$

$$F''_Y(y \mid \beta) = \begin{cases} 0, & y < 2R\cos\left(\frac{\psi_1}{2}\right) \\ 1 - \dfrac{2\mathrm{arco}\cos\left(\frac{y}{2R}\right) - \psi_1}{\psi_2 - \psi_1}, & 2R\cos\left(\frac{\psi_2}{2}\right) \le y \le 2R \\ 1, & y > 2R \end{cases} \tag{11.73}$$

The PDF of y is

$$f_Y(y \mid \beta) = f'_Y(y \mid \beta) + f''_Y(y \mid \beta) \tag{11.74}$$

$$f'_Y(y \mid \beta) = \begin{cases} \dfrac{-1}{(\psi_2 - \psi_1)\sqrt{R^2 - \left(\frac{y}{2}\right)^2}}, & 2R\cos\left(\frac{\psi_1}{2}\right) \le y \le 2R \\ 0, & \text{elsewhere} \end{cases} \tag{11.75}$$

$$f'_Y(y \mid \beta) = \begin{cases} \dfrac{1}{(\psi_2 - \psi_1)\sqrt{R^2 - \left(\frac{y}{2}\right)^2}}, & 2R\cos\left(\frac{\psi_2}{2}\right) \le y \le 2R \\ 0, & \text{elsewhere} \end{cases} \tag{11.76}$$

The mean distance $E\left[Y|\beta\right]$ is

$$E[Y \mid \beta] = \begin{cases} \displaystyle\int_{2R\cos(\psi_1/2)}^{2R} y \cdot \frac{-1}{(\psi_2 - \psi_1)\sqrt{R^2 - \left(\frac{y}{2}\right)^2}}\,\mathrm{d}y \\ \displaystyle + \int_{2R\cos(\psi_2/2)}^{2R} y \cdot \frac{1}{(\psi_2 - \psi_1)\sqrt{R^2 - \left(\frac{y}{2}\right)^2}}\,\mathrm{d}y \end{cases} \tag{11.77}$$

$$= \left\{ \frac{4R}{\psi_2 - \psi_1}\left[\sin\left(\frac{\psi_2}{2}\right) - \sin\left(\frac{\psi_1}{2}\right)\right] \right. \tag{11.78}$$

The mean distance $E[Y]$ for cell $id(m, i, j)$ for a mobile path entering cell i from cell m and exiting cell i to cell j is

$$\begin{aligned} d(m, i, j) = E[Y] &= \int_{\beta_1}^{\beta_2} E[Y \mid \beta] f_\beta(\beta)\,\mathrm{d}\beta \\ &= \left\{ \int_{\beta_1}^{\beta_2} \frac{1}{\beta_2 - \beta_1} \cdot \frac{4R}{\psi_2 - \psi_1}\left[\sin\left(\frac{\psi_2}{2}\right) - \sin\left(\frac{\psi_1}{2}\right)\right] \right. \end{aligned} \tag{11.79}$$

$$
= \begin{cases} \dfrac{8R}{(\omega_2 - \omega_1)(\beta_2 - \beta_1)} \left[\sin\left(\dfrac{\beta_2 - \omega_2}{2}\right) - \sin\left(\dfrac{\beta_1 - \omega_1}{2}\right) \right. \\ \qquad \left. - \sin\left(\dfrac{\beta_2 - \omega_1}{2}\right) + \sin\left(\dfrac{\beta_1 - \omega_2}{2}\right) \right] \quad \beta_1 \geq \omega_2 \\ \dfrac{8R}{(\omega_2 - \omega_1)(\beta_2 - \beta_1)} \left[\sin\left(\dfrac{\omega_1 - \beta_2}{2}\right) - \sin\left(\dfrac{\omega_1 - \beta_1}{2}\right) \right. \\ \qquad \left. - \sin\left(\dfrac{\omega_2 - \beta_1}{2}\right) + \sin\left(\dfrac{\omega_2 - \beta_1}{2}\right) \right] \quad \beta_2 \leq \omega_1 \end{cases} \tag{11.80}
$$

REFERENCES

[1] A. Acampora, Wireless ATM: A perspective on issues and prospects, *IEEE Person. Commun.*, vol. 3, 1996, pp. 8–17.

[2] A. Acampora, An architecture and methodology for mobile-executed handoff in cellular ATM networks, *IEEE J. Select. Areas Commun.*, vol. 12, 1994, pp. 1365–1375.

[3] A. Acharya, J. Li, F. Ansari and D. Raychaudhuri, Mobility support for IP over wireless ATM, *IEEE Commun. Mag.*, vol. 36, 1998, pp. 84–88.

[4] A. Acharya, J. Li, B. Rajagopalan and D. Raychaudhuri, Mo-bility management in wireless ATM networks, *IEEE Commun. Mag.*, vol. 35, 1997, pp. 100–109.

[5] I.F. Akyildiz, J. McNair, J.S.M. Ho, H. Uzunalioglu and W. Wang, Mobility management in current and future communication networks, *IEEE Network Mag.*, vol. 12, 1998, pp. 39–49.

[6] I.F. Akyildiz and J.S.M. Ho, On location management for personal communications networks, *IEEE Commun. Mag.*, vol. 34, 1996, pp. 138–145.

[7] I.F. Akyildiz, J.S.M. Ho and Y.B. Lin, Movement-based location update and selective paging for PCS networks, *IEEE/ACM Trans. Networking*, vol. 4, 1996, pp. 629–636.

[8] I.F. Akyildiz and J.S.M. Ho, Dynamic mobile user location update for wireless PCS networks, *ACM-Baltzer J. Wireless Networks*, vol. 1, no. 2, 1995, pp. 187–196.

[9] B. Akyol and D. Cox, Re-routing for handoff in a wireless ATM network, *IEEE Personal Commun.*, vol. 3, 1996, pp. 26–33.

[10] V. Anantharam, M.L. Honig, U. Madhow and V.K. Wei, Optimization of a database hierarchy for mobility tracking in a personal communications network, *Performance Eval.*, vol. 20, no. 1–3, 1994, pp. 287–300.

[11] E. Ayanoglu, K. Eng and M. Karol, Wireless ATM: Limits, challenges, and proposals, *IEEE Personal Commun.*, vol. 3, 1996, pp. 19–34.

[12] A. Bar-Noy, I. Kessler and M. Sidi, Topology-based tracking strategies for personal communication networks, *ACM-Baltzer J. Mobile Networks and Applications (MONET)*, vol. 1, no. 1, 1996, pp. 49–56.

[13] A. Bar-Noy, I. Kessler, and M. Sidi, Mobile users: to update or not to update? *ACM-Baltzer J. Wireless Networks*, vol. 1, no. 2, 1995, pp. 175–186.

[14] S. Dolev, D.K. Pradhan and J.L. Welch, Modified tree structure for location management in mobile environments, *Comput. Commun.*, vol. 19, no. 4, 1996, pp. 335–345.

[15] F. Dosiere, T. Zein, G. Maral and J.P. Boutes, A model for the handover traffic in low earth-orbiting (LEO) satellite networks for personal communications, *Int. J. Satellite Commun.*, vol. 11, 1993, pp. 145–149.

[16] N. Efthymiou, Y.F. Hu and R. Sheriff, Performance of inter-segment handover protocols in an integrated space/terrestrial-UMTS environment, *IEEE Trans. Veh. Technol.*, vol. 47, 1998, pp. 1179–1199.

[17] E. Guarene, P. Fasano and V. Vercellone, IP and ATM integration perspectives, *IEEE Commun. Mag.*, vol. 36, 1998, pp. 74–80.

[18] J.S.M. Ho and I.F. Akyildiz, Dynamic hierarchical data-base architecture for location management in PCS networks, *IEEE/ACM Trans. Networking*, vol. 5, no. 5,1997, pp. 646–661.

[19] J.S.M. Ho and I.F. Akyildiz Local anchor scheme for reducing signaling cost in personal communication networks, *IEEE/ACM Trans. Networking*, vol. 4, no. 5, 1996, pp. 709–726.

[20] J.S.M. Ho and I.F. Akyildiz, A mobile user location update and paging mechanism under delay constraints, *ACM-Baltzer J. Wireless Networks*, vol. 1, no. 4, 1995, pp. 413–425.

[21] D. Hong and S. Rappaport, Traffic model and performance analysis for cellular mobile radio telephone systems with prioritized and nonprioritized handoff procedures, *IEEE Trans. Veh. Technol.*, vol. 35, 1986, pp. 77–92.

[22] L.-R. Hu and S. Rappaport, Adaptive location management scheme for global personal communications, in *Proc. IEEE Communications*, vol. 144, no. 1, 1997, pp. 54–60.

[23] C.-L. I, G.P. Pollini and R.D. Gitlin, PCS mobility manage-ment using the reverse virtual call setup algorithm, *IEEE/ACM Trans. Networking*, vol. 5, 1997, pp. 13–24.

[24] R. Jain and Y.B. Lin, An auxiliary user location strategy employing forwarding pointers to reduce network impact of PCS, *ACM-Baltzer J. Wireless Networks*, vol. 1, no. 2, 1995, pp. 197–210.

[25] R. Jain, Y.B. Lin and S. Mohan, A caching strategy to reduce network impacts of PCS, *IEEE J. Select. Areas Commun.*, vol. 12, 1994, pp. 1434–1444.

[26] D. Johnson and D. Maltz, Protocols for adaptive wireless and mobile networking, *IEEE Personal Commun.*, vol. 3, 1996, pp. 34–42.

[27] S.J. Kim and C.Y. Lee, Modeling and analysis of the dynamic location registration and paging in microcellular systems, *IEEE Trans. Veh. Technol.*, vol. 45, 1996, pp. 82–89.

[28] P. Krishna, N. Vaidya, and D.K. Pradhan, Static and adaptive location management in mobile wireless networks, *Comput. Commun.*, vol. 19, no. 4, 1996, pp. 321–334.

[29] B. Li, S. Jiang and D. Tsang, Subscriber-assisted handoff support in multimedia PCS, *Mobile Comput. Commun. Rev.*, vol. 1, no. 3, 1997, pp. 29–36.

[30] Y.B. Lin Paging systems: network architectures and inter-faces, *IEEE Network*, vol. 11, 1997, pp. 56–61.

[31] Y.B. Lin, Reducing location update cost in a PCS network, *IEEE/ACM Trans. Networking*, vol. 5, 1997, pp. 25–33.

[32] Y.-B. Lin and I. Chlamtac, Heterogeneous personal communication services: Integration of PCS systems, *IEEE Commun. Mag.*, vol. 34, 1996, pp. 106–113.

[33] Y.B. Lin, F.C. Li, A. Noerpel and I.P. Kun, Performance modeling of multitier PCS system, *Int. J. Wireless Information Networks*, vol. 3, no. 2, 1996, pp. 67–78.

[34] Y.B. Lin and S.K. DeVries, PCS network signaling using SS7, *IEEE Commun. Mag.*, vol. 33, 1995, pp. 44–55.

[35] Y.B. Lin, Determining the user locations for personal communications services networks, *IEEE Trans. Veh. Technol.*, vol. 43, 1994, pp. 466–473.

[36] J. Markoulidakis, G. Lyberopoulos, D. Tsirkas and E. Sykas, Mobility modeling in third-generation mobile telecommunications systems, *IEEE Personal Commun.*, vol. 4, 1997, pp. 41–56.

[37] M. Marsan, C.-F. Chiasserini, R. Lo Cigno, M. Munafo and A. Fumagalli, Local and global handovers for mobility management in wireless ATM networks, *IEEE Personal Commun.*, vol. 4, 1997, pp. 16–24.

[38] A.R. Modarressi and R.A. Skoog, Signaling system 7: a tutorial, *IEEE Commun. Mag.*, vol. 28, 1990, pp. 19–35.

[39] S. Mohan and R. Jain, Two user location strategies for personal communications services, *IEEE Personal Commun.*, vol. 1, 1994, pp. 42–50.

[40] R. Pandya, D. Grillo, E. Lycksell, P. Mieybegue, H. Okinaka and M. Yabusaki, IMT-2000 standards: Network aspects, *IEEE Personal Commun.*, 1997, pp. 20–29.

[41] C.E. Perkins, *Mobile IP: Design Principles and Practices*, Addison-Wesley Wireless Communications Series. Reading, MA: Addison Wesley, 1998.

[42] C.E. Perkins, IP mobility support version 2, Internet Engineering Task Force, Internet draft, draft-ietf-mobileip-v2-00.text, November 1997.

[43] C. Perkins, Mobile IP, *IEEE Commun. Mag.*, vol. 35, 1997, pp. 84–99.

[44] C. Perkins, Mobile-IP local registration with hierarchical foreign agents, Internet Engineering Task Force, Internet draft; draft-perkins-mobileip-hierfa-00.txt, February 1996.

[45] B. Rajagopalan, An overview of ATM forum's wireless ATM standards activities, *ACM Mobile Comput. Commun. Rev.*, vol. 1, no. 3, 1997.

[46] B. Rajagopalan, Mobility management in integrated wireless-ATM networks, *ACM-Baltzer J. Mobile Networks Applicat.* (*MONET*), vol. 1, no. 3, 1996, pp. 273–285.

[47] E. del Re, A coordinated European effort for the definition of a satellite integrated environment for future mobile communications, *IEEE Commun. Mag.*, vol. 34, 1996, pp. 98–104.

[48] C. Rose, State-based paging/registration: A greedy technique, *IEEE Trans. Veh. Technol.*, vol. 48, 1999, pp. 166–173.

[49] C. Rose and R. Yates, Ensemble polling strategies for in-creased paging capacity in mobile communication networks, *ACM/Baltzer Wireless Networks J.*, vol. 3, no. 2, 1997, pp. 159–167.

[50] C. Rose and R. Yates, Location uncertainty in mobile networks: a theoretical framework, *IEEE Commun. Mag.*, vol. 35, 1997, pp. 94–101.

[51] C. Rose, Minimizing the averagecost of paging and reg-istration: a timer-based method, *ACM-Baltzer J. Wireless Networks*, vol. 2, no. 2, 1996, pp. 109–116.

[52] C. Rose and R. Yates, Minimizing the average cost of paging under delay constraints, *ACM-Baltzer J. Wireless Networks*, vol. 1, no. 2, 1995, pp. 211–219.

[53] S. Tabbane, Location management methods for 3rd generation mobile systems, *IEEE Commun. Mag.*, vol. 35, 1997, pp. 72–78.

[54] C.-K. Toh, A unifying methodology for handovers of het-erogeneous connections in wireless ATM networks, *ACM SIGCOMM Comput. Commun. Rev.*, vol. 27, no. 1, 1997, pp. 12–30.

[55] C.-K. Toh, A hybrid handover protocol for local area wireless ATM networks, *ACM-Baltzer J. Mobile Networks Applicat.* (*MONET*), vol. 1, no. 3, 1996, pp. 313–334.

[56] M. Veeraraghavan and G. Dommetry, Mobile location man-agement in ATM networks, *IEEE J. Select. Areas Commun.*, vol. 15, 1997, pp. 1437–1454.

[57] M. Veeraraghavan, M. Karol, and K. Eng, Mobility and connection management in a wireless ATM LAN, *IEEE J. Select. Areas Commun.*, vol. 15, 1997, pp. 50–68.

[58] J.Z. Wang, A fully distributed location registration strategy for universal personal communication systems, *IEEE J. Select. Areas Commun.*, vol. 11, 1993, pp. 850–860.

[59] M. Werner, C. Delucchi, H.-J. Vogel, G. Maral, and J.-J. De Ridder, ATM-based routing in LEO/MEO satellite networks with intersatellite links, *IEEE J. Select. Areas Commun.*, vol. 15, 1997, pp. 69–82.

[60] M. Werner, A. Jahn, E. Lutz, and A. Bottcher, Analysis of system parameters for LEO/ICO-satellite communication net-works, *IEEE J. Select. Areas Commun.*, vol. 13, 1995, pp. 371–381.

[61] R. Yates, C. Rose, B. Rajagopalan and B. Badrinath, Analysis of a mobile-assisted adaptive location management strategy, *ACM-Baltzer J. Mobile Networks Applicat. (MONET)*, vol. 1, no. 2, 1996, pp. 105–112.

[62] A. Yenerand C. Rose, Highly mobile users and paging: Optimal polling strategies, *IEEE Trans. Veh. Technol.*, vol. 47, 1998, pp. 1251–1257.

[63] D.R. Wilson, Signaling system no. 7, IS-41 and cellular telephony networking, *Proc. IEEE*, vol. 80, 1992, pp. 664–652.

[64] C. Perkins and D. Johnson, Route optimization in mobile IP, Internet Engineering Task Force, Internet draft; draft-ietf-mobileip-optom-07.txt, 20 November 1997.

[65] P. Calhoun and C. Perkins, Tunnel establishment protocol, Internet Engineering Task Force, Internet draft; draft-ietfmobileip- calhoun-tep-00.txt, 21 November 1997.

[66] G. Troxel and L. Sanchez, Rapid authentication for mobile IP, Internet Engineering Task Force, Internet draft; draft-ietf-mobileip-ra-00.txt, December 1997.

[67] R. Yuan, S.K. Biswas, L.J. French, J. Li, and D. Raychaudhuri, A signaling and control architecture for mobility support, *ACM-Baltzer J, Mobile Networks Applicat. (MONET)*, vol. 1, no. 3, 1996, pp. 287–298.

[68] M. Johnsson, Simple mobile IP, Internet Engineering Task Force, Internet-draft, Ericsson; draft-ietf-mobileip-simple-00.txt, March 1999.

[69] C.B. Becker, B. Patil and E. Qaddoura, IP mobility architecture framework, Internet Engineering Task Force, Internet draft; draft-ietf-mobileip-ipm-arch-00.txt, March 1999.

[70] J.M. Benedetto, Economy-class ion-defying IC's in orbit, *IEEE Spectrum*, vol. 35, 1998, pp. 36–41.

[71] E. Lutz, Issues in satellite personal communication systems, *ACM J. Wireless Networks*, vol. 4, no. 2, 1998, pp. 109–124.

[72] B. Miller, Satellite free mobile phone, *IEEE Spectrum*, vol. 35, 1998, pp. 26–35.

[73] F. Ananasso and F.D. Priscoli, Issues on the evolution toward satellite personal communication networks, in *Proc. GLOBECOM'95*, London, pp. 541–545.

[74] E. del Re, R. Fantacci and G. Giambene, Call blocking performance for dynamic channel allocation technique in future mobile satellite systems, *Proc. Inst. Elect. Eng., Commun.*, vol. 143, no. 5, 1996, pp. 289–296.

[75] D. Hong and S.S. Rappaport, Traffic model and performance analysis for cellular mobile radio telephone systems with prioritized and non-prioritized handoff procedures, *IEEE Trans. Vehic. Technol.,* vol. VT- 35, no. 3, 1986, pp. 77–92.

[76] CEAS Technical Report no. 773, 1 June 1999, College of Engineering and Applied Sciences, State University of New York, Stony Brook, NY, USA.

[77] A. Acampora and M. Naghshineh, Control and quality-of-service provisioning in high speed microcellular networks, *IEEE Person. Commun.*, vol. 1, 1994, pp. 36–42.

[78] D. Levine, I. Akyildiz and M. Naghshineh, Resource estimation and call admission algorithm for wireless multimedia using the shadow cluster concept, *IEEE/ACM Trans. Networking*, vol. 5, no. 1, 1997, pp. 1–12.

[79] V. Bharghavan and J. Mysore, Profile based next-cell prediction in in-door wireless LANs, in *Proc. IEEE Singapore Int. Conf. Networking*, April 1997, pp. 147–152.

[80] G. Liu and G.Q. Maguire Jr, Transmit activity and intermodal route planner, Technical Report, Royal Institute of Technology, Stockholm, February 1995.

[81] P. Bahl, T. Liu and I. Chlamtac, Mobility modeling, location tracking, and trajectory prediction in wireless ATM networks, *IEEE J. Select. Areas Commun.*, vol. 16, 1998, pp. 922–937.

[82] A. Aljadhai and T. Znati, A framework for call admission control and QoS support in wirelessnetworks, in *Proc. INFOCOM99*, vol. 3, New York, March 1999, pp. 1014–1026.

[83] A. Aljadhai and T. Znati, A predictive bandwidth allocation scheme for multimedia wireless networks, in *Proc. Conf. Communication Networks and Distributed Systems Modeling and Simulation,* Phoenix, AZ, January 1997, pp. 95–100.

[84] A.R. Aljadhai and T.F. Znati. Predictive mobility support for QoS provisioning in mobile wireless environments *IEEE J. Selected Areas Commun.*, vol. 19, no. 10, 2001, pp. 1915–1931.

[85] A. Talukdar, B.R. Badrinath, and A. Acharya, On accommodating mo-bile hosts in an integrated services packet network, in *Proc. IEEE IN-FOCOM*, vol. 3, Kobe, Japan, April 1997, pp. 1046–1053.

[86] S.E. Dreyfus, An appraisal of some shortest-path algorithms, *Opns. Res.*, vol. 17, 1969, pp. 395–412.

12

Adaptive Resource Management

12.1 CHANNEL ASSIGNMENT SCHEMES

A given radio spectrum (or bandwidth) can be divided into a set of disjoint or noninterfering radio channels. All such channels can be used simultaneously while maintaining an acceptable received radio signal. In order to divide a given radio spectrum into such channels, many techniques such as frequency division (FDMA/OFDMA), time division (TDMA/TH UWB), or code division (CDMA/MC CDMA) can be used, as discussed in Chapter 2. In FDMA, the spectrum is divided into disjoint frequency bands, whereas in TDMA the channel separation is achieved by dividing the usage of the channel into disjoint time periods called time slots. In CDMA, the channel separation is achieved by using different spreading codes. The major criteria in determining the number of channels with a certain quality that can be used for a given wireless spectrum is the level of received signal quality that can be achieved in each channel.

If $Si(k)$ is the set (i) of wireless terminals that communicate with each other using the same channel k, then due to the physical characteristics of the radio environment, the same channel k can be reused simultaneously by another set j if the members of sets i and j are spaced sufficiently apart. All such sets which use the same channel are referred to as co-channel sets or simply co-channels. The minimum distance at which co-channels can be reused with acceptable interference is called the 'co-channel reuse distance' D. For illustration see Figure 12.1.

This is possible because, due to propagation path loss in the radio environment, the average power received from a transmitter at distance d is proportional to $P_{\mathrm{T}}d^{-\alpha}$ where α is a number in the range 3–5 depending on the physical environment, and P_{T} is the average transmitter power. Thus, by adjusting the transmitter power level and/or the distance d

Advanced Wireless Networks: 4G Technologies Savo G. Glisic
© 2006 John Wiley & Sons, Ltd.

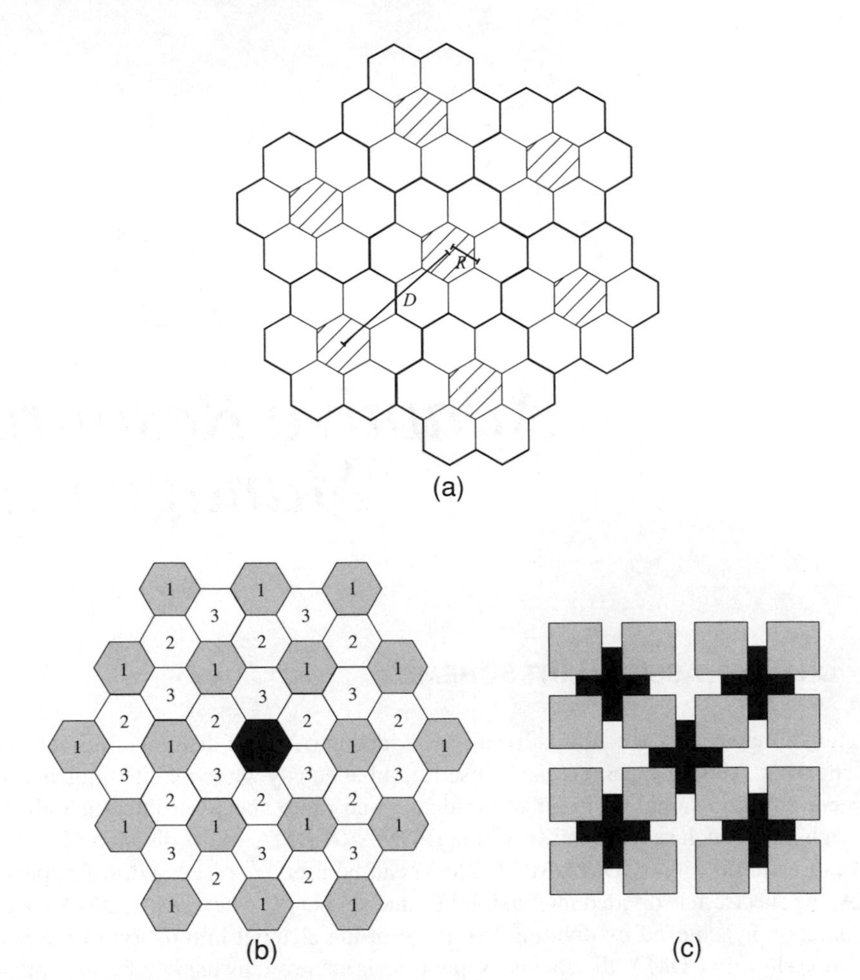

Figure 12.1 Examples of frequency reuse factor. (a) Uniform hexagonal cellular layout with reuse 7. (b) Macrocell layout with reuse 3. (c) Street microcell layout with reuse 2.

between co-channels, a channel can be reused by a number of co-channels if the CIR (carrier-to-interference ratio) in each co-channel is above the required value $CIR_{\min}$. In general, for a wireless station R at distance d from a transmitter T, using the same reference radio channel as the set of other transmitters T_i at distances d_i from R, we have

$$CIR = \frac{S}{I + N_0} = \frac{Pd^{-\alpha}}{\sum_i P_i d_i^{-\alpha} + N_0} \tag{12.1}$$

where N_0 represents the background noise. To achieve a certain level of CIR at the reference station R, different methods can be used.

In general we can represent the residual interference signal power as

$$I \rightarrow I_r(d, \theta, f, t) = (1 - C_f)(1 - C_p)(1 - C_\theta)(1 - C_t)I(d, \theta, f, t) \tag{12.1a}$$

where C_f, C_p, C_θ and C_t are frequency, propagation (distance + shadowing + fading), angle (space) and time isolation coefficients, respectively. $I(d, \theta, f, t)$ is the interference signal power without any suppression techniques [1]. For perfect isolation at least one of these coefficients is equal to one and the interference has no influence on the received signal. In practice, it is rather difficult and economically impractical to reach the point where $C_i = 1$. Instead, the product $(1 - C_r)\,(1 - C_\theta)\,(1 - C_t)$ depending on these coefficients should be kept as low as possible with an affordable effort measured by cost, power consumption and physical size of the hardware required for the solution.

Coefficient C_f is related to frequency assignment in the cellular network while coefficient C_p is related to the propagation conditions. $C_f = 1$ if the interfering signal frequency is different from the frequency of the useful signal. $C_p = 1$ if, due to propagation losses, the interfering signal cannot reach the site of the useful reference signal. In general, the same frequency can be used in two cells only if the propagation losses between the two cells are high enough so that the interfering signals are attenuated to the acceptable level. Coefficient C_θ is related to antenna beamforming and the possibilities of reducing the interference level by spatial filtering. Finally, interference cancellation and equalization in time domain can also be used to reduce the level of interference.

In this chapter we will focus on two methods for reducing the interference level. First, the distance between interfering stations using the co-channel and the reference station R can be increased to reducc the co-channel interference level. Many channel allocation schemes are based on this idea of physical separation. Another solution to reduce the CIR at R is to reduce the interfering powers transmitted from interfering stations and/or to increase the desired signal's power level P. This is the idea behind power control schemes. These two methods present the underlying concept for channel assignment algorithms in cellular systems. Each of these algorithms uses a different method to achieve a $CIR_{\min}$ at each mobile terminal by separating co-channels and/or by adjusting the transmitter power.

12.1.1 Different channel allocation schemes

Channel allocation schemes can be divided into a number of different categories depending on the comparison basis. For example, when channel assignment algorithms are compared based on the way in which co-channels are separated, they can be divided into fixed channel allocation (FCA), dynamic channel allocation (DCA), and hybrid channel allocation (HCA).

In *FCA schemes*, the area is partitioned into a number of cells, and a number of channels are assigned to each cell according to some reuse pattern, depending on the desired signal quality. In *DCA*, all channels are placed in a pool and are assigned to new calls as needed such that the criterion is satisfied. At the cost of higher complexity, DCA schemes provide flexibility and traffic adaptability. However, DCA strategies are less efficient than FCA under high load conditions. To overcome this drawback, HCA techniques were designed by combining FCA and DCA schemes. Channel assignment schemes can be implemented in many different ways. For example, a channel can be assigned to a radio cell based on the coverage area of the radio cell and its adjacent cells such that the $CIR_{\min}$ is maintained with high probability in all radio cells. Channels could be also assigned by taking the local CIR measurements of the mobile's and base station's receiver into account. That is, instead of allocating a channel blindly to a cell based on worst-case conditions (such as letting co-channels be located at the closest boundary), a channel can be allocated to a mobile based on its local CIR measurements [2, 3].

Channel assignment schemes can be implemented in centralized or distributed fashion. In the centralized schemes the channel is assigned by a central controller, whereas in distributed schemes a channel is selected either by the local base station of the cell from which the call is initiated or selected autonomously by the mobile. In a system with cell-based control, each base station keeps information about the current available channels in its vicinity. Here the channel availability information is updated by exchange of status information between base stations. In autonomously organized distributed schemes, the mobile chooses a channel based on its local CIR measurements without the involvement of a central call assignment entity. This scheme is simpler but less efficient. The channel assignment based on local assignment can be done for both FCA and DCA schemes.

12.1.2 Fixed channel allocation

In the FCA strategy a set of nominal channels is permanently allocated to each cell for its exclusive use. Here a definite relationship is assumed between each channel and each cell, in accordance with co-channel reuse constraints [4–8].

The total number of available channels in the system C is divided into sets, and the minimum number of channel sets N (reuse factor) required to serve the entire coverage area is related to the reuse distance s as follows [5]: $N = \sigma^2/3$, for hexagonal cells.

Here σ is defined as D/R, where R is the radius of the cell and D is the physical distance between the two cell centres [4]. N can assume only the integer values 3, 4, 7, $9, \ldots,$ as generally presented by the series, $(i + j)2 - ij$, with i and j being integers [4]. Figure 12.1(a) and 4(b) gives the allocation of channel sets to cells for $N = 3$ and $N = 7$, respectively.

In the simple FCA strategy, the same number of nominal channels is allocated to each cell. This uniform channel distribution is efficient if the traffic distribution of the system is also uniform. In that case, the overall average blocking probability of the mobile system is the same as the call blocking probability in a cell. Because traffic in cellular systems can be nonuniform with temporal and spatial fluctuations, a uniform allocation of channels to cells may result in high blocking in some cells, while others might have a sizeable number of spare channels. This could result in poor channel utilization. So, the number of channels in a cell can match the load in it by nonuniform channel allocation [9, 10] or static borrowing [11, 12].

In nonuniform channel allocation the number of nominal channels allocated to each cell depends on the expected traffic profile in that cell. Thus, heavily loaded cells are assigned more channels than lightly loaded ones. In Zhang and Yum [9] an algorithm, called *nonuniform compact pattern allocation*, is proposed for allocating channels to cells according to their traffic distributions. The technique attempts to allocate channels to cells in such a way that the average blocking probability in the entire system is minimized. A similar technique for nonuniform channel allocation is also employed in the algorithms proposed in Oh *et al.* [10].

Simulation results in Zhang and Yum [9] show that the blocking probability using nonuniform compact pattern allocation is always lower than the blocking probability of uniform channel allocation. Also, for the same blocking probability, the system can carry on average 10 % (maximum 22 %) more traffic with the use of the nonuniform pattern allocation [9].

In the static borrowing schemes proposed in References [11, 12], unused channels from lightly loaded cells are reassigned to heavily loaded ones at distances greater than the minimum reuse distance σ. Although in static borrowing schemes channels are permanently assigned to cells, the number of nominal channels assigned in each cell may be reassigned periodically according to spatial inequities in the load. This can be done in a scheduled or predictive manner, with changes in traffic known in advance or based on measurements, respectively.

12.1.3 Channel borrowing schemes

In a channel borrowing scheme, an acceptor cell that has used all its nominal channels can borrow free channels from its neighboring cells (donors) to accommodate new calls. A channel can be borrowed by a cell if the borrowed channel does not interfere with existing calls. When a channel is borrowed, several other cells are prohibited from using it. This is called channel locking. In contrast to static borrowing, channel borrowing strategies deal with short-term allocation of borrowed channels to cells: once a call is completed, the borrowed channel is returned to its nominal cell. The proposed channel borrowing schemes differ in the way a free channel is selected trom a donor cell to be borrowed by an acceptor cell.

The channel borrowing schemes can be divided into simple and hybrid. In simple channel borrowing schemes, any nominal channel in a cell can be borrowed by a neighboring cell for temporary use. In hybrid channel borrowing strategies, the set of channels assigned to each cell is divided into two subsets, A (standard or local channels) and B (nonstandard or channels that can be borrowed). Subset A is for use only in the nominally assigned cell, while subset B is allowed to be lent to neighboring cells.

The channel borrowing schemes can be categorized as:

Simple channel borrowing:

- simple borrowing (SB);
- borrow from the richest (SBR);
- basic algorithm (BA);
- basic algorithm with reassignment (BAR);
- borrow first available (BFA).

Hybrid channel borrowing:

- simple hybrid borrowing scheme (SHCB);
- borrowing with channel ordering (BCO);
- borrowing with directional channel locking (BDCL);
- sharing with bias (SHB);
- channel assignment with borrowing and reassignment (CABR);
- ordered dynamic channel assignment with rearrangement (ODCA).

In the next two subsections we discuss the simple and hybrid borrowing schemes in detail.

12.1.3.1 Simple channel borrowing schemes

In the simple borrowing (SB) strategy [11–14], a nominal channel set is assigned to a cell, as in the FCA case. After all nominal channels are used, an available channel from a neighboring cell is borrowed. To be available for borrowing, the channel most not interfere with existing calls. Although channel borrowing can reduce call blocking, it can cause interference in the donor cells from which the channel is borrowed and prevent future calls in these cells from being completed [15].

As shown in Singh *et al.* [14], the SB strategy gives lower blocking probability than static FCA under light and moderate traffic, but static FCA performs better in heavy traffic conditions. This is due to the fact that, in light and moderate traffic conditions, borrowing of channels provides a means to serve the fluctuations of offered traffic, and as long as the traffic intensity is low, the number of donor cells is small. In heavy traffic, the channel borrowing may proliferate to such an extent, due to channel locking, that the channel usage efficiency drops drastically, causing an increase in blocking probability and a decrease in channel utilization [16].

Because the set of channels that can be borrowed in a cell may contain more than one candidate channel, the way a channel is selected from the set plays an important role in the performance of a channel borrowing scheme. The objective of all the schemes is to reduce the number of locked channels caused by channel borrowing. The difference between them is the specific algorithm used for selecting one of the candidate channels for borrowing. Along these lines, several variations of the SB strategy have been proposed where channels are borrowed from nonadjacent cells [9, 11–14]. In the following, we discuss briefly each of the proposed schemes.

12.1.3.2 Borrow from the richest (SBR)

In this scheme, channels that are candidates for borrowing are available channels nominally assigned to one of the adjacent cells of the acceptor cell [11]. If more than one adjacent cell has channels available for borrowing, a channel is borrowed from the cell with the greatest number of channels available for borrowing. As discussed earlier, channel borrowing can cause channel locking. The SBR scheme does not take channel locking into account when choosing a candidate channel for borrowing.

The *basic algorithm* (BA) is an improved version of the SBR strategy which takes channel locking into account when selecting a candidate channel for borrowing [11, 12]. This scheme tries to minimize the future call blocking probability in the cell that is most affected by the channel borrowing. As in the SBR, channels that are candidates for borrowing are available channels nominally assigned to one of the adjacent cells of the acceptor cell. The algorithm chooses the candidate channel that maximizes the number of available nominal channels in the worst-case nominal cell in distance σ to the acceptor cell.

12.1.3.3 Basic algorithm with reassignment (BAR)

This scheme [12] provides for the transfer of a call from a borrowed channel to a nominal channel whenever a nominal channel becomes available. The choice of the particular borrowed channel to be freed is again made in a manner that minimizes the maximum probability of future call blocking in the cell most affected by the borrowing, as in the BA scheme.

12.1.3.4 Borrow first available (BFA)

Instead of trying to optimize when borrowing, this algorithm selects the first candidate channel it finds [11]. Here, the philosophy of the nominal channel assignment is also different. Instead of assigning channels directly to cells, the channels are divided into sets, and then each set is assigned to cells at reuse distance σ. These sets are numbered in sequence. When setting up a call, channel sets are searched in a prescribed sequence to find a candidate channel.

A general conclusion reached by most studies on the performance comparison of the previous schemes is that adopting a simple test for borrowing (e.g. borrowing the first available channel that satisfies the σ constraint) yields performance results quite comparable to systems which perform an exhaustive and complex search method to find a candidate channel [9, 11–13]. SBR, BA and BFA were evaluated by simulation in Anderson [11] using a two-dimensional hexagonal cell layout with 360 service channels. The offered load was adjusted for an average blocking of 0.02. The results show that all three schemes exhibit nearly the same average blocking probability versus load. The BFA has an advantage over the other two in that its computing effort and complexity are significantly less. Here the complexity of each algorithm is determined based un the average number of channel tests per call while searching for a candidate channel to borrow. In Anderson [11], simulation results showed a large variation in the complexity of these algorithms depending on network load. For example, for a 20 % increase in the traffic, SBR requires 50 %, and the BA 100 %, more channel tests compared with BFA.

12.1.4 Hybrid channel borrowing schemes

12.1.4.1 Simple hybrid channel borrowing strategy (SHCB)

In the SHCB strategy [4, 9, 13] the set of channels assigned to each cell is divided into two subsets, A (standard) and B (borrowable) channels. Subset A is nominally assigned in each cell, while subset B is allowed to be lent to neighboring cells. The ratio A:B is determined *a priori*, depending on an estimation of the traffic conditions, and can be adapted dynamically in a scheduled or predictive manner [13].

12.1.4.2 Borrowing with channel ordering (BCO)

The BCO [9, 13,14] outperforms SHCB by dynamically varying the A:B channel ratio according to changing traffic conditions [13, 14]. In the BCO strategy, all nominal channels are ordered such that the first channel has the highest priority for being assigned to the next local call, and the last channel is given the highest priority for being borrowed by the neighboring cells. A variation of the BCO strategy, called BCO with reassignment, allows intercellular handoff, that is, immediate reallocation of a released high-rank channel to a call existing in a lower-rank channel in order to minimize the channel locking effect.

12.1.4.3 Borrowing with directional channel locking (BDCL)

In this strategy, a channel is suitable for borrowing only if it is simultaneously free in three nearby co-channel cells. This requirement is too stringent and decreases the number of channels available for borrowing. In the BDCL strategy, the channel locking in the co-channel cells is restricted to those directions affected by the borrowing. Thus, the number of channels available for borrowing is greater than that in the BCO strategy. To determine in which case a 'locked' channel can he borrowed, 'lock directions' are specified for each locked channel. The scheme also incorporates reallocation of calls from borrowed to nominal channels and between borrowed channels in order to minimize the channel borrowing of future calls, especially the multiple-channel borrowing observed during heavy traffic.

It was shown in Zhang and Yum [9] by simulation that BDCL gives the lowest blocking probability, followed by HCO and FCA, for both uniform and nonuniform traffic. The reduction of the blocking probability for BDCL and BCO over FCA for the system in Zhang and Yum [9] is almost uniformly 0.04 and 0.03, respectively, for the range of traffic load tested.

The nonuniform pattern allocation FCA scheme, discussed in the previous section, can also be applied in the case of the hybrid channel borrowing strategies. With the use of nonuniform pattern allocation the relative performance of the BDCL, BCO and uniform FCA schemes remains the same as before, but the traffic-carrying capacity of a system can be increased by about 10 %. This advantage is in addition to those gained from the channel borrowing strategies [9].

12.1.4.4 Sharing with Bias (SHB)

In Yum and Wong [17] a scheme of channel borrowing with coordinated sectoring is described. The SHB strategy is similar to the join-biased queue rule, which is a simple but effective way to balance the load of servers in the presence of unbalanced traffic. Each cell in the system is divided in three sectors, X, Y, Z, as shown in Figure 12.2. Only calls initiated in one of these sectors can borrow channels from the two adjacent cells neighboring it (donor cells). In addition, the nominal channels in donor cells are divided into two subsets, A and B, as in the SHCB ease. Channels from set A can only be used inside the donor cell, while channels in set B can be loaned to an acceptor cell.

For the example shown in Figure 12.2 a call initiated in sector X of cell number 3 can only borrow a channel from set A of the cells numbered 1 and 2.

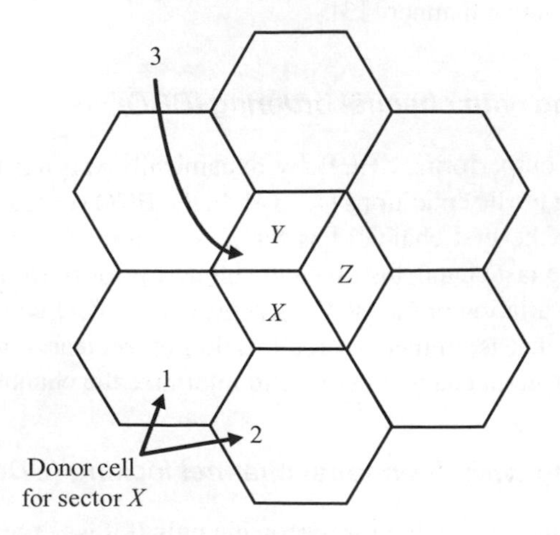

Figure 12.2 Sharing with bias.

12.1.4.5 Channel assignment with borrowing and reassignment (CARB)

The scheme was proposed in Engel and Peritsky [12] and is statistically optimum in a certain min–max sense. Here channels are borrowed on the basis of causing the least harm to neighboring cells in terms of future call blocking probability. Likewise, reassignment of borrowed channels is done in such a way as to cause maximum relief to neighboring cells.

12.1.4.6 Ordered channel assignment scheme with rearrangement (ODCA)

This scheme, proposed in Kuek [18], combines the merits of CARB and BCO with improvements to yield higher performance. In ODCA, when a call requests service, the base station of the cell checks to see if there are any nominal channels available. If there are channels available, the user will be assigned one on an ordered basis, as in BCO. Here all channels are numbered in predetermined order according to the same criterion as in the CARB scheme, and the lowest-numbered available idle channel is always selected. If all nominal channels are busy, the cell may borrow a nonstandard channel from a neighboring cell. Once a nonstandard channel is assigned, the availability lists of all affected cells where the assigned channel can cause interference are updated. Whenever a channel is no longer required, the availability lists of the affected cells are updated accordingly. Whenever a standard channel is available, the channel reassignment procedure is initiated to ensure efficient utilization. If there is a nonstandard channel in use in the cell, the call served by that channel is switched to the newly freed standard channel; the necessary availability lists are also updated. If no nonstandard channels are used in the cell, a call served by a standard channel with lower priority than the newly freed one is switched to the newly freed channel [18].

The performance of ODCA was studied in Kuek [18] for a highway microcellular environment with nonuniform teletraffic load. Performance comparison with the FCA and CARB shows significant improvement. The ODCA scheme exhibits better channel utilization compared with the CARE and FCA; the OUCA scheme also performs better than CARE and FCA at blocking probabilities below 0.1. For example, at a blocking probability of 0.05, ODCA is capable of supporting 4 % more traffic than CARE and 35 % more traffic than FCA [181]. However, the ODCA scheme incurs a higher computational overhead in assigning and reassigning channels, and more frequent switching of channels due to the reassignment propagation effect.

12.1.5 Dynamic channel allocation

Owing to short-term temporal and spatial variations of traffic in cellular systems, FCA schemes are not able to attain high channel efficiency. To overcome this, DCA schemes have been studied during the past 20 years. In contrast to FCA, there is no fixed relationship between channels and cells in DCA. All channels are kept in a central pool and are assigned dynamically to radio cells as new calls arrive in the system [19]. After a call is completed, its channel is returned to the central pool.

In DCA, a channel is eligible for use in any cell provided that signal interference constraints are satisfied. Because, in general, more than one channel might be available in the central pool to be assigned to a cell that requires a channel, some strategy must be applied to select the assigned channel [8]. The main idea of all DCA schemes is to evaluate the cost of using each candidate channel, and select the one with the minimum cost provided that certain interference constraints are satisfied. The selection of the cost function is what differentiates DCA schemes [8].

The selected cost function might depend on the future blocking probability in the vicinity of the cell, the usage frequency of the candidate channel, the reuse distance, channel occupancy distribution under current traffic conditions, radio channel measurements of individual mobile users or the average blocking probability of the system [16].

Based on information used for channel assignment, DCA strategies could be classified either as call-by-call DCA or adaptive DCA schemes [20, 23, 27]. In the call-by-call DCA, the channel assignment is based only on current channel usage conditions in the service area, while in adaptive DCA the channel assignment is adaptively carried out using information on the previous as well as present channel usage conditions [20]. DCA schemes can be also divided into centralized and distributed schemes with respect to the type of control they employ. In general the schemes can be classified into the following groups:

(1) *Centralized DCA*
- First available (FA)
- Locally optimized dynamic assignment (LODA)
- Selection with maximum usage on the reuse ring (RING)
- Mean square (MSQ)
- Nearest neighbor (NN)
- Nearest neighbor + 1 (NN + 1)
- 1-clique

(2) *Distributed DCA*
- Locally packing distributed DCA (LP-DDCA)
- LP-DDCA with ACI constraint
- Moving direction (MD)

(3) *CIR measurement DCA schemes*
- Sequential channel search (SCS)
- MSIR
- Dynamic channel selection (DCS)
- Channel segregation

(4) *One-dimensional Systems*
- MINMAX
- Minimum interference (MI)
- Random minimum interference (RMI)
- Random minimum interference with reassignment (RMIR)
- Sequential minimum interference SMI

12.1.6 Centralized DCA schemes

In centralized DCA schemes, a channel from the central pool is assigned to a call for temporary use by a centralized controller. The difference between these schemes is the specific cost function used for selecting one of the candidate channels for assignment. *First available (FA)* is the simplest of the DCA schemes strategy. In FA the first available channel within the reuse distance found during a channel search is assigned to the call. The FA strategy minimizes the system computational time; and, as shown by simulation in Cox and Reudink [8] for a linear cellular mobile system, it provides an increase of 20 % in the total handled traffic compared with FCA for low and moderate traffic loads.

Locally optimized dynamic assignment uses cost function based on the future blocking probability in the vicinity of the cell in which a call is initiated [9, 13].

12.1.6.1 Channel reuse optimization schemes

The objective of any mobile system is to maximize the efficiency of the system. Maximum efficiency is equivalent to maximum utilization of every channel in the system. It is obvious that the shorter the channel reuse distance, the greater the channel reuse over the whole service area. The cost functions selected in the following schemes attempt to maximize the efficiency of the system by optimizing the reuse of a channel in the system area.

Selection with maximum usage on the reuse ring (RING) selects a candidate channel which is in use in the most cells in the co-channel set. If more than one channel has this maximum usage, an arbitrary selection among such channels is made to serve the call. If none is available, the selection is made based on the FA scheme.

12.1.6.2 Mean square, nearest neighbor, nearest neighbor plus one

The MSQ scheme selects the available channel that minimizes the mean square of the distance among the cells using the same channel. The NN strategy selects the available channel occupied in the nearest cell in distance $\geq \sigma$, while the NN + 1 scheme selects an eligible channel occupied in the nearest cell within distance $\geq \sigma + 1$ or distance σ if an available channel is not found in distance $\sigma + 1$ [8].

Computer simulations of FCA, MSQ, NN and NN + 1 strategies show that, under light traffic conditions, NN exhibits the lowest blocking rate, followed by MSO, FA and NN + 1 [20]. When applied to a microcellular system the NN + 1 strategy leads to lower forced call termination and channel changing because the mobile unit is more likely to keep the same channel when it moves to an adjacent cell [21]. In addition, simulation results of FA, RING and NN [8, 22] show that, for both one- and two-dimensional mobile systems, all of the above schemes operate at very low blocking rates until the offered traffic reaches some threshold. A small increase in the offered traffic above this threshold produces a considerable increase in the blocking probability of new calls and results in very little increase in the traffic carried by the system. The load at which blocking begins to occur in one-dimensional systems [22] is somewhat greater than that in two-dimensional systems [8]. Finally, the simulation results in Cox and Reudink [22] show that strategies like RING and NN, which use a channel reuse optimization approach, are able to carry 5 % more traffic at a given blocking rate of 3 % compared with a channel assignment strategy like FA, which does not employ any channel reuse optimization.

12.1.6.3 1-Clique

All four previous schemes employ local channel reuse optimization schemes. A global channel reuse optimization approach is used in the *1-clique* strategy. The 1-clique scheme uses a set of graphs, one for each channel, expressing the non-co-channel interference structure over the whole service area for that channel. In each graph a vertex represents a cell, and cells without co-channel interference are connected with edges. So, each graph reflects the results of a possible channel assignment. A channel is assigned from the several possibilities such that as many vertices as possible still remain available after the assignment. This scheme shows a low probability of blocking, but when there are a lot of cells the required computational time makes quick channel selection difficult [19].

12.1.6.4 Schemes with channel rearrangement

Compared with FCA schemes, DCA schemes do not carry as much traffic at high blocking rates because they are not able to maximize channel reuse as they serve the randomly offered call attempts. In order to improve the performance of DCA schemes at high traffic density, channel reassignment techniques have been suggested [5, 8]. The basic goal of channel reassignment is to switch calls already in process, whenever possible, from the channels these calls are using to other channels, with the objective of keeping the distance between cells using the same channel simultaneously to a minimum. Thus, channel reuse is more concentrated, and more traffic can be carried per channel at a given blocking rate.

12.1.6.5 Distributed DCA schemes

Microcellular systems have shown great potential for capacity improvement in high-density personal communication networks. However, propagation characteristics will be less predictable and network control requirements more complex than in the other systems. Centralized DCA schemes can produce near-optimum channel allocation, but at the expense of a high centralization overhead. Distributed schemes are therefore more attractive for implementation in microcellular systems, owing to the simplicity of the assignment algorithm in each base station.

The proposed distributed DCA schemes use either local information about the current available channels in the cell's vicinity (cell-based) [24] or signal strength measurements [25]. In cell-based schemes a channel is allocated to a call by the base station at which the call is initiated. The difference with the centralized approach is that each base station keeps information about the current available channels in its vicinity. The channel pattern information is updated by exchanging status information between base stations. The cell-based scheme provides near-optimum channel allocation at the expense of excessive exchange of status information between base stations, especially under heavy traffic loads.

Particularly appealing are the DCA interference adaptation schemes that rely on signal strength measurements where a base station uses only local information, without the need to communicate with any other base station in the network. Thus, the system is self-organizing, and channels can be placed or added everywhere, as needed, to increase capacity or to improve radio coverage in a distributed fashion. These schemes allow fast real-time processing and maximal channel packing at the expense of increased co-channel interference probability with respect to ongoing calls in adjacent cells, which may lead to undesirable effects such as interruption, deadlock, and instability.

Table 12.1 ACO matrix at base station i

Base station number	Channel number								Number of assignable channels
	1	2	3	4	5	6	...	M	
i		×					...		0
i_1	×			×			...		0
i_2		×					...		2
⋮	⋮	⋮	⋮	⋮	⋮	⋮	⋮	⋮	⋮
i_{k_i}			×		×				4

12.1.7 Cell-based distributed DCA schemes

Local packing dynamic distributed channel assignment (LP DDCA) is a scheme where each base station assigns channels to calls using the augmented channel occupancy (ACO) matrix, which contains necessary and sufficient local information for the base station to make a channel assignment decision. For M available channels in the system and k neighboring cells to cell i within the co-channel interference distance, the ACO matrix is shown in Table 12.1. It has $M + 1$ columns and $k_i + 1$ rows. The first M columns correspond to the M channels. The first row indicates the channel occupancy in cell i and the remaining k_i rows indicate the channel occupancy pattern in the neighborhood of i, as obtained from neighboring base stations. The last column of the matrix corresponds to the number of current available channels for each of the $k_i + 1$ co-channel cells. So, an empty column indicates an idle channel which can be assigned to cell i. When a call requests service from cell i, its base station will use the ACO matrix to assign the first channel with an empty column. The content of the ACO table is updated by collecting channel occupancy information from interfering cells. Whenever a change of channel occupancy happens in one cell, the base station of the cell informs the base stations of all the interfering cells about the change in order to update the data in the local ACO matrices.

In addition to constraining co-channel interference, the design of a wireless cellular systems must also include measures to limit adjacent channel interference (ACT). In Cox and Reudink [8] a modified version of the LP-DDCA scheme was proposed that incorporates the ACI constraint.

The *Moving direction* (MD) strategy was proposed in Okada and Kubota [24] for one-dimensional microcellular systems. In these systems, forced call termination and channel changing occur frequently because of their small cell size [24]. The MD strategy uses information on moving directions of the mobile units to decrease both the forced call termination blocking probability and the channel changing. An available channel is selected among those assigned to mobile units that are elsewhere in the service area and moving in the same direction as the mobile in question. The search for such a channel starts from the nearest noninterfering cell to the one where the new call was initiated, and stops at the cell that is α reuse distances away, where α is a parameter.

A channel assignment example is given in Figure 12.3 where b, c, d and e are the available channels, and DR is the minimum reuse distance. For this example the parameter α is set to

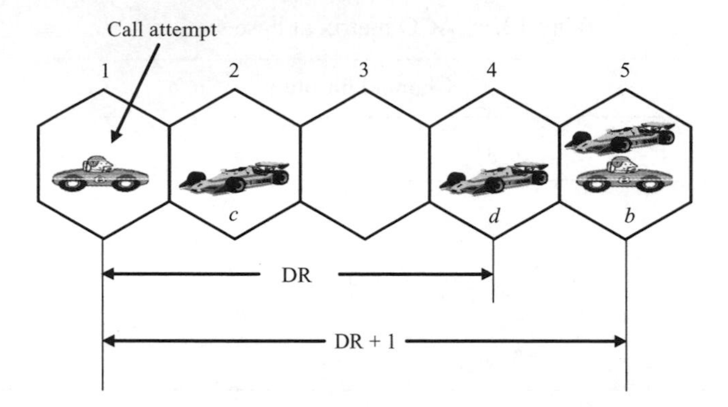

Figure 12.3 Moving direction strategy illustration.

one. The new call attempt is assigned channel *b* because the mobile requesting the channel is moving in the same direction as the mobile in cell number 5.

The sets of mobiles moving in the same direction and assigned the same channel are thus formed. Thus, when a mobile of a set crosses a cell boundary it is likely that a same set of mobiles has already crossed out of its cell to the next cell. In this manner a mobile can use the same channel after handoff with higher probability. This lowers the probability of both changing channels and forced call termination. The strategy is efficient in systems where mobiles move at nearly the same speed through the cells laid along a road or a highway and for one-dimensional microcellular systems. The simulation results in Okada and Kubota [24] for a one-dimensional system show that the MD strategy provides lower probability of forced call termination compared with the NN, NN + 1 and FCA strategies. Although the MD scheme has attractive features, it is not obvious how it could be expanded to a two-dimensional system.

12.1.8 Signal strength measurement-based distributed DCA schemes

A large body of research has been published on the performance analysis of traditional channel allocation schemes, both FCA and DCA [2, 4, 26], in which knowledge of the mobile's locations is not taken into account. In all of these schemes, channels are allocated to cells based on the assumption that the mobile may be located anywhere within the boundary of the cell. Thus, the packing of channels is not maximal. These schemes suffer from the fact that the selected fixed reusability distance might be too pessimistic. In the interference adaptation schemes, mobiles measure the amount of co-channel interference to determine the reusability of the channel. If a mechanism is assumed to exist by which mobiles and base stations can measure the amount of interference, then maximal channel packing could be achieved. An example of a system based on this principle is the Digital European Cordless Telecommunications (DECT) standard. However, local decisions can lead to suboptimal allocation. In interference adaptation DCA schemes, mobiles and base stations estimate CIR and allocate a channel to a call when predicted CIRs are above a threshold. It is possible that this allocation will cause the CIR of established calls to deteriorate, in which case a service interrupt occurs. If the interrupted call cannot find an acceptable new channel immediately, the result is a premature service termination, referred to as deadlock. Even if the interrupted call finds an acceptable channel, setting up a link using the new channel can cause interruption of another established link. These successive interruptions are referred to as instability. If no channel is available for the initial call request, the call is blocked.

12.1.8.1 Sequential channel search

The simplest scheme among the interference adaptation DCA schemes is the SCS strategy; where all mobile–base station pairs examine channels in the same order and choose the first available with acceptable CIR. It is expected that SCS will support a volume of traffic by suboptimal channel packing at the expense of causing many interruptions.

Dynamic channel selection (DCS) is a fully distributed algorithm for flexible mobile cellular radio resource sharing based on the assumption that mobiles are able to measure the amount of interference they experience in each channel. In DCS, each mobile station estimates the interference probability and selects the base station which minimizes its value. The interference probability is a function of a number of parameters, such as the received signal power from base stations, the availability of channels, and co-channel interference. In order to evaluate the interference probability, specific models for each of the above parameters should be developed. In Hong and Rappaport [28], models are developed to calculate probabilities of channel availability, desired carrier power and the CIR for constant traffic load.

The *Channel segregation* strategy was proposed in Akaiwa and Andoh [25] as a self-organized dynamic channel assignment scheme. By scanning all channels, each cell selects a vacant channel with an acceptable co-channel interference level. The scanning order is formed independently for each cell in accordance with the probability of channel selectability, $P(i)$, which is renewed by learning. For every channel i in the system, each cell keeps the current value of $P(i)$. When a call request arrives at the base station, the base station channel with the highest value of $P(i)$ under observation is selected. Subsequently, the received power level of the selected channel is measured in order to determine whether the channel is used or not. If the measured power level is below (or above) a threshold value, the channel is considered to be idle (or busy). If the channel is idle, the base station starts communication using the channel, and its priority is increased. It the channel is busy, the priority of the channel is decreased and the next-highest-priority channel tried. If all channels are busy, the call is blocked. The value of $P(i)$ and the update mechanism determine the performance of the algorithm. In Furuya and Akaiwa [29], $P(i)$ is updated to show the successful transmission probability on channel i as follows:

$$\begin{aligned} P(i) &= [P(i)N(i) + 1]/[N(i) + 1] \text{ and} \\ N(i) &= N(i) + 1 \text{ if the channel is idle} \\ P(i) &= [P(i)N(i)]/[N(i) + 1] \text{ and} \\ N(i) &= N(i) + 1 \text{ if the channel is busy} \end{aligned}$$

Here $N(i)$ is the number of times channel i is accessed. In Akaiwa and Andoh [25] the update mechanism for $P(i)$ is defined as $F(i) = N\mathrm{s}(i)/N(i)$ where $N\mathrm{s}(i)$ is the number of successful uses of channel i. Simulation results show that interference due to carrier sense error is reduced by 1:10 to 1:100 with channel segregation [29]. Also, the blocking probability is greatly reduced compared with FCA and DCA schemes.

12.1.9 One-dimensional cellular systems

All the DDCA schemes described in this section are applicable for one-dimensional cellular mobile systems. One-dimensional structures can be identified in cases such as streets with tall buildings shielding interference on either side, see Figure 12.1(c).

The *minimum interference* (MI) scheme is well known and among the simplest for one-dimensional cellular systems. It is incorporated in the enhanced cordless telephone (CT-2) and DECT systems. In an MI scheme, a mobile signals its need for a channel to its nearest base station. The base station then measures the interfering signal power on all channels not already assigned to other mobiles. The mobile is assigned the channel with the minimum interference. The order in which mobiles are assigned channels affects the efficiency of channel reuse. Taking into consideration the order of service, we discuss three variations of the MI scheme:

- *Random minimum interference* (RMI) is the scheme where the mobiles are served according to the MI scheme in a random order or, equivalently, in the order in which calls arrive in the system.
- *Random minimum interference with reassignment* (RMIR) is a modification where mobiles are first served according to the RMI scheme. Each mobile is then reassigned a channel by its base station according to the MI scheme. Those mobiles denied service by the initial RMI scheme also try to obtain a channel again. The order in which mobiles are reassigned is random. The number of times this procedure is carried out is the number of reassignments, R.
- *Sequential minimum interference* (SAI) is a modification where mobiles are assigned channels according to the MI scheme in a sequential order. The sequence followed is such that any mobile is served only after all the mobiles that are ahead of it have had a chance to be served. This procedure would require some coordination between base stations because of the sequential order of service.

In general, there is a trade-off between quality of service, the implementation complexity of the channel allocation algorithms and spectrum utilization efficiency. Simulation [4, 7, 8] results show that, under low traffic intensity, DCA strategies performs better. However, FCA schemes become superior at high offered traffic, especially in the case of uniform traffic. In the ease of nonuniform traffic and light to moderate loads, it is believed that the DCA scheme will perform better due to the fact that, under low traffic intensity, DCA uses channels more efficiently than FCA.

In DCA, the assignment control is made independently in each cell by selecting a vacant channel among those allocated to that cell in advance. In DCA the knowledge of occupied channels in other cells as well as in the cell in question is necessary. The amount of control is different in each DCA strategy. If the DCA requires a lot of processing and complete knowledge of the state of the entire system, the call setup delay would be significantly long without high-speed computing and signaling. The implementation complexity of the DCA is discussed in Vucetic [30].

Hybrid channel assignment schemes are a mixture of the FCA and DCA techniques. In HCA, the total number of channels available for service is divided into fixed and dynamic sets. The fixed set contains a number of nominal channels that are assigned to cells as in the FCA schemes and in all cases are to be preferred for use in their respective cells. The second set of channels is shared by all users in the system to increase flexibility. When a call requires service from a cell and all of its nominal channels are busy, a channel from the dynamic set is assigned to the call. The channel assignment procedure from the dynamic set follows any of the DCA strategies described in the previous section. For example, in the studies presented in References [4, 31], the FA and RING strategies are used, respectively, for DCA. Variations of the main RCA schemes include HCA with channel reordering [31] and HCA schemes where calls that cannot find an available channel are queued instead of blocked [5]. The call blocking probability for an HCA scheme is defined as the probability that a call arriving at a cell finds both the fixed and dynamic channels busy. Performance evaluation results of different HCA schemes have been presented in References [4–6, 32]. In Kahwa and Georganas [4], a study is presented for an HCA scheme with Erlang-b service discipline for cells of uniform size and shape where traffic is uniformly distributed over the whole system. The measure of interest is the probability of blocking as the load increases for different ratios of fixed to dynamic cells. As shown in Kahwa and Georganas [4], for a system with fixed to dynamic channel ratio 3:1, the HCA gives a better grade of service than FCA for load increases up to 50 %. Beyond this load HCA performs better in all cases studied in Kahwa and Georganas [4]. A similar pattern of behaviour is obtained for an HCA scheme with an FA DCA scheme and Ertang-c service discipline (calls that cannot find an available channel are queued instead of blocked). In addition, the HCA scheme with Erlang-c service discipline [5] has a lower probability of blocking than the HCA scheme with Erlang-b service discipline [4]. This phenomenon is expected because in the former case calls are allowed to be queued until they can be served.

In Yue [32], two different approximating models were presented. In the first model the traffic offered in the dynamic channels is modeled as an interrupted Poisson process, while the second used GI/M/m(m). The blocking probability vs the arrival rate for both models presents the same pattern of behavior as the simulation results of References [4, 5].

Flexible channel allocation (FlCA) schemes divided the set of available channels into fixed and flexible sets. Each cell is assigned a set of fixed channels that typically suffices under a light traffic load. The flexible channels are assigned to those cells whose channels have become inadequate under increasing traffic loads. The assignment of these emergency channels among the cells is done in either a scheduled or predictive manner [33]. In the literature, proposed F1CA techniques differ according to the time at which and the basis on which additional channels are assigned.

In the predictive strategy, the traffic intensity or, equivalently, the blocking probability is constantly measured at every cell site so that the reallocation of the flexible channels can be carried at any point in time [16]. Fixed and flexible channels are determined and assigned (or released) to (or from) each cell according to the change in traffic intensity or blocking probability measured in each cell. The number of dynamic channels required in a cell is determined according to the increase in measured traffic intensity. The acquired flexible channels can be used in a manner identical to the fixed channels in a cell as long as the cell possesses the channels. As long as a cell has several free fixed channels, no flexible channels are assigned to it if the traffic intensity is below a certain threshold [33].

If the flexible channels are assigned on a scheduled basis, it is assumed that the variation of traffic, such as the movement of traffic peaks in time and space, are estimated *a priori*. The change in assignment of flexible channels is then made at the predetermined peaks of traffic change [16]. Flexible assignment strategies use centralized control and require the

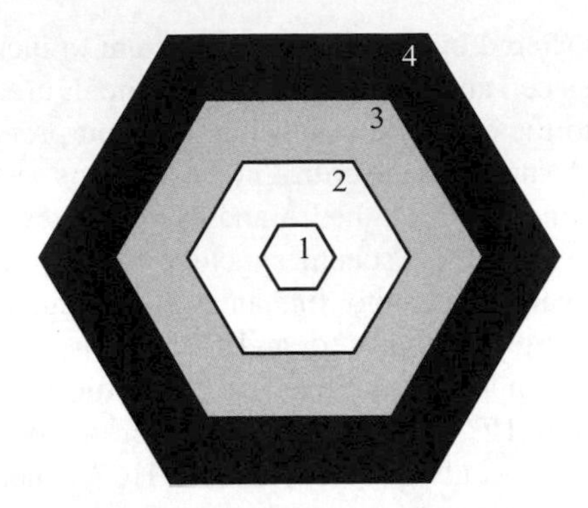

Figure 12.4 Concentric sub-cells.

central controller to have up-to-date information about the traffic pattern in its area in order to manage the assignment of the flexible channels [16].

Reuse partitioning (RUP) is an effective concept to obtain high spectrum efficiency in cellular systems. In RUP, as shown in Figure 12.4, each cell in the system is divided into two or more concentric subcells (zones). Because the inner zones are closer to the base station located at the center of the cell, the power level required to achieve a desired CIR in the inner zones can be much lower compared with the outer zones. Thus, the channel reuse distance (i.e. the distance between cells using the same channel) can be smaller for the inner zones than for the outer ones, resulting in higher spectrum efficiency. Reuse partitioning schemes could be divided into fixed and adaptive.

Fixed reuse partitioning

- Simple reuse partitioning.
- Simple sorting channel assignment algorithm.

Adaptive reuse partitioning

- Autonomous reuse partitioning ARP.
- Flexible reuse FRU.
- DDCA.
- All channel con-centric allocation ACCA.
- Self organized reuse partitioning SPRP.

12.1.10 Fixed reuse partitioning

Simple reuse partitioning was introduced in Halpern [35]. In this scheme, available channels are split among several overlaid cell plans with different reuse distances. The underlying principle behind RUP is to reduce signal-to-interference ratio (SIR) for those units that already have more than adequate transmission quality while offering greater protection to those units that require it. The goal is to produce an overall SIR distribution that satisfies system quality objectives while bringing about a general increase in system capacity. For the same SIR objective, those partitioning have the potential to obtain a significant increase in system capacity when compared with a system that uses only a single reuse factor [35].

Simple RUP can be implemented by dividing the spectrum allocation into two [35] or more [34] groups of mutually exclusive channels. Channel assignment within the ith group is then determined by the reuse factor N for that group. Mobile units with the best received signal quality will be assigned to the group of channels with the smallest reuse factor value, while those with the poorest received signal quality will be assigned to the group of channels with the largest reuse factor value. As the received signal quality for a mobile unit changes, it can be handed off to

(1) a channel that belongs to a different reuse group on the same zone at the same cell;

(2) a channel that belongs to the same or to a different group on another zone at the same cell;

(3) a channel belonging to the same or a different group at another cell.

Typically, the mobile units closer to a cell site will be served by channels from a group having a small value of N [35].

There are two main design issues related to the simple RUP concept. The first issue is the capacity allocation problem, which is to decide how many channels should he assigned to each zone. The second issue is the actual assignment of channels to calls. In Zande and Frodigh [34] the performance limits of the RUP concept were explored, and methods for allocating capacity to the different cell zones as well as optimum real-time channel assignment schemes have been presented.

The simple sorting channel assignment algorithm also divides each cell into a number of cocentric zones and assignes a number of channels, as in simple RUP. For each mobile in the cell, the base station measures the level of SIR and places the measurements in a descending order. Then it assigns channels to the set of at most M mobiles with the largest values of SIR, where M is the number of available channels in the entire cell. The mobile in the set with the smallest value of SIR is assigned a channel from the outer cell zone. The assignment of mobile channels according to ascending values of SIR continues until all channels from the outer zone are used. The base station continues to assign channels in the next zone, and so on, until all mobiles in the set have been assigned channels.

The simple sorting channel algorithm achieves almost optimum performance. It also allows 1.4–3 times more traffic than the FCA scheme. An important remaining issue is that the sorting scheme only determines which cell plan each mobile should use. It does not assign actual channels, which must be done with some care. In addition, if all cells using a certain channel group started the channel assignment by using the first channel in the group, we would get an exceptionally high interference level on that particular channel. A random selection procedure would be one way to solve this problem.

12.1.11 Adaptive channel allocation reuse partitioning (ACA RUP)

In this case any channel in the system can be used by any base station, as long as the required CIR is maintained. *Autonomous reuse partitioning* is based on the RUP concept and real-time CIR measurements. In this technique, all the channels are viewed in the same order by all base stations, and the first channel which satisfies the threshold condition is allocated to the mobile attempting the call. Therefore, each channel is reused at a minimum distance with respect to the strength of the received desired signal. In ARP base stations conduct their allocations independent of one another, and no cooperative control is necessary.

As compared with simple FCA, ARP doubles the traffic handling capacity of the system and decreases the co-channel interference by 1:4 [36]. ARP improves the traffic handling at the cost of the SIR margin in each channel. This creates problems for fast moving mobile stations such as car-mounted units, which suffer from rapid fluctuations in signal level.

The *flexible reuse* (FRU) scheme is based on the principle that, whenever a call requests service, the channel with the smallest CIR margin among those available is selected. If there is no available channel, the call is blocked. Simulations in Onoe and Yasuda [37] showed that FRU can effectively improve system capacity, especially for users with portable units. More specifically, a capacity gain of 2.3–2.7 of FRU over FCA was observed. However, the FRU strategy requires a large number of CIR measurements, which makes it virtually impractical for high-density microcellular systems.

The *self-organized reuse partitioning scheme* (SORP) uses the method where each base station has a table in which average power measurements for each channel in its cell and the surrounding cells are stored. When a call arrives, the base station measures the received power of the calling mobile station (in order to define at which subcell the mobile station is located) and selects a channel, which shows the average power closest to the measured power. The channel is used if available, otherwise the second closest candidate is tried. The content of the table for the chosen channel is updated with the average value of the measured power and the power of the mobile stations using the same channel. The power level of the other mobile stations is broadcast by their base station. As a consequence of this procedure, in each base station channels that correspond to the same power are grouped autonomously for self-organized partitioning.

In Chawla and Qiu [85], a performance comparison is made between SORP, conventional ARP and random DCA schemes. The simulation analysis showed that SORP and ARP show almost the same performance, which is far superior to random DCA. SORP can reduce the occurrence of intracell handoff and can reach a desired channel quickly, while achieving high traffic capacity. The essential difference between ARP and SORP is that ARP always senses the channels in the same order until one is available, while SORP learns which channel is proper for the calling mobile, so it can find a desired channel more quickly.

All channel concentric allocation (ACCA) is an extension of the RUP concept. All radio channels of a system are now allocated nominally in the same manner for each cell, as in Figure 12.5. Each cell is divided into N concentric regions. Each region has its own channel allocation. Here, each channel is assigned a mobile belonging to the concentric region in which that channel is allocated, and has a specific desired signal level corresponding to the channel location. Therefore, each channel has its own reuse distance determined from the desired signal level. Thus, ACCA accomplishes effective channel allocation in a global sense, although it is a self-organizing distributed control algorithm. Computer simulations have shown that the system capacity at a blocking rate of 3 % is improved by a factor of 2.5 compared with the FCA [38]. If, in addition, a transmitter power control is implemented on top of ACCA, the system accomplishes a capacity 3.4 times greater than FCA.

Distributed control channel allocation (DCCA) is a dynamic channel allocation scheme based on the ARP concept. In this scheme all cells are identical, and channels are ordered in the same manner, starting with channel number one, by all the base stations in the

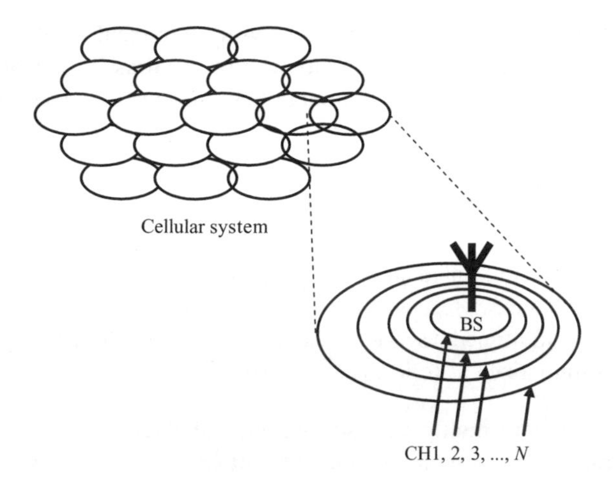

Figure 12.5 Principle of the all-channel concentric allocation.

Figure 12.6 DCCA cell structure.

network. The decision to allocate a channel is made locally, based on CIR measurements. The architecture of a cell in DCCA is shown in Figure 12.6. It consists of an omnidirectional central station connected to six symmetrically oriented substations [39, 40]. The substations are simple transceivers, and can be switched on and off under the control of the main station. When the traffic density of the cell is low, all the substations are off and the only operating station is the main station, at the center of the cell covering the entire cell area. Gradually, as call traffic increases, forced call blocking will occur due to an unacceptable level of co-channel interference or the unavailability of resources. In this case, the main base station switches on the nearest substation to the mobile unit demanding access. This in effect relocates the main base station closer to the mobile requesting service: therefore, CIR measurements will now be higher, thus improving the probability of finding an acceptable channel. If the traffic is reduced, the main station switches off a number of substations. The system therefore automatically adapts itself to time-variant cell traffic density. As a result, an improvement in both system efficiency and traffic capacity can be achieved. As discussed in Madani and Aghvami [40], the DCCA system results in lower probability of forced termination of calls. Computer simulation showed a drastic reduction in the number of handoffs and almost 50 % less forced termination of calls compared with the ARP scheme.

12.2 RESOURCE MANAGEMENT IN 4G

Based on the discussion presented so far in Chapters 11 and 12 an integrated bandwidth management approach is now presented that can be implemented in next generation wireless networks that support multimedia services (data, voice, video, etc.). The main principles of the approach are summarized as follows:

(1) The system supports multiple classes of service that require various levels of quality that may range from strict to flexible and soft QoS requirements.

(2) Advanced bandwidth reservation is performed to the cell a mobile user is moving towards in order to assist and support a seamless handoff process.

(3) User mobility is introduced into the reservation scheme and process in order to optimize the efficiency of handoff mechanisms and minimize, if not eliminate, the unnecessary reservation of resources and, therefore, improve the system capacity and throughput.

(4) Bandwidth reconfiguration processes are used that may allow the efficient resource redistribution in a cell to satisfy the QoS requirements of all the mobile users in the cell, especially when users with flexible QoS requirements are supported in the system.

(5) A mobile agent based framework is used to facilitate the efficient implementation of the above integrated approach.

One of the elements in future resource management is mobile locationing and tracking. Position location technology can be loosely classified into two major categories: mobile-based solution and network-based solution. The global positioning system (GPS) is a worldwide radio-navigation system formed from a constellation of 24 satellites and their ground stations. As an alternative to mobile-based approaches, cellular networks can be used as the sole means of providing location services, where the MSs are located by measuring the signals traveling to and from a set of fixed cellular BSs. The signal measurements are used, for example, to determine the length and/or direction of the individual radio paths, and then the MS position is computed from geometric relationships. Basically, radiolocation systems can be implemented based on either signal strength, angle of arrival (AOA), or time of arrival (TOA) measurements or their combinations [41, 42].

However, there are several problems associated with the existing network-based geolocation systems which limit and prevent the use of location information of mobile users for network management purposes. First, the triangulation measurement data collecting procedure is quite complicated. At least two or three BSs should be involved in locating the position of a mobile user. Each BS will perform some triangulation measurement. Then, all this data has to be sent to some device to carry out the calculation. This procedure requires some specialized protocol to support the information exchange. This also would waste some bandwidth in data transmission and introduce some delay to the overall location algorithm. Nowadays, for some practical systems, it may take several seconds to locate a mobile user. So, if the user moves with high speed, we cannot obtain an accurate position information in real time. Furthermore, performing the calculation at BS instead of MS raises some difficulties in obtaining accurate radio parameters of an MS that may vary with time and position. Finally, if all the mobile users in the network need geolocation information, the computational load at one BS becomes very heavy. Such an overload condition may lead to high processing delays or failures of the location module in the BSs.

By moving the measurement and calculation functionality to the mobile station, all the information needed for the geolocation can be gathered almost at one time and then be treated locally. In order to allow the geolocation system to have more flexibility so that the MS can smoothly roam among different networks with different geolocation approaches, a mobile agent-based system framework can be used. The use of an approach based on code-on-demand increases the flexibility of the system and maintains agents simple and small at the same time. This is an element of a broader concept of *active networks* that will be discussed later in Chapter 16. In this section, one of the basic ideas introduced is that, when the MS is turned on or roaming into a network with different geolocation approach, it sends a message agent (MA) to the BS. Then, the BS will generate a geolocation agent (GA) and send it to the mobile station. The GA should contain the signal processing algorithm for signal measurement and the triangulation algorithm. By using this framework, mobile stations can obtain all the data for triangulation calculation almost at the same time and carry out the geolocation calculation locally.

12.3 MOBILE AGENT-BASED RESOURCE MANAGEMENT

In general, a major incentive for an agent-based approach is that policies can be implemented dynamically, allowing for a resource-state-based call admission, reservation and network management strategy. Agents are used to discover resources available inside the network and claim resources on behalf of customers according to some 'figures of merit' [43], which represent tradeoffs between bandwidth claimed and loss risk incurred due to high utilization. Agents are able to trigger adaptation of applications inside the network on behalf of customers. This allows for an immediate response to resource shortages, decreases the amount of useless data transported, and reduces signaling overhead. Mobile agents provide the highest possible degree of flexibility and can carry application-specific knowledge into the network to locations where it is needed. In this section, we provide a detailed description of the mobility-assisted handoff/bandwidth reservation/call admission control scheme which can provide a flexible QoS management strategy in wireless networks. We also present a mobile agent-based framework and demonstrate its use in the implementation of the integrated strategy.

The agents used in the system and their corresponding roles are as follows:

- *Message agents* (MAs) are used to exchange information and management data among agents and managers. They are created and received by message managers (MMs). MM can also forward messages contained by MAs to the corresponding managers or agents in the same network element.

- *Geolocation agents* (GAs) are created by geolocation managers (GM) in base stations and are sent to mobile stations. They contain the signal processing algorithm for signal measurement and the triangulation algorithm used in the current networks.
- *Bandwidth reservation agents* (BRAs) are created by bandwidth reservation managers (BRMs) and are sent to mobile stations. They contain the bandwidth reservation algorithm used in the current networks.
- *Call admission control agents* (CACAs) are created by call admission control managers (CACMs) embedded in base stations and are sent to mobile stations. The CAC strategy can be flexibly deployed in this way. CACAs will collect customer requirement and send necessary information by MAs to CACMs. The CAC algorithm is carried by CACMs.

When a mobile station is turned on, an MA is sent out to the base station containing initial information about the mobile station. MM in the base station relays the information in MA to CACM, GM and BRM. CACA will create a new CACA and send it back to the mobile station. CACA will collect user information about the QoS requirement and send it with call-in requirement back to CACM by MA. The CAC algorithm embedded in CACM will make the decision of accepting the new call or not. Once the new call is accepted, GM and BRM also will create a new GA and a new BRA, respectively, and send them back to the mobile station. Within the mobile station, GA can output the position of mobile station itself periodically and this position information is input into the position information processing module. This module then provides mobility information about mobile stations to BRA. BRA can use mobility information to calculate the bandwidth required for handoff purposes. The bandwidth reservation is also sent to the BS by MA. Finally, BRM at base station will summarize the bandwidth needed by each of its neighboring cells and send each neighboring cell an MA containing the corresponding bandwidth reservation requirements. BRM at base station in the neighboring cell can use this result to perform the bandwidth reservation process. In the case of handoff, a handoff requirement and also the QoS requirement will be sent by MA to the CACM embedded in the target base station. The pool used by bandwidth reconfiguration procedure is maintained by CACM. After bandwidth reallocation, CACM will send out message agents to MSs informing them about the reconfiguration results.

For the mobility predictive resource management scheme in a centralized management architecture, the following two modules are required: geolocation module and bandwidth reservation computation module. In CS (traditional centralized *client–server*) approach, the management primitives are often low level and fixed, and no semantic compression is allowed to be performed. As a result, these two modules work independently. The only entity that these two module can operate with is the *management information base* (MIB) embedded in the mobile station. The geolocation module periodically updates the position information in MIB, while the bandwidth reservation module will access the MIB for position information to carry out the prediction and calculate the bandwidth required by the mobile devices under consideration. In the mobile agent based approach, the agent has the ability to perform the semantic compression. This means that all the simple steps that should be processed one by one in CS mode now can be integrated into some agent that can carry out a more complicated task autonomously. The agents move to the mobile station and carry out the functionality of geolocation, bandwidth prediction locally. As a result, we should expect that the interactive traffic between mobile station and network management station can be reduced dramatically.

The various steps involved in the centralized management approach and the corresponding control messages involved are summarized as follows:

(1) *Geolocation* – whenever a mobile station needs to update its position information in the MIB, it sends out a request I_{mn} to the corresponding NMS (network management station), which in turn sends out requests I_{nb} to β different base stations near the mobile station (the number of β depends on the geolocation technology). After each one of these base stations obtains the corresponding measurement results, they return the measurements R_{bn} to the NMS. The NMS then uses the results to perform the triangulation computation and estimate the position of the mobile station. Finally, the position result R_{nm} is returned to the mobile station for MIB update. The traffic generated in this phase is: $C_{\text{geo}} = I_{mn} + \beta I_{nb} + \beta R_{bn} + R_{nm}$.

(2) *Position prediction and bandwidth reservation calculation* – when the NMS wants to predict the mobility of each mobile station, the current position and a few historical positions of MS are required. Since in the CS approach no semantic compression is allowed to be performed, every request for the position information (current or historical) of mobile stations will generate an inquiry I_{nm} to mobile station to search its MIB. The mobile station then sends back the results R_{mn} one by one to the NMS. Finally, the NMS can use this information to do the prediction and calculate the bandwidth needing to be reserved in neighboring cells.

So in this phase, the traffic generated by CS mode management is: $C_{\text{pos}} = \alpha I_{nm} + \alpha R_{mn}$, where α is the total new updates (recent positions and current position in the MIB of mobile station) required for prediction. Please note here that the prediction and reservation interval maybe different from the position calculation/update interval and, as a result, its is possible that some recent positions (stored in the MIB of the mobile node) along with the current one may have to be sent to the NMS as required by the bandwidth reservation algorithm.

Assuming that the average call life time is T_{l} and the geolocation/prediction update interval is t_{u}, then in the CS mode the total control traffic generated for the resource management is:

$$
\begin{aligned}
C(cs) &= \frac{T_{\text{l}}}{t_{\text{u}}}(C_{\text{geo}} + C_{\text{pos}}) \\
&= \frac{T_{\text{l}}}{t_{\text{u}}}(I_{mn} + 3I_{nb} + 3R_{bn} + R_{nm} + \alpha I_{nm} + \alpha R_{nm})
\end{aligned}
$$

Similarly, the various steps involved in the mobile agent-based approach and the corresponding control messages involved can be summarized as follows:

- *Initial phase – agent(s) transport*. When a mobile station is initialized, or whenever the mobile station enters a new network domain that supports different geolocation method or bandwidth calculation/reservation method, the resource management agents (including geolocation agent A_{geo}, bandwidth reservation agent A_{bwrsv} and call admission control agent A_{cac}) will be downloaded from the NMS. The total traffic introduced in this initial phase is equal to: $A_{\text{geo}} + A_{\text{bwrsv}} + A_{\text{cac}}$.
- *Phase 1 – geolocation phase*. Since the geolocation module is included in the agent and is executed locally, no traffic is generated in the network. The geolocation results can be written into the MIB locally.

- *Phase 2 – prediction phase.* Since the prediction module has already been included in the agent code, the prediction can be made locally without generating any traffic in the network.
- *Phase 3 – bandwidth reservation.* The bandwidth reservation agent calculates the bandwidth needed to be reserved in a certain cell and sends the reservation result R_{mb} via the message agents to BS.

Therefore, after initialization phase of agent download, the total communications control traffic generated for resource management purposes is only: $C(ma) = T_1 R_{mb}/t_u$. We should note here that, once the agents are embedded (downloaded) in the mobile station, unless a new resource management strategy is needed (when a management system is upgraded or the mobile stations travels to a new foreign domain), we do not need to transport the code of the agent again. Since all the involved agents are, in general, very small in size (up to a few kbytes), except for the graphical interface which, however, can vary from implementation to implementation and in any case does not need to migrate, it is expected that the migration times (e.g. message agents) and corresponding consumed energy are very low as well. Based on the above discussion and arguing that the size of R_{mb} is of the same order of message R_{mn}, we can easily conclude that, after the initialization phase, the control traffic generated between the mobile stations and base stations is reduced significantly by taking advantage of the mobile agent technology.

Security is another crucial aspect of agent systems. An adequate security model is required in order to insure a high level of protection to the agents and the nodes, so that nodes can be protected from the attacks of unsafe agents, and the agents can run on the nodes of the network without being damaged [44]. These issues are discussed in Chapter 15.

12.3.1 Advanced resource management system

In this section we discuss performance of an advanced resource management system (*arm*) using:

(1) The predictive mobility-based bandwidth reservation scheme (PMBBR), as described in Chapter 11; and

(2) QoS management with flexible bandwidth sharing (FBS).

For the system with two classes of traffic, FBS operates as follows:

- *Class 1 (higher priority)* – the desired bandwidth for this kind of traffic is BW_1^u. If class 1 traffic cannot obtain the desired bandwidth, it may have the option to continue at a lower bandwidth requirement BW_1^l. For instance, this can be achieved by adjusting the coding rate so that the video/audio quality is still acceptable (i.e., real-time traffic).
- *Class 2 (lower priority)* – the desired bandwidth for this kind of traffic is BW_2^u and there are no strict QoS requirements. However, some flexible QoS requirements are defined for such a service. The user could specify a set of acceptable QoS levels that correspond to bandwidth requirements that range from a lower bound bandwidth requirement BW_2^l to a maximum bandwidth requirement BW_2^u and expect a QoS varying in the specified range (i.e. nonreal-time traffic).

Such an *arm* system is compared with the corresponding results of a conventional system where the fixed bandwidth reservation is implemented, in terms of achievable new call (P_{nb}) and handoff P_{hb} call blocking probabilities. Instead of two or more classes of users, the

Table 12.2 Simulation parameters

Parameter	Value	Description
BW_{total}	1000 kbs	Total bandwidth capacity of a cell
BW_1^{u}	30 kbs	Desired bandwidth requirement for class 1 users
BW_1^{l}	25 kbs	Lower bound bandwidth requirement for class 1 users
BW_2^{u}	50 kbs	Desired bandwidth requirement for class 2 users
BW_2^{l}	5 kbs	Lower bound bandwidth requirement for class 2 users

concept of flexible bandwidth sharing will be extended in 4G to different systems or even different operators. The wireless network used in this study is composed of 37 hexagonal cells. The cell radius is set to be 1000 m. In order to approximate the performance of a large cellular system, the cells are wrapped around to eliminate the border effect. The arrival of new calls initiated in each cell forms a Poisson process with rate λ. The life time of each call is exponentially distributed with mean 240 s.

Additional system and traffic parameters are summarized in Table 12.2 [122]. It should be noted here that the *arm* framework aims to suggest a general approach that supports the seamless operation and provides flexibility for the resource management in the next-generation wireless networks that support multimedia services. The bandwidth values for the different classes of service were chosen for illustration purposes. New voice coding technologies and emerging data applications may bring different bandwidth requirements to the wireless networks, which, however, can be easily fit into the framework. Throughout the simulation study, it was assumed that the desired bandwidth requirement for class 1 (e.g. voice) users is 30 kbs and that for class 2 (data) users is 50 kbs.

The mobility model used this study is as follows: when a new call is initiated, the corresponding MS is assigned with a random initial position inside the cell, a random moving direction and an initial moving speed which is chosen according to a uniform distribution in the interval [0; $V_{\max}$] mile/h. The speed (v) and direction (θ) are updated to v' and θ' every time interval Δt according to the following model:

$$v' = \begin{cases} \min\{\max[v + \Delta v, 0], V_{\max}\}, & p \le 0.9 \\ 0 & \text{otherwise} \end{cases}$$

$$\theta' = \theta + \Delta\theta$$

where Δv models the acceleration/deceleration of the mobile user and is a uniformly distributed random variable over the interval [−5 mile/h, 5 mile/h], $\Delta\theta$ characterizes the user's change in moving direction and is a uniformly distributed variable over the interval $[-\Delta\theta_{\max}, \Delta\theta_{\max}]$ and p is a uniformly distributed random variable over the interval [0, 1]. The use of variable p allows us to simulate the situation where a mobile user may stop occasionally during the course of moving. Based on this mobility model, two mobility patterns are considered in this study: high-speed pattern and low-speed pattern. In the high-speed pattern, the mobility parameters are $V_{\max} = 60\,\text{mile/h}$, $\Delta\theta_{\max} = \pi/4$, which correspond to highly directional, fast-moving traffic (e.g. highway traffic); for the low-speed pattern, $V_{\max} = 30\,\text{mile/h}$, $\Delta\theta_{\max} = \pi/2$ which corresponds to a less directional, slow-moving traffic (e.g. downtown traffic). The mobility update interval (Δt) is chosen to be 10 s.

Numerical results are presented for two different traffic profiles (TP) regarding the composition of class 1 and class 2 traffic. In profile TP1, 10 % of the new call attempts are class 2 calls, while, in profile TP2, 50 % of the new calls are class 2 calls. Note that in the proposed CAC and resource reconfiguration scheme only class 2 calls can lend bandwidth to handoff calls and, therefore, the number of ongoing class 2 calls in a cell will influence the performance of the system.

Figure 12.7 compares the handoff call and new call blocking probabilities of the *arm* system with the corresponding results of the conventional system vs λ, under TP1, with users moving in the high-speed pattern. In the figures, 'c1' and 'c2' stand for class 1 and class

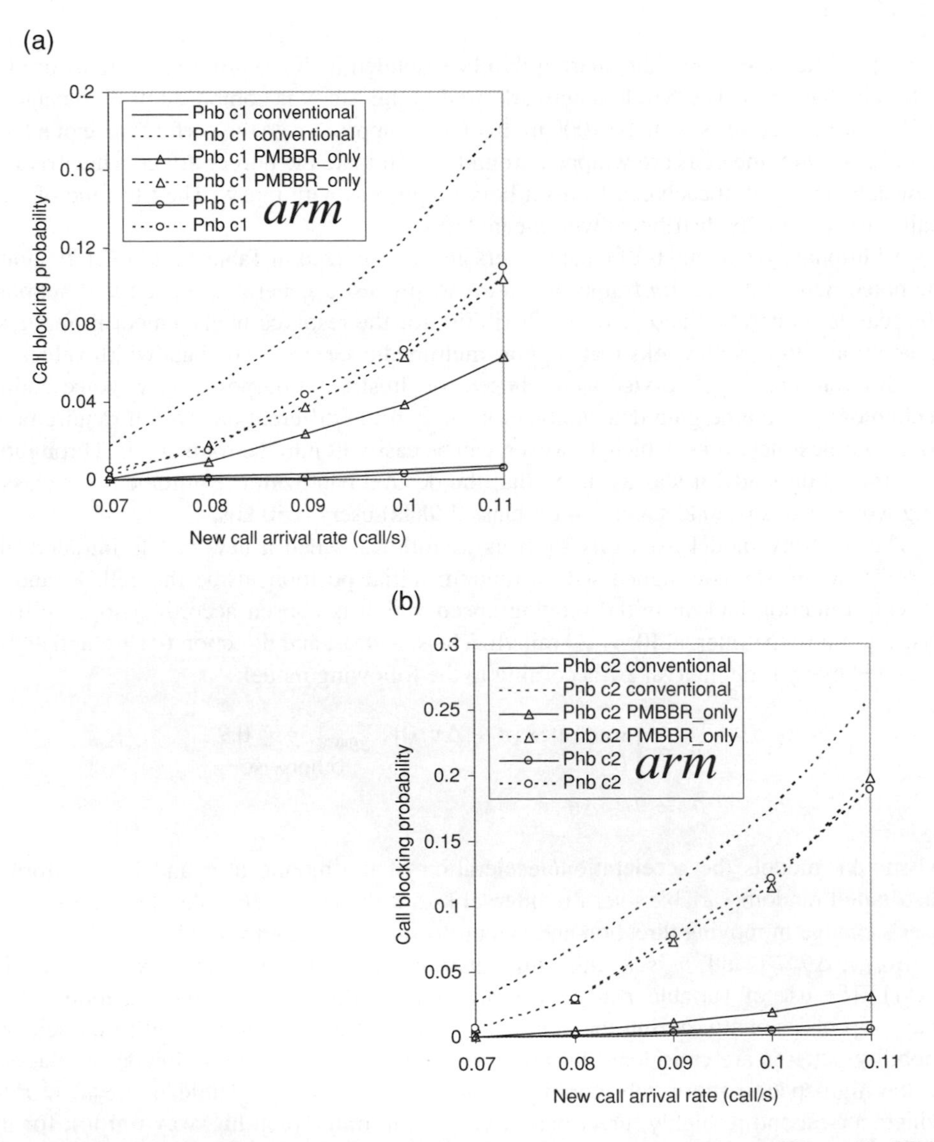

Figure 12.7 Call blocking probabilities for high mobility pattern under traffic profile TP1. (a) Call blocking probabilities for class 1 users. (b) Call blocking probabilities for class 2 users.

Table 12.3 Average bandwidth used by each class 2 user

λ	0.05	0.06	0.07	0.08	0.09	0.10	0.11
arm (kbs)	50.00	49.60	49.00	48.10	46.85	45.56	42.50
Conventional (kbs)	50.00	50.00	49.90	49.77	49.75	49.67	49.42

2 users, respectively. The conventional system uses the fixed bandwidth reservation, where the reservation value represents a fixed percentage of the total capacity of the cell. In the following study, the corresponding reservation value of the conventional system for class 1 users is $BW_{\text{res1}} = 30$ kps and for class 2 users is $BW_{\text{res2}} = 50$ kps. The reservation values are selected based on experimentation with the objective of keeping similar handoff blocking probability for both the *arm* system and the conventional system. The call admission control procedure for a conventional system is the same except that the bandwidth reconfiguration is not used. From the figure, we observe that, for the given parameters, the *arm* system and the conventional system have similar handoff call blocking probabilities.

However, the *arm* system can significantly decrease the new call blocking probabilities for both user classes, which demonstrates that it can admit more users than the conventional system while still guaranteeing the same level of QoS for handoff calls. One should bear in mind that the connection level QoS improvement achieved by the *arm* system is obtained at the cost of slightly decreasing the bandwidth actually used by class 2 users. Table 12.3 lists the average used bandwidth by each class 2 user.

Figure 12.8 presents the corresponding numerical results for TP2 where 50 % of the new calls belong to class 2 traffic. Under this traffic configuration, the arm system can achieve an even better performance improvement.

Figure 12.9 shows the corresponding blocking probabilities that can be achieved by the *arm* system for different mobility patterns (high vs low) under TP1. From the figure, we observe that the *arm* scheme is capable of achieving good performance even when the users move in the low-speed pattern. The *arm* scheme achieves very low handoff failure rates for both user classes, which is similar to the results obtained under the high-speed mobility pattern.

12.4 CDMA CELLULAR MULTIMEDIA WIRELESS NETWORKS

In cellular segments of wireless IP networks, there is a need for simple radio resource and teletraffic management solutions that allow operators to customize various network and user profiles, optimize the grade of service (GoS) while fulfilling the QoS requirements. In such solutions, CAC for real-time (RT) services and packet access control for nonreal-time (NRT) services play key roles. This section

(1) presents a comparative study of CAC policies in the uplink (UL) of WCDMA cellular networks used as segments of the wireless IP networks;

(2) introduces a simple, effective and robust soft-decision CAC policy that exploits the UL interference distributions to compensate fluctuations of the local average SIRs dominating QoS while expanding call-admissible region to maximize the system capacity; the method requires neither complex measurements nor mutual exchange of information between adjacent CDMA cells about the state of the network; this is therefore especially applicable in wireless IP networks, consisting of number of

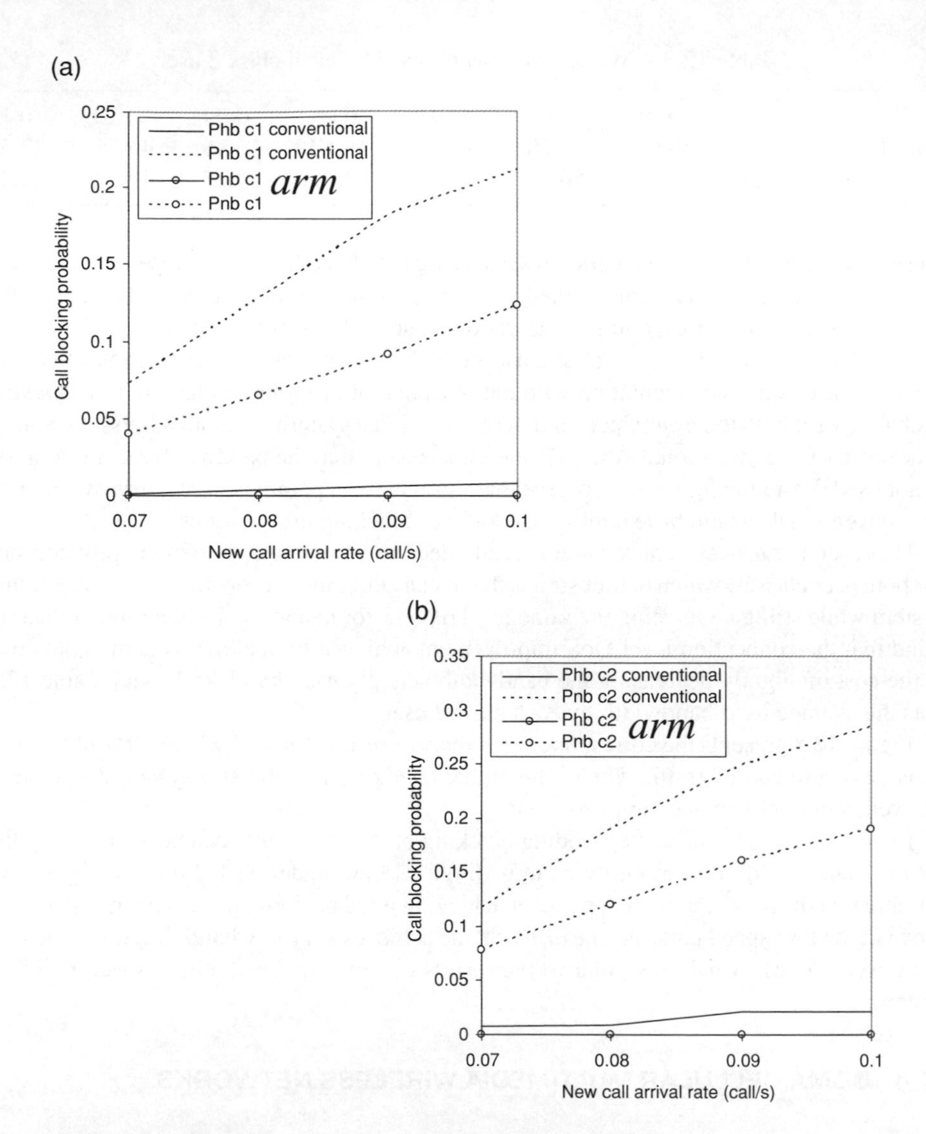

Figure 12.8 Call blocking probabilities with high mobility pattern under TP2. (a) Call blocking probabilities for class 1 users. (b) Call blocking probabilities for class 2 users.

segments or domains, where the simplicity and the scalability requirements favor localized radio resource control;

(3) introduces multiple call-blocking thresholds or load-based fracturing factors for different types of calls, services and subscription classes to provide operators a flexible tool for QoS differentiation and GoS management;

(4) presents a reliable dimensioning of the free capacity for NRT packet-data access and estimation of the upper-limit UL data throughput for different packet-lengths and data-rates based on asymptotic and quasi-stationary analysis of the RT traffic;

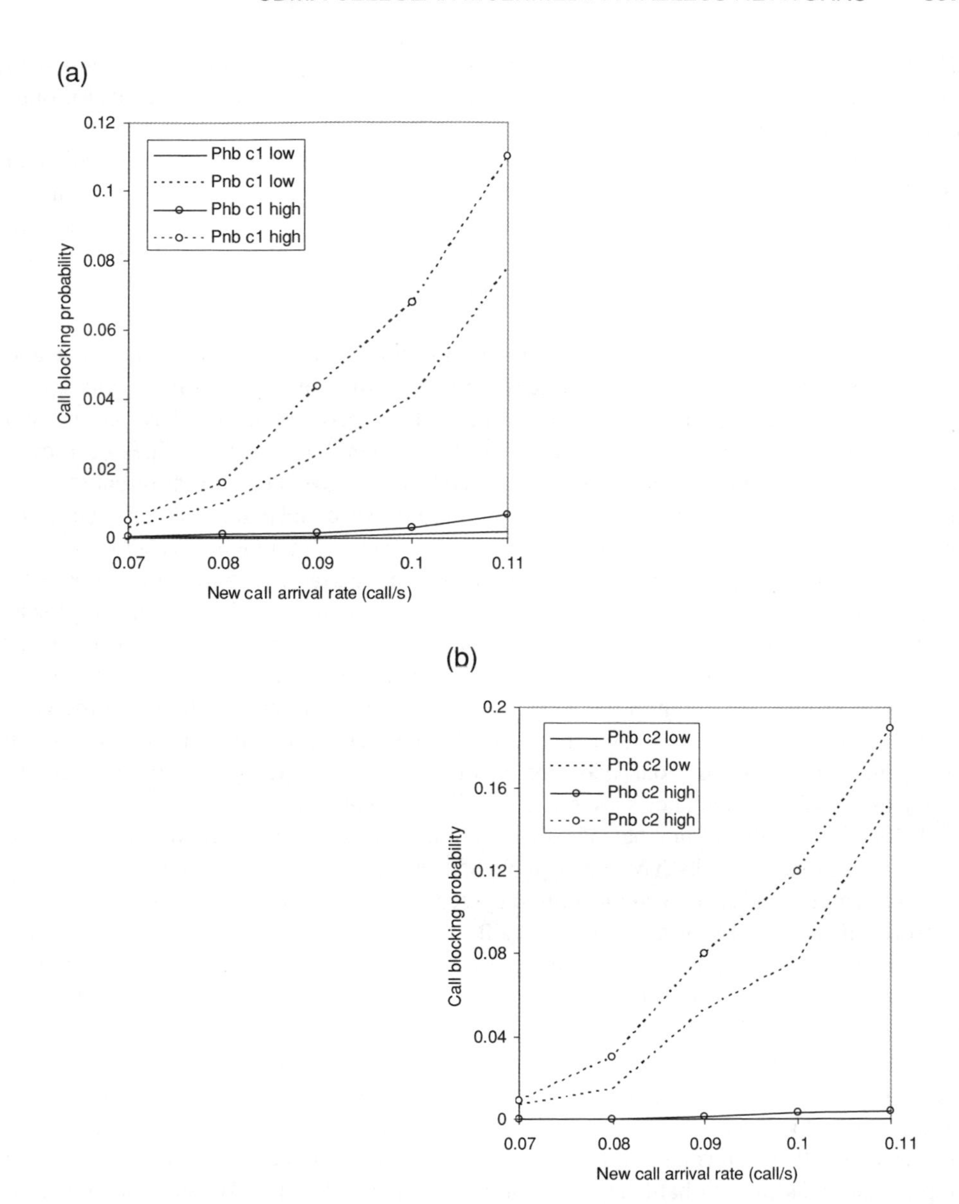

Figure 12.9 Call blocking probabilities in *arm* system for different mobility patterns under TP1. (a) Call blocking probabilities for class 1 users. (b) Call blocking probabilities for class 2 users.

(5) introduces a dynamic feedback packet access control scheme that adapts to the UL free capacity left by the RT traffic with optimal throughput-delay tradeoff;

(6) provides a simple and accurate analytical method for evaluating the teletraffic performance of the system with different CAC policies and QoS differentiation, taking into account priority handoffs and user mobility equilibrium.

The results are important for optimizing the overall Erlang capacity and throughputs with differentiated services in WCDMA segments or domains of wireless IP networks for multimedia applications.

A variety of emerging technologies and protocol enhancements has been investigated and designed to provide IP-based multimedia services to mobile users. These research and standardization efforts are summarized in References [45–47]. By using the radio resource management framework, in this section we address some simple solutions for future network operators to customize and optimize the performance of their multimedia networks serving various user and network profiles.

The traffic-engineering framework for wireless multimedia networks must be simple and flexible for supporting a variety of services with QoS differentiation to satisfy wide range of customer demands and to maximize revenues. The operators need to have an in-hand tool to manage their network operation more or less similarly to the airlines customer-service systems where passengers are categorized into classes receiving different QoS on any required routes, but all sharing expensive limited space of planes. The limitation of extremely expensive radio resources is also the bottleneck of future wireless networks. The key in providing the QoS for users, while efficiently sharing radio resources to optimize the GoS for networks, is the access control at the air interface acting as ticket-selling and check-in processes. The access control herein includes CAC for the RT connection-oriented service domain and packet access control for the NRT connectionless service domain. Hence, by using intelligent access control strategies, we can have an effective solution to the two-fold objective: (1) to optimize the network utility; and (2) to provide operators freedom in controlling pay-load and group behavior of traffic classes regarding required services as well as subscriber potential classes, e.g. gold, silver and bronze.

To date, a significant number of papers and standards dealing with various issues of CDMA wireless networks have been published. Major efforts have been paid to study detailed features and specific algorithms, e.g. receivers, power control, handoff, etc. Investigations of the CDMA system capacity, radio resource and teletraffic management have been mostly based on single-service 'worst case' scenarios, or high-cost simulations [50]. In more sophisticated scenarios, the simplicity requirement is usually ignored despite the fact that only this condition can ensure system efficiency and guarantee fast response to multiple QoS requirements from the users [52]. Examples are References [44, 47, 59], where valuable results are presented for measurement-based adaptive admission control and bandwidth management schemes. However, it has been pointed out, e.g. in Reference [47], that the price of better performance offered by these schemes is an increased complexity of the implementations in both hardware and software. The results on CAC and packet access control in CDMA wireless networks are summarized extensively in References [49, 50, 51, 52, 54].

This section emphasizes practical operating aspects of multimedia CDMA mobile cellular networks, used as segments of the future wireless IP, and the need for simple and effective radio resource and traffic handling mechanisms. This is investigated in contexts of the access control, which is mainly viewed as an optimal interference management problem in CDMA systems. Various CAC policies are considered simultaneously and an effective soft-decision mechanism, namely SCAC, is presented. The motivation behind SCAC is to use probabilistic functions, adapted to inter-cell interference distributions, for making decision upon receipt of a call request in order to optimize the radio resource utilization (RRU) of interference-limited CDMA systems, while keeping overall communication quality under acceptable outage levels. This method requires neither complex measurements nor mutual exchange of information between the adjacent cells about the state of the network. Thus, the complexity and the cost of implementations can be significantly reduced. This is of importance in the wireless IP networks where a number of segments or domains are involved. The scalability also favors distributed or localized control of radio resources. The normalized load factors, which essentially characterize spread spectrum CDMA transmissions, are used to represent the resource consumption, similarly to the well-known effective bandwidth concepts [50, 68]. Because attributes of the required service and adequate transmit power control (TPC) schemes can provide and insure *a priori* knowledge and stationary behavior of the resource consumption [41–44, 46, 53], SCAC can be promising in providing good multiplexing gain and robustness. Multiple call-blocking thresholds or load-based fracturing factors are introduced that give operators a simple and flexible means of differentiating and/or prioritizing different types of calls (e.g. new call and handoff call), services and/or subscription classes for desired serving customs. The performance of CAC policies is evaluated with respect to the following:

(1) effectiveness represented by tradeoffs of the simplicity in implementations and the ability in autoconfiguring with various users QoS and networks GoS profiles for maximizing system capacity or revenues; performance is measured by the probabilities of call blocking and handoff failure or some additional cost functions;

(2) reliability measured by the probability of losing communication quality during services, which is also known as the outage probability when the received interference exceeds the tolerable level;

(3) robustness seen against a need for redesign due to uncertainty and changes of the system parameters.

With respect to controlling NRT packet radio access in UL, using the free capacity left by the RT traffic , this section provides

(1) a precise dimensioning of the UL radio resources available for packet access;

(2) estimates of the average upper-limit UL data throughput for different bit-rates and packet-lengths that can be used for tradeoff of the design parameters for packet access control schemes;

(3) the dynamic feedback information-based multiple access scheme (DFIMA), which is suitable for QoS differentiation in background services.

These issues are increasingly important for the efficiency of the RRU in serving mobile messaging services. Examples are short message service or wireless e-mail, which have been taking the consumer market by storm recently.

The analytical models are presented based on modified product-form loss network models and results of the stochastic knapsack problem [64–67]. The priority handoff and the cell-condition of user mobility equilibrium are considered. The UL dynamic resources and upper-limit throughput for NRT packet transmissions are estimated using asymptotic and quasi-stationary analysis of the RT traffic. This provides a reliable and flexible tool for designers and operators in teletraffic engineering of the networks. Only the UL direction is chosen to present the results. The traffic asymmetry between UL and DL is therefore not taken into consideration. The RT services are of CBR classes, whereas the NRT packet services are of background classes having a lower priority that can tolerate moderate delays. The VBR and the ABR interactive services, pre-emption in QoS differentiation, are not considered. Despite these limitations, such a practical and comprehensive system model and analysis for radio resource and teletraffic management of multimedia CDMA cellular networks, with discussion of simple and effective SCAC and DFIMA, should be of interest to system designers. Moreover, this section lays the groundwork for investigating more advanced control mechanisms as well as more sophisticated traffic models for future wireless IP networks.

12.4.1 Principles of SCAC

An overview of publications on QoS and resource management including CAC techniques in 3G multimedia networks can be found in References [52, 54]. In general, CAC policies can be divided into two categories: static and dynamic [59]. From the implementation point of view, these can be further divided into modeling-based and measurement-based policies [57]. Under the static CAC category, a call request is accepted only when sufficient resources are available to meet its required QoS and to maintain the QoS of ongoing calls. The admission controller must decide which resource combinations can be accepted into the network. In FDMA/TDMA cellular systems, the fixed channel assignment-based policy [55] is often adopted because of its simplicity and capacity maximization in line with a fixed threshold of the number of channels (cell-capacity). In CDMA cellular systems, such a fixed threshold is hard to determine efficiently because the qualitative and quantitative features of CDMA systems appear to be heavily inter-twined and mutually supportive. Under the dynamic CAC category, a call request can be rejected even if a set of new calls may meet their QoS requirements. On the other hand, a new call can be accepted into the network even if the instantaneous QoS may be violated, but when averaged over states, the service quality is met.

Therefore, the network utility can be increased and various cost-effective and traffic-shaping policies can be adopted. The reinforcement learning-based CAC, described by a semi-Markov decision process, is believed to be a good solution for a class of adaptive policies [56, 59]. Later in this section, a much simpler and more effective dynamic SCAC is presented by exploiting the soft-capacity feature of CDMA radios.

Consider the UL of a cell in a CDMA cellular system consisting of multiple, identical and independent cells using omni-directional antennas in BS. This network is supporting M independent connection-oriented RT service classes, each with a constant bit-rate and a target value of signal-to-interference ratio (SIR) for meeting its QoS requirements. Reference [53] describes the spectral efficiency of an interference-limited CDMA air-interface in term of load factors.

The starting point in this concept is the target signal to noise ratio for a given user k defined as

$$(E_b/N_0)_k = \frac{G_k P_k}{\rho_k(I_{\text{tot}} - P_k)} = \frac{W P_k}{\rho_k r_k(I_{\text{tot}} - P_k)} = \gamma_{\text{target\_}k} = \gamma_k$$

where ρ_k is the voice activity factor, r_k data rate, P_k received power, W signal bandwidth and I_{tot} includes overall interference, from its own cell, other cells and noise. The previous relation can be also represented as

$$P_k = \frac{1}{1 + \dfrac{W}{\gamma_k r_k \rho_k}} I_{\text{total}} = w_k I_{\text{total}}$$

with

$$w_k = \frac{1}{1 + \dfrac{W}{\gamma_k r_k \rho_k}}$$

where w_k is referred to as the load factor.

Let $\mathbf{K} = \{1, 2, \ldots, M\}$. Define: $\mathbf{n} = (n_k,\ k \in \mathbf{K})$ is the occupancy vector, where n_k is the number of class-k calls in progress; $\mathbf{w} = (w_k,\ k \in \mathbf{K})$ is the load vector, where w_k is the average load factor of a class-k connection. For all $k \in \mathbf{K}$, let r_k be the bit-rate, γ_{target_k} the target SIR and ρ_k the activity factor of class-k data-sources. The most common way to define the average capacity of a cell in CDMA cellular systems is based on the load equation in the boundary condition of the cell tolerable (required) interference level, $I_{0\text{req}}$. Let C_{a} be an average capacity of a cell. This represents the spectral efficiency and can be expressed as follows:

$$\lim_{I_{\text{total}} \to I_{0\text{req}}} \left(\frac{I_{\text{own}}}{I_{\text{total}}} \right)_{\text{avrg}} = \overline{\sup} \sum_{k \in \mathbf{K}} n_k w_k \equiv C_a \tag{12.2}$$

where I_{own} is the own-cell interference level, i.e. the received power caused by signals of active users in the cell, and I_{total} is the total interference level, i.e. the total received power in UL including thermal noise and interference caused by other cells. It is shown in References [46, 53] that the mean and the variance of other-cell interference can be approximated by own-cell interference multiplied by a constant coefficient f. Let η be the fraction of the thermal noise density with respect to $I_{0\text{req}}$. The left side of Equation (12.2) can be rewritten as [46]:

$$\lim_{I_{\text{total}} \to I_{0\text{req}}} \left(\frac{I_{\text{own}}}{I_{\text{total}}} \right)_{\text{avrg}} = \frac{1 - \eta}{1 + f} \tag{12.3}$$

Table 12.4 summarizes the parameters needed to calculate w_k and C_{a}.

For example, with f equal to 40 % and η equal to 10 %, C_{a} is around 65 % and w_k for a voice-service connection is around 0.5 %, which means the system can serve at-most $n_k = C_{\text{a}}/w_k = 130$ voice calls simultaneously. It has been shown theoretically and by simulations [41–46] that the values of SIR and f can vary significantly due to the changes of propagation parameters, TPC inaccuracy, traffic distributions, etc. For instance, SIR has been found in References [42, 46] as a log–normal random variable with total error standard deviation of the order of 2 dB. The parameter f varies between 25 and 55 % in a microcellular environment with omnidirectional antennas [53]. The fluctuations of local average SIR and other-cell interference distribution have essential impacts on system capacity resulting in the 'soft-capacity' feature of CDMA systems.

Based on the knowledge of local average SIR and interference distributions, their bounds, means and variances, the resource consumption of connection- and cell-basis can be modeled as follows. For all $k \in \mathbf{K}$, the resource consumption of a class-k connection is represented by average load factor of its service class, which is a stationary, independent and bounded

Table 12.4 Parameters for microcellular multiservice WCDMA system

Parameter	Definition	Values
W	WCDMA chip-rate	3.84 Mcs
f	Coefficient of the equilibrium other-to-own cell interference	$40 \pm 15\,\%$ in microcellular
η	Coefficient of the thermal noise density and $I_{0\mathrm{req}}$	-10 dB
C_{a}	Mean of stationary system capacity (SSC)	0.6429
C_{l}	Lower limit of SSC	0.5806
C_{u}	Upper limit of SSC	0.7200
Σ^2	Variance of SSC	9e-03
L	Cell perimeter	200 m (microcellular)
A	Cell area with hexagon shape	$2.5981 \times 10^4 \mathrm{m}^2$
$E[v]$	Mean of users' velocity	1 m/s
r_1	Bit-rate for class 1	12.2 kbs (EFR voice)
$\gamma_{\mathrm{target}\text{-}1}$	SIR target to meet class 1 QoS (BER)	6 ± 1 dB for BER $= 10 \times 10^{-3}$
ρ_1	Voice activity factor during the call	0.4
r_2	Bit-rate for class 2	64 kbs (video)
$\gamma_{\mathrm{target}\text{-}2}$	SIR target to meet class 2 QoS	2.75 ± 0.75 dB for BER $= 10 \times 10^{-5}$
r_3	Bit-rate for class 3	144 kbs (multimedia)
$\gamma_{\mathrm{target}\text{-}3}$	SIR target to meet class 3 QoS	2 ± 0.5 dB for BER $= 10 \times 10^5$
w_1	Mean of class-1 connection-basis effective load factor (CELF)	0.0050
$w_{\mathrm{l},1}$	Lower limit of class 1 CELF	0.0040
$w_{\mathrm{u},1}$	Upper limit of class 1 CELF	0.0063
σ_1^2	Variance of class 1 CELF	2.69×10^{-6}
w_2	Mean of class 2 CELF	0.0304
$w_{\mathrm{l},2}$	Lower limit of class 2 CELF	0.0257
$w_{\mathrm{u},2}$	Upper limit of class 2 CELF	0.0360
σ_2^2	Variance of class 2 CELF	3×10^{-5}
w_3	Mean of class-3 CELF	0.0561
$w_{\mathrm{l},3}$	Lower limit of class 3 CELF	0.0503
$w_{\mathrm{u},3}$	Upper limit of class 3 CELF	0.0625
σ_3^2	Variance of class 3 CELF	7×10^{-5}
$1/\mu_1$	Mean call-holding time	120 s
ξ	Empirical factor in $\Theta(c)$	0

normal-distributed random variable having mean w_k and falling in between a max- and min-guaranteed resource consumption $w_{u,k}$ and$w_{l,k}$. Similar to the well-known effective bandwidth concepts, this can be easily extended for modeling resource consumption of a VBR or ABR connection based on extensive reported results in analysis of, e.g., ATM networks. The cell-capacity is modeled accordingly with a bounded Gaussian random variable having mean C_a and lying in between C_l and C_u. The mean and boundary values of resource consumption and cell-capacity can be estimated using Equations (12.1) and (12.3) together with mean and boundary values of local average SIR target and f. Further, C_l and C_u constraints can be more relaxed to determine. For example C_u can be set up to $(1 - \eta)$ in theory.

The most straightforward CAC scheme is to share C_a units of the cell resources completely and statically among calls of different service classes, where a class-k call consumes w_k units of resources for its service. This leads to product-formed loss network model with modeling-based static complete-sharing CAC (MdCAC), where a call request for class-k service is accepted immediately if at most $(C_a - w_k)$ units of resources are occupied, otherwise rejected. Such MdCAC in CDMA cellular systems may face the following problems:

(1) the QoS guaranteed scenario results in the 'worst case' of system capacity and RRU;

(2) there is a need for redesign of the system-capacity constraint C_a for the changes of CDMA radio channel parameters and traffic statistics;

(3) the advantages of CDMA techniques, such as the 'soft capacity' feature, cannot be exploited.

The results of recent research have emphasized that robust CAC policies may require online measurements [57, 59], resulting in far more complex and high-cost SW/HW implementations [47, 49].

The principle of measurement-based CAC (MsCAC) is that a new call is accepted immediately into the system if at least the required resource consumption of its service class is available, otherwise rejected. The decision is made based on online measurements of the related statistics. Analysis and simulation results show that, when the UL traffic is less bursty, the measurement-based CAC has no capacity-gain over the modeling-based CAC described above. It may even suffer degradation due to measurement errors, which agrees with References [49, 57, 58].

To harmonize the advantages and overcome the problems of MdCAC and MsCAC with a simple and effective SCAC policy, let us first reconsider Equations (12.2) and (12.3) in a slightly different way. Owing to the interference-limited nature of CDMA cellular networks, **n**-forming active-user combinations in the cell shall meet their QoS constraints if:

$$c_{\text{other}} \leq (1 - \eta) - c_{\text{own}} \tag{12.4}$$

where c_{other} represents the load factor produced by transforming the other-cell interference; $c_{\text{own}} = \mathbf{nw}$ represents load factor generated by **n**-forming active-user combinations in the cell, **n** is the occupancy vector and **w** is the load vector defined above. Because it has been shown that c_{other} is well modeled with a Gaussian random variable [41–43, 46], we have:

$$\Pr\{c_{\text{other}} \leq 1 - \eta - c_{\text{own}}\} = 1 - Q\left(\frac{1 - \eta - c_{\text{own}} - E[c_{\text{other}}]}{\sqrt{Var[c_{\text{other}}]}}\right) \tag{12.5}$$

where $Q(x)$ is the standard normal integral function and

$$E[c_{\text{other}}] = fE[c_{\text{own}}] \tag{12.6}$$

$$Var[c_{\text{other}}] = f\,Var[c_{\text{own}}] \tag{12.7}$$

The above relations imply that, as UL traffic increases beyond the average capacity, whether users experience a degradation of the communication quality depends on the other-cell interference level. Therefore, an optimal way to benefit from this soft-capacity feature is to use a dynamic CAC mechanism having decision functions adapted to interference distributions to compensate for the fluctuation of local average SIR while expanding the admissible region. SCAC is such a solution that works as follows. Instead of using an average cell capacityC_{a}, use an upper- and a lower-limit denoted by C_{u} and C_{l}, respectively. This improves the system capacity, at the same time enhances the robustness. In principle, the mechanism of SCAC is defined as follows:

(1) a new call of class-k shall be admitted immediately into the system if the current UL load factor is at most $(C_l - w_k)$;

(2) else if the current own-cell UL load factor is at most $(C_u - w_k)$, a new call of class-k shall be admitted with a permission probability based on Equation (12.5), which is given in detail later;

(3) otherwise, a new call of class-k shall be rejected immediately;

(4) nonpre-emptive priority is given to handoff calls, i.e. handoff failure of class-k calls occurs only when more than $(C_{\text{u}} - w_k)$ resources have been occupied already.

SCAC policy is supposed to give advantages over both MdCAC and MsCAC policies, i.e. simple, robust and improved capacity utilization. Soft decisions can also be combined with measurement-based techniques to reduce the complexity of estimators, to compensate for bias and to enhance performance. The implementation issues will be addressed later. Moreover, one can notice that SCAC provides better traffic shaping gain than MdCAC and MsCAC. It gives more chance for calls of high resource-consuming classes to access the system when traffic intensity of lower classes is heavy that cannot be improved with the other two.

12.4.2 QoS differentiation paradigms

To regulate the operation of multimedia systems in the way mentioned above, the access control has to consider not only characteristics of requested services but also subscription profiles of user classes. This will provide fair decisions in resource allocation for agreed QoS upon accommodating a new call request. For example, a gold-class customer request for any services shall be served immediately as long as there are enough network resources for that. On the other hand, operators might reject all bronze-class requests as well as resource-consuming requests from silver-class if the current load has already exceeded a certain level, and so forth. However, once a call request is accepted, QoS in terms of BER for that RT connection is assured independently of its associated user class. QoS differentiation paradigms for CBR services specify such serving traffic patterns depending on offered traffic intensity distributions among classes during different busy periods of the day for shaping traffic and maximizing revenues. These therefore need to be simple and easy to reconfigure and extend for a variety of user and network profile characteristics. QoS differentiation can be viewed as traffic prioritization in CAC. Herein, we use multiple thresholds or load-based fracturing factors for hard or softer blocking of call requests. The methods are similar to the well-known threshold dropping or weighted fair queueing mechanisms, but no actual queues are allowed in this system. Thus, system resources are shared noncompletely among RT traffic classes in a blocked-call-cleared fashion. Define: $c = \mathbf{nw}$ is the stationary system load state, which is identical to c_{own} above. Suppose there are J different user classes sorted in the decreasing order of priority, e.g. 1 is gold, 2 is silver, etc. Let $\mathbf{J} = \{1, 2, \ldots, J\}$. Taking M different service classes into account, which are sorted in the increasing order of resource consumption, we introduce a $J \times M$ table of prioritized admission probability given in matrix-form as follows:

$$\mathbf{A} = [a_{jk}(c)]\mathrm{j} \in \mathbf{J},\ \mathrm{k} \in \mathbf{K}, 0 \le a_{jk}(c) \le 1 \tag{12.8}$$

where $a_{jk}(c)$ is the admission probability of a call request for class-k service from a class-j user, which depends on system load state and needs further modification when combining with SCAC described above. However, it can be considered that there are $J \times M$ prioritized traffic classes, which form group behaviors according to user class or service class.

Note that hereafter we use term 'traffic class' or class-(j, k) when QoS differentiation is applied to distinguish from service class-k alone in complete-sharing scenarios. In hard threshold blocking case, straightforward, $J \times M$ thresholds may be needed, each corresponding to a traffic class. Define a threshold table given in matrix-form as follows:

$$\mathbf{L} = [l_{jk}]j \in \mathbf{J}, k \in \mathbf{K}, l_{jk} \ge l_{uv} \text{ if } j \ge u \text{ and } k \ge v$$

Thus, not taking effects of SCAC into account, $a_{jk}(c)$ can be determined as:

$$a_{jk}(c) = \begin{cases} 1 & \text{if } c + w_k \le l_{jk} \\ 0 & \text{otherwise} \end{cases} \tag{12.9}$$

The number of needed thresholds is significantly reduced if only group behavior levels are of interest. That is, J or M thresholds may be needed instead of $J \times M$. For example, the first priority user class has a single threshold up to upper bound C_{u} of the system capacity regardless of required services, whereas the lowest priority user class has a threshold as lower bound C_{l} of the system capacity for any services. The access of the lowest priority user class would not be granted for k to M services if system load state exceeded $l_{Jk} \le C_{\text{l}}$. In fractional (soft) blocking case, $a_{jk}(c)$ may not take a hard value of 1 or 0, but in between depending on system load state and priority of traffic class, resulting in load-based fracturing factors for each traffic class or group behavior. As a simple example, let say a call request from a class-j user for class-k service is admitted immediately into the system if the load state after that does not exceed the l_{jk} threshold; else it is accepted with a probability of 0.8 if the system load state after that does not exceed the lower bound C_{l} of the system capacity; else it is accepted with a probability of 0.2 as long as there are enough resources left for that. Generally, this QoS differentiation paradigm on the one hand gives operators better flexibility to customize and to tune their network performance; on the other hand it allows unified analysis of a class of guard resource schemes for QoS differentiation and CAC. The hard threshold blocking case described previously is in fact a special variation of this fractional rule.

Up to this point, handoff calls have been assumed to have either the highest priority regardless of their associated traffic classes or equal priority as of new calls regarding their associated traffic classes. If not so, prioritization of handoff calls can be handled similarly to new calls.

QoS differentiation of the background NRT packet-switched services can be done not only in blocking or dropping fashion, but also in granting throughput-delay of a connection basis. For instance, the transmissions of certain customers should be guaranteed with higher data-rate and lower delay. Detailed paradigms and mechanisms are discussed later.

12.4.3 Traffic model

The traffic model is based on the following assumptions:

(1) Sources are of the ON–OFF nature. Unless stated otherwise, $\rho_k = 1$ for all $k \in \mathbf{K}$. The impacts of employing user activity detection are precisely studied in the example of a single-class system.

(2) TPC is sufficient to ensure that the standard deviation of the local average SIR is always in control. The cell-capacity as well as the resource consumption of each class-k connection for all $k \in \mathbf{K}$ is a bounded Gaussian random variable lying in $[C_{\mathrm{l}}, C_{\mathrm{u}}]$ and $[w_{\mathrm{l},k}, w_{\mathrm{u},k}]$ respectively with means and variances given in Table 12.1.

(3a) Without QoS differentiation users and calls are served on complete-sharing FCFS (first come first served) basis. For all $k \in \mathbf{K}$, class-k calls arrive at the corresponding cell according to Poisson process with rates $\lambda_{\mathrm{l},k}$ and $\lambda_{\mathrm{hl},k}$ for the new and handoff calls respectively.

(3b) With QoS differentiation, for all $j \in \mathbf{J}$ and $k \in \mathbf{K}$, let $\lambda_{\mathrm{l},jk}$ and $\lambda_{\mathrm{hl},jk}$ be Poisson arrival rates for new and handoff calls respectively of traffic class-(j, k).

(4) The mean call-holding time given that there are no handoff failures is $1/\mu_1$, and the outgoing handoff rate per cell per call is μ_2, which is commonly valid for all classes of calls. These are the exponentially distributed random variables. Therefore, the service time in the cell, i.e. cell-resident time, is also an exponentially distributed random variable having a mean equal to $1/\mu = 1/(\mu_1 + \mu_2)$. In the condition of user mobility equilibrium, i.e. the mean number of incoming mobile terminals equal to the mean number of outgoing mobile terminals per time-unit per cell, there are two simple models to determine the value for μ_2 [60]. Denote: v as the speed of a mobile terminal, L as the length of cell-perimeter and A as the cell-area. In a macrocellular environment, terminals usually move with high speed along the cell in one direction. The linear model can be used to approximate μ_2 as follows:

$$\mu_2 = E[v]/L \tag{12.10}$$

In a microcellular environment represented by a two-dimensional model with user random movement, μ_2 is given by:

$$\mu_2 = E[v]L/\pi A \tag{12.11}$$

With assumption (3a), the arrival rate of handoff calls can be approximated also as in [61] and [62]:

$$\lambda_{hl,k} = \lambda_{l,k}(1 - B_k)(\mu_1/\mu_2 + F_k)^{-1} \tag{12.12}$$

where B_k is the new-call blocking probability and F_k is the handoff failure probability of class-k. With assumption (3b), we have:

$$\lambda_{\text{hl},jk} = \lambda_{1,jk}(1 - B_{jk})(\mu_1/\mu_2 + F_{jk})^{-1} \tag{12.13}$$

where B_{jk} is the new-call blocking probability and F_{jk} is the handoff failure probability of class-(j, k).

(5) Three different CAC policies are adopted alternatively for the performance comparison purposes. The first and the second one are the modeling-based MdCAC and the measurement-based MsCAC of static complete-sharing policies. The third one is the SCAC policy, which is discussed first in complete-sharing and then in QoS differentiation scenarios.

For the complete-sharing SCAC, define: $\pi_k(c) \equiv$ the admission probability of a new call for class-k service when the system is in state c. From Equations (12.45), (12.46) and (12.47), $\pi_k(c)$ can be given by:

$$\pi_k(c) = \begin{cases} 1 \text{ , if } c + w_k \leq C_l \\ 1 - Q\left(\dfrac{1 - \eta - c - w_k - fE[c]}{\sqrt{fVar[c]}}\right) \text{, if } C_l < c + w_k \leq C_u \\ 0 \text{ , otherwise} \end{cases} \tag{12.14}$$

To avoid the overestimate of C_u that may increase the system outage probability, some adjustments of the mean and variance, e.g. by increasing them slightly, or other functions, such as the incomplete Gamma, can be used. For the later option, $\pi_k(c)$ becomes:

$$\pi_k(c) = 1 - \Gamma[\nu, \alpha(1 - \eta - c - w_k)] \text{ for } C_l < c + w_k \leq C_u \tag{12.15}$$

where $\Gamma(\nu, x)$ is the incomplete Gamma distribution function. The α and ν parameters have the following relations:

$$fE[c] = \nu/\alpha \text{ and } f\text{Var}[c] = \nu/\alpha^2$$

In general, $E[c]$ and Var$[c]$ can be calculated by using steady-state solutions and inversion techniques. To reduce the computation complexity, the corresponding mean and variance values of MdCAC system can be reused. This is reliable because, with assumptions (1) and (2) above, the loss network model can be expected to represent well the stationary behavior of those systems. For SCAC with QoS differentiation, by invoking Equation (12.8) we have:

$$a_{jk}(c) = a_{0\_jk}(c)\pi_k(c) \quad \text{for all } j \in \mathbf{J} \text{ and } k \in \mathbf{K} \tag{12.16}$$

where the $a_{0\_jk}(c)$ factor is the admission probability of traffic class-(j, k) determined by QoS differentiation paradigms; the $\pi_k(c)$ factor comes from optimal interference management of SCAC given by Equation (12.14) or (12.15). The connection-oriented RT traffic has higher priority over the NRT traffic, which is served as background services. For clarity, additional modeling issues and assumptions for NRT packet access and services are discussed later.

12.4.4 Performance evaluation

The performance measures of interests, which characterize the GoS, are the probabilities of new-call blocking and handoff failure. In optimization framework, a linear combination of these probabilities is usually used to define the GoS as a cost-function. The weighting factors are usually decided based on the 'importance' of calls considering new-call or handoff, user classes, cost of service classes, etc. For example, GoS for a single-class system in Zander Kim [50] has been defined as: $GoS \equiv (10F + B)$, where F is the handoff failure probability and B is the new-call blocking probability. The other important performance measure is the equilibrium QoS loss probability, i.e. the probability of losing communication quality in the system outage state. The Erlang capacity of the system can be determined for given percentages of these loss probabilities. In complete-sharing systems, for all $k \in \mathbf{K}$ and in steady-state condition, denote: B_k as the new-call blocking probability of class-k calls; and F_k as the handoff failure probability of class-k calls. In systems with QoS differentiation, for all $j \in \mathbf{J}$ and $k \in \mathbf{K}$ and in steady-state condition, denote: B_{jk} as the new-call blocking probability of traffic class-(j,k); and F_{jk} as the handoff failure probability of traffic class-(j, k).

In addition, operators may be interested in group behavior. Thus, for all $j \in \mathbf{J}$ and in steady-state condition, denote: B_j as the new-call blocking probability of class-j users; and F_j as the handoff failure probability of class-j users. Finally, denote: P_{loss} as the equilibrium QoS loss probability.

12.4.5 Related results

In this subsection, the results of extensive studies on product-form loss networks, stochastic knapsack problems and effective-bandwidth allocation [63–68], are summarized, which supports analytical methods used in this section. The single-link with no buffering and complete-sharing loss system is considered. Let the system capacity be C and the offered traffic intensity of class-k is $\alpha_k = \lambda_k/\mu_k$, where λ_k is the Poisson arrival rate and $1/\mu_k$ is the mean service time of class-k call. Define the set of all possible system states:

$$\Omega = \left\{\mathbf{n} : \sum_{k \in \mathbf{K}} n_k w_k \leq C\right\} \tag{12.17}$$

The steady-state probability of system being in state $\mathbf{n}$ has a product-form:

$$p(\mathbf{n}) = \frac{1}{G} \prod_{k \in \mathbf{K}} \frac{\alpha_k^{n_k}}{n_k!} \quad \mathbf{n} \in \Omega \tag{12.18}$$

where

$$G = \sum_{\mathbf{n} \in \Omega} \prod_{k \in \mathbf{K}} \frac{\alpha_k^{n_k}}{n_k!} \tag{12.19}$$

The blocking probability of class-k calls can be determined theoretically by:

$$B_k = \sum_{\mathbf{n} \in \Omega : \mathbf{n}\mathbf{w} + w_k \geq C} p(\mathbf{n}) \tag{12.20}$$

The cost of computation with the above formula can be prohibitively high for larger-size state set, i.e. for large $\mathbf{K}$ and C/w_k. This problem has been considered by many authors, resulting in elegant and efficient recursion techniques for the calculation of blocking probabilities. For practical evaluation purposes of this section, i.e. keeping it accurate, but as simple as possible without loss of generality, the stochastic-knapsack approximation described in References [63, 66] is chosen and believed to suite best investigating various load-based and cost-effective CAC policies.

Define the set of all feasible system load states:

$$\Psi = \{c : c = \mathbf{nw}, \mathbf{n} \in \Omega\} \tag{12.21}$$

The steady-state probability of the load state c is given by:

$$s(c) = \frac{q(c)}{\sum\limits_{c \in \Psi} q(c)} \tag{12.22}$$

where $q(c)$ is given in a recursive form as follows:

$$q(c) = \frac{1}{c} \sum_{k \in \mathbf{K}} w_k \alpha_k q(c - w_k); c \in \Psi^+, q(0) = 1 \text{ and } q(-) = 0 \tag{12.23}$$

The blocking probability of class-k calls now can be determined by:

$$B_k = \sum_{c \in \Psi : c > C - w_k} s(c) \tag{12.24}$$

12.4.6 Modeling-based static complete-sharing MdCAC system

The above results can be applied directly for analyzing this system with the parameters as follows: the system capacity is C_a instead of C; $\alpha_k = (\lambda_{1,k} + \lambda_{hl,k})/(\mu_1 + \mu_2)$. Then, B_k is obtained by using Equation (12.24):

$$F_k = B_k \tag{12.25}$$

$$P_{loss} = \frac{\sum\limits_{c \in \Psi} cs(c) Q\left(\dfrac{(1 - \eta - c) - fE[c]}{\sqrt{fVar[c]}}\right)}{\sum\limits_{c \in \Psi} cs(c)} \tag{12.26}$$

where

$$E[c] = \sum_{c \in \Psi} cs(c) \tag{12.27}$$

$$\mathrm{Var}[c] = \sum_{c \in \Psi} c^2 s(c) - E^2[c] \tag{12.28}$$

Given the offered traffic intensity of each service class, the performance of the system strictly depends on C_a and $\mathbf{w}$, which are needed a *priori*. Overestimates or underestimates of these parameters may result in wasting or insufficiently allocating resources, thereby reducing the GoS or lowering the perceived QoS, respectively. Furthermore, in the presence of additional uncertainty causing the changes in traffic descriptor parameters, the system cannot provide appropriate QoS to the users and therefore C_a and $\mathbf{w}$ need to be redesigned.

However, with a reliable modeling, good overall statistical multiplexing gain and performance can be expected. From the TDMA/GSM experiences, static channel allocations in some circumstances can even provide better capacity gain than dynamic ones. Note that C_a and $\mathbf{w}$ can be improved, i.e. maximizing C_a while keeping $\mathbf{w}$ as low as possible for given QoS requirements, by using techniques like sectorization, multiuser detection, smart antennas, etc.

12.4.7 Measurement-based complete-sharing MsCAC system

MsCAC should overcome the nonrobustness problem of MdCAC described above. The network side attempts to learn statistics of the traffic by online measurements. Under the assumption (2) of the traffic model, stationary w_k and C_a are bounded Gaussian random variables having mean w_k and C_a, and variance σ_k^2 and Σ^2 respectively. Here, C_a and w_k denote the random variables and also their mean values. The variances can be estimated by using corresponding boundary and mean values. Let β be the acceptable equilibrium outage probability of the system.

The goal is to quantify and to illustrate the system behavior with its sensitivity to estimation errors, and to lay the groundwork for studying the scaled aggregate UL load fluctuations of the traffic with more sophisticated models. It does not aim to study details of any specific measurement-based CAC algorithms, neither to provide the parameter estimation framework. Some valuable results for measurement-based CAC in single-class systems can be found in Dziong *et al.* [57], whereas Sampath and Holtzman [58] investigate measurement-based CAC algorithms by simulations. Herein, the multivariate Gaussian approximation is used to dimension the admissible region. In the MsCAC system, stationary constraints of the system capacity can be defined by:

$$\sum_{k\in\mathbf{K}} n_k w_k + Q^{-1}(\beta)\left(\sum_{k\in\mathbf{K}} n_k \sigma_k^2\right)^{1/2} \leq C_a + Q^{-1}(\beta)\Sigma \tag{12.29}$$

The components of Equation (12.29) are obtained by online estimations. The additive uncertainty due to measurement errors may have significant impacts on the system performance. In addition, the need for reliable estimations of mean and variance values for various service classes and cell-basis may result in far more complex hardware/software implementations compared with the modeling-based policies. The performance characteristics of the MsCAC system will be obtained by simulations of a simple memoryless and an auto-regressive measurement-based system. The over-bounding hyperplane of the admissible region can be determined by intersection of the axes in M-dimensional Euclidean space. The intersection point $(0, 0, \ldots, c_k, 0, \ldots, 0)$ corresponds to the single-class capacity region of class-k that is given in estimation by:

$$c_k = \frac{\hat{C}_\mathrm{a}}{\hat{w}_k} + Q^{-1}(\beta)\left(\hat{\Sigma} - \frac{\hat{\sigma}_k}{\hat{w}_k}\sqrt{\frac{\hat{C}_\mathrm{a}}{\hat{w}_k}}\right) \tag{12.30}$$

For the Gaussian estimation errors, c_k can be quantified by:

$$c_k \approx \frac{C_\mathrm{a}}{w_k} + Q^{-1}(\beta)\left(\Sigma - \frac{\sigma_k}{w_k}\sqrt{\frac{C_\mathrm{a}}{w_k}}\right)\varepsilon - \delta\sqrt{\frac{C_\mathrm{a}}{w_k}} \tag{12.31}$$

where ε, δ are $N(0, 1)$ normal random variables [57] representing impacts of measurement errors. It has been shown in Sampath and Holtzman [58] by simulations that measurement-based CAC policies are best suited for serving quite bursty traffic. Their performance is almost the same, and none of them are capable of meeting the loss targets accurately. One can reach the same conclusion from the above analysis.

12.4.8 Complete-sharing dynamic SCAC system

It has been shown [64, 66, 67] that for larger C/w_k systems, which is true in WCDMA cellular systems, the stationary behavior of system load states c approaches the one-dimensional Markov process. Thus, for analysis of the SCAC system we invoke Equations (12.21)–(12.23) with the following modifications. The system capacity is now C_u. The behavior of the system when the system load state is less than C_l is exactly the same as that of the complete-sharing system. When the system load state is between C_l and C_u, a soft decision is used for the acceptance of a new call. The highest priority is given to a handoff call. Equation (12.23) now becomes:

$$q(c) = \frac{1}{c} \sum_{k \in \mathbf{K}} w_k \alpha_k (c - w_k) q(c - w_k); c \in \psi^+, q(0) = 1 \text{ and } q(-) = 0 \tag{12.32}$$

where

$$\alpha_k(c) = \frac{\lambda_{hl,k} + \lambda_{l,k} \pi_k(c)}{(\mu_1 + \mu_2)[1 + \Theta(c)]} \tag{12.33}$$

$\pi_k(c)$ is the admission probability of a new call given by Equation (12.11) or (12.12); $\Theta(c)$ is the function of system load state representing the call-drop rate due to loss of the communication quality. For instance, a linear model for $\Theta(c)$ with a constant empirical weight-factor ξ can be defined as follows:

$$\Theta(c) = \begin{cases} 0 \text{ if } c \leq C_l \\ \xi(c - C_l)/(C_u - C_l) \in [0, 1] \text{ if } C_l < c \leq C_u \end{cases} \tag{12.34}$$

The performance measures are obtained as follows:

$$B_k = \sum_{c \in \Psi : c > C_l - w_k} s(c)[1 - \pi_k(c)] \tag{12.35}$$

$$F_k = \sum_{c \in \Psi : c > C_u - w_k} s(c) \tag{12.36}$$

$$P_{\text{loss}} = \frac{\sum_{c \in \Psi} c s(c) Q \left(\dfrac{(1 - \eta - c) - f E[c]}{\sqrt{f} \operatorname{Var}[c]} \right)}{\sum_{c \in \Psi} c s(c)} \tag{12.37}$$

With this SCAC, GoS including new-call blocking probability and handoff failure probability can be significantly improved over MdCAC and MsCAC. However, in complete-sharing systems, beyond a certain load state, higher resource-consuming users hardly can gain access into the system if the traffic intensity of lower resource-consuming users is relatively heavy. This can be overcome by using QoS differentiation combined with SCAC for meeting potential customer demands.

12.4.9 Dynamic SCAC system with QoS differentiation

Assume that handoff calls have the highest priority regardless of their user class association. For characterization of the steady system load states, we invoke Equations (12.17), (12.21), (12.22), (12.32) and (12.34) and the assumption ($c2$). Equation (12.32) is now modified as follows:

$$\alpha_k(c) = \frac{\sum_{j\in \mathbf{J}} \lambda_{hl,jk} + \lambda_{l,jk} a_{jk}(c)}{(\mu_1 + \mu_2)\,[1 + \Theta(c)]} \tag{12.38}$$

The performance measures of interests for this system are given as:

$$B_{jk} = \sum_{c\in\Psi} s(c)\left[1 - a_{jk}(c)\right] \tag{12.39}$$

$$F_{jk} = \sum_{c\in\Psi: c>C_u - w_k} s(c) \tag{12.40}$$

By summing up B_{jk} or F_{jk} over $\mathbf{J}$ or $\mathbf{K}$ set, we can obtain loss probabilities for group behaviors. Thus:

$$B_j = \sum_{k\in\mathbf{K}} B_{jk} \tag{12.41}$$

$$F_j = \sum_{k\in\mathbf{K}} F_{jk} \tag{12.42}$$

The equilibrium QoS loss probability can be obtained by using Equation (12.37). By updating admission probability table, $\mathbf{A} = [a_{jk}(c)] j \in \mathbf{J}, k \in \mathbf{K}, 0 \le a_{jk}(c) \le 1$, as defined in Equation (12.8), operators can easily control and optimize the GoS of their serving networks.

12.4.10 Example of a single-class system

This section provides a simple probabilistic interpretation of the SCAC analysis presented above based on a single-class system. In this case, the analysis as well as the implementation can be very much simplified and presented by numbers. However, the insight into system behavior is still preserved and well represented. Let the call arrival rates at the corresponding cell be λ_{l} and λ_{h} for new and handoff calls respectively, r the bit-rate and γ_{target} the SIR target of the service. Let ρ be the probability that source is active during its conversation, i.e. actually transmits data independently from other sources. Denote n as the number of users occupying the cell, i.e. the number of calls being served in the system and j as the number of active users among n users. The average capacity, i.e. average upper limit of the number of admissible calls in the cell, can be expressed as follows:

$$c_{\mathrm{avrg}} = \left\lfloor \frac{(W/r)(1-\eta)}{\gamma_{\mathrm{target}}(1+f)\rho} \right\rfloor \tag{12.43}$$

where $\lfloor x \rfloor$ is the maximum integer not exceeding the argument. For the fixed channel assignment-based CAC system, Erlang's B-formula with the number of servers equal to c_{avrg} can be used to derive the performance measures. For the SCAC system, the other-cell interference impacts can be expressed by:

$$j + m \le \frac{(W/r)(1-\eta)}{\gamma_{\text{target}}} = c_0 \tag{12.44}$$

where m is a nonnegative number obtained by transforming other-cell interference into an equivalent non-integral number of active users. Since the other-cell interference is modeled as a Gaussian random variable, we have:

$$\Pr\{m \le c_0 - j\} = 1 - Q\left(\frac{(c_0 - j) - E[m]}{\sqrt{Var[m]}}\right) \tag{12.45}$$

where $E[m]$ and Var$[m]$ can be obtained as:

$$E[m] = fE[j] \tag{12.46}$$

$$\text{Var}[m] = f\,\text{Var}[j] \tag{12.47}$$

The lower and upper limits of the cell-capacity are given by c_{l} and c_{u}, respectively. The permission probability of a new call at state n is defined as follows:

$$\pi(n) = \begin{cases} 1 \text{ if } n < c_l \\ 1 - \sum\limits_{j=0}^{n} b(j;n,\rho) Q\left(\dfrac{(c_0 - j) - fE[j]}{\sqrt{f\text{Var}[j]}}\right) \text{ if } c_{\text{l}} \le n < c_{\text{u}} \\ 0 \quad \text{otherwise} \end{cases} \tag{12.48}$$

where $b(j;\, n,\, \rho)$ is the binomial distribution:

$$b(j;n,\rho) = \binom{n}{j} \rho^j (1-\rho)^{n-j} \tag{12.49}$$

The handoff request is accepted right away if fewer than c_{u} calls are being served in the cell, otherwise it is rejected. Denote p_n as the steady-state probability of the system being in state n. The general steady-state solution of the birth–death process has a form:

$$p_n = p_0 \prod_{i=1}^{n} \frac{\lambda_{i-1}}{\mu_i} \text{ for } n \ge 1 \tag{12.50}$$

where λ_n and μ_n are the birth and death rates at state n, respectively. In our system model:

$$\lambda_n = \begin{cases} \lambda_{\text{h}} + \pi(n)\lambda_l \ \text{ if } 0 \le n < c_{\text{u}} \\ 0 \text{ otherwise} \end{cases} \tag{12.51}$$

$$\mu_n = n(\mu_1 + \mu_2) \text{ for } 1 \le n \le c_{\text{u}} \tag{12.52}$$

Therefore, the steady-state probabilities can be determined using Equations (12.50)–(12.52) and the following condition:

$$\sum_{n=0}^{c_{\text{u}}} p_n = 1 \tag{12.53}$$

Define s_j as the steady-state probability that there are exact j active users in the cell:

$$s_j = \sum_{n=j}^{c_{\text{u}}} b(j;n,\rho) p_n \tag{12.54}$$

where $b(j; n, \rho)$ is given by Equation (12.49). $E[j]$ and $\mathrm{Var}[j]$ in Equation (12.48) are given by:

$$E[j] = \sum_{j=0}^{c_u} j s_j \tag{12.55}$$

$$\mathrm{Var}[j] = \sum_{j=0}^{c_u} j^2 s_j - E^2[j] \tag{12.56}$$

Closed-form solution of the performance measures are obtained as follows:

$$B = \sum_{n=c_l}^{c_u} p_n[1 - \pi(n)] \tag{12.57}$$

$$F = p_{c_u} \tag{12.58}$$

$$P_{\text{loss}} = \frac{\sum_{j=0}^{c_u} j s_j Q\left(\frac{(c_0 - j) - fE[j]}{\sqrt{f\mathrm{Var}[j]}}\right)}{\sum_{j=0}^{c_u} j s_j} \tag{12.59}$$

Figure 12.10 depicts the above characteristics for a voice-only system with voice-source activity factor $\rho = 0.4$. In order to ease the computations, the system parameters given in Table 12.5 correspond to the IS-95 system [48, 49]. It shows in comparison with Erlang's loss system that the SCAC system gains significant improvement in GoS. Overall, the SCAC system offers much better Erlang capacity and performance.

12.4.11 NRT packet access control

As mentioned before, NRT packet access in UL needs to be controlled so that QoS as well as GoS of the RT traffic is not affected by optimal throughput-delay tradeoff. For

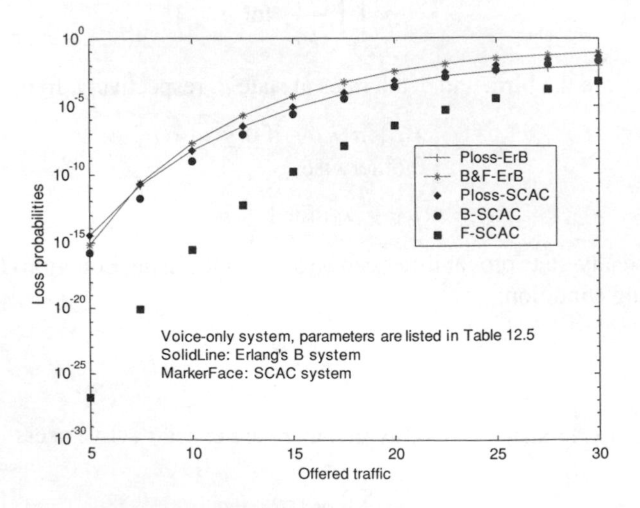

Figure 12.10 Benefits of SCAC in a voice-only system.

Table 12.5 Parameter for single-class voice-only CDMA system

	Definition	Values
W	CDMA chip rate	1.25 Mcs
r	Bit-rate of the users	9.6 kbs
γ_{target}	SIR to meet the required QoS, i.e. BER target	7 ± 1 dB for BER $= 10 \times 10^{-3}$
f	Coefficient of the equilibrium other-to-own cell interference	40 ± 15 %
η	Constant coefficient of the thermal noise density and $I_{0\text{req}}$	-10 dB
ρ	Voice-source activity factor during the call	0.4
c_0	Maximum number of active users in an isolated cell	23
c_{avrg}	Average system capacity	32
c_{l}	Lower limit of the system capacity	25
c_{u}	Upper limit of the system capacity	45

design of such dynamic control mechanisms, there is a need for precise understanding of the dynamics of available UL radio resources for NRT packet transmissions while RT traffic is being served adequately. The main purpose of this section is to provide a reliable method based on asymptotic and quasi-stationary analysis of the RT traffic, to predict free capacity and average upper limits of UL data throughput. Based on the prediction results, a simple and effective DFIMA scheme is proposed for NRT packet access control.

12.4.12 Assumptions

Denote R_{p} as the bit-rate for packet-data transmission in UL, L_{p} as the packet-length in bits, γ_{p} as the SIR target to meet the QoS requirements of packet transmission and T_{p} as the transmit time interval (TTI) needed for a packet at the air interface, $T_{\text{p}} = L_{\text{p}}/R_{\text{p}}$. The following assumptions are made in addition to those made earlier in this section.

(1) RT services have higher priority over NRT messaging services. Resource consumption of the NRT packet-service domain should never exceed the free resources left by the RT traffic.

(2) Users are sharing common channel for NRT packet access in UL using different DS/CDMA code sequences. The amount of UL resources consumed by a packet transmission can be determined similarly to the resource consumption of an RT connection that is given in Equation (12.1). The parameters needed for packet transmissions can be predefined or decided by the access control.

(3) The time axis is divided into T_{s}-long time-slots for packet transmissions, which can be equal to one or multiple radio-frame duration of, e.g., 10 ms as defined in 3GPP standards. Packet transmissions are synchronized starting at the beginning of a time-slot.

(4a) Without QoS differentiation, T_p is set equal to T_s. Thus, packet transmissions are synchronized starting at the beginning and finishing at the end of the same time-slot. For a given time-slot, all transmissions have the same bit-rate and SIR target that may be changed on slot-by-slot basis.

(4b) For different user classes, QoS differentiation should guarantee different maximum delays and minimum data rates when sending messages, e.g. emails, pictures, etc. Therefore, packet transmissions may have different parameters (bit-rate, SIR target, TTI) and resource consumption, which are controlled by the access control taking into account QoS differentiation paradigms.

12.4.13 Estimation of average upper-limit (UL) data throughput

Let free capacity of the system left by the RT traffic at time t be $z(t)$. The stationary distribution and quasi-stationary behavior of $z(t)$ over a sustained period of time are of interest. Let $c(t)$ be the RT system load state at time t and $C(t)$ is the system 'soft' capacity at time t. Hence:

$$\begin{aligned} z(t) &= C(t) - c(t) \\ E[\lim_{t\to\infty} z(t)] &= E\{\lim_{t\to\infty}[C(t) - c(t)]\} \\ E[z] &= C_a - E[c] \end{aligned} \tag{12.60}$$

where z and c are the free capacity and the RT system load state in equilibrium condition. The z and c process itself is not Markovian in general, but for larger C_a/w_k it behaves as an approximate Markov process. The quasi-stationarity of c over time interval τ is approximated by an exponential function [64, 69].

For instance, to obtain the equivalent one-dimensional birth–death process for z and c in complete-sharing systems, the Pascal approximation can be used as in Grossglauser [67]. First, we need to scale the system capacity and resource consumption of each RT service class into integers. To do so we first assume $\min\{w_k, k \in \mathbf{K}\} \equiv w_1$. Then define the scaled load vector, the scaled system state and the scaled average system capacity as follows:

$$\mathbf{w}^* = \{w^*{}_k, k \in \mathbf{K}\} \equiv \{\lfloor w_k/w_1 \rfloor, k \in \mathbf{K}\}; c^* \equiv \lfloor c/w_1 \rfloor; \text{ and } C^*{}_a \equiv \lfloor C_a/w_1 \rfloor \text{ respectively.}$$

The set of equivalent system states is then given by:

$$\psi^* \equiv \left\{c^* : c^* = 0, 1, 2, \ldots, C_a^*\right\}$$

Normalize the mean service-time of a RT call: $\mu \equiv 1$. The equivalent birth–death process being in steady state c^* has a death rate of c^* and a birth rate of:

$$\lambda(c^*) = \upsilon^2/\omega^2 + c^*(1 - \upsilon/\omega^2) \tag{12.61}$$

where υ and ω^2 are given by:

$$\upsilon = \sum_{k\in\mathbf{K}} w_k^* \alpha_k \text{ and } \omega^2 = \sum_{k\in\mathbf{K}} (w_k^*)^2 \alpha_k \tag{12.62}$$

with $\alpha_k = (\lambda_{l,k} + \lambda_{hl,k})/(\mu_1 + \mu_2)$ or its normalized value $\alpha_k \equiv (\lambda_{l,k} + \lambda_{hl,k})$. The steady-state probability $s(c^*)$ for $c^* \in \psi^*$ satisfies:

$$c^* s(c^*) = \lambda(c^* - 1)s(c^* - 1) \text{ for } \mu \equiv 1, c^* \geq 1 \text{ and } \sum_{0}^{C_a^*} s(c^*) = 1 \tag{12.63}$$

The quasi-stationary probability of RT traffic being in load state c^* over period of time τ can be defined by:

$$p(c^*, \tau) = \lim_{t \to \infty} Pr\{c^*(t + \tau) = c^* | c^*(t) = c^*\} \tag{12.64}$$

Using the exponential property of quasi-stationary probability in loss networks mentioned above, Equation (12.64) can be approached by:

$$p(c^*, \tau) = \mathrm{e}^{-\tau[\lambda(c^*)+c^*]} \tag{12.65}$$

Equation (12.65) for $\tau = T_\mathrm{p}$ provides the equilibrium probability that there are maximum $z^* = C^*{}_\mathrm{a} - c^*$ units of resources available for packet transmissions. Let $S(c^*)$ be the upper-limit number of packets that can be transmitted successfully in a given time-slot, for a given load state of the RT service domain $c^*.S(c^*)$ can be calculated by:

$$S(c^*) = p(c^*, T_\mathrm{p}) \frac{(C_\mathrm{a}^* - c^*)w_1}{R_\mathrm{p}\gamma_\mathrm{p}/W} \tag{12.66}$$

The above equation can be used to study tradeoffs of the parameters for optimal packet access. Let us assume that there are N data users, each generating data-packets according to the Poisson process with rate Λ per time-slot, $\Lambda \leq 1$. Hence, there are $D = \lfloor N\Lambda \rfloor$ average active data-sources per slot. If each of them attempts to transmit their packets at the beginning of a given time-slot with the probability of $\min[1, S(c^*)/D]$, the probability of successful packet transmission, i.e. having less than $S(c^*)$ initiated transmissions, is at least $1/2$. This is in accordance with the binomial distribution. The average upper-limit of UL packet-data throughput denoted by S is given by:

$$S = (1 - P_\mathrm{loss}) \sum_{c^* \in \Psi^*} s(c^*)S(c^*) \tag{12.67}$$

where P_loss is the QoS loss probability. By using Little's formula, the average lower limit of packet delay is given by S/D time-slots.

12.4.14 DFIMA, dynamic feedback information-based access control

The NRT packet access in UL needs to adapt to a quasi-stationary stochastic process of free capacity left by the RT traffic. DFIMA scheme is a promising candidate to optimize UL packet transmission characteristics and RRU. The motivation behind this scheme is as follows. Data terminals attempt to transmit their packets with a transmit permission probability (TPP), bit-rate and TTI changing dynamically on slot-by-slot basis according to the feedback information from the network. Similar access control schemes have appeared in different contexts. For instance, Rappaport [71] investigated the feedback channel state information-based carrier sensing MAC protocols for a centralized asynchronous CDMA packet radio system. Thomas *et al.* [70] proposed a real-time access control scheme for a data-only system and Viterbi [48] proposed a 1-bit feedback scheme in an integrated voice/data system. The implementations might vary from one to another, but they all agreed that the feedback schemes should be simple and effective. Based on the results from the previous subsection, TPP and optimized bit-rate for packet transmissions in the next time-slot can be predicted. In the case of QoS differentiation, TTI or packet-length can be changed as well. These parameters can be transmitted in the content of feedback information. For instance, let the service criteria be to serve as many users simultaneously as possible with optimal throughput-delay tradeoff and no QoS differentiation. The number of active data users in next time-slot, $D = \lfloor N\Lambda \rfloor$, and load state c^* of the RT traffic at the end of current time-slot are supposed to be known. Based on Equations (12.65) and (12.66) together with assumption (4a), a bit-rate can be chosen so that the $S(c^*)/D$ factor is as close to 1 as possible. TPP for users attempting to transmit their packet in next time-slot is set equal to min[1, $S(c^*)/D$]. Thus, the maximum number of users can be served with optimum UL throughput and reasonable average packet-delay of S/D, where S is given by Equation (12.67). However, one can notice that, if a terminal has a packet to send, it needs to wait until next time-slot to attempt its transmission. Therefore, access delay to the first attempt per packet can be up to 1 time-slot. A packet transmission can be synchronized to the beginning of a mini-slot (e.g. a 10 ms radio frame can be further divided into 15 mini-slots as in 3GPP standards) to reduce the delay. This results in a spread-slotted asynchronous multiple access system with feedback control, which is similar to the schemes reported in References [71, 75–79] for a data-only CDMA radio system. DFIMA can be extended to provide flexible means of QoS differentiation. This can be done through setting different TPP or/and TTI or/and bit-rate for different user classes in feedback information depending on load state of the RT traffic and offered load of the NRT traffic. DFIMA can be expected to outperform the well-known ALOHA and CSMA in throughput-delay tradeoff to the same extend as shown in Rappaport [71]. Also, this approach should overcome the hidden terminal problem using feedback, and be simpler and more flexible than PRMA.

12.4.15 Performance examples

A number of numerical and simulation results are presented in this section in order:

- to quantify the system performance and the benefits of using the described SCAC and the QoS differentiation;
- to study the effects and tradeoffs of the design parameters in the access control on the system behaviors and the performance characteristics through various system scenarios;
- to demonstrate the flexibility and the accuracy of the analytical methods used to evaluate the teletraffic performance of WCDMA cellular systems in this section.

The results are presented for a cellular system supporting three RT service classes: class 1 is 12.2 kbs voice, class 2 is 64 kbs video and class 3 is 144 kbs multimedia calls. The parameters are given in Table 12.4, which are in agreement with Ariyavisitakul [53]. Owing to limited space, we present results for offered traffic intensities of the three service classes in following proportions: 7:2:1, 5:4:1 and 4:3:3 of class 1, class 2 and class 3, respectively. These proportions are called multimedia traffic intensity profiles, MTIP. In words, let λ be a so-called common divisor of the three-class offered traffic; 7:2:1 MTIP, for instance, means

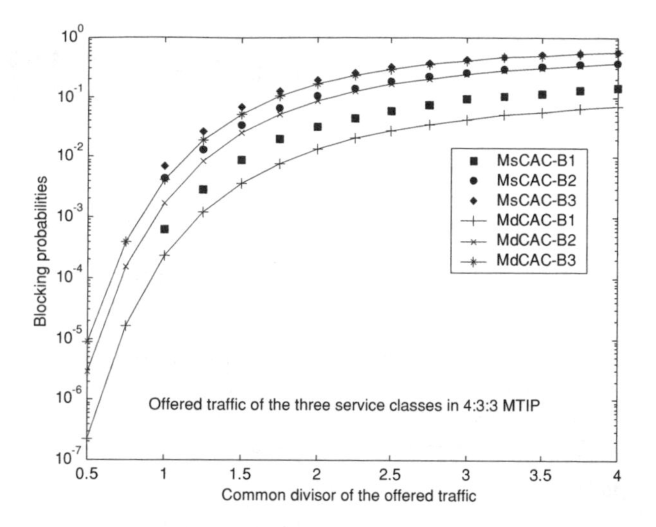

Figure 12.11 MdCAC vs MsCAC.

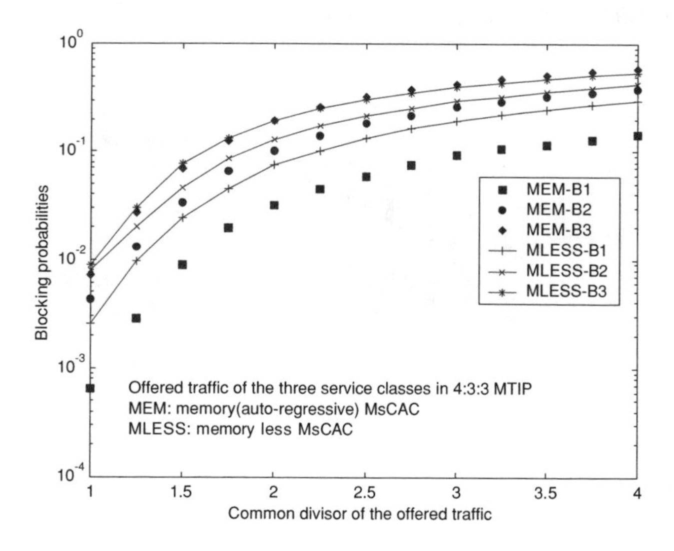

Figure 12.12 Memory vs memoryless MsCAC systems.

that the offered traffic of class 1 is 7λ Erlangs, class 2 is 2λ Erlangs and class 3 is λ Erlangs. Thus, the total offered traffic is 10 Erlangs if $\lambda = 1$.

Figure 12.11 presents the new-call blocking as well as the handoff failure probability of each service class in MdCAC and MsCAC systems vs the common divisor of the offered traffic intensities of the service classes in 4:3:3 MTIP. This figure confirms that there is no capacity gain using MsCAC for serving CBR services. Figure 12.12 shows the need for stable and reliable measurements in MsCAC systems. The performance characteristics of a memoryless measurement-based and a memory (auto-regressive) measurement-based system are presented also for 4:3:3 MTIP. Estimations with the help of auto-regressive filters may results in better performance, but more complex hardware/software is needed. Figures 12.13–12.15 present the handoff failure and new-call blocking probabilities of the

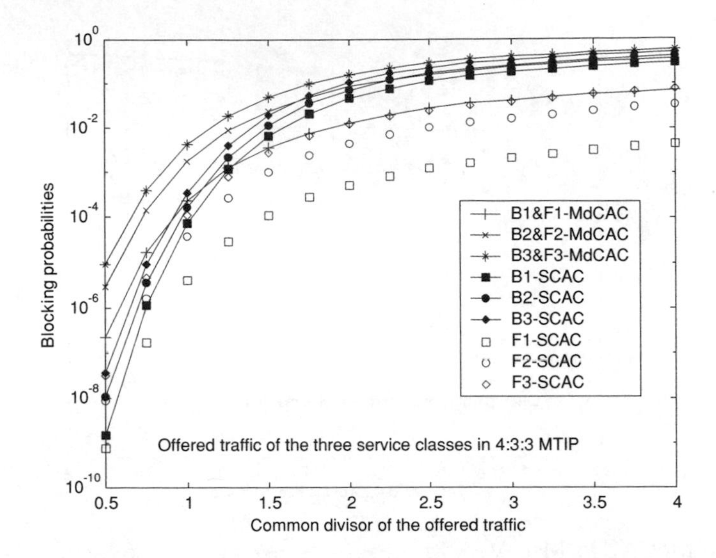

Figure 12.13 SCAC vs MdCAC in 4:3:3 MTIP.

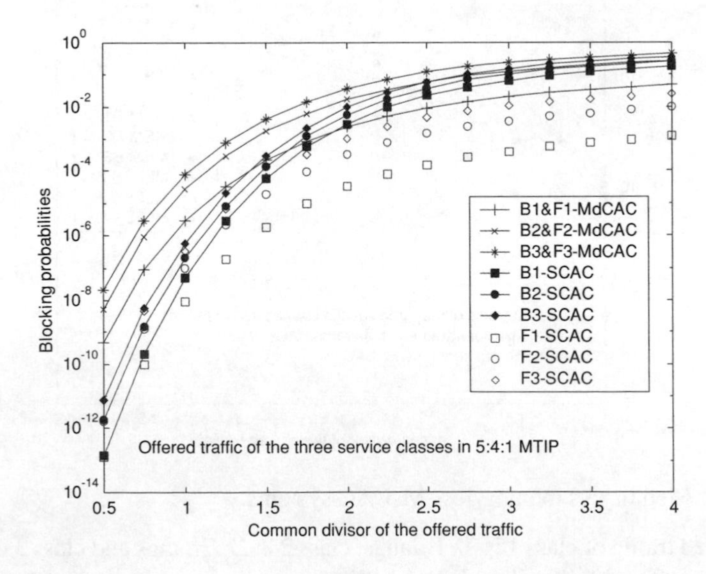

Figure 12.14 SCAC vs MdCAC in 5:4:1 MTIP.

SCAC system in comparison with the MdCAC system for 4:3:3, 5:4:1 and 7:2:1 MTIPs respectively. Figure 12.16 demonstrates that the SCAC system with the incomplete Gamma decision function offers better average communication quality, but slightly worse equilibrium blocking and dropping characteristics compared with the Gaussian function. The freedom of choosing such decision functions to fulfill the performance requirements increases the flexibility of the SCAC policy. The 4:3:3 MTIP is used in Figure 12.16. The QoS loss probability vs the offered traffic in 7:2:1 MTIP are listed in Table 12.6 for different

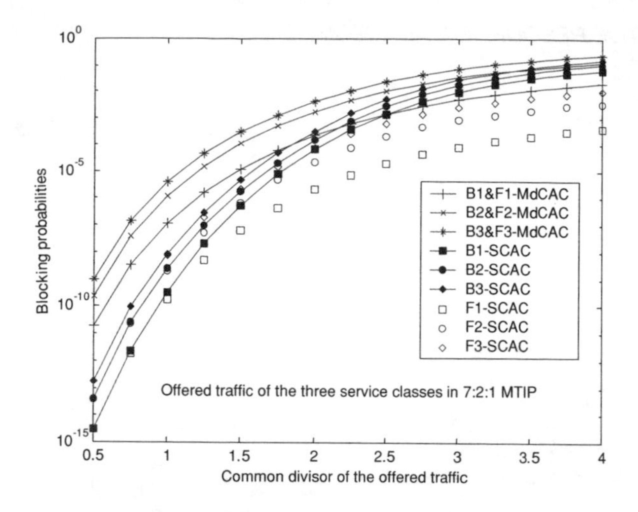

Figure 12.15 SCAC vs MdCAC in 7:2:1 MTIP.

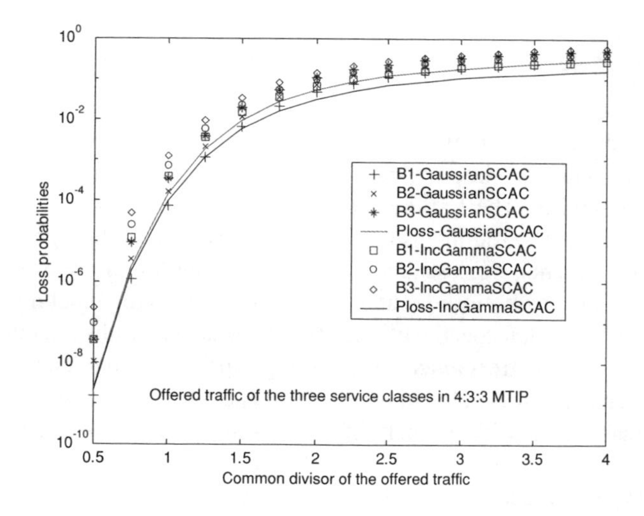

Figure 12.16 Flexibility of choosing decision function in the SCAC system.

Table 12.6 QoS loss probability in comparison

The three-class RT offered traffic, e.g. in 7:2:1 MTIP					
	Common divisor				
Traffic	1	1.5	2	2.5	3
MdCAC	0	0	0	0.0001	0.0007
MsCAC	0	0.0002	0.0024	0.0114	0.0315
GaussianSCAC	0	0	0.0001	0.0020	0.0119
IncGammaSCAC	0	0	0.0001	0.0013	0.0072
SCAC with threshold QoSDiff	0	0	0	0.0001	0.0005
SCAC with fracturing QoSDiff	0	0	0	0.0003	0.0021

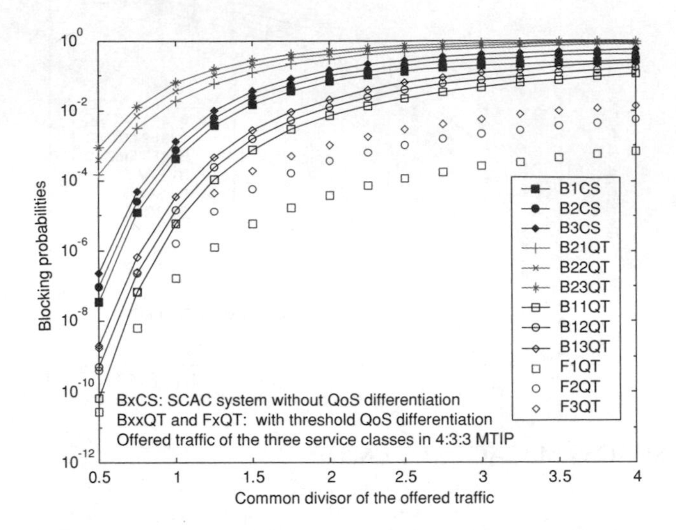

Figure 12.17 SCAC system with threshold hard blocking QoS differentiation.

CAC systems. Although the SCAC system suffers slight degradation of the communication quality, it yields significant improvements in the handoff failure probability and in the call blocking probability. The traffic shaping gain of the SCAC is clearly illustrated in Figure 12.15, where the traffic intensity of the voice calls is really high. For a predefined performance requirements, e.g. less than 0.1 and 0.5 % handoff failure probability for voice and other calls respectively, less than 1, 5 and 10 % new-call blocking probability for class 1, class 2 and class 3, respectively, and 10 % allowable equilibrium outage probability, the SCAC system overall offers much better Erlang capacity. Moreover, there is no need for redesign of the capacity thresholds in the SCAC system as long as the range of allowable uncertainty is maintained with the help of other control mechanisms such as TPC, link-adaptation, etc. Thus, the robustness is also well improved over the MdCAC system. SCAC has demonstrated an efficient RRU and capacity enhancement.

With QoS differentiation, operators can customize the operation of serving networks. Figure 12.17 presents the performance characteristics of the following simple scenario. Users are divided into two user classes: business ($j = 1$) and economy class ($j = 2$). Requests of the business users for any RT services are served immediately as long as there are enough resources for accommodating them. On the other hand, requests of the economy users are served only if less than 70 % of effective resources are occupied by RT traffic, i.e. system load state c less than 0.5. Assume that demands for services of user classes are equal. Thus, arrival rates of new call requests from user classes for each service class are equal. Invoke assumption (3b) with $\lambda_{l,1k} = \lambda_{l,2k}$ for $k = 1, 2, 3$. The offered traffic of e.g. 4:3:3 MTIP above can be split for each user class resulting in 2:2:1.5:1.5:1.5:1.5 MTIP of six traffic classes. The factor $a_{0\_jk}(c)$ of admission probability in (12.56) can be determined by (12.49), where the blocking threshold of business class l_{1k} is C_u and of economic class l_{2k} is 0.5 for all k. This numerical example clearly demonstrates the effects of QoS differentiation on performance characteristics. The business class not only experiences much better GoS, but also better communication quality during the calls.

Figure 12.18 illustrates the QoS differentiation with load-based fracturing factors for soft-blocking of new calls of the economic class. This is based on a simple scenario as follows. Again we assume handoff calls have the highest priority regardless of associated user class. The business class is served as long as resources are available. The economy class can share the resources equally with the business class if less than 65 % of effective resources are occupied, i.e. c is less than 0.47. Otherwise, if c is less than C_l, invoke (12.56) with: $a_{0\_21}(c) = 0.8$, $a_{0\_22}(c) = a_{0\_23}(c) = 0.6$. If c is less than C_u, $a_{0\_21}(c) = 0.4$, $a_{0\_22}(c) = a_{0\_23}(c) = 0.3$. Otherwise, $a_{0\_21}(c) = a_{0\_22}(c) = a_{0\_23}(c) = 0$.

The offered traffic is the same as in the previous scenario. Figures 12.17 and 12.18 show that the performance characteristics of the system can easily be tuned by using either threshold-based hard blocking or fracturing factor-based soft blocking paradigms. For NRT packet radio access, additional parameters are given in Table 12.7.

The average upper-limit UL data throughput for packet transmissions with 64 kbs and 10 ms TTI is presented in Figure 12.19 vs different offered traffic intensities of the three RT service classes, which are in 7:2:1, 5:4:1 and 4:3:3 MTIPs. The impacts of bit-rates on average upper-limit throughput with constant T_p duration of 10 ms are presented in Figure 12.20. Figure 12.21 illustrates the effects of T_p with a constant bit-rate of 64 kbs. Table 12.8 summarizes the mean values of aggregate RT traffic and quasi-stationary free

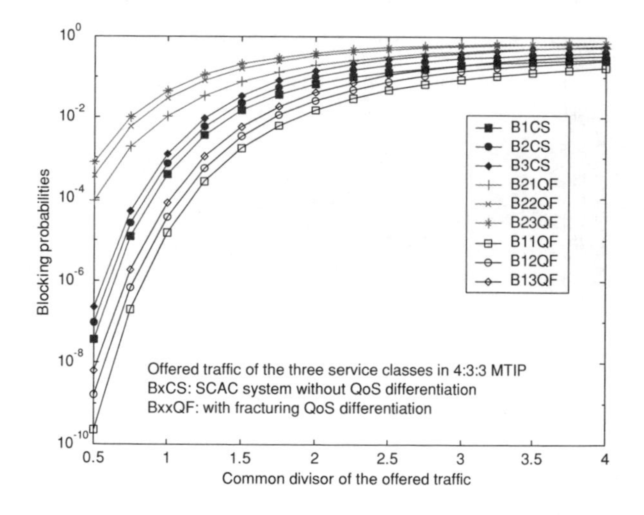

Figure 12.18 SCAC system with fracturing soft blocking QoS differentiation.

Table 12.7 Parameter summary for packet radio access

	Definition	Values
T_p	Packet transmit duration	10, 20, 30 or 40ms
R	Bit-rate for packet transmissions	32, 64, 144 or 384 kbs
γ_p	SIR target	3, 2, 1.5 or 1 dB for the above bit-rates, respectively

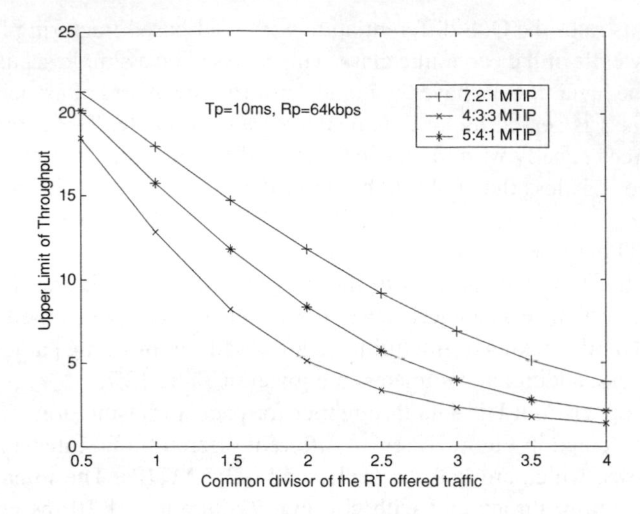

Figure 12.19 Average upper-limit UL data throughput in different RT MTIPs.

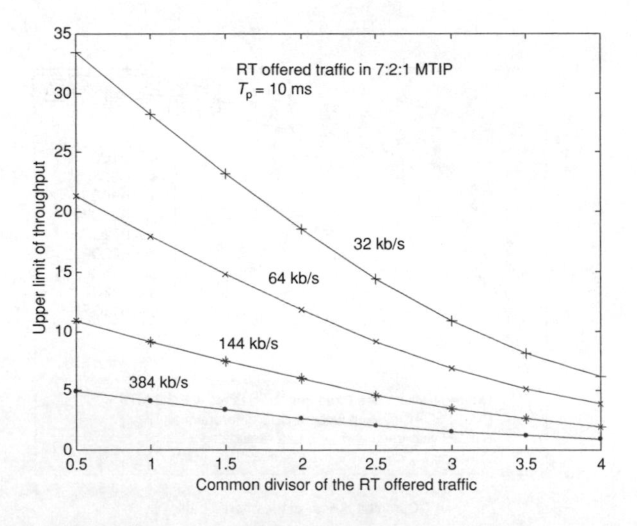

Figure 12.20 Effects of the bit-rates to the throughput.

capacity over T_p time-interval of 10 ms. Figure 12.20 and 12.21 and Table 12.8 are for 7:2:1 MTIP. One can see that throughput characteristics are affected significantly by the dynamic of RT traffic as well as the packet-transmission parameters. The results give valuable quantitative merits for studying the design parameters and the performance tradeoffs of packet access control schemes. This explains the motivations of using DFIMA scheme presented above, where content of feedback information provides 1:1 mapping of optimal transport format combination (including TPP, bit-rate, packet-length or TTI) for packet transmission in the next time-slot based on feasible free resource predictions. For example, assume 50 % cell capacity is occupied by the RT traffic of the 7:2:1 MTIP at the end of a given time slot of 10 ms. The possible bit-rates for packet transmissions are 32, 64 and 144 kbs. Consider

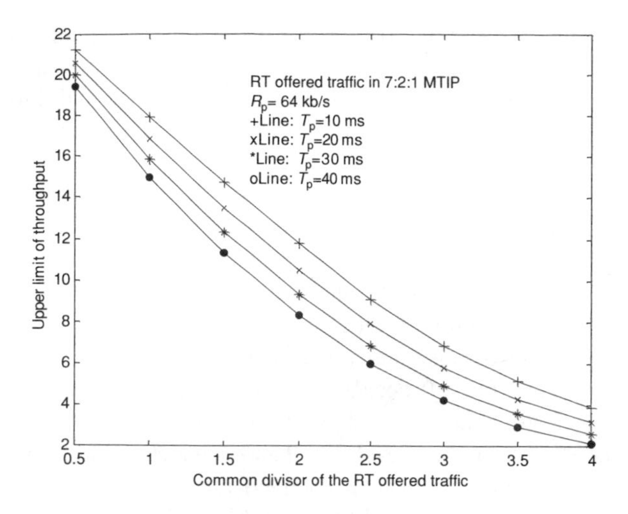

Figure 12.21 Effects of the packet transmission durations.

Table 12.8 Means of stationary RT aggregate traffic and quasi-stationary NRT resource availability

The three-class RT offered traffic, e.g. in 7:2:1 MTIP					
	Common divisor				
Traffic	1	1.5	2	2.5	3
$E[c]$	0.1500	0.2249	0.2988	0.3690	0.4305
$E[(z; T_p = 10\text{ ms})]$	0.4609	0.3793	0.3032	0.2347	0.1772

two cases of the NRT offered traffic: 10 or 50 active data users in the next time-slot. Using Equation (12.65) and (12.66), one can predict the maximum numbers of successful packet transmissions for each possible bit-rate, e.g. in our examples 16.91 packets of 32 kbs, 10.76 packets of 64 kbs and 5.53 packets of 144 kbs. Therefore, in the case of having 10 active users in the next time slot, the feedback information should tell them to transmit their packet immediately with bit-rate of 64 kbs. If 50 users want to transmit their packet, they have to attempt with TPP of 16/50 and bit-rate of 32 kbs at the beginning of the next time slot. The expected throughput in this case is about eight packets since the free capacity is successfully utilized for packet transmissions with the probability of 1/2. The performance of DFIMA can be optimized with respects to TPP, bit-rate, TTI and QoS differentiation paradigms, which is flexible and effective.

12.4.16 Implementation issues

Although the MdCAC and the SCAC policies are simple to implement without need for any special software and hardware, they may face a problem because the modeling parameters are required *a priori*. Owing to the diverse nature of different traffic sources and their often-complex statistics, some of the parameters may be hard to determine without which the modeling-based CAC policies cannot operate. The soft-decision solutions are believed to give more flexibility in determining the modeling parameters, and thus are quite suited to achieving good multiplexing gain and robustness. On the other hand, implementations of the MsCAC policy require advanced hardware and software to ensure the reliability of measurements. For this reason, it is not cost-effective. Moreover, estimation errors in some circumstances may cause significant degradations of the system performance. However, the advantage of MsCAC is that it seems 'insensitive' to the traffic nature and the operation is robust. The network can learn and adapt to the statistics of traffic even when the burstiness of traffic is considered as out of control for the modeling-based systems. To gain tradeoff of all design criteria, a hybrid soft-decision/measurement-based implementation is a reasonable choice. Parameters needed for soft-decision functions, i.e. means and variances, can rely on auto-regressive measurements. For such solution, parameters and constraints can simply be thresholds of the UL interference level of cell and connection basis, an allowable outage probability, estimates of the current total received interference level with its mean and variance, etc. These are anyhow needed for the TPC mechanism of CDMA systems. For implementations of the DFIMA scheme, measurements or estimations of RT system load state for CAC can be reused. The NRT offered traffic needs to be measured or estimated for prediction of optimal parameters (e.g. TPP, bit-rate, TTI) that are used as the content of feedback information. Look-up tables for transport format combinations of UL packet transmission can be implemented or configured in both mobile and access network sides in order to minimize the size of feedback information. Eight bit feedback is enough to ensure sufficient exchange of control information in DFIMA, even with QoS differentiation.

12.5 JOINT DATA RATE AND POWER MANAGEMENT

As already seen from the previous discussion the radio resource manager (RRM) contains a number of sub-blocks like the connection admission controller, the traffic classifier, the radio resource scheduler and the interference and noise measurements. The main role of the RRM is to manage the different available resources to achieve a list of target QoS. The radio resource scheduler (RRS) is an essential part of the RRM. The RRS has two important radio resources to control: MS transmitting power and transmitted data rate. The RRS uses those two resources to achieve different objectives like maximizing the number of simultaneous users, reducing the total transmitting power, or increasing the total throughput. The conventional way to achieve these objectives is to select one of them as a target to optimize and use other objectives as constraints. More sophisticated algorithms based on multiobjective (MO) optimization and Kalman filter techniques have been also proposed. Here we address the problem of how to combine the power and rate in an optimum way.

Even Shannon's equation shows that the achievable information rate in a radio channel is an increasing function of the signal-to-interference and noise ratio. Increasing the information rate in data communication systems is restricted by the SINR. Increasing the SINR can be done in two ways. The first way is by reducing the total interference and noise affecting that user. This depends on some characteristics of the noise and the interference. For example, if the structure of the interference from other users is known at the receiver then, by applying one of the multi-user detection methods, that interference can be reduced. Also if the users are spatially distributed then the interference can be reduced by using a multi-antenna system (see Section 12.1). If the users concurrently use the channel (as in DS-CDMA) then the interference can be reduced using power control techniques. From previous studies we can see that some characteristics of the interference are assumed to be known or can be controlled. There are many sources of interference and noises that cannot be reduced by the first way such as thermal noise, interference from other cells etc. The second way of increasing the SINR is simply by increasing the transmitted power. In a single user communication (point-to-point) or in broadcasting, this can be an acceptable solution and the main disadvantages are the cost and the nonlinearities in the power amplifiers. However, in a multiuser communication environment, increasing the transmitted power means more co-channel and cross-channel interference problems.

Therefore, a joint control of data rate as well as the transmitted power is an important topic in modem communication systems. The modern communication systems (3G or 4G) are supporting the multirate data communication because they are designed not only for voice communication but also for data and multimedia communication.

An efficient combining algorithm for the power control and the rate control is required for these systems. The term 'efficient' here refers to optimization of the transmitted power and data rate to meet the required specifications. There are many proposed combining algorithms for power and rate control in the literature. The objectives of those algorithms are quite varied. Some algorithms suggest maximizing the throughput; others minimizing the packet delay or minimizing the total power consumption.

The 3G/4G mobile communication systems based on WCDMA support the multi-rate transmission. There are mainly two methods to achieve the multi-rate transmission, the multicode (MC) scheme and the variable-spreading length (VSL) scheme. In the MC-CDMA system, all the data signals over the radio channel are transmitted at a basic rate, R_b. Any connection can only transmit at rates mR_b, referred to as m-rate, where m is a positive integer. When a terminal needs to transmit at m-rate, it converts its data stream, serial-to-parallel, into m basic-rate streams. Then each stream is spread using different and orthogonal codes. In a VSL-CDMA system, the chip rate is fixed at a specified value (3.84 Mb/s for UMTS) and the data rate can take different values. This means that the processing gain (PG) is variable. The processing gain can be defined as the number of chips per symbol.

12.5.1 Centralized minimum total transmitted power (CMTTP) algorithm

The mathematical formulation of the CMTTP problem is *find the power vector* $\mathbf{P} = [P_1, \ldots, P_Q]^T$ *and the rate vector* $\mathbf{R} = [R_1, \ldots, R_Q]^T$ *minimizing the cost function*:

$$J(\mathbf{P}) = \mathbf{1}^T\mathbf{P} = \sum_{i=1}^{Q} P_i \tag{12.68}$$

given that the required signal-to-noise ratio is guaranteed to each user

$$\frac{R_s}{R_i}\frac{P_i G_{ki}}{\sum\limits_{\substack{j=1\\j\neq i}}^{Q} P_j G_{kj} + N_i} \geq \delta_i^*, \quad \forall i = 1, \ldots, Q \tag{12.69}$$

and

$$P_{\min} \le P_i \le P_{\max},\ R_i \ge r_i, \quad \forall i = 1, \ldots, Q \tag{12.70}$$

where δ_i^* is the minimum required SINR for user i, r_i is the minimum rate limit for i. The problem presented in Equations (12.68)–(12.70) can be reduced to a system of linear equations. If the constraints Equations (12.69)–(12.70) cannot be achieved, then the problem is called infeasible. In this case either some user should be dropped from this link or some of the constraints should be relaxed. At the optimal solution, all QoS constraints are met with equality. Also, the optimal power vector is the one that achieves all rate constraints with equality. So, the optimum rate vector is $\mathbf{R}^* = [r_1, \ldots, r_Q]^{\mathrm{T}}$. The corresponding power vector can be obtained by solving the QoS equation. This is a system of linear equations in power. From Equation (12.69) we have

$$\frac{R_s}{r_i}\frac{P_j G_{kj}}{\sum\limits_{\substack{j=1 \\ j\neq Q}}^{Q} P_j G_{kj} + N_i}\delta_i^{\mathrm{T}}, \quad \forall i = 1, \ldots, Q \tag{12.71}$$

where δ_i^{T} is the target SINR for user i. Let $\tilde{r}_i = \delta_i^{\mathrm{T}} r_i / R_s$ and substitute it into Equation (12.71), to obtain

$$P_i = \tilde{r}_i \left[\sum_{\substack{j=1 \\ j\neq i}}^{Q} \frac{G_{kj}}{G_{ki}} P_j + \frac{N_i}{G_{ki}} \right] \tag{12.72}$$

In matrix form

$$\mathbf{P} = \mathbf{rHP} + \mathbf{ru} \tag{12.73}$$

where

$$H_{ij} = \begin{cases} 0 & i = j \\ \dfrac{G_{kj}}{G_{ki}} > 0 & i \neq j \end{cases} \qquad u_i = \frac{N_i}{G_{ki}} \tag{12.74}$$

$$\mathbf{r} = \mathrm{diag}\{\tilde{r}_1 \ldots \tilde{r}_Q\} \tag{12.75}$$

Then the optimum power vector is

$$\mathbf{P}^* = [\mathbf{I} - \mathbf{rH}]^{-1}\mathbf{ru} \tag{12.76}$$

In order to obtain a nonnegative solution of Equation (12.77), the following condition should hold:

$$\rho(\mathbf{rH}) < 1$$

where $\rho(\mathbf{A})$ is the spectral radius of matrix $\mathbf{A}$.

12.5.2 Maximum throughput power control (MTPC)

This algorithm has been suggested in Chawla and Qiu [85]. The algorithm is based on the maximization of the total throughput in a cellular system. There is no need to generate all solutions in this method. Since the gain links and the interference of other users are needed to calculate the transmitted power of each user, the MTPC algorithm is a centralized algorithm. The throughput of user i can be approximated when M-QAM modulation is used by

$$T_i = \Theta + \log_2(\Gamma_i)$$

where T_i is the throughput of user i, Θ is a constant, and Γ_i is the CIR of user i, and it is given in general by

$$\Gamma_{ki} = \frac{P_i G_{ki}}{\sum\limits_{\substack{j=1 \\ j \neq i}} P_j G_{kj} + N_i} \geq \Gamma_{\min}, i = 1, \ldots, Q, k = 1, \ldots, M$$

where Q = number of mobile stations, M = number of base stations and G_{kj} = channel gain between mobile station j and base-station k. The total throughput T is given by

$$T = \sum_{i=1}^{Q} T_i = Q\Theta + \log_2\left(\prod_{i=1}^{Q} \Gamma_i\right) \quad (12.77)$$

where Q is the number of users.

Now the problem can be defined as follows: *given the link gains G_{ij} of the users, what is the power vector $\mathbf{P} = [P_1, P_2, \ldots, P_Q]'$ which maximizes the total throughput*? Since the first term in Equation (12.77) is constant and the logarithmic function is an increasing function, then maximizing the multiplicative term

$$\prod_{i=1}^{Q} \Gamma_i$$

will lead to maximizing the total throughput T. The problem considered in Chawla and Qiu [85] is

$$\max_P \left[\prod_{i=1}^{Q} \Gamma_i(P)\right] \text{ s.t. } P \in \Omega \quad (12.77a)$$

where $\Omega = \{P | P_{\min} \leq P_i \leq P_{\max}, i = 1, \ldots, Q\} \subset \Re^Q$. The MTPC algorithm to solve Equation (12.77a) is given by

$$P_k(t+1) = \frac{1}{\sum\limits_{r \neq k}^{Q} \dfrac{G_{rk}}{\left(\sum\limits_{j \neq r}^{Q} G_{rj} P_j(t) + n\right)}} \qquad t = 0, 1, \ldots, \quad k = 1, \ldots, Q \quad (12.78)$$

$$P_{\min} \leq P_k(t+1) \leq P_{\max}$$

where G_{ij} is the channel gain between user j and base station i. Without loss of generality user i is assumed to be assigned to base station i. Starting from any initial vector $\mathbf{P}(0) \in \Omega$, the iteration specified by Equation(12.78) converges to a unique point $\mathbf{P}^* \in \Omega$, which achieves the global maximum [85].

12.5.3 Statistically distributed multirate power control (SDMPC)

A distributed solution of the optimization problem given by Equations (12.68)–(12.70) is proposed for one cell case in Morikawa *et al.* [80]. It is assumed that every user has two states, active ON or passive OFF. The transition probabilities of the ith user from idle to active state at any packet slot is υ_i, and from active to idle state is ζ_i. The durations of the active and idle periods are geometrically distributed with a mean of $1/\zeta_i$ and $1/\upsilon_i$ (in packet slots), respectively. The optimization problem Equations (12.68)–(12.70) is slightly modified as

Find

$$\min_{\mathbf{P}} J[\mathbf{P}(t)] = \sum_{i=1}^{Q} \beta_i(t) P_i(t) \tag{12.79}$$

subject to

$$\frac{R_s}{R_i} \frac{P_i G_{ki}}{\sum\limits_{\substack{j=1 \\ j\neq Q}}^{Q} P_j \beta_j(t) G_{kj} + N_i} \geq \delta_i^*, \quad \forall i = 1, \ldots, Q \tag{12.80}$$

$$P_{\min} \leq P_i \leq P_{\max}, \; R_i = r_i, \quad \forall i = 1, \ldots, Q, \tag{12.81}$$

One parameter has been added to original optimization problem which is the indicator function $\beta_j(t)$. The indicator function is equal to one if the jth user is currently active, and zero otherwise. It is assumed in Morikawa *et al.* [80] that the random process $\hat{\beta}(t)$ has Markovian property since geometric distribution is memoryless over the duration of traffic. The centralized solution (if the system is feasible) is given by

$$P_i(t) = \frac{\beta_i(t)\gamma_i}{G_{ki}} \times \frac{N_i}{1 - \sum\limits_{j=1}^{Q} \beta_j(t)\gamma_j} \tag{12.82}$$

where

$$\gamma_i = \frac{\delta_i^{\mathrm{T}}}{\delta_i^{\tau} + R_s/R_i} \tag{12.83}$$

The main idea behind the SDMPC algorithm is to estimate the other users' information part. Therefore the term $(\sum_{j=1}^{Q} \beta_j(t)\gamma_j)$ is estimated. The Markovian property of the random process $\beta_j(t)$ has been exploited to obtain a good estimate of the other users' information part. The SDMPC algorithm is given by

$$P_i(t) = \frac{\beta_i(t)\gamma_i}{G_{ki}} \times \frac{N_i}{1 - \hat{\beta}(t)} \tag{12.84}$$

where $\hat{\beta}(t)$ is the estimation of $\sum\limits_{j=1}^{Q} \beta_j(t)\gamma_j$.

The estimated parameter $\hat{\beta}(t)$ has been derived in Morikawa *et al.* [80] for two cases: (i) there is no 'collision' at t; and (ii) a 'collision' occurs at t. There are at least three drawbacks of this algorithm. First of all in the cellular CDMA system there is a control channel always active (when the mobile phone is ON). Then, in the SDMPC algorithm, the channel gain and the average power of the additive noise are assumed to be known. In reality they should be estimated as well. Good estimation of the channel gain and the noise variance is usually difficult. In practice it is easer to estimate CIR or SINR because they have direct impact on BER. Finally, it was assumed that the durations of active and idle periods are geometrically distributed. This assumption is oversimplified and far from reality.

12.5.4 Lagrangian multiplier power control (LRPC)

As mentioned previously, the data rates which can be achieved belong to a set of integers. In the formulation of the optimization problem, to maximize the data rate we assume that the date rate is continuous. This assumption can be relaxed in the simulation by rounding the optimum data rate to the nearest floor of the data rate set. It can be proven that the solution of the optimization problem with continuity assumption is not necessarily the same as the solution of the actual discrete problem. The advantage of the LRPC algorithm is that the optimization problem has been formulated without the continuity assumption of the data rates [81]. It has been assumed that each user has a set of m transmission rates $\mathbf{M} = \{r_1, r_2, \ldots, r_m\}$ to choose from. Let the rates be ordered in as ascending way, i.e. $r_1 < r_2 < \ldots < r_m$. To properly receive messages at transmission rate r_k, mobile i is expected to attain $\Gamma_i(\mathbf{P}) \geq \Gamma^{\mathrm{T}}_{i,k}$, where $\Gamma^{\mathrm{T}}_{i,k}$ is the target CIR.

Define $\mathbf{Y} = \left[y_i^k\right]$ to be a 0–1 matrix such that, for every mobile i and rate r_k

$$y_i^k = \begin{cases} 1, & \text{if mobile i is transmitting with rate } \mathrm{r_k} \\ 0, & \text{otherwise} \end{cases} \tag{12.85}$$

The combined rate and power control is formulated as the following optimization problem [81]

$$\mathbf{R} \triangleq \max_{Y,P} \sum_{i=1}^{Q} \sum_{k=1}^{m} r_k y_i^k \tag{12.86}$$

subject to the following constraints

$$\sum_{k=1}^{m} y_i^k \leq 1,\ y_i^k \in \{0, 1\},\ \text{and } 0 \leq P_i \leq P_{\max} \tag{12.87}$$

$$P_i + (1 - y_i^k) B_i^k \geq \frac{P_i \Gamma^{\mathrm{T}}_{i,k}}{\Gamma_i(\mathbf{P})} \tag{12.88}$$

where B_i^k is an arbitrary large number satisfying

$$B_i^k \geq \max_{P} \frac{P_i \Gamma^{\mathrm{T}}_{i,k}}{\Gamma_i(\mathbf{P})} \tag{12.89}$$

The above optimization problem is solved using Lagrangian multiplier method. The main goal of LRPC algorithm is to maximize the total throughput of the system. Although the LRPC improves the system throughput, its power consumption for supported users as well as the outage probability is rather high. Therefore it is not recommended to be used in the systems where the fairness is an important issue.

12.5.5 Selective power control (SPC)

The SPC algorithm has been suggested in Kim *et al.* [81]. The SPC algorithm is a logical extension of the DCPC algorithm [82]. The main idea of the SPC algorithm is to adapt the target CIR of each user to utilize any available resources. The suggested SPC algorithm is given by

$$P_i(t+1) = \max_k \left\{ \frac{P_i(t)\Gamma_{i,k}^{\mathrm{T}}}{\Gamma_i(P)} \times \chi\left(\frac{P_i(t)\Gamma_{i,k}^{\mathrm{T}}}{\Gamma_i(P)} \le P_{\max}\right)\right\}$$

$$t = 0, 1, \ldots, \quad i = 1, \ldots, Q$$

where $\chi(E)$ is the indicator function of the event E. Although the SPC algorithm improves the outage probability compared with LRPC algorithm, its outage is still high.

Jäntti suggested an improved version of the SPC algorithm. It is called selective power control with active link protection (SPC-ALP) Algorithm [83]. The SPC-ALP algorithm has less outage probability and better performance than the SPC algorithm. The main idea of the SPC-ALP algorithm is to admit new users into the network with at least the minimum data rate and also if possible allow old users to choose higher data rates. This is done by defining three different modes of operation for each user:

- Standard mode, where the user updates its power using SPC algorithm. In this mode the rate cannot be increased but it can be decreased if needed. If there are more resources to be utilized by increasing the rate, the used mode is changed to the transition mode.
- Transition mode, where the user updates its power using ALP algorithm. Also the rate is adapted to the maximum rate that can be supported.
- Passive mode, where the user stops its transmission. More details about the SPC-ALP algorithm can be found in Jäntti and Kim [83].

12.5.6 RRM in multiobjective (MO) framework

The QoS can be defined for a set of factors. In this Section we will consider only the BER and the user data rate in the uplink. The objectives of the RRS could be defined as

(1) Minimize the total transmitting power.

(2) Achieve the target SINR in order to achieve a certain BER level (depends on the application).

(3) Maximize the fairness between the users. In our definition, the system is fair as long as each user is supported by at least its minimum required QoS. In this sense, minimizing the outage probability leads to maximizing the fairness.

(4) Maximize the total transmitted data rate or at least achieve the minimum required data rate.

It is clear that objective (1) is totally conflicting with objective (4) and partially conflicting with objective (2). Objective (3) is totally incompatible with objective (4). Objective (2) is partially contradictory with objective (4).

So far, the RRM problem was formulated as a single objective (SO) optimization problem considering the other parameters as constraints. Solving the objectives (1)–(4) at the same time as using MO optimization technique, leads to a more general solution than the conventional methods. In this section we discuss an MO optimization method to solve the RRM problem. In subsequent subsections we will discuss some radio resource scheduler algorithms based on MO optimization. The field is very wide and many different algorithms and methods can be derived based on MO optimization. One formulation of the RRS optimization problem can be defined as:

$$\min_{P_i, R_i} \left\{ \sum_{i=1}^{Q} P_i, -\sum_{i=1}^{Q} \psi(R_i), O\_P \right\}, i = 1, \ldots, Q \tag{12.90}$$

subject to

$$P_{\min} \leq P_i \leq P_{\max}, R_{i,\min} \leq R_i \leq R_{i,\max} \tag{12.91}$$

where O_P is the outage probability. The outage probability is defined as the probability that a user cannot achieve at least the minimum required QoS. We can see that the O_P reflects the fairness situation in the communication system. The minus sign associated with the sum of the rate function in Equation (12.90) refers to the maximization process of the total utility functions.

Defining the objectives and the constraints is the first step. Selecting the proper MO optimization method to solve the problem is the second step. Then the (weakly) Pareto optimal set of solutions is generated, where every solution is optimal in a different sense. Finally, the decision maker selects the optimum solution from the optimal set which best achieves the required specifications. In this section we discuss a framework to use the MO optimization techniques in RRM.

12.5.7 Multiobjective distributed power and rate control (MODPRC)

The algorithm is based on minimizing a multi-objective definition of an error function. In this algorithm we defined three objectives: (1) minimize the transmitted power; (2) achieve at least the minimum CIR, which is defined at the minimum data rate; and (3) achieve the maximum CIR, which is defined at maximum data rate. An optimized solution can be obtained using an MO optimization.

The derivations of the algorithms are based on a VSL-CDMA communication system. After the dispreading process at the receiver, the SINR is

$$\delta_i(t) = \frac{R_s}{R_i(t)} \Gamma_i(t), t = 0, 1, \ldots \tag{12.92}$$

where $\delta_i(t)$ is the SINR of user i at t, R_s is the fixed chip rate (= 3.84 Mb/s for UMTS), $R_i(t)$ is the data rate for user i at t, and $\Gamma_i(t)$ is the CIR of user i at t. In wireless and digital communication, it is well known that the BER is a decreasing function in the SINR. In case of coherent binary PSK, the BER can be approximated by (when the interference is assumed Gaussian)

$$\mathrm{BER}_{\mathrm{PSK}} = \frac{1}{2}\mathrm{erfc}\big(\sqrt{\delta}\big) \tag{12.93}$$

For example, if the BER should not be more than 10^{-4} then the target SINR is obtained from Equation (12.93) as $\delta^{\mathrm{T}} \geq 8.3$ dB. In the case of fixed data rate power control there is one target CIR that corresponds to the target SINR, because we have only one spreading factor value. In a case of multirate services there are different target CIR values corresponding to the target SINR. From Equation (12.92) it is clear that, in case of constant target SINR, maximizing CIR leads to maximizing data rate as follows:

$$R_i(t) = \frac{R_{\mathrm{s}}}{\delta_i^{\mathrm{T}}}\Gamma_i(t), t = 0, 1, \ldots \tag{12.94}$$

Trying to achieve the maximum CIR for all users will end up in high outage probability. If there is a reasonable dropping algorithm then only one or few users will be supported [84]. To reduce the outage probability, we will define the target CIR at the minimum transmitted rate as

$$\Gamma_{i,\min} = \frac{R_{i,\min}}{R_s}\delta_i^{\mathrm{T}} \tag{12.95}$$

Also we will define the maximum CIR which is defined at the maximum transmitted rate as

$$\Gamma_{i,\max} = \frac{R_{i,\max}}{R_s}\delta_i^{\mathrm{T}} \tag{12.96}$$

The target SINR, the minimum/maximum CIR, and the minimum/maximum data rate are time-dependent, but we dropped the time symbol (t) for simplicity. In UMTS specifications the power is updated on slot-by-slot basis. The data rate is updated on a frame-by-frame basis. To generalize the analysis, we use the same time symbol for power and rate.

To increase the fairness, the users should achieve *at least* the minimum target CIR, which corresponds to the minimum transmitted rate (e.g. 15 kb/s in UMTS). The multirate power control problem is defined as: given the target SINR vector $\boldsymbol{\delta} = [\delta_1^{\mathrm{T}}, \delta_2^{\mathrm{T}}, \ldots, \delta_Q^{\mathrm{T}}]'$, the minimum requested data rate vector $\mathbf{R}_{\min} = [R_{1,\min}, R_{2,\min}, \ldots, R_{Q,\min}]'$, and without loss of generality, assuming the maximum allowed data rate $R_{\max}$ to be the same for all users, find the optimum power vector $\mathbf{P} = [P_1, P_2, \ldots, P_Q]'$ and the optimum rate vector $\mathbf{R} = [R_1, R_2, \ldots, R_Q]'$ that minimize the following cost function

$$J(\mathbf{P}) = \left[\sum_{i=1}^{Q}\sum_{t=1}^{N} \gamma^{N-t} e_i^2(t)\right], t = 1, \ldots, N, \tag{12.97}$$

subject to

$$P_{\min} \leq P_i \leq P_{\max}, i = 1, \ldots, Q \tag{12.98}$$

N is the optimization time window, γ is a real-valued constant adaptation factor. The notation $()'$ is used for transposed. The error $e_i(t)$ has been defined according to the *weighted metrics method*

$$e_i(t) = \lambda_{i,1}|P_i(t) - P_{\min}| + \lambda_{i,2}|\Gamma_i(t) - \Gamma_{i,\min}| + \lambda_{i,3}|\Gamma_i(t) - \Gamma_{i,\max}| \tag{12.99}$$

where $0 \leq \lambda_{i,1}, \lambda_{i,2}, \lambda_{i,3} \leq 1$ are real-valued, constant tradeoff factors, $\sum_{k=1}^{3} \lambda_{i,k} = 1$. The advantages of joining the weighting metrics method with the least square formula of

Equation (12.97) are:

- The least squares method is well known and its derivation is straightforward.
- A general solution is obtained using Equation (12.97), minimizing over all users and for time window N.

The error function Equation (12.100) is the mathematical interpretation of the RRM objectives given in (a)-(d). The first term of Equation (12.100) is to keep the transmitted power $P_i(t)$ as close as possible to $P_{\min}$, so we try to achieve objective (1). Objectives (2) and (3) will be achieved in the second part of the error function. In this part, the transmitted power is selected to obtain CIR very close to the minimum required CIR. Achieving the minimum required QoS for every user maximizes the fairness in the cell. The third term in Equation (12.100) represents the objective (4), where the users will try to obtain the maximum allowed QoS if possible.

By solving Equations (12.97) and (12.100) (using same procedure of MODPC algorithm) for a one-dimensional ($N = 1$) case we obtain for $i = 1, \ldots, Q$:

$$P_i(t+1) = \frac{\lambda_{i,1} P_{\min} + \lambda_{i,2}\Gamma_{i,\min} + \lambda_{i,3}\Gamma_{i,\max}}{\lambda_{i,1} P_i(t) + (\lambda_{i,2} + \lambda_{i,3})\Gamma_i(t)} P_i(t),\ t = 0, 1, \ldots, \qquad (12.100)$$

and as before

$$R_i(t+1) = \frac{R_s}{\delta_i^{\mathrm{T}}} \Gamma_i(t) \qquad (12.101)$$

$$P_{\min} \le P_i(t) \le P_{\max};\ R_{i,\min} \le R_i(t) \le R_{\max} \qquad (12.102)$$

If the minimum solution places such demands on some users that they cannot be achieved, then dropping or the handoff process should be applied [84]. The multirate power control algorithm given by Equations (12.100)–(12.103) has some interesting characteristics. By changing the values of the tradeoff factors λ_i, different solutions with different meanings are obtained. For example, when $\lambda_{i,1} = 1$, $\lambda_{i,2} = 0$, and $\lambda_{i,3} = 0$, it is clear that Equation (12.100) will be reduced to a fixed level (no) power control and user i will send at minimum power. For $\lambda_{i,1} = 0$, $\lambda_{i,2} = 1$ and $\lambda_{i,3} = 0$, Equation (12.100) becomes the distributed power control (DPC) algorithm of Grandhi and Zander [82]. In this case, the fairness is maximized. When $\lambda_{i,1} = 0$, $\lambda_{i,2} = 0$, and $\lambda_{i,3} = 1$, algorithm Equation (12.100) will maximize the average transmitted rate (with using reasonable dropping algorithm for nonsupported users). In this case one or a few users will be supported, so the outage probability will be high. From previous extreme conditions, one can make a tradeoff between these objectives to get the best performance according to the required specifications. The selection of the tradeoff values should be based on the communication link condition as well as the network and the user requirements. A wide range of different solutions can be obtained by changing the values of tradeoff factors. The selection of one solution is a job for the decision maker.

12.5.8 Multiobjective totally distributed power and rate control (MOTDPRC)

In this section, we discuss a slight modification of the MODPRC algorithm, the totally distributed algorithm. The MODPRC algorithm, Equations (12.100)–(12.103), assumes the availability of the actual CIR value. In the existing and near-future cellular systems, only an up–down command of the power is available at the MS. The estimated CIR is used with the MOTDPRC algorithm. The CIR (in dB) could be estimated as

$$\tilde{\Gamma}_i(t)\_\mathrm{dB} = \Gamma_i^{\mathrm{T}}(t)\_\mathrm{dB} - \tilde{e}_i(t),\ t = 0, 1, \ldots \tag{12.103}$$

where $\tilde{e}_i(t)$ is power control error, $\Gamma_i^{\mathrm{T}}(t)$ is the target CIR, and $\tilde{\Gamma}_i(t)$ is the estimated CIR. Using the estimated CIR in the MODPRC algorithm we obtain

$$P_i(t) = \frac{\lambda_{i,1} P_{\min} + \lambda_{i,2}\Gamma_{\min} + \lambda_{i,3}\Gamma_{\max}}{\lambda_{i,1} P_i(t-1) + (\lambda_{i,2} + \lambda_{i,3})\tilde{\Gamma}_i(t)} P_i(t-1),\ t = 0, 1, \ldots \tag{12.104}$$

$$R_i(t) = \frac{R_{\mathrm{s}}}{\delta^{\mathrm{T}}}\tilde{\Gamma}_i(t) \tag{12.105}$$

$$P_{\min} \le P_i(t) \le P_{\max};\ R_{i,\min} \le R_i(t) \le R_{\max} \tag{12.106}$$

12.5.9 Throughput maximization/power minimization (MTMPC)

Another application of the MO optimization in the RRM can be achieved by modifying the maximum throughput power control algorithm. The algorithm was based on maximizing the throughput and ignoring the transmitted power levels. In practice, reducing the transmitted power is very desirable. In this section we will formulate the cost function with two objectives. The first objective is the maximization of the total throughput as in Chawla and Qiu [85]. The second objective is to minimize the total transmitted power. The approach is treating the total throughput maximization and the total power minimization simultaneously using multiobjective optimization techniques.

The problem is defined as follows: *given the link gains of the users find the power vector which increases the total throughput as much as possible) and at the same time reduces the total transmitted power (as much as possible)*. The problem can be represented mathematically as

$$\max\{O_1(\mathbf{P}), -O_2(\mathbf{P})\}\ \text{s.t.} P \in \Omega \tag{12.107}$$

where $\mathbf{P} = [P_1, P_2, \ldots, P_Q]^{\mathrm{T}}$ is the power vector, the objective functions

$$O_1(\mathbf{P}) = \prod_{i=1}^{Q} \Gamma_i,\ \text{and}\ O_2(\mathbf{P}) = \sum_{i=1}^{Q} p_i$$

and the admissible power set $\Omega = \{\mathbf{P} | P_{\min} \le P_i \le P_{\max}, i = 1, \ldots, Q\}$. The minus sign is used to minimize the second objective. We will use the weighting method to solve the multiobjective optimization problem. The idea of the weighting method is to associate each objective function with a tradeoff factor (weighting coefficient) and maximizes (or minimizes) the weighted sum of the objectives [86]. Applying the weighting method in our problem we obtain,

$$\max_{P}\{O(\mathbf{P})\}\ \text{s.t.}\ \mathbf{P} \in \Omega, \tag{12.108}$$

where

$$O(\mathbf{P}) = \lambda_1 \prod_{i=1}^{Q} \Gamma_i(\mathbf{P}) - \lambda_2 \mathbf{1}'\mathbf{P} \tag{12.109}$$

is the multiobjective function, $\mathbf{1} = [1, 1, \ldots, 1]'$, and the tradeoff factors are real numbers, $0 \leq \lambda_1 \leq 1$, and $\lambda_2 = 1 - \lambda_1$.

Necessary conditions for solving the problem Equation (12.109) are

$$\nabla O(\mathbf{P}) = \mathbf{0} \tag{12.110}$$

where $\nabla O(P) = [\partial O/\partial P_1, \partial O/\partial P_2, \cdots, \partial O/\partial P_Q]$ is the gradient of O. Substituting the CIR expression into Equation (12.109) we obtain

$$O[P(t)] = \lambda_1 \prod_{i=1}^{Q} \frac{P_i(t)G_{ii}}{\sum_{\substack{j=1 \\ J\neq i}}^{Q} P_j(t)G_{ij} + N_i} - \lambda_2 \sum_{i=1}^{Q} P_i(t) \tag{12.111}$$

To maximize the reward functions, Equation (12.111), we find the power vector $\mathbf{P}$ which satisfies Equation (12.110). Since the obtained equations are nonlinear, it will be very complicated to get an analytical solution. An iterative solution for $k = 1, \ldots, Q$ will be formulated (we will drop the iteration argument t for simplicity)

$$\frac{\partial O}{\partial P_k} = \lambda_1 \frac{\left(G_{kk} \prod_{i\neq k}^{Q} G_{ii} P_i\right) \prod_{i=1}^{Q} \left(\sum_{j\neq i}^{Q} G_{ij} P_j + n\right) - \left(\prod_{i=1}^{Q} G_{ii} P_i\right) \left[\sum_{r\neq k}^{Q} G_{rk} \prod_{i\neq r}^{Q} \left(\sum_{j\neq i}^{Q} G_{ij} P_j + n\right)\right]}{\left[\prod_{i=1}^{Q} \left(\sum_{j\neq i}^{Q} G_{ij} P_j + n\right)\right]^2} - \lambda_2 = 0 \tag{12.112}$$

After simplification,

$$\lambda_1 \frac{\left(G_{kk} \prod_{i\neq k}^{Q} G_{ii} P_i\right) - \left(\prod_{i=1}^{Q} G_{ii} P_i\right) \sum_{r\neq k}^{Q} \frac{G_{rk}}{\left(\sum_{j\neq r}^{Q} G_{rj} P_j + n\right)}}{\prod_{i=1}^{Q} \left(\sum_{j\neq r}^{Q} G_{ij} P_j + n\right)} - \lambda_2 = 0 \tag{12.113}$$

which can be rewritten as

$$\lambda_1 G_{kk} \prod_{i\neq k}^{Q} G_{ii} P_i - \lambda_1 \left(\prod_{i=1}^{Q} G_{ii} p_i\right) \sum_{r\neq k}^{Q} \frac{G_{rk}}{\left(\sum_{j\neq r}^{Q} G_{rj} P_j + n\right)} = \lambda_2 \prod_{i=1}^{Q} \left(\sum_{j\neq i}^{Q} G_{ij} P_j + n\right) \tag{12.114}$$

or

$$\lambda_1 G_{kk} - \lambda_1 G_{kk} P_k \sum_{r\neq k}^{Q} \frac{G_{rk}}{\left(\sum_{j\neq r}^{Q} G_{rj} P_j + n\right)} = \frac{\lambda_2}{\prod_{i\neq k}^{Q} G_{ii} P_i} \prod_{i=1}^{Q} \left(\sum_{j\neq i}^{Q} G_{ij} P_j + n\right) \tag{12.115}$$

Solving for P_k leads to

$$P_k = \left[\lambda_1 G_{kk} - \frac{\lambda_2}{\prod_{i \neq k}^{Q} G_{ii} P_i} \prod_{i=1}^{Q} \left(\sum_{j \neq i}^{Q} G_{ij} P_j + n \right) \right] \frac{1}{\lambda_1 G_{kk} \sum_{r \neq k}^{Q} \frac{G_{rk}}{\left(\sum_{j \neq r}^{Q} G_{rj} P_j + n \right)}} \tag{12.116}$$

and further to

$$P_k(t+1) = \frac{1}{\sum_{r \neq k}^{Q} \frac{G_{rk}}{\left(\sum_{j \neq r}^{Q} G_{rj} P_j(t) + n \right)}} - \frac{\lambda_2}{\lambda_1 G_{kk} \prod_{i \neq k}^{Q} G_{ii} P_i(t)} \frac{\prod_{i=1}^{Q} \left(\sum_{j \neq i}^{Q} G_{ij} P_j(t) + n \right)}{\sum_{r \neq k}^{Q} \frac{G_{rk}}{\left(\sum_{j \neq r}^{Q} G_{rj} P_j(t) + n \right)}} \tag{12.117}$$

For $\lambda_2 = 0$, the problem is reduced to maximizing the throughput and Equation (12.118) reduces to Equation (12.78). Without power constraints Equation (12.78) is rewritten as

$$\widehat{P}_k(t+1) = \frac{1}{\sum_{r \neq k}^{Q} \frac{G_{rk}}{\left(\sum_{j \neq r}^{Q} G_{rj} P_j(t) + n \right)}} \tag{12.118}$$

and Equation (12.118) can be rewritten in a more compact form as

$$P_k(t+1) = \frac{\widehat{P}_k(t+1)}{\left[1 + \frac{\lambda_2}{\lambda_1} \frac{\widehat{P}_k(t+1)}{\prod_{i=1}^{Q} \Gamma_i(t)} \right]}, \quad t = 0, 1, \ldots \tag{12.119}$$

where $P_k \in [P_{\min}, P_{\max}]$, $k = 1, \ldots, Q$. From Equation (12.119), the new transmitted power is a scaled value of the transmitted power in the case of the maximum throughput algorithm. To compare the two-algorithms, the numerical example from References [85, 87] is used. Consider the system with $Q = 5$ users and the path gain matrix, G, shown below.

$$G(\text{dB}) = \begin{Bmatrix} -5.8 & -18.2 & -55.3 & -20.3 & -33.6 \\ -36.0 & -9.7 & -43.5 & -22.2 & -15.9 \\ -41.6 & -30.9 & -9.3 & -38.6 & -36.5 \\ -14.2 & -20.6 & -38.5 & -6.8 & -36.6 \\ -22.6 & -23.9 & -20.1 & -16.4 & -10.8 \end{Bmatrix}$$

The tradeoff factors have been set to $\{\lambda_1, \lambda_2\} = \{0.9999, 0.0001\}$. In this case we penalize power usage. From Table 12.9, we can see that the summation of the SINR (dB) of the users [which is related to the total throughput as in Equation (12.78)] has not changed very much in both schemes (only 0.04 %), but the power has been reduced by more than 98 % in the case of MTMPC method. Additional information on the topic can be found in References [88–90].

Table 12.9 Comparison of MTPC and MTMPC algorithms

	MTMPC algorithm $\lambda_1 = 0.0001$ and $\lambda_2 = 0.9999$		MTPC algorithm $\lambda_1 = 1$ and $\lambda_2 = 0$	
User	P(dBw)	SINR(dB)	$\bar{p}$(dBw)	SINR(dB)
1	−13.9789	16.8345	−0.5580	16.9295
2	−16.8918	−0.8548	−6.6072	−0.9234
3	−4.6187	36.0956	13.4264	36.8300
4	−12.8725	8.6383	1.2111	8.5561
5	−14.3460	−5.9278	−1.5289	−6.5922
	Average power (W) = 0.1	Sum[SINR(dB)] = 54.78	Average power (W) = 5	Sum[SINR(dB)] = 54.80

12.6 DYNAMIC SPECTRA SHARING IN WIRELESS NETWORKS

In this section we present schemes for interference suppression in UWB-based WPAM systems when sharing the same band as other communications networks. The scheme can be used to significantly improve the performance of UWB systems in the presence of interference from mobile communication systems such as GSM and WCDMA. It is also effective in the presence of WLAN systems which are nowadays based on OFDMA technology or in military communications where the interference is generated by intentional jamming. The section also discuss the effectiveness of the scheme to suppress MC CDMA, which is a candidate technology for 4G mobile communications. In order to demonstrate the relevancy of these results we first provide a systematic review of the existing work in this field and then present specific scheme. The results show that significant suppression gain up to 40 dB can be achieved in the presence of OFDM, WCDMA and MC CDMA, enabling coexistence of different networks in the same frequency bandwidth. The effectiveness decreases if the number of subcarriers is increased.

The online source [91] gives historical perspective to UWB technologies. It lists down the early UWB references and patents from the 1960s and 1970s. In [92] a comprehensive overview of UWB wireless systems is given. It discusses the FCC allocation of 7.5 GHz (3.1–10.6 GHz) unlicensed band for the UWB devices. Potential UWB modulation schemes, multiple access issues, single vs multiband implementation and link budgets are also discussed. Paper [93] is a very frequently referenced one giving a brief introduction to the basics of impulse radio systems. It describes the characteristics of impulse radio and gives analytical estimates of the multiaccess capability under idealistic channel conditions.

12.6.1 Channel capacity

Some new channel capacity results for M-ary pulse position modulation (M-PPM) time hopping UWB systems are presented in [94]. It is demonstrated that the previous results based on the 'pure PPM model' have overestimated the real UWB capacity. The proposed model is extended with the correlator and soft decision decoding. The capacity is evaluated in the single-user case and with asynchronous multiple user interference (MUI) when the inputs are equiprobable. It is found that larger M leads to increased capacity only at the high bit SNR region. Furthermore, optimal time offset values for each M are independent of the bit SNR. The MUI influence is detrimental for the capacity, especially at high bit SNRs.

12.6.2 Channel models

Paper [95] focuses on the UWB indoor channel modeling issues. The measurement data is collected from an extensive campaign in a typical modern office building with a 2 ns delay resolution. The model is formulated as a stochastic tapped delay line (STDL). The energy statistics due to small-scale effects seem to follow a Gamma distribution for all bins. Large-scale parameters can be modeled as stochastic parameters that can change, e.g., from room to room. UWB propagation channels are also discussed in [96]. Based on the modified CLEAN algorithm, estimates of time-of-arrival, angle-of-arrival, and waveform shape are derived. Key parameters of the model are intercluster decay rate, intracluster decay rate, cluster arrival rate, ray arrival rate and standard deviation of the relative azimuth arrival angles. Intercluster signal decay rate is generally determined by the architecture of the building. Intracluster decay rate, on the other hand, depends on the objects close to the receive antenna (e.g. furniture). Relative azimuth arrival angles were best fit to a Laplacian probability density function. Saleh and Valenzuela [97] present a model that has become a frequently referenced and adopted source in indoor multipath propagation channel modeling. They propose a statistical indoor radio channel model that (1) has flexibility to reasonable fit with the measured data, (2) is simple enough to be used in simulation and analysis, and (3) can be extended by adjusting parameters to represent various buildings. In the developed statistical model the rays of the received signal arrive in clusters.

The ray amplitudes are independent Rayleigh random variables with exponentially decaying variances with respect to the cluster delay and the ray delay. The clusters and the rays within the cluster form Poisson arrival processes with different, fixed rates. Paper [98] characterizes measurement-based UWB wireless indoor channels from the communications theoretic viewpoint. The bandwidth of the signal used in the measurement is over 1 GHz, resulting in the less than 1 ns time resolution. Robustness of the UWB signal to multipath fading is quantitatively evaluated through histograms and cumulative distributions. Two rake structures are introduced: the all rake serves as the best-case (benchmark) receiver and the maximum-energy-capture selective rake is a realistic sub-optimal approach. Multipath components of the measured waveforms are detected using a maximum-likelihood estimator based on a separable specular multipath channel model.

12.6.3 Diversity reception

Performance of PPM and on–off keying (OOK) binary block-coded modulation formats using a maximal ratio combining rake receiver is studied analytically in [99]. The trade-off between receiver complexity and performance is examined. Several suboptimal receivers in indoor multipath AWGN channels have been employed. Results indicate the robust performance may require an increase in rake complexity. This implies allocation of more rake fingers and tracking of the strongest multipaths to help in the selection combining. Rake performance for a pulse-based high data rate UWB system in an Intel Labs indoor channel model is addressed in [100]. It is noted that, at low input SNR values (0–10 dB) and small number of rake fingers, it is more beneficial to add rake taps for energy capture rather than for intersymbol interference (ISI) mitigation. In the presence of channel estimation errors, equal gain combining can be more robust than maximal ratio combining, and therefore yield better performance. In order to quantify the trade-off between rake receiver energy capture and diversity order [101] presents partly quasi-analytical and partly experimental analysis suited to dense multipath propagation environments. Numerical results show that a diversity level of less than 50 is adequate in typical indoor office conditions.

12.6.4 Performance evaluation

In [102] a method to evaluate the BER performance of time hopping (TH) PPM in the presence of multiuser interference and AWGN channel is proposed. Gaussian quadrature rules are used in this approach. Paper [103] concentrates on the signal design for binary UWB communications in dense multipath channels. The aim is to find signals with good distance properties leading to good BER performance that both depend on the time shift parameter τ. Performance of UWB correlation receivers for equal mean power Gaussian monocyles is studied in [104]. Channel conditions vary among ideal single user AWGN, nonideal synchronous, multipath fading and multiple access interference. It is shown that the pulse shape has a notable impact on the correlation receiver performance. The effects can be seen in the autocorrelation function, especially in the mainlobe. The autocorrelation is highly related to the SNR gain of the output and to interference resistance properties. Special characteristics of the Gaussian monocycles include: (1) higher order derivatives have higher SNR gain in single user and asynchronous multiple access channel but are less robust to interference than lower order derivatives; (2) narrower pulses have higher SNR gain in asynchronous multiple access channel at the cost of inferior interference resistance ability. Exact bit error rate performance of TH-PPM UWB systems in the presence of multiple access interference (MAI) is analyzed and simulated in [105]. Furthermore, it is shown that, with a moderate number of MAI sources, the standard Gaussian approximation becomes inaccurate at high SNRs.

12.6.5 Multiple access techniques and user capacity

The main principles for multiaccess in UWB systems are discussed in [106]. A functional medium access, radio link and radio resource control architecture is proposed and open issues for future activities are addressed. Numerical throughput and delay performance results in radio resource sharing are shown. Uncoded and coded performance analysis for TH-UWB systems is covered in [107]. A practical low-rate error correcting coding scheme is presented that requires no bandwidth expansion. Gaussian assumption for multiuser interference is shown to be invalid for high uncoded data rates. The user capacity is shown to increase radically with the proposed coding scheme. M-ary signals ranging from $M = 2$ up to $M = 256$ are used in [108] together with PPM signal formats in the UWB multiple access capacity analysis. Performance is analyzed in free-space propagation conditions. The number of supported users is dependent on the given bit error rate, SNR, transmission rate and modulation alphabet size. Performance and receiver complexity trade-off is discussed. Upper bounds are derived for the combinations of user capacity and total transmission rate. According to the numerical examples it is possible to achieve a system with high capacity and data rate, low bit error probability, and yet only moderate receiver complexity. Reference [109] is one of the first public and widely cited papers outlining the potential of time hopping impulse radio multiple access communications. It describes the basic building blocks of the impulse transmitter and receiver and their mathematical formulations. It also shows an example for the bit error vs user capacity estimate at variable data rates. Finally, some drawbacks of the high time-resolution impulse radio systems are mentioned: (1) the need for up to thousands of rake fingers in the multipath receiver; and (2) complex initial clock acquisition. In [110] a comprehensive overall description of the time-hopping UWB system physical layer issues is given. Achievable transmission rates and multiple access capacities are estimated for analog and digital modulation formats. Numerical results indicate that the digital implementation has the potential for nearly one order of magnitude higher user densities than the analog one.

12.6.6 Multiuser detection

Reference [111] is focused on the multiuser detection (MUD) possibilities for direct sequence UWB systems. It is demonstrated that the adaptive minimum mean squared error (MMSE) MUD receiver outperforms the rake receiver both in energy capture and in interference rejection sense. Studied interference sources are narrowband IEEE 802.11a interference and wideband multiuser UWB interference. Ideally, MMSE receiver can achieve AWGN bit error rate within a 1–2 dB margin even in dense multipath channels. In heavily loaded conditions the penalty of 6 dB is experienced, but at the same time the rake receivers suffer from unbearable error floors. Iterative partial parallel multiuser interference cancellation (PIC) is applied to the UWB multiuser system in [112]. Matched filter, maximum-likelihood, and linear minimum mean squared error receivers are also used in the performance comparison. In this paper, multiuser detection is combined with error control coding. The UWB system includes only one pulse per symbol and AWGN channel is assumed. Numerical results show that it is possible to attain the coded single user BER bound for eight to 15 users in a heavily loaded system without any processing gain. As the number of users increases and the bandwidth to pulse repetition frequency decreases, MAI is expected to adversely affect system capacity and performance. As a consequence, a framework for the design of multiuser detectors for UWB multiple access communications systems is presented in [113]. An optimum multiuser detector is also proposed.

12.6.7 Interference and coexistence

Coexistence of UWB system with some other radio systems is studied in [114]. This means the evaluation of interference caused by the UWB system to the other radio systems and vice versa. The coexisting radio concepts are GSM900, UMTS/WCDMA and GPS. Several short Gaussian-based UWB pulses are employed. According to the numerical results, convenient selection of pulse waveform and width leads to interference resistance up to a certain limit. The pulse shape is in interaction with the data rate. High-pass filtered waveforms are preferred in the case of short UWB pulses, whereas generic Gaussian ones are favorable if long pulses are utilized. Interference caused by narrowband systems is the most detrimental to UWB if it is located at the UWB system's nominal center frequency. In the GPS band the DS based UWB system interfered less than the time hopping system.

12.6.8 Channel estimation/imperfections

Channel estimation for time hopping UWB communications is dealt with in [115]. Multipath propagation and MAI are taken into consideration. Maximum-likelihood estimation is applied in data-aided and nondata-aided scenarios. Numerical results show that the performance is reasonable if the number of simultaneous users is below 20. The impact of the timing jitter and tracking to the impulse radio system performance is investigated in [116]. Binary and 4-ary modulations are used. According to these studies both modulations suffer from the jitter, however 4-ary is better. Timing jitter is also discussed in [117]. Static and Rayleigh fading channels are assumed. Orthogonal PPM, optimum PPM, OOK and BPSK modulations are compared in performance evaluation. Similar performance degradation is noted for BPSK and PPM schemes, while OOK is more susceptible to large jitter. The probability density function of timing jitter due to rake finger estimation is simulated. The results depend on the pulse shape and SNR. Worst case distribution is shown to provide an upper bound for BER performance. The above survey of issues in UWB communications indicates a need for special attention to interference avoidance or interference suppression due to extremely wide signal bandwidth and the possibility of interference with other systems operating in the same bandwidth. One way to deal with the problem is to design the pulse shape in such a way that the signal has no significant spectral component in the occupied frequency bands. Pulse shapes respecting the FCC spectral mask were proposed in References [118–120]. The drawback of such a solution is the need for over sampling and lack of flexibility in the case that the interfering signal is not stationary like in military applications.

Another approach is to use adaptive interference suppression like the schemes summarized in the previous sub-section entitled *interference and co-existence.* The solution discussed in this section belongs to the latter category. We will demonstrate the advantages of this approach with a number of numerical results. The solution is adaptive and can be implemented with no over-sampling, unlike other schemes.

12.6.9 Signal and interference model

In general the signal transmitted by the desired user is modeled as:

$$s(t) = \sum_i b[t - iNT_f - (1 - a_i)\Delta] \cos \omega_c t \tag{12.120}$$

where

$$b(t) = \sum_{n=0}^{N-1} g[t - nT_f - h(n)T_c] \tag{12.121}$$

The signal can also be transmitted in the baseband with no carrier. In Equation (12.121), $g(t)$ represents the basic pulse shape (monocycle pulse) and T_f represents frame duration during which there is only one pulse T_c seconds wide. The sequence $h(n)$ is the user's time-hopping code and its elements are integers taking values in the range $0 \le h(n) \le N - 1$. The parameter T_c is the duration of an addressable time bin within a frame. In other words, the right-hand side of Equation (12.121) consists of a block of N time-hopped monocycles. a_i represents information bits (0,1). Equation (12.121) says that, if a_i were all zero, the signal would be a repetition of $b(t)$-shaped blocks with period NT_f. Δ may be viewed as the time shift impressed by a unit data symbol on the monocycles of a block. It is clear that the choice of Δ affects the detection process and can be exploited to optimize system performance. In summary, the transmitted signal consists of a sequence of $b(t)$-shaped position-modulated blocks.

The code sequence restarts at every data symbol. This 'short-code' assumption is made here for the sake of simplicity and is in keeping with some trends in the design of third-generation CDMA cellular systems. Longer codes are conceivable and perhaps more attractive but lead to more complex channel estimation schemes. The OFDM interference, generated for example by a WLAN user, is modeled as

$$j(t) = \sum_{i=0}^{N-1} d_{i+\left\lfloor \frac{t}{T_j} \right\rfloor} J_i \cos\left[2\pi\left(f_c + \Delta f_c + \frac{i}{T_j}\right)t + \varphi_i\right] \tag{12.122}$$

where N is the number of channels, d_i is the FDM interference information bits, $f_c + \Delta f_c$ is the first channel carrier frequency, J_i is the OFDM interference amplitudes, φ_i is the channel phase at the receiver input and T_j is the bit interval.

The MC CDMA interference can be modeled as

$$j(t) = \sum_{k=1}^{K}\sum_{i=0}^{N-1} d_k(t)\, c_k(i)\, J_{k,i} \cos\left[2\pi\left(f_c + \Delta f_c + \frac{i}{T_j}\right)t + \varphi_{k,i}\right] \tag{12.123}$$

where N is the number of channels, $d_k(t)$ is the kth user interference bit, $c_k(i)$ is the PN sequence of the kth user and the ith channel, K is the number of users, $f_c + \Delta f_c$ is the first channel carrier frequency, $J_{k,i}$ is the interference amplitude, $\varphi_{k,i}$ is the channel phase at the receiver input and T_j is the bit interval.

12.6.10 Receiver structure

The receiver block diagram is shown in Figure 12.22. There are two interference rejection circuits, A and B . The first one (A) processes signal if logic one is sent, and the other (B) processes signal for logic zero. When several time-hopping signals are simultaneously received over a channel with L_c paths, the composite waveform at the output of the receiver antenna may be written as:

$$r(t) = \sum_{l=1}^{L_c} [\gamma I_l s(t-\tau_l)\cos\omega_c t + \gamma Q_l s(t-\tau_l)\sin\omega_c t] + n(t) + j(t) \tag{12.124}$$

where $n(t)$ is noise, and $j(t)$ is total interference, $\gamma_l = \gamma I_l + j\gamma Q_l$ is the complex attenuation and τ_l is the delay in the lth path. If we consider signal sampled at chip interval T_c we have:

$$r(k) = rI(k) + jrQ(k) \tag{12.125}$$

where

$$rI(t) = r(t)\cos\omega_c t \tag{12.126a}$$

$$rQ(t) = r(t)\sin\omega_c t \tag{12.126b}$$

$$k = \frac{t}{T_c} \tag{12.127}$$

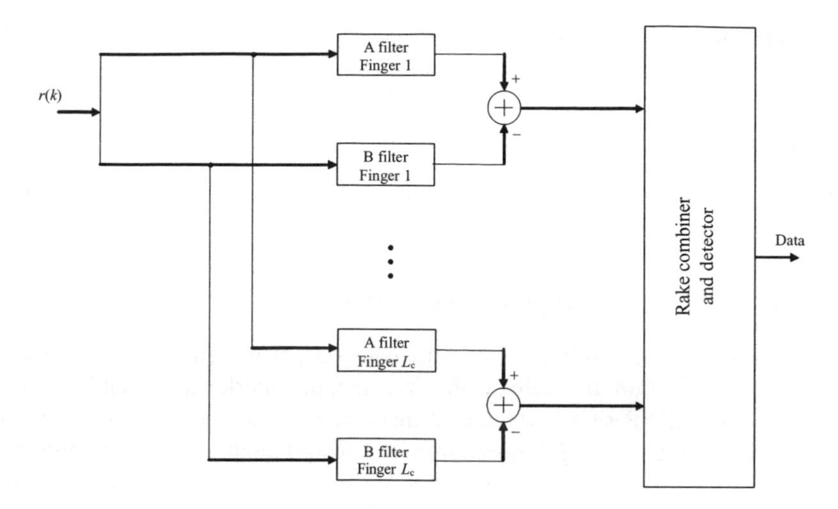

Figure 12.22 Receiver block diagram.

The detection variable in the lth rake receiver finger is:

$$DI^l(i) = DAI^l(i) - DBI^l(i) \tag{12.128}$$

$$DQ^l(i) = DAQ^l(i) - DBQ^l(i) \tag{12.129}$$

$$DAI^l(i) = \sum_{k=iN\frac{T_f}{T_c}-\frac{\tau_l}{T_c}}^{(i+1)N\frac{T_f}{T_c}-\frac{\tau_l}{T_c}} rI(k)b\left(k - iN\frac{T_f}{T_c} - \frac{\tau_l}{T_c}\right) \tag{12.130}$$

$$DAQ^l(i) = \sum_{k=iN\frac{T_f}{T_c}-\frac{\tau_l}{T_c}}^{(i+1)N\frac{T_f}{T_c}-\frac{\tau_l}{T_c}} rQ(k)b\left(k - iN\frac{T_f}{T_c} - \frac{\tau_l}{T_c}\right) \tag{12.131}$$

$$DBI^l(i) = \sum_{k=iN\frac{T_f}{T_c}-\frac{\tau_l}{T_c}}^{(i+1)N\frac{T_f}{T_c}-\frac{\tau_l}{T_c}} rI(k)b\left(k - iN\frac{T_f}{T_c} - \frac{\tau_l}{T_c} - \frac{\Delta}{T_c}\right) \tag{12.132}$$

$$DBQ^l(i) = \sum_{k=iN\frac{T_f}{T_c}-\frac{\tau_l}{T_c}}^{(i+1)N\frac{T_f}{T_c}-\frac{\tau_l}{T_c}} rQ(k)b\left(k - iN\frac{T_f}{T_c} - \frac{\tau_l}{T_c} - \frac{\Delta}{T_c}\right) \tag{12.133}$$

where i represents the bit index. Now, the signal at the output of RAKE combiner is:

$$d(i) = \sum_{l=1}^{L_c} \left[DI^l(i)TI^l(i) + DQ^l(i)TQ^l(i)\right] \tag{12.134}$$

The weight of the I branch in the lth finger is:

$$TI^l(i) = \overline{DAI^l(i) + DBI^l(i)} \tag{12.135}$$

and the weight of the Q branch in the lth finger is:

$$TQ^l(i) = \overline{DAQ^l(i) + DBQ^l(i)} \tag{12.136}$$

12.6.11 Interference rejection circuit model

Interference rejection at UWB radio system may be performed by a transversal filter employing LMS algorithm. Basically, in the first step, the interfering signal is estimated in the presence of the UWB signal which is at that stage considered as an additional noise. The estimated interference $\hat{j}$ is subtracted from the overall input signal r, creating the input signal $r' = r - \hat{j} = s + n + j - \hat{j} = s + n + \Delta j$ to the standard UWB receiver. In order to predict the interference signal, sampling is performed at frame rate, and the adaptation of filter weights using LMS algorithm is performed at bit rate. It is already known that the changes of the symbol in the interfering signal will disrupt the estimation process. Curve 3, in Figure 12.23, shows the detection variable at the output of the transversal filter when there

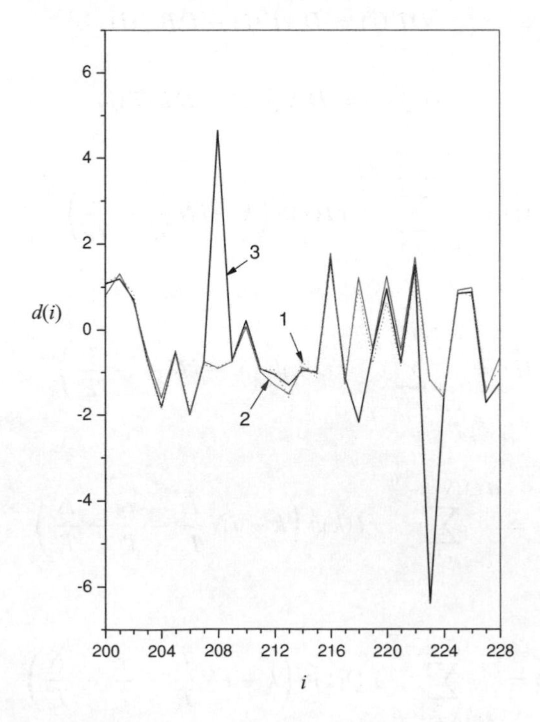

Figure 12.23 Detection variable: 1, no interference; 2, interference with $J{:}S = 40$ dB, with interference rejection circuit; 3, interference with $J : S = 40$ dB, with transversal filter using classical LMS algorithm, PSK interference, $M = 4$, $\Delta = 5$ ns, $\upsilon_{bJ} = 100$ Msymbol/s, $\upsilon_{bTH} = 5$ Mbt/s, $T_{frame} = 10$ ns, $\Delta f_c = 800$ MHz.

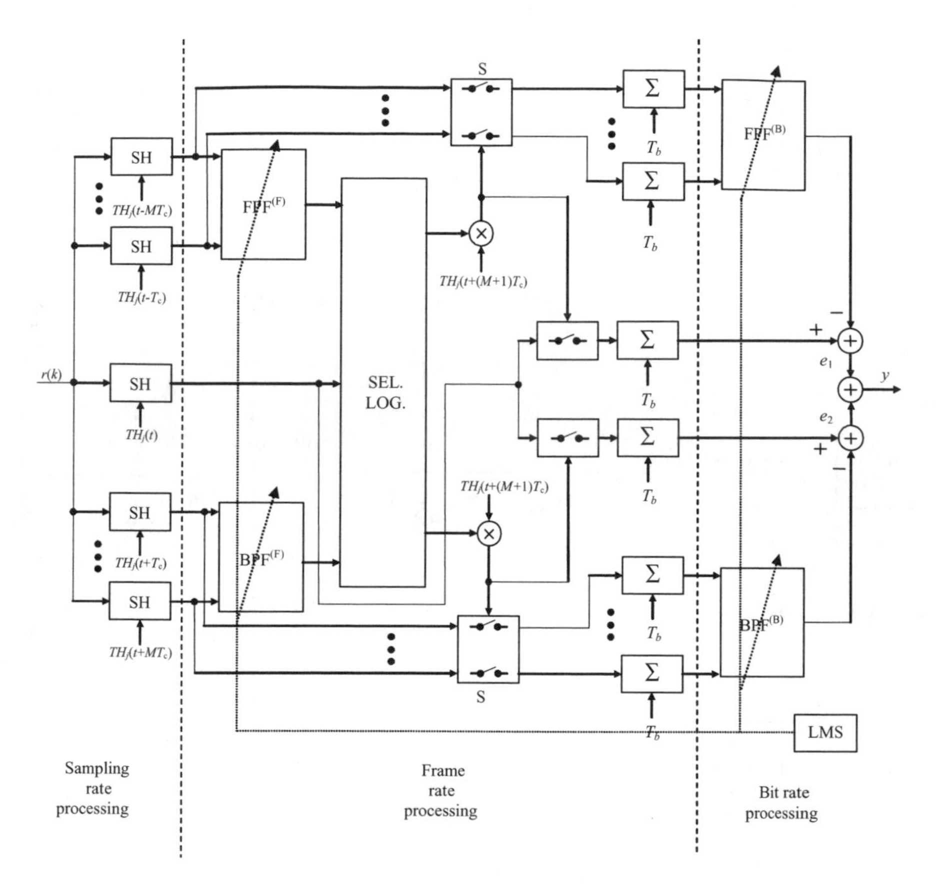

Figure 12.24 Interference rejection circuit block diagram. $TH_j(t)$ = time hopping sequence [$TH_j(t) = b(t - iNT_f - j\Delta)$, $j = 1$ for filter A, $j = 0$ for filter B]; SH = sample and hold circuit; $BPF^{(F)}$, $BPF^{(B)}$ = backward prediction filter; $FPF^{(F)}$, $FPF^{(B)}$ = forward prediction filter; SEL. LOG. = selection logic; A_i = amplifier, i = 1, 2; and C_i = comparator, i = 1, 2.

is PSK interference (with interference to signal ratio of 40 dB) at the input of the receiver together with the useful UWB signal and Gaussian noise. The presence of an impulse interference may be seen in the figure. The appearance of the impulse interference is very similar to the one at DSSS system using transversal filter for interference rejection [121]. A 'U' structure, that successfully rejects this impulse interference, was proposed in the mentioned reference. Similarly, in this section we discuss a structure, shown in Figure 12.24, for the interference rejection, which is based on the 'U' structure and has been modified to match the UWB radio environment.

Curve 2 in Figure 12.23 shows the signal at the output of the interference rejection structure in the case when there is PSK interference with interference-to-signal ratio of 40 dB at the input of the receiver. It can be noted that the interference is rejected and the detection variable is very similar to the one when there is no interference (curve 1).

More details on operating principle of the interference rejection circuit is shown in Figures 12.24 and 12.25. We consider the case of a bit interval having 20 frames with five

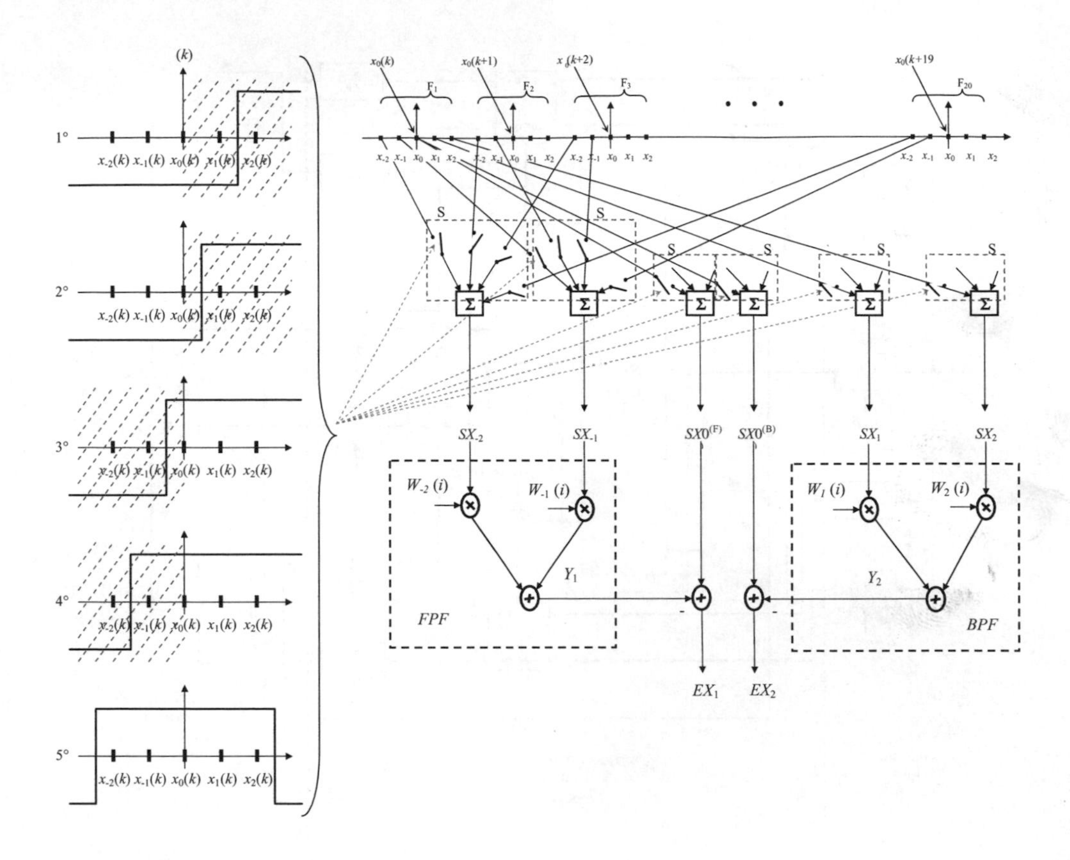

Figure 12.25 Signal processing.

samples per frame ($M = 2$). Central samples, within each frame, carry the same information about the useful signal and the interference symbols, and, since each sample belongs to a different frame, all those samples originate from different instances of time. Similarly, samples from different frames equally distant from the central sample carry the correlated interference signal. Therefore, an equivalent signal may be formed in the following way: Mth equivalent signal sample is the sum of Mth samples from each frame. Adaptation of filter weights using LMS algorithm and interference prediction is performed using the equivalent signal samples.

As already mentioned, the changes of the symbol in the interfering signal will disrupt the estimation process so that a forward and backward prediction are used simultaneously. When the symbol change occurs, the filter with less disruption (smaller error at its output) is used to deliver the estimates. This process is described in the following in more detail. For additional insight into the problem the reader is also referred to Cox and Reudink [31]. Possible moments of the transition to happen are shown in Figure 12.25 and are denoted with 1°, 2°, 3° and 4°. Therefore, at frame rate $BPF^{(F)}$ (backward prediction filter) and $FPF^{(F)}$ (forward prediction filter) filters are operating with weights being forwarded from the LMS algorithm adapted by the equivalent signal. $BPF^{(F)}$ and $FPF^{(F)}$ filters have the same weights during the useful signal bit interval, i.e. within all the frames belonging to the same bit interval. So the interference is predicted using the same weights computed using the equivalent signal. If, in cases 1°, and 2°, we discard samples belonging to the BPF$^{(F)}$ filter we will also discard the interference transition influence on the prediction. For cases 3° and 4°, samples that belong to the FPF$^{(F)}$ filter should be discarded. This sample discarding is performed in the selector S at the outputs of BPF$^{(F)}$ and FPF$^{(F)}$ filters, based on the error signal.

Therefore, if there is no interference signal transition during the sampling within one frame, the equivalent signal will be formed using all the samples from that frame. Also, the same equivalent central sample (SX0$^{(B)}$ = SX0$^{(F)}$) is passed to both BPF$^{(B)}$ and FPF$^{(B)}$ LMS algorithms. On the other hand, if there is interference signal transition during the sampling within one frame, the equivalent signal will be formed using samples from FPF$^{(F)}$ (cases 1° and 2°) or BPF$^{(F)}$ (cases 3° and 4°), and there will be two different equivalent central samples.

For the described Interference rejection circuit we have: *the first part of the interference rejection circuit* processes data at frame level, and at each frame the following input signal processing is performed. At filter A, the signal is sampled very close in time to the useful signal. For $-M \leq m \leq M$ we have:

$$A_m^l(n) = AI_m^l(n) + jAQ_m^l(n) \tag{12.137}$$

$$AI_m^l(n) = \sum_{k=n\frac{T_f}{T_c}-\frac{\tau_l}{T_c}}^{(n+1)\frac{T_f}{T_c}-\frac{\tau_l}{T_c}} rI(k)g\left[k - n\frac{T_f}{T_c} - \frac{\tau_l}{T_c} - h(n) - m\right] \tag{12.138}$$

$$AQ_m^l(n) = \sum_{k=n\frac{T_f}{T_c}-\frac{\tau_l}{T_c}}^{(n+1)\frac{T_f}{T_c}-\frac{\tau_l}{T_c}} rQ(k)g\left[k - n\frac{T_f}{T_c} - \frac{\tau_l}{T_c} - h(n) - m\right] \tag{12.139}$$

For filter B we have:

$$B_m^l(n) = BI_m^l(n) + jBQ_m^l(n) \tag{12.140}$$

$$BI_m^l(n) = \sum_{k=n\frac{T_f}{T_c}-\frac{\tau_l}{T_c}}^{(n+1)\frac{T_f}{T_c}-\frac{\tau_l}{T_c}} rI(k)g\left[k - n\frac{T_f}{T_c} - \frac{\tau_l}{T_c} - h(n) - m - \frac{\Delta}{T_c}\right] \tag{12.141}$$

$$BQ_m^l(n) = \sum_{k=n\frac{T_f}{T_c}-\frac{\tau_l}{T_c}}^{(n+1)\frac{T_f}{T_c}-\frac{\tau_l}{T_c}} rQ(k)g\left[k - n\frac{T_f}{T_c} - \frac{\tau_l}{T_c} - h(n) - m - \frac{\Delta}{T_c}\right] \tag{12.142}$$

After that, variables $C1$ and $C2$, which the operation of one side of filter X (A and B) is based on, are determined:

$$CX1^l(n) = \begin{cases} 0, \left(\sum_{m=-M}^{-1} X_m^l(n)W_m(i)]\right)^2 \geq A_1\left(\sum_{m=1}^{M} X_m^l(n)W_m(i)]\right)^2 \\ 1, \left(\sum_{m=-M}^{-1} X_m^l(n)W_m(i)]\right)^2 < A_1\left(\sum_{m=1}^{M} X_m^l(n)W_m(i)]\right)^2 \end{cases} \tag{12.143}$$

$$CX2^l(n) = \begin{cases} 0, \left(\sum_{m=1}^{M} X_m^l(n)W_m(i))\right)^2 \geq A_2\left(\sum_{m=-M}^{-1} X_m^l(n)W_m(i))\right)^2 \\ 1, \left(\sum_{m=1}^{M} X_m^l(n)W_m(i))\right)^2 < A_2\left(\sum_{m=-M}^{-1} X_m^l(n)W_m(i))\right)^2 \end{cases} \tag{12.144}$$

where $A_i(i = 1, 2)$ are constants (for $A_i \to \infty$, the selector selects all the samples and the structure operates as a traditional LMS algorithm). These gains are introduced because of the decrease of noise influence on the irregular selections.

The second part of the interference rejection circuit operates at bit interval level $T_b = NT_f$, and for each ith bit we have the following signals:

$$SX_m^l(i) = \sum_{n=iN}^{(i+1)N} X_m^l(n)CX1^l(n) - M \leq m \leq -1 \tag{12.145}$$

$$SX_m^l(i) = \sum_{n=iN}^{(i+1)N} X_m^l(n)CX2^l(n)1 \leq m \leq M \tag{12.146}$$

$$SX1^l(i) = \sum_{n=iN}^{(i+1)N} X_0^l(n)CX1^l(n) \tag{12.147}$$

$$SX2^l(i) = \sum_{n=iN}^{(i+1)N} X_0^l(n)CX2^l(n) \tag{12.148}$$

The error signal used for w coefficients adaptation is:

$$EX1^l(i) = SX1^l(i) - \sum_{m=-M}^{-1} SX_m^l(i)W_m(i) \tag{12.149}$$

$$EX2^l(i) = SX2^l(i) - \sum_{m=1}^{M} SX_m^l(i)W_m(i) \tag{12.150}$$

The filter output signal, being led to the RAKE combiner, is:

$$DX^l(i) = EX1^l(i) + EX2^l(i) \tag{12.151}$$

The adaptation algorithm is defined as:

$$W_m(i+1) = W_m(i) + \frac{\mu EX1^l(i)\left[SX_m^l(i)\right]^*}{\sum_{j=-M}^{-1}\left[SX_j^l(i)\right]^2}, \quad -M \le m \le -1 \tag{12.152}$$

$$W_m(i+1) = W_m(i) + \frac{\mu EX2^l(i)\left[SX_m^l(i)\right]^*}{\sum_{j=1}^{M}\left[SX_j^l(i)\right]^2}, \quad 1 \le m \le M \tag{12.153}$$

Finally, the interference rejection circuit is made symmetrical in the following way:

$$W_m(i+1) = \frac{\left[W_m(i+1) + W_{-m}(i+1)^*\right]}{2}, \quad 1 \le m \le M \tag{12.154}$$

$$W_m(i+1) = W_{-m}(i+1)^*, \quad -1 \le m \le -M \tag{12.155}$$

12.6.12 Performance analysis

The error probability per bit is:

$$Pe = \frac{1}{N_i}\sum_{i=1}^{N_i} Pe\,(i) \tag{12.156}$$

where

$$Pe\,(i) = \begin{cases} \dfrac{1}{2}\text{erfc}\left(\sqrt{\dfrac{SNR(i)}{2}}\right), & a_i \sum_{j=1}^{N_a} d^j\,(i) \ge 0 \\ 1 - \dfrac{1}{2}\text{erfc}\left(\sqrt{\dfrac{SNR(i)}{2}}\right), & a_i \sum_{j=1}^{N_a} d^j\,(i) < 0 \end{cases} \tag{12.157}$$

N_i is the bit ensemble size (measured in number of information bits) used for averaging the result and N_a is the number of ensemble members.

Estimated signal-to-noise ratio per bit is:

$$SNR\,(i) = \frac{\left[\dfrac{1}{N_a}\sum_{j=1}^{N_a} d^j\,(i)\right]^2}{\dfrac{1}{N_a}\sum_{j=1}^{N_a}\left[d^j\,(i)\right]^2 - \left[\dfrac{1}{N_a}\sum_{j=1}^{N_a} d^j\,(i)\right]^2} \tag{12.158}$$

where $d^j(i)$ is the jth ensemble member.

12.6.13 Performance examples

Figure 12.26 presents the results for BER as a function of signal to noise ratio SNR, in the presence of different PSK/QAM type interfering signals. Additional parameters of the signals are: filter length $M = 4$, $\Delta = 5$ ns, $\upsilon_{bJ} = 100$ Msymbol/s, $\upsilon_{bTH} = 5$ Mbt/s,

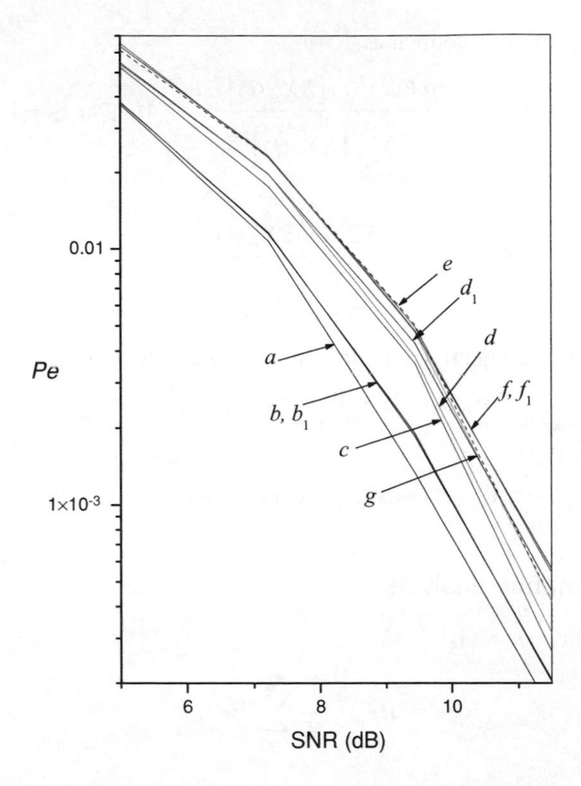

Figure 12.26 Error probability as a function of signal-to-noise ratio. Error probability based on Monte-Carlo simulation: *a*, no interference, without interference rejection filter; *b*, no interference, with interference rejection filter; *c*, PSK interference, $J{:}S = 40$ dB, with interference rejection filter; *d*, QPSK interference, $J{:}S = 40$ dB, with interference rejection filter; *e*, 16QAM interference, $J{:}S = 40$ dB, with interference rejection filter; *f*, 64QAM interference, $J{:}S = 40$ dB, with interference rejection filter; *g*, 256QAM interference, $J{:}S = 40$ dB, with interference rejection filter; Error probability based on estimated detection variable signal to noise ratio: b_1, the same parameters as *b*; d_1, the same parameters as *d*; f_1, the same parameters as *f*; $\Delta f_c = 800$ MHz, $M = 4$, $\Delta = 5$ ns, $v_{bJ} = 100$ Msymbol/s, $v_{bTH} = 5$ Mbt/s, $T_{frame} = 10$ ns.

$\Delta f_c = 800$ MHz and $T_{frame} = 10$ ns. One can see: (1) fair agreement of simulation and numerical results; (2) the performance results are close to no interference case, although interference with the level of 40 dB above the UWB signal is present; (3) there is also a slight degradation of the performance when the interfering signal constellation size is increased.

Figure 12.27 presents the results for BER as a function of interference to signal ratio *J:S*, in the presence of different PSK/QAM-type interfering signals. Additional parameters of the signals are: filter length $M = 4$, SNR=7 dB, $\Delta = 5$ ns, $v_{bJ} = 100$ Msymbol/s, $v_{bTH} = 5$ Mbt/s, $\Delta f_c = 800$ MHz and $T_{frame} = 10$ ns. One can see that, when *J:S* becomes larger than zero (5 dB), the BER increases rapidly if there is now interference suppression (curves A). The performance is very similar if a standard LMS algorithm is used (curves C). On the other hand the U-type filter is performing significantly better (curves B). There is again a slight

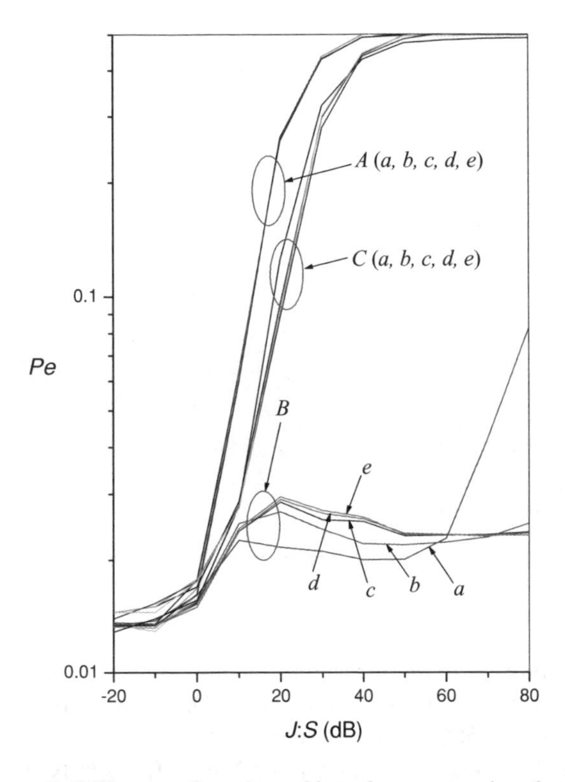

Figure 12.27 Error probability as a function of interference-to-signal ratio. A, without interference rejection; B, with interference rejection circuit; C, with classical LMS interference rejection filter; a, PSK interference; b, QPSK interference; c, 16QAM interference; d, 64QAM interference; e, 256QAM interference; $\Delta f_c = 800$ MHz, $M = 4$, SNR = 7 dB, $\Delta = 5$ ns; $\upsilon_{bJ} = 100$ Msymbol/s, $\upsilon_{bTH} = 5$ Mbt/s, $T_{frame} = 10$ ns.

degradation of the performance when the interfering signal constellation size is increased. Figure 12.28 presents the results for BER as a function of interference symbol duration T_j/T_c in the presence of different PSK/QAM-type interfering signals. Additional parameters of the signals are J: $S = 30$ dB, SNR = 7 dB, $M = 4$, $\Delta = 5$ ns, $\upsilon_{bTH} = 5$ Mbt/s, $\Delta f_c = 800$ MHz and $T_{frame} = 10$ ns. One can see that BER decreases when T_j/T_c increases. There is again a slight degradation of the performance when the interfering signal constellation size is increased. Figure 12.29 presents the results for BER as a function of interference symbol duration T_j/T_c and the number of subcarriers N in the presence of OFDM-type interfering signals. Additional parameters of the signals are J: $S = 30$ dB, SNR = 7 dB, $M = 4$, $\Delta = 5$ ns, $\upsilon_{bTH} = 5$ Mbt/s and $T_{frame} = 10$ ns, $\Delta f_c = 800$ MHz and 16QAM per subcarrier. One can see that BER decreases when T_j/T_c increases up to $T_j/T_c \approx 200$. Beyond that point there is no significant reduction in BER if T_j/T_c is further increased. There is a significant degradation of the performance when the number of subcarriers in the OFDM signal is increased.

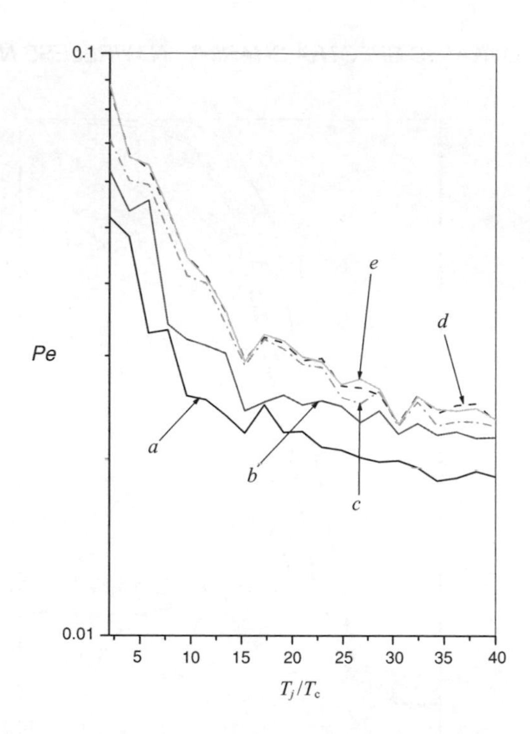

Figure 12.28 Error probability as a function of interference bit duration. a, PSK interference; b, QPSK interference; c, 16QAM interference; d, 64QAM interference; e, 256QAM interference; $\Delta f_c = 800$ MHz; $J{:}S = 30$ dB; SNR = 7 dB; $M = 4$; $\Delta = 5$ ns; $v_{bTH} = 5$ Mbt/s; $T_{frame} = 10$ ns.

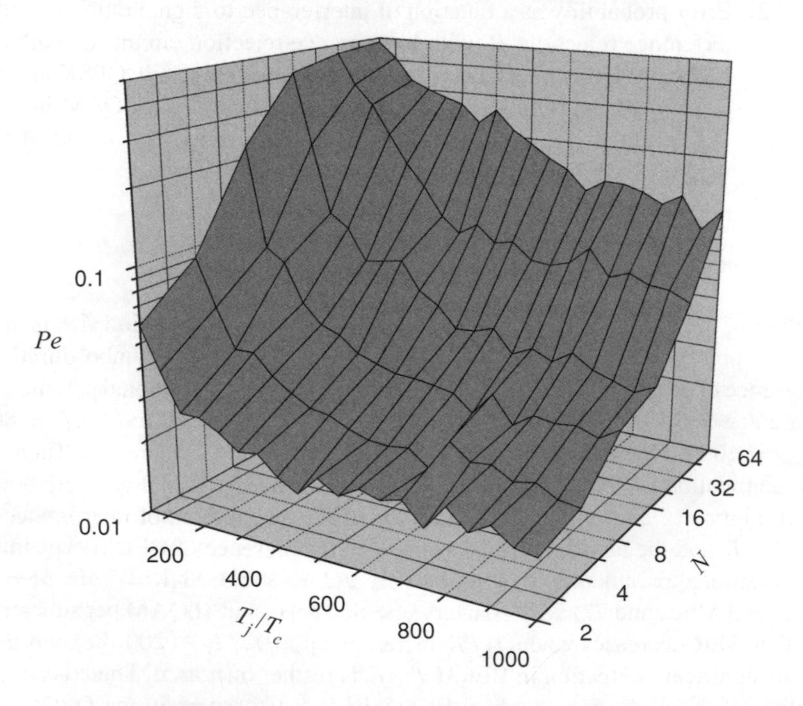

Figure 12.29 Error probability as a function of OFDM interference bit duration and the number of subcarriers. OFDM/16QAM interference; $\Delta f_c = 800$ MHz; $J : S = 30$ dB; SNR = 7 dB; $M = 4$; $\Delta = 5$ ns; $v_{bTH} = 5$ Mbt/s; $T_{frame} = 10$ ns.

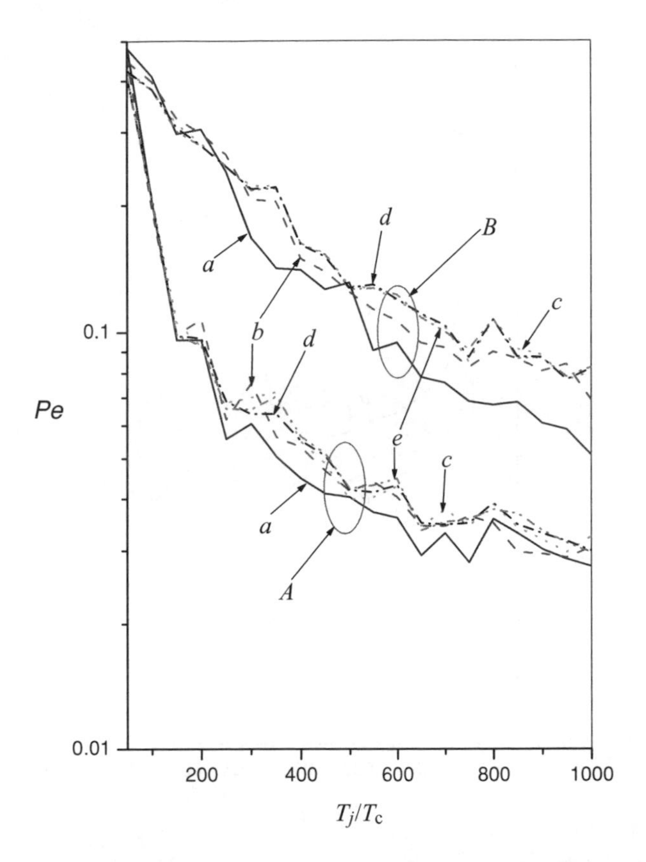

Figure 12.30 Error probability as a function of OFDM interference bit duration. A, interference rejection circuit; B, classical LMS interference rejection filter; a, OFDM/PSK interference; b, OFDM/QPSK interference; c, OFDM/16QAM interference; d, OFDM/ 64QAM interference; e, OFDM/256QAM interference; $N = 16, \Delta f_c = 800$ MHz, $J{:}S = 30$ dB, SNR = 7 dB, $M = 4$, $\Delta = 5$ ns; $\upsilon_{bTH} = 5$ Mbt/s, $T_{frame} = 10$ ns.

Figure 12.30 presents the results for BER as a function of interference symbol duration T_j/T_c in the presence of OFDM-type interfering signals. Additional parameters of the signals are $J{:}S = 30$ dB, SNR $= 7$ dB, $M = 4$, $\Delta = 5$ ns, $\upsilon_{bTH} = 5$ Mbt/s and $T_{frame} = 10$ ns , $\Delta f_c = 800$ MHz and $N = 16$. One can see again that BER decreases when T_j/T_c increases. There is again a slight degradation of the performance when the interfering signal constellation size is increased. Once again, the U-type filter performs much better than the classical LMS algorithm. Figure 12.31 presents the results for BER as a function of interference symbol duration T_j/T_c in the presence of MC CDMA-type interfering signals for different number of subcarriers N and number of users K. Additional parameters of the signals are $J{:}S = 30$ dB, SNR $= 7$ dB, $M = 4$, $\Delta = 5$ ns, $\upsilon_{bTH} = 5$ Mbt/s and $T_{frame} = 10$ ns and $\Delta f_c = 800$ MHz. One can see again that BER decreases when T_j/T_c increases. The performance are improved if the number of subcarriers is decreased.

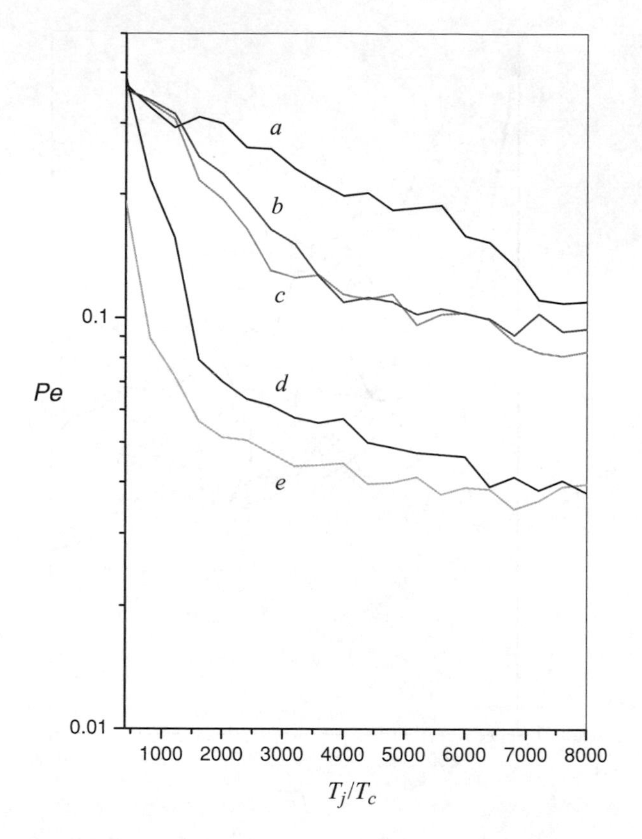

Figure 12.31 Error probability as a function of MC CDMA interference bit duration. a, $N = 64$, $K = 10$; b, $N = 32$, $K = 5$; c, $N = 32$, $K = 10$; d, $N = 32$, $K = 20$; e, $N = 16$, $K = 10$. MC CDMA interference; $J{:}S = 30$ dB, SNR $= 7$ dB, $M = 4$, $\Delta = 5$ ns; $v_{\mathrm{bTH}} = 5$ Mbt/s, $T_{\mathrm{frame}} = 10$ ns, $\Delta f_{\mathrm{c}} = 800$ MHz.

The performance is improved if the number of users is increased for the same overall power of the interfering signal. This can be explained by the fact that a sum of MC CDMA signals will create an equivalent multicarrier signal with fewer dominant subcarriers which are suppressed more effectively by the filter because the LMS algorithm better adjusts the filter weights. This is demonstrated in Figures 12.29 and 12.30 for OFDM signal.

In this section we presented a U-type estimation filter based scheme for interference suppression in UWB systems and discussed its performance. It was shown that the scheme can be used to significantly improve the performance of UWB systems in the presence of interference from mobile communication systems such as GSM and WCDMA. It is also effective in the presence of WLAN systems which are based on OFDMA technology or in military communications where the interference is generated by intentional jamming. The section also discusses the effectiveness of the scheme to suppress MC CDMA, which is a candidate technology for 4G mobile communications. The results show that significant suppression gain up to 40 dB can be achieved in the presence of OFDM, WCDMA and MC CDMA. The effectiveness decreases if the size of the number of subcarriers is increased.

REFERENCES

[1] S.G. Glisic and P. Pirinen, *Co-Channel Interference in Digital Cellular TDMA Networks*, *John Wiley Encyclopedia of Telecommunications*, ed. J. Proakis. John Wiley & Sons Ltd: Chichester, 2003.

[2] J. Zander, Asymptotic bounds on the performance of a class of dynamic channel assignment algorithms, *IEEE JSAC*, vol. 11, 1993, pp. 926–933.

[3] J. C.-I. Chuang, Performance issues and algorithms for dynamic channel assignment, *IEEE JSAC*, vol. 11, 1993, p. 6.

[4] T.J. Kahwa and N. Georganas, A hybrid channel assignment scheme in large scale cellular-structured mobile communication systems, *IEEE Trans. Commun.*, vol. COM 26, 1978, pp. 432–438.

[5] J. Sin and N. Georganas, A simulation study of a hybrid channel assignment scheme for cellular land mobile radio systems with Erlang C service, *IEEE Trans. on Commun.*, vol. COM-9, 1981, pp. 143–147.

[6] D. Cox and D.O. Reudink, Increasing channel occupancy in large scale mobile radio systems: dynamic channel reassignment, *IEEE Trans. Vehicular Technol.*, vol.VT-22, 1973, pp. 218–222.

[7] D. Cox and D.O. Reudink, A comparison of some non-uniform spatial demand profiles on mobile radio system performance, *IEEE Trans. Commun.*, vol. COM 20, 1972, pp. 190–195.

[8] D.C. Cox and D.O. Reudink, Dynamic channel assignment in two dimensional large-scale mobile radio systems, *Bell Syst. Tech. J.*, vol. 51, 1972, pp. 1611–1628.

[9] M. Zhang and T.-S. Yum, The non-uniform compact pattern allocation algorithm for cellular mobile systems, *IEEE Trans. Vehicular Technol.* vol. VT-40, 1991, pp. 387–391.

[10] S.-H. Oh and D.W. Tcha, Prioritized channel assignment in a cellular radio network, *IEEE Trans. Commun.*, vol. 40, 1992, pp. 1259–1269.

[11] T. Anderson, A simulation study of sole dynamic channel assignment algorthms in high capacity mobile telecommunications system, *IEEE Trans. Vehicular Technol.*, vol. VT-22, 1973, p. 210.

[12] J.S. Engel and M. Peritsky, Statistically optimum dynamic server assignment in systems with interfering servers, *IEEE Trans. Vehicular Technol.*, vol. VT-22, 1973, pp. 203–209.

[13] M. Zhang, Comparisons of channel assignment strategies in cellular mobile telephone systems, *IEEE Trans. Vehicular Technol.*, vol. VT38, 1989, pp. 211–215.

[14] R. Singh, S.M. Elnoubi and C. Gupta, A new frequency channel assignment algorithm in high capacity mobile communications systems, *IEEE Trans. Vehicular Tech.*, vol. VT-31, 1982.

[15] P. John, An insight into dynamic channel assignment in cellular mobile communication systems, *Eur. J. Opnl. Res.*, vol. 74, 1994, pp. 70–77.

[16] S. Tekinay and B. Jabbari, Handover and channel assignment in mobile cellular networks, *IEEE Commun. Mag.* vol. 29, 1991, pp. 42–46.

[17] T.-S. P. Yum and W.-S. Wong, Hot spot traffic relief in cellular systems, *IEEE JSAC*, vol. 11, 1993, pp. 934–940.

[18] S. Kuek, Ordered dynamic channel assignment scheme with reassignment in highway microcell, *IEEE Trans. Vehicular Technol.*, vol. 41, 1992, pp. 271–277.

[19] K. Okada and F. Kubota, On dynamic channel assignment in cellular mobile radio systems, *Proc. IEEE Int. Symp. Circuits and Systems*, vol. 2, 1991, pp. 938–941.
[20] K. Okada and F. Kubota, On dynamic channel assignment strategies in cellular mobile radio systems, *IEICE Trans. Fundamentals*, vol. 75, 1992, pp. 1634–1641.
[21] D. Cox and D. Reudink, A comparison of some channel assignment strategies in large mobile communication systems, *IEEE Trans. Commun.*, vol. 20, 1972, pp. 190–195.
[22] D. Cox and D. Reudink, Dynamic channel assignment in high capacity mobile communications systems, *Bell Syst. Tech. J.*, vol . 50, 1971, pp. 1833–1857.
[23] A. Gamst, Some lower bounds for a class of frequency assignment problems, *IEEE Trans. Vehicular Technol.*, vol. 35, 1986.
[24] K. Okada and F. Kubota, A proposal of a dynamic channel assignment strategy with information of moving directions, *IEICE Trans. Fund.*, vol. E75-a, 1992, pp. 1667–1673.
[25] V. Akaiwa and H. Andoh, Channel segregation-A self organized dynamic allocation method: application to TDMA/FDMA microcellular system, *JSAC*, vol. 11, 1993, pp. 949–954.
[26] V. Prabhl and S.S. Rappaport. Approximate analysis of dynamic channel assignment in large systems with cellular structure, *IEEE Trans. Commun.*, vol. 22, 1974, pp. 1715–1720.
[27] J.B. Punt and D. Sparreboom, Mathematical models for the analysis of dynamic channel selection or indoor mobile wireless communications systems, *PIMRC*, vol. E6.5, 1994, pp. 1081–1085.
[28] D. Hong and S. Rappaport, Traffic modelling and performance analysis for cellular mobile radio telephone systems with prioritized and nonprioritized handoff procedures, *IEEE Trans. Vehicular Technol.*, vol. VT 35, 1986, pp. 77–92.
[29] Y. Furuya and V. Akaiwa, Channel segregation, A distributed channel allocation scheme for mobile communication systems, *IEICE Trans.*, vol. 74, 1991, pp. 1531–1537.
[30] J. Vucetic, A hardware implementation of channel allocation algorithm based on a space-bandwidth model of a cellular network, *IEEE Trans. Vehicular Technol.*, vol. 42, 1993, pp. 444–455.
[31] D. Cox and D. Reudink, Increasing channel occupancy in large scale mobile radio systems: dynamic channel reassignment, *IEEE Trans. Commun.*, vol. 21, 1973, pp. 1302–1306.
[32] W. Yue, Analytical methods to calculate the performance of a cellular mobile radio communication system with hybrid channel assignment, *IEEE Trans. Vehicular Technol.*, vol VT-40, 1991, pp. 453–459.
[33] J. Tajime and K. Imamura, A strategy for flexible channel assignment in mobile communication systems, *IEEE Trans. Vehicular Technol.*, vol. VT-37, 1988, pp. 92–103.
[34] J. Zande and J. Frodigh, Capacity allocation and channel assignment in cellular radio systems using reuse partitioning, *Electron. Lett.*, vol. 28, 1991.
[35] S. W. Halpern, Reuse partitioning in cellular systems, *IEEE Trans. Vehicular Technol.*, 1983, pp. 322–327.
[36] T. Kanai, Autonomous reuse partitioning in cellular systems, *IEEE VTC*, 1992 pp. 782–785.

[37] S. Onoe and S. Yasuda, Flexible re-use for dynamic channel assignment in mobile radio systems, *Proc. IEEE ICC*, 1989, pp. 472–476.

[38] T. Takenaka, T. Nakamura and V. Tajima, All-channel concentric allocation in cellular systems, *IEEE ICC'93*, Geneva, 23–26 May pp. 920–924.

[39] K. Madani and A.H. Aghvami, Performance of distributed control channel allocation (DCCA) under uniform and non-uniform traffic conditions in microcellular radio communications. *IEEE, SUPERCOMM/ICC'94*, Dallas, TX, 1–5 May 1994.

[40] K. Madani and A.H. Aghvami, Investigation of handover in distributed control channel allocation (DCCA) for microcellular radio systems, *Person. Indoor Mobile Radio Commun.*, 1994, p. 82.1.

[41] S. Glisic, *Advanced Wireless Communications: 4G Technology*. John Wiley and Sons Ltd: Chichester, 2004.

[42] J.J. Caffery Jr and G.L. Stuber, Overview of radiolocation in CDMA cellular systems, *IEEE Commun. Mag.*, 1998, pp. 38–45.

[43] H. De Meer, A. La Corte, A. Puliafito and O. Tomarchio, Programmable agents for flexible QoS management in IP networks, *IEEE J. Selected Areas Commun.*, vol. 18, no. 2, 2000.

[44] M.S. Greenberg, J.C. Byington and T. Holding, Mobile agents and security, *IEEE Commun. Mag.*, vol. 7, no. 31, 1998, pp. 76–85.

[45] C. Perkins, Mobile networking through mobile IP, *IEEE Internet Comput.*, Tutorial, January–February 1998.

[46] J.P. Macker, V.D. Park and M.S. Corson, Mobile and wireless internet services: putting the pieces together, *IEEE Comm. Mag.*, June 2001, pp. 148–155.

[47] The book of visions 2000—visions of wireless world 2000, *EC IST-WSI*, 2000.

[48] A.J. Viterbi, Capacity of a simple and stable protocol for short message service over CDMA network, in *Communication and Cryptography*, ed. R. Blahut, pp. 423–429, 1994.

[49] S. Glisic and V.V. Phan, Sensitivity function of soft decision carrier sense MAC protocols for wireless CDMA networks with specified QoS, invited paper, *IEEE 11th PIMRC Conf. Proc.*, September 2000, London, pp. 205–211.

[50] J. Zander and S.L. Kim, *Radio Resource Management for Wireless Networks*. Artech House: London, 2001.

[51] A.M. Viterbi and A.J. Viterbi, Erlang capacity of a power controlled CDMA system, *IEEE J. Select. Areas Comm.*, vol. 11, 1993, pp. 892–900.

[52] A.A. Abu-Dayya and N.C. Beaulieu, Outage probabilities in the presence of correlated lognormal interferers, *IEEE Trans. Vehicle Technol.*, vol. 43, 1994, pp. 164–173.

[53] S. Ariyavisitakul, Signal and interference statistics of a CDMA system with feedback power control II, *IEEE Trans. Commun.*, vol. 42, 1994, pp. 597–605.

[54] Z.Liu and M.E.Zarki, SIR-based call admission control for DS-CDMA cellular systems, *IEEE J. Select. Areas Commun.*, vol. 12, 1994, pp. 638–644.

[55] W.B. Yang and E. Geraniotis, Admission policies for integrated voice and data traffic in CDMA packet radio networks, *IEEE J. Select. Areas Commun.*, vol. 12, 1994, pp. 654–664.

[56] A.J. Viterbi, *Principle of Spread Spectrum Communication*. Addison-Wesley: Reading, MA: 1995.

[57] Z. Dziong, J. Ming and P. Mermelstein, Adaptive traffic admission for integrated services in CDMA wireless-access networks, *IEEE J. Select. Areas Commun.*, vol. 14, 1996, pp. 1737–1747.

[58] A. Sampath and J.M. Holtzman, Access control of data in integrated voice/data CDMA systems: benefits and tradeoffs, *IEEE Select. Areas Commun.,* vol. 15, 1997, pp. 1511–1526.

[59] Y. Ishikawa and N. Umeda, Capacity design and performance of call admission control in cellular CDMA systems, *IEEE J. Select. Areas Commun.*, vol. 15, 1997, pp. 1627–1635.

[60] J.S. Evans and D. Everitt, Effective bandwidth-based admission control for multiservice CDMA cellular networks, *IEEE Trans. Vehicle Technol.*, vol. 48, 1999, pp. 36–46.

[61] S.M. Shin, C.H. Cho and D.K. Sung, Interference-based channel assignment for DS-CDMA cellular systems, *IEEE Trans. Vehicle Technol.*, vol. 48, 1999, pp. 233–239.

[62] N. Dimitriou, R. Tafazolli and G. Sfikas, Quality of service for multimedia CDMA, *IEEE Commun. Mag.*, 2000, pp. 88–94.

[63] H. Holma and A. Toskala, *WCDMA for UMTS Radio Access for Third Generation Mobile Communications*. John Wiley & Sons Inc.: New York, 2000.

[64] S. Dixit, Y. Gou and Z. Antoniou, Resource management and quality of service in third generation wireless networks, *IEEE Commun. Mag.*, 2001, pp. 125–133.

[65] R. Ramjee, R. Nagarajan and D. Towsley, On optimal call admission control in cellular networks, *IEEE InfoCom1996 Conf.*, vol. 1, 1996, pp. 43–50.

[66] D. Mitra, M.I. Reiman and J. Wang, Robust dynamic admission control for unified cell and call QoS in statistical multiplexer, *IEEE J. Select. Areas Commun.*, vol. 16, 1998, pp. 692–707.

[67] M. Grossglauser and D. Tse, A framework for robust measurement-based admission control, *IEEE ACM Trans. Networks*, vol. 7, 1999, pp. 293–309.

[68] L. Breslau, S. Jamin and S. Shenker, Comments on the performance of measurement-based admission control algorithms, *IEEE InfoCom2000 Conf.*, 26–30 March 2000, pp. 1233–1242.

[69] H. Tong and T.X. Brown, Adaptive call admission control under quality of service constraints: a reinforcement learning solution, *IEEE J. Select. Areas Commun.*, vol. 18, 2000, pp. 209–221.

[70] R. Thomas, H. Gilbert and G. Mazziotto, Influence of the movement of the mobile station on the performance of a radio cellular network, in *3rd Nordic Seminar*, Copenhagen, September 1988.

[71] S.S. Rappaport, Modeling the hand-off problem in personal communications networks, in *41st IEEE Vehicle Technology Conference*, 19–22 May 1991, pp. 517–523.

[72] Y.B. Lin, S. Mohan and A. Noerpel, Queueing priority channel assignment strategies for PCS handoff and initial access, *IEEE Trans. Vehicle Technol.*, vol. 43, no. 3, 1994, pp. 704–712.

[73] J.S. Kaufman, Blocking in a Shared Resource Environment, *IEEE Trans. Commun.*, vol. 29, no. 10, 1981, pp. 1474–1481.

[74] J.W. Roberts, A service system with heterogeneous user requirements, *Perform. Data Commun. Syst. Applic.*, 1981, pp. 423–431.

[75] F.P. Kelly, Blocking probabilities in larger circuit-switched networks, *Adv. Appl. Prob.*, vol. 18, 1986, pp. 473–505.

[76] K.W. Ross and D.H.K. Tsang, The stochastic knapsack problem, *IEEE Trans. Commun.*, vol. 37, no. 7, 1989, pp. 740–747.

[77] S.P. Chung and K.W. Ross, Reduced load approximations for multirate loss networks, *IEEE Trans. Commun.*, vol. 41, no. 8, 1993, pp. 1222–1231.

[78] D. Mitra, J.A. Morrison and K.G. Ramakrishnan, ATM network design and optimization: a multirate loss network framework, *IEEE ACM Trans. Networkes*, vol. 4, no. 4, 1996, pp. 531–543.

[79] N.G. Bean, R.J. Gibbens and S. Zachary, Asymptotic analysis of single resource loss systems in heavy traffic, with applications to integrated networks, *Adv. Appl. Prob.*, vol. 27, 1995, pp. 273–292.

[80] H. Morikawa, T. Kajiya, T. Aoyama and A. Campbell, Distributed power control for various QoS in a CDMA wireless system, in *Proc. IEEE PIMRC*, Helsinki, September 1997, pp. 903–907.

[81] S. Kim, Z. Rosberg and J. Zander, Combined power control and transmission rate selection in cellular networks, in *Proc. IEEE Vehicular Technology Conf.*, Amsterdam, September 1999, pp. 1653–1657.

[82] S. Grandhi and J. Zander, Constrained power control in cellular radio systems, in *Proc. IEEE Vehicular Technology Conf.*, vol. 2, Piscataway, NJ, June 1994, pp. 824–828.

[83] R. Jäntti and S. Kim, Selective power control and active link protection for combined rate and power management, in *Proc. IEEE Vehicular Technology Conf.*, Tokyo, May 2000, pp. 1960–1964.

[84] G. Janssen and J. Zander, Power control and stepwise removal algorithms for a narrowband multiuser detector in a cellular system," in *Proc. IEEE ICC*, New York, May 2002, pp. 345–350.

[85] K. Chawla and X. Qiu, Throughput performance of adaptive modulation in cellular systems, in *Proc. IEEE Int. Conf. Universal Personal Communications*, Florence, October 1998, pp. 945–950.

[86] K. Miettinen, *Nonlinear Multiobjective Optimization.* Kluwer Academic: Boston, MA, 1998.

[87] M. Elmusrati, Radio resource scheduler and smart antenna in cellular communication systems, PhD thesis, HUT, Helsinki, 2004.

[88] J. Zander, Distributed cochannel interference control in cellular radio systems, *IEEE Trans. Vehicle Technol.*, vol. 41, 1992, pp. 305–311.

[89] S. Grandhi, R. Vijayan and D. Goodman, Distributed power control in cellular radio systems, *IEEE Trans. Commun.*, vol. 42, 1994, pp. 226–228.

[90] T. Lee and J. Lin, A fully distributed power control algorithm for cellular mobile systems, *IEEE Trans. Commun.*, vol. 14, 1996, pp. 692–697.

[91] Multispectral Solutions Inc., History of UWB technology; www.multispectral.com/history.html

[92] Aiello GR and Rogerson Ultra-wideband wireless systems, *IEEE Microwave Mag.* vol. 4, no. 2, 2003, pp. 36–47.

[93] Win MZ and Scholtz RA, Impulse radio: how it works, *IEEE Commun.* Lett., vol. 2, 1998, pp. 36–38.

[94] Zhang J, Kennedy RA and Abhayapala TD, New results on the capacity of M-ary PPM ultra-wideband systems, in *Proc. IEEE Int. Conf. Communications (ICC)*, Anchorage, AK, 2003, vol. 4, pp. 2867–2871.

[95] Cassioli D, Win MZ and Molisch AF, The ultra-wideband indoor channel: from statistical model to simulations, *IEEE J. Selected Areas Commun.*, vol. 20, no. 6, 2002, pp. 1247–1257.

[96] Cramer RJ-M, Scholtz RA and Win MZ, Evaluation of an ultra-wide-band propagation channel, *IEEE Trans. Antennas Propagation*, vol. 50, 2002, pp. 561–570.

[97] Saleh AAM and Valenzuela RA, A statistical model for indoor multipath propagation, *IEEE J. Selected Areas Commun.*, vol. 5, 1987, pp. 128–137.

[98] Win MZ and Scholtz RA, Characterization of ultra-wide bandwidth wireless indoor channels: a communication-theoretic view, *IEEE J. Selected Areas Commun.*, vol. 20, 2002, pp. 1613–1627.

[99] Choi JD and Stark WE, Performance of ultra-wideband communications with sub-optimal receivers in multipath channels, *IEEE J. Selected Areas Commun.*, vol. 20, 2002, pp. 1754–1766.

[100] Rajeswaran A, Somayazulu VS and Foerster JR, Rake performance for a pulse based UWB system in a realistic UWB indoor channel, in *Proc. IEEE Int. Conf. Communications* (*ICC*), Anchorage, AK, vol. 4, 2003, pp. 2879–2883.

[101] Win MZ and Scholtz RA, On the energy capture of ultrawide bandwidth signals in dense multipath environments, *IEEE Commun. Lett.*, vol. 2, 1998, pp. 245–247.

[102] Durisi G and Benedetto S, Performance evaluation of TH-PPM UWB systems in the presence of multiuser interference, *IEEE Commun. Lett.*, vol. 7, 2003, pp. 224–226.

[103] Ramírez-Mireles F, Signal design for ultra-wide-band communications in dense multipath, *IEEE Trans. Vehicular Technol.*, vol. 51, 2002, pp. 1517–1521.

[104] Zhang J, Abhayapala TD and Kennedy RA, Performance of ultra- wideband correlator receiver using Gaussian monocycles, in *Proc. IEEE Int. Conf. Communications* (*ICC*), Anchorage, AK, vol. 3, 2003, pp. 2192–2196.

[105] Hu B and Beaulieu NC, Exact bit error rate analysis of TH-PPM UWB systems in the presence of multiple-access interference, *IEEE Commun. Lett.*, vol. 7(12), 2003, pp. 572–574.

[106] Cuomo F, Martello C, Baiocchi A and Capriotti F, Radio resource sharing for *ad hoc* networking with UWB, *IEEE J. Selected Areas Commun.*, vol. 20, 2002, pp. 1722–1732.

[107] Forouzan AR, Nasiri-Kenari M and Salehi JA, Performance analysis of time-hopping spread-spectrum multiple-access systems: uncoded and coded schemes, *IEEE Trans. Wireless Commun.*, vol. 1, 2002, pp. 671–681.

[108] Ramírez-Mireles F, Performance of ultrawideband SSMA using time hopping and M-ary PPM, *IEEE J. Selected Areas Commun.* vol. 19, 2001, pp. 1186–1196.

[109] Scholtz RA, Multiple access with time-hopping impulse modulation, in *Proc. IEEE Military Communications Conf.* (*MILCOM*), Boston, MA, vol. 2, 1993, pp. 447–450.

[110] Win MZ and Scholtz RA, Ultra-wide bandwidth time-hopping spread-spectrum impulse radio for wireless multiple-access communications, *IEEE Trans. Commun.*, vol. 48, 2000, pp. 679–691.

[111] Li Q and Rusch LA, Multiuser detection for DS-CDMA UWB in the home environment, *IEEE J. Selected Areas Commun.*, vol. 20, 2002, pp. 1701–1711.

[112] Trichard LGF and Kohno R, Iterative multiuser partial parallel interference cancellation for UWB-MA systems, in *Proc. 2003 Int. Workshop on Ultra Wideband Systems* (*IWUWBS*), Oulu, Finland, 2003.

[113] Yoon YC and Kohno R, Optimum multi-user detection in ultra-wideband (UWB) multiple-access communication systems, in *Proc. IEEE Int. Conf. Communications (ICC)*, New York, vol. 2, 2002, pp. 812–816.

[114] Hämäläinen M, Hovinen V, Tesi R, Iinatti JHJ and Latva-aho M, On the UWB system coexistence with GSM900, UMTS/WCDMA, and GPS, *IEEE J. Selected Areas Commun.*, vol. 20, 2002, pp. 1712–1721.

[115] Lottici V, D'Andrea A and Mengali U, Channel estimation for ultra-wideband communications, *IEEE J. Selected Areas Commun.*, vol. 20(9), 2002, pp. 1638–1645.

[116] Lovelace WM and Townsend JK, The effects of timing jitter and tracking on the performance of impulse radio, *IEEE J. Selected Areas Commun.*, vol. 20, 2002, pp. 1646–1651.

[117] Güvenç İ and Arslan H, Performance evaluation of UWB systems in the presence of timing jitter, in *IEEE Conf. Ultra Wideband Systems and Technologies (UWBST)*, Reston, VA, 2003.

[118] X. Luo, L. Yang and G. Giannakis, Designing optimal pulse-shapers for ultra-wideband radios, *J. Commun. Networks*, vol. 5, no. 4, 2003, pp. 344–354.

[119] B. Parr, B. Chao and Z. Ding, A new UWB pulse generator gor FCC spectral mask, in *Proc. Vehicular Technology Conf.*, vol. 3, April 2003, pp. 1664–1666.

[120] B. Parr *et al.*, A novel ultra wide-band pulse design algorithm, *IEEE Commun. Lett.*, vol. 7, no. 5, 2003, pp. 219–221.

[121] Glisic S, Nikolic Z and Dimitrijevic B, Adaptive self-reconfigurable interference suppression schemes for CDMA wireless networks, *IEEE Trans. Commun.*, vol. 47, 1999, pp. 598–607.

[122] J. Ye, J. Hou and S. Papavassiliou, A comprehensive resource management framework for next generation wireless networks, *IEEE Trans. Mobile Computing*, vol. 1, no. 4, 2002, pp. 249–265.

13

Ad Hoc *Networks*

13.1 ROUTING PROTOCOLS

Self-organizing wireless networks composed of mobile nodes and requiring no fixed infrastructure will be referred to as mobile *ad hoc* networks (MANET). These networks are characterized by dynamic topology. In other words, nodes can join or leave the network as well as being able to change the range of their transmissions. Each node acts as an independent router. Because of the wireless mode of communication, links will be bandwidth-constrained and of variable capacity. In addition, there will be limited transmitter range, energy-limitations and limited physical security. MAC and network protocols are of a distributed nature. Also, complex routing protocols with large transmission overheads and large processing loads on each node will be used.

Dynamic topology is illustrated in Figure 13.1. Node mobility creates a dynamic topology, i.e. changes in the connectivity between the nodes as a direct function of the distance between each other, the characteristics of the area where they are deployed and, of course, the power of their transmitters. Node mobility and architecture reconfigurability in *ad hoc* networks have had a major effect on the design of routing protocols. MANETs are used for military applications and rescue and recovery scenarios, in local areas like offices, building WLANs, home networks, robot networks, sensor networks and personal Area networking, also for interconnection of wireless devices like games, and habitat, wildlife and micro climate monitoring.

Based on the above characteristics of MANET the routing protocols should be distributed, based on both uni- and bi-directional links, energy efficient and secure. Performance metrics include:

(1) data throughput, i.e. how well the network delivers packets from the source to destination;

(2) delay, e.g. end-to-end delay of packet delivery;

© 2006 John Wiley & Sons, Ltd.

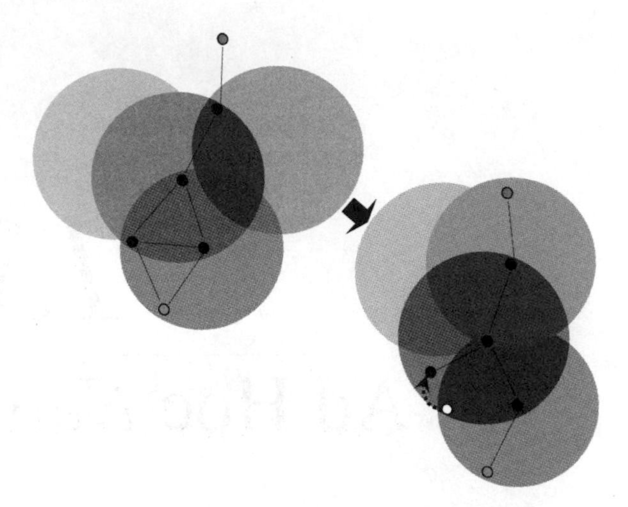

Figure 13.1 Illustration of dynamic topology.

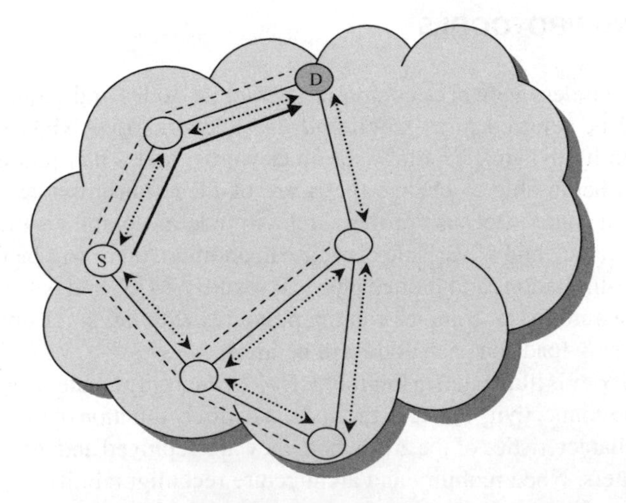

Figure 13.2 Proactive/timer: route updates (←⋯→), routes (------) and data (⟶).

(3) overhead costs, i.e. average of control packets produced per node;

(4) power consumption, i.e. average power used by each node.

Two classes of routing protocols are considered for these applications, proactive and reactive routing. Proactive routing, illustrated in Figure 13.2, maintains routes to every other node in the network. This is a table-driven protocol where regular routing updates impose large overhead. On the other hand, there is no latency in route discovery, i.e. data can be sent immediately. The drawback is that most routing information might never be used. These protocols are suitable for high traffic networks and most often are based on Bellman–Ford type algorithms. Reactive routing, illustrated in Figure 13.3 maintains routes to only those nodes which are needed. The cost of finding routes is expensive since flooding is involved. Owing to the nature of the protocol, there might be a delay before transmitting data. The

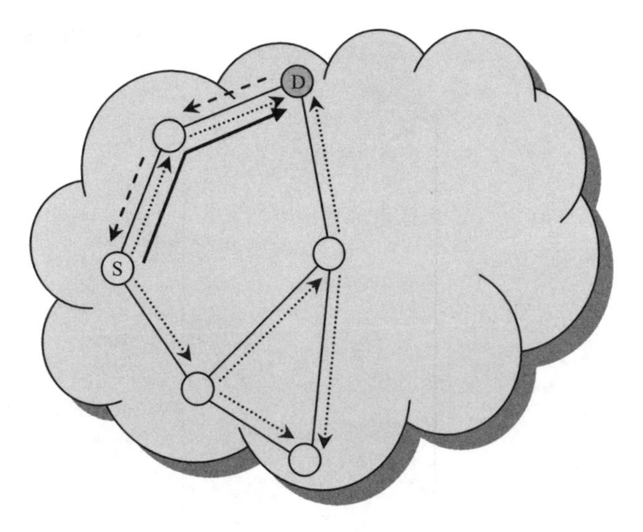

Figure 13.3 Reactive/on demand: route REQ (←·······→), route REP (------) and data flow (——→).

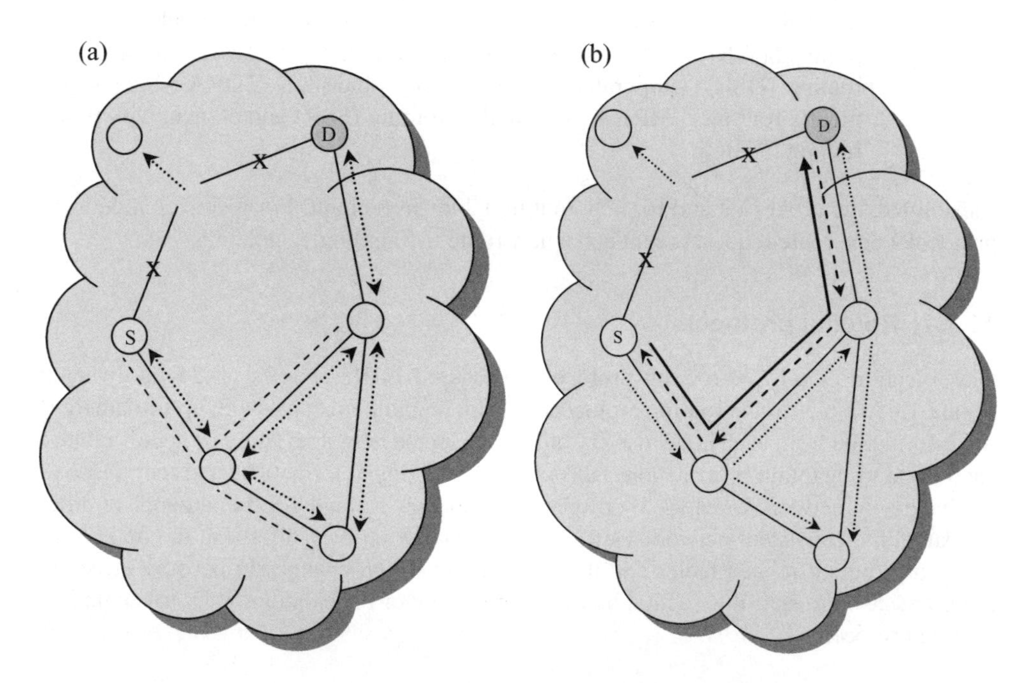

Figure 13.4 (a) Proactive/timer protocol operation when the topology is changed.
(b) Reactive/on demand protocol operation when the topology is changed.

protocol may not be appropriate for real-time applications but is good for low/medium traffic networks.

When the network topology is changed, the two protocols will behave differently, as can be seen in Figure 13.4(a, b). Proactive/timer-based protocol must wait for the updates to be

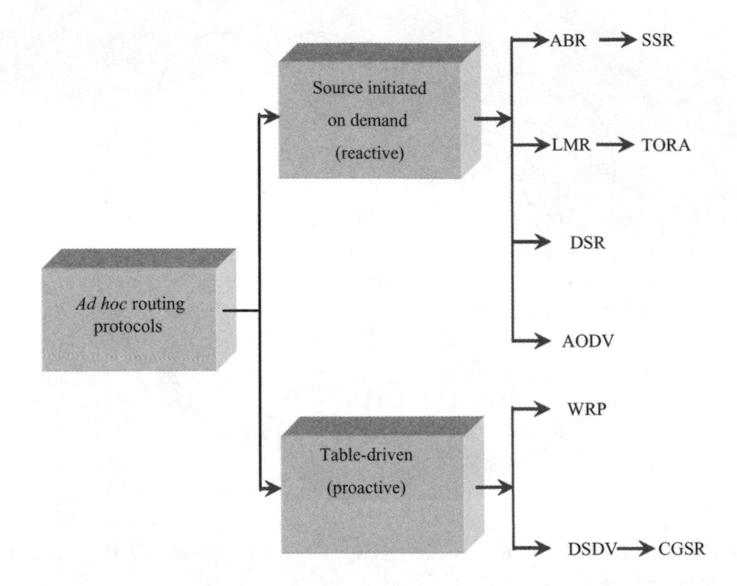

Figure 13.5 Classification of *ad hoc* routing protocols [destination sequenced distance vector (DSDV), clusterhead gateway switch routing (CGSR), wireless routing protocol (WRP), *ad hoc* on-demand distance vector (AODV), dynamic source routing (DSR), temporally ordered routing algorithm (TORA), lightweight mobile routing (LMR), signal stability routing (SSR) and associativity based routing (ABR)].

transmitted and processed and for new routing tables to be built. For reactive/on demand protocol a new route request is sent and a new route is found independently.

13.1.1 Routing protocols

Classification of *ad hoc* routing protocols, discussed in References [1–24] is given in Figure 13.5. Table-driven routing protocols attempt to maintain consistent, up-to-date routing information from each node to every other node in the network. These protocols require each node to maintain one or more tables to store routing information, and they respond to changes in network topology by propagating updates throughout the network in order to maintain a consistent network view. The areas in which they differ are the number of necessary routing-related tables and the methods by which changes in network structure are broadcast. We start this section by discussing some of the existing table-driven *ad hoc* routing protocols.

13.1.1.1 Destination sequenced distance vector, DSDV

Traditional distance vector algorithms are based on the classical Bellman–Ford (DBF) algorithm discussed in Chapter 7. This algorithm has the drawback that routing loops can occur. To eliminate or minimize the formation of loops the nodes are required to often coordinate and communicate among themselves. The problem is caused by frequent topological changes. The RIP is based on this type of algorithm. The application of RIP to *ad hoc* networks is limited since it was not designed for such an environment. The objective of DSDV protocols is to preserve the simplicity of RIP and avoid the looping problem in a mobile wireless environment. The main features of DSDV are:

- routing tables have entries with number of hops to each network destination;
- each route table entry is tagged with a sequence number originated by the destination;
- nodes periodically communicate their routing table to their neighbors and when there is a significant new information available;
- routing information is normally transmitted using a broadcasting or multicasting mode.

Route tables entry consists of:

(1) destination address;

(2) number of hops required to reach the destination;

(3) sequence number of the information received regarding that destination, as originally stamped by the destination.

Within the headers of the packet, the transmitter route tables usually carry the node hardware address, the network address and the route transmission sequence number. Receiving a transmission from a neighbor does not indicate immediately the existence of a bi-directional link with that neighbor. A node does not allow routing through a neighbor until that neighbor shows that it also has the node as a neighbor. This means that DSDV algorithms use only bi-directional links.

An important parameter value is the time between broadcasting routing information packets. However when any new and substantial route information is received by a mobile node, the updated route information is transmitted as soon as possible.

Broken links are a normal occurrence in *ad hoc* networks. They are detected by the MAC layer or inferred if no transmissions are received from a neighbor for some time. A broken link is given a metric of ∞ and an updated sequence number. A broken link translates into a substantial route change and thus this new routing information is immediately broadcasted to all neighbors. To propagate new routing information, particularly the one generated by broken links, and to avoid large transmission overheads, two types of routing packets are used: (1) full dump – carries all the available information; and (2) incremental – carries only information changed after the last dump. When a node receives a new routing packet the information is compared with the one already available at the node. Routes with a more recent sequence number are always used. Newly computed routes are scheduled for immediate advertisement. The route updating process is illustrated in Figure 13.6.

When A moves and it is detected as a routing neighbor by G and H, it causes these nodes to advertise their updated routing information (incremental update). Upon reception of this update, F updates its own routing tables and broadcasts the new information. D receives this update and carries out an update of its routing table. The steps are illustrated in Figure 13.7.

13.1.1.2 Clusterhead gateway switch routing

The clusterhead gateway switch routing (CGSR) protocol differs from the previous protocol in the type of addressing and network organization scheme employed. Instead of a 'flat' network, CGSR is a clustered multihop mobile wireless network with several heuristic routing schemes [4]. In the next section we will see that, by having a cluster head controlling a

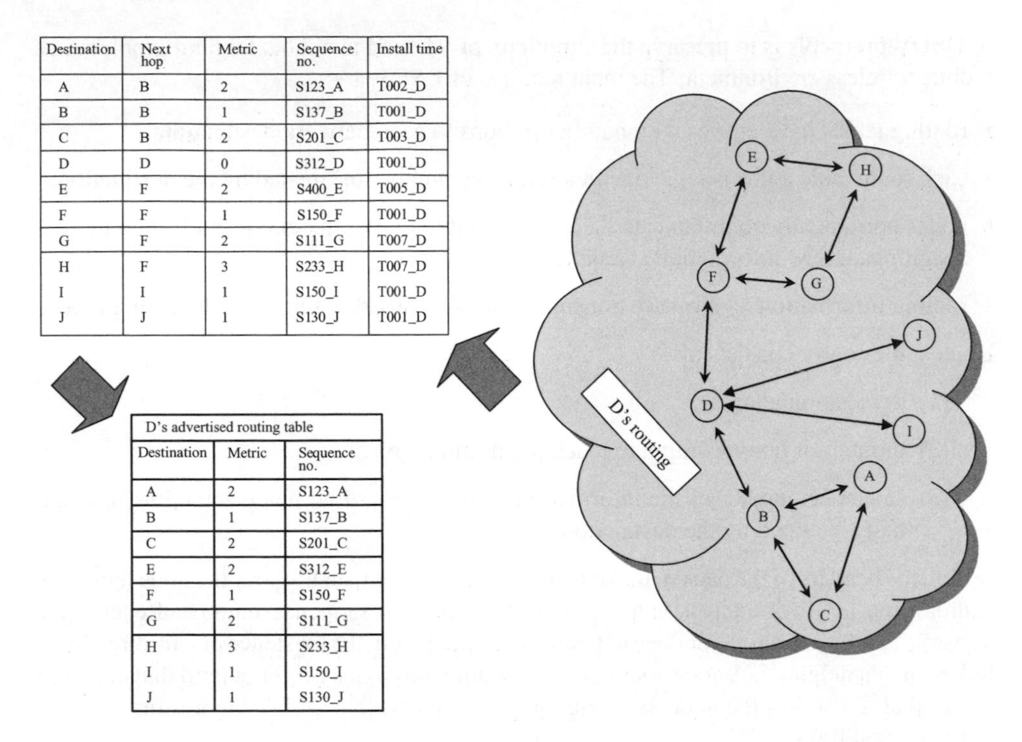

Destination	Next hop	Metric	Sequence no.	Install time
A	B	2	S123_A	T002_D
B	B	1	S137_B	T001_D
C	B	2	S201_C	T003_D
D	D	0	S312_D	T001_D
E	F	2	S400_E	T005_D
F	F	1	S150_F	T001_D
G	F	2	S111_G	T007_D
H	F	3	S233_H	T007_D
I	I	1	S150_I	T001_D
J	J	1	S130_J	T001_D

D's advertised routing table

Destination	Metric	Sequence no.
A	2	S123_A
B	1	S137_B
C	2	S201_C
E	2	S312_E
F	1	S150_F
G	2	S111_G
H	3	S233_H
I	1	S150_I
J	1	S130_J

Figure 13.6 Illustration of route updating process. Metric = number of hops to reach destination; sequence number = the freshness of received route, used to avoid routing loops; install time = when a received route was installed, used for damping route fluctuations.

group of *ad hoc* nodes, a framework for code separation (among clusters), channel access, routing and bandwidth allocation can be achieved. A cluster head selection algorithm is utilized to elect a node as the cluster head using a distributed algorithm within the cluster. The disadvantage of having a cluster head scheme is that frequent cluster head changes can adversely affect routing protocol performance since nodes are busy in cluster head selection rather than packet relaying. Hence, instead of invoking cluster head reselection every time the cluster membership changes, a least cluster change (LCC) clustering algorithm is introduced. Using LCC, cluster heads only change when two cluster heads come into contact, or when a node moves out of contact of all other cluster heads.

CGSR uses DSDV as the underlying routing scheme, and hence has much of the same overhead as DSDV. However, it modifies DSDV by using a hierarchical cluster-head-to-gate way routing approach to route traffic from source to destination. Gateway nodes are nodes that are within a communication range of two or more cluster heads. A packet sent by a node is first routed to its cluster head, and then the packet is routed from the cluster head to a gateway to another cluster head, and so on until the cluster head of the destination node is reached. The packet is then transmitted to the destination. Using this method, each node must keep a "cluster member table" where it stores the destination cluster head for each mobile node in the network. These cluster member tables are broadcast by each node periodically using the DSDV algorithm. Nodes update their cluster member tables on reception of such a table from a neighbor.

D's updated routing table				
Destination	Next hop	Metric	Sequence no.	Install time
A	**F**	**4**	**S412_A**	**T509_D**
B	B	1	S137_B	T001_D
C	B	2	S201_C	T003_D
D	D	0	S312_D	T001_D
E	F	2	S400_E	T005_D
F	F	1	S150_F	T001_D
G	F	2	S111_G	T007_D
H	F	3	S233_H	T007_D
I	I	1	S150_I	T001_D
J	J	1	S130-J	T001_D

D's updated advertised routing table		
Destination	Metric	Sequence no.
A	**3**	**S412_A**
B	1	S137_B
C	2	S201_C
E	2	S400_E
F	1	S150_F
G	2	S111_G
H	3	S233_H
I	1	S150_I
J	1	S130_J

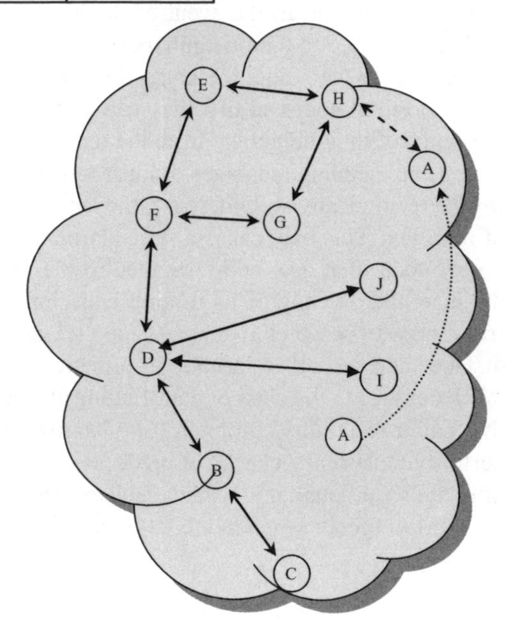

Figure 13.7 Illustration of route updating process after A moves.

In addition to the cluster member table, each node must also maintain a routing table which is used to determine the next hop in order to reach the destination. On receiving a packet, a node will consult its cluster member table and routing table to determine the nearest cluster head along the route to the destination. Next, the node will check its routing table to determine the next hop used to reach the selected cluster head. It then transmits the packet to this node.

13.1.1.3 The wireless routing protocol

The wireless routing protocol (WRP) described in Murthy and Garcia-Luna-Aceves [5] is a table-based protocol with the goal of maintaining routing information among all nodes in the network. Each node in the network is responsible for maintaining four tables:

(1) distance table; (2) routing table; (3) link-cost table; and (4) message retransmission list (MRL) table.

Each entry of the MRL contains the sequence number of the update message, a retransmission counter, an acknowledgment-required flag vector with one entry per neighbor, and a list of updates sent in the update message. The MRL records which updates in an update message need to be retransmitted and which neighbors should acknowledge the retransmission [5].

Mobiles inform each other of link changes through the use of update messages. An update message is sent only between neighboring nodes and contains a list of updates (the destination, the distance to the destination and the predecessor of the destination), as well as a list of responses indicating which mobiles should acknowledge (ACK) the update. Mobiles send update messages after processing updates from neighbors or detecting a change in a link to a neighbor. In the event of the loss of a link between two nodes, the nodes send update messages to their neighbors. The neighbors then modify their distance table entries and check for new possible paths through other nodes. Any new paths are relayed back to the original nodes so that they can update their tables accordingly. Nodes learn of the existence of their neighbors from the receipt of acknowledgments and other messages. If a node is not sending messages, it must send a hello message within a specified time period to ensure connectivity. Otherwise, the lack of messages from the node indicates the failure of that link. This may cause a false alarm. When a mobile receives a hello message from a new node, that new node is added to the mobile's routing table, and the mobile sends the new node a copy of its routing table information. Part of the novelty of WRP stems from the way in which it achieves loop freedom. In WRP, routing nodes communicate the distance and second-to-last hop information for each destination in the wireless networks. WRP belongs to the class of path-finding algorithms with an important exception. It avoids the 'count-to-infinity' problem [6], also discussed in Chapter 7, by forcing each node to perform consistency checks of predecessor information reported by all its neighbors. This ultimately (although not instantaneously) eliminates looping situations and provides faster route convergence when a link failure event occurs.

13.1.2 Reactive protocols

13.1.2.1 Dynamic source routing (DSR)

In this case every packet carries the routing sequence. Intermediate nodes may learn routes from 'heard' traffic (RREQ, RREP, DATA). No periodic sending of routing packets occurs. The system may piggyback route requests on route replies and must use link layer feedback to find broken links. To send a packet the sender constructs a *source route* in the packet's header.

The source route has the address of every host through which the packet should be forwarded to reach its destination. Each host in the *ad hoc* network maintains a *route cache* in which it stores source routes it has learned. Each entry in the route cache has an expiration period, after which it will be deleted. If the sender does not have a route to a destination it then attempts to find out by using a *routing discovery* process. This process is illustrated in Figures 13.8–13.10. While waiting for the routing discovery to complete, the sender continues sending and receiving packets with other hosts. Each host uses a *route maintenance* procedure to monitor the correct operation of a route.

Usually the data link layer has a mechanism to detect a link failure. When a link failure is detected the host sends a *route error packet* to the original sender of the packet. The route error packet has the address of the host who detected the failure and the host to which it was attempting to transmit the packet. When a route error packet is received by a host, the hop in error is removed from the host's route cache, and all routes which contain this hop are truncated at that point.

To return the route error packet the host uses a route in its route cache to the sender of the original packet. If the host does not have a route it can reverse the route information carried in the packet that could not be forwarded because of the link error. The later assumes that only bidirectional links are being used for routing. Another option to return route error packets is to perform a route discovery process to find a route to the original sender and use that route. Several optimizations are possible to reduce the amount of overhead traffic.

13.1.2.2 Route cache

During the process of route discovery and maintenance a host receives, directly or indirectly, information about routes to other hosts, thus minimizing the need to search for that

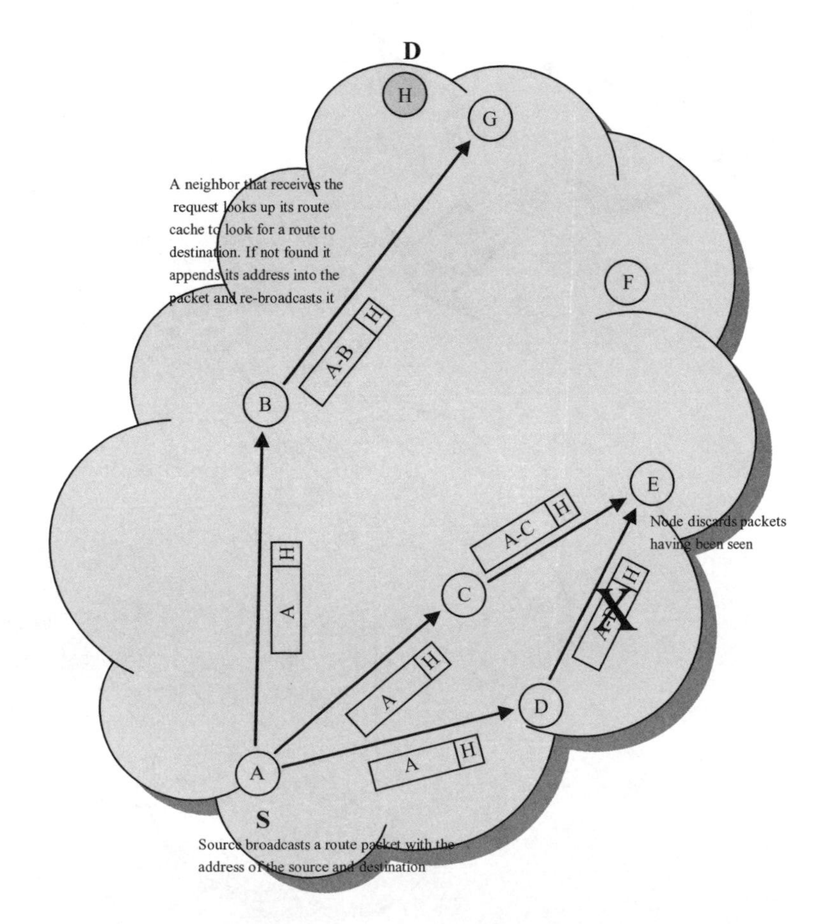

Figure 13.8 DSR – route discovery.

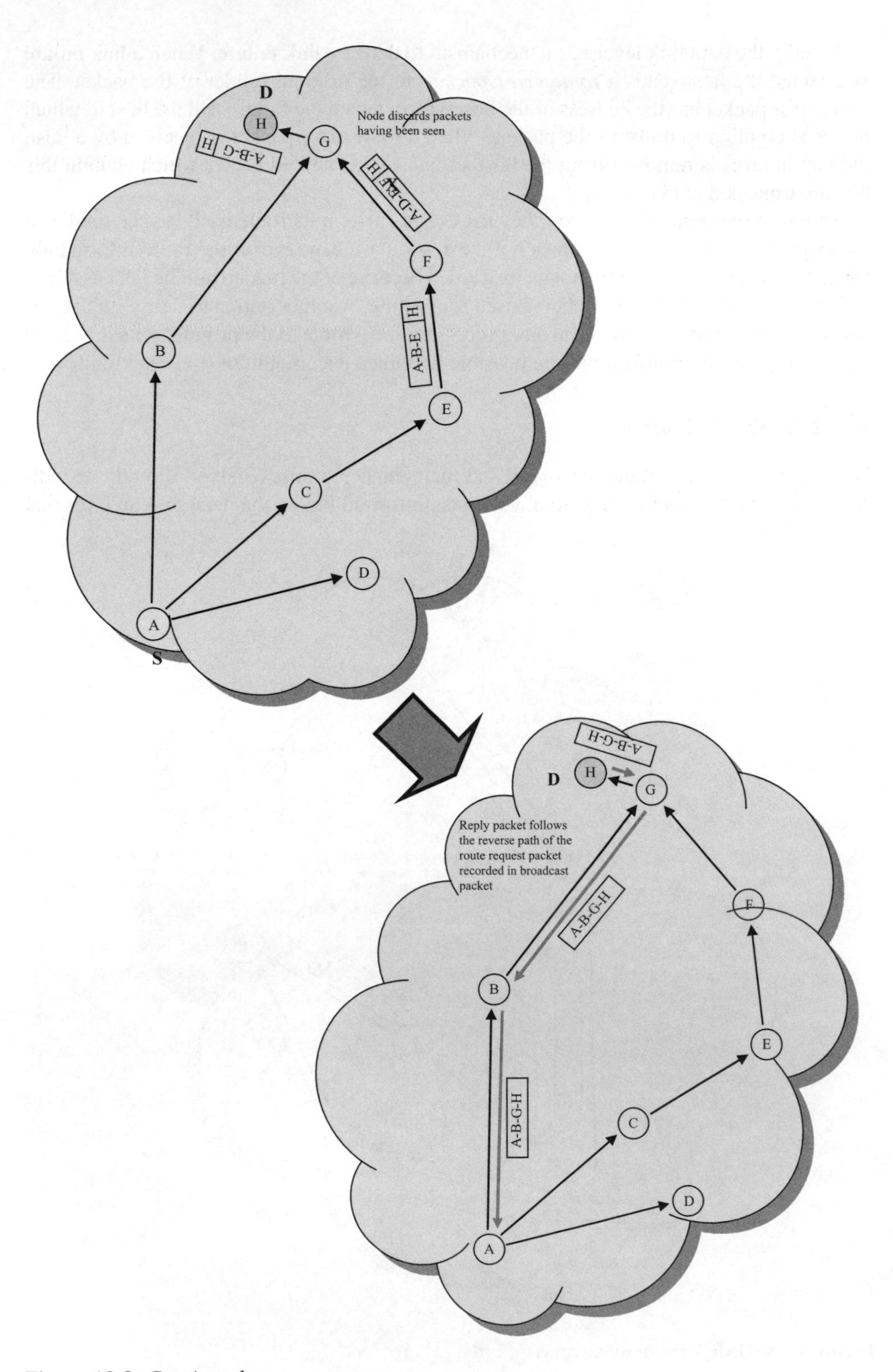

Figure 13.8 *Continued.*

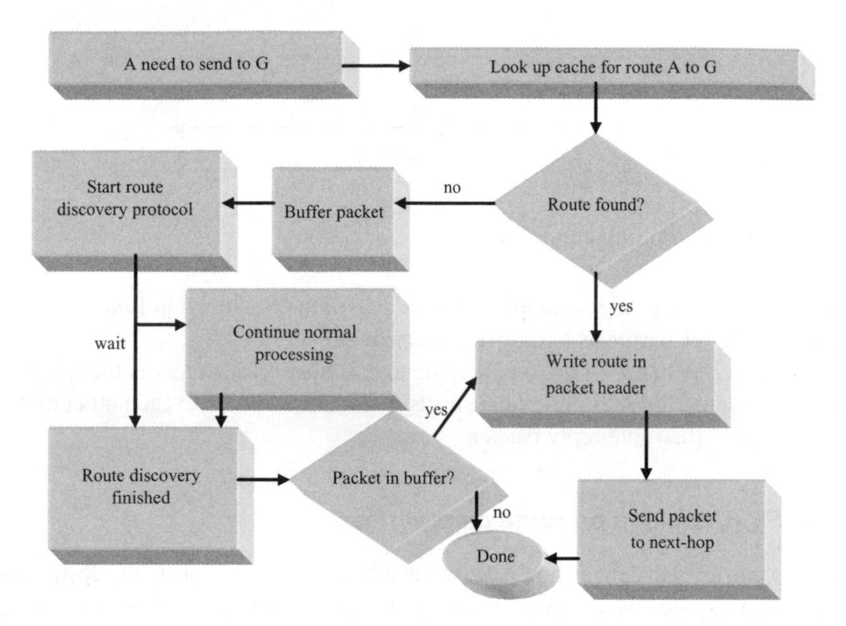

Figure 13.9 DSR – route discovery decision process at source A.

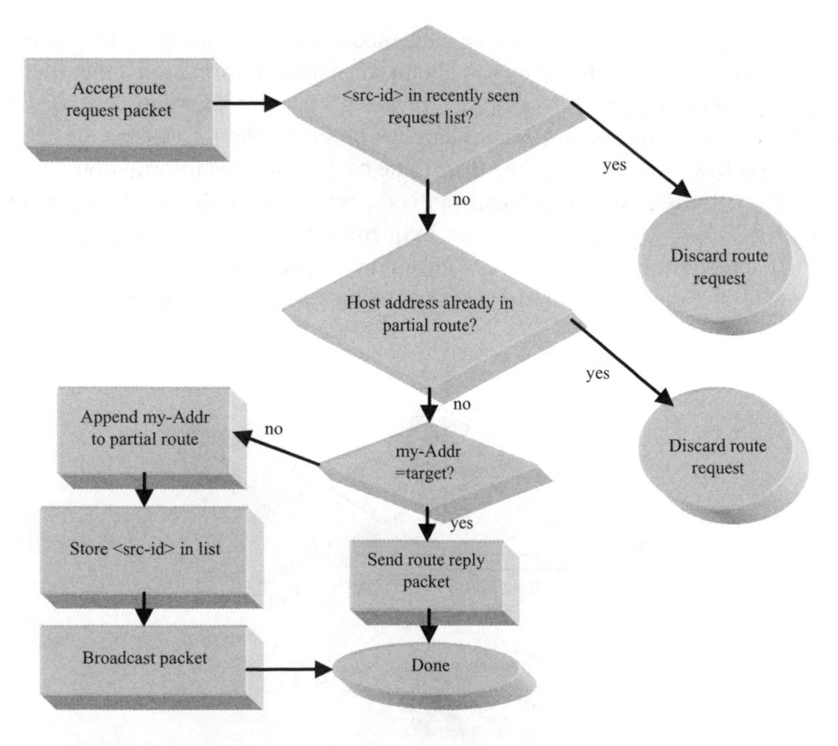

Figure 13.10 DSR – route discovery decision process at an intermediate node.

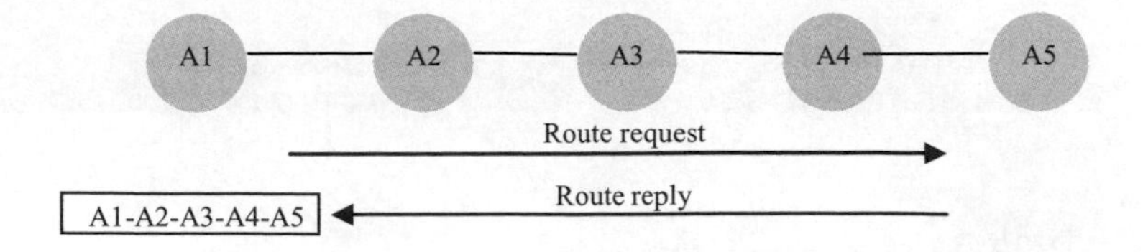

Figure 13.11 DSR – optimizations.

information in the future. For example in the *ad hoc* network shown in Figure 13.11 let us assume that node A1 performs a route discovery to A5.

Since hosts A2, A3 and A4 are on the route to A5, host A1 also learns the routes to A2, A3 and A4. Likewise, these 'intermediate hosts' learn about routes to each other by looking into the content of the route reply packet.

13.1.2.3 Piggybacking on route discoveries

To minimize the delay in delivering a data packet when there is no route to the destination and a route discovery process is needed, one can piggyback the data on the route request packets.

13.1.2.4 Learning by 'listening'

If the host operate in promiscuous receiving mode, i.e. they receive and process every transmission in their range, then they can obtain substantial information about routing, e.g. in the network in Figure 13.12. Nodes A2, A3 and A5 listen to the route error packet from A4 to A1. Since the route error packet identifies precisely the hop where the failure was detected, hosts B, C and D can update their route cache with this information.

In summary, SDR is an on-demand protocol, with potentially zero control message overhead if the topology does not change frequently. Packet delays/jitters are associated with on-demand routing. It can work with unidirectional links as well as bidirectional links. Route caching is used to minimize the route discovery overhead. Promiscuous mode

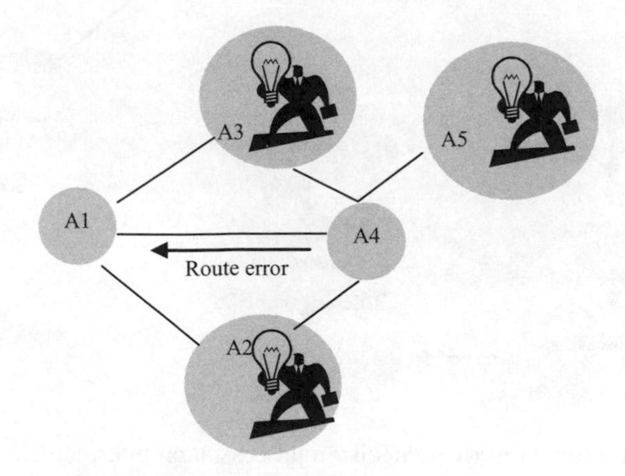

Figure 13.12 Learning about the routs by 'listening'.

operations can translate in excessive use of power. This is not easily scalable to large networks since the protocol design assumes a small network diameter. The need to place the entire route in the route replies and data packets translates in large on-the-air packet overheads. Protocol allows for the possibility to keep multiple routes to a destination in the route cache. CPU and memory use demands on each host are high, since the routes have to be continuously maintained and updated.

13.1.2.5 Ad-hoc on demand distance vector (AODV)

The protocol uses 'traditional' routing tables. Hello messages are sent periodically to identify neighbors. Sequence numbers guarantee freshness. Route requests are sent in the reverse direction, i.e. they only use bi-directional links. The system may use link layer feedback to find broken links. The protocol is based on the destination-sequenced distance-vector (DSDV) algorithm. It provides on-demand route acquisition. Nodes maintain the route cache and use a destination sequence number for each route entry. The protocol does nothing when connection between end points is still valid. The route discovery mechanism, illustrated in Figure 13.13, is initiated by broadcasting a route request packet (RREQ), when a route to new destination is needed.

The neighbors forward the request to their neighbors until either the destination or an intermediate node with a 'fresh enough' route to the destination is located. Route reply packets are transmitted upstream of the path taken by the route request packet to inform the original sender (an intermediate nodes) of the route finding. Route error packets (RERR) are used to erase broken links.

13.1.2.6 AODV – path maintenance

Periodic hello messages can be used to ensure symmetric links and to detect link failures. Hello messages include a list of nodes that the host has heard of. Once the next hop becomes unavailable the host upstream of the link failure propagates an unsolicited RREP with a hop count of ∞ to all active upstream neighbors. Upon receiving notification of a broken link, a source node can restart the discovery process if they still need a route to destination.

13.1.2.7 Temporal order routing algorithm (TORA)

TORA is also a distributed protocol. It provides loop-free routes and multiple routes. It is source-initiated and creates routes to destination only when desired.

The protocol minimizes reaction to topological changes, by localizing reaction to a very small group of nodes near the change and provides fast recovery upon route failures. Also, it detects a network partition when there is one and erases all invalid routes within a finite time.

The protocol has three basic functions: route creation, route maintenance and route erasure. During the route creation, nodes use a 'height' metric to build a directed acyclic graph (DAG) rooted at the destination. Links are assigned that have a direction (upstream or downstream) based on the relative height metric of neighboring nodes, as illustrated in Figure 13.14.

From Chapter 7, $G(V, E)$ represents the network, where V is the set of nodes and E is the set of links. Each node $i \in V$ is assumed to have a unique identifier (ID), and each

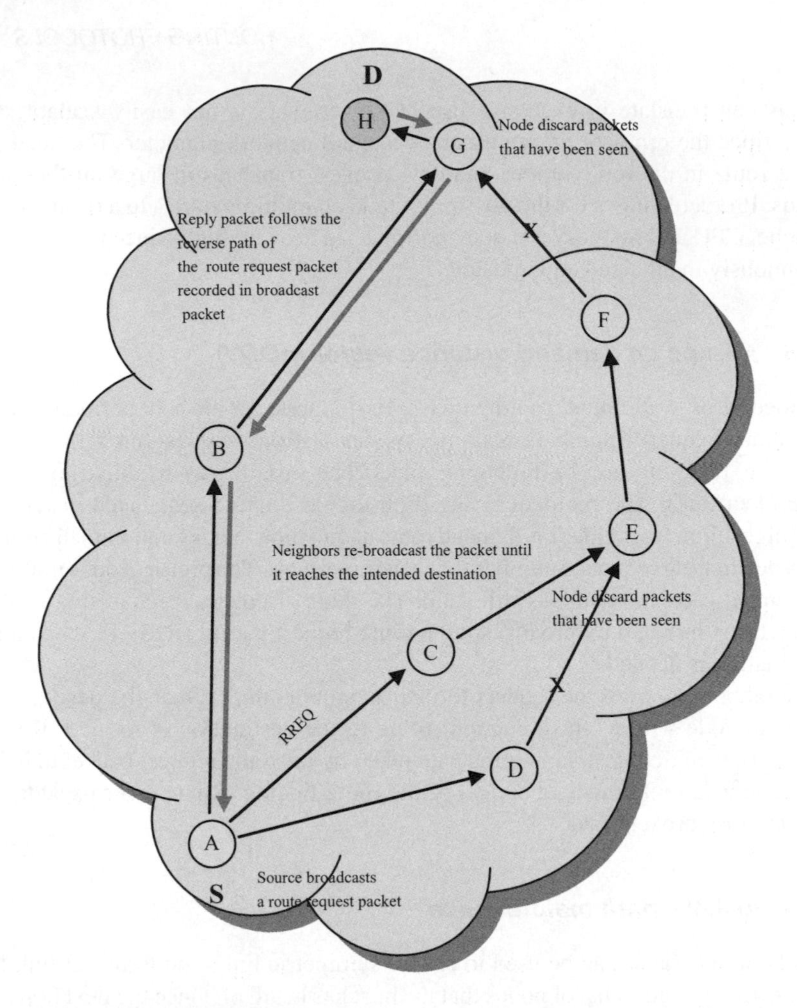

Figure 13.13 AODV – path finding.

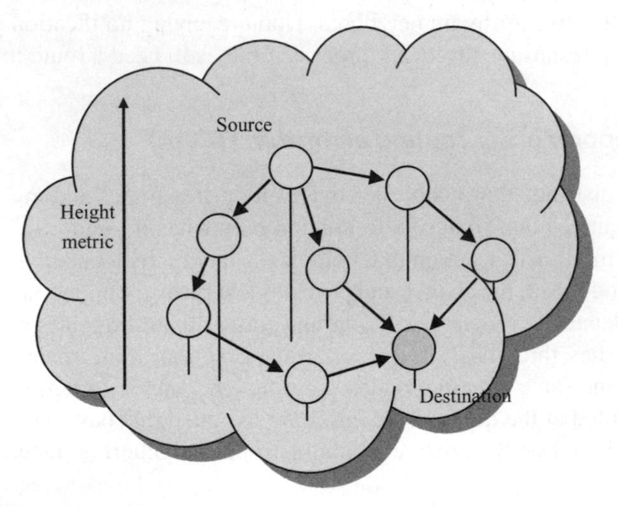

Figure 13.14 TORA.

link $(i, j) \in E$, is assumed to be bidirectional, i.e. it allows two-way communications. The height metric associated with each node is of the form $Hi = (\tau i, oidi, ri, \delta i, i)$ where:

- τi is the logical time of a link failure, and defines a new reference level;
- $oidi$ is the unique ID of the node that defined the reference level;
- ri denotes two unique sub-levels of a reference level;
- δi is used to order the nodes with respect to a common reference level; and
- i is the unique ID of the node itself.

Each node i (other than the destination) maintains its height Hi. Initially the height is set to NULL, $Hi = [-, -, -, -, i]$. The height of the destination is always ZERO, $H_{dID} = [0, 0, 0, 0, dID]$. At each node there is a height array with an entry $NH_{i,j}$ for each neighbor $i, \in V$. Initially the height of each neighbor is set to NULL; if a neighbor is a destination its corresponding entry in the height array is set to ZERO.

The operation of TORA protocols is illustrated in Figures 13.15–13.18. The protocols are well suited for networks with large dense populations of nodes. They posses the ability to detect network partitions and erase all invalid routes within a finite time.

Protocol quickly creates and maintains routes for a destination for which routing is desired. It minimizes the number of nodes reacting to topological changes. It needs a synchronization mechanism to achieve a temporal order of events.

13.1.2.8 Associativity-based routing

The associativity based routing (ABR) protocol [12] is free from loops, deadlock and packet duplicates, and defines a different routing metric for *ad hoc* mobile networks. This metric is known as the degree of association stability. In ABR, a route is selected based on the degree of association stability of mobile nodes. Each node periodically generates a beacon to signify its existence. When received by neighboring nodes, this beacon causes their associativity tables to be updated. For each beacon received, the associativity tick of the current node with respect to the beaconing node is incremented. Association stability is defined by connection stability of one node with respect to another node over time and space. A high degree of association stability may indicate a low state of node mobility, while a low degree may indicate a high state of node mobility. Associativity ticks are reset when the neighbors of a node or the node itself move out of proximity. A fundamental objective of ABR is to derive longer-lived routes for *ad hoc* mobile networks.

The three phases of ABR are: (1) route discovery; (2) route reconstruction (RRC); and (3) route deletion. The route discovery phase is accomplished by a broadcast query and await-reply (BQ-REPLY) cycle. A node desiring a route broadcasts a BQ message in search of mobiles that have a route to the destination. All nodes receiving the query (that are not the destination) append their addresses and their associativity ticks with their neighbors along with QoS information to the query packet. A successor node erases its upstream node neighbors' associativity tick entries and retains only the entry concerned with itself and its upstream node. In this way, each resultant packet arriving at the destination will contain the associativity ticks of the nodes along the route to the destination. The destination is then able to select the best route by examining the associativity ticks along each of the paths.

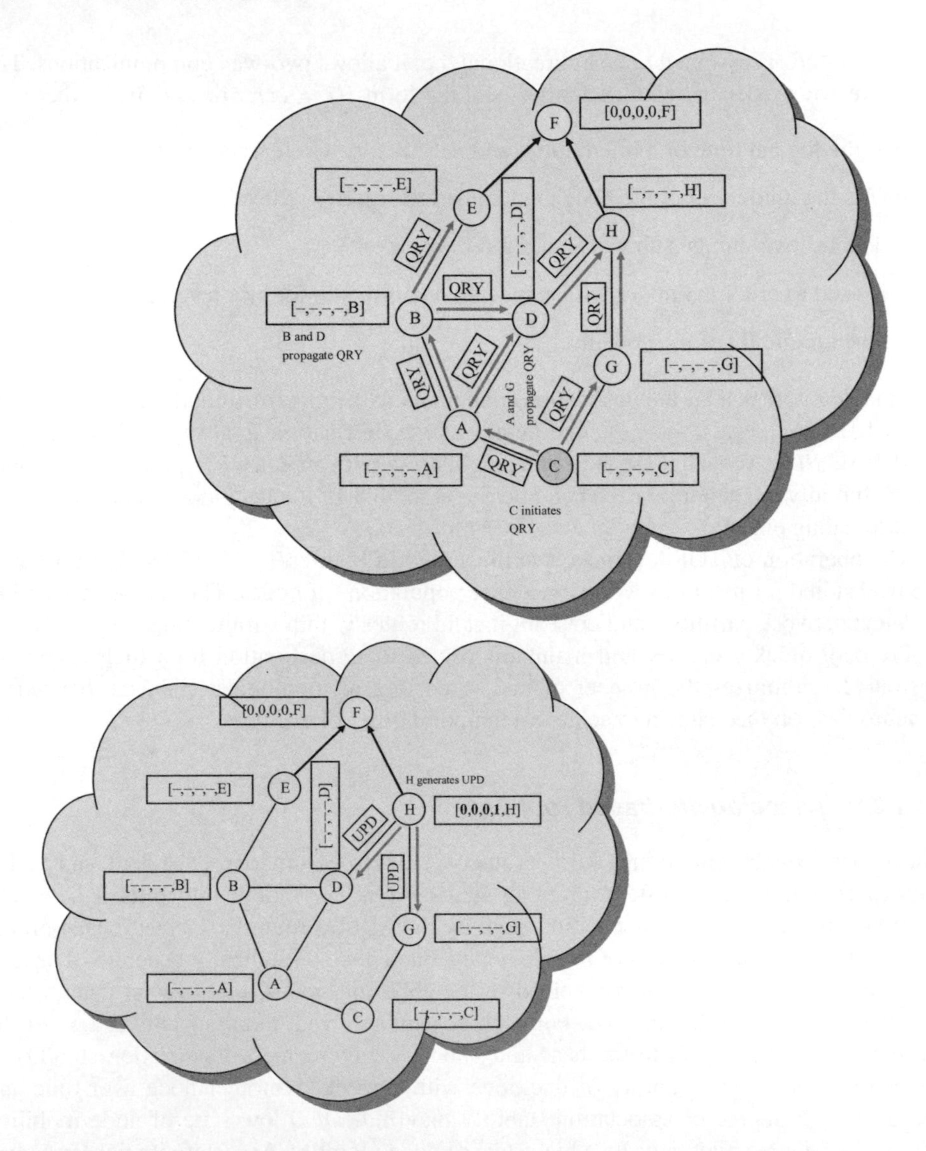

Figure 13.15 TORA – route creation.

When multiple paths have the same overall degree of association stability, the route with the minimum number of hops is selected. The destination then sends a REPLY packet back to the source along this path. Nodes propagating the REPLY mark their routes as valid. All other routes remain inactive, and the possibility of duplicate packets arriving at the destination is avoided. RRC may consist of partial route discovery, invalid route erasure, valid route updates and new route discovery, depending on which node(s) along the route move. Movement by the source results in a new BQ-REPLY process. The RN message is a route notification used to erase the route entries associated with downstream nodes. When the destination moves, the immediate upstream node erases its route and determines if the node is still reachable by a localized query (LQ[H]) process, where H refers to the hop count from the upstream node to the destination. If the destination receives the LQ

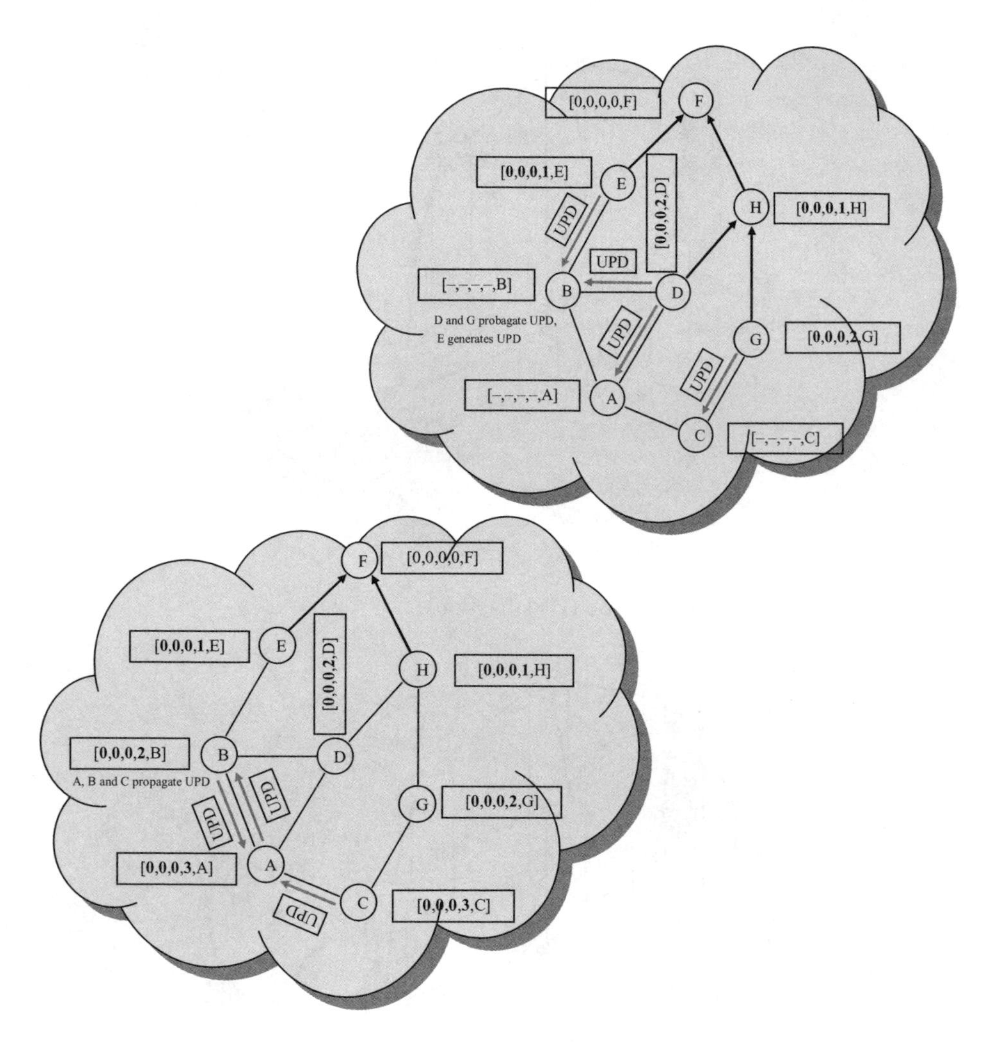

Figure 13.15 *Continued.*

packet, it REPLYs with the best partial route; otherwise, the initiating node times out and the process backtracks to the next upstream node. Here an RN message is sent to the next upstream node to erase the invalid route and inform this node that it should invoke the LQ process. If this process results in backtracking more than halfway to the source, the LO process is discontinued and a new BQ process is initiated at the source. When a discovered route is no longer desired, the source node initiates a route delete (RD) broadcast so that all nodes along the route update their routing tables. The RD message is propagated by a full broadcast, as opposed to a directed broadcast, because the source node may not be aware of any route node changes that occurred during RRCs.

13.1.2.9 Signal stability routing

Another on-demand protocol is the signal stability-based adaptive routing protocol (SSR) presented in Dube *et al.* [13]. Unlike the algorithms described so far, SSR selects routes

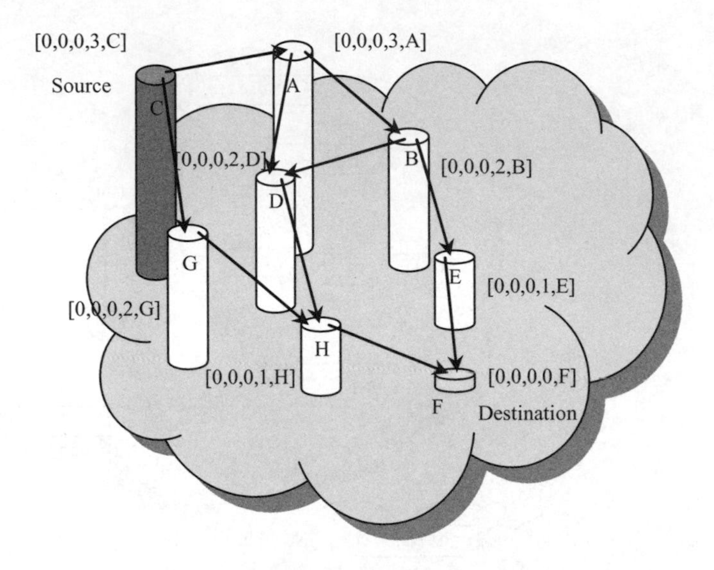

Figure 13.16 TORA – route creation (visualization).

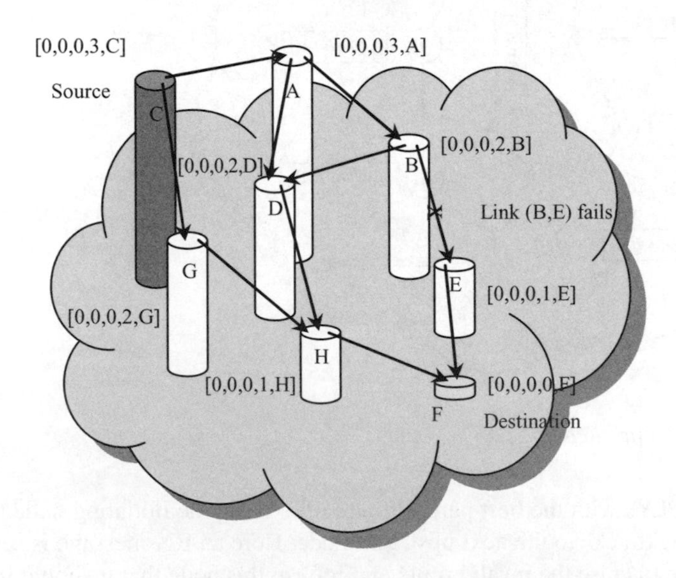

Figure 13.17 TORA – maintaining routes link failure with no reaction.

based on the signal strength between nodes and a node's location stability. This route selection criteria has the effect of choosing routes that have 'stronger' connectivity. SSR can be divided into two cooperative protocols: the dynamic routing protocol (DRP) and the static routing protocol (SRP). The DRP is responsible for the maintenance of the signal stability table (SST) and routing table (RT). The SST records the signal strength of neighboring nodes, which is obtained by periodic beacons from the link layer of each neighboring node. Signal strength may be recorded as either a strong or weak channel. All transmissions are received by, and processed in, the DRP. After updating all appropriate table entries, the DRP passes a received packet to the SRP.

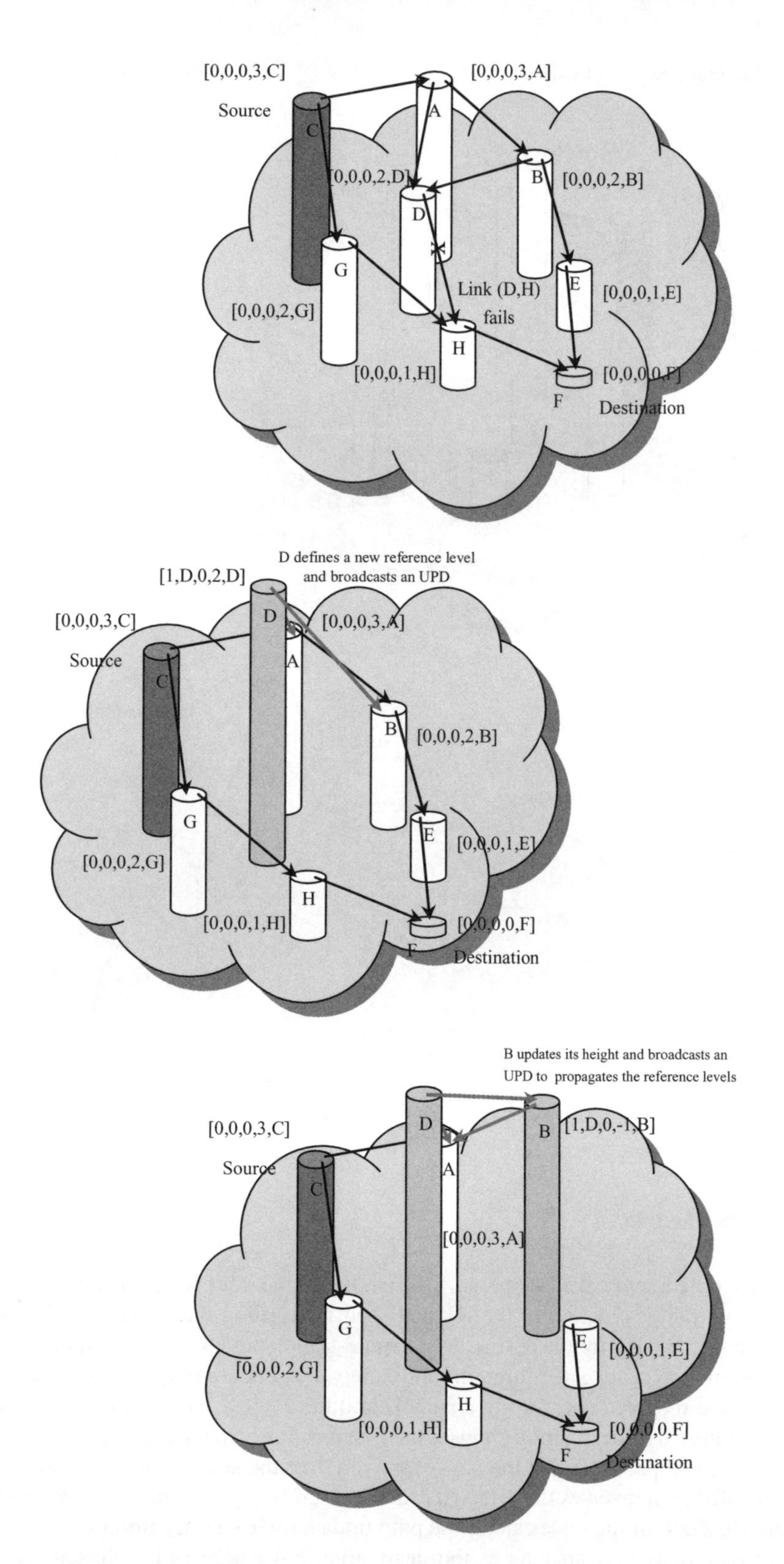

Figure 13.18 TORA – re-establishing routes after link failure of last downstream link.

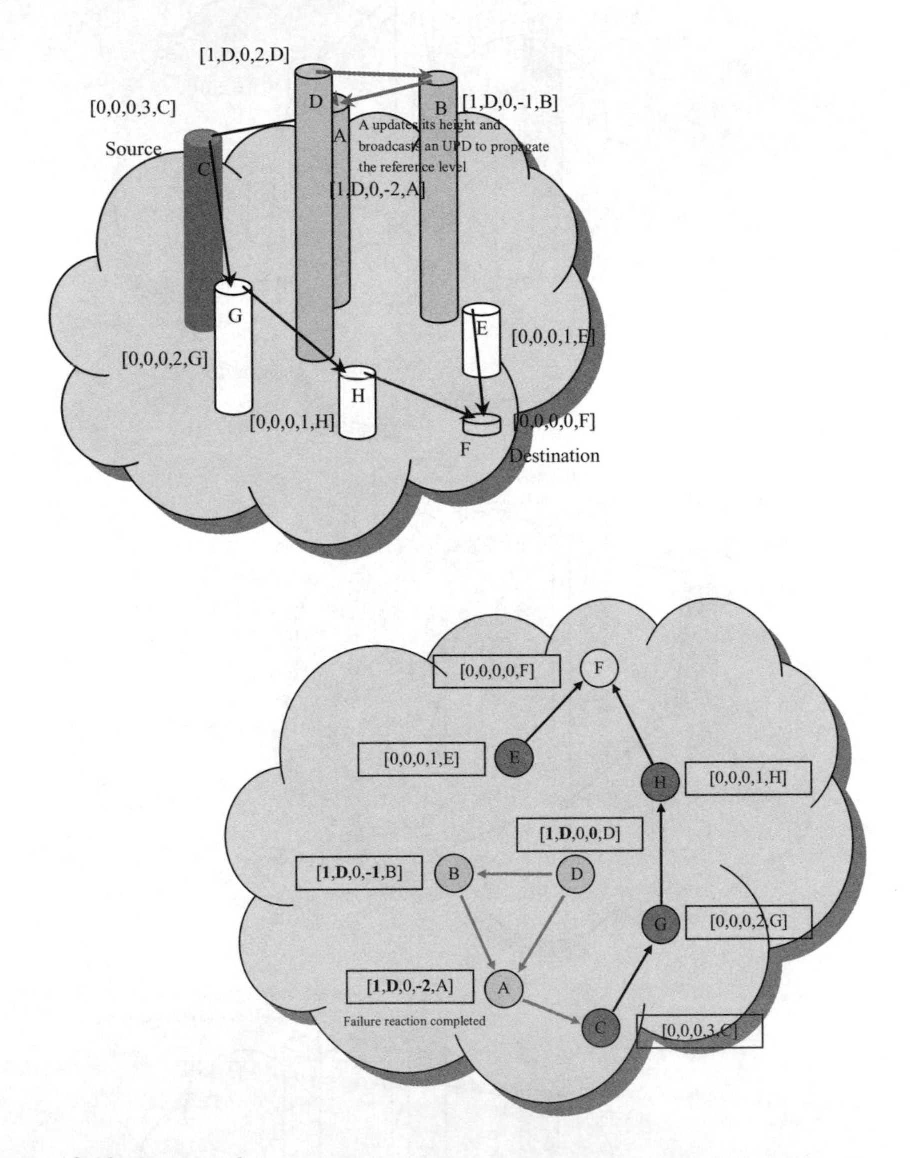

Figure 13.18 *Continued.*

The SRP processes packets by passing the packet up the stack if it is the intended receiver or looking up the destination in the RT and then forwarding the packet if it is not. If no entry is found in the RT for the destination, a route-search process is initiated to find a route. Route requests are propagated throughout the network, but are only forwarded to the next hop if they are received over strong channels and have not been previously processed (to prevent looping). The destination chooses the first arriving route-search packet to send back because it is most probable that the packet arrived over the shortest and/or least congested path. The DRP then reverses the selected route and sends a route-reply message back to the initiator. The DRP of the nodes along the path update their RTs accordingly.

Route-search packets arriving at the destination have necessarily chosen the path of strongest signal stability, since the packets are dropped at a node if they have arrived over a weak channel. If there is no route-reply message received at the source within a specific timeout period, the source changes the PREF field in the header to indicate that weak channels are acceptable, since these may be the only links over which the packet can be propagated. When a failed link is detected within the network, the intermediate nodes send an error message to the source indicating which channel has failed. The source then initiates another route-search process to find a new path to the destination. The source also sends an erase message to notify all nodes of the broken link.

13.2 HYBRID ROUTING PROTOCOL

The zone routing protocol (ZRP) is a hybrid routing protocol that proactively maintains routes within a local region of the network (referred to as the routing zone). Knowledge of this routing zone topology is leveraged by the ZRP to improve the efficiency of a reactive route query/reply mechanism. The ZRP can be configured for a particular network through adjustment of a single parameter, the routing zone radius. A routing zone of radius r is defined for each node and includes the nodes whose minimum distance in hops from a given node is at most r hops. An example of a routing zone (for node S) of radius two hops is shown in Figure 13.19. Nodes within the circle, are said to be within the routing zone of the central node S. Nodes outside the circle are said to be outside S's routing zone. Peripheral nodes are nodes whose minimum distance to S is exactly equal to the zone radius. The remaining nodes are categorized as interior nodes.

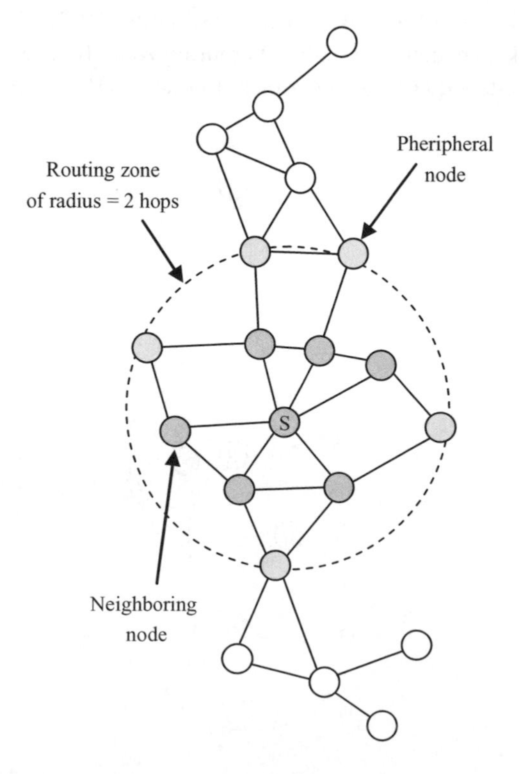

Figure 13.19 A routing zone of radius $r = 2$ hops.

For a routing zone of radius r, the number of routing zone nodes can be regulated through adjustments in each node's transmitter power. Subject to the local propagation conditions and receiver sensitivity, the transmission power determines the set of neighbor nodes, i.e. those nodes that are in direct communication with a node. To provide adequate network reachability, it is important that a node be connected to a sufficient number of neighbors. However, more is not necessarily better. As the transmitters' coverage areas grow larger, so does the membership of the routing zones. This can result in an excessive amount of route update traffic.

Each node is assumed to continuously (proactively) maintain routing information only to those nodes that are within its routing zone. Because the updates are only propagated locally, the amount of update traffic required to maintain a routing zone does not depend on the total number of network nodes (which can be quite large). This is referred to as the *intrazone routing protocol* (IARP). The *interzone routing protocol (IERP)* is responsible for reactively discovering routes to destinations located beyond a node's routing zone. The IERP operates as follows: the source node first checks whether the destination is within its zone. If so, the path to the destination is known, and no further route discovery processing is required. If the destination is not within the source's routing zone, the source broadcasts a route request (referred to as request) to all its peripheral nodes. Now, in turn, all the peripheral nodes execute the same algorithm: they check whether the destination is within their zone. If so, a route reply (referred to as reply) is sent back to the source indicating the route to the destination. If not, the peripheral node forwards the query to its peripheral nodes, which in turn execute the same procedure.

An example of this route discovery procedure is demonstrated in Figure 13.20. The source node S needs to send a packet to the destination D. To find a route within the network, S first checks whether D is within its routing zone. If so, S knows the route to D. Otherwise, S broadcasts a query to its peripheral nodes; that is, S sends a query to nodes

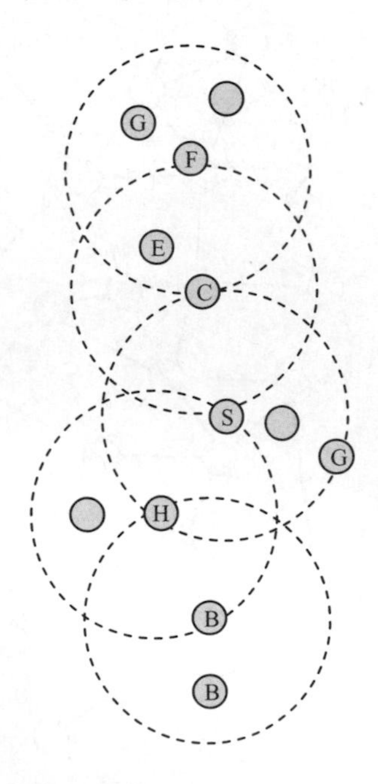

Figure 13.20 Illustration of IERP operation.

H, G and C. Now, in turn, after verifying that D is not in its routing zone, each one of these nodes forwards the query by broadcasting the query to its peripheral nodes. In particular, H sends the query to B, which recognizes D as being in its routing zone and responds to the query, indicating the forwarding path: S–H–B–D. As indicated by this example, a route can be specified by a sequence of nodes that have received the successful IERP query thread. The manner in which this information is collected and distributed is specified by a route accumulation procedure. In the basic route accumulation, a node appends its ID to a received query packet. When a node finds the destination in its zone, the accumulated sequence of IDs specifies a route between querying source and destination. By reversing the accumulated route, a path is provided back to the query source. This information can be used to return the route reply through source routing.

The intuition behind the ZRP is that querying can be performed more efficiently by broadcasting queries to the periphery of a routing zone rather than flooding the queries over the same area. However, problems can arise once the query leaves the initial routing zone. Because the routing zones heavily overlap, a node can be a member of many routing zones. It is very possible that the query will be forwarded to all the network nodes, effectively flooding the network.

Yet a more disappointing result is that the IERP can result in much more traffic than the flooding itself, due to the fact that broadcasting involves sending the query along a path equal to the zone radius. Excess route query traffic can be regarded as a result of overlapping query threads (i.e. overlapping queried routing zones). Thus, the design objective of query control mechanisms should be to reduce the amount of route query traffic by steering threads outward from the source's routing zone and away from each other, as indicated in Figure 13.21. This problem is addressed primarily through appropriate mechanisms of query detection and query termination.

13.2.1 Loop-back termination

The query is terminated when the accumulated route (excluding the previous node) contains the host which lies in routing zone, e.g. for route = {S→A→B→C} in Figure 13.22, C terminates the query, because S is in C's routing zone.

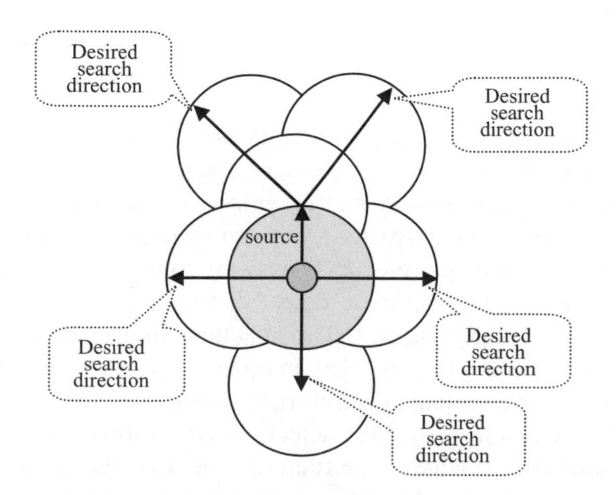

Figure 13.21 Guiding the search in desirable directions.

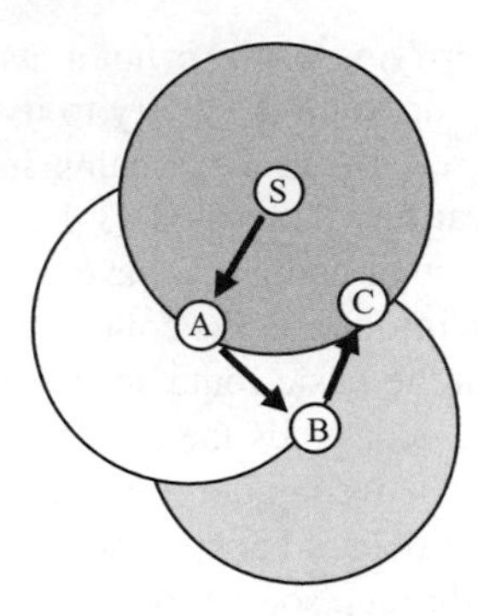

Figure 13.22 Loop-back termination.

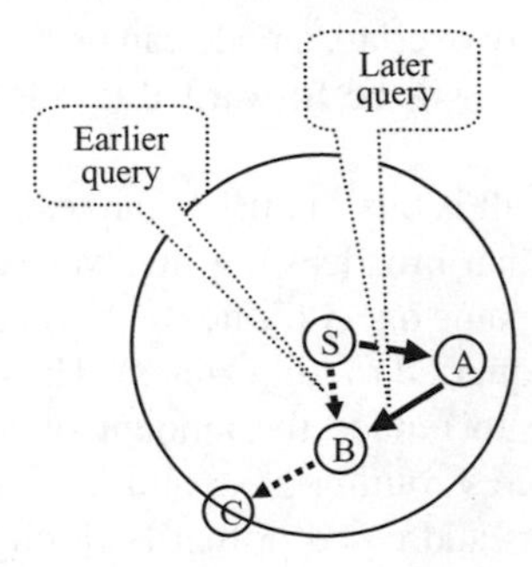

Figure 13.23 Early termination.

13.2.2 Early termination

When the ability to terminate route query threads is limited to peripheral nodes, threads are allowed to penetrate into previously covered areas, which generates unnecessary control traffic. This excess traffic can be eliminated by extending the thread termination capability to the intermediate nodes that relay the thread. This approach is referred to as early termination (ET). Figure 13.23 illustrates the operation of the ET mechanism. Node S broadcasts a route query with node C as one of the intended recipients. Intermediate node A passes along the query to B. Instead of delivering the query to node C, node B terminates the thread because a different thread of this query was previously detected. Intermediate nodes may terminate existing queries but are restricted from issuing new queries. Otherwise, the ZRP would degenerate into a flooding protocol. The ability to terminate an overlapping query thread depends on the ability of nodes to detect that a routing zone they belong to has been previously queried. Clearly, the central node in the routing zone (which processed the query) is aware that its zone has been queried. In order to notify the remaining routing zone nodes without introducing additional control traffic, some form of 'eavesdropping' needs to be implemented. The first level of query detection (QD1) allows the intermediate nodes, which transport queries to the edge of the routing zone, to detect these queries. In single channel networks, it may be possible for queries to be detected by any node within the

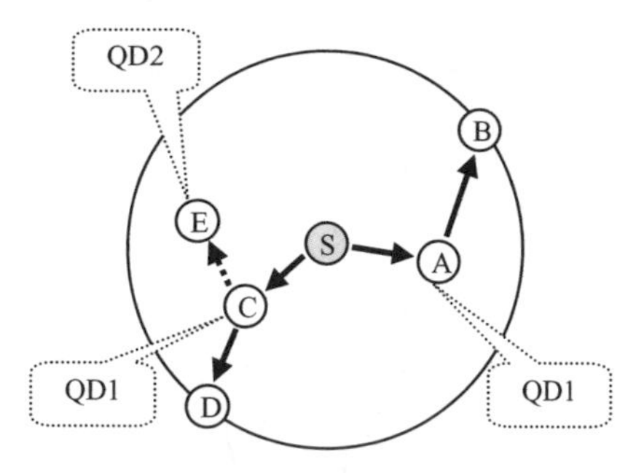

Figure 13.24 Query detection (QD1/QD2).

range of a query-transmitting node. This extended query detection capability (QD2) can be implemented by using IP broadcasts to send route queries. Figure 13.24 illustrates both levels of advanced query detection. In this example, node S broadcasts to two peripheral nodes, B and D. The intermediate nodes A and C are able to detect passing threads using QD1. If QD2 is implemented, node E will be able to 'eavesdrop' on A's transmissions and record the query as well.

The techniques just discussed improve the efficiency of the IERP by significantly reducing the cost of propagating a single query. Further improvements in IERP performance can be achieved by reducing the frequency of route queries, initiating a global route discovery procedure only when there is a substantial change in the network topology. More specifically, active routes are cached by nodes: the communicating end nodes and intermediate nodes. Upon a change in the network topology, such that a link within an active path is broken, a local path repair procedure is initiated. The path repair procedure substitutes a broken link by a minipath between the ends of the broken link. A path update is then generated and sent to the end points of the path. Path repair procedures tend to reduce the path optimality (e.g. increase the length for shortest path routing). Thus, after some number of repairs, the path endpoints may initiate a new route discovery procedure to replace the path with a new optimal one.

13.2.3 Selective broadcasting (SBC)

Rather than broadcast queries to all peripheral nodes, the same coverage can be provided by broadcasting to a chosen subset of peripheral nodes. This requires IARP to provide network topology information for an extended zone that is twice the radius of the routing zone.

A node will first determine the subset of other peripheral nodes covered by its assigned inner peripheral nodes. The node will then broadcast to this subset of assigned inner peripheral nodes which forms the minimal partitioning set of the outer peripheral nodes.

This is illustrated in Figure 13.25. S's inner peripheral nodes are A, B and C. Its outer peripheral nodes are F, G, H, X, Y and Z. Two inner peripheral nodes of B (H and X) are also inner peripheral nodes of A and C. S can then choose to eliminate B from its list of broadcast recipients since A can provide coverage to H and C can cover X.

The position of the routing functions in the protocol stack are illustrated in Figure 13.26. Route updates are triggered by the MAC-level neighbor discovery protocol (NDP). IARP is notified when a link to a neighbor is established or broken. IERP reactively acquires routes

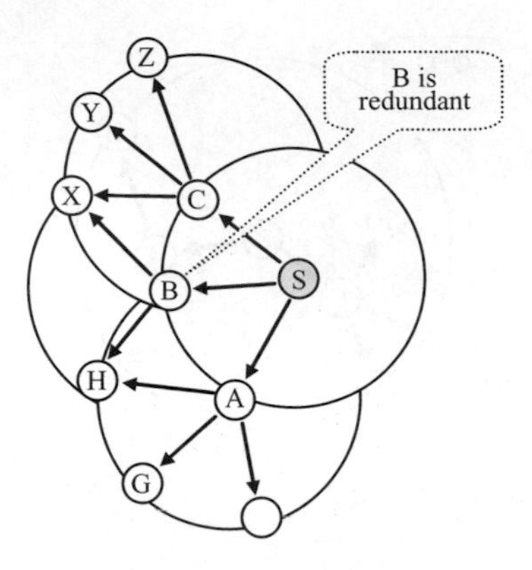

Figure 13.25 Selective broadcasting.

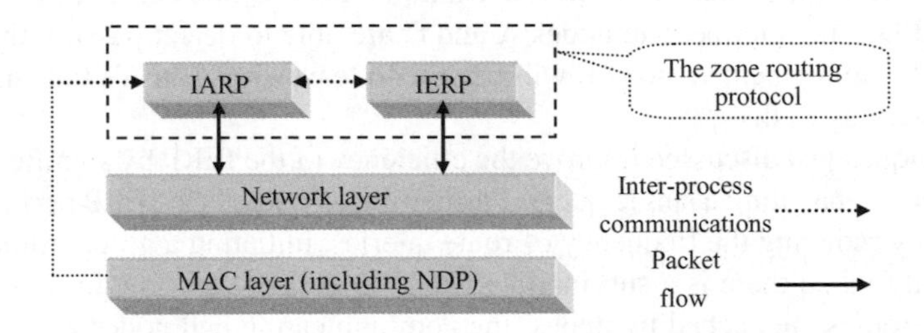

Figure 13.26 The position of the routing functions in the protocol stack.

Table 13.1 Simulation parameters

Parameter	Symbol	Values	Default
Zone radius [hops]	ρ	1–8	—
Node density [neighbors/node]	δ	3–9	6
Relative node speed [neighbors/s]	V	0.1–2.0	1.0
Number of nodes [nodes]	N	200–1000	500

to nodes beyond the routing zone. IERP forwards queries to its peripheral nodes (BRP) keeping track of the peripheral nodes through the routing topology information provided by IARP.

Pearlman and Haas [38] present the performance evaluation of the hybrid protocol, described above for the simulation set of parameters given in Table 13.1. From Figure 13.27 one can see that the IARP control traffic per node is increased as the radius of the zone is increased.

At the same time IERP traffic generated per zone would be reduced. So, the total ZRP traffic has a minimum value for some zone diameter r, which is demonstrated in Figure 13.28.

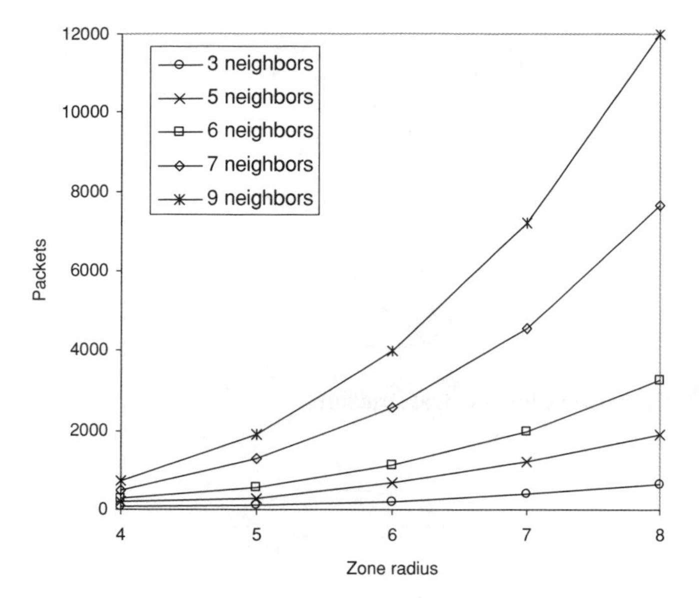

Figure 13.27 IARP traffic generated per neighbor.

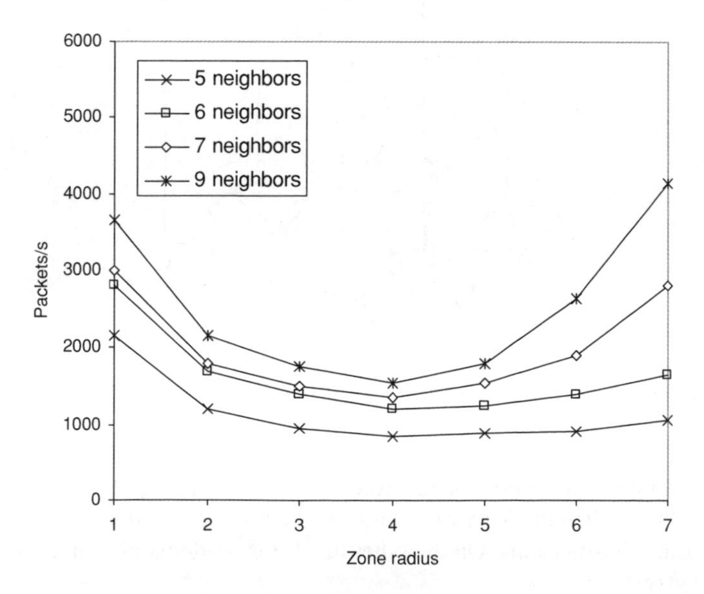

Figure 13.28 ZRP traffic per node ($N = 1000$ nodes, $v = 0{:}5$ neighbors/s).

13.3 SCALABLE ROUTING STRATEGIES

13.3.1 Hierarchical routing protocols

A hierarchical approach to routing, often referred to as *hierarchical state routing* (HSR), has been a traditional option when the network has a large number of nodes. The approach has a lot in common with routing with aggregation presented in Chapter 7. Common

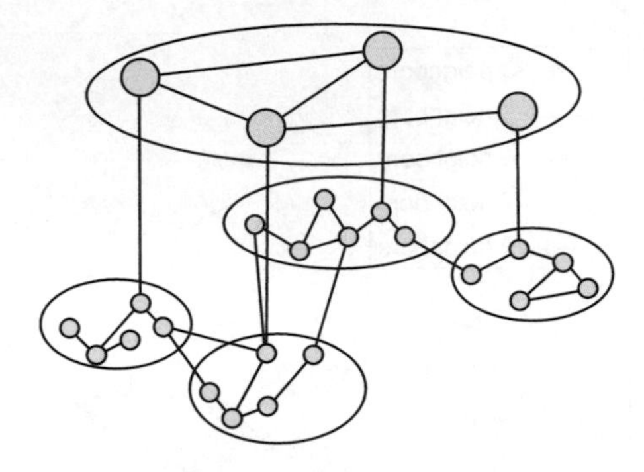

Figure 13.29 The network hierarchical structure.

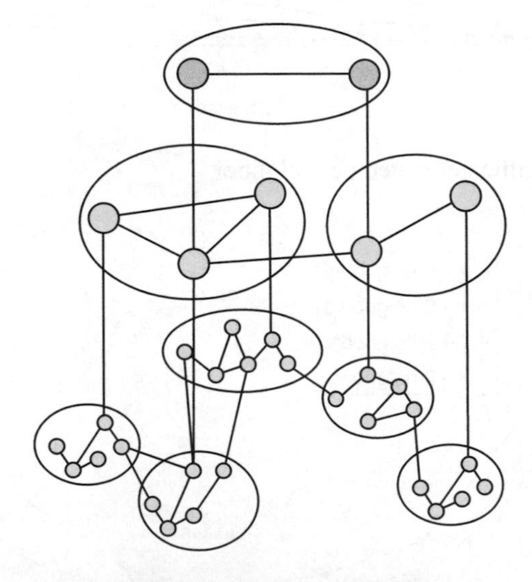

Figure 13.30 Three layers network hierarchical structure.

table-driven protocols and on-demand protocols are for flat topologies and thus have a *scalability* problem when the network is large. For table-driven protocols there is high volume of overhead transmissions. On the other hand, for on-demand protocols there is large discovery latency.

The experience gained in wired networks suggests the use of a hierarchical structure to address the scalability problem. The use of hierarchical routing protocol in *ad-hoc* networks reduces overhead traffic and discovery latency but it has drawbacks, such as: suboptimal routes and complex management of the network hierarchical structure due to its dynamic nature. The basic idea is to divide the network into clusters or domains, as illustrated in Figures 13.29 and 13.30.

The mobile nodes are grouped into regions, regions are grouped into super-regions, and so on, as shown in Figure 13.30. A specific mobile host is chosen as the clusterhead for each region. In hierarchical routing, mobile nodes know how to route packets to their destination

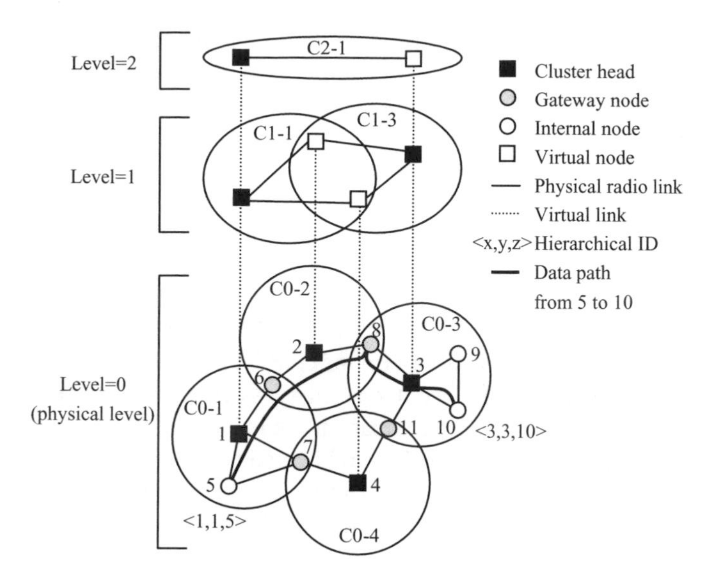

Figure 13.31 Physical multilevel clustering.

within its own region, but do not know the route outside of its own region. Clusterheads know how to reach other regions.

Figure 13.31 shows an example of physical clustering in more detail. At level $l = 0$, we have four physical level clusters, C0–1, C0–2, C0–3 and C0–4. Level 1 and level 2 clusters are generated by recursively selecting cluster heads. Different clustering algorithms can be used for the dynamic creation of clusters and the election of cluster heads [39, 40]. At level 0 clustering, spread-spectrum radios and CDMA can be introduced for spatial reuse across clusters. Within a level 0 cluster, the MAC layer can be implemented by using a variety of different schemes (polling, MACA, CSMA, TDMA, etc.).

Generally, as in ZRP, there are three kinds of nodes in a cluster, namely, the cluster head node (e.g. nodes 1–4), gateway node (e.g. nodes 6–8, and 11) and internal nodes (e.g. nodes 5, 9 and 10). The cluster head node acts as a local coordinator of transmissions within the cluster. The node IDs shown in Figure 13.31 (at level $l = 0$) are physical (e.g. MAC layer) addresses. They are hardwired and are unique to each node.

Within a physical cluster, each node monitors the state of the link to each neighbor (i.e. up/down state and possibly QoS parameters such as bandwidth) and broadcasts it within the cluster. The cluster head summarizes LS information within its cluster and propagates it to the neighbor cluster heads (via the gateways). The knowledge of connectivity between neighbor cluster heads leads to the formation of level 1 clusters. For example, as shown in Figure 13.31, neighbor cluster heads 1 and 2 become members of the level 1 cluster C1–1. To carry out LS routing at level 1, an LS parameter of the 'virtual' link in C1–1 between nodes 1 and 2 (which are neighbor cluster heads) is calculated from the LS parameters of the physical path from cluster head 1 to next cluster head 2 through gateway 6. More precisely, gateway 6 passes the LS update for link (6–2) to cluster head 1. Cluster head 1 estimates the parameters for the path (1–6–2) using its local estimate for (1–6) and the estimate for (6–2) it just received from gateway 6. The result becomes the LS parameter of the 'virtual link' between nodes 1 and 2 in C1–1. This is equivalent to the aggregation process discussed in

Chapter 7. The virtual link can be viewed as a 'tunnel' implemented through lower level nodes.

Applying the aforementioned clustering procedure (aggregation) recursively, new cluster heads are elected at each level and become members of the higher level cluster (e.g. node 1 is elected as a cluster head at level 1 and becomes a member of level 2 cluster C2–1).

Nodes within a cluster exchange virtual LS information as well as summarized lower-level cluster information. After obtaining the LS information at this level, each virtual node floods it down to nodes within the lower level cluster. As a result, each physical node has a 'hierarchical' topology information, as opposed to a full topology view as in flat LS schemes. The hierarchy so developed requires a new address for each node, the hierarchical address. There are many possible solutions for the choice of the hierarchical address scheme. In hierarchical state routing (HSR), we define the hierarchical ID (HID) of a node as the sequence of the MAC addresses of the nodes on path from the top hierarchy to the node itself. For example, in Figure 13.31 the hierarchical address of node 6 [called HID(6)], is 3, 2, 6. In this example, node 3 is a member of the top hierarchical cluster (level 2). It is also the cluster head of C1–3. Node 2 is member of C1–3 and is the cluster head of C0–2. Node 6 is a member of C0–2 and can be reached directly from node 2. The advantage of this hierarchical address scheme is that each node can dynamically and locally update its own HID upon receiving the routing updates from the nodes higher up in the hierarchy. The hierarchical address is sufficient to deliver a packet to its destination from anywhere in the network using HSR tables.

Referring to Figure 13.31, consider for example the delivery of a packet from node 5 to node 10. Note that HID(5) =<1,1,5> and HID(10) =<3,3,10>. The packet is forwarded upwards (to node 1) to the top hierarchy by node 5. Node 1 delivers the packet to node 3, which is the top hierarchy node for destination 10. Node 1 has a 'virtual link', i.e. a tunnel, to node 3, namely, the path (1, 6, 2, 8, 3). It thus delivers the packet to node 3 along this path. Finally, node 3 delivers the packet to node 10 along the downwards hierarchical path, which in this case reduces to a single hop.

Gateways nodes can communicate with multiple cluster heads and thus can be reached from the top hierarchy via multiple paths. Consequently, a gateway has multiple hierarchical addresses, similar to a router in the wired Internet (see Chapter 1), equipped with multiple subnet addresses.

13.3.2 Performance examples

Performance analysis of the system described above can be found in Iwata *et al.* [41]. In most experiments, the network consistes of 100 mobile hosts roaming randomly in all directions at a predefined average speed in a 1000×1000 m area(i.e. no group mobility models are used). A reflecting boundary is assumed. The radio transmission range is 120 m. A free space propagation channel model is assumed. The data rate is 2 Mb/s. The packet length is 10 kb for data, 2 kb for a cluster head neighbor list broadcast, and 500 b for MAC control packets. Transmission time is 5 ms for a data packet, 1 ms for a neighboring list, and 0.25 ms for a control packet. The buffer size at each node is 15 packets. Figures 13.32 and 13.33 illustrate the tradeoffs between throughput and control overhead (O/H) in HSR when the route refresh rate is varied.

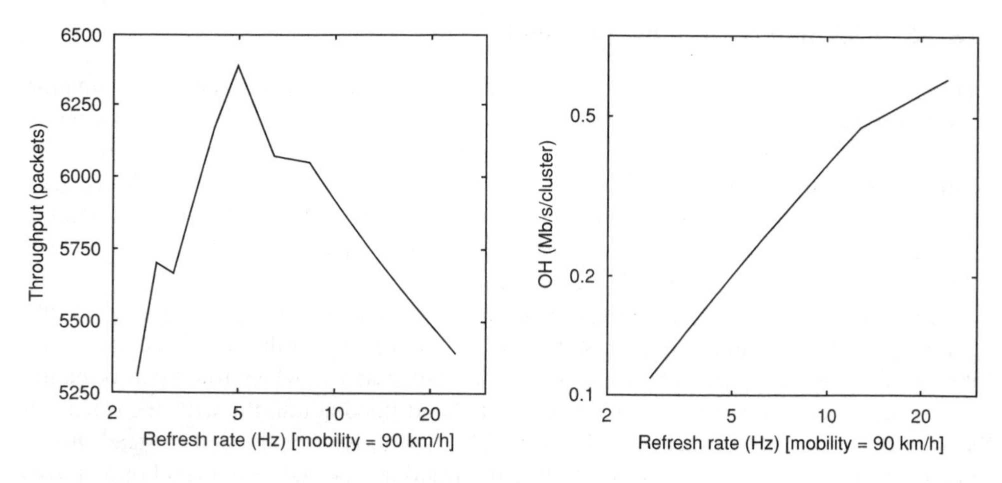

Figure 13.32 System performance vs the routing information refresh rate with $v =$ 90 km/h.

In Figure 13.32 (at 90 km/h), we note that the O/H increases linearly with refresh rate until the network becomes saturated with control packets and starts dropping them. The data throughput first increases rapidly with the refresh rate, owing to more accurate routes and lower packet drops due to the lack of a route. Eventually, throughput peaks and then starts decreasing as the network becomes saturated, and data packets are dropped because of buffer overflow. Figure 13.33 reports the 'optimal' HSR refresh rate as a function of speed.

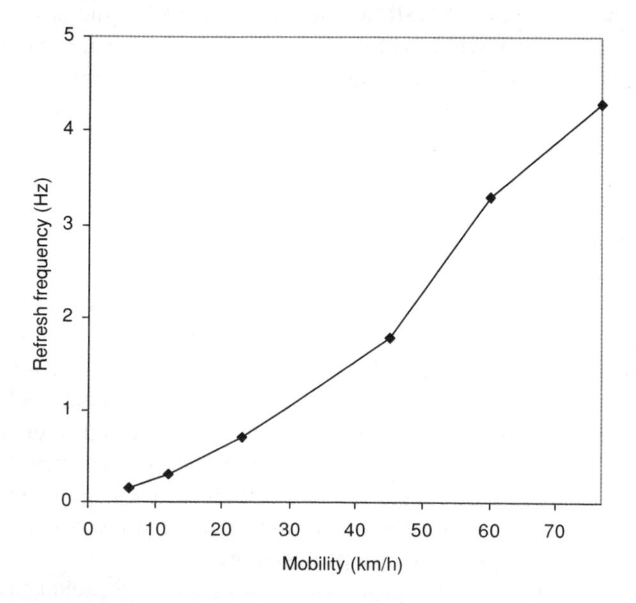

Figure 13.33 Optimum routing information refresh rate vs mobility.

13.3.3 FSR (fisheye routing) protocol

This protocol represents a different way to reduce (aggregate) the amount of information used for routing purposes. In Kleinrock and Stevens [42], a 'fisheye' technique was proposed to reduce the size of information required to represent graphical data. The eye of a fish captures with high detail the pixels near the focal point. The detail decreases as the distance from the focal point increases. In routing, the fisheye approach translates to maintaining accurate distance and path quality information about the immediate neighborhood of a node, with progressively less detail as the distance increases.

The FSR scheme presented in Iwata *et al.* [41] is built on top of another routing scheme called 'global state routing' (GSR) [43]. GSR is functionally similar to LS routing in that it maintains a topology map at each node. The key is the way in which routing information is disseminated. In LS, LS packets are generated and flooded into the network whenever a node detects a topology change. In GSR, LS packets are not flooded. Instead, nodes maintain an LS table based on the up-to-date information received from neighboring nodes and periodically exchange it with their local neighbors only (no flooding).

Through this exchange process, the table entries with larger sequence numbers replace the ones with smaller sequence numbers. The GSR periodic table exchange resembles the DSDV, discussed earlier in this chapter, where the distances are updated according to the time stamp or sequence number assigned by the node originating the update. In GSR (like in LS), LSs are propagated, a full topology map is kept at each node, and shortest paths are computed using this map.

In a wireless environment, a radio link between mobile nodes may experience frequent disconnects and reconnects. The LS protocol releases an LS update for each such change, which floods the network and causes excessive overhead. GSR avoids this problem by using periodic exchange of the entire topology map, greatly reducing the control message overhead [43]. The drawbacks of GSR are the large size update message that consumes a considerable amount of bandwidth and the latency of the LS change propagation, which depends on the update period. This is where the fisheye technique comes to help, by reducing the size of update messages without seriously affecting routing accuracy.

Figure 13.34 illustrates the application of fisheye in a mobile wireless network. The circles with different shades of gray define the fisheye scopes with respect to the centre node (node 11). The scope is defined as the set of nodes that can be reached within a given number of hops. In our case, three scopes are shown for one, two and three hops, respectively. Nodes are color-coded as black, gray and white, accordingly. The reduction of update message size is obtained by using different exchange periods for different entries in the table. More precisely, entries corresponding to nodes within the smaller scope are propagated to the neighbors with the highest frequency. Referring to Figure 13.35, entries in bold are exchanged most frequently. The rest of the entries are sent out at a lower frequency. As a result, a considerable fraction of LS entries are suppressed, thus reducing the message size. This strategy produces timely updates from near stations, but creates large latencies from stations that are far away. However, the imprecise knowledge of the best path to a distant destination is compensated for by the fact that the route becomes progressively more accurate as the packet gets closer to its destination.

In summary, FSR scales well to large networks, by keeping LS exchange overhead (O/H) low without compromising route computation accuracy when the destination is near. By retaining a routing entry for each destination, FSR avoids the extra work of 'finding' the

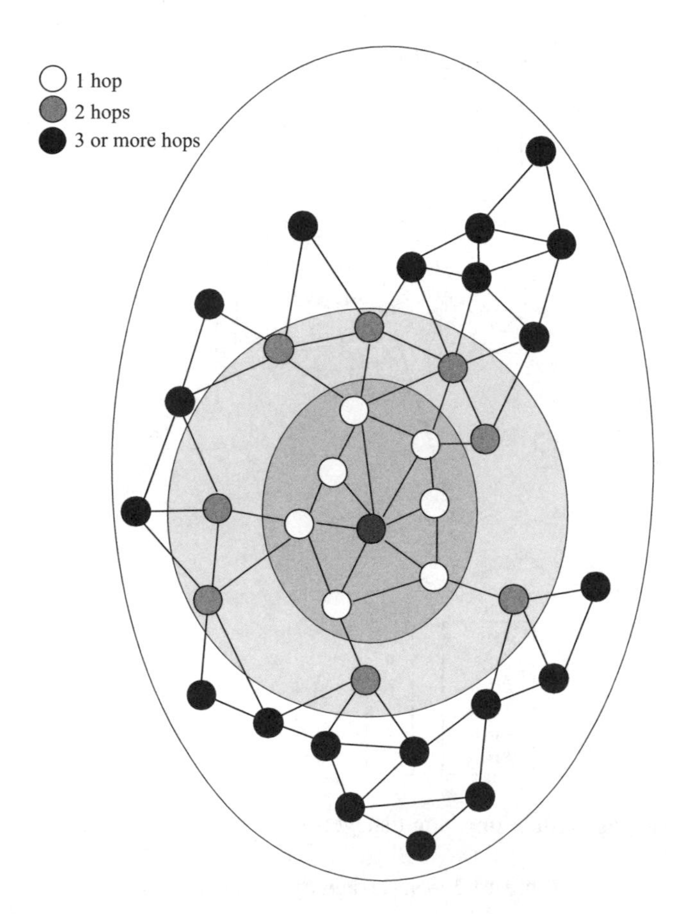

Figure 13.34 Illustration of a fisheye.

destination (as in on-demand routing) and thus maintains low-single packet transmission latency. As mobility increases, routes to remote destinations become less accurate. However, when a packet approaches its destination, it finds increasingly accurate routing instructions as it enters sectors with a higher refresh rate.

Figure 13.36 shows the increase in the control O/H as a function of number of nodes. Geographical node density is kept the same for all runs, as shown in Table 13.2 [41]. One can see that as network size grows larger, the fisheye technique aggressively reduces the O/H.

13.4 MULTIPATH ROUTING

A routing scheme that uses multiple paths simultaneously by splitting the information between a multitude of paths, so as to increase the probability that the essential portion of the information is received at the destination without incurring excessive delay is referred to as *multipath routing*. Such a scheme is needed to mitigate the instability of the topology (e.g. failure of links) in an *ad hoc* network due to nodal mobility and changes in wireless propagation conditions. The scheme works by adding an overhead to each packet, which

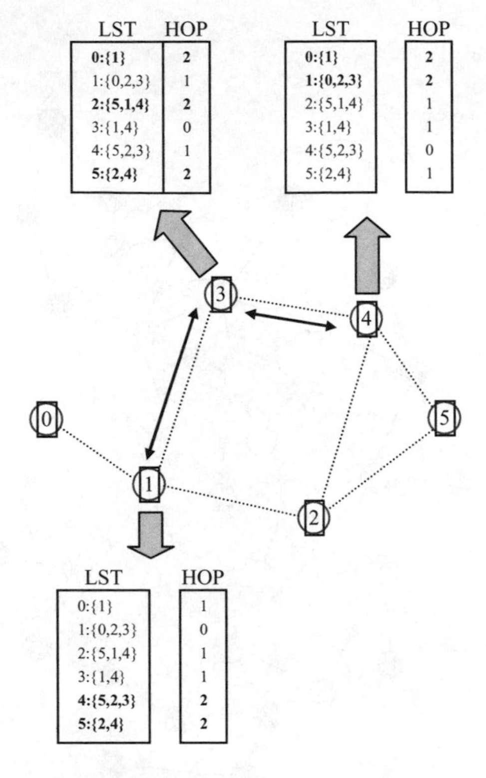

Figure 13.35 Message reduction using fisheye.

Table 13.2 Node density (nodes vs area)

Number of nodes	Simulation area
25	500 × 500
49	700 × 700
100	1000 × 1000
225	1500 × 1500
324	1800 × 1800
400	2000 × 2000

is calculated as a linear function of the original packet bits. The process has its analogy in coding theory. The resulting packet (information and overhead) is fragmented into smaller blocks and distributed over the available paths. The probability of reconstructing the original information at the destination is increased as the number of used paths is increased.

A lot of research has been done in the area of multipath routing in wired networks. One of the initial approaches to this problem was the dispersity routing [44]. In order to achieve self-healing and fault tolerance in digital communication networks, diversity coding is suggested in Ayanoglu *et al.* [45]. In Krishnan and Silvester [46], a per-packet allocation granularity for multipath source routing schemes was shown to perform better than a per-connection allocation.

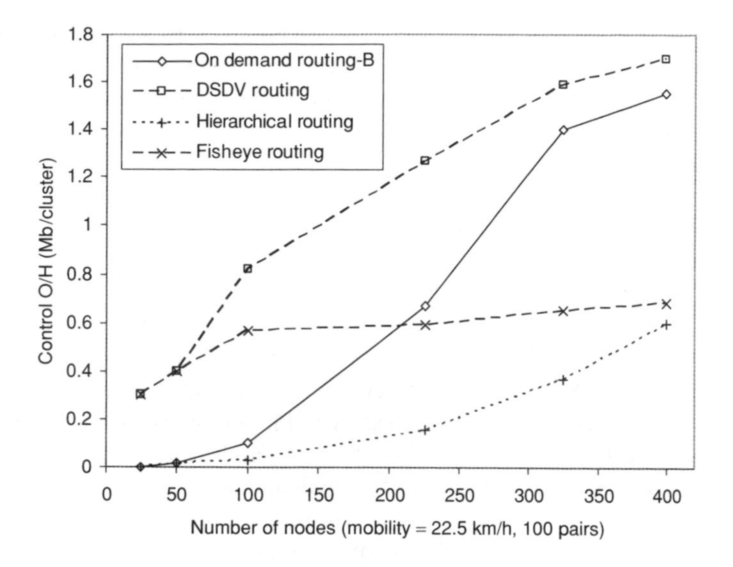

Figure 13.36 Control O/H vs number of nodes.

An exhaustive simulation of the various tradeoffs associated with dispersity routing is presented in Banerjea [47]. The inherent capability of this routing method to provide a large variety of services was pointed out. Owing to this fact, numerous schemes employing multipath routing have been proposed for wired networks in order to perform QoS routing [48–55]. All these protocols are based on proactive routing, since they maintain tables that reflect the state of the entire network. For this reason, owing to the unreliability of the wireless infrastructure and the nodal mobility, which can trigger an excessive amount of updates in the state tables, they cannot be successfully applied to mobile networks.

The application of multipath techniques in mobile *ad hoc* networks seems natural, as multipath routing allows reduction of the effect of unreliable wireless links and the constantly changing topology. The on-demand multipath routing scheme is presented in Nasipuri and Das [56] as a multipath extension of dynamic source routing (DSR) [57], described in Section 13.1. The alternate routes are maintained, so that they can be utilized when the primary one fails. TORA [58], routing on demand acyclic multipath (ROAM) [59] and *ad hoc* on-demand distance vector-backup routing (AODV-BR) [60], which is based on the *ad hoc* on-demand distance vector (AODV) protocol [61], are also examples of schemes that maintain multiple routes and utilize them only when the primary root fails. However, these protocols do not distribute the traffic into the multiple available paths.

Another extension of DSR, multiple source routing (MSR) [62], proposes a weighted round-robin heuristic-based scheduling strategy among multiple paths in order to distribute load, but provides no analytical modeling of its performance. The split multipath routing (SMR), proposed in Lee and Gerla [63], focuses on building and maintaining maximally disjoint paths; however, the load is distributed only in two routes per session. In Papadimitratos *et al.* [64], the authors propose a novel and nearly linear heuristic for constructing a highly reliable path set. In Pearlman *et al.* [65], the effect of alternate path routing (APR) on load balancing and end-to-end delay in mobile *ad hoc* networks has been explored. It was argued, however, that the network topology and channel characteristics (e.g. route coupling) can severely limit the gain offered by APR strategies. In an interesting application [66], multi-path path transport (MPT) is combined with multiple description coding (MDC) in order to send video and image information in a multihop mobile radio network. In this section , we discuss a multipath scheme for mobile *ad hoc* networks based on diversity coding [45]. Data load is distributed over multiple paths in order to minimize the packet drop rate and achieve load balancing in a constantly changing environment.

Suppose that $n_{\max}$ paths are available for the transmission of data packets from a source to a destination. Any of the multipath schemes mentioned in the introduction can be employed in order to acquire these paths. No paths have nodes in common (mutually disjoint).

Each path, indexed as $i, i = 1, \ldots n_{\max}$, is either down at the time that the source attempts to transmit with probability of failure p_i or the information is received correctly with probability $1 - p_i$. Since there are no common nodes among the paths, they are considered independent in the sense that success or failure of one path cannot imply success or failure of another. It should be noted here that, in wireless *ad hoc* networks, nodes are sharing a single channel for transmission, so node disjointness does not guarantee the independence of the paths. Taking this into account, the paths are ideally considered independent as an approximation of a realistic *ad hoc* wireless network. For a more realistic modeling of the paths in a wireless network, one may refer to Tsirigos and Haas [67], where path correlation is included in the analysis. The failure probabilities of the available paths are organized in the probability vector $\mathbf{p} = \{p_i\}$, in such a way that $p_i \leq p_{i+1}$. The vector of success probabilities is defined as $\mathbf{q}\{q_i\} = 1 - \mathbf{p} = \{1 - p_i\}$.

Let us now suppose that we have to send a packet of D data bits utilizing the set of available independent paths in such a way as to maximize the probability that these bits are successfully communicated to the destination. This probability is denoted as P. In order to achieve this goal, we employ a coding scheme in which C extra bits are added as overhead. The resulting $B = D + C$ bits are treated as one network-layer packet. The extra bits are calculated as a function of the information bits in such a way that, when splitting the B-bit packet into multiple equal-size nonoverlapping blocks, the initial D-bit packet can be reconstructed, given any subset of these blocks with a total size of D or more bits. First, we define the overhead factor $r = B/D = b/d$ where b and d take integer values and the fraction b/d cannot be further simplified. One should note that $1/r$ would be equivalent to coding gain in channel coding theory.

Next we define the vector $\mathbf{v} = \{v_i\}$, where v_i is the number of equal-size blocks that is allocated to path i. Some of the paths may demonstrate such a poor performance that there is no point in using them at all. This means that we might require using only some of the available paths. If n is the number of the paths we have to use in order to maximize P, it would be preferable to define the block allocation vector $\mathbf{v} = \{v_i\}$ as a vector of a variable size n, instead of fixing its size to the number of available paths $n_{\max}$.

Given the fact that the probability failure vector is ordered from the best path to the worst one, a decision to use n paths implies that these paths will be the first n ones. Based on these observations, the allocation vector $\mathbf{v} = \{v_i\}$ has the following form: $\mathbf{v} = \{v_1, v_2, \ldots, v_n\}$, $n \leq n_{\max}$.

If the block size is w, then $w \sum_{i=1}^{n} v_i = B = rD$. Therefore, the total number of blocks that the B-bit packet is fragmented into is $a = \sum_{i=1}^{n} v_i = rD/w$. From $p_i \leq p_{i+1}$ it follows that $v_i \geq v_{i+1}$, because a path with higher failure probability cannot be assigned fewer blocks than a path with a lower failure probability. The original D-bit packet is fragmented into Nw-size blocks, $d_1, d_2, d_3, \ldots, d_N$, and the C-bit overhead packet into M w-size blocks, $c_1, c_2, c_3, \ldots, c_M$. Based on this we have $N = D/w = a/r$ and $M = C/w = (r-1)N = (r-1)a/r$. Path 1 will be assigned the first v_1 blocks of the B-bit sequence, path two will receive the next v_2 blocks, and so on. Thus, path i will be assigned v_i blocks, each block of size w. Like parity check bits in error correcting $(N+M, M)$ block coding, the overhead symbols are generated as linear combination of the original packets as

$$c_j = \sum_{i=1}^{N} \beta_{ij} d_i; \quad 1 \leq j \leq M \tag{13.1}$$

where multiplication and summation are performed in Galois Fields $GF(2^m)$. The relations between probability of successful packet transmission P, parameters N and M and link failure probabilities are available from coding theory and will not be repeated here. One of the important results from that theory is that the block size has to satisfy the following inequality, so that the original information can be recovered [45]:

$$w \geq \lceil \log_2(N+M+1) \rceil \geq \log_2(a+1) \tag{13.2}$$

By incorporating the previous definitions in Equation (13.2), we obtain an inequality for the number of blocks, into which we can split the B-bit packet

$$B \geq a \log_2(a+1) \equiv B_{\min} \tag{13.3}$$

13.5 CLUSTERING PROTOCOLS

13.5.1 Introduction

In dynamic cluster-based routing, described so far in this chapter, the network is dynamically organized into partitions called clusters, with the objective of maintaining a relatively stable effective topology [70]. The membership in each cluster changes over time in response to node mobility and is determined by the criteria specified in the clustering algorithm. In order to limit far-reaching reactions to topology dynamics, complete routing information is maintained only for intracluster routing. Intercluster routing is achieved by hiding the topology details within a cluster from external nodes and using hierarchical aggregation, reactive routing or a combination of both techniques. The argument made against dynamic clustering is that the rearrangement of the clusters and the assignment of nodes to clusters may require excessive processing and communications overhead, which outweigh its potential benefits. If the clustering algorithm is complex or cannot quantify a measure of cluster stability, these obstacles may be difficult to overcome. A desirable design objective for an architectural framework capable of supporting routing in large *ad hoc* networks subject to high rates of node mobility incorporates the advantages of cluster-based routing and balances the tradeoff between reactive and proactive routing while minimizing the shortcomings of each. Furthermore, the consequences of node mobility suggest the need to include a quantitative measure of mobility directly in the network organization or path selection process.

Specifically, a strategy capable of evaluating the probability of path availability over time and of basing clustering or routing decisions on this metric can help minimize the reaction to topological changes. Such a strategy can limit the propagation of far-reaching control information while supporting higher quality routing in highly mobile environments.

In this section we present the (*c*,*t*) cluster framework, which defines a strategy for dynamically organizing the topology of an *ad hoc* network in order to adaptively balance the tradeoff between proactive and on demand-based routing by clustering nodes according to node mobility. This is achieved by specifying a distributed asynchronous clustering algorithm that maintains clusters which satisfy the (*c*,*t*) criteria that there is a probabilistic bound c on the mutual availability of paths between all nodes in the cluster over a specified interval of time t. In order to evaluate the (*c*,*t*) criteria, a mobility model is used that characterizes the movement of nodes in large *ad hoc* networks. It is shown how this model is used to determine the probability of path availability when links are subject to failure due to node mobility.

Based on the (*c*,*t*) cluster framework, intracluster routing requires a proactive strategy, whereas intercluster routing is demand-based. Consequently, the framework specifies an adaptive-hybrid scheme whose balance is dynamically determined by node mobility. In networks with low rates of mobility, (*c*,*t*) clustering provides an infrastructure that is more proactive. This enables more optimal routing by increasing the distribution of topology information when the rate of change is low. When mobility rates become very high, cluster size will be diminished and reactive routing will dominate. The (*c*,*t*) cluster framework decouples the routing algorithm specification from the clustering algorithm, and thus, it is flexible enough to support evolving *ad hoc* network routing strategies described so far in both the intra- and intercluster domains.

Several dynamic clustering strategies have been proposed in the literature [70–73]. These strategies differ in the criteria used to organize the clusters and the implementation of the distributed clustering algorithms. McDonald and Znati [74] use prediction of node mobility as a criteria for cluster organization. Clustering decisions in [70–73] are based on static views of the network at the time of each topology change. Consequently, they do not provide for a quantitative measure of cluster stability. In contrast, the (c, t) cluster strategy [74] forms the cluster topology using criteria based directly on node mobility. According to Ramanathan and Steenstrup [73], the ability to predict the future state of an *ad hoc* network comprising highly mobile nodes is essential if the network control algorithms are expected to maintain any substantive QoS guarantees to real-time connections. The multimedia support for wireless network (MMWN) system proposed by Ramanathan and Steenstrup [73] is based upon a hybrid architecture that includes the characteristics of *ad hoc* and cellular networks. Their framework uses hierarchical routing over dynamic clusters that are organized according to a set of system parameters that control the size of each cluster and the number of hierarchical levels. Aggregation of routing information is used to achieve scalability and limit the propagation of topological change information. A multilevel strategy is used to repair virtual circuit (VC) connections that have been disturbed due to node mobility. MMWN does not predict node movement. Consequently, it is unable to provide a quantitative bound on the stability of its cluster organization.

Vaidya *et al.* [72] proposed a scheme that dynamically organizes the topology into k clusters, where nodes in a cluster are mutually reachable via k-hop paths. The algorithm considers $k = 1$ and reduces to finding cliques in the physical topology. Using a first-fit heuristic, the algorithm attempts to find the largest cliques possible. Although the algorithm does not form optimal clusters, it still requires a three-pass operation each time a topology change occurs: one for finding a set of feasible clusters, a second for choosing the largest of the feasible clusters that are essential to maintain cluster connectivity, and a third to eliminate any existing clusters that are made superfluous by the new clusters.

The objective of the scheme proposed by Lin and Gerla [70] differs significantly from the previous examples. Rather than using clustering to minimize the network's reaction to topological changes, their scheme is intended to provide controlled access to the bandwidth and scheduling of the nodes in each cluster in order to provide QoS support. Hierarchical routing and path maintenance were a secondary concern. The proposed algorithm is very simple and uses node ID numbers to deterministically build clusters of nodes that are reachable by two-hop paths.

The zone routing protocol (ZRP), described in Section 13.2 is a hybrid strategy that attempts to balance the tradeoff between proactive and reactive routing. The objective of ZRP is to maintain proactive routing within a zone and to use a query–response mechanism to achieve interzone routing. In ZRP, each node maintains its own hop-count constrained routing zone; consequently, zones do not reflect a quantitative measure of stability, and the zone topology overlaps arbitrarily. These characteristics differ from (c,t) clusters, which are determined by node mobility and do not overlap. Both strategies assume a proactive routing protocol for intrazone/cluster routing, and each organizes its topology based upon information maintained by that protocol. ZRP also defines the query control scheme to achieve interzone routing. Although ZRP is not a clustering algorithm and the (c,t) cluster framework is not a routing protocol, the comparison demonstrates a close relationship that could be leveraged by incorporating the (c,t) cluster into ZRP. The use of (c,t) clusters in ZRP could achieve more efficient and adaptive hybrid routing without significantly increasing its complexity.

13.5.2 Clustering algorithm

The objective of the clustering algorithm is to partition the network into several clusters. Optimal cluster size is dictated by the tradeoff between spatial reuse of the channel (which drives toward small sizes) and delay minimization (which drives toward large sizes). Other constraints also apply, such as power consumption and geographical layout. Cluster size is controlled through the radio transmission power. For the cluster algorithm, we assume that transmission power is fixed and is uniform across the network. Within each cluster, nodes can communicate with each other in at most two hops. The clusters can be constructed based on node ID.

The following algorithm partitions the multihop network into some nonoverlapping clusters. The following operational assumptions underlying the construction of the algorithm in a radio network are made. These assumptions are common to most radio data link protocols [75–78]:

(A1) Every node has a unique ID and knows the IDs of its one-hop neighbors. This can be provided by a physical layer for mutual location and identification of radio nodes.

(A2) A message sent by a node is received correctly within a finite time by all of its one-hop neighbors.

(A3) Network topology does not change during the algorithm execution.

The distributed clustering algorithm is shown in Figure 13.37

```
Γ: the set of ID's of my one-hop neighbors and myself
{
        if (my_id == min(Γ))
        {
                my_cid = my_id;
                broadcast cluster(my_id,my_cid);
                Γ = Γ - {my_id};
        }

        for (:;)
        {
                on receiving cluster(id, cid)
                {
                        set the cluster ID of node id to cid;
                        if (id==cid and (my_cid==UNKNOWN or my_cid>cid))
                                my_cid = cid;
                        Γ = Γ - {id};
                        if (my_id == min(Γ))
                        {
                                if (my_cid==UNKNOWN) my_cid = my_id;
                                broadcast cluster(my_id,my_cid);
                                Γ = Γ - {my_id};
                        }
                }
                if (Γ==Ø) stop;
        }
}
```

Figure 13.37 Distributed clustering algorithm (cluster ID – cid). (Reproduced by permission of IEEE [70].)

As an example, the topology from Figure 13.38 after clustering is given in Figure 13.39. From Figures 13.37–13.39 one can see that the cluster ID of each node is equal to either its node ID or the lowest cluster ID of its neighbors. Every node must have its cluster ID once it becomes the lowest ID node in its locality. This cluster ID will be broadcast at this time, and will not be changed before the algorithm stops. Hence, every node can determine its cluster and only one cluster.

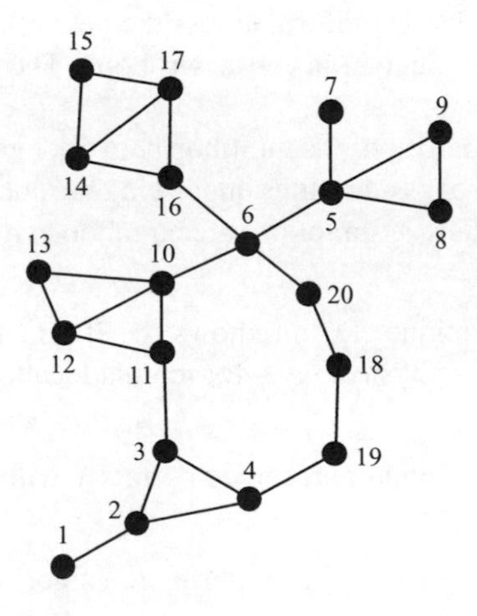

Figure 13.38 System topology.

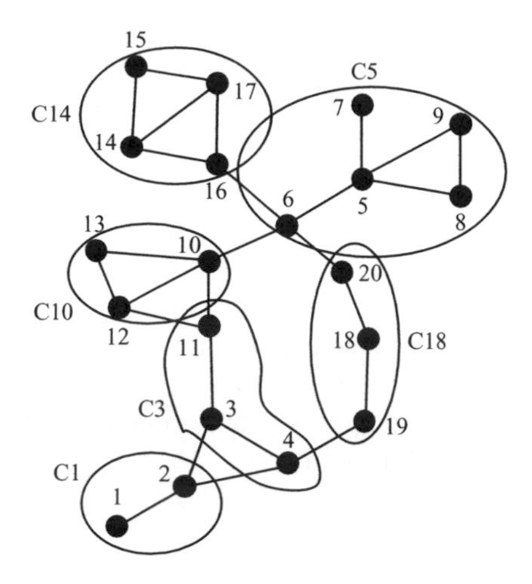

Figure 13.39 Clustering.

13.5.3 Clustering with prediction

13.5.3.1 (c,t) Cluster framework

The objective of the (*c,t*) cluster framework is to maintain an effective topology that adapts to node mobility so that routing can be more responsive and optimal when mobility rates are low and more efficient when they are high. This is accomplished by a simple distributed clustering algorithm using a probability model for path availability as the basis for clustering decisions. The algorithm dynamically organizes the nodes of an *ad hoc* network into clusters where probabilistic bounds can be maintained on the availability of paths to cluster destinations over a specified interval of time.

The (*c,t*) cluster framework can also be used as the basis for the development of adaptive schemes for probabilistic QoS guarantees in *ad hoc* networks. Specifically, support for QoS in time-varying networks requires addressing: (1) connection-level issues related to path establishment and management to ensure the existence of a connection between the source and the destination; and (2) packet-level performance issues in terms of delay bounds, throughput and acceptable error rates.

Ideally, it is desirable to guarantee that the QoS requirements of ongoing connections are preserved for their entire duration. Unfortunately, this is not possible in a time-varying network environment as connections may fail randomly due to user mobility.

A more realistic and practical approach is to provide some form of probabilistic QoS guarantees by keeping connection failures below a prespecified threshold value and by ensuring with high probability that a minimum level of bandwidth is always available to ongoing connections.

The basic idea of the (*c,t*) cluster strategy is to partition the network into clusters of nodes that are mutually reachable along cluster internal paths that are expected to be available for a period of time t with a probability of at least c. The union of the clusters in a network must cover all the nodes in the network. Assume, without loss of generality, that t is identical at every node in a cluster. If the cluster's topology remains stable over the interval of length t, then routing will be deterministic during this interval, and standard assumptions permit the *ad hoc* network to be modeled as a network of Jackson queues. Assuming that path availability is an ergodic process, c represents the average proportion of time a (c,t) path is available to carry data. Consequently, c places a lower bound on the effective capacity of the path over an interval of length t.

Let the link capacity be C b/s and the mean packet length $1/\mu$ b. The effective packet service rate μ_{eff} over the interval t can be determined based upon the path availability according to Equation (13.4). Based on the Jackson model, each node can be treated as an independent M/M/1 queue. Using knowledge of the current aggregate arrival rate λ and the effective service rate μ_{eff}, the M/M/1 results can be applied to find the mean total packet delay T. Since this delay must be less than t, this approach establishes a lower bound on the path availability, as shown in Equation (13.7)

$$\mu_{\text{eff}} = cC\mu \tag{13.4}$$

$$T = \frac{1}{\mu_{\text{eff}} - \lambda} \tag{13.5}$$

$$t \geq \frac{1}{cC\mu - \lambda} \tag{13.6}$$

$$c \geq \frac{1 + \lambda t}{\mu t C} \tag{13.7}$$

An effective adaptive strategy for determining the value of c controls the minimum level of cluster stability required to support the traffic load and QoS requirements of established connections. The choice of the parameter t is a system design decision that determines the maximum cluster size achievable for different rates of mobility when no traffic exists in the network.

13.5.3.2 (c,t) Cluster algorithm

There are five events which drive the (c,t) cluster algorithm, namely, node activation, link activation, link failure, expiration of the timer and node deactivation.

Node activation

The primary objective of an activating node is to discover an adjacent node and join its cluster. In order to accomplish this, it must be able to obtain topology information for the cluster from its neighbor and execute its routing algorithm to determine the (c,t) availability of all the destination nodes in that cluster. The source node can join a cluster if and only if all the destinations are reachable via (c,t) paths. The first step upon node activation is the initialization of the source node's CID (cluster ID) to a predefined value that indicates its unclustered status. The network-interface layer protocol is required to advertise the node's CID as part of the neighbor greeting protocol [79] and in the header of the encapsulation protocol. This enables nodes to easily identify the cluster status and membership of neighboring nodes and of the source of the routing updates – a necessary function to control the dissemination of routing information. When its network-interface layer protocol identifies one or more neighboring nodes, the source node performs the following actions. First, the source node identifies the CIDs associated with each neighbor. Next, it evaluates the link availability associated with each neighbor according to either a system default mobility profile or mobility information obtained through the network-interface layer protocol or physical-layer sensing. The precise methodology and the information required for the evaluation of link availability is described later in this section.

Finally, the neighbors, having discovered the unclustered status of the source node, automatically generate and transmit complete cluster topology information, which they have stored locally as a result of participating in the cluster's intracluster routing protocol. This topology synchronization function is a standard feature of typical proactive routing protocols when a router discovers the activation of a link to a new router. The source node does not immediately send its topology information to any of the neighbors.

Link activation

A link activation detected by a clustered node that is not an orphan is treated as an intracluster routing event. Hence, the topology update will be disseminated throughout the cluster. Unlike reactive routing that responds after path failure, the dissemination of link activation updates is a key factor to an (c,t) cluster node's ability to find new (c,t) paths in anticipation of future link failures or the expiration of the timer.

Link failure

The objective of a node detecting a link failure is to determine if the link failure has caused the loss of any (c,t) paths to destinations in the cluster. A node's response to a link failure event is twofold. First, each node must update its view of the cluster topology and re-evaluate the path availability to each of the cluster destinations remaining in the node's routing table. Second, each node forwards information regarding the link failure to the remaining cluster destinations.

Expiration of c timer

The c timer controls cluster maintenance through periodic execution of the intracluster routing algorithm at each node in a cluster. Using the topology information available at each node, the current link availability information is estimated and maximum availability paths are calculated to each destination node in the cluster. If any of the paths are not (c,t) paths, then the node leaves the cluster.

Node deactivation

The event of node deactivation encompasses four related events, namely, graceful deactivation, sudden failure, cluster disconnection and voluntary departure from the cluster. In general, each of these events triggers a response by the routing protocol. As a result, nodes determine that the node that has deactivated is no longer reachable.

13.5.3.3 **Ad hoc** *mobility model*

The *random ad hoc mobility model* used in this section is a continuous-time stochastic process, which characterizes the movement of nodes in a two-dimensional space. Based on

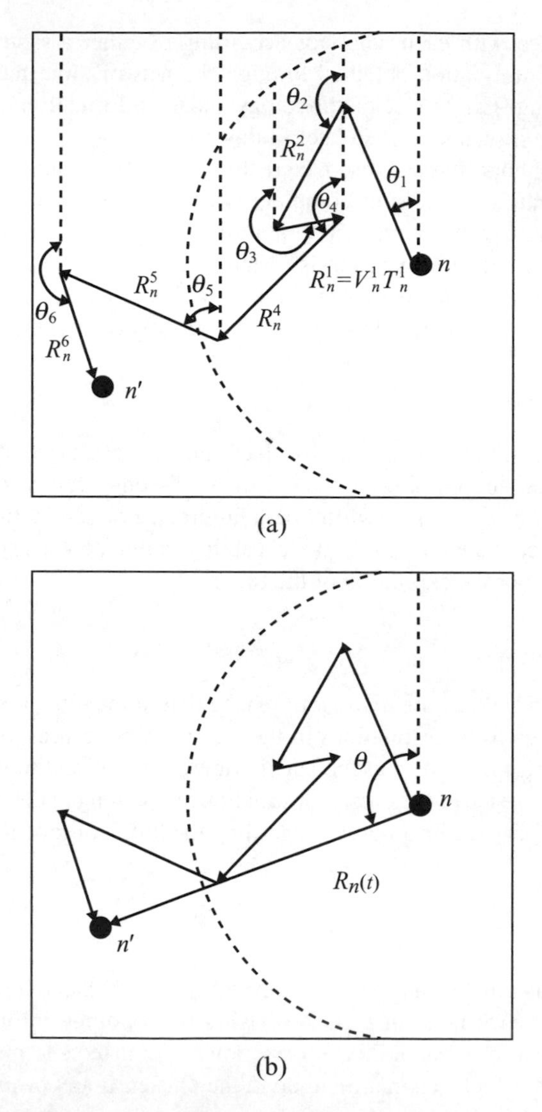

Figure 13.40 *Ad hoc* mobility model node movement: (a) epoch random mobility vectors; (b) *ad hoc* mobility model node movement.

the random *ad hoc* mobility model, each node's movement consists of a sequence of random length intervals called mobility epochs during which a node moves in a constant direction at a constant speed. The speed and direction of each node varies randomly from epoch to epoch. Consequently, during epoch i of duration T_n^i, node n moves a distance of $V_n^i T_n^i$ in a straight line at an angle of θ_n^i. The number of epochs during an interval of length t is the discrete random process $N_n(t)$. Figure 13.40(a) illustrates the movement of the node over six mobility epochs, each of which is characterized by its direction, θ_n^i, and distance $V_n^i T_n^i$. The mobility profile of node n moving according to the random *ad hoc* mobility model requires three parameters: λ_n, μ_n and σ_n^2. The following list defines these parameters and states the assumptions made in developing this model:

(1) The epoch lengths are identically, independently distributed (i.i.d.) exponentially with mean $1/\lambda_n$.

(2) The direction of the mobile node during each epoch is i.i.d. uniformly distributed over $(0, 2\pi)$ and remains constant only for the duration of the epoch.

(3) The speed during each epoch is an i.i.d. distributed random variable (e.g. i.i.d. normal, i.i.d. uniform) with mean μ_n and variance σ_n^2 and remains constant only for the duration of the epoch.

(4) Speed, direction and epoch length are uncorrelated.

(5) Mobility is uncorrelated among the nodes of a network, and links fail independently.

Nodes with limited transmission range are assumed to experience frequent random changes in speed and direction with respect to the length of time a link remains active between two nodes. Furthermore, it is assumed that the distributions of each node's mobility characteristics change slowly relative to the rate of link failure. Consequently, the distribution of the number of mobility epochs is stationary and relatively large while a link is active. Since the epoch lengths are i.i.d. exponentially distributed, $N_n(t)$ is a Poisson process with rate λ_n. Hence, the expected number of epochs experienced by node n during the interval $(0,t)$ while a link is active is $\lambda_n t \triangleq 1$.

These assumptions reflect a network environment in which there are a large number of heterogeneous nodes operating autonomously in an *ad hoc* fashion, which conceptually reflects the environment considered in the design of the (c,t) cluster framework. In order to characterize the availability of a link between two nodes over a period of time $(t_0, t_0 + 1)$, the distribution of the mobility of one node with respect to the other must be determined. To characterize this distribution, it is first necessary to derive the mobility distribution of a single node in isolation. The single node distribution is extended to derive the joint mobility distribution that accounts for the mobility of one node with respect to the other. Using this joint mobility distribution, the link availability distribution is derived.

The random mobility vector can be expressed as a random sum of the epoch random mobility vectors

$$\mathbf{R}_n(t) = \sum_{i=1}^{N_n(t)} \mathbf{R}_n^i$$

as shown in Figure 13.40(b). Let be $\mathbf{R}_n(t)$ the resulting random mobility vector of a mobile node which is located at position $[X(t_0), Y(t_0)]$ at time t_0 and moves according to a random *ad hoc* mobility profile, $\langle \lambda_n, \mu_n, \sigma_n^2 \rangle$. The phase of the resultant vector $\mathbf{R}_n(t)$ is uniformly distributed over $(0, 2\pi)$ and its magnitude represents the aggregate distance moved by the node and is approximately Raleigh distributed with parameter

$$\alpha_n = (2t/\lambda_n)(\sigma_n^2 + \mu_n^2)$$

$$\Pr(\theta_n \le \phi) = \phi/2\pi, \quad 0 \le \phi \le 2\pi \tag{13.8}$$

$$\Pr[R_n(t) \le r] \approx 1 - \exp(-r^2/\alpha_n), \quad 0 \le r \le \infty \tag{13.9}$$

The derivation of these distributions is an application of the classic theory of uniform random phasor sums [80] that applies central limit theorem to a large number of i.i.d. variables.

Joint node mobility
Based on the assumption of random link failures, we can consider the mobility of two nodes at a time by fixing the frame of reference of one node with respect to the other. This transformation is accomplished by treating one of the nodes as if it were the base station of a cell, keeping it at a fixed position. For each movement of this node, the other node is translated an equal distance in the opposite direction. So, the vector $\mathbf{R}_{m,n}(t) = \mathbf{R}_m(t) - \mathbf{R}_n(t)$, representing the equivalent random mobility vector of node m with respect to node n, is obtained by fixing m's frame of reference to n's position and moving m relative to that point. Its phase is uniformly distributed over $(0, 2\pi)$ and its magnitude has Raleigh distribution with parameter $\alpha_{m,n} = \alpha_m + \alpha_n$.

Random ad hoc *link availability*
If $L_{m,n}(t) = 1$ denotes an active and $L_{m,n}(t) = 0$ an inactive link, then for nodes n and m, link availability is defined as

$$A_{m,n}(t) \equiv \Pr[L_{m,n}(t_0 + t) = 1 \mid L_{m,n}(t_0) = 1] \tag{13.10}$$

Note that a link is still considered available at time t even if it experienced failures during one or more intervals (t_i, t_j); $t_0 < t_i < t_j < t_0 + t$. By definition, if m lies within the circular region of radius R centered at n, the link between the two nodes is considered to be active.

Depending on the initial status and location of nodes m and n, two distinct cases of link availability can be identified.

(1) Node activation – node m becomes active at time t_0, and it is assumed to be at a random location within range of node n. In this case we have

$$\begin{aligned} A_{m,n}(t) &\approx 1 - \Phi\left\{\frac{1}{2}, 2, -R^2/\alpha_{m,n}\right\} \\ \Phi\left\{\frac{1}{2}, 2, z\right\} &= e^{z/2}\left[I_0(z/2) - I_1(z/2)\right] \\ \alpha_{m,n} &= 2t\left\{\frac{\sigma_m^2 + \mu_m^2}{\lambda_m} + \frac{\sigma_n^2 + \mu_n^2}{\lambda_n}\right\} \end{aligned} \tag{13.11}$$

(2) Link activation: node m moves within range of node n at time t_0 by reaching the boundary defined by R, and it is assumed to be located at a random point around the boundary. In this case we have

$$A_{m,n}(t) = \frac{1}{2}\{1 - I_0(-2R^2/\alpha_{m,n})\exp(-2R^2/\alpha_{m,n})\} \tag{13.12}$$

Random ad hoc *path availability*
Let $P^k_{m,n}(t)$ indicate the status of path k from node n to node m at time t. $P^k_{m,n}(t) = 1$ if all the links in the path are active at time t, and $P^k_{m,n}(t) = 0$ if one or more links in the path are inactive at time t. The path availability $\pi^k_{m,n}(t)$ between two nodes n and m at time $t \geq t_0$ is given by the following probability

$$\pi^k_{m,n}(t) \equiv \Pr\left\{P^k_{m,n}(t_0 + t) = 1 \,\middle|\, P^k_{m,n}(t_0 = 1)\right\} = \prod_{(i,j)\in k} A_{i,j}(t_0 + t) \tag{13.13}$$

If $\pi^k_{m,n}(t)$ is the path availability of path k from node n to node m at time t, then path k is defined as an (c,t) path if and only if

$$\pi^k_{m,n}(t) \geq c \tag{13.14}$$

Node n and node m are (c,t) available if they are mutually reachable over (c,t) paths. An (c,t) cluster is a set of (c,t) available nodes. This definition states that every node in an (c,t) cluster has a path to every other node in the cluster that will be available at time $t_0 + t$ with a probability $\geq c$.

Path availability cost calculation
The above discussion demonstrates how the link availability can be calculated, thereby providing a link metric that represents a probabilistic measure of path availability. This metric can be used by the routing algorithm in order to construct paths that support a lower bound c on availability of a path over an interval of length t. The availabilities of each of the links along a path are used by the (c,t) cluster protocol to determine if the path is an (c,t) path, and consequently, if a cluster satisfies the (c,t) criteria. In order to support this functionality in an *ad hoc* network, the routing protocol must maintain and disseminate the following status information for each link:

(1) the initial link activation time, t_0;

(2) the mobility profiles for each of the adjacent nodes $\langle \lambda_i, \mu_i, \sigma_i^2 \rangle$, $i = m, n$;

(3) the transmission range of each of the adjacent nodes, R;

(4) the event which activated the link: (a) node activation at time t_0 or (b) nodes moving into range of each other at time t_0.

Based on this information, any node in an (c,t) cluster can estimate, at any time τ, the availability of a link at time $t + \tau$. This can be achieved because each node knows the initial link activation time t_0; hence, link availability is evaluated over the interval $(t_0, t + \tau)$. Nodes can use conditional probability to evaluate the availability of their own links because they have direct knowledge of such a link's status at time τ, whereas remote nodes do not. Specifically, for an incident link that activated at time t_0, a node will evaluate the availability at time t, given that it is available at time $\tau \geq t_0$.

13.5.3.4 Performance example

A range of node mobility with mean speeds between 5.0 and 25.0 km/h was simulated in McDonald and Znati [74]. The speeds during each mobility epoch were normally distributed, and the direction was uniformly distributed over $(0, 2\pi)$. A node activation rate of 250 nodes/h was used. The mean time to node deactivation was 1 h. Nodes were initially randomly activated within a bounded region of 5×5 km. Nodes that moved beyond this boundary were no longer considered to be part of the *ad hoc* network and were effectively deactivated. (c,t) path availability was evaluated using Dijkstra's algorithm.

For each simulation run, data was collected by sampling the network status once per second over an observation interval of 1 h. The first 2 h of each run were discarded to eliminate transient effects, and each simulation was rerun 10 times with new random seeds.

Simulation results for cluster size and cluster survival times are given in Figures 13.40 and and 13.41. Finally logical relationships among MANET network-layer entities is given in Figure 13.42.

13.6 CASHING SCHEMES FOR ROUTING

A large class of routing protocols for MANETs, namely reactive protocols, employ some form of caching to reduce the number of route discoveries. The simplest form of caching is based on timeouts associated with cache entries. When an entry is cached, a timer starts. When the timeout elapses, the entry is removed from the cache. Each time the entry is used, the timer restarts. Therefore, the effectiveness of such a scheme depends on the timeout

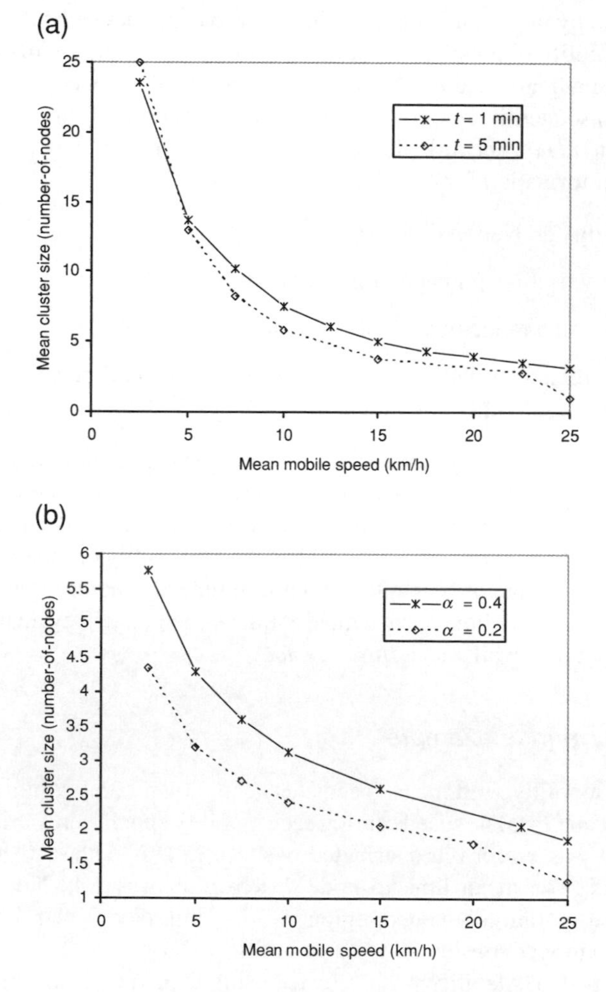

Figure 13.41 Simulation results: (a) cluster size ($R = 1000$ m); (b) cluster size ($R = 500$ m); (c) cluster survival ($R = 1000$ m); and (d) cluster survival ($R = 500$ m). (Reproduced by permission of IEEE [74].)

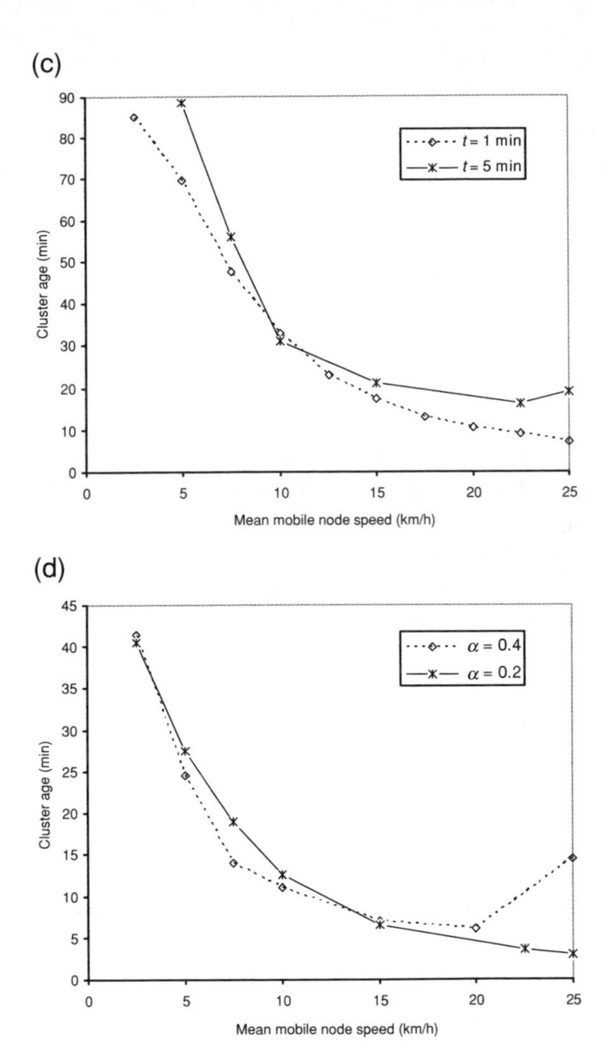

Figure 13.41 *Continued.*

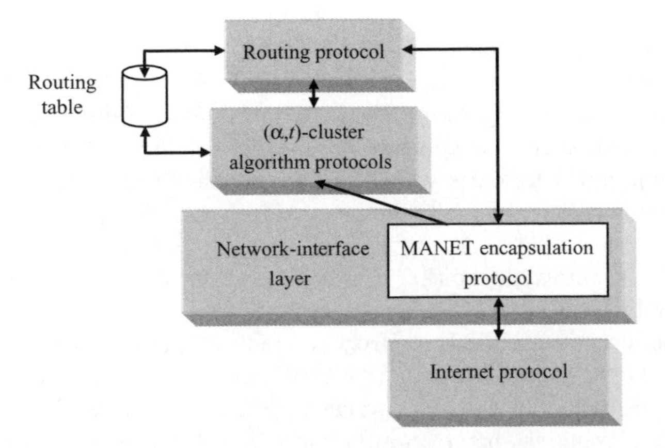

Figure 13.42 Logical relationships among MANET network-layer entities.

value associated with a cached route. If the timeout is well-tuned, the protocol performance increases; otherwise, a severe degradation arises as entries are removed either prematurely or too late from the cache.

13.6.1 Cache management

A cache scheme is characterized by the following set of design choices that specify cache management in terms of space (cache structure) and time (i.e. when to read/add/ delete a cache entry): store policy, read policy, writing policy and deletion policy.

The *store policy* determines the structure of the route cache. Recently, two different cache structures were studied [81], namely link cache and path cache, and applied to DSR. In a link cache structure, each individual link in the routes returned in RREP packets is added to a unified graph data structure, managed at each node, that reflects the node's current view of the network topology. In so doing, new paths can be calculated by merging route information gained from different packets. In the path cache, however, each node stores a set of complete paths starting from itself. The implementation of the latter structure is easier compared with the former, but it does not permit inference of new routes and exploitation of all topology information available at a node.

The *reading policy* determines rules of using a cache entry. Besides the straightforward use from the source node when sending a new message, several other strategies are possible. For example, DSR defines the following policies:

- *cache reply* – an intermediate node can reply to a route request with information stored in its own cache;
- *salvaging* – an intermediate node can use a path from its own cache when a data packet meets a broken link on its source route;
- *gratuitous reply* – a node runs the interface in the promiscuous mode and it listens for packets not directed to itself. If the node has a better route to the destination node of a packet, it sends a gratuitous reply to the source node with this new better route.

The *writing policy* determines when and which information has to be cached. Owing to the broadcast nature of radio transmissions, it is quite easy for a node to learn about new paths by running its radio interface in the promiscuous mode. The main problem for the writing policy is indeed to cache valid paths. Negative caches are a technique proposed in Johnson and Maltz [82] and adapted in Marina and Das [83] to filter the writing of cache entries in DSR out. A node stores negative caches for broken links seen either via the route error control packets or link layer for a period of time of δt s. Within this time interval, the writing of a new route cache that contains a cached broken link is disabled.

The *deletion policy* determines which information has to be removed from the cache and when. Deletion policy is actually the most critical part of a cache scheme. Two kinds of 'errors' can occur, owing to an imprecise erasure: (1) *early deletion*, a cached route is removed when it is still valid; and (2) *late deletion*, a cached route is not removed even if it is no longer valid.

The visible effect of these kinds of errors is a reduction in the packet delivery fraction and an increase in the routing overhead (the total number of overhead packets) [84]. Late deletions create the potential risk of an avalanche effect, especially at high load. If a node replies with a stale route, the incorrect information may be cached by other nodes and, in turn, used as a reply to a discovery. Thus, cache 'pollution' can propagate fairly quickly [83].

Caching schemes in DSR

All such schemes rely on a local timer-based deletion policy [81, 84]. The only exception has been proposed in Marina and Das [83]. They introduce a reactive caching deletion policy, namely, the wider error notification, that propagates route errors to all the nodes, forcing each node to delete stale entries from its cache.

Simulation results reported in References [81, 83] show that performance of a timer-based caching deletion policy is highly affected by the choice of the timeout associated with each entry. In the path cache, for a value of timeout lower than the optimal one (i.e. early deletion), the packet delivery fraction and routing overhead are worse than caching schemes that do not use any timeout. In the link cache, late deletion errors increase the routing overhead while the packet delivery fraction falls sharply.

The cache timeout can obviously be tuned dynamically. However, adaptive timer-based deletion policies have their own drawbacks. This policy suffers from late or early deletion errors during the transition time from one optimal timeout value to the successive one. So, the more the network and the data load are variable, the worse the performance will be.

To reduce the effect of such imprecise deletions, the adaptive timer-based cache scheme has been combined with the wide error notification deletion technique and studied for DSR in Perkins and Royer [17]. According to such a combined scheme, routes that become stale before their timeout expiration are removed reactively from all the sources using that route. In this combined technique, however, two more points remain unresolved: (1) Owing to the reactive nature of the deletions, if a cache entry is not used, it remains in the cache, even if no longer valid, thus it can be used as a reply to a path discovery and (2) the effect of early deletions cannot be avoided.

Caching schemes in ZRP

The caching zone with radius k^* for a cache leader n is defined as the set of nodes at a distance of at most k^* hops from n. An active path is created as a result of the discover phase and it is composed of a set of nodes, referred to as active nodes, forming a path from a source node S to a destination node D. Cache leader nodes are a subset of the active nodes.

The key consideration is to avoid the possibility that nodes can cache route information autonomously. Therefore, a cache leader n is the only node that is authorized to advertise route information inside its caching zone which is written into caches. On receiving the advertising message, a node proactively maintains a path to n so that it can be used as the next-hop node to any of the advertised routes. A cache leader is responsible for the validity of the advertised routes. Thus, it monitors such routes and forces each node in its caching zone to remove a route as soon as it becomes stale, so the deletion policy is proactive. Let us note that, if we consider $k^* = k$ and each node of a ZRP interzone path as a cache leader, we get the same underlying zone structure of ZRP (this implies that each active node is a cache leader). However, more generally, a cache leader can decide to advertise paths only to those nodes located at a distance $k^* < k$, and not all active nodes need to be cache leaders.

Implementation of C-ZRP

For simplicity, the implementation assumes:

(1) $k = k^*$.

(2) All active nodes act as cache leader nodes and vice versa.

(3) Only paths to active nodes are advertised as external routes.

(4) Caches are managed using explicit injection/deletion messages.

(5) To stop redundant query threads, LT (loop back termination), QD2 (query detection) and ET (early termination) redundant filtering rules are used which have been described earlier in this chapter.

When a node S, in Figure 13.43, executes a route request for a node D, an interzone path from S to D is identified. A node B_i belonging to an interzone path is an active node for the caching scheme. In Figure 13.43 an interzone path between S and D is formed by nodes b, e, p and t. Thus, an interzone path is also an active path. An interzone path is stored according to a distributed next-hop fashion, where the next-hop node is an active node. B_i stores B_{i+1} as the next-hop active node for all the downstream nodes from B_{i+2} to B_{M+1} and B_{i-1} as the next-hop active node for all the upstream nodes from B_0 to B_{i-2}. These two

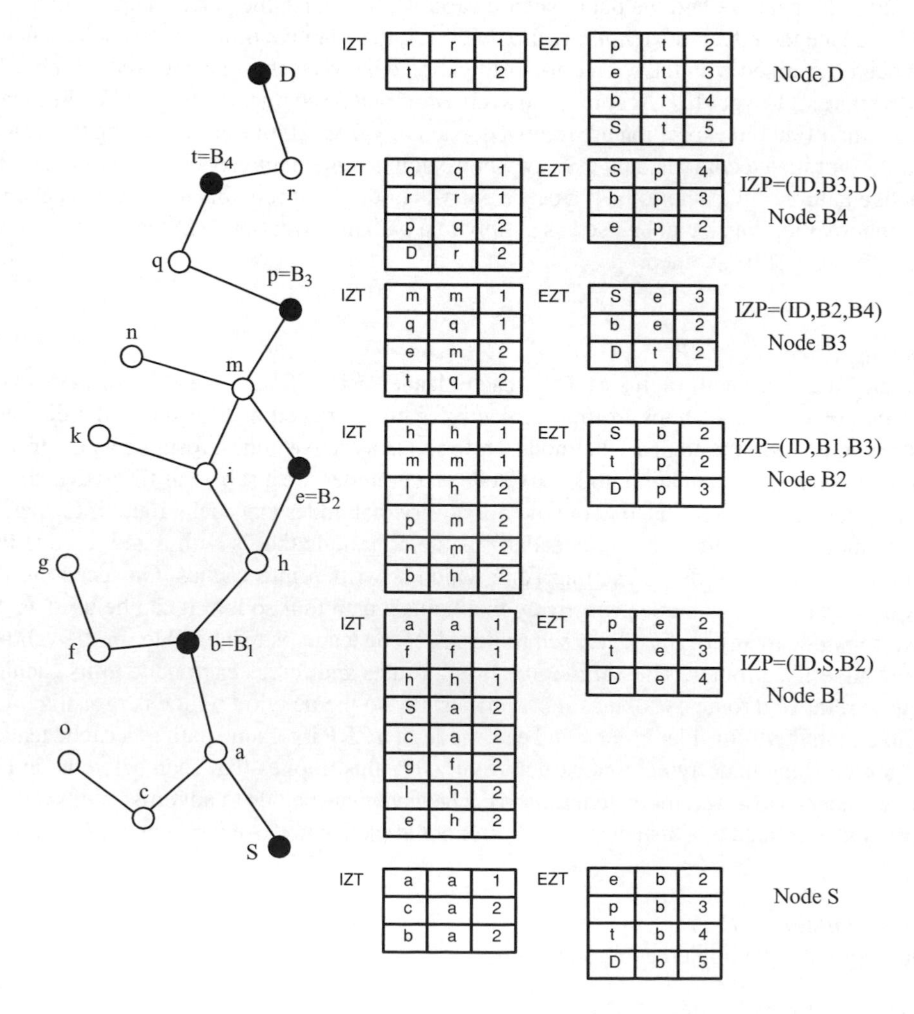

Figure 13.43 An example of values of data structures used in C-ZRP, $k = 2$.

active nodes will be referred to as companion nodes (as an example, the companion nodes of node b, with respect to the interzone path from S to D are S and e).

All routing information concerning nodes belonging to an interzone path is advertised inside the caching zone of each member of the path, which thus acts as cache leader for that information. Such routes are then maintained proactively by the IARP. If a new node joins B_i's zone, it acquires, by means of the IARP, all previously advertised routing information by B_i. Since a node may belong to more than one overlapping zone, it can acquire more than a single path to the same destination.

When a node, say B_{i+1}, leaves B_i's routing zone, not all the routing information gathered during the route request/reply is lost. Roughly speaking, two active paths from S to B_{i-1} and from B_{i+1} to D are still up. Hence, all the routing information concerning these subpaths is still valid. However, nodes $B_0, \ldots, B_{i-1}$ ($B_{i-1}, \ldots, B_{M+1}$) notify the nodes inside their own zones, using a delete control message, that the destinations $B_{i-1}, \ldots, B_{M+1}$ ($B_0, \ldots, B_i$) are no longer reachable.

Data structures

Each node X uses the following local data structures:

- *Internal zone routing table (IZT)* – an entry of IZT is a triple (d, n, #*h*), where d is the destination node, n is the next-hop node (located in X's transmission range), and #*h* is the path cost in number of hops.
- *External zone routing table (EZT)* – a row of EZT is a triple (d, n, #*z*), where d is the destination node, n is the next-hop active node (n belongs to X's routing zone and is not restricted to be in its transmission range), and #*z* is the cost of the path from X to d, given as the number of active nodes that have to be traversed. For example, in Figure 13.43, node b sets node e as the next-hop active node for p with cost two (nodes e and p).
- *Interzone path table (IZP)* – an interzone path corresponds to an entry in X's IZP table provided that X is an active node and ($X \neq S,D$). In this case, let the path id be ID and $X = B_i$. The entry is the triple (ID, B_{i-1}, B_{i+1}).
- *Reachable nodes (RN) list* – this is a sequence of pairs (d,#*z*), where d is an active node belonging to an interzone path and #*z* is the cost of the path from X expressed as number of active nodes that must be traversed to reach d. A node X advertises RN to nodes belonging to Z_k(X). RN includes the projection of EZT along the first and third components. For example, node *b* of Figure 13.43 will include the pairs (p, 2), (t, 3), and (D, 4) in RN.
- *Unreachable nodes (UN) set* – this set of nodes is used to advertise destinations that become unreachable.

Interzone path creation

A single interzone path from S to D is created during a route request/reply cycle by allowing only the destination D to send a single reply for a given request. The path is tagged with a unique identifier ID, for example, obtained by using increasing sequence numbers generated by the requesting node. When S triggers a new route discovery for a node D, it bordercasts a query message to all its border nodes. The message contains the identifier ID and a route accumulation vector AV[ ], initialized with AV[0] = S. Let M be the number of active nodes (not including S and D).

(1) When a border node X $\neq$ D receives a query message, if the message is received for the first time and the redundant query filter rules are satisfied:
  (a) It adds its own identification into the accumulation vector. As an example, if the node X corresponds to node B_j in the interzone path, then AV $[j] = X$.
  (b) If D belongs to X's routing zone, then the latter unicasts the query message to D. Otherwise, it executes a bordercast.

(2) When the destination node D receives a query message with identifier ID for the first time:
  (a) It stores the tuples (AV[i], AV[M], $M + 1 - i$, for $0 \leq i \leq M - 1$ in EZT.
  (b) It prepares the list RN = (AV[i], $M + 1 - i$)], for $0 \leq i \leq M$.
  (c) It sets AV[$M + 1$] = D.
  (d) It sends a reply message to AV[M]. The message contains the AV vector accumulated in the query message.

  An example of path creation is given in Figure 13.44(a).

(3) When a border node B_j receives a reply message:
  (a) If $B_j \neq$ S, then it stores the triple (ID,AV[$j - 1$], AV [$j + 1$]) in the IZP table, thus becoming an active node.
  (b) It stores the following tuples in EZT: (AV [i], AV[$j - 1$], $j - i$), for $0 \leq i \leq j - 2$; (AV [i], AV[$j + 1$], $j - i$), for $j + 2 \leq i \leq M + 1$.

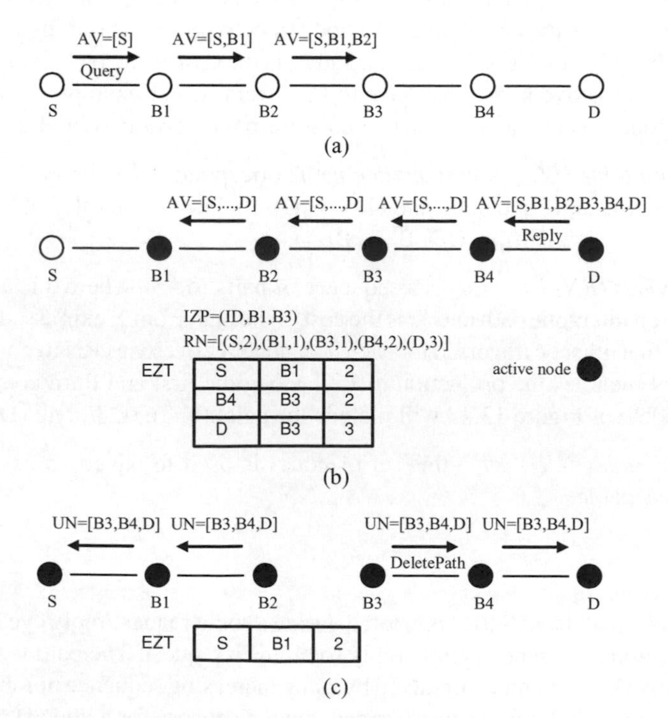

Figure 13.44 An example of interzone path creation and deletion.

(c) It prepares RN = [(AV[$j+i$], $|i|$)] for $-j \leq i \leq M+1$.
(d) If $B_j \neq S$, then it forwards the reply message to the node AV [$j-1$].

Figure 13.43(b) shows the state at node B2 after the reception of the reply message with AV [S,B1,B2,B3,B4,D] that caused the execution of the following actions:

(1) B_2 becomes a member of an interzone path [it stores the triple (ID,B_1,B_3) in IZP].

(2) B_2 adds the entries (S,B_1,2), (B_4,B_3, 2), (D,B_3, 3) in EZT.

(3) B_2 prepares the list of reachable nodes, RN = [(S,2),(B1,1),(B3,1),(B4, 2),(D,3)].

(4) B_2 forwards the reply to B1.

Interzone path deletion
An interzone path is broken at node B_j when B_{j-1} (or B_{j+1}) is no longer in B_j's routing zone. In this case, the path is divided in two subpaths and the source node is notified with an error message. An active node B_j executes the following actions (in the following notation '–' means any):

(1) Deletes the entry (–,B_{j-1},–) or (–,B_{j+1},–) from EZT.

(2) Checks for the companion node B_{j+1} or B_{j-1} in the IZP table.

(3) If the companion node is found, then it prepares the following list of unreachable nodes: N = [B_0,B_1,...,B_{j-1}] (UN = [B_{j+1},B_{j+2},...,B_{M+1}]); and sends a Delete_Path message, containing UN and the path identifier ID, to the companion node.

(4) Deletes the entry (ID,B_{j-1},B_{j+1}) from IZP after the successful transmission of the message.

When an active path is broken, the source node either receives the Delete_Path message from B_1 [if the link is broken between (B_j,B_{j+1}), with $j > 0$], or is able to detect the break autonomously via IARP. The source node thus triggers a new route discovery if required to send other packets, while the two subpaths (B_0,B_1,...,B_{j-1} and B_{j+1},B_{j+2},...,B_{M+1}) remain active. Figure 13.44(c) shows the case when the 'link' between B_2 and B_3 is broken (i.e. their distance becomes higher than k). Two interzone subpaths, (S,B_1,B_2) and (B_3,B_4,D), are generated. In the figure, B_2's EZT data structure is also shown.

When an active node receives a Delete_Path message from one of its companion nodes X, it deletes the entries stored in the UN list from EZT and forwards the message to the other companion node. If the receiving node has some another route to a node stored in UN, then it does not include such a node when forwarding UN.

Cache management
In order to allow all the nodes of B_j's routing zone to use the acquired information, B_j broadcasts RN inside its zone. Such a message is referred to as *the inject message*. On receiving an inject message carrying the reachable node list RN from a node $X = B_j$, a node Y creates a set of entries (RN[i].d,X,RN[i].#z) into its own EZT, $0 \leq i \leq$ |RN|, where RN[i].d is the first component (destination node) of the ith pair of RN, RN[i],#z, the second component (i.e. the length), and |RN| is the number of elements of RN. Figure 13.44(a)

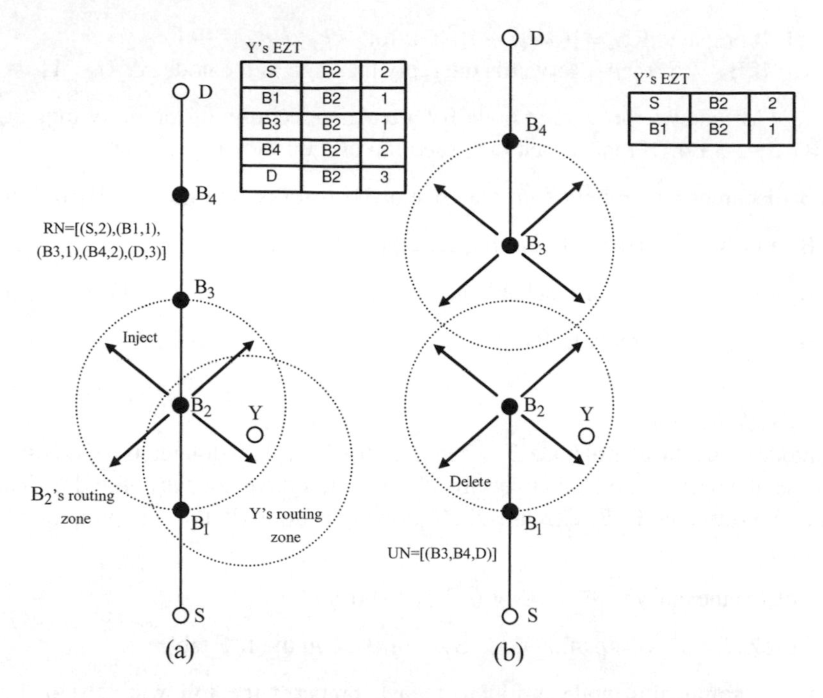

Figure 13.45 An example of (a) injection and (b) deletion of external nodes.

shows node B_2 injecting the external routes to nodes S,B_1,B_3,B_4,D into its zone. Note that Y now has two routes to node B_1 since such a node is in Y' routing zone.

Deleting external routes

When a node B_j either detects a path breakage or receives a Delete_Path message, it broadcasts a Delete message into its zone containing the list of unreachable nodes UN. When an internal node receives a Delete message it deletes all the matching entries from EZT. Figure 13.45(b) shows the delete mechanism on node Y .

13.7 DISTRIBUTED QoS ROUTING

In this section a distributed QoS routing scheme for *ad hoc* networks is discussed. Two routing problems are presented, *delay constrained least cost routing* (DCLC) and *bandwidth constrained least cost routing* (BCLC). As before, the path that satisfies the delay (or bandwidth) constraint is called a feasible path. The algorithms can tolerate the imprecision of the available state information. Good routing performance in terms of success ratio, message overhead and average path cost is achieved even when the degree of information imprecision is high. Note that the problem of information imprecision exists only for QoS routing; all best-effort routing algorithms, such as DSR and ABR, do not consider this problem because they do not need the QoS state in the first place. Multipath parallel routing is used to increase the probability of finding a feasible path. In contrast to the flooding-based path discovery algorithms, these algorithms search only a small number of paths, which limits the routing overhead. In order to maximize the chance of finding a feasible path, the state information at the intermediate nodes is collectively utilized to make intelligent hop-by-hop path selection. The logic behind this is very much equivalent to using the Viterbi instead of ML (maximum likelihood) algorithm in trellis-based demodulation processes.

The algorithms consider not only the QoS requirements, but also the optimality of the routing path. Low-cost paths are given preference in order to improve the overall network performance. In order to reduce the level of QoS disruption, fault-tolerance techniques are brought in for the maintenance of the established paths. Different levels of redundancy provide tradeoff between the reliability and the overhead. The dynamic path repairing algorithm repairs the path at the breaking point shifts the traffic to a neighbor node, and reconfigures the path around the breaking point without rerouting the connection along a completely new path. Rerouting is needed in two cases. One case is when the primary path and all secondary paths are broken. The other case is when the cost of the path grows large and hence it becomes beneficial to route the traffic to another path with a lower cost.

13.7.1 Wireless links reliability

One element of the cost function will be reliability of the wireless links. The links between the stationary or slowly moving nodes are likely to exist continuously. Such links are called *stationary links*. The links between the fast moving nodes are likely to exist only for a short period of time. Such links are called *transient links*. A routing path should use stationary links whenever possible in order to reduce the probability of a path breaking when the network topology changes. A *stationary neighbor* is connected to a node with a stationary link. As in Chapter 7, delay of path P between two nodes equals the sum of the link delays on the path between the two nodes and will be denoted as *delay(P)*. Similarly *bandwidth(P)* equals the minimum link bandwidth on the path P, and *cost(P)* equals the sum of the link costs.

13.7.2 Routing

Given a source node s, a destination node t, and a delay requirement D, the problem of delay-constrained routing is to find a feasible path P from s to t such that $delay(P) \leq D$. When there are multiple feasible paths, we want to select the one with the least cost. Another problem is bandwidth-constrained routing, i.e. finding a path P such that $bandwidth(P) \geq B$, where B is the bandwidth requirement. When there are multiple such paths, the one with the least cost is selected. Finding a feasible path is actually the first part of the problem. The second part is to maintain the path when the network topology changes.

13.7.3 Routing information

The following end-to-end state information is required to be maintained at every node i for every possible destination t. The information is updated periodically by a distance-vector protocol discussed in Section 13.1:

(1) Delay variation – $\Delta D_i(t)$ keeps the estimated maximum change of $D_i(t)$ before the next update. That is, based on the recent state history, the actual minimum end-to-end delay from i to t is expected to be between $D_i(t) - \Delta D_i(t)$ and $D_i(t) + \Delta D_i(t)$ in the next update period.

(2) Bandwidth variation – $\Delta B_i(t)$ keeps the estimated maximum change of $B_i(t)$ before the next update. The actual maximum bandwidth from i to t is expected to be between $B_i(t) - \Delta B_i(t)$ and $B_i(t) + \Delta B_i(t)$ in the next period.

(3) The cost metric $C_i(t)$ is used for optimization, in contrast to the delay and bandwidth metrics used in QoS constraints.

Consider an arbitrary update of $\Delta D_i(t)$ and $D_i(t)$. Let $\Delta D_i(t)$ and $\Delta D_i^{'}(t)$ be the values of $\Delta D_i(t)$ before and after the update, respectively. Similarly, let $D_i(t)$ and $D_i^{'}(t)$ be the values of $D_i(t)$ before and after the update, respectively. $D_i^{'}(t)$ is provided by a distance-vector protocol. $\Delta D_i^{'}(t)$ is calculated as follows:

$$\Delta D_i^{'}(t) = \alpha \Delta D_i(t) + (1-\alpha)\left|D_i^{'}(t) - D_i(t)\right| \tag{13.15}$$

The factor $\alpha(<1)$ determines how fast the history information $\Delta D_i(t)$ is forgotten, and $1-\alpha$ determines how fast $\Delta D_i^{'}(t)$ converges to $|D_i^{'}(t) - D_i(t)|$. By the previous formula, it is still possible for the actual delay to be out of the range $[D_i(t) - \Delta D_i(t), D_i(t) + \Delta D_i(t)]$.

One way to make such probability sufficiently small is to enlarge $\Delta D_i(t)$. Hence, we shall modify the formula and introduce another factor $\beta(>1)$:

$$\Delta D_i^{'}(t) = \alpha \Delta D_i(t) + (1-\alpha)\beta\left|D_i^{'}(t) - D_i(t)\right| \tag{13.16}$$

$\Delta D_i^{'}(t)$converges to $\beta\left|D_i^{'}(t) - D_i(t)\right|$at a speed determined by $1-\alpha$.

13.7.4 Token-based routing

There are numerous paths from s to t. We shall not randomly pick several paths to search. Instead, we want to make an intelligent hop-by-hop path selection to guide the search along the best candidate paths. This is what, for example, the Viterbi algorithm would be doing in trying to avoid search through all possible trajectories of a trellis (ML approach) in a signal demodulation/decoding process.

The basic idea of token-based probing is outlined below. A token is the permission to search one path. The source node issues a number of tokens based on the available state information. One guideline is that more tokens are issued for the connections with tighter requirements. Probes (routing messages) are sent from the source toward the destination to search for a low-cost path that satisfies the QoS requirement. Each probe is required to carry at least one token. At an intermediate node, a probe with more than one token is allowed to be split into multiple ones, each searching a different downstream subpath. The maximum number of probes at any time is bounded by the total number of tokens. Since each probe searches a path, the maximum number of paths searched is also bounded by the number of tokens. See Figure 13.46 for an example. Upon receipt of a probe, an intermediate node decides, based on its state: (1) whether the received probe should be split; and (2) to which neighbor nodes the probe(s) should be forwarded. The goal is to collectively utilize the state information at the intermediate nodes to guide the limited tickets (the probes carrying them) along the best paths to the destination, so that the probability of finding a low-cost feasible path is maximized.

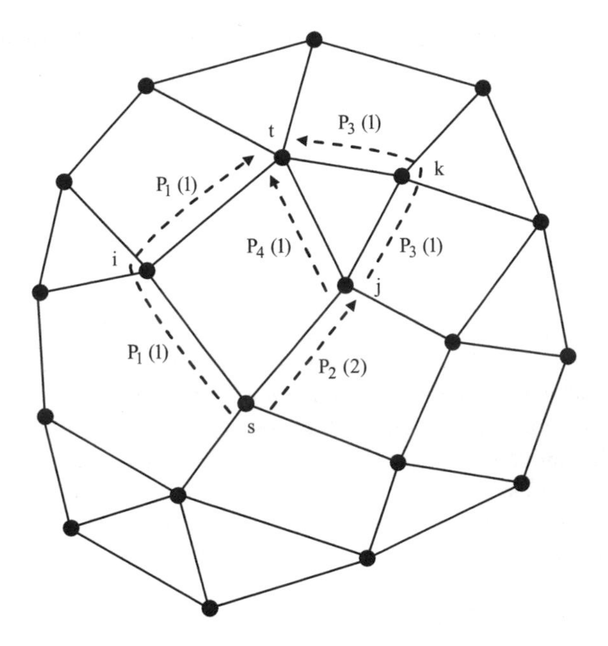

Figure 13.46 Generation of probes p (number of tokens).

13.7.5 Delay-constrained routing

When a connection request arrives at the source node, a certain number N_0 of tokens are generated, and probes are sent toward the destination t. Each probe carries one or more tokens. Since no new tokens are allowed to be created by the intermediate nodes, the total number of tokens is always N_0, and the number of probes is at most N_0 at any time. When a node receives a probe p with $N(p)$ tokens, it makes at most $N(p)$ copies of p, distributes the received tickets among the new probes, and then forwards them along to selected outgoing links toward t. Each probe accumulates the delay of the path it has traversed so far. A probe can proceed only when the accumulated delay does not violate the delay requirement. Hence, any probe arriving at the destination detects a feasible path, which is the one it has traversed.

There are two basic guidelines for how to determine N_0 and how to distribute the tokens in a received probe among the new probes:

(1) Different numbers of tokens are assigned to different connections based on their 'needs'. For a connection whose delay requirement is large and can be easily satisfied, one token is issued to search a single path; for a connection whose delay requirement is smaller, more tokens are issued to increase the chance of finding a feasible path; for a connection whose delay requirement is too small to be satisfied, no tokens are issued, and the connection is immediately rejected.

(2) When a node i forwards the received tokens to its neighbors, the tokens are distributed unevenly among the neighbors, depending on their chances of leading to reliable low-cost feasible paths. A neighbor having a smaller end-to-end delay to the destination should receive more tickets than a neighbor having a larger delay; a neighbor

having a smaller end-to-end cost to the destination should receive more tokens than a neighbor having a larger cost; a neighbor having a stationary link to should be given preference over a neighbor having a transient link to i. Note that some neighbors may not receive any tokens because i may have only a few or just one token to forward.

13.7.6 Tokens

The two types of tokens, constraint (limitation) tokens (CT) and optimization tokens (OT) have different purposes. CT tokens prefer paths with smaller delays, so that the chance of satisfying a given delay (constraints) requirement is higher. OT tokens prefer the paths with smaller costs. The overall strategy is to use the more aggressive OT tokens to find a low-cost feasible path with relatively low success probability and to use the CT tokens as a backup to guarantee a high success probability of finding a feasible path. The number of CT tokens L, and OT tokens O is determined based on the delay requirement D. If D is very large and can definitely be satisfied, a single CT token will be sufficient to find a feasible path. If D is too small to be possibly satisfied, no CT token is necessary, and the connection is rejected. Otherwise, more than one CT token is issued to search multiple paths for a feasible one. Based on the previous guideline, a linear token curve shown in Figure 13.47(a) was chosen in Chen and Nahrstedt [85] for simplicity and efficient computation.

Parameter Φ is a system parameter specifying the maximum allowable number of CT. It shows that more CT are assigned for smaller D. Number of optimization tokens O is also determined based on the delay requirement D as shown in Figure 13.47(b).

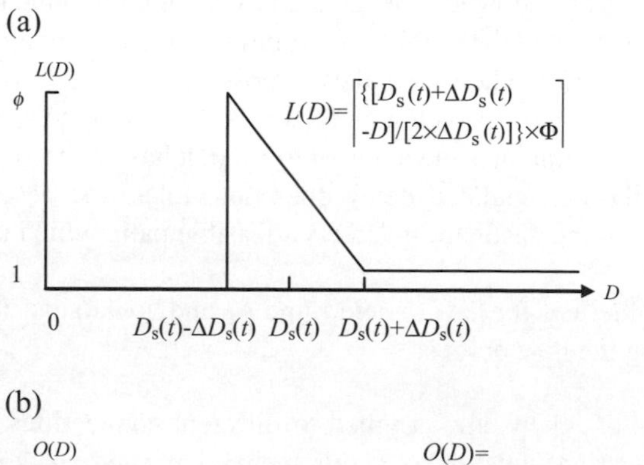

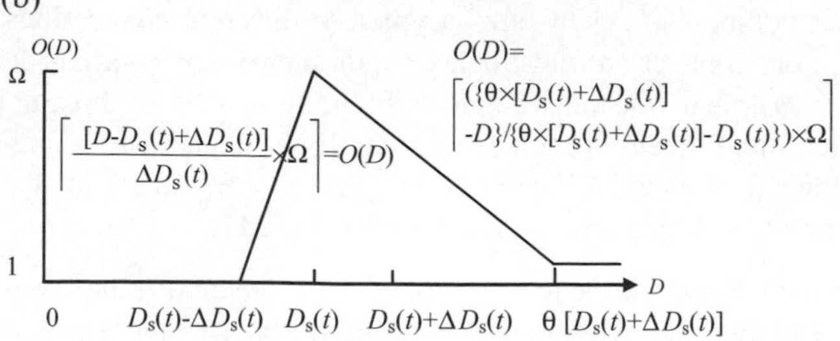

Figure 13.47 Token curves.

13.7.7 Forwarding the received tokens

13.7.7.1 Candidate neighbors

If $L + O = 0$, the connection request is rejected. Otherwise, probes carrying the tokens are sent from s to t. A probe proceeds only when the path has a delay of no more than D. Hence, once a probe reaches t, it detects a delay-constrained path. Each probe accumulates the delay of the path it has traversed so far. A data field, denoted as *delay(p)*, is defined in a probe p. Initially, *delay(p)*$\Rightarrow$ 0; whenever p proceeds for another *link (i,j)*, *delay(p)* $\Rightarrow$ *delay(p)* + *delay(i,j)*. Suppose a node i receives a probe p with $L(p)$ constraint tokens and $O(p)$ optimization tokens. Suppose k is the sender of the probe p. The set $R_i^p(t)$ of candidate neighbors, to which i will forward the received tokens, is determined as follows. We first consider only the stationary neighbors (V_i^s) of i. Let

$$R_i^p(t) = \left\{ j \,\middle|\, delay(p) + delay(i, j) + D_j(t) - \Delta D_j(t) \leq D,\ j \in V_i^s - \{k\} \right\}$$

$R_i^p(t)$is the set of neighbors to which the tickets should be forwarded. If $R_i^p(t) = \{0\}$, we take the transient neighbors into consideration and redefine $R_i^p(t)$ to be

$$R_i^p(t) = \left\{ j \,\middle|\, delay(p) + delay(i, j) + D_j(t) - \Delta D_j(t) \leq D,\ j \in V_i - \{k\} \right\}$$

If we still have $R_i^p(t) = \{0\}$, all received tokens are invalidated and discarded. If $R_i^p(t) \neq \{0\}$, then for every $j \in R_i^p(t)$, i makes a copy of p, denoted as p_j. Let p_j have $L(p_j)$ constraint tokens and $O(p_j)$ optimization tokens. These parameters are calculated as

$$L(p_j) = \frac{\left[delay(i, j) + D_j(t)\right]^{-1}}{\sum\limits_{j' \in R_i^p(t)} \left[delay(i, j') + D_{j'}(t)\right]^{-1}} \times L(p)$$

$$O(p_j) = \frac{\left[\cos t(i, j) + C_j(t)\right]^{-1}}{\sum\limits_{j' \in R_i^p(t)} \left[\cos t(i, j') + C_{j'}(t)\right]^{-1}} \times O(p) \tag{13.17}$$

These numbers will be rounded to the closest integer.

The *data structure* carried by a probe p is shown in Table 13.3. The last six fields, *k, path, L(p), O(p), delay(p)* and *cost(p)*, are modified as the probe traverses. Tokens are logical entities, and only the number of tokens is important: there can be at most $L(p) + O(p)$ new probes descending from p, among which probes with constraint tokens choose paths based on delay, probes with optimization tokens choose paths based on cost, and probes with both types of tokens choose paths based on both delay and cost.

13.7.8 Bandwidth-constrained routing

The algorithm shares the same computational structure with the delay-constrained routing algorithm. The differences are the metric-dependent token curves and token distribution formulas. The token curves are given in Figure 13.48.

Table 13.3

id	System-wide unique identification for the connection request
s	Source node
t	Destination node
D	Delay requirement
$L + O$	Total number of tokens
k	Sender of *p*
path	Path *p* has traversed so far
$L(p)$	Number of constrained tokens carried by *p*
$O(p)$	Number of optimization tokens carried by *p*
delay(p)	Accumulated delay of the path traversed so far
cost(p)	Accumulated cost of the path traversed so far

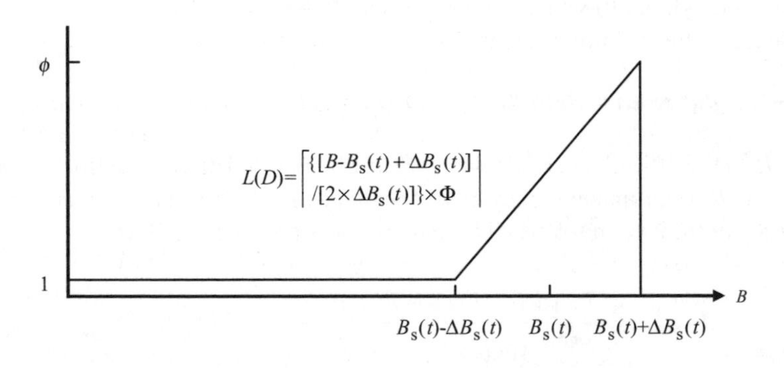

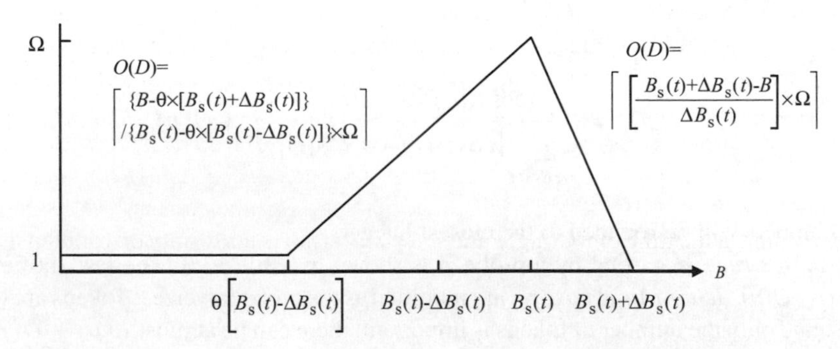

Figure 13.48 Token curves as functions of B.

13.7.9 Forwarding the received tickets

Suppose a node i receives a probe p with $L(p)$ constraint tokens and $O(p)$ optimization tokens. Suppose k is the sender of the probe p. The set $R_i^p(t)$ of candidate neighbours, to which i will forward the received tokens, is determined as follows. Define

$$R_i^p(t) = \left\{ j \,\middle|\, bandwidth(i, j) \geq B \wedge B_j(t) + \Delta B_j(t) \geq B,\ j \in V_i^s - \{k\} \right\}$$

If $R_i^p(t) = \{0\}$, we take the transient neighbours into consideration and redefine $R_i^p(t)$ to be

$$R_i^p(t) = \left\{ j \mid bandwidth(i, j) \geq B \wedge B_j(t) + \Delta B_j(t) \geq B,\ j \in V_i - \{k\} \right\}$$

If we still have $R_i^p(t) = \{0\}$ all received tokens are invalidated and discarded. If $R_i^p(t) \neq \{0\}$, then for every $j \in R_i^p(t)$, i makes a copy of p, denoted as p_j. Let p_j have $L(p_j)$ constraint tokens and $O(p_j)$ optimization tokens. $L(p_j)$ is determined based on the observation that a probe sent toward the direction with a larger residual bandwidth should have more L tokens. These parameters are now calculated as

$$L(p_j) = \frac{\min\left[bandwidth(i, j), B_j(t)\right]}{\sum\limits_{j' \in R_i^p(t)} \min\left[bandwidth(i, j'), B_{j'}(t)\right]} \times L(p)$$

$$O(p_j) = \frac{\left[\cos t(i, j) + C_j(t)\right]^{-1}}{\sum\limits_{j' \in R_i^p(t)} \left[\cos t(i, j') + C_{j'}(t)\right]^{-1}} \times O(p) \tag{13.18}$$

13.7.10 Performance example

Three performance metrics: (1) success ratio = number of connections accepted/total number of connection requests; (2) avg msg overhead = total number of messages sent/total number of connection requests; and (3) avg path cost = total cost of all established paths/number of established paths, are analysed by simulation in Cher and Nahrstedt [85].

Sending a probe over a link is counted as one message. Hence, for a probe that has traversed a path of l hops, l messages are counted. The network topology used in simulation is randomly generated. Forty nodes are placed randomly within a 15×15 m area. The transmission range of a node is bounded by a circle with a radius of 3 m. A link is added between two nodes that are in the transmission range of each other. The average degree of a node is 3.4.

The source node, the destination node, and parameter D of each connection request are randomly generated. D is uniformly distributed in the range of [30, 160 ms]. The cost of each link is uniformly distributed in [0, 200]. Each link (j, k) is associated with two delay values: delay-old denoted as *delay(j,k)* and delay-new denoted as *delay'(j,k); delay(j,k)* is the last delay value advertised by the link to the network. Parameter *delay'(j,k)* is the actual delay of the link at the time of routing. Parameter *delay (j,k)* is uniformly distributed in [0, 50 ms], while *delay'(j,k)* is uniformly distributed in $[(1 - \xi)delay(j, k), (1 + \xi)delay(j, k)]$, where ξ is a simulation parameter, called the imprecision rate, defined as

$$\xi = \sup\left\{\left[delay'(j, k) - delay(j, k)\right] / delay(j, k)\right\}$$

Three algorithms are simulated: the flooding algorithm, the token-based probing algorithm (TBP) and the shortest-path algorithm (SP). The flooding algorithm is equivalent to TBP with infinite number of constraint tokens and zero optimization tokens. It floods routing messages from the source to the destination. Each routing message accumulates the delay of the path it has traversed, and the message proceeds only if the accumulated delay does not exceed the delay bound. The system parameters of the TBP algorithm are $\Phi = 4$, $\theta = 1.5$, $\Omega = 3$. The values are obtained by extensive simulation runs. The SP algorithm maintains a state vector at each node I by a distance-vector protocol.

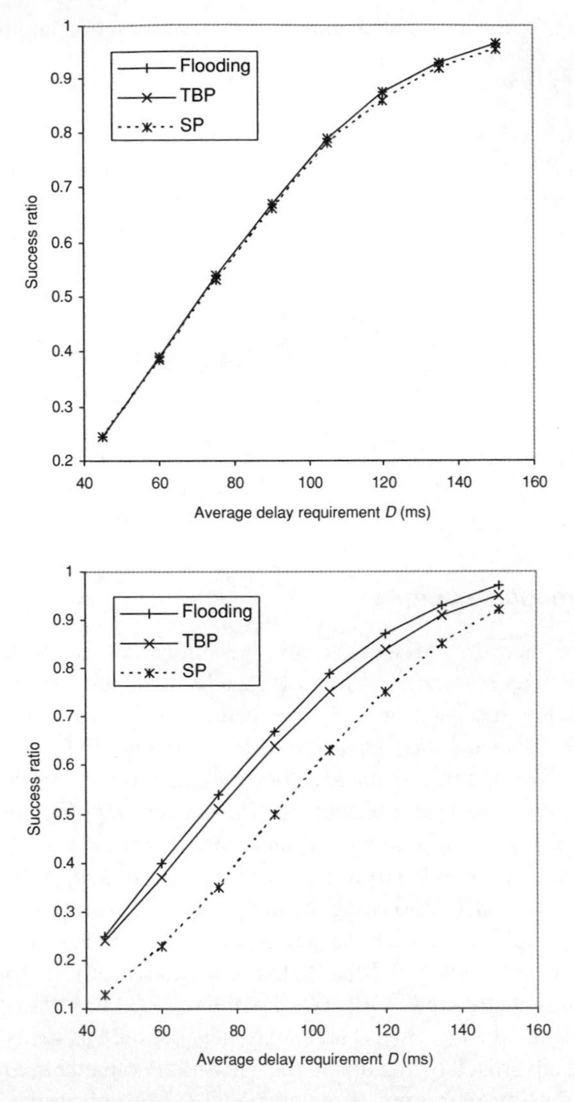

Figure 13.49 Success ratio [imprecision rate: (a) 5 % and (b) 50 %]. (Reproduced by permission of IEEE [85].)

Comparison of the performance results shown in Figures 13.49–13.52 demonstrates advantages of the TBP. The protocol presented in this section is a brief interpretation of Ticket Based Protocol represented in Chen and Nahrstedt [85] with slightly modified terminology adjusted to that of the rest of the book. The problem of QoS routing in wireline and wireless networks has been attracting much attention in both academia and industry. For more information see References [86–88]. A comprehensive overview of the literature is given in Chen and Nahrstedt [88]. The problem will be revisited again in Chapter 21 in more detail.

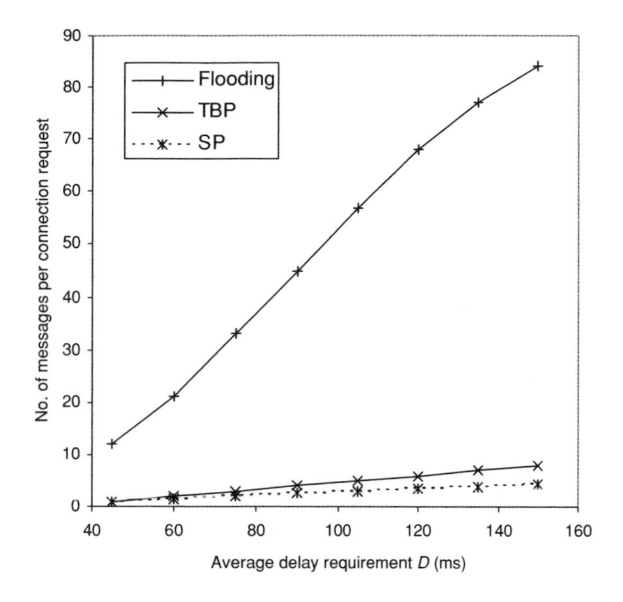

Figure 13.50 Messages overhead (imprecision rate: 10 %). (Reproduced by permission of IEEE [85].)

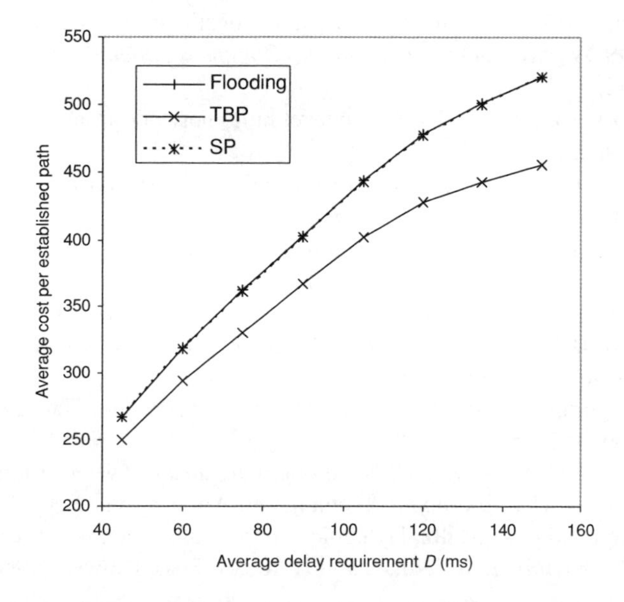

Figure 13.51 Cost per established path (imprecision rate: 5 %). (Reproduced by permission of IEEE [85].)

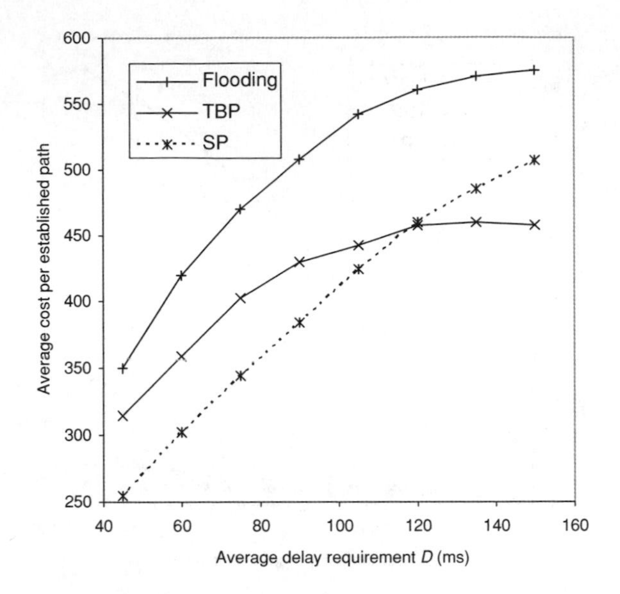

Figure 13.52 Cost per established path (imprecision rate: 50 %). (Reproduced by permission of IEEE [85].)

REFERENCES

[1] C.E. Perkins and P. Bhagwat, Highly dynamic destination-sequenced distance-vector routing (DSDV) for mobile computers, *Comput. Commun. Rev.*, October 1994, pp. 234–244.

[2] J. Jubin and J. Tornow, The DARPA packet radio network protocols, *Proc. IEEE*, vol. 75, no. 1, 1987, pp. 21–32.

[3] E. Royer and C.K. Toh, A review of current routing protocols for *ad hoc* mobile wireless networks, *IEEE Person. Commun.*, April 1999, pp. 46–54.

[4] C.-C. Chiang, Routing in clustered multihop, mobile wireless networks with fading channel, in *Proc. IEEE SICON* '97, April 1997, pp. 197–211.

[5] S. Murthy and I.I. Garcia-Luna-Aceves, An efficient routing protocol for wireless networks, *ACM Mobile Networks App.* J., (Special Issue on Routing in Mobile Communication Networks), October 1996, pp. 183–197.

[6] A.S. Tanenbaum, *Computer Networks*, 3rd edn, Ch. 5, Prentice Hall: Englewood Cliffs, NJ, 1996, pp. 357–358.

[7] C.E. Perkins and E.M. Royet, *Ad-hoc* on-demand distance vector routing, in *Proc 2nd IEEE Workshop. Mobile Comput*. Technol. and Applications, Feb. 1999, pp. 90–100.

[8] D.B. Johnson and O.A. Maltz, Dynamic source routing in *ad-hoc* wireless networks, in *Mobile Computing, L.* Irnielinski and H. Korth (Eds). Kluwer: Norwell, MA, 1996, pp. 153–181.

[9] J. Broch, O.B. Johnson, and D.A. Maltz, The dynamic source routing protocol for mobile *ad hoc* networks, IETF Internet draft, draft-ietf manet-dsr-01.txt, December 1998 (work in progress).

[10] V.D. Park and M.S. Corson, A highly adaptive distributed routing algorithm for mobile wireless networks, in *Proc. INFOCOM* '97, April 1997.

[11] M.S. Corson and A. Ephremides, A distributed routing algorithm for mobile wireless networks, *ACM/Baltzer Wireless Networks J.*, vol. 1, no. 1, 1995, pp. 61–81.

[12] C.-K. Toh, A novel distributed routing protocol to support *ad-hoc* mobile computing, in *Proc. 1996 IEEE 15th Annual Int. Phoenix Conf. Computing and Communications*, March 1996, pp. 480–486.

[13] R. Dube *et al.*, Signal stability based adaptive routing (SSA) for *ad-hoc* mobile networks, *IEEE Person. Commun.*, February 1997, pp. 36–45.

[14] C.-K. Toh, Associativity-based routing for *ad-hoc* mobile networks, *Wireless Person. Commun.*, vol. 4, no. 2, March 1997, pp. 1–36.

[15] S. Murthy and I.I. Garcia-Luna-Aceves, Loop-free internet routing using hierarchical routing trees, in *Proc. INFOCOM '97*, 7–11, April 1997.

[16] C-C. Chiang, M. Gerla and S. Zhang, Adaptive shared tree multicast in mobile wirelesa networks, in *Proc. GLOBECOM '98*, November 1998, pp. 1817–1822.

[17] C.E. Perkins and E.M. Royer, *Ad hoc* on demand distance vector (AODV) routing, IETF Internet draft, draft-ietf-manet-aodv-02.txt, November 1998 (work in progress),

[18] L. Ji and M.S. Corson, A lightweight adaptive multicast algorithm, in *Proc. GLOBECOM '98*, November 1998, pp. 1036–1042.

[19] C.-K. Toh and G. Sin, Implementing associativity-based routing for *ad hoc* mobile wireless networks, Unpublished article, March 1998.

[20] D. Baker *et al.*, Flat vs. hierarchical network control architecture, in *ARO/DARPA Workshop mobile ad-hoc Networking*; www.isr.umd. edu, Mar, 1997.

[21] M. Gerla, C-C. Chiang and L. Zhang, Tree multicast strategies in mobile, multihop wireless networks, *ACM/Baltzer Mobile Networks Applic.* J., 1998.

[22] S. Singh, T. Woo and C.S. Raghavendra, Power-aware routing in mobile *ad hoc* networks, in *proc. ACM/IEEE MO6'ICOM '98*, October 1998.

[23] Y. Ko and N.H. Vaidya, Location-aided routing (LAR) in mobile *ad hoc* networks, in *Proc. ACM/IEEE MCIBCOM '98*, October 1998.

[24] C.R. Sin and M. Gerla, MACNPR: an asynchronous multimedia multi- hop wireless network, in *Proc. IEEE INFOCOM '97*, March 1997.

[25] Z.J. Haas and M.R. Pearlman, The performance of a new routing protocol for the reconfigurable wireless networks, in *Proc. ICC'98*, pp. 156–160.

[26] Z.J. Haas and M.R. Pearlman, Evaluation of the *ad-hoc* connectivity with the reconfigurable wireless networks, in *Virginia Tech's Eighth Symp. Wireless Personal Communications*, 1998, pp. 156–160.

[27] Z.J. Haas and M.R. Pearlman, The performance of query control schemes for the zone routing protocol, in *Proc. SIGCOMM'98*, pp. 167–177.

[28] P. Jacquet, P. Muhlethaler and A. Qayyum, Optimized link state routing protocol, *IETF MANET,* Internet Draft, November 1998. http://www.ietf.cnri.reston.va.us

[29] D.B. Johnson and D.A. Maltz, Dynamic source routing in ad hoc wireless networking, in *Mobile Computing,* T. Imielinski and H. Korth (Eds). Kluwer: Norwell, MA, 1996.

[30] J. Moy, OSPF version 2, RFC 2178, March 1997.

[31] S. Murthy and J.J. Garcia-Luna-Aceves, A routing protocol for packet radio networks, in *Proc. ACM Mobile Computing and Networking Conf. (MOBICOM'95)*, pp. 86–94.

[32] An efficient routing protocol for wireless networks, *MONET*, vol. 1, 1986, pp. 183–197.

[33] V.D. Park and M.S. Corson, A highly adaptive distributed routing algorithm for mobile wireless networks, in *Proc. IEEE INFOCOM '97*, Kobe, pp. 1405–1413.

[34] C.E. Perkins and P. Bhagwat, Highly dynamic destination-sequenced distance-vector routing (DSDV) for mobile computers, in *Proc. ACM SIGCOMM,* vol. 24, no. 4, October 1994, pp. 234–244.

[35] C.E. Perkins and E.M. Royer, *Ad hoc* on-demand distance vector routing, in *Proc. IEEE WMCSA'99,* vol. 3, New Orleans, LA, 1999, pp. 90–100.

[36] J. Sharony, A mobile radio network architecture with dynamically changing topology using virtual subnets, *MONET*, vol. 1, 1997, pp. 75–86.

[37] P.F. Tsuchiya, The landmark hierarchy: a new hierarchy for routing in very large networks, *ACM Comput. Commun. Rev.*, vol. 18, 1988, pp. 35–42.

[38] M.R. Pearlman and Z.J. Haas, Determining the optimal configuration for the zone routing protocol, *IEEE J. Selected Areas Commun.*, vol. 17, no. 8, 1999, pp. 1395–1414.

[39] C.-C. Chiang, H.-K. Wu, W. Liu and M. Gerla, Routing in clustered multihop, mobile wireless networks, in *Proc. IEEE Singapore Int. Conf. Networks*, 1997, pp. 197–211.

[40] M. Gerla and J. Tsai, Multicluster, mobile, multimedia radio network, *ACM-Baltzer J. Wireless Networks*, vol. 1, no. 3, 1995, pp. 255–265.

[41] A. Iwata, C.-C. Chiang, G. Pei, M. Gerla and T.-W. Chen, Scalable routing strategies for *ad hoc* wireless networks, *IEEE J. Selected Areas Commun.* vol. 17, no. 8, 1999, pp. 1369–1379.

[42] L. Kleinrock and K. Stevens, Fisheye: a lenslike computer display transformation, Computer Science Department, University of California, Los Angeles, CA, Technical Report, 1971.

[43] T.-W. Chen and M. Gerla, Global state routing: a new routing schemes for *ad-hoc* wireless networks, in *Proc. IEEE ICC'98*, 7–11 June 1998, pp. 171–175.

[44] N.F. Maxemchuk, Dispersity routing, in *Proc. IEEE ICC '75*, June 1975, pp. 41.10–41.13.

[45] E. Ayanoglu, C.-L. I, R.D. Gitlin and J.E. Mazo, Diversity coding for transparent self-healing and fault-tolerant communication networks, *IEEE Trans. Commun.*, vol. 41, 1993, pp. 1677–1686.

[46] R. Krishnan and J.A. Silvester, Choice of allocation granularity in multipath source routing schemes, in *Proc. IEEE INFOCOM '93*, vol. 1, 1993, pp. 322–329.

[47] A. Banerjea, Simulation study of the capacity effects of dispersity routing for fault-tolerant realtime channels, in *Proc. ACM SIGCOMM'96*, vol. 26, 1996, pp. 194–205.

[48] D. Sidhu, R. Nair and S. Abdallah, Finding disjoint paths in networks, in *Proc. ACM SIGCOMM '91*, 1991, pp. 43–51.

[49] R. Ogier, V. Rutemburg and N. Shacham, Distributed algorithms for computing shortest pairs of disjoint paths, *IEEE Trans. Inform. Theory*, vol. 39, 1993, pp. 443–455.

[50] S. Murthy and J.J. Garcia-Luna-Aceves, Congestion-oriented shortest multipath routing, in *Proc. IEEE INFOCOM '96*, March 1996, pp. 1028–1036.

[51] W.T. Zaumen and J.J. Garcia-Luna-Aceves, Loop-free multipath routing using generalized diffusing computations, in *Proc. IEEE INFOCOM '98*, March 1998, pp. 1408–1417.

[52] J. Chen, P. Druschel and D. Subramanian, An efficient multipath forwarding method, in *Proc. IEEE INFOCOM '98*, 1998, pp. 1418–1425.
[53] N. Taft-Plotkin, B. Bellur and R. Ogier, Quality-of-service routing using maximally disjoint paths, in *Proc. 7th Int. Workshop Quality of Service (IWQoS'99)*, June 1999, pp. 119–128.
[54] S. Vutukury and J.J. Garcia-Luna-Aceves, An algorithm for multipath computation using distance vectors with predecessor information, in *Proc. IEEE ICCCN '99*, October 1999, pp. 534–539.
[55] I. Cidon, R. Rom, and Y. Shavitt, Analysis of multi-path routing, *IEEE/ACM Trans. Networking*, vol. 7, 1999, pp. 885–896.
[56] A. Nasipuri and S.R. Das, On-demand multipath routing for mobile *ad hoc* networks, in *Proc. IEEE ICCCN*, October 1999, pp. 64–70.
[57] D. Johnson and D. Maltz, Dynamic source routing in *ad hoc* wireless networks, in *Mobile Computing*, T. Imielinski and H. Korth (Eds). Kluwer: Norwell, MA, 1996.
[58] V.D. Park and M.S. Corson, A highly adaptive distributed routing algorithm for mobile wireless networks, in *Proc. IEEE INFOCOM '99*, 1999, pp. 1405–1413.
[59] J. Raju and J.J. Garcia-Luna-Aceves, A new approach to on-demand loop-free multipath routing, in *Proc. IEEE ICCCN*, October 1999, pp. 522–527.
[60] S.J. Lee and M. Gerla, AODV-BR: backup routing in *ad hoc* networks, in *Proc. IEEE WCNC*, 2000, pp. 1311–1316.
[61] C.E. Perkins and E.M. Royer, *Ad-hoc* on-demand distance vector routing, in *Proc. IEEE WMCSA*, New Orleans, LA, February 1999, pp. 90–100.
[62] L. Wang, L. Zhang, Y. Shu and M. Dong, Multipath source routing in wireless *ad hoc* networks, in *Proc. Canadian Conf. Electrical and Computer Engineering*, vol. 1, 2000, pp. 479–483.
[63] S.J. Lee and M. Gerla, Split multipath routing with maximally disjoint paths in *ad hoc* networks, in *Proc. ICC 2001*, vol. 10, June 2001, pp. 3201–3205.
[64] P. Papadimitratos, Z.J. Haas and E.G. Sirer, Path set selection in mobile *ad hoc* networks, in *Proc. ACM MobiHOC 2002*, Lausanne, 9–11, June 2002, pp. 160–170.
[65] M.R. Pearlman, Z.J. Haas, P. Sholander and S.S. Tabrizi, On the impact of alternate path routing for load balancing in mobile *ad hoc* networks, in *Proc. MobiHOC*, 2000, pp. 150–310.
[66] N. Gogate, D. Chung, S. Panwar and Y. Wang, Supporting video/image applications in a mobile multihop radio environment using route diversity, in *Proc. IEEE ICC '99*, vol. 3, June 1999, pp. 1701–1706.
[67] A. Tsirigos and Z.J. Haas, Analysis of Multipath Routing: part 2 – mitigation of the effects of frequently changing network topologies, *IEEE Trans. Wireless Commun.*, vol. 3, No. 2, March 2004, pp. 500–511.
[68] A.B. McDonald and T. Znati, A path availability model for wireless ad hoc networks, in *Proc. IEEE WCNC*, vol. 1, 1999, pp. 35–40.
[69] A. Tsirigos and Z.J. Haas, Analysis of multipath routing: part 1 – the effect on packet delivery ratio, *IEEE Trans. Wireless Commun.*, vol. 3, no. 1, 2004, pp. 138–146.
[70] C.R. Lin and M. Gerla, Adaptive clustering for mobile wireless networks,"*IEEE J. Selected Areas Commun.*, vol. 15, no. 7, 1997, pp. 1265–1275.
[71] M. Gerla and J.T. Tsai, Multicluster, mobile, multimedia radio network, *Wireless Networks,* vol. 1, 1995, pp. 255–265.

[72] N.H. Vaidya, P. Krishna, M. Chatterjee and D.K. Pradhan, A cluster-based approach for routing in dynamic networks, *ACM Comput. Commun. Rev.*, vol. 27, no. 2, 1997.

[73] R. Ramanathan and M. Steenstrup, Hierarchically-organized, multihop mobile wireless networks for quality-of-service support, *Mobile Networks Applic.*, vol. 3, no. 1, 1998, pp. 101–119.

[74] A.B. McDonald and T.F. Znati, A Mobility-based framework for adaptive clustering in wireless *ad hoc* Networks, *IEEE J. Selected Areas Commun.*, vol. 17, No. 8, 1999, pp. 1466–1488.

[75] D.J. Baker and A. Ephremides, The architectural organization of a mobile radio network via a distributed algorithm, *IEEE Trans. Commun.*, 1981, pp. 1694–1701.

[76] D.J. Baker, J. Wieselthier and A. Ephremides, A distributed algorithm for scheduling the activation of links in a self-organizing, mobile, radio network, in *Proc. IEEE ICC'82*, pp. 2F.6.1–2F.6.5.

[77] I. Chlamtac and S.S. Pinter, Distributed nodes organization algorithm for channel access in a multihop dynamic radio network, *IEEE Trans. Comput.*, 1987, pp. 728–737.

[78] M. Gerla and J. T.-C. Tsai, Multicluster, mobile, multimedia radio network, *ACM-Baltzer J. Wireless Networks*, vol. 1, no. 3, 1995, pp. 255–265.

[79] A.B. McDonald and T. Znati, Performance evaluation of neighbour greeting protocols: ARP versus ES-IS, *Comput. J.,* vol. 39, no. 10, 1996, pp. 854–867.

[80] P. Beckmann, *Probability in Communication Engineering.* Harcourt Brace World: New York, 1967.

[81] Y. Hu and D.B. Johnson, Caching strategies in on-demand routing protocols for wireless *ad hoc* networks, in *Proc. MobiCom 2000*, August 2000, pp. 231–242.

[82] D.B. Johnson and D.A. Maltz, Dynamic source routing in *ad hoc* wireless networks, www.ietf.org/internet-drafts/ draft-ietf-manet-dsr-03.txt, IETF Internet Draft (work in progress), October 1999.

[83] M.K. Marina and S.R. Das, Performance of route cache strategies in dynamic source routing: *Proc. Second Wireless Networking and Mobile Computing (WNMC)*, April 2001.

[84] D. Maltz, J. Broch, J. Jetcheva and D.B. Johnson, The effects of on-demand behavior in routing protocols for multi-hop wireless *ad hoc* networks, *IEEE J. Selected Areas in Commun.*, Special Issue on Mobile and Wireless Networks, August 1999.

[85] S. Chen and K. Nahrstedt, Distributed quality-of-service routing in *ad hoc* Networks, *IEEE J. Selected Areas Commun.*, vol. 17, no. 8, August 1999, pp. 1488–1505.

[86] H.F. Soloma, S. Reeves and Y. Viniotis, Evaluation of multicast routing algorithms for real-time communication on high-speed networks, *IEEE J. Selected Area Commun.* vol. 15, no. 3, 1997, pp. 332–345.

[87] I. Cidon, R. Ram and V. Shavitt, Multi-path routing combined with resource reservation,: *IEEE INFOCOM '97*, Japan, April 1997, pp. 92–100.

[88] S. Chen and K. Nahrstedt, An overview of quality-of-service routing for the next generation high-speed networks: problems and solutions, *IEEE Networks*, Special Issue on Transmission and Distribution of Digital Video, November/December 1998, pp. 64–79.

14

Sensor Networks

14.1 INTRODUCTION

A sensor network is composed of a large number of sensor nodes, which are densely deployed either inside the phenomenon they are observing, or very close to it. Most of the time the nodes are randomly deployed in inaccessible terrains or disaster relief operations. This also means that sensor network protocols and algorithms must possess self-organizing capabilities. Another unique feature of sensor networks is the cooperative effort of sensor nodes. Sensor nodes are fitted with an on-board processor. Instead of sending the raw data to the nodes responsible for the fusion, sensor nodes use their processing abilities to locally carry out simple computations and transmit only the required and partially processed data.

The above described features ensure a wide range of applications for sensor networks. Some of the application areas are health, military and security. For example, the physiological data about a patient can be monitored remotely by a doctor. While this is more convenient for the patient, it also allows the doctor to better understand the patient's current condition. Sensor networks can also be used to detect foreign chemical agents in the air and the water. They can help to identify the type, concentration and location of pollutants. In essence, sensor networks will provide the end user with intelligence and a better understanding of the environment. It is expected that, in future, wireless sensor networks will be an integral part of our lives, even more than the present-day personal computers.

Realization of these and other sensor network applications require wireless *ad hoc* networking techniques. Although many protocols and algorithms have been proposed for traditional wireless *ad hoc* networks, as described in Chapter 13, they are not well suited for the unique features and application requirements of sensor networks. To illustrate this point, the differences between sensor networks and *ad hoc* networks (see Chapter 13, and

Advanced Wireless Networks: 4G Technologies Savo G. Glisic
© 2006 John Wiley & Sons, Ltd.

references therein) are outlined below [1–5]:

- sensor nodes are densely deployed;
- sensor nodes are prone to failure;
- the number of sensor nodes in a sensor network can be several orders of magnitude higher than the nodes in an *ad hoc* network;
- the topology of a sensor network changes very frequently;
- sensor nodes mainly use broadcast communication paradigms whereas most *ad hoc* networks are based on point-to-point communications;
- sensor nodes are limited in power, computational capacities and memory;
- sensor nodes may not have global identification (ID) because of the large amount of overhead and large number of sensors.

One of the most important constraints on sensor nodes is the low power consumption requirement. Sensor nodes carry limited, quite often irreplaceable, power sources. Therefore, while traditional networks aim to achieve high QoS provisions, sensor network protocols must focus primarily on power conservation. They must have inbuilt trade-off mechanisms that give the end user the option of prolonging network lifetime at the cost of lower throughput or higher transmission delay. This problem will be the focus of this chapter.

Sensor networks may consist of many different types of sensors such as seismic, low sampling rate magnetic, thermal, visual, infrared, acoustic and radar. These sensors are able to monitor a wide variety of ambient conditions that include the current characteristics such as speed, direction and size of an object, temperature, humidity, vehicular movement, lightning condition, pressure, soil makeup, noise levels, the presence or absence of certain kinds of objects, mechanical stress levels on attached objects etc.

Sensor nodes can be used for continuous sensing, event detection, event ID, location sensing, and local control of actuators. The concept of micro-sensing and wireless connection of these nodes promises many new application areas. Usually these applications are categorized into military, environment, health, home and other commercial areas. It is possible to expand this classification with more categories such as space exploration, chemical processing and disaster relief.

Wireless sensor networks can be an integral part of military command, control, communications, computing, intelligence, surveillance, reconnaissance and targeting (C4ISRT) systems. They are used for monitoring friendly forces, equipment and ammunition and battlefield surveillance (Figure 14.1). Sensor networks can be deployed in critical terrains, and some valuable, detailed and timely intelligence about the opposing forces and terrain can be gathered within minutes before the opposing forces can intercept them. Sensor networks can also be incorporated into the guidance systems of intelligent ammunition.

Sensor networks deployed in the friendly region and used as a chemical or biological warning system can provide friendly forces with critical reaction time, which decreases casualty numbers drastically. Environmental applications of sensor networks include tracking the movements of birds, small animals and insects; monitoring environmental conditions that affect crops and livestock; irrigation; macroinstruments for large-scale Earth monitoring and planetary exploration; chemical/biological detection; precision agriculture; biological, Earth and environmental monitoring in marine, soil and atmospheric contexts; forest fire

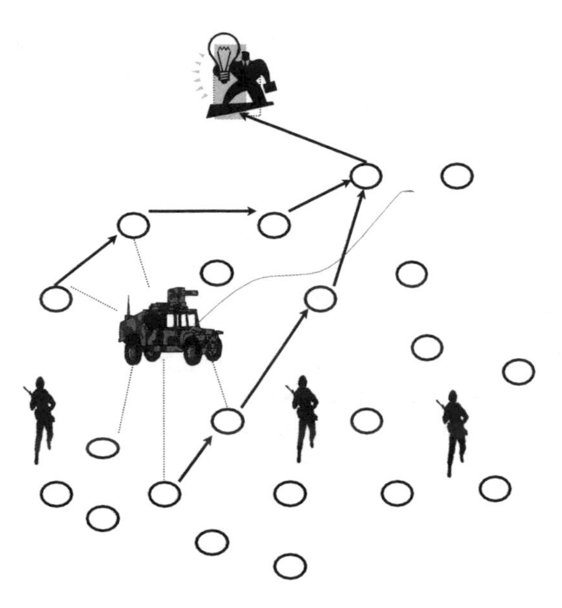

Figure 14.1 Battlefield surveillance.

detection; meteorological or geophysical research; flood detection; bio-complexity mapping of the environment; and pollution study.

The health applications for sensor networks are providing interfaces for the disabled, integrated patient monitoring, diagnostics, drug administration in hospitals, monitoring the movements and internal processes of insects or other small animals, telemonitoring of human physiological data, and tracking and monitoring doctors and patients inside a hospital. More details on sensor networks applications can be found in References [6–64].

14.2 SENSOR NETWORKS PARAMETERS

A sensor network design is influenced by many parameters, which include fault tolerance, scalability, production costs, operating environment, sensor network topology, hardware constraints, transmission media and power consumption. The failure of sensor nodes should not affect the overall task of the sensor network. This is the reliability or fault tolerance issue. Fault tolerance is the ability to sustain sensor network functionalities without any interruption due to sensor node failures [25, 49].

The protocols and algorithms may be designed to address the level of fault tolerance required by the sensor networks. If the environment where the sensor nodes are deployed has little interference, then the protocols can be more relaxed. For example, if sensor nodes are being deployed in a house to keep track of humidity and temperature levels, the fault tolerance requirement may be low since this kind of sensor networks is not easily damaged or interfered with by environmental noise. On the other hand, if sensor nodes are being deployed in a battlefield for surveillance and detection, then the fault tolerance has to be high because the sensed data are critical and sensor nodes can be destroyed by hostile actions. As a result, the fault tolerance level depends on the application of the sensor networks, and the schemes must be developed with this in mind. The number of sensor nodes deployed in studying a phenomenon may be in the order of hundreds or thousands. The networks must be able to work with this number of nodes. The density can range from few sensor nodes to few hundred sensor nodes in a region, which can be less than 10 m in diameter [12]. The node density depends on the application in which the sensor nodes are deployed. For machine diagnosis application, the node density is around 300 sensor nodes in a 5×5 m^2 region, and the density for the vehicle tracking application is around 10 sensor nodes per region [51]. In some cases, the density can be as high as 20 sensor nodes/m^3 [51]. A home may contain around two dozen home appliances containing sensor nodes [42], but this number will grow if sensor nodes are embedded into furniture and other miscellaneous items. For habitat monitoring application, the number of sensor nodes ranges from 25 to 100 per region. The density will be extremely high when a person normally containing hundreds of sensor nodes, which are embedded in eye glasses, clothing, shoes, watch, jewelry and the human body, is sitting inside a stadium watching a basketball, football or baseball game.

As a consequence, the cost of each sensor node has to be kept low. The state-of-the-art technology allows a Bluetooth radio system to be less than US$10 [46]. Also, the price of a PicoNode is targeted to be less than US$1 [45]. The cost of a sensor node should be much less than US$1 in order for the sensor network to be feasible [45]. The cost of a Bluetooth radio, which is known to be a low-cost device, is 10 times more expensive than the targeted price for a sensor node. Note that a sensor node also has some additional units such as sensing and processing units. In addition, it may be equipped with a location-finding system, mobilizer or power generator depending on the applications of the sensor networks. As a result, the cost of a sensor node is a very challenging issue given the amount of functionality with a price of much less than a dollar.

An illustration of sensor network topology Is shown in Figure 14.2. Deploying a high number of nodes densely requires careful handling of topology maintenance. Issues related to topology maintenance and change can be classified in three phases [66–77].

14.2.1 Pre-deployment and deployment phase

Sensor nodes can be either thrown in mass or placed one by one in the sensor field. They can be deployed by dropping from a plane, delivering in an artillery shell, rocket or missile, throwing by a catapult (from a ship, etc.), placing in factory or placing one by one either by a human or a robot. Although the sheer number of sensors and their unattended deployment usually preclude placing them according to a carefully engineered deployment plan, the schemes for initial deployment must reduce the installation cost, eliminate the need for any preorganization and preplanning, increase the flexibility of arrangement and promote self-organization and fault tolerance.

14.2.2 Post-deployment phase

After deployment, topology changes are due to change in sensor nodes' [27, 34] position, reachability (due to jamming, noise, moving obstacles, etc.), available energy, malfunctioning and task details. Sensor nodes may be statically deployed. However, device failure is a regular or common event due to energy depletion or destruction. It is also possible to have sensor networks with highly mobile nodes. In addition, sensor nodes and the network experience varying task dynamics, and they may be a target for deliberate jamming. Therefore, all these factors cause frequent changes in sensor network topologies after deployment.

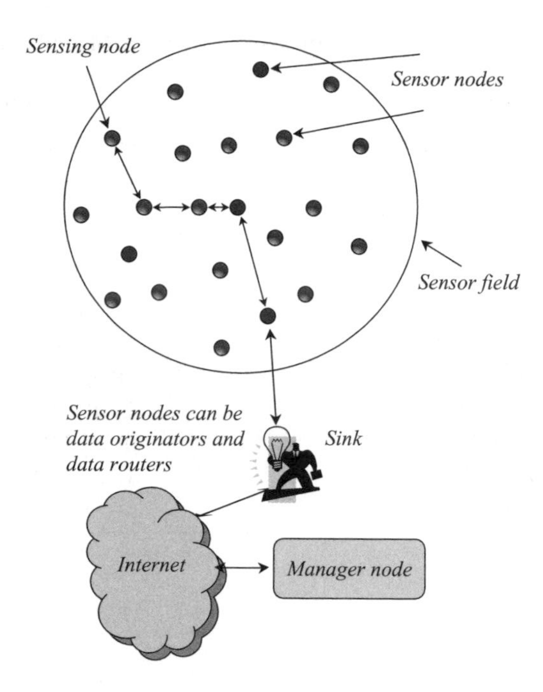

Figure 14.2 Sensor network topology.

14.2.3 Re-deployment of additional nodes phase

Additional sensor nodes can be re-deployed at any time to replace the malfunctioning nodes or due to changes in task dynamics. Addition of new nodes poses a need to re-organize the network. Coping with frequent topology changes in an *ad hoc* network that has myriad nodes and very stringent power consumption constraints requires special routing protocols.

In a multihop sensor network, communicating nodes are linked by a wireless medium. One option for radio links is the use of industrial, scientific and medical (ISM) bands, listed in Table 14.1, which offer license-free communication in most countries.

Some of these frequency bands are already being used for communication in cordless phone systems and wireless local area networks (WLANs). For sensor networks, a small-sized, low-cost, ultralow power transceiver is required. According to Porret *et al.* [43], certain hardware constraints and the trade-off between antenna efficiency and power consumption limit the choice of a carrier frequency for such transceivers to the ultrahigh frequency range. They also propose the use of the 433 MHz ISM band in Europe and the 915 MHz ISM band in North America. The transceiver design issues in these two bands are addressed in References [16, 35].

14.3 SENSOR NETWORKS ARCHITECTURE

The sensor nodes are usually scattered in a *sensor field* as shown in Figure 14.2. Each of these scattered sensor nodes has the capabilities to collect data and route data back to the *sink* and the end users. Data are routed back to the end user by a multihop infrastructureless

Table 14.1 Frequency bands available for ISM applications

Frequency band	Center frequency
6765–6795 kHz	6780 kHz
13 553–13 567 kHz	13,560 kHz
26 957–27 283 kHz	27,120 kHz
40.66–40.70 MHz	40.68 MHz
433.05–434.79 MHz	433.92 MHz
902–928 MHz	915 MHz
2400–2500 MHz	2450 MHz
5725–5875 MHz	5800 MHz
24–24.25 GHz	24.125 GHz
61–61.5 GHz	61.25 GHz
122–123 GHz	122.5 GHz
244–246 GHz	245 GHz

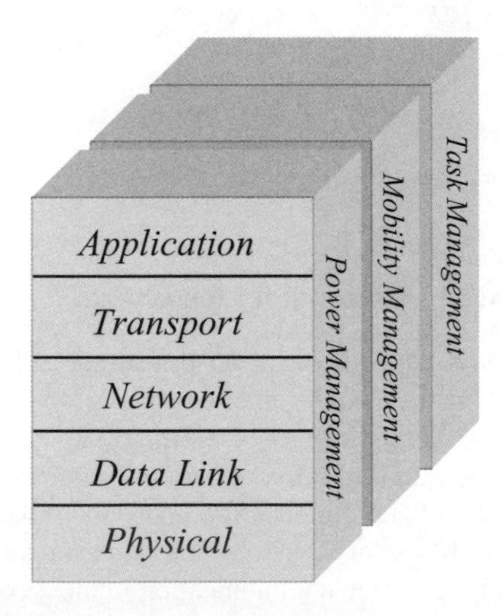

Figure 14.3 The sensor networks protocol stack.

architecture through the sink, as shown in Figure 14.2. The sink may communicate with the *task manager node* via Internet or satellite.

The protocol stack used by the sink and all sensor nodes is given in Figure 14.3. This protocol stack combines power and routing awareness, integrates data with networking protocols, communicates power efficiently through the wireless medium and promotes cooperative efforts of sensor nodes. The protocol stack consists of the *application layer*, *transport layer*, *network layer*, *data link layer*, *physical layer*, *power management plane*, *mobility management plane* and *task management plane* [81–94]. Depending on the sensing tasks, different types of application software can be built and used on the application layer. The transport layer helps to maintain the flow of data if the sensor networks application requires it. The network layer takes care of routing the data supplied by the transport layer. Since the environment is noisy and sensor nodes can be mobile, the MAC protocol must be power aware and able to minimize collision with neighbors' broadcast. The physical layer addresses the needs of a simple but robust modulation, transmission and receiving techniques. In addition, the power, mobility and task management planes monitor the power, movement and task distribution among the sensor nodes. These planes help the sensor nodes coordinate the sensing task and lower the overall power consumption.

14.3.1 Physical layer

The physical layer is responsible for frequency selection, carrier frequency generation, signal detection, modulation and data encryption. The choice of a good modulation scheme is critical for reliable communication in a sensor network. Binary and M-ary modulation schemes are compared in Shih *et al.* [51]. While an M-ary scheme can reduce the transmit on-time by sending multiple bits per symbol, it results in complex circuitry and increased radio power consumption. These tradeoff parameters are formulated in Shih *et al.* [51] and it is concluded that under start-up power dominant conditions, the binary modulation scheme is more energy-efficient. Hence, M-ary modulation gains are significant only for low start-up power systems.

14.3.2 Data link layer

The data link layer is responsible for the multiplexing of data streams, data frame detection, medium access and error control.

14.3.2.1 Medium access control

The MAC protocol in a wireless multihop self-organizing sensor network must achieve two goals [61, 62, 138]. The first is the creation of the network infrastructure. Since thousands of sensor nodes are densely scattered in a sensor field, the MAC scheme must establish communication links for data transfer. This forms the basic infrastructure needed for wireless communication hop by hop and gives the sensor network self-organizing ability. The second objective is to fairly and efficiently share communication resources between sensor nodes. Traditional MAC schemes, described in Chapter 5, are summarized in Figure 14.4 and Table 14.2.

A lot of modifications are needed in MAC protocols when applied in sensor networks. In a cellular system, the base stations form a wired backbone. A mobile node is only a single hop away from the nearest base station. The primary goal of the MAC protocol in

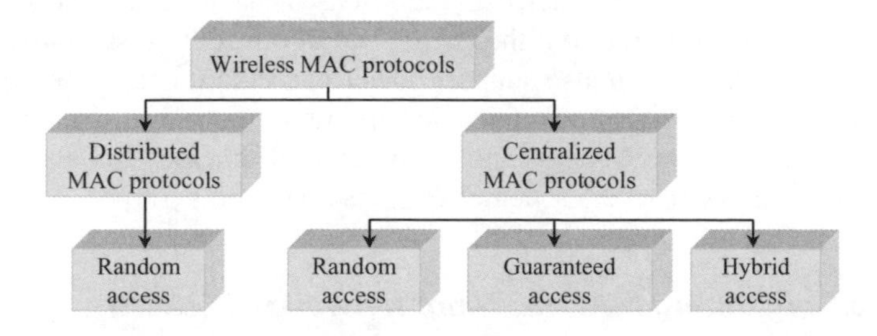

Figure 14.4 Classification of MAC schemes.

Table 14.2 Categorization of MAC protocols

Category	Resource sharing mode	Application domain	Disadvantages
Dedicated assignment or fixed allocation	Pre-determined fixed allocation	Continuous traffic/provides bounded delay	Inefficient for bursty traffic
Demand based	On demand or user request	Variable rate and multimedia traffic	Overhead and delay due to reservation process
Random access or contention-based	Contention when transmission packets are available	Bursty traffic	Inefficient for delay-sensitive traffic

such systems is the provision of high QoS and bandwidth efficiency. Power conservation assumes only secondary importance as base stations have unlimited power supply and the mobile user can replenish exhausted batteries in the handset. Hence, medium access is focused on dedicated resource assignment strategy. Such an access scheme is impractical for sensor networks as there is no central controlling agent like the base station.

The mobile *ad hoc* networks (MANET), discussed in Chapter 13, are probably the closest peers to the sensor networks. The MAC protocol in a MANET has the task of forming the network infrastructure and maintaining it in the face of mobility. Hence, the primary goal is the provision of high QoS under mobile conditions. Although the nodes are portable battery-powered devices, they can be replaced by the user and hence power consumption is only of secondary importance.

Therefore, the MAC protocol for sensor networks must have built-in power conservation, mobility management and failure recovery strategies. Although many schemes for medium access have been proposed for MANETs (see Chapter 13), the design of an efficient MAC scheme for the new regime of sensor networks is still an open research issue. The *fixed allocation* and *random access* versions of medium access have been discussed in References [54, 60]. *Demand-based* MAC schemes may be unsuitable for sensor networks due their large messaging overhead and link set-up delay. Power conservation is achieved by the use of power saving operation modes and by preferring time-outs to acknowledgements, wherever possible.

Since radios must be turned off during idling for precious power savings, the MAC scheme should include a variant of TDMA [39]. Such a medium access mechanism is presented in Sohrabi *et al.* [54]. Further, contention-based channel access is deemed unsuitable due to its requirement to monitor the channel at all times. It must be noted, however, that random medium access can also support power conservation, as in the IEEE 802.11 standard for WLANs, by turning off radios depending on the status of the net allocation vector. Constant listening times and adaptive rate control schemes can also help achieve energy efficiency in random access schemes for sensor networks [60].

14.3.2.2 Self-organizing MAC for senor networks (SMACS)

The SMACS protocol [54] achieves network start-up and link-layer organization, and the *eavesdrop-and-register* (*EAR*) algorithm enables seamless connection of mobile nodes in a sensor network. SMACS is a distributed infrastructure-building protocol which enables nodes to discover their neighbors and establish transmission/reception schedules for communication without the need for any local or global master nodes. In this protocol, the neighbor discovery and channel assignment phases are combined so that, by the time nodes hear all their neighbors, they will have formed a connected network. A communication link consists of a pair of time slots operating at a randomly chosen, but fixed frequency (or frequency hopping sequence). This is a feasible option in sensor networks, since the available bandwidth can be expected to be much higher than the maximum data rate for sensor nodes. Such a scheme avoids the necessity for network-wide synchronization, although communicating neighbors in a subnet need to be time-synchronized. *Power conservation* is achieved using a random wake-up schedule during the connection phase and by turning the radio off during idle time slots. The process is based on using an ultralow power radio to wake-up the neighbors. This second radio uses much less power via either a low duty cycle or hardware design. Usually this second radio can only transmit a busy tone. This broadcast tone should not disrupt any on-going data transmission, e.g. use a different channel.

The amount of time and power needed to wake-up (start-up) a radio is not negligible and thus just turning off the radio whenever it is not being used is not necessarily efficient. The energy characteristics of the start-up time should also be taken into account when designing the size of the data link packets.

The EAR protocol [49] attempts to offer continuous service to the mobile nodes under both mobile and stationary conditions. Here, the mobile nodes assume full control of the connection process and also decide when to drop connections, thereby minimizing messaging overhead. The EAR is transparent to the SMACS, so that the SMACS is functional until the introduction of mobile nodes into the network. In this model, the network is assumed to be mainly static, i.e. any mobile node has a number of stationary nodes in its vicinity. A drawback of such a time-slot assignment scheme is the possibility that members already belonging to different subnets might never get connected. For more details see Sohrabi *et al.* [54]. *Carrier sense media access* (CSMA)-based MAC scheme for sensor networks is presented in Woo and Culler [60]. Traditional-CSMA based schemes are deemed inappropriate as they all make the fundamental assumption of stochastically distributed traffic and tend to support independent point-to-point flows. On the contrary, the MAC protocol for sensor networks must be able to support variable, but highly correlated and dominantly periodic traffic. Any CSMA-based medium access scheme has two important components, the *listening mechanism (sensing)* and the *backoff scheme*. Based on simulations in References [60–64], the constant listen periods are energy-efficient and the introduction of random delay (*p-persistance*) provides robustness against repeated collisions. Fixed window and binary exponential decrease backoff schemes are recommended to maintain proportional fairness in the network. A phase change at the application level is also advocated to get over any capturing effects. It is proposed in this work that the energy consumed/throughput can serve as a good indicator of *energy efficiency*.

14.3.3 Network layer

The *ad hoc* routing techniques, already discussed in Chapter 13, do not usually fit the requirements of the sensor networks. The networking layer of sensor networks is usually designed according to the following principles. First of all, power efficiency is always an important design parameter, see Figure 14.5. Sensor networks are mostly data centric. Data

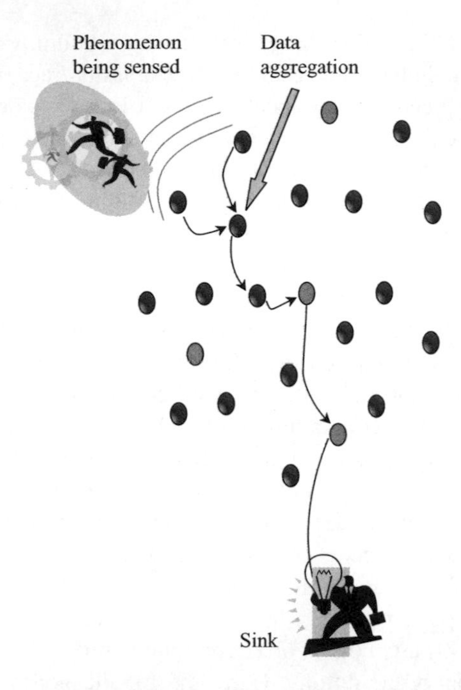

Figure 14.5 Multihop routing due to limited transmission range.

aggregation is useful only when it does not hinder the collaborative effort of the sensor nodes. An ideal sensor network has attribute-based addressing and location awareness. One of the following approaches can be used to select an energy efficient route.

(1) *Maximum available power (PA) route* – the route that has maximum total available power is preferred.

(2) *Minimum energy (ME) route* – the route that consumes ME to transmit the data packets between the sink and the sensor node is the ME route.

(3) *Minimum hop (MH) route* – the route that makes the MH to reach the sink is preferred. Note that the ME scheme selects the same route as the MH when the same amount of energy is used on every link. Therefore, when nodes broadcast with same power level without any power control, MH is then equivalent to ME.

(4) *Maximum minimum PA node route* – the route along which the minimum PA is larger than the minimum PAs of the other routes is preferred. This scheme precludes the risk of using up a sensor node with low PA much earlier than the others because they are on a route with nodes which have very high PAs.

14.3.3.1 Data centric routing

In data-centric routing, the interest dissemination is performed to assign the sensing tasks to the sensor nodes. There are two approaches used for interest dissemination: sinks broadcast the interest [27], and sensor nodes broadcast an advertisement for the available data [23] and wait for a request from the interested sinks. For an illustration see Figure 14.6.

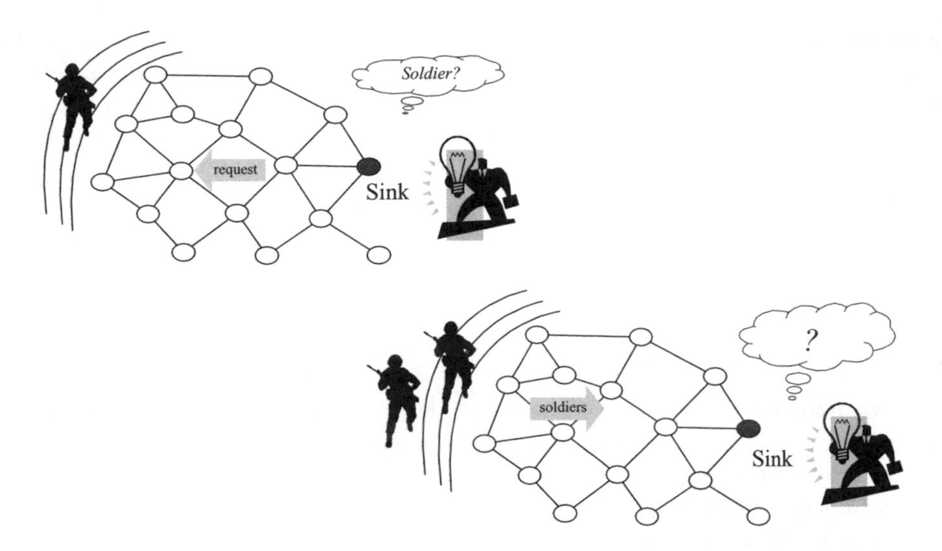

Figure 14.6 Broadcasting the interest (are there any soldiers in the area?) and advertising (there are soldiers in the area).

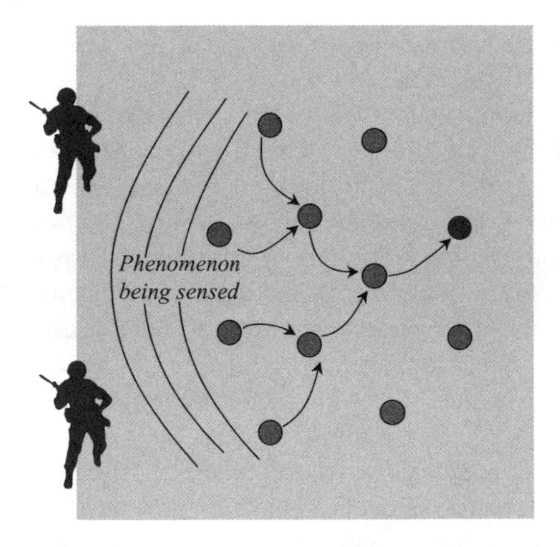

Figure 14.7 The data aggregation.

The data-centric routing requires attribute-based naming [49, 65]. For attribute-based naming, the users are more interested in querying an attribute of the phenomenon, rather than querying an individual node. For instance 'the areas where the moisture is over 70 %' is a more common query than 'the moisture read by a certain node'. The attribute-based naming is used to carry out queries using the attributes of the phenomenon. The attribute-based naming also makes broadcasting, attribute-based multicasting, geo-casting and any-casting important for sensor networks.

Data aggregation is a technique used to solve the implosion and overlap problems in data-centric routing [23]. In this technique, a sensor network is usually perceived as a reverse multicast tree, as shown in Figure 14.7, where the sink asks the sensor nodes to report the ambient condition of the phenomena. Data coming from multiple sensor nodes are aggregated as if they are about the same attribute of the phenomenon when they reach the same routing node on the way back to the sink. Data aggregation can be perceived as a set of automated methods of combining the data that comes from many sensor nodes into a set of meaningful information [22]. With this respect, data aggregation is known as data fusion [23]. Also, care must be taken when aggregating data, because the specifics of the data, e.g. the locations of reporting sensor nodes, should not be left out. Such specifics may be needed by certain applications.

14.3.3.2 Internetworking

One other important function of the network layer is to provide internetworking with external networks such as other sensor networks, command and control systems and the Internet. In one scenario, the sink nodes can be used as a gateway to other networks. Another option is creating a backbone by connecting sink nodes together and making this backbone access other networks via a gateway.

14.3.3.3 Flooding and gossiping

Flooding has already been described in Chapters 7 and 13 as a technique used to disseminate information across a network. The drawbacks are [23]: (1) *implosion*, when duplicated messages are sent to the same node; (2) *overlap* – when two or more nodes share the same observing region, they may sense the same stimuli at the same time, and as a result, neighbor nodes receive duplicated messages; and (3) *resource blindness*, not taking into account the available energy resources. Control of the energy consumption is of paramount importance in WSNs; a promiscuous routing technique such as flooding wastes energy unnecessarily.

Gossiping is a variation of flooding attempting to correct some of its drawbacks [21]. Nodes do not indiscriminately broadcast but instead send a packet to a randomly selected neighbor who, once it receives the packet, repeats the process. It is not as simple to implement as the flooding mechanism and it takes longer for the propagation of messages across the network.

Data funneling by data aggregation concentrates, e.g. funnels, the packet flow into a single stream from the group of sensors to the sink. It reduces (compresses) the data by taking advantage of the fact that the destination is not that interested in a particular order of how the data packets arrive.

In the setup phase, the controller divides the sensing area into regions and performs a directional flood towards each region. When the packet reaches the region the first receiving node becomes a border node and modifies the packet (add fields) for route cost estimations within the region. Border nodes flood the region with modified packets. Sensor nodes in the region use cost information to schedule which border nodes to use.

In the data communication phase, when a sensor has data it uses the schedule to choose the border node that is to be used. It then waits for a time inversely proportional to the number of hops from the border. Along the way to the border node, the data packets joined together until they reach the border node. The border node collects all packets and then sends one packet with all the data back to the controller These steps are illustrated in Figure 14.8.

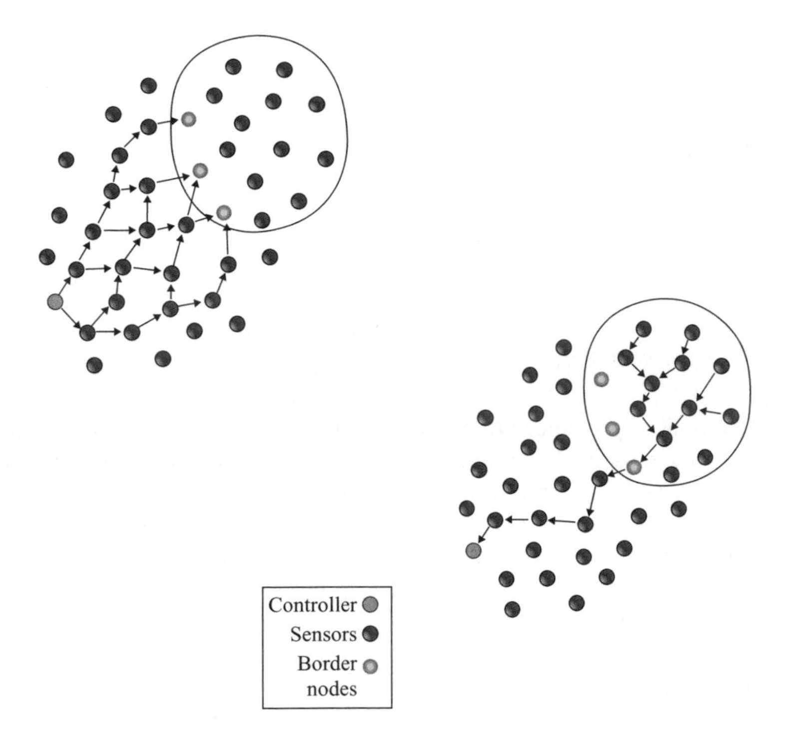

Figure 14.8 Data funneling: (a) setup phase; (b) data communication phase.

14.3.3.4 Sensor protocols for information via negotiation (SPIN)

A family of adaptive protocols called SPIN [23] is designed to address the deficiencies of classic flooding by negotiation and resource adaptation. The SPIN family of protocols is designed based on two basic ideas: sensor nodes operate more efficiently and conserve energy by sending data that describe the sensor data instead of sending the whole data, e.g. image, and sensor nodes must monitor the changes in their energy resources.

The *sequential assignment routing* (*SAR*) [54] algorithm creates multiple trees where the root of each tree is a one-hop neighbor from the sink. Each tree grows outward from the sink while avoiding nodes with very low QoS (i.e. low-throughput/high delay) and energy reserves. At the end of this procedure, most nodes belong to multiple trees. This allows a sensor node to choose a tree to relay its information back to the sink.

The *low-energy adaptive clustering hierarchy* (*LEACH*) [22] is a clustering-based protocol that minimizes energy dissipation in sensor networks. The purpose of LEACH is to randomly select sensor nodes as cluster-heads, so the high-energy dissipation in communicating with the base station is spread to all sensor nodes in the sensor network. During the set-up phase, a sensor node chooses a random number between 0 and 1. If this random number is less than the threshold $T(n)$, the node is a cluster-head. $T(n)$ is calculated as

$$T(n) = \begin{cases} \dfrac{P}{1 - P[r mod(1/P)]} & \text{if } n \in G \\ 0 & \text{otherwise,} \end{cases}$$

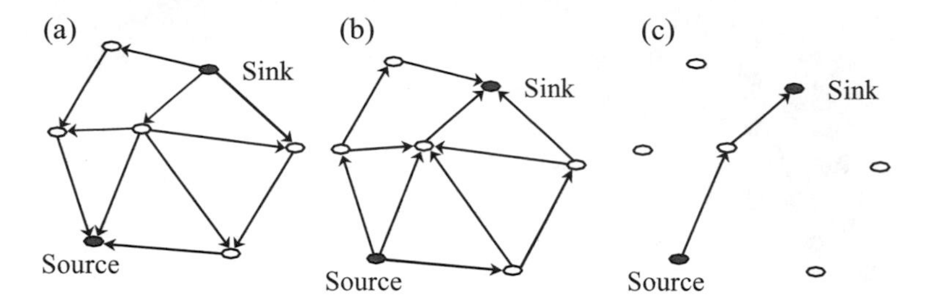

Figure 14.9 An example of directed diffusion [27]: (a) propagate interest, (b) set up gradient and (c) send data.

where P is the desired percentage to become a cluster-head; r, the current round; and G, the set of nodes that have not being selected as a cluster-head in the last $1/P$ rounds. After the cluster-heads are selected, the cluster-heads advertise to all sensor nodes in the network that they are the new cluster-heads. During the steady phase, the sensor nodes can begin sensing and transmitting data to the cluster-heads. After a certain period of time spent on the steady phase, the network goes into the set-up phase again and entering into another round of selecting the cluster-heads.

14.3.3.5 Directed diffusion

The directed diffusion data dissemination is discussed in Intanagonwiwat *et al.* [27]. The sink sends out interest, which is a task description, to all sensors, as shown in Figure 14.9(a). The task descriptors are named by assigning attribute-value pairs that describe the task. Each sensor node then stores the interest entry in its cache. The interest entry contains a timestamp field and several gradient fields. As the interest is propagated throughout the sensor network, the gradients from the source back to the sink are set up as shown in Figure 14.9(b). When the source has data for the interest, the source sends the data along the interest's gradient path as shown in Figure 14.9(c). The interest and data propagation and aggregation are determined locally. Also, the sink must refresh and reinforce the interest when it starts to receive data from the source. The directed diffusion is based on data-centric routing where the sink broadcasts the interest.

14.3.4 Transport layer

The need for transport layer is discussed in References [44, 46]. This layer is especially needed when access to the system is planned through Internet or other external networks. TCP with its current transmission window mechanisms, as discussed in Chapter 9, does match to the extreme characteristics of the sensor network environment. In Bakre and Badrinath [62] an approach called TCP splitting is considered to make sensor networks interact with other networks such as the Internet. In this approach, TCP connections are ended at sink nodes, and a special transport layer protocol can handle the communications between the sink node and sensor nodes. As a result, the communication between the user and the sink node is by UDP or TCP via the Internet or Satellite. Owing to sensor node limited memory, the communication between the sink and sensor nodes may be purely by UDP type protocols.

Unlike protocols such as TCP, the end-to-end communication schemes in sensor networks are not based on global addressing. These schemes must consider that attribute-based naming is used to indicate the destinations of the data packets.

As a conclusion, TCP variants developed in Chapter 9 for the traditional wireless networks are not suitable for WSNs where the notion of *end-to-end reliability* has to be redefined due to the specific nature of the sensor network which comes with features such as: (1) multiple senders, the sensors and one destination, the sink, which creates a reverse multicast type of data flow; (2) for the same event there is a high level of redundancy or correlation in the data collected by the sensors and thus there is no need for end-to-end reliability between individual sensors and the sink, but instead between the event and the sink; (3) on the other hand there is need of end-to-end reliability between the sink and individual nodes for situations such as re-tasking or reprogramming; (4) the protocols developed should be energy aware and simple enough to be implemented in the low-end type of hardware and software of many WSN applications.

Pump slowly, fetch quickly (PFSQ) is designed to distribute data from a source node by pacing the injection of packets into the network at relatively low speed (*pump slowly*) which allows nodes that experience data loss to aggressively recover missing data from their neighbors (*fetch quickly*). The goals of this protocols are: (1) to ensure that all data segments are delivered to the intended destinations with minimum especial requirements on the nature of the lower layers; (2) to minimize the number of transmissions to recover lost information; (3) to operate correctly even in situations where the quality of the wireless links is very poor; and (4) to (5) provide loose delay bounds for data delivery to all intended receivers. PFSQ has been designed to guarantee sensor-to-sensor delivery and to provide end-to-end reliability for control management distribution from the control node (sink) to the sensors. It does not address congestion control.

Event-to-sink reliable transport (ESRT), illustrated in Figure 14.10, is designed to achieve reliable event detection (at the sink node) with a protocol that is energy aware and has congestion control mechanisms.

The protocol provides self-configuration, even in the case of a dynamic topology. For energy awareness sensor nodes are notified to decrease their frequency of reporting if the reliability level at the sink node is above the minimum. Congestion control takes advantage of the high level of correlation between the data flows corresponding to the same event. Sink is only interested in the collective information from a group of sensors, not in their

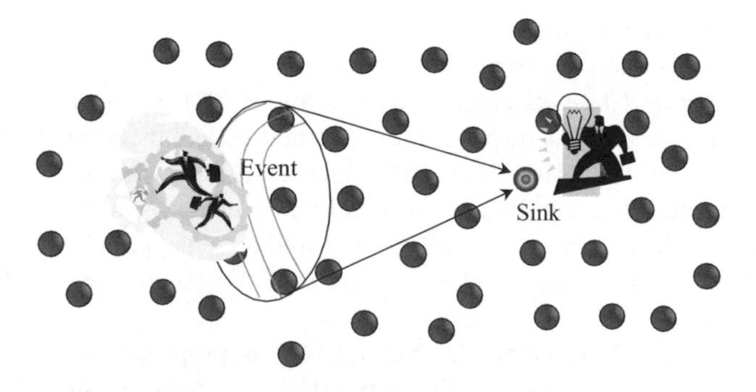

Figure 14.10 Event-to-sink reliable transport (ESRT).

individual report. Most of the complexity of the protocol falls on the sink node, minimizing the requirements on the sensor nodes.

14.3.5 Application layer

In this section, we discuss three application layer protocols, i.e. sensor management protocol (SMP), task assignment and data advertisement protocol (TADAP), and sensor query and data dissemination protocol (SQDDP). Additional work is expected in this segment of sensor networks.

14.3.5.1 Sensor management protocol

Sensor networks have many different application areas, and accessing them through networks such as the Internet is an option [44]. Designing an application layer management protocol has several advantages. It makes the hardware and software of the lower layers transparent to the sensor network management applications. System administrators interact with sensor networks using the sensor management protocol (SMP). Unlike many other networks, sensor networks consist of nodes that do not have global IDs, and they are usually infrastructureless. Therefore, SMP needs to access the nodes by using attribute-based naming and location-based addressing, which are explained in detail earlier in this section. SMP is a management protocol that provides the software operations needed to perform the following administrative tasks: introducing the rules related to data aggregation, attribute-based naming and clustering to the sensor nodes, exchanging data related to the location-finding algorithms, time synchronization of the sensor nodes, moving sensor nodes, turning sensor nodes on and off, querying the sensor network configuration and the status of nodes, re-configuring the sensor network, and authentication, key distribution and security in data communications. Discussion of these issues is given in References [15, 41, 48, 49, 65, 95–97].

The *task assignment and data advertisement protocol* controls interest dissemination in sensor networks. As discussed earlier in this section (Figure 14.6), users send their interest to a sensor node, a subset of the nodes or whole network. This interest may be about a certain attribute of the phenomenon or a triggering event. Another approach is the advertisement of available data in which the sensor nodes advertise the available data to the users, and the users query the data they are interested in. An application layer protocol that provides the user software with efficient interfaces for interest dissemination is useful for lower layer operations, such as routing, as explained in Figure 14.6.

The *sensor query and data dissemination protocol* (SQDDP) provides user applications with interfaces to issue queries, respond to queries and collect incoming replies. Note that these queries are generally not issued to particular nodes. Instead, attribute-based or location-based naming is preferred. For instance, 'the locations of the nodes in the supermarket where the selling items stock is bellow a threshold and should be re-supplied' is an attribute-based query. Similarly, 'the selling item stock size at the node A' is an example for location-based naming.

Sensor query and tasking language (SQTL) [49] is proposed as an application that provides even a larger set of services. SQTL supports three types of events, which are defined by the keywords *receive*, *every* and *expire*. Receive keyword defines events generated by a sensor node when the sensor node receives a message; every keyword defines events occurring periodically due to a timer time-out; and the expire keyword defines the events occurring when a timer is expired. If a sensor node receives a message that is intended for it and contains a script, the sensor node then executes the script. Although SQTL is proposed, different types of SQDDP can be developed for various applications. The use of SQDDPs may be unique to each application.

14.4 MOBILE SENSOR NETWORKS DEPLOYMENT

A mobile sensor network is composed of a distributed collection of *nodes*, each of which in addition to sensing, computation and communication also has locomotion capabilities. Locomotion facilitates a number of useful network capabilities, including the ability to self-deploy and self-repair. The use of mobile sensor networks includes applications ranging from urban combat scenarios to search-and-rescue operations and emergency environment monitoring. An example is a scenario involving a hazardous materials leak in an urban environment. In general, we would like to be able to throw a number of sensor nodes into a building through a window or doorway. The nodes are equipped with chemical sensors that allow them to detect the relevant hazardous material, and deploy themselves throughout the building in such a way that they maximize the area 'covered' by these sensors. Data from the nodes are transmitted to a base station located safely outside the building, where they are assembled to form a live map showing the concentration of hazardous compounds within the building.

For the sensor network to be useful in this scenario, the location of each node must be determined. In urban environments, accurate localization using GPS is generally not possible (due to occlusions or multipath effects), while landmark-based approaches require prior models of the environment that may be either unavailable (destroyed), incomplete or inaccurate. This is particularly true in disaster scenarios, were the environment may have undergone recent (and unplanned) structural modifications. Therefore it is of interest to determine the location of network nodes using the nodes *themselves* as landmarks. This particular technique, however, requires that nodes maintain line-of-sight relationships with one another. This condition also enables establishment of line-of-sight communications links which operate with minimum energy consumption. An additional demand is that nodes should deploy in such a way that they maximize the area 'covered' by the network, while simultaneously ensuring that each node can be seen by at least one other node.

The concept of *coverage* as a paradigm for evaluating *multi-robot* [98–114] systems was introduced in Gage [93]. Three basic types of coverage were defined: *blanket coverage*, where the objective is to achieve a static arrangement of nodes that maximizes the total detection area; *barrier coverage*, where the objective is to minimize the probability of undetected penetration through the barrier; and *sweep coverage*, which is more or less equivalent to a moving barrier. The problem of exploration and map-building by a single robot in an unknown environment has been considered by a number of authors [113–115]. The frontier-based approach described in References [113,114] proceeds by incrementally building a global occupancy map of the environment, which is then analyzed to find the 'frontiers' between free and unknown space. The robot is directed to the nearest such frontier. The network deployment algorithm described in Howard *et al.* [98] shares a number of similarities with [113]. It also builds a global occupancy grid of the environment and direct nodes to the frontier between free and unknown space. However, in this deployment algorithm the map is built entirely from live, rather than stored, sensory data. It also satisfies an additional constraint: that each node must be visible to at least one other node.

Multi-robot exploration and map-building has been explored by a number of authors [100,106,109,111] who use a variety of techniques ranging from topological matching [100] to fuzzy inference [106] and particle filters [111]. Once again, there are two key differences between these earlier works and the work described in Howard *et al.* [98], where maps are built entirely from live, not stored, sensory data, and the deployment algorithm must satisfy an additional constraint (i.e. line-of-sight visibility).

A distributed algorithm for the deployment of mobile robot teams has been described in [108], where the concept of 'virtual pheromones' is introduced. These are localized messages that are emitted by one robot and detected by nearby robots. Virtual pheromones can be used to generate either 'gas expansion' or 'guided growth' deployment models. The key advantage of this approach is that the deployment algorithm is entirely distributed, and has the potential to respond dynamically to changes in the environment. This algorithm does, however, lead to relatively slow deployment; it is also unclear, from the published results, how effective this algorithm is at producing good area coverage. A somewhat similar algorithm based on artificial potential fields is described in Howard *et al.* [104].

The algorithm elaborated in this section is an *incremental* deployment algorithm [98], where nodes are deployed one at a time, with each node making use of information gathered by the previously deployed nodes to determine its ideal deployment location. The algorithm aims to maximize the total network *coverage*, i.e. the total area that can be 'seen' by the network. At the same time, the algorithm must ensure that the *visibility constraint* is satisfied, i.e. each node must be visible to at least one other node. The algorithm relies on a number of key assumptions:

(1) *Homogeneous nodes* – all nodes are assumed to be identical. We also assume that each node is equipped with a range sensor, a broadcast communications device, and is mounted on some form of mobile platform.

(2) *Static environment* – the environment is assumed to be static, at least to the extent that gross topology remains unchanged while the network is deploying.

(3) *Model-free* – this algorithm is intended for applications in which environment models are unavailable; indeed, a key task for the network may be to *generate* such models.

(4) *Full communication* – all nodes in the network can communicate with some remote base station on which the deployment algorithm is executed. Note that this does not automatically imply that *all* nodes must be within radio range of the base station; the nodes may, for example, form an *ad hoc* multihop network.

(5) *Localization* – the position of each node is known in some arbitrary global coordinate system. This technique does not require external landmarks or prior models of the environment, but does require that each node is visible to at least one other node. It is this requirement that gives rise to the *visibility constraint*, i.e. each node must be visible to at least one other node at its deployed location.

Two performance metrics are of interest: *coverage*, i.e. the total area visible to the network's sensors; and *time*, i.e. the total deployment time, including both the time taken to perform the necessary computations and the time taken to physically move the nodes. The objective is to maximize the coverage while minimizing the deployment time.

The algorithm has four phases: *initialization, selection, assignment* and *execution*.

- *Initialization* – nodes are assigned one of three states: *waiting, active* or *deployed*. As the names suggest, a *waiting* node is waiting to be deployed, an *active* node is in the process of deploying, and a *deployed* node has already been deployed. Initially, the state of all nodes is set to *waiting*, with the exception of a single node that is set to *deployed*. This node provides a starting point, or 'anchor', for the network, and is not subject to the visibility constraint.
- *Selection* – sensor data from the deployed nodes is combined to form a common map of the environment (*occupancy grid*). This map is analyzed to select the deployment location, or goal, for the next node. Each cell in this grid is assigned one of three states: *free*, *occupied* or *unknown*. A cell is *free* if it is known to contain no obstacles, *occupied* if it is known to contain one or more obstacles, and *unknown* otherwise. In the combined occupancy grid, any cell that can be seen by one or more nodes will be marked as either free or occupied; only those cells that cannot be seen by *any* node will marked as unknown. We can therefore ensure that the visibility constraint is satisfied by always selecting goals that lie somewhere in free space.
- *Assignment* – in the simplest case, the selected goal is assigned to a waiting node, and the node's state is changed from *waiting* to *active*. More commonly, assignment is complicated by the fact that deployed nodes tend to obstruct waiting nodes, necessitating a more complex assignment algorithm. That is, the algorithm may have to re-assign the goals of a number of previously deployed nodes, changing their state from *deployed* to *active*.
- *Execution* – active nodes are deployed sequentially to their goal locations. The state of each node is changed from *active* to *deployed* upon arrival at the goal. The algorithm iterates through the selection, assignment and execution phases, terminating only when all nodes have been deployed. Performance examples of the algorithm can be found in Howard *et al.* [98].

14.5 DIRECTED DIFFUSION

As already indicated in Section 14.3.3, directed diffusion consists of several elements. Data is *named* using attribute-value pairs. A sensing task (or a subtask thereof) is disseminated throughout the sensor network as an *interest* for named data. This dissemination sets up *gradients* within the network designed to 'draw' events (i.e. data matching the interest). Events start flowing towards the originators of interests along multiple paths. The sensor network *reinforces* one, or a small number of these paths as illustrated in Figure 14.9. In this section we elaborate these elements in more detail.

In directed diffusion, task descriptions are *named* by, for example, a list of attribute-value pairs that describe a task. For example a surveillance system (military or civil application) which is expected to report an intrusion in a given area might be described as

type = human	*// detect location*
interval = 20 ms	*// send back events every 20 ms*
duration = 10 s	*// .. for the next 10 s*
rect = [−100, 100, 200, 400]	*// from sensors within rectangle*

For simplicity, we choose the subregion representation to be a rectangle defined on some coordinate system; in practice, this might be based on GPS coordinates. The task description specifies an interest for data matching the attributes. For this reason, such a task description is called an *interest*. The data sent in response to interests are also named using a similar naming scheme. Thus, for example, a sensor that detects an intrusion might generate the following data:

type = human *// type of intruder seen*
instance = military *// instance of this type*
location = [125, 220] *// node location*
intensity = 0.6 *// signal amplitude measure*
confidence = 0.85 *// confidence in the match*
timestamp = 01:20:40 *// event generation time*

Given our choice of naming scheme, we now describe how interests are *diffused* through the sensor network. Suppose that a task, with a specified *type* and *rect*, a *duration* of 10 min and an *interval* of 10 ms, is instantiated at a particular node in the network. The interval parameter specifies an event data rate; thus, in our example, the specified data rate is 100 events per second. This sink node records the task; the task state is purged from the node after the time indicated by the duration attribute.

For each active task, the sink periodically *broadcasts* an interest message to each of its neighbors. This initial interest contains the specified *rect* and *duration* attributes, but contains a much larger *interval* attribute. Intuitively, this initial interest may be thought of as exploratory; it tries to determine if there indeed are any sensor nodes that detect the human intrusion. To do this, the initial interest specifies a low data rate (in our example, one event per second). Then, the initial interest takes the following form:

type = human
interval = 1 s
rect = [−100, 200, 200, 400]
timestamp = 01:20:40
expiresAt = 01:30:40

The interest is periodically *refreshed* by the sink. To do this, the sink simply re-sends the same interest with a monotonically increasing timestamp attribute. Every node maintains an interest cache. Each item in the cache corresponds to a *distinct* interest. Two interests are distinct if their *type*, *interval* or *rect* attributes are different. Interest entries in the cache *do not contain information about the sink*. Thus, interest state scales with the number of distinct active interests. The definition of distinct interests also allows interest *aggregation*. Two interests I_1 and I_2, with identical types, completely overlapping *rect* attributes, can, in some situations, be represented by a single interest entry.

An entry in the interest cache has several fields. A *timestamp* field indicates the timestamp of the last received matching interest. The interest entry also contains several *gradient fields*, up to one per neighbor. Each gradient contains a *data rate* field requested by the specified neighbor, derived from the *interval* attribute of the interest. It also contains a *duration* field, derived from the *timestamp* and *expiresAt* attributes of the interest, and indicating the approximate lifetime of the interest.

When a node receives an interest, it checks to see if the interest exists in the cache. If no matching entry exists (where a match is determined by the definition of distinct interests specified above), the node creates an interest entry. The parameters of the interest entry are instantiated from the received interest. This entry has a single gradient towards the neighbor from which the interest was received, with the specified event data rate. In the above example, a neighbor of the sink will set up an interest entry with a gradient of one event per second towards the sink. For this, it must be possible to distinguish individual neighbors. Any locally unique neighbor identifier may be used for this purpose.

If there exists an interest entry, but no gradient for the sender of the interest, the node adds a gradient with the specified value. It also updates the entry's timestamp and duration fields appropriately. Finally, if there exist both an entry *and* a gradient, the node simply updates the timestamp and duration fields.

After receiving an interest, a node may decide to re-send the interest to some subset of its neighbors. To its neighbors, this interest *appears to originate from the sending node*, although it might have come from a distant sink. This is an example of a *local interaction*. In this manner, interests *diffuse* throughout the network. Not all received interests are re-sent. A node may suppress a received interest if it recently re-sent a matching interest.

Generally speaking, there are several possible choices for neighbors, as presented in Figure 14.11. The simplest alternative is to *re-broadcast* the interest to all neighbors. It may also be possible to perform geographic routing, using some of the techniques described in Chapters 7 and 13. This can limit the topological scope for interest diffusion, thereby resulting in energy savings. Finally, in an immobile sensor network, a node might use cached

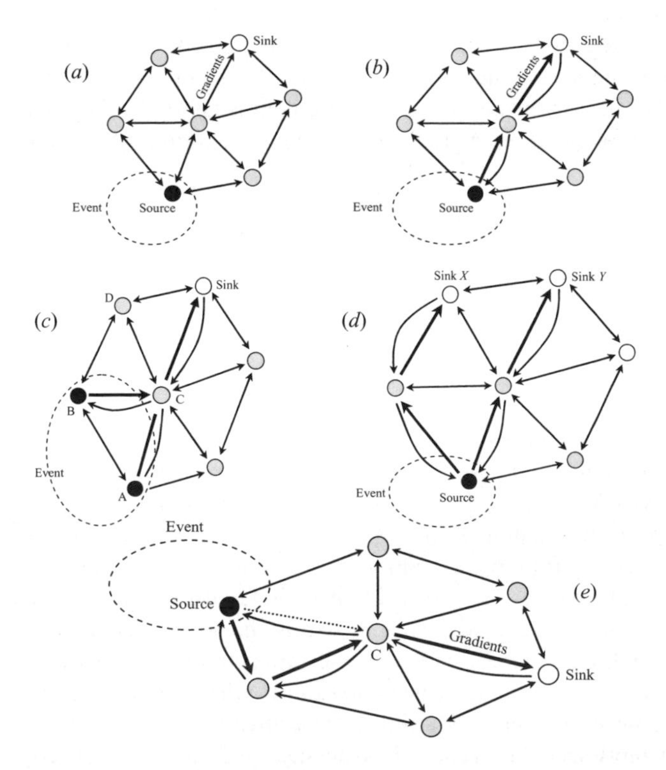

Figure 14.11 Diffusion: (a) gradient establishment; (b) reinforcement; (c) multiple sources; (d) multiple siniks; and (e) repair.

data to direct interests. For example, if in response to an earlier interest, a node heard from some neighbor *A* data sent by some sensor within the region specified by the *rect* attribute, it can direct this interest to *A*, rather than broadcasting to all neighbors.

Figure 14.11(a) shows the gradients established in the case where interests are flooded through a sensor field. Unlike the simplified description in Figure 14.9(b), notice that every pair of neighboring nodes establishes a gradient towards each other. This is a consequence of local interactions.

For a sensor network, a gradient specifies both a data rate and a direction in which to send events. More generally, a gradient specifies a *value* and a direction. Figure 14.9(c) implicitly depicts binary valued gradients. In the sensor networks presented in this section, gradients have two values that determine the event reporting rate. In other sensor networks, gradient values might be used to, for example, probabilistically forward data along different paths, achieving some measure of load balancing.

14.5.1 Data propagation

A sensor node that is within the specified *rect* processes interests as described in the previous section. In addition, the node tasks its local sensors to begin collecting samples. A sensor node that detects a target searches its interest cache for a matching interest entry. In this case, a matching entry is one whose *rect* encompasses the sensor location, and the type of the entry matches the detected target type. When it finds one, it computes the highest requested event rate among all its outgoing gradients. The node tasks its sensor subsystem to generate event samples at this highest data rate. In the previous example, this data rate is initially one event per second (until reinforcement is applied). The source then sends to each neighbor for whom it has a gradient, an event description every second of the form:

```
type = human                // type of intruder seen
instance = military        // instance of this type
location = [125, 220]      // node location
intensity = 0.6            // signal amplitude measure
confidence = 0.85          // confidence in the match
timestamp = 01:20:40       // local time when event was generated
```

This *data* message is unicast individually to the relevant neighbors. A node that receives a data message from its neighbors attempts to find a matching interest entry in its cache. The matching rule is as described in the previous paragraph. If no match exists, the data message is silently dropped. If a match exists, the node checks the *data cache* associated with the matching interest entry for loop prevention. If a received data message has a matching data cache entry, the data message is silently dropped. Otherwise, the received message is added to the data cache and the data message is re-sent to the node's neighbors.

By examining its data cache, a node can determine the data rate of received events. To re-send a received data message, a node needs to examine the matching interest entry's gradient list. If all gradients have a data rate that is greater than or equal to the rate of incoming events, the node may simply send the received data message to the appropriate neighbors. However, if some gradients have a lower data rate than others (caused by selectively reinforcing paths, then the node may *downconvert* to the appropriate gradient.

14.5.2 Reinforcement

The sink initially diffuses an interest for a low event-rate notification (one event per second). Once sources detect a matching target, they send low-rate events, possibly along multiple paths, towards the sink. After the sink starts receiving these low data rate events, it *reinforces* one particular neighbor in order to 'draw down' higher quality (higher data rate) events. In general, this feature of directed diffusion is achieved by *data driven* local rules. One example of such a rule is to reinforce any neighbor from which a node receives a previously unseen event. To reinforce this neighbor, the sink re-sends the original interest message but with a smaller interval (higher data rate). The reinforcement propagates back, and one way to chose reinforced path is to pick up one with low delay, as shown in Figure 14.11(b). Similar mechanisms are used to handle multiple source and multiple sinks scenarios, as shown in Figure 14.11(c) and (d).

In directed diffusion, *intermediate* nodes on a previously reinforced path can also apply the reinforcement rules. This is useful to enable *local repair* of failed or degraded paths as indicated in Figure 14.11(e).

14.6 AGGREGATION IN WIRELESS SENSOR NETWORKS

Data fusion or aggregation is an important concept in sensor networks. The key idea is to combine data from different sensors to eliminate redundant transmissions, and still provide a rich, multidimensional view of the environment being monitored. This concepts shifts the focus from address-centric approaches (finding routes between pairs of end nodes) to a more data-centric approach (finding routes from multiple sources to a destination that allows in-network consolidation of data).

Consider a network of n sensor nodes $1, 2, \ldots, n$ and a sink node t labeled $n+1$ distributed over a region. The locations of the sensors and the sink are fixed and known apriori. Each sensor produces some information as it monitors its vicinity. We assume that each sensor generates one data packet per time unit to be transmitted to the base station. For simplicity, we refer to each time unit as a transmission cycle or simply *cycle*. We assume that all data packets have size k bits. The information from all the sensors needs to be gathered at each cycle and sent to the sink for processing. We assume that each sensor has the ability to transmit its packet to any other sensor in the network or directly to the sink. Further, each sensor i has a battery with finite, non-replenishable energy E_i. Whenever a sensor transmits or receives a data packet, it consumes some energy from its battery. The sink has an unlimited amount of energy available to it. Typical assumptions used in the modeling of energy consumption are that sensor consumes $\epsilon_{\text{elec}} = 50$ nJ/b to run the transmitter or receiver circuitry and $\epsilon_{\text{amp}} = 100$ pJ/bit/m^2 for the transmitter amplifier. Thus, the energy consumed by a sensor i in receiving a k-bit data packet is given by, $Rx_i = \epsilon_{\text{elec}} \times k$, while the energy consumed in transmitting a data packet to sensor j is given by, $Tx_{i,j} = \epsilon_{\text{elec}} \times k + \epsilon_{\text{amp}} \times d_{i,j}^2 \times k$, where $d_{i,j}$ is the distance between nodes i and j.

We define the *lifetime* T of the system to be the number of cycles until the first sensor is drained of its energy. A *data gathering schedule* specifies, for each cycle, how the data packets from all the sensors are collected and transmitted to the base station. A schedule can be thought of as a collection of T directed trees, each rooted at the base station and spanning all the sensors, i.e. a schedule has one tree for each round. The lifetime of a schedule equals the lifetime of the system under that schedule. The objective is to find a schedule that maximizes the system lifetime T.

Data aggregation performs in-network fusion of data packets, coming from different sensors enroute to the sink, in an attempt to minimize the number and size of data transmissions and thus save sensor energies. Such aggregation can be performed when the data from different sensors are highly correlated. As usual, we make the simplistic assumption that an intermediate sensor can aggregate multiple incoming packets into a single outgoing packet.

The problem is to find a data gathering schedule with maximum lifetime for a given collection of sensors and a sink, with known locations and the energy of each sensor, where sensors are permitted to aggregate incoming data packets.

Consider a schedule S with lifetime T cycles. Let $f_{i,j}$ be the total number of packets that node i (a sensor) transmits to node j (a sensor or sink) in S. The energy constraints at each sensor, impose

$$\sum_{j=1}^{n+1} f_{i,j} \cdot Tx_{i,j} + \sum_{j=1}^{n} f_{j,i} \cdot Rx_i \leq E_i, i = 1, 2, \ldots, n.$$

The schedule S induces a flow network $G = (V, E)$. The flow network G is a directed graph having as nodes all the sensors and the sink, and edges (i, j) with capacity $f_{i,j}$ whenever $f_{i,j} > 0$.

If S is a schedule with lifetime T, and G is the flow network induced by S, then, for each sensor s, the maximum flow from s *to the sink t in G is $\geq T$.* This is due to the fact that each data packet transmitted from a sensor must reach the base station. The packets from s could possibly be aggregated with one or more packets from other sensors in the network. Intuitively, we need to guarantee that each of the T values from s *influences* the final value(s) received at the sink. In terms of network flows, this implies that sensor s must have a maximum $s - t$ flow of size $\geq T$ to the sink in the flow network G. Thus, a necessary condition for a schedule to have lifetime T is that each node in the induced flow network can push flow T to the sink.

Now, we consider the problem of finding a flow network G with maximum T, that allows each sensor to push flow T to the base station, while respecting the energy constraints at all the sensors. What needs to be found are the capacities of the edges of G. Such a flow network G will be referred to as an *admissible* flow network with lifetime T. An admissible flow network with maximum lifetime is called an *optimal admissible* flow network.

An optimal admissible flow network can be found using the integer program with linear constraints. If for each sensor $k = 1, 2, \ldots, n$, $\pi_{i,j}^{(k)}$ is a flow variable indicating the flow that k sends to the sink t over the edge (i, j), the integer program is given by:

Maximize T with

$$\sum_{j=1}^{n+1} f_{i,j} \cdot Tx_{i,j} + \sum_{j=1}^{n} f_{j,i} \cdot Rx_i \leq E_i, i = 1, 2, \ldots, n$$

$$\sum_{j=1}^{n} \pi_{j,i}^{(k)} = \sum_{j=1}^{n+1} \pi_{i,j}^{(k)}, \text{ for all } i = 1, 2, \ldots, n \text{ and } i \neq k,$$

$$T + \sum_{j=1}^{n} \pi_{j,k}^{(k)} = \sum_{j=1}^{n+1} \pi_{k,j}^{(k)},$$
$$0 \le \pi_{i,j}^{(k)} \le f_{i,j}, \text{ for all } i = 1, 2, \ldots, n \text{ and } j = 1, 2, \ldots, n+1$$
$$\sum_{i=1}^{n} \pi_{i,n+1}^{(k)} = T, \; k = 1, 2, \ldots, n,$$

The first line imposes the energy constraint per node; the next two lines enforce the flow conservation principle at a sensor; the next line ensures that the capacity constraints on the edges of the flow network are respected and the last line ensures that T flow from sensor k reaches the sink.

Now we can get a schedule from an admissible flow network. A schedule is a collection of directed trees rooted at the sink that span all the sensors, with one such tree for each cycle. Each such tree specifies how data packets are gathered and transmitted to the sink. These trees are referred to as *aggregation trees*. An aggregation tree may be used for one or more cycles. The number of cycles f for which an aggregation tree is used is indicated by associating the value f with each one of its edges. In the following f is referred to as the lifetime of the aggregation tree. The *depth* of a sensor v is the average of its depths in each of the aggregation trees, and the depth of the schedule is be $\max\{depth(v) : v \in V\}$.

Figure 14.12 shows an admissible flow network G with lifetime $T = 50$ and two aggregation trees A_1 and A_2, with lifetimes 30 and 20, respectively. By looking at one of these trees, say A_1, we see that, for each one of 30 cycles, sensor 2 transmits one packet to sensor 1, which in turn aggregates it with its own data packet and then sends one data packet to the base station. Given an admissible flow network G with lifetime T and a directed tree

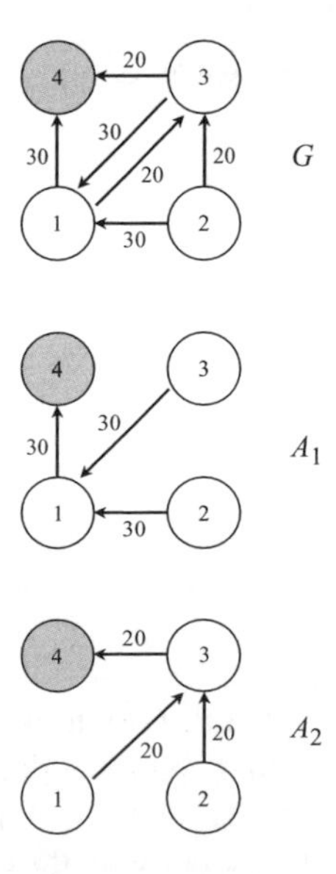

Figure 14.12 An admissible flow network G with lifetime 50 rounds and two aggregation trees A_1 and A_2 with lifetimes 30 and 20 rounds, respectively. The depth of the schedule with aggregation trees A_1 and A_2 is 2.

A rooted at the sink t with lifetime f, we define the (A, f)-*reduction* G' *of* G to be the flow network that results from G after reducing the capacities of all of its edges, that are also in A, by f. We call G' the (A, f)-*reduced* G. An (A, f)-reduction G' of G is *feasible* if the maximum flow from v to the sink t in G' is $\geq T - f$ for each vertex v in G'. Note that A does not have to span all the vertices of G, and thus it is not necessarily an aggregation tree. Moreover, if A is an aggregation tree, with lifetime f, for an admissible flow network G with lifetime T, and the (A, f)-reduction of G is feasible, then the (A, f)-reduced flow network G' of G is an admissible flow network with lifetime $T - f$. Therefore, we can devise a simple iterative algorithm, to construct a schedule for an admissible flow network G with lifetime T, provided we can find such an aggregation tree A.

Aggretree (G, T, t)

1 initialize $f \leftarrow 1$

2 let $A = (V_o, E_o)$ where $V_o = \{t\}$ and $E_o = \emptyset$

3 while A does not span all the nodes of G do

4   for each edge $e = (i, j) \in G$ such that $i \notin V_o$ and $j \in V_o$ do

5   let A' be A together with the edge e

6   // check if the $(A', 1)$-reduction of G is feasible

7   let G_r be the $(A', 1)$-reduction of G

8   if MAXFLOW$(v, t, G_r) \geq T - 1$ for all nodes v of G

9   // replace A with A'

10             $V_o \leftarrow V_o \cup \{i\}$, $E_o \leftarrow E_o \cup \{e\}$

11   break

12 let $c_{\min}$ be the minimum capacity of the edges in A

13 let G_r be the $(A, c_{\min})$-reduction of G

14 if MAXFLOW$(v, t, G_r) \geq T - c_{\min}$ for all nodes v of G

15     $f \leftarrow c_{\min}$

16 replace G with the (A, f)-reduction of G

17 return f, G, A

The aggretree (G, T, t) algorithm can be used to obtain an aggregation tree A with lifetime f from an admissible flow network G with lifetime $T \geq f$. Tree A is formed as follows. Initially A contains just the sink t. While A does not span all the sensors, we find and add to A an edge $e = (i, j)$, where $i \notin A$ and $j \in A$, provided that the (A', f)-reduction of G is feasible – here A' is the tree A together with the edge e and f is the minimum of the capacities of the edges in A'. Given a flow network G and sink t such that each sensor has a minimum $s - t$ cut of size $\geq T$ (i.e. the maximum flow from s to t in G is $\geq T$), we can prove that it is always possible to find a sequence of aggregation trees, via the algorithm, that can be used to aggregate T data packets from each of the sensors. The proof of correctness is based on a minimax theorems in graph theory [116, 117].

Experimental results show [118] that, for a network with 60 nodes, the above algorithm can improve the network lifetime by a factor of more than 20.

14.7 BOUNDARY ESTIMATION

An important problem in sensor networking applications is *boundary estimation* [79, 80, 127, 129]. Consider a network sensing a field composed of two or more regions of distinct behavior (e.g. differing mean values for the sensor measurements). An example of such a field is depicted in Figure 14.13(a). In practice this may represent the bound of the area under the fire or contaminated area. Boundary estimation is the process of determining the delineation between homogeneous regions. By transmitting to the sink only the information about the boundary instead of the transmission from each sensor, a significant aggregation effect can be achieved. There are two fundamental limitations in the boundary estimation problem. First, the accuracy of a boundary estimate is limited by the spatial density of sensors in the network and by the amount of noise associated with the measurement process. Second, energy constraints may limit the complexity of the boundary estimate that is ultimately transmitted to a desired destination.

The objective is to consider measurements from a collection of sensors and determine the boundary between two fields of relatively homogeneous measurements.

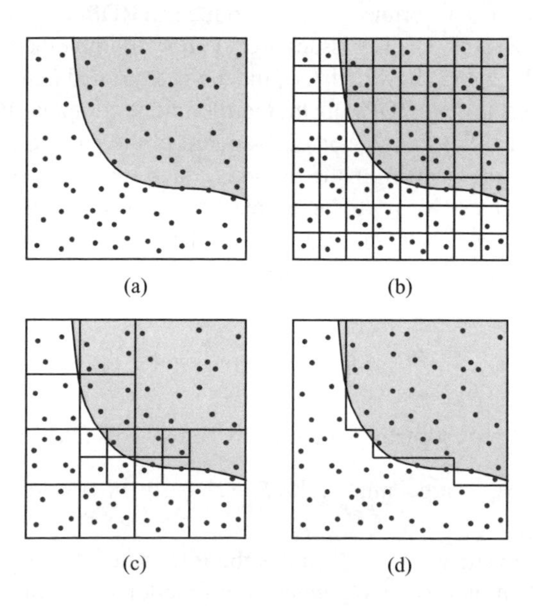

Figure 14.13 Sensing an inhomogeneous field. (a) Points are sensor locations. The environment has two conditions indicated by the gray and white regions of the square. (b) The sensor network domain is partitioned into square cells. (c) Sensors within the network operate collaboratively to determine a pruned partition that matches the boundary. (d) Final approximation to the boundary between the two regions which is transmitted to a remote point.

We presume a hierarchical structure of 'clusterheads' which manage measurements from nodes below them in the hierarchy. Thus, the nodes in each square of the partition communicate their measurements to a clusterhead in the square. Index the squares at the finest scale by row and column (i, j). The clusterhead in square (i, j) computes the average of these measurements to obtain a value $x_{i,j}$: $N(\mu_{i,j}, \sigma^2/m_{i,j})$, where $\mu_{i,j}$ is the mean value, σ^2 is the noise variance for each sensor measurement, and $m_{i,j}$ is the number of nodes in square (i, j). Thus we assume sensor measurements that have a Gaussian distribution. For simplicity, we assume $m_{i,j} = 1$. The random distribution is to account for noise in the system as well as for the small probability of node failure.

A possible approach to the boundary estimation problem is to devise a hierarchical processing strategy that enables the nodes to collaboratively determine a nonuniform rectangular partition of the sensor domain that is adapted to the boundaries [119–125]. The partition will have high, fine resolution along the boundary, and low, coarse resolution in homogeneous regions of the field, as depicted in Figure 14.13. The partition effectively provides a 'staircase'-like approximation to the boundary.

The estimation process partitions the sensor domain of a normalized unit square [0, 1] into n sub-squares of sidelength $1/\sqrt{n}$, as shown in Figure 14.13(b). The sidelength $1/\sqrt{n}$ is the finest resolution in the analysis. In principle, this initial partition can be generated by a *recursive dyadic partition* (RDP). First divide the domain into four sub-squares of equal size. Repeat this process again on each sub-square. Repeat this $(1/2)\log_2 n = J$ times. This gives rise to a *complete* RDP of resolution $1/\sqrt{n}$ (the rectangular partition of the sensing domain shown in Figure 14.13(b). The RDP process can be represented with a quadtree structure. The quadtree can be pruned back to produce an RDP with nonuniform resolution as shown in Figure 14.13(c). The key issues are: (1) how to implement the pruning process in the sensor network; and (2) how to determine the best pruned tree.

Let P_n denote the set of all RDPs, including the initial complete RDP and all possible prunings. For a certain RDP $P \in P_n$, on each square of the partition, the estimator of the field averages the measurements from the sensors in that square and sets the estimate of the field to that average value. This results in a piecewise constant estimate, denoted by θ, of the field. This estimator will be compared with the data $x = \{x_{i,j}\}$. The empirical measure of performance is the sum-of-squared errors between $\theta = \theta(P)$ and the data $x = \{x_{i,j}\}$.

$$\Delta(\theta, x) = \sum_{i,j=1}^{\sqrt{n}} [\theta(i, j) - x_{i,j}]^2 \tag{14.1}$$

The complexity penalized estimator is defined by References [119–125]:

$$\hat{\theta}_n = \arg \min_{\theta(P)P \in P_n} \Delta[\theta(P), x] + 2\sigma^2 p(n) N_{\theta(P)} \tag{14.2}$$

where σ^2 is the noise variance, $N_{\theta(P)}$ denotes the total number of squares in the partition P, and $p(n)$ is a certain monotonically increasing function of n that discourages unnecessarily high-resolution partitions [appropriate choices of $p(n)$ will be discussed below]. The optimization in Equation (14.2) can be solved using a bottom-up tree pruning algorithm in $O(n)$ operations [122, 126, 128,]. At each level of the hierarchy, the clusterhead receives the best sub-partition/subtree estimates from the four clusterheads below it, and compares the total cost of these estimates with the cost of the estimate equal to the average of all sensors in that cluster to make the decision on pruning.

14.7.1 Number of RDPs in *P*

Set P of RDPs consists of all RDPs resulting from pruning P_J, the uniform partition of the unit square into n squares of sidelength $1/\sqrt{n}$. We need to determine how many RDPs there are in P or, more specifically, we need to know how many partitions there are with exactly ℓ squares/leafs. Since the RDP is based on recursive splits into four, the number of leafs in every partition in P is of the form $\ell = 3m + 1$, for some integer $0 \leq m \leq (n-1)/3$. The integer m corresponds to the number of recursive splits. For each RDP having $3m + 1$ leafs there is a corresponding partially ordered sequence of m split points (at dyadic positions in the plane). In general, there are

$$\binom{n}{m} \equiv \frac{n!}{(n-m)!m!}$$

possible selections of m points from n (n corresponding to the vertices of the finest resolution partition, P_J). This number is an upper bound on the number of partitions in P with $\ell = 3m + 1$ leafs (since RDPs can only have dyadic split points).

14.7.2 Kraft inequality

Let Θ_n denote the set of all possible models of the field. This set contains piecewise constant models (constant on the dyadic squares corresponding to one of the partitions in P_n). The constant values are in a prescribed range $[-R, R]$, and are quantized to k bits. The range corresponds to the upper and lower limits of the amplitude range of the sensors. The set Θ_n consists of a finite number of models derived in the previous section. Here we show that with the number of bits k employed per transmission and $p(n)$ properly calibrated, we have

$$\sum_{\theta \in \Theta_n} \mathrm{e}^{-p(n)|\theta|} \leq 1 \tag{14.3}$$

where for simplicity notation $N_{\theta(P)} = |\theta|$ is used. If $\Theta_n^{(m)}$ denotes the subset of Θ_n consisting of models based on $\ell = 3m + 1$ leaf partitions, then we have

$$\begin{aligned}
\sum_{\theta \in \Theta_n} \mathrm{e}^{-p(n)|\theta|} &= \sum_{m=0}^{(n-1)/3} \sum_{\theta \in \Theta_n^{(m)}} \mathrm{e}^{-(3m+1)p(n)} \leq \sum_{m=0}^{(n-1)/3} \binom{n}{m} (2^k)^{3m+1} \mathrm{e}^{-(3m+1)p(n)} \\
&\leq \sum_{m=0}^{(n-1)/3} \frac{n^m}{m!} (2^k)^{3m+1} \mathrm{e}^{-(3m+1)p(n)} \\
&= \sum_{m=0}^{(n-1)/3} \frac{1}{m!} \mathrm{e}^{-[m \log n + (3m+1)\log(2^k) - (3m+1)p(n)]}
\end{aligned}$$

If $A \equiv m \log n + (3m+1)\log(2^k) - (3m+1)p(n) < -1$ (then $\mathrm{e}^A < \mathrm{e}^{-1}$), then we have

$$\sum_{\theta \in \Theta_n} \mathrm{e}^{-p(n)|\theta|} \leq 1/\mathrm{e} \sum_{m=0}^{(n-1)/3} \frac{1}{m!} \leq 1 \tag{14.4}$$

To guarantee $A < -1$, we must have $p(n)$ growing at least like $\log n$. Therefore, set $p(n) = \gamma \log n$, for some $\gamma > 0$. Also, as we will see later in the next section, to guarantee that the quantization of our models is sufficiently fine to contribute a negligible amount to the overall error we must select 2^k: $n^{1/4}$. With these calibrations we have $A = [(7/4 - 3\gamma)\, m + (1/4 - \gamma)] \log n$. In order to guarantee that the MSE converges to zero, we will see in the next section that m must be a monotonically increasing function of n. Therefore, for n sufficiently large, the term involving $(\frac{1}{4} - \gamma)$ is negligible, and the condition $A < -1$ is satisfied by $\gamma > 7/12$. In References [119–125] $\gamma = 2/3$ is used.

14.7.3 Upper bounds on achievable accuracy

Assume that $p(n)$ satisfies the condition defined by Equation (14.4) where again $|\theta|$ denotes the number of squares (alternatively we shall call this the number of leafs in the pruned tree description of the boundary) in the partition θ. It is shown in the above section that $p(n) \leq \gamma \log n$ satisfies Equation (14.4). Let $\hat{\theta}_n$ denote the solution to

$$\hat{\theta}_n = \arg\min_{\theta \in \Theta_n} \Delta(\theta, x) + 2\sigma^2 p(n)|\theta| \tag{14.5}$$

where, as before, x denotes the array of measurements at the finest scale $\{x_{i,j}\}$, and $|\theta|$ denotes the number of squares in the partition associated with θ. This is essentially the same estimator as defined in Equation (14.2) except that the values of the estimate are quantized in this case.

If θ_n^* denote the true value of the field at resolution $1/\sqrt{n}$ [i.e. $\theta_n^*(i, j) = E[x_{i,j}]$] then, applying Theorem 7 in References [119, 124], the MSE of the estimator $\hat{\theta}_n$ is bounded above as

$$\frac{1}{n} \sum_{i,j=1}^{\sqrt{n}} E\{[\hat{\theta}_n(i, j) - \theta_n^*(i, j)]^2\} \leq \min_{\theta \in \Theta_n} \frac{1}{n} \left\{ 2 \sum_{i,j=1}^{\sqrt{n}} [\theta(i, j) - \theta_n^*(i, j)]^2 + 8\sigma^2 p(n)|\theta| \right\} \tag{14.6}$$

The upper bound involves two terms. The first term, $2\sum_{i,j=1}^{\sqrt{n}} [\theta(i, j) - \theta_n^*(i, j)]^2$, is a bound on the bias or approximation error. The second term, $8\sigma^2 p(n)\,|\theta|$, is a bound on the variance or estimation error. The bias term, which measures the squared error between the best possible model in the class and the true field, is generally unknown. However, if we make certain assumptions on the smoothness of the boundary, then the rate at which this term decays as function of the partition size $|\theta|$ can be determined.

If the field being sensed is composed of homogeneous regions separated by a one-dimensional boundary and if the boundary is a Lipschitz function [122, 128], then by carefully calibrating quantization and penalization [taking k : $1/4 \log n$ and setting $p(n) = 2/3 \log n$] we have [119,125]

$$\frac{1}{n} \sum_{i,j=1}^{\sqrt{n}} E\{[\hat{\theta}_n(i, j) - \theta_n^*(i, j)]^2\} \leq O\left[\sqrt{(\log n)/n}\right] \tag{14.7}$$

This result shows that the MSE decays to zero at a rate of $\sqrt{[(\log n)/n]}$.

14.7.4 System optimization

The system optimization includes energy-accuracy trade-off. Energy consumption is defined by two communication costs: the cost of communication due to the construction of the tree (*processing cost*) and the cost of communicating the final boundary estimate (*communication cost*). We will show that the expected number of leafs produced by the algorithm is $O(\sqrt{n})$, and that the *processing* and *communication* energy consumption is proportional to this number. Having in mind MSE : $\sqrt{[(\log n)/n]}$ and ignoring the logarithmic factor, the accuracy-energy trade-off required to achieve this optimal MSE is roughly MSE : 1/energy. If each of the n sensors transmits its data, directly or by multiple hops, to an external point, the processing and communication energy costs are $O(n)$, which leads to the trade-off MSE : $1/\sqrt{\text{energy}}$, since we know that no estimator exists that can result in an MSE decaying faster than $O(1/\sqrt{n})$. Thus, the hierarchical boundary estimation method offers substantial savings over the naive approach while optimizing the tradeoff between accuracy and complexity of the estimate.

Communication cost is proportional to the final description of the boundary, thus it is of interest to compute the expected size of the tree, or $E[|\hat{\theta}|]$. We construct an upperbound for $E[|\hat{\theta}|]$ under the assumption of a homogeneous field with no boundary. Let P denote the tree-structured partition associated with $\hat{\theta}$. Note that, because P is an RDP, it can have $d + 1$ leafs (pieces in the partition), where $d = 3m, m = 0, \ldots, (n - 1)/3$. Therefore, the expected number of leafs is given by

$$E[|\hat{\theta}|] = \sum_{m=0}^{(n-1)/3} (3m + 1)\Pr(|\hat{\theta}| = 3m + 1)$$

The probability Pr $(|\hat{\theta}| = 3m + 1)$ can be bounded from above by the probability that one of the possible partitions with $3m + 1$ leafs, $m > 0$, is chosen in favor of the trivial partition with just a single leaf. That is, the event that one of the partitions with $3m + 1$ leafs is selected implies that partitions of all other sizes were not selected, including the trivial partition, from which the upper bound follows. This upper bound allows us to bound the expected number of leafs as follows:

$$E[|\hat{\theta}|] \leq \sum_{m=0}^{(n-1)/3} (3m + 1)N_m P_m$$

where N_m denotes the number of different $(3m + 1)$-leaf partitions, and p_m denotes the probability that a particular $(3m + 1)$-leaf partition is chosen in favor of the trivial partition (under the homogeneous assumption). The number N_m can be bounded above by $\binom{n}{m}$, just as in the verification of the Kraft inequality. The probability p_m can be bounded as follows. Note this is the probability of a particular outcome of a comparison of two models. The comparison is made between their respective sum-of-squared errors plus complexity penalty, as given by Equation (14.2). The single leaf model has a single degree of freedom (mean value of the entire region), and the alternate model, based on the $(3m + 1)$-leaf has $3m + 1$ degrees of freedom. Thus, under the assumption that the data are i.i.d. zero-mean Gaussian distributed with variance σ^2, it is easy to verify that the difference between the sum-of-squared errors of the models [single-leaf model sum-of-squares minus $(3m + 1)$-leaf model sum-of-squares] is distributed as $\sigma^2 W_{3m}$, where W_{3m} is a chi-square distributed random variable with $3m$ degrees of freedom (precisely the difference between the degrees of freedom in the two models). This follows from the fact that the difference of the sum-of-squared errors is equal to the sum-of-squares of an orthogonal projection of the data onto a $3m$-dimensional subspace.

The single-leaf model is rejected if $\sigma^2 W_{3m}$ is greater than the difference between the complexity penalties associated with the two models; that is, if $\sigma^2 W_{3m} > (3m + 1)2\sigma^2$

$p(n) - 2\sigma^2 p(n) = 6m\sigma^2 p(n)$, where $2\sigma^2 p(n)$ is the penalty associated with each additional leaf in P. According to the MSE analysis in the previous section, we require $p(n) = \gamma \log n$, with $\gamma > 7/12$. In References [119–125] $\gamma = 2/3$, in which case the rejection of the single-leaf model is equivalent to $W_{3m} > 4m \log n$. The probability of this condition, $p_m = \Pr(W_{3m} > 4m \log n)$, is bounded from above using Lemma 1 of Laurent and Massart [130]: 'If W_d is chi-square distributed with d degrees of freedom, then for $s > 0$ $\Pr(W_d \geq d + s\sqrt{2d} + s^2) \leq \mathrm{e}^{-s^2/2}$'. Making the identification $d + g\sqrt{2d} + s^2 = 4m \log n$ produces the bound

$$p_m = \Pr(W_{3m} > 4m \log n) \leq \mathrm{e}^{-2m \log n + m\sqrt{[3/2(4 \log n - 3/2)]}}$$

Combining the upper bounds above, we have

$$\begin{aligned} E[|\hat{\theta}|] &\leq \sum_{m=0}^{(n-1)/3} (3m+1) \binom{n}{m} \mathrm{e}^{-2m \log n + m\sqrt{3/2(4 \log n - 3/2)}} \\ &= \sum_{m=0}^{(n-1)/3} (3m+1) \binom{n}{m} n^{-m} \mathrm{e}^{-m \log n + m\sqrt{3/2(4 \log n - 3/2)}} \end{aligned} \tag{14.8}$$

For $n \geq 270$ the exponent $-\log n + \sqrt{3/2(4 \log n - 3/2)} < 0$ and therefore

$$E[|\hat{\theta}|] \leq \sum_{m=0}^{(n-1)/3} (3m+1) \binom{n}{m} n^{-m} \leq \sum_{m=0}^{(n-1)/3} (3m+1) \frac{n^m}{m!} n^{-m} \leq \sum_{m=0}^{(n-1)/3} (3m+1)/m! < 11$$

Furthermore, note that, as $n \to \infty$, the exponent $-\log n + \sqrt{[3/2(4 \log n - 3/2)]} \to \infty$. This fact implies that the factor $\mathrm{e}^{-m \log n + m\sqrt{[3/2(4 \log n - 3/2)]}}$ tends to zero when $m > 0$. Therefore, the expected number of leafs $E[|\hat{\theta}|] \to 1$ as $n \to \infty$.

Thus, for large sensor networks, the expected number of leafs (partition pieces) in the case where there is no boundary (simply a homogeneous field) is one. To consider the inhomogeneous case where a boundary does exist, if the boundary is a Lipschitz function or has a box counting dimension of 1, there exists a pruned RDP with at most $C'\sqrt{n}$ squares (leafs) that includes the $O(\sqrt{n})$ squares of sidelength $1/\sqrt{n}$ that the boundary passes through. Thus an upper bound on the number of leafs required to describe the boundary in the noiseless case is given by $C'\sqrt{n}$.

In the presence of noise, we can use the results above for the homogeneous case to bound the number of spurious leafs due to noise (zero as n grows); as a result, for large sensor networks, we can expect at most $C'\sqrt{n}$ leafs in total. Thus, the expected energy required to transmit the final boundary description is energy $= O(\sqrt{n})$.

The *processing cost* is intimately tied to the expected size of the final tree, as this value determines how much pruning will occur. We have seen above that the communication cost is proportional to $\sqrt{n}$ and herein we shall show that the processing cost is also $O(\sqrt{n})$. At each scale $2^j/\sqrt{n}$, $j = 0, \ldots, 1/2 \log_2 n - 1$, the hierarchical algorithm passes a certain number of data or averages, n_j, corresponding to the number of squares in the best partition (up to that scale), up the tree to the next scale. We assume that a constant number of bits k is transmitted per measurement. These $k\, n_j$ bits must be transmitted approximately $2^j/\sqrt{n}$ meters (assuming the sensor domain is normalized to 1 square meter). Thus, the total in-network communication energy in bit-meters is:

$$\varepsilon = k \sum_{j=0}^{1/2 \log_2 n - 1} n_j 2^j / \sqrt{n}$$

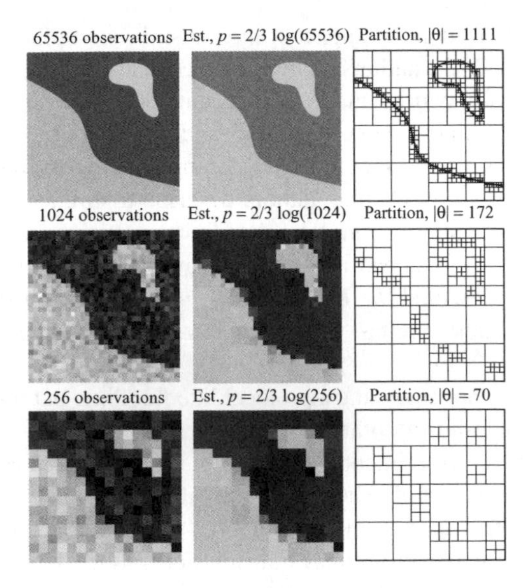

Figure 14.14 Effect of sensor network density (resolution) on boundary estimation. Column 1, noisy set of measurements; column 2, estimated boundary; and column 3, associated partition. (Reproduced by permission of IEEE [121].)

In the naive approach, $n_j = n$ for all j, and therefore $\varepsilon \approx kn$. In the hierarchical approach, first consider the case when there is no boundary. We have already seen that in such cases the tree will be pruned at each stage with high probability. Therefore, $n_j = n/4^j$ and $\varepsilon \approx 2k\sqrt{n}$. Now if a boundary of length $C\sqrt{n}$ is present, then $n_j \leq n/4^j + C\sqrt{n}$. This produces $\varepsilon \leq k(C+2)\sqrt{n}$. Thus, we see that the hierarchical algorithm results in $\varepsilon = O(\sqrt{n})$.

Finally, a performance example is shown in Figure 14.14 [121].

14.8 OPTIMAL TRANSMISSION RADIUS IN SENSOR NETWORKS

In this section we discuss the problem of finding an optimal transmission radius for flooding in sensor networks. On one hand, a large transmission radius implies that fewer retransmissions will be needed to reach the outlying nodes in the network; therefore, the message will be heard by all nodes in less time. On the other hand, a larger transmission radius involves a higher number of neighbors competing to access the medium, and therefore each node has a longer contention delay for packet transmissions. In this section we discuss this tradeoff in CSMA/CA wireless MAC protocols.

Even though flooding has some unique advantages – it maximizes the probability that all reachable nodes inside a network will receive the packet – it has several disadvantages as well. Several works have proposed mechanisms to improve flooding efficiency. The broadcast storm paper by Ni *et al.* [131] suggests a way to improve flooding by trading robustness. The authors propose to limit the number of nodes that transmit the flooded packet. The main idea is to have some nodes refrain from forwarding their packet if its transmission will not contribute to a larger coverage. Nevertheless, the basic flooding technique is in wide use for a number of querying techniques for sensor networks (in large part because of its guarantee of maximal robustness), and in this section we focus on analyzing its MAC-layer effects and improving its performance by minimizing the settling time of flooding.

Other studies have looked at the impact of the transmission radius in wireless networks. In Gupta and Kumar [132] the authors analyzed the critical transmission range to maintain connectivity in wireless networks and present a statistical analysis of the probability of connectivity. On the same line of work, Kleinrock and Silvester [133] analyze the minimum number of neighbors that a node should have to keep the network connected.

In Takagi and Kleinrock [134], the authors describe a similar tradeoff for increasing the transmission radius: a shorter range implies fewer collisions and a longer range implies moving a packet further ahead in one hop. However, in that work the authors want to maximize a parameter called *the expected one-hop progress in the desired direction*, which essentially measures how fast a packet can reach its destination in point-to-point transmissions.

All these studies were not analyzing a protocol like flooding, but instead trying to obtain an optimal transmission radius for other metrics such as connectivity, throughput or energy. In Ganesan *et al.* [135] an experimental testbed of 150 Berkeley motes [136] run flooding as the routing protocol. The study showed empirical relations between the reception and settling times – parameters used in this section – for different transmission ranges.

In this section we discuss an optimal transmission radius. However, in this case the important metric is the amount of time that a flooded packet captures the transmission medium. To accomplish the goal of minimizing the settling time, the tradeoff between reception and contention times is studied including the interaction between the MAC-layer and network-level behavior of an information dissemination scheme in wireless networks.

The network model is based on the following assumptions:

(1) The MAC protocol is based on a CSMA/CA scheme.

(2) All the nodes have the same transmission radius R.

(3) The area of the network can be approximated as a square.

(4) No mobility is considered.

(5) The nodes are deployed in either a grid or uniform topology. In a uniform topology, the physical terrain is divided into a number of cells based on the number of nodes in the network, and each node is placed randomly within each cell.

The analytical model is described by the following terms:

(1) Reception time (T_R) – average time when all the nodes in the network have received the flooded packet.

(2) Contention time (T_C) – average time between reception and transmission of a packet by all the nodes in the network.

(3) Settling time (T_S) – average time when all the nodes in the network have transmitted the flooded packet and signals the end of the flooding event.

From these definitions we observe that $T_S = T_R + T_C$. If the transmission radius of the nodes is not carefully chosen, the flooded packet may take too long to be transmitted by all the nodes in the network, impacting overall network throughput. *The more time the channel is captured by a flooding event, the fewer queries can be disseminated, and the less time the channel is available for other packet transmissions*. We can state the relation between settling time and throughput Th in sensor networks as $Th \propto 1/T_S$. So, the goal is to minimize the settling time T_S. Since the settling time is the sum of the reception and contention times, the remainder of this section will analyze the relationships between T_R and T_C with respect to the range of the transmission radius.

The reception time T_R represents the average time at which nodes received the packet. If the transmission radius of each node is increased, the reception time in the network will decrease, because there are fewer hops needed to reach outlying nodes. Therefore, the reception time T_R is directly proportional to the maximum distance between any two nodes in the network, and inversely proportional to the transmission radius. Owing to the kind of topologies considered here (grid or uniform), the maximum distance between the nodes is the diagonal of the network area. If R is transmission radius (m) and S the length of the side of the square area (m) then $T_R = cS/R$, where c is a constant.

If a node increases its transmission radius, it will increase its number of neighbors, which will cause an increase in the contention time. If we consider the area covered by the network as S^2 then the expected number of neighbors of a given node is described by $m = \pi R^2 n / S^2$, where n is the total number of nodes in the network. However, the contention time is not directly proportional to above equation. There are two phenomena that influence T_c, the *edge phenomenon* and the *back-off phenomenon*.

The *edge phenomenon* can be described as follows: nodes close to the edges of the network area will not increase their number of neighbors proportionally to the square of the radius. The reason is that only a fraction of the area covered by its transmission radius intersects the area of the network. This phenomenon is illustrated in Figure 14.15, which shows a square topology with a given node (black point). In this figure, we can observe three regions as the transmission radius is increased:

- Region 1 – when R ranges from 0 to the edge of the network (R_e).
- Region 2 – when R ranges from R_e until it covers the entire network (R_w).
- Region 3 – when R is greater than R_w.

Each of these regions will have a different expression for the number of neighbors. For the first region, the number of nodes inside the transmission radius is directly proportional to

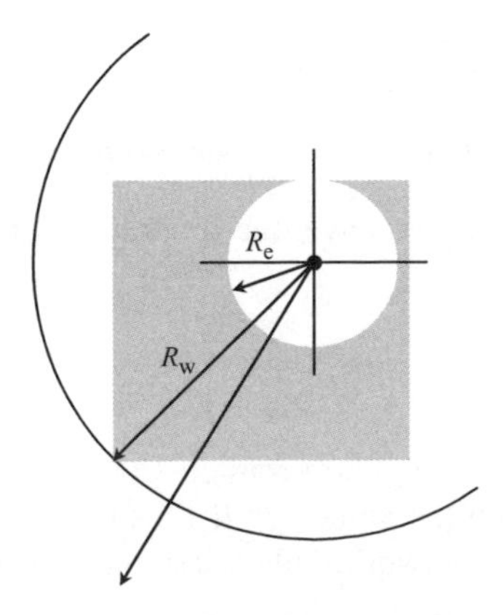

Figure 14.15 Different regions to calculate the number of neighbors versus the transmission radius of the node.

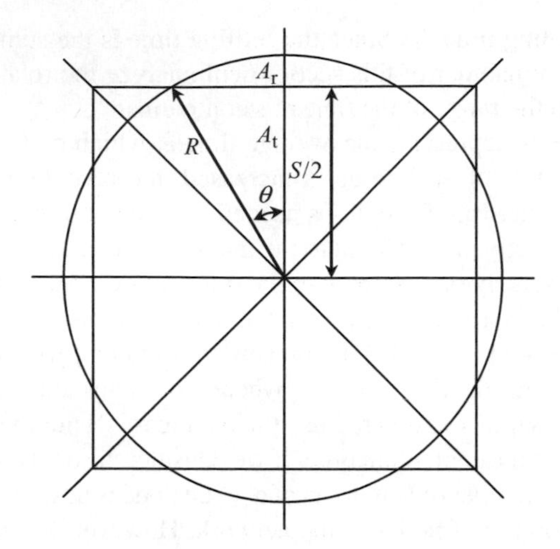

Figure 14.16 The overlapping area between the transmission radius coverage and the effective area of the network.

the square of the radius. In the second region, the number of neighbors increases proportionally to the overlapping area between the transmission range and the network area. The overlapping area (A_O) is shown in Figure 14.16

Defining A_r as the residual area beyond $S/2$, because of symmetry, the total overlapping area is given by

$$A_O = 8\left(\frac{\pi R^2}{8} - A_r\right)$$

where $A_\theta = A_r + A_t$. Since $\theta = \arccos(S/2R)$, we have $A_\theta = \theta(R^2/2)$ and $A_t = R^2 \sin(\theta)\cos(\theta)$.

As a consequence $A_r = \theta(R^2/2) - R^2 \sin(\theta)\cos(\theta)$ and we get for the center (+) and corner position ($\angle$)

$$A_{O+} = 8\left[\frac{\pi R^2}{8} - \theta\frac{R^2}{2} - R^2 \sin(\theta)\cos(\theta)\right]$$
$$A_{O+} = R^2[\pi - 4\theta - 4\sin(\theta)\cos(\theta)]$$

In the case of the lower bound, we have one quarter of a circle and ϕ is given by $\phi = \arccos(S/R)$ and

$$A_{O\angle} = 2\left[\frac{\pi R^2}{8} - \theta\frac{R^2}{2} - R^2 \sin(\theta)\cos(\theta)\right], A_{O\angle} = R^2\left[\frac{\pi}{4} - \theta - \sin(\theta)\cos(\theta)\right]$$

In the third region, the number of neighbors remains constant and is equal to the total number of nodes in the network. Since the values of R for the three different regions depend on the position of the node in the network, only the bounds of the edge phenomenon will be analyzed.

The node closest to the center of the network is the one increasing its number of neighbors most aggressively, hence it represents the upper bound. For this node, the second region begins when R is greater than $S/2$ and the third region begins when R is greater than $(S/2)\sqrt{2}$. The following equation shows the upper bound $\lceil m \rceil$ for the number of neighbors of this node:

$$\lceil m \rceil = \begin{cases} \pi \dfrac{R^2}{S^2} n & 0 < R < \dfrac{S}{2} \\ R^2[\pi - 4\theta + 4\cos(\theta)\sin(\theta)] & \dfrac{S}{2} < R \dfrac{S}{2}\sqrt{2} \\ n & \dfrac{S}{2}\sqrt{2} < R \end{cases}$$

where $\theta = \arccos(S/2R)$. The lower bound is given by nodes located on the corners of the network. In this case, there is no region 1 as such but rather the number of nodes increases as $\pi(R^2/4)$ and finishes when R equals S. The second region finishes when R equals $S\sqrt{2}$. The next equation represents the lower bound for the number of neighbors $\lfloor m \rfloor$:

$$\lfloor m \rfloor = \begin{cases} \pi \dfrac{R^2}{4S^2} n & 0 < R < S \\ R^2 \left[\dfrac{\pi}{4}\pi - \theta + \cos(\theta)\sin(\theta)\right] & S < R < S\sqrt{2} \\ n & S\sqrt{2} < R \end{cases} \quad \text{where } \theta = \arccos\left(\frac{S}{R}\right)$$

14.8.1 Back-off phenomenon

In CSMA/CA protocols, a node checks if the medium is clear before sending a packet; when the medium is clear for a small period of time, the node transmits the packet. If the channel becomes busy during this waiting period, it chooses a random time in the future to transmit the packet. This mechanism leads to a non linear relationship between the contention time and the number of neighbors. This nonlinear relationship is referred to as the back-off phenomenon.

By simulating a one-hop network with a varying number of nodes the nonlinear relationship between the number of neighbors and the contention time can be numerically approximated by [137] $f(m) = J\log^3(m)$, where m is the number of neighbors and J is a constant. If we incorporate the edge and back-off phenomena explained above, we obtain that the upper bound for the contention time T_C is given by:

$$\lceil T_C \rceil = \begin{cases} Kf\left(\pi \dfrac{R^2}{S^2} n\right) & 0 < R < \dfrac{S}{2} \\ Lf\{R^2[\pi - 4\theta + 4\cos(\theta)\sin(\theta)]\} & \dfrac{S}{2} < R < \dfrac{S}{2}\sqrt{2} \\ Mf(n) & \dfrac{S}{2}\sqrt{2} < R \end{cases}$$

and the lower bound is:

$$\lfloor T_C \rfloor = \begin{cases} Kf\left(\pi \dfrac{R^2}{4S^2} n\right) & 0 < R < S \\ Lf\left\{R^2\left[\dfrac{\pi}{4} - \theta - \cos(\theta)\sin(\theta)\right]\right\} & S < R < S\sqrt{2} \\ Mf(n) & S\sqrt{2} < R \end{cases}$$

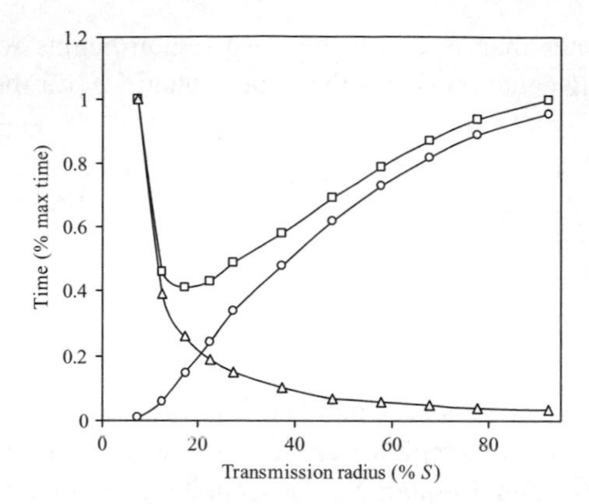

Figure 14.17 Analytical curve for the settling time, which is the sum of the reception and contention times.

where K, L and M are constants and $f(.)$ is the function described above. The settling time is the sum of the reception and contention times. Figure 14.17 shows the analytical settling time, for a network with four hundred nodes. For illustration purposes, the analytical reception and contention times are also plotted. The settling time curve shows a minimum, as expected.

14.9 DATA FUNNELING

In this section, a combination of two methods is discussed which improve the lifetime of energy constrained networks by reducing the amount of communication needed to send readings from a set of sensors to a controller. The first scheme is a packet aggregation technique, which has been already discussed, while the second scheme performs data compression. While either of the two schemes can be used separately, using them together provides maximum gain.

The main idea behind the algorithm, called data funneling [139], is the following. The controller breaks up the space into different regions (e.g. cuboids) and sends interest packets to each region, as shown in Figure 14.8(a). Upon receiving the interest packet, each node in the region will start periodically sending its readings back to the controller at an interval specified in the interest packet, usually every few minutes. Since many or all of the nodes within the region will be sending their readings back to the controller at the same time, it would be much more efficient to combine these readings into a single packet, so that only one packet with only one header travels from the region to the controller. The question is, how can all these reading be collected at a single point and combined into a single packet?

As already discussed in Section 14.3.3, the data funneling algorithm works as follows. The interest packets are sent toward the region using directional flooding. Each node that receives the interest packet checks if it is in the target region. If it is not, it computes its cost for communicating back to the controller, updates the cost field within the interest packet, and sends it on toward the specified region. This is the *directional flooding phase*.

When a node that is in the target region receives the interest packet from a neighbor node that lies outside the target region, the directional flooding phase concludes. The node realizes that it is on the border of the region and designates itself to be a border node, as shown in Figure 14.8(b). Each border node computes its cost for communicating with the controller in the same manner as was done by the nodes outside the region during the directional flooding phase. It then floods the entire region with a modified version of the interest packet. The 'cost to reach the controller' field is reset to zero and becomes the 'cost to reach the border node field.' Within the region, each node only keeps track of its cost for communicating with the border node, not its cost for communicating with the controller. Intuitively, it is as if the border node becomes the controller of the specified region. It is at one of the border nodes that all the readings from within the region will be collated into a single packet.

In addition, two new fields are added to the modified interest packet. One field keeps track of the number of hops that have been traversed between the border node and the node currently processing the packet. The other field specifies the border node's cost for communicating with the controller, and this field, once defined by the border node, does not change as the packet travels from one node to another.

Once the nodes within the region receive the modified interest packet from the border nodes, they will then route their readings to the controller via each of the border nodes in turn. Since there are several border nodes within the region, maximizing aggregation of sensor readings requires all the nodes within the region to agree to route their data via the same border node during every given round of reporting back to the controller. This is accomplished by having every node compute an identical schedule of which border node to use during each round of reporting. This is achieved by each node in the region applying the same deterministic function to the vector of costs to reach the controller seen by each border node. Since all the nodes apply the same function to the same inputs, they will all compute the same schedule, allowing them to collect all of their data at one border node during each round of reporting. The function used to compute the schedule can be similar to the function used to compute the probabilities for selecting different paths in probabilistic routing. This allows border nodes with a low cost for communicating to the controller to be used more frequently than the ones with a high cost.

As data flows within the region from the sensors to the border nodes it can be aggregated along the way, as shown in Figure 14.8(b). When the time comes to send a new round of observations back to the controller, the sensor nodes do not immediately start sending their packets. Instead, they wait an amount of time inversely proportional to their distance (in number of hops) to the border node that will be used in that round of reporting before sending their readings toward that border node. This allows the nodes that are far away from the border node to send their data earlier than the nodes that are closer to the border node. This way, nodes close to the border will first receive the readings from upstream nodes and bundle those readings with their own. In the end, all of the data to be sent out by all the nodes within the region will be collated at one border node and sent back to the controller in a single packet, as shown in Figure 14.8(b).

If α is the ratio of bits in a packet header to the total number of bits in a packet containing the header and a single sensor reading for a particular application, and m is the average number of sensor readings per transmitted packet when data funneling is employed, then the total energy expended by the network on communication is reduced by $\alpha \times (m - 1/m) \times 100\,\%$ due to data funneling if no compression of the sensor readings is done at the aggregation points. Performing compression on the sensor readings at the aggregation points within a region, as discussed in the sequel, would result in even greater energy savings. For this purpose *coding by ordering* is used.

The main idea behind 'coding by ordering' is that, when transmitting many unique pieces of data, and the order in which the data is sent is not important to the application (i.e. the transmitter may choose the order in which to send those pieces of data), then the choice of the order in which those pieces of data are sent can be used to convey additional information to the receiver. In fact it is possible to avoid explicitly transmitting some of those pieces of data, and use the ordering of the other information to convey the information contained in the pieces of data that were not sent.

Consider the case of the data funneling algorithm. In each round of reporting, the border node receives the packets containing sensor readings from n sensors in its region. It then places each node's packet (containing the node ID, which may be just the node's position, and payload) into a large superpacket containing the data of all the nodes and sends the superpacket to the controller. The border node has to include the ID of each node, which is unique, along with the node's sensor reading so as to make it clear which payload corresponds to which node. Since all of the sensor readings from the region will reach the controller at the same time and the ordering of the packets within the superpacket does not affect the application, the border node has the freedom to choose the ordering of the packets within the superpacket. This allows the border node to choose to 'suppress' some of the packets (i.e. choose not to include them in the superpacket), and order the other packets within the super-packet in such a way as to indicate the values contained within the suppressed packets.

For example, consider the case when there are four nodes with IDs 1, 2, 3 and 4 in the region. Each of the four sensors generates an independent reading, which is a value from the set $\{0, \ldots, 5\}$. The border node can choose to suppress the packet from node 4 and, instead, choose the appropriate ordering among the $3! = 6$ possible orderings of the packets from nodes 1, 2 and 3 to indicate the value generated by node 4. Note that in this case the border node need not encode the ID of the suppressed node because that information can be recovered from the fact that there are only four nodes and the packets of three of them were explicitly given in the superpacket. The question is, how many packets can be suppressed? Let n be the number of packets present at the encoder, k be the range of possible values generated by each sensor (e.g. if each sensor generates a 4-b value, then $k = 2^4$), and d be the range of node IDs of the sensor nodes. Given n, k and d, what is the largest number of packets, l, that can be suppressed?

One strategy is to have the encoder (located at the border node) throw away any l packets and appropriately order the remaining $n - l$ packets to indicate what values were contained in the suppressed packets. A total of $(n - l)!$ values can be indexed by ordering $n - l$ distinct objects. Each of the suppressed packets contains a payload that can take on any of the k possible values and an ID, which can be any value from the set of d valid IDs except for the ones that belong to the packets included in the super packet. The values contained within the suppressed packets can be regarded as symbols from a $(d - n + l) \times k$ ary alphabet, giving $(d - n + l)^l \times k^l$ possible values for the suppressed packets. In order for it to be possible to suppress l out of n packets in this manner, the following relationship must be satisfied $(n - l)! \geq (d - n + l)^l k^l$; or by using approximation, $n! = \sqrt{2\pi n}(n/e)^n$ we have

$$(n - l)[\ln(n - i) - 1] + 0.5 \ln[2\pi(n - l)] - l \ln k - l \ln(d - n + l) \geq 0$$

If this inequality is satisfied, then it is possible to suppress l packets. The suppressed packets can contain identical values. While their payloads may be identical, each packet has to have a unique ID. Since each packet has to be identified with a unique ID from among the d possible IDs, and $n - l$ of the possible IDs are taken up by the transmitted packets, there are $\binom{d-n+l}{l}$ possible combinations of IDs that the l suppressed packets can take on; therefore, when enumerating the possible values contained within the suppressed packets, the $(d - n + l)^l$ term should be replaced by $\binom{d-n+l}{l}$ giving $(n - l)! \geq \binom{d-n+l}{l}k^l$ as the relationship that must be satisfied in order for it to be possible to suppress l out of n packets. Again, approximation can be used to convert the inequality to the following equivalent relationship with more manageable terms:

$$\ln(2\pi) + l + (l + 0.5)\ln l + (n - l + 0.5)\ln(n - l) + (d - n + 0.5)\ln(d - n) - l\ln k \\ - (d - n + l + 0.5)\ln(d - n + l) - n \geq 0$$

The two schemes presented above assume that the encoder will suppress l packets without giving much consideration to which l packets are suppressed; however, since the encoder has the freedom to choose which l packets to suppress, the number of values that may be indexed by dropping l out of n packets and ordering the remaining $n - l$ packets increases by a factor of $\binom{n}{l}$. Combining this with the previous condition gives the following relationship, which must be satisfied if it is to be possible to suppress l out of n packets:

$$\frac{n!}{l!} \geq \binom{d-n+l}{l}k^l$$

As before, applying approximation and some manipulation can reduce the inequality to an equivalent one:

$$(n + 0.5)\ln n + (d - n + 0.5)\ln(d - n) + 0.5\ln(2\pi) - d - 0.5\ln(d - n + l) \\ + (d - n + l)[\ln(d - n + l) - 1] + l\ln k \geq 0$$

For example when $n = 30$, using the low-complexity scheme allows the encoder to suppress $l = 6$ packets, a 20 % saving in energy spent on transmitting sensor data. The bound on the number of packets that can be suppressed at $n = 30$ is 10. As n grows, the savings also increase. When $n = 100$, the low-complexity scheme provides 32 % savings, the higher-complexity scheme guarantees 44 % savings, while the bound is 53 %.

14.10 EQUIVALENT TRANSPORT CONTROL PROTOCOL IN SENSOR NETWORKS

The need for a transport layer for data delivery in WSN can be questioned under the premise that data flows from source to sink are generally loss tolerant. While the need for end-to-end reliability may not exist due to the sheer amount of correlated data flows, an event in the sensor field still needs to be tracked with a certain accuracy at the sink. So, instead of a traditional TCP layer, the sensor network paradigm necessitates an *event-to-sink* reliability notion at the transport layer that will be referred to as equivalent TCP (ETCP) [143–148]. Such a notion of collective identification of data flows from the event to the sink was illustrated earlier in Figure 14.10. An example of ETCP is event-to-sink reliable transport

(ESRT) protocol for WSN, discussed in Sankarasubramaniam *et al.* [123]. Some of its features are

(1) *Self-configuration* – ESRT is self-configuring and achieves flexibility under dynamic topologies by self-adjusting the operating point.

(2) *Energy awareness* – if reliability levels at the sink are found to be in excess of that required, the source nodes can conserve energy by reducing their reporting rate.

(3) *Congestion control* – required event detection accuracy may be attained even in the presence of packet loss due to network congestion. In such cases, however, a suitable congestion control mechanism can help conserve energy while maintaining desired accuracy levels at the sink. This is done by conservatively reducing the reporting rate.

(4) *Collective identification* – ESRT does not require individual node IDs for operation. This is also in tune with ESRT model rather than the traditional end-to-end model. More importantly, this can ease implementation costs and reduce overhead.

(5) *Biased Implementation* – the algorithms of ESRT mainly run on the sink with minimum functionalities required at sensor nodes.

In another example [141, 142], the PSFQ mechanism is used for reliable retasking/reprogramming in WSN. PSFQ is based on slowly injecting packets into the network, but performing aggressive hop-by-hop recovery in case of packet loss. The pump operation in PSFQ simply performs controlled flooding and requires each intermediate node to create and maintain a data cache to be used for local loss recovery and in-sequence data delivery. Although this is an important transport layer solution for WSN, it is applicable only for strict sensor-to-sensor reliability and for purposes of control and management in the reverse direction from the sink to sensor nodes. Event detection/tracking in the forward direction does not require guaranteed end-to-end data delivery as in PSFQ. Individual data flows are correlated and loss-tolerant to the extent that desired event features are collectively and reliably informed to the sink. Hence, the use of PSFQ for the forward direction can lead to a waste of valuable resources. In addition to this, PSFQ does not address packet loss due to congestion. For this reason in the sequel we elaborate more ESRT algorithm.

The operation of the algorithm is based on the notion of *observed event reliability*, r_i (the number of received data packets in decision interval i at the sink), and *desired event reliability*, R (the number of data packets required for reliable event detection). R depends on the application. If the observed event reliability, r_i, is greater than the desired reliability, R, then the event is deemed to be reliably detected. Else, appropriate action needs to be taken to achieve the desired reliability, R. With the above definition, r_i can be computed by stamping source data packets with an event ID and incrementing the received packet count at the sink each time the ID is detected in decision interval i. Note that this does not require individual identification of sensor nodes. Further, we model any increase in source information about the event features as a corresponding increase in the reporting rate, f, of sensor nodes. The reporting rate of a sensor node is defined as the number of packets sent out per unit time by that node. The transport problem in WSN is to *configure the reporting*

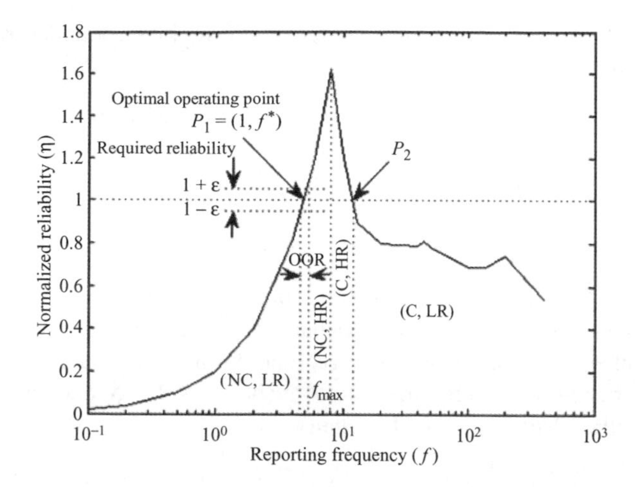

Figure 14.18 The five characteristic regions in the normalized reliability, η, vs reporting frequency, f, behavior. Number of source nodes-81. (Reproduced by permission of IEEE [140].)

rate, f, of source nodes so as to achieve the required event detection reliability, R, at the sink with minimum resource utilization.

In order to study the relationship between the observed reliability at the sink, r, and the reporting frequency, f, of sensor nodes, Sankarasubramaniam *et al.* [140] developed an evaluation environment using *ns*-2 . The parameters used in the study are:

Area of sensor field: 100 × 100 m^2
Number of sensor nodes: 200
Radio range of a sensor node: 40 m
Packet length: 30 bytes
IFQ length: 65 packets
Transmit power: 0.660 W
Receive power: 0.395 W
Decision interval (τ): 10 s

Event centers (X_{ev}, Y_{ev}) were randomly chosen and all sensor nodes within the event radius behave as sources for that event. Let the desired reliability as laid down by the application be R. Hence, a normalized measure of reliability is $\eta = r/R$. Parameter, η_i denotes the normalized reliability at the end of decision interval i. The results of the above experiment are shown in Figure 14.18.

The aim is to operate as close to $\eta = 1$ as possible, while utilizing minimum network resources (f close to f^* in Figure 14.18). We call this the *optimal operating point*, marked as P_1 in Figure 14.18. For practical purposes, we define a tolerance zone of width 2ε around P_1. Here, ε is a protocol parameter to be optimized. Although the event is reliably detected at P_2 too, the network is congested and some source data packets are lost. Event reliability is achieved only because the high reporting frequency of source nodes compensates for this congestion loss. However, this is a waste of limited energy reserves and hence is not the operating point of interest. Similar reasoning holds for $\eta > 1 + \varepsilon$.

From Figure 14.25, we identify five characteristic regions (bounded by dotted lines) using the following decision boundaries:

(NC,LR): $f < f_{\max}$ and $\eta < 1 - \varepsilon$ (no congestion, low reliability)
(NC,HR): $f \leq f_{\max}$ and $\eta > 1 + \varepsilon$ (no congestion, high reliability)
(C,HR): $f > f_{\max}$ and $\eta > 1$ (congestion, high reliability)
(C,LR): $f > f_{\max}$ and $\eta \leq 1$ (congestion, low reliability)
OOR: $f < f_{\max}$ and $1 - \varepsilon \leq \eta \leq 1 + \varepsilon$ (optimal operating region)

As seen earlier, the sink derives a reliability indicator η_i at the end of decision interval i. Coupled with a congestion detection mechanism (to determine $f \gtrless f_{\max}$), this can help the sink determine in which of the above regions the network currently resides. Hence, these characteristic regions identify the state of the network. Let S_i denote the network state variable at the end of decision interval i. Then,

$$S_i \in \{(NC, LR), (NC, HR), (C, HR), (C, LR), OOR\}$$

The operation of ESRT is closely tied to the current network state S_i. The ESRT protocol state model and transitions are shown in Figure 14.19. ESRT identifies the current state S_i from: (1) reliability indicator η_i computed by the sink for decision interval i; (2) a congestion detection mechanism, using the decision boundaries as defined above. Depending on the

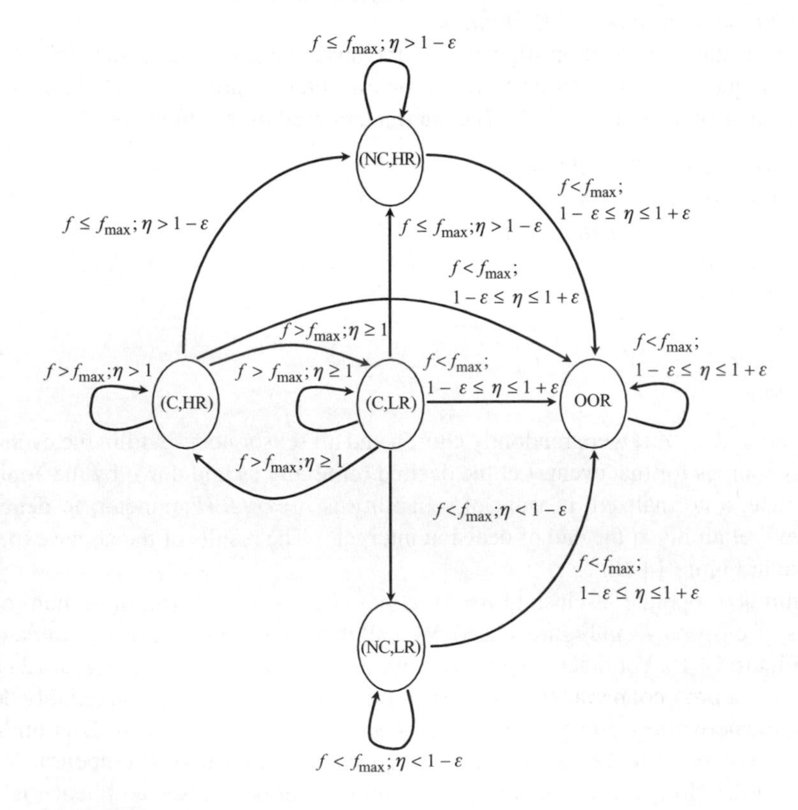

Figure 14.19 ESRT protocol state model and transitions. (Reproduced by permission of IEEE [27].)

current state S_i, and the values of f_i and η_i, ESRT then calculates the updated reporting frequency f_{i+1} to be broadcast to the source nodes. At the end of the next decision interval, the sink derives a new reliability indicator η_{i+1} corresponding to the updated reporting frequency f_{i+1} of source nodes. In conjunction with any congestion reports, ESRT then determines the new network state S_{i+1}. This process is repeated until the optimal operating region (state OOR) is reached. The state model of the ESRT protocol and state transitions are shown in Figure 14.19. The following reporting rate updating rules are used [140]:

(NC,LR) $$f_{i+1} = \frac{f_i}{\eta_i}$$

(NC,HR) $$f_{i+1} = \frac{f_i}{2}\left(1 + \frac{1}{\eta_i}\right)$$

(C,HR) $$f_{i+1} = \frac{f_i}{\eta_i}$$

(C,LR) $$f_{i+1} = f_i^{(\eta_i/k)}$$

where k denotes the number of successive decision intervals for which the network has remained in state (C,LR) including the current decision interval

OOR $$f_{i+1} = f_i$$

In order to determine the current network state S_i in ESRT, the sink must be able to detect congestion in the network. ESRT uses a congestion detection mechanism based on local buffer level monitoring in sensor nodes. Any sensor node whose routing buffer overflows due to excessive incoming packets is said to be congested and it informs the sink of the same. For more on system performance see Sankarasubramaniam *et al.* [140].

REFERENCES

[1] I.F. Akyildiz, W. Su, Y. Sankarasubramaniam and E. Cayirci, A survey on sensor networks, *Comput. Networks*, 2002, pp. 393–422.

[2] G.J. Pottie and W.J. Kaiser, Wireless integrated network sensors, *Commun. ACM*, vol. 43, no. 5, 2000, pp. 51–58.

[3] J. Rabaey, M.J. Ammer, J.L. da Silva Jr, D. Patel and S. Roundy, Picoradio supports *ad hoc* ultra-low power wireless networking, *Comput. Mag.*, July 2000, pp. 42–48.

[4] S. Tilak, N. Abu-Ghazaleh and W. Heinzelman, A taxonomy of wireless micro-sensor network models, *ACM Mobile Comput. Commun. Rev. (MC2R)*, vol. 6, no. 2, April 2002, pp. 28–36.

[5] A. Mainwaring, J. Polastre, R. Szewczyk, D. Culler and J. Anderson, Wireless sensor networks for habitat monitoring, in *1st Workshop on Sensor Networks and Applications*, Atlanta, GA, October 2002, pp. 88–97.

[6] G.D. Abowd and J.P.G. Sterbenz, Final report on the interagency workshop on research issues for smart environments, *IEEE Person. Commun.*, October 2000, pp. 36–40.

[7] J. Agre and L. Clare, An integrated architecture for cooperative sensing networks, *IEEE Comput. Mag.*, May 2000, pp. 106–108.

[8] A. Bakre and B.R. Badrinath, I-TCP: indirect TCP for mobile hosts, in *Proc. 15th Int. Conf. Distributed Computing Systems*, Vancouver, BC, May 1995, pp. 136–143.

[9] P. Bonnet, J. Gehrke and P. Seshadri, Querying the physical world, *IEEE Person. Commun.*, October 2000, pp. 10–15.

[10] B.G. Celler, T. Hesketh, W. Earnshaw and E. Ilsar, An instrumentation system for the remote monitoring of changes in functional health status of the elderly, in *Int. Conf. IEEE-EMBS*, NewYork, 1994, pp. 908–909.

[11] A. Chandrakasan, R. Amirtharajah, S. Cho, J. Goodman, G. Konduri, J. Kulik, W. Rabiner and A. Wang, Design considerations for distributed micro-sensor systems, in *Proc. IEEE 1999 Custom Integrated Circuits Conf.*, San Diego, CA, May 1999, pp. 279–286.

[12] S. Cho and A. Chandrakasan, Energy-efficient protocols for low duty cycle wireless microsensor, in *Proc. 33rd Annual Hawaii Int. Conf. System Sciences*, Maui, HI, vol. 2, 2000, p. 10.

[13] G. Coyle *et al.*, Home telecare for the elderly, *J. Telemed. Telecare*, vol. 1, 1995, pp. 183–184.

[14] I.A. Essa, Ubiquitous sensing for smart and aware environments, *IEEE Person. Commun.*, October 2000, pp. 47–49.

[15] D. Estrin, R. Govindan, J. Heidemann and S. Kumar, Next century challenges: scalable coordination in sensor networks, in *ACM MobiCom'99*, Washingtion, DC, 1999, pp. 263–270.

[16] P. Favre, N. Joehl, A. Vouilloz, P. Deval, C. Dehollain and M. J. Declerz, A 2 V, 600 A, 1 GHz BiCMOS super regenerative receiver for ISM applications, *IEEE J. Solid St. Circuits*, vol. 33, 1998, pp. 2186–2196.

[17] K. Govil, E. Chan and H. Wasserman, Comparing algorithms for dynamic speed-setting of a low-power CPU, in *Proc. ACM MobiCom'95*, Berkeley, CA, November 1995, pp. 13–25.

[18] M.P. Hamilton and M. Flaxman, Scientific data visualization and biological diversity: new tools for spatializing multimedia observations of species and ecosystems, *Landscape Urban Plann.*, vol. 21, 1992, pp. 285–297.

[19] M.P. Hamilton and Hummercams, robots, and the virtual reserve, Directors Notebook, 6 February 2000; available from www.jamesreserve.edu/news.html

[20] B. Halweil, Study finds modern farming is costly, *World Watch*, vol. 14, no. 1, 2001, pp. 9–10.

[21] S. Hedetniemi and A. Liestman, A survey of gossiping and broadcasting in communication networks, *Networks*, vol. 18, no. 4, 1988, pp. 319–349.

[22] W.R. Heinzelman, A. Chandrakasan and H. Balakrishnan, Energy-efficient communication protocol for wireless microsensor networks, in *IEEE Proc. Hawaii Int. Conf. System Sciences*, January 2000, pp. 1–10.

[23] W.R. Heinzelman, J. Kulik and H. Balakrishnan, Adaptive protocols for information dissemination in wireless sensor networks, *Proc. ACM MobiCom'99*, Seattle, WA, 1999, pp. 174–185.

[24] C. Herring and S. Kaplan, Component-based software systems for smart environments, *IEEE Person. Commun.*, October 2000, pp. 60–61.

[25] G. Hoblos, M. Staroswiecki and A. Aitouche, Optimal design of fault tolerant sensor networks, *IEEE Int. Conf. Control Applications*, Anchorage, AK, September 2000, pp. 467–472.

[26] T. Imielinski and S. Goel, DataSpace: querying and monitoring deeply networked collections in physical space, in *ACM Int. Workshop on Data Engineering for Wireless and Mobile Access MobiDE 1999*, Seattle, WA, 1999, pp. 44–51.

[27] C. Intanagonwiwat, R. Govindan and D. Estrin, Directed diffusion: a scalable and robust communication paradigm for sensor networks, in *Proc. ACM Mobi- Com'00, Boston,* MA, 2000, pp. 56–67.

[28] P. Johnson *et al.*, Remote continuous physiological monitoring in the home, *J. Telemed. Telecare*, vol. 2, no. 2, 1996, pp. 107–113.

[29] J.M. Kahn, R.H. Katz and K.S.J. Pister, Next century challenges: mobile networking for smart dust, in *Proc. ACM MobiCom'99*, Washington, DC, 1999, pp. 271–278.

[30] T.H. Keitt, D.L. Urban and B.T. Milne, Detecting critical scales in fragmented landscapes, *Conserv. Ecol.*, vol. 1, no. 1, 1997, p. 4. Available from www.consecolo.org/vol1/iss1/art4

[31] R. Kravets, K. Schwan and K. Calvert, Power-aware communication for mobile computers, in *Proc. Mo-MUC'99*, San Diego, CA, November 1999, pp. 64–73.

[32] H. Lee, B. Han, Y. Shin and S. Im, Multipath characteristics of impulse radio channels, in *Proc. IEEE Vehicular Technology Conf.*, Tokyo, vol. 3, 2000, pp. 2487–2491.

[33] P. Letteri and M.B. Srivastava, Adaptive frame length control for improving wireless link throughput, range and energy efficiency, in *Proc. IEEE INFOCOM'98*, San Francisco, CA, March 1998, pp. 564–571.

[34] S. Meguerdichian, F. Koushanfar, G. Qu and M. Potkonjak, Exposure in wireless *ad-hoc* sensor networks, in *Proc. ACM MobiCom'01*, Rome, 2001, pp. 139–150.

[35] T. Melly, A. Porret, C.C. Enz and E.A. Vittoz, A 1.2 V, 430 MHz, 4 dBm power amplifier and a 250 W frontend, using a standard digital CMOS process, in *IEEE Int. Symp. Low Power Electronics and Design Conf.*, San Diego, CA, August 1999, pp. 233–237.

[36] F.R. Mireles and R.A. Scholtz, Performance of equicorrelated ultra-wideband pulse-position-modulated signals in the indoor wireless impulse radio channel, in *IEEE Conf. Communications, Computers and Signal Processing*, vol. 2, 1997, pp. 640–644.

[37] Y.H. Nam, Z. Halm, Y.J. Chee and K.S. Park, Development of remote diagnosis system integrating digital telemetry for medicine, in *Int. Conf. IEEE-EMBS*, Hong Kong, 1998, pp. 1170–1173.

[38] N. Noury, T. Herve, V. Rialle, G. Virone, E. Mercier, G. Morey, A. Moro and T. Porcheron, Monitoring behavior in home using a smart fall sensor, in *IEEE-EMBS Special Topic Conf. Microtechnologies in Medicine and Biology*, October 2000, pp. 607–610.

[39] M. Ogawa, T. Tamura and T. Togawa, Fully automated biosignal acquisition in daily routine through 1 month, in *Int. Conf. IEEE-EMBS*, Hong Kong, 1998, pp. 1947–1950.

[40] N. Priyantha, A. Chakraborty and H. Balakrishnan, The cricket location-support system, in *Proc. ACM MobiCom'00*, August 2000, pp. 32–43.

[41] A. Perrig, R. Szewczyk, V. Wen, D. Culler and J.D. Tygar, SPINS: security protocols for sensor networks, in *Proc. ACM MobiCom'01*, Rome, 2001, pp. 189–199.

[42] E.M. Petriu, N.D. Georganas, D.C. Petriu, D. Makrakis and V.Z. Groza, Sensor-based information appliances, in *IEEE Instrum. Msmt. Mag.*, December 2000, pp. 31–35.

[43] A. Porret, T. Melly, C.C. Enz and E.A. Vittoz, A low-power low-voltage transceiver architecture suitable for wireless distributed sensors network, in *IEEE Int. Symp. Circuits and Systems'00*, Geneva, vol. 1, 2000, pp. 56–59.

[44] G.J. Pottie and W.J. Kaiser, Wireless integrated network sensors, *Commun. ACM*, vol. 43, no. 5, 2000, pp. 551–558.

[45] J. Rabaey, J. Ammer, J.L. da Silva Jr and D. Patel, Pico-radio: *ad-hoc* wireless networking of ubiquitous low-energy sensor/monitor nodes, in *Proc. IEEE Computer Society Annual Workshop on VLSI (WVLSI'00)*, Orlanda, FL, April 2000, pp. 9–12.

[46] J.M. Rabaey, M.J. Ammer, J.L. da Silva Jr, D. Patel and S. Roundy, Picoradio supports *ad hoc* ultra-low power wireless networking, *IEEE Comput. Mag.*, 2000, pp. 42–48.

[47] V. Rodoplu and T.H. Meng, Minimum energy mobile wireless networks, *IEEE J. Selected Areas Commun.*, vol. 17, no. 8, 1999, pp. 1333–1344.

[48] A. Savvides, C. Han and M. Srivastava, Dynamic fine-grained localization in ad-hoc networks of sensors, in *Proc. ACM MobiCom'01*, Rome, July 2001, pp. 166–179.

[49] C. Shen, C. Srisathapornphat and C. Jaikaeo, Sensor information networking architecture and applications, *IEEE Person. Commun.*, August 2001, pp. 52–59.

[50] E. Shih, B.H. Calhoun, S. Cho and A. Chandrakasan, Energy-efficient link layer for wireless microsensor networks, in *Proc. IEEE Computer Society Workshop on VLSI 2001*, Orlando, FL, April 2001, pp. 16–21.

[51] E. Shih, S. Cho, N. Ickes, R. Min, A. Sinha, A. Wang and A. Chandrakasan, Physical layer driven protocol and algorithm design for energy-efficient wireless sensor networks, in *Proc. ACM MobiCom'01*, Rome, July 2001, pp. 272–286.

[52] B. Sibbald, Use computerized systems to cut adverse drug events: report, *CMAJ. Can. Med. Assoc. J.*, vol. 164, no. 13, 2001, p. 1878.

[53] S. Singh, M. Woo and C.S. Raghavendra, Power-aware routing in mobile *ad hoc* networks, in *Proc. ACM MobiCom'98*, Dallas, TX, 1998, pp. 181–190.

[54] K. Sohrabi, J. Gao, V. Ailawadhi and G.J. Pottie, Protocols for self-organization of a wireless sensor network, *IEEE Person. Commun.*, October 2000, pp. 16–27.

[55] Y. Tseng, S. Wu, C. Lin and J. Sheu, A multi-channel MAC protocol with power control for multi-hop mobile *ad hoc* networks, in *IEEE Int. Conf. Distributed Computing Systems*, Mesa, AZ, April 2001, pp. 419–424.

[56] B. Walker and W. Steffen, An overview of the implications of global change of natural and managed terrestrial ecosystems, *Conserv. Ecol.*, vol. 1, no. 2, 1997. Available from www.consecol.org/vol1/iss2/art2

[57] B. Warneke, B. Liebowitz and K.S.J. Pister, Smart dust: communicating with a cubic-millimeter computer, *IEEE Comput.*, January 2001, pp. 2–9.

[58] www.fao.org/sd/EIdirect/EIre0074.htm

[59] www.alertsystems.org

[60] A. Woo and D. Culler, A transmission control scheme for media access in sensor networks, in *Proc. ACM MobiCom'01*, Rome, July 2001, pp. 221–235.

[61] S. Wu, C. Lin, Y. Tseng and J. Sheu, A new multi channel MAC protocol with on-demand channel assignment for multihop mobile *ad hoc* networks, *Int. Symp. Parallel Architectures, Algorithms, and Networks, I-SPAN 2000*, Dallas, TX, 2000, pp. 232–237.

[62] S. Wu, Y. Tseng and J. Sheu, Intelligent medium access for mobile *ad hoc* networks with busy tones and power control, *IEEE J. Selected Areas Commun.*, September 2000, pp. 1647–1657.
[63] Y. Xu, J. Heidemann and D. Estrin, Geography-informed energy conservation for *ad hoc* routing, in *Proc. ACM MobiCom'2001*, Rome, July 2001.
[64] M. Zorzi and R. Rao, Error control and energy consumption in communications for nomadic computing, *IEEE Trans. Computers*, vol. 46, no. 3, 1997, pp. 279–289.
[65] J. Elson and D. Estrin, Random, ephemeral transaction identifiers in dynamic sensor networks, in *Proc. 21st Int. Conf. Distributed Computing Systems*, Mesa, AZ, April 2001, pp. 459–468.
[66] A. Howard, M. J. Mataric and G. S. Sukhatme, An incremental self-deployment algorithm for mobile sensor networks, *Autonomous Robots*, Special Issue on Intelligent Embedded Systems, vol. 13, 2002, pp. 113–126.
[67] T. Clouqueur, V. Phipatanasuphorn, P. Ramanathan and K. Saluja, Sensor deployment strategy for target detection, in *1st Workshop on Sensor Networks and Applications*, Atlanta, GA, October 2002, pp. 42–48.
[68] W. Rabiner Heinzelman, A. Chandrakasan and H. Balakrishnan, Energy-efficient communication protocol for wireless microsensor networks, in *Proc. 33rd Int. Conf. System Sciences (HICSS '00)*, January 2000, pp. 1–10.
[69] S. Lindsey and C. S. Raghavendra, PEGASIS: power efficient gathering in sensor information systems, in *2002 IEEE Aerospace Conf.*, March 2002, pp. 1–6.
[70] B. Krishnamachari, S. Wicker, R. Bejar and Marc Pearlman, Critical density thresholds in distributed wireless networks, in *Communications, Information and Network Security*, H. Bhargava, H.V. Poor, V. Tarokh and S. Yoon (eds). Kluwer: Norwell, MA, 2002, pp. 1–15.
[71] C. Intanagonwiwat, R. Govindan and D. Estrin, Directed diffusion: a scalable and robust communication paradigm for sensor networks, *Proc. ACM Mobicom*, Boston MA, August 2000, pp. 1–12.
[72] D. Braginsky and D. Estrin, Rumor routing algorithm for sensor networks, in *First Workshop on Sensor Networks and Applications (WSNA)*, Atlanta, GA, October 2002, pp. 1–12.
[73] D. Petrovic, R.C. Shah, K. Ramchandran and J. Rabaey, Data funneling: routing with aggregation and compression for wireless sensor networks, in *Proc. 1st IEEE Int. Workshop Sensor Network Protocols and Applications*, Anchorage, AK, May 2003, pp. 1–7.
[74] S. Verdu, Recent advances on the capacity of wideband channels in the low-power regime, *IEEE Wireless Commun.*, August 2002, pp. 40–45.
[75] E. J. Duarte-Melo and M. Liu, Data-gathering wireless sensor networks: organization and capacity, *Comput. Networks*, November 2003, pp. 519–537.
[76] K. Kalpakis, K. Dasgupta and P. Namjoshi, Maximum Lifetime Data Gathering and Aggregation in Wireless Sensor Networks, in *Proc. 2002 IEEE Int. Conf. Networking (ICN'02)*, Atlanta, GA, 26–29 August 2002. pp. 685–696.
[77] M. Bhardwaj, T. Garnett and A.P. Chandrakasan, Upper bounds on the lifetime of sensor networks, *Proc. ICC 2001*, Helsinki, June 2001, pp. 1–6.
[78] K. Kant Chintalapudi and R. Govindan, Localized edge detection in a sensor field, *SNPA, 2003*, pp. 1–11.

[79] R. Nowak and U. Mitra, Boundary estimation in sensor networks: theory and methods, in *2nd Int. Workshop on Information Processing in Sensor Networks*, Palo Alto, CA, 22–23 April 2003, pp. 1–16.

[80] D. Li, K. Wong, Y.H. Hu and A. Sayeed, Detection, classification and tracking of targets in distributed sensor networks, *IEEE Signal Proc. Mag.*, vol. 19, no. 2, March, 2002 pp. 1–23.

[81] C.-Y. Wan, A.T. Campbell and L. Krishnamurthy, PSFQ: a reliable transport protocol for wireless sensor networks, in *First ACM International Workshop on Wireless Sensor Networks and Applications (WSNA 2002)*, Atlanta, GA, 28 September 2002, pp. 1–11.

[82] W. Ye, J. Heidemann and D. Estrin, An energy-efficient MAC protocol for wireless sensor networks, in *Proc. 21st Int. Annual Joint Conf. IEEE Computer and Communications Societies (INFOCOM 2002)*, New York, June, 2002 pp. 1–10.

[83] Y. Sankarasubramaniam, O.B. Akan and I.F. Akyildiz, ESRT: event-to-sink reliable transport in wireless sensor networks, in *Proc. ACM MobiHoc'03*, Annapolis, MD, June 2003, pp. 177–188.

[84] F. Stann and J. Heidemann, RMST: reliable data transport in sensor networks, in *Proc. 1st IEEE Int. Workshop Sensor Networks Protocols and Applications*, Anchorage, AK, May 2003, pp. 1–11.

[85] M. Zuniga and B. Krishnamachari, Optimal transmission radius for flooding in large scale sensor networks, in *Workshop on Mobile and Wireless Networks, MWN 2003*, held in conjunction with the *23rd IEEE Int. Conf. Distributed Computing Systems (ICDCS)*, May 2003, pp. 1–29.

[86] G. Lu, B. Krishnamachari and C.S. Raghavendra, An adaptive energy efficient and low-latency MAC for data gathering in sensor networks, in *4th Int. Workshop on Algorithms for Wireless, Mobile, Ad Hoc and Sensor Networks (WMAN 04)*, held in conjunction with the *IEEE IPDPS Conf. 18th Int. Parallel and Distributed Processing Symp.*, April 2004, pp. 1–12.

[87] A. Depedri, A. Zanella and R. Verdone, An energy efficient protocol for wireless sensor networks, in *Autonomous Intelligent Networks and Systems (AINS 2003)*, Menlo Park, CA, 30 June to 1 July 2003, pp. 1–6.

[88] A. P. Chandrakasan, R. Min, M. Bhardwaj, S. Cho and A. Wang, Power aware wireless microsensor systems, Keynote Paper ESSCIRC, Florence, September 2002, pp. 1–8.

[89] P. Chen, B. O'Dea and E. Callaway, Energy efficient system design with optimum transmission range for wireless *ad hoc* networks, in *IEEE Int. Conf. Communication (ICC 2002)*, vol. 2, 28 April to 2 May 2002, pp. 945–952.

[90] S. Banerjee and A. Misra, Energy efficient reliable communication for multi-hop wireless networks, *J. Wireless Networks*, pp. 1–23.

[91] N. Lay, C. Cheetham, H. Mojaradi and J. Neal, Developing low-power transceiver technologies for *in situ* communication applications, IPN Progress Report 42–147, 15 November 2001, pp. 1–22.

[92] Y. Prakash and S.K.S Gupta, Energy efficient source coding and modulation for wireless applications, *IEEE Wireless Commun. Networking Conf.*, 2003, vol. 1, 16–20 March 2003, pp. 212–217.

[93] M. Khan, G. Pandurangan and B.Bhargava, Energy-efficient routing schemes for

wireless sensor networks, Technical Report CSD TR 03-013, Department of Computer Science, Purdue University, 2003, pp. 1–12.
[94] E. Shih, S-H Cho, N. Ickes, R. Min, A. Sinha, A. Wang and A. Chandrakasan, Physical layer driven protocol and algorithm design for energy-efficient wireless sensor networks, in *Proc. 7th Annual Int. Conf. Mobile Computing and Networking*, Rome, 2001, pp. 272–287.
[95] D. Ganesan, R. Govindan, S. Shenker and D. Estrin, Highly resilient, energy efficient multipath routing in wireless sensor networks, *Mobile Comput. Commun. Rev. (MC2R)*, vol. 1, no. 2, 2002, pp. 10–24.
[96] B. Krishnamachari and S. Iyengar, Distributed bayesian algorithms for fault-tolerant event region detection in wireless sensor networks, *IEEE Trans. Comput.*, vol. 53, 2004, pp. 241–250.
[97] C. Karlof and D. Wagner, Secure routing in sensor networks: attacks and countermeasures, in *Proc. 1st IEEE Int. Workshop Sensor Networks Protocol and Applications*, Anchorage, AK, May 2003, pp. 1–15.
[98] A. Howard, M. Matari and G. Sukhatme, An incremental self-deployment algorithm for mobile sensor networks, in *Autonomous Robots, Special Issue on Intelligent Embedded Systems*, vol. 13(2). Kluwer Academic: Norwell, MA, 2002, pp. 113–126.
[99] T. Balch and M. Hybinette: Behavior-based coordination of large-scale robot formations, in *Proc. Fourth Int. Conf. Multiagent Systems (ICMAS')*, Boston, MA, 2000, pp. 363–364.
[100] G. Dedeoglu and G.S. Sukhatme, Landmark-based matching algorithms for cooperative mapping by autonomous robots, in *Distributed Autonomous Robotics Systems*, L.E. Parker, G.W. Bekey and J. Barhen (eds), vol. 4. Springer: Berlin, 2000, pp. 251–260.
[101] A. Elfes, Sonar-based real-world mapping and navigation, in *IEEE J. Robot. Autom.*, vol. RA-3, 1987, pp. 249–265.
[102] D.W. Gage, Command control for many-robot systems, in *AUVS-, the Nineteenth Annual AUVS Technical Symp.*, Hunstville AB, 1992, pp. 22–24. Reprinted in *Unmanned Syst. Mag.*, vol. 10, no. 4, Fall 1992, pp. 28–34.
[103] B.P. Gerkey, R.T. Vaughan, K. Stoy, A. Howard, G.S. Sukhatme and M. J. Matarić, Most valuable player: a robot device server for distributed control, in *Proc. of the IEEE/RSJ Int. Conf. on Intelligent Robots and Systems (IROSOl)*, Wailea, HJ, 2001, pp. 1226–1231.
[104] A. Howard, M.J. Matarić and G.S. Sukhatme, Mobile sensor network deployment using potential fields: a distributed, scalable solution to the area coverage problem, in *Distributed Autonomous Robotic Systems 5: Proc. 6th Int. Conf. Distributed Autonomous Robotic Systems (DARS02)*, Fukuoka, 2002, pp. 299–308.
[105] C. Intanagonwiwat, R. Govindan and D. Estrin, Directed diffusion: a scalable and robust communication paradigm for sensor networks, in *Proc. Sixth Annual International Conf. on Mobile Computing and Networks (MobiCOM 2000)*, Boston, MA, 2000, pp. 56–67.
[106] M. López-Sánchez, F. Esteva, R.L. de Mántaras, C. Sierra and J. Amat, Map generation by cooperative low-cost robots in structured unknown environments, *Autonomous Robots*, vol. 5, 1998, pp. 53–61.

[107] T. Lozano-Perez and M. Mason, Automatic synthesis of fine-motion strategies for robots, *Int. J. Robot. Res.*, vol. 3, 1984, pp. 3–24.
[108] D. Payton, M. Daily, R. Estkowski, M. Howard and C. Lee, Pheromone robotics, *Autonomous Robots*, vol. 11, 2001, pp. 319–324.
[109] I.M. Rekleitis, G. Dudek and E.E. Milios, Graph-based exploration using multiple robots, in L.E. Parker, G.W. Bekey and J. Barhen (eds.), *Distributed Autonomous Robotics Systems*, vol. 4, Springer: Berlin, 2000, pp. 241–250.
[110] F.E. Scheider, D. Wildermuth and H.-L. Wolf, Motion coordination in formations of multiple mobile robots using a potential field approach, in L.E. Parker, G.W. Bekey and J. Barhen (eds). *Distributed Autonomous Robotics Systems*, vol. 4, Springer: Berlin, 2000, pp. 305–314.
[111] S. Thrun, D. Fox, W. Burgard and F. Dellaert, Robust Monte Carlo localization for mobile robots, *Artif. Intell. J.* vol. 128, 2001, pp. 99–141.
[112] A.F. Winfield, Distributed sensing and data collection via broken *ad hoc* wireless connected networks of mobile robots, in L.E. Parker, G.W. Bekey and J. Barhen (eds). *Distributed Autonomous Robotics Systems*, vol. 4. Springer: Berlin, 2000, pp. 273–282.
[113] B. Yamauchi, Frontier-based approach for autonomous exploration, in *Proc. IEEE Int. Symp. Computational Intelligence, Robotics and Automation*, 1997, pp. 146–151.
[114] B. Yamauchi, A. Shultz and W. Adams, Mobile robot exploration and map-building with continuous localization, in *Proc. 1998 IEEE/RSJ Int. Conf. Robotics and Automation*, vol. 4, San Francisco, CA, 1998, pp. 3175–3720.
[115] A. Zelinksy, A mobile robot exploration algorithm, *IEEE Trans. Robot. Autom.*, vol. 8, 1992, pp. 707–717.
[116] J. Edmonds, Edge-disjoint branchings, in *Combinatorial Algorithms*. Academic Press: London, 1973.
[117] L. Lovász, On two minimax theorems in graph theory, *J. Combin. Theory Ser. B*, 1976.
[118] K. Dasgupta, K. Kalpakis and P. Namjoshi, An efficient clustering-based heuristic for data gathering and aggregation in sensor networks, in *IEEE Wireless Communications and Networking, WCNC 2003*, vol. 3, 16–20 March 2003, pp. 1948–1953.
[119] R.D. Nowak, E.D. Kolaczyk, Multiscale maximum penalized likelihood estimators, *IEEE Int. Symp. Information Theory*, Lozana, 2002, p. 156.
[120] R.D. Nowak and E.D. Kolaczyk, A statistical multiscale framework for Poisson inverse problems, *IEEE Trans. Inform. Theory*, vol. 46, no. 5, August 2000, pp. 1811–1825.
[121] R. Nowak, U. Mitra and R. Willett, Estimating inhomogeneous fields using wireless sensor networks, *IEEE J. Select. Areas Commun.*, vol. 22, no. 6, August 2004, pp. 999–1006.
[122] R.M. Willett and R.D. Nowak, Platelets: a multiscale approach for recovering edges and surfaces in photon-limited medical imaging, *IEEE Trans. Medical Imag.*, vol. 22, no. 3, March 2003, pp. 332–350.
[123] R.M. Willett, A.M. Martin and R.D. Nowak, Adaptive sampling for wireless sensor networks, in *Int. Symp. Information Theory, ISIT*, 27 June to 2 July 2004, p. 519.
[124] www.ece.rice.edu/~nowak/pubs.html (www.ece.rice.edu/~nowak/msla.pdf).
[125] R. Nowak and U. Mitra, *Boundary Estimation in Sensor Networks: Theory and*

Methods, 2nd International Workshop on Information Processing in Sensor Networks, Palo Alto, CA, 22–23 April 2003, pp. 1–15.

[126] L. Breiman, J. Friedman, R. Olshen and C. J. Stone. *Classification and Regression Trees*. Wadsworth: Belmont, CA, 1983.

[127] K. Chintalapudi and R. Govindan. Localized edge detection in sensor fields. University of Southern California, Computer Science Department, Technical Report, 2002; available at www.cs.usc.*edu/tech-reports/techniathrmca1-reports.html*

[128] D. Donoho and Wedgelets: nearly minimax estimation of edges, *Ann. Statist.*, vol. 27, 1999, pp. 859–897.

[129] A.P. Korostelev and A. B. Tsybakov. *Minimax Theory of Image Reconstruction*. Springer: New York, 1993.

[130] B. Laurent and P. Massart. Adaptive estimation of a quadratic functional by model selection. *Ann. Stat.*, vol. 5, October 2000, pp. 37–52.

[131] S. Ni, Y. Tseng, Y. Chen and J. Chen, The broadcast storm problem in a mobile *ad hoc* network, in *Annual ACM/IEEE Int. Conf. Mobile Computing and Networking (MOBICOM)*, August 1999, pp. 151–162.

[132] P. Gupta and P.R. Kumar, Critical power for asymptotic connectivity in wireless networks, in *Stochastic Analysis, Control, Optimization and Applications*, W.M. McEneany *et al.* (eds). Birkhauser: Boston, MA, 1998, pp. 547–566.

[133] L. Kleinrock and J.A. Silvester. Optimum transmission radii for packet radio networks or why six is a magic number, in *Record National Telecommun. Conf.*, December 1978, pp. 4.3.1–4.3.5.

[134] H. Takagi and L. Kleinrock, Optimal transmission ranges for randomly distributed packet radio Terminals, *IEEE Trans. Commun.*, vol. COM-32, no. 3, March 1984, pp. 246–257.

[135] D. Ganesan, B. Krishnamachari, A. Woo, D. Culler, D. Estrin and S. Wicker, Complex behavior at scale: an experimental study of low-power wireless sensor networks, UCLA Computer Science Technical Report UCLA/CSD-TR 02-0013.

[136] TinyOS Homepage; http://webs.cs.berkeley.edu/tos/

[137] Z.M. Zuniga and B. Krishnamachari, Optimal transmission radius for flooding in large scale wireless sensor networks, in *Int. Workshop on Mobile and Wireless Networks*, Providence, RI, May 2003.

[138] K. Sohrabi, J. Gao, V. Ailawadhi and G.J. Pottie. Protocols for self-organization of a wireless sensor network, *IEEE Person. Commun.*, vol. 7, no. 5, October 2000, pp. 16–27.

[139] D. Petrovic, R.C. Shah, K. Ramchandran and J. Rabaey, Data funneling: routing with aggregation and compression for wireless sensor networks, *IEEE Int. Workshop on Sensor Network Protocols and Applications*, 11 May 2003, pp. 156–162.

[140] Y. Sankarasubramaniam, Ö. B. Akan and I.F. Akyildiz, ESRT: event-to-sink reliable transport in wireless sensor networks, in *MobiHoc'03,* Annapolis, MD, 1–3 June 2003, pp. 177–188.

[141] C.-Y. Wan, A.T. Campbell and L. Krishnamurthy, Pump-slowly, fetch-quickly (PSFQ): a reliable transport protocol for sensor networks, *IEEE J. Select. Areas Commun.*, volume 23, no. 4, 2005, pp. 862–872.

[142] C.-Y. Wan, A.T. Campbell and L. Krishnamurthy, PSFQ: a reliable transport protocol for wireless sensor networks, in *Proc. 1st ACM Int. Workshop Wireless Sensor Network Applications (WSNA)*, Atlanta, GA, 28 September 2002, pp. 1–11.

[143] S. Floyd, V. Jacobson, C. Liu, S. Macanne and L. Zhang, A reliable multicast framework for lightweight session and application layer framing, *IEEE/ACM Trans. Networks*, vol. 5, no. 2, 1997, pp. 784–803.

[144] R. Stann and J. Heidemann, RMST: reliable data transport in sensor networks, in *Proc. 1st IEEE Int. Workshop Sensor Network Protocols Applications*, Anchorage, AK, May 2003, pp. 102–112.

[145] C.-Y. Wan, S.B. Eisenman and A.T. Campbell, CODA: congestion detection and avoidance in sensor networks, in *Proc. 1st ACM Conf. Embedded Networked Sensor Syst. (SenSys)*, Los Angeles, CA, 5–7 November 2003, pp. 266–279.

[146] C.-Y. Wan, A resilient transport system for wireless sensor networks, Ph.D. dissertation, Department of Electronic Engineering, Columbia University, New York, 2005. Available at: http://comet.columbia.edu/armstrong/wan-2005.pdf

[147] H. Balakrishnan, V.N. Padmanabhan, S. Seshan and R.H. Katz, A comparison of mechanisms for improving TCP performance over wireless links, *IEEE/ACM Trans. Networking*, vol. 5, no. 6, 1997, pp. 756–769.

[148] B. Langen, G. Lober and W. Herzig, Reflection and transmission behavior of building materials at 60 GHz, in *Proc. IEEE PIMRC'94*, The Hague, September 1994, pp. 505–509.

15

Security

A necessary component of network security is the ability to reliably authenticate communication partners and other network entities.For this reason we will start this chapter by discussing authentication protocols [1–19]. The focus will be on a systematic approach to the design of the protocols rather then describing a specific protocol used in practice. The following session will discuss the security architectures. Principles of key distribution will be covered in the third section followed up by some specific solutions in *ad hoc* networks, sensor networks, GSM and UMTS systems.

15.1 AUTHENTICATION

Many designs dealing with authentication in networks or distributed systems combine the issues of authentication with those of key distribution. These designs typically assume that all network parties share a key with a common trusted entity, a key distribution center (KDC), from which they can get pair-wise shared keys to carry out mutual authentication protocols. These protocols are called three-party authentication protocols, and have been studied extensively [5, 6, 10–12, 15, 16]. Most of the corresponding implementations [10] require the exchange of long messages, which is possible for application layer protocols, but makes them unsuitable for use in lower-layer networking protocols where limited packet sizes are an important consideration. Some require synchronized clocks or counters [10] that pose system management and initialization issues, as will be discussed below.

Two-party authentication protocols are used in many networks. Some of these use public key cryptography [1–3, 18]. With a public key cryptographic system, each party only has to know and verify the public key of the other party, and there is no need to share secret keys. The authentication method most widely used in network environments today consists of asking users to prove their identity by demonstrating knowledge of a secret they know, *a password*. This widespread but old technique has several weaknesses.

Advanced Wireless Networks: 4G Technologies Savo G. Glisic
© 2006 John Wiley & Sons, Ltd.

In most password systems, the password typed by a user is sent in cleartext over the network to the computer that requested it. This means that an intruder can spy on passing traffic to collect passwords. Since users need to memorize their passwords, they typically select passwords that they can easily remember. So, such passwords are selected from within a relatively small vocabulary, and are thus easily guessed by potential intruders [4].

Of all authentication techniques that circumvent the drawbacks of passwords, the most promising ones are those using cryptographic means. With such techniques, users are typically provided with smart cards or chip cards equipped with a processor capable of cryptographic operations. Authentication is based upon using the card to compute or verify cryptographic messages exchanged with the computer.

Cryptographic authentication consists of challenging the user or communicating party being authenticated to prove its identity by demonstrating its ability to encipher or decipher some item with a secret or private key known to be stored inside the smart card. Of course, since the secret or private key stored on the card or secure device changes infrequently, the item to be enciphered or deciphered with it must change for every execution of the authentication protocol, otherwise, even though the secret never flows in cleartext over the network, an intruder could still tap a line, record the cryptic message that flows over it, and play that recording back at a later time without even knowing what it means.

To guarantee that the item that gets enciphered (or deciphered), called the *challenge*, is different for every execution of the authentication protocol, three techniques are used. The challenge may be derived either from a real-time clock reading, in which case it is called a *timestamp*, from a *counter* that is incremented for every protocol execution, or from a random number generator, in which case it is called a *nonce*. In any case, a new challenge (timestamp, counter value or nonce) is generated for each protocol run.

With timestamps, the user being authenticated, A, enciphers the current reading of its clock and sends the result to the party requesting the authentication B. B then deciphers the received message, and verifies that the timestamp corresponds to the current realtime. The drawback of timestamps is thus immediately apparent. A and B must have synchronized real-time clocks for the verification to be possible. However, since clocks can never be perfectly synchronized and messages take time to travel across networks anyway, B cannot expect that the deciphered timestamp received from A will ever be equal to its own real-time clock reading. A's time-stamp at best can (and normally should) be within some limited time window of B's real-time clock. However, as soon as a time window of tolerance is defined, a potential intruder could exploit it to impersonate A by replaying one of A's recent authentication messages within that time window. Preventing this requires putting a limit to the number of authentication protocol runs allowable during the time window, and having B save all the authentication messages within the window. Thus, both efficiency and security require pretty good clock synchronization. Achieving such synchronization is often difficult, and making it secure would itself require using some other authentication method not depending on time.

With counting challenges, A and B maintain synchronized counters of the number of times they have authenticated one another. Every time A wants to communicate with B, it enciphers its counter and sends the result to B, who deciphers it and verifies that it matches its own counter value, whereupon both parties increment their respective counters for the next time. The drawback of such counting challenges is that both parties must maintain synchronized counters, which poses stable storage and counter management problems in the long run. The counters must be long enough to prevent an attacker from waiting for a (deterministic) counter wraparound; using counters complicates resolution of conflicts when both parties want to initiate protocol executions at the same time; after detecting loss of synchronization between two counters, some other method of authentication must be used to securely resynchronize them; and an undetected error in the counter values may have catastrophic effects.

Therefore, both timestamps and counter techniques, while useful, are not a complete solution, especially for protocols in the lower layers of a network architecture. The best technique for this purpose, and also the simplest one to implement, consists of using nonces. The price for this simplicity is an extra network message. While A can authenticate itself to B with a single message if timestamps or counters are used, two messages are necessary with nonces. Indeed, it is B and not A who needs to be sure that the nonce has never been used before. Thus, it is B and not A who must generate it. B must encipher it and send it to A, then A must decipher it and send it back in cleartext for B to authenticate it, which costs a total of two messages. The cost of one more message is usually tolerable, especially since the extra message can often be piggy-backed onto regular traffic or its cost can be amortized over many subsequent messages authenticated with sequence counters. Given these advantages of nonces over timestamps and counters, the rest of this section focuses on nonce-based authentication protocols.

Figure 15.1 is a simple representation of the nonce-based one-way authentication protocol, where N is a nonce generated by B and $Ea(N)$ is its value enciphered under some key Ka that B knows is associated with A. It may be a secret key shared between A and B or the public key of A. The protocol allows B to authenticate A. A number of modifications are shown in the same figure.

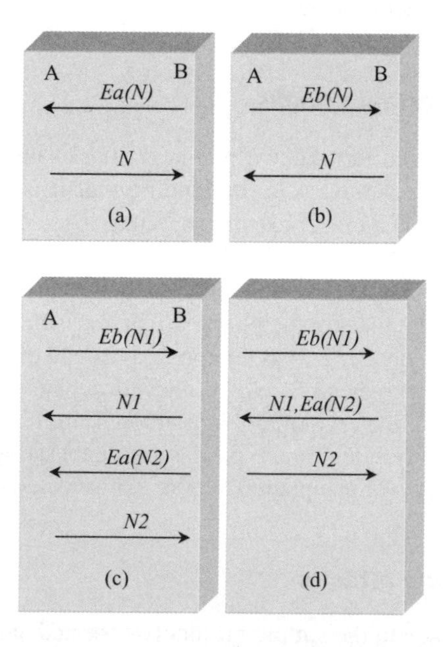

Figure 15.1 Authentication protocols: (a) one-way authentication of A; (b) one-way authentication of B; (c) two-way authentication; and (d) two-way authentication with three messages.

The simple and elegant two-way authentication protocol, which was derived most naturally by combining two instances of the simplest possible one-way protocol, presents a number of undesirable characteristics. In particular, we will show that an intruder may break this protocol by interleaving messages from different executions of it, if the same shared key is used by both parties *or* if the same key is always used by the party who starts the protocol. One could still argue that the protocol is safe if used with four different keys; however, this creates additional overhead and key management issues, as well as migration problems for existing designs.

15.1.1 Attacks on simple cryptographic authentication

The simple two-way authentication protocol illustrated in Figure 15.1(d) suffers from a number of defects.

15.1.1.1 Known plaintext attacks

A first weakness of the protocol is its openness to known plaintext attacks. Every enciphered message flowing between A and B is the ciphertext of a bit string (the nonce) that flows in plaintext in a subsequent message between A and B. This enables a passive wiretapping intruder to collect two cleartext–ciphertext pairs every time the protocol is run, which at the very least helps it accumulate encryption tables in the long run, and may even help it break the scheme and discover the encryption key, depending on the quality of the encryption algorithm used. It is in general desirable that the plaintext of exchanged enciphered messages not be known or derivable by intruders.

15.1.1.2 Chosen ciphertext attacks

A potential intruder may even turn known plaintext attacks into selected text attacks by playing an active instead of passive role. Pretending that it is A or B, an intruder may send the other party (B or A) a ciphertext message that it selected itself and wait for that other party to reply with the deciphered value of that text. Of course, the intruder, not knowing the right key, may not be able to complete the third protocol flow. However, it can accumulate knowledge about cleartext–ciphertext pairs *of which it selected the ciphertext itself* (or the cleartext, if the challenge was enciphered instead of deciphered). So it may try specific ciphertext strings such as all zeros, all ones, or whatever else might help it break the key faster, depending on the cryptographic method used. It is in general desirable that an intruder not be able to trick legitimate parties into generating deciphered versions of selected ciphertext messages (or enciphered versions of selected cleartext).

15.1.1.3 Oracle session attacks

If A and B use the same key in the simple protocol suggested earlier, the intruder actually *can* break the authentication without even breaking the key. This is illustrated in Figure 15.2, where the intruder, noted X and posing as A, starts a session with B by sending

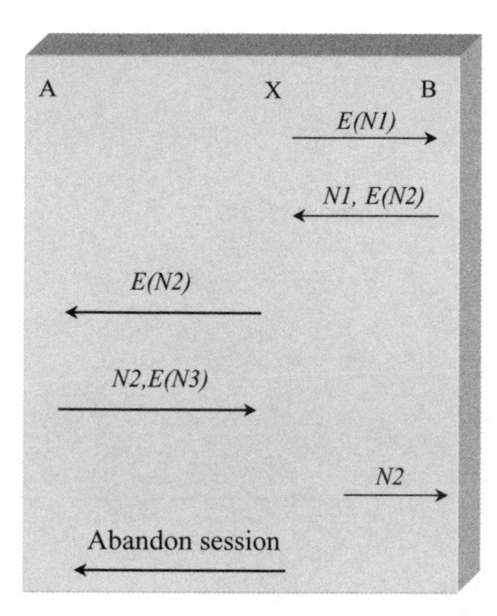

Figure 15.2 An oracle session attack.

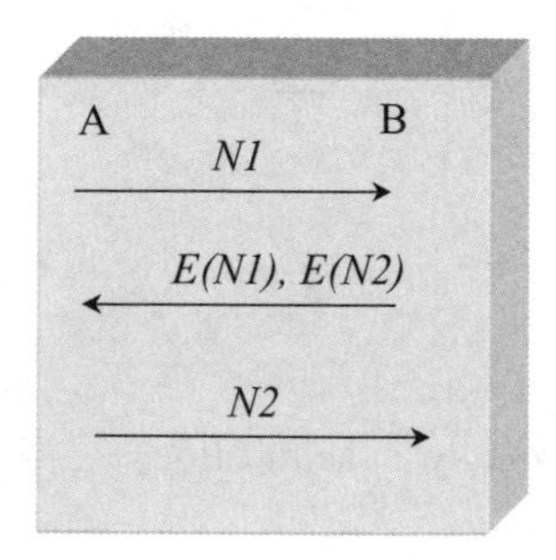

Figure 15.3 Improved two-way protocol.

B some enciphered nonce $E(N1)$. B replies with the deciphered value $N1$ and its own enciphered challenge $E(N2)$. X cannot decipher $N2$, but it can take advantage of a selected ciphertext attack on A, using A as an oracle who will provide the necessary deciphered value $N2.X$ accomplishes this by posing now as B to start a separate 'oracle' session with A, sending $E(N2)$ as the initial message on that session. A will reply with the needed value $N2$ and some own enciphered nonce $E(N3)$. X then drops the oracle session with A since it cannot and does not care about deciphering $N3$, but it then turns around and successfully assumes its faked identity A with respect to B by sending $N2$ to B. This may be prevented by modifying protocol as shown in Figure 15.3. This is an ISO SC27 protocol [19].

15.1.1.4 Parallel session attacks

Another defect commonly found in simple protocols, such as those seen above, is depicted in Figure 15.4, where the intruder assumes a passive role instead of an active one. It intercepts

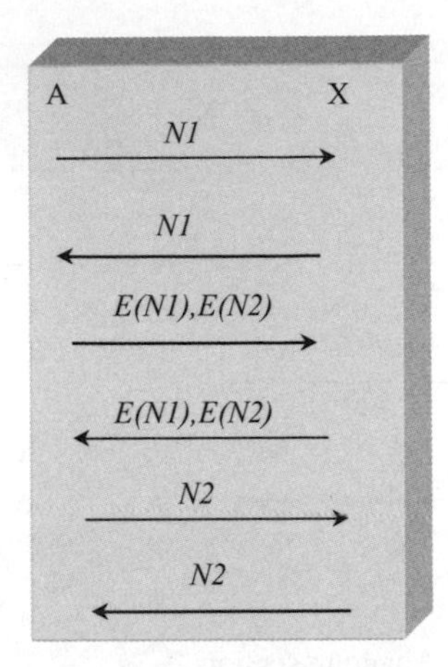

Figure 15.4 A parallel session attack.

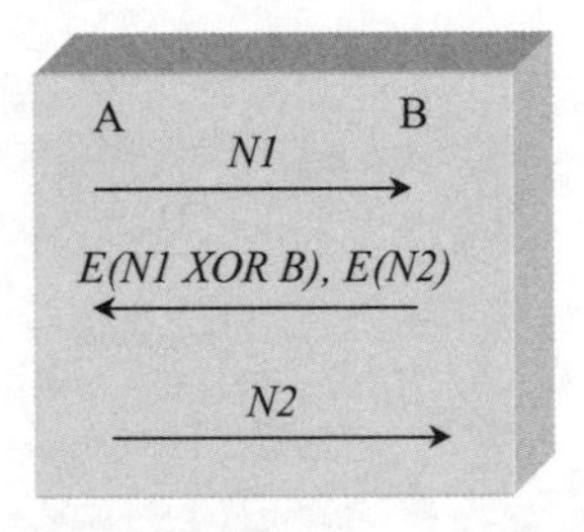

Figure 15.5 An asymmetric two-way authentication.

a call from A to B with challenge $N1$ and since it cannot answer the challenge $N1$ by replying $E(N1)$, it simply pretends that it is B trying to start a *parallel session* with A. Of course, it selects to use just the same $N1$ as the first challenge on the original session, thus causing A to provide it with precisely the answer $E(N1)$ needed to complete the first authentication. The remaining communication shown in the figure will complete the second authentication as well.

Parallel session attacks illustrate another fundamental flaw of many simple authentication protocols: the cryptographic expression used in the second message must be asymmetric (i.e. direction-dependent) so that its value in a protocol run initiated by A cannot be used in a protocol run initiated by B. Based on the above observations, i.e. that the cryptographic message in the second flow must be asymmetric (direction-dependent) and different from the one in the third flow, one may think that a protocol such as the one in Figure 15.5 should be secure. Unfortunately, in addition to that the fact that known- and selected-text attacks are still possible, the problem now is that the simple function XOR has not fixed anything since an intruder can still resort to a parallel session attack. It merely needs to use proper offsetting in addition, as illustrated in Figure 15.6.

The simple and apparently sound protocol of Figure 15.1(d) is thus broken in many ways. In the following, oracle-session attacks, parallel-session attacks, offset attacks and/or other

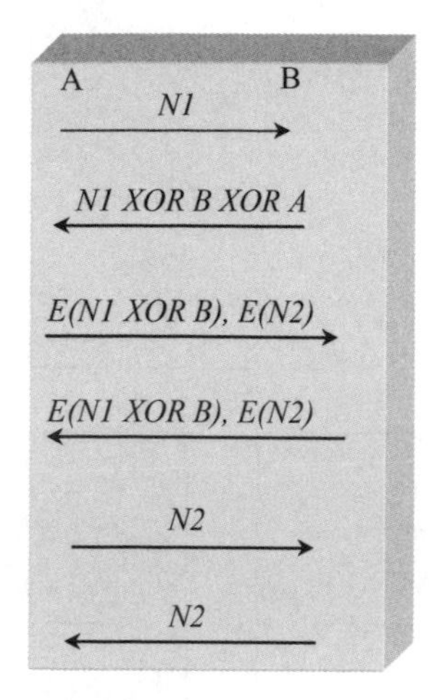

Figure 15.6 An offset attack (through a parallel session).

types or combinations of attacks that involve replaying messages and functions of challenges observed in other runs of the protocol will be collectively referred to as *interleaving* attacks. Based on the above observations we can specify the following design requirements for authentication protocols:

(1) nonce-based (avoid synchronization and stable counter storage);

(2) resistant to common attacks;

(3) usable at any layer of any network architecture (small messages);

(4) usable on any processing base (few computations);

(5) using any cryptographic algorithm, either symmetric ones, such as DES [20] or asymmetric ones such as RSA [7];

(6) exportable – the chance to receive proper licensing for a technology is larger if it provides only message authentication code (MAC) functions and does not require full-fledged encryption and decryption of large messages;

(7) extendable – it should support the sharing of secret keys between multiple users and should allow additional fields to be carried in the messages, which would thus be authenticated as part of the protocol.

15.1.2 Canonical authentication protocol

A number of protocols meeting the above requirements have been developed [13, 14]. Some of these protocols are summarized in Figure 15.7. Given that the focus is on authentication protocols using nonces, the most general canonical form for all such protocols is depicted in Figure 15.7 (P1). Here, A sends some nonce $N1$ to $B.B$ authenticates itself by sending back a function $u(\,)$ of several parameters, including at least its secret $K1$ and the challenge $N1$ it received from A. It also sends a challenge $N2$ to A. A finally completes the protocol by authenticating itself with a function $v(\,)$ of several parameters including at least its secret

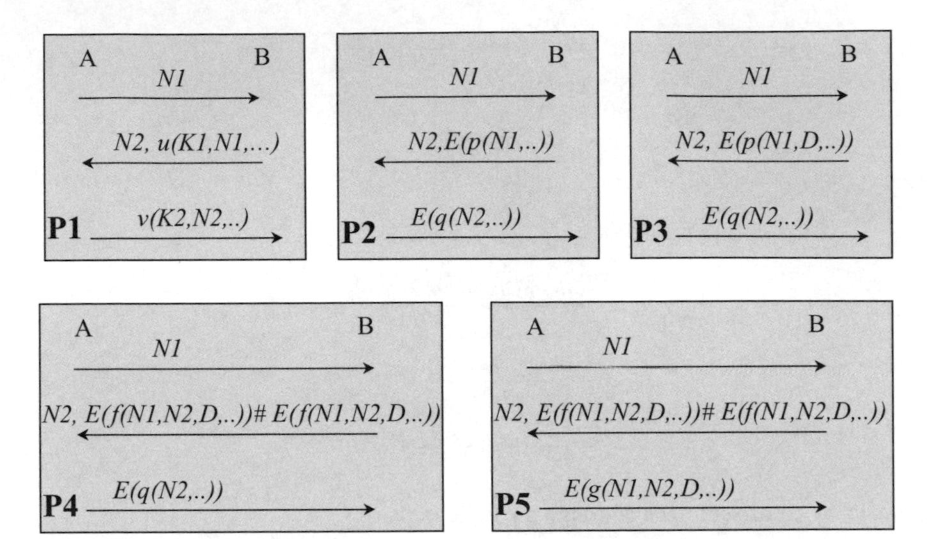

Figure 15.7 P1, resistance to replay attacks through use of nonces; P2, use of symmetric cryptography; P3, resistance to parallel session attacks; P4, resistance to selected-text and offset attacks; P5, minimal number of encryptions.

$K2$ and the challenge $N2$ from B. As indicated in the earlier section on attacks, in order for this protocol to resist oracle attacks, the functions $u(\ )$ and $v(\ )$ must be different from one another, meaning that a cryptographic message from flow (2) can never be used to derive the necessary message for flow (3).

The protocols must work with either symmetric or asymmetric cryptographic systems. With the latter, the secrets $K1$ and $K2$ would be the private keys of B and A, respectively. However, with symmetric systems, $K1$ and $K2$ would be shared secret keys. In fact, in many typical scenarios, they would be the same shared secret key. In the sequel we will continue under that assumption because it opens the door to attacks that would not be possible otherwise, yet it has advantages of efficiency and ease of migration of existing systems over solutions which use different keys for A and B. Any protocol that resists interleaving attacks using equal symmetric keys is also safe with asymmetric ones while the opposite is not true.

Under the assumption of symmetric cryptography with a single shared secret key K protocol P1 becomes as depicted in Figure 15.7 (P2), where the functions $u(K1$, decryption) are represented by operations $E(\ )$ under the same key K of two different functions $p(\ )$ and $q(\ )$ of the remaining parameters. As indicated earlier in the section on attacks, the prevention of parallel session attacks suggests that function $p(\ )$ must be asymmetric, i.e. direction-dependent. In other words, the arguments to $p(\ )$ must be different depending on whether A or B starts the communication session. This is depicted in Figure 15.7 (P3) by the addition of a parameter D in $p(\ )$, which stands for anything indicating or tied to the direction of the flow (e.g. the name or address of one of the parties).

As shown in Figure 15.7 (P3), the complete protocol P3 requires three messages. In any networking environment, these messages need not travel in packets of their own. In practice, they can and typically would be piggy-backed onto other network packets, such as, for instance, connection requests and confirmations at whatever layer entities are being authenticated (e.g. link, network, transport, application, etc.). Within packets at any such layer, the authentication fields should take as little space as possible. As it stands presently, the canonical protocol requires a nonce to be carried on the first flow, a nonce and a cryptographic expression on the second one, and a cryptographic expression on the third one. The cleartext of the directional parameter, e.g. party name or address, is implicitly known or already present in the enclosing packet headers. The required size of nonces is given by the amount of security desired: the larger they are, the lower the probability will be that a given nonce gets reused. The size of the cryptographic expressions depends on the encryption algorithm used and on the amount of information being encrypted.

As an example, if the DES is used, the smallest cryptographic expression will be 64 b (key 56 b) long. Since the security of the whole authentication protocol rests on the unforgeability of the cryptographic expressions, if expressions of 64 b are used, the statistical security, ss, of the expressions is $ss = 1/p(\text{expression}) = 2^{64}$, the inverse of the probability that an intruder can successfully guess a cryptographic expression. For many environments, such a degree of security is already quite satisfactory, although longer cryptographic expressions are already being used. Triple DEA (ANSI X9.17, DEA 1999) uses three keys and three executions of DEA with effective key length 112 and 168 b. Advanced encryption standard (AES, 1997) uses blocks of length 128 b and key lengths 128, 192 and 256 b. If 64 b nonces are used, $p(\,)$ and $q(\,)$ may be restricted to 64 b functions of their parameters without compromising the degree of security. We further make this simplifying assumption.

Restricting $p(\,)$ and $q(\,)$ to 64 b functions to limit message size has its own implications. Many simple and attractive 64 b functions of only $N1$ and D (e.g. XOR, rotations, conditional permutations, etc.) could be subjected to selected-plaintext and offset attacks by an intruder. By including inside function $p(\,)$ an additional internal encrypted function of the same parameters $N1$ and D, one can separate these parameters cryptographically and thus bar the way to offset attacks. By including $N2$ inside function $p(\,)$, one bars the way to a potential intruder using B as an oracle to get some selected plaintext enciphered in the second flow because the cleartext of that flow then depends on nonce $N2$ which is not under the control of the intruder initiating a protocol run. These further conditions on $p(\,)$ are represented in Figure 15.7 (P4), where $p(\,)$ would thus be a 64 b function (noted #) of two operands, which themselves include functions $f(\,)$ and $g(\,)$ of $N1$, N2 and D.

The different fields in the message should be cryptographically separated, so that the attacker should not be able to control one field through another field. This separation should be ensured by an appropriate selection of the functions $f(\,)$ and $g(\,)$. We will later give exact conditions on $f(\,)$ and $g(\,)$ to prevent interleaving attacks, and give specific examples for these functions.

Any peer authentication protocol requires at least two cryptographic expressions. In Figure 15.7 (P5) a third cryptographic operation is used to protect against offset attacks. To keep the protocol simple, by putting additional conditions on $q(\,)$, we can return to a protocol requiring only two cryptographic block operations. This can be achieved by imposing that the functions $g(\,)$ and $q(\,)$ actually be the same, as represented in Figure 15.7 (P5). Indeed, in this case, the innermost cryptographic expression required to produce or verify the second flow is the same as the cryptographic expression of the third flow, and can thus be computed only once and saved to minimize computation.

(a)

Key generation
Select p,q (p,q prime)
Calculate n=pq
Calculate $\phi(n)=(p-1)(q-1)$
Select integer e (gcd $(\phi(n),e)=1,\ 1<e<\phi(n)$)
Calculate d ($d=e^{-1} \bmod \phi(n)$)
Public key KU=[e,n]
Private key KR=[d,n]
Encription
Plaintext M<n
Ciphertext $C= M^e \pmod n$
Decription
Ciphertext C
Plaintext $M=C^d \pmod n$

(b)

p=17
q=11
n=pq=187
$\phi(n)=(p-1)(q-1)=160$
e=7
d=23 (*de* mod 160=1)
KU=[7,187]
KR=[23,187]
M=88
Ciphertext
$C = 88^7 \bmod 187 = 11$
Plain text
$M = 11^{23} \bmod 187 = 88$

Figure 15.8 (a) Operation of a public key encryption system; (b) numerical example of operation of a public key encryption system.

Finally, it may be instructive at this point to describe the operation of public key encryption system used for authentication. In this case user generates pair of keys and places one key in the public domain. To send a message the user encrypts using the public key and decrypts using a private key as illustrated in Figure 15.8(a); a numerical example is given in Figure 15.8(b).

15.2 SECURITY ARCHITECTURE

Cryptography provides only two fundamental services, namely, *confidentiality* and *authentication.* All communications security services are concerned with the identity of the senders or recipients of information. Cryptography allows users to be identified by allocating secret keys to them. There are only two ways that cryptography can work, and these define the fundamental services [22, 23]:

- *Confidentiality* – only the user (or set of users) in possession of the secret key can read the message;
- *Authentication* – only the user (or set of users) in possession of the secret key can write the message.

A *secure channel* can be thought of as a relationship between two system users which provides some security services.

In this section, two basic types of secure channel are considered, namely, confidentiality channels and authentication channels. In addition, we define symmetric channels, each of which may coincide with a confidentiality channel, an authentication channel or both.

In a conventional symmetric cipher, a pair of users share the same key. This symmetric channel typically provides both a confidentiality and an authentication channel between a specific pair of users. Symmetric ciphers can also be used to provide authentication alone, as when a message authentication code is used [25].

A public key cryptosystem, as described above, usually has one public key and a corresponding secret key. For certain algorithms, such as RSA [31], the public key may be used for confidentiality and the secret key may be used for authentication. However, for other algorithms, it may be that only one channel is provided. For example, signature schemes provide authentication but not confidentiality,while schemes such as the McEliece algorithm [25] are suitable for authentication but not confidentiality.

There are three components that make up a *security architecture* in the model of this paper: (1) *users*; (2) *trusted users*, including information on who trusts whom; and (3) *secure channels*, which may provide confidentiality authentication or both.

The *formal model* is a relatively simple specification and defines a state-based sequential system as described in Spivey [32]. The first line of the specification defines the abstract types which will be used. These are users and keys [*User, Key*] and they are fundamental values in the model. Other components, such as secure channels and trusted users, will be defined in terms of these sets.

Next, some global, or axiomatic, definitions are made. These are things that are not expected to change within the model and so can be excluded from the rest of the system state. Five sets of keys are defined concerning two separate properties. A key must be in exactly one of the sets *public*, *secret*, *shared*; in other words, these sets partition (⊵) the keys. In addition, a key must be a confidentiality key (in the set *Conf*) or an authentication key (in the set *Auth*) or both. The dual, or inverse, is defined for each key. Taking the dual of a key is a self-inverse operation. The dual of a confidentiality key is still a confidentiality key and similarly for authentication keys. Secret and public keys are interchanged under the dual map, while the dual of a shared key is still shared. The trusted users are defined by a map which defines those users trusted by each user. For a formal representation of the

Table 15.1 Special Z symbols

Symbol	Meaning
$f: X \rightarrow Y$	Function between X and Y
$f: X \leftrightarrow Y$	Relation between X and Y
id X	The identity function on X
dom f	The domain of f
ran f	The range of f
$f(\lvert X \rvert)$	Image of the set X under the function f
$f \oplus g$	Function which takes values of the function f except on the domain of g, where it takes the values of g
$f \circ g$	Functional composition where the domain of g must equal the range of f
$X \setminus Y$	Set difference of X and Y
$\mathbb{P} X$	the set of sets with elements of X as values

system components, notation from Table 15.1 known as Z notation [32] is used, resulting in [21]:

$$
\begin{array}{l}
\hline
\textit{Shared, Public, Secret, Auth, Conf} : \mathbb{P}\ \textit{Key} \\
\textit{dual} : \textit{Key} \rightarrow \textit{Key} \\
\textit{Trusted} : \textit{User} \rightarrow \mathbb{P}\ \textit{User} \\
\hline
\textit{(Shared, Public, Secret) partition Key} \\
\textit{Auth} \cup \textit{Conf} = \textit{Key} \\
\textit{dual} \circ \textit{dual} = \textit{id Key} \\
\textit{dual}(\lvert \textit{Conf} \rvert) = \textit{Conf} \\
\textit{dual}(\lvert \textit{Auth} \rvert) = \textit{Auth} \\
\textit{dual}(\lvert \textit{Shared} \rvert) = \textit{Shared} \\
\textit{dual}(\lvert \textit{Public} \rvert) = \textit{Secret} \\
\textit{dual}(\lvert \textit{Secret} \rvert) = \textit{Public}
\end{array}
$$

The first ordinary schema defines the variable that records what keys are known by each user, and with whom they are associated.

$$
\begin{array}{|l|}
\hline
\textit{Keys} \\
\hline
\textit{keys} : \textit{User} \rightarrow \mathbb{P}\ (\textit{User} \times \textit{Key}) \\
\hline
\end{array}
$$

A set of (*user, key*) pairs is associated to each user, where 'x maps to (y, k)' means that x knows key k and uses it in communications with user y. The following three schemas define the state space of the model by giving formal definitions of secure channels in terms of possession of keys.

Confidentiality channels

$$
\begin{array}{l}
\mathit{Keys} \\
\mathit{ConfChannels} : \mathit{User} \rightarrow \mathit{User} \\
\hline
\forall x, y : \mathit{User} \bullet (x, y) \in \mathit{ConfChannels} \Leftrightarrow \\
\{\exists k : \mathit{Conf} \setminus \mathit{Secret};\ z : \\
\qquad \mathit{User} \bullet (y, k) \in \mathit{keys}(x) \wedge [z, \mathit{dual}(k)] \in \mathit{keys}(y)\}
\end{array}
$$

Confidentiality channels define relations between pairs of users. These are ordered pairs as the channel may be in only one direction. It will be noticed that the definition is not symmetrical with respect to x and y. The predicate states that, for a confidentiality channel to exist from x to y, there must be a key whose use includes confidentiality and is either shared or public; x must associate this key with y; y must know the dual of the key. In the model, this means that y must associate it with some user(s), but it is not of any concern to y exactly which users know the key. To see why this is reasonable, consider a public key system providing confidentiality to y. Any user z who knows the public key of y has a confidentiality channel to y, and it is important to z that it is only y who has the secret dual key; y must of course know the secret key, but it does not matter to y who knows the public key. This corresponds to the viewpoint that confidentiality is a service provided to the sender of information.

Authentication channels

$$
\begin{array}{l}
\mathit{Keys} \\
\mathit{AuthChannels} : \mathit{User} \rightarrow \mathit{User} \\
\hline
\forall x, y : \mathit{User} \bullet (x, y) \in \mathit{AuthChannels} \Leftrightarrow \\
\{\exists k : \mathit{Auth} \setminus \mathit{Public};\ z : \mathit{User} \bullet \\
\quad (z, k) \in \mathit{keys}(x) \wedge [x, \mathit{dual}(k)] \in \mathit{keys}(y)\}
\end{array}
$$

Authentication channels also define relations between pairs of users. The definition of authentication channels is dual to the definition of confidentiality channels and corresponds to the viewpoint that authentication is a service provided to the receiver of information.

Symmetric channels

$$
\begin{array}{l}
\mathit{Keys} \\
\mathit{SymmChannels} : \mathbb{P}\{\mathbb{P}(\mathit{User})\} \\
\hline
\mathit{SymmChannels} \subseteq \{x, y : \mathit{User} | x \neq y \bullet \{x, y\}\} \\
\forall x, y : \mathit{User} \bullet (x, y) \in \mathit{SymmChannels} \Leftrightarrow \\
\{(\exists k : \mathit{Shared} \bullet (y, k) \in \mathit{keys}(x) \wedge [x, \mathit{dual}(k)] \in \mathit{keys}(y)\}
\end{array}
$$

Symmetric channels are in both directions and so are defined as sets of two different users. They correspond to the situation where neither key is public, and in practice the key and its dual are usually equal.

State

$$\begin{array}{|l|}\hline ConfidentialityChannels \\ AuthenticationChannels \\ SymmetricChannels \\ \hline\end{array}$$

The system state is defined exactly by what keys are known by each user, thereby defining what secure channels exist.

Transfer

$$\begin{array}{|l|}\hline \Delta State \\ orig?, dest?, recip?, sender? : User \\ k? : Key \\ \hline (orig?, k?) \in keys\ (sender?) \\ keys' = keys \oplus \{recip? \rightarrow keys(recip?) \cup (orig?, k?)\} \\ \hline\end{array}$$

This schema says that if a key $k?$ is sent from one user to another, then the keys known to the recipient are updated to associate the key sent with the originator. In this model, the only state changes are those which happen as a result of passing keys from one user to another. Such a key exchange may or may not result in new channels being formed. The key passes from the sender to the recipient. It may be that the users between whom communication is intended (originator and destination) are different from the sender and recipient involved in a particular exchange. This is the situation if the sender is a key server. The recipient will therefore associate the received key with the originator, who may or may not be the sender.

SecureTransfer

$$\begin{array}{|l|}\hline Transfer \\ \hline k? \in Secret \Rightarrow (sender?, recip?) \in ConfChannels \\ k? \in Public \Rightarrow (sender?, recip?) \in AuthChannels \\ k? \in Shared \Rightarrow (sender?, recip?) \in ConfChannels \cap \\ \qquad\qquad\qquad\qquad\qquad AuthChannels \\ orig? \neq sender? \wedge k? \in Public \cup Shared \Rightarrow \\ \qquad sender? \in Trusted(recip?) \\ recip? \neq dest? \wedge k? \in Secret \cup Shared \Rightarrow \\ \qquad recip? \in Trusted(sender?) \\ \hline\end{array}$$

This schema specifies the conditions which must exist if a key exchange should be performed securely: (1) secret keys may only be transferred over a confidentiality channel; (2) public keys must be transferred over authentication channels; (3) if the key is to be shared, the channel should provide both confidentiality and authentication since both users need to associate the key only with each other; (4) if the key is public or shared then the recipient must trust the sender, unless the sender is the originator; this is because the key must be correctly assigned to the originator; and (5) if the key is secret or shared, then the sender must trust the recipient not to reveal it, unless the recipient is the destination. More details on formal protocol presentation can be found in References [21–32].

15.3 KEY MANAGEMENT

Information protection mechanisms, discussed so far in this chapter, assume cryptographic keys to be distributed to the communicating parties prior to secure communications. The secure management of these keys is one of the most critical elements when integrating cryptographic functions into a system, since even the most elaborate security concept will be ineffective if the key management is weak.

An automatic distribution of keys typically employs different types of messages. A transaction usually is initiated by requesting a key from some central facility (e.g. a key distribution center, KDC), or from the entity a key is to be exchanged with. Cryptographic service messages (CSMs) are exchanged between communicating parties for the transmission of keying material, or for authentication purposes. CSMs may contain keys, or other keying material, such as the distinguished names of entities, key-IDs, counters or random values. CSMs have to be protected depending on their contents and on the security requirements. Generic requirements include the following:

(1) *Data confidentiality* should be provided while secret keys and possibly other data are being transmitted or stored.

(2) *Modification detection* prevents the active threat of unauthorized modification of data items. In most environments, all cryptographic service messages have to be protected against modification.

(3) *Replay detection* is to counter unauthorized duplication of data items.

(4) *Timeliness* requires that the response to a challenge message is prompt and does not allow for playback of some authentic response message by an impersonator.

(5) *Entity authentication* is to corroborate that an entity is the one claimed.

(6) *Data origin authentication* (*proof/nonrepudiation of origin*) is to make certain that the source of a message is the one claimed.

(7) *Proof/nonrepudiation of reception* shows the sender of a message that the message has been received by its legitimate receiver correctly.

(8) *Notarization* is the registration of messages to attest at a later stage its content, origin, destination or time of issue.

The correctness of key management protocols requires more than the existence of secure communication channels between entities and key management servers. For example, it critically depends on the capability of those servers to reliably follow the protocols. Therefore, each entity has to base its deductions not only on the protocol elements sent and received, but also on its trust in the server which, for that reason, often is called the *trusted party*.

Key management is facilitated by the key management services, which include entity registration, key generation, certification, authentication, key distribution and key maintenance.

Entity registration is a procedure by which an individual or a device is authenticated to the system. An absolute identification is provided if a link between an ID (e.g. a distinguished name or a device-ID) and some physical representation of the identified subject (e.g. a person or a device) can be established. An identification can be carried out manually or automatically. Absolute identification always requires at least one initial manual identification (e.g. by showing a passport or a device-ID).

Mutual authentication is usually based on the exchange of certificates. In any system, an entity is represented by some public data, called its (public) credentials (e.g. ID and address). Besides that, an entity may own secret credentials (e.g. testimonials) that may or may not be known by some trusted party. Whenever an entity is registered, a certificate based upon its credentials is issued as a proof of registration. This may involve various procedures, from a protected entry in a specific file to a signature by the certification authority on the credentials.

Key generation refers to procedures by which keys or pairs of keys of good cryptographic quality are securely and unpredictably generated. This implies the use of a random or pseudorandom process involving random seeds, which cannot be manipulated. The requirements are that certain elements of the key space are not more probable than others, and that it is not possible for the unauthorized to gain knowledge about keys.

Certificates are issued for authentication purposes. A credential containing identifying data together with other information (e.g. public keys) is rendered unforgeable by some certifying information (e.g. a digital signature provided by the key certification center). Certification may be an on-line service where some certification authority provides interactive support and is actively involved in key distribution processes, or it may be an off-line service so that certificates are issued to each entity only at some initial stage.

Authentication/Verification may be either (1) entity authentication or identification, (2) message content authentication and (3) message origin authentication. The term verification refers to the process of checking the appropriate claims, i.e. the correct identity of an entity, the unaltered message content or the correct source of a message. The validity of a certificate can be verified using some public information (e.g. a public key of the key certification center), and can be carried out without the assistance of the certification authority, so that the trusted party is only needed for issuing the certificates.

Key distribution refers to procedures by which keys are securely provided to parties legitimately asking for them. The fundamental problem of key exchange or distribution is to establish keying material to be used in symmetric mechanisms whose origin, integrity and confidentiality can be guaranteed. As a result of varied design decisions appropriate to different circumstances, a large variety of key distribution protocols exist [37, 39]. The basic elements of a key distribution protocol are as follows.

15.3.1 Encipherment

The confidentiality of a data item D can be ensured by enciphering D with an appropriate key K which is denoted $eK(D)$. Depending on whether a secret key algorithm or a public key algorithm is used for the enciphering process, D will be enciphered with a secret key K shared between the sender and the legitimate recipient of the message, or with the legitimate recipient B's public key K_{Bp}. Encipherment with the sender A's private key K_{As}, may be used to authenticate the origin of data item D, or to identify A. Encipherment with a secret key provides modification detection if B has some means to check the validity of D (e.g. if B knows D beforehand, or if D contains suitable redundancy).

15.3.2 Modification detection codes

To detect the modification of a data item D, one can add some redundancy that has to be calculated using a collision-free function, i.e. it must be infeasible to find two different values of D that render the same result. Moreover, this process has to involve a secret parameter K in order to prevent forgery. Appropriate combination of K and D also allows for data origin authentication. Examples of suitable building blocks are message authentication codes, or hash functions combined with encipherment. The generic form of this building block is $D||mdcK(D)$. Modification detection codes (*mdc*) enable the legitimate recipient to detect unauthorized modification of the transmitted data immediately after receipt. The correctness of distributed keying material can also be checked if the sender confirms his knowledge of the key in a second step (see below).

15.3.3 Replay detection codes

To detect the replay of a message and to check its timeliness, some explicit or implicit challenge-and-response mechanism has to be used, since the recipient has to be able to decide on the acceptance. In most applications, the inclusion of a replay detection code denoted by $D||rdc$ (e.g. a timestamp *TD*, a counter *CT*, or a random number *R*) will only make sense if it is protected by modification detection. With symmetric cryptographic mechanisms, key modification, i.e. some combination (e.g. XOR) of the secret key with an *rdc*, can be used to detect the replay of a message. A special case is the process of key offsetting used to protect keying material enciphered for distribution where the key used for encipherment is XORed with a count value.

15.3.4 Proof of knowledge of a key

Authentication can be implemented by showing knowledge of a secret (e.g. a secret key). Nevertheless, a building block that proves the knowledge of a key K can also be useful, when K is public. There are several ways for A to prove to B the knowledge of a key that are all based on the principle of challenge and response in order to prevent a replay attack. Depending on the challenge which may be a data item in cleartext or in ciphertext, A has to process the key K and the *rdc* in an appropriate way (e.g. by encipherment, or by calculating a message authentication code), or A has to perform a deciphering operation. The challenge may explicitly be provided by B (e.g. a random number R) or implicitly be given by a synchronized parameter (e.g. a timestamp TD or a counter CT). For some building blocks, the latter case requires only one pass to prove knowledge of K; its tradeoff is the necessary synchronization. If B provides a challenge enciphered with a key K^*, the enciphered data item has to be unpredictable (e.g. a random number R, or a key K^{**}). The generic form of this building block is *authK(A to B)*.

15.3.5 Point-to-point key distribution

This is the basic mechanism of every key distribution scheme. If based on symmetric cryptographic techniques, point-to-point key distribution requires that the two parties involved already share a key that can be used to protcct the keying material to be distributed. If based on asymmetric techniques, point-to-point key distribution requires that each of the two parties has a public key with its associated secret key, and the certificate of the public key produced by a certification authority known to the other party. General assumptions are: (1) the initiator A is able to generate or otherwise acquire a secret key K^*; (2) security requirements are confidentiality of K^*, modification and replay detection, mutual authentication of A and B, and a proof of delivery for K^*. For point-to-point key distribution protocols based on symmetric cryptographic techniques we additionally assume: (3) a key K_{AB} is already shared by A and B. In generic form, a point-to-point key distribution protocol meeting those requirements can be described as shown in Table 15.2 [33].

Table 15.3 is an example of a specific point-to-point key distribution protocol [33] (N denotes a nonrepeating number, R a random number).

For a point-to-point key distribution protocol based on public key techniques, we make the following supplementary assumptions: (4) there is no shared key known to A and B before the key exchange process starts; (5) there is a trusted third party C, where A can receive a certificate that contains the distinguished names of A and C, A',s public key K_{Ap}, and the certificate's expiration date *TE*. The integrity of the certificate is protected by C's signature. As an example, A's certificate can look like $ID_C || ID_A || K_{Ap} || TE || eK_{Cs}[h(ID_C || ID_A || K_{Ap} || TE)]$.

The exchange of certificates can be performed off-line and is not shown in the following protocol. In this protocol, A sends a message (often referred to as *token*) to B that consists of a

Table 15.2 Generic point-to-point key distribution

A		B
(1) →	$eK_{AB}(K^* \| rdc)$	
(2) →	$authK^*(A\ to\ B)$	
	$authK^*(B\ to\ A)$	← (3)

Table 15.3 Point-to-point key distribution

A	(ISO/IEC CD 9798-2)	B
(1) →	$eK_{AB}(K^* \| N)$	
	$eK^*(N \| R)$	← (2)
(3) →	R	

Table 15.4 Point-to-point key distribution

A	(ISO/IEC CD 9798-2)	B
(1) $\rightarrow$	$eK_{Bp}(K^*)\|\| rdc\|\| eK_{As}(h(eK_{Bp}(K^*)\|\| rdc))$	
	$eK^*(rdc)$	$\leftarrow$ (2)

secret key K^* enciphered with B's public key and an appended *rdc*. The integrity of the token is protected by A's signature. This guarantees modification and replay detection, as well as data origin authentication. B responds with the enciphered *rdc*, thereby acknowledging that it has received the key K^*, as shown in Table 15.4.

Key maintenance includes procedures for key activation, key storage, key replacement, key translation, key recovery, black listing of compromised keys, key deactivation and key deletion. Some of the issues of key maintenance are addressed below.

Storage of keying material refers to a key storage facility which provides secure storage of keys for future use, e.g. confidentiality and integrity for secret keying material, or integrity for public keys. Secret keying material must be protected by physical security (e.g. by storing it within a cryptographic device) or enciphered by keys that have physical security. For all keying material, unauthorized modification must be detectable by suitable authentication mechanisms.

Key archival refers to procedures by which keys for notarization or nonrepudiation services can be securely archived. Archived keys may need to be retrieved at a much later date to prove or disprove certain claims.

Key replacement enables parties to securely update their keying material. A key shall be replaced when its compromise is known or suspected. A key shall also be replaced within the time deemed feasible to determine it by an exhaustive attack. A replaced key shall not be reused. The replacement key shall not be a variant or any nonsecret transformation of the original key.

Key recovery refers to cryptographic keys which may become lost due to human error, software bugs or hardware malfunction. In communication security, a simple handshake at session initiation can ensure that both entities are using the same key. Also, message authentication techniques can be used for testing that plaintext has been recovered using the proper key. Key authentication techniques permit keys to be validated prior to their use. In the case where a key was lost, it still may be possible to recover that key by searching part of the key space. This approach may be successful, if the number of likely candidates is small enough.

Key deletion refers to procedures by which parties are assured of the secure destruction of keys that are no longer needed. Destroying a key means eliminating all records of this key, such that no information remaining after the deletion provides any feasibly usable information about the destroyed key.

More information on key management can be found in References [33–39].

15.4 SECURITY MANAGEMENT IN GSM NETWORKS

The *authentication* generic process is shown in Figure 15.9 [41,42]. The authentication algorithm (called A3 in the GSM specifications) computes from a random number RAND,

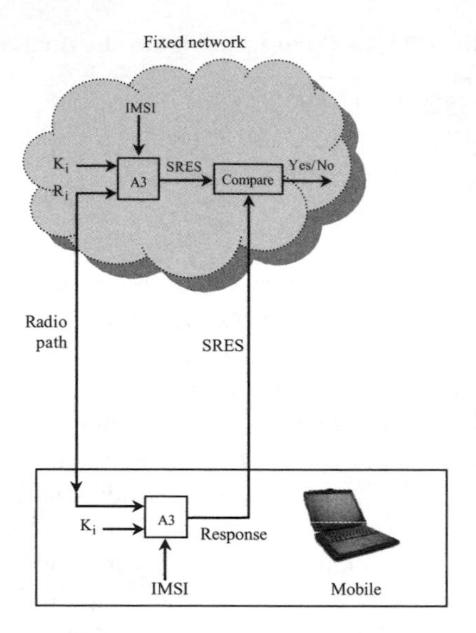

Figure 15.9 Generic authentication process.

both at the MS (mobile station) and at the AuC, a signed response (SRES), using an individual secret key K_i attached to the mobile subscriber. The number RAND, whose value is drawn randomly between 0 and $2^{128} - 1$, is used to generate the response by the mobile as well as by the fixed part of the network. The authentication process is carried out both at the mobile and at the MSC simultaneously. The BSS remains transparent to this process. The mobile only receives the random number over the radio path and in turn returns the signed response to the network. Thus, an air interface mobile designation is not disclosed. At subscription time, the subscriber authentication key K_i is allocated to the subscriber together with its IMSI (international mobile station identity). The key K_i is stored in the AuC and used to generate a triplet (K_c, signed response, RAND) within the GSM system. As stated above, the same K_i is also stored at the mobile in the subscriber ID (SIM). In the AuC, the following steps are carried out in order to produce one triplet. A nonpredictable RAND is produced. RAND and K_i are used to calculate the signed response and the ciphering key (K_c) using two different algorithms (A3, A8). This triplet (RAND, signed response, and K_c) is generated for each and every user and is then delivered to the HLR. This procedure is shown in Figure 15.10 [41–43].

The AuC begins the authentication and cipher key generation procedures after receiving an identification of the subscriber from the MSC/VLR. The AuC first queries the HLR for the subscriber's authentication key K_i. It then generates a 128 b RAND for use as a challenge (nonce), to be sent to the MS for verification of the MS's authenticity. RAND is also used by the AuC, with K_i in the algorithm A3 for authentication, to calculate the expected correct signed response from the MS. RAND and K_i are also used in the AuC to calculate the cipher key K_c with algorithm A8. The signed response is a 32 b number, and K_c is a 64 b number.

The values of RAND, signed response and K_c are transmitted to the MSC/VLR for interaction with the MS. Algorithms A3 and A8 are not fully standardized by GSM and

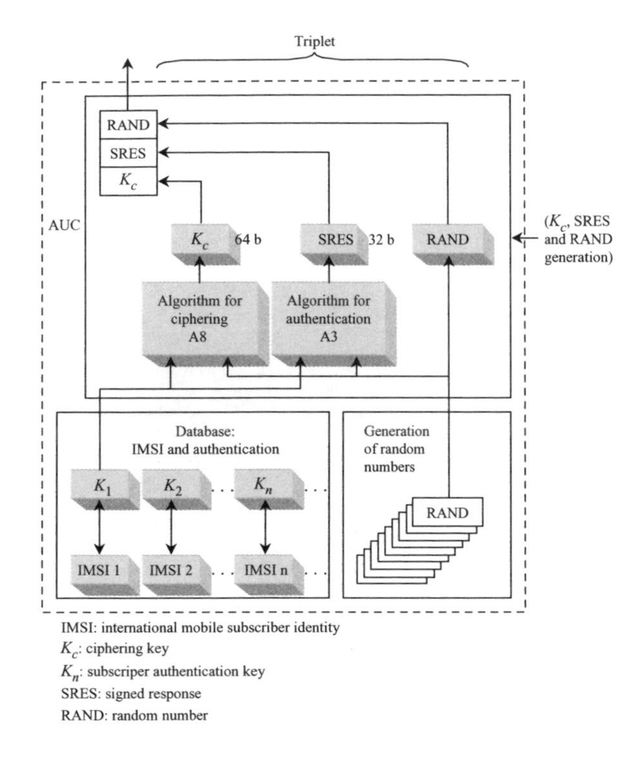

Figure 15.10 Generation of K_c, signed response, and RAND at the AuC.

may be specified at the direction of PLMN operators. Different PLMNs may use different and proprietary versions of these algorithms. Also, to protect the secrecy of the user, the authentication key K_i is not sent to the MSC/VLR. Based on the discretion of the PLMN operator, K_i can be of any format and length. The MSC/VLR forwards the value of the RAND to the MS, which also has the correct K_i and algorithm A3, which is stored in its SIM. The SIM then uses RAND and K_c in these algorithms to calculate the authentication SRESc and the cipher key, K_c. The MS sends the calculated response, SRESc, back to the MSC/VLR, which compares it with the value signed response received from the HLR/AuC. If the SRESc and the signed response agree, the subscriber access to the system is granted, and the cipher key K_c is transferred to the BTS for use in encrypting and decrypting messages to and from the MS. If the SRESc (computed signed response at the mobile) and the signed response disagree, the subscriber access to the system is denied. In summary, the VLR initiates authentication toward the MS and checks the authentication result. The complete process is shown in Figure 15.11.

The *ciphering/deciphering* algorithm (called A5) uses a cipher key K_c, which is allocated to each mobile subscriber during the authentication procedures. The key K_c is computed from the RAND by an algorithm (called A8) driven by the mobile subscriber authentication key K_i. Algorithm A8 is common to all GSMs. Figure 15.10 has shown the process of generating K_c. For the authentication procedure, when a signed response is being calculated at the mobile, the ciphering key (K_c) is also calculated using another algorithm (A8), as shown in Figure 15.11. This key is set in the fixed system and in the MS. At the ciphering

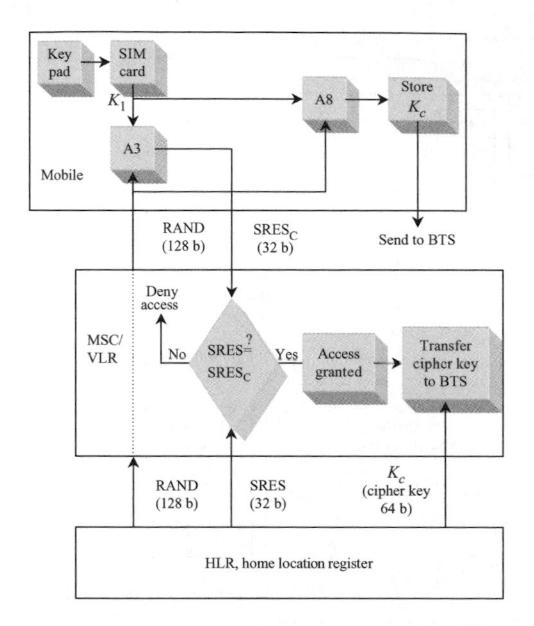

Figure 15.11 Authentication process in GSM system.

start command (from VLR to BSS), K_c is used by the MS and the BTS in order to cipher and decipher the bitstream that is sent over the radio path. In addition to the authentication procedures, a key setting may be initiated by the network as often as the network operator wishes. The command to use the encryption key is sent over the logical channel and dedicated control channel, as soon as the identity of the mobile subscriber is known by the network.

The key K_c must be agreed upon by the mobile station and the network prior to the start of encryption. The choice in GSM is to compute the key K_c independently from the effective start of encryption during the authentication process. K_c is then stored in a nonvolatile memory inside the SIM so as to be remembered even after a switched-off phase. This key is also stored in the visited MSC/VLR on the network side and is ready to be used for the start of encryption. The actual encryption/decryption of user data takes place within the mobile station and the BSS. For this purpose, the encryption key is downloaded from the MSC to the BTS via the BSC. After authentication, the transmission is ciphered, and K_c is used for ciphering/deciphering. This process is shown in Figure 15.12. Data flow on the radio path is obtained by a bit-per-bit binary addition of the user data flow and ciphering bitstream generated by the GSM algorithm A5 using a ciphering key (K_c). This exact process of encryption/decryption at the mobile and at BTS is shown in Figure 15.13. Code words S_1 and S_2 for downlinks and uplinks are changed at every frame. When modulo 2 is added with plain text, S_1 outputs cipher text. On the other side, the cipher text, when modulo 2 is added with S_1, outputs the plain text. The ciphering/deciphering function is placed on the transmission chain between the interleaver and the modulator. Since A3 and A8 are always running together, these two are implemented as a single algorithm in most cases. The algorithm A3 is standardized in the whole of GSM.

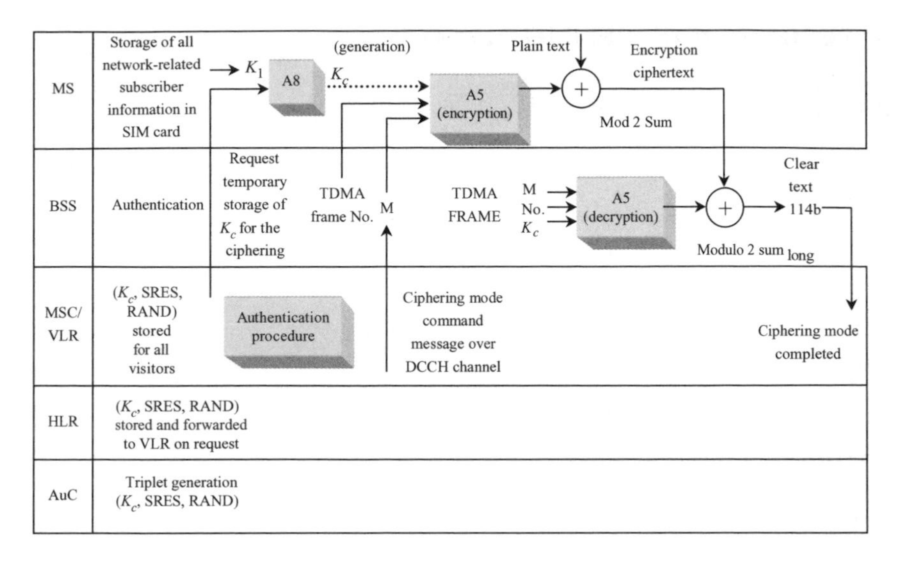

Figure 15.12 Sequential steps for encryption and decryption process.

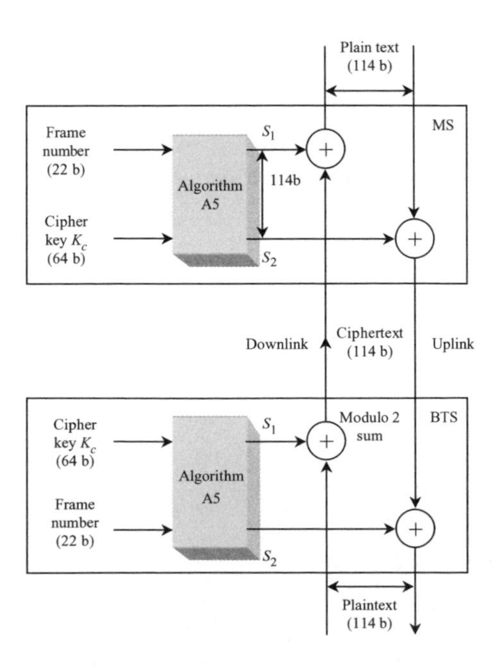

Figure 15.13 Encryption/decryption process.

15.5 SECURITY MANAGEMENT IN UMTS

Cryptographic functions used in UMTS are specified below:

Algorithm: Purpose/usage
O,S (operator specific, fully standardized)/location in the network

---

f0 : random challenge generating function
O /AuC
f1 : network authentication function
O – (MILENAGE) /USIM and AuC
f1 : resynchronization message authentication function*
O – (MILENAGE) /USIM and AuC
f2 : user challenge-response authentication function
O – (MILENAGE) /USIM and AuC
f3 : cipher key derivation function
O – (MILENAGE) /USIM and AuC
f4 : integrity key derivation function
O – (MILENAGE) /USIM and AuC
f5 : anonymity key derivation function for normal operation
O – (MILENAGE) /USIM and AuC
f5 : anonymity key derivation function for resynchronization*
O – (MILENAGE) /USIM and AuC
f6 : MAP encryption algorithm
S/MAP nodes
f7 : MAP integrity algorithm
S/MAP nodes
f8 : UMTS encryption algorithm
S – (KASUMI)/ MS and RNC
f9 : UMTS integrity algorithm
S – (KASUMI) / MS and RNC

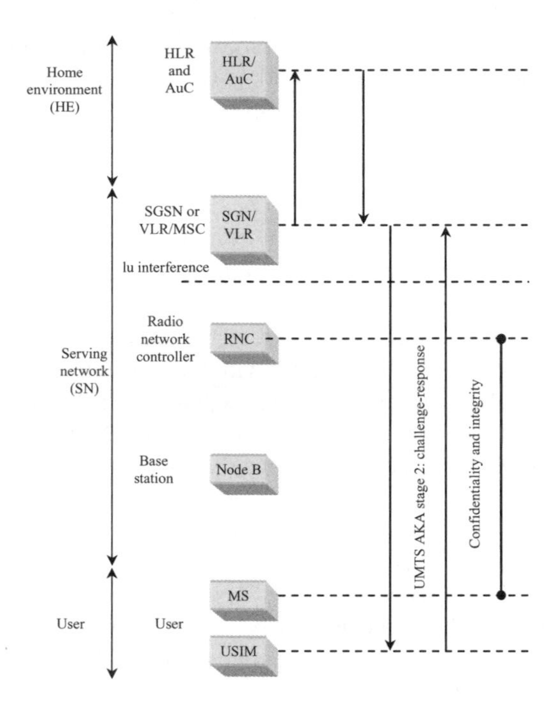

Figure 15.14 Simplified UMTS security architecture.

More details on these functions can be found in References [46–60] especially [52, 53, 56, 57]. A simplified UMTS security architecture is shown in Figure 15.14.

The security architecture is designed around a mutual authentication procedure that is executed between the user (USIM) and the SGSN/VLR3 at the network end. The procedure is called the UMTS Authentication and Key Agreement (AKA) since, in addition to providing authentication services, it also includes generation of session keys for confidentiality and integrity protection at the user end. The cryptographic algorithms/functions are defined in a requirements specification [49]. The AKA procedure is executed in two stages, as shown in Figure 15.14. The first stage involves transfer of security credentials (authentication vector, AV) from the home environment (HE) to the serving network (SN). The HE mainly consists of the home location register (HLR) and authentication center (AuC); the SN consists of the parts of the core network that are directly involved in setting up connections. With respect to access security, the SN network elements of interest are the SGSN, which handles packet-switched traffic, and the circuit-switched GSM nodes VLR/MSC (mobile switching center). An operator with a physical access infrastructure will normally have both HE and SN nodes.

The authentication vectors contain sensitive data like *challenge-response* authentication data and cryptographic keys. It is therefore clear that the transfer of authentication vectors between the HLR/AuC and the SGSN/VLR needs to be secured against eavesdropping and modification (i.e. both the transfer's confidentiality and integrity must be protected). The actual transfer mechanism for the AVs is the SS7-based mobile application part (MAP) protocol. The MAP protocol itself contains no security functionality, but a security extension to MAP called MAPsec [50] has been developed by the 3G Partnership Project (3GPP). The MAPsec protocol belongs to the Network Domain Security (NDS) work area in 3GPP. NDS covers both the MAPsec specification and specifications for how to protect IP connections on the control plane of the UMTS core network [51].

The second AKA stage is where the SGSN/ VLR executes the one-pass challenge-response procedure to achieve mutual entity authentication between the USIM and the network (SN, HE). A point to be made is that, in a two-staged AKA approach, the HE delegates responsibility for executing the security procedures to the SN. There must therefore be a trust relationship between the HE and the SN in this matter. In the GSM environment, this trust relationship is regulated through roaming agreements; the same model should be applicable to UMTS. The cryptographic functions ($f0 - f5^*$) used in the AKA procedure are implemented exclusively in the USIM and AuC. UMTS operators are free to choose any algorithm they want provided it complies with the function input/output specification given in Reference [49]. However, 3GPP has developed the MILENAGE [51, 52] algorithm set to provide the AKA functions. The formal status of MILENAGE is that it is provided as an example algorithm set, but in practice it is the recommended algorithm set for the AKA functions.

MILENAGE itself is built around the symmetric block cipher Rijndael. A consequence of having mutual authentication is that the USIM is now an active entity. In GSM, the user could not authenticate the network; hence, the UE could not reject the network. In UMTS, the USIM will attempt to authenticate the network and it is now possible that the USIM will reject the network. Details on the implementation of different security services in UMTS can be found in References [46–60].

15.6 SECURITY ARCHITECTURE FOR UMTS/WLAN INTERWORKING

As specified in Chapter 1, 4G will integrate different wireless systems and provide intertechnology roaming. Security architectures will depend on the type (tight or loose) of interworking. A tight UMTS/WLAN interworking solution would mandate the full 3GPP security architecture and require the 3GPP protocol stacks and interfaces to be present in the WLAN system [64] . The loose interworking options merely require the 3GPP authentication method to be implemented. To avoid link layer modifications, the authentication protocol is allowed to run at the link layer using Internet protocols – extensible authentication protocol (EAP) [63] and authentication, authorization and accounting (AAA) – as transport mechanisms. When used in WLAN the protocol EAP will be referred to as EAPOL. The main 3GPP-WLAN interworking architecture is defined in 3GPP TS 23.234 [61]; the security architecture is found in 3GPP TS 33.234 [62].

A benefit of the loose interworking approach is that the 3GPP-WLAN architecture is a fairly simple architecture. The architecture contains the WLAN access network and a UMTS core network in addition to glue technology to connect the two systems. Figure 15.15 gives an overview of the proposed architecture [64]. The two key glue components of the interworking solution are the AAA and EAP technologies. These are used to execute the UMTS AKA protocol from the 3G system's home domain toward the WLAN user equipment. The AAA architecture and the RADIUS and/or Diameter protocol are to be used as the bridge between the 3GPP system and the WLAN access network. Both Diameter and RADIUS are generic protocols and are intended to provide support for a diverse set of AAA applications, including network access, IP mobility and interoperator roaming. The EAP-AKA [67] protocol allows the UMTS AKA security protocol, which was originally

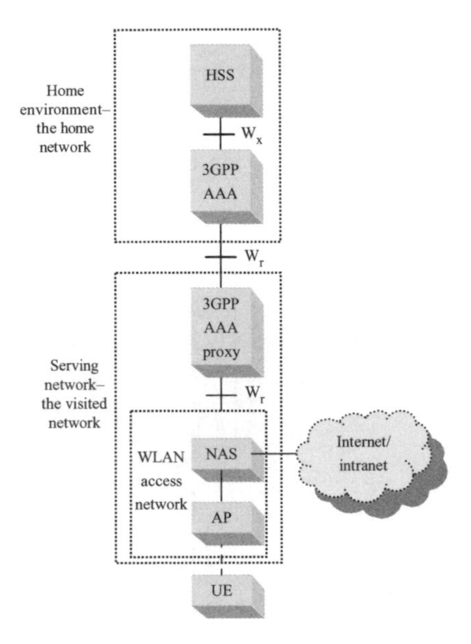

Figure 15.15 3GPP-WLAN architecture.

designed for execution over UTRAN, to be executed over the WLAN access toward the user equipment. EAP [63] is a key element in the 3GPP-WLAN security architecture. EAP provides, in essence, a generic peer-to-peer based *request–response* transaction environment for authentication dialogs, and supports multiple authentication mechanisms. EAP typically runs directly over the link layer without requiring IP. EAP has its own flow control mechanisms, and is capable of removing duplicate messages and retransmitting lost messages. EAP can be used over different link layer protocols including the IEEE WLAN link layer. The necessary EAP encapsulation is described in the EAP-over-LAN specification [68].

The EAP protocol does not natively provide much in terms of authentication mechanisms. Instead, its power lies in its generic mechanism to support existing authentication methods through specialized EAP methods. EAP contains a negotiation sequence where the authenticator requests information about which authentication method to use. The EAP architecture does not require the authenticator to support all authentication methods. Instead, the authenticator can request assistance from a backend authentication server to complete the authentication processing. The specific authentication methods supported are defined in separate specifications detailing how the EAP framework is to be used to run the target authentication methods. For 3GPPWLAN interworking, primary interest is in the EAP-AKA [64, 67] methods. An execution of the EAP-AKA procedure specific to the IEEE 802.11 WLAN is illustrated in Figure 15.16.

15.7 SECURITY IN *AD HOC* NETWORKS

Traditional security mechanisms, such as authentication protocols, digital signature and encryption, still play important roles in achieving confidentiality, integrity, authentication and nonrepudiation of communication in *ad hoc* networks. However, these mechanisms are not sufficient by themselves.

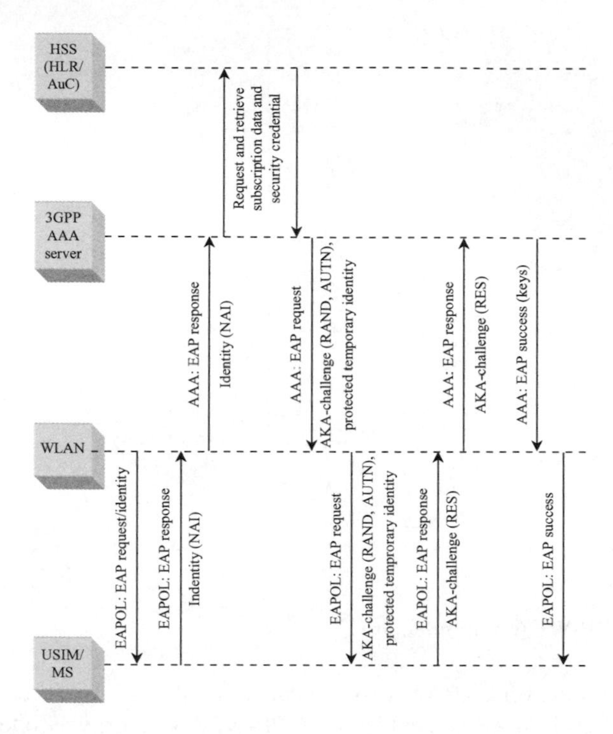

Figure 15.16 UMTS AKA procedure between a 3GPP network and an 802.11 WLAN MS.

In Zhou and Haas [69], two additional principles are discussed. First, the *redundancies* in the network topology (i.e. multiple routes between nodes) are exploited to achieve availability. The second principle is *distribution of trust.* Although no single node is trustworthy in an *ad hoc* network because of low physical security and availability, the trust can be distributed to an aggregation of nodes. Assuming that any $t + 1$ nodes will be unlikely to all be compromised, consensus of at least $t + 1$ nodes is trustworthy.

Although certain physical layer countermeasures, such as spread spectrum and coding, are possible [70, 79, 83, 84], we will only focus on how to defend against denial of service attacks towards routing protocols.

In most routing protocols, discussed in Chapter 13, routers exchange information on the topology of the network and this information could become a target for malicious adversaries who intend to bring the network down.

There are two sources of threats to routing protocols. The first comes from external attackers. By injecting erroneous routing information replaying old routing information or distorting routing information, an attacker could successfully partition a network or introduce excessive traffic load into the network by causing retransmission and inefficient routing. The second and also the more severe kind of threats comes from compromised nodes, which might advertise incorrect routing information to other nodes. Detection of such incorrect information is difficult. Merely requiring routing information to be signed by each node would not work, because compromised nodes are able to generate valid signatures using their private keys.

In the first case, nodes can protect routing information in the same way they protect data traffic, i.e. through the use of cryptographic schemes such as digital signature. However, this defense is ineffective against attacks from compromised servers. Worse yet, we cannot neglect the possibility of nodes being compromised in an *ad hoc* network. Detection of compromised nodes through routing information is also difficult in an *ad hoc* network because of its dynamically changing topology. When a piece of routing information is found to be invalid, the information could be generated by a compromised node, or, it could have become invalid as a result of topology changes. It is difficult to distinguish between the two cases.

On the other hand, certain properties of *ad hoc* networks can be exploited to achieve secure routing. For example, the routing protocols for *ad hoc* networks must handle outdated routing information to accommodate the dynamically changing topology. False routing information generated by compromised nodes could, to some extent, be considered outdated information. As long as there are sufficiently many correct nodes, the routing protocol should be able to find routes that go around these compromised nodes. Such capability of the routing protocols usually relies on the inherent redundancies due to multiple, possibly disjoint, routes between nodes in *ad hoc* networks. If routing protocols can discover multiple routes (e.g. protocols in ZRP, DSR, TORA and AODV, discussed in Chapter 13; all can achieve this), nodes can switch to an alternative route when the primary route appears to have failed.

Diversity coding, discussed in Chapter 7, takes advantage of multiple paths in an efficient way without message retransmission. The basic idea, discussed in Chapter 7, is to transmit redundant information through additional routes for error detection and correction. For example, if there are n disjoint routes between two nodes, then we can use $n - r$ channels to transmit data and use the other r channels to transmit redundant information. Even if certain routes are compromised, the receiver may still be able to validate messages and to recover messages from errors using the redundant information from the additional r channels.

Key management service using a single CA (*certification authority*) in *ad hoc* networks is problematic. The CA, responsible for the security of the entire network, is a vulnerable point of the network. If the CA is unavailable, nodes cannot obtain the current public keys of other nodes or to establish secure communication with others. If the CA is compromised and leaks its private key to an adversary, the adversary can then sign any erroneous certificate using this private key to impersonate any node or to revoke any certificate.

A standard approach to improve availability of a service is replication. However, a simple replication of the CA makes the service more vulnerable. Compromise of any single replica, which possesses the service private key, could lead to collapse of the entire system. To solve this problem, the trust can be distributed to a set of nodes with a collective key management responsibility [69] .

In such a system, the service as a whole has a public/private key pair. All nodes in the system know the public key of the service and trust any certificates signed using the corresponding private key. Nodes, as clients, can submit *query* requests to get other clients' public keys or submit *update* requests to change their own public keys.

The key management service, with an $(n, t + 1)$ configuration $(n \geq 3t + 1)$, consists of n special nodes, called *servers*, located within an *ad hoc* network. Each server also has its own key pair and stores the public keys of all the nodes in the network. In particular, each server knows the public keys of other servers. Thus, servers can establish secure links among them. It is assumed that the adversary can compromise up to t servers in any period of time with a certain duration.

If a server is compromised, then the adversary has access to all the secret information stored on the server. A compromised server might be unavailable or exhibit Byzantine behavior (i.e. it can deviate arbitrarily from its protocols). It is also assumed that the adversary lacks the computational power to break the cryptographic schemes employed. The service is correct if *robustness* and *confidentiality* are preserved. Robustness assumes that the service is always able to process *query* and *update* requests from clients. Every *query* always returns the last updated public key associated with the requested client, assuming no concurrent *updates* on this entry. Confidentiality assumes that the private key of the service is never disclosed to an adversary so that an adversary is never able to issue certificates, signed by the service private key, for erroneous bindings.

Threshold cryptography is used to accomplish distribution of trust in the key management service [71, 72]. An $(n, t+1)$ threshold cryptography scheme allows n parties to share the ability to perform a cryptographic operation (e.g. creating a digital signature), so that any $t+1$ parties can perform this operation jointly, whereas it is infeasible for at most t parties to do so, even by collusion.

In this case, the n servers of the key management service share the ability to sign certificates. For the service to tolerate t compromised servers, an $(n, t+1)$ threshold cryptography scheme is employed and the private key k of the service is divided into n shares $(s_1, s_2, \ldots, s_n)$, by assigning one share to each server. In this context $(s_1, s_2, \ldots, s_n)$ is referred to as an $(n, t+1)$ *sharing* of k. The service is configured as illustrated in Figure 15.17.

For the service to sign a certificate, each server generates a partial signature for the certificate using its private key share and submits the partial signature to a combiner. With $t+1$ correct partial signatures, the combiner is able to compute the signature for the certificate as illustrated in Figure 15.18, where servers generate a signature using a (3, 2) threshold signature scheme. The compromised servers (there are at most t of them) cannot generate correctly signed certificates by themselves, because they can generate at most t partial signatures.

If K/k is the public/private key pair of the service then by using a (3, 2) threshold cryptography scheme in Figure 15.18, each server i gets a share si of the private key k. For a message m, server i can generate a partial signature $PS(m, s_i)$ using its share s_i. Correct servers 1 and 3 both generate partial signatures and forward the signatures to a combiner c. Even though server 2 fails to submit a partial signature, c is able to generate the signature $(m)_k$ of m signed by service private key k [71, 72].

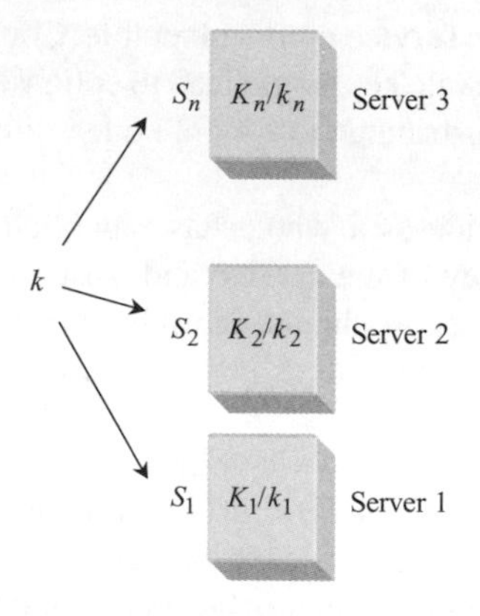

Figure 15.17 The configuration of a key management service.

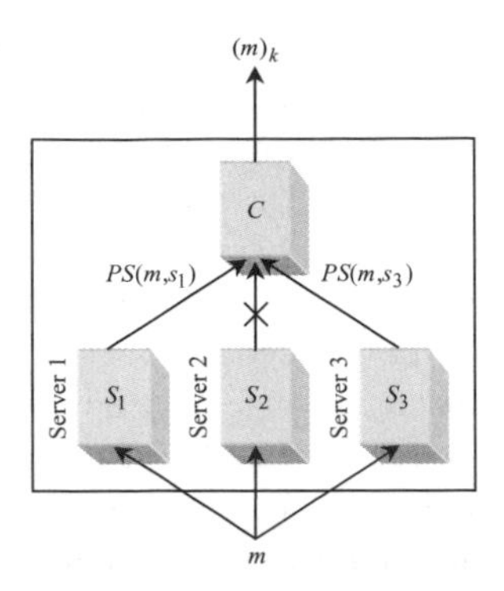

Figure 15.18 Threshold signature with three servers.

A compromised server could generate an incorrect partial signature, and use of this partial signature would yield an invalid signature. Fortunately, a combiner can verify the validity of a computed signature using the service public key. In case verification fails, the combiner tries another set of $t + 1$ partial signatures. This process continues until the combiner constructs the correct signature from $t + 1$ correct partial signatures. More efficient robust combining schemes are proposed [73, 74]. These schemes exploit the inherent redundancies in the partial signatures (note that any $t + 1$ correct partial signatures contain all the information of the final signature) and use error correction codes to mask incorrect partial signatures. In Gennaro *et al.* [73], a robust threshold DSS (digital signature standard) scheme is proposed. The process of computing a signature from partial signatures is essentially an interpolation. The authors uses the Berlekamp and Welch decoder, so that the interpolation still yields a correct signature despite a small portion (fewer than one-quarter) of partial signatures being missing or incorrect.

Even if the compromised servers are detected and excluded from the service, the adversary could still gather more than t shares of the private key from compromised servers over time. This would allow the adversary to generate any valid certificates signed by the private key.

Proactive schemes [75, 76] are proposed as a countermeasure to mobile adversaries. A proactive threshold cryptography scheme uses share refreshing, which enables servers to compute new shares from old ones in collaboration without disclosing the service private key to any server. The new shares constitute a new $(n, t + 1)$ sharing of the service private key. After refreshing, servers remove the old shares and use the new ones to generate partial signatures. Because the new shares are independent of the old ones, the adversary cannot combine old shares with new shares to recover the private key of the service. Thus, the adversary is challenged to compromise $t + 1$ servers between periodic refreshing.

Share refreshing must tolerate missing newly generated shares (called subshares) and erroneous subshares from compromised servers. A compromised server may not send any subshares. However, as long as correct servers agree on the set of subshares to use, they can generate new shares using only subshares generated from $t + 1$ servers. For servers to detect incorrect subshares, verifiable secret sharing schemes are used, for example, those in Pedersen [77]. A verifiable secret sharing scheme generates extra public information for each (sub)share using a one-way function. The public information can testify the correctness of the corresponding (sub)shares without disclosing the (sub)shares.

A variation of share refreshing also allows the key management service to change its configuration from $(n, t+1)$ to $(n', t'+1)$. This way, the key management service can adapt itself, on the fly, to changes in the network. If a compromised server is detected, the service should exclude the compromised server and refresh the exposed share. If a server is no longer available or if a new server is added, the service should change its configuration accordingly. For example, a key management service may start with the (7, 3) configuration. If, after some time, one server is detected to be compromised and another server is no longer available, then the service could change its setting to the (5, 2) configuration. If two new servers are added later, the service could change its configuration back to (7, 3) with the new set of servers.

15.7.1 Self-organized key management

In this section we discuss a *self-organizing public-key management system* that allows users to create, store, distribute and revoke their public keys without the help of any trusted

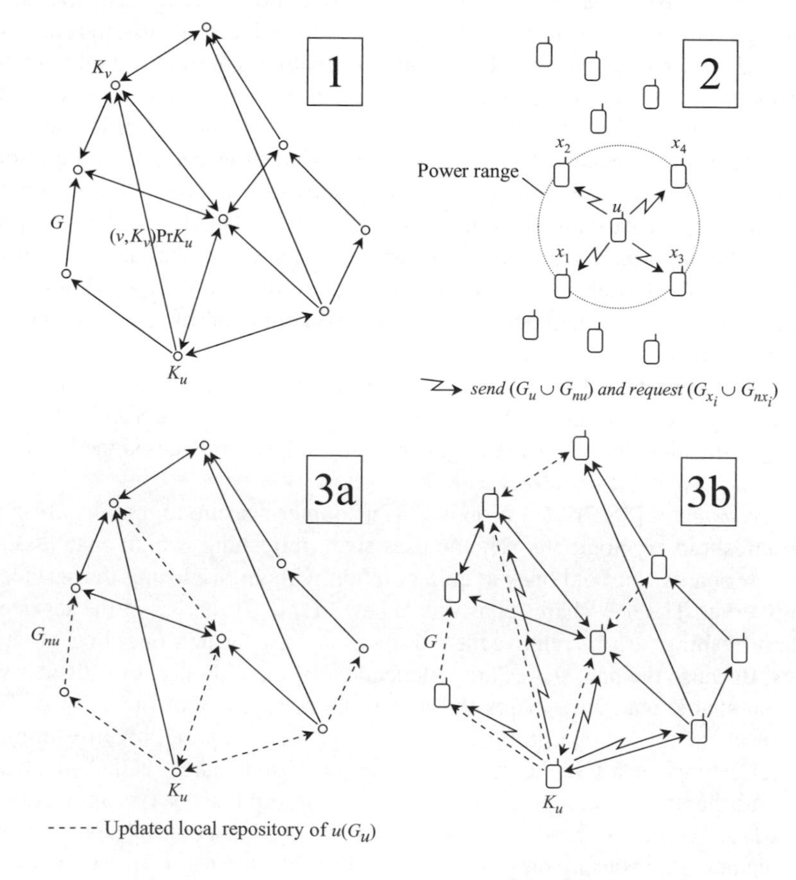

Figure 15.19 The creation of nodes' updated and nonupdated repositories (where $G_u^N = G_{nu}$).

authority or fixed server [85]. In the system model, the public keys and the certificates of the system are represented as a directed graph $G(V, E)$, where V and E stand for the set of vertices and the set of edges, respectively. This graph will be referred to as the certificate graph. The vertices of the certificate graph represent public keys and the edges represent certificates. More precisely, there is a directed edge from vertex Ku to vertex Kw if there is a certficate signed with the private key of u that binds Kw to an identity. A certificate chain from a public key Ku to another public key Kv is represented by a directed path from vertex Ku to vertex Kv in G. Thus, the existence of a certificate chain from Ku to Kv means that vertex Kv is reachable from vertex Ku in G (denoted below $K_{u \to G} K_v$). In the following, the certificate graph G designates the graph comprising only the valid (not expired) certificates of the whole network. In the model, the updated and the nonupdated certificate repositories of user u are represented by the certificate graphs Gu and Gnu, respectively. Therefore, for any u, Gu is a subgraph of G, but Gnu is not necessarily a subgraph of G, as it may also contain some implicitly revoked certificates.

As shown in Figure 15.19, the initial phase of the scheme is executed in four steps: the creation of public/private key pairs, the issuing of certificates, the certificate exchange, and the creation of nodes' updated certificate repositories.

In step 0, *the user creates her own public/private key pair.*

In step 1, *she issues public-key certificates based on her knowledge about other users' public keys. The issuing of public-key certificates also continues when the system is fully operational (i.e. when the updated and nonupdated repositories are already constructed) as users get more information about other users' public keys. During this process, the certificate graph G is created. The speed of the creation of a usable (i.e. sufficiently connected) certificate graph heavily depends on the motivation of the users to issue certificates.*

In step 2, *the node performs the certificate exchange. During this step, the node collects certificates and thus creates its nonupdated certificate repository. Along with the creation of new certificates, the certificate exchange also continues even when the system is fully operational. This means that nodes' nonupdated repositories will be continuously upgraded with new certificates. The nodes' mobility determines the speed at which certificates are accumulated by the nodes themselves.*

In step 3, *the node constructs its updated certificate repository. The node can perform this operation in two ways, either by communicating with its certificate graph neighbors (**Step 3a**), or by applying the repository construction algorithm on the nonupdated certificate repository (**Step 3b**). When the node constructed its updated certificate repository, it is ready to perform authentication.*

The *authentication* is performed in such a way that, when a user u wants to verify the authenticity of the public key Kv of another user v, u tries to find a directed path from Ku to Kv in $Gu \cup Gv$. The certificates on this path are then used by u to authenticate Kv. An example of a certificate graph with updated local repositories of the users is shown in Figure 15.20. If there is no path from Ku to Kv in $Gu \cup Gv$, u tries to find a path from Ku

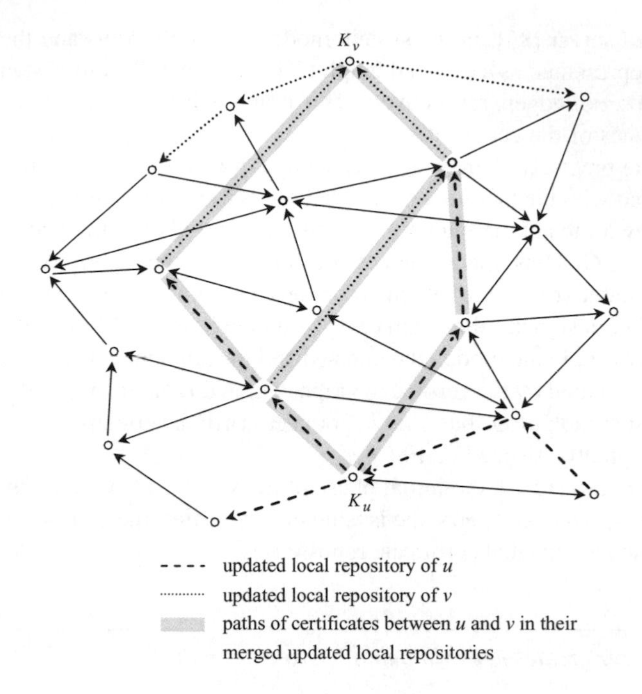

Figure 15.20 A certificate graph and paths of certificates between users u and v in their merged updated local repositories.

to Kv in $Gu \cup Gnv$. If such a path is found, u updates the expired certificates, checks their correctness and performs authentication. If there is no path from Ku to Kv in $Gu \cup Gnv$, u fails to authenticate Kv. A detailed description of each operation can be found in Capkun *et al.* [85].

15.8 SECURITY IN SENSOR NETWORKS

Owing to its inherent simplicity, sensor network routing protocols are sometimes even more susceptible to attacks against general *ad-hoc* routing protocols. Most network layer attacks against sensor networks fall into one of the following categories: *spoofed, altered or replayed routing information, selective forwarding, sinkhole attacks, sybil attacks, wormholes, HELLO flood attacks or acknowledgement spoofing. Spoofing, altering or replaying* routing information may be used by adversaries to create routing loops, attract or repel network traffic, extend or shorten source routes, generate false error messages, partition the network, increase end-to-end latency, etc.

Selective forwarding attack refers to the case where malicious nodes refuse to forward certain messages and simply drop them, ensuring that they are not propagated any further. Selective forwarding attacks are typically most effective when the attacker is explicitly included on the path of a data flow. In the next two sections, we discuss sinkhole attacks and the Sybil attack, two mechanisms by which an adversary can efficiently include herself on the path of the targeted data flow.

Sinkhole attacks typically work by making a compromised node look especially attractive to surrounding nodes with respect to the routing algorithm. For instance, an adversary could spoof or replay an advertisement for an extremely high quality route to a base station.

One motivation for mounting a sinkhole attack is that it makes selective forwarding trivial. By ensuring that all traffic in the targeted area flows through a compromised node, an adversary can selectively suppress or modify packets originating from any node in the area.

Sybil attack refers to the case where a single node presents multiple identities to other nodes in the network. Sybil attacks pose a significant threat to geographic routing protocols. Location-aware routing often requires nodes to exchange coordinate information with their neighbors to efficiently route geographically addressed packets. It is only reasonable to expect a node to accept but a single set of coordinates from each of its neighbors, but by using the Sybil attack an adversary can 'be in more than one place at once'.

The *wormhole attack* refers to the case where an adversary tunnels messages received in one part of the network over a low latency link and replays them in a different part. Wormhole attacks most commonly involve two distant malicious nodes colluding to understate their distance from each other by relaying packets along an out-of-bounds channel available only to the attacker. An adversary situated close to a base station may be able to completely disrupt routing by creating a well-placed wormhole. An adversary could convince nodes who would normally be multiple hops from a base station that they are only one or two hops away via the wormhole. This can create a sinkhole: since the adversary on the other side of the wormhole can artificially provide a high-quality route to the base station, potentially all traffic in the surrounding area will be drawn through it if alternate routes are significantly less attractive. This will most likely always be the case when the endpoint of the wormhole is relatively far from a base station. Figure 15.21 shows an example of a wormhole being used to create a sinkhole. Wormholes can also be used simply to convince two distant nodes that they are neighbors by relaying packets between the two of them. Wormhole attacks would probably be used in combination with selective forwarding or eavesdropping. Detection is potentially difficult when used in conjunction with the Sybil attack.

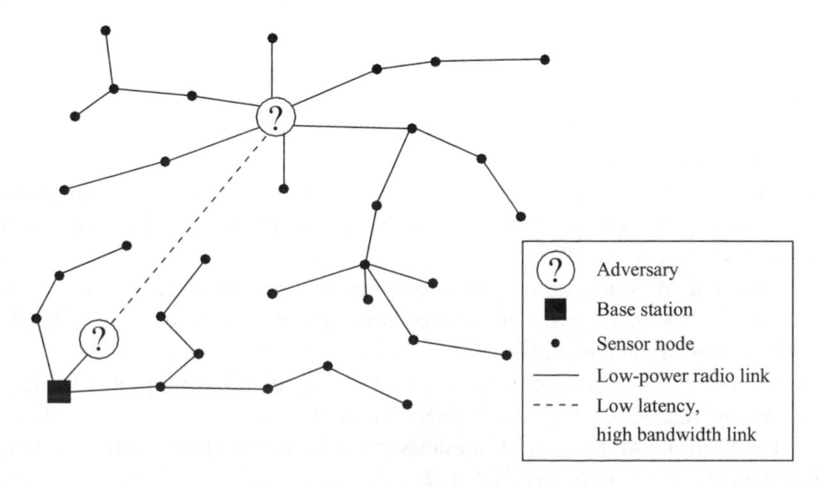

Figure 15.21 A laptop-class adversary using a wormhole to create a sinkhole in TinyOS beaconing.

HELLO floods can be thought of as one-way, broadcast wormholes. Many protocols require nodes to broadcast HELLO packets to announce themselves to their neighbors, and a node receiving such a packet may assume that it is within (normal) radio range of the sender. This assumption may be false: a laptop-class attacker broadcasting routing or other information with large enough transmission power could convince every node in the network that the adversary is its neighbor.

An adversary does not necessarily need to be able to construct legitimate traffic in order to use the HELLO flood attack. It can simply re-broadcast overheard packets with enough power to be received by every node in the network.

Acknowledgement spoofing refers to the case where an adversary can spoof link layer acknowledgments for 'overheard' packets addressed to neighboring nodes. Goals include convincing the sender that a weak link is strong or that a dead or disabled node is alive. For example, a routing protocol may select the next hop in a path using link reliability. Artificially reinforcing a weak or dead link is a subtle way of manipulating such a scheme. Since packets sent along weak or dead links are lost, an adversary can effectively mount a selective forwarding attack using acknowledgement spoofing by encouraging the target node to transmit packets on those links.

Attacks on specific sensor networks routing protocol and possible countermeasures are discussed in References [86, 87].

REFERENCES

[1] *OSI Directory – Part 8: Authentication Framework.* ISO 9594-8. ISO: Geneva, 1988.

[2] J.J. Jueneman, S.M. Matyas and C.H. Meyer, Message authentication, *IEEE Commun. Mag.*, 1985, pp. 29–40.

[3] C.H. Meyer and S.M. Matyas, *Cryptography: a New Dimension in Computer Data Security.* Wiley: New York, 1982.

[4] R. Morris and K. Thompson, Password security: a case history, *CACM*, vol. 22, no. 11, 1979, pp. 594–597.

[5] R.M. Needham and M.D. Schroeder, Using encryption for authentication in large networks of computers, *CACM*, vol. 21, no. 12, 1978, pp. 993–998.

[6] D. Otway and O. Rees, Efficient and timely mutual authentication, *ACM OSR*, vol. 21, no. 1, 1987, pp. 8–10.

[7] R.L. Rivest, A. Shamir and L. Adlcman, A method for obtaining digital signatures and public-key crypto-systems, *CACM,* vol. 21, no. 2, 1978, pp. 120–126; *CACM*, vol. 26, no. 1, 1983, pp. 96–99.

[8] R. Rivest, The MD4 message digest algorithm, *Internet RFC*, vol. 1320, April 1992.

[9] R. Rivest, The MD5 message digest algorithm, *Internet RFS,* vol. 1321, April 1992.

[10] *Banking – key Management (Wholesale).* ISO 8732. ISO: Geneva, 1988.

[11] R.K. Bauer, T.A. Berson and R.J. Freiertag, A key distribution protocol using event markers, *ACM TOCS,* vol. 1, no. 3, 1983, pp. 249–255.

[12] S.M. Bellovin and M. Merritt, Limitations of the Kerberos authentication system, *ACM CCR,* vol. 20, no. 5, 1990, pp. 119–132.

[13] R. Bird, I. Gopal, A. Herzberg, P.A. Janson, S. Kutten, R. Mowa and M. Yung, Systematic design of a family of attack-resistant authentication protocols, *IEEE J. Select. Area Commun.*, vol. 11. no. 5, 1993, pp. 679–692.

[14] R. Bird, I. Gopal, A. Herzberg, P.A. Janson, S. Kutten, R. Mowa and M. Yung, Systematic design of two-party authentication protocols, in *Proc. Crypto* 91,Santa Barbara, CA, Aug. 1991, pp. 44–61, available as *Advances in Cryptology, Lecture Notes in Computer Science* 576, J. Feigenbaum (ed.). Springer: New York, 1991.

[15] M. Burrows, M. Abadi and R.M. Needham, A logic of authentication, in *Proc. 12th ACM SOSP, ACM OSR,* vol. 23, no. 5, 1989, pp. 1–13.

[16] D.E. Denning and G.M. Sacco, Timestamps in key distribution systems, *CACM,* vol. 24, no. 8, 1981, pp. 533–536.

[17] L. Gong, Using one-way functions for authentication, *ACM CCR,* vol. 19, no. 5, 1989, pp. 8–11.

[18] C. I'Anson and C. Mitchell, Security defects in CCITT Recommendation X.509 – The Directory Authentication Framework, *ACM CR,* vol. 20, no. 2, 1990, pp. 30–34.

[19] *Entity Authentication Using Symmetric Techniques.* ISO-IEC JTC1.27.02.2 (20.03.1.2). ISO: Geneva, June 1990.

[20] *Data Encryption Standard,* FIPS 46, NBS, January 1977.

[21] C. Boyd, Security architectures using formal methods, *IEEE J. Select. Areas Commun.*, vol. 11, no. 5, June 1993, pp. 694–701.

[22] C.A. Boyd, Hidden assumptions in cryptographic protocols, *Proc. IEE, Part E,* vol. 137, 1990, pp. 433–436.

[23] M. Burrows, M. Abadi and R. Needham, A logic of authentication, *ACM Trans. Comput. Syst.,* vol. 8, no. 1, 1990, pp. 18–36.

[24] *CCITT, The Directory, Part 8: Authentication Framework, Recommendation X.509.* ISO: Geneva, 1989.

[25] D.W. Davies and W.L. Price, *Security in Computer Networks.* Wiley: New York, 1989.

[26] L. Gong, R. Needham and R. Yahalom, Reasoning about belief in cryptographic protocols, in *Proc. 1990 IEEE Computer Soc. Symp. Security Privacy.* IEEE Computer Society Press, 1990, pp. 234–248.

[27] *Information Processing Systems – Open Systems Interconnection Reference Model, Security Architecture,* ISO 7498-2. ISO: Geneva, 1988.

[28] R. Kailar and V.D. Gligor, On belief evolution in authentication protocols, in *Proc. Computer Security Foundations Workshop IV.* IEEE Press: New York, 1991, pp. 103–116.

[29] C. Meadows, A system for the specification and analysis of key management protocols, in *Proc. 1991 IEEE Computer Soc. Symp. Security Privacy.* IEEE Computer Society Press: New York, 1991, pp. 182–195.

[30] R.M. Needham and M.D. Schroeder, Using encryption for authentication in large networks of computers, *Commun. ACM,* vol. 21, no. 12, 1978, pp. 993–999.

[31] R. Rivest, A Shamir and L. Adelman, A method for obtaining digital signatures and public key cryptosystems, *Commun. ACM,* vol. 21, no. 2, 1978, pp. 120–126.

[32] J.M. Spivey, *The Z Notation.* Prentice-Hall: Englewood Cliffs, NJ, 1989.

[33] W. Fumy and P. Landrock, Principles of key management, *IEEE J. Select. Areas Commun.*, vol. 11, no. 5, 1993, pp. 785–793.

[34] D.M. Balenson, Automated distribution of cryptographic keys using the financial institution key management standard. *IEEE Commun. Mag.,* July 1985, pp. 41–46.

[35] W. Diffie and M.E. Hellman, New directions in cryptography, *IEEE Trans. Inform. Theory,* vol. 22, 1976, pp. 644–654.

[36] S.M. Matyas, Key handling with control vectors, *IBM Syst. J.,* vol. 30, no. 2, 1991, pp. 151–174.

[37] R.M. Needham and M.D. Schroeder, Using encryption for authentication in large networks of computers, *Commun. ACM,* vol. 21, 1978, pp. 993–999.
[38] E. Okamoto, Proposal for identity-based key distribution systems, *Electron. Lett.,* vol. 22, 1986, pp. 1283–1284.
[39] D. Otway and O. Rees, Efficient and timely mutual authentication. *Opns. Syst. Rev.*, vol. 21, 1987, pp. 8–10.
[40] A. Mehrotra and L.S. Golding, Mobility and security management in the GSM system and some proposed future improvements, *Proc. IEEE*, vol. 86, no. 7, 1998, pp. 1480–1486.
[41] A. Mehrotra, *GSM System Engineering*. Artech House: Norwood, MA, 1997.
[42] A. Mehrotra, *Cellular Radio Analog and Digital Systems*. Artech House: Norwood, MA, 1994, section 7.5.2.4, pp. 305–309.
[43] S.M. Redl, M.K. Weber and M.W. Oliphant, Security parameter, in *An Introduction to GSM*. Artech House: Norwood, MA, 1995, section 3.8, pp. 44–48.
[44] European Telecommunication Standard Institute/Global System for Mobility, ETSI/GSM specification vol. 2.17, section 3, Jan. 1993.
[45] G. Koien, An Introduction to Access Security in UMTS, *IEEE Wireless Commun.*, February 2004, pp. 8–18.
[46] 3G TS 33.120, *3G Security; Security Principles and Objectives.*
[47] 3G TS 21.133, *3G Security; Security Threats and Requirements.*
[48] 3G TS 33.102, *3G Security; Security Architecture.*
[49] 3G TS 33.105, *3G Security; Cryptographic Algorithm Requirements.*
[50] 3G TS 33.200, *3G Security; Network Domain Security; MAP Application Layer Security.*
[51] 3G TS 33.210, *3G Security; Network Domain Security; IP Network Layer Security.*
[52] 3G TS 35.205, *3G Security; Specification of the MILENAGE Algorithm Set: An Example Algorithm Set for the 3GPP Authentication and Key Generation Functions f1, f1*, f2, f3, f4, f5, and f5*; Document 1: General.*
[53] 3G TS 35.206, *3G Security; Specification of the MILENAGE Algorithm Set: An Example Algorithm Set for the 3GPP Authentication and Key Generation Functions f1, f1*, f2, f3, f4, f5, and f5*; Document 2: Algorithm Specification.*
[54] *Information Technology – Security Techniques – Entity Authentication - Part 4: Mechanisms Using a Cryptographic Check Function.* ISO/IEC 9798-4. ISO: Geneva.
[55] National Institute of Standards and Technology. *FIPS-197, Advanced Encryption Standard (AES)* (FIPS PUB 197). NIST: Gaithersburg, MD, 26 November 2001.
[56] 3G TS 35.201, *3G Security; Specification of the 3GPP Confidentiality and Integrity Algorithms; Document 1: f8 and f9 Specification.*
[57] 3G TS 35.202, *3G Security; Specification of the 3GPP Confidentiality and Integrity Algorithms; Document 2: KASUMI specification.*
[58] 3GPP, *Document TSGS#14(01)0622, Work Item Description: Support for Subscriber Certificates, SA#14*, Tokyo, Japan, Dec. 2001.
[59] S. Murphy and M. J. B. Robshaw, Essential algebraic structure within the AES: in *Proc. Crypto2002, LNCS*, vol. 2442. Springer: Berlin, 2002, pp. 1–16.
[60] 3G TS 33.904, *3GPP, SAGE; Report on the Evaluation of 3GPP Standard Confidentiality and Integrity Algorithms (SAGE v. 2.0).*
[61] 3G TS 23.234, *3GPP System to Wireless Local Area Network (WLAN) Interworking System Description, Release 6*, work in progress.

[62] 3G TS 33.234 v050, *3G Security; Wireless Local Area Network (WLAN) Interworking Security, Release 6*, work in progress.

[63] L. Blunk *et al.*, Extensible Authentication Protocol (EAP), Internet draft, draft-ietf-eap-rfc2284bis-04.txt, June 2003, work in progress.

[64] G. Koien and T. Haslestad, Security Aspects of'3G-WLAN Interworking, *IEEE Commun. Mag.*, November 2003, pp. 82–88.

[65] ETSI TR 101 957, *Broadband Radio Access Networks (BRAN); HIPERLAN Type 2; Requirements and Architectures for Interworking between HIPERLAN/2nd and 3rd Generation Cellular Systems.*

[66] IEEE Std 802.11i/D4.0, *Draft Amendment to Standard for Telecommunications and Information Exchange Between Systems – LAN/MAN Specific Requirements – Part 11: Wireless Medium Access Control (MAC) and Physical Layer (PHY) specifications: Medium Access Control (MAC) Security Enhancements*, May 2003, work in progress.

[67] J. Arkko and H. Haverinen, EAP AKA Authentication, Internet Draft: draft-arkko-pppext-eap-aka-10.txt, June 2003, work in progress.

[68] IEEE Std 802.1X-2001, *IEEE Standard for Local and Metropolitan Area Networks – Port-Based Network Access Control*, July 2001.

[69] L. Zhou and Z. Haas, Securing *ad hoc* Networks, *IEEE Networks*, November/ December 1999, pp. 24–30.

[70] E. Ayanoglu, C.-L. I, R. D. Gitlin and J. E. Mazo, Diversity coding for transparent self-healing and fault-tolerant communication networks. *IEEE Trans. Commun.*, vol. 41, 1993, pp. 1677–1686.

[71] Y. Desmedt and Y. Frankel, Threshold cryptosystems. In G. Brassard (ed.), *Advances in Cryptology – Crypto'89, Proc. 9th Annual International Cryptology Conference*, Santa Barbara, CA A, 20–24 August 1989, vol. 435 of Lecture Notes in Computer Science. Springer: Berlin, 1990, pp. 307–315.

[72] Y. Desmedt, Threshold cryptography. *Eur. Trans. Telecommun.*, vol. 5, 1994, pp. 449–457.

[73] R. Gennaro, S. Jarecki, H. Krawczyk and T. Rabin, Robust threshold DSS signatures. In U. M. Maurer (ed.), *Advances in Cryptology – Proc. Eurocrypt'96, International Conference on the Theory and Application of Cryptographic Techniques*, Saragossa, 12–16 May 1996, vol. 1233 of Lecture Notes in Computer Science. Springer: Berlin, 1996, pp. 354–371.

[74] R. Gennaro, S. Jarecki, H. Krawczyk and T. Rabin, Robust and efficient sharing of RSA functions. In N. Koblitz (ed.), *Advances in Cryptology – Proc. Crypto'96, the 16th Annual International Cryptology Conference*, Santa Barbara, CA, 18–22 August 1996, vol. 1109 of Lecture Notes in Computer Science. Springer: Berlin, 1996, pp. 157–172.

[75] A. Herzberg, S. Jarecki, H. Krawczyk and M. Yung, Proactive secret sharing or: How to cope with perpetual leakage. In D. Coppersmith (ed.), *Advances in Cryptology – Proc. Crypto'95, the 15th Annual International Cryptology Conference*, Santa Barbara, CA USA, 27–31 August 1995, vol. 963 of Lecture Notes in Computer Science. Springer: Berlin, 1995, pp. 457–469.

[76] Y. Frankel, P. Gemmell, P. MacKenzie and M. Yung, Proactive RSA. In B. S. Kaliski Jr. (ed.), *Advances in Cryptology – Proc. Crypto'97, the 17th Annual International Cryptology Conference*, Santa Barbara, CA, 17–21 August 1997, vol. 1294 of Lecture Notes in Computer Science. Springer: Berlin, 1997, pp. 440–454.

[77] T. Pedersen, Non-interactive and information-theoretic secure verifiable secret sharing. In J. Feigen-baum (ed.), *Advances in Cryptology – Proc. Crypto'91, the 11th Annual International Cryptology Conference*, Santa Barbara, CA, 11–15 August 1991, vol. 576 of Lecture Notes in Computer Science. Springer: Berlin, 1992, pp. 129–140.
[78] B. Kumar, Integration of security in network routing protocols. *SIGSAC Rev.,* vol. 11, 1993, pp. 18–25.
[79] K.E. Sirois and S.T. Kent, Securing the Nimrod routing architecture, in *Proc. Symp. Network and Distributed System Security*, Los Alamitos, CA, February 1997. The Internet Society, IEEE Computer Society Press, 1997, pp. 74–84.
[80] B.R. Smith, S. Murphy and J.J. Garcia-Luna-aceves, Securing distance-vector routing protocols. In *Proc. Symp. Network and Distributed System Security,* Los Alamitos, CA, February 1997. The Internet Society, IEEE Computer Society Press, 1997, pp. 85–92.
[81] R. Hauser, T. Przygienda and G. Tsudik, Lowering security overhead in link state routing. *Comput. Networks,* vol. 31, 1999, pp. 885–894.
[82] M.K. Reiter, Distributing trust with the Rampart toolkit, *Commun. ACM,* vol. 39, 1996, pp. 71–74.
[83] M.K. Reiter, M.K. Franklin, J.B. Lacy and R.N. Wright. The Ω key management service, *J. Comput. Security,* vol. 4, 1996, pp. 267–297.
[84] L. Gong, Increasing availability and security of an authentication service. *IEEE J. Select. Areas Commun.*, vol. 11, 1993, pp. 657–662.
[85] S. Capkun, L. Buttyán and J.-P. Hubaux, Self-organized public-key management for mobile *ad hoc* networks, *IEEE Trans. Mobile Comput.*, vol. 2, no. 1, 2003, pp. 52–64.
[86] C. Karlof and D. Wagner, Secure routing in wireless sensor networks: attacks and countermeasures, *Proc. First IEEE Int. Workshop on Sensor Network Protocols and Applications*, 11 May 2003, pp. 113–127.
[87] J. Kulik, W. Heinzelman and H. Balabishnan, Negotiation-based protocols for disseminating information in wireless sensor networks, *Wireless Networks*, vol. 8, nos 2–3, 2002, pp. 169–185.

16

Active Networks

16.1 INTRODUCTION

The basic goals of active networking (AN) are to create networking technologies that, in contrast to current networks, are easy to evolve and which allow application specific customization. To achieve these goals, AN uses a simple idea, that the network would be easier to change and customize if it were programmable [1–6]. While AN has the high-level goals of improving evolvability and customizabilty, there are a number of low-level concerns that must be balanced to achieve these goals. The first of these concerns is flexibility. AN systems aim to significantly improve the flexibility with which we can build networks. The second concern is safety and security. It is crucial that, while adding flexibility, we do not compromise the safety or security of the resulting system. The third concern is performance. If adding flexibility results in a system that cannot achieve its performance goals, it will be pointless. The final concern is usability. It is important that the resulting system not be so complex as to be unusable.

One of the basic techniques of AN is that code (program) is moved to the node at which it should execute. One place this idea first arose was in distributed systems supporting process migration. Another significant early influence was in generalization of remote procedure call (RPC) to support remote evaluation (RE). Both of these techniques will be discussed in detail in applications for network management in Chapter 18. The next important step in this direction was a DARPA proposal on the topic of 'Protocol boosters for distributed computing systems'. The idea was to *dynamically construct* protocols using protocol element insertion and deletion on an as-needed basis, to respond to network dynamics. Protocols were constructed *optimistically*; that is, ideal operating conditions (no errors, low delays, adequate throughput, etc.) were assumed, and protocol elements (such as error detection and correction mechanisms) were inserted into protocols on-demand, as conditions were encountered that deviated from the best case where the protocol element would not be needed.

Advanced Wireless Networks: 4G Technologies Savo G. Glisic
© 2006 John Wiley & Sons, Ltd.

On the network level the previous concepts resulted in middleboxes [7]. Domain-specific 'middleboxes' [7] are appearing such as firewalls, network address translators (NATs) and intrusion detection systems (IDSs). Examples of such middleboxes include NATs [8–10], NAT with protocol translator (NAT-PT) [11], SOCKS gateway [12], Packet classifiers, markers and schedulers [13], TCP performance enhancing proxies [14], IP firewalls [15, 16], gatekeepers/session control boxes [17, 18], transcoders, proxies [17, 19]. These various application-driven *ad hoc* examples of STF (store, translate and forward) functionality can be unified in a common framework, and embedded in a common programming model reinforced with the safety and security.

The IETF's forwarding and control element separation (ForCES) working group [20] models router architectures in a way that allows the introduction of programmable and active technologies into the Internet control plane, in the style of base building block (BBN)s FIRE [21].

The principal observation here is that forwarding and routing are distinct activities. Routing is part of the 'control plane' of the IP Internet, while forwarding is the 'transport plane'. These activities were consolidated in traditional IP routers, but are now recognized as logically separate (viz. MPLS). The separation permits more general network elements to replace IP routers in performing control plane activities, allowing distributed routing with improved performance, slightly more global optimization, and perhaps surprisingly, an increase in security. In addition, the separation of routing and forwarding functions permits decoupling of their performance, allowing better tracking of technology improvements in both forwarder mechanics and routing algorithms.

Active router control uses fast forwarders as a virtual link layer, managed by specialized active router controllers (ARCs), as shown in Figure 16.1. Using a set of routers as a 'router in a room' is not uncommon; one could simply reduce their autonomy and specialize them to IP forwarding, making way for source routing over all optical networks or routing/router control internal to the network. The use of general purpose network elements permits this separation of concerns, while offering the *potential* for improvement of the Internet on a number of axes.

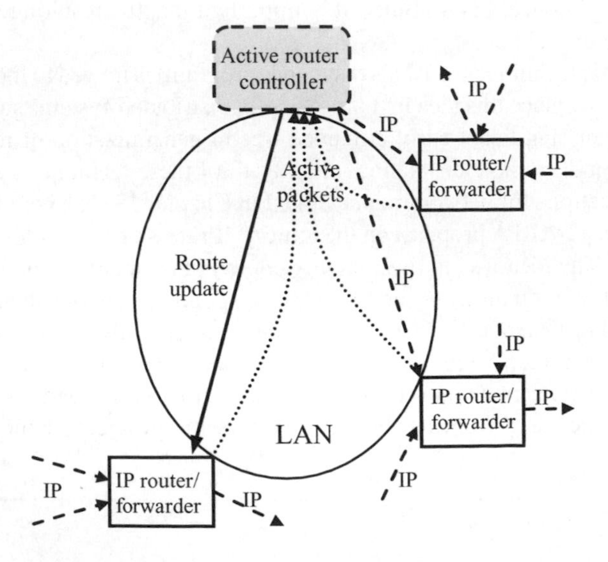

Figure 16.1 Active router controller managing a set of forwarder/routers.

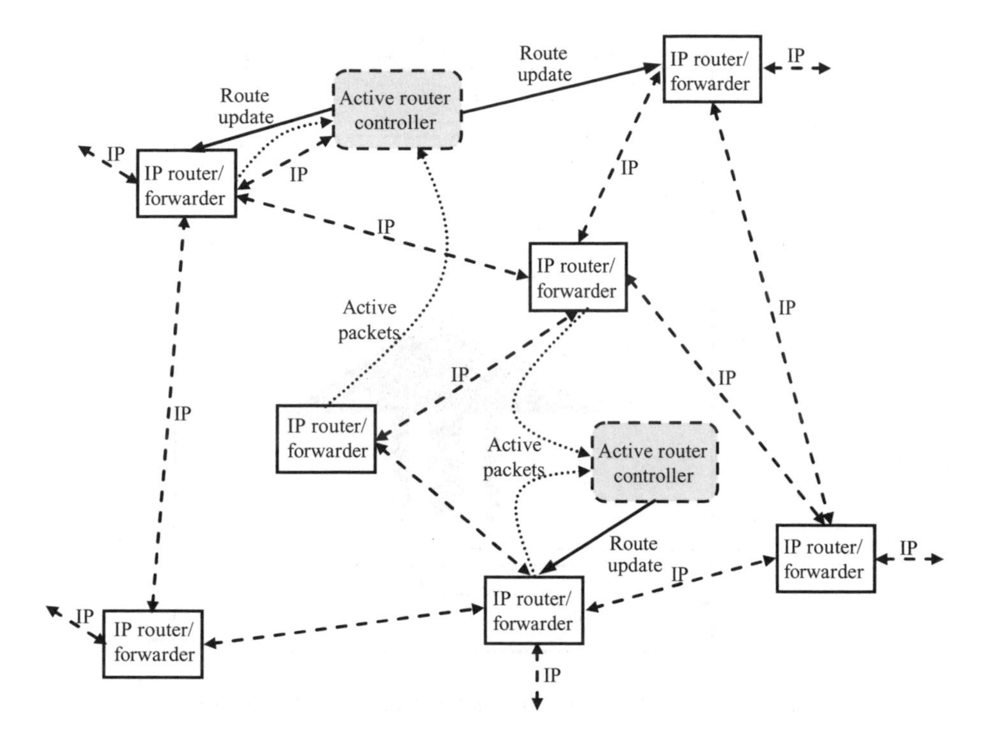

Figure 16.2 ARCs distributed throughout an Internet.

In this configuration, choices of routes are made by the ARC. A significant advantage is that specialized forwarding tables can be loaded into each forwarder; these tables can be small since entries must exist only for adjacent nodes.

A key advantage of the ARC model is that, for computationally centered tasks (routing, or more general computations if the programmability of ARCs is exploited), a computer which tracks computer technology trend exponentials (faster CPU, larger RAM) is used, while the forwarders independently track networking technology trend exponentials such as bandwidth improvements.

The basic distributed control architecture of Figure 16.1 can be replicated throughout an Internetwork, with the active elements using the managed Internet routes as link layers. This is shown in Figure 16.2, where a set of active nodes has been grafted into a larger collection of forwarders to create an active Internet.

16.2 PROGRAMABLE NETWORKS REFERENCE MODELS

IEEE P1520 is formalized by the IEEE Project 1520 standards initiative for programmable network interfaces and its corresponding reference model [22]. The IEEE P1520 RM, depicted in Figure 16.3, defines the following four interfaces.

(1) CCM interface – the connection control and management interface is a collection of protocols that enable the exchange of state and control information at a very low level between the NE and an external agent.

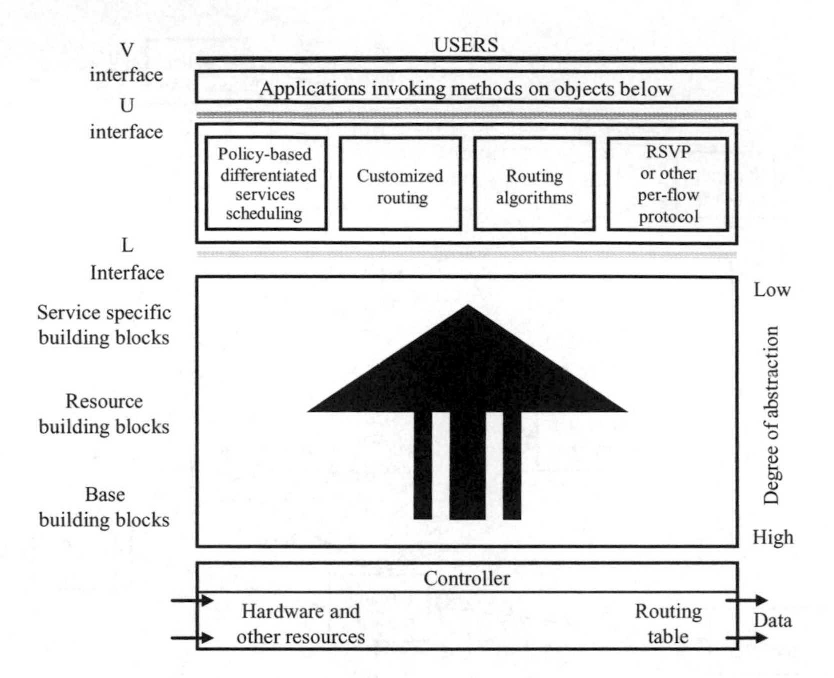

Figure 16.3 P1520 reference model and the L-interface abstraction model.

(2) L-interface – this defines an application program interface (API) that consists of methods for manipulating local network resources abstracted as objects. This abstraction isolates upper layers from hardware dependencies or other proprietary interfaces.

(3) U-interface – this mainly provides an API that deals with connection setup issues. The U-interface isolates the diversity of connection set-up requests from the actual algorithms that implement them.

(4) V-interface – it provides a rich set of APIs to write highly customized software, often in the form of value-added services.

CCM and L-interfaces fall under the category of NE (network element) interfaces, whereas U- and V-interfaces constitute network-wide interfaces. Initial efforts through the ATM sub-working group (P1520.2) focused on telecommunication networks based on ATM and introduced programmability in the control plane [23]. Later, the IP Sub-working Group extended these principles to IP networks and routers.

16.2.1 IETF ForCES

Recently, a working group of IETF, called forwarding and control element separation (ForCES) was formed with a similar objective to that of P1520 [24, 25]. Document [25] classifies the NE as a collection of components of two types: control elements (CE) and forwarding elements (FE) operating in the control and forwarding (transport) planes, respectively. CE's host controls functionality, like routing and signaling protocols, whereas FEs perform operations on packets, like header processing, metering, scheduling, etc., when

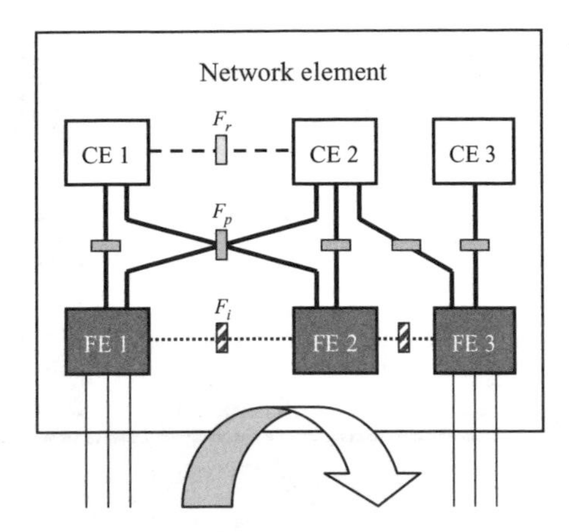

Figure 16.4 ForCES architectural representation of NE.

passing through them. CEs and FEs may be interconnected with each other in every possible combination (CE-CE, CE-FE, FE-FE), thus forming arbitrary types of logical topologies (see Figure 16.4). Every distinct combination defines a reference point, namely, F_r, F_p and F_i. Each one of these reference points may define a protocol or a collection thereof, but ForCES protocol is only defined for the F_p reference point.

16.2.2 Active networks reference architecture

Active networks transform the store-and-forward network into store-compute-and-forward network. The difference here is that packets are no longer passive but rather active in the sense that they carry executable code together with their data payload. This code is dispatched and executed at designated (active) nodes performing operations on the packet data as well as changing the current state of the node to be found by the packets that follow. Two approaches are possible based on whether programs and data are carried discretely, within program and data packets (out-of-band) or in an integrated manner, i.e. in-band.

In the discrete case, injecting code into the node and processing packets are two separated jobs. The user or network operator first injects his customized code into the routers along a path. Then the data packet arrives, its header is examined and the appropriate preinstalled code is loaded to operate on its contents [26, 27]. Separate mechanisms for loading and executing may be required for the control thereof. This separation enables network operators to dynamically download code to extend a node's capabilities, which in turn become available to customers through execution.

At the other extreme lies the integrated approach where code and data are carried by the same packet [28]. In this context, when a packet arrives at a node, code and data are separated, and the code is loaded to operate on the packet or change the state of the node. A hybrid approach has also been proposed [29].

The *active networks reference architecture* model [30, 31] is shown in Figure 16.5. An active network is a mixture of active and legacy (nonactive) nodes. The active nodes run the node operating system (NodeOS) – not necessarily the same, while a number of execution environments (EE) coexist at the same node. Finally a number of active applications (AA) make use of services offered by the EEs.

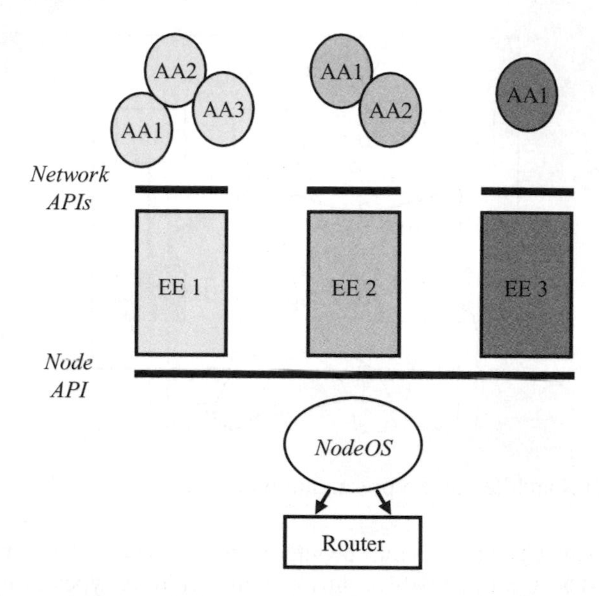

Figure 16.5 Active node architecture.

The NodeOS simultaneously supports multiple EEs. Its major functionality is to provide isolation among EEs through resource allocation and control mechanisms, and to provide security mechanisms to protect EEs from each other. It may also provide other basic facilities like caching or code distribution that EEs may use to build higher abstractions to be presented to their AAs. All these capabilities are encapsulated by the node interface through which EEs interact with the NodeOS. This is the minimal fixed point at which interoperability is achieved [31]. In contrast EEs implement a very broad definition of a network API ranging from programming languages to virtual machines, to static APIs in the form of a simple list of fixed-size parameters, etc. [32]. EE takes the form of a middleware toolkit for creating, composing and deploying services.

The AN reference architecture [30] is designed for simultaneously supporting a multiplicity of EEs at a node. Only EEs of the same type are allowed to communicate with each other, whereas EEs of different type are kept isolated from each other. A thorough analysis and comparison of programmable networks may be found in References [33, 34].

The *FAIN (future active IP network) NE* reference architecture is depicted in Figure 16.6. It describes how the ingredients identified previously may synergistically be combined in building next-generation NEs capable of seamlessly incorporating new functionality or dynamically configured to change their behavior according to new service requirements. One of the key concepts defined by the FAIN architecture is the EE. In FAIN, drawing from an analogy based on the concepts of class and object in object-oriented systems, we distinguish EEs between the *EE type* and the *EE instances* thereof.

An EE type is characterized by the programming methodology and the programming environment that is created as a result of the methodology used. The EE type is free of any implementation details. In contrast, an EE instance represents the realization of the EE type in the form of a runtime environment by using specific implementation technology, e.g. programming language and binding mechanisms to maintain operation of the runtime environment. For details see References [21, 35].

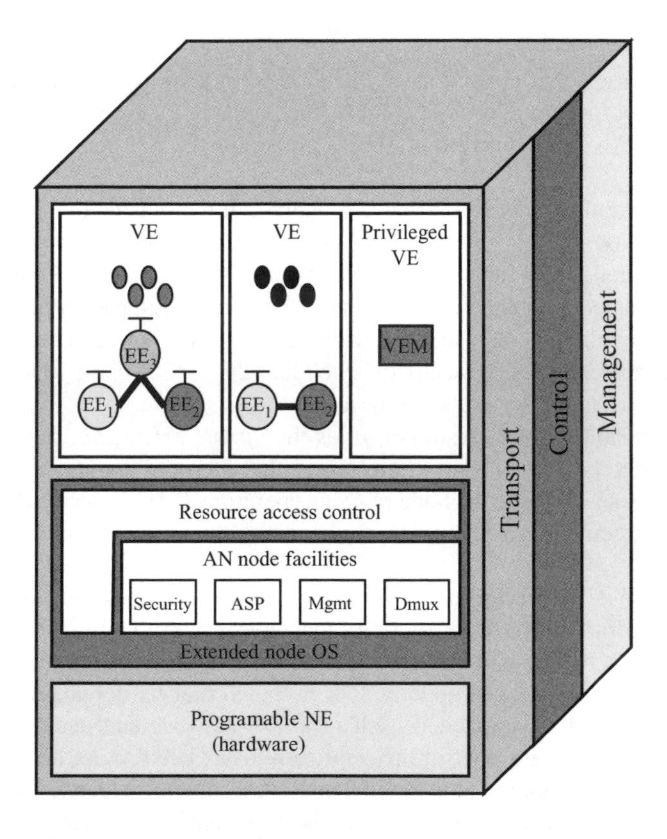

Figure 16.6 FAIN NE reference architecture.

16.3 EVOLUTION TO 4G WIRELESS NETWORKS

Given the volume of investments in existing networks infrastructure, a possible way to 4G might be evolution rather than any revolution. By 'network evolution' we mean any incremental change to a network that modifies or enhances existing functionality or adds new functionality. In the context of active networking, evolution should be able to occur at remote nodes while the network is operational with only minimal disruption to existing services. AN achieves evolution by changing the programs that operate the network. Thus the ways in which we can evolve the network are dictated by the programmability mechanisms that are available to make such changes.

Active packets (AP) are perhaps the most radical AN technology for evolution and they are the only mechanism that, at a high-level, are specific to AN. Such packets carry (or literally are) programs that execute as they pass through the nodes of network. A packet can perform management actions on the nodes, effect its own routing, or form the basis of larger protocols between distributed nodes, e.g. routing protocols. Such packets (ANTS [38], Smart Packets [39] and PLAN [40]) are like conventional packets, but with qualitatively more power and flexibility.

Active packet evolution does not require changes to the nodes of the network. Instead, it functions solely by the execution of APs utilizing standard services. The disadvantage of this approach is that taking advantage of new functionality requires the use of new packet programs. This means that at some level the applications using the functionality must be aware that the new functionality exists. This is the kind of evolution facilitated by pure AP systems, such as ANTS [38] and in essence it embodies the AN goal of application-level customization.

The programmability mechanism that is broadly familiar outside the AN community is the *plug-in extension*. Plug-in extensions can be downloaded and dynamically linked into a node to add new node-level functionality. For this new functionality to be used, it must be callable from some prebuilt and known interface. For example, a packet program will have a standard way of calling node resident services. If it is possible to add a plug-in extension to the set of callable servicces (typically by extending the service name space) then such an extension 'plugs in' to the service call interface.

The programmability mechanism known as the *update extensions* may also be downloaded, but they go beyond plug-in extensions in that they can update or modify existing code and can do so even while the node remains operational. Thus, such extension can add to or modify a system's functionality even when there does not exist an interface for it to hook into.

An example of AP evolution program written in a special-purpose language, PLAN (packet language for AN) [40], for mobility management protocol can be found in Song *et al.* [36]. Before a mobile node leaves its home network, it must identify the router that serves as its home agent. For simplicity, it is assumed that its default router serves this purpose. Once the node has attached itself to a new network and has a unique address, it sends an AP containing a control program to register itself to its home agent. When executed, this program simply adds the information to the home agent's soft-state keyed by the mobile node's home network address. Both the application aware and transparent versions share the same soft-state entries, allowing them to use the same control program and to co-exist.

Figure 16.7 illustrates how we detect that a packet is at the home agent of a mobile host and how the packet is then tunneled to the unique address. The PLAN code for the packet

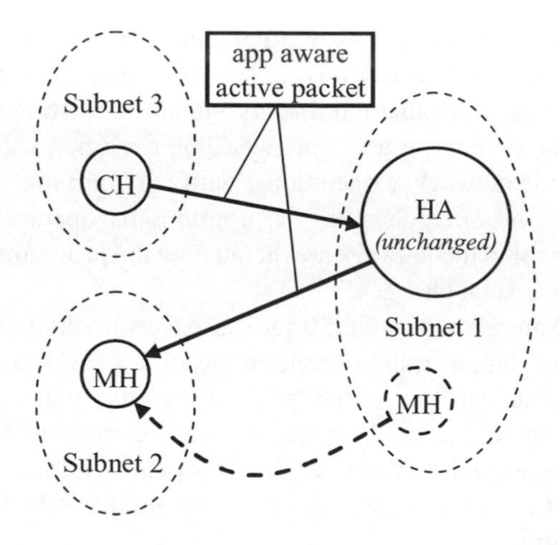

Figure 16.7 Active packets for mobile-IP (MH, mobile host; HA, home agent; CH, corresponding host.)

that must be sent by the application is [36]:

```
fun getToAgent(dest, payload, port) =
  try
    let val fagent = lookupTuple(dest) in
      OnRemote(|FoundFA|(dest, payload, port), fagent, getRB( ), defRoute)
    end
  handle NotFound =>
    if (thisHostIs(dest)) then deliver(payload, port)
    else
      let val next = defaultRoute(dest) in
        OnNeighbor(|getToAgent|(dest, payload, port), #1 next, getRB( ),#2 next))
      end
fun FoundFA(dest, payload, port) =
  let val hop = defaultRoute(dest) in
    OnNeighbor(|deliver|(payload, port), #1 hop, getRB(), #2 hop))
  end
```

GetToAgent is the main function and, when it executes, it first looks up *dest* in the soft-store using *lookupTuple*. If that succeeds, it has found the home agent and it uses *OnRemote* to send a new packet, the tunnel, that will execute *FoundFA* at the foreign agent with the same arguments as *getToAgent*.

OnRemote provides multihop transmission of the packet without execution until it reaches the foreign agent. If the lookup fails the handle will execute. If we have actually reached the host then we deliver the packet. Otherwise, it looks up the next hop toward *dest*. It then uses *OnNeighbor*, which only transmits a packet one hop, to send the packet. Thus the packet travels hop-by-hop looking for the home agent. *FoundFA* function executes on the foreign agent, which in this case is the mobile host, but might be some other node on the same sub-net. It sends a packet to the *dest* that does the delivery. This is where the original packet is removed from the tunnel. Notice that all of that functionality is encoded in the tunnel packet program itself; the foreign agent does not need to have any knowledge of its role as a tunnel endpoint – it just has to support PLAN.

If a node defines an extensible interface, new extensions can be loaded and plugged into this interface to provide extended or enhanced functionality. This is the essence of plug-in extension evolution. As an illustration consider the pseudocode shown below for a simple AP that implements route discovery [36]:

```
Route Discover(Simple DSR)
procedure Route Request(Target,RouteRec)
  if Duplicate Request Packet Or My Address Already In Route Rec
    then Discard
         { if This Host is Target
         {   then Route Reply(RouteRec)
    else {
         {        { Append My Address To Route
         { else   { Flood To All Neighbours
```

The packet itself embodies many key aspects of the protocol directly. In particular, it does duplicate elimination, tests for routing loops, detects termination and sends a reply, and performs flooding. In general, since many protocols have relatively simple control flow, simple APs can implement key aspects of the protocol directly. This algorithm benefits from noderesident services that are specific to the protocol, particularly to detect if this is a duplicate request or if the current node is already in the route record.

16.4 PROGRAMMABLE 4G MOBILE NETWORK ARCHITECTURE

In this section, we introduce a possible high-level architecture for future mobile systems with the focus on programmability [37, 41]. To address the creation and provisioning of unanticipated services, the whole system has to bc designed to be as flexible as possible. Openness and configurability, not only at the service level but also with respect to the whole system architecture, will attract third-party vendors to evolve the system as it unfolds and is therefore the key to viable solutions. As pointed out in Chapter 1, 4G systems are expected to provide an integration platform to seamlessly combine heterogeneous environments. Core abstraction layers cover hardware platform, network platform, middleware platform and applications. The abstraction layers can interact with each other using well-defined interfaces. Besides their regular cooperation in an operational setting, each layer can be configured separately and independently via configuration interfaces.

The generic architecture for the network elements of a mobile network is shown in Figure 16.8 (excluding applications). In this architecture, the following abstraction layers are considered, each programmable with configurable components:

- Middleware platform, typically with a virtual execution environment for platform-independent distributed processing.
- The computing platform serves as a general purpose platform for processing stateful protocols, e.g. routing, QoS signaling or connection management.
- The forwarding engine is in the data path of a network node and it connects the interface modules, e.g. by a switch matrix. This engine can be implemented as dedicated

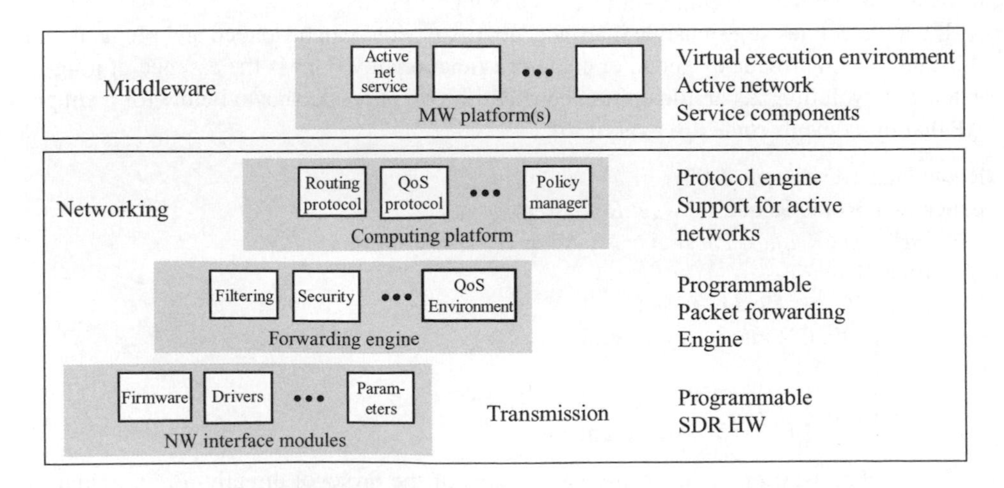

Figure 16.8 Programmable network element model.

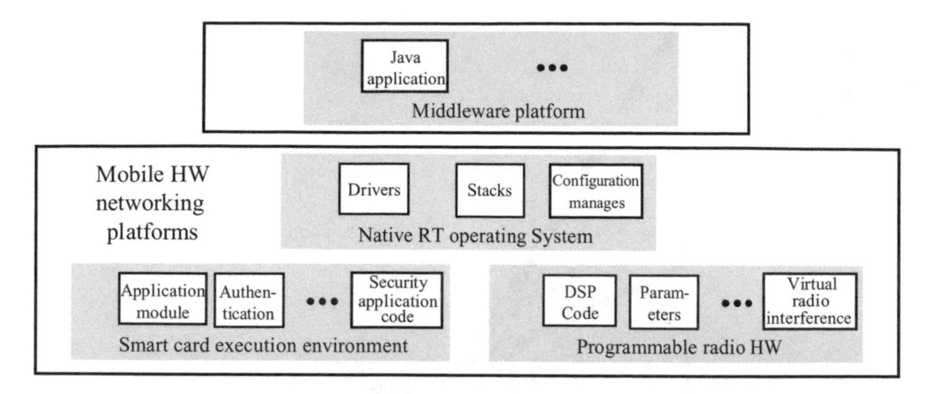

Figure 16.9 Mobile terminal architecture.

hardware or as a kernel module of common operating systems. The forwarding engine is programmable for performance-critical tasks which are performed on a per-packet basis.

- The interface modules are medium specific for different wireless or wired standards. They can be configured or programmed to adapt to new physical layer protocols or for triggering events in higher layers.

There are different approaches to active networking in this architecture. Some approaches offer a virtual execution environment, e.g. a Java virtual machine, on the middleware layer. Some options also include native code in the computing platform, e.g. for flexible signaling. Others employ programmable hardware for forwarding.

A key ingredient in these approaches is interfaces to the lower layers and programmable filters for identifying the packets to be processed in the active networking environment. The terminal architecture is shown in Figure 16.9. It consists of:

- Middleware platform, typically with a limited virtual execution environment.
- Smart Card, e.g. USIM for UMTS, which includes subscriber identities and also a small, but highly secure execution environment. This can be used ideally for personal services like electronic wallet.
- Programmable hardware, which is designed for one or more radio standards.
- Native operating system which provides real-time support, needed for stacks and certain critical applications, e.g. multimedia codecs.

Compared with network elements, the SmartCard is a new, programmable component. Owing to resource limitations, the forwarding engine and computation platform just collapse to one operating system platform. Also, the middleware layers are typically quite restricted. Service deployment and control of reconfigurations are complex since there is a split of responsibility between operator and manufacturer. For instance, the manufacturer has to provide or digitally sign appropriate low-level code for reconfigurations. On the other hand, the operator is interested in controlling the configuration to fit the user and network needs.

There are a number of examples to show the importance of flexible networking protocols and the need of dynamic cross layer interfaces in future mobile networks. As an example we show that hand-over can be optimized by information about the user context [37, 42, 43]

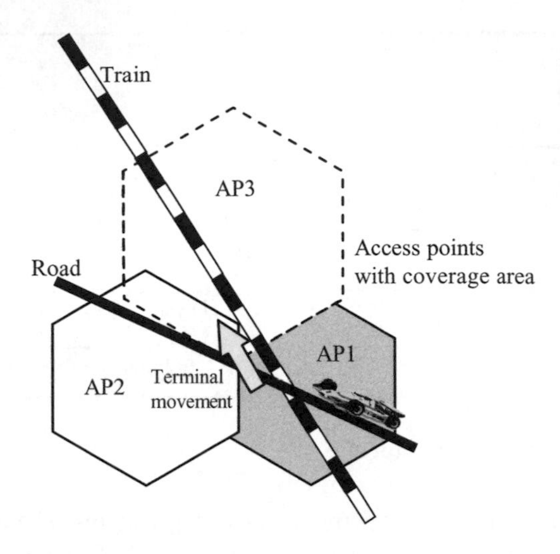

Figure 16.10 Context-aware hand-over prediction.

as shown in Figure 16.10. In this scenario, the terminal moves into an area covered by both AP2 and AP3. The problem is to decide which access point (AP) to choose. State of the art are many algorithms based on signal strength analysis or on available radio resources. Even if one AP is slightly better regarding these local measurements, the decision may not be the best. For instance, if the terminal in Figure 16.10 is on the train, it is obviously better to hand over to AP3, even if AP2 is more reachable for a short period of time. In many cases, the hand-over can be optimized by knowledge of terminal movement and user preferences. For instance, if the terminal is in a car or train, its route may be constrained to certain areas. Also, the terminal profile may contain the information that the terminal is built in a car. Alternatively, the movement pattern of a terminal may suggest that the user is in a train. A main problem is that hand-over decisions have to be executed fast. However, the terminal profile and location information is often available on a central server in the core network. Retrieving this information may be too slow for hand-over decisions.

Furthermore, the radio conditions during hand-over may be poor and hence limit such information exchange. The idea of the solution is to proactively deploy a context-aware decision algorithm onto the terminal which can be used to assist hand-over decisions. A typical example is the information about the current movement pattern, e.g. by knowledge of train or road routes. For implementation, different algorithms can be deployed by the network on the terminal, depending on the context information. This implementation needs a cross layer interface, which collects the context information from different layers and makes an optimized decision about deploying a decision algorithm.

16.5 COGNITIVE PACKET NETWORKS

We now discuss packet switching networks in which intelligence is incorporated into the packets, rather than at the nodes or in the protocols. Networks which contain such packets are referred to as cognitive packet networks (CPN). Cognitive packets in CPN route themselves, and learn to avoid congestion and being lost or destroyed. Cognitive packets learn from their own observations about the network and from the experience of other packets. They rely

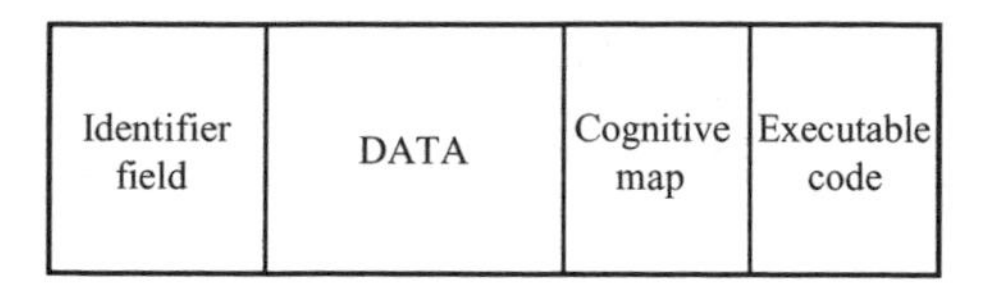

Figure 16.11 Format of cognitive packet.

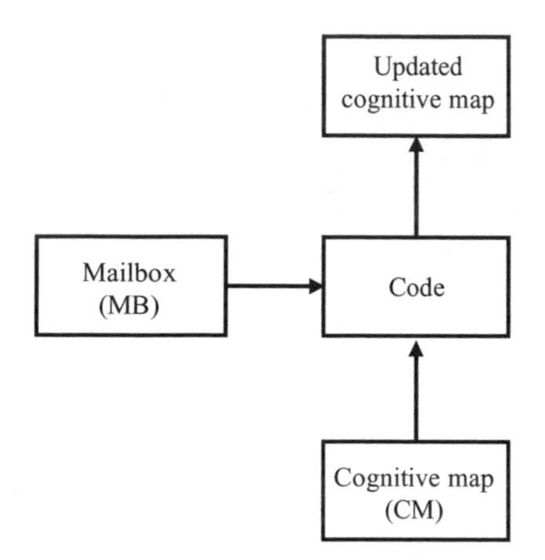

Figure 16.12 CP update by a CPN node.

minimally on routers. Each cognitive packet progressively refines its own model of the network as it travels through the network, and uses the model to make routing decisions. Cognitive packets (CPs) store information in their private cognitive map (CM) and update the CM and make their routing decisions using the code which is in each packet. This code may include neural networks or other adaptive algorithms. Figure 16.11 presents the format of a cognitive packet. The manner in which cognitive memory at a node is updated by the node's processor is shown in Figure 16.12. In a CPN, the packets use nodes as 'rented space', where they make decisions about their routes. They also use nodes as places where they can read their mailboxes. *Mailboxes may be filled by the node, or by other packets which pass through the node.* Packets also use nodes as processors which execute their code to update their CM and then execute their routing decisions. As a result of code execution, certain information may be moved from the CP to certain mailboxes. The nodes may execute the code of CPs in some order of priority between classes of CPs, for instance as a function of QoS requirements which are contained in the identification field).

As a routing decision, a CP will be placed in some output queue, in some order of priority, determined by the CP code execution. A CPN and a CPN node are schematically represented in Figures 16.13 and 16.14. CPs use units of information 'signals' to communicate with each other via mailboxes (MBs) in the nodes. These signals can also emanate from the environment (nodes, existing end-to-end protocols) toward the CPs. CP, shown in Figure 16.11, contain the following fields [44]:

(1) The *identifier field* (IF), which provides a unique identifier for the CP, as well as information about the class of packets it may belong to, such as its quality of service (QoS) requirements.

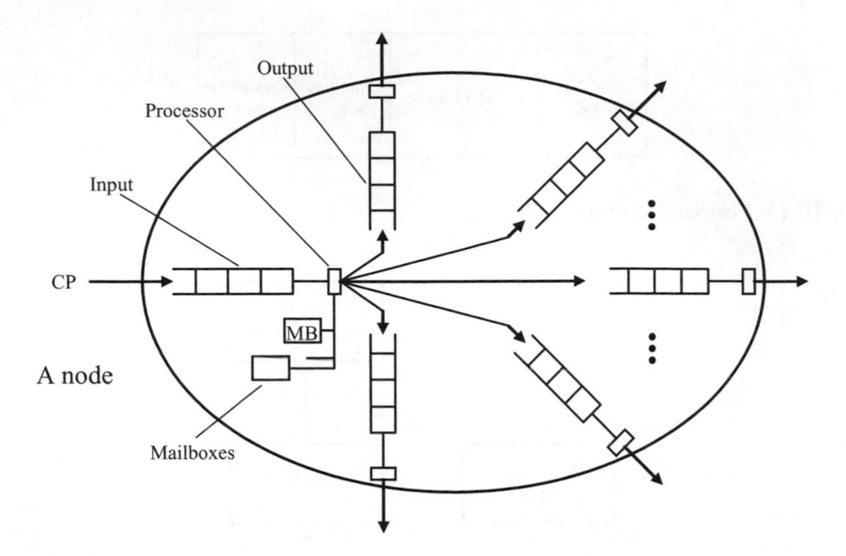

Figure 16.13 CPN node model.

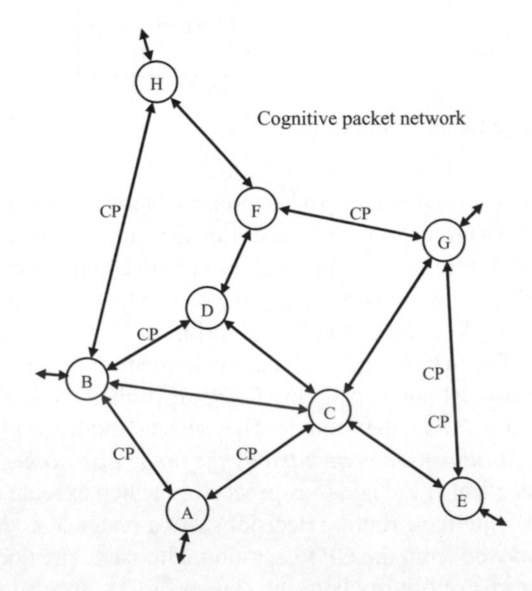

Figure 16.14 CPN example.

(2) The *data field* containing the ordinary data it is transporting.

(3) A *cognitive map* (CM), which contains the usual source and destination (S-D) information, as well as a map showing where the packet currently 'thinks' it is, the packet's view of the state of the network, and information about where it wants to go next; the S-D information may also be stored in the IF.

(4) *Executable code* that the CP uses to update its CM. This code will contain learning algorithms for updating the CM, and decision algorithms which use the CM.

A node in the CPN provides a storage area for CPs and for mailboxes which are used to exchange data between CPs, and between CPs and the node. It has an input buffer for CPs arriving from the input links, a set of mailboxes, and a set of output buffers which are associated with output links. Nodes in a CPN carry out the following functions [44]:

(1) A node receives packets via a finite set of ports and stores them in an input buffer.

(2) It transmits packets to other nodes via a set of output buffers. Once a CP is placed in an output buffer, it is transmitted to another destination node with some priority indicated in the output buffer.

(3) A node receives information from CPs which it stores in MBs. MBs may be reserved for certain classes of CPs, or may be specialized by classes of CPs . For instance, there may be different MBs for packets identified by different source-destination (S-D) pairs.

(4) A node executes the code for each CP in the input buffer. During the execution of the CPs code, the CP may ask the node to decline its identity, and to provide information about its local connectivity (i.e. this is node A, and I am connected to nodes B, C, D via output buffers) while executing its code. In some cases, the CP may already have this information in its CM as a result of the initial information it received at its source, and as a result of its own memory of the sequence of moves it has made.

As a result of this execution:

(1) The CMs of the packets in the input buffer are updated.

(2) Certain information is moved from CPs to certain MBs.

(3) A CP which has made the decision to be moved to an output buffer is transfered there, with the priority it may have requested.

We have been already discussing some networks which offer users the capability of adding network executable code to their packets. Additional material can be found in References [44–68].

16.5.1 Adaptation by cognitive packets

Each cognitive packet starts with an initial representation of the network from which it then progressively constructs its own cognitive map of the network state and uses it to make routing decisions. Learning paradigms are used by CPs to update their CM and reach decisions using the packet's prior experience and the input provided via mailboxes. In the adaptive approach for CPs, each packet entering the network is assigned a goal before it enters the network, and the CP uses the goal to determine its course of action each time it has to make a decision [44]. For instance if the CP contains part of a telephone conversation, a typical goal assigned to the packet should reflect the concern about the delay. A more sophisticated goal in this case could be delay and sequenced (in order) delivery. On the other hand, if these were data packets, the goal may simply be packet loss rate. These goals are translated into numerical quantities (e.g. delay values, loss probabilities and weighted combinations of such numerical quantities), which are then used directly in the adaptation.

An example is where all the packets were assigned a common goal, which was to minimize a weighted combination of delay (W) and loss (L) as

$$G = \alpha W + \beta L \tag{16.1}$$

A simple approach to adaptation is to respond in the sense of the most recently available data. Here the CP's cognitive memory contains data which is updated from the contents of the node's mailbox. After this update is made, the CP makes the decision which is most advantageous (lowest cost or highest reward) simply based on this information. This approach is referred to as the Bang-Bang algorithm. Instead some other learning paradigms for CPs can be also used like *learning feedforward random neural networks* (LFRNN) [57, 68] or *random neural networks with reinforcement learning* (RNNRL) [57]. Adaptive stochastic finite-state machines (ASFSM) [55, 56, 58] are another class of adaptive models which could be used for these purposes.

16.5.2 The random neural networks-based algorithms

For the reinforcement learning approach to CP adaptation, as well the feed-forward neural network predictor, the RNN [57] was used in Gelenbe *et al.* [44]. IT is an analytically tractable model whose mathematical structure is akin to that of queueing networks. It has product form just like many useful queueing network models, although it is based on nonlinear mathematics. The state qi of the ith neuron in the network is the probability that it is excited. *Each neuron i is associated with a distinct outgoing link at a node.* These quantities satisfy the following system of nonlinear equations:

$$q_i = \lambda^+(i) / \left[r(i) + \lambda^-(i) \right] \tag{16.2}$$

with

$$\lambda^+(i) = \sum_j q_j w_{ji}^+ + \Lambda_i, \ \lambda^-(i) = \sum_j q_j w_{ji}^- + \lambda_i \tag{16.3}$$

Here w_{ij}^+ is the rate at which neuron i sends 'excitation spikes' to neuron j when i is excited, and w_{ij}^- is the rate at which neuron i sends "inhibition spikes" to neuron j when i is excited and $r(i)$ is the total firing rate from the neuron i. For an n neuron network, the network parameters are these n by n 'weight matrices' $\mathbf{W}^+ = \|w_{ij}^+\|$ and $\mathbf{W}^- = \|w_{ij}^-\|$ which need to be 'learned' from input data. Various techniques for learning may be applied to the RNN. These include reinforcement learning and gradient-based learning, which are used in the following.

Given some goal G that the CP has to achieve as a function to be to be minimized [i.e. transit delay or probability of loss, or a weighted combination of the two as in Equation (16.1)], a reward R is formulated which is simply $R = 1/G$. Successive measured values of the R are denoted by $Rl, l = 1, 2, \ldots$ These are first used to compute a decision threshold:

$$T_l = aT_{l-1} + (1 - a)R_l \tag{16.4}$$

where a is some constant $0 < a < 1$, typically close to 1. Now an RNN with (at least) as many nodes as the decision outcomes is constructed. Let the neurons be numbered $1, \ldots, n$.

Thus for any decision i, there is some neuron i. Decisions in this RL algorithm with the RNN are taken by selecting the decision j for which the corresponding neuron is the most excited, i.e. the one with has the largest value of q_j. Note that the lth decision may not have contributed directly to the lth observed reward because of time delays between cause and effect. Suppose that we have now taken the lth decision which corresponds to neuron j, and that we have measured the lth reward Rl. Let us denote by r_i the firing rates of the neurons before the update takes place. We first determine whether the most recent value of the reward is larger than the previous 'smoothed' value of the reward, which is referred to as the threshold T_{l-1}. If that is the case, then we increase very significantly the excitatory weights going into the neuron that was the previous winner (in order to reward it for its new success), and make a small increase of the inhibitory weights leading to other neurons. If the new reward is not better than the previously observed smoothed reward (the threshold), then we simply increase moderately all excitatory weights leading to all neurons, except for the previous winner, and increase significantly the inhibitory weights leading to the previous winning neuron (in order to punish it for not being very successful this time). This is detailed in the algorithm given below. We compute T_{l-1} and then update the network weights as follows for all neurons $i \neq j$:

$$
\begin{aligned}
&\text{If} \quad T_{l-1} \leq R_l \\
&\quad - w^+(i,j) \leftarrow w^+(i,j) + R_l \\
&\quad - w^-(i,k) \leftarrow w^-(i,k) + R_l/(n-2), \qquad \text{if} \quad k \neq j \\
&\text{Else} \\
&\quad - w^+(i,j) \leftarrow w^+(i,j) + R_l/(n-2), \qquad \text{if} \quad k \neq j \\
&\quad - w^-(i,k) \leftarrow w^-(i,k) + R_l
\end{aligned}
$$

Then we re-normalize all the weights by carrying out the following operations, to avoid obtaining weights which indefinitely increase in size. First for each i we compute:

$$
\bar{r}_i = \sum_{m=1}^{n} [w^+(i,m) + w^-(i,m)] \tag{16.5}
$$

and then renormalize the weights with

$$
\begin{aligned}
w^+(i,j) &\leftarrow w^+(i,j) r_i / \bar{r}_i \\
w^-(i,j) &\leftarrow w^-(i,j) r_i / \bar{r}_i
\end{aligned}
$$

The probabilities qi are computed using the nonlinear iterations in Equations (16.2) and (16.3), leading to a new decision based on the neuron with the highest probability of being excited.

16.5.2.1 Performance examples

Simulation results are generated in a scenario as in Gelenbe *et al.* [44]. The CPN included $10 \times 10 = 100$ nodes, interconnected within rectangular grid topology. All link speeds were normalized to 1, and packets were allowed to enter and leave the network either from the top 10 nodes or the bottom 10 nodes in the grid. All the packets were assigned a common goal, which was to minimize a weighted combination of delay (W) and loss (L) as defined

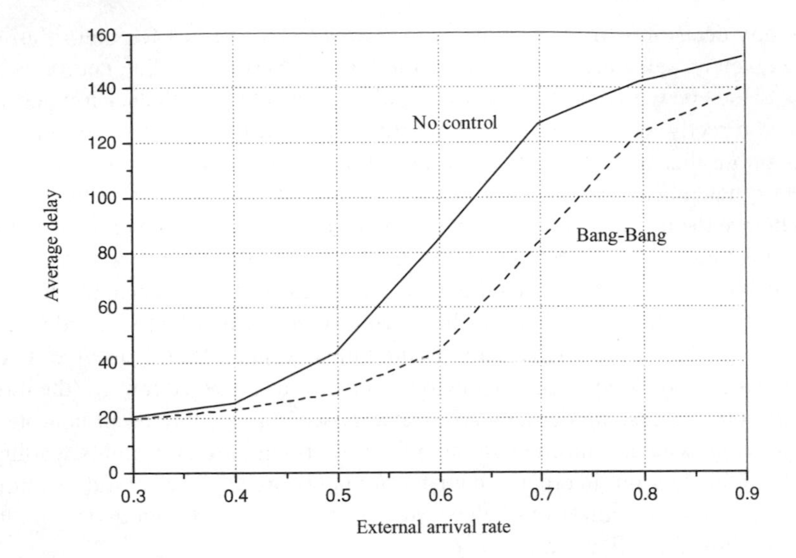

Figure 16.15 Comparison of average delay through the CPN with Bang-Bang control based on estimated delay.

by Equation (16.1). All the algorithms are allowed to use four items of information which are deposited in the nodes' mailboxes:

(1) the length of the local queues in the node;

(2) recent values of the downstream delays experienced by packets which have previously gone through the output links and reached their destinations;

(3) the loss rate of packets which have passed through the same node and gone through the output links;

(4) estimates made by the most recent CPs which have used the output links headed for some destination d of its estimated delay D_d and loss L_d from this node to its destination.

The value D_d is updated by each successive CP passing through the node and whose destination is d. Samples of the simulation results are given in Figures 16.15–16.18.

16.6 GAME THEORY MODELS IN COGNITIVE RADIO NETWORKS

Game theory is a set of mathematical tools used to analyze interactive decision makers. We will show how these tools can be used in the analysis of cognitive networks [71–77]. The fundamental component of game theory is the concept of a game, formally expressed in *normal form* as [69, 70]:

$$G = \langle N, A, \{u_i\} \rangle$$

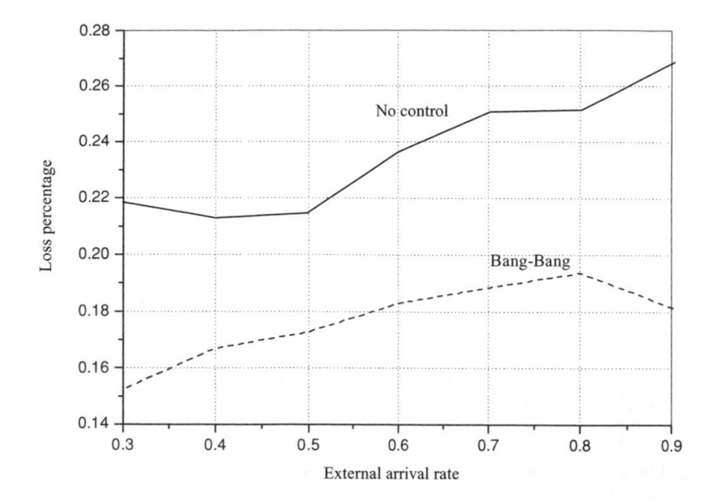

Figure 16.16 Comparison of average loss through the CPN with Bang-Bang control based on delay and loss.

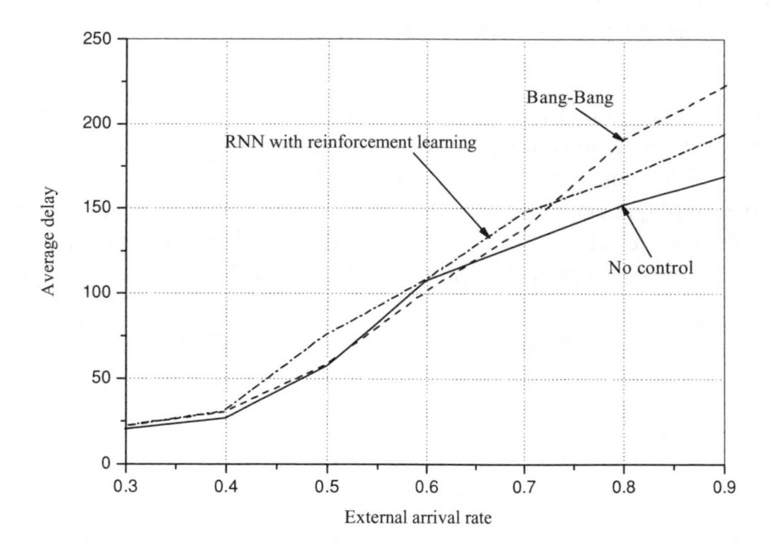

Figure 16.17 Comparison of average delay through the CPN with reinforcement learning-based control using delay and loss as the goal.

Where

(1) G is a particular game;

(2) $N = \{1, 2, \ldots, n\}$ is a finite set of players (decision makers);

(3) A_i is the set of actions available to player i;

(4) $A = A_1 \times A_2 \times \cdots \times A_n$ is the action space; and

(5) $\{u_i\} = \{u_1, u_2, \cdots, u_n\}$ is the set of utility (objective) functions that the players wish to maximize.

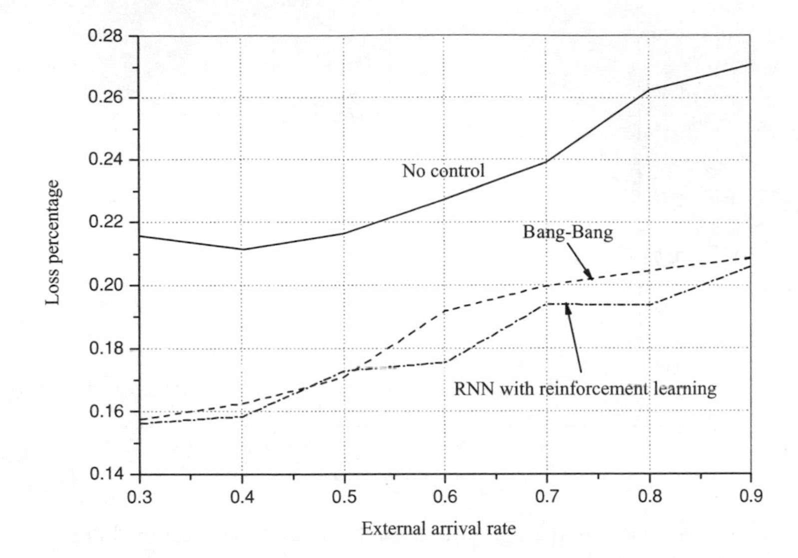

Figure 16.18 Comparison of average loss through the CPN with reinforcement learning-based control using delay and loss as the goal.

Each player's objective function, u_i, is a function of the particular action chosen by player i, a_i and the particular actions chosen by all of the other players in the game, a_{-i}, and yields a real number. Other games may include additional components, such as the information available to each player and communication mechanisms. In a *repeated game*, players are allowed to observe the actions of the other players, remember past actions, and attempt to predict future actions of players. An action vector a is said to be a *Nash equilibrium* (NE) if

$$u_i(a) \geq u_i(b_i, a_{-i}) \forall i \in N, b_i \in A_i \tag{16.6}$$

Restated, an NE is an action vector from which no player can profitably unilaterally deviate. NE correspond to the steady-states of the game and are then predicted as the most probable outcomes of the game.

A *repeated game* is a sequence of stages where each stage is the same normal form game. When the game has an infinite number of stages, the game is said to be an infinite horizon game. Players choose strategies (actions at each stage), based on their knowledge of the game-past actions, future expectations and current observations. These strategies can be fixed, contingent on the actions of other players or adaptive. These strategies can be also designed to punish players who deviate from agreed upon behavior. When punishment occurs, players choose their actions to minimize the payoff of the offending player. However, i is still able to achieve some payoff v_i. Thus there is a limit to the how much a player can be punished.

As estimations of future values of u_i are uncertain, many repeated games modify the original objective functions by discounting the expected payoffs in future stages by δ, where $0 < \delta < 1$ such that the anticipated value in stage k to player i is given by

$$u_{i,k}(a) = \delta^k u_i(a) \tag{16.7}$$

It can be shown that, in a repeated game with an infinite horizon and discounting, for every feasible payoff vector $v > v_i$ for all $i \in N$, there exists a $\delta < 1$ such that for all $\delta \in (\delta, 1)$ there is an NE with payoffs v (*Folk theorem* [69]). To generalize the Folk theorem, given a discounted infinite horizon repeated game, nearly any behavior can be designed to be the steady-state through the proper choice of punishment strategies and δ. Thus convergent behavior of a repeated game can be achieved nearly independent of the objective function.

A *myopic game* is defined as a repeated game in which there is no communication between the players, memory of past events or speculation of future events. Any adaptation by a player can still be based on knowledge of the current state of the game. As players have no consideration of future payoffs, the Folk theorem does not hold for myopic games and the convergence to steady-state behavior must occur through other means. Two convergence dynamics possible in a myopic game are the *best response dynamic* and the *better response dynamic*. Both dynamics require additional structure in the stage game to ensure convergence. Best response dynamic [70] refers to the case where, at each stage, one player $i \in N$ is permitted to deviate from a_i to some randomly selected action $b_i \in A_i$ if

$$u_i(b_i, a_{-i}) \geq u_i(c_i, a_{-i}) \forall c_i \neq b_i \in A_i \text{ and } u_i(b_i, a_{-i}) > u_i(a)$$

Better response dynamic [70] refers to the case where, at each stage, one player $i \in N$ is permitted to deviate from a_i to some randomly selected action $b_i \in A_i$ if $u_i(b_i, a_{-i}) > u_i(a_i, a_{-i})$.

An *S-modular game* restricts $\{u_j\}$ such that, for all $i \in N$, either Equation (16.8) or (16.9) is satisfied.

$$\frac{\partial^2 u_i(a)}{\partial a_i \partial a_j} \geq 0 \forall j \neq i \in N \tag{16.8}$$

$$\frac{\partial^2 u_i(a)}{\partial a_i \partial a_j} \leq 0 \forall j \neq i \in N \tag{16.9}$$

In the former case, the game is said to be *supermodular*; in the latter the game is said to be *submodular*. Myopic games whose stages are S-modular games with a unique NE and follow a best response dynamic converge to the NE when the NE is unique [71].

A *potential game* is a special type of game where $\{u_j\}$ are such that the change in value seen by a unilaterally deviating player is reflected in a function $P\colon A \to \Re$. All myopic games where the stages are the same potential game converge to a NE when decisions are updated according to a better response dynamic [70].

An *exact potential game* (EPG) is the game where there exists some function (EPF) $P\colon A \to \Re$ such that $\forall i \in N, \forall a \in A$

$$u_i(a_i, a_{-i}) - u_i(b_i, a_{-i}) = P(a_i, a_{-i}) - P(b_i, a_{-i}) \tag{16.10}$$

A necessary and sufficient condition for a game to be an exact potential game is [72]

$$\frac{\partial^2 u_i(a)}{\partial a_i \partial a_j} = \frac{\partial^2 u_j(a)}{\partial a_j \partial a_i} \forall i, j \in N, a \in A \tag{16.11}$$

Coordination-dummy game [72] is a composite of a coordination game with identical interest function V and a dummy game with dummy function D_i whose value is solely dependent on the actions of the other players and can be expressed as

$$u_i(a) = V(a) + D_i(a_{-i}) \tag{16.12}$$

An EPF for this game can be written as

$$P(a) = V(a) \tag{16.13}$$

Self-motivated games' utility functions are a function solely of that player's actions.

$$u_i(a) = h_i(a) \tag{16.14}$$

A self-motivated game can be shown to be an EPG by introducing the EPF as

$$P(a) = \sum_{i \in N} h_i(a_i) \tag{16.15}$$

Bilateral symmetric interaction (BSI) game [73] refers to the case where every player's objective function can be characterized by

$$u_i(a) = \sum_{j \in N \backslash \{i\}} w_{ij}(a_i, a_j) - h_i(a_i) \tag{16.16}$$

where $w_{ij}: A_i \times A_j \to \Re$ and $h_i: A_i \to \Re$ such that for every $(a_i, a_j) \in A_i \times A_j$, $w_{ij}(a_i, a_j) = w_{ji}(a_j, a_i)$. An EPF for a BSI game is given by

$$P(a) = \sum_{i \in N} \sum_{j=1}^{i-1} w_{ij}(a_i, a_j) - \sum_{i \in N} h_i(a_i) \tag{16.17}$$

Ordinal potential games (OPG) refer to the case where there exists some function (OPF) $P: A \to \Re$ such that, in addition to Equation (16.10), we also have

$$u_i(a_i, a_{-i}) > u_i(b_i, a_{-i}) \Leftrightarrow P(a_i, a_{-i}) > P(b_i, a_{-i}), \forall i \in N, \forall a \in A \tag{16.18}$$

All EPG are also OPG.

16.6.1 Cognitive radio networks as a game

The cognitive radios in the network form the game's set of decision makers. The set of physical layer parameters which a radio is permitted to alter forms the player's action set. From these action sets, the action space is formed. Preference relations over the action space are formed by an exhaustive evaluation of the adaptation algorithms. Objective functions are then formed by mapping the preference relations to the real number line so that preferable action vectors are larger than less preferable action vectors.

16.6.1.1 Distributed power control

As an example we examine distributed power control algorithms within the context of CRNs [71, 74, 77]. Below, the following notation is used:

- *N*, the set of decision-making (cognitive) radios in the network;
- *i*, *j*, two different cognitive radios, $i, j \in N$;
- P_j, the set of power levels available to radio j; this is presumed to be a segment of the real number line $\Re$;

- p_j, a power level chosen by j from P_j;
- $\mathbf{P}$, the power space ($\Re^n$) formed from the Cartesian product of all P_j; $\mathbf{P} = P_1 \times P_2 \times \cdots P_n$;
- $\mathbf{p}$, a power profile (vector) from $\mathbf{P}$ formed as $\mathbf{p} = \{p_1, p_2, \cdots p_n\}$;
- $u_j(p)$, the utility that j receives from $\mathbf{p}$; this is the specific objective function that j is looking to maximize.

Based on these conventions, a power control game G can be formulated as $G = \langle N, P, \{u_j\} \rangle$.

16.6.1.2 Repeated power games

MacKenzie and Wicker [74] consider a discounted repeated power control game implemented on a packet-based network wherein the original objective function for each radio j is the modified function of throughput given as

$$u_j(\mathbf{p}) = R\left[1 - 2BER(\mathbf{p})\right]^L / p_j \tag{16.19}$$

where R is the transmission rate, L is the packet length and BER is the bit error rate which is a function of the SINR seen by j. As the modeled game has an infinite horizon, a CRN implemented in this manner will exhibit convergent behavior if:

(1) some mechanism exists for broadcasting the desired operating vector, the discount factor, and the punishment strategy;

(2) there is full knowledge of the environment so the radios can differentiate deviant behavior from fades and jammers external to the CRN;

(3) there is knowledge of the action chosen by each radio at each stage.

16.6.1.3 S-modular games

Altman and Altman [71] examines the application of super-modular games to distributed power control. Thus the objective functions for this game are characterized by Equations (16.8) and (16.9). Herein each game follows a general updating algorithm (GUA), which is actually a best response dynamic. Thus if these network games have a unique NE, behavior converges to the NE from any initial p. For a CRN which satisfies this characterization, the conditions for convergence are:

(1) The adaptation algorithms must incorporate perfect knowledge of the objective function.

(2) The network must have a unique steady state.

(3) Some method must exist for measuring current performance and for sensing relevant environmental parameters. Depending on the particular adaptation algorithm, it might be necessary to know the number and type of other radios in the network.

Altman and Altman [71] assert that the classes of power control algorithms considered in by Yates [75] are S-modular games. Yates examines power control in the context of the uplink of a cellular system under five scenarios:

(1) fixed assignment where each mobile is assigned to a particular base station;

(2) minimum power assignment where each mobile is assigned to the base station where its SINR is maximized;

(3) macro diversity where all base stations combine the signals of the mobiles;

(4) limited diversity where a subset of the base stations combine the signals of the mobiles; and

(5) multiple connection reception where the target SINR must be maintained at a number of base stations.

In each scenario, each mobile, j, tries to achieve a target SINR γ_j as measured by a function, $I(\mathbf{p})$. $I(\mathbf{p})$ is the standard effective interference function which has the following properties: (1) *positivity*, e.g. $I(\mathbf{p}) > 0$; (2) *monotonicity*, e.g. if $\mathbf{p} \geq \mathbf{p}^*$, then $I(\mathbf{p}) \geq I(p^*)$; (3) *scalability*, e.g. for all $\alpha > 1$, $\alpha I(\mathbf{p}) > I(\alpha \mathbf{p})$, where the convention that $\mathbf{p} > \mathbf{p}^*$ means that $\mathbf{p}_i > \mathbf{p}_i{}^* \forall i \in N$.

Single-cell target SINR games are also ordinal potential games. To prove it, consider a modified game where the action sets are the received power levels at the base station. A received power level for mobile j, r_j, is the product of its path loss to base station k, $h_{j,k}$, and its transmitted power level. Then the objective functions of a target SINR game can be expressed as

$$u_j(p) = 1 - \left\{ -\gamma_j + \left[r_j - \left(\sum_{i \in N \setminus j} r_i \right) - \sigma_k \right] \right\}^2 \tag{16.20}$$

where σ_k is the noise power at k. Expanding the above expression gives

$$\begin{aligned} u_j(\mathbf{p}) = {} & 1 - \gamma_j^2 - \sigma_k^2 - 2\gamma_j\sigma_k - \left(\sum_{i \in N \setminus j} r_i \right)^2 - 2\gamma_j \left(\sum_{i \in N \setminus j} r_i \right) - 2\sigma_k \left(\sum_{i \in N \setminus j} r_i \right) \\ & - r_j^2 + 2\gamma_j r_j + 2\sigma_k r_j + 2r_j \left(\sum_{i \in N \setminus j} r_i \right) \end{aligned} \tag{16.21}$$

The first line of Equation (16.21) is a dummy game; the following three terms in Equation (16.21) are a self-motivated game, EPG. The final term is also an EPG as it is a BSI game. Since a composite game formed from a linear combination of two EPGs is itself an EPG, it is seen that the target SINR game is an EPG. As other forms of target SINR games are ordinal transformations (OT) of Equation (16.21), all target SINR games are OPG, since OT of an OPG is an OPG. It can be also shown that all target throughput power control games are OPG (all throughput maximization power control games are OPG).

16.6.1.4 Nonlinear group pricing

Goodman and Mandayam [76] consider a scenario wherein mobile devices are trying to maximize a function of their throughput. Note that the only NE of games whose utility functions are pure throughput function is the power vector where all mobiles transmit at maximum power. To avoid this problem, Goodman and Mandayam [76] introduce the modified throughput function as

$$u_j(\mathbf{p}) = \frac{BT(\mathbf{p})}{p_j} - cRp_j \tag{16.22}$$

where R is the transmission rate, c is a constant, B is the battery life, and T is the throughput function. This can neither be shown to be an OPG nor an OPG. The left-most term is generally only an OPG, whereas the pricing function is an EPG as it is a self-motivated game.

Goodman and Mandayam [76] use a best response dynamic in their experiments, which converge to a NE. Although not stated in Goodman and Mandayam [76], Saraydar *et al.* [78] show that Equation (16.22) is indeed a supermodular game. Also note that, without the cost function, Equation (16.19) is a particular instance of Equation (16.22). So a CRN implementing the repeated games of References [74], [76] or [78] is a supermodular game and will exhibit convergent behavior if the properties specified above for S-modular games are satisfied. However, it should be noted that, when left as a repeated game, there is greater flexibility in dynamically selecting the steady-state.

16.6.1.5 Nonlinear group pricing

Sung and Wong [77] consider another single cell distributed power control game with pricing characterized by

$$u_j(\mathbf{p}) = R_j T(\mathbf{p}) - \frac{\lambda h_j p_j}{\sum_{i \in N} h_i p_i} \tag{16.23}$$

where λ is a constant and the base station subscript has been dropped in this case as all mobiles are communicating with the same base station. The pricing function is to reflect that the damage caused by p_j to the other devices in the network is a function of the relative difference in powers rather than absolute power level. Note that Equation (16.23) is just a composite game of throughput maximization and a cost function. As we have indicated that throughput maximization is an OPG, Equation (16.23) can only be guaranteed to be an OPG only if, though not necessarily if, the pricing function has an EPF which can only be true if Equation (16.11) is satisfied. Differentiating the price function twice yields:

$$\frac{\partial^2 C_i}{\partial p_i \partial p_j} = \frac{\left(\sum_{k \in N} h_k p_k\right)^2 \lambda h_i h_j - 2h_j \left(\sum_{k \in N} h_k p_k\right) g_i(\mathbf{p})}{\left(\sum_{k \in N} h_k p_k\right)^4}$$

$$\frac{\partial^2 C_j}{\partial p_j \partial p_i} = \frac{\left(\sum_{k \in N} h_k p_k\right)^2 \lambda h_i h_j - 2h_i \left(\sum_{k \in N} h_k p_k\right) g_j(\mathbf{p})}{\left(\sum_{k \in N} h_k p_k\right)^4}$$

where $g_i(\mathbf{p}) = \lambda h_i \sum_{k \in N} h_k p_k - \lambda h_i^2 p_i$. Further evaluation leads to the result that this price function has an EPF if $h_j p_j = h_i p_i$. Note that this is only satisfied when the received powers are identical, which will generally not be true. Thus the cost function does not have an EPF and the nonlinear group priced game is not an OPG. Also note that Equation (16.23) cannot be guaranteed to be a supermodular game either as properly chosen power vectors p and p^* will yield different signs for the second partial derivative of the cost functions.

Therefore neither the better response dynamic nor the best response dynamic will assuredly converge, and a more complex convergence dynamic is required. As the repeated game dynamic is guaranteed for convergence, it can still be used. Thus this modification of the pricing function significantly complicated the network.

Additional examples of game theory applications in resource allocation modeling can be found in References [116–127] and routing in [79–115].

16.7 BIOLOGICALLY INSPIRED NETWORKS

Current Internet protocols were never planned for the emerging pervasive environments where the amount of information will be enormous. The communications requirements placed by these protocols on the low cost sensor and tag nodes are in direct contradiction to the fundamental goals of these nodes, being small, inexpensive and maintenance-free. Biological systems provide insights into principles which can be adopted to completely redefine the basic concepts of control, structure, interaction and function of the emerging pervasive environments. The study of the rules of genetics and evolution combined with mobility leads to the definition of service-oriented communication systems which are autonomous and autonomously self-adaptive. Based on References [128–131] in this section we discuss how this paradigm shift, which views a network only as a randomly self-organizing by-product of a collection of self-optimizing services, may become the enabler of the new world of omnipresent low cost pervasive environments of the future.

16.7.1 Bio-analogies

In Carreras *et al.* [131] the depicted scenario services are associated with living organisms. Service is defined by chromosomes. In this way service evolves and adapts to the environment constantly and autonomously. By analogy with living organisms, chromosomes are collections of *genes* that are the smallest service (related) data unit and inborn intelligence/instincts and thus represent all the information needed for the organism to function and service to be executed. Like in nature, this concept defines a complete life-cycle of the organisms and therefore of services. The life cycle starts from the birth of an organism, goes through the reproduction and ends with the death. *Reproduction and evolution* occur, applying evolution rules inherited from nature. Fitness is measuring the correspondence of the organism's genetic information with the environment and determines the exchange of information (genetic information). Therefore no-end-to-end communication concept exists in these systems. Information is only exchanged as needed, locally, between mating organisms.

Environment is determining the *natural selection* based on the *fitness* of the organisms, with the environment leading to the best services possible as a function of the environment. In this concept *the service is the organism.* A scenario is envisioned where users will be more and more interested in a service able to provide reliable localized information. The role of the service will be, for instance, to provide answers to questions like *How is the weather around the train station?* or *Where will I find a free parking space around there?* Services will be hosted on users' devices and will go around through the physical movement of the users. Each service is constituted by a program and its related data that is organized into *chromosomes*. Each chromosome consists of:

(1) data that is the genetic information of the organism, organized in *genes*;

(2) a *plugin* that stores a syntax notation describing the actions dictated by the chromosome and the fitness (degree of attraction) operator yielding *natural selection* through preferred *mating*;

Genes are a tuple of information and consist of: (a) *value*; (b) *timing information*; (c) *information source ID/location information*; and (d) *other data depending on the service*.

Organisms are diploid, which means that there is always two homologous chromosomes associated with each service. The two homologous chromosomes have the same genes but in different forms. Alleles are different forms of the same gene and correspond to different values in the tuple of gene information. They may differ in timing information or in the source value information. Each allele may be dominant or recessive depending on the service behavior. Having two chromosomes allows us to estimate the reliability of the data. We would probably always choose the youngest data value to be the actual one if it is a parking lot, but we might also average the sensor data if it represents temperature.

The choice of the preferred value among the two reflects the concept of dominant and recessive genes. As in nature, recessive information enables the service to survive in different environments, providing the service with higher resilience against data corruption and fraud, and may even allow for additional features. Analogy with the *life cycle* is also possible. In this concept service is born when the user downloads the chromosome onto his device. From that moment on, the user is able to interact with the other organisms (i.e. users carrying chromosomes) and with the environment where the users are physically moving. When gathering information from the environment the service *grows*. While growing, the service improves its functionalities, meaning that it becomes able to increase performance. When a user meets another user while moving, services may reproduce and produce offspring. It is in this phase that evolution and natural selection occur. In order to be able to reproduce the service must satisfy some fitness requirements, since it is not willing to spread useless information. We assume a service to be dead when it cannot reproduce anymore.

Analogies with *birth*, *growth*, *reproduction* and *death* are even further elaborated in this concept. As we have already said, the service is born when the user obtains (downloads) an empty chromosome of a certain service that consists only of the plugin. From that instant on the user can read data from sensors and use the syntax definition from the plugin. At this stage the user is haploid, i.e. it has only one chromosome per service. When reading sensor data, the user fills the chromosomes both at the same time. This information is in any case more reliable than the previous information on the chromosomes.

The concept of mating is performed the following way: in *meiosis*, the diploid cell splits into two haploid reproductive cells (the eggs or sperm). In real life, four haploid cells would appear because of one step of copying the chromosomes, including cross-over. Each two copies would be identical. This means that the chromosome pair is being split and copied. Two packets are sent out one after another containing one chromosome each. In the best case when all the sent packets reach the receiver, it has four different combinations of the chromosome pairs. This is the best case; the user may for energy reasons decide to send out only one chromosome, or receive only one because of packet loss. A selection has to take place to decide which of these combinations survives. This selection could be influenced by the quality of the plugin (by the version number). Having such a selection can help to spread new versions of plugins. It also may help to repair broken chromosomes (that were damaged during the wireless transmission). In a sense, we allow for truly spontaneous mutations and we may even find that mutations spread. It remains to be seen if this is any good. The selection occurs as a consequence of localization and age. The fitness of a chromosome is defined as the average of the timing information of the genes weighted with the localization information. In this sense the environment participates in the selection, since the survival of a service also depends on where the user is (but not only) when he mates with another user.

If the sensor data in the chromosome is too old, it is very likely useless. Thus we forbid sending out the chromosome after a certain threshold (coded in the plugin). This way the chromosome can die, but it may be reborn through reading new sensor data or receiving a fresh chromosome. The service is considered *alive* as long as it is able to answer to questions. Death is therefore a consequence of outdated chromosomal information (sensor-gathered information is aging). It is in the interest of the user to exchange information and to gather sensor information and this same interest drives the service *instinct to survive*. This defines a complete life cycle of the service.

As an example of the environmental monitoring applications, Carreras *et al.* [132] describe a parking lot finding application as a possible application scenario of the above concept for the nomadic sensor network:

Each parking spot in a city is equipped with a sensor that is capable of notifying whether the parking lot is free or not together with its geographical position (or ID). Each user subscribing to the service is equipped with a hand-held device that communicates both with the sensor nodes and with other users who subscribed to the service. The users, while moving around the city, collect information on the status (free/not free) of the parking lots they encounter on their way. This information is stored in their devices, together with a time stamp of when the information has been collected. Whenever two users come into communication range, they exchange their knowledge of the parking lot status (and update their own if the information is 'fresh' enough). The basic service of the system to the user would be the assistance in finding the nearest parking lot. The user queries his device (which might be some PDA), asking for the nearest free parking lot in a specific area of the city. The device will provide this information based on the data that has stored so far. Of course, the information cannot be considered real-time.

16.7.2 Bionet architecture

The bio-networking architecture [130–134] applies key concepts and mechanisms described above to design network applications. One of the key concepts in biological systems is emergence. In biological systems, beneficial system properties (e.g. adaptability) often emerge through the simple and autonomous interactions among diverse biological entities (see the above example). The bio-networking architecture applies the concept of emergence by implementing network applications as a group of distributed, autonomous and diverse objects called cyber-entities (CEs). In Itao *et al.* [134], analogy to a bee colony (a network application) consisting of multiple bees (CEs) is used. Each CE implements a functional service related to the application and follows simple behaviors similar to biological entities, such as *reproduction, death, migration and environment sensing*, as discussed in the previous section.

In the bio-networking architecture, CEs are designed based on the three principles described below in order to interact and collectively provide network applications that are autonomous, scalable, adaptive, and simple.

(1) *CEs are decentralized* – there are no central entities to control and coordinate CEs (i.e. no directory servers and no resource managers). Decentralization allows network applications to be scalable and simple by avoiding a single point of performance bottleneck and failure and by avoiding any central coordination in developing and deploying CEs.

(2) *CEs are autonomous* – CEs monitor their local network environments, and based on the monitored environmental conditions, they autonomously interact without any intervention from human users or from other controlling entities.

(3) *CEs are adaptive* to dynamically changing environnemental conditions (e.g. user demandes, user locations and resource availability) over the short- and long-term.

The bionet platform described in Suzuki *et al.* [130], and presented in Figure 16.19, provides an execution environment for CEs. It consists of two types of software components. The first type of components, *supporting components*, abstracts low-level operating and networking details (e.g. network I/O and concurrency control for executing CEs). The second type of components, *runtime components*, provides runtime services that CEs use to perform their services and behaviours. The bionet platform is implemented in Java, and each bionet platform runs on a Java virtual machine (JVM). Each CE is implemented as a Java object and runs on a bionet platform.

As shown in Figure 16.20, a CE consists of three main segment: *attributes, body* and *behaviours. Attributes* carry descriptive information regarding the CE (e.g. CE ID and description of a service it provides). The *body* implements a service that the CE provides

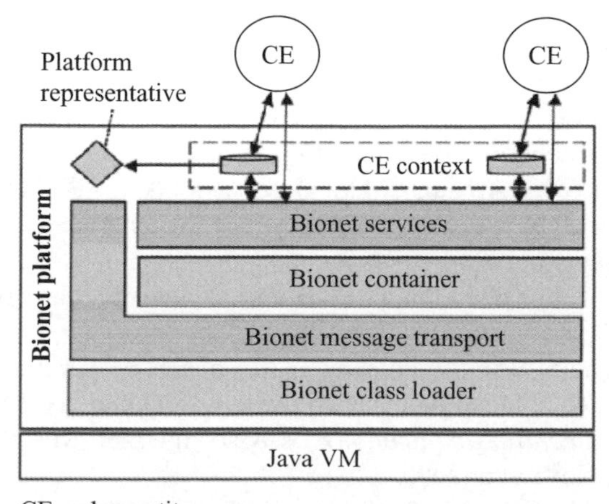

Figure 16.19 Bionet platform architecture. (Reproduced by permission of IEEE [130].)

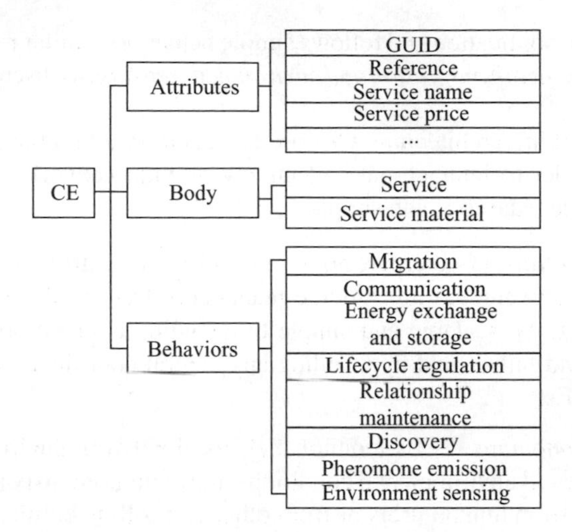

Figure 16.20 Design of a CE.

and contains materials relevant to the service (e.g. data, application code or user profiles). CE *behaviors* implement nonservice-related actions that are inherent to all CEs. Examples of behavior include migration, reproduction and energy exchange. More information on different segments of the architecture can be found in Suzuki *et al.* [130].

REFERENCES

[1] D.S. Alexander, M. Shaw, S. Nettles and J. Smith, Active bridging, in *Proc. ACM SIGCOMM Conf.*, 1997, pp. 101–111.

[2] B. Schwartz, A.W. Jackson, W.T. Strayer, W. Zhou, D. Rockwell and C. Partridge, Smart packets for active networks, *ACM Trans. Comput. Syst.*, vol. 18, no. 1, 2000, pp. 67–88.

[3] D. Tennenhouse and D. Wetherall, Toward an active network architecture, *Comput. Commun. Rev.*, vol. 26, no. 2, 1996.

[4] D.L. Tennenhouse, J.M. Smith, W.D. Sincoskie, D.J. Wetherall and G.J. Minden, A survey of active network research, *IEEE Commun.*, vol. 35, 1997, pp. 80–86.

[5] D.S. Alexander, W.A. Arbaugh, M.W. Hicks, P. Kakkar, A.D. Keromytis, J.T. Moore, C.A. Gunter, S.M. Nettles and J.M. Smith, The SwitchWare active network architecture, *IEEE Network*, 1998, pp. 29–36.

[6] *FAIN – Future Active IP Networks*. Available at: www.ist-fain.org/publications/ publications.html

[7] B. Carpenter and S. Brim, *Middleboxes: Taxonomy and Issues*. Internet Engineering Task Force, RFC 3234. Available at: www.iets.org, February 2002.

[8] T. Hain, *Architectural Implications of NAT*. Internet RFC 2993, Available at: www.iets.org, November 2000.

[9] P. Srisuresh and K. Egevang, *Traditional IP Network Address Translator* (*Traditional NAT*). Internet RFC 3022. Available at: www.iets.org, January 2001.

[10] M. Holdrege and P. Srisuresh, *Protocol Complications With the IP Network Address Translator*. Internet RFC 3027. Available at: www.iets.org, January 2001.

[11] G. Tsirtsis and P. Srisuresh, *Network Address Translation – Protocol Translation (NAT-PT)*. Internet RFC 2766. Available at: www.iets.org, February 2000.

[12] M. Leech, M. Ganis, Y. Lee, R. Kuris, D. Koblas and L. Jones, *SOCKS Protocol Version 5*. Internet RFC 1928. Available at: www.iets.org, March 1996.

[13] S. Blake, D. Black, M. Carlson, E. Davies, Z. Wang and W. Weiss, *An Architecture for Differentiated Service*. Internet RFC 2475. Available at: www.iets.org, December 1998.

[14] J. Border, M. Kojo, J. Griner, G. Montenegro and Z. Shelby, *Performance Enhancing Proxies Intended to Mitigate Link-Related Degradations*. Internet RFC 3135. Available at: www.iets.org, June 2001.

[15] N. Freed, *Behavior of and Requirements for Internet Fire-Walls*. Internet RFC 2979. Available at: www.iets.org, October 2000.

[16] B. Cheswick and S. Bellovin, *Firewalls and Internet Security: Repelling the Wily Hacker*. Addison-Wesley: Reading, MA, 1994.

[17] M. Handley, H. Schulzrinne, E. Schooler and J. Rosenberg, *SIP: Session Initiation Protocol*. Internet RFC 2543. Available at: www.iets.org, March 1999.

[18] F. Cuervo, N. Greene, A. Rayhan, C. Huitema, B. Rosen and J. Segers, *Megaco Protocol 1.0*. Internet RFC 3015. Available at: www.iets.org, November 2000.

[19] R. Fielding, J. Gettys, J. Mogul, H. Frystyk, L. Masinter, P. Leach and T. Berners-Lee. *Hypertext Transfer Protocol* – HTTP/1.1. Internet RFC 2616. Available at: www.iets.org, June 1999.

[20] IETF Forwarding Control Element Separation Working Group Home Page; www.ietf.org/html.charters/forces-charter.html

[21] C. Partridge, A.C. Snoeren, W.T. Strayer, B. Schwartz, M. Condell and I. Castineyra, FIRE: flexible intra-AS routing environment, *IEEE J. Select. Areas Commun.*, vol. 19, no. 3, 2001, pp. 410–425.

[22] J. Biswas, A.A. Lazer, J.-F. Huard, K. Lim, H. Mahjoub, L.-F. Pau, M. Suzuki, S. Torstensson, W. Wang and S. Weinstein, The IEEE P1520 standards initiative for programmable network interfaces, *IEEE Commun. Mag.*, vol. 36, no.10, 1998, pp. 64–70.

[23] IEEE P1520.2, *Draft 2.2, Standard for Application Programming Interfaces for ATM Networks*. Available at: www.ieee-pin.org/pin-atm/intro.html

[24] IETF ForCES; www.ietf.org/html.charters/ forces-charter.html

[25] IETF For CES, *Draft-IETF-Forces-Framework-04.txt*. Available at: www.ietf.org/internet-drafts/draft-ietf-forces-framework-04.txt, 2002.

[26] D.J. Wetherall, J.V. Guttag and D.L. Tennenhouse, *ANTS: a Toolkit For Building and Dynamically Deploying Network Protocols IEEE Openarch*. Available at: ftp://ftp.tns.lcs.mit.edu/pub/papers/openarch98.ps.gz, 1998.

[27] D. Decasper, G. Parulkar, S. Choi, J. DeHart, T. Wolf and B. Plattner, *A Scalable, High Performance Active Network Node IEEE Network*. Available at: www.tik.ee.ethz.ch/~dan/papers/ieee_ann_1.pdf

[28] B. Schwartz, A.W. Jackson, W.T. Strayer, W. Zhou, D. Rockwell and C. Partridge, *Smart Packets for Active Networks OPENARCH'99*. Available at: ftp://ftp.bbn.com/pub/AIR/smart.ps, 1999.

[29] D. Scott Alexander, W.A. Arbaugh, M.W. Hicks, P. Kakkar, A.D. Keromytis, J.T. Moore, C.A. Gunter, S.M. Nettles and J.M. Smith, The switchware active network architecture, *IEEE Network*, vol. 12, 1998, pp. 29–36.

[30] K.L. Calvert (ed.), *Architectural Framework for Active Networks Draft Version 1.0*. Available at: protocols.netlab.uky.edu/~calvert/arch-latest.ps, 1999.

[31] L. Peterson (ed.), *Node OS interface specification AN node OS working group*. Available at: www.cs.princeton.edu/nsg/papers/nodeos-02.ps, 2001.

[32] K. Calvert, S. Bhattacharjee, E. Zegura and J. Sterbenz, Directions in active networks, *IEEE Commun.*, vol. 36, 1998, pp. 72–78.

[33] A.T. Campbell, H. De Meer, M.E. Kounavis, K. Miki, J. Vicente and D. Villela, A survey of programmable networks . *ACM Comput. Commun. Rev*. April, 1999.

[34] *Initial Active Network and Active Node Architecture FAIN Project Deliverable* 2. Available at: www.ist-fain.org/deliverables/del2/d2.pdf

[35] *Revised Active Network and Active Node Architecture FAIN Project Deliverable* 4. Available at: www.ist-fain.org/deliverables/del4/d4.pdf

[36] S.-K. Song *et al.*, *Evolution in Action: Using Active Networking to Evolve Network Support for Mobility*, J. Sterbenz *et al.* (eds). IWAN 2002. LNCS 2546, Springer: Berlin, 2002, pp. 146–161.

[37] C. Prehofer and Q. Wei, *Active Networks for 4G Mobile Communication:Motivation, Architecture, and Application Scenarios*, J. Sterbenz *et al*. (eds). IWAN 2002, LNCS 2546. Springer: Berlin, 2002, pp. 132–145.

[38] D.J. Wetherall, J. Guttag and D.L. Tennenhouse. ANTS: a toolkit for building and dynamically deploying network protocols, in *IEEE OPENARCH*, April 1998, pp. 117–129.

[39] B. Schwartz, A.W. Jackson, W.T. Strayer, W. Zhou, R.D. Rockwell and C. Partridge. Smart packets: Applying active networks to network management, *ACM Trans. Comput. Systems*, vol. 18, no. 1, February 2000.

[40] M. Hicks, P. Kakkar, J.T. Moore, C.A. Gunter and S. Nettles. PLAN: a packet language for active networks, in *Proc. Third ACM SIGPLAN Int. Conf. Functional Programming*. ACM, 1998, pp. 86–93.

[41] WWRF, Wireless World Research Forum (WWRF); www.wireless-worldresearch.org/

[42] M. Kounavis and A. Campbell, *Design, Implementation and Evaluation of Programmable Handoff in Mobile Networks, Mobile Networks and Applications*, vol. 6. Kluwer Academic: Norwell, MA, 2001, pp. 443–461.

[43] H.J. Wang, R.H. Katz and J. Giese, Policy-enabled handoffs across heterogeneous wireless networks, in *WMCSA* 99, New Orleans, LA, 25–26 February 1999, pp. 51–60.

[44] E. Gelenbe, Z. Xu and E. Seref, Cognitive packet networks, *11th IEEE Int. Conf. Tools with Artificial Intelligence*, 9–11 November 1999, pp. 47–54.

[45] E. Gelenbe, R. Lent and A. Nunez, Self-aware networks and QoS, *Proc. IEEE*, vol. 92, no. 9, 2004, pp. 1478–1489.

[46] R.E. Ramos and K. Madani, A novel generic distributed intelligent re-configurable mobile network architecture, in *IEEE VTS 53rd Vehicular Technology Conf.* vol. 3, 6–9 May 2001, pp. 1927–1931.

[47] T. Kocak and J. Seeber, Smart packet processor design for the cognitive packet network router, in *The 2002 45th Midwest Symp. Circuits and Systems, MWSCAS-2002*, vol. 2, 4–7 August 2002, pp. II-513–II-516.

[48] X. Hu, A.N. Zincir-Heywood and M.I. Heywood, Testing a cognitive packet concept on a LAN, in *IEEE CCECE 2002. Canadian Conf. Electrical and Computer Engineering*, vol. 3, 12–15 May 2002, pp. 1577–1582.

[49] Y. Miao, Z.-Q. Liu, C.K. Siew and C.Y. Miao, Dynamical cognitive network – an extension of fuzzy cognitive map, *IEEE Trans. Fuzzy Syst.*, vol. 9, no. 5, 2001, pp. 760–770.

[50] J. Neel, R.M. Buehrer, B.H. Reed and R.P. Gilles, Game theoretic analysis of a network of cognitive radios, in *The 2002 45th Midwest Symp. Circuits and Systems. MWSCAS-2002*, vol. 3, 4–7 August 2002, pp. III-409–III-412.

[51] E. Gelenbe, M. Gellman, R. Lent, P. Liu and P. Su, Autonomous smart routing for network QoS, in *Int. Conf. Autonomic Computing*, 17–18 May 2004, pp. 232–239.

[52] W.-R. Zhang, S.-S. Chen, W. Wang and R.S. King, A cognitive-map-based approach to the coordination of distributed cooperative agents, *IEEE Trans. Syst. Man Cybernet.*, vol. 22, no. 1, 1992, pp. 103–114.

[53] J.O. Neel, J.H. Reed and R.P. Gilles, Convergence of cognitive radio networks, in *2004 IEEE Wireless Communications and Networking Conf., WCNC*, vol. 4, 21–25 March 2004, pp. 2250–2255.

[54] P. Mahonen, Cognitive trends in making: future of networks, in *15th IEEE Int. Symp. Personal, Indoor and Mobile Radio Communications, PIMRC 2004*, vol. 2, 5–8 September 2004, pp. 1449–1454.

[55] R. Viswanathan and K.S. Narendra, Comparison of expedient and optimal reinforcement schemes for learning systems, *J. Cybernet.* vol. 2, 1972, pp. 21–37.

[56] K.S. Narendra and P. Mars, The use of learning algorithms in telephone traffic routing – a methodology, *Automatica*, vol. 19, 1983, pp. 495–502.

[57] E. Gelenbe, Learning in the recurrent random neural network, *Neural Comput.*, vol. 5, no. 1, 1993, pp. 154–164.

[58] P. Mars, J.R. Chen and R. Nambiar, *Learning Algorithms: Theory and Applications in Signal Processing, Control and Communications*. CRC Press: Boca Raton, FL, 1996.

[59] D.L. Tennenhouse, J.M. Smith, D.W. Sincoskie, D.J. Wetherall and G.J. Minden, A survey of active network research, *IEEE Commun. Mag.*, vol. 35, no. 1, 1997, pp. 80–86.

[60] M. Bregust and T. Magedanz, Mobile agents-enabling technology for active intelligent network implementation, *IEEE Network Mag.*, vol. 12, no. 3, 1998, pp. 53–60.

[61] T. Faber, ACC: using active networking to enhance feedback congestion control mechanisms, *IEEE Network Mag.*, vol. 12, no. 3, 1998, pp. 61–65.

[62] S. Rooney, J.E. van der Merwe, S.A. Crosby and I.M. Leslie, The tempest: a framework for safe, resource-assured, programmable networks, *IEEE Commun. Mag.* vol. 36, no. 10, 1998, pp. 42–53.

[63] J.-F. Huard and A.A. Lazar, A programmable transport architecture with QoS guarantee, *IEEE Commun.*, vol. 36, no. 10, 1998, pp. 54–63.

[64] J. Biswas, A.A. Lazar, S. Mahjoub, L.-F. Pau, M. Suzuki, S. Torstensson, W. Wang and S. Weinstein, The IEEE P1520 standards initiative for programmable network interface, *IEEE Commun.*, vol. 36, no. 10, 1998, pp. 64–72.

[65] K.L. Calvert, S. Bhattacharjee, E. Zegura and J. Sterbenz, Directions in active networks, *IEEE Commun.*, vol. 36, no. 10, 1998, pp. 64–72,

[66] W. Marcus, I. Hadzic, A.J. McAuley and J.M. Smith Protocol boosters: applying programmability to network infrastructures, *IEEE Commun.*, vol. 36, no. 10, 1998, pp. 79–83.

[67] D.S. Alexander, W.A. Arbaugh, A.D. Keromytis and J.M. Smith, Safety and security of programmable networks infrastructures, *IEEE Commun.*, vol. 36, no. 10, 1998, pp. 84–92.

[68] E. Gelenbe, Zhi-Hong Mao and Y. Da-Li, Function approximation with spiked random networks, *IEEE Trans. Neural Networks*, vol. 10, no. 1, 1999, pp. 3–9.
[69] D. Fudenberg and J. Tirole, *Game Theory*. MIT Press: Cambridge, MA, 1991.
[70] J. Friedman and C. Mezzetti, Learning in games by random sampling, *J. Econ. Theory*, vol. 98, 2001, pp. 55–84.
[71] E. Altman and Z. Altman. *S*-modular games and power control in wireless networks, *IEEE Trans. Autom. Control*, vol. 48, 2003, pp. 839–842.
[72] D. Monderer and L. Shapley, Potential games, *Games Econ. Behav.*, vol. 14, 1996, pp. 124–143.
[73] T. Ui, A shapley value representation of potential games, *Games Econ. Behav.*, vol. 14, 2000, pp. 121–135.
[74] A. MacKenzie and S. Wicker, Game theory in communications: motivation, explanation, and application to power control, in *Globecom 2001*, pp. 821–825.
[75] R. Yates, A framework for uplink power control in cellular radio systems, *IEEE J. Select. Areas Commun.*, vol. 13, no. 7, 1995, pp. 1341–1347.
[76] D. Goodman and N. Mandayam. Power control for wireless data, *IEEE Person. Commun.*, April 2000, pp. 48–54.
[77] C. Sung and W. Wong, A noncooperative power control game for multirate CDMA data networks, *IEEE Trans. Wireless Commun.*, vol. 2, no. 1, 2003, pp. 186–219.
[78] C. Saraydar, N. Mandayam and D. Goodman, Pareto efficiency of pricing-based power control in wireless data networks, *Wireless Commun. Networking Conf.*, 21–24 September 1999, pp. 231–235.
[79] J.P. Hespanha and S. Bohacek, Preliminary results in routing games, in *Proc. 2001 American Control Conf.*, vol. 3, 25–27 June 2001, pp. 1904–1909.
[80] R. Kannan and S.S. Iyengar, Game-theoretic models for reliable path-length and energy-constrained routing with data aggregation in wireless sensor networks, *IEEE J. Select. Areas Commun.*, vol. 22, no. 6, 2004, pp. 1141–1150.
[81] V. Anantharam, On the Nash dynamics of congestion games with player-specific utility, in *43rd IEEE Conf. Decision and Control, CDC*, vol. 5, 14–17 December 2004, pp. 4673–4678.
[82] A.A. Economides and J.A. Silvester, A game theory approach to cooperative and non-cooperative routing problems, in *SBT/IEEE Int. Telecommunications Symp. ITS '90 Symposium Record*, 3–6 September 1990, pp. 597–601.
[83] A.A. Economides and J.A. Silvester, Multi-objective routing in integrated services networks: a game theory approach, in *INFOCOM '91. Tenth Annual Joint Conf. IEEE Computer and Communications Societies. Networking in the 90s*, 7–11 April 1991, vol. 3. IEEE: New York, 1991, pp. 1220–1227.
[84] J. Cai and U. Pooch, Play alone or together – truthful and efficient routing in wireless *ad hoc* networks with selfish nodes, in *2004 IEEE Int. Conf. Mobile Ad-hoc and Sensor Systems*, 25–27 October 2004, pp. 457–465.
[85] I. Sahin and M.A. Simaan, A game theoretic flow and routing control policy for two-node parallel link communication networks with multiple users, in *15th IEEE Int. Symp. Personal, Indoor and Mobile Radio Communications, PIMRC 2004*, vol. 4, 5–8 September 2004, pp. 2478–2482.
[86] E. Altman, T. Basar and R. Srikant, Nash equilibria for combined flow control and routing in networks: asymptotic behavior for a large number of users, *IEEE Trans. Autom. Control*, vol. 47, no. 6, 2002, pp. 917–930.

[87] T. Boulogne, E. Altman, H. Kameda and O. Pourtallier, Mixed equilibrium (ME) for multiclass routing games, *IEEE Trans. Autom. Control*, vol. 47, no. 6, 2002, pp. 903–916.

[88] A. Orda, R. Rom and N. Shimkin, Competitive routing in multi-user communication networks, in *INFOCOM '93. IEEE Twelfth Annual Joint Conf. IEEE Computer and Communications Societies. Networking: Foundation for the Future*, 28 March to 1 April 1993, vol. 3, pp. 964–971.

[89] C.-H. Yeh and E.A. Varvarigos, A mathematical game and its applications to the design of interconnection networks, in *Int. Conf. Parallel Processing*, 3–7 September 2001, pp. 21–30.

[90] E. Altman and H. Kameda, Equilibria for multiclass routing in multi-agent networks, in *IEEE Conf. Decision and Control*, vol. 1, 4–7 December 2001, pp. 604–609.

[91] V. Marbukh, QoS routing under adversarial binary uncertainty, in *IEEE Int. Conf. Communications, ICC 2002*, vol. 4, 28 April to 2 May 2002, pp. 2268–2272.

[92] R.J. La and V. Anantharam, Optimal routing control: repeated game approach, *IEEE Trans. Autom. Control*, vol. 47, no. 3, 2002, pp. 437–450.

[93] R.J. La and V. Anantharam, Optimal routing control: game theoretic approach, in *IEEE Conf. Decision and Control*, vol. 3, 10–12 December 1997, pp. 2910–2915.

[94] Y.A. Korilis, A.A. Lazar and A. Orda, Capacity allocation under noncooperative routing, *IEEE Trans. Autom. Control*, vol. 42, no. 3, 1997, pp. 309–325.

[95] W. Wang, X.-Y. Li and O. Frieder, *k*-Anycast game in selfish networks, in *Int. Conf. Computer Communications and Networks*, 11–13 October 2004, pp. 289–294.

[96] V. Marbukh, Minimum cost routing: robustness through randomization, in *IEEE Int. Symp. Information Theory*, 2002, p. 127.

[97] R.E. Azouzi, E. Altman and O. Pourtallier, Properties of equilibria in competitive routing with several user types, in *IEEE Conf. Decision and Control*, vol. 4, 10–13 December 2002, pp. 3646–3651.

[98] O. Kabranov, A. Yassine and D. Makrakis, Game theoretic pricing and optimal routing in optical networks, in *Int. Conf. Communication Technology Proc.*, vol. 1, 9–11 April 2003, pp. 604–607.

[99] M. Kodialam and T.V. Lakshman, Detecting network intrusions via sampling: a game theoretic approach, in *IEEE Joint Conf. IEEE Computer and Communications Societies, INFOCOM 2003*. vol. 3, 30 March to 3 April 2003, pp. 1880–1889.

[100] J. Zander, Jamming in slotted ALOHA multihop packet radio networks, *IEEE Trans. Commun.*, vol. 39, no. 10, 1991, pp. 1525–1531.

[101] O. Ercetin and L. Tassiulas, Market-based resource allocation for content delivery in the Internet, *IEEE Trans. Comput.*, vol. 52, no. 12, 2003, pp. 1573–1585.

[102] Y.A. Korilis, A.A. Lazar and A. Orda, Achieving network optima using Stackelberg routing strategies, *IEEE/ACM Trans. Networking*, vol. 5, no. 1, 1997, pp. 161–173.

[103] J. Zander, Jamming games in slotted Aloha packet radio networks, *IEEE Military Communications Conf., MILCOM '90, 'A New Era'*, vol. 2, 30 September to 3 October 1990, pp. 830–834.

[104] K. Yamamoto and S. Yoshida, Analysis of distributed route selection scheme in wireless *ad hoc* networks, in *IEEE Int. Symp. Personal, Indoor and Mobile Radio Communications*, vol. 1, 5–8 September 2004, pp. 584–588.

[105] Y.A. Korilis and A. Orda, Incentive compatible pricing strategies for QoS routing, in *INFOCOM '99*, vol. 2, 21–25 March 1999, pp. 891–899.

[106] K. Yamamoto and S. Yoshida, Stability of selfish route selections in wireless *ad hoc* networks, in *Int. Symp. Multi-Dimensional Mobile Communications, 2004 and Joint Conf. 10th Asia-Pacific Conf. Communications*, vol. 2, 29 August to 1 September 2004, pp. 853–857.
[107] R. Kannan and S.S. Iyengar, Game-theoretic models for reliable path-length and energy-constrained routing with data aggregation in wireless sensor networks, *IEEE J. Select. Areas Commun.*, vol. 22, no. 6, August 2004, pp. 1141–1150.
[108] J. Cai and U. Pooch, Allocate fair payoff for cooperation in wireless *ad hoc* networks using Shapley value, in *Int. Parallel and Distributed Processing Symp.*, 26–30 April 2004, p. 219.
[109] M. Alanyali, Learning automata in games with memory with application to circuit-switched routing, in *IEEE Conf. Decision and Control*, vol. 5, 14–17 December 2004, pp. 4850–4855.
[110] R. Atar and P. Dupuis, Characterization of the value function for a differential game formulation of a queueing network optimization problem, in *IEEE Conf. Decision and Control*, vol. 1, 7–10 December 1999, pp. 131–136.
[111] E. Altman, T. Basar and R. Srikant, Nash equilibria for combined flow control and routing in networks: asymptotic behaviour for a large number of users, in *IEEE Conference on Decision and Control*, vol. 4, 7–10 December 1999, pp. 4002–4007.
[112] A. Orda, R. Rom and N. Shimkin, Competitive routing in multiuser communication networks, *IEEE/ACM Trans. Networking*, vol. 1, no. 5, 1993, pp. 510–521.
[113] S. Irani and Y. Rabani, On the value of information in coordination games, in *Annual Symp. Foundations of Computer Science*, 1993, 3–5 November 1993, pp. 12–21.
[114] K. Loja, J. Szigeti and T. Cinkler, Inter-domain routing in multiprovider optical networks: game theory and simulations, in *Next Generation Internet Networks*, 18–20 April 2005, pp. 157–164.
[115] J.A. Almendral, L.L. Fernandez, V. Cholvi and M.A.F. Sanjuan, Oblivious router policies and Nash equilibrium, in *Int. Symp. Computers and Communications*, vol. 2, 28 June to 1 July 2004, pp. 736–741.
[116] J. Virapanicharoen and W. Benjapolakul, Fair-efficient guard bandwidth coefficients selection in call admission control for mobile multimedia communications using game theoretic framework, in *Int. Conf. Communications*, vol. 1, 20–24 June 2004, pp. 80–84.
[117] L. Berlemann, B. Walke and S. Mangold, Behavior based strategies in radio resource sharing games, in *IEEE Int. Symp. Personal, Indoor and Mobile Radio Communications*, vol. 2, 5–8 September 2004, pp. 840–846.
[118] N. Feng, S.-C. Mau and N.B. Mandayam, Pricing and power control for joint network-centric and user-centric radio resource management, *IEEE Trans. Commun.*, vol. 52, no. 9, September 2004, pp. 1547–1557.
[119] R. Kannan, S. Sarangi, S.S. Iyengar and L. Ray, Sensor-centric quality of routing in sensor networks, in *IEEE Joint Conf. IEEE Computer and Communications Societies*, vol. 1, 30 March to 3 April 2003, pp. 692–701.
[120] C.U. Saraydar, N.B. Mandayam and D.J. Goodman, Efficient power control via pricing in wireless data networks, *IEEE Trans. Commun.*, vol. 50, no. 2, 2002, pp. 291–303.
[121] V. Anantharam, On the Nash dynamics of congestion games with player-specific utility, in *Conf. on Decision and Control, CDC*, vol. 5, 14–17 December 2004, pp. 4673–4678.

[122] Y. Zheng and Z. Feng, Evolutionary game and resources competition in the Internet, in *The IEEE-Siberian Workshop of Students and Young Researchers Modern Communication Technologies*, 28–29 November 2001, pp. 51–54.

[123] A. Aresti, B.M. Ninan and M. Devetsikiotis, Resource allocation games in connection-oriented networks under imperfect information, in *IEEE Int. Conf. Communications*, vol. 2, 20–24 June 2004, pp. 1060–1064.

[124] L. Libman and A. Orda, Atomic resource sharing in noncooperative networks, in *INFOCOM '97*, vol. 3, 7–11 April 1997, pp. 1006–1013.

[124] T. Heikkinen, On distributed resource allocation of a multimedia network, in *IEEE Vehicular Technology Conf.*, vol. 4, 19–22 September 1999, pp. 2116–2118.

[125] P. Fuzesi and A. Vidacs, Game theoretic analysis of network dimensioning strategies in differentiated services networks, in *IEEE Int. Conf. Communications*, vol. 2, 28 April to 2 May 2002, pp. 1069–1073.

[126] T. Alpcan and T. Basar, A game-theoretic framework for congestion control in general topology networks, in *IEEE Conf. Decision and Control*, vol. 2, 10–13 December 2002, pp. 1218–1224.

[127] M. Chatterjee, Haitao Lin, S.K. Das and K. Basu, A game theoretic approach for utility maximization in CDMA systems, *IEEE Int. Conf. Commun., ICC '03*, vol. 1, 11–15 May 2003, pp. 412–416.

[128] I. Chlamtac, I. Carreras and H. Woesner, *From Internets to BIONETS: Biological Kinetic Service Oriented Networks*. Springer Science: Berlin, 2005, pp. 75–95.

[129] T. Nakano and T. Suda, Adaptive and evolvable network services, in K. Deb *et al.* (eds). *Genetic and Evolutionary Computation GECCO 2004*, vol. 3102. Springer: Heidelberg, 2004, pp. 151–162.

[130] J. Suzuki and T. Suda, A middleware platform for a biologically inspired network architecture supporting autonomous and adaptive applications, *IEEE J. Select. Areas Commun.*, vol. 23, no. 2, February 2005, pp. 249–260.

[131] I. Carreras, I. Chlamtac, H. Woesner and C. Kiraly, BIONETS: bio-inspired next generation networks, private communication, January 2005.

[132] I. Carreras, I. Chlamtac, H. Woesner and H. Zhang, Nomadic sensor networks, private communication, January 2005.

[133] T. Suda, T. Itao and M. Matsuo, The bio-networking architecture: the biologically inspired approach to the design of scalable, adaptive, and survivable/available network applications, in *The Internet as a Large-Scale Complex System*, K. Park (ed.). Princeton University Press: Princeton, NJ, 2005.

[134] T. Itao, S. Tanaka, T. Suda and T. Aoyama, A framework for adaptive UbiComp applications based on the jack-in-the-net architecture, *Kluwer/ACM Wireless Network J.*, vol. 10, no. 3, 2004, pp. 287–299.

17

Network Deployment

17.1 CELLULAR SYSTEMS WITH OVERLAPPING COVERAGE

The concept of cellular communication systems is based on the assumption that a mobile user is served by the base station that provides the best link quality [1, 2]. Spectrum allocation strategies, discussed in Chapter 12, are based on this assumption. In many cases, however, a mobile user can establish a communication link of acceptable quality with more than one base. Therefore, at many locations there is overlapping coverage, usually by nearby base stations [3].

The *coverage overlap* can be used to improve the system performance. Several schemes that consider this have been suggested [4–8]. Generalized fixed channel assignment (GFCA), a scheme that allows a call to be served by any of several nearby base stations, was considered in Choudhury and Rappaport [4]. Directed retry, discussed in References [5, 6], allows a new call that cannot be served at one base to attempt access via a nearby alternative base. Load sharing is an enhancement of directed retry that allows calls in congested cells to be served by neighboring base stations. In Chu and Rappaport [7] overlapping coverage for highway microcells was considered. The use of overlapping coverage with channel rearrangement was discussed in Chu and Rappaport [8].

Reuse partitioning [9–11] can also improve traffic performance of fixed channel assignment (FCA). The method divides the channels into several disjoint partitions. These partitions are associated with different cluster sizes (or reuse factors). Channels are allocated to base stations according to these cluster sizes. To meet the same signal quality requirement, channels corresponding to smaller cluster sizes are used within a smaller area than that for channels associated with larger cluster sizes [9]. Since channels are reused more often for a smaller cluster size, there may be more channels available at a base in reuse partitioning than in FCA. Therefore, improved traffic performance can be obtained. Because there is a fixed relationship between channels and base stations, reuse partitioning is a fixed channel assignment scheme. In this section the acronym FCA is used only to refer

Advanced Wireless Networks: 4G Technologies Savo G. Glisic
© 2006 John Wiley & Sons, Ltd.

fixed channel assignment without utilizing overlapping coverage or reuse partitioning. That is, in FCA all channels are allocated using a single cluster size and overlapping coverage is not exploited.

When overlapping coverage exists in a system and is being used to provide enhanced access, users in overlapping areas may benefit. However, this may be at the expense of users in nonoverlapping areas who may encounter increased blocking or handoff failure because of the higher channel occupancy of the system. Generally, overlapping areas tend to occur at the periphery of cells (i.e. distant from bases). In reuse partitioning, on the other hand, calls that are close to base stations can access channels from both the smaller cluster and larger cluster partitions. Those calls distant from bases cannot use channels from partitions of smaller cluster sizes. Such calls may in fact do more poorly than in FCA. When *both* overlapping coverage and reuse partitioning are used, they can complement one another. Such a combination will be discussed in this section.

The *system layout* is shown in Figure 17.1. The *cell radius*, r, is defined as the distance from a base to a vertex of its own cell. The *coverage* of a base is the area in which users can establish a link with acceptable signal quality with that base. This area can be modeled by a circle with the center at the corresponding base station. The *coverage radius*, R, is defined as the distance from a base to its coverage boundary.

The coverage of a base is overlapped with coverages of neighboring bases. Therefore, calls can potentially access one, two, three or even more base stations depending on the platform location and the ratio of the coverage radius to the cell radius (R/r). For the

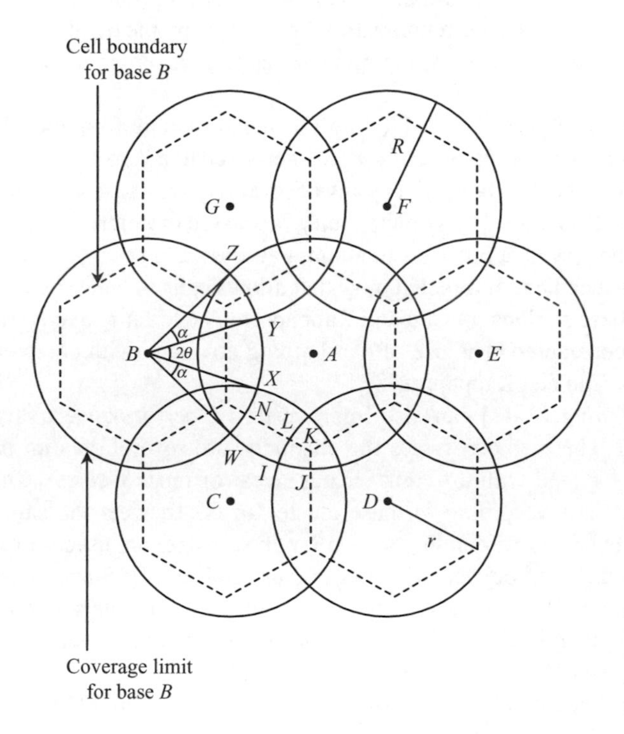

Figure 17.1 System layout for overlapping coverage.

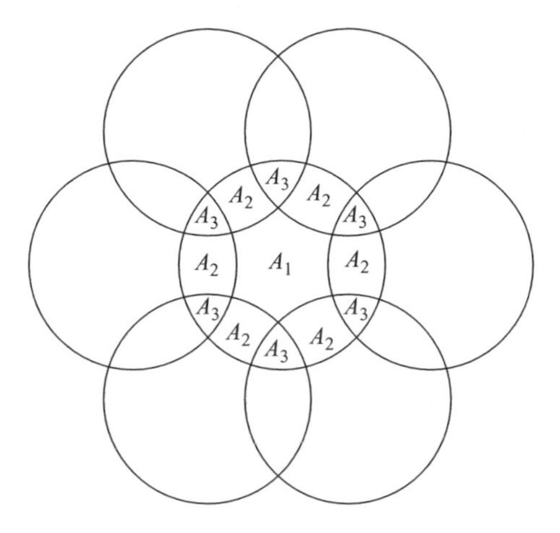

Figure 17.2 Three kinds of regions in the coverage of a base.

interesting range of R/r, there are three kinds of regions in which calls have potential access to one, two or three base stations. These regions in Figure 17.2 are denoted A_1, A_2 and A_3, respectively. A_1 is the nonoverlapping region while both A_2 and A_3 are overlapping regions. We consider (without loss of generality) the cell radius, r, to be normalized to unity. The coverage radius, R, is determined by the requirement of the link quality. Once R is found, the percentage of a cell that belongs to region A_1, A_2 or A_3 can be calculated. They are denoted by p_1, p_2 and p_3, respectively. These relationships are discussed letter in the section.

Reuse partitioning that has two partitions of channels, denoted a and b, is used. Systems with more partitions can be considered similarly. Channels of these two partitions are referred to as a- and b-type channels, respectively. These two partitions are used with different cluster sizes, N_a and N_b, to allocate channels to base stations. Channels of type a are equally divided into N_a groups each with C_a channels. Similarly, b-type channels are equally divided into N_b groups each with C_b channels. Every base station is assigned one group of channels from each partition in such a way that co-channel interference is minimized. As a result, there are totally $C_a + C_b$ channels available at a base. Let C_T denote the total number of available channels and f denote the fraction of channels that are assigned to partition a. So,

$$f = N_a \cdot C_a / C_T \quad \text{and} \quad 1 - f = N_b \cdot C_b / C_T \tag{1}$$

Since N_a and N_b are different, N_a is assumed to be the smaller one. Consequently a-type channels must be used in a smaller area. Channels of type a are intended for use by users who can access only one base, which corresponds to users in region A_1. Channels of type b can be used by all users. Figure 17.3 shows the channel assignment and the available channel groups in a specific region for such a system with $N_a = 3$ and $N_b = 7$. The three groups of a-type channels are labeled a_1, a_2 and a_3, and the seven groups of b-type channels are labeled $b_1, b_2, \ldots, b_7$. Because of overlapping coverage, users in region A_2 can access two groups of channels from partition b. Similarly, users in A_3 can access three groups of b-type channels.

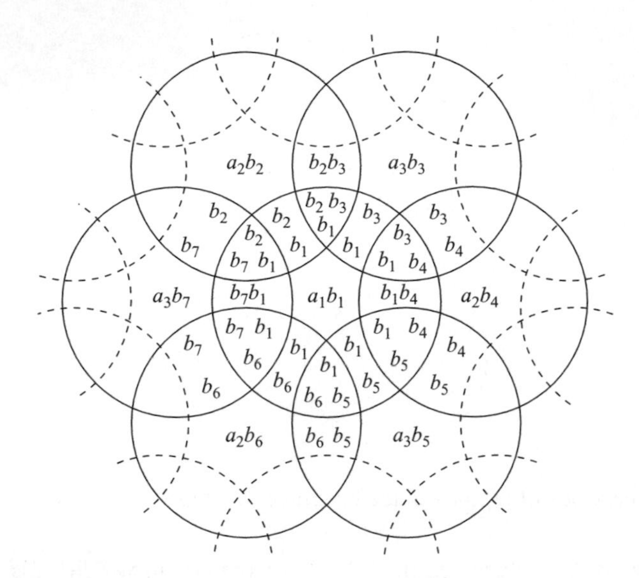

Figure 17.3 The relationship between channel groups and various regions for $N_a = 3$ and $N_b = 7$.

Priority to handoff calls is given at each base, by allocating C_{ha} of the C_a a-type channels to handoffs. Similarly, C_{hb} of the C_b b-type channels are reserved at a base. Specific channels are not reserved, only the numbers, C_{ha} and C_{hb}. As a result, new calls that arise in region A_1 can access $C_a - C_{ha}$ a-type channels and $C_b - C_{hb}$ b-type channels at the corresponding base. New calls that arise in region A_2 or A_3 can access $C_b - C_{hb}$ channels of b-type at each potential base. New calls that arise in region A_1 are served by b-type channels only when no a-type channel is available. When more than one base has a channel available to serve a new call that arises in region A_2 or A_3, the one with the best link quality is chosen. Most likely, it is the nearest one under uniform propagation and flat terrain conditions.

The possible types of handoffs are shown in Figure 17.3. The received signal power is inversely proportional to the distance raised to an exponent, γ, which is called path loss exponent. The coverage radius, R, is determined such that the worst case SIR of a-type calls is equal to the worst case SIR of b-type calls. Therefore, the desired R is the one that satisfies $\mathrm{SIR}_a(R, N_a, \gamma) = \mathrm{SIR}_b(R, N_b, \gamma)$. Mobile users are assumed to be uniformly distributed throughout the service area. Call arrivals follow a Poisson point process with a rate Λ_n per cell, and this rate is independent of the number of calls in progress. Session duration, T, is a random variable with a negative exponential probability density function with mean $\bar{T}(= \mu^{-1})$. An a (or b)-type call resides within the pertaining region during the dwell time T_a (or T_d).

These are random variables with a negative exponential distribution of mean $\bar{T}_a(= \mu_a^{-1})$ (or $\bar{T}_d(= \mu_d^{-1})$. $\bar{T}_a$ and $\bar{T}_d$ are assumed to be proportional to the radii of the corresponding areas. Although region A_1 is not circular, an equivalent radius (which is the radius of a circle that has the same area as A_1) can be used. *System modeling* is based on using two variables to specify the state of a base. One is the number of a-type channels in use; the other is the number of b-type channels in use. A complete state representation for the whole system will be a string of base station states with two variables for each base. This state representation can keep track of all events that occur in the system. Unfortunately, the huge number of system states precludes using this approach for most cases of interest. A simplified approach is to decouple a base from others using average teletraffic demands related to neighboring bases. This is similar to the approach used in References [13, 14]. As a result, the state, s, is characterized from a given base by $a(s)$, $b(s)$, where $a(s)$ is the number of a-type channels in use in state s, and $b(s)$ is the number of b-type channels in use in state s with constraint $0 \leq a(s) \leq C_a$ and $0 \leq b(s) \leq C_b$. The state probabilities, $P(s)$, in statistical equilibrium are needed for determining the performance measures of interest. To calculate state probabilities, the state transitions and the corresponding transition rates must be identified and calculated. Owing to overlapping coverage and reuse partitioning, state transitions result from six driving processes:

(1) new call arrivals;

(2) call completions;

(3) handoff departures of a type; calls that use a-type channels are a-type calls and those that use b-type channels are b-type calls. An a-type call initiates a handoff (a-type handoff) when it leaves region A_1. An a-type handoff may be an inter-base or an intra-base handoff depending on which base continues its service (see Figure 17.3);

(4) hand-off arrivals of a type;

(5) handoff departures of b type; a b-type call initiates a handoff (b-type handoff) only when it leaves the coverage of the serving base. A b-type handoff that enters region A_1 of a neighboring base (out of the coverage of the serving base) can access channels from both a-type and b-type (a-type channels first) at that target base. A b-type handoff that enters region A_2 can access all channels of b-type at each potential target base; and

(6) hand-off arrivals of b type.

The transition rates from a current state s to next state s_n due to these driving processes are denoted by $r_n(s, s_n)$, $r_c(s, s_n)$, $r_{da}(s, s_n)$, $r_{ha}(s, s_n)$, $r_{db}(s, s_n)$, and $r_{hb}(s, s_n)$, respectively. The details for state transition rate derivation and solving the system of state probabilities equations is based on the general principles discussed in Chapter 6. Some specific details can be found also in References [7, 8, 13, 14–16] and especially [17].

Some examples of performance results with $C_T = 180$, $N_a = 3$, $N_b = 9$, $\bar{T} = 100$ s, $\bar{T}_d = 60$ s, $C_{ha} = 0$, $C_{hb} = 0$ and $\gamma = 4$, are shown in Figures 17.4–17.6.

17.2 IMBEDDED MICROCELL IN CDMA MACROCELL NETWORK

Combining a macrocell and microcell with a two-tier overlaying and underlaying structure for a CDMA system is to become an increasingly popular trend in 4G wireless networks. However, the cochannel interference from a microcell to macrocell and vice versa in this two-tier structure is different from that in the homogeneous structure (i.e. consisting of only macrocells or only microcells).

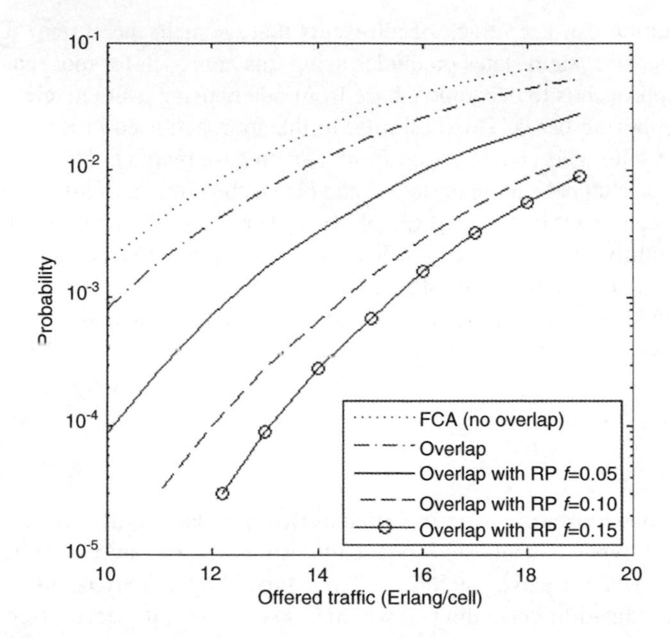

Figure 17.4 Overall blocking probability for FCA, coverage overlap and coverage overlap with reuse partitioning.

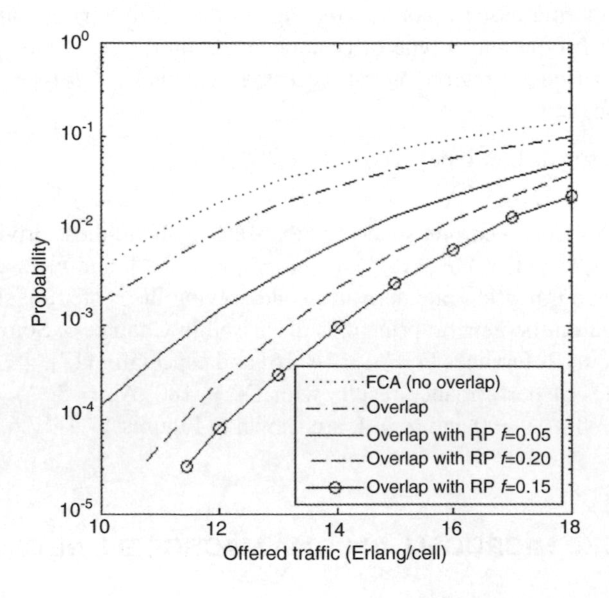

Figure 17.5 Forced termination probability for FCA, coverage overlap, and coverage overlap with reuse partitioning.

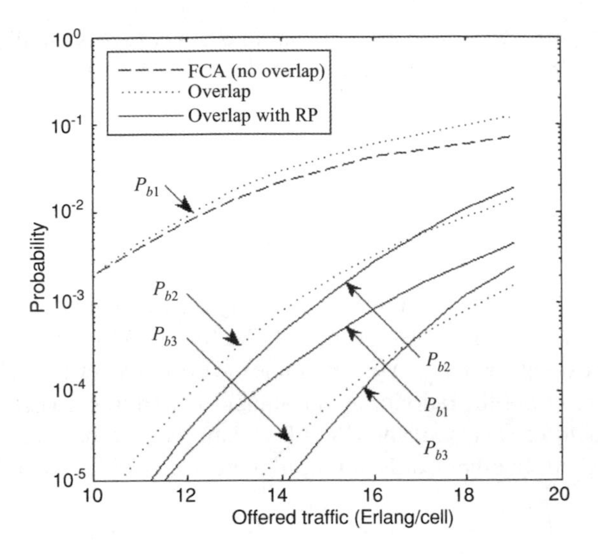

Figure 17.6 Blocking probability in areas A_1, A_2 and A_3 for FCA, coverage overlap and coverage overlap with reuse partitioning.

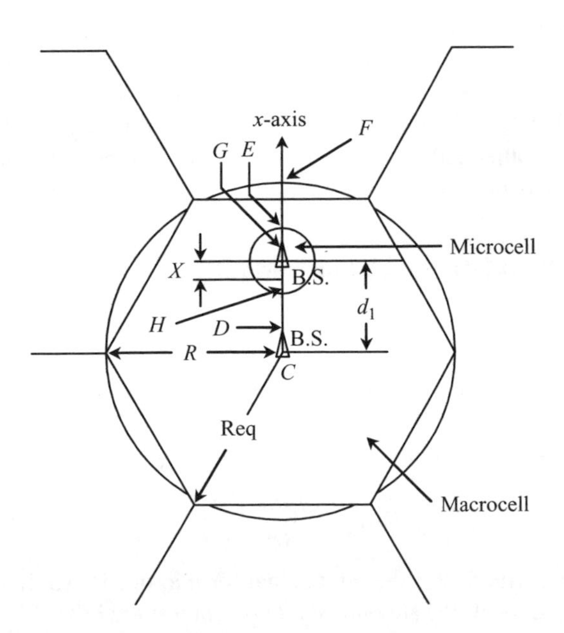

Figure 17.7 Macrocell/microcell overlayed/underlayed structure.

A possibility of embedding a CDMA micro cell into a TDMA macrocell is discussed in References [18, 19] and a CDMA micro cell into a CDMA macrocell in Wu *et al.* [20]. Figure 17.7 presents the geometry of overlayed and underlayed CDMA macrocell (radius R)/CDMA microcell (radius r) structures where a microcell is located at distance d_1 from the BS of the central macrocell. Propagation loss before breakpoint (bp) is assumed to be proportional to d^2, otherwise, it is proportional to d^4, where d is the distance away from the transmitter. The received power is represented as

$$P_l = \begin{cases} \dfrac{(\lambda/4\pi)^2 P_t}{d^2}, & d \leq bp = \dfrac{4\pi h_m h_b}{\lambda} \\ \dfrac{(h_m h_b)^2 P_t}{d^4}, & \text{otherwise} \end{cases} \tag{17.1}$$

where P_t is the power of the transmitter, h_m and h_b are the height of the mobile and BS antennas respectively, and bp is the breakpoint [17–20]. The MS will access a cell BS with the strongest pilot signal. After an MS determines to access a certain cell BS, the cell BS will verify whether or not the ratio for uplink signal to interference (C/I) from a mobile is less than a threshold of $(C/I)_{tu}$. If the C/I is less than the threshold of $(C/I)_{tu}$, the MS is inhibited and blocked. In other words, for a call to be accepted we should have

$$\left(\frac{C}{I}\right)_u = \frac{P_{rj}}{\sum\limits_{i=1,i\neq j}^{M} P_{ri} + \sum\limits_{k=1}^{N} P'_{tk} \times (\beta)'_k} \geq \left(\frac{C}{I}\right)_{tu} \tag{17.2}$$

where P_{rj} is the power received from the desired jth MS by a dedicated BS, P'_{tk} is the power transmitted from the kth MS belonging to other BSs, and $(\beta)'_k$ is the path loss from the kth MS to the dedicated BS. The first and second terms in the denominator represent the interference from other MSs in the same BS and from MSs belonging to other BSs, respectively. Thermal noise is rather small relative to the cells' interference, so it is neglected. If P_r is increased, the interference from other cells is relatively smaller so that the C/I requirement for newly active users can be satisfied.

17.2.1 Macrocell and microcell link budget

For the handoff analysis we want to find the point where the strength of pilot signal emitted from a microcell BS is equal to that from a macrocell at the microcell boundary. The microcell boundary is assumed here to be distant from the macrocell bp. The distance from the microcell BS to the center of microcell is x. Therefore, the pilot power (emitted from microcell BS) received at boundary point D shown in Figure 17.7 is represented as

$$P_{rsD} = \frac{(h_m h_{bs})^2 \alpha_s P_{ts}}{(r-x)^4} \tag{17.3}$$

where α_s is the power fraction of the pilot signal for microcells. Similarly, the pilot power received from the macrocell BS at boundary D is represented by

$$P_{rlD} = \frac{(h_m h_{bl})^2 \alpha_l P_{tl}}{(d_l - x)^4} \tag{17.4}$$

where α_l is the ratio of pilot power to the BS's downlink power and d_l is the distance between the centers of macrocells and microcells. The above equation is similar to (17.1). The subscripts xs (or xl) of P_{xs} (or P_{xl}) denote symbols for the microcell (*small cell*) or for the macrocell (*large cell*). Since the strengths of the pilot signals emitted from both the microcell BS and macrocell BS at D point are equal (i.e. $P_{rsD} = P_{rlD}$), the transmission power emitted from both BSs must satisfy the following formula:

$$P_{\mathrm{ts}} = \frac{h_{\mathrm{bl}}^2 (r - x)^4 \alpha_{\mathrm{l}}}{h_{\mathrm{bs}}^2 (d_1 - x)^4 \alpha_{\mathrm{s}}} P_{\mathrm{tl}} \tag{17.5}$$

Similarly, it holds for point E which is on distance $r + x$ from the BS as shown in Figure 17.7. Thus, we obtain

$$P_{\mathrm{rsE}} = \frac{(h_{\mathrm{m}} h_{\mathrm{bs}})^2 \, \alpha_{\mathrm{s}} P_{\mathrm{ts}}}{(r + x)^4} \quad \text{and} \quad P_{\mathrm{rlE}} = \frac{(h_{\mathrm{m}} h_{\mathrm{bl}})^2 \, \alpha_{\mathrm{l}} P_{\mathrm{tl}}}{(d_1 + r)^4} \tag{17.6}$$

Equating P_{rsE} and P_{rlE} yields

$$P_{\mathrm{ts}} = P_{\mathrm{tl}} \left(\frac{h_{\mathrm{bl}}}{h_{\mathrm{bs}}} \right)^2 \left(\frac{r + x}{d_1 + r} \right)^4 \left(\frac{\alpha_{\mathrm{l}}}{\alpha_{\mathrm{s}}} \right) \tag{17.7}$$

Dividing Equation (17.5) by Equation (17.7) gives $x = r^2/d_1$, which means that the microcell BS must shift from its center by x distance. When d_1 is getting smaller, the value x is getting larger. This fact will cause an asymmetric environment and make the site engineering more difficult. Substituting $x = r^2/d_1$ into Equation (17.5) gives the power ratio of

$$\frac{P_{\mathrm{tl}}}{P_{\mathrm{ts}}} = \frac{h_{\mathrm{bs}}^2 \, d_1^4 \, \alpha_{\mathrm{s}}}{h_{\mathrm{bl}}^2 \, r^4 \, \alpha_{\mathrm{l}}} \tag{17.8}$$

The *downlink C/I*s for the central macrocell and for the microcell can be expressed as [1]:

$$\left(\frac{C}{I} \right)_{1} = \frac{\dfrac{P_{\mathrm{tl}}(1 - \alpha_{\mathrm{l}})}{N} \times \beta}{\left[1 - \dfrac{(1 - \alpha_{\mathrm{l}})}{N} \right] P_{\mathrm{tl}} \beta + P_{\mathrm{ts}} \beta' + \sum_{i=1}^{6} P_{\mathrm{tl}} \times \beta_i} \tag{17.9}$$

$$\left(\frac{C}{I} \right)_{\mathrm{s}} = \frac{\dfrac{P_{\mathrm{ts}}(1 - \alpha_{\mathrm{s}})}{M} \beta'}{\left[1 - \dfrac{(1 - \alpha_{\mathrm{s}})}{M} \right] P_{\mathrm{ts}} \beta' + P_{\mathrm{ts}} \beta + \sum_{i=1}^{6} P_{\mathrm{tl}} \beta_i} \tag{17.10}$$

where N and M denote the number of active users in the central macrocell BS and microcell BS, β, β' and β_i are path loss for macrocell, microcell and adjacent macrocell, respectively, and the third term in the denominator represents the interference from the six adjacent macrocells. The above formulas are the decision rules for the downlink to accept a newly active MS. For a given threshold *C/I* they determined maximum N and M (*capacity*).

The *uplink C/I*s measured by both the macrocell and microcell BSs are derived as

$$\left(\frac{C}{I} \right)_{1} = \frac{P_{\mathrm{r}}}{(N - 1) P_{\mathrm{r}} + \sum_{i=1}^{M} P_{\mathrm{r}}' \left(\frac{\mathrm{r_i}}{\mathrm{d_i}} \right)^4 \left(\frac{\mathrm{h_{bl}}}{\mathrm{h_{bs}}} \right)^2 + I_{ll}} \tag{17.11}$$

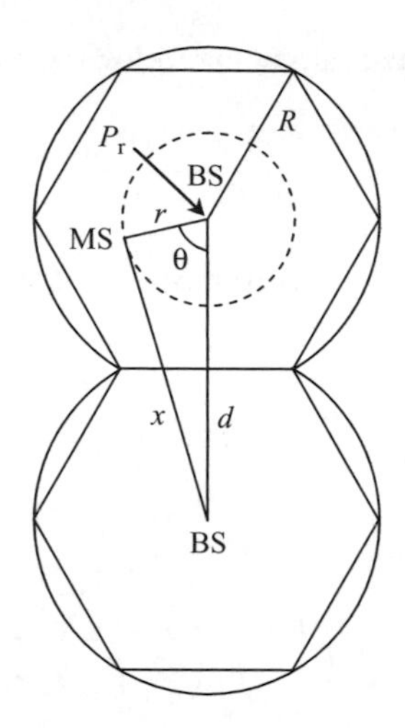

Figure 17.8 Cell geometry for calculation of interference.

and

$$\left(\frac{C}{I}\right)_s = \frac{P_r'}{(M-1)P_r' + \sum_{i=1}^{N} P_r \left(\frac{d_i}{r_i}\right)^4 \left(\frac{h_{bs}}{h_{bl}}\right)^2 + I_{ls}} \tag{17.12}$$

respectively, where P_r and P_r' are mobile powers received by the macrocell's BS and microcell's BS, respectively, M and N denote the number of active mobiles in each microcell and macrocell, respectively. The first term and the second term in the denominator represent the interference from other MSs in the same BS and from all MSs in the other kind of BS, and r_i and d_i are the distance from the ith mobile to its own BS and to the other kind of BS, respectively. Finally, the third term I_{ll} (or I_{ls}) represents the interference from adjacent macrocells to the central macrocell BS (or microcell BS). The above formulas are the decision rules for the uplink to accept a newly active MS. To find expression for I_{ll} (or I_{ls}), N mobile units are assumed to be uniformly distributed in an equivalent disc of radius R shown in Figure 17.8. The density of mobile unit is $\rho = N/\pi R^2$ [1, 20].

The power received in the BS from each mobile is assumed here to be P_r with a path loss proportional to the fourth power of distance. The interference caused by mobiles in an adjacent macrocell is

$$P(d) = \frac{2NP_r}{\pi R^2} \int_0^R r^5 \, \mathrm{d}r \int_0^{\pi} \frac{\mathrm{d}\theta}{(d^2 + r^2 - 2\,\mathrm{d}r\cos\theta)^2} \tag{17.13}$$

where r is the distance between the mobile and the adjacent macrocell BS, $x = \surd(d^2 + r^2 - 2\mathrm{d}r\cos\theta)$ is the distance between the mobile in the adjacent macrocell and the interfered BS, and d is the distance between two BSs. Therefore, the interference to the central macrocell BS caused by the six adjacent macrocells can be calculated as

$$I_{ll} = \sum_{i=1}^{6} P(d_i = 2R) = 0.378NP_r \tag{17.14}$$

Similarly, the interference to the microcell BS caused by the six adjacent macrocells is expressed as

$$I_{ls} = \sum_{i=1}^{6} P(d_i)(h_{bs}/h_{bl})^2 = i_s N P_r \tag{17.15}$$

where

$$i_s = \sum_{i=1}^{6} 2(h_{bs}/h_{bl})^2 \left[2(d_i^2/R^2) \ln \frac{d_i^2}{d_i^2 - R^2} - \left(\frac{4d_i^4 - 6d_i^2 R^2 + R^4}{2(d_i^2 - R^2)^2} \right) \right] \tag{17.16}$$

depending on the location of microcell.

17.2.2 Performance example

A detailed analysis of the above system by simulation is given in Wu *et al.* [20]. Four different power control schemes are analyzed:

(1) both the macrocell and microcell use the signal-strength-based power control for the uplink and there is no power control for the downlink;

(2) (a) [or scheme 2(b)], both the macrocells and microcells use the signal-strength-based power control for the uplink, and downlink power control is also applied for the microcell (or applied for macrocells and microcells);

(3) both the macrocells and microcells use the *C/I* power control for the uplink and there is no power control for the downlink;

(4) both the macrocells and microcells apply the *C/I* power control for uplink, however, but only the microcell uses power control for the downlink.

In addition, antenna heights of macrocells and microcells are 45 and 6 m, respectively. The radius of the microcell is $r = 1000$ m, the radius of the macrocell is R, and $R = 10r$. The bps are 342 m for the microcell and 2569 m for the macrocell.

The arrival process for users in a cell is Poisson distributed with a mean arrival rate of λ, and the holding time for a call is exponentially distributed with the mean of $1/\mu$. The required E_b/N_0 for uplink and downlink in a CDMA system is around 7 and 5 dB, respectively, if the bit error rate is less than 10^{-3}. The major performance we are concerned with is the capacity. Under this maximum loading condition, the background noise is very little relative to interference, so that *C/I* is the only index to survey. The processing gain for a CDMA system is defined as the ratio of bandwidth over data rate, that is, 21.0 dB (i.e. 1.2288 MHz/9.6 kbs), so that the threshold of $(C/I)_{tu}$ is 14 dB for uplink and the threshold $(C/I)_{td}$ is 16 dB for the downlink.For this set of parameters the blocking probabilities for the two cells are given in Figures 17.9 and 17.10.

17.3 MULTITIER WIRELESS CELLULAR NETWORKS

In this section we further extend the previous problem to a general cell-design methodology in multitier wireless cellular network. Multitier networks may be a solution when there are a

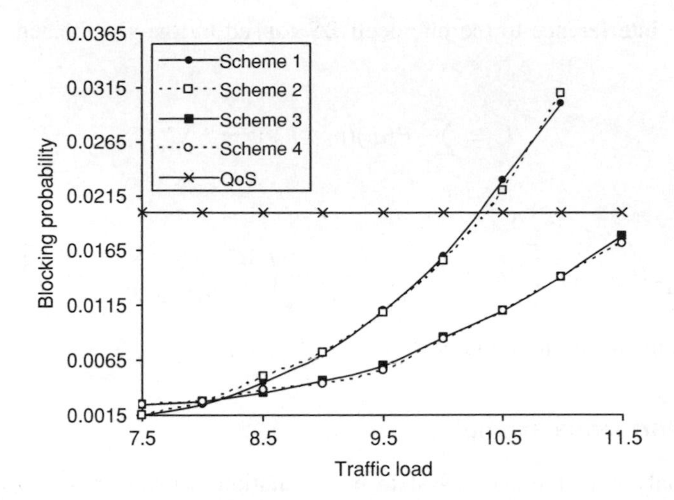

Figure 17.9 Blocking probability of macrocell. (Reproduced by permission of IEEE [20].)

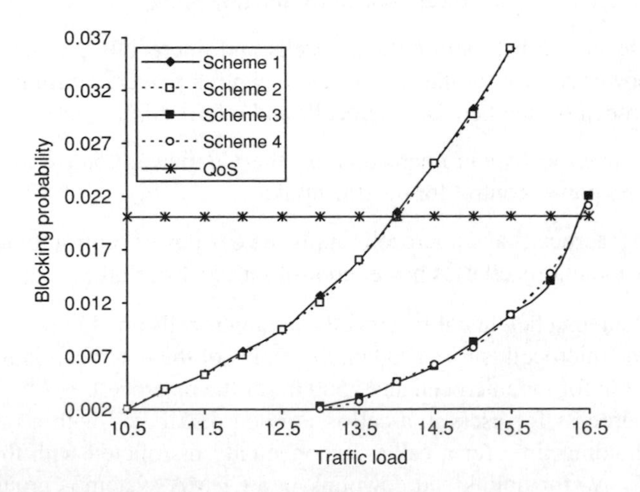

Figure 17.10 Blocking probability in microcell. (Reproduced by permission of IEEE [20].)

multitude of traffic types with drastically different parameters and/or different requirements, such as different mobility parameters or quality-of-service requirements. In such situations, it may be cost-effective to build a multitude of cellular infrastructures, each serving a particular traffic type. The network resources (channels) are then partitioned among the multitude of tiers. In general terms, we are interested in quantifying the cost reduction due to the *multitier* network design, as opposed to a single-tier network. 4G networks are expected to serve different mobility platforms and support multimedia applications through a newly deployed infrastructure based on the multitier approach.

The cellular systems for the 4G of wireless multimedia networks will rely on cells that are smaller than those used today. In particular, in the proposed microcellular systems, the cell radius can shrink to as small as 50 m (compared with 0.5–10 mile radius range for today's macrocellular systems). Smaller cell radii are also possible for systems with smaller coverage, such as pico- and nanocells in a local-area environment. The size of the cells is closely related to the expected speed of the mobiles that the system is to support. In other words the faster the mobiles move, the larger the cells should be. This will keep the complexity of handoffs at a manageable level. The cell size is also dependent on the expected system load (Erlangs per area unit). The larger the system load, the smaller the cell should be. This is the direct outcome of the limited wireless resources assigned to handle cell traffic. Smaller cells result in channels being reused a greater number of times and thus greater total system capacity. Finally, the cell size also depends on the required QoS level. This includes the probability of call blocking, call dropping or call completion. In general, the more stringent the QoS, the smaller the cells need to be, since smaller cells increase the total system capacity. Unfortunately, smaller cells require more base stations, and hence higher costs.

In this section, we discuss the multitier concept to show how to optimize the design of a system with differing traffic requirements and mobile characteristics. We examine how to lay out a multitier cellular system, given a total number of channels, the area to be covered, the average speed of mobiles in a tier, call arrival and duration statistics for each tier, and a constraint on the QoS (i.e. blocking and dropping probabilities). We discuss how to design a multitier cellular system in terms of the number of tier-i cells (e.g. of macrocells and microcells in a two-tier system) and the number of channels allocated to each tier so that the total system cost is minimized.

17.3.1 The network model

For each tier, the total area of the system (*coverage* area) is partitioned into cells. All cells of the same tier are of equal size. The network resources are also partitioned among the network tiers. Channels allocated to a particular tier are then reused based on the reuse factor determined for the mobiles of that tier (i.e. within each tier, channels are divided into channel sets). One such set is then allocated to each cell of that tier. For simplicity, FCA for the allocation of channels both among the tiers and within a tier is used.

In this section, connection-oriented traffic is considered. Each tier is identified by several parameters: *call arrival rate*, *call duration time*, *call data rate*, *average speed of mobiles*, *performance factors* and *QoS factors*. Additional assumptions are that:

- handoffs and new calls are served from the same pool of available channels which means that, for a given tier, the probability of blocking equals the probability of dropping;
- the speed of each class of mobiles is constant for each tier;
- new call arrivals and call terminations are independent of the handoff traffic;
- the traffic generation is spatially uniformly distributed;
- call arrivals follow a Poisson process;
- the spatial and time distribution of call arrivals are the same over all cells in a given tier, but can be different from tier to tier;
- call duration for a given tier is exponentially distributed, and may vary from tier to tier.

In the presentation the following notation is used [21]:

- λ_i^0, rater of tier i calls initiated per unit area, calls/(s m^2);
- λ_i, call initiation rate in a tier i cell (calls/s);
- $1\mu_i$, average call duration of a tier i cell (s call);
- h_i, mean number of handoffs for a tier i call (handoffs/p call);
- γ_i, the relative cost of a tier i base station to the cost of a tier S base-station ($\gamma_S \equiv 1$);
- A, total area to be covered (m^2);
- S, number of tiers;
- C, total number of available network channels;
- C_i, number of channels allocated to tier i;
- N_i, number of channels allocated to a tier i cell;
- f_i, number of tier i cells in the frequency reuse cluster;
- m_i, number of tier i cells contained in a tier (i – 1) cell; tier 0 defines the total coverage area; $m_1 \stackrel{def}{=}$ total number of tier 1 cells;
- R_i, radius of a tier i cell (m);
- V_i, average speed of tier i mobile users (ms);
- $P_{\rm B}(i)$, actual blocking probability for a tier i cell;
- $P_{\rm D}(i)$, actual dropping probability for a tier i cell;
- $P_{\rm B}^{\rm max}(i)$, the maximum acceptable blocking probability for tier i calls;
- $P_{\rm D}^{\rm max}(i)$, the maximum acceptable dropping probability for tier i calls;
- $P_{\rm L}(i)$, loss probability of tier i;
- PLT, overall system loss probability, weighted by the amount of traffic of each tier;
- PLT$_{\rm max}$, the maximum acceptable weighted system loss probability;
- TSC, total system cost;
- TSC$_{\rm max}$, maximum allowable total system cost; input design parameter tabbing.

Cost optimization is supposed to minimize the cost of the multitier infrastructure. In our model, we assume that the major part of the total system cost is the cost of base station deployment, which in our model is proportional to the total number of base stations. Thus, the total system cost, TSC, is:

$$C = m_1\{\gamma_1 + m_2[\gamma_2 + m_3(\cdots)]\} = \sum_{i=1}^{S} \gamma_i \cdot \prod_{j=1}^{i} m_j, \quad S \geq 1 \tag{17.17}$$

equation, we assume that cells are arbitrary split so that the location of tier i is independent of the location of the base stations of the other tiers. In a more

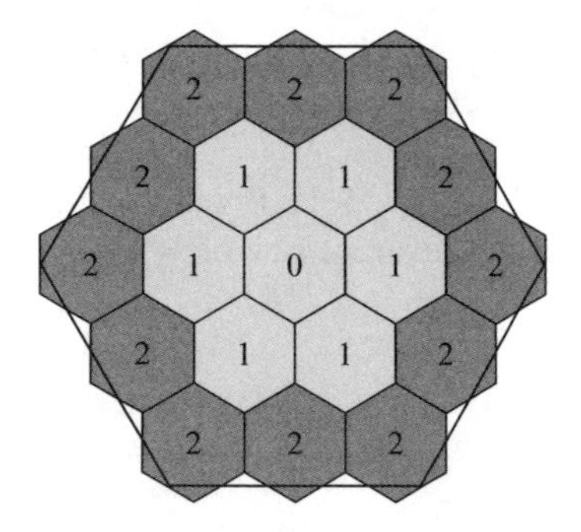

Figure 17.11 Cell splitting in multitier network.

structured splitting cells of tier i form rings within a cell of tier $(i-1)$ as shown for the two-tier case in Figure 17.11. Thus, if a base station already exists in a cell for tier i, the cost of placing an additional base station for a higher tier is negligible. Therefore, for example, in the case of a two-tier network the total cost can be reduced by the cost of the tier 2 base stations. In such a case, the total cost is

$$\mathrm{TSC} = \sum_{i=1}^{S} \gamma_i \left(\prod_{j=1}^{i} m_j - \prod_{j=1}^{i-1} m_j \right) = \gamma_1 m_1 + \sum_{i=2}^{S} \gamma_i (m_i - 1) \prod_{j=1}^{i-1} m_j, \quad S \geq 1 \tag{17.18}$$

It will be further assumed that the set of all the available channels to tier i, C_i, is equally divided among the f_i cells in the frequency reuse cluster, which means no channel sharing between the cells. There is no channel sharing among the tiers either. It is also assumed that no channels are put aside for handling handoffs. Therefore, the blocking and dropping probabilities are equal, and will be referred to as the probability of loss, $P_\mathrm{L}(i)$; that is,

$$\forall i, \; P_\mathrm{L}(i) = P_\mathrm{B}(i) = P_\mathrm{D}(i) \text{ and } P_\mathrm{L}^{\max}(i) = \min \left\{ P_\mathrm{B}^{\max}(i), P_\mathrm{D}^{\max}(i) \right\}$$

The overall system loss probability, PLT, is given by:

$$\mathrm{PLT} = \frac{\sum_{i=1}^{S} P_\mathrm{L}(i) \lambda_i^0}{\sum_{i=1}^{S} \lambda_i^0} \tag{17.19}$$

The layout design problem is to minimize TSC, when the total number of available channels is at most C, and subject to the following QoS constraints: $\mathrm{PLT} \leq \mathrm{PLT}_{\max}$ and $\bar{P}_\mathrm{L} \leq \bar{P}_\mathrm{L}^{\max}$. In other words: Find $\{m_i\}$ minimizing TSC with $\sum_{j=1}^{S} C_j \leq C$ and $\mathrm{PLT} \leq \mathrm{PLT}_{\max}$ and $\bar{P}_L \leq \bar{P}_L^{\max}$ [21].

The average number of handoffs, h_i, that a call of tier i will undergo during its lifetime is

$$h_i = (3 + 2\sqrt{3}) V_i / 9 \mu_i R_i, \quad i = 1, 2, \ldots, S \text{ [22]}$$

Now assume that there are a total of M_i cells in tier-i ($M_i = \prod_{j=1}^{i} m_j$). Then the total average number of handoffs per second in all the cells is $h_i\lambda_i M_i$. The total arrival rate of handoff and initial calls in all tier i cells is $\lambda_i M_i\,(h_i + 1)$, and the average total (handoffs and initial calls) rate per cell is $\lambda_i^{\text{total}} = \lambda_i(h_i + 1)$. It was assumed that the total arrival process to a cell is still Markovian, with the average rate of λ_i^{total}. Since the area of a hexagon with radius R is $3\sqrt{3}/2R^2$, we obtain $\lambda_i = 3\sqrt{2}\lambda_i^0 R_i^2/2$. This leads to

$$\lambda_i^{\text{total}} = \lambda_i^0 R_i \frac{\left[(2+\sqrt{3})V_i + 3\sqrt{3}\mu_i R_i\right]}{2\mu_i} \tag{17.20}$$

The average call termination rate is [22] $\mu_i^{\text{total}} = \mu_i(1 + 9h_i)$. We can use the Erlang-B formula (see Chapter 6) extended to *M/G/c/c* systems. The probability of loss of a tier i call is therefore

$$P_{\text{L}}(i) = \frac{\left(\lambda_i^{\text{total}}/\mu_i^{\text{total}}\right)^{N_i}/N_i!}{\sum_{j=0}^{C_i}\left(\lambda_i^{\text{total}}/\mu_i^{\text{total}}\right)^j/j!} \tag{17.21}$$

$$= \frac{\{[\lambda_i(1+h_i)]/[\mu_i(1+9h_i)]\}^{N_i}/N_i!}{\sum_{j=0}^{C_i}\{[(\lambda_i(1+h_i)]/[\mu_i(1+9h_i)]\}^j/j!}, \quad i = 1, 2, \ldots, S$$

where λ_i^{total}, μ_i^{total}, λ_i and h_i are given above, and $N_i = \lfloor C_i/f \rfloor$, $i = 1, 2, \ldots, S$.

For a two-tier system ($S = 2$), the cost function simplifies to TSC $= \lambda_1 \cdot m_1 + \lambda_2 \cdot m_1 \cdot (m_2 - 1)$. Parameters m_1, m_2 and R_2 as a function of the area, A and the tier 1 cell radius, R_1, can be expressed as

$$m_1 = \lceil A/A_1 \rceil = \left\lceil 2\sqrt{3}A/9R_1^2 \right\rceil \tag{17.22}$$

where $A1$ is area of tier 1 cell. We assume that both tier 1 and tier 2 cells are hexagonally shaped and that tier 2 cells are obtained by suitably splitting the tier 1 cells. Each tier 1 cell will contain k layers of tier 2 cells. Figure 17.11 shows an example of a tier 1 cell which contains three layers (0, 1 and 2). From the geometry of a hexagon, the number of tier 2 cells in a tier 1 cell is given by $m_2 = 1 + 6 + \cdots + 6k = 1 + 3k(k+1)$, $k = 0, 1, \ldots,$ where $(k+1)$ denotes the number of 'circular layers' in the cell splitting. Also, the radius of a tier-2 cell is given by $R_2 = R_1/\sqrt{m_2}$.

The *constrained optimization problem* outlined above consists of finding optimal values of R_1, R_2, C_1, C_2, m_1 and m_2, which are the system design parameters. The optimal values will be referred to as R_1^*, R_2^*, C_1^*, C_2^*, m_1^* and m_2^*. The parameters m_1, m_2, C_1 and C_2 are integers, while R_1 and R_2 are continuous variables. However, we only need to consider a discrete subset of the possible values of R_1. From Equation (17.22), it is sufficient to consider just those values of R_1 for which $R_1 = \sqrt{(2\sqrt{3}A/9\ell)}$ where $\ell = m_{\text{min}1}, \ldots, m_{\text{max}1}$; $m_{\text{min}1}$ and $m_{\text{max}1}$ are the lower and upper bounds, respectively, of the number of tier 1 cells in the system. Thus, R_1 can only assume one of $m_{\text{max}1} - m_{\text{min}1} + 1$ values.

The lower bound, $m_{\text{min}\,1}$, is usually assumed to be equal to 1, while the upper bound, $m_{\text{max}\,1}$, can be estimated from TSC$_{\text{max}}$, which is given to the designer, as $m_{\text{max}\,1} = \lceil \text{TSC}_{\text{max}}/\gamma_1 \rceil$. The following search procedure is used in Ganz *et al.* [21] to find the design parameters.

$$\begin{aligned}
\mathrm{TSC}^* &= \mathrm{TSC}_{\max} \\
m_{\max 1} &= \lceil \mathrm{TSC}_{\max}/\gamma_1 \rceil \\
a_1 &= 3\sqrt{2}\lambda_1^0/2 \\
a_2 &= 3\sqrt{2}\lambda_2^0/2 \\
b_1 &= (3+2\sqrt{3})V_1/9\mu_1 \\
b_2 &= (3+2\sqrt{3})V_2/9\mu_2 \\
c &= \sqrt{2\sqrt{3}A/9}
\end{aligned}$$

`while` $(m_{\min 1} \leq m_1 \leq m_{\max 1})$ `do`

$$\begin{aligned}
R_1 &= c/\sqrt{m_1} \\
\lambda_1 &= a_1 R_1^2 \\
h_1 &= b_1/R_1
\end{aligned}$$

`while` $(1 \leq C_1 < C)$ `do`

$$\begin{aligned}
N_1 &= \lfloor C_1/f_1 \rfloor \\
N_2 &= \lfloor (C - C_1)/f_2 \rfloor \\
P_L(1) &= \frac{\{[\lambda_1(1+h_1)]/[\mu_1(1+9h_1)]\}^{N_1}/N_1!}{\sum_{j=0}^{N_i}\{[\lambda_1(1+h_1)]/[\mu_1(1+9h_1)]\}^j/j!}
\end{aligned}$$

`for` $(k = 1;;$ k++$)$ `do`

$$\begin{aligned}
m_2 &= 1 + 3k(k+1) \\
\mathrm{TSC} &= \gamma_1 m_1 + \gamma_2 m_1(m_2 - 1)
\end{aligned}$$

`if` $(\mathrm{TSC} > \mathrm{TSC}^*)$ `then break`

$$\begin{aligned}
R_2 &= R_1/\sqrt{m_2} \\
\lambda_2 &= a_2 R_2^2 \\
h_2 &= b_2/R_2 \\
P_L(2) &= \frac{\{[\lambda_2(1+h_2)]/[\mu_2(1+9h_2)]\}^{N_2}/N_2!}{\sum_{j=0}^{N_2}\{[\lambda_2(1+h_2)]/[\mu_2(1+9h_2)]\}^j/j!} \\
\mathrm{PLT} &= \frac{\lambda_1^0 P_L(1) + \lambda_2^0 P_L(2)}{\lambda_1^0 + \lambda_2^0}
\end{aligned}$$

`if` $(\mathrm{PLT} < \mathrm{PLT}_{\max})$ `and` $[P_L(1) < P_L^{\max}(1)]$ `and` $[P_L(2) < P_L^{\max}(1)]$ `then break`

`end for`

```
if (TSC < TSC*) then
```

$$\hat{k} = k$$

$$m_1^* = m_1$$

$$m_2^* = m_2$$

$$R_1^* = R_1$$

$$R_2^* = R_1/\sqrt{m_2}$$

$$C_1^* = C_1$$

$$C_2^* = C - C_1$$

```
    end if
  end while
end while
```

The outputs are: TSC*, m_1^*, m_2^*, R_1^*, R_2^*, C_1^* and C_2^*.

17.3.2 Performance example

To generate numerical results, a system with the parameters shown in Table 17.1 is used [21]. The performance of the two-tier system is compared with that of a one-tier system. To obtain results for a one-tier system the optimization algorithm with $R_1 = R_2$ was run. This results in both tiers sharing the same cells and $m_2 = 1$. The total cost is then computed as TSC $= m_1\gamma_1$.

Table 17.1 Example system parameters

Parameter	Value/range	Units
A	100	km^2
C	90	Channels
S	2	
λ_1^0	0.23.0	Calls/(min km^2)
λ_2^0	5.040.0	Calls/(min km^2)
μ_1, μ_2	0.33	Calls/min
γ_1	10.0	$ (in 1000 s)/base
γ_2	1.0	$ (in 1000 s)/base
V_1	30540	km/h
V_2	1.512.0	km/h
f_1, f_2	3	
TSC$_{max}$	10 000–20 000	$ (in 1000 s)
PLT$_{max}$	0.01	
$P_B^{max}(*)$	0.01	
$P_D^{max}(*)$	0.01	

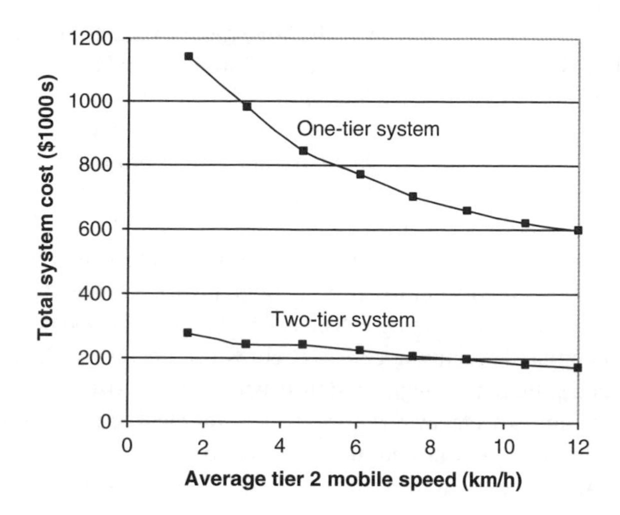

Figure 17.12 Comparison of the total system costs between the one-tier and two-tier systems, $\lambda_1^0 = 1$[calls/(min km^2)] and $\lambda_2^0 = 20$[calls/(min km^2).]

Figure 17.12 depicts the total system cost as a function of tier 2 mobile speed, while tier 1 mobile speed is fixed for one-tier and a two-tier systems. The tier 1 mobile speeds considered are 30, 90, 180, 270, 360 and 540 km/h.

The main conclusion from Figure 17.12 and many other similar runs [21] is that, for the parameter ranges used in this study, the two-tier system outperforms the single-tier system for all the values of the slower and faster mobile speeds.

17.4 LOCAL MULTIPOINT DISTRIBUTION SERVICE

Wireless systems can establish area-wide coverage with the deployment of a single base station. The local multipoint distribution service (LMDS) offers a wireless method of access to broadband interactive services. The system architecture is considered point-to-multipoint since a centralized hub, or base station, simultaneously communicates with many fixed subscribers in the vicinity of the hub. Multiple hubs are required to provide coverage over areas larger than a single cell. Because of the fragile propagation environment at 28 GHz, LMDS systems have small cells with a coverage radius on the order of a few kilometers. Digital LMDS systems can flexibly allocate bandwidth across a wide range of bi-directional broadband services including telephony and high-speed data access.

Multiple LMDS hubs are arranged in a cellular fashion to reuse the frequency spectrum many times in the service area. Complete frequency reuse in each cell of the system is attempted with alternating polarization in either adjacent cells or adjacent hub antenna sectors within the same cell. Subscriber antennas are highly directional with roughly a 9 inch diameter (30–35 dBi) to provide additional isolation from transmissions in adjacent cells and to reduce the received amount of multipath propagation that may cause signal degradation. Since cells are small and the entire spectrum is reused many times, the overall system capacity is quite high, and backhaul requirements can be large. Backhaul networks will probably be a combination of fiber-optics and point-to-point radio links.

The system capacity comes mainly from the huge radio frequency (RF) bandwidth available: block A, 1150 MHz (27.50–28.35, 29.100–29.250 and 31.075–31.225 GHz); and block B, 150 MHz (31.000–31.075 and 31.225–31.300 GHz). For the purpose of frequency planning for two-way usage it is essential to solve the problem of LMDS spectrum partitioning for the upstream and downstream. The standard duplexing options, frequency-division duplex (FDD) and time-division duplex (TDD), are applicable in LMDS too and will not be discussed here in any more detail. Instead, as a wireless point-to-multipoint system, the deployment of LMDS will be assessed by the basic parameters of cell size and capacity. Obviously, an operator would want to cover as large an area as possible with a minimum number of cell sites, hence maximizing cell size. The modulation and coding, and the ensuing E_b/N_0, are factors in determining cell size. However, because of high rain attenuation at LMDS frequencies, there is a major trade-off between cell size and system availability determined by the rain expectancy for a given geographical area. In most of the United States, for the forthright aim of providing 'wireless fiber' availability of 0.9999–0.99999, the cells would only be between 0.3 and 2 miles [23]. Consequently, the deployment of LMDS will extensively involve multicell scenarios. In the following we review the methods of optimizing system capacity in these scenarios through frequency reuse. The main problem related to frequency reuse is the interference between different segments of the system. Therefore, patterns that create bound and predictable values of S/I are essential to device deployment. Frequency reuse and capacity in interference-limited systems have been discussed in previous sections for mobile cellular and personal communications services (PCS) systems. As the basic methods and principles also apply to fixed systems, there are several major differences in the treatment of fixed broadband wireless systems such as LMDS.

Unlike mobile cellular, in LMDS the subscriber antennae are highly directional and point toward one specific base station. This, together with the nonmoving nature of the subscriber, results in much lower link dynamics. The channel can mostly be described as Rician (with a strong main ray) and not Rayleigh-like in mobile cellular. Since the subscriber employs directive antennae, it is compelled to communicate with only one base station, which excludes the use of macro diversity (or cell diversity) – a very beneficial method with mobile cellular, but one that complicates the interference and frequency reuse analysis.

Also, mobile cellular service was originally intended for voice; therefore, it is designed for symmetric loading, while broadband wireless services generally have more downstream traffic than upstream. This reflects in the main issue of concern for interference study, which in PCS and mobile cellular is upstream interference because the base station receiver has the hard task of receiving from a multitude of mobiles transmitting through nondirectional antennae and suffering different fast fading.

In LMDS the upstream is usually lower-capacity and employs lower-order modulation, which offers better immunity. Also, the slower fading environment allows a closed-loop transmission power control system to operate relatively accurately. Consequently, upstream most subscribers transmit at a power lower than nominal, unlike the downstream transmission. Also, the narrow beam of their antennae is an interference-limiting factor. Therefore, downstream interference is the most problematic, as opposed to the mobile cellular case.

Another parallel with mobile cellular refers to cellular patterns. Owing to the mobile nature of its subscribers, a mobile cellular system has to provide service from the start to a whole area (town, region), with a large number of adjacent cells. The number of sectors is relatively low (2–4); otherwise, the mobile would go through frequent handoffs.

In LMDS, in the long run, owing to the small cell size, the ever increasing hunger for bandwidth and the availability of radio and networking technologies, deployment is also expected to be blanket coverage. However, the economics of LMDS do not require this from the beginning. Operators will probably first offer the service in clusters of business clients with high data bandwidth demands and with the financial readiness for the new service, and later will gradually expand the service to larger areas. Second, the sectorization will be in denser patterns determined by the increasing demand for bandwidth and not limited by the requirement for handoff. The narrower sectors employing higher antenna gain in the base station also provide for larger cell size.

Frequency reuse in one cell is illustrated in Figure 17.13 with three examples of frequency reuse by sectorization. The first step in frequency planning is to assume the division of the available spectrum into subbands, so adjacent sectors will operate on different subbands. Also, the sector structure has to be designed as a function of the capacity required and the S/I specification of the modem used. The need to divide into subbands is a function of base station antenna quality, specifically the steepness of the rolloff from the main lobe to sidelobes in the horizontal antenna pattern. If the antenna sidelobes roll off very steeply after the main lobe, it is possible to reuse the frequency every two sectors, resulting in patterns of the type A, B, A, B, ..., as in Figure 17.13(b). In this context the reuse factor, F_R, is the number of times the whole band is used in the cell, which for the simplified regular patterns is the number of sectors divided by the number of subbands; in this case $F_R = 3$. In a more conservative deployment the frequency is reused only in back-to-back sectors as in Figure 17.13(a), which has six sectors with reuse pattern A, B, C, A, B, C. Figure 17.13(c) shows a higher-capacity cell where the pattern A, B, C, ..., is repeated in 12 sectors, resulting in $F_R = 4$. Obviously, to keep the equipment cost low, we want a low number of sectors; schemes where we divide the spectrum in as few subbands as possible (two is best).

With present antenna technology at 28 GHz it is possible to achieve sidelobes under -33 dB, but the radiation pattern of the deployed antenna is different. The sidelobes are significantly higher because of scattering effects such as diffraction, reflection or dispersion caused by foliage. The very narrow beamwidth of the subscriber antenna (which at 28 GHz can be reasonably made 3° or lower) helps limit the scattering effect. Obviously, the sum of such effects depends on the millimeter-wave effects in the specific buildings and terrain in the area, and have to be estimated through simulation and measurements for each particular case. Another uncertain factor in the estimation of S/I is the fading encountered by the direct signal. As a starting assessment, we shall consider the above-mentioned effects accountable for increasing the equivalent antenna sidelobe radiation to -25 dB.

Based on this, the following estimations of S/I are only orientative. For each particular deployment the worst case has to be estimated considering the particular conditions.

17.4.1 Interference estimations

To allow A, B, C, ..., type sectorization, in Figure 17.13(a) and (c) the base station antenna sidelobe is $\alpha = -25$ dB at an angle more than 2.5 B_{3dB} from boresight, where B_{3dB} is the -3dB beamwidth (the main lobe is between ± 0.5 B_{3dB}). In Figure 17.13(a) and (b),

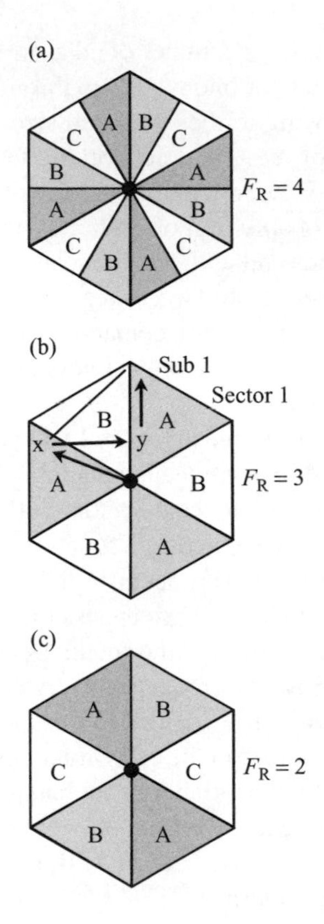

Figure 17.13 Cells with reuse factors of 2, 3 and 4.

$B_{3dB} = 60°$, while in Figure 17.13(c) $B_{3dB} = 30°$. In Figure 17.13(b), better-quality antennas are considered, which would achieve the same sidelobe rejection at 1.5 B_{3dB}, driving frequency reuse higher for the same number of sectors.

17.4.2 Alternating polarization

Figure 17.14(a) shows how a high reuse factor of 6 can be achieved in 12 sectors by alternating the polarity in sectors. The lines' orientations show the polarization, horizontal (H) or vertical (V). The amount of discrimination that can be achieved depends on the environment. Although the antenna technology may provide for polarization discrimination of 30–40 dB, we shall consider that the combination of depolarization effects raises the cross-polarization level to $p = -7$ dB.

The interference is reduced, so the same frequency at the same polarization comes only in the fourth sector. The problem is that a deployment has to allow for gradual sectorization – start by deploying minimum equipment, then split into more sectors as demand grows. However, if alternating polarization is employed in sectors, the operator would have to visit the subscriber sites in order to reorient the antennas. A more conservative approach would be to set two large areas of different polarity, as in Figure 17.14(b). This does not reduce the close-in sidelobes but reduces overall interference and also helps in the multiplecell design.

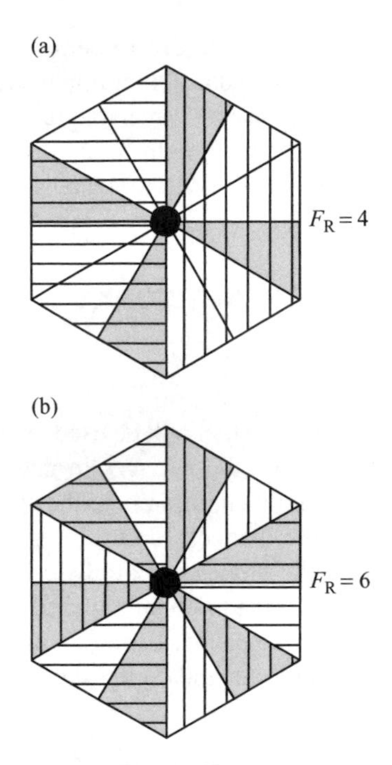

Figure 17.14 Twelve-sector cells with cross-polarization.

Following is an approximation for the S/I:

$$\frac{S}{I} = \frac{S}{\sum_i I_i + N_R} = \frac{S}{\sum_i \alpha_i S_i + \sum_j p_j \alpha_j S_j + N_R} \tag{17.23}$$

where S_i are the transmission powers in other sectors using the same subband, I_i the interferers, and N_R the receiver input noise. Later the case will be considered where all transmission powers in sectors are equal to $S_i = S$, and, N_R is neglected. N_1 and N_2 are the number of sectors with the same polarization and the cross-polarized ones, respectively. As well, the worst case values α and p will be taken for the sidelobe gains and cross-polarization:

$$\frac{S}{I} = \frac{1}{\alpha(N_1 + N_2 p)}. \tag{17.24}$$

Thus, in a first approximation [24], by applying Equation (17.24), the S/I level (or co-channel interference, CCI) at the subscriber receiver is given in Table 17.2. The type of modulation and the subsequent modem S/I specification have to be specification have to be

Table 17.2 The S/I level at the subscriber receiver

	Figure 17.13(a)	Figure 17.13(b)	Figure 17.13(c)	Figure 17.14(a)	Figure 17.14(b)
S/I in dB	25	22	20	22	25

specified with sufficient margin. A modem receiver operates in a complex environment of challenges of which the S/I or CCI is only one. Other impairments such as equalizer errors, adjacent channel interference, phase noise and inter-modulation induced by the RF chain limit the S/I with which the modem can work in the real operation environment. The above technique can be extended to multiple cell scenario. Details can be found in Roman [24].

17.5 SELF-ORGANIZATION IN 4G NETWORKS

17.5.1 Motivation

Self-organisation is an emerging principle that will be used to organise 4G cellular networks [25–40]. It is a functionality that allows the network to detect changes, make intelligent decisions based upon these inputs, and then implement the appropriate action, either minimizing or maximizing the effect of the changes. Figure 17.15 illustrates a multitier scenario where numerous self-organizing technologies potentially could be applied.

Frequency planning discussed so far in this chapter was performed by choosing a suitable reuse pattern. Individual frequencies are then assigned to different base stations according to propagation predictions based on terrain and clutter databases.

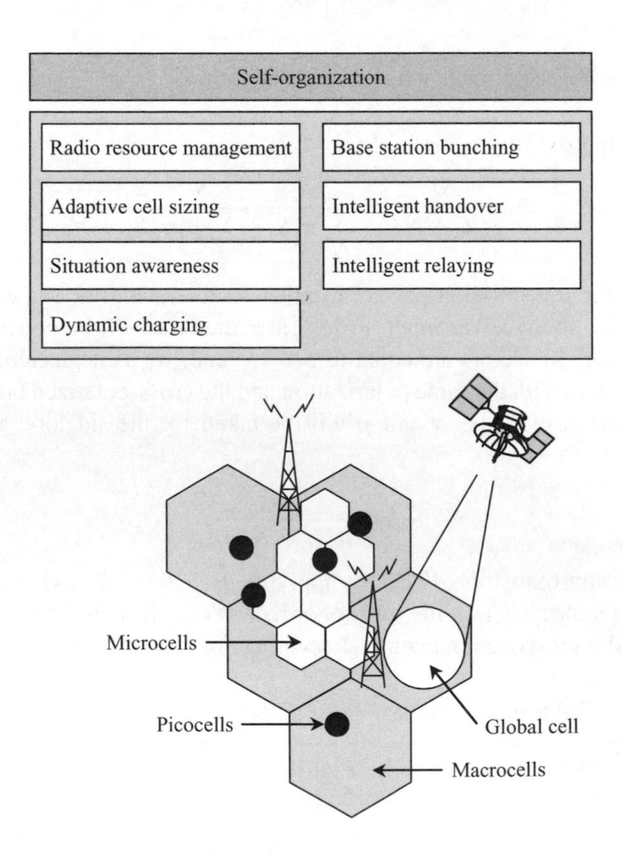

Figure 17.15 Multitier scenario with self-organising technologies.

The need to move away from this type of frequency planning has been expressed in the literature [28] as well as being emphasized by ETSI in the selection criteria for the UMTS air–interface technique. The main reason for this departure is the need for very small cell sizes in urban areas with highly varying morphology, making traditional frequency planning difficult. Another reason lies in the difficulty associated with the addition of new base stations to the network, which currently requires extensive reconfiguration. In view of these two arguments, a desirable solution would require the use of unconfigured base transceiver stations (BTSs) at all sites; these BTSs are installed without a predefined set of parameters and select their operating characteristics on the basis of information achieved from runtime data. For instance, they may operate at all the available carriers and select their operating frequencies to minimize mutual interference with other BTSs [28].

The increasing demand for data services means that the next generation of communication networks must be able to support a wide variety of services. The systems must be location- and situation-aware, and must take advantage of this information to dynamically configure themselves in a distributed fashion [29].

There will be no central control strategy in 4G, and all devices will be able to adapt to changes imposed by the environment. The devices are intelligent and clearly employ some form of self-organization [30, 31]. So far in our previous discussion, coverage and capacity have been the two most important factors in cellular planning. To have good coverage in both rural and urban environments is important so as to enable customers to use their terminals wherever they go. Coverage gaps mean loss of revenue and can also lead to customers moving to a different operator (which they believe is covering this area better). On the other hand, some areas may not be economically viable to cover from the operator's point of view due to a low population density. Other locations, for instance a sports stadium, may only require coverage at certain times of the day or even week.

Capacity is equally important. Without adequate capacity, users will not be able to enter the network even though there might be suitable coverage in the area. Providing the correct capacity in the correct location is essential to minimize the amount of infrastructure, while ensuring a high utilization of the hardware that has actually been installed. Based on the traffic distribution over the duration of a day reported in Lam *et al.* [32], an average transceiver utilization of only 35 % has been estimated. Improving this utilization is therefore of great interest.

A flexible architecture is essential to enable the wide variety of services and terminals expected in 4G to co-exist seamlessly in the same bandwidth. In addition, future upgrades and reconfigurations should require minimum effort and cost. The initial investment, the running costs and the cost of future upgrades are expected to be the three most important components in determining the total cost of the system.

17.5.2 Networks self-organizing technologies

Capacity, both in terms of bandwidth and hardware, will always be limited in a practical communication system. When a cell becomes congested, different actions are possible. The cell could borrow resources, bandwidth or hardware, from a neighboring cell. It could also make a service handover request to a neighbor in order to minimize the congestion. Thirdly, a service handover request could be made to a cell in a layer above or below in the hierarchical cell structure. Finally, the cell could try to reduce the path loss to the mobile terminal to minimise the impact of other cell interference. If neighboring cells are unable to 'assist' the congested cell, the options left for the cell are to degrade the users' service quality (if it is interference limited) or to try and influence the users' behavior. This can be achieved through service pricing strategies. The pricing scheme can be regarded as a protection mechanism for the network. Since it cannot create capacity, only utilise what it already has, it needs to force the users to adapt their behavioral pattern until the network is upgraded or there is more capacity available. Self-organizing technologies fall into one of these categories.

17.5.2.1 Bunching of base stations

In a micro- and picocellular environment there will be severe fluctuations in traffic demand, user mobility and traffic types. This highly complex environment will require advanced radio resource management (RRM) algorithms and it will be beneficial to have a central intelligent unit that can maximize the resource utilization. The bunch concept has been proposed as a means to deal with this issue. It involves a central unit (CU) that controls a set of remote antennas or base stations (which have very little intelligence). The central units will deal with all decisions on channel allocation, service request and handover. Algorithms for layers 1 and 2 (such as power control) may be controlled by the remote unit itself. The bunch concept can be viewed simply as a very advanced base station with a number of small antennas for remote sensing. The central unit will therefore have complete control over all the traffic in its coverage area and will be able to maximize the resource utilization for the current traffic. This provides opportunities for uplink diversity and avoids intercell handovers in its coverage area. The bunch approach will typically be deployed in city centres, large buildings or even a single building floor.

17.5.2.2 Dynamic charging

The operators need to encourage users to utilize the network more efficiently, something that can be achieved through a well thought-out pricing strategy. Pricing becomes particularly important for data services such as e-mail and file transfer as these may require considerable resources but may not be time-critical. A large portion of e-mails (which are not time-critical) could, for example, be sent during off-peak hours, hence improving the resource utilization. In this area two main approaches are used, *user utility method* and *maximum revenue method*.

User utility algorithms assume that the user associates a value to each service level that can be obtained. The service level is often referred to as the user's *utility function* and it can be interpreted as the amount the user is willing to pay for a given quality of service. It is assumed that the user acts 'selfishly', always trying to maximize their own utility (or service). The whole point with a pricing strategy is to enable the operator to predict how users will react to it, something which is not trivial. Current proposals do not try to determine the exact user's utility function, but rather to postulate a utility function which is based on the characteristics of the application or service. Two prime examples are voice and data services, which exhibit very different characteristics. Although speech applications are very sensitive to time delays, they are relatively insensitive to data errors. Similarly, although data services are relatively insensitive to time delays, they are very sensitive to data errors.

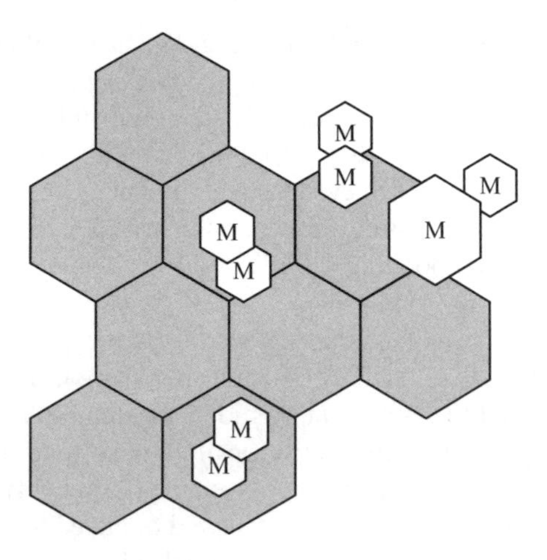

Figure 17.16 An intelligent-relaying overlay.

The *maximum revenue method* suggested in Bharghavan *et al.* [36] is based on letting the network optimize its revenue by allocating resources to users in a manner which is beneficial for the network. The two main principles are to maximize the resources allocated to static flows and to minimize the variance in resources allocated to mobile flows. These rules are based on the assumption that, whereas a static user's preference is for maximum data rate, mobile users are more concerned with the variance in service quality as they move from cell to cell. The actual revenue model is based on a 4-tuple, $<A, T, C_a, F>$, namely admission fee, termination credit, adaptation credit and revenue function. The network uses these parameters to optimize its revenue. Maximum revenue is calculated according to the following rules: (1) the flow pays an admission fee once it is granted its minimum requested resource allocation; (2) if a flow is prematurely terminated by the network, the latter pays the flow a termination credit; (3) the network pays the flow an adaptation credit if the resource allocation is changed during the transmission, regardless of whether the allocation is increased or reduced; (4) the flow pays a positive but decreasing marginal revenue for each extra unit granted by the network. The flow does not pay for resources allocated above its maximum requested resource allocation.

Intelligent relaying is a technique that can minimize the amount of planning and the number of base stations required in a cellular network. A network employing intelligent relaying includes mobiles that are capable of passing data directly to other mobiles, rather then directly to a base station, as for a conventional cellular network. In order to plan a network incorporating intelligent relaying, it is convenient to consider each mobile as a 'virtual cell', acting as a base station at its center. The coverage area of this virtual cell, as seen in Figure 17.16, can be varied according to the circumstances, as the mobile changes its transmit power and according to the mobility of the user. The mobile will set the radius of its virtual cell according to the number of other mobiles in the vicinity available to relay data; the size of the virtual cell will be minimized to improve frequency reuse.

Context awareness in 4G will, for example, be a scenario in which devices such as personal digital assistants (PDAs) should be able to communicate at short range with a number of other devices. These devices might be a fax machine, a computer, a mobile phone, a printer or a photocopier. Assuming all these devices have a low-power radio, then the PDA would be able to engage any of the devices. The task could be printing documents, downloading documents, faxing a message or uploading data for storage. To the user, the interdevice communication will be transparent. The user will only be concerned with the task they are performing, not the devices they are connected with. This sort of communication between electronic devices can be achieved through transmission of beacon signals, which provide relevant information such as device capability and identification. As manufacturers throughout the world are incorporating the Bluetooth technology into their products, this concept is about to materialize [41].

The context or situation awareness concept can also be exploited in cellular networks. In current cellular systems, base stations transmit information on their broadcast control channel, which can be used to implement this idea. Assuming this information includes $I = [id, x\_lat, y\_long, Tx]$ (base station identity, position in latitude and longitude, broadcast control channel transmit power), then this would enable the network to reconfigure its base station coverage areas when a base station is removed or added to the network.

Dynamic cell sizing is another emerging technique that has received considerable interest in the literature [40]. By dynamically adjusting the coverage areas of the cells, optimum network performance can be achieved under any traffic conditions. When a single cell is heavily occupied, whilst the surrounding cells are lightly loaded, it is possible with this scheme to reduce the cell size of the loaded cell and to increase the size of the surrounding cells. In this manner, more users can be accommodated in the centre cell. This can be implemented in such a way that the base station controls its attachment area by increasing or decreasing its beacon transmit power, hence increasing or decreasing the area in which mobiles will connect to the base station. Under congestion the cell will contract such as to limit its service area and reduce the inter-base-station interference. This will enable it to serve more users closer to the base station. In the limit, the cell is so small that the interference contribution of neighboring base stations can be ignored and maximum capacity is achieved. In light traffic conditions it will expand and hence improve the network coverage with cells overlapping the same area. The cell to which the user connects will therefore be a function not only of the path loss between the base station and the user but also of the beacon transmit power.

Intelligent handover (IH) techniques can be made more intelligent, also considering parameters such as resource utilization. A fast-moving mobile user who is currently served by a cell may run into problems in the next cell because it is fully congested. An intelligent handover algorithm would be able to recognize this problem and try to solve it by handing the user to the microcellular layer. The user will stay in this layer until the blocked cell has been passed, upon which it will be handed back to the macro layer. Similarly, if there is no coverage on the user's home network in the area in which they are moving, then the IH algorithm should seek to maintain the connection by handing over to a competitor's network or a private network offering capacity to external users. Performance evaluation of the above techniques can be found in Spilling *et al.* [42].

REFERENCES

[1] S.G. Glisic, *Adaptive WCDMA, Theory and Practice*. Wiley: Chichester, 2003.

[2] S.G. Glisic and P. Pirinen, Co-channel interference in digital cellular TDMA networks, in *Encyclopedia of Telecommunications*, ed. J. Proakis. Wiley: Chichester, 2003.

[3] N. Srivastava and S.S. Rappaport, Models for overlapping coverage areas in cellular and micro-cellular communication systems, in *IEEE GLOBECOM '91*, Phoenix, AZ, 2–5 December 1991, pp. 26.3.1–26.3.5.

[4] G.L. Choudhury and S.S. Rappaport, Cellular communication schemes using generalized fixed channel assignment and collision type request channels, *IEEE Trans. Vehicular Technol.*, vol. VT-31, 1982, pp. 53–65.

[5] B. Eklundh, Channel utilization and blocking probability in a cellular mobile telephone system with directed retry, *IEEE Trans. Commun.*, vol. COM-34, 1986, pp. 329–337.

[6] J. Karlsson and B. Eklundh, A cellular mobile telephone system with load sharing – an enhancement of directed retry, *IEEE Trans. Commun.*, vol. COM-37, 1989, pp. 530–535.

[7] T.-P. Chu and S.S. Rappaport, Generalized fixed channel assignment in microcellular communication systems, *IEEE Trans. Vehicular Technol.*, vol. 43, 1994, pp. 713–721.

[8] T.-P. Chu and S.S. Rappaport, Overlapping coverage and channel rearrangement in microcel-lular communication systems, *IEEE Proc. Comm.*, vol. 142, no. 5, 1995, pp. 323–332.

[9] S.W. Halpern, Reuse partitioning in cellular systems, in *IEEE Vehicular Technology Conf., VTC '83*, Toronto, 25–27 May 1983, pp. 322–327.

[10] K. Sallberg, B. Stavenow and B. Eklundh, Hybrid channel assignment and reuse partitioning in a cellular mobile telephone system, in *IEEE Vehicular Technology Conf. VTC '87*, Tampa, FL 1–3 June 1987, pp. 405–411.

[11] J. Zander and M. Frodigh, Capacity allocation and channel assignment in cellular radio systems using reuse partitioning, *Electron. Lett.*, vol. 28, no. 5, 1992, pp. 438–440.

[12] D. Hong and S.S. Rappaport, Traffic model and performance analysis for cellular mobile radio telephone systems with prioritized and non-prioritized handoff procedures, *IEEE Trans. Vehicular Technol.*, vol. VT-35, 1986, pp. 77–92.

[13] S.S. Rappaport, The multiple-call handoff problem in high-capacity cellular communications systems, *IEEE Trans. Vehicular Technol.*, vol. 40, 1991, pp. 546–557.

[14] S.S. Rappaport, Blocking, handoff and traffic performance for cellular communication systems with mixed platforms, *IEEE Proc. I*, vol. 140, no. 5, 1993, pp. 389–401.

[15] L.P.A. Robichaud, M. Boisvert and J. Robert, *Signal Flow Graphs and Applications*. Prentice-Hall: Englewood Cliffs, NJ, 1962.

[16] H. Jiang and S.S. Rappaport, Handoff analysis for CBWL schemes in cellular communications, College of Engineering and Applied Science, State University of New York, Stony Brook, CEAS Technical Report no. 683, 10 August 1993.

[17] T.-P. Chu and S. Rappaport, Overlapping coverage with reuse partitioning in cellular communication systems, *IEEE Trans. Vehicular Technol.*, vol. 46, no. 1, 1997, pp. 41–54.

[18] D. Grieco, The capacity achievable with a broadband CDMA microcell underlying to an existing cellular macrosystem, *IEEE J. Select. Areas Commun.*, vol. 12, no. 4, 1994, pp. 744–750.

[19] S. Glisic and B. Vucetic, *Spread Spectrum CDMA Systems for Wireless Communications*. Artech House: Norwood, MA, 1997.

[20] J.-S. Wu, J.-K. Chung and Y.-C. Yang, Performance study for a microcell hot spot embedded in CDMA macrocell systems, *IEEE Trans. Vehicular Technol.*, vol. 48, no. 1, 1999 pp. 47–59.

[21] A. Ganz, C.M. Krishna, D. Tang and Z.J. Haas, On optimal design of multitier wireless cellular systems, *IEEE Commun. Mag.*, vol. 35, 1997, pp. 88–94.

[22] R. Guérin, Channel occupancy time distribution in a cellular radio system, *IEEE Trans. Vehicular Technol.*, vol. VT-35, no. 3, 1987, pp. 627–635.

[23] P.B. Papazian, G.A. Hufford, R.J. Achate and R. Hoffman, Study of the local multipoint distribution service radio channel, *IEEE Trans. Broadcasting*, vol. 43, no. 2, 1997, pp. 175–184.

[24] V.I. Roman, Frequency reuse and system deployment in local multipoint distribution service, *IEEE Person. Commun.*, December 1999, pp. 20–28.

[25] M. Schwartz, Network management and control issues in multimedia wireless networks, *IEEE Person. Commun.*, June 1995, pp. 8–16.

[26] A.O. Mahajan, A.J. Dadej and K.V. Lever, Modelling and evaluation network formation functions in self-organising radio networks, in *Proc. IEEE Global Telecommunications Conf., GLOBECOM*, London, 1995, pp. 1507–1511.

[27] R.W Nettleton and G.R. Schloemar, Selforganizing channel assignment for wireless systems, *IEEE Commun. Mag.*, August 1997, pp. 46–51.

[28] M. Frullone, G. Rira, P. Grazioso and G. Fabciasecca, Advanced planning criteria for cellular systems, *IEEE Person. Commun.*, December 1996, pp. 10–15.

[29] R. Katz, Adaptation and mobility in wireless information systems, *IEEE Person. Commun.*, vol. 1, 1994, pp. 6–17.

[30] M. Flament, F. Gessler, F. Lagergen, O. Queseth, R. Stridh, M. Unbedaun, J. Wu and J. Zander, An approach to 4th Generation wireless infrastructures – scenarios and key research issues, *IEEE 49th Vehicular Technology Conf.*, Houston, TX, 16–20 May 1999, vol. 2, pp. 1742–1746.

[31] M. Flament, F. Gessler, F. Lagergen, O. Queseth, R. Stridh, M. Unbedaun, J. Wu and J. Zander, Telecom scenarios 2010 – a wireless infrastructure perspeclive. A PCC report is available at: www.s3,kth.se/radio/4GW/publk7Papers/ScenarioRcport.pdf

[32] D. Lam, D.C. Cox and J. Widom, Teletraffic modelling for personal communications services, *IEEE Commun. Mag.*, vol. 35, no. 2, 1997, pp. 79–87.

[33] E.K. Tameh and A.R. Nix, The use of measurement data to analyse the performance of rooftop diffraction and foliage loss algorithms in 3-D integrated urban/rural propagation model, in *Proc. IEEE 48th Vehicular Technology Conf.*, Ottawa, vol. 1, May 1998, pp. 303–307.

[34] First European initiative on re-configurable radio systems and networks, European Commission Green Paper version 1.0.

[35] A.G. Spilling, A.R. Nix, M.P. Fitton and C. Van Eijl, Adaptive networks for UMTS, in *Proc. 49th IEEE Vehicular Technology Conf.*, Houston, TX, vol. 1, 16–18 May 1999, pp. 556–560.

[36] V. Bharghavan, K.-W. Lee, S. Lu, S. Ha, J.-R. Li and D. Dwyer, The TIMELY adaptive resource management architecture, *IEEE Person. Commun.*, August 1998, pp. 20–31.

[37] T.J. Harrold and A.R. Nix, Intelligent relaying for future personal communication systems, in *IRK Colloquium on Capacity and Range Enhancement Techniques for Third Generation Mobile Communications and Beyond*, February 2000, IEEE Colloquium Digest no. 00/003, pp. 9/l–5.

[38] N. Bambos, Toward power-sensitive network architectures in wireless communications: concepts, issues and design aspects, *IEEE Person. Commun.*, June 1998, pp. 50–59.

[39] A.G. Spilling and A.R. Nix, aspects of self-organisation in cellular networks, in *9th IEEE Symp. Personal Indoor and Mobile Radio Communications*, Boston, MA, 1998, pp. 682–686.

[40] T. Togo, I. Yoshii and R. Kohro, Dynamic cell-size control according to geographical mobile distribution in a DS/CDMA cellular system, in *9th IEEE Symp. Personal Indoor and Mobile Radio Communications*, Boston, MA, 1998, pp. 677–681.

[41] J.C. Tiaartson, The Bluetooth radio system, *IEEE Person. Commun.*, vol. 7, no. 1, February 2000, pp. 28–36.

[42] A.G. Spilling, A.R. Nix, M.A. Beach and T.J. Harrold, Self-organisation in future mobile communications, *Electron. Commun. Engng. J.*, June 2000, pp. 133–147.

18
Network Management

18.1 THE SIMPLE NETWORK MANAGEMENT PROTOCOL

Back in 1988, the simple network management protocol (SNMP), was designed to provide an easily implemented, low-overhead foundation for multivendor network management of different network resources. The SNMP specification: (1) defines a protocol for exchanging information between one or more management systems and a number of agents; (2) provides a framework for formatting and storing management information; and (3) defines a number of general-purpose management information variables, or objects. The model of network management that is used for SNMP includes the following key elements: management station, management agent, management information base and network management protocol.

The *management station* will have, at least: (1) a set of management applications for data analysis, fault recovery, and so on; (2) an interface by which the network manager may monitor and control the network by communicating with the managed elements of the network; (3) a protocol by which the management station and managed entities exchange control and management information; (4) The management station maintains at least a summary of the management information maintained at each of the managed elements in the network in its own database of information. Only the last two elements are the subject of SNMP standardization.

The *management agent* is a piece of SNMP software in key platforms, such as hosts, bridges, routers and hubs, by which they may be managed from a management station. The management agent responds to requests for information from a management station, responds to requests for actions from the management station, and may asynchronously provide the management station with important but unsolicited information. In order to manage the resources in a network, these resources are represented as objects. Each object is essentially a data variable that represents one aspect of the managed system. The collection of objects is referred to as a management information base (MIB). The MIB functions as

Advanced Wireless Networks: 4G Technologies Savo G. Glisic
© 2006 John Wiley & Sons, Ltd.

a collection of access points at the agent for the management station; the agent software maintains the MIB. These objects are standardized across systems of a particular class (e.g. bridges all support the same management objects). A management station performs the monitoring function by retrieving the value of MIB objects. A management station can cause an action to take place at an agent or can change the configuration settings of an agent by modifying the value of specific variables. The management station and agents are linked by a network management protocol, which includes the following key capabilities:

(1) Get – used by the management station to retrieve the values of objects at the agent.

(2) Set – used by the management station to set the values of objects at the agent.

(3) Trap – used by an agent to notify the management station of significant events.

SNMP was designed to be an application-level protocol that is part of the TCP/IP protocol suite and typically operates over the user datagram protocol (UDP), although it may also operate over TCP. A manager process controls access to the central MIB at the management station and provides an interface to the network manager. The manager process achieves network management by using SNMP, which is implemented on top of UDP and IP.

Each agent must also implement SNMP, UDP and IP. In addition, there is an agent process that interprets the SNMP messages and controls remote access to the agent's MIB. For an agent device that supports other applications, such as FTP, TCP as well as UDP is required. From a management station, three types of SNMP messages are issued on behalf of a management application: *GetRequest*, *GetNextRequest* and *SetRequest.* The first two are variations of the *get* function. All three messages are acknowledged by the agent in the form of a *GetResponse* message, which is passed up to the management application. In addition, an agent may issue a *trap* message in response to an event that affects the MIB and the underlying managed resources. SNMP relies on UDP, which is a connectionless protocol, and SNMP is itself connectionless. No ongoing connections are maintained between a management station and its agents. Instead, each exchange is a separate transaction between a management station and an agent.

Trap-directed polling technique is used if a management station is responsible for a large number of agents, and if each agent maintains a large number of objects. In this case it becomes impractical for the management station to regularly poll all agents for all of their readable object data. Instead, at initialization time, and perhaps at infrequent intervals, such as once a day, a management station can poll all the agents it knows of for some key information, such as interface characteristics, and perhaps some baseline performance statistics, such as average number of packets sent and received over each interface over a given period of time. Once this baseline is established, the management station refrains from polling. Instead, each agent is responsible for notifying the management station of any unusual event. Examples are if the agent crashes and is rebooted, the failure of a link or an overload condition as defined by the packet load crossing some threshold. These events are communicated in SNMP messages known as traps.

Once a management station is alerted to an exception condition, it may choose to take some action. At this point, the management station may direct polls to the agent reporting the event and perhaps to some nearby agents in order to diagnose any problem and to gain more specific information about the exception condition. Having in mind that traps are communicated via UDP and are therefore delivered unreliably, a management station may wish to infrequently poll agents. All these options should be used in such a way as to provide a reliable network management with minimum overhead traffic generated in the network.

The use of SNMP requires that all agents, as well as management stations, must support UDP and IP. This limits direct management to such devices and excludes other devices, such as some bridges and modems, that do not support any part of the TCP/IP protocol suite. Furthermore, there may be numerous small systems (personal computers, workstations, programmable controllers), that do implement TCP/IP to support their applications, but for which it is not desirable to add the additional burden of SNMP, agent logic and MIB maintenance.

To accommodate devices that do not implement SNMP, the concept of proxy was developed. In this scheme an SNMP agent acts as a proxy for one or more other devices; that is, the SNMP agent acts on behalf of the proxied devices. Figure 18.1 indicates the type of protocol architecture that is often involved. The management station sends queries concerning a device to its proxy agent. The proxy agent converts each query into the management protocol that is used by the device. When a reply to a query is received by the agent, it passes that reply back to the management station. Similarly, if an event notification of some sort from the device is transmitted to the proxy, the proxy sends that on to the management station in the form of a trap message. The format of the SNMPv1 message is given in Figure 18.1. More data for versions $v2 - v3$ and any additional detail on SNMP protocols can be found in References [1–6].

In a traditional centralized network management scheme, one host in the configuration has the role of a network management station; there may possibly be one or two other management stations in a backup role. The remainder of the devices on the network contain agent software and an MIB, to allow monitoring and control from the management station. As networks grow in size and traffic load, such a centralized system is unworkable. Too much burden is placed on the management station, and there is too much traffic, with reports from every single agent having to wend their way across the entire network to headquarters. In such circumstances, a decentralized, distributed approach works better. In a decentralized network management scheme, there may be multiple top-level management stations, which might be referred to as management servers. Each such server might directly manage a portion of the total pool of agents. However, for many of the agents, the management server delegates responsibility to an intermediate manager. The intermediate manager plays the role of manager to monitor and control the agents under its responsibility. It also plays an agent role to provide information and accept control from a higher-level management server. This type of architecture spreads the processing burden and reduces total network traffic and will be discussed in more detail in the next section.

The *common management information protocol* (CMIP) [8, 9] was proposed as a standard to supercede SNMP. The standards are exhaustive in their specifications and include almost every possible detail for network management functions. CMIP was specified to work over the OSI [Chapter 1, 10] protocol stack. The drawbacks of CMIP are: (1) it is complex to use and implement; (2) it requires many resources to execute; (3) it has high overhead; and (4) few networks use the OSI protocol suites.

The *LAN man management protocol* (LMMP) [8] was developed as a network management solution for LANs. LMMP was built over the IEEE 802 logical link layer (LLC). Therefore, it is independent of the network layer protocol. Functionally, it is equivalent to common management information service over IEEE 802 LLC (CMOL). The advantages of LMMP are that it is easier to implement and protocol-independent. The disadvantages

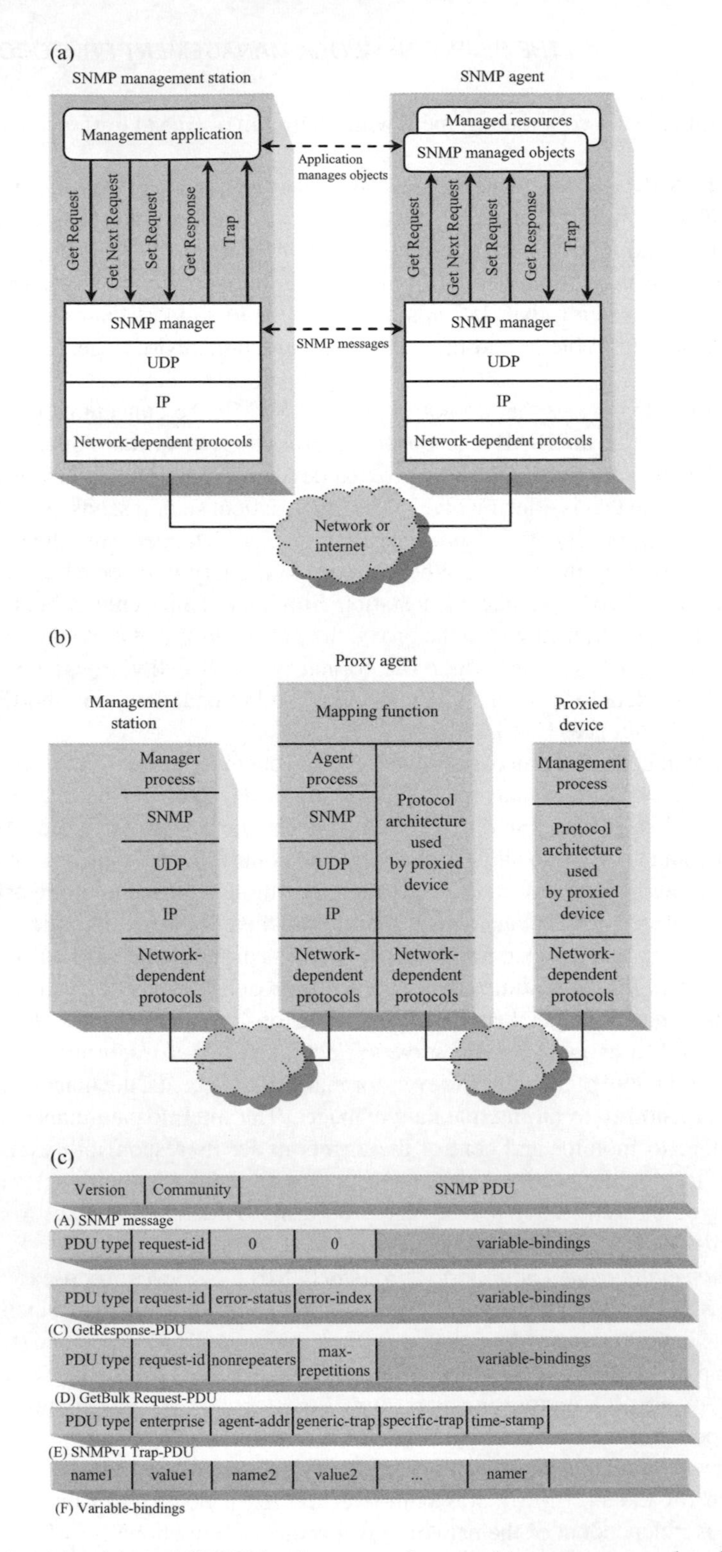

Figure 18.1 (a) Position of SNMP in the protocol stack; (b) proxy protocol architecture; (c) SNMP formats (reproduced by permission of IEEE [47]); (d) SNMP message and PDU fields (reproduced by permission of IEEE [47]).

(d)

Field	Description
Version	SNMP version; RFC 1157 is version 1
Community	A pairing of an SNMP agent with some arbitrary set of SNMP application entities. The name of the community functions as a password to authenticate the SNMP message
Request-id	Used to distinquish among outstanding request by providing each request with unique ID
Error-status	Used to indicate that an expection occured while processing a request. Values are: noError (0), tooBig (1), noSuchname (2), badValue (3), readOnly (4), genErr (5)
Error-index	When error-status is nonzero, error-index may provide additional information by indicating which variable in a list caused the exception. A variable is an instance of a managed object
Variable-bindings	A list of variable names and corresponding values. In some cases (e.g. GetRequest-PDU), the values are null
Enterprise	Type of object generating trap; based on sysObjectID
Agent-addr	Address of object generating trap
Generic-trap	Generic trap type. Values are: coldStart (0), warmStart (1), linkDown (2), linkUp (3), authenticationFailure (4), egpNeighborLoss (5), enterpriseSpecific (6)
Specific-trap	Specific trap code
Time-stamp	Time elapsed between the last (re)initialization of the network entity and the generation of the trap; contains the value of sysUpTime
Non-repeaters	Indicates how many listed variables are to return just one value each
Max-repetitions	Indicates number of values to be returned for each of the remaining variables

Figure 18.1 *(Continued.)*

are that LMMP messages cannot go beyond a LAN boundary because LMMP does not use any network layer facilities. LMMP agents are also not dynamically extensible.

The *telecommunications management network* (TMN) [11] was built over the OSI reference model, but it includes support for signaling system number 7, TCP/IP, ISDN, X.25 and 802.3 LAN-based networks. TMN has some advantages over the existing standards. For instance, TMN supports a larger number of protocols suites, incorporates features from existing management protocols, has wider domain of acceptance, and is 'future proof'. The disadvantages are that large amounts of processing are required, and agents are not extensible at run time.

18.2 DISTRIBUTED NETWORK MANAGEMENT

As already indicated in Section 18.1, centralized network management systems (NMSs) are clients to management agents (servers) residing permanently in each managed network element (NE). Although adequate for most practical management applications, the limitations of SNMP, such as the potential processing and traffic bottleneck at the NMS, have been recognized for many years.

The inefficiency can be reduced by distributing network management functions to a hierarchy of mid-level managers, each responsible for managing a portion of the entire network, as in SNMPv2. Subnetworks can be managed in parallel, reducing the traffic and processing burden on the highest level NMS. Typically, the distribution of network management functions and the organization of managers are fairly static. It has been observed that decentralizing network management functions may achieve several benefits [12–22].

(1) network traffic and processing load in the NMS can be both reduced by performing data processing closer to the NEs;

(2) scalability to large networks is improved;

(3) searches can be performed closer to the data, improving speed and efficiency; and

(4) distributed network management is inherently more robust without depending on continuous communications between the NMS and NEs.

In general in this context, the network management approaches will be classified as shown in Figure 18.2. In Figure 18.2(a), the *client–server* (CS) model represents the traditional SNMP paradigm where a centralized NMS polls a network of network elements. The communications between the NMS (*client*) and agents (*server*) is characterized by pairs of query–response messages for every interaction. Figure 18.2(b) represents a *hierarchical static* (HS) approach modeled as mid-level managers, each managing a separate subnetwork of network elements.

A two-level hierarchy is considered here, although multiple hierarchical layers may be possible. Each subnetwork is managed in a client–server manner and mid-level managers may communicate with a centralized high-level NMS as needed.

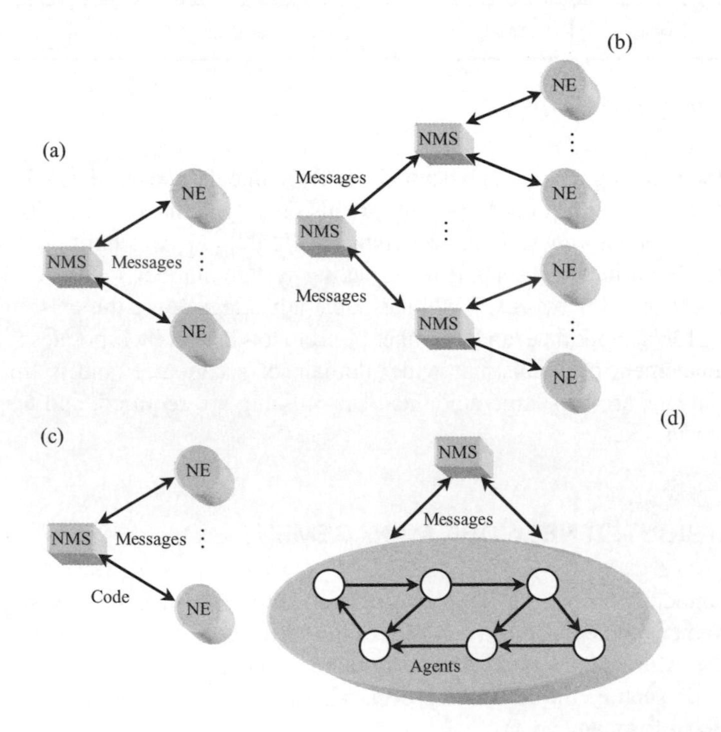

Figure 18.2 Centralized and distributed network management approaches. (a) Client–server; (b) hierarchical static; (c) weak mobility; and (d) strong mobility.

In the *weak mobility* (WM) approach, the NMS distributes code to specific NEs where the code is executed, as shown in Figure 18.2(c). After performing its specific task, the code will typically report results back to the NMS and expire at the NE. During execution, the code does not have a capability for autonomous migration to other NEs. In the *strong mobility* (SM) approach, the NMS dispatches one or more agents to carry out a specific task, as shown in Figure 18.2(d). Agents have the capability to autonomously travel (execution state and code) among different NEs to complete their tasks. The route may be predetermined or chosen dynamically, depending on the results at each NE.

18.3 MOBILE AGENT-BASED NETWORK MANAGEMENT

As already discussed in Chapters 1 and 16, 4G itself will become a competitive marketplace with a diversity of vendors, operators and customers. In this environment, users will be able to choose from a wide range of network services, which provide different bandwidth requirements and various QoS, and which will operate under different pricing schemes. Technical and market forces are driving the evolution of 4G toward the traded resource service architecture (TRSA) model, whereby network resources such as bandwidth, buffer space and computational power will be traded in the same manner as existing commodities.

In such complicated environments, users, whether buyers or sellers, need tools that facilitate expertise brokering activities such as buying or selling the right products, at the right price and at the right time. The brokering process will be guided by user preferences, which need to be processed according to some customized logic. The mobile agent technology seems to be a feasible option for dealing with these issues [23–34]. In other words, an *ad hoc* software sent across the network may allow network resources to be used more effectively, as they can be directly controlled at the application level.

In order to provide each customer with the opportunity of implementing his/her own policy, trading costs with service quality according to his/her own preferences and capacities, network service must be open, flexible and configurable upon demand. A first step toward the realization of a programmable network is that of using an agent-based approach, in order to obtain a faster time scale of service deployment. A major incentive for an agent-based approach is that policies can be implemented dynamically, allowing for a resource-state-based admission and reservation strategy. Agents are used to discover about resources available inside the network and claim resources on behalf of customers according to some 'figures of merit' [25], which represent tradeoffs between bandwidth claimed and loss risk incurred due to high utilization. Different customers may pay for resources in a different way, negotiating the costs for obtaining a certain 'figure of merit'. Agents are able to trigger adaptation of applications inside the network on behalf of customers. This allows for an immediate response to resource shortages, decreases the amount of useless data transported and reduces the signaling overhead. Mobile agents [26] provide the highest possible degree of flexibility and can carry application-specific knowledge into the network to locations where it is needed, following an approach similar to the one shown in some other programmable network projects discussed in Chapter 16 [23, 27].

As already indicated in Section 18.2, by adopting mobile code technology in network management it is possible to overcome some limitations of the centralized management approach. In general, such technology is based on the idea of reversing the logic according to which the data produced by network devices are periodically transferred to the central management station. Management applications can then be *moved* to the network devices, thus, performing (locally) some micromanagement operations, and reducing the workload for the management station and the traffic overhead in the network.

The *code on demand* paradigm, used in this concept, relies on a client which can dynamically download and execute code modules from a remote code server. The client has no need for unnecessary code installed, except for the runtime system allowing these mechanisms. Java applets are a very common example of such type of technology. The use of an approach based on code on demand increases the flexibility of the system, and maintains agents simple and *small* at the same time.

The *remote evaluation* paradigm, is based on the idea that a client sends a procedure to be executed on a remote host, where it is run up to the end; the results are, therefore, returned to the client host which sent the code. This paradigm allows the issue of the bandwidth waste that occurs in a centralized system when micromanagement operations are needed to be dealt with. In a traditional system, no matter which processing has to be performed on the data of the device, they have to be moved from the SNMP-agent to the NMS, where they are processed. Conversely, thanks to the mechanism of remote evaluation, the actions to be performed can be developed and sent to the remote device, where they will be executed without generating traffic in the network (except the initial one for sending the code to be executed). A typical example is the computation of the so-called *health functions* [24]. In general, by health function we mean an expression consisting of several elementary management variables. Each variable provides a particular measure concerning the device to be controlled. By using a technology based on remote evaluation, we can compute these functions directly on the devices, and only when necessary. The code that performs such computations will be sent by the NMS and dynamically executed. This approach allows obtaining what is called '*semantic compression of data*'. In general the manager of a network is not interested in the single value of an MIB variable, but in aggregate values containing higher-level '*information*'. We can, therefore, develop a system in which the manager writes its management functions (or they might be already available), and then invokes their remote execution on specific devices, when necessary.

The *mobile agent* adds more benefits to the ones that can be obtained with remote evaluation. In fact, in this case the action on a device is always expressly started by the NMS. Conversely, in the case of mobile agents, the ability to store the state during the movements allows applications to be developed in which the agent moves from one device to another, performing the management functions required. In addition, the agent can be delegated the task of deciding where and when to migrate according to its current state, thus reducing the interaction with the NMS, and making processing more distributed.

18.3.1 Mobile agent platform

Different platforms have been designed for the development and the management of mobile agents that give all the primitives needed for their creation, execution, communication, migration, etc. [24, 32, 35, 36]. The following agents are used:

The *browser agent* (BA) collects some MIB variables from a set of nodes specified by the user. Both the variables of the MIB-tree to be collected and the servers to be visited are selected through an appropriate dialog box of the application '*SNMP-monitoring*'. After being started, the agent reaches the first node to be visited, opens an SNMP local communication session, builds a PDU-request containing the MIB variables to be searched, waits for the reply from the SNMP daemon, and saves the data obtained in an internal structure. Then, if other nodes need to be visited, it reaches them, and repeats the procedure mentioned above. Otherwise, it returns to the platform from which it has been launched, where the results of the research are presented.

The *daemon agent* (DA) monitors a 'health function' defined by the user. For starting the agent, the function to be computed and the node of the network (where the computation has to be done) must be provided. Then this agent moves to the node in question, where it records the value of the function: if the value is higher than a fixed threshold (defined by the user), a notification message is sent to the server from which the agent has departed. The two agents described before directly interact with the SNMP daemon present in the different nodes.

The *messenger agent* (MA), during its migration through the nodes of the network, interacts with other agents for collecting specific information produced by them. During the configuration, the agents to be contacted and the servers where they have to be searched need to be selected and (if necessary) also the number of times the agent has to contact such agents. Thus, the MA performs operations at a higher abstraction level than the mere retrieval of MIB variables. In fact, since DAs can perform the computation of any function on the different nodes of the network, the messenger allows collection of such information, thus obtaining a general description about the state of the network.

The *verifier agent* does not perform an actual network management action. Its task is that of collecting important information, which might be useful for further operations of network management, from remote network devices to the starting host. It visits the nodes selected during the configuration, and computes a function whose purpose is the evaluation of the presence of a specific property of the system visited (for example, a software version, or the available space on disk, or the verification of some log files, etc.). The verifier agent then reports to the server, from which it departed, the list of the nodes that satisfy this property.

18.3.2 Mobile agents in multioperator networks

In 4G networks, many types of providers will be usually involved in order to complete the end-to-end service offering. Specifically, the SP (service provider) is responsible for the definition end delivery of the service characteristics, while the NP (network provider) provides the network infrastructure (i.e. high-speed network). In such an arrangement, the SP is essentially a customer of the NP, while the SP provides the service to its own customers or end-users (usually multiple customers with small to medium size). As a means of competition, many different NPs offer access to a remote CP (content provider) which provides multimedia services such as voice, data and video. Similarly, the SP is capable of accessing many different NPs to request service. The various network operators (providers) will then be competing to sell their network links to clients through a representative agent host, the *network access broker* (NAB)/the *internetwork access broker* (INAB).

A scenario that involves three competing NPs and two SPs is illustrated in Figure 18.3. The three networks are owned by three different operators, and each one of them is responsible for resource allocation and pricing strategies within its own environment.

Moreover, the three networks have some interconnection points with each other, therefore allowing traffic to flow among the different networks, as expected in an open marketplace.

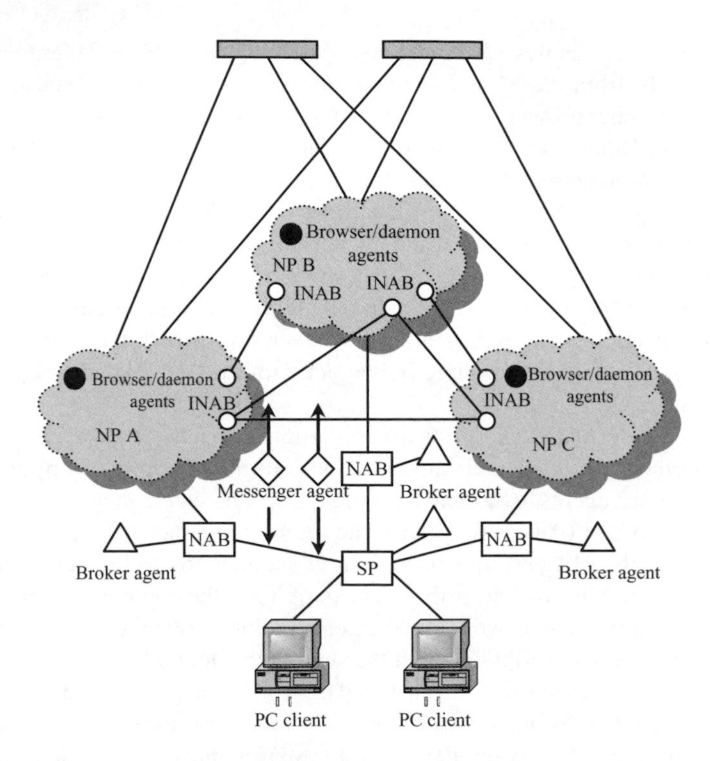

Figure 18.3 Network model for agent brokering environment.

The SP is informed periodically about the cost changes associated with each of those interconnection points. As a means of competition, all three networks offer the same access to the remote CP. The SP serving a particular client is responsible for identifying the best/optimal connection path (routing) per request of the customer.

In Chapter 7, QoS routing with multiconstraint has been shown to be an NP-hard problem [34]. For this reason, in this section we additionally consider a genetic algorithm (GA), which presents a good method to handle multiconstraint optimization problems and does not depend on the properties of the problem itself [37, 38]. The remaining issue is how the underlying agent architecture may interact with the genetic algorithm itself. It is assumed that an SP may host a particular kind of agent [named *broker agent* (BrkA)] which is in charge of identifying the optimal path to manage a specific connection request. The interaction between a BrkA and the algorithm may occur according to the following strategies:

(1) The BrkA is able to execute the algorithm in run-time upon the request from the PC client.

(2) The BrkA sends a request to a network node where the genetic algorithm can be executed. The optimal path is then sent back to the BrkA, which activates the setup procedure.

(3) A set of optimal paths for different pairs of PC client and CPs is stored in a database (eventually distributed), which is accessed from the BrkA to retrieve the more convenient path to satisfy the specific request. Once the connection is established, the genetic algorithm can be re-executed in order to identify a more convenient path, if the case.

If network performance and reliability change for some reason, monitoring agents distributed in the system will promptly react to pass the genetic algorithm the new data to recompute the new optimal paths.

18.3.3 Integration of routing algorithm and mobile agents

When network conditions change, SP always wants to find a good route for its customers in real time. That is, with certain QoS constraints, SP wants to find a cost-reasonable route for its costumers. The routing algorithm in SP needs to know:

(1) traffic from SP to CP;

(2) QoS requirements/constraints (i.e. time delay constraint from SP to CP);

(3) connectivity of nodes of NPs,

(4) bandwidth available for each link between nodes;

(5) cost for the traffic to pass through each link; and

(6) time delay for the traffic to pass through each link.

The use of mobile agents provides the complementary underlying structure in order to obtain this information in a distributed and efficient manner. In Section 18.3.1, we provided a brief description of how the agents can be used to deal with the collection of information [31, 33] about the state of the network and the monitoring of health functions that can be defined by the SP. The definition and creation of BAs and DAs serve the purpose of collecting specific variables from a set of nodes (e.g. NABs and INABs) specified by the SP. Then, MAs, during their migration through specific nodes of the network, interact with the other agents for collecting specific information produced by them and, therefore, obtain a general description about the state of the network. Moreover due to the mechanisms of remote evaluation and mobile agents described in the previous sections, the actions to be performed can be developed by the BrkA of a specific SP and sent to other remote devices where they will be executed on behalf of the BrkA, therefore limiting the generated traffic in the network and distributing the computational load and effort. Once the data are collected the routing algorithms can be executed.

A number of these algorithms were discussed in Chapter 6. For this reason in this section we additionally discuss only the GA. In general, GA is a stochastic algorithm searching process in the solution space by emulating biological selection and reproduction. It has been used to solve various optimization problems [38]. In this section, we discuss a genetic-based algorithm in order to address the problem described in the previous sections. In the following, we describe the different features and phases of the algorithm [35].

Encoding is used to map the search space of the specific problem into the space where GA can operate. In the literature, related work in using GA as an optimization tool to solve the famous Traveling Salesman Problem [37] has been reported. There are several encoding approaches mentioned in the literature, such as adjacency representation, ordinal representation and path representation. Some encoding approaches are not suitable for GA operation (crossover and mutation), and some other encoding approaches are inefficient in searching the solution space. In this section, *path representation approach* is used to naturally encode a route due to its easy implementation. That is, all the nodes that the route passes through are listed in sequence. In order to encode the route in a fixed data structure, 'zeroes' are filled into the empty space of the code.

For instance, in a scenario with 10 nodes in addition to the CP and SP, an array with 10 elements can be used to represent the route. Once the number of nodes that the route passes is less than 10, the corresponding element will be zero. For example [1 2 3 4 0 0 0 0 0 0] is used to represent the SP–1–2–3–4–CP route.

Population initialization is used to start GA computation. In the population initialization process, the number of nodes the route will pass through, which node will be in the route and the sequence of the nodes of the route can be randomly determined. However, there will be some solutions that may violate constraints of delay or interconnectivity. The 'penalty method' is used to deal with these constraints, as follows:

(1) for those links that do not exist, a very large delay value is assigned; and

(2) for those routes that violate the delay constraint, a penalty to their cost is added.

In the algorithm, the following expression is used to evaluate the weighted cost of those illegal routes:

$$C'(r) = C(r)[\alpha + D(r)/D_{\max}]$$

where $C'(r)$is the weighted cost of route r, $C(r)$ is the function that evaluates the total cost of the links that the route may pass through, $D(r)$ is the function that evaluates the time delay of the route, $D_{\max}$ is the upper bound of the time delay constraint, and α is the penalty constant (in the algorithm it is set to 2).

The *fitness* $F(r)$ of the solutions is proportional to their survivability during the algorithm computation. The selection operation, defined below, will use these values to keep 'good' solutions and discard the 'bad' solutions. For simplicity, we can use the cost of each route as the fitness of each solution. In order to prevent those solutions with very low fitness from being discarded by selection operation of GA at the first several computation loops of GA, the value of fitness of those 'bad' solutions is increased. In this way, some 'bad' solutions still have chances to survive at the beginning of the evolution of GA and the 'good' parts in them will have chances to transfer to new generations. For convenience, we normalize the fitness of solutions to [0,1] by the following expression:

$$F(r) = \frac{[C_{\max} - C(r)]C_{\min}}{(C_{\max} - C_{\min})C(r)}$$

where $C_{\max}$ and $C_{\min}$ and are maximum and minimum cost of routes in each generation of population, respectively.

Selection operation keeps 'good' solutions while discard 'bad' solutions at the same time. This operation tries to simulate 'natural selection' in real life. Those readers with background in communications theory will easily recognize the similarities between GA and the Viterbi algorithm (VA) used to approximate maximum likelihood (ML) demodulators.

Two selection operators are used in the algorithm. The first one is based on the 'fitness proportional model'. It is also called 'Monte Carlo selection'. The algorithm is as follows:

(1) add the fitness of all solutions;

(2) randomly generate a number between zero and the sum of fitness; this number is called the pointer of the Monte Carlo wheel; and

(3) add the fitness of each solution one by one until the value is greater than the pointer. Then the last solution is being selected.

Using this algorithm, the higher the fitness value, the bigger the chance of that solution being chosen. The second selection operator used in the algorithm is the 'best solution reservation'. The best solution in the population will always survive and several duplicated copies will be generated for mutation operation. In this way, the GA will always converge to a certain 'good' solution. Moreover, there will be a good chance to find a better solution on the base of the best solution of each generation.

Crossover operation enables any pair of solutions in the population to have a chance to exchange part of their solution information with others. Therefore, those 'good' parts from different solutions may be combined together to create a new, better solution. The two original solutions are the 'parents' of the new solutions generated. There are many crossover operators designed to solve different problems. In this section, we use traditional one-point crossover method. That is, we find a certain point of the array and swap the part before and after the cross point to generate two new solutions. However, because the number of nodes the route may pass through is not fixed, it is difficult to determine a fixed crossover point. In Papavassiliou *et al.* [35] the crossover point is determined by $\lfloor(A + B)/4\rfloor$, where A and B are the numbers of nodes that the two routes will pass through, respectively, and the operator '$\lfloor\,\rfloor$' is a rounding function, e.g. if it is before the crossover operation, there are two routes: [1 2 3 4 5 0 0 0 0 0] and [6 7 8 0 0 0 0 0 0 0]. According to this procedure ($\lfloor(A + B)/4\rfloor = 2$), after the crossover we get: [1 2 8 0 0 0 0 0 0 0] and [6 7 3 4 5 0 0 0 0 0]. Note that only part of the population will experience crossover operation; this rate is called the crossover rate.

The *mutation operation* randomly chooses a solution in the population and then slightly change it to generate a new solution. In this way, there is a chance to find better solution that cannot be found by only crossover operation. In the algorithm, four mutation operators are used:

(1) randomly delete a node from a route;

(2) randomly add a node into a route;

(3) randomly delete two nodes from a route; and

(4) randomly add two nodes into a route.

Only part of population will experience the mutation operation (this is characterized by mutation rate). In order to enhance the local searching ability, those copies of the best solution of each generation are all treated by mutation operator. In this way, the GA may find a better solution that is 'close' to the best solution of each generation.

In the *repair operation*, during crossover and mutation operation, illegal representation of route may be generated because duplicated elements (nodes) may appear in the same route. In the algorithm, those duplicated nodes that would bring high cost to the route are deleted. For example, there may be a route like [1 2 3 4 5 3 7 0 0 0], where node '3' is duplicated. We can evaluate the cost of strings (2 3 4), (5 3 7) and the delay of strings (2 3 4), (5 3 7). If the weighted cost of string (2 3 4) is less than (5 3 7), node 3 in (5 3 7) will be discarded.

The *computation efficiency* can be improved if after a minimum number of trails (MinTrails), when the algorithm has found a feasible solution and has made no more improvement for a specific period of time, the computation process is stopped. In the algorithm, the 'improvement' is presented by the average cost change rate of the best solution of certain generation. This change rate is evaluated by the following expression:

$$\frac{\Delta C}{\Delta t} = \Delta R(k) = \frac{C(i) - C(i+1)}{k - i}$$

where we assume the cost of the best solution of that generation changes at ith step and ChangeRate(k) $= \Delta R(k)$is the average change rate of cost at kth step ($k > i$). Once this value is less than a certain lower-bound MinChangeRate, GA computation may be stopped.

The *updating process* may be improved by assuming that the price of each link and the congestion of the network will change gradually. So when new traffic comes, the SP will recompute routes for its customers. During the dynamic operation of the system, in order to improve the efficiency of thc algorithm, the results of the last computation can be partly reused. One possibility is to mix certain 'training genes' into the initial population of the new route computation. However, this may lead to premature discards and prevent GA from finding better solutions. Instead of mixing the past solution into the initial solution of GA, they may be mixed into population after, for example, 70 % of MinTrails of GA loops. In this way, we can still take advantage of the results of the last computation and prevent premature discards at the same time. If the network conditions change smoothly, we can take advantage of the past best solution of last computation. If the network conditions change dramatically at a certain time and the optimal route may totally differ from the past solution, GA will not take advantage of the past best solution by mixing the past solution into the population during the GA computation. The flow chart in Figure 18.4 summarizes the operation of the algorithm. The integration of the mobile agents into the routing algorithm is presented in Figure 18.5.

As shown in the figure, BrkAs, MAs, BAs and DAs are used to migrate among different network elements to implement the proposed routing algorithm. Once the PC client needs a connection to the CP, an MA will be sent from PC client to SP containing information about the upper bound of setup time delay of the connection and the corresponding QoS

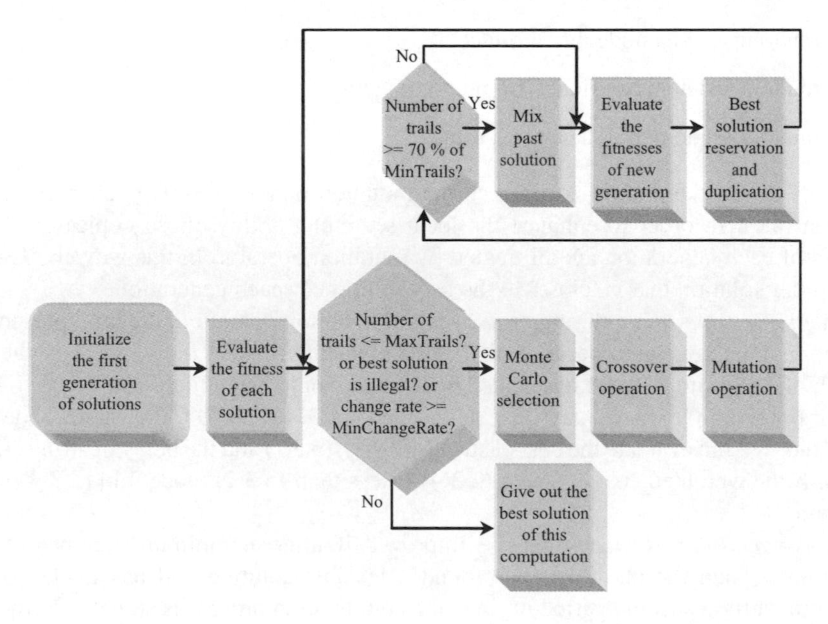

Figure 18.4 Flow chart of GA.

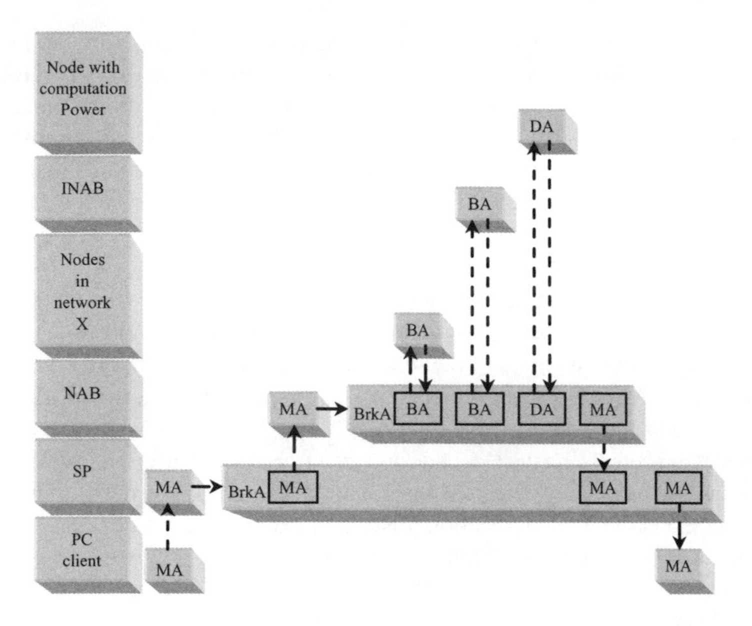

Figure 18.5 Agents used to implement routing algorithms based on GA.

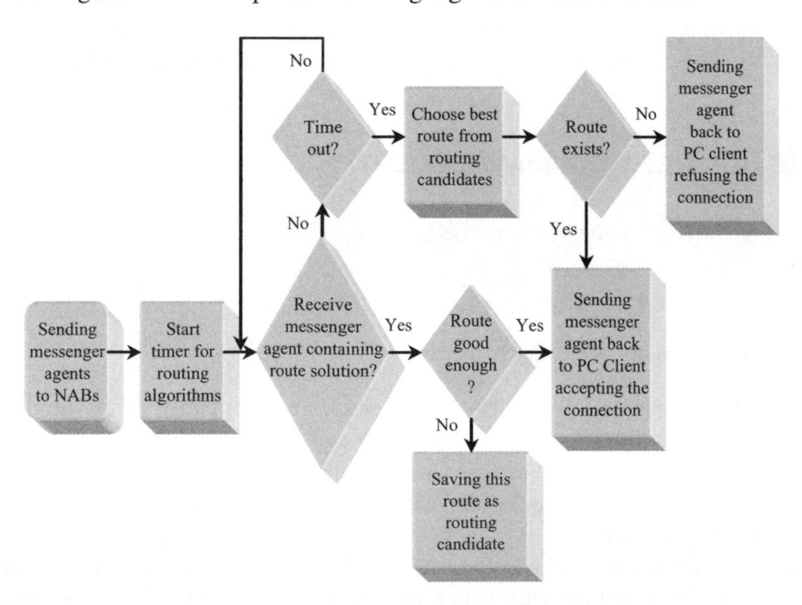

Figure 18.6 Algorithm for BrkAs in SP to choose a route for its client.

requirements. After receiving the MAs from PC client, the SP creates a BrkA to deal with this connection requirement.

This BrkA creates MAs containing source and destination information, as well as QoS requirements, and multicasts the agents to each NAB that it is connected with. Then the BrkA in SP waits for the agents from NABs to obtain the routing solution according to the scheme depicted in Figure 18.6.

As seen by the flow chart, in order to control the connection setup time, a timer is used to determine the deadline of the route searching procedure. If the BrkA receives an MA from NAB with satisfactory routing solution before the expiration of the timer, the route searching process stops and this solution is selected. Otherwise, when the timer expires the agent chooses the best route among the route candidates found until that time. Each NAB also creates a BrkAs to deal with the connection when it receives the MAs with the corresponding connection request from the SP. Then, three kinds of agents are used to implement the routing algorithm as follows.

A *browser* agent will be created and sent to nodes inside the individual private network that the NAB belongs to. These agents will collect resource information such as available bandwidth, delay of the link, price of the link, etc. In a similar way, the BrkA in each NAB will also send out BAs to INABs to see if it can take advantage of network resources from other NPs.

A *daemon* agent containing the GA code and resource-related information will be created after collecting the necessary resource information, to implement the routing algorithm described in detail in the previous section. Instead of executing the algorithm in each NAB, the BrkA sends DAs to the most suitable nodes inside its private network (e.g. nodes with enough computation resources such as CPU, memory, etc). In this way, we can balance the computation load among nodes in the private networks, if needed.

A *messenger agent* will be used to bring results back to the BrkA from DAs after the genetic-based route computation. This agent will be forwarded to the BrkA in the SP. Performance results for the algorithm can be found in Papavassiliou *et al.* [35].

18.4 *AD HOC* NETWORK MANAGEMENT

We start by identifying some of the properties of *ad hoc* networks that make them difficult to manage.

18.4.1 Heterogeneous environments

First of all, nodes of an *ad hoc* network can range in complexity from simple sensors located in the field to fully functional computers such as laptops. An implication of this diversity is that not all nodes will be able to contribute equally to the management task. For instance, it is likely that sensors and small personal digital assistant (PDA)-type devices will contribute minimally to the task of management, while more powerful machines will need to take on responsibilities such as collecting data before forwarding it to the management station, tracking other mobiles in the neighborhood as they move, etc. Thus, the management protocol needs to function in very heterogeneous environments.

18.4.2 Time varying topology

One mission of a network management protocol is to present the topology of the network to the network manager. In wireline networks, this is a very simple task because changes to the topology are very infrequent (e.g. a new node gets added, failure of a node, or addition/deletion of a subnetwork, etc.). In mobile networks, on the other hand, the topology changes very frequently because the nodes move about constantly. Thus, the management station needs to collect connectivity information from nodes periodically. An implication of this is an increased message overhead in collecting topology information.

18.4.3 Energy constraints

Most nodes in *ad hoc* networks run on batteries. Thus, we need to ensure that the network management overhead is kept to a minimum so that energy is conserved. Different energy is consumed by a radio when a packet is transmitted or received. In addition, the CPU expends energy in processing these packets. Thus, we need to reduce the number of packets transmitted/received/processed at each node. This requirement is contradictory to the need for topology update messages previously discussed.

18.4.4 Network partitioning

Energy constraints and mobility can result in the network becoming partitioned frequently. For instance, nodes may power themselves off to conserve energy resulting in partitions, or a node may move out of transmission range of other nodes. Similarly, a node may die when its battery runs out of power. In all these cases, the partitioned subnetworks need to continue running independently, and the management protocol must be robust enough to adapt. For instance, when the network gets partitioned, the management protocol entities must quickly learn that the partition has occurred and reconfigure the subnetwork(s) to function autonomously. Furthermore, when partitions merge, the management protocol must be capable of updating the network view without too much of an overhead.

18.4.5 Variation of signal quality

Signal quality can vary quite dramatically in wireless environments. Thus, fading and jamming may result in a link going down periodically. An effect of this is that the network topology from a graph theoretic point of view changes. However, the physical layout of the network may not change at all. The management protocol must be able to distinguish this case from the case when node moves cause topology changes, because in the case of changing link quality/connectivity, it may not be necessary to exchange topology update messages at all. In order to be able to do this, the management protocol entity (which resides at the application layer) must be able to query the physical layer. This obviously violates the layering concept of OSI, but it results in enormous savings.

18.4.6 Eavesdropping

Ad hoc networks are frequently set up in hostile environments (e.g. battlesite networks) and are therefore subject to eavesdropping, destruction and possibly penetration (i.e. a node is captured and used to snoop). Thus, the management protocol needs to incorporate encryption as well as sophisticated authentication procedures.

18.4.7 *Ad hoc* network management protocol functions

In this section we will discuss some main functions of *ad hoc* network management protocol (ANMP). *Data collection* is a necessary function in ANMP where the management entities collect node and network information from the network. SNMP specifies a large list of information items that can be collected from each node. However, this list does not include some crucial data items that are specific to the *ad hoc* environment like the status of the battery power (expected remaining lifetime), link quality, location (longitude and latitude), speed and direction, etc. All this information needs to be collected as (and when) it changes 'significantly'. For example, the location of a mobile node changes continuously, but there is little point in updating it constantly because the overhead in message complexity is high. The best solution is to update this information when some other aspect of the node changes. For instance, if the node's connectivity changes as a result of the motion, then we may need to update its location.

One problem that arises in *ad hoc* networks in relation to data collection is the message overhead. *Ad hoc* networks have limited bandwidth (whose quality is variable), and we must ensure that the process of management does not consume significant amounts of this resource. Since network management runs at the application layer, the simplest way to implement data collection (at the manager station) is to poll each node individually. This method, unfortunately, results in a very high message overhead. A more efficient method of data collection is to use a spanning tree rooted in the manager station.

Configuration/fault management is needed because nodes in *ad hoc* networks die, move or power themselves off to save energy. In all of these cases, the network topology changes, and the manager station needs to know the fate of these nodes. In cases when a node is unavailable for the reasons just stated, the manager records that fact in its database. However, even in the case when a manager knows that a node is dead, the entry for that node is not removed from the database because the node may be resurrected (e.g. we put in a new battery) or, keeping in mind that *ad hoc* networks are temporary, we may need a complete history of the network's behavior to effect a redesign of protocols, evaluate security breaches, etc.

New nodes may join a network periodically, and these nodes must be incorporated seamlessly into the network. A network may also be partitioned periodically. In this case, we need to ensure that each partition selects its own manager. However, when these partitions merge, one common manager needs to be chosen. Manager selection must be done based on the hardware and software capabilities of nodes and the available battery power. We may also have geographically coexisting but independent networks. An example is a battlefield, where a naval unit may be physically collocated with an infantry unit, each of which is using its own *ad hoc* network. In this case, the two networks may decide to be managed together or continue being managed independently but exchange information (such as link quality, presence of jamming, etc.) with each other as an aid to better deploy the network resources. It is also possible that a node may belong to two different networks and be managed by two (or more) managers. An example is a disaster relief model, where a police officer may remain on a police (secure) *ad hoc* network but simultaneously be connected (and managed) to an *ad hoc* network of medical relief teams. In 4G management, protocols for *ad hoc* networks must be able to operate in all these scenarios.

Security management deals with security threats [42–53, 55]. *Ad hoc* networks are very vulnerable to security threats because the nodes (e.g. the unmanned nodes) can easily be tampered with, and signals can be intercepted, jammed or faked. Current protocols, such as SNMPv3 [41, 56, 62, 63], do provide us with some mechanisms to guard against eavesdropping and replay attacks using secure unicast, which is not efficient for all incoming network architectures.

ANMP designing should address the issues raised above. It should be also compatible with SNMP. This is necessary because: (1) SNMP is a widely used management protocol today; and (2) *ad hoc* networks can be viewed as extensions of today's networks that are used to cover areas lacking a fixed infrastructure. In operation, it is quite likely that an *ad hoc* network would be connected to a local area wireline network (using a gateway). In such cases, ANMP manager should be designed to be viewed either as a peer of the SNMP manager (which is managing the wireline network) or as an agent of the SNMP manager. This flexibility is a major strength of ANMP. Obviously ANMP can operate in isolated *ad hoc* networks as well. In the next section, we provide an overview of the possible design choices in ANMP with the following constraints [39]:

(1) the PDU structure used is identical to SNMP's PDU structure;

(2) UDP is the transport protocol used for transmitting ANMP messages;

(3) lost data is not retransmitted by ANMP because information is periodically updated anyway. Furthermore, if the application sitting on top of ANMP wishes to obtain the lost information, it can request the ANMP manager to solicit that information again.

18.4.8 ANMP architecture

In order to have a protocol that is message-efficient, a hierarchical model for data collection is appropriate, since intermediate levels of the hierarchy can collect data (possibly producing a digest) before forwarding it to upper layers of the hierarchy. A problem, however, with utilizing a hierarchical approach in *ad hoc* networks is the cost of maintaining a hierarchy in the face of node mobility. A good tradeoff is to use a three-level hierarchical architecture for ANMP. Figure 18.7 illustrates this architecture. The lowest level of this architecture consists of individual managed nodes called agents. Several agents (that are close to one another) are grouped into clusters and are managed by a cluster head (the nodes with squares around them in the figure). The cluster heads in turn are managed by the network manager. It is important to emphasize that: (a) clustering for management (in ANMP) is very different from clustering for routing, as we discuss in Chapter 13; and (b) a manager is frequently more than one hop away from the cluster heads (Figure 18.7 is a logical view and not a physical view).

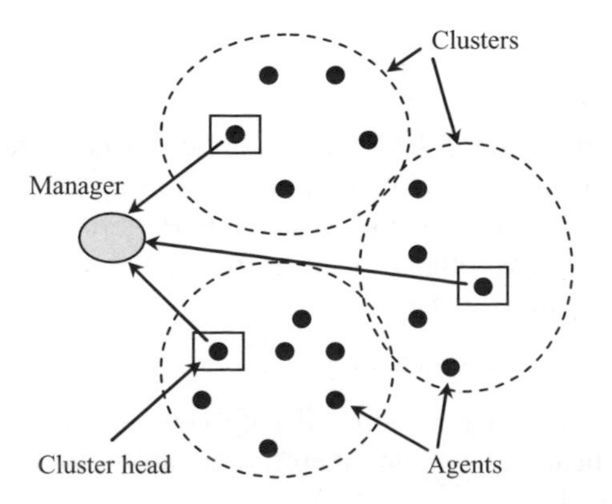

Figure 18.7 ANMP's hierarchical architecture.

The structure of the clusters is dynamic. Thus, as nodes move about, the number and composition of the clusters change. Similarly, the nodes serving as cluster heads also change over time. Different algorithms for forming and maintaining clusters will be discussed below. The clusters should have the following properties:

(1) The clusters are neither too large nor too small. The message overhead of collecting data within large clusters will be high. Likewise, if we have very small clusters, then there will be many cluster heads, all of which will be controlled by the manager. Thus, the message overhead in transactions between the cluster heads and the manager is high.

(2) The clusters are formed such that node mobility does not result in frequent recomputation of the clusters. This property is necessary if we are to reduce the message overhead of maintaining clusters.

(3) Sometimes nodes move out of one cluster and into another but are not incorporated into the new cluster immediately. This is because cluster maintenance algorithms only run periodically and not continuously. The effect of this is that some percentage of nodes may be unmanaged by cluster heads for short periods of time. This is not really a problem (except for the message overhead in data collection) because these nodes are still in communication with the overall manager, and they can be directly managed by the manager.

It is important to make a distinction between the use of clusters for management vs clusters for routing. Clustering in ANMP is used to logically divide the network into a three-level hierarchy in order to reduce the message overhead of network management. Since ANMP is an application-layer protocol, ANMP presupposes the existence of an underlying routing protocol. Thus, the manager node can always reach any of the nodes in the *ad hoc* network (so long as they both lie in the same partition) and can manage them directly. Clustering simply introduces intermediate nodes called cluster heads for the purpose of reducing the message overhead. Thus, clustering algorithms used for management serve a weaker and different objective when compared with clustering algorithms used for routing.

In cluster-based routing [40], neighboring nodes form clusters and select a cluster head. Cluster heads have the responsibility of routing packets sent to nodes outside the cluster. It is easy to see that clustering here serves a fundamental goal of maintaining routes. Finally, we note that if the underlying routing protocol is cluster-based, ANMP could simply use these clusters for management as well. However, if the routing protocol is not cluster-based [41], the two clustering algorithms describe here form clusters and rely on routing support to exchange control messages.

Graph-based clustering is described first as a basic concept. After that the maintenance algorithm that deals with node mobility after clusters have been formed will be discussed. A node in the graph represents a mobile host in the network. There is an undirected link between two nodes if they are within transmission range of each other. For the purpose of clustering we assume the following.

(1) each node in the network has a unique ID;

(2) each node in the network maintains a list of its one-hop neighbors (recall that ANMP runs at the application layer, and therefore it can obtain this information from the network or MAC layer);

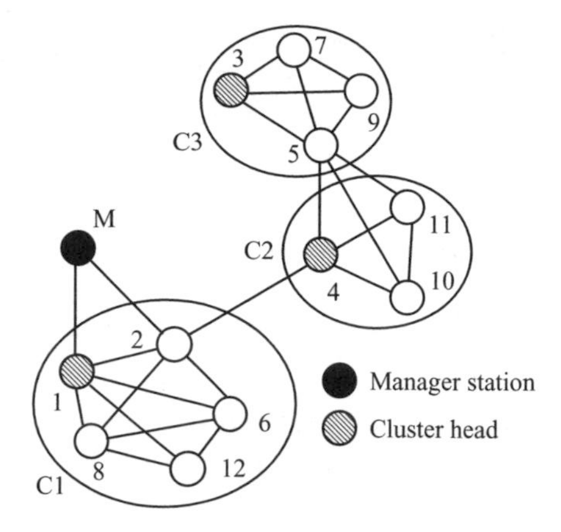

Figure 18.8 Clusters formed using graphical clustering.

(3) Messages sent by a node are received correctly by all its neighbors within a finite amount of time.

In the algorithm, the node with minimum ID among its neighbors (which have not joined any other cluster) forms a cluster and becomes the cluster head. Upon hearing from a cluster head, each node that has not yet joined any cluster declares that it has joined a cluster. If any node has been included in more than one cluster, then all but one cluster head prunes the node entry from their list when they do not hear any information from that node.

Figure 18.8 illustrates cluster formation in a simple *ad hoc* network. The node with the minimum ID forms the cluster and becomes the cluster head. Here node 1 is the minimum ID node among its neighbors; therefore, it forms a cluster C1. Node 4 does not initiate cluster formation because it is not the minimum ID node among its neighbors. When node 2 broadcasts a message that it has joined cluster C1, node 4 realizes that it is now the minimum ID node. Since the cluster formation runs in a distributed way, it is possible that, by the time node 4 receives the broadcast message from node 2, node 3 has already initiated cluster formation, and node 5 also sends a message that it has joined some cluster C3. Thus, when node 4 starts cluster formation, it only includes the remaining nodes among its neighbors and from C2.

Because the node with minimum ID considers only its one-hop neighbors while forming the cluster, the nodes in a cluster are one hop away from the cluster head and at most two hops away from any other cluster mate when the cluster is formed. The information maintained by each node after clusters are formed is:

- a neighbor list – a list of nodes that are one hop away;
- a cluster list – a list of all nodes that are its cluster mates;
- a ping counter – a counter that indicates the time steps elapsed since it last heard from its cluster head.

Cluster maintenance for graph-based clustering deals with the changes in the network topology. As a result, the clusters and cluster membership have to be updated. Changes in cluster membership are triggered when a node moves out of a cluster (and into another)

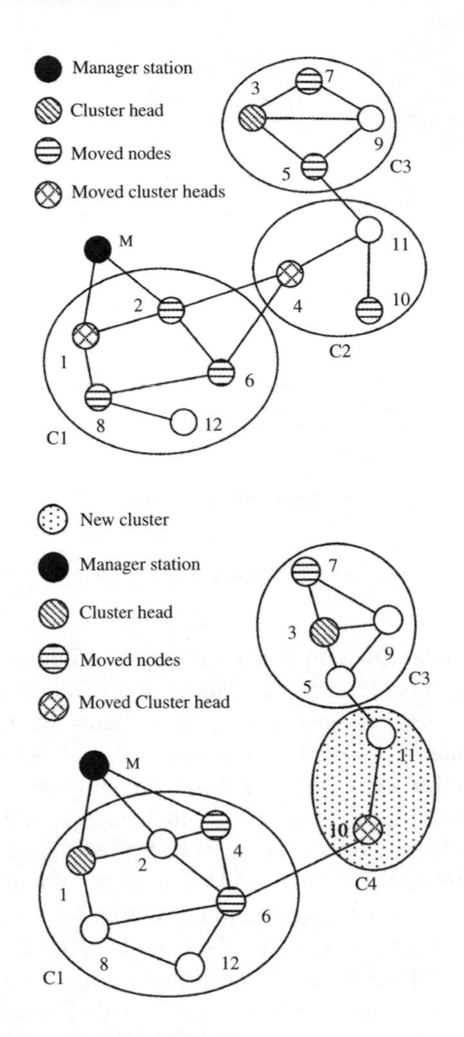

Figure 18.9 Effect of node mobility on clusters.

or when the cluster head itself moves out of a cluster (or, relatively speaking, cluster members move away from the cluster head). Figure 18.9 presents some illustrative cases. Figure 18.9(a) shows a situation where nodes move about but are still connected to at least one of their cluster mates (see Figure 18.8). Here, there is no need to recompute clusters. Figure 18.9(b) shows two scenarios: (1) when a node moves across the cluster boundary; and (2) when the cluster head gets disconnected. It can be seen that node 4 gets disconnected from all the members of its previous cluster. Since node 4 is two hops away from the cluster head of cluster C1, it sends a join request to node 2. On receiving such event from node 4, node 2 adds node 4 to its cluster list and broadcasts it to all the members. Meanwhile, nodes 10 and 11 discover that their cluster head has moved away, and they initiate cluster formation and form a new cluster C4. The previous example indicates an important property of the maintenance algorithm.

When new clusters are formed, all nodes in the cluster are one hop away from the cluster head. However, as nodes move about, we allow nodes to join clusters even if they are two hops away from the cluster head of an existing cluster. This flexibility drastically reduces the message overhead of the algorithm.

18.4.8.1 Performance of cluster maintenance algorithm

If a node is connected to at least one node in its cluster list, it is still considered to be part of that cluster. If a node detects that it has no links to any of its previous cluster mates, it either forms a new cluster by itself or joins another cluster. The algorithm should be able to differentiate between node movements within the cluster and movements across the cluster boundary. Four types of events that a node can detect as a result of mobility are identified. A node can detect: (1) a new neighbor, who is also a cluster mate; (2) that a previous neighbor and cluster mate has moved; (3) that it was previously directly connected to the cluster head but is no longer directly connected, or it was previously not directly connected but is now directly connected; (4) a new neighbor, who wants to join the cluster.

At every fixed interval, called a time step, each node locally creates a list of events that it has observed and sends it to its cluster head. The cluster heads collects these events and recomputes the cluster list. If there is any change in the cluster membership, the cluster head broadcasts the new list. Thus, whenever a node moves out of a cluster or joins a cluster, the message exchange is restricted to within that cluster. In order to minimize the number of cluster changes, a node is allowed to join a cluster if the cluster head is two hops away. The restriction of two hops (as opposed to, say, three hops) has been enforced to avoid the creation of big clusters.

In such a division of the network, the cluster head plays an important role. That is why a major event that can occur is the movement of the cluster head. To determine if the cluster head has moved away, a ping counter is used at each node. This counter gets incremented at every time step. If the counter at the cluster head crosses some threshold value, then the cluster head sends a ping message to all its cluster mates, indicating that it is still alive. If the cluster mates do not hear a ping after their ping counters cross a threshold, they assume that the cluster head is either dead or has moved out. Once a node detects such an event, it cannot be certain about its cluster list. New cluster(s) are formed with one-hop neighbors in the same way as the clusters were initially formed. It is easy to see that the frequency at which these pings are sent plays an important role in maintaining the clusters. If the ping frequency is small, then between consecutive pings, some nodes may be unmanaged by a cluster head (i.e. they do not belong to any cluster). Unfortunately, even though a high frequency of pings minimizes the number of nodes unmanaged by cluster heads, it results in a higher message overhead.

One method of reducing the number of ping messages while simultaneously keeping the fraction of nodes unmanaged by clusterheads small is to exploit information available at the MAC layer. The MAC layer in wireless networks periodically transmits a beacon to announce itself to its neighbors. Thus, the MAC layer keeps an updated list of its one-hop neighbors. If nodes transmit this list to their neighbors, changes in the cluster membership can be detected quickly. If the cluster head moves away, its departure will be noticed by the MAC layer of its one-hop neighbors. These nodes can act on this information to quickly reform clusters.

Another characteristic of *ad hoc* networks are partitions. If a subnetwork gets partitioned from the main network, it is treated as any other cluster because the protocol does not require

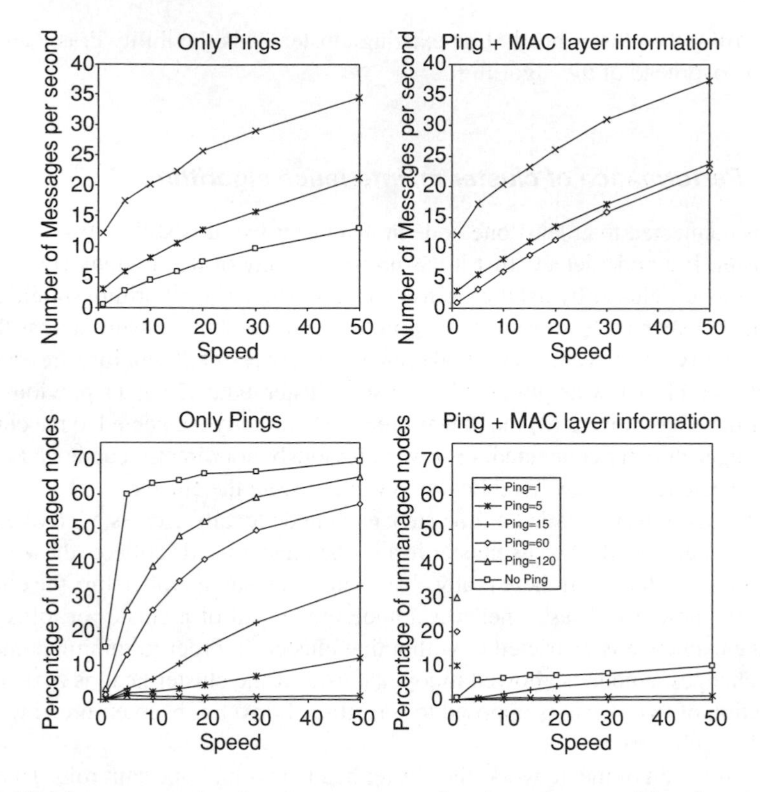

Figure 18.10 Message overhead and percentage of nodes unmanaged by cluster heads.

any information exchange between clusters. If a single node gets partitioned, it forms a cluster by itself. However, when it gets reconnected, it tries to join another cluster because the clusters of too small or too large a size are both inefficient from the point of view of network management.

To study the performance of this clustering algorithm, an ad hoc network was simulated in Chen *et al.* [39] in which the 30 nodes move randomly within a 1500×1500 unit box (for 60 nodes, the area of the playground was twice that, and so on). Each node selects a direction randomly uniformly in the range [0, 360] degrees. It moves along that direction for a length of time that is exponentially distributed. At the end of this time, it recomputes the direction and traveling time. When a node hits the edge of the box, it wraps around and emerges from the opposite edge. In the simulation, the same transmission power was used for all the nodes in the network and the *ad hoc* network was represented as an undirected graph with links between two nodes if they are within transmission range of each other. The average speed of nodes is 1–50 unit/s in different runs. The transmission range of a node is fixed at 450 units. Each reading is an average of 10 readings, and the simulation time is 1000 s. Finally, a packet loss probability of 10^{-3} in all simulations was assumed.

Figure 18.10 (upper part) shows the message overhead of the protocol for the case when only pings to detect topology changes (graph on the left) and when pings along with MAC-layer information (the graph on the right) are used. The x-axis indicates the speed in units per second, and the y-axis shows the number of messages exchanged per second for cluster maintenance. Each curve in the graph depicts the message overhead incurred for different ping intervals, which vary from no ping (equivalent to infinite time steps) down to one ping message every time step. It may be noted that message overhead increases almost linearly with increase in speed. The message overhead when we use MAC-layer information shows similar behavior, but there are more messages exchanged. This is obvious because of the fact that nodes do not wait for the ping message from the cluster head. The clustering algorithm is triggered as soon as the nodes detect cluster head movement.

The same figure (lower part) shows the percentage of nodes unmanaged by cluster heads. The plot on the left is for the case when only pings are used to detect topology changes, while the graph on the right is for the case when pings as well as MAC-layer information to detect topology changes are used. It can be seen from the graph that at higher speeds the percentage of nodes unmanaged by cluster heads goes up. This is mainly because of frequent disconnections and partitions generated in the network. Interestingly, the percentage of nodes unmanaged by cluster heads is as high as 50–70 % when we only use pings, but stays below 10 % when we use MAC-layer connectivity information as well.

REFERENCES

[1] www.snmp.com/

[2] W. Stallings, *SNMP, SNMPv2, and RMON: Practical Network Management*, 2nd edn., Addison-Wesley: Reading, MA, 1996.

[3] M. Rose, *The Simple Book: an Introduction to Network Management*, 3rd edn. Prentice Hall: Upper Saddle River, NJ, 1996.

[4] http://netman.cit.buffalo.edu/index.html

[5] J. Case, M. Fedor, M. Schoffstall and J. Davin, a simple network management protocol (SNMP), IETF, RFC 1157, 1990.

[6] *Information Technology – Open Systems Interconnection-Common Management Information Protocol Definition*, ITU-T, Geneva, ITU Recommendation X.711, 1992.

[7] W. Stallings, SNMP and SNMPv2: the infrastructure for network management, *IEEE Commun. Mag.,* vol. 36, 1998, pp. 37–43.

[8] A. Leinwand and K. Fang, *Network Management: A Practical Perspective.* Addison-Wesley: Reading, MA, 1993.

[9] W. Stallings, *SNMP, SNMPv2, and CMIP: the Practical Guide to Network Management Standards,* 1st edn. Addison-Wesley: Reading, MA, 1993.

[10] L. Raman, OSI systems and network management, *IEEE Commun. Mag.*, vol. 36, 1998, pp. 46–53.

[11] M. Kahani and H.W.P. Beadle, Decentralised approach for network management, *ACM SIGCOM Comput. Commun. Rev.*, 1997, pp. 36–47.

[12] G. Goldszmidt and Y. Yemini, Delegated agents for network management, *IEEE Commun. Mag.*, vol. 36, 1998, pp. 66–70.

[13] M. Kahani and H. Beadle, Decentralized approaches for network management, *Computer Commun. Rev.*, vol. 27, 1997, pp. 36–47.

[14] S. Vinoski, CORBA: integrating diverse applications within distributed heterogeneous environments, *IEEE Commun. Mag.*, vol. 35, 1997, pp. 46–55.

[15] M. Greenberg, J. Byington and D. Harper, Mobile agents and security, *IEEE Commun. Mag.*, vol. 36, 1998, pp. 76–85.

[16] V. Pham and A. Karmouch, Mobile software agents: an overview, *IEEE Commun. Mag.*, vol. 36, 1998, pp. 26–37.
[17] D. Milojicic, F. Douglis and R. Wheeler, *Mobility Processes, Computers and Agents*, D. Milojicic, F. Douglis and R. Wheeler (eds). ACM Press,1999.
[18] A. Fuggetta, G. Picco and G. Vigna, Understanding code mobility, *IEEE Trans. Software Engng*, vol. 24, 1998, pp. 342–361.
[19] M. Baldi and G. Picco, Evaluating the tradeoffs of mobile code design paradigms in network management applications, in *Proc. ICSE'98*, Kyoto, 19–25 April 1998, pp. 146–155.
[20] M. Siegl and G. Trausmuth, Hierarchical network management: A concept and its prototype in SNMPv2, *Comput. Networks ISDN Syst.*, vol. 28, 1996, pp. 441–452.
[21] J. Stamos and D. Gifford, Remote evaluation, *ACMTrans. Prog. Lang. and Syst.*, vol. 12, 1990, pp. 537–565.
[22] A. Bieszczad, B. Pagurek and T. White, Mobile agents for network management, *IEEE Commun. Surv.*, vol. 1, no. 4Q, 1998, pp. 2–9.
[23] A.T. Campbell, H.G. de Meer, M.E. Kounavis, K. Miki, J. Vicente and D. Villela, A survey of programmable networks, *ACM Comput. Commun. Rev.*, vol. 29, 1999, pp. 7–23.
[24] A. Puliafito, O. Tomarchio and L. Vita, MAP: design and implementation of a mobile agent platform, *J. Syst. Architect.*, vol. 46, 2000, pp. 256–267.
[25] H. De Meer, A. La Corte, A. Puliafito and O. Tomarchio, Programmable agents for flexible QoS management in IP networks, *IEEE J. Select. Areas Commun.*, vol. 18, 2000, pp. 145–162.
[26] M.R. Genesereth and S.P. Ketchpel, Software agents, *Commun. ACM*, vol. 37, 1994, pp. 48–53.
[27] D.S. Alexander, W.A. Arbaugh, A.D. Keromytis and J.M. Smith, The switchware active network architecture, *IEEE Network*, vol. 12, 1998, pp. 29–36.
[28] D. Tennenhouse, S.M. Smith W.D. Sincoskie, D.J. Weatheall and G.J. Minden, A survey of active networks research, *IEEE Commun. Mag.*, vol. 35, 1997, pp. 80–85.
[29] *IEEE Networks*, Special Issue on Programmable Networks, vol. 12, May/June 1998.
[30] *AdventNet Management APIs*. Available at: www.adventnet.com
[31] A. Bivens, L. Gao, M.F. Hulber and B. Szymanski, Agent-based network monitoring, in *Proc. Autonomous Agents99 Conf., Workshop 1, Agent Based High Performance Computing: Problem Solving Applications and Practical Deployment*, Seattle, WA, May 1999, pp. 41–53.
[32] M.S. Greenberg, J.C. Byington and T. Holding, Mobile agents and security, *IEEE Commun. Mag.*, vol. 7, 1998 pp. 76–85.
[33] A. Puliafito and O. Tomarchio, Using mobile agents to implement flexible network management strategies, *Comput. Commun. J.*, vol. 23, 2000, pp. 708–719.
[34] Z. Wang and J. Crowcroft, Quality-of-service routing for supporting multimedia applications, *IEEE J. Select. Areas Commun.*, vol. 14, 1996, pp. 1228–1234.
[35] S. Papavassiliou, A. Puliafito, O. Tomarchio and J. Ye, Mobile agent-based approach for efficient network management and resource allocation: framework and applications, *IEEE J. Select. Areas Commun.*, vol. 20, no. 4, 2002, pp. 858–872.
[36] R. Zahavi and T.J. Mowbray, *The Essential CORBA: Systems Integration Using Distributed Objects*. Wiley: New York, 1995.

[37] D.E. Goldberg, *Genetic Algorithms in Search, Optimization and Machine Learning*. Addison-Wesley: Reading, MA, 1989.

[38] Z. Michalewicz, *Genetic Algorithms + Data Structure = Evolution Programs*. Springer: New York, 1992.

[39] W. Chen, N. Jain and S. Singh, ANMP: *ad hoc* network management protocol, *IEEE J. Select. Areas Commun.*, vol. 17 no. 8, 1999, pp. 1506–1531.

[40] M. Gerla and T.J. Tzu-Chich, Multicluster, mobile, multimedia radio network, *ACM-Baltzer J. Wireless Networks,* vol. 1, 1995, pp. 255–266.

[41] D.B. Johnson and D.A. Maltz, Dynamic source routing in ad hoc wireless networks, in *Mobile Computing*, T. Imielinski and H. F. Korth (eds). Kluwer: Norwell, MA, 1996, pp. 153–181.

[42] K. Terplan, *Communication Network Management.* Prentice-Hall: Englewood Cliffs, NJ, 1989.

[43] Mobile MIB Taskforce, *System Components*, Available at: www.epilogue.com/mmtf/

[44] J. Pavon and J. Tomas, CORBA for network and service management in the TINA framework, *IEEE Trans. Commun.,* vol. 36, 1998, pp. 72–79.

[45] G. Goldszmidt and Y. Yemini, Delegated agents for network management, *IEEE Trans. Commun.,* vol. 36, 1998, pp. 66–70.

[46] V.S. Acosta, OSF/DME *(Distributed Management Environment).* Available at: www.frontiernet.net/vsa184/papers/osf_dme.htm, 1998.

[47] OMG. *CORBA Overview.* Available at: www.infosys.tuwien.ac.at/Research/Corba/OMG/arch2.htm#446864, 1998.

[48] U. Blumenthal and B. Wijnen, User-based security model (USM) for version 3 of the simple network management protocol (SNMPv3), RFC 2274, January 1998.

[49] B. Wijnen, R. Presuhn and K. McCloghrie, View-based access control for the simple network management protocol (SNMP), RFC 2275, January 1998.

[50] C.P. Pfleeger, *Security in Computing,* 2nd edn. Prentice-Hall: Englewood Cliffs, NJ, 1997.

[51] G.-H. Chiou and W.-T. Chen, Secure broadcasting using secure lock, *IEEE Trans. Software Engng*, vol. 15, 1989, pp. 929–933.

[52] L. Gong and N. Shacham, Multicast security and its extension to a mobile environment, *Wireless Networks,* vol. 1, August 1995, pp. 281–295.

[53] J. McLean, The specification an modeling of computer security, *IEEE Comput.* vol. 3, 1990, pp. 9–17.

[54] *Network Management Server.* Available at: http://netman.cit.buffalo.edu/index.html, 1998.

[55] U. Blumenthal and B. Wijnen, User-based security model (USM) for version 3 of the simple network management protocol (SNMPv3), RFC 2274, January 1998.

[56] B. Wijnen, R. Presuhn and K. McCloghrie, View-based access control for the simple network management protocol (SNMP), RFC 2275, January 1998.

[57] R. Lin Chunhung and M. Gerla, Adaptive clustering for mobile wireless networks, *IEEE J. Select. Areas Commun.*, vol. 15, 1997, pp. 1265–1274.

[58] SNMP Research Group, *The EMANATE run-time extensible agent system, SNMP Version 3 Charter*. Available at: www.snmp.com/emanateintro.html, 1998.

[59] *SNMP Version3 Charter.* Available at: www.ieft.org/html.charters/snmpv3-charter.html, 1998.

[60] D.J. Sidor, TMN Standards: Satisfying today's needs while preparing for tomorrow, *IEEE Commun. Mag.*, vol. 36, 1998, pp. 54–64.

[61] Mobile Management Task Force, *Mobile MIB Draft 2.0*, Available at: www.epilogue.com/mmtf/, 1998.

[62] C.E. Perkins and P. Bhagwat, Routing over multi-hop wireless network of mobile computers, in *Mobile Computing,* T. Imielinski and H. F. Korth (eds). Kluwer: Norwell, MA, 1996, pp. 183–205.

[63] C.-K. Toh, The Cambridge ad hoc mobile routing protocol, in *Wireless ATM and Ad Hoc Networks*, Chap. 9. Kluwer: Reading, MA, 1997.

19

Network Information Theory

Information theory has made a significant contribution to the development of communication theory and practice. This is especially true in the domain of physical layer including channel capacity issues, coding and modulation. Unfortunately, information theory has not yet made a comparable mark in the field of communication networks, which is today the center of activity and attention in most information technology areas. The principal reason for this is twofold. First, by focusing on the classical point-to-point, source–channel–destination model of communication, information theory has ignored the bursty nature of real sources. In advanced networks, source burstiness is the central phenomenon that underlies the process of resource sharing for communication. Secondly, by focusing on the asymptotic limits of the tradeoff between accuracy and rate of communication, information theory ignored the role of delay as a parameter that may affect this tradeoff. In networking, delay is a fundamental quantity, not only as a performance measure, but also as a parameter that may control and affect the fundamental limits of the rate–accuracy tradeoff.

A comprehensive survey of information theory contributions to the study of different network layers is given in Ephremides and Hajek [1]. The effective capacity of communications links including time-varying channel capacity, adaptive coding and modulation and queueing has already been discussed in Chapters 4 and 8. In this chapter we additionally discus the effective capacity of advanced cellular network, transport capacity of wireless networks and network coding.

19.1 EFFECTIVE CAPACITY OF ADVANCED CELLULAR NETWORKS

In this section we discuss effective capacity and capacity losses in an advanced CDMA network due to imperfections in the operation of the system components. In addition to the standard WCDMA technology both base stations and mobile units use antenna beam forming and self-steering to track the incoming (and transmitted) signal direction. By using high

Advanced Wireless Networks: 4G Technologies Savo G. Glisic
© 2006 John Wiley & Sons, Ltd.

directivity antennas and antenna pointer tracking, the level of multiple access interference (MAI) and the required transmitted power are reduced. In order to exploit the available propagation diversity signals arriving from different directions (azimuth ψ, elevation φ) and delay τ are combined in a three-dimensional (ψ, φ, τ) RAKE receiver. This is expected to significantly improve the system performance. In this section we present a systematic mathematical framework for capacity evaluation of such a CDMA network in the presence of implementation imperfections and a fading channel. The theory is general and some examples of a practical set of channel and system parameters are used as illustrations. As an example, it was shown that, in the case of voice applications and a two-dimensional (4 antennas$\times$4 multipaths) RAKE receiver, up to 90 % of the system capacity can be lost due to the system imperfections. Further elaboration of these results, including extensive numerical analysis based on the offered analytical framework, would provide enough background for understanding of possible evolution of advanced W-CDMA and MC CDMA towards the fourth generation of mobile cellular communication networks.

The physical layer of the third generation of mobile communication system (3G) is based on wideband CDMA. The CDMA capacity analysis is covered in a number of papers and recently has become a subject in standard textbooks [2–8].

The effect of more sophisticated receiver structures (like multiuser detectors MUD or joint detectors) on CDMA or hybrid systems capacity has been examined in [9–13]. The results in Hämäläinen *et al.* [9] show roughly twofold increase in capacity with MUD efficiency 65 % compared with conventional receivers. The effect of the fractional cell load on the coverage of the system is presented in [10]. The coverage of MUD-CDMA uplink was less affected by the variation in cell loading than in conventional systems. References [11] and [12] describe a CDMA system where joint data estimation is used with coherent receiver antenna diversity. This system can be used as hybrid multiple access scheme with TDMA and FDMA component. In Manji and Mandayam [13] significant capacity gains are reported when zero forcing multiuser detectors are used instead of conventional single-user receivers.

In most of the references it has been assumed that the service of interest is low rate speech. In next generation systems (4G), however, mixed services including high rate data have to be taken into account. This has been done in Huang and Bhargava [14], where the performance of integrated voice/data system is presented. It is also anticipated that 4G will be using adaptive antennas to further reduce the MAI. The effects of adaptive base station antenna arrays on CDMA capacity have been studied [6, 7]. The results show that significant capacity gains can be achieved with quite simple techniques.

One conventional way to improve cellular system capacity, used in 3G systems, is cell splitting, i.e. sub-dividing the coverage area of one base station to be covered by several base stations (smaller cells) [15]. Another simple and widely applied technique to reduce interference spatially in 3G is to divide cells into sectors, e.g. three 120° sectors. These sectors are covered by one or several directional antenna elements. The effects of sectorization to spectrum efficiency are studied in Chan [16]. The conclusion in Chan [16] is that sectorization reduces co-channel interference and improves the signal-to-noise ratio of the desired link at the given cluster size. However, at the same time the trunking efficiency is decreased [17]. Owing to the improved link quality, a tighter frequency reuse satisfies the performance criterion in comparison to the omnicellular case. Therefore, the net effect of sectorization is positive at least for large cells and high traffic densities.

By using M-element antenna arrays at the base station the spatial filtering effect can be further improved. The multiple beam adaptive array would not reduce the network trunking efficiency, unlike sectorization and cell splitting [18]. These adaptive or smart antenna techniques can be divided into switched-beam, phased array and pure adaptive antenna systems. Advanced adaptive systems are also called spatial division multiple access (SDMA) systems. Advanced SDMA systems maximize the gain towards the desired mobile user and minimize the gain towards interfering signals in real time.

According to Winters [19], by applying a four-element adaptive array at the TDMA, uplink frequencies can be reused in every cell (three-sector system) and sevenfold capacity increase is achieved. Correspondingly, a four-beam antenna leads to reuse of three or four and doubled capacity at small angular spread.

Some practical examples of the impact of the use of advanced antenna techniques on the existing cellular standards are described in References [20, 21]. In Petrus *et al.* [20] the reference system is AMPS and in Mogensen *et al.* [21] it is GSM. The analysis in Kudoh and Matsumoto [22] uses ideal and flat-top beamformers. The main lobe of the ideal beamformer is flat and there are no sidelobes, whereas the flat-top beamformer has a fixed sidelobe level. The ideal beamformer can be seen as a realization of the underloaded system, i.e. there are less interferers than there are elements in the array. The overloaded case is better modeled by the flat-top beamformer because all interferers cannot be nulled and the sidelobe level is increased. Performance results show that reuse factor of one is not feasible in AMPS, but reuses four and three can be achieved with uniform linear arrays (ULA) with five and eight elements, respectively. Paper [21] concentrates on the design and performance of the frequency hopping GSM network using conventional beamforming. Most of the results are based on the simulated and measured data of eight-element ULA. The simulated C/I improvement follows closely the theoretical gain at low azimuth spreads. In urban macrocells the C/I gain is reduced from the theoretical value 9 dB down to approximately 5.5–7.5 dB. The designed direction of arrival (DoA) algorithm is shown to be very robust to co-channel interference. The potential capacity enhancement is reported to be threefold in a $1/3$ reuse FH-GSM network for an array size of $M = 4$–6. A number of papers [23–30] present the analysis of capacity improvements using spatial filtering.

Solutions in 4G will go even beyond the above options and assume that both base station and the mobile unit are using beam forming and self-steering to continuously track transmitter–receiver direction (two side beam pointer tracking 2SBPT). Owing to user mobility and tracking imperfections there will be always tracking error that will result in lower received signal level, causing performance degradation. In this chapter we provide a general framework for performance analysis of a network using this technology. It is anticipated that this technology will be used in 4G systems.

19.1.1 4G cellular network system model

Although the general theory of MIMO system modeling is applicable for the system description, performance analysis will require more details and a slightly different approach will be used. This model will explicitly present signal parameters sensitive to implementation imperfections.

19.1.1.1 Transmitted signal

The complex envelope of the signal transmitted by user $k \in \{1, 2, \ldots, K\}$ in the nth symbol interval $t \in [nT, (n+1)T]$ is

$$s_k = A_k T_k(\psi, \varphi) e^{j\phi_{k0}} S_k^{(n)}(t - \tau_k) \tag{19.1}$$

where A_k is the transmitted signal amplitude of user k, $T_k(\psi, \varphi)$is the transmitting antenna gain pattern as a function of azimuth ψ and elevation angle φ, τ_k is the signal delay, ϕ_{k0} is the transmitted signal carrier phase, and $S_k^{(n)}(t)$ can be represented as

$$S_k^{(n)}(t) = S_k^{(n)} = S_k = S_{ik} + jS_{qk} = d_{ik}c_{ik} + jd_{qk}c_{qk} \tag{19.2}$$

In this equation d_{ik} and d_{qk} are two information bits in the I- and Q-channels, respectively, and $c_{ikm}^{(n)}$ and $c_{qkm}^{(n)}$ are the mth chips of the kth user PN codes in the I- and Q-channel respectively. Equations (19.58) and (19.63) are general and different combinations of the signal parameters over most of the signal formats of practical interest. In practical systems the codes will be a combination of a number of component codes [2].

19.1.1.2 Channel model

The channel impulse responses consist of discrete multipath components represented as

$$h_k^{(n)}(\psi, \varphi, t) = \sum_{l=1}^{L} h_{kl}^{(n)}(\psi, \varphi)\delta\left(t - \tau_{kl}^{(n)}\right) = \sum_{l=1}^{L} H_{kl}^{(n)}(\psi, \varphi)e^{j\phi_{kl}}\delta(t - \tau_{kl}^{(n)})$$

If antenna lobes are narrow we can use a discrete approximation of this functions in spatial domain too and implement 3D RAKE receive as follows:

$$\begin{aligned} h_k^{(n)}(\psi, \varphi, t) &= \sum_{l=1}^{L} h_{kl}^{(n)}\delta\left(\psi - \psi_{kl}, \varphi - \varphi_{kl}, t - \tau_{kl}^{(n)}\right) \\ &= \sum_{l=1}^{L} H_{kl}^{(n)}e^{j\phi_{kl}}\delta\left(\psi - \psi_{kl}, \varphi - \varphi_{kl}, t - \tau_{kl}^{(n)}\right) \end{aligned} \tag{19.3a}$$

$$h_{kl}^{(n)} = H_{kl}^{(n)}e^{j\phi_{kl}} \tag{19.3b}$$

where L is the overall number of spatial-delay multipath components of the channel. Each path is characterized by a specific angle of arrival $(\psi, \varphi)$and delay τ. Parameter $h_{kl}^{(n)}$ is the complex coefficient (gain) of the lth path of user k at symbol interval with index n, $\tau_{kl}^{(n)} \in [0, T_m)$ is the delay of the lth path component of user k in symbol interval n and $\delta(t)$ is the Dirac delta function. We assume that T_m is the delay spread of the channel. In what follows, indices n will be dropped whenever this does not produce any ambiguity. It is also assumed that $T_m < T$.

19.1.2 The received signal

The base station receiver block diagram is shown in Figure 19.1 [31]. The overall received signal at the base station site during N_b symbol intervals can be represented as

$$\begin{aligned} r(t) &= \mathrm{Re}\left\{e^{j\omega_0 t}\sum_{n=0}^{N_b-1}\sum_{k=1}^{K} s_k^{(n)}(t) * h_k^{(n)}(t)\right\} + \mathrm{Re}\left\{z(t)e^{j\omega_0 t}\right\} \\ &= \mathrm{Re}\left\{e^{j\omega_0 t}\sum_{n=0}^{N_b-1}\sum_{k}\sum_{l} a_{kl}S_k^{(n)}(t - nT - \tau_k - \tau_{kl})\right\} + \mathrm{Re}\left\{z(t)e^{j\omega_0 t}\right\} \end{aligned} \tag{19.4}$$

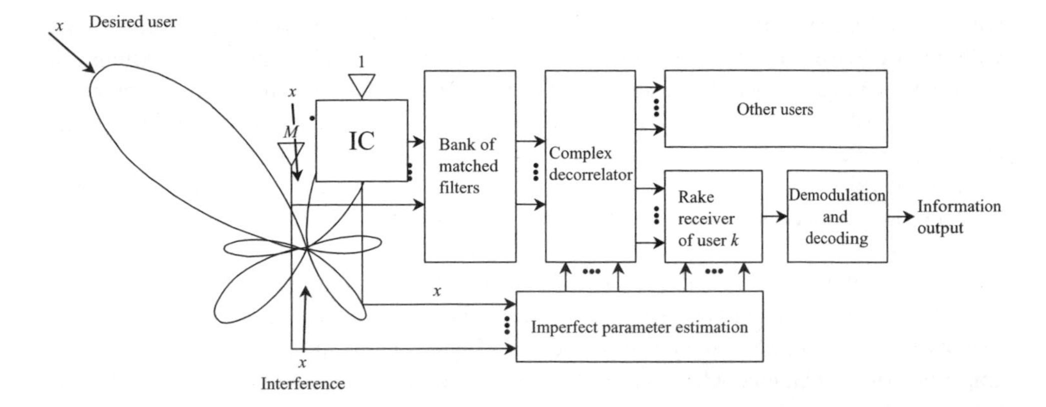

Figure 19.1 Receiver block diagram.

where $a_{kl} = A_k T_k(\psi, \varphi) H_{kl}^{(n)} \mathrm{e}^{j\Phi_{kl}} = A'_{kl} \mathrm{e}^{j\Phi_{kl}}$, $A_k T_k(\psi, \varphi) H_{kl}^{(n)} = A'_{kl}$, $\Phi_{kl} = \phi_0 + \phi_{k0} - \phi_{kl}$, ϕ_0 is the frequency down-conversion phase error and $z(t)$ is a complex zero mean additive white Gaussian noise process with two-sided power spectral density σ^2 and ω_0 is the carrier frequency. In general, in the sequel we will refer to A'_{kl} as received signal amplitude. This amplitude will be further modified by the receiver antenna gain pattern. The complex matched filter of user k with receiver antenna pattern $R_k(\psi, \varphi)$ will create two correlation functions for each path:

$$
\begin{aligned}
y_{ikl}^{(n)} &= \int_{nT+\tau_k+\tau_{kl}}^{(n+1)T+\tau_k+\tau_{kl}} r(t) R_k(\psi, \varphi) c_{ik}(t - nT - \tau_k + \tau_{kl}) \cos(\omega_0 t + \tilde{\Phi}_{kl}) \, \mathrm{d}t \qquad (19.5) \\
&= \sum_{k'} \sum_{l'} A_{k'l'} [d_{ik'} \rho_{ik'l',ikl} \cos \varepsilon_{k'l',kl} + d_{qk'} \rho_{qk'l',ikl} \sin \varepsilon_{k'l',kl}] = \sum_{k'} \sum_{l'} y_{ikl}(k'l')
\end{aligned}
$$

where $A_{k'l'} = A'_{k'l'} R_k(\psi, \varphi)$, parameter $\tilde{\Phi}_{kl}$ is the estimate of Φ_{kl} and

$$
\begin{aligned}
y_{ikl}(k'l') &= y_{iikl}(k'l') + y_{iqkl}(k'l') \\
&= A_{k'l'} [d_{ik'} \rho_{ik'l',ikl} \cos \varepsilon_{k'l',kl} + d_{qk'} \rho_{qk'l',ikl} \sin \varepsilon_{k'l',kl}]
\end{aligned}
$$

$$
\begin{aligned}
y_{qkl}^{(n)} &= \int_{nT+\tau_k+\tau_{kl}}^{(n+1)T+\tau_k+\tau_{kl}} r(t) R_k(\psi, \varphi) c_{qk}(t - nT - \tau_k + \tau_{kl}) \sin(\omega_0 t + \tilde{\Phi}_{kl}) \, \mathrm{d}t \\
&= \sum_{k'} \sum_{l'} A_{k'l'} [d_{qk'} \rho_{qk'l',qkl} \cos \varepsilon_{k'l',kl} - d_{ik'} \rho_{ik'l',qkl} \sin \varepsilon_{k'l',kl}] \\
&= \sum_{k'} \sum_{l'} y_{qkl}(k'l') \qquad (19.7a)
\end{aligned}
$$

$$
\begin{aligned}
y_{qkl}(k'l') &= y_{qqkl}(k'l') + y_{qikl}(k'l') \\
&= A_{k'l'} [d_{qk'} \rho_{qk'l',qkl} \cos \varepsilon_{k'l',kl} - d_{ik'} \rho_{ik'l',qkl} \sin \varepsilon_{k'l',kl}] \qquad (19.7b)
\end{aligned}
$$

where $\rho_{x,y}$ are cross-correlation functions between the corresponding code components x and y. Each of these components is defined with three indices. Parameter $\varepsilon_{a,b} = \Phi_a - \tilde{\Phi}_b$

where a and b are defined with two indices each. In order to receive the incoming signal without any losses, the receiving antenna should be directing (pointing) the maximum of its radiation diagram towards the angle of arrival of the incoming signal. In this segment we will use the following terminology. The direction of the signal arrival is characterized by the pointer $\boldsymbol{p}_a = (\psi_a, \varphi_a)$. In order to receive the maximum signal available, the receiver antenna pointer should be $\boldsymbol{p}_r = (\psi_r, \varphi_r) = (\psi_a + \pi, \varphi_a + \pi)$. Owing to mobility, the receiver will be tracking the incoming signal pointer and the pointer tracking error will be defined as $\Delta\boldsymbol{p} = \boldsymbol{p}_r - \boldsymbol{p}_a = (\psi_r - \psi_a, \varphi_r - \varphi_a) = (\Delta\psi, \Delta\varphi)$. Owing to this error the amplitude of the received signal will be reduced by ε_p with respect to the maximum value. These issues will be elaborated later. When necessary, we should make a distinction between the amplitude seen by the receiver *(asr)* for a given pointer tracking error $\Delta\boldsymbol{p}$ and the maximum available amplitude *(maa)* obtained when $\Delta\boldsymbol{p} = 0$. Let the vectors $\Im(\,)$ of MF output samples for the nth symbol interval be defined as

$$\boldsymbol{y}_{ik}^{(n)} = \Im^L\left(y_{ikl}^{(n)}\right) = \left(y_{ik1}^{(n)}, y_{ik2}^{(n)}, \ldots, y_{ikL}^{(n)}\right), \quad \in C^L \tag{19.8a}$$

$$\boldsymbol{y}_{qk}^{(n)} = \Im^L\left(y_{qkl}^{(n)}\right) = \left(y_{qk1}^{(n)}, y_{qk2}^{(n)}, \ldots, y_{qkL}^{(n)}\right), \quad \in C^L \tag{19.8b}$$

$$\mathrm{y}_{k}^{(n)} = \mathrm{y}_{ik}^{(n)} + j\mathrm{y}_{qk}^{(n)}; \quad \boldsymbol{y}^{(n)} = \Im^K\left(\boldsymbol{y}_k^{(n)}\right), \quad \in \mathrm{C}^{KL}; \quad \boldsymbol{y} = \Im^{N_b}(\boldsymbol{y}^{(n)}), \quad \in C^{N_bKL} \tag{19.8c}$$

19.1.2.1 CDMA system capacity

The starting point in the evaluation of CDMA system capacity is the parameter $Y_m = E_{bm}/N_0$, the received signal energy per symbol per overall noise density in a given reference receiver with index m. For the purpose of this analysis we can represent this parameter in general case as

$$Y_m = \frac{E_{bm}}{N_0} = \frac{ST}{I_{oc} + I_{oic} + I_{oin} + \eta_{th}} \tag{19.9}$$

where I_{oc}, I_{oic} and I_{oin} are power densities of intracell-, intercell- and overlay-type internetwork interference, respectively, and η_{th} is thermal noise power density. S is the overall received power of the useful signal and $T = 1/R_b$ is the information bit interval. Contributions of I_{oic} and I_{oin} to N_0 have been discussed in a number of papers [2]. In order to minimize repetition in our analysis, we will parameterize this contribution by introducing $\eta_0 = I_{oic} + I_{oin} + \eta_{th}$ and concentrate on the analysis of the intracell interference in the CDMA network based on advanced receivers using imperfect rake and MAI cancellation. A general block diagram of the receiver is shown in Figure 19.1. An extension of the analysis, to both intercell and internetwork interference, is straightforward.

19.1.3 Multipath channel: near–far effect and power control

We start with the rejection combiner, which will choose the first multipath signal component and reject (suppress) the others. In this case, Equation (19.9) for the I-channel becomes

$$\begin{aligned} Y_{ibm} &= \frac{\alpha_{iim1}(m1)S/R_b}{\{\alpha_{iqm1}(m1) + I(k'l')\}S/R_b + \eta_0} \\ &= \frac{\alpha_{iim1}(m1)}{\alpha_{iqm1}(m1) + I(k'l') + \eta_0 R_b/S} \end{aligned} \tag{19.10}$$

where $\alpha_x(z)$, (for $x = iim1, iqm1, im1$ and $z = m1, k'l'$) is the power coefficient defined as $\alpha_x(z) = E_\varepsilon\{|y_x(z)|^2\}/S$, S is normalized power level of the received signal and parameters $y_x(z)$ are in general defined by Equation (19.6).

$$I(k'l') = \sum_{k'=1}^{K} \sum_{\substack{l' = 1(k' \neq m) \\ l' = 2(k' = m)}}^{L} \alpha_{im1}(k'l')$$

is the equivalent MAI. $E_\varepsilon\{\ \}$ stands for averaging with respect to corresponding phases $\varepsilon_{a,b}$ defined by Equation (19.7). Based on this, we have

$$\begin{aligned} \alpha_{im1}(k'l') &= E_\varepsilon\left\{y_{im1}^2(k'l')\right\} \\ &= A_{k'l'}^2\rho_{im1}^2(k'l')/2 \Rightarrow A_{k'l'}^2/2 \Rightarrow \{A_{k'}H_{k'l'}T_{k'}(\psi,\varphi)R_m(\psi,\varphi)\}^2 \end{aligned} \quad (19.11)$$

where

$$\rho_{im1}^2(k'l') = \rho_{ik'l',im1}^2 + \rho_{qk'l',im1}^2, \ \rho^2 = E_\rho\left\{\rho_{ik'l',im1}^2 + \rho_{qk'l',im1}^2\right\}$$

and normalization $A_{k'l'}^2\rho^2/2 \Rightarrow A_{k'l'}^2/2$.

A similar equation can be obtained for the Q-channel too. It has been assumed that all 'interference per path' components are independent. In what follows we will simplify the notation by dropping all indices $im1$ so that $\alpha_{im1}(k'l') \Rightarrow \alpha_{kl}$. With no power control (*npc*) α_{kl} will depend only on the channel characteristics. In partial power control (*ppc*) only the first multipath component of the signal is measured and used in a power control (open or closed) loop. Full power control (*fpc*) will normalize all components of the received signal and rake power control (*rpc*) will normalize only those components combined in the rake receiver. The (*rpc*) control seems to be more feasible because these components are already available. These concepts for ideal operation are defined by the following equations

$$\begin{aligned} &npc \Rightarrow \alpha_{kl} = \alpha_{kl}, \quad \forall k, l; \quad ppc \Rightarrow \alpha_{k1} = 1, \quad \forall k \\ &fpc \Rightarrow \sum_{l=1}^{L} \alpha_{kl} = 1, \quad \forall k; \quad rpc \Rightarrow \sum_{l=1}^{L_0} \alpha_{kl} = 1, \quad \forall k \end{aligned} \quad (19.12)$$

where L_0 is the number of fingers in the rake receiver. The contemporary theory in this field does not recognize these options which causes a lot of misunderstanding and misconceptions in the interpretation of the power control problem in the CDMA network. Although *fpc* is not practically feasible, the analysis including *fpc* should provide the reference results for the comparison with other, less efficient options. Another problem in the interpretation of the results in the analysis of the power control imperfections is caused by the assumption that all users in the network have the same problem with power control. Hence, the imperfect power control is characterized with the same variance of the power control error. This is more than pessimistic assumption and yet it has been used very often in analyses published so far. The above discussion is based on the signal amplitude seen by the receiver (asr). System losses due to difference between the *asr* and *maa* will be discussed in the next section. If we now introduce matrix $\boldsymbol{\alpha}_m$ with coefficients $\|\alpha_{kl}\|$, $\forall k, l$ except for $\alpha_{m1} = 0$ and use notation $\mathbf{1}$ for vector of all ones, Equation (19.10) becomes

$$Y_{bm} = \frac{\alpha_{m1}}{\mathbf{1} \cdot \boldsymbol{\alpha}_m \cdot \mathbf{1}^T + \eta_0 R_b/S} \quad (19.13)$$

Compared with Equation (19.10), the index i is dropped in order to indicate that the same form of equation is valid for both, the I- and Q-channel defined by Equation (19.8).

19.1.4 Multipath channel: pointer tracking error, rake receiver and interference canceling

If L_0-fingers rake receiver ($L_0 \leq$ L) with combiner coefficients w_{mr} ($r = 1, 2, \ldots, L_0$) and interference canceller are used, the signal-to-noise ratio will become

$$Y_{bm} = \frac{r_m^{(L_0)}}{f(m, \boldsymbol{\alpha}, \mathbf{c}, \mathbf{r})K + \varsigma_0 \eta R_b/S} \tag{19.14}$$

where

$$\varsigma_0 = \sum_{r=1}^{L_0} w_{mr}^2 = \mathbf{w}_m \mathbf{w}_m^{\mathrm{T}}; \quad \mathbf{w}_m = (w_{m1}, w_{m2}, \ldots, w_{mn}) \tag{19.15}$$

is due to Gaussian noise processing in the rake receiver, and noise density η_0 becomes η due to additional signal processing. Also we have

$$\begin{aligned} f(m, \boldsymbol{\alpha}, \mathbf{c}, \mathbf{r}) &= \frac{1}{K} \sum_{\substack{k=1 \\ k \neq m}}^{K} \sum_{r=1}^{L_0} \sum_{l=1}^{L} w_{mr}^2 \bar{\alpha}_{kl}(1 - C_{kl}) + \frac{1}{K} \sum_{r=1}^{L_0} \sum_{\substack{l=1 \\ l \neq r}}^{L} w_{mr}^2 \bar{\alpha}_{ml}(1 - C_{ml}) \\ &= \frac{1}{K} \left\{ \mathbf{w}_m \left(\mathbf{1} \cdot \boldsymbol{\alpha}_{cmr} \cdot \mathbf{1}^{\mathrm{T}} \right) \mathbf{w}_m^{\mathrm{T}} \right\} \end{aligned} \tag{19.16}$$

with $\boldsymbol{\alpha}_{cmr}$ being a matrix of size $K \times L$ with coefficients $\|\bar{\alpha}_{kl}(1 - C_{kl})\|$ except for $\bar{\alpha}_{mr}(1 - C_{mr}) = 0$ and C_{ml} is efficiency of the canceller. The parameter $\bar{\alpha}_{kl} = E_{p(\phi,\psi)}\{\alpha_{kl}\}$ is the average value of the interfering signal power α_{kl} coming from direction (azimuth, elevation) $(\psi_{kl}, \varphi_{kl}) = (\psi_m + \Delta\psi_{kl}, \varphi_m + \Delta\varphi_{kl})$ with respect to the reference pointer $p(\psi_m, \varphi_m)$of the useful signal. Formally this can be represented as

$$\bar{\alpha}_{kl} = \iint_{\phi,\psi} \iint_{\Delta\psi,\Delta\varphi} \alpha_{kl}(\psi, \varphi) \mathrm{PDF}(\Delta\psi_{kl}, \Delta\varphi_{kl})\, \mathrm{d}\psi\, \mathrm{d}\varphi\, \mathrm{d}(\Delta\psi_{kl})\, \mathrm{d}(\Delta\varphi_{kl}) \tag{19.16a}$$

were $\mathrm{PDF}(\Delta\psi_{kl}, \Delta\varphi_{kl})$ is probability density function of the arguments. For the first insight into the system performance a uniform distribution of $\Delta\psi_{kl}$ and $\Delta\varphi_{kl}$ can be assumed along with a rectangular shape of $T(\psi, \varphi)$ and $R(\psi, \varphi)$ in the range $\psi \in \psi_0 + \Psi$, $\varphi \in \varphi_0 + \Phi$, so that evaluation of Equation (19.16a) becomes trivial. For the amplitudes A_{kl} seen by the receiver *asr*, the parameter $r_m^{(L_0)}$ in Equation (19.14), called rake receiver efficiency, is given as

$$r_m^{(L_0)} = \left(\sum_{r=1}^{L_0} w_{mr} \cos \varepsilon_{\theta mmr} \sqrt{\alpha_{mr}} \right)^2 = \left(\mathrm{W}_m \cdot \boldsymbol{\alpha}_{mm\surd} \right)^2 \tag{19.17}$$

with

$$\boldsymbol{\alpha}_{mm\surd} = \left(\cos \varepsilon_{\theta mm1} \sqrt{\alpha_{m1}}, \cos \varepsilon_{\theta mm2} \sqrt{\alpha_{m2}}, \ldots \right)^{\mathrm{T}}$$

Parameter $\varepsilon_{\theta mmr} = \theta_{mmr} - \hat{\theta}_{mmr}$ is the carrier phase synchronization error in receiver m for signal of user m in path r. We will drop index *mkl* whenever it does not result into any ambiguity. In the sequel we will use the following notation: $\alpha_{kl} = A_{kl}^2/2$, $\hat{A}_{mkl}$ is the estimation of A_{kl} by the receiver m, $\varepsilon_a = \Delta A_{mkl}/A_{kl} = (A_{kl} - \hat{A}_{mkl})/A_{kl}$ is the relative amplitude estimation error, $\varepsilon_m = BER =$ bit error rate that can be represented as $\overline{m\hat{m}} = 1 - 2BER$, and $\varepsilon_\theta =$ carrier phase estimation error.

For the equal gain combiner (EGC) the combiner coefficients are given as $w_{mr} = 1$. Having in mind the notation used so far in the sequel, we will drop index m for simplicity. For the maximal ratio combiner (MRC) the combiner coefficients are based on estimates as

$$\hat{w}_r = \frac{\cos\varepsilon_{\theta r}}{\cos\varepsilon_{\theta 1}}\cdot\frac{\hat{A}_r}{\hat{A}_l} \cong \frac{\left(1-\varepsilon_{\theta r}^2/2\right)}{\left(1-\varepsilon_{\theta l}^2/2\right)}\cdot\frac{A_r(1-\varepsilon_{ar})}{A_l(1-\varepsilon_{al})} \tag{19.18}$$

$$E\left\{\hat{w}_r\right\} = w_r\left(1-\sigma_{\theta r}^2\right)\left(1+\sigma_{\theta 1}^2\right)(1-\varepsilon_{ar})(1+\varepsilon_{a1}) \tag{19.19}$$

$$E\left\{\hat{w}_r^2\right\} = w_r^2\left(1-2\sigma_{\theta r}^2+3\sigma_{\theta r}^4\right)\left(1+2\sigma_{\theta l}^2-3\sigma_{\theta l}^4\right)(1-\varepsilon_{ar})^2(1+\varepsilon_{a1})^2$$

Averaging Equation (19.17) gives for EGC

$$\begin{aligned} E\left\{r^{(L_0)}\right\} &= E\left\{\left(\sum_{r=1}^{L_0}\cos\varepsilon_{\theta r}\sqrt{\alpha_r}\right)^2\right\} = E\left\{\left(\sum_{r=1}^{L_0}\left(1-\varepsilon_{\theta r}^2/2\right)\sqrt{\alpha_r}\right)^2\right\} \\ &= \sum_r\sum_{\substack{l\\ l\neq r}}\left(1-\sigma_{\theta r}^2\right)\left(1-\sigma_{\theta l}^2\right)\sqrt{\alpha_r\alpha_l} + \sum_r \alpha_r\left(1-2\sigma_{\theta r}^2+3\sigma_{\theta r}^4\right) \end{aligned} \tag{19.20}$$

For MRC the same relation becomes

$$\begin{aligned} E\left\{r^{(L_0)}\right\} &= E\left\{\left(\sum_{r=1}^{L_0}\frac{\alpha_r}{\sqrt{\alpha_1}}\frac{\left(1-\varepsilon_{\theta r}^2/2\right)^2}{\left(1-\varepsilon_{\theta l}^2/2\right)}\frac{(1-\varepsilon_{ar})}{(1-\varepsilon_{al})}\right)^2\right\} \\ &= E\left\{\sum_{r=1}^{L_0}\frac{\alpha_r^2}{\alpha_l}\frac{\left(1-\varepsilon_{\theta r}^2/2\right)^4}{\left(1-\varepsilon_{\theta l}^2/2\right)^2}\frac{(1-\varepsilon_{ar})^2}{(1-\varepsilon_{a1})^2}\right\} \\ &\quad + \sum_r\sum_{\substack{l\\ l\neq r}}\frac{\alpha_r\alpha_l}{\alpha_l}\left(1-2\sigma_{\theta r}^2+3\sigma_{\theta r}^4\right)\left(1-2\sigma_{\theta l}^2+3\sigma_{\theta l}^4\right) \\ &\quad \times\left(1+2\sigma_{\theta 1}^2-3\sigma_{\theta 1}^4\right)(1-\varepsilon_{ar})(1-\varepsilon_{al})(1+\varepsilon_{al})^2 \end{aligned} \tag{19.21}$$

In order to evaluate the first term we use limits. For the upper limit we have $\varepsilon_{\theta r}^2 \Rightarrow \varepsilon_{\theta l}^2$. By using this we have

$$\left(1-\varepsilon_{\theta r}^2/2\right)^4/\left(1-\varepsilon_{\theta l}^2/2\right)^2 \Rightarrow \left(1-\varepsilon_{\theta l}^2/2\right)^2$$

and the first term becomes

$$\sum_{r=l}^{L_0}\frac{\alpha_r^2}{\alpha_l}\left(1-2\sigma_{\theta 1}^2+3\sigma_{\theta 1}^4\right)(1-\varepsilon_{ar})^2/(1-\varepsilon_{a1})^2$$

For the lower limit we use $\varepsilon_{\theta l}^2 \Rightarrow \varepsilon_{\theta r}^2$ and the first term becomes

$$\sum_{r=l}^{L_0}\frac{\alpha_r^2}{\alpha_l}\left(1-2\sigma_{\theta r}^2+3\sigma_{\theta r}^4\right)(1-\varepsilon_{ar})^2/(1-\varepsilon_{al})^2$$

For a signal with the I- and Q-component the parameter $\cos\varepsilon_{\theta r}$ should be replaced by $\cos\varepsilon_{\theta r} \Rightarrow \cos\varepsilon_{\theta r} + m\rho\sin\varepsilon_{\theta r}$, where m is the information in the interfering channel (I or Q), and ρ is the cross-correlation between the codes used in the I- and Q-channel. For small tracking errors this term can be replaced as $\cos\varepsilon_{\theta r} + m\rho\sin\varepsilon_{\theta r} \approx 1 + m\rho\varepsilon - \varepsilon^2/2$, where the notation is further simplified by dropping the subscript $(\,)_{\theta r}$. Similar expressions can be derived for the complex signal format. In the above discussion the signal amplitude seen by the receiver *(asr)* is used for A_r. In the presence of pointer estimation error this is related to the (maa) as $A_r \Rightarrow A_r - \varepsilon_p$. So, to account for the losses due to ε_p, parameter α_r in the above equation should be replaced by $\alpha_r \Rightarrow E(A_r - \varepsilon_p)^2 = \alpha_r - 2\bar{\varepsilon}_p\sqrt{\alpha_r} + \sigma_p^2$ where $\bar{\varepsilon}_p$ and $\sigma_p^2 = E(\varepsilon_p^2)$ are the mean and variance of the pointer tracking error. The power control will compensate for the pointer tracking losses by increasing the signal power by the amount equal to the losses. This means that the level of interference will be increased which can be taken into account in our model by modifying the parameters α_{kl} as follows: $\alpha_{kl} \Rightarrow \alpha_{kl} + 2\bar{\varepsilon}_p\sqrt{\alpha_{kl}} - \sigma_{pkl}^2$

19.1.5 Interference canceler modeling: nonlinear multiuser detectors

For the system performance evaluation a model for the canceller efficiency is needed. Linear multiuser structure might not be of much interest in the next generation of the mobile communication systems where the use of long codes will be attractive. An alternative approach is nonlinear (multistage) multiuser detection, that would include channel estimation parameters too. This would be based on interference estimation and cancellation schemes (OKI standard-IS-665/ITU recommendation M.1073 or UMTS defined by ETSI)

In general if the estimates of Equation (19.8) are denoted $\hat{\mathbf{y}}_i$ and $\hat{\mathbf{y}}_q$, then the residual interference after cancellation can be expressed as

$$\Delta\mathbf{y}_i = \mathbf{y}_i - \hat{\mathbf{y}}_i, \quad \Delta\mathbf{y}_q = \mathbf{y}_q - \hat{\mathbf{y}}_q, \quad \Delta\mathbf{y} = \Delta\mathbf{y}_i + j\Delta\mathbf{y}_q = \mathrm{Vec}\left\{\Delta y_\varsigma\right\} \tag{19.22}$$

where index $\varsigma \Rightarrow k, l$ spans all combinations of k and l. By using Equation (19.22), each component $\alpha_{kl}(1 - C_{kl})$ in Equation (19.16) can be obtained as a corresponding entry of $\mathrm{Vec}\{|\Delta y_\varsigma|^2\}$.

To further elaborate these components we will use a simplified notation and analysis. After frequency downconversion and despreading, the signal from user k, received through path l at the receiver m, would have the form

$$\hat{S}_{mkl} = \hat{A}_{mkl}\hat{m}_k\cos\hat{\theta}_{mkl} = (A_{mkl} + \Delta A_{mkl})\hat{m}_k\cos(\theta_{mkl} + \varepsilon_{\theta mkl}) \tag{19.23}$$

for a single signal component and

$$\begin{aligned} \hat{S}_{mkl}^{i} &= \hat{A}_{mkl}\hat{m}_{ki}\cos\theta_{mkl} + \hat{A}_{mkl}\hat{m}_{kq}\sin\theta_{mkl}; \\ \hat{S}_{mkl}^{q} &= -\hat{A}_{mkl}\hat{m}_{ki}\sin\theta_{mkl} + \hat{A}_{mkl}\hat{m}_{kq}\cos\theta_{mkl} \end{aligned} \tag{19.24}$$

for a complex (I&Q) signal structure. In a given receiver m, components $\hat{S}_{mkl}^{i}$ and $\hat{S}_{mkl}^{q}$ correspond to Δy_{ikl} and Δy_{qkl}. Parameter A_{mkl} includes both amplitude and correlation function. In Equation (19.23) ΔA_{mkl} and $\varepsilon_{\theta mkl}$ are amplitude and phase estimation errors. The canceller would create $S_{mkl} - \hat{S}_{mkl} = \Delta S_{mkl}$ and the power of this residual error (with index m dropped for simplicity) would be

$$E_\theta\left[(\Delta S_{kl})^2\right] = E_\theta\left[A_{kl}m_k\cos\theta_{kl} - (A_{kl} + \Delta A_{kl})\hat{m}_k\cos(\theta_{kl} + \varepsilon_{\theta k})\right]^2 \tag{19.25}$$

where $E_\theta\,[\,]$ stands for averaging with respect to θ_{kl} and m_k. Parameter $(\Delta S_{kl})^2$ corresponds to $|\Delta y_n|^2$. This can be represented as

$$E_\theta\left[(\Delta S_{kl})^2\right] = \alpha_{kl}\left[1 + (1+\varepsilon_a)^2 - 2\,(1+\varepsilon_a)\cdot(1-2\varepsilon_m)\cos\varepsilon_\theta\right] \tag{19.26}$$

From this equation we have $1 - C_{kl} = E_\theta\lfloor(\Delta S_{kl})^2\rfloor/\alpha_{kl}$ and $C_{kl} = 2\,(1+\varepsilon_a)(1-2\varepsilon_m)\cos\varepsilon_\theta - (1+\varepsilon_a)^2$. Expanding $\cos\varepsilon_\theta$ as $1 - \varepsilon_\theta^2/2$ and averaging gives

$$C_{kl} = 2\,(1+\varepsilon_a)(1-2\varepsilon_m)\left(1-\sigma_\theta^2\right) - (1+\varepsilon_a)^2 \tag{19.27}$$

For zero mean ε_θ , $\sigma_\theta^2 = E\lfloor\varepsilon_\theta^2/2\rfloor$ is the carrier phase tracking error variance. For the complex (I&Q) signal structure cancellation efficiencies in I- and Q-channels can be represented as

$$\begin{aligned} C_{kl}^i &= 4\,(1+\varepsilon_a)(1-2\varepsilon_m)\left(1+\sigma_\theta^2\right) - 2\,(1+\varepsilon_a)^2 - 1 \\ C_{kl}^q &= 4\,(1+\varepsilon_a)(1-2\varepsilon_m)\left(1+\sigma_\theta^2\right) - 2\,(1+\varepsilon_a)^2 - 1 \end{aligned} \tag{19.28}$$

So, in this case the canceller efficiency is expressed in terms of amplitude, phase and data estimation errors. These results should be now used for analysis of the impact of large-scale channel estimators on overall CDMA network sensitivity. The performance measure of any estimator is parameter estimation error variance that should be directly used in Equations (19.28) for cancellation efficiency and Equations (19.19)–(19.22) for the rake receiver. If joint parameter estimation is used, based on ML criterion, then the Cramer–Rao bound could be used for these purposes. For the Kalman type estimator, the error covariance matrix is available for each iteration of estimation. If each parameter is estimated independently, then for carrier phase and code delay estimation error a simple relation $\sigma_{\theta,\tau}^2 = 1/SNR_{\mathrm{L}}$ can be used where SNR_{L} is the signal-to-noise ratio in the tracking loop. For the evaluation of this SNR_{L} the noise power is in general given as $N = B_{\mathrm{L}}N_0$. For this case, the noise density N_0 is approximated as a ratio of the overall interference plus noise power divided by the signal bandwidth. The loop bandwidth will be proportional to f_{D} where f_{D} is the fading rate (Doppler). The higher the f_{D}, the higher the loop noise bandwith, the higher the equivalent noise power ($N_0 f_{\mathrm{D}}$). If interference cancellation is performed prior to parameter estimation, N_0 is obtained from $f(\,)$ defined by Equation (9.16). If parameter estimation is used without interference cancellation the same $f(\,)$ is used with $C_{kl} = 0$. In addition to this

$$\varepsilon_a \Rightarrow \frac{A - \hat{A}(1-\varepsilon_\tau)}{A} = \frac{\Delta A + \hat{A}\varepsilon_\tau}{A}\;;\quad \varepsilon_a \Rightarrow \varepsilon_A + \varepsilon_\tau(1-\varepsilon_A) \tag{19.29}$$

where ε_τ is the code delay estimation error and $\varepsilon_A = (A - \hat{A})/A = 1 - \hat{A}/A$. For noncoherent estimation we have

$$\frac{\hat{A}}{A} = \left(\frac{\pi}{4y}\right)^{1/2}\exp\left(-\frac{y}{2}\right)\left\{1 + yI_0\left(\frac{y}{2}\right) + yI_1\left(\frac{y}{2}\right)\right\} \tag{19.30}$$

where $I_0(\,)$ and $I_1(\,)$ are the zeroth- and first- order Bessel functions, respectively, and y is the signal-to-noise ratio.

19.1.6 Approximations

If we assume that the channel estimation is perfect ($\varepsilon_a = \varepsilon_\theta = 0$) the parameter C_{mkl} becomes $C_{kl} = 2(1 - 2\varepsilon_m) - 1 = 1 - 4\varepsilon_m$. For DPSK modulation ε_m = (1/2)exp($-y$/2) where y is signal-to-noise ratio and for CPSK $\varepsilon_m = (1/2)\text{erfc}(\sqrt{y})$. So $C_{kl} = 1 - 2\text{e}^{-y}$ for DPSK, and $C_{kl} = 1 - 2\,\text{erfc}(\sqrt{y})$ for CPSK. For large y, $C_{kl} \Rightarrow 1$ and for small y in DPSK system we have $\text{e}^{-y} \cong 1 - y$ and $C_{kl} \cong 2y - 1$. This can be presented as $C_{kl} = 2Y_b - 1$, where Y_b is given by Equation (19.14). Bearing in mind that Y_b depends on C_{kl} the whole equation can be solved through an iterative procedure starting up with an initial value of $C_{kl} = 0, \forall m, k, l$. Similar approximations can be obtained for σ_θ^2 and ε_a. From practical point of view an attractive solution could be a scheme that would estimate and cancel only the strongest interference (e.g. successive interference cancellation schemes [2]).

19.1.7 Outage probability

The previous section already completely defines the simulation scenario for the system performance analysis. For the numerical analysis further assumptions and specifications are necessary. First of all we need the channel model. The exponential multipath intensity profile (MIP) is widely used analytical model realized as a tapped delay line [2]. It is very flexible in modeling different propagation scenarios. The decay of the profile and the number of taps in the model can vary. Averaged power coefficients in the multipath intensity profile are

$$\overline{\alpha_l} = \overline{\alpha_0}\text{e}^{-\lambda l} \quad l, \lambda \geq 0 \tag{19.31}$$

where λ is the decay parameter of the profile. Power coefficients should be normalized as $\sum_{l=0}^{L-1} \overline{\alpha_0}\text{e}^{-\lambda l} = 1$. For $\lambda = 0$ the profile will be flat. The number of resolvable paths depends on the channel chip rate and this must be taken into account. We will start from Equation (19.14) and look for the average system performance for $\rho^2 = 1/G$ where $G = W/R_b$ is the system processing gain and W is the system bandwidth (chip rate). The average signal to noise ratio will be expressed as

$$\overline{Y_b} = \frac{r^{(L_0)}G}{f(\alpha)K + \varsigma_0\eta' W/S} \tag{19.32}$$

Now, if we accept some quality of transmission, BER = 10^{-e}, that can be achieved with the given SNR = Y_0, then with the equivalent average interference density $\eta_0 = I_{\text{oic}} + I_{\text{oin}} + \eta_{\text{th}}$ SNR will be

$$Y_0 = \frac{r^{(L_0)}G}{\eta_0} \tag{19.33}$$

To evaluate the outage probability P_{out} we need to evaluate [2]

$$\begin{aligned} P_{\text{out}} &= \Pr\left(\text{BER} > 10^{-e}\right) = \Pr\left(\text{MAI} + \frac{\eta W}{S} > \eta_0\right) \\ &= \Pr\left(\text{MAI} > \eta_0 - \frac{\eta W}{S}\right) = \Pr\left(\text{MAI} > \delta\right) \end{aligned} \tag{19.34}$$

where δ is given as

$$\delta = \frac{r^{(L_0)}G}{Y_0} - \frac{\eta W}{S}$$

It can be shown that this outage probability can be represented by Gaussian integral

$$P_{\text{out}} = Q\left(\frac{\delta - m_g}{\sigma_g}\right) \tag{19.35}$$

where m_g and σ_g are the mean value and the standard variance respectively of the overall interference. From Equation (19.32) we have for the system capacity K with ideal system components

$$K_{\max} = \frac{r_0^{(L_0)}G}{Y_0 f_0(\alpha)} - \varsigma_0 \eta W / S f_0(\alpha) \tag{19.36}$$

Owing to imperfections in the operation of the rake receiver and interference canceller, this capacity will be reduced to

$$K' = \frac{r^{(L_0)}G}{Y_0 f(\alpha)} - \varsigma_0 \eta' W / S f(\alpha) \tag{19.37}$$

where $r_0^{(L_0)}$ and $f_0(\alpha)$ are now replaced by real parameters $r^{(L_0)}$ and $f(\alpha)$ that take into account those imperfections. The system sensitivity function is defined as

$$\Re = \frac{K_{\max} - K'}{K_{\max}} = \frac{1}{K_{\max}}\left\{\frac{\Delta r^{(L_0)}G}{Y_0 f(\alpha) f_0(\alpha)} - \frac{\varsigma_0 \eta' W \Delta f(\alpha)}{S f(\alpha) f_0(\alpha)}\right\} \tag{19.38}$$

where $\Delta r^{(L_0)} = r_0^{(L_0)} f(\alpha) - r^{(L_0)} f_0(\alpha)$ and $\Delta f(\alpha) = f(\alpha) - f_0(\alpha)$.

19.1.7.1 Performance example

In this section we present some numerical results for illustration purposes. The results are obtained for a channel with double exponential (space and delay) profile with decay factors λ_s and λ_t. Graphical results are presented with: solid line, 4 × 4 rake; dashed line, 4 × 1 rake. In the case of a *3 dB approximation of the real antenna*, the beam forming is approximated by the rectangular shape of the antenna pattern in the range $\varphi_{3\text{dB}} = 30°$ (for $\rho_c = 1$), and uniform distribution of $\Delta\psi_{kl}$ and $\Delta\varphi_{kl}$ ($[0, \pi]$, $[0, 2\pi]$, respectively) , $\lambda_s = \lambda_t = 0$ if not specified otherwise, $Y_0 = 2$, $L = 4$, interference margin $\text{SNR} = (20 * \alpha_0/alfa\_mean)^{-1}$.

The users are uniformly distributed in the sphere (φ, ψ) with $\varphi, \psi \in (0, 360°)$. The results can be easily scaled down for a more realistic scenario where $\varphi \in (0, 360°)$ and $\psi \in (\psi_{\min}, \psi_{\max})$. The canceling efficiency for maximum capacity is calculated by assuming no estimation errors in Equation (19.27). When estimation errors are included, canceling efficiencies are given by Equation (19.27). Carrier phase tracking error variance is assumed to be $\sigma_\theta^2 = 1/\text{SNR}_\text{L}$. For MRC the amplitude estimation error is approximated from Equation (19.30) to follow $\varepsilon_a = 1/4\ \text{SNR}_\text{L}$. For *real antennas* the amplitude patterns are specified in Figure 19.2. The base station amplitude antenna pattern is given as

$$A(\varphi, \psi) = \frac{1}{N}\sum_{n=1}^{N} \exp\left\{-\pi\left[\rho_c \sin\psi \cos\left(\varphi - 2\pi\frac{n}{N}\right)\right]^2\right\}$$

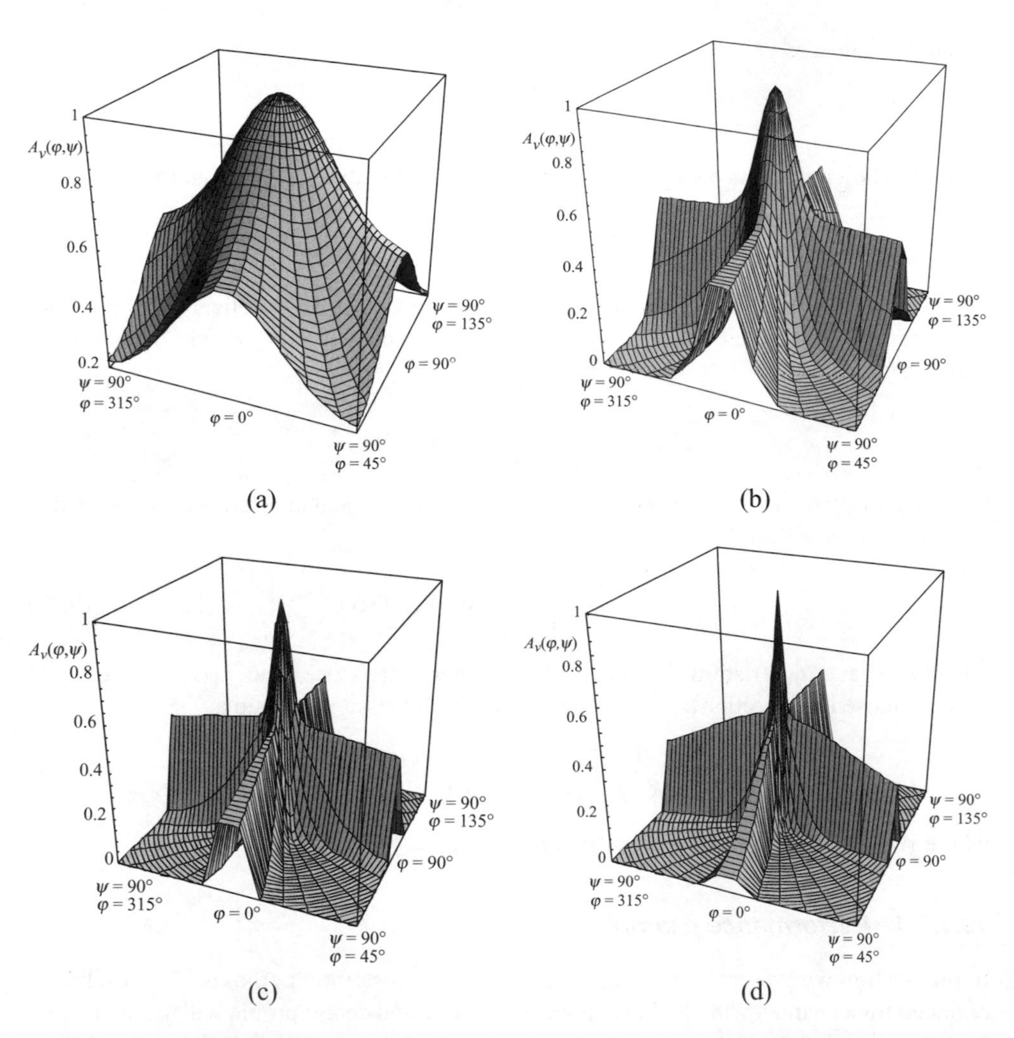

Figure 19.2 Peak-amplitude pattern $A(\varphi, \psi)$ for nonsinusoidal Gaussian pulses received by the circular array with $N = 4$ elements and (a) $\rho_c = 1$; (b) $\rho_c = 3$; (c) $\rho_c = 6$; (d) $\rho_c = 12$.

and four examples cover a relatively wide range of shapes, from a wide lobe for $\rho_c = 1$ to a rather narrow lobe for $\rho_c = 12$. At mobile stations an omnidirectional antenna pattern $A(\varphi, \psi) = 1$ is assumed. Capacity curves, defined as the number of users with data rate $R =$ chip rate/$G = 4.096/G$, for different antenna patterns, are shown in Figure 19.3. The highest capacity is obtained for $\rho_c = 12$ because with the narrowest lobe the spatial division multiple access effect is the most effective. With increased receiver velocity the capacity will be reduced due to the increased effect of imperfections. A comparison of the systems using ideal and real antennas is shown in Figure 19.4. Figure 19.4 presents the system capacity vs G. In general higher G means more users in the network and more MAI resulting in more impact of imperfections. A 4×4 rake performs better for lower G but for higher G (more users) it deteriorates faster. The degradation is more severe for higher receiver velocities.

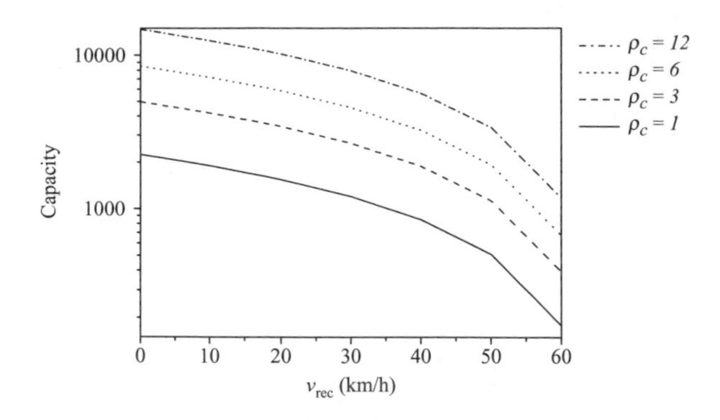

Figure 19.3 Capacity vs receiver velocity for EGC for different antenna patterns $A(\varphi, \psi)$. $N = 4$, 4×4 rake, $G = 256$, $\lambda_t = \lambda_s = 0$, $Y_0 = 2$, $L = 4$, SNR $= (20^* \alpha_0/\text{alpha\_mean})^{-1}$.

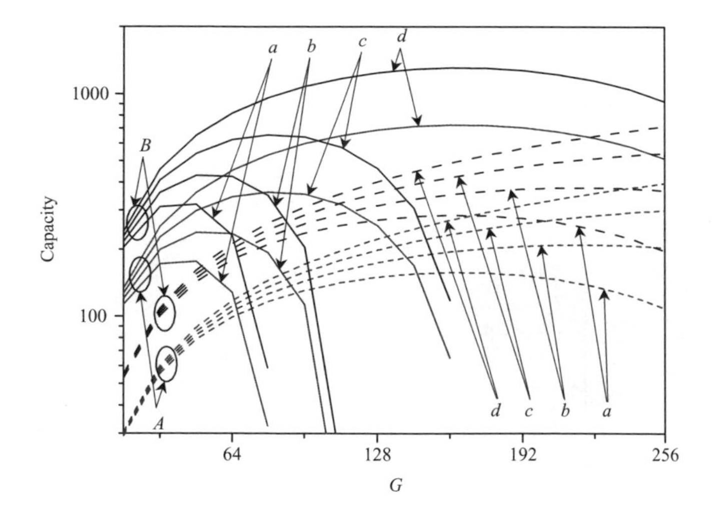

Figure 19.4 Capacity vs processing gain for EGC. a, $v_{\text{rec}} = 200$ km/h, b, $v_{\text{rec}} = 150$ km/h, c, $v_{\text{rec}} = 100$ km/h; d, $v_{\text{rec}} = 50$ km/h. Solid line: 4×4 rake; dashed line: 4×1 rake; A real antenna pattern of circular array at base station; B, 3 dB aproximation ($\psi_{3\text{dB}} = 30°$) of the real antenna pattern; $\rho_c = 1$; $N = 4$, $\lambda_s = \lambda_t = 0$, $Y_0 = 2$, $L = 4$, SNR $= (20^* \alpha_0/\text{alpha\_mean})^{-1}$.

Figure 19.5 represents the same results as a function of the receiver velocity. The system sensitivity function defined by Equation (19.38) is shown in Figure 19.6. Sensitivity equal to 1 means that all capacity has been lost due to imperfections. Figure 19.6 demonstrates that very high values for the system sensitivity, even in the range close to 0.9, can be expected if a large number of users (low data rate corresponding to high G) are in the network.

In this section we have presented a systematic analytical framework for the capacity evaluation of an advanced CDMA network. This approach provides a relatively simple way to specify the required quality of a number of system components. This includes multiple access interference canceller and rake receiver, taking into account all their imperfections. The

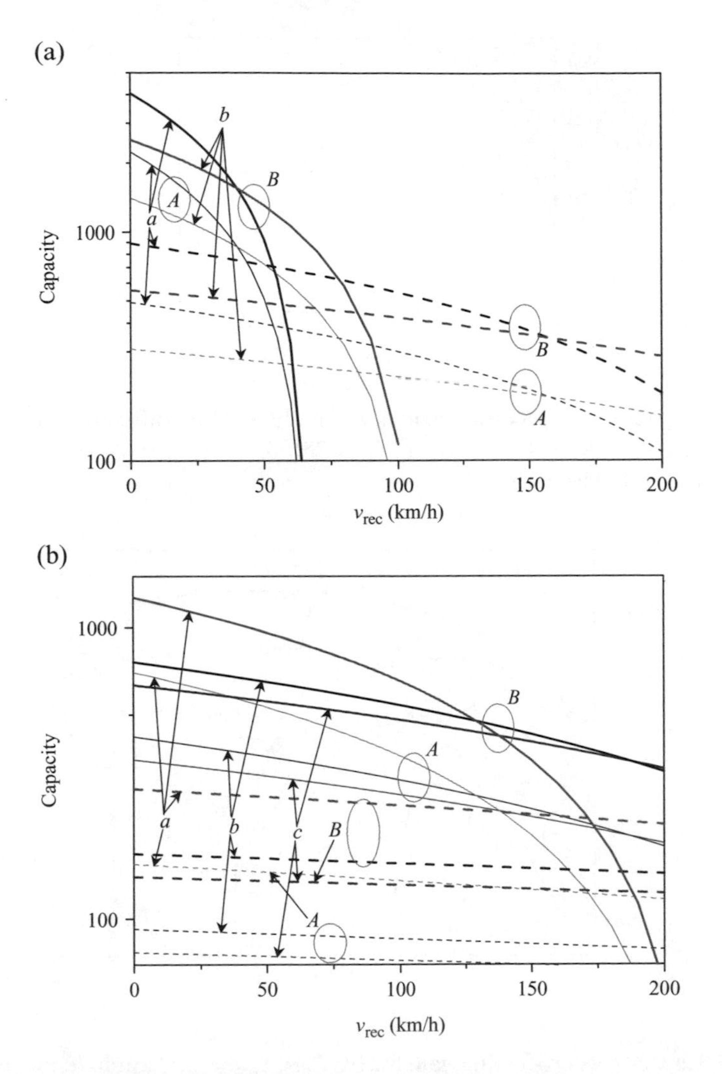

Figure 19.5 Capacity vs the receiver velocity for EGC; solid lines, 4 × 4 rake; dashed lines, 4 × 1 rake. *A*, real antenna; *B*, 3 dB approximation of the real antenna. $\rho_c = 1; N = 4, \lambda_t = \lambda_s = 0, Y_0 = 2, L = 4$, SNR = $(20^*\alpha_0/\text{alpha\_mean})^{-1}$. (a) $a, G = 256; b, G = 160$; (b) $a, G = 80; b, G = 48; c, G = 40$.

system performance measure is the network sensitivity function representing the relative losses in capacity due to all imperfections in the system implementation. Some numerical examples are presented for illustration purposes. These results are obtained for a channel with double exponential (space and delay) profile. It was shown that for the receiver velocity 100–200 km/h as much as 70–90 % of the system capacity can be lost due to the imperfections of the three-dimensional rake receiver and interference cancellation operation. variety of results are presented for different channel decay factor, fading rate and number of rake fingers. In general, under ideal conditions, the system capacity is increased if the number of fingers is increased. At the same time one should be aware that the system sensitivity

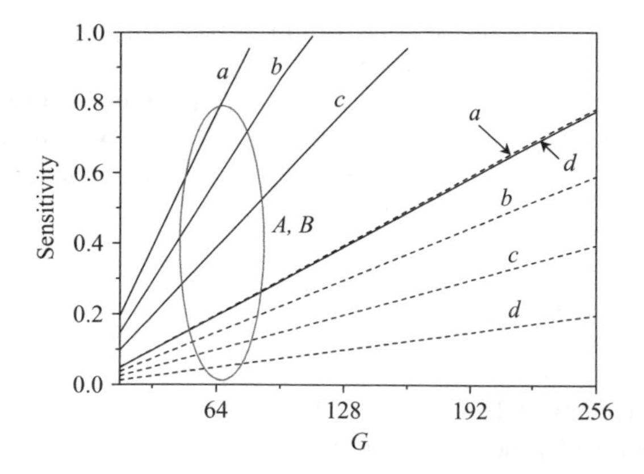

Figure 19.6 Sensitivity vs processing gain for EGC. a, $v_{\text{rec}} = 200$ km/h; b, $v_{\text{rec}} = 150$ km/h; c, $v_{\text{rec}} = 100$ km/h; d, $v_{\text{rec}} = 50$ km/h. Solid line, 4×4 Rake; dashed line; 4×1 rake. A, real antenna pattern of circular array at base station; B, 3 dB aproximation ($\psi_{3\text{dB}} = 30°$) of the real antenna pattern. $\rho_c = 1; N = 4, \lambda_s = \lambda_t = 0, Y_0 = 2, L = 4$, SNR $= (20^*\alpha_0$ /alpha_mean$)^{-1}$.

is also increased if the fading rate and number of rake fingers are higher. The results and methodology presented in this section offer enough tools and data for the careful choice of the system parameters in realistic environments which are characterized by imperfections.

19.2 CAPACITY OF *AD HOC* NETWORKS

In this section we now discuss the capacity of wireless networks. The discussion is based on concepts presented in References [32, 33]. In an *ad hoc* network, it is supposed that n nodes are located in a region of area 1 m^2. Each node can transmit at W b/s over a common wireless channel. The channel in general may be broken up into several subchannels of capacity $W_1, W_2, \ldots, W_M$ b/s. This will be immaterial for the final results as long as $\sum_{m=1}^{M} W_m = W$. Packets are sent from node to node in a multihop fashion until they reach their final destination. They can be buffered at intermediate nodes while awaiting transmission. Owing to spatial separation, several nodes can make wireless transmissions simultaneously, provided there is no excessive interference from others. In the sequel we will discuss the conditions under which a wireless transmission over a subchannel is received successfully by its intended recipient.

Two types of networks are considered, *Arbitrary networks*, where the node locations, destinations of sources, and traffic demands, are all arbitrary, and *Random Networks*, where the nodes and their destinations are randomly chosen.

19.2.1 Arbitrary networks

In the arbitrary setting we suppose that n nodes are arbitrarily located in a disk of unit area in the plane. Each node chooses an arbitrary destination to which it wishes to send traffic at an arbitrary rate, so that the traffic pattern is arbitrary too. Each node can choose an arbitrary range or power level for each transmission.

To define when a transmission is received successfully by its intended recipient we will allow for two possible models for successful reception of a transmission over one hop, called the *protocol model* and the *physical model*. Let X_i denote the location of a node; we will also use X_i to refer to the node itself.

19.2.1.1 The protocol model

Suppose node X_i transmits over the mth subchannel to a node X_j. In this case transmission is successfully received by node X_j if $|X_k - X_j| \geq (1 + \Delta)|X_i - X_j|$. In this case a guard zone $\Delta > 0$ is specified by the protocol to prevent a neighboring node from transmitting on the same subchannel at the same time.

19.2.1.2 The physical model

Let $\{X_k; k \in T\}$ be the subset of nodes simultaneously transmitting at some time instant over a certain subchannel. Let P_k be the power level chosen by node X_k for $k \in T$. In this case the transmission from a node X_i, $i \in T$, is successfully received by a node X_j if $S/I \geq \beta$, where

$$S = P/\left|X_i - X_j\right|^\alpha$$

and

$$I = N + \sum_{\substack{k \in T \\ k \neq i}} P/\left|X_i - X_j\right|^\alpha$$

This models a situation where a minimum signal-to-interference ratio (SIR) of β is necessary for successful receptions, the ambient noise power level is N, and signal power decays with distance r as $1/r^\alpha$. For a model outside a small neighborhood of the transmitter, $\alpha > 2$.

19.2.1.3 The transport capacity of arbitrary networks

In this contest we say that the network transports one *bit-meter* (b-m) when 1 b has been transported a distance of 1 m towards its destination. (We do not give multiple credit for the same bit carried from one source to several different destinations as in the multicast or broadcast cases.) This sum of products of bits and the distances over which they are carried is a valuable indicator of a network's *transport capacity* C_T. (It should be noted that, when the area of the domain is A square meters rather than the normalized 1 m^2, then all the transport capacity results presented below should be scaled by $\sqrt{A}$.) By using the notation: $f(n) = \Theta[g(n)]$ when $f(n) = O[g(n)]$ as well as $g(n) = O[f(n)]$, we will show later in the section that the transport capacity of an *arbitrary network* under the *protocol model* is $C_\mathrm{T} = \Theta(W\sqrt{n})$ b-m/s if the nodes are optimally placed, the traffic pattern is optimally chosen, and if the range of each transmission is chosen optimally. An upper bound is $C_T = \sqrt{(8/\pi)}(W/\Delta)\sqrt{n}$ b-m/s for every arbitrary network for all spatial and temporal scheduling strategies, while $C_T = Wn/[(1 + 2\Delta)(\sqrt{n} + \sqrt{8\pi})]$ bit-meters per second (for

n a multiple of four) can be achieved when the nodes and traffic patterns are appropriately chosen, and the ranges and schedules of transmissions are appropriately chosen.

If this transport capacity were to be equally divided between all the n nodes, then each node would obtain $\Theta(W/\sqrt{n})$ b-m/s. If, further, each source has its destination about the same distance of 1 m away, then each node would obtain a *throughput capacity* of $\Theta(W/\sqrt{n})$ b/s.

The upper bound on transport capacity does not depend on the transmissions being omnidirectional, as implied by Equation (19.1), but only on the presence of some dispersion in the neighborhood of the receiver. It will be shown later in the section that, for the physical model, $cW\sqrt{n}$ b-m/s is feasible, while $c'Wn^{\alpha-1/\alpha}$ b-m/s is not, for appropriate c, c'. Specifically,

$$C_{\mathrm{T}} = Wn/\left\{\left(\sqrt{n}+\sqrt{8\pi}\right)\left\{16\beta\left[2^{\frac{\alpha}{2}}+6^{\alpha-2}/(\alpha-2)\right]\right\}^{1/\alpha}\right\} \text{ b-m/s (for } n \text{ a multiple of 4)}$$

is feasible when the network is appropriately designed, while an upper bound is

$$C_{\mathrm{T}} = \left[(2\beta+2)/\beta\right]^{\frac{1}{\alpha}} W n^{\frac{\alpha-1}{\alpha}}/\sqrt{\pi} \text{ b-m/s}$$

It is suspected that that an upper bound of order $\Theta(W\sqrt{n})$ b-m/s may actually hold. In the special case where the ratio $P_{\max}/P_{\min}$ between the maximum and minimum powers that transmitters can employ is bounded above by β, then an upper bound is in fact

$$C_{\mathrm{T}} = (W\sqrt{8n/\pi})/\left[(\beta P_{\min}/P_{\max})^{\frac{1}{\alpha}}-1\right] \text{ b-m/s}$$

Both bounds suggest that transport capacity improves when α is larger, i.e. when the signal power decays more rapidly with distance.

19.2.2 Random networks

In this case, n nodes are randomly located, i.e. independently and uniformly distributed, either on the surface S^2 of a three-dimensional sphere of area 1 m^2, or in a disk of area 1 m^2 in the plane. The purpose in studying S^2 is to separate edge effects from other phenomena. Each node has a randomly chosen destination to which it wishes to send $\lambda(n)$ b/s. The destination for each node is independently chosen as the node nearest to a randomly located point, i.e. uniformly and independently distributed. (Thus destinations are on the order of 1 m away on average.) All transmissions employ the same nominal range or power (homogeneous nodes). As for arbitrary networks, both a protocol model and a physical model are considered.

19.2.2.1 The protocol model

All nodes employ a common *range* r for all their transmissions. When node X_i transmits to a node X_j over the mth subchannel, this transmission is successfully received by X_j if $|X_i - X_j| \leq r$ and for every other node X_k simultaneously transmitting over the same subchannel $|X_k - X_j| \geq (1+\Delta)r$.

19.2.2.2 The physical model

All nodes choose a common power level P for all their transmissions. Let $\{X_k; k \in T\}$ be the subset of nodes simultaneously transmitting at some time instant over a certain subchannel. A transmission from a node X_i, $i \in T$, is successfully received by a node X_j if $S/I \geq \beta$, where

$$S = P/|X_i - X_j|^\alpha$$

and

$$I = N + \sum_{\substack{k \in T \\ k \neq i}} P/|X_i - X_j|^\alpha$$

19.2.2.3 The throughput capacity of random networks

The throughput is defined in the usual manner as the time average of the number of bits per second that can be transmitted by every node to its destination.

19.2.2.4 Feasible throughput

A throughput of $\lambda(n)$ b/s for each node is *feasible* if there is a spatial and temporal scheme for scheduling transmissions, such that by operating the network in a multihop fashion and buffering at intermediate nodes when awaiting transmission, every node can send $\lambda(n)$ b/s on average to its chosen destination node. That is, there is a $T < \infty$ such that in every time interval $[(i-1)T, iT]$ every node can send $T\lambda(n)$ b to its corresponding destination node.

19.2.2.5 The throughput capacity of random wireless networks

We say that the *throughput capacity* of the class of random networks is of order $\Theta[f(n)]$ b/s if there are deterministic constants $c > 0$ and $c' < +\infty$ such that

$$\lim_{n \to \infty} \text{Prob}\,[\lambda(n) = cf(n) \text{ is feasible}] = 1$$

$$\liminf_{n \to \infty} \text{Prob}\,[\lambda(n) = c'f(n) \text{ is feasible}] < 1$$

It will be shown in the next section that, in the case of both the surface of the sphere and a planar disk, the order of the throughput capacity is $\lambda(n) = \Theta[W/\surd(n \log n)]$ b/s for the protocol model. For the upper bound for some c',

$$\lim_{n \to \infty} \text{Prob}\left[\lambda(n) = c'W/\sqrt{n \log n} \text{ is feasible 1e}\right] = 0$$

Specifically, there are deterministic constants c'' and c''' not depending on n, Δ or W, such that $\lambda(n) = c''W/\left[(1+\Delta)^2\surd(n \log n)\right]$ b/s is feasible, and $\lambda(n) = c'''W/\left[\Delta^2\surd(n \log n)\right]$ b/s is infeasible, both with probability approaching 1 as $n \to \infty$.

It will be also shown that, for the physical model, a throughput of $\lambda(n) = cW/\surd(n \log n)$ b/s is feasible, while $\lambda(n) = c'W/\surd n$ bits per second is not, for appropriate c, c', both with probability approaching 1 as $n \to \infty$. Specifically, there are deterministic constants c'' and

c''' not depending on n, N, α, β or W, such that

$$\lambda(n) = c''W / \left[\sqrt{n \log n} \left(2 \left\{ c'''\beta \left[3 + 1/(\alpha - 1) + 2/(\alpha - 2) \right] \right\}^{\frac{1}{\alpha}} - 1 \right)^2 \right] \text{ b/s}$$

is feasible with probability approaching one as $n \to \infty$. If $\bar{L}$ is the mean distance between two points independently and uniformly distributed in the domain (either surface of sphere or planar disk of unit area), then there is a deterministic sequence $\varepsilon(n) \to 0$, not depending on N, α, β or W, such that $\left(\sqrt{8/\pi n}\right) W \left[1 + \varepsilon(n)\right] / \left[\bar{L}(\beta^{\frac{1}{\alpha}} - 1) \right]$ b-m/s is infeasible with probability approaching 1 as $n \to \infty$.

19.2.3 Arbitrary networks: an upper bound on transport capacity

The following assumptions for the setting on a planar disk of unit area are used [32, 33]:

(a1) There are n nodes arbitrarily located in a disk of unit area on the plane.

(a2) The network transports $\lambda n T$ b over T s.

(a3) The average distance between the source and destination of a bit is $\bar{L}$. Together with (a2), this implies that a transport capacity of $\lambda n \bar{L}$ b-m/s is achieved.

(a4) Each node can transmit over any subset of M subchannels with capacities W_m b/s, $1 \le m \le M$, where $\sum_{m=1}^{M} W_m = W$.

(a5) Transmissions are slotted into synchronized slots of length τ s. (This assumption can be eliminated, but makes the exposition easier.)

(a6) Definitions of physical model and protocol model from the previous section are used.

While retaining the restriction for the case of the physical model, we can either retain the restriction in the protocol model or consider an alternate restriction as follows: if a node X_i transmits to another node X_j located at a distance of r units on a certain subchannel in a certain slot, then there can be no other receiver within a radius of Δr around X_j on the same subchannel in the same slot. This alternative restriction addresses situations where the transmissions are not omnidirectional, but there is still some dispersion in the neighborhood of the receiver. Under the above assumptions the following results for the transport capacity have been obtained [32, 33]:

(r1) In the *protocol model*, the transport capacity $\lambda n \bar{L}$ is bounded as: $\lambda n \bar{L} \le W \sqrt{8n/\pi}$ Δ b-m/s.

(r2) In the *physical model*, $\lambda n \bar{L} \le W n^{\alpha - 1/\alpha} \left[(2\beta + 2) / \beta \right]^{1/\alpha} / \sqrt{\pi}$ b-m/s.

(r3) If the ratio $P_{\max}/P_{\min}$ between the maximum and minimum powers that transmitters can employ is strictly bounded above by β, then $\lambda n \bar{L} \le W \sqrt{(8n/\pi)}$ $\left[(\beta P_{\min}/P_{\max})^{1/\alpha} - 1 \right]^{-1}$ b-m/s.

(r4) When the domain is of A square meters rather than 1 m^2, then all the upper bounds above are scaled by $\sqrt{A}$.

To prove the above results, consider bit b, where $1 \leq b \leq \lambda nT$. Let us suppose that it moves from its origin to its destination in a sequence of $h(b)$ hops, where the hth hop traverses a distance of r_b^h. Then from (a3) we have

$$\sum_{b=1}^{\lambda nT} \sum_{h=1}^{h(b)} r_b^b \geq \lambda nT \bar{L} \tag{19.39}$$

Having in mind that in any slot at most $n/2$ nodes can transmit for any subchannel m and any slot s we have

$$\sum_{b=1}^{\lambda nT} \sum_{h=1}^{h(b)} 1 \text{ (the } h\text{th hop of bit } b \text{ is over subchannel } m \text{ in slot } s \text{)} \leq \frac{W_m \tau n}{2}$$

Summing over the subchannels and the slots, and noting that there can be no more than T/τ slots in T s yields

$$H := \sum_{b=1}^{\lambda nT} h(b) \leq \frac{WTn}{2} \tag{19.40}$$

From the triangle inequality and (a6), for the *protocol model*, where X_j is receiving a transmission from X_i over the mth subchannel and at the same time X_ℓ is receiving a transmission from X_k over the same subchannel, we have

$$|X_j - X_\ell| \geq |X_j - X_k| - |X_\ell - X_k| \geq (1 + \Delta)|X_i - X_j| - |X_\ell - X_k|$$

Similarly,

$$|X_\ell - X_j| \geq (1 + \Delta)|X_k - X_\ell| - |X_j - X_i|$$

Adding the two inequalities, we obtain

$$|X_\ell - X_j| \geq \frac{\Delta}{2}(|X_k - X_\ell| + |X_i - X_j|)$$

This means that disks of radius $\Delta/2$ times the lengths of hops centered at the receivers over the same subchannel in the same slot are essentially disjoint. This conclusion also directly follows from the alternative restriction in (a6). Allowing for edge effects where a node is near the periphery of the domain, and noting that a range greater than the diameter of the domain is unnecessary, we see that at least a quarter of such a disk is within the domain. Having in mind that at most $W_m \tau$ bits can be carried in slot from a receiver to a transmitter over the mth subchannel, we have

$$\sum_{b=1}^{\lambda nT} \sum_{h=1}^{h(b)} 1 \text{ (the } h\text{th hop of bit } b \text{ is over subchannel } m \text{ in slot s)} \times \left(\frac{1}{4}\right) \pi \left(\frac{\Delta}{2}\right)^2 (r_b^h)^2 \leq W_m \tau \tag{19.41}$$

Summing over the subchannels and the slots gives

$$\sum_{b=1}^{\lambda nT} \sum_{h=1}^{h(b)} \frac{\pi \Delta^2}{16} (r_b^h)^2 \leq WT$$

or equivalently

$$\sum_{b=1}^{\lambda nT}\sum_{h=1}^{h(b)}\frac{1}{H}(r_b^h)^2 \le \frac{16WT}{\pi\Delta^2 H} \tag{19.42}$$

Since the quadratic function is convex we have

$$\left(\sum_{b=1}^{\lambda nT}\sum_{h=1}^{h(b)}\frac{1}{H}r_b^h\right)^2 \le \sum_{b=1}^{\lambda nT}\sum_{h=1}^{h(b)}\frac{1}{H}\left(r_b^h\right)^2 \tag{19.43}$$

Finally combining Equations (19.64) and (19.65) yields

$$\sum_{b=1}^{\lambda nT}\sum_{b=1}^{h(b)} r_b^h \le \sqrt{\frac{16WTH}{\pi\Delta^2}} \tag{19.44}$$

and substituting Equation (19.40) in Equation (19.66) gives

$$\lambda nT\bar{L} \le \sqrt{\frac{16WTH}{\pi\Delta^2}} \tag{19.45}$$

Substituting Equation (19.63) in Equation (19.67) yields the result (r 1).

For the *physical model* suppose X_i is transmitting to $X_{j(i)}$ over the mth subchannel at power level P_i at some time, and let $\Im$ denote the set of all simultaneous transmitters over the mth subchannel at that time. The initial constraint introduced by (a6) can be represented as

$$\frac{S}{I} = \frac{\frac{P_i}{|X_i - X_j|^\alpha}}{N + \sum\limits_{\substack{k\in\Im\\ k\ne i}}\frac{P_k}{|X_k - X_j|^\alpha}} \ge \beta \tag{19.46}$$

By also including the signal power of X_i in the denominator, the signal-to-interference requirement for $X_{j(i)}$ can be written as

$$\frac{S}{I} = \frac{\frac{P_i}{|X_i - X_{j(i)}|^\alpha}}{N + \sum\limits_{k\in\Im}\frac{P_k}{|X_k - X_{j(i)}|^\alpha}} \ge \frac{\beta}{\beta+1}$$

which results in

$$|X_i - X_{j(i)}|^\alpha \le \frac{\beta+1}{\beta}\frac{P_i}{N + \sum\limits_{k\in\Im}\frac{P_k}{|X_k - X_{j(i)}|^\alpha}} \le \frac{\beta+1}{\beta}\frac{P_i}{N + (\frac{\pi}{4})^{\alpha/2}\sum\limits_{k\in\Im}P_k}$$

(since $|X_k - X_{j(i)}| \le 2/\sqrt{\pi}$).

Summing over all transmitter–receiver pairs,

$$\sum_{i \in \Im} |X_i - X_{j(i)}|^\alpha \le \frac{\beta+1}{\beta} \frac{\sum_{i \in \Im} P_i}{N + (\frac{\pi}{4})^{\alpha/2} \sum_{i \in \Im} P_k} \le 2^\alpha \pi^{-(\alpha/2)} \frac{\beta+1}{\beta}$$

Summing over all slots and subchannels gives

$$\sum_{b=1}^{\lambda n T} \sum_{h=1}^{h(b)} r^\alpha(h,b) \le 2^\alpha \ \pi^{-\frac{\alpha}{2}} \frac{\beta+1}{\beta} WT$$

The rest of the proof proceeds along lines similar to the *protocol model*, invoking the convexity of r^α instead of r^2. For the consideration of the special case where $P_{\max}/P_{\min} < \beta$, we start with Equation (19.46). From it, it follows that if X_i is transmitting to X_j at the same time that X_k is transmitting to X_ℓ, both over the same subchannel, then

$$\frac{\frac{P_i}{|X_i - X_j|^\alpha}}{\frac{P_k}{|X_k - X_j|^\alpha}} \ge \beta$$

Thus

$$|X_k - X_j| \ge (\beta P_{\min}/P_{\max})^{\frac{1}{\alpha}} \ |X_i - X_j| = (1+\Delta)|X_i - X_j|$$

where $\Delta := (\beta P_{\min}/P_{\max})^{\frac{1}{\alpha}} - 1$. Thus the same upper bound as for the *protocol model* carries over with Δ defined as above.

19.2.4 Arbitrary networks: lower bound on transport capacity

There is a placement of nodes and an assignment of traffic patterns such that the network can achieve

$$Wn/\left\{(1+2\Delta)\left(\sqrt{n} + \sqrt{8\pi}\right)\right\} \text{ b-m/s}$$

under the *protocol model*,

$$Wn/\left\{\left(\sqrt{n} + \sqrt{8\pi}\right)\left[16\beta\left(2^{\frac{\alpha}{2}} + \frac{6^{\alpha-2}}{\alpha-2}\right)\right]^{-1/\alpha}\right\} \text{ b-m/s}$$

under the *physical model*, both whenever n is a multiple of 4. To prove it, consider the *protocol model* and define

$$r := 1/\left\{(1+2\Delta)\left(\sqrt{n/4} + \sqrt{2\pi}\right)\right\}$$

Recall that the domain is a disk of unit area, i.e. of radius $1/\sqrt{\pi}$ in the plane. With the center of the disk located at the origin, place transmitters at locations $[j(1+2\Delta)r \pm \Delta r, k(1+2\Delta)r]$ and $[j(1+2\Delta)r, k(1+2\Delta)r \pm \Delta r]$ where $|j+k|$ is even. Also place receivers at $[j(1+2\Delta)r \pm \Delta r, k(1+2\Delta)r]$ and $[j(1+2\Delta)r, k(1+2\Delta)r \pm \Delta r]$ where $|j+k|$ is odd. Each transmitter can transmit to its nearest receiver, which is at a distance r away, without interference from any other transmitter–receiver pair. It can be verified that there are at least $n/2$ transmitter–receiver pairs, all located within the domain. This is based on the fact that for a tessellation of the plane by squares of side s, all squares intersecting a disk of radius $R - \sqrt{2}s$ are entirely contained within a larger concentric disk of radius R. The number of such squares is greater than $\pi(R - \sqrt{2}s)^2/s^2$. This proves the above statement for $s = (1 + 2\Delta)r$ and $R = 1/\sqrt{\pi}$. Restricting attention to just these pairs, there are a total of $n/2$ simultaneous transmissions, each of range r, and each at W b/s. This achieves the transport capacity indicated.

For the physical model, a calculation of the SIR shows that it is lower-bounded at all receivers by

$$(1 + 2\Delta)^\alpha \left[16\left(2^{\frac{\alpha}{2}} + \frac{6^{\alpha-2}}{\alpha - 2}\right)\right]^{-1}$$

Choosing Δ to make this lower bound equal to β yields the result.

In the *protocol model*, there is a placement of nodes and an assignment of traffic patterns such that the network can achieve

$2W/\sqrt{\pi}$ *b-m/s for* $n \geq 2$
$4W\left[\sqrt{\pi}(1 + \Delta)\right]^{-1}$ *b-m/s, for* $n \geq 8$
$Wn\left[(1 + 2\Delta)\left(\sqrt{n} + \sqrt{8\pi}\right)\right]^{-1}$ *b-m/s, for* $n = 2, 3, 4, \cdots, 19, 20, 21$ *and*
$4\lfloor n/4 \rfloor W\left\{(1 + 2\Delta)\left(\sqrt{4\lfloor n/4 \rfloor} + \sqrt{8\pi}\right)\right\}^{-1}$ *b-m/s, for all n*

With at least two nodes, clearly $2W/\sqrt{\pi}$ b-m/s can be achieved by placing two nodes at diametrically opposite locations. This verifies the formula for the bound for $n \leq 8$. With at least eight nodes, four transmitters can be placed at the opposite ends of perpendicular diameters, and each can transmit toward its receiver located at a distance $1/\left[\sqrt{\pi}(2 + 2\Delta)\right]$ towards the center of the domain. This yields $4W/\left[\sqrt{\pi}(1 + \Delta)\right]$ b-m/s, verifying the formula up to $n = 21$.

19.2.5 Random networks: lower bound on throughput capacity

In this section we show that one can spatially and temporally schedule transmissions in a random graph so that, when each randomly located node has a randomly chosen destination, each source–destination pair can indeed be guaranteed a 'virtual channel' of capacity $cW\left[(1 + \Delta)^2\sqrt{n \log n}\right]^{-1}$ b/s with probability approaching 1 as $n \to \infty$, for an appropriate constant $c > 0$. We will show how to route traffic efficiently through the random graph so that no node is overloaded.

19.2.5.1 Spatial tessellation

In the following a Voronoi tessellation of the surface S^2 of the sphere is used. For a set of p points $\{a_1, a_2, \cdots, a_p\}$ on S^2 the Voronoi cell $V(a_i)$ is the set of all points which are closer to a_i than to any of the other a_js, i.e. [34],

$$V(a_i) := \{x \in S^2 : |x - a_i| = \min_{1 \leq j \leq p} |x - a_j|\}$$

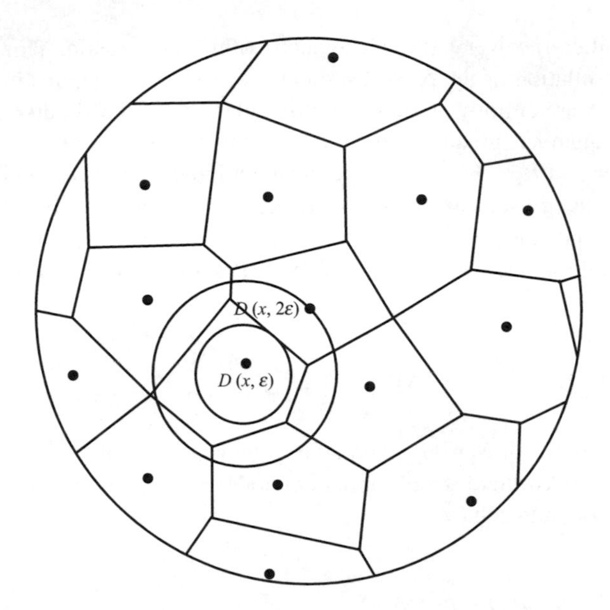

Figure 19.7 A tessellation of the surface S^2 of the sphere.

Above and throughout, distances are measured on the surface S^2 of the sphere by segments of great circles connecting two points. The point a_i is called the generator of the Voronoi cell $V(a_i)$. The surface of the sphere does not allow any regular tessellation where all cells look the same. For our application Voronoi tessellations will also need to be not too eccentrically shaped. Therefore, the properties of Voronoi tessellations needed in this section can be summarized as:

(v1) For every $\varepsilon > 0$, there is a Voronoi tessellation of S^2 with the property that every Voronoi cell contains a disk of radius ε and is contained in a disk of radius 2ε (see illustration in Figure 19.7).

To prove this we denote by $D(x, \varepsilon)$ a disk of radius ε centered at x. Choose a_1 as any point in S^2. Suppose that $a_1, \cdots, a_p$ have already been chosen such that the distance between any two a_js is at least 2ε. There are two cases to consider. Suppose there is a point x such that $D(x, \varepsilon)$ does not intersect any $D(a_i, \varepsilon)$. Then x can be added to the collection: Define $a_{p+1} := x$. Otherwise, we stop. This procedure has to terminate in a finite number of steps since the addition of each a_i removes the area of a disk of radius $\varepsilon > 0$ from S^2. When we stop we will have a set of generators such that they are at least 2ε units apart, and such that all other points on S^2 are within a distance of 2ε from one of the generators. The Voronoi tessellation obtained in this way has the desired properties.

In the sequel we will use a Voronoi tessellation V_n for which

(v2) Every Voronoi cell contains a disk of area $100 \log n/n$.

Let $\rho(n) :=$ radius of a disk of area $(100 \log n)/n$ on S^2.

(v3) Every Voronoi cell is contained in a disk of radius $2\rho(n)$.

We will refer to each Voronoi cell $V \in V_n$ as simply a 'cell'.

19.2.5.2 Adjacency and interference

By definition, two cells are *adjacent*, if they share a common point. If we choose the range $r(n)$ of each transmission so that $r(n) = 8\rho(n)$, this range will allow direct communication within a cell and between adjacent cells.

(v4) Every node in a cell is within a distance $r(n)$ from every node in its own cell or adjacent cell.

To prove it let we notice that the diameter of cells is bounded by $4\rho(n)$, see (v3). The range of a transmission is $8\rho(n)$. Thus the area covered by the transmission of a node includes adjacent cells.

19.2.5.3 Interfering neighbors

By definition we say that two cells are *interfering neighbors* if there is a point in one cell which is within a distance $(2 + \Delta)r(n)$ of some point in the other cell. In other words if two cells are not interfering neighbors, then in the protocol model a transmission from one cell cannot collide with a transmission from the other cell.

19.2.5.4 Bound on the number of interfering neighbors of a cell

An important property of the constructed Voronoi tessellation V_n is that the number of interfering neighbors of a cell is uniformly bounded. This will be exploited in the next section in constructing a spatial transmission schedule which allows for a high degree of spatial concurrency and thus frequency reuse.

(v5) Every cell in V_n has no more than c_1 interfering neighbors were c_1 depends only on Δ and grows no faster than linearly in $(1 + \Delta)^2$.

To prove it, Let V be a Voronoi cell. If V' is an interfering neighboring Voronoi cell, there must be two points, one in V and the other in V', which are no more than $(2 + \Delta)r(n)$ units apart. From (v3), the diameter of a cell is bounded by $4\rho(n)$. Hence V', and similarly every other interfering neighbor in the protocol model, must be contained within a common large disk D of radius $6\rho(n) + (2 + \Delta)r(n)$. Such a disk D cannot contain more than $c_2\{[6\rho(n) + (2+ \stackrel{\Delta}{=})r(n)]^2/\rho^2(n)\}$ disks of radius $\rho(n)$. From (v2), there can therefore be no more than this number of cells within D. This then is an upper bound on the number of interfering neighbors of the cell V. The result follows from the chosen magnitudes of $\rho(n)$ and $r(n)$.

19.2.5.5 A bound on the length of an all-cell inclusive transmission schedule

The bounded number of interfering neighbors for each cell allows the construction of a schedule of bounded length which allows one opportunity for each cell in the tessellation V_n to transmit.

(v6) In the Protocol Model there is a schedule for transmitting packets such that in every $(1 + c_1)$ slots, each cell in the tessellation V_n gets one slot in which to transmit, and such that all transmissions are successfully received within a distance $r(n)$ from their transmitters.

(v7) There is a deterministic constant c not depending on n, N, α, β, or W such that if Δ is chosen to satisfy

$$(1+\Delta)^2 > \left(2\left\{c\beta\left[3+(\alpha-1)^{-1}+2(\alpha-2)^{-1}\right]\right\}^{\frac{1}{\alpha}}-1\right)^2$$

then for a large enough common power level P, the above result (v6) holds even for the physical model.

To prove it we show first the result for the protocol model. This follows from a well-known fact about vertex coloring of graphs of bounded degree. A graph of degree no more than c_1 can have its vertices colored by using no more than $(1+c_1)$ colors, with no two neighboring vertices having the same color [35]. One can therefore color the cells with no more than $(1+c_1)$ colors such that no two interfering neighbors have the same color. This gives a schedule of length at most $(1+c_1)$, where one can transmit one packet from each cell of the same color in a slot.

For the physical model one can show that, under the same schedule as above, the required SIR of β is obtained if each transmitter chooses an identical power level P that is high enough, and Δ is large enough. From the previous discussion we know that any two nodes transmitting simultaneously are separated by a distance of at least $(2+\Delta)r(n)$ and disks of radius $(1+\Delta/2)r(n)$ around each transmitter are disjoint. The area of each such disk is at least $c_3\pi(1+\Delta/2)^2r^2(n)$. (In the case of disks on the plane $c_3=1$, but it is smaller for disks on the surface of the sphere.)

Consider a node X_i transmitting to a node X_j at a distance less than $r(n)$. The signal power received at X_j is at least $Pr^{-\alpha}(n)$. Now we look at the interference power due to all the other simultaneous transmissions. Consider the annulus of all points lying within a distance between a and b from X_j. A transmitter within this annulus has the disk centered at itself and of radius $(1+\Delta/2)r(n)$ entirely contained within a larger annulus of all points lying between a distance $a-(1+\Delta/2)r(n)$ and $b+(1+\Delta/2)r(n)$. The area of this larger annulus is no more than

$$c_4\pi\left\{[b+(1+\Delta/2)r(n)]^2-[a-(1+\Delta/2)r(n)]^2\right\}$$

Each transmitter above 'consumes' an area of at least $c_3\pi\,(1+\Delta/2)^2\,r^2(n)$, as noted earlier. Hence the annulus of points at a distance between a and b from the receiver X_j cannot contain more than

$$c_4\pi\{[b+(1+\frac{\Delta}{2})r(n)]^2-[a-(1+\frac{\Delta}{2})r(n)]^2\}\left[c_3\pi(1+\frac{\Delta}{2})^2r^2(n)\right]^{-1}$$

transmitters. Also, the received power at X_j from each such transmission is at most P/a^α. Noting that there can be no other simultaneous transmitter within a distance $(1+\Delta)r(n)$ of X_j, and taking $a=k(1+\Delta/2)r(n)$ and $b=(k+1)(1+\frac{\Delta}{2})r(n)$ for $k=1,2,3,\cdots$, we see that the SIR at X_j is lower-bounded by

$$\mathrm{Pr}^{-\alpha}(n)\left\{N+\sum_{k=1}^{+\infty}c_4P\left[(k+2)^2-(k-1)^2\right]\left[c_3k^\alpha(1+\frac{\square}{2})^\alpha r^\alpha(n)\right]^{-1}\right\}^{-1}$$

$$=\frac{P}{N}\left\{r^\alpha(n)+\frac{c_4}{c_3(1+\Delta/2)^\alpha}\frac{P}{N}\sum_{k=1}^{+\infty}\frac{6k+3}{k^\alpha}\right\}^{-1}$$

Since $\alpha > 2$, the sum in the denominator converges, and is smaller than $9 + 3(\alpha - 1)^{-1} + 6(\alpha - 2)^{-1}$. When Δ is as specified and $P \to \infty$, the lower bound on the SIR converges to a value greater than β. Using similar arguments one can show that [32]

(v8) Each cell contains at least one node.

(v9) The mean number of routes served by each cell $\leq c_{10}\sqrt{(n \log n)}$.

(v10) The actual traffic served by each cell $\leq c_5\lambda(n)\sqrt{(n \log n \text{ whp})}$.

19.2.5.6 Lower bound on throughput capacity of random networks

From (v6) we know that there exists a schedule for transmitting packets such that in every $(1 + c_1)$ slots, each cell in the tessellation V_n gets one slot to transmit, and such that each transmission is received within a range $r(n)$ of the transmitter. Thus the rate at which each cell gets to transmit is $W/(1 + c_1)$ b/s. On the other hand, the rate at which each cell *needs* to transmit is with high probability (*whp*) less than $c_5\lambda(n)\sqrt{n \log n}$ [see (v10)]. With high probability, this rate can be accommodated by all cells if it is less than the rate available, i.e. if $c_5\lambda(n)\sqrt{(n \log n)} \leq W(1 + c_1)^{-1}$. Moreover, within a cell, the traffic to be handled by the entire cell can be handled by any one node in the cell, since each node can transmit at rate W b/s whenever necessary. In fact, one can even designate one node in each cell as a 'relay' node. This node can handle all the traffic needing to be relayed. The other nodes can simply serve as sources or sinks.

We have proved the following theorem, noting the linear growth of c_1 in $(1 + \Delta)^2$ in (v5), and the choice of Δ in (v6) for the physical model.

(v11) For random networks on S^2 in the protocol model, there is a deterministic constant $c > 0$ not depending on n, Δ, or W, such that $\lambda(n) = cW\left\{(1 + \Delta)^2\sqrt{(n \log n)}\right\}^{-1}$ b/s is feasible whp.

(v12) For random networks on S^2 in the physical model, there are deterministic constants c' and c'' not depending on n, N, α, β, or W, such that

$$\lambda(n) = c'W\left\{\left(2\left\{c''\beta\left[3 + (\alpha - 1)^{-1} + 2(\alpha - 2)^{-1}\right]\right\}^{\frac{1}{\alpha}} - 1\right)^2 \sqrt{n \log n}\right\} \text{ b/s}$$

is feasible whp. These throughput levels have been attained without subdividing the wireless channel into subchannels of smaller capacity.

19.3 INFORMATION THEORY AND NETWORK ARCHITECTURES

19.3.1 Network architecture

For the optimization of *ad hoc* and sensor networks we will discuss some performance measure for the architectures shown in Figures 19.8–19.11. We will mainly focus on transport capacity $C_{\mathrm{T}} := \sup \sum_{l=1}^{m} R_l \rho_l$, where the supremum is taken over m, and vectors $(R_1, R_2, \ldots, R_m)$ of feasible rates for m source-destination pairs, and ρ_l is the distance

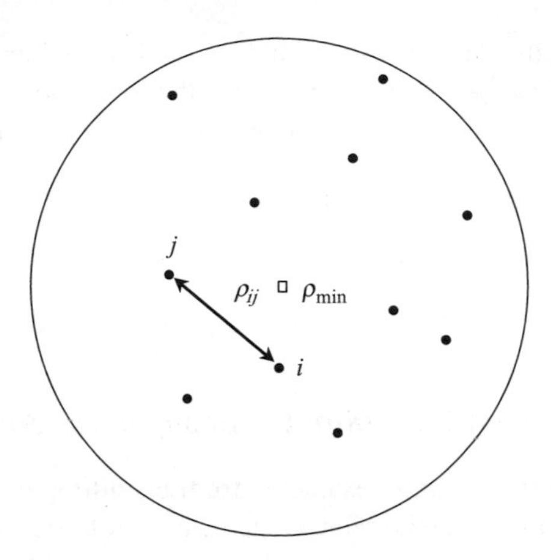

Figure 19.8 A planar network: n nodes located on a two-dimensional plane, with minimum separation distance $\rho_{\min}$.

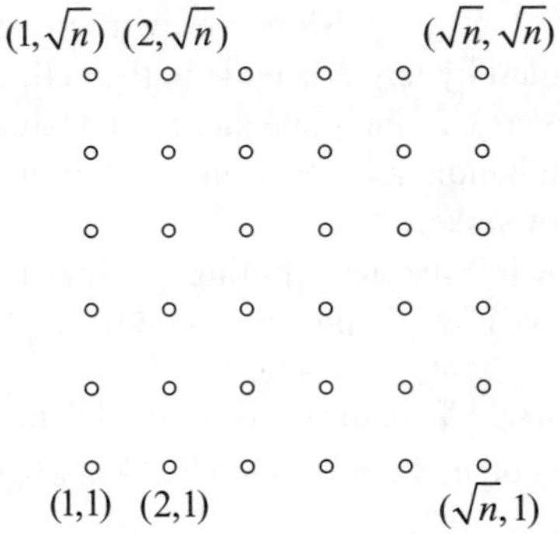

$(1, \sqrt{n})$ $(2, \sqrt{n})$ $(\sqrt{n}, \sqrt{n})$

$(1,1)$ $(2,1)$ $(\sqrt{n}, 1)$

Figure 19.9 A regular planar network: n nodes located on a plane at (i, j) with $1 \leq i, j \leq \sqrt{n}$. (Reproduced by permission of IEEE [33].)

between the lth source and its destination. For the planar network from Figure 19.8 we assume [33]:

(1) There is a finite set N of n nodes located on a plane.

(2) There is a minimum positive separation distance $\rho_{\min}$ between nodes, i.e. $\rho_{\min} := \min_{i \neq j} \rho_{ij} > 0$, where ρ_{ij} is the distance between nodes $i, j \in N$.

(3) Every node has a receiver and a transmitter. At time instants $t = 1, 2, \ldots,$ node $i \in N$ sends $X_i(t)$, and receives $Y_i(t)$ with

$$Y_i(t) = \sum_{j \neq i} \frac{e^{\gamma \rho_{ij}} X_j(t)}{\rho_{ij}^{\delta}} + Z_i(t)$$

where $Z_i(t), i \in N, t = 1, 2, \ldots$ are Gaussian independent and identically distributed (i.i.d.) random variables with mean zero and variance σ^2. The constant $\delta > 0$ is referred to as the path loss exponent, while $\gamma \geq 0$ will be called the

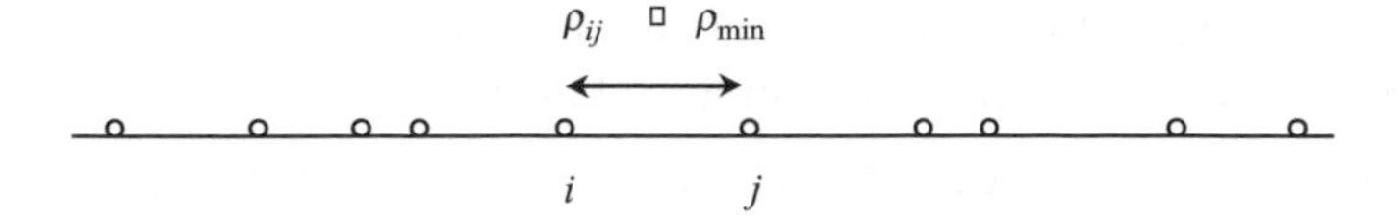

Figure 19.10 A linear network: n nodes located on a line, with minimum separation distance $\rho_{\min}$. (Reproduced by permission of IEEE [33].)

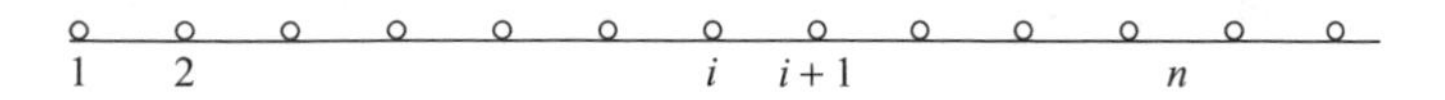

Figure 19.11 A regular linear network: n nodes located on a line at $1, 2, \ldots, n$. (Reproduced by permission of IEEE [33].)

absorption constant. A positive γ generally prevails except for transmission in a vacuum, and corresponds to a loss of $20\gamma \log_{10} e$ decibel per meter.

(4) Denote by $P_i \geq 0$ the power used by node i. Two separate constraints on $\{P_1, P_2, \ldots, P_n\}$ are studied: total power constraint $P_{\text{total}} : \sum_{i=1}^{n} P_i \leq P_{\text{total}}$ or individual power constraint $P_{\text{ind}} : P_i \leq P_{\text{ind}}$, for $i = 1, 2, \ldots, n$.

(5) The network can have several source–destination pairs $(s_\ell, d_\ell), \ell = 1, \ldots, m$, where s_ℓ, d_ℓ are nodes in N with $s_\ell \neq d_\ell$, and $(s_\ell, d_\ell) \neq (s_j, d_j)$ for $\ell \neq j$. If $m = 1$, then there is only a single source–destination pair, which we will simply denote (s, d).

A special case is a regular planar network where the n nodes are located at the points (i, j) for $1 \leq i, j \leq \sqrt{n}$; see Figure 19.9. This setting will be used mainly to exhibit achievability of some capacities, i.e. inner bounds. Another special case is a *linear network* where the n nodes are located on a straight line, again with minimum separation distance $\rho_{\min}$; see Figure 19.10. The main reason for considering linear networks is that the proofs are easier to state and comprehend than in the planar case, and can be generalized to the planar case. Also, the linear case may have some utility for, say, networks of cars on a highway, since its scaling laws are different.

A special case of a linear network is a regular linear network where the n nodes are located at the positions $1, 2, \ldots, n$; see Figure 19.11. This setting will also be used mainly to exhibit achievability results.

19.3.2 Definition of feasible rate vectors

Definition 1

Consider a wireless network with multiple source–destination pairs $(s_\ell, d_\ell), \ell = 1, \ldots, m$, with $s_\ell \neq d_\ell$, and $(s_\ell, d_\ell) \neq (s_j, d_j)$ for $\ell \neq j$. Let $S := \{s_\ell, \ell = 1, \ldots, m\}$ denote the set of source nodes. The number of nodes in S may be less than m, since we allow a node to have originating traffic for several destinations. Then a $[(2^{TR_1}, \ldots, 2^{TR_m}), T, P_e^{(T)}]$ code with total power constraint P_{total} consists of the following:

(1) m independent random variables W_ℓ (transmitted words TR_l bits long) with $P(W_l = k_l) = 1/2^{TR_l}$, for any $k_\ell \in \{1, 2, \ldots, 2^{TR_\ell}\}$, $\ell = 1, \ldots, m$. For any $i \in S$, let $\bar{W}_i := \{W_\ell : s_\ell = i\}$ and $\bar{R}_i := \sum_{\{\ell : s_\ell = i\}} R_\ell$

(2) Functions $f_{i,t} : \mathbb{R}^{t-1} \times \{1, 2, \ldots, 2^{T\bar{R}_i}\} \to \mathbb{R}$, $\quad t = 1, 2, \ldots, T$, for the source nodes $i \in S$ and $f_{j,t} : \mathbb{R}^{t-1} \to \mathbb{R}$, $t = 2, \ldots, T$, for all the other nodes $j \notin S$, such that

$$X_i(t) = f_{i,t}(Y_i(1), \ldots, Y_i(t-1), \bar{W}_i), \quad t = 1, 2, \ldots, T$$
$$X_j(1) = 0, \qquad X_j(t) = f_{j,t}(Y_j(1), \ldots, Y_j(t-1)), \quad t = 2, 3, \ldots, T$$

such that the following total power constraint holds:

$$\frac{1}{T} \sum_{t=1}^{T} \sum_{i \in N} X_i^2(t) \le P_{\text{total}} \tag{19.47}$$

(3) m decoding functions

$$g_\ell : \mathbb{R}^T \times \{1, 2, \ldots, |\bar{W}_{d_\ell}|\} \to \{1, 2, \ldots, 2^{TR_\ell}\}$$

for the destination nodes of the m source-destination pairs $\{(s_\ell, d_\ell), \ell = 1, \ldots, m\}$, where $|\bar{W}_{d_\ell}|$ is the number of different values $\bar{W}_{d_\ell}$ can take. Note that W_{d_ℓ} may be empty.

(4) The average probability of error:

$$P_e^{(T)} := \text{Prob}\left[(\hat{W}_1, \hat{W}_2, \ldots, \hat{W}_m) \ne (W_1, W_2, \ldots, W_m)\right] \tag{19.48}$$

where $\hat{W}_\ell := g_\ell(Y_{d_\ell}^T, \bar{W}_{d_\ell})$, with $Y_{d_\ell}^T := \left[Y_{d_\ell}(1), Y_{d_\ell}(2), \ldots, Y_{d_\ell}(T)\right]$.

Definition 2
A rate vector $(R_1, \ldots, R_m)$ is said to be *feasible* for the m source-destination pairs (s_ℓ, d_ℓ), $\ell = 1, \ldots, m$, with total power constraint P_{total}, if there exists a sequence of $[(2^{TR_1}, \ldots, 2^{TR_m}), T, P_e^{(T)}]$ codes satisfying the total power constraint P_{total}, such that $P_e^{(T)} \to 0$ as $T \to \infty$.

The preceding definitions are presented with total power constraint P_{total}. If an individual power constraint P_{ind} is placed on each node, then Equation (19.47) should be modified:

$$\frac{1}{T} \sum_{t=1}^{T} X_i^2(t) \le P_{\text{ind}}, \quad \text{for } i \in N \tag{19.49}$$

and correspondingly modify the rest of the definitions to define the set of feasible rate vectors under an individual power constraint.

19.3.3 The transport capacity

The capacity *region*, is the closure of the set of all such feasible vector rates. As in Section 19.2 we will, focus mainly on the distance-weighted sum of rates.

Definition 3
As in Section 19.2, the network's *transport capacity* C_{T} is

$$C_{\mathrm{T}} := \sup_{(R_1,\ldots,R_m)\text{ feasible}} \sum_{\ell=1}^{m} R_\ell \cdot \rho_\ell$$

where $\rho_\ell := \rho_{s_\ell d_\ell}$ is the distance between s_ℓ and d_ℓ, and $R_\ell := R_{s_\ell d_\ell}$. In the following, due to limited space, a number of results from information theory will be presented without the formal proof. For more details the reader is referred to Xie and Kumar [33].

19.3.4 Upper bounds under high attenuation

(r1) The transport capacity is bounded by the network's total transmission power in media with $\gamma > 0$ or $\delta > 3$.

For a single link (s, d), the *rate* R is bounded by the *received* power at d. In wireless networks, owing to mutual interference, the transport capacity is upper-bounded by the total transmitted power P_{total} used by the entire network.

In any planar network, with either positive absorption, i.e. $\gamma > 0$, or with path loss exponent $\delta > 3$

$$C_{\mathrm{T}} \leq \frac{c_1(\gamma, \delta, \rho_{\min})}{\sigma^2} \cdot P_{\mathrm{total}} \tag{19.50}$$

where

$$c_1(\gamma, \delta, \rho_{\min}) := \begin{cases} \dfrac{2^{2\delta+7}\log \mathrm{e}\,\mathrm{e}^{-\gamma\rho_{\min}/2}(2-\mathrm{e}^{-\gamma\rho_{\min}/2})}{\delta^2 \rho_{\min}{}^{2\delta+1}\ (1-\mathrm{e}^{-\gamma\rho_{\min}/2})}, & \text{if } \gamma > 0 \\ \dfrac{2^{2\delta+5}(3\delta-8)\log \mathrm{e}}{(\delta-2)^2(\delta-3)\rho_{\min}{}^{2\delta-1}}, & \text{if } \gamma = 0 \text{ and } \delta > 3 \end{cases} \tag{19.51}$$

(r2) The transport capacity follows an $O(n)$ scaling law under the individual power constraint, in media with $\gamma > 0$ or $\delta > 3$.

Consider any planar network under the individual power constraint P_{ind}. Suppose that either there is some absorption in the medium, i.e. $\gamma > 0$, or there is no absorption at all but the path loss exponent $\delta > 3$. Then its transport capacity is upper-bounded as follows:

$$C_{\mathrm{T}} \leq \frac{c_1(\gamma, \delta, \rho_{\min})P_{\mathrm{ind}}}{\sigma^2} \cdot n \tag{19.52}$$

where $c_1(\gamma, \delta, \rho_{\min})$ is given by Equation (19.51).

As in the previous section we use notation:

$$f = O(g) \text{ if } \limsup_{n\to+\infty} [f(n)/g(n)] < +\infty$$

$f = \Omega(g)$ if $g = O(f)$; $f = \Theta(g)$ if $f = O(g)$ as well as $g = O(f)$. Thus, all $O(\cdot)$ results are upper bounds, all $\Omega(\cdot)$ results me lower bounds, and all $\Theta(\cdot)$ results are sharp order estimates for the transport capacity. For n nodes located in an area of A square meters, it is shown in Section 19.2 that the transport capacity is of order $O[\surd(An)]$ under a noninformation theoretic protocol model. If A itself grows like n, i.e. $A = \Theta(n)$, then the scaling law is $O[\surd(An)] = O(n)$, which coincides with the information-theoretic scaling law here. In fact, A must grow at least this rate since nodes are separated by a minimum distance $\rho_{\min} > 0$, i.e. $A = \Omega(n)$, and so the $O(n)$ result here is slightly stronger than the $O[\surd(An)]$ result in the previous section.

(r3) If either $\gamma > 0$ or $\delta > 2$ in any linear network, then

$$C_{\mathrm{T}} \leq \frac{c_2(\gamma, \delta, \rho_{\min})}{\sigma^2} \cdot P_{\text{total}} \tag{19.53}$$

where

$$c_2(\gamma, \delta, \rho_{\min}) := \begin{cases} \dfrac{2\mathrm{e}^{-\gamma\rho\min}\log \mathrm{e}}{(1-\mathrm{e}^{-\gamma\rho\min})^2(1-\mathrm{e}^{-2\gamma\rho\min})\rho_{\min}^{2\delta-1}}, & \text{if } \gamma > 0 \\ \dfrac{2\delta(\delta^2-\delta-1)\log \mathrm{e}}{(\delta-1)^2(\delta-2)\rho_{\min}^{2\delta-1}}, & \text{if } \gamma = 0 \text{ and } \delta > 2 \end{cases} \tag{19.54}$$

(r4) For any linear network, if either $\gamma > 0$ or $\delta > 2$, then the transport capacity is upper-bounded as follows:

$$C_{\mathrm{T}} \leq \frac{c_2(\gamma, \delta, \rho_{\min}) P_{\text{ind}}}{\sigma^2} \cdot n$$

where $c_2(\gamma, \delta, \rho_{\min})$ is as in Equation (19.44).

19.3.5 Multihop and feasible lower bounds under high attenuation

The $O(n)$ upper bound on transport capacity is tight for regular planar networks in media with $\gamma > 0$ or $\delta > 3$, and it is achieved by multihop. The 'multihop strategy' is defined as the following. Let Π_ℓ denote the set of all paths from source s_ℓ to destination d_ℓ, where by such a path π we mean a sequence $(s_\ell = j_0, j_1, \ldots, \ldots, j_z = d_\ell)$ with $j_q \neq j_r$ for $q \neq r$. The total traffic rate R_ℓ to be provided to the source destination pair (s_ℓ, d_ℓ) is split over the paths in Π_ℓ in such a way that if traffic rate $\lambda_\pi \geq 0$ is to be carried over path π, then $\sum_{\pi \in P_\ell} \lambda_\pi = R_\ell$. On each path π, packets are relayed from node to next node. On each such hop, each packet is fully decoded, treating all interference as noise. Thus, only point-to-point coding is used, and no network coding or multiuser estimation is employed. Such a strategy is of great interest and it is currently the object of much protocol development activity.

The following result implies that, when $\gamma > 0$ or $\delta > 3$, the sharp order of the transport capacity for a regular planar network is $\Theta(n)$, and that it can be attained by multihop.

(r5) In a regular planar network with either $\gamma > 0$ or $\delta > 1$, and individual power constraint P_{ind}

$$C_{\mathrm{T}} \geq S\left(\frac{\mathrm{e}^{-2\gamma} P_{\text{ind}}}{c_3(\gamma, \delta) P_{\text{ind}} + \sigma^2}\right) \cdot n$$

where

$$c_3(\gamma, \delta) := \begin{cases} \dfrac{4(1+4\gamma)\mathrm{e}^{-2\gamma} - 4\mathrm{e}^{-4\gamma}}{2\gamma(1-\mathrm{e}^{-2\gamma})}, & \text{if } \gamma > 0; \\ \dfrac{16\delta^2 + (2\pi - 16)\delta - \pi}{(\delta - 1)(2\delta - 1)}, & \text{if } \gamma = 0 \text{ and } \delta > 1 \end{cases}$$

and $S(x)$ denotes the Shannon function $S(x) := \left[\log(1+x)\right]/2$.

This order of distance weighted sum of rates is achievable by multihop. Multihop is order-optimal in a random scenario over a regular planar network in media with $\gamma > 0$ or $\delta > 3$, providing some theoretical justification for its use in situations where traffic is diffused over the network.

Consider a regular planar network with either $\gamma > 0$ or $\delta > 1$, and individual power constraint P_{ind}. The n source–destination pairs are randomly chosen as follows: every source s_ℓ is chosen as the node nearest to a randomly (uniformly i.i.d.) chosen point in the domain, and similarly for every destination d_ℓ. Then $\lim_{n\to\infty} \text{Prob}\,[\, R_\ell = c/\sqrt{(n \log n)}$ is feasible for every $\ell \in \{1, 2, \ldots, n\}] = 1$ for some $c > 0$. Consequently, a distance weighted sum of rates $C_{\text{T}} = \Omega[n/\sqrt{(\log n)}]$ is supported with probability approaching one as $\eta \to \infty$. This is within a factor $l/\sqrt{\log n}$ of the transport capacity $\Theta(n)$ possible when $\delta > 3$.

(r6) A vector of rates $(R_1, R_2, \ldots, R_m)$ can be supported by multihop in a planar network in media with $\gamma > 0$ or $\delta > 1$, if the traffic can be load balanced such that no node is overloaded and no hop is too long.

This is a fairly straightforward result saying nothing about order optimality, and is provided only in support of the above theme that multihop is an appropriate architecture for balanceable scenarios.

(r7) A set of rates $(R_1, R_2, \ldots R_m)$ for a planar network can be supported by multihop if no hop is longer than a distance $\bar{\rho}$, and for every $1 \le i \le n$, the traffic to be relayed by node i

$$\sum_{\ell=1}^{m} \sum_{\{\pi\in\Pi_\ell:\ \text{node } i \text{ belongs to } \pi\}} \lambda_\pi < S\left\{\frac{\mathrm{e}^{-2\gamma\bar{\rho}} P_{\text{ind}}}{\bar{\rho}^{2\delta}[c_4(\gamma, \delta, \rho_{\min})P_{\text{ind}} + \sigma^2]}\right\}$$

where

$$c_4(\gamma, \delta, \rho_{\min}) := \begin{cases} \dfrac{2^{3+2\delta}\mathrm{e}^{-\gamma\rho_{\min}}}{\gamma\rho_{\min}^{1+2\delta}}, & \text{if } \gamma > 0 \\ \dfrac{2^{2+2\delta}}{\rho_{\min}^{2\delta}(\delta - 1)}, & \text{if } \gamma = 0 \text{ and } \delta > 1 \end{cases}$$

19.3.6 The low-attenuation regime

In this scenario no absorption, i.e. $\gamma = 0$, and small path loss exponent are assumed. In this case coherent relaying with interference subtraction (CRIS) is considered an interesting strategy for information transmission in the following scenarios. For a source–destination pair (s, d), the nodes are divided into groups, with the first group containing only the source, and the last group containing only the destination d. Call the higher numbered groups 'downstream' groups, although they need not actually be closer to the destination. Nodes in group i, for $1 \leq i \leq k-1$, dedicate a portion P_{ik} of their power to coherently transmit for the benefit of node k and its downstream nodes. Each node k employs interference subtraction during decoding to subtract out the known portion of its received signal being transmitted by its downstream nodes.

(r8i) If there is no absorption, i.e. $\gamma = 0$, and the path loss exponent $\delta < 3/2$, then even with a fixed total power P_{total}, any arbitrarily large transport capacity can be supported by CRIS in a regular planar network with a large enough number of nodes n.

(r8ii) If $\gamma = 0$ and $\delta < 1$, then even with a fixed total power P_{total}, CRIS can support a fixed rate $R_{\min} > 0$ for any single source–destination pair in any regular planar network, irrespective of the distance between them.

A similar result exists for the regular linear networks.

(r9i) If $\gamma = 0$ and $\delta < 1$, then even with a fixed total power P_{total}, any arbitrarily large transport capacity can be supported by CRIS in a regular linear network with a large enough number of nodes n.

(r9ii) If $\gamma = 0$ and $\delta < 1/2$, then even with a fixed total power P_{total}, CRIS can support a fixed rate $R_{\min} > 0$ for any single source–destination pair in any regular linear network, irrespective of the distance between them.

A superlinear $\Theta(n^{\theta})$ scaling law with $1 < \theta < 2$ is feasible for some linear networks when $\gamma = 0$ and $\delta < 1$.

(r10) For $\gamma = 0$ and individual power constraint P_{ind} for every $0.5 < \delta < 1$, and $1 < \theta < 1/\delta$, there is a family of linear networks for which the transport capacity is $C_T = \Theta(n^{\theta})$. This order optimal transport capacity is attained in these networks by CRIS.

19.3.7 The Gaussian multiple-relay channel

The results for the low-attenuation regime rely on the following results for the Gaussian multiple-relay channel. An example of a four-node network with two parallel relays as shown in Figure 19.12. Consider a network of n nodes with α_{ij} the attenuation from node i to node j and i.i.d. additive $N(0, \sigma^2)$ noise at each receiver. Each node has an upper bound on the power available to it, which may differ from node to node. Suppose there is a single source-destination pair (s, d). We call this the Gaussian multiple relay channel.

The first result addresses the case where each relaying group consists of only one node. The strategy used is CRIS. Consider the Gaussian multiple-relay channel with coherent multistage relaying and interference subtraction. Consider $M + 1$ nodes, sequentially denoted

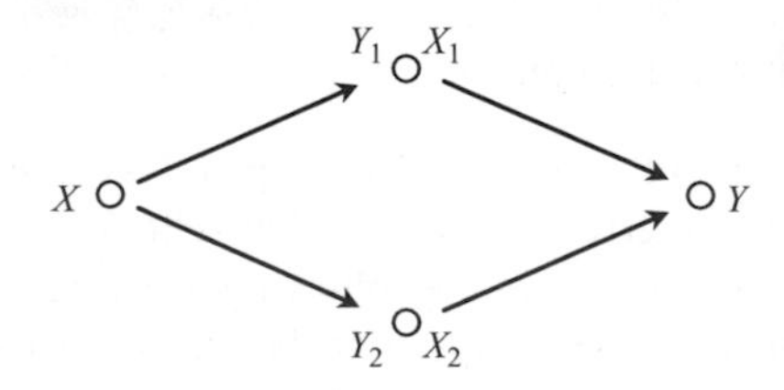

Figure 19.12 A four-node network with two parallel relays [33].

by 0, 1, ..., M, with 0 as the source, M as the destination, and the other $M-1$ nodes serving as $M-1$ stages of relaying.

(r11) Any rate R satisfying the following inequality is achievable from 0 to M:

$$R < \min_{1\le j\le M} S\left[\frac{1}{\sigma^2}\sum_{k=1}^{j}\left(\sum_{i=0}^{k-1}\alpha_{ij}\sqrt{P_{ik}}\right)^2\right]$$

where $P_{ik} \ge 0$ satisfies $\sum_{k=i+1}^{M} P_{ik} \le P_i$.

For the network setting in (r11), Theorem 3.1 in Gupta and Kumar [36] shows that a rate R_0 is achievable if there exist some $\{R_1, R_2, \ldots, R_{M-1}\}$ such that

$$R_{M-1} < S\left[P^R_{M,M-1}\left(\sigma^2 + \sum_{\ell=0}^{M-2} P^R_{M,\ell}\right)^{-1}\right]$$

and

$$R_m < \min\left\{S\left(\frac{P^R_{m+1,m}}{\sigma^2 + \sum_{\ell=0}^{m-1} P^R_{m+1,\ell}}\right),\quad R_{m+1} + \min_{m+2\le k\le M} S\left(\frac{P^R_{k,m}}{\sigma^2 + \sum_{\ell=0}^{m-1} P^R_{k,\ell}}\right)\right\}$$

for each $m = 0, 1, \ldots, M-2$, where

$$P^R_{k,\ell} := \left(\sum_{i=0}^{\ell}\alpha_{ik}\sqrt{P_{i,\ell+1}}\right)^2, \quad 0 \le \ell < k \le M.$$

From the above, recursively for $m = M-2, M-1, \ldots, 0$, it is easy to prove that

$$R_m < \min_{m+1\le j\le M} S\left[\left(\sigma^2 + \sum_{\ell=0}^{m-1} P^R_{j,\ell}\right)^{-1} \sum_{k=m}^{j-1} P^R_{j,k}\right]$$

For $m = 0$, this inequality is exactly (r11), showing a higher achievable rate. The right-hand side (RHS) in (r11) can be maximized over the choice of order of the $M-1$ intermediate nodes. The relaying can also be done by groups, and the next result addresses this. As above, maximization can be done over the assignment of nodes to the groups.

Consider again the Gaussian multiple-relay channel using coherent multistage relaying with interference subtraction. Consider any $M+1$ groups of nodes sequentially denoted by $N_0, N_1, \ldots N_M$ with $N_0 = \{s\}$ as the source, $N_M = \{d\}$ as the destination, and the other $M-1$ groups as $M-1$ stages of relay. Let n_i be the number of nodes in group N_i, $i \in \{0, 1, \ldots, M\}$. Let the power constraint for each node in group N_i be $P_i/n_i \ge 0$.

(r12) Any rate R satisfying the following inequality is achievable from s to d:

$$R < \min_{1\le j\le M} S\left[\frac{1}{\sigma^2}\sum_{k=1}^{j}\left(\sum_{i=0}^{k-1}\alpha_{N_iN_j}\sqrt{P_{ik}/n_i}\cdot n_i\right)^2\right]$$

where $P_{ik} \ge 0$ satisfies $\sum_{k=i+1}^{M} P_{ik} \le P_i$, and $\alpha N_{\dot{x}} N_j := \min\{\alpha_{k\ell} : k \in Ni, \ell \in Nj\}$, $i, j \in \{0, 1, \ldots, M\}$.

As pointed out earlier, more on the results (r1)–(r12) can be found in Xie and Kumar [33].

19.4 COOPERATIVE TRANSMISSION IN WIRELESS MULTIHOP *AD HOC* NETWORKS

The technique discussed in this section allows us to transmit reliably to far destinations that the individual nodes are not able to reach without rapidly consuming their own battery resources, even when using multihop links discussed so far. The results are of interest in both *ad hoc* and sensor networks. The key idea is to have the nodes simply echo the source's (leader) transmission operating as active scatterers while using adaptive receivers that acquire the equivalent network signatures corresponding to the echoed symbols. The active nodes in the network operate either as regenerative or nonregenerative relays. The intuition is that each of the waveforms will be enhanced by the accumulation of power due to the aggregate transmission of all the nodes while, if kept properly under control, the random errors or the receiver noise that propagate together with the useful signals will cause limited deterioration in the performance. The avalanche of signals triggered by the network leaders forms the so-called opportunistic large array (OLA).

In contrast to Sections 19.2 and 19.3, we are interested in this section in a method that utilizes the network as a *distributed modem*, where one or few sources are effectively transmitting data and all the other users are operating as repeaters. A fresh look into the concept of repeaters as a form of cooperative transmission came recently from References [37–40].

We will assume that, in a network of N nodes transmitting over a shared medium, each node is part of a multiple stage relay of a single source transmitting toward a remote receiver whose position is unknown to all the nodes. If no node in the network is powerful enough to communicate reliably with the remote receiver, the problem is referred to as the *reach-back problem*. Coordination among nodes in a large network is an extremely difficult task. In a cooperative transmission mechanism for which cooperation is obtained in a distributed fashion the source *(leader)* transmits a pulse with complex envelope $p_m(t)$ out of an M-ary set of waveforms. The resulting signal at the ith receiver is $r_i(t) = s_{i,m}(t) + n_i(t)$ where $s_{i,m}(t)$ is the network-generated signature of the mth symbol. If N nodes echo *exactly the same symbol*,

$$s_{i,m}(t) = \sum_{n=1}^{N} A_{i,n}(t)p_m\left[t - \tau_{i,n}(t)\right], m = 0, \ldots, M-1$$

where $n_i(t)$ is the ith receiver AWGN with variance N_0, $\tau_{i,n}(t)$ is the delay of the link between the ith and the nth node, including the asynchronism of the beginning of transmission for each node n, and $A_{i,n}(t)$ is the product of a complex fading coefficient $\omega_{i,n}(t)$, the transmit power P_t and the channel average gain, e.g. $\propto (1 + d_{i,n})^{-\alpha_{i,n}}$(log–normal fading), where $d_{i,n}$ is the distance, and $\alpha_{i,n}$ the decay constant between the ith and nth nodes.

The following assumptions are used:

(a1) $A_{i,n}(t)$ and $\tau_{i,n}(t)$ are constant over multiple symbol durations T_s; the nodes are quasi-stationary for a time much greater than T_s.

(a2) The delays are $\tau_{i,1} < \tau_{i,2} \leq \cdots \leq \tau_{i,N}$, where the minimum delay $\tau_{i,1}$ corresponds to the leader. To avoid ISI, the upper bound for the effective symbol rate is $R_s = 1/T_s \leq 1/\Delta\tau$, where $\Delta\tau$ is the maximum *delay spread* of $s_{i,m}(t)$ for all i. The delay spread for node i is defined as

$$\sigma_{\tau i} = \sqrt{\frac{\int_{-\infty}^{\infty} (t - \bar{\tau}_i)^2 \cdot \left|S_{i,m}(t)\right|^2 \mathrm{d}t}{\int_{-\infty}^{\infty} \left|S_{i,m}(t)\right|^2 \mathrm{d}t}}$$

where the average delay

$$\bar{\tau}_i = \frac{\int_{-\infty}^{\infty} t \cdot \left|S_{i,m}(t)\right|^2 \mathrm{d}t}{\int_{-\infty}^{\infty} \left|S_{i,m}(t)\right|^2 \mathrm{d}t}$$

and thus, $\Delta\tau = \max_i \sigma_{\tau_i}$. Echoes that come from farther away are strongly attenuated (by $\approx d^{-\alpha}$); therefore, the echoes received at node i are nonnegligible only for those coming from nodes within a certain distance Δd, which essentially depends on the transmit power and path loss. Hence, R_s can be increased by lowering the transmit power, capitalizing on spatial bandwidth reuse. In the reach back problem, however, the delay spread is $\Delta\tau \approx \sup_i[\tau_{i,N} - \tau_{i,1}]$ because the receiver is roughly at the same distance from all nodes.

(a3) T_s is fixed for all nodes to $c_1\Delta\tau$, where c_1 is a constant taken to satisfy the ISI constraint. With (a3), we guarantee that no ambiguity will occur at the nodes in timing their responses. The transmission activity of the node is solely dependent on the signal that the node receives. Based on the evolution of $s_{i,m}(t)$, we can distinguish two phases: (1) the earlier *receive phase*, when the upstream waves of signals approach the node, and (2) the period after the *firing instant*, which we call the *rest phase*, where the node hears the echoes of the downstream wave of signals fading away (for the regenerative case, the firing instant occurs shortly after the time when the node has accumulated enough energy to detect the signal). The switching between the two modes can be viewed as a very elementary form of time-division duplex (TDD).

(a4) The leader (and also the nodes in the regenerative case) transmits pulses with complex envelope $p_m(t)$ having limited double-sided bandwidth W and, approximately, duration T_p. By sampling at the Nyquist rate, $N_p = T_p W$ is the approximate length of the sequence $\{p_m(k/W)\}$ of samples. Multipath propagation can be simply included in the model by increasing the number of terms in the summation in $s_{i,m}(t)$; therefore, it does not require special attention. In fact, when we neglect the propagation of errors and noise that occurs in the case of regenerative and nonregenerative repeaters, respectively, the OLA itself is equivalent to a multipath channel, created by a set of active scatterers. In the regenerative case, the ideal OLA response is

$$g_i(\tau) = \sum_{n=1}^{N} A_{i,n}\delta(\tau - \tau_{i,n}) \tag{19.55}$$

The nonregenerative OLA scattering model is more complex due to the feedback effect, which implies that not one but several signal contributions are scattered by each source.

The received OLA response is

$$g_i(\tau) = \sum_{n=1}^{N'} A_{i,n'}\delta(\tau - \tau_{i,n'}) \tag{19.55a}$$

For every possible path in the network, there is a contribution to the summation in Equation (19.55a) that has an amplitude equal to the product of all the path link gains traveled so far and a delay equal to the sum of all the path delays. Theoretically, the number of reflections $N' \to \infty$ because the signals and their amplified versions keep cycling in the network and adding up. If properly controlled, the contributions will keep adding up and then opportunistically serve the purpose of enhancing the signal. Hence, the key for the nonregenerative design is to control the noise that accompanies the useful signal. In both regenerative and nonregenerative cases, the received signal can be rewritten as the following convolution:

$$r_i(t) = g_i(t)^* p_m(t) + n_i(t) \tag{19.56}$$

where $g_i(t)$ is the network impulse response, which is analogous to that of a multipath channel. Based on Equation (19.46), the idea is to let the nodes operate as regenerative and nonregenerative repeaters and avoid any complex coordination procedure to forward their signals at the network layer and share the bandwidth at the MAC layer. In addition, no channel state information is used. The information flow is carried forward using receivers that are capable of tracking the *unknown* network response $g_i(t)$ or, directly, the signature waveforms $s_{i,m}(t) \triangleq g_i(t)^* p_m(t)$. We should expect that the OLA behaves as a frequency-selective channel. Nodes' mobility causes changes of the response $g_i(t)$ over time. If most of the network is stationary and N is large, the inertia of the system will be such that mobile nodes will cause small changes in $g_i(t)$.

Since the transmission channel is bandlimited with passband bandwidth W, the signature waveform $p_m(t)$ will have to be bandlimited and, therefore, uniquely expressible through its samples taken at the Nyquist rate $1/T_c$, where $T_c = 1/W$. In general, $p_m(t)$ corresponds to a finite number of samples N_p and is approximately time limited with duration $T_p \approx N_p/W$. Introducing the vectors $\mathbf{p}_m$, $\mathbf{g}_i$ and $\mathbf{r}_i$ such that

$$\begin{aligned}
&\{\mathbf{p}_m\}_k = p_m(kT_c), k = 0, \ldots N_p - 1 \\
&\{\mathbf{g}_i\}_k = \int \operatorname{sinc}(\pi W\tau) g_i(kT_c + l_i T_C - \tau) d\tau, \quad k = 0, \ldots N_i - 1 \\
&\{\mathbf{r}_i\}_k = r_i(kT_c + l_i T_c) k = 0, \ldots, N_i + N_p - 2 \\
&\{\mathbf{n}_i\}_k = n_i(kT_c + l_i T_c), k = 0, \ldots, N_i + N_p - 2
\end{aligned}$$

where N_i is the number of samples needed to represent $g_i(t)$, we have

$$\{\mathbf{r}_i\}_k = \sum_{n=0}^{N_p-1} \{\mathbf{p}_m\}_n \{\mathbf{g}_i\}_{k-n} + \{\mathbf{n}_i\}_k$$

By using the following Toeplitz convolution matrix:

$$\{\mathbf{G}_i\}_{k,n} = \{\mathbf{g}_i\}_{k-n}, \quad n = 0, \ldots, N_p - 1; k = 0, \ldots, N_i + N_p - 2$$

we obtain

$$\mathbf{r}_i = \mathbf{G}_i \mathbf{p}_m + \mathbf{n}_i \tag{19.57}$$

19.4.1 Transmission strategy and error propagation

The transmission of the OLA is led by a predetermined source node in the network. All the other nodes form multiple stages of relays to either flood the network with the information from the source, or just to pass the information to a remote receiver. The intermediate nodes in OLA have a choice of whether to relay or not, depending on the performance at that node. In the *regenerative scheme*, the OLA nodes has the choice of retransmitting its detected symbol or staying silent. Only nodes that are *connected* actively reply, where connectivity is defined as follows.

Definition 1
The ith regenerative node is connected if, based on its estimates of all possible signatures $G_i p_m$ and receiver noise variance, the pairwise symbol error probability of the ith receiver not considering error propagation) is below *a* fixed upper-bound ε , i.e. $\max_m Pr\{m \rightarrow \mu\} \leq \varepsilon, \forall \mu \neq m, m = 0, \ldots, M-1$.

In the N_s samples contained in each symbol period, the time instant selected for the detection and subsequent echo is the first sample $\bar{N}_i \leq N_s$ at which the node is connected. If there is no such sample, the node will never echo the signal (but it may obviously detect the information at his own risk.)

In the *nonregenerative scheme*, every node that achieves the SNR constraint amplifies the signal coming from the other nodes as well as their receiver noise. Hence, the noise n_i has a rather complex structure since it includes the noise that comes from every node that has transmitted previously and all its subsequent amplifications along with the signal. Since the geographical area in an *ad hoc* or sensor network is limited, the delay spread of each node response will also be limited, as far as the signal-to-noise contribution is concerned. Considerations on the SNR can be deduced by considering the inherent recursive structure of the signal composition. Details can found in Scaglione and Hong [40].

Definition 2
The ith nonregenerative node is said to be connected if the signal to noise ratio at the node ξ_i is above *a* fixed threshold $\xi_i > \bar{\xi}$.

19.4.2 OLA flooding algorithm

In this section, we compare numerically OLA with the more traditional ways of flooding the network described in Chapter 13. The flooding algorithm in Royer and Toh [41] is commonly indicated as one of the simplest ways of distributing information in the network or of searching for a path to the desired destination to initialize table-driven protocols. Some interesting alternatives are the probabilistic scheme [42] and the scalable broadcast algorithm [43]. Most of these methods require the MAC and physical layer to provide virtual transmission pipes (obtained typically with contention) that connect each pair of nodes, which imitates wired networks. The approach is legitimate but obviously inefficient. In fact, in solving the network broadcast problem, it is natural to utilize and integrate in the design the fact that wireless devices are physically broadcasting. This is precisely what happens in the OLA, where, contrary to the networking approaches, the transmission protocol and cooperation strategy operate at the physical layer. In OLA framework, the receiver has to solve an equivalent point-to-point communication problem without requiring higher layers' interventions (MAC or network layer). The benefits from eliminating two layers are higher connectivity and faster flooding. However, the OLA flooding algorithm cannot be used for some broadcasting applications such as route discovery due to the fact that the higher layer information is eliminated.

Each node in the OLA is assumed to have identical transmission resources; therefore, any node has the ability of assuming the role of a leader. The leader can be chosen to be the leader of a troop, the clusterheads in clustering algorithms, or simply some node that has information to send.

19.4.3 Simulation environment

Various network broadcasting methods were analyzed using the ns2 network simulator. The simulation parameters are specified in Table 19.1 [40, 44]. Physical layer resources of the IEEE 802.11 DSSS PHY specifications [45] for the 2.4 GHz carrier were used. Each node-to-node transmission is assumed to experience independent small-scale fading with Rayleigh coefficients of variance 1. The large-scale fading is deterministic, and the path loss model is based on the model used in *ns2* [46], where the *free space model* is used for distance $d < d_c$ (the *cross-over distance*) and the *two-ray ground reflection model* is used for $d > d_c$, where $d_c = 4\pi/\lambda$. The position of the 'leader' is randomly selected, and the OLA is regenerative.

There are three parameters that define the simulation setting. The first is the *point-to-point average SNR* (averaged over the small scale fading), which is defined as $SNR_{p2p}(d) \triangleq P_t/(N_0 d^\alpha)$ where d^α is the path loss. The second is the *transmission radius* $d_{p2p} \triangleq [P_t/(N_0\xi)]^{-\alpha}$, which is equal to the distance at which the $SNR_{p2p}(d) = \xi$ using the specified path loss model. The exact *SNR* at each node is different due to the accumulation of signals. Therefore, we define a third parameter, which is *the node SNR at the detection level* $\mathrm{SNR}_{\mathrm{det}} \triangleq ||g_i||^2/E\{||n_i||^2\}$.The value of SNR_{det} can be mapped one-to-one into the node error rate if error propagation is neglected and provides a criterion equivalent to the one in definition 1 to establish whether a node is connected or not. In all cases, the threshold on $\mathrm{SNR}_{\mathrm{det}}$ is the same as the required point-to-point $\mathrm{SNR}_{p2p}(d)$ used to define network links in conventional networks, and let this value be ξ.

To simplify the network simulations, it is assumed that the transmission propagates through the network approximately in a multiple ring structure shown in Figure 19.13. In

Table 19.1 Simulation parameters

Simulation Paramaters	Values
Network area	350 × 350 m
Radius of TX	100 m
Payload	64 bytes/packet
Number of trials	100
Modulation	BPSK
Bandwidth (IEEE 802.11)	83.5 Mbs

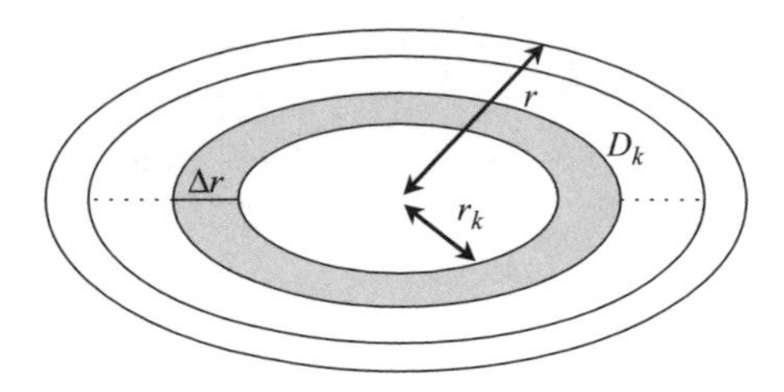

Figure 19.13 Nodes in the ring Δr transmit approximately simultaneously.

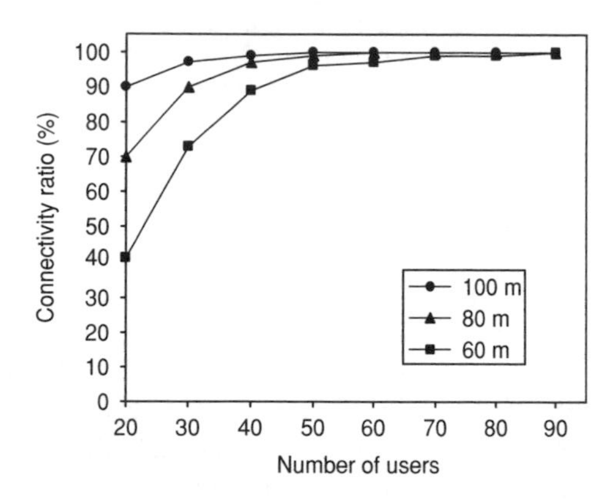

Figure 19.14 Connectivity ratio vs number of nodes in the network.

each ring, we prune away the nodes that do not have strong enough SNR_{det}, but we do not detect and retransmit at the exact time when the SNR_{det} reaches the threshold. Because we just partition the network geographically, we can expect, in general, nonuniform and lower error rates than the ones prescribed by the threshold on SNR_{det}. In the experiments, it was assumed that the signal space is perfectly estimated at each relaying node. This assumption is practical, because the network is static, and when the number of training symbols is sufficiently large, the noise variance caused by the contribution of the estimation error can be neglected. Figure 19.14 shows to what degree the network is connected according to definition 1 when the threshold for $(\text{SNR}_{\text{det}})_{\text{dB}}$ is 10 dB. Specifically, the *connectivity ratio* (CR) is shown, which is defined as *the number of nodes that are 'connected', over the total number of nodes in the network.* The nodes' transmit power and thermal noise are constant and are fixed so that $(\text{SNR}_{p2p})_{\text{dB}} = 10$ at the distances $d_{p2p} = 100$, 80 and 60 m, representing the transmission radii.

For $d_{p2p} = 100$ m, the CR is 100 % even at very low node density. As we shorten the radius of transmission, the connectivity of the network will decrease. Figure 19.15 plots the *delivery ratio* (DR), which is defined as the ratio between the average number of nodes that receive packets using a specific flooding algorithm over the number of nodes that are *connected in multiple hops*, i.e. nodes for which there exists a path from the leader that is formed with point-to-point links having SNR_{p2p} above a fixed threshold. The only cause of packet loss considered in Williams and Camp [44] is the fact that the packet is not delivered because it is dropped by the intermediate relays' queues to reduce the nodes congestion. *Routing, MAC and physical layer errors and their possible propagation are ignored* in the

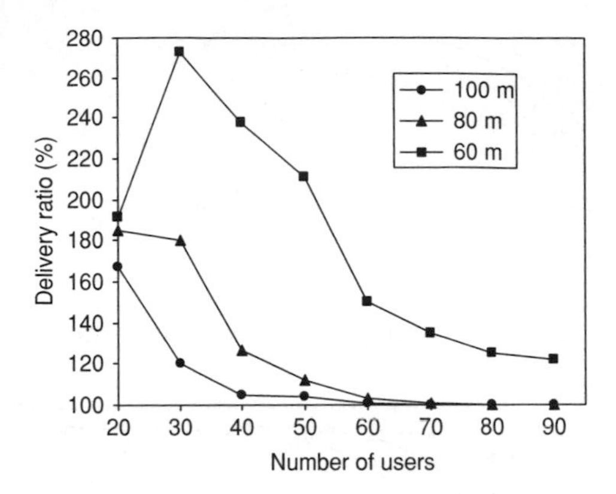

Figure 19.15 Delivery ratio vs number of nodes in the network.

definition of DR. DR essentially shows how routing and MAC problems can reduce the probability of successfully reaching the nodes. Hence, according to Williams and Camp [44], the simple flooding algorithm achieves 100 % DR, even if it might create longer delays and instability due to the increased level of traffic. In the OLA, the accumulation of signal energy may still allow extra nodes (besides the ones that have a multihop route) to receive the broadcasted packets reliably. Therefore, if we calculate the ratio between the number of nodes connected in the OLA according to definition 1 and the number of nodes that are connected through multihop point-to-point links, we must be able to achieve more than 100 % DR. Using the parameters in Table 19.1, Figure 19.15 plots the DR vs the number of nodes. It is shown that there can be remarkable gains in connectivity over any scheme operating solely on point-to-point links.

The *end-to-end delay* is the time required to broadcast the packet to the entire network. In the OLA flooding algorithm, there is no channel contention, and therefore, the overhead necessary for carrier sensing and collision avoidance used in IEEE 802.11 is eliminated. With the reduction of overheads and the time saved by avoiding channel contention, it is clear that the speed of flooding will be much higher than the traditional broadcasting methods. Figure 19.16 shows the end-to-end delay vs the number of nodes in the network. The end-to-end delay is only in the order of milliseconds for a packet payload of 64 bytes, which coincides with the symbol period T_s times the number of bits in a packet.

19.5 NETWORK CODING

In this section we consider a communication network in which certain source nodes multicast information to other nodes on the network in multihops where every node can pass on any of its received data to others. The question is how the data should be sent and how fast *each* node can receive the complete information. Allowing a node to encode its received data before passing it on, the question involves optimization of the multicast mechanisms at the nodes. Among the simplest coding schemes is linear coding, which regards a block of data as a vector over a certain base field and allows a node to apply a linear transformation to a vector before passing it on.

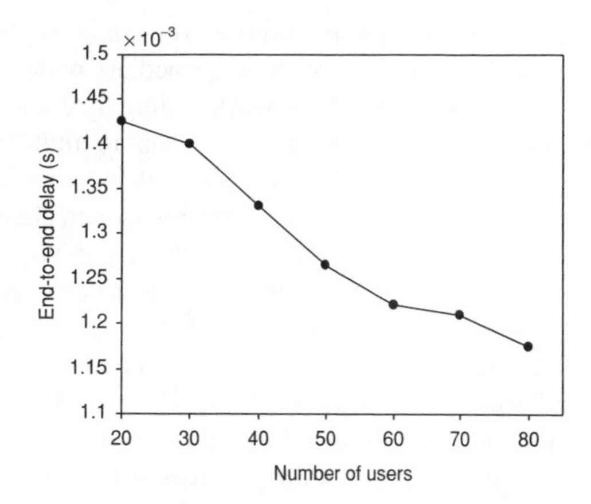

Figure 19.16 End-to-end delay vs number of nodes in the network.

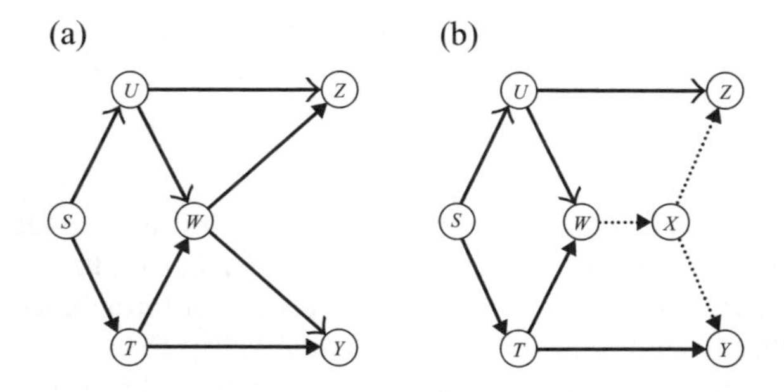

Figure 19.17 Illustration of network coding.

As in Chapter 7, we define a *communication network* as a pair (G, S), where G is a finite *directed* multigraph and S (source) is the unique node in G without any incoming edges. A directed edge in G is called a *channel* in the communication network (G, S). A channel in graph G represents a noiseless communication link on which one unit of information (e.g. a bit) can be transmitted per unit time. The multiplicity of the channels from a node X to another node Y represents the *capacity* of direct transmission from X to Y. We assume that, every single channel has unit capacity.

At the source S, information is generated and multicast to other nodes on the network in the multihop fashion where every node can pass on any of its received data to other nodes. At each nonsource node which serves as a sink, the complete information generated at S is recovered. Now the question is how fast each sink node can receive the complete information. As an example, consider the multicast of two data bits, b_1 and b_2, from the source S in the communication network depicted by Figure 19.17(a) as both nodes Y and Z. One solution is to let the channels ST, TY, TW and WZ carry the bit b_1 and channels SU, UZ, UW and WY carry the bit b_2. Note that, in this scheme, an intermediate node sends out a data bit only if it receives the same bit from another node. For example, the node T receives the bit b_1 and sends a copy on each of the two channels TY and TW. Similarly, the node U receives the bit b_2 and sends a copy into each of the two channels UW and UZ. We assume that there is no processing delay at the intermediate nodes.

Unlike a conserved physical commodity, information can be replicated or coded. The notion of network coding refers to coding at the intermediate nodes when information is multicast in a network. Let us now illustrate network coding by considering the communication network depicted in Figure 19.17(b). Again, we want to multicast two bits b_1 and b_2 from the source S to both the nodes Y and Z. A solution is to let the channels ST, TW and TY carry the bit b_1, channels SU, UW and UZ carry the bit b_2, and channels WX, XY and XZ carry the exclusive-OR $b_1 \oplus b_2$. Then, the node Y receives b_1 and $b_1 \oplus b_2$, from which the bit $b_2 = b_1 \oplus (b_1 \oplus b_2)$ can be decoded. Similarly, the node Z can decode the bit b_1 from b_2 and $b_1 \oplus b_2$ as $b_1 = b_2 \oplus (b_1 \oplus b_2)$. The coding/decoding scheme is assumed to have been agreed upon beforehand. In order to discuss this issue in more detail, in this section we first introduce the notion of a *linear-code multicast* (LCM). Then we show that, with a 'generic' LCM, every node can simultaneously receive information from the source at rate equal to its max-flow bound. After that, we describe the physical implementation of an LCM, first when the network is acyclic and then when the network is cyclic followed by a presentation of a greedy algorithm for constructing a generic LCM for an acyclic network. The same algorithm can be applied to a cyclic network by expanding the network into an acyclic network. This results in a 'time-varying' LCM, which, however, requires high complexity in implementation. After that, we introduce the time-invariant LCM (TILCM).

Definition 1
Over a communication network a *flow* from the source to a nonsource node T is a collection of channels, to be called the *busy* channels in the flow, such that: (1) the subnetwork defined by the busy channels is acyclic, i.e. the busy channels do not form directed cycles; (2) for any node other than S and T, the number of incoming busy channels equals the number of outgoing busy channels; (3) the number of outgoing busy channels from S equals the number of incoming busy channels to T. The number of outgoing busy channels from S will be called the *volume* of the flow. The node T is called the *sink* of the flow. All the channels on the communication network that are not busy channels of the flow are called the *idle* channels with respect to the flow.

Definition 2
For every nonsource node T on a network (G, S), the maximum volume of a flow from the source to T is denoted $max\ flow_G(T)$, or simply *mf(T)* when there is no ambiguity.

Definition 3
A *cut* on a communication network (G, S) between the source and a nonsource node T means a collection C of nodes which includes S but not T. A channel XY is said to be *in* the cut C if $X \in C$ and $Y \notin C$. The number of channels in a cut is called the *value* of the cut.

19.5.1 Max-flow min-cut theorem (mfmcT)

For every nonsource node T, the minimum value of a cut between the source and a node T is equal to *mf(T)*. Let d be the maximum of *mf(T)* over all T. In the sequel, the symbol Ω will denote a fixed d-dimensional vector space over a sufficiently large base field. The information unit is taken as a symbol in the base field. In other words, one symbol in the base field can be transmitted on a channel every unit time.

Definition 4

An LCM υ on a communication network *(G, S)* is an assignment of a vector *space* $\upsilon(X)$ to every node X and a vector $\upsilon(XY)$ to every channel *XY* such that (1) $\upsilon(S) = \Omega$; (2) $\upsilon(XY) \in \upsilon(X)$ for every channel XY; and (3) for any collection $\wp$ of nonsource nodes in the network $\langle\{\upsilon(T) : T \in \wp\}\rangle = \langle\{\upsilon(XY) : X \notin \wp Y \in \wp\}\rangle$. The notation $\langle\cdot\rangle$ is for *linear span*. Condition (3) says that the vector spaces $\upsilon(T)$ on all nodes T inside $\wp$ together have the same linear span as the vectors $\upsilon(XY)$ on all channels XY to nodes in $\wp$ from outside $\wp$.

LCM v data transmission: *The information to be transmitted from S is encoded as a d-dimensional row vector, referred to as an information vector. Under the transmission mechanism prescribed by the LCM v, the data flowing on a channel XY is the matrix product of the information (row) vector with the (column) vector v(XY). In this way, the vector v(XY) acts as the kernel in the linear encoder for the channel XY. As a direct consequence of the definition of an LCM, the vector assigned to an outgoing channel from a node X is a linear combination of the vectors assigned to the incoming channels to X. Consequently, the data sent on an outgoing channel from a node X is a linear combination of the data sent on the incoming channels to X. Under this mechanism, the amount of information reaching a node T is given by the dimension of the vector space v(T) when the LCM v is used.*

Coding in Figure 19.17(b) is achieved with the LCM υ specified by

$$\begin{aligned} \upsilon(ST) &= \upsilon(TW) = \upsilon(TY) = (1\ \ 0)^{\mathrm{T}} \\ \upsilon(SU) &= \upsilon(UW) = \upsilon(UZ) = (0\ \ 1)^{\mathrm{T}} \end{aligned} \tag{19.58}$$

and

$$\upsilon(WX) = \upsilon(XY) = \upsilon(XZ) = (1\ \ 1)^{\mathrm{T}}$$

where $(\,)^{\mathrm{T}}$ stands for transposed vector. The data sent on a channel is the matrix product of the *row* vector $(b_1\ b_2)$ with the *column* vector assigned to that channel by υ. For instance, the data sent on the channel *WX* is

$$(b_1, b_2)\upsilon(WX) = (b_1, b_2)\,(1\ \ 1)^{\mathrm{T}} = b_1 + b_2$$

Note that, in the special case when the base field of Ω is *GF*(2), the vector $b_1 + b_2$ reduces to the exclusive-OR $b_1 \oplus b_2$ in an earlier example.

Proposition P1

For every LCM υ *on a network, for all nodes T* $\dim[\upsilon(T)] \leq mf\ (T)$. *To prove it fix a nonsource node T and any cut C between the source and T* $\upsilon(T) \subset \langle \upsilon(Z) : Z \notin C\rangle = \langle \upsilon(YZ) : Y \in C \text{ and } Z \notin C\rangle$. *Hence,* $\dim[\upsilon(T)] \leq \dim(\langle \upsilon(YZ) : Y \in C \text{ and } Z \notin C\rangle)$, *which is at most equal to the value of the cut. In particular,* $\dim[\upsilon(T)]$ *is upper-bounded by the minimum value of a cut between S and T, which by the max-flow min-cut theorem is equal to mf(T).*

This means that *mf (T)* is an upper bound on the amount of information received by T when an LCM v is used.

19.5.2 Achieving the max-flow bound through a generic LCM

In this section, we derive a sufficient condition for an LCM v to achieve the max-flow bound on $\dim[v(T)]$ in Proposition 1.

Definition

An LCM v on a communication network is said to be generic if the following condition holds for any collection of channels $X_1Y_1, X_2Y_2, \ldots, X_mY_m$ for $1 \le m \le d$: $(*)$ $v(X_k) \not\subset \langle\{v(X_jY_j) : j \ne k\}\rangle$ for $1 \le k \le m$ if and only if the vectors $v(X_1Y_1)$, $v(X_2Y_2)$, $\ldots$, $v(X_mY_m)$ are linearly independent. If $v(X_1Y_1)$, $v(X_2Y_2)$, $\ldots$, $v(X_mY_m)$ are linearly independent, then $v(X_k) \not\subset \langle\{v(X_jY_j) : j \ne k\}\rangle$ since $v(X_kY_k) \in v(X_k)$. A generic LCM requires that the converse is also true. In this sense, a generic LCM assigns vectors which are as linearly independent as possible to the channels.

With respect to the communication network in Figure 19.17(b), the LCM v defined by Equation (19.58) is a generic *LCM*. However, the *LCM* u defined by

$$u(ST) = u(TW) = u(TY) = (1\ 0)^{\mathrm{T}}$$
$$u(SU) = u(UW) = u(UZ) = (0\ 1)^{\mathrm{T}} \qquad (19.59)$$

and

$$u(WX) = u(XY) = u(XZ) = (1\ 0)^{\mathrm{T}}$$

is not generic. This is seen by considering the set of channels $\{ST, WX\}$ where

$$u(S) = u(W) = \left\langle \begin{pmatrix} 1 \\ 0 \end{pmatrix}, \begin{pmatrix} 0 \\ 1 \end{pmatrix} \right\rangle$$

Then $u(S) \not\subset \langle u(WX)\rangle$ and $u(W) \not\subset \langle u(ST)\rangle$, but $u(ST)$ and $u(WX)$ are not linearly independent. Therefore, u is not generic. Therefore, in a generic LCM v any collection of channels $XY_1, XY_2, \ldots, XY_m$ from a node X with $m \le \dim[v(X)]$ must be assigned linearly independent vectors by v.

Theorem T1

If v is a generic LCM on a communication network, then for all nodes T, $\dim[v(T)] =$ *mf*(T). To prove it, consider a node T not equal to S. Let f be the common value of *mf*(T) and the minimum value of a cut between S and T. The inequality $\dim[v(T)] \le f$ follows from Proposition 1. So, we only have to show that $\dim[v(T)] \ge f$. To do so, let $\dim(C) = \dim(\langle v(X, Y) : X \in C \text{ and } Y \not\in C\rangle)$ for any cut C between S and T. We will show that $\dim[v(T)] \ge f$ by contradiction. Assume $\dim[v(T)] < f$ and let A be the collection of cuts U between S and T such that $\dim(U) < f$. Since $\dim[v(T)] < f$ implies $V\backslash\{T\} \in A$, where v is the set of all the nodes in G, A is nonempty.

By the assumption that v is a generic LCM, the number of edges out of S is at least d, and $\dim(\{S\}) = d \geq f$. Therefore, $\{S\} \notin$ A. Then there must exist a minimal member $U \in A$ in the sense that for any $Z \in U\backslash\{S\} \neq \phi, U\backslash\{Z\} \notin$ A. Clearly, $U \neq \{S\}$ because $\{S\} \notin A$. Let K be the set of channels in cut U and B be the set of boundary nodes of U, i.e. $Z \in B$ if and only if $Z \in U$ and there is a channel (Z, Y) such that $Y \notin U$. Then for all $W \in B$, $v(W) \not\subset \langle v(X, Y) : (X, Y) \in K\rangle$ which can be seen as follows. The set of channels in cut $U\backslash\{W\}$ but not in K is given by $\{(X, W) : X \in U\backslash\{W\}\}$. Since v is an LCM $\langle v(X, W) : X \in U\backslash\{W\}\rangle \subset v(W)$. If $v(W) \subset \langle v(X, Y) : (X, Y) \in K\rangle$, then $\langle v(X', Y') : X' \in U\backslash\{W\}, Y' \notin U\backslash\{W\}\rangle$ the subspace spanned by the channels in cut $U\backslash\{W\}$, is contained by $\langle v(X, Y) : (X, Y) \in K\rangle$. This implies that $\dim(U\backslash\{W\}) \leq \dim(U) < f$ is a contradiction. Therefore, for all $W \in B$, $v(W) \not\subset \langle v(X, Y) : (X, Y) \in K\rangle$. For all $(W, Y) \in K$, since $\langle v(X, Z) : (X, Z) \in K\backslash\{(W, Y)\}\rangle \subset \langle v(X, Y) : (X, Y) \in K\rangle$, $v(W) \not\subset \langle v(X, Y) : (X, Y) \in K\rangle$ implies that $v(W) \not\subset \langle v(X, Z) : (X, Z) \in K\backslash\{(W, Y)\}\rangle$.

Then, by the definition of a generic LCM $\{v(XY) : (X, Y) \in K\}$ is a collection of vectors such that $\dim(U) = \min(|K|, d)$. Finally, by the max-flow min-cut theorem, $|K| \geq f$, and since $d \geq f$, $\dim(U) \geq f$. This is a contradiction to the assumption that $U \in A$. The theorem is proved.

An LCM for which $\dim[v(T)] = mf(T)$ for all T provides a way for broadcasting a message generated at the source S for which every nonsource node T receives the message at rate equal to $mf(T)$. This is illustrated by the next example, which is based upon the assumption that the base field of Ω is an infinite field or a sufficiently large finite field. In this example, we employ a technique which is justified by the following arguments.

Lemma 1

Let X, Y and Z be nodes such that mf(X) = i, mf(Y) = j, and mf(Z) = k, where $i \leq j$ and $i > k$. By removing any edge UX in the graph, mf(X) and mf(Y) are reduced by at most 1, and mf(Z) remains unchanged.

To prove it we note that, by removing an edge UX, the value of a cut C between the source S and node X (respectively, node Y) is reduced by 1 if edge UX is in C, otherwise, the value of C is unchanged. By the *mfmcT*, we see that *mf(X)* and *mf(Y)* are reduced by at most 1 when edge UX is removed from the graph. Now consider the value of a cut C between the source S and node Z. If C contains node X, then edge UX is not in C, and, therefore, the value of C remains unchanged upon the removal of edge UX. If C does not contain node X, then C is a cut between the source S and node X. By the *mfmcT*, the value of C is at least i. Then, upon the removal of edge UX, the value of C is lower-bounded by $i - 1 \geq k$. Hence, by the *mfmcT*, *mf(Z)* remains to be k upon the removal of edge UX.

Example E1

Consider a communication network for which $mf(T) = 4, 3$ or 1 for nodes T in the network. The source S is to broadcast 12 symbols $a_1, \ldots, a_{12}$ taken from a sufficiently large base field F. (Note that 12 is the least common multiple of 4, 3 and 1.) Define the set $\mathbf{T}_i = \{T : mf(T) = i\}$, for $i = 4, 3, 1$. For simplicity, we use the second as the time unit. We now describe how $a_1, \ldots, a_{12}$ can be broadcast to the nodes in $\mathbf{T}_4$, $\mathbf{T}_3$, $\mathbf{T}_1$, in 3, 4 and 12 s, respectively, assuming the existence of an LCM on the network for $d = 4, 3, 1$.

(1) Let v_1 be an *LCM* on the network with d $= 4$. Let $\alpha_1 = (a_1\ a_2\ a_3\ a_4), \alpha_2 = (a_5\ a_6\ a_7\ a_8)$ and $\alpha_3 = (a_9\ a_{10}\ a_{11}\ a_{12})$. In the first second, transmit α_1 as the information vector using v_1, in the second second, transmit α_2, and in the third second, transmit α_3. After 3 s, after neglecting delay in transmissions and computations all the nodes in $\mathbf{T}_4$ can recover α_1, α_2 and α_3.

(2) Let r be a vector in F^4 such that $\langle\{r\}\rangle$ intersects trivially with $v_1(T)$ for all T in $\mathbf{T}_3$, $i.e. \langle\{r, v_1(T)\}\rangle = F^4$ for all T in $\mathbf{T}_3$. Such a vector r can be found when F is sufficiently large because there are a finite number of nodes in $\mathbf{T}_3$. Define $b_i = \alpha_i r$ for $i = 1, 2, 3$. Now remove incoming edges of nodes in $\mathbf{T}_4$, if necessary, so that *mf*(T) becomes 3 if T is in $\mathbf{T}_4$, otherwise, *mf*(T) remains unchanged. This is based on Lemma 1). Let v_2 be an *LCM* on the resulting network with $d = 3$. Let $\beta = (b_1\ b_2\ b_3)$ and transmit β as the information vector using v_2 in the fourth second. Then all the nodes in $\mathbf{T}_3$ can recover β and hence α_1, α_2 *and* α_3.

(3) Let s_1 and s_2 be two vectors in F^3 such that $\langle\{s_1, s_2\}\rangle$ intersects with $v_2(T)$ trivially for all T in $\mathbf{T}_1$, $i.e. \langle\{s_1, s_2, v_2(T)\}\rangle = F^3$ for all T in $\mathbf{T}_1$. Define $\gamma_i = \beta s_i$ for $i = 1, 2$. Now remove incoming edges of nodes in $\mathbf{T}_4$ and $\mathbf{T}_3$, if necessary, so that *mf*(T) becomes 1 if T is in $\mathbf{T}_4$ or $\mathbf{T}_3$, otherwise, *mf*(T) remains unchanged. Again, this is based on Lemma 1). Now let v_3 be an *LCM* on the resulting network with $d = 1$. In the fifth and the sixth seconds, transmit γ_1 and γ_2 as the information vectors using v_3. Then all the nodes in $\mathbf{T}_1$ can recover β.

(4) Let t_1 and t_2 be two vectors in F^4 such that $\langle\{t_1, t_2\}\rangle$ intersects with $\langle\{r, v_1(T)\}\rangle$ trivially for all T in $\mathbf{T}_1$, $i.e.$ $\langle\{t_1, t_2, r, v_1(T)\}\rangle = F^4$ for all T in $\mathbf{T}_1$. Define $\delta_1 = \alpha_1 t_1$ and $\delta_2 = \alpha_1 t_2$. In the seventh and eighth seconds, transmit δ_1 and δ_2 as the information vectors using v_3. Since all the nodes in $\mathbf{T}_1$ already know b_1, upon receiving δ_1 and δ_2, α_1 can then be recovered.

(5) Define $\delta_3 = \alpha_2 t_1$ and $\delta_4 = \alpha_2 t_2$. In the ninth and tenth seconds, transmit δ_3 and δ_4 as the information vectors using v_3. Then α_2 can be recovered by all the nodes in $\mathbf{T}_1$.

(6) Define $\delta_5 = \alpha_3 t_1$ and $\delta_6 = \alpha_0 t_2$. In the eleventh and twelveth seconds, transmit δ_5 and δ_6 as the information vectors using v_3. Then α_3 can be recovered by all the nodes in $\mathbf{T}_1$.

So, in the ith second for $i = 1, 2, 3$, via the generic LCM v_1, each node in $\mathbf{T}_4$ receives all four dimensions of α_i, each node in $\mathbf{T}_3$ receives three dimensions of α_i, and each node in T_1 receives one dimension of α_i. In the fourth second, via the generic LCM v_2, each node in $\mathbf{T}_3$ receives the vector β, which provides the three missing dimensions of α_1, α_2 and α_3 (one dimension for each) during the first 3 s of multicast by v_1. At the same time, each node in $\mathbf{T}_1$ receives one dimension of β. Now, in order to recover β, each node in $\mathbf{T}_1$ needs to receive the two missing dimensions of β during the fourth second. This is achieved by the generic LCM v_3 in the fifth and sixth seconds. So far, each node in $\mathbf{T}_1$ has received one dimension of α_i for $i = 1, 2, 3$ via v_1 during the first 3 s, and one dimension of α_i for $i = 1, 2, 3$ from β via v_2 and v_3 during the fourth to sixth seconds. Thus, it remains to provide the six missing dimensions of α_1, α_2 and α_3 (two dimensions for each) to each node in $\mathbf{T}_1$, and this is achieved in the seventh to the twelfth seconds via the generic LCM v_3. The previous scheme can be generalized to arbitrary sets of max-flow values.

19.5.3 The transmission scheme associated with an LCM

Let v be an LCM on a communication network (G, S), where the vectors $v(SX)$ assigned to outgoing channels SX linearly span a d-dimensional space. As before, the vector $v(XY)$ assigned to a channel XY is identified with a d-dimensional column vector over the base field of Ω by means of the choice of a basis. On the other hand, the total information to be transmitted from the source to the rest of the network is represented by a d-dimensional row vector, called the information vector. Under the transmission scheme prescribed by the LCM v, the data flowing over a channel XY is the matrix product of the information vector with the column vector $v(XY)$. We now consider the physical realization of this transmission scheme associated with an LCM.

A communication network (G, S) is said to be *acyclic* if the directed multigraph G does not contain a directed cycle. The nodes on an acyclic communication network can be sequentially indexed such that every channel is from a smaller indexed node to a larger indexed node. On an acyclic network, a straightforward realization of the above transmission scheme is as follows. Take one node at a time according to the sequential indexing. For each node, 'wait' until data is received from every incoming channel before performing the linear encoding. Then send the appropriate data on each outgoing channel. This physical realization of an LCM over an acyclic network, however, does not apply to a network that contains a directed cycle. This is illustrated by the following example.

Example 2

Let p, q, and r be vectors in Ω, where p and q are linearly independent. Define $v(SX) = p$, $v(SY) = q$ and $v(WX) = v(XY) = v(YW) = r$.

This specifies an *LCM* v on the network illustrated in Figure 19.18 if the vector r is a linear combination of p and q. Otherwise, the function v gives an example in which the law of information flow is observed for every single node but not observed for every set of nodes. Specifically, the law of information flow is observed for each of the nodes X, Y and W, but not for the set of nodes $\{X, Y, W\}$. Now, assume that $p = (1\,0)^T$, $q = (0\,1)^T$ and $r = (1\,1)^T$.

Then, v is an LCM. Write the information vector as $(b_1\ b_2)$, where b_1 and b_2 belong to the base field of Ω. According to the transmission scheme associated with the LCM, all three channels on the directed cycle transmit the same data $b_1 + b_2$. This leads to the logical problem of how any of these cyclic channels acquires the data $b_1 + b_2$ in the first place.

In order to discuss the transmission scheme associated with an LCM over a network containing a directed cycle, we need to introduce the parameter of time into the scheme. Instead of transmitting a single data symbol (i.e. an element of the base field of Ω) through

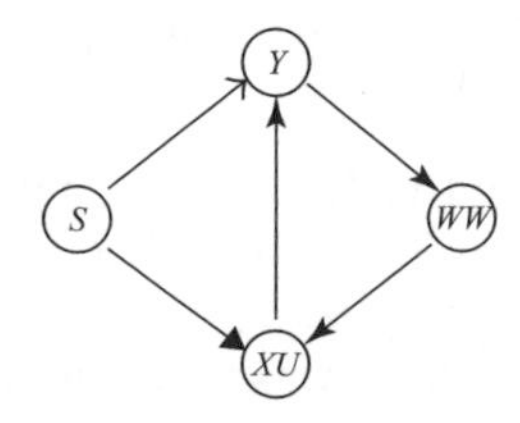

Figure 19.18 An LCM on a cyclic network.

each channel, we shall transmit a time-parameterized stream of symbols. In other words, the channel will be time-slotted. As a consequence, the operation of coding at a node will be time-slotted as well.

19.5.4 Memoryless communication network

Given a communication network (G, S) and a positive integer τ, the associated *memoryless* communication network denoted as $(G^{(\tau)}, S)$ is defined as follows. The set of nodes in $G^{(\tau)}$ includes the node S and all the pairs of the type $[X, t]$, where X is a nonsource node in G and t ranges through integers $1 - \tau$. The channels in the network $(G^{(\tau)}, S)$ belong to one of the three types listed below. For any nonsource nodes X and Y in (G, S):

(1) for $t \leq \tau$, the multiplicity of the channel from S to $[X, t]$ is the same as that of the channel SX in the network (G, S);

(2) for $t < \tau$, the multiplicity of the channel from $[X, t]$ to $[Y, t+1]$ is the same as that of the channel XY in the network (G, S);

(3) for $t < \tau$, the multiplicity of the channel from $[X, t]$ to $[X, \tau]$ is equal to $max\ flow_G(X) = mfG(X)$.

Lemma 2
The memoryless communication network $(G^{(\tau)}, S)$ is acyclic.

Lemma 3
There exists a fixed number ε, independent of τ, such that for all nonsource nodes X in (G, S), the maximum volume of a flow from S to the node $[X, \tau]$ in $(G^{(\tau)}, S)$ is at least $\tau - \varepsilon$ times *mfG*(X). For proof see Li *et al.* [47].

Transmission of data symbols over the network $(G^{(\tau)}, S)$ may be interpreted as '*memoryless*' transmission of data *streams* over the network (G, S) as follows:

(1) A symbol sent from S to $[X, t]$ in $(G^{(\tau)}, S)$ corresponds to the symbol sent on the channel SX in (G, S) during the time slot t.

(2) A symbol sent from $[X, t]$ to $[Y, t+1]$ in $(G^{(\tau)}, S)$ corresponds to the symbol sent on the channel XY in (G, S) during the time slot $t+1$. This symbol is a linear combination of symbols received by X during the time slot t and is unrelated to symbols received earlier by X.

(3) The channels from $[X, t]$ to $[X, t]$ for $t < \tau$ signify the accumulation of received information by the node X in (G, S) over time.

Since this is an acyclic network, the LCM on the network $(G^{(\tau)}, S)$ can be physically realized in the way mentioned above. The physical realization can then be interpreted as a *memoryless* transmission of data streams over the original network (G, S).

19.5.5 Network with memory

In this case we have to slightly modify the associated acyclic network.

Definition 1
Given a communication network $(G,\ S)$ and a positive integer τ, the associated communication network *with memory*, denoted as $(G^{[\tau]}, S)$, is defined as follows. The set of nodes in $G^{[\tau]}$ includes the node S and all pairs of the type $[X,\ t]$, where X is a nonsource node in G and t ranges through integers 1 to τ. Channels in the network $(G^{[\tau]},\ S)$ belong to one of the three types listed below. For any nonsource nodes X and Y in $(G,\ S)$;

(1) for $t \leq \tau$, the multiplicity of the channel from S to $[X,\ t]$ is the same as that of the channel SX in the network $(G,\ S)$;

(2) for $t < \tau$, the multiplicity of the channel from $[X, t]$ to $[Y, t+1]$ is the same as that of the channel XY in the network (G, S);

(3) for $t < \tau$, the multiplicity of channels from $[X, t]$ to $[X, t+1]$ is equal to $t \times mfG(X)$.

(4) The communication network $(G^{[\tau]},\ S)$ is acyclic.

(5) Every flow from the source to the node X in the network $(G^{(\tau)},\ S)$ corresponds to a flow with the same volume from the source to the node $[X,\ t]$ in the network $(G^{[\tau]},\ S)$.

(6) Every *LCM* v on the network $(G^{(\tau)},\ S)$ corresponds to an *LCM* u on the network $(G^{[\tau]},\ S)$ such that for all nodes X in G: $\dim[u([X, \tau])] = \dim[v(X)]$.

19.5.6 Construction of a generic LCM on an acyclic network

Let the nodes in the acyclic network be sequentially indexed as $X_0 = S, X_1, X_2, \ldots, X_n$ such that every channel is from a smaller indexed node to a larger indexed node. The following procedure constructs an LCM by assigning a vector $v(XY)$ to each channel XY, one channel at a time.

```
{
  for all channels XY
     v(XY) = the zero vector; // initialization
  for (j = 0;   j ≤ n;   j ++)
{
     arrange all outgoing channels X_jY from X_j
      in an arbitrary order;
     take one outgoing channel from X_j at a time
     {
        let the channel taken be X_jY;
        choose a vector w in the space v(X_j) such
        that w ∉ ⟨v(UZ) : UZ ∈ ξ⟩ for any collection
        ξ of at most d − 1 channels with
```

$v(X_j) \not\subset \langle v(UZ) : UZ \in \xi \rangle$;
$v(X_jY) = w$;
}
$v(X_{j+1})$ = *the linear span by vectors*
$v(XX_{j+1})$ *on all incoming channels* XX_{j+1}
to X_{j+1};
}
}

The essence of the above procedure is to construct the generic LCM iteratively and make sure that in each step the partially constructed LCM is generic.

19.5.7 Time-invariant LCM and heuristic construction

In order to handle delays, we can use an element $a(z)$ of $F[(z)]$ to represent the z-transform of a stream of symbols $a_0, a_1, a_2, \ldots, a_t, \ldots$ that are sent on a channel, one symbol at a time. The formal variable z is interpreted as a unit-time shift. In particular, the vector assigned to an outgoing channel from a node is *z times* a linear combination of the vectors assigned to incoming channels to the same node. Hence a TILCM is completely determined by the vectors that it assigns to channels. On the communication network illustrated in Figure 19.18, define the TILCM v as

$$v(SX) = (1\ 0)^{\mathrm{T}},\ v(SY) = (0\ 1)^{\mathrm{T}},\ v(XY) = (z\ z^3)^{\mathrm{T}}$$
$$v(YW) = (z^2\ z)^{\mathrm{T}} \text{ and } v(WX) = (z^3\ z^2)$$

By formal transformation we have for example

$$\begin{aligned} v(XY) = (z\ z^3) &= z(1 - z^3)(1\ 0)^{\mathrm{T}} + z(z^3\ z^2)^{\mathrm{T}} \\ &= z[(1 - z^3)v(SX) + v(WX)] \end{aligned}$$

Thus, $v(XY)$ is equal to z times the linear combination of $v(SX)$ and $v(WX)$ with coefficients $1 - z^3$ and 1, respectively. This specifies an encoding process for the channel XY that does not change with time. It can be seen that the same is true for the encoding process of every other channel in the network. This explains the terminology '*time-invariant*' for an LCM.

To obtain further insight into the physical process write the information vector as $[a(z)\quad b(z)]$, where

$$a(z) = \sum_{j\geq 0} a_j z^j \text{ and } b(z) = \sum_{j\geq 0} b_j z^j$$

belong to $F[(z)]$. The product of the information (row) vector with the (column) vector assigned to that channel represents the data stream transmitted over a channel

$$\begin{aligned} [a(z)\quad b(z)] \cdot v(SX) &= [a(z)\quad b(z)] \cdot (1\ 0)^{\mathrm{T}} \\ &= a(z) \to (a_0, a_1, a_2, a_3, a_4, a_5, \ldots, a_t, \ldots)[a(z)\quad b(z)] \cdot v(SY) \\ &= b(z) \to (b_0, b_1, b_2, b_3, b_4, b_5, \ldots, b_t, \ldots)[a(z)\quad b(z)] \cdot v(XY) \\ &= za(z) + z^3 b(z) \to (0, a_0, a_1, a_2 + b_0, a_3 + b_1, a_4 + b_2, \ldots, \\ &\qquad a_{t-1} + b_{t-3}, \ldots)[a(z)\quad b(z)] \cdot v(YW) \\ &= z^2 a(z) + zb(z) \to (0, b_0, a_0 + b_1, a_1 + b_2, a_2 + b_3, a_3 + b_4, \ldots, \end{aligned}$$

$$\begin{aligned} & a_{t-2} + b_{t-1}, \ldots)[a(z) \quad b(z)] \cdot v(WX) \\ & = z^3 a(z) + z^2 b(z) \rightarrow (0, 0, b_0, a_0 + b_1, a_1 + b_2, a_2 + b_3, \ldots, \\ & a_{t-3} + b_{t-2}, \ldots) \end{aligned}$$

Adopt the convention that $a_t = b_t = 0$ for all $t < 0$. Then the data symbol flowing over the channel XY, for example, at the time slot t is $a_{t-1} + b_{t-3}$ for all $t \geq 0$. If infinite loops are allowed, then the previous definition of TILCM v is modified as follows:

$$v(SX) = (1\ 0)^{\mathrm{T}}, \quad v(SY) = (0\ 1)^{\mathrm{T}}$$
$$v(XY) = (1 - z^3)^{-1}(z\ z^3)^{\mathrm{T}}, \quad v(YW) = (1 - z^3)^{-1}(z^2\ z)^{\mathrm{T}}$$

and

$$v(WX) = (1 - z^3)^{-1}(z^3\ z^2)$$

The data stream transmitted over the channel XY, for instance, is represented by

$$\left(\sum_{j\geq 0} a_j z^j \sum_{j\geq 0} b_j z^j\right) \cdot v(XY) = \sum_{j\geq 0} (a_j z^{j+1} + b_j z^{j+3})/(1 - z^3)$$
$$= \left[\sum_{j\geq 0} \left(a_j z^{j+1} + b_j z^{j+3}\right)\right] \sum_{i\geq 0} z^{3i} = \sum_{j\geq 0} \sum_{j\geq 0} \left(a_j z^{3i+j+1} + b_j z^{3i+j+3}\right)$$
$$= \sum_{t\geq 1} z^t (a_{t-1} + a_{t-4} + a_{t-7} + \cdots + b_{t-3} + b_{t-6} + b_{t-9} + \cdots)$$

That is, the data symbol $a_{t-1} + a_{t-4} + a_{t-7} + \cdots + b_{t-3} + b_{t-6} + b_{t-9} + \cdots$ is sent on the channel XY at the time slot t. This TILCM v, besides being time invariant in nature, is a 'memoryless' one because the following linear equations allows an encoding mechanism that requires no memory:

$$\begin{aligned} v(XY) &= zv(SX) + zv(WX) \\ v(YW) &= zv(SY) + zv(XY) \end{aligned} \tag{19.60}$$

and

$$v(WX) = zv(YW)$$

19.5.7.1 TILCM construction

There are potentially various ways to define a *generic* TILCM and, to establish desirable dimensions of the module assigned to every node. In this section we present a heuristic construction procedure based on graph-theoretical *block decomposition* of the network. For the sake of computational efficiency, the procedure will first remove '*redundant*' channels from the network before identifying the 'blocks' so that the 'blocks' are smaller. A channel in a communication network is said to be *irredundant* if it is on a simple path starting at the source. otherwise, it is said to be *redundant*. Moreover, a communication network is said to be *irredundant* if it contains no redundant channels. In the network illustrated in Figure 19.19, the channels ZX, TX and TZ are redundant.

The deletion of a redundant channel from a network results in a subnetwork with the same set of irredundant channels. Consequently, the irredundant channels in a network define an

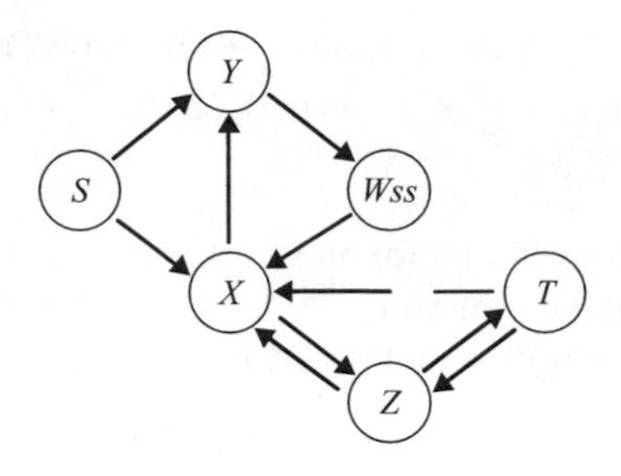

Figure 19.19 Redundant channels in a network.

irredundant subnetwork. It can be also shown that, if v is an LCM (respectively, a TILCM) on a network, then v also defines an LCM (respectively, a TILCM) on the subnetwork that results from the deletion of any redundant channel. In addition we say that two nodes are equivalent if there exists a directed path leading from one node to the other and vice versa. An equivalence class under this relationship is called a *block* in the graph. The source node by itself always forms a block. When every block '*contracts*' into a single node, the resulting graph is acyclic. In other words, the blocks can be sequentially indexed so that every interblock channel is from a smaller indexed block to a larger indexed block.

For the construction of a 'good' TILCM, smaller sizes of blocks tend to facilitate the computation. The extreme favorable case of the block decomposition of a network is when the network is acyclic, which implies that every block consists of a single node. The opposite extreme is when all nonsource nodes form a single block exemplified by the network illustrated in Figure 19.19. The removal of redundant channels sometimes serves for the purpose of breaking up a block into pieces. For the network illustrated in Figure 19.19, the removal of the three redundant channels breaks the block $\{T, W, X, Y, Z\}$ into the three blocks $\{T\}$, $\{W, X, Y\}$ and $\{Z\}$.

In the construction of a 'good' LCM on an *acyclic* network, as before, the procedure takes one node at a time according to the acyclic ordering of nodes and assigns vectors to outgoing channels from the taken node. For a general network, we can start with the trivial TILCM v on the network consisting of just the source and then expand it to a 'good' TILCM v that covers one more block at a time.

The sequential choices of blocks are according to the acyclic order in the block decomposition of the network. Thus, the expansion of the 'good' TILCM v at each step involves only *incoming* channels to nodes in the new block. A heuristic algorithm for assigning vectors $v(XY)$ to such channels XY is for $v(XY)$ to be z times an *arbitrary convenient* linear combination of vectors assigned to incoming channels to X. In this way, a system of linear equations of the form of $Ax = b$ is set up, where A is a square matrix with the dimension equal to the total number of channels in the network and x is the unknown column vector whose entries are $v(XY)$ for all channels XY. The elements of A and b are polynomials in z. In particular, the elements of A are either ± 1, 0, or a polynomial in z containing the factor z. Therefore, the determinant of A is a formal power series with the constant term (the zeroth power of z) being ± 1, and, hence is invertible in $F(z)$. According to Cramer's rule, a unique solution exists. This is consistent with the physical intuition because the whole network is completely determined once the encoding process for each channel is specified. If this unique solution does not happen to satisfy the requirement for being a 'good' TILCM, then the heuristic algorithm calls for adjustments on the coefficients of the linear equations on the trial-and-error basis.

After a 'good' TILCM is constructed on the subnetwork formed by irredundant channels in a given network, we may simply assign the zero vectors to all redundant channels.

Example

After the removal of redundant channels, the network depicted by Figure 19.19 consists of four blocks in the order of $\{S\}$, $\{W, X, Y\}$, $\{Z\}$ and $\{T\}$. The subnetwork consisting of the first two blocks is the same as the network in Figure 19.18. When we expand the trivial TILCM on the network consisting of just the source to cover the block $\{W, X, Y\}$, a heuristic trial would be

$$v(SX) = (1\ 0)^T \quad v(SY) = (0\ 1)^T$$

together with the following linear equations:

$$v(XY) = zv(SX) + zv(WX)$$
$$v(YW) = zv(SY) + zv(XY)$$

and

$$v(WX) = zv(YW).$$

The result is the memoryless TILCM v in the preceding example. This TILCM can be further expanded to cover the block $\{Z\}$ and then the block $\{T\}$.

19.6 CAPACITY OF WIRELESS NETWORKS USING MIMO TECHNOLOGY

In this section an information theoretic network objective function is formulated, which takes full advantage of multiple input multiple output (MIMO) channels. The demand for efficient data communications has fueled a tremendous amount of research into maximizing the performance of wireless networks. Much of that work has gone into enhancing the design of the receiver, however considerable gains can be achieved by optimizing the transmitter.

A key tool for optimizing transmitter performance is the use of channel reciprocity. Exploitation of channel reciprocity makes the design of optimal transmit weights far simpler . It has been recognized that network objective functions that can be characterized as functions of the output signal-to-interference noise ratio (SINR) are well suited to the exploitation of reciprocity, since it permits us to relate the uplink and downlink objective functions [48–50]. It is desirable, however, to relate the network objective function to information theory directly, rather than *ad-hoc* SINR formulations, due to the promise of obtaining optimal channel capacity. This section demonstrates a link between the information theoretic, Gaussian interference channel [51] and a practical network objective function that can be optimized in a simple fashion and can exploit channel reciprocity. The formulation of the objective function permits a water filling [52] optimal power control solution, that can exploit multipath modes, MIMO channels and multipoint networks.

Consider the network of transceivers suggested by Figure 19.20. Each transceiver is labeled by a node number. The nodes are divided into two groups. The group 1 nodes transmit in a given time-slot, while the group 2 nodes receive. In alternate time-slots, group

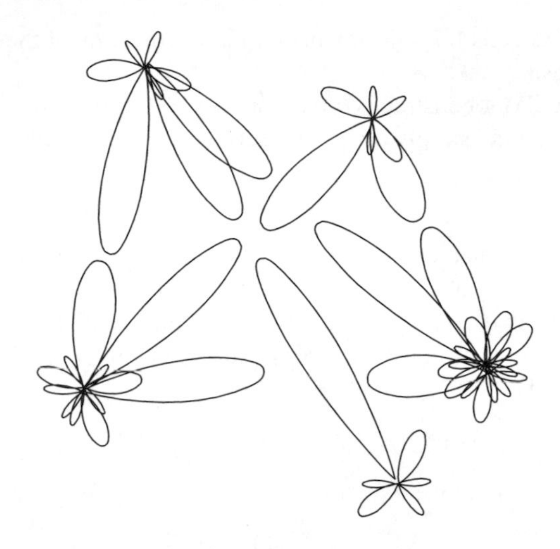

Figure 19.20 Network layout.

2 nodes transmit, while group 1 nodes receive, according to a time division duplex (TDD) transmission scheme.

The channel is assumed to be channelized into several narrow band frequency channels, whose bandwidth is small enough so that all multipath is assumed to add coherently. This assumption allows the channel between any transmit antenna and any receive antenna to be modeled as a single complex gain and is consistent with an orthogonal frequency division multiplexing (OFDM) modulation scheme. All transceivers are assumed to have either multiple antennas or polarization diversity, for both transmit and receive processing. The link connecting nodes may experience multipath reflections. For this analysis we assume that the receiver synchronizes to a given transmitter and removes any propagation delay offsets. The channel, for a given narrow band, between the transmitter and receiver arrays is therefore modeled as a complex matrix multiply.

Let ℓ_1 be an index into the group 1 transceivers, and ℓ_2 an index into the group 2 transceivers. When node ℓ_2 is transmitting during the uplink, we model the channel between the two nodes as a complex matrix $\mathbf{H}_{12}(\ell_1, \ell_2)$ of dimension $M_1(\ell_1) \times M_2(\ell_2)$, where $M_1(\ell_1)$ is the size of the antenna array at node ℓ_1, and $M_2(\ell_2)$ is the size of the array at node ℓ_2. Polarization diversity is treated like additional antennae. In the next TDD time slot, during the downlink, node ℓ_1 will transmit and the channel from ℓ_1 to ℓ_2 is described by the complex $M_2(\ell_1) \times M_1(\ell_2)$ matrix $\mathbf{H}_{21}(\ell_2, \ell_1)$. For every node pair (ℓ_1, ℓ_2) that forms a communications link in the network, we assign a MIMO channel link number, indexed by k or m, in order to label all such connections established by the network. Obviously not every node pair is necessarily assigned a link number, only those that actually communicate. We also define the mapping from the MIMO link number to the associated group 1 node, $\ell_1(k)$ and the associated group 2 node $\ell_2(k)$ by the association of k with the link $[\ell_1(k), \ell_2(k)]$. Because each channel is MIMO, a given node will transmit multiple symbols over possibly more than one transmission mode. The set of all transmission modes over the entire network is indexed by q and p. This index represents all the symbol streams that are transmitted from one node to another and therefore represents a lower level link number. The low level link numbers will map to its MIMO channel link, via the mapping $k(q)$. Because our network will exploit channel reciprocity, we assume that every uplink symbol stream indexed by q has an associated downlink symbol stream assigned to the same index. The q'th uplink symbol stream is spread by an $M_1(q) \times 1$ complex vector $g_1(q)$, where $M_1(q) \equiv M_1\{\ell_1[k(q)]\}$. Similarly $g_2(q)$ is the associated transmit vector for the downlink.

For each node, we group the transmit vectors into a single matrix,

$$\mathbf{G}_2(k) \equiv \left[g_2(q_1), g_2(q_2), \ldots, g_2(q_{M_c(k)})\right] \tag{19.61}$$

and

$$\mathbf{G}_1(k) \equiv \left[g_1(q_1), g_1(q_2), \ldots, g_1(q_{M_c(k)})\right] \tag{19.62}$$

where $k(q_i) = k$ and there are $M_c(k)$ transmission modes associated with MIMO link k. With these conventions the signal model can be written as:

$$\mathbf{x}_1(n;k) = \mathbf{i}_1(n;k) + \mathbf{H}_{12}(k,k)\mathbf{G}_2(k)\mathbf{d}_2(n;k) \tag{19.63}$$

$$\mathbf{x}_2(n;k) = \mathbf{i}_2(n;k) + \mathbf{H}_{21}(k,k)\mathbf{G}_1(k)\mathbf{d}_1(n;k) \tag{19.64}$$

where $\mathbf{x}_1(n;k)$ is the received complex data vector at sample n, and node $\ell_1(k)$, $\mathbf{H}_{12}(k,m) \equiv \mathrm{H}_{12}[\ell_1(k), \ell_2(m)]$ is the $M_1[\ell_1(k)] \times M_2[\ell_2(m)]$ complex MIMO channel matrix for downlink transmission, n is a time/frequency index, that represents an independent reuse of the channel, either due to adjacent frequency channels (e.g. adjacent OFDM channels) or due to multiple independent time samples, $\mathbf{i}_1(n;k)$ is the interference vector seen at node $\ell_1(k)$ due to the other transmitting nodes as well as due to background radiation, and $\mathbf{d}_2(n;k)$ is the $M_c(k) \times 1$ downlink information symbol vector, transmitted for sample n. The analogous model for the uplink case is shown in (4). The interference vector can be written as

$$\mathbf{i}_r(n;k) = \sum_{m \neq k} \mathbf{H}_{rt}(k,m)G_t(m)\mathbf{d}_t(n;m) + \varepsilon_r(n,k) \tag{19.65}$$

where $\varepsilon_r(n,k), r = \{1,2\}$, is the background radiation noise vector seen by the receiving node $\ell_r(k)$ and $t = \{2,1\}$ is the transmission timeslot indicator. The convention is adopted that $t = 2$ if $r = 1$, otherwise $t = 1$ when $r = 2$.

19.6.1 Capacity metrics

In this section we will be interested in the asymptotic maximum achievable throughput under certain network constraints:

(1) We will assume that signal energy from all the nodes in the network, other than the one from the link being processed, is treated as interference. This is due to difficulty of synchronizing a node to multiple interference sources and reduces the complexity of the receiver.

(2) The interference $\mathbf{i}_r(n;k)$ is modeled as complex circular Gaussian noise. The superposition of many distant interferers will appear Gaussian by the central limit theorem.

Considering therefore the channel model described by Equations (19.63) and (19.65), we can write the mutual information between the source data vector and the received data vector as [52, 53]:

$$I[\mathbf{x}_r(k); \mathbf{d}_t(k)] = \log_2 \left| \mathbf{I} + \mathbf{R}_{\underline{i}_r \underline{i}_r}^{-1}(k)\mathbf{R}_{s_t s_t}(k) \right| \triangleq C_{rt}(k; \mathbf{G}_t) \tag{19.66}$$

where, $\mathbf{G}_t \leftrightarrow \{g_t(q)\}$ represents all transmit weights stacked into a single parameter vector and where we neglect the n dependency for the random vectors $\mathbf{x}_r(n; k)$ and $\mathbf{d}_t(n; k)$. The interference and the signal covariance matrices are defined as:

$$\mathbf{R}_{i_r i_r}(k) \triangleq \sum_{m \neq k} \mathbf{A}_{rt}(k, m) A_{rt}^H(k, m) + \mathbf{R}_{\varepsilon_r \varepsilon_r}(k) \tag{19.67}$$

$$\mathbf{A}_{rt}(k, m) \triangleq \mathbf{H}_{rt}(k, m)\mathbf{G}_t(m) \tag{19.68}$$

$$\mathbf{R}_{s_t s_t}(k) \triangleq \mathbf{A}_{rt}(k, k)\mathbf{A}_{rt}^H(k, k), \tag{19.69}$$

and where $\mathbf{R}_{\varepsilon_r \varepsilon_r}(k)$ is the covariance of the background noise vector $\varepsilon_r(n, k)$. The covariance of the source statistics is assumed governed by the transmit weights $\mathbf{G}_t(k)$, therefore the covariance of $\mathbf{d}_t(k)$ is assumed to be the identity matrix.

Let us now consider the introduction of complex linear beam-forming weights at the receiver. For each low level link q, we assign a receive weight vector $\mathbf{w}_r(q), r \in \{1, 2\}$ for use at receiver $\ell_r[\mathbf{k}(q)]$. The weights, in theory, are used to copy the transmitted information symbol,

$$\hat{d}_t(n; q) \triangleq w_r^{\mathrm{H}}(q) x_r(n; k) \tag{19.70}$$

$$= \mathbf{w}_r^{\mathrm{H}}(q)\mathbf{i}_r(n; k) + \mathbf{w}_r^{\mathrm{H}}(q)\mathbf{A}_{rt}(k, k)\mathbf{d}_t(n; k) \tag{19.71}$$

$k(q) = k$ where,

$$\mathbf{d}_t(n; k) \equiv \left[d_t(q_1), d_t(q_2), \ldots d_t(q_{M_c(k)})\right]^{\mathrm{T}} \tag{19.72}$$

and $k(q_j) = k$. The vector version of this can be written as:

$$\hat{\mathbf{d}}_t(n; k) = \mathbf{W}_r^{\mathrm{H}}(k)\mathbf{i}_r(n; k) \; + \; \mathbf{W}_r^{\mathrm{H}}(k)\mathbf{A}_{rt}(k, k)\mathbf{d}_t(n; k) \tag{19.73}$$

where

$$\mathbf{W}_r(k) \equiv \left[\mathbf{w}_r(q_1), \mathbf{w}_r(q_2), \cdots \mathbf{w}_r(q_{M_c(k)})\right] \tag{19.74}$$

and $k(q_j) = k$. The structure of the transceiver channel model is illustrated in Figure 19.21 for the uplink case, highlighting the first $M_c(k)$ low-level links, all assumed to be associated with the same MIMO-channel link. The MIMO channel link index k and sample index n are omitted for simplicity. The downlink case is the same after the 1 and 2 indices are swapped.

Owing to the additional processing [52], the mutual information either remains the same or is reduced by the application of the linear receive weights,

$$I[\hat{\mathbf{d}}_t(k); \mathbf{d}_t(k)] \leq I[\mathbf{x}_r(k); \mathbf{d}_t(k)] \tag{19.75}$$

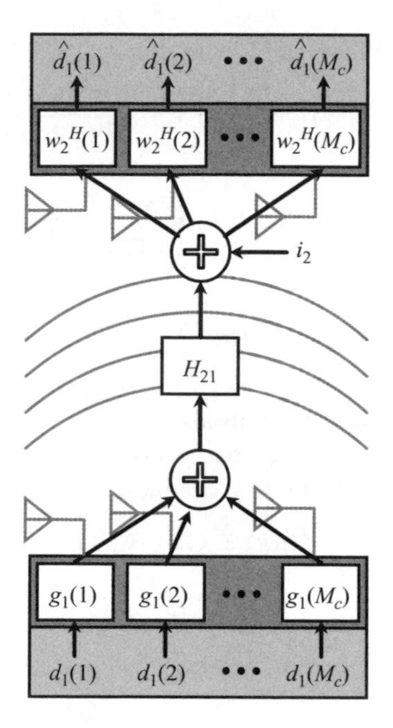

Figure 19.21 Uplink transceiver channel model.

In addition to inequality, Equation (19.75), it can be also shown [54, 55] that, if the source symbols $d_t(q)$ are mutually independent then,

$$I\left[\hat{\mathbf{d}}_t(k);\mathbf{d}_t(k)\right] \geq \sum_{q:\,k(q)=k} I\left[\hat{d}_t(q);d_t(q)\right] \tag{19.76}$$

The lower bound of in Equation (19.76) is referred to as the decoupled capacity and can be written as:

$$D_{rt}(k;\mathbf{W}_r,\mathbf{G}_t) \equiv \sum_{q:\,k(q)=k} I[\hat{d}_t(q);d_t(q)] \tag{19.77}$$

where the collection of all receive weights is stacked into the parameter vector $\mathbf{W}_r \leftrightarrow \{\mathbf{w}_r(q)\}$. Equations (19.75) and (19.76) demonstrate that

$$C_{rt}(k;\mathbf{G}_t) \geq D_{rt}(k;\mathbf{W}_r,\mathbf{G}_t) \tag{19.78}$$

The decoupled capacity is implicitly a function of the *transmit weights* $\mathbf{g}_t(q)$ and the *receive weights* $\mathbf{w}_r(q)$. Optimizing over these weights can always acheive the full capacity of the Gaussian interference channel and hence the upper bounds in Equation (19.78).

First we write $\hat{d}_t(n;q)$ as,

$$\hat{d}_t(n;q) = \mathbf{w}_r^{\mathrm{H}}(q)\mathbf{a}_r(q) + \mathbf{w}_r^{\mathrm{H}}(q)\mathbf{i}_r(q),$$

$$i_r(n;q) \equiv \sum_{p\neq q} \mathbf{H}_{rt}[k,k(p)]\mathbf{g}_t(p)d_t(n;p) + \varepsilon_r(n,k)$$

$$\mathbf{a}_r(q) \equiv \mathbf{H}_{rt}(k,k)\mathbf{g}_t(q),\ k = k(q) \tag{19.79}$$

Now the mutual information on the right-hand side of Equation (19.55a) can be written as,

$$I\left[\hat{d}_t(q); d_t(q)\right] = \log_2\left(1 + \frac{|\mathbf{w}_r^{\mathrm{H}}(q)\mathbf{a}_r(q)|^2}{\mathbf{w}_r^{\mathrm{H}}(q)\mathbf{R}_{i_r i_r}(q)\mathbf{w}_r}\right) \tag{19.80}$$

$$R_{i_r i_r}(q) = R_{i_r i_r}(k) + \cdots \quad H_{rt}(k,k)\left(\sum_{\substack{p:k(p)=k \\ p\neq q}} = g_t(p)g_t^{\mathrm{H}}(p)\right) H_{rt}^{\mathrm{H}}(k,k)$$

$$= \quad R_{i_r i_r}(k) + \sum_{\substack{p:k(p)=k \\ p\neq q}} a_r(p)d_r^{\mathrm{H}}(p) \tag{19.81}$$

where $k = k(q)$. In the following we will also use the fact that scrambling the transmit weights by an orthonormal scrambling matrix does not change the values of the mutual informations for the Gaussian interference channel. In other words, if

$$\tilde{\mathbf{G}}_t(k) \triangleq \mathbf{G}_t(k)\mathbf{V}_t(k) \tag{19.82}$$

$$\mathbf{V}_t(k)\mathbf{V}_t^{\mathrm{H}}(k) = \mathbf{I}$$

then for all k,

$$C_{rt}(k;\tilde{\mathbf{G}}) = C_{rt}(k;\mathbf{G}) \tag{19.83}$$

where $\tilde{\mathbf{G}} \leftrightarrow \{\tilde{\mathbf{g}}_t(q)\}$ is the parameter vector of all stacked, scrambled transmit weights, $\tilde{\mathbf{g}}_t(q)$ drawn from the columns of $\tilde{\mathbf{G}}_{\mathbf{t}}(k)$. The proof is based on the fact that the mutual information is completely determined by the statistics, $\mathbf{R}_{\mathbf{i_r i_r}}(k)$ and $\mathbf{R}_{\mathbf{s_t s_t}}(k)$. From Equations (19.67) and (19.47), these depend only on the outer products $\mathbf{G}_t(m)\mathbf{G}_t^{\mathrm{H}}(m)$, which are invariant with respect to orthonormal scrambling because $\tilde{\mathbf{G}}_t(m)\,\tilde{\mathbf{G}}_t^{\mathrm{H}}(m) = \mathbf{G}_t(m)\mathbf{G}_t^{\mathrm{H}}(m)$.Therefore replacing $\mathbf{G}_t(m)$ with $\tilde{\mathbf{G}}_t(m)$ does not change the mutual information. Based on this we have the following relation. For any set of network wide transmit weights $\mathbf{G}$, there exists a set of receive weights $\hat{\mathbf{w}}_{\mathbf{r}}(q)$ and transmit weights $\tilde{\mathbf{g}}_t(q)$ such that for all k,

$$C_{rt}(k;\mathbf{G}_t) = C_{rt}(k;\tilde{\mathbf{G}}_t) = D_{rt}(k;\hat{\mathbf{W}}_r,\tilde{\mathbf{G}}_t) \tag{19.84}$$

where $\hat{\mathbf{W}}_r \leftrightarrow \{\hat{\mathbf{w}}_{\mathbf{r}}(q)\}$ and $\tilde{\mathbf{G}}_{\mathbf{t}} \leftrightarrow \{\tilde{\mathbf{g}}_t(q)\}$. Given the transmit weights $\tilde{\mathbf{G}}_t$, the receive weights can be found from

$$\hat{\mathbf{w}}_r(q) = \arg\max_{\mathbf{w}_r(q)} I[\hat{d}_t(q); d_t(q)] = \arg\max_{\mathbf{w}_r(q)} \gamma_r(q) = \mathbf{R}_{\mathbf{i_r i_r}}^{-1}(q)\mathbf{a}_r(q) \tag{19.85a}$$

where $\gamma_r(q)$ is the output signal to interference noise ratio (SINR) given by,

$$\gamma_r(q) \triangleq \frac{|\mathbf{w}_r^{\mathrm{H}}(q)\mathbf{a}_r(q)|^2}{\mathbf{w}_r^{\mathrm{H}}(q)\mathbf{R}_{\mathbf{i_r i_r}}(q)\mathbf{w}_r(q)} \tag{19.85b}$$

For the formal proof see References [54–57]. Equations (19.84)–(19.85) are significant because they allow us to optimize the Gaussian interference channel using decoupled capacity. The decoupled capacity can be shown to obey the reciprocity theorem [54, 55], which permits us to relate uplink network metrics to downlink network metrics. We can either consider minimizing total power subject to a channel capacity constraint, or maximizing channel capacity subject to a transmit power constraint.

In References [54, 55] a technique called locally enabled global optimization (LEGO) is designed to fully exploit the reciprocity theorem. This technique transforms the optimization over the transmit powers to one over the set of achievable output SINRs in Equation (19.85). The LEGO algorithm can be efficiently implemented using local information, and requires only an estimate of the post beamforming interference power, $y_r(q) \equiv \mathbf{w}_r^{\mathrm{H}}(q)\mathbf{R}_{\mathbf{i}_r\mathbf{i}_r}\mathbf{w}_r(q)$, and an estimate of the post-beamforming channel gain, $T_{rt}(q,q) \equiv |\mathbf{w}_r^{\mathrm{H}}(q)\mathbf{H}_{rt}[k(q),k(q)]\mathbf{g}_t(q)|^2$. If a collection of links through which we desire to compute a channel capacity is $Q(m) \triangleq \{q : k(q) = m\}$, then the algorithm can be summarized as below.

Algorithm 1: LEGO

(1) Update the remote receiver weights for every link during downlink transmission:

$$\mathbf{w}'_1(p) = \arg\max_{\mathbf{w}_1(p)} \gamma_1(p)$$

(2) Set the remote transmit weights to a scaled version of the conjugated receive weights,

$$\mathbf{g}_1(p) = \mathbf{w}_1^*(p)\sqrt{\pi_l(p)/[\mathbf{w}_1^H(p)\mathbf{R}_{\varepsilon_r\varepsilon_r}\mathbf{w}_1(p)]}$$

(3) For each link p, estimate the associated post-beamforming interference power $y_1(p)$ and relay this information back to the base station.

(4) Update the base receiver weights for every link during uplink transmission:

$$\mathbf{w}'_2(p) = \arg\max_{\mathbf{w}_2(p)} \gamma_2(p).$$

(5) For each link p, estimate the associated post-beamforming interference power $y_2(p)$.

(6) Update the target SINR for each link, by optimizing the linearized model over each aggregate set m:

$$\gamma(q) = \arg\min_{\gamma(q)} \sum_{q\in Q(m)} \gamma(q)\frac{y_1(q)y_2(q)}{T_{12}(q,q)} s.t. c(m) \quad \leq \quad \sum_{q\subset Q(m)} \log[1+\gamma(q)]$$

for $q \in Q(m)$,

$$\gamma(p)' = \alpha[\hat{\gamma}(p) - \gamma(p)] + \gamma(p), \textit{for some } 0 < \alpha \leq 1 (\alpha \textit{ is initially set to 1}).$$

(7) Update the downlink transmit powers $\pi_2(q)$:

$$\pi'_2(q) = \gamma(q)y_1(q)/T_{12}(q,q)$$

(8) Update the uplink transmit powers $\pi_1(q)$ and relay them back to the remote units. $\pi'_1(q) = \gamma(q)y_2(q)/T_{21}(q,q)$.

(9) Set the base transmit weights to a scaled version of the conjugated receive weights,

$$g_2(p) = \mathbf{w}_2^*(p)\sqrt{\pi_2(p)/\left[\mathbf{w}_2^H(p)\mathbf{R}_{\varepsilon_r\varepsilon_r}\mathbf{w}_2(p)\right]}$$

(10) Return to step 1.

References [54, 55] provide a numerical example of a four-cell wireless network as illustrated in Figure 19.22. Each 1 km radius cell contains two remote units (RU), each of which communicates to two base stations. The base stations have eight antennae and the RUs two antennae. There are eight independent 50 kHz channels and thus each RU has 16 transmit modes [i.e. each $Q(m)$ contains 16 links]. The performance of LEGO is compared with a standard power management algorithm, that seeks to transmit a constant power for each link from each base station, and a single antenna network, that uses frequency division multiplexing to isolate each RU into a separate channel. Background radiation is assumed to be thermal white noise at room temperature, with an added 10 dB noise figure. As can be seen from the figure, the performance improvement of the LEGO algorithm is significant.

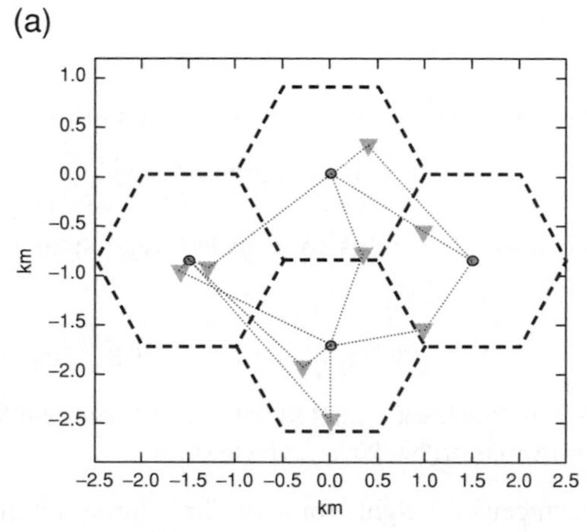

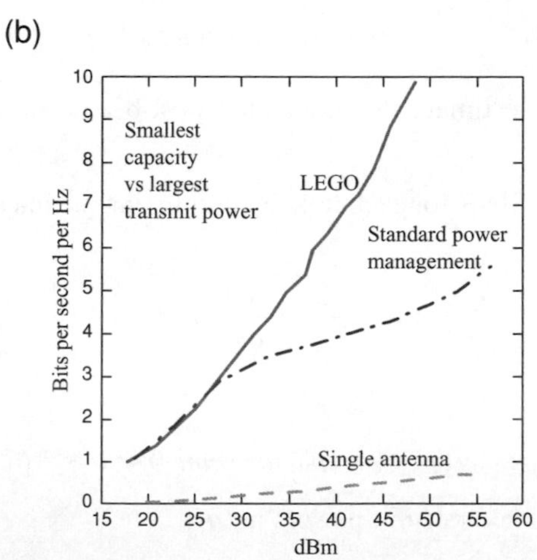

Figure 19.22 (a) Cell network geometry; (b) LEGO performance: worst case capacity vs worst transmit power. (Reproduced by permission of IEEE [55].)

19.7 CAPACITY OF SENSOR NETWORKS WITH MANY-TO-ONE TRANSMISSIONS

In sensor networks the many-to-one throughput capacity is the *per source data throughput*, when all or many of the sources are transmitting to a single fixed receiver or sink [58–63]. Earlier in this chapter we have shown that the achievable per node throughput in a wireless network is $\theta\left[W/(n\log n)\right]$, where W is the transmission capacity and n is the total number of nodes in the network.

The result was based on the assumption that communications are one-to-one, and that sources and destinations are randomly chosen. It does not apply to scenarios where there are communication hot spots in the network. Since many-to-one communication causes the sink to become a point of traffic concentration, the throughput achievable per source node in this case is reduced. In this section we are only interested in the case where every source gets an equal (on average) amount of original data (not including relayed data) across to the sink. This is because otherwise throughput can be maximized by having only the sensors closest to the destination transmit. Equal share of throughput from every sensor is desired for applications like imaging where each sensor represents a certain region of the whole field and data from each part are equally important. When distributed data compression is used, this is again approximately the case. However, when conditional coding is used this may no longer be true, since the amount of processed data can vary from source to source. In order to achieve the above goal we use the following assumptions about the network architecture.

19.7.1 Network architecture

(1) The network is deployed in a field of circular shape. There are n nodes, sources (we will use nodes, sources and sensors interchangeably in subsequent discussions) deployed in a network. A sink/destination is located at the center of the network/circle. Each node is not only a source of data, but also a relay for some other sources to reach the sink.

(2) A network where the nodes are randomly placed following a uniform distribution will be referred to as a randomly deployed network or a random network. In such a network we have no direct control over the exact location of the nodes. A network where we can determine the exact locations of the nodes will be referred to a as an arbitrary network.

(3) Two network organizational architectures are considered: (a) a flat architecture where nodes communicate with the sink via possibly multi-hop routes by using peer nodes as relays; and (b) hierarchical architecture, where clusters are formed so that sources within a cluster send their data (via a single hop or multihop depending on the size of the cluster) to a designated node known as the clusterhead. The clusterhead can potentially perform data aggregation and processing and then forward data to the sink. In this study, we will assume that the clusterheads serve as simple relays and no data aggregation is performed. We will also assume that the communication between nodes and clusterheads and communication between clusterheads and the sink are on separate frequency channels so that the two layers do not interfere.

(4) Throughout the section we will assume that the sources transmit following a schedule that consists of time slots.

(5) To simplify the resulting expressions, we assume the field has an area of 1. Nodes share a common wireless channel using omnidirectional antennas. We assume nodes use a fixed transmission power and achieve a fixed transmission range. We adopt the commonly used interference model, as earlier in this chapter. Let X_i and X_j be two

sources with distance $d_{i,j}$ between them. Then the transmission from X_i to X_j will be successful if and only if

$$d_{i,j} \leq r \text{ and } d_{k,j} > r + \Delta, \Delta \geq 0 \tag{19.76}$$

for any source X_k that is simultaneously transmitting. In the following r will be referred to as the transmission range. There are two interference concepts here. A node may interfere with another node that is transmitting if it is within distance $2r + \Delta$ of that node. To see this consider two transmitting nodes. If one node is within $2r + \Delta$ of the other node there will be an overlap between a circle of radius r around the first transmitting node and a circle of radius $r + \Delta$ around the second transmitting node. If the intended receiver is located within the overlapping area the transmission will fail because of interference. Therefore the two nodes need to be at least $2r + \Delta$ apart. On the other hand a node will interfere with another node that is receiving if it is within distance $r + \Delta$ for obvious reasons. In the following both will be generally referred to as interference range. The distinction will be clear from the context. Also note that this interference model, Equation (19.76), essentially implies that no nodes can receive more than one transmission at a time. We will also assume that no node can transmit and receive at the same time.

(6) The network scenario is depicted in Figure 19.23. The sink is placed at the center of this field. It receives all data generated by sources in the network.

(7) In the following W refers to the transmission capacity of the channel in a flat network. In a hierarchical network W refers to the transmission capacity of the channel used within clusters. W' refers to the transmission capacity of the channel used from the heads to the sink. The capacity is derived as a function of the transmission range, assuming the transmission range can provide connectivity.

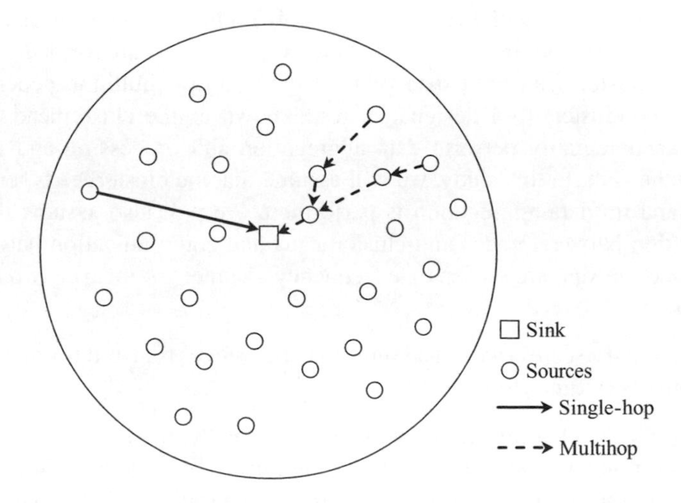

Figure 19.23 Many-to-one network scenario.

19.7.2 Capacity results

In this section we summarize capacity results for the *network defined above*. For formal proofs of the results the reader is referred to References [58–63].

19.7.2.1 Capacity in a flat network

(1) The maximum per node throughput in the network is upper bounded by W/n.

(2) $\lambda = W/n$ can be achieved when every source can directly reach the sink.

(3) $\lambda = W/n$ is not achievable if not every source can directly reach the destination and $\Delta > r$.

(4) $\lambda = W/n$ may be achieved in an arbitrary network when not every source can directly reach the destination and $\Delta < r$.

When the sink cannot directly receive from every source in the network, and assuming that the channel allocation does not take into account difference in traffic load, then $\lambda = W/n$ is not achievable with high probability regardless of the value of Δ. In this case we will use the upper bound on throughput by deriving the maximum number of simultaneous transmissions.

Denote by A_r the area of a circle of radius r, i.e. $A_r = \pi r^2$. Let random variable V_r denote the number of nodes within an area of size A_r and assume a total area of 1. We then have [58–63]:

(5) In a randomly deployed network with n nodes,

$$Pr\left(nA_r - \sqrt{\alpha_n n} \le V_r \le nA_r + \sqrt{\alpha_n n}\right) \to 1 \text{ as } n \to \infty$$

where the sequence $\{\alpha_n\}$ is such that $\lim_{n\to\infty} \alpha n/n = \varepsilon$, ε being positive but arbitrarily small.

(6) If a network has randomly deployed sources and the transmission range r is such that not all sources can directly reach the sink, then with high probability the throughput upper bound $\lambda = W/n$ is not achievable.

(7) A randomly deployed network using multihop transmission for many-to-one communication can achieve throughput

$$\lambda \ge W\left(\pi r^2 - \sqrt{\varepsilon}\right) / \left\{n\left(4\pi r^2 + 4\pi r\Delta + \pi\Delta^2 + \sqrt{\varepsilon}\right)\right\}$$

with high probability, when no knowledge of the traffic load is assumed and ε is as given in (1).

In the following we will use a concept of *virtual sources*. As an example consider a simple network consisting of three sources and a sink, shown in Figure 19.24(a). The distance between adjacent nodes is r. Regardless of the value of Δ, when one source transmits, it interferes with all other sources in this network. Therefore only one source can transmit at a time. The number of interfering neighbors for any of the sources is two, which is the highest degree of the graph that represents the interference relationship in this network. Thus a schedule of length 3 allows all sources to transmit once during the schedule. The

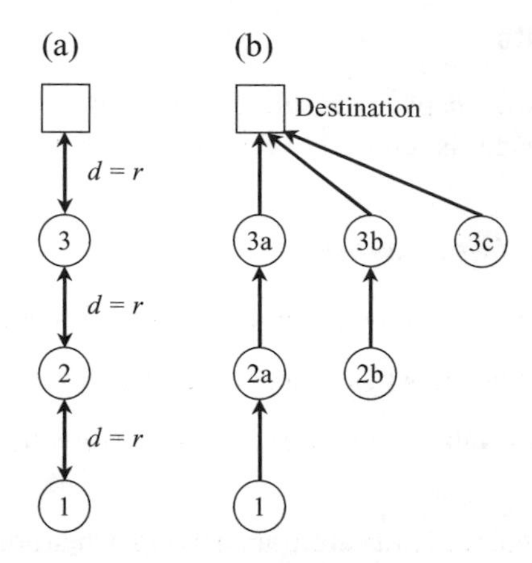

Figure 19.24 (a) Chain network (b) virtual sources.

load on the source closest to the sink, source 3, is 3λ, since it carries the traffic of all three sources. The achievable throughput is then calculated as $3\lambda = W/3$, thus $\lambda = W/9$.

The way the schedule was calculated previously assigned the same share of the resources (time) to all the sources. Since we used the source with the highest need of resource (the one carrying the most traffic) to calculate the amount of resource needed, every other source is wasting resource. In our example we are giving every source the possibility of making three transmissions. Source 3 does indeed need all three transmissions, but source 2 only needs two and source 1 only needs one, hence a total of six transmissions. Now consider a similar network, only this time we have three sources that can reach the sink, shown in Figure 19.24(b). We create a schedule where each one of the sources gets to transmit once and once only. However this time not all sources generate data. Using labels shown in Figure 19.24(b), source 1 generates a packet and transmits it to source $2a$. Source $2a$ relays the packet to source $3a$, which then relays it to the sink. Then source $2b$ generates a packet and transmits it to source $3b$, which relays it to the sink. Finally source $3c$ generates and transmits a packet to the destination. We can view each raw of sources in this network as an equivalent of a single source in the previous example, i.e. $2a$ and $2b$ combined are equivalent to 2 in Figure 19.24(a); $3a$, $3b$ and $3c$ combined are equivalent to 3, in terms of interference and traffic load. We will define sources $2a$, $2b$ and $3a$, $3c$ as *virtual sources* in the sense that they each represent one actual source in the network but they are co-located in one physical source. Adopting this concept, in this network the highest number of interfering neighbors is five (with a total of six virtual sources all in one interference area) and therefore there exists a schedule of length 6 that enables every virtual source to transmit once. Since the traffic load is the same for all virtual sources, the resources will be shared equally and no source will be wasting its share. In this case we get $\lambda = W/6$. Note that this is the largest λ that could be obtained for the example in Figure 19.24(a). This concept allows us to define a 'traffic load-aware' schedule in the following way.

(1) For each source node, create one virtual source for every source node whose traffic goes through this node, including itself.

(2) Counting all the virtual sources we can determine the number of interfering neighbors (virtual sources) k. The new maximum degree of the interference graph is then $k - 1$.

(3) A schedule of length $s \leq k$ exists which is equally shared among virtual sources.

(4) The achievable throughput per node is simply the share obtained by any virtual source in the network, i.e. $\lambda = W/s \geq W/k$.

The concept of virtual source is used in References [58–63] to prove the following theorems:

(8) A randomly deployed network using multi-hop transmission for many-to-one communication can achieve $\lambda \geq W/\Sigma_{h=1}^{h=\lceil 2+\frac{\Delta}{r} \rceil} l_h^+ n_h^+$ with high probability, when knowledge of the traffic load is assumed. l_h^+ and n_h^+ are the upper bounds on the number of virtual sources per actual source and the number of actual sources respectively, that are h hops away from the sink with high probability.

(9) A randomly deployed network using multihop transmission for many-to-one communication can achieve a throughput arbitrarily close to $W/\{n(2 - \pi r^2)\}$, when knowledge of the traffic load is assumed and $\Delta = 0$.

In a *hierarchical network* with H clusters (heads) we assume that each cluster head creates a cluster containing the sources closest to it. Within each cluster the communication is either via a single hop or via multihop, while the communication from clusterheads to the sink is assumed to be done via a single hop on a different channel. We assume that cluster heads cannot transmit and receive simultaneously. In order to avoid boundary problems, we will assume there is at least a distance of $2(2r + \Delta)$ between any two clusterheads. We will also assume that each cluster covers an area of same size, as though not necessarily the same shape. Following these two assumptions and using result (5), we have with high probability that the number of nodes in each cluster is within $(\alpha_n n)$ of n/H, where α_n is such that $\lim_{n\to\infty} \alpha_n/n = \varepsilon$. Therefore the clusters essentially form a Voronoi tessellation of the field, where every cluster (or Voronoi cell) contains a circle of radius $2r + \Delta$. Consequently sources located near the boundary between two clusters will not have a higher number of interfering neighbors (in terms of virtual sources), due to low traffic load, than the ones closer to the clusterheads. Thus previous results are directly applicable and we do not have to be concerned with the boundary.

The question of interest is whether there exists an appropriate number of clusters H that would allow the network to achieve $\lambda = W/n$ with high probability using clustering, when clusterheads have the same transmission capacity W as the sources. That W/n remains to be the upper bound is again obvious considering the fact that the sink cannot receive from more than one node (at rate W), and that there are n sources in the network. In References [58–63] the following results are proven:

(10) In a network using clustering, where cluster heads have the same transmission capacity W as the sources, there exists an appropriate number of clusters H and an appropriate range of transmission r that would allow the network to achieve $\lambda = W/n$ with high probability as $n \to \infty$. The range of transmission r must satisfy

$$\frac{20r^4 + 36\Delta r^3 + 25\Delta^2 r^2 + 8\Delta^3 r + \Delta^4}{r^2 - \sqrt{\epsilon}\left(4r^2 + 4r\Delta + \Delta^2 - \frac{1}{\pi}\right)} \leq \frac{1}{\pi} \text{ and } H \geq 3 - \pi r^2$$

In the case of $\Delta = 0$ and letting $\sqrt{\varepsilon} \approx 0$, we need $r < \sqrt{1/20\pi}$.

(11) In a network using clustering, where cluster heads have transmission capacity W', there exists an appropriate number of clusters H and an appropriate range of transmission r, as $n \rightarrow \infty$, that allows the network to achieve $\lambda = W'/n$ with high probability. W'/n is also the upper bound on throughput in this scenario. The condition is $H \geq \left(3 - \pi r^2\right) W'/W$.

REFERENCES

[1] A. Ephremides and B. Hajek, Information theory and communication networks: an unconsummated union, *IEEE Trans. Inform. Theory*, vol. 44, no. 6, 1998, pp. 2416–2434.

[2] S. Glisic, *Adaptive WCDMA, Theory and Practice*. Wiley: Chichester 2003.

[3] A. Baiocchi *et al.*, Effects of user mobility on the capacity of a CDMA cellular network, *Eur. Trans. Telecommun.*, vol. 7, no. 4, 1996, pp. 305–314.

[4] A.M. Viterbi and A.J. Viterbi, Erlang capacity of a power controlled CDMA system, *IEEE J. Select. Areas Commun.*, vol. 11, no. 6, 1993, pp. 892–899.

[5] M.A. Landolsi, V.V. Veeravalli and N. Jain, New results on the reverse link capacity of CDMA cellular networks, *Proc. IEEE VTC'96*, 1996, pp. 1462–1466.

[6] J.C. Liberti Jr and T.S. Rappaport, Analytical results for capacity improvements in CDMA, *IEEE Trans. Vehicular Technol.*, vol. 43, no. 3, 1994, pp. 680–690.

[7] A.F. Naguib, A. Paulraj and T. Kailath, Capacity improvement with base-station antenna arrays in cellular CDMA, *IEEE Trans. Vehicular Technol.*, vol. 43, no. 3, 1994, pp. 691–698.

[8] L. Tomba, Outage probability in CDMA cellular systems with discontinuous transmission, *Proc. IEEE ISSSTA'96*, 1996, pp. 481–485.

[9] S. Hämäläinen, H. Holma and A. Toskala, Capacity evaluation of a cellular CDMA uplink with multiuser detection, *Proc. IEEE ISSSTA'96*, 1996, pp. 339–343.

[10] H. Holma, A. Toskala and T. Ojanperä, Cellular coverage analysis of wideband MUD-CDMA system, *Proc. IEEE PIMRC'97*, 1997, pp. 549–553.

[11] J. Blanz, A. Klein, M. Naβhan and A. Steil, Capacity of a cellular mobile radio system applying joint detection, COST 231 TD94 002, 1994.

[12] J. Blanz, A. Klein, M. Naβhan and A. Steil, Performance of a cellular hybrid C/TDMA mobile radio system applying Joint detection and coherent receiver antenna diversity, *IEEE J. Select. Areas Commun.*, vol. 12, no. 4, 1994, pp. 568–579.

[13] S. Manji and N.B. Mandayam, Outage probability for a zero forcing multiuser detector with random signature sequences, *Proc. IEEE VTC'98*, 1998, pp. 174–178.

[14] W. Huang and V.K. Bhargava, Performance evaluation of a DS/CDMA cellular system with voice and data services, *Proc. IEEE PIMRC'96*, 1996, pp. 588–592.

[15] V.H. MacDonald, The cellular concept, *Bell Syst. Tech. J.*, vol., 58, 1979, pp. 15–41.

[16] G.K. Chan, Effects of sectorization on the spectrum efficiency of cellular radio systems, *IEEE Trans. Vehicular Technol.*, vol. 41, 1992, 217–225.

[17] S.G. Glisic and P. Pirinen, Co-channel interference in digital cellular TDMA networks, in *Encyclopedia of Telecommunications*, J. Proakis (ed.). Wiley: Chichester, 2003.

[18] S.C. Swales, M. A. Beach, D.J. Edwards and J.P. McGeehan, The performance enhancement of multibeam adaptive base station antennas for cellular land mobile radio systems, *IEEE Trans. Vehicular Technol.*, vol. 39, 1990, pp. 56–67.
[19] J.H. Winters, Smart antennas for wireless systems, *IEEE Person. Commun.*, vol. 5, 1998, pp. 23–27.
[20] P. Petrus, R.B. Ertel and J.H. Reed, Capacity enhancement using adaptive arrays in an AMPS system, *IEEE Trans. Vehicular Technol.*, vol. 47, 1998, pp. 717–727.
[21] P.E. Mogensen, P.L. Espensen, P. Zetterberg, K.I. Pedersen and F. Frederiksen, Performance of adaptive antennas in FH-GSM using conventional beamforming, *Wireless Person. Commun.*, vol. 14, 2000, pp. 255-274.
[22] E. Kudoh and T. Matsumoto, Effect of transmitter power control imperfections on capacity in DS/CDMA cellular mobile radios, *Proc. IEEE ICC'92*, 1992, pp. 237–242.
[23] A.J. Paulraj and C.B. Papadias, Space–time processing for wireless communications, *IEEE Signal Process. Mag.*, vol. 14, 1997, pp. 49–83.
[24] L.C. Godara, Applications of antenna arrays to mobile communications, part I: performance improvement, feasibility, and system considerations, *Proc. IEEE*, vol. 85, 1997, pp. 1031–1060.
[25] J. Litva and T. K.-Y. Lo, *Digital Beamforming in Wireless Communications*. Artech House: Boston, MA, 1996.
[26] J.H. Winters, Optimum combining in digital mobile radio with cochannel interference, *IEEE Trans. Vehicular Technol.*, vol. VT-33, 1984, pp. 144–155.
[27] P. Zetterberg, A comparison of two systems for downlink communication with base station antenna arrays, *IEEE Trans. Vehicular Technol.*, vol. 48, 1999, pp. 1356–1370.
[28] P. Zetterberg and B. Ottersten, The spectrum efficiency of a base station antenna array system for spatially selective transmission, *IEEE Trans. Vehicular Technol.*, vol. 44, 1995, pp. 651–660.
[29] W.S. Au, R.D. Murch and C.T. Lea, Comparison between the spectral efficiency of SDMA systems and sectorized systems, *Wireless Personal Commun.*, vol. 16, 2001, pp. 15–67.
[30] L.-C. Wang, K. Chawla and L.J. Greenstein, Performance studies of narrow-beam trisector cellular systems, *Int. J. Wireless Inform. Networks*, vol. 5, 1998, pp. 89–102.
[31] S. Glisic *et al.* Effective capacity of advanced wireless cellular networks (invited paper), in *PIMRC2005*, Berlin, 11–14 September 2005.
[32] P.Gupta and P.R. Kumar, The capacity of wireless networks, *IEEE Trans. Inform. Theory*, vol. 46, no. 2, 2000 , pp. 388–404.
[33] L.-L. Xie and P.R. Kumar, A network information theory for wireless communication: scaling laws and optimal operation, *IEEE Trans. Inform. Theory*, vol. 50, no. 5, 2004 , pp. 748–767.
[34] A. Okabe, B. Boots and K. Sugihara, *Spatial Tessellations Concepts and Applications of Voronoi Diagrams*. Wiley: New York, 1992.
[35] J.A. Bondy and U. Murthy, *Graph Theory with Applications*. Elsevier: New York, 1976.
[36] P. Gupta and P.R. Kumar, Toward an information theory of large networks: an achievable rate region, *IEEE Trans. Inform. Theory*, vol. 49, 2003, pp. 1877–1894.

[37] A. Sendonaris, E. Erkip and B. Aazhang, Increasing uplink capacity via user cooperation diversity, *Proc. IEEE Int. Symp. Inform. Theory*, 2001, p. 156.
[38] J. Laneman and G. Wornell, Energy-efficient antenna sharing and relaying for wireless networks, *Proc. IEEE Wireless Commun. Networking Conf.*, 2000, p. 294.
[39] J. Laneman, G.Wornell and D. Tse, An efficient protocol for realizing cooperative diversity in wireless networks, *Proc. IEEE Int. Symp. Inform. Theory*, 2001, p. 294.
[40] A. Scaglione and Y.W. Hong, Opportunistic large arrays: cooperative transmission in wireless multihop *ad hoc* networks to reach far distances, *IEEE Trans. on Signal Process.*, vol. 51, no. 8, 2003, pp. 2082–2093.
[41] E.M. Royer and C.K. Toh, A review of current routing protocols for *ad hoc* mobile wireless networks, *IEEE Person. Commun. Mag.*, vol. 6, 1999, pp. 46–55.
[42] Y.-C. Tseng, S.-Y. Ni, Y.-S. Chen and J.-P. Sheu, The broadcast storm problem in a mobile *ad hoc* network, *ACM Wireless Networks*, vol. 8, 2002, pp. 153–167.
[43] W. Peng and X.-C. Lu, On the reduction of broadcast redundancy in mobile *ad hoc* networks, in *Proc. IEEE/ACM Mobile Ad Hoc Networking Computing*, November 2000, pp. 129–130.
[44] B. Williams and T. Camp, Comparison of broadcasting techniques for mobile *ad hoc* networks, in *Proc. ACM Int. Symp. Mobile Ad Hoc Networking Computing*, June 2002.
[45] ANSI/IEEE Std 802.11, 1999 ed., Available at: http://standards.ieee.org/getieee802 /download /802.11–1999.pdf, 1999.
[46] Network Simulator – ns2. Available at: www.isi.edu/nsnam/ns/
[47] S.-Y. R. Li, R.W. Yeung and N. Cai, Linear network coding, *IEEE Trans. Inform. Theory*, vol. 49, no. 2, February 2003, pp. 371–381.
[48] J.-H Chang, L. Tassiulas and F Rashid-Farrokhi, Joint transmitter receiver diversity for efficient space division multiaccess, *IEEE Trans. Wireless Commun.*, vol. 1, 2002, pp. 16–27.
[49] F. Rashid-Fwrokhi, K.J.R. Liu, and L. Tassiulas, Transmit beamforming and power control for cellulw wireless systems, *IEEE J. Select. Areas Commun.*, vol. 16, 1988, pp. 1437–1450.
[50] F. Rashid-Farrokhi, L. Tassiulas and K.J.R. Liu, Joint optimal power control and beamforming in wireless networks using antenna wrays, *IEEE Trans. Commun.*, vol. 46, 1998, pp. 1313–1324.
[51] A. Cwleial, A case where interference does not reduce capacity, *IEEE Trans. Inform. Theory*, vol. IT-21, 1975, pp. 569–570.
[52] T. Cover and J. Thomas, *Elements of Information Theory*. Wiley: New York, 1991.
[53] *G.G.* Raleigh and *V*K. Jones, Multivariate modulation and coding for wireless communication, *IEEE J. Select. Areas Commun.*, vol. 17, no. 5, 1999, pp. 851–866.
[54] M.C. Bromberg, Optimizing MIMO multipoint wireless networks assuming Gaussian other-user interference, IEEE Trans. Inform. Theory, vol. 49, no. 10, 2003, pp. 2352–2362.
[55] M.C. Bromberg and B.G.Agee, Optimization of spatially adaptive reciprocal multipoint communication networks, *IEEE Trans. Commun.*, vol. 51, no. 8, 2003, pp. 1254–1257.
[56] J.-H Chang, L. Tassiulas and F Rashid-Farrokhi, Joint transmitter receiver diversity for efficient space division multiaccess, *IEEE Trans. Wireless Commun.*, vol. 1, 2002, pp. 16–27.

[57] F. Rashid-Farrokhi, L. Tassiulas and K.J.R. Liu, Joint optimal power control and beamforming in wireless networks using antenna wrays, *IEEE Trans. Commun.*, vol. 46, 1998, pp. 1313–1324.

[58] H.E. Gamal, On the transport capacity of the many-to-one dense wireless network, *IEEE Vehicular Technol. Conf., VTC 2003*, vol. 5, 6–9 October 2003, pp. 2881–2885.

[59] A.D. Murugan, P.K. Gopala and H.E.Gamal, Correlated sources over wireless channels: cooperative source-channel coding, *IEEE J. Select. Areas Commun.*, vol. 22, no. 6, 2004, pp. 988–998.

[60] T. J. Kwon, M.Gerla, V.K. Varma, M. Barton and T.R. Hsing. Efficient flooding with passive clustering-an overhead-free selective forward mechanism for *ad hoc*/sensor networks *Proc. IEEE*, vol. 91, no. 8, 2003, pp. 1210–1220.

[61] A. Sabharwal, On capacity of relay-assisted communication, in *IEEE GLOBECOM'02*, vol. 2, 17–21 November 2002, pp. 1244–1248.

[62] H.E. Gamal, On the scaling laws of dense wireless sensor networks: the data gathering channel, *IEEE Trans. Inform. Theory*, vol. 51, no. 3, 2005, pp. 1229–1234.

[63] D. Marco, E.J. Duarte-Melo, M. Liu and D.L. Neuhoff, On the many-to-one transport capacity of a dense wireless sensor network and the compressibility of its data, in *Int. Workshop on Information Processing in Sensor Networks*, Berkeley, CA, April 2003, pp.104–109.

[64] L. Tomba, Computation of the outage probability in rice fading radio channels, *Eur. Trans. Telecommun.*, vol. 8, no. 2, 1997, pp. 127–134.

[65] I. Howitt and Y.M. Hawwar, Evaluation of outage probability due to cochannel interference in fading for a TDMA system with a beamformer, *Proc. IEEE Vehicular Technology Conf.*, 1998, pp. 520–524.

20

Energy-efficient Wireless Networks

20.1 ENERGY COST FUNCTION

In Chapter 5 we discussed the impact of MAC layer protocols on energy efficiency, including TCP controlled retransmissions. In this chapter, we extend this analysis to the network layer and focus on routing algorithms. We discuss how the error rate associated with a link affects the overall probability of reliable delivery, and consequently the energy associated with the reliable transmission of a single packet. For any particular link $\langle i, j \rangle$ between a transmitting node i and a receiving node j, let $T_{i,j}$ denote the transmission power and $p_{i,j}$ represent the packet error probability. Assuming that all packets are of a constant size, the energy involved in a packet transmission, $E_{i,j}$, is simply a fixed multiple of $T_{i,j}$.

Any signal transmitted is affected by two different factors: attenuation due to the medium, and interference with ambient noise at the receiver. The attenuation is proportional to D^K, where D is the distance between the receiver and the transmitter. The bit error rate associated with a particular link is essentially a function of the ratio of the received signal power to the ambient noise. In the constant-power scenario, $T_{i,j}$ is independent of the characteristics of the link $\langle i, j \rangle$ and is a constant. In this case, a receiver located further away from a transmitter will suffer greater signal attenuation (proportional to D^K) and will, accordingly, be subject to a larger bit-error rate. In the variable-power scenario, a transmitter node adjusts $T_{i,j}$ to ensure that the strength of the (attenuated) signal received by the receiver is *independent of* D and is above a certain threshold level Th. The minimum transmission power associated with a link of distance D in the variable-power scenario is $T_m = Th \times \gamma \times D^K$, where γ is a constant and K is the coefficient of channel attenuation ($K \geq 2$). Since Th is typically a technology-specific constant, we can see that the minimum transmission energy over such a link varies as $E_m(D) \propto D^K$.

Advanced Wireless Networks: 4G Technologies Savo G. Glisic
© 2006 John Wiley & Sons, Ltd.

If links are considered error-free, then minimum hop paths are the most energy-efficient for the fixed-power case. Similarly, in the absence of transmission errors, paths with a large number of small hops are typically more energy efficient in the variable power case. However in the presence of link errors, none of the above choices may give optimal energy efficient paths. We now analyze the consequences of this behavior for the variable-power scenario and end-to-end (EER) and hop-by-hop (HHR) packet retransmission techniques. The analysis for the fixed-power scenario is simpler, and is a special case of the variable-power scenario. Energy consumption for additional signal processing in the transmitter/receiver (modulation/demodulation) will be neglected. Modification of the models to include these losses is straightforward.

In the EER case, a transmission error on any link leads to an end-to-end retransmission over the path. Given the variable-power formulation of E_m, it is easy to see why breaking up a link of distance D into two shorter links of distance D_1 and D_2 such that $D_1 + D_2 = D$ always reduces the total E_m. To elaborate on this, let us consider communication between a sender (S) and a receiver (R) separated by a distance D. Let N represent the total number of hops between S and R, so that $N - 1$ represents the number of forwarding nodes i: $i = \{2, \ldots, N\}$, with node i referring to the $(i - 1)$th intermediate hop in the forwarding path. Node 1 refers to S and node $N + 1$ refers to R. In this case, the total energy spent in simply transmitting a packet once (without considering whether or not the packet was reliably received) from the sender to the receiver over the $N - 1$ forwarding nodes is:

$$E_{\mathrm{t}} = \sum_{i=1}^{N} E_m^{i,i+1} = \sum_{i=1}^{N} \alpha D_{i,i+1}^K \tag{20.1}$$

where $D_{i,j}$ refers to the distance between nodes i and j and α is a proportionality constant. To understand the tradeoffs associated with the choice of $N - 1$, we compute the lowest possible value of E_{t} for any given layout of $N - 1$. Using very simple symmetry arguments, it is easy to see that the minimum transmission energy case occurs when each of the hops are of equal length D/N. In that case, E_{t} is given by

$$E_{\mathrm{t}} = \sum_{i=1}^{N} \alpha D^K/N^K = \alpha D^K/N^{K-1}$$

We now consider how the choice of N affects the probability of transmission errors and the consequent need for retransmissions. Clearly, increasing the number of intermediate hops increases the likelihood of transmission errors over the entire path. Assuming that each of the N links has an independent packet error rate of p_{link}, the probability of a transmission error over the entire path, denoted by p, is given by $p = 1 - (1 - p_{\mathrm{link}})^N$.

The number of transmissions (including retransmissions) necessary to ensure the successful transfer of a packet between S and D is then a geometrically distributed random variable X, such that $\Pr\{X = k\} = p^{k-1} \times (1 - p)$, $\forall k$. The *mean* number of individual packet transmissions for the successful transfer of a single packet is thus $1/(1 - p)$. Since each such transmission uses total energy E_{t} given above, the total expected energy required in the reliable transmission of a single packet is given by:

$$E_{\mathrm{t}}(EER) = \alpha \frac{D^K}{N^{K-1}} \cdot \frac{1}{1-p} = \frac{\alpha D^K}{N^{K-1}(1 - p_{\mathrm{link}})^N} \tag{20.2}$$

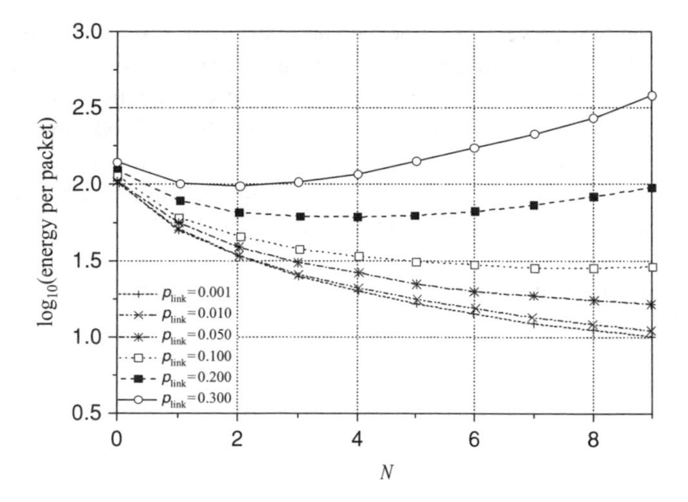

Figure 20.1 Total energy costs (K = 2, EER).

By treating N as a continuous variable and differentiating, it follows that the optimal value of the number of hops, N_{opt} is given by:

$$N_{\text{opt}} = -(K - 1)/\log(1 - p_{\text{link}}) \tag{20.3}$$

The existence of the optimum value is demonstrated in Figure 20.1.

In the case of the HHR model, the number of transmissions on each link is *independent of the other links* and is geometrically distributed. The total energy cost for the HHR case with N intermediate nodes, with each hop being of distance D/N and having a link packet error rate of p_{link}, is

$$E_t(HHR) = \sum_{i=1}^{N} \alpha \frac{D_{i,i+1}^k}{1 - p_{i,i+1}} = \alpha \frac{D^K}{N^{K-1} \cdot (1 - p_{\text{link}})} \tag{20.4}$$

In this case, it is easy to see that the total energy required always decreases with increasing N. One should be aware that in a practical system at some point when N is sufficiently large, the signal processing energy will become comparable with the energy spent for transmissions.

20.2 MINIMUM ENERGY ROUTING

Energy-aware routing protocols typically compute the shortest-cost path, where the cost associated with each link is some function of the transmission (and/or reception) energy associated with the corresponding nodes. To adapt such minimum cost route determination algorithms (such as Dijkstra's or the Bellman–Ford algorithm) for energy-efficient reliable routing, the link cost must now be a function of not just the associated transmission energy, but the link error rates as well. A link is assumed to exist between node pair $\{i, j\}$ as long as node j lies within the transmission range of node i. This transmission range is uniquely defined for the constant-power case. For the variable-power case, this range is really the *maximum permissible range* corresponding to the maximum transmission power of a sender. Let $E_{i,j}$ be the energy associated with the transmission of a packet over link $l_{i,j}$, and $p_{i,j}$ be the link packet error probability associated with that link. In the fixed-power scenario, $E_{i,j}$ is independent of the link characteristics; in the variable-power scenario, $E_{i,j}$ is a function of the distance between nodes i and j. Now, the routing algorithm's job is to compute the shortest path from a source to the destination that minimizes the sum of the energy costs over each constituent link.

Choosing path P for communication between S and D implies that the total energy cost for HHR is

$$E_P = \sum_{i=1}^{N} \frac{E_{i,i+1}}{1 - p_{i,i+1}} \tag{20.5}$$

Choosing a minimum-cost path from node 1 to node $N + 1$ is thus equivalent to choosing the path P that minimizes Equation (20.5). It is thus easy to see that the corresponding link cost for link $L_{i,j}$, denoted $C_{i,j}$, is given by $C_{i,j} = E_{i,j}/(1 - p_{i,j})$. *Ad-hoc* routing protocols, discussed in Chapter 13, such as AODV, DSR and TORA, can use this link cost to compute the appropriate energy-efficient routes. Some of the existing energy-efficient routing techniques, e.g. PARO, can also be easily adapted to use this new link cost formulation to compute minimum-energy routes. Thus, in such a modified version of the PARO algorithm, an intermediate node C would offer to interject itself between two nodes A and B if the sum of the link cost $C_{A,C} + C_{C,B}$ was less than the 'direct' link cost $C_{A,B}$.

In *end-to-end retransmissions*, the total energy cost along a path contains a multiplicative term involving the packet error probabilities of the individual constituent links. In fact, assuming that transmission errors on a link do not stop downstream nodes from relaying the packet, the total energy cost can be now expressed as:

$$E_P = \frac{\sum_{i=1}^{N} E_{i,i+1}}{\prod_{i=1}^{N} (1 - p_{i,i+1})} \tag{20.5a}$$

Given this form, the total cost of the path cannot be expressed as a linear sum of individual link costs, thereby making the exact formulation inappropriate for traditional minimum-cost path computation algorithms. In [1] a heuristic cost function for a link, was suggested as

$$C_{i,j} = \frac{E_{i,j}}{(1 - p_{i,j})^L} \tag{20.6}$$

where $L = 2, 3, \ldots,$ and is chosen to be identical for all links. Clearly, if the exact path length is known and all nodes on the path have identical link error rates and transmission costs, L should be chosen equal to that path length. However, in accordance with current routing schemes, we require that a link should associate only a single link cost with itself, irrespective of the lengths of specific routing paths that pass through it. Therefore, we need to fix the value of L independent of the different paths that cross a given link. If better knowledge of the network paths is available, then *L should be chosen to be the average path length of this network.* Higher values of L impose progressively stiffer penalties on links with non-zero error probabilities. Given this formulation of the link cost, the minimum-cost path computation effectively computes the path with the minimum "approximate" energy cost given by:

$$E_P \sim \sum_{i=1}^{N} \frac{E_{i,i+1}}{(1 - p_{i,i+1})^L} \tag{20.7}$$

As before, protocols like AODV, DSR, TORA and PARO can use this new link cost function to make their routing decisions.

20.3 MAXIMIZING NETWORK LIFETIME

We now discuss how we can include the retransmission-aware formulation of the link cost in an algorithm, that attempts to increase the operational lifetime of multihop wireless networks. Unlike previous protocols, *maximum reliable packet carrying capacity* (MRPC) considers both the node characteristics (residual battery energy at the transmitting node) and the link characteristics (link distance and link error rates), while evaluating the suitability of alternative paths. Given the current battery power levels at the different nodes, MRPC selects a route that has the *maximum reliable packet carrying capacity* among all possible paths, assuming no other cross-traffic passes through the nodes on that path.

To formalize the algorithm, let us assume that the residual battery power at a certain instance of time at node i is B_i. As before, let the transmission energy required by node i to transmit a packet over link $\langle i, j \rangle$ to node j be $E_{i,j}$. Let the source and destination nodes for a specific session (route) be S and D respectively. If the route-selection algorithm then selects a path P from S to D that includes the link $\langle i, j \rangle$, then the maximum number of packets that node i can forward over this link is clearly $B_i/E_{i,j}$. Accordingly, we can define a node-link metric, $M_{i,j}$ for the link $\langle i, j \rangle$ as:

$$M_{i,j} = \frac{B_i}{E_{i,j}} \tag{20.8}$$

The key point in this formulation is that the cost metric includes both a node-specific parameter (the battery power) and a link-specific parameter (the packet transmission energy for reliable communication across the link). Clearly, the 'lifetime' of the chosen path P, defined by the maximum number of packets that may be potentially forwarded between S and D using path P, is determined by the *weakest intermediate node* – one with the smallest value of $M_{i,j}$. Accordingly, the 'lifetime' $\Im$ associated with route P is:

$$\Im_P = \min_{(i,j) \in P} \{M_{i,j}\} \tag{20.9}$$

The MRPC algorithm then selects the candidate route P_c that maximizes the 'lifetime' of communication between S and D. Formally, the chosen route is such that:

$$P_c = \arg\max\{\Im_P \mid P \in \textit{all possible routes}\} \tag{20.10}$$

Given the cost and lifetime formulations for MRPC it is then easy to use a modified version of Dijkstra's minimum cost algorithm for decentralized route computation.

To apply Dijkstra's algorithm for determining the minimum-cost path, the distance metric from any node to the given destination should be defined as the value of $\Im_P$ over the optimal path from that node to D. Now consider a node A that sees advertisements from its neighbors,

$\{X, Y, Z, \ldots\}$, with corresponding distance metrics $\Im_X, \Im_Y, \Im_Z, \ldots$ for a given destination D. Node A can then compute the best path to D (using its optimal neighbor) by using the following simple algorithm:

(1) For each of the neighboring nodes ($j \in \{X, Y, Z, \ldots\}$), compute the link cost $M_{A,j}$ using Equation (20.8).

(2) For each of the neighboring nodes ($j \in \{X, Y, Z, \ldots\}$) compute the potential new value of $\Im_{\text{pot}}$ using $\Im_{\text{pot}}(A, j) = \min\{M_{A,j}, \Im_j\}$.

(3) Select as the next-hop neighbor towards D the node which results in the maximum value of $\Im_{\text{pot}}$, i.e. choose node k such that $k = \arg\max_{j\varepsilon\{X,Y,Z\}}\{\Im_{\text{pot}}(A, j)\}$ and assign $\Im_A = \Im_{\text{pot}}(A, k)$.

Using this recursive formulation allows all nodes in the *ad hoc* network to iteratively build their optimal route towards a specific destination D. The distance-vector formulation presented here can easily be incorporated in protocols, such as AODV, DSR and TORA, that are specifically designed for *ad hoc* mobile environments.

The basic MRPC formulation for power-aware routing does not need to specify the value of the transmission energy cost associated with a specific link. Note that Equation (20.8) is expressed as a function of a generic link cost $M_{i,j}$. Accordingly, by specifying different forms of $M_{i,j}$ it is possible to tailor the MRPC mechanism for specific technologies and/or scenarios.

For the fixed-power scenario, the energy involved in a single packet transmission attempt, $E_{i,j}$, is a constant for all $\langle i, j\rangle$ and is independent of the distance between neighboring nodes i and j. For the variable-power scenario, $E_{i,j}$ will typically be $\propto D_{i,j}^K$, where $D_{i,j}$ is the distance between nodes i and j. A routing algorithm for reliable packet transfer should include the link's packet error probability in formulating the transmission energy cost. By ignoring the packet error probability, the link cost concentrates (wrongly) only on the energy spent in transmitting a single packet. The correct metric is the *effective* packet transmission energy for reliable transmission, which includes the energy spent in one or more re-transmissions that might be necessary in the face of link errors. A transmission energy metric of the form $C_{i,j} = E_{i,j}/(1 - p_{i,j})^L$ was suggested, where $p_{i,j}$ is the link's packet error probability, $L \geq 1$. For hop-by-hop re-transmissions L should be chosen to be 1. In the absence of hop-by-hop re-transmissions (i.e. re-transmissions are only performed end-to-end), the transmission cost is well approximated by $L \in [3, 5]$. Power-aware routing has been studied in a number of papers [1–42].

The *conditional MRPC* (CMRPC) algorithm is the MRPC equivalent of the *conditional min–max minimum battery cost routing* (CMMBCR) algorithm presented in Toh *et al.* [2]. The CMMBCR algorithm is based on the observation that using residual battery energy as the sole metric throughout the lifetime of the *ad hoc* network can actually lower the overall lifetime, since it never attempts to minimize the total energy consumption. Accordingly, the CMMBCR algorithm uses regular minimum-energy routing as long as there is even one candidate path, where the remaining battery power level in all the constituent nodes lies above a specified threshold γ. When no such path exists, CMMBCR switches to MMBCR, i.e. it picks the path with the maximum residual capacity on the ‘critical node’.

The CMRPC algorithm differs from CMMBCR in that the cost-functions at all times include the link-specific parameters (e.g. error rates) as defined earlier in this chapter. The algorithm can thus be specified as follows. Let Ψ be the set of all possible paths between the source S and destination D and let Ω represent the set of paths such that: for any route $Q \in \Omega$, $\Im_Q \geq \gamma$. In other words Ω represents the set of paths whose most critical nodes have a lifetime greater than a specified threshold. The routing scheme thus consists of the following actions:

(1) If $\Omega \neq \emptyset$ (there are one or more paths with $\Im > \gamma$, the algorithm selects a path $\bar{Q} \in \Omega$ that minimizes the total transmission energy for reliable transfer, i.e.

$$\bar{Q} = \arg\min_{Q \in \Omega} \left\{ \sum_{(i,j) \in Q} M_{i,j} \right\} \tag{20.11}$$

(2) Otherwise, switch to the MRPC algorithm, i.e. select $\bar{Q}$ such that

$$\bar{Q} = \arg\max_{Q \in \Omega} \{\Im_Q | Q \in \Psi\}$$

The threshold γ is a parameter of the CMRPC algorithm. A lower value of γ implies a smaller protection margin for nodes nearing battery power exhaustion. Accordingly, the performance of the CMRPC algorithm will be a function of γ.

The *performance example* is based on network topology shown in Figure 20.2. The corner nodes and the mid-points of each side of the rectangular grid were chosen as traffic sources and destinations; the bold lines in the figure show the session end-points [1].

Each (source, destination) pair had two simultaneous sessions activated in the opposite direction, giving rise to a total of 16 different sessions. For the results reported here, each session consisted of a UDP traffic generated by a CBR source whose inter-packet gap was distributed uniformly between 0.1 and 0.2 s. The error rate on each link was independently distributed uniformly between (0.05, $p_{\max}$). Varying values of $p_{\max}$ were used. Routes were recomputed at 2 s intervals in these simulations to capture the effect of changes in the residual packet capacity on the link metrics.

Whenever nodes died (when its battery power gets completely drained) during the course of a simulation, the simulation code would check whether the graph became partitioned. The simulations were run until each of the 16 sessions failed to find any route from their source to the corresponding destination. To avoid the termination of a simulation due to battery power exhaustion at source or destination nodes, all source and sink nodes were configured

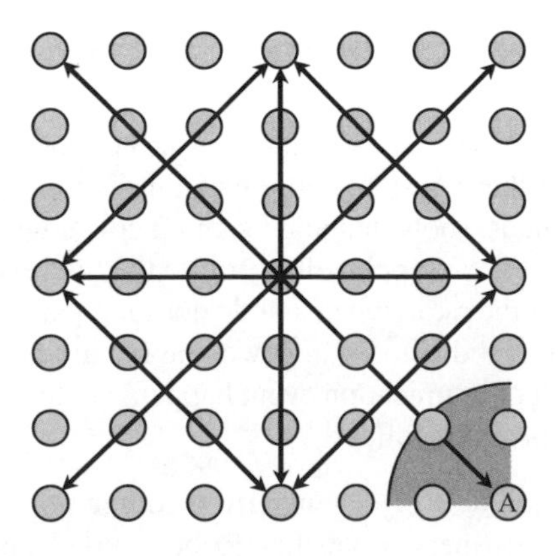

Figure 20.2 Simulation scenario.

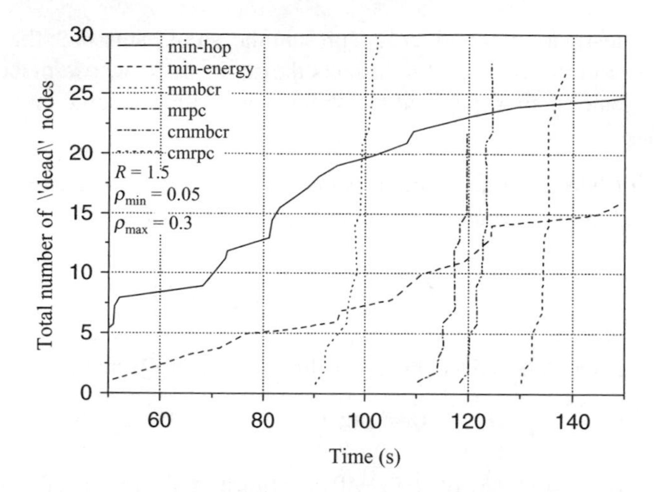

Figure 20.3 Expiration sequence for different algorithms, $R = 1.5$. (Reproduced by permission of IEEE [1].)

to have 'infinite' power resources. All the other 'intermediate' nodes were configured with identical initial battery power levels.

To study the performance of the various algorithms, experiments were performed where the maximum transmission radius, R, of each node was varied. Figure 20.2 shows the set of neighboring nodes for a corner node when the transmission radius is set to 1.5. The expiration sequence, as well as the node expiry times were noted, for each simulation. The *expiration sequence* (sorted in ascending order of the expiration times) provides a useful indicator of how each algorithm affects the lifetime of the individual nodes, and the entire network. In addition to the expiration sequence, the *total packet throughput* was also calculated by counting the total number of packets successfully received at the destination nodes, and the *energy costs per packet* by dividing the total energy expenditure by the total packet throughput. Except for the expiration sequences, all other metrics were obtained by averaging over multiple runs. The results are shown in Figures 20.3–20.5. From these results one can see that CMRPC/MRPC outperforms other options.

20.4 ENERGY-EFFICIENT MAC IN SENSOR NETWORKS

Among the requirements for MACs in wireless sensor networks, energy efficiency is typically the primary goal. In these systems, idle listening is identified as a major source of energy wastage. Measurements show that idle listening consumes nearly the same power as receiving. Since in sensor network applications traffic load is very light most of the time, it is often desirable to turn off the radio when a node does not participate in any data delivery. Some schemes put (scheduled) idle nodes in power-saving mode (SMAC) and switch nodes to full active mode when a communication event happens. Although a low duty cycle MAC is energy-efficient, it has three side-effects.

(1) It increases the packet delivery latency. At a source node, a sampling reading may occur during the sleep period and has to be queued until the active period. An intermediate node may have to wait until the receiver wakes up before it can forward

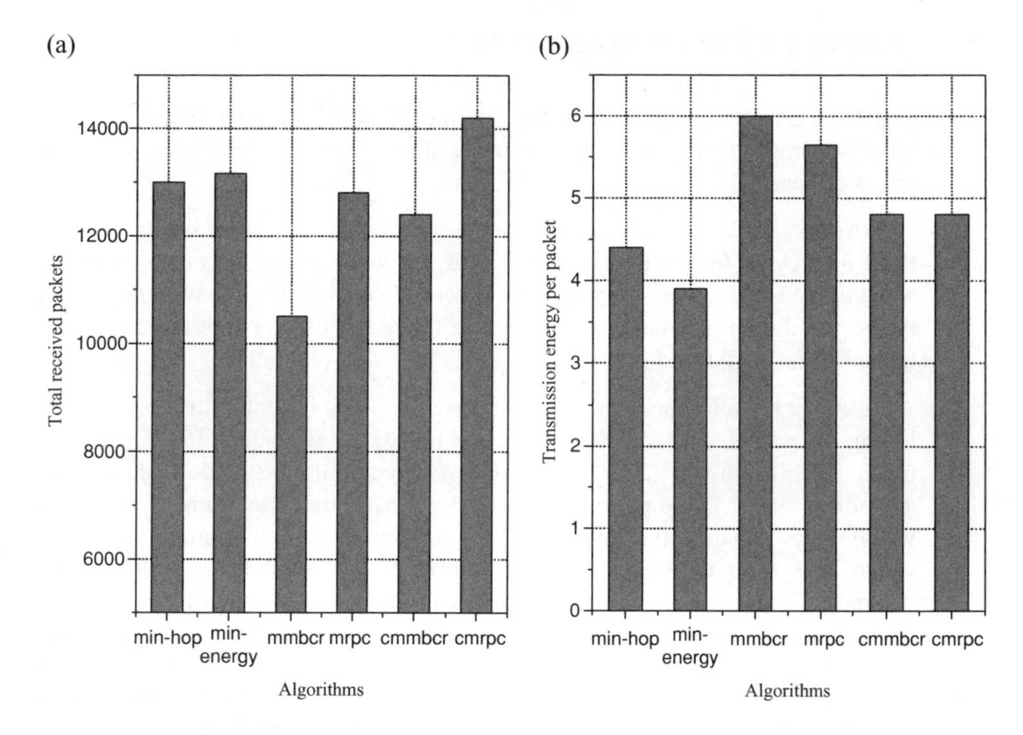

Figure 20.4 (a) Total packet throughput; (b) average transmission energy per received packet (UDP sources), $R = 1.5$. (Reproduced by permission of IEEE [1].)

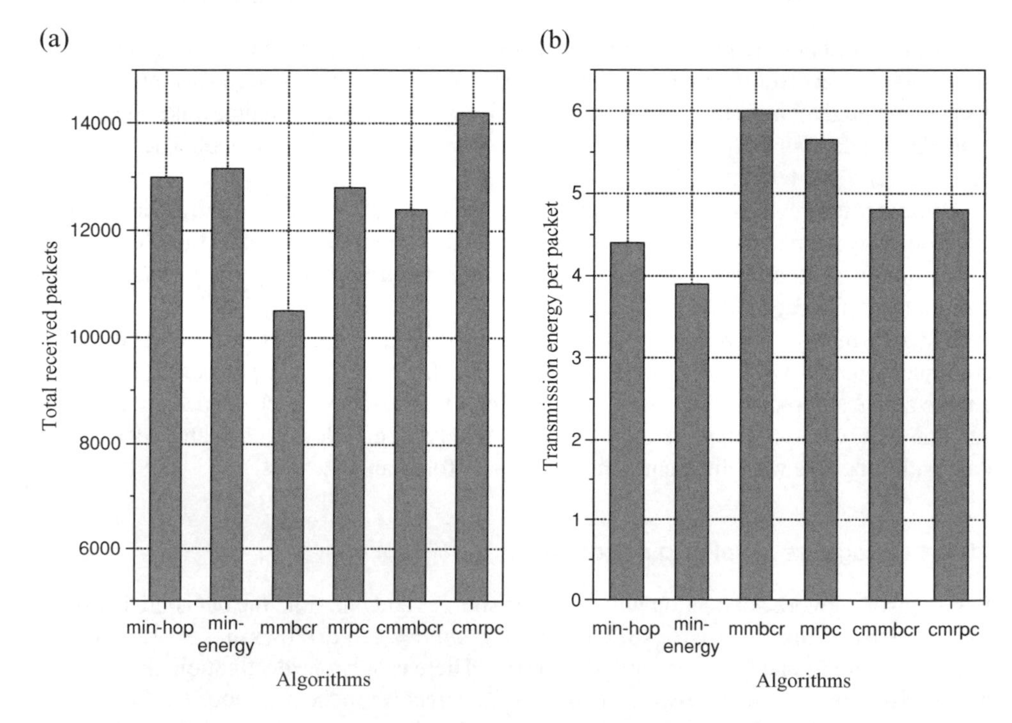

Figure 20.5 CMRPC: (a) total packet throughput; (b) average. transmission energy per received packet vs the protection threshold. (Reproduced by permission of IEEE [1].)

a packet received from its previous hop. This is called *sleep latency* in SMAC, and it increases proportionally with hop length by a slope of schedule length (active period plus sleep period).

(2) A fixed duty cycle does not adapt to the varying traffic rate in sensor network. A fixed duty cycle for the highest traffic load results in significant energy wastage when traffic is low while a duty cycle for low traffic load results in low message data delivery and long queuing delay. Therefore it is desirable to adapt the duty cycle under variant traffic load.

(3) A fixed synchronous duty cycle may increase the possibility of collision. If neighboring nodes turn to active state at the same time, all may contend for the channel, making a collision very likely. There are several possibilities to reduce sleep delay and adjust duty cycle to the traffic load. Those mechanisms are either implicit, in which nodes remain active on overhearing an ongoing transmission or explicit, in which there are direct duty cycle adjusting messages. In adaptive listening, a node that overhears its neighbor's transmission wakes up for a short period of time at the end of the transmission, so that if it is the next hop of its neighbor, it can receive the message without waiting for its scheduled active time. A node also can keep listening and potentially transmitting as long as it is in an active period. An active period ends when no activation event has occurred for a certain time. The activation time events include reception of any data, the sensing of communication on the radio, the end-of-transmission of a node's own data packet or acknowledgement, etc.

If the number of buffered packets for an intended receiver exceeds a threshold L, the sender can signal the receiver to remain on for the next slot. A node requested to stay awake sends an acknowledgement to the sender, indicating its willingness to remain awake in the next slot. The sender can then send a packet to the receiver in the following slot. The request is renewed on a slot-by-slot basis.

However, in previous mechanisms (whether explicit or implicit), not all nodes beyond one hop away from the receiver can overhear the data communication, and therefore packet forwarding will stop after a few hops. This *data forwarding interruption problem* causes sleep latency for packet delivery.

DMAC employs a *staggered active/sleep schedule* to solve this problem and enable continuous data forwarding on the multihop path. In DMAC, *data prediction* is used to enable active slot request when multiple children of a node have packets to send in a same sending slot, while the *more to send packet* is used when nodes on the same level of the data gathering tree with different parents compete for channel access.

20.4.1 Staggered wakeup schedule

For a sensor network application with multiple sources and one sink, the data delivery paths from sources to sink are in a tree structure, a *data gathering tree*. Flows in the data gathering tree are unidirectional from sensor nodes to sink. There is only one destination, the sink. All nodes except the sink will forward any packets they receive to the next hop. The key insight in designing a MAC for such a tree is that it is feasible to stagger the wake-up scheme so that packets flow continuously from sensor nodes to the sink. *DMAC is designed to deliver data along the data gathering tree*, aiming at both energy efficiency and low latency.

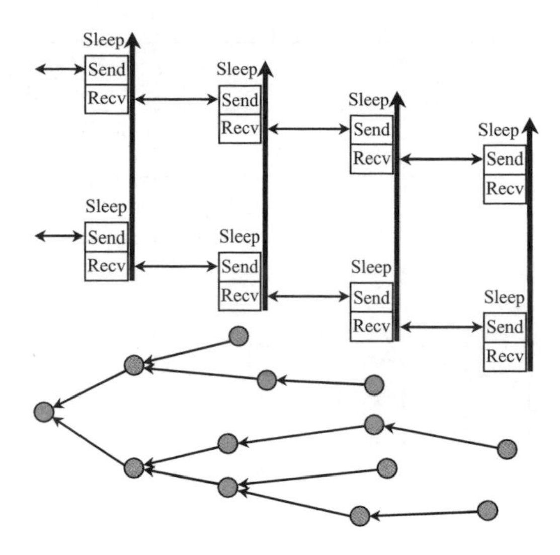

Figure 20.6 DMAC in a data gathering tree.

In DMAC, the activity schedule of nodes on the multihop path is staggered to wake up sequentially like a chain reaction. Figure 20.6 shows a data gathering tree and the staggered wake-up scheme. An interval is divided into receiving, sending and sleep periods. In receiving state, a node is expected to receive a packet and send an ACK packet back to the sender. In the sending state, a node will try to send a packet to its next hop and receive an ACK packet. In sleep state, nodes will turn off radio to save energy. The receiving and sending periods have the same length of μ, which is enough for one packet transmission and reception. Depending on its depth d in the data gathering tree, a node skews its wake-up scheme $d\mu$ ahead from the schedule of the sink. In this structure, data delivery can only be done in one direction towards the root. Intermediate nodes have a sending slot immediately after the receiving slot.

A staggered wake-up schedule has four advantages: (1) since nodes on the path wake up sequentially to forward a packet to next hop, sleep delay is eliminated if there is no packet loss due to channel error or collision; (2) a request for longer active period can be propagated all the way down to the sink, so that all nodes on the multihop path can increase their duty cycle promptly to avoid data stuck in intermediate nodes; (3) since the active periods are now separated, contention is reduced; and (4) only nodes on the multihop path need to increase their duty cycle, while the other nodes can still operate on the basic low duty cycle to save energy. The simulation results for the three different protocols are shown in Figure 20.7. DMAC demonstrates good performance. In the simulation the following parameters were used, as in Lu *et al.* [43]: Radio bandwidth 100 kbps, radio transmission range 250 m, radio interference range 550 m, packet length 100 bytes, transmit power 0.66 W, receive power 0.395 W and idle power 0.35 W. The sleeping power consumption is set to 0. An MTS (more to send) packet is 3 bytes long. According to the parameters of the radio and packet length, the receiving and sending slot μ is set to 10ms for DMAC and 11 ms for DMAC/MTS. The active period is set to 20 ms for SMAC with adaptive listening. All schemes have the basic duty cycle of 10 %. This means a sleep period of 180 ms for DMAC and SMAC, 198 ms for DMAC/MTS.

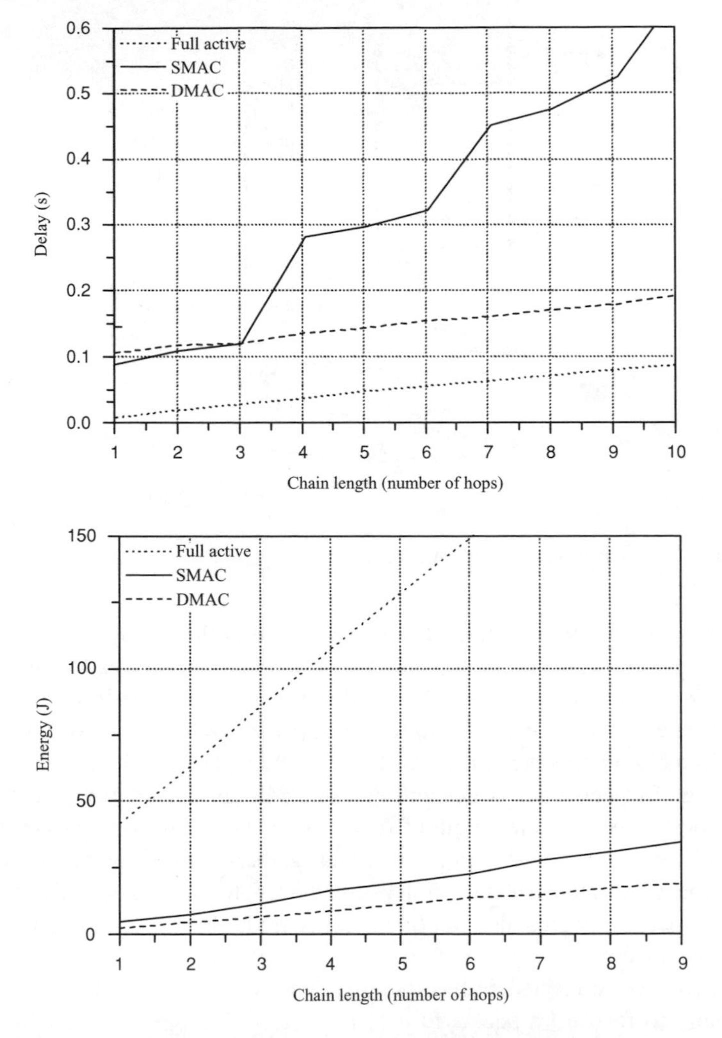

Figure 20.7 (a) Mean packet latency on each hop under low traffic load; (b) total energy consumption on each hop under low traffic load.

REFERENCES

[1] A. Misra and S. Banerjee, MRPC: maximizing network lifetime for reliable routing in wireless environments, in *Wireless Communications and Networking Conf., WCNC 2002*, vol. 2, 17–21 March 2002. IEEE: New York, pp. 800–806.

[2] C.-K.,Toh, H. Cobb and D.A. Scott, Performance evaluation of battery-life-aware routing schemes for wireless *ad hoc* networks, in *IEEE Int. Conf. Communications, ICC 2001*, Helsinki, vol. 9, 11–14 June 2001, pp. 2824–2829.

[3] I. Stojmenovic and X. Lin, Power-aware localized routing in wireless networks, *IEEE Trans. Parallel and Distributed Systems*, vol. 12, no. 11, November 2001, pp. 1122–1133.

[4] J.-C. Cano and D. Kim, Investigating performance of power-aware routing protocols for mobile ad-hoc networks, in *International Mobility and Wireless Access Workshop, MobiWac 2002*, 12 October 2002, pp. 80–86.
[5] C.-K. Toh, Maximum battery life routing to support ubiquitous mobile computing in wireless *ad hoc* networks, *IEEE Commun. Mag.*, vol. 39, no. 6, 2001, pp. 138–147.
[6] J.-H. Chang and L. Tassiulas, Maximum lifetime routing in wireless sensor networks, *IEEE/ACM Trans. Networking*, vol. 12, no. 4, 2004, pp. 609–619.
[7] M. Tarique, K.E. Tepe and M. Naserian, Energy saving dynamic source routing for *ad hoc* wireless networks, in *Third Int. Symp. Modeling and Optimization in Mobile, Ad Hoc, and Wireless Networks, WIOPT 2005*, 4–6 April 2005, pp. 305–310.
[8] J. Gomez, A.T. Campbell, M. Naghshineh and C. Bisdikian, Power-aware routing in wireless packet networks, in *IEEE Int. Workshop on Mobile Multimedia Communications (MoMuC '99)*, 15–17 November 1999, pp. 380–383.
[9] Y. Xue and B. Li, A location-aided power-aware routing protocol in mobile *ad hoc* networks, in *IEEE Global Telecommun. Conf., GLOBECOM '01*, vol. 5, 25–29 November 2001, pp. 2837–2841.
[10] A. Avudainayagam, Y. Fang and W. Lou, DEAR: a device and energy aware routing protocol for mobile *ad hoc* networks, in *Proc. MILCOM 2002*, vol. 1, 7–10 October 2002, pp. 483–488.
[11] R. Ranjan and A. Saad, Generic architecture for power aware routing in wireless sensor networks, *29th Annual IEEE Int. Conf. Local Computer Networks*, 16–18 November 2004, pp. 575–576.
[12] B. Zhang and H.T. Mouftah, Localized power-aware routing for wireless *ad hoc* networks, in *IEEE Int. Conf. Communications*, vol. 6, 20–24 June 2004, pp. 3754–3758.
[13] J. Shen and J. Harms, Position-based routing with a power-aware weighted forwarding function in MANETs, in *IEEE Int. Conf. Performance, Computing, and Communications*, 2004, pp. 347–355.
[14] A. Safwat, H. Hassanein and H. Moufta, A MAC-based performance study of energy-aware routing schemes in wireless *ad hoc* networks, in *IEEE Global Telecommunications Conf., GLOBECOM '02*. vol. 1, 17–21 November 2002, pp. 47–51.
[15] L. De Nardis, G. Giancola and M.-G. Di Benedetto, A position based routing strategy for UWB networks, in *IEEE Conf. Ultra Wideband Systems and Technologies*, 16–19 November 2003, pp. 200–204.
[16] J. Nie and Z. Zhou, An energy based power-aware routing protocol in *ad hoc* networks, *IEEE Int. Symp. Communications and Information Technology, ISCIT*, vol. 1, 26–29 October 2004, pp. 280–285.
[17] J.-P. Sheu, C.-W. Lai and C.-M. Chao, Power-aware routing for energy conserving and balance in *ad hoc* networks, in *IEEE Int. Conf. Networking, Sensing and Control*, vol. 1, 21–23 March 2004, pp. 468–473.
[18] S.-H. Lee, E. Choi and D.-H. Cho, Timer-based broadcasting for power-aware routing in power-controlled wireless *ad hoc* networks, *IEEE Commun. Lett.*, vol. 9, no. 3, 2005, pp. 222–224.
[19] A. Helmy, Contact-extended zone-based transactions routing for energy-constrained wireless *ad hoc* networks, *IEEE Trans. Vehicular Technol.*, vol. 54, no. 1, 2005, pp. 307–319.

[20] M. Maleki, K. Dantu and M. Pedram, Power-aware source routing protocol for mobile *ad hoc* networks, in *Proc. Int. Symp. Low Power Electronics and Design, ISLPED '02*, 2002, pp. 72–75.
[21] R.K. Guha, C.A. Gunter and S. Sarkar, Fair coalitions for power-aware routing in wireless networks, *43rd IEEE Conf. Decision and Control, CDC*, vol. 3, 14–17 December 2004, pp. 3271–3276.
[22] L. De Nardis, G. Giancola and M.-G. Di Benedetto, Power-aware design of MAC and routing for UWB networks, in *IEEE Global Telecommunications Conf. Workshops*, 29 November to 3 December 2004, pp. 235–239.
[23] Q. Li, J. Aslam and D. Rus, Distributed energy-conserving routing protocols, in *Proc. 36th Annual Hawaii Int. Conf. System Sciences*, 6–9 January 2003, p. 10.
[24] J. Gomez, A.T. Campbell, M. Naghshineh and C. Bisdikian, Conserving transmission power in wireless *ad hoc* networks, in *Ninth Int. Conf. Network Protocols*, 11–14 November 2001, pp. 24–34.
[25] J.-E. Garcia, A. Kallel, K. Kyamakya, K. Jobmann, J.C. Cano and P. Manzoni, A novel DSR-based energy-efficient routing algorithm for mobile *ad-hoc* networks, *IEEE Vehicular Technol. Conf., VTC 2003*, vol. 5, 6–9 October, pp. 2849–2854.
[26] N. Gemelli, P. LaMonica, P. Petzke and J. Spina, Capabilities aware routing for dynamic *ad hoc* networks, *Int. Conf. Integration of Knowledge Intensive Multi-Agent Systems*, 30 September to 4 October 2003, pp. 585–590.
[27] M. Krunz, A. Muqattash and S.-J. Lee, Transmission power control in wireless *ad hoc* networks: challenges, solutions and open issues, *IEEE Networks*, vol. 18, no. 5, 2004, pp. 8–14.
[28] J. Schiller, A. Liers, H. Ritter, R. Winter and T. Voigt, ScatterWeb – low power sensor nodes and energy aware routing, in *Proc. 38th Annual Hawaii Int. Conf. System Sciences*, 3–6 January 2005, p. 286c.
[29] B. Zhang and H. Mouftah, Adaptive energy-aware routing protocols for wireless *ad hoc* networks, in *First Int. Conf. Quality of Service in Heterogeneous Wired/Wireless Networks, QSHINE*, 18–20 October 2004, pp. 252–259.
[30] Y. Zhou, D.I. Laurenson and S. McLaughlin, High survival probability routing in power-aware mobile *ad hoc* networks, *Electron. Lett.*, vol. 40, no. 22, 2004, pp. 1424–1426.
[31] S. Agarwal, A. Ahuja, J.P. Singh and R. Shorey, Route-lifetime assessment based routing (RABR) protocol for mobile *ad-hoc* networks, in *IEEE Int. Conf. Commun., ICC*, vol. 3, 18–22 June 2000, pp. 1697–1701.
[32] A. Safwat, H. Hassanein and H. Mouftah, Power-aware fair infrastructure formation for wireless mobile *ad hoc* communications, in *IEEE Global Telecommun. Conf., GLOBECOM '01*, vol. 5, 25–29 November 2001, pp. 2832–2836.
[33] S. Guo and O. Yang, An optimal TDMA-based MAC scheduling for the minimum energy multicast in wireless *ad hoc* networks, *IEEE Int. Conf. Mobile Ad-hoc and Sensor Systems*, 25–27 October 2004, pp. 552–554.
[34] K. Wang, Y.-L. Xu, G.-L. Chen and Y.-F. Wu, Power-aware on-demand routing protocol for MANET, in *Proc. 24th Int. Conf. Distributed Computing Systems Workshops*, 23–24 March 2004, pp. 723–728.
[35] R. Min and A. Chandrakasan, A framework for energy-scalable communication in high-density wireless networks, in *Proc. Int. Symp. Low Power Electronics and Design, ISLPED '02*, 2002, pp. 36–41.

[36] D. Shin and J. Kim, Power-aware communication optimization for networks-on-chips with voltage scalable links, in *Int. Conf. Hardware/Software Codesign and System Synthesis, CODES + ISSS*, 8–10 September 2004, pp. 170–175.

[37] L. Hughes and Y. Zhang, Self-limiting, adaptive protocols for controlled flooding in *ad hoc* networks, in *Proc. Second Annual Conf. Communication Networks and Services Research*, 19–21 May 2004, pp. 33–38.

[38] Y. Liu and P.X. Liu, A two-hop energy-efficient mesh protocol for wireless sensor networks, in *Proc. of IEEE/RSJ Int. Conf. Intelligent Robots and Systems (IROS 2004)*, vol. 2, 28 September to 2 October 2004, pp. 1786–1791.

[39] A. Safwat, H. Hassanein and H. Mouftah, Energy-efficient infrastructure formation in MANETs, in *Proc. IEEE Conf. Local Computer Networks*, 14–16 November 2001, pp. 542–549.

[40] G. Dimitroulakos, A. Milidonis, M.D. Galaris, G. Theodoridis, C.E. Gontis and F. Catthoor, Power aware data type refinement on the HIPERLAN/2, in *IEEE Int. Conf. Electronics, Circuits and Systems, ICECS,* vol. 1, 14–17 December 2003, pp. 216–219.

[41] S. Jayashree, B.S. Manoj and C.S.R. Murthy, Next step in MAC evolution: battery awareness?, in *IEEE Global Telecommunications Conf., GLOBECOM '04*, vol. 5, 29 November to 3 December 2004, pp. 2786–2790.

[42] L. Zhao, X. Hong and Q. Liang, Energy-efficient self-organization for wireless sensor networks: a fully distributed approach, in *IEEE Global Telecommunications Conf., GLOBECOM '04*, vol. 5, 29 November to 3 December 2004, pp. 2728–2732.

[43] G. Lu, B. Krishnamachari and C.S. Raghavendra, An adaptive energy-efficient and low-latency MAC for data gathering in wireless sensor networks, in *Proc. Int. Parallel and Distributed Processing Symp.*, 26–30 April 2004, p. 224.

21

Quality-of-Service Management

QoS has been the main criterion in the analysis of the schemes presented so far in the book. However, in the last chapter we present some additional solutions that will be of interest in 4G networks.

21.1 BLIND QoS ASSESSMENT SYSTEM

In this section we present a method to blindly estimate the quality of a multimedia communication link using digital fragile watermarking. Data hiding by digital watermarking is usually employed for multimedia copyright protection, authenticity verification or similar purposes. However, watermarking is here adopted as a technique to provide a blind measure of the quality of service in multimedia communications [1–28]. The watermark embedding procedure is sketched in Figure 21.1. It consists of embedding a watermark sequence, which is usually binary, into host data by means of a key. In the detection phase, the key is used to verify the presence of the embedded sequence. With regard to the domain where the watermark embedding occurs, we can distinguish methods operating in the spatial domain [15], in the discrete cosine transform DCT domain [16–19], in the Fourier transform domain [20], and in the wavelet transform domain [1– 5].

When unwanted modifications of the watermarked data affect even the extracted watermark, the embedding scheme is known as fragile. Fragile watermarking [6–8] can be used to obtain information about the tampering process. In fact, it indicates whether or not the data has been altered and supplies localization information as to where the data was altered.

Here, an unconventional use of a fragile watermark to evaluate the QoS in multimedia mobile communications is presented. Specifically, a known watermark is superimposed onto the host data. The rationale behind this approach is that, by transmitting the watermarked

Advanced Wireless Networks: 4G Technologies Savo G. Glisic
© 2006 John Wiley & Sons, Ltd.

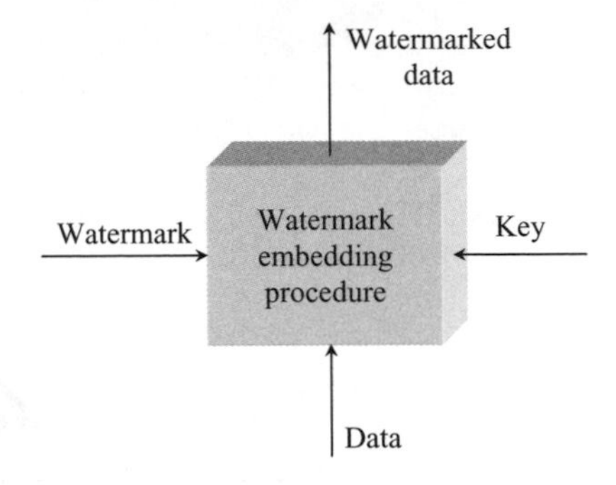

Figure 21.1 Watermark embedding process.

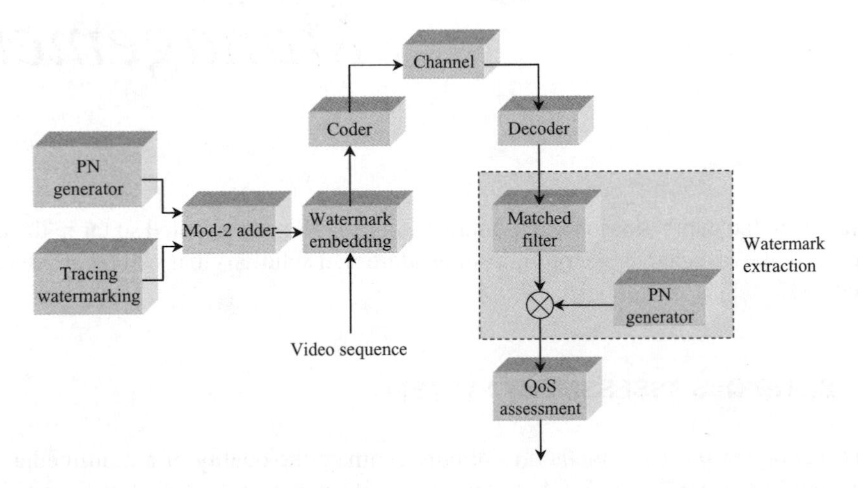

Figure 21.2 Principle scheme of tracing watermarking for coder-channel quality assessment in multimedia communications.

data onto a channel, the mark undergoes the same alterations as suffered by the data. At the receiving side, the watermark is estimated and compared with the original. Since the alterations endured by the watermark are also likely to be suffered by the entire data, as they follow the same communication link, the watermark degradation can be used to estimate the overall alterations endured by the data.

The tracing watermarking procedure for coder-channel quality assessment is given in Figure 21.2. The watermark embedding is performed by resorting to the spread-spectrum technique proposed in Cox *et al.* [17] for still images and applied to video sequences in Hartung and Girod [18]. The watermark is a narrowband low-energy signal. It is then spread so that it has a larger bandwidth. Consequently, the watermark energy contribution for each host frequency bin is negligible, which makes the watermark imperceptible.

In the application addressed here, a system embedded into the data stream that is able to *trace* the degradations introduced by the transmission system composed by the coder-channel cascade and not perceptually affecting the data themselves is described.

A set of uncorrelated pseudorandom noise (PN) matrices is multiplied by the reference watermark. Both the PN matrices and the embedded watermark are known at the receiving side. The watermark is the same for each video sequence frame, whereas the PN matrices are different for each frame. This insures that the spatial localization of the mark is different frame-by-frame so that the watermark visual persistency is negligible.

After the randomization of the watermark by the PN matrices, the embedding of the tracing marks is performed in the DCT domain. The watermark is embedded in the DCT middle-band frequencies of the whole image. After the inverse DCT (IDCT) has been performed on each frame, the whole sequence is coded by a video coder and, finally, transmitted.

At the receiving side, the video is first decoded; then, from the DCT of each received frame of the sequence, a matched filter extracts the spread watermark, which is finally despread using the known PN matrices. After having extracted the received watermark, it is matched to the reference one, which is known at the receiving side, and the mean-square error (MSE), between the original mark and the received one, is used as an index of the degradation affecting the received watermark.

21.1.1 System modeling

A two-dimensional video sequence will be represented as $\{f_i[n_1, n_2], i = 1, 2, \ldots M\}$. The sequence is composed of M frames $f_i[n_1, n_2]$ of $N_1 \times N_2$ pixels. Let $\omega[k_1, k_2]$ be the employed watermark, having dimensions $K_1 \times K_2$ and $\{p_i[k_1, k_2], i = 1, 2, \ldots M\}$ the M PN matrices of dimensions $K_1 \times K_2$. The watermark $\omega[k_1, k_2]$ is a visual pattern, like a logo, consisting of binary pixels.

The PN matrices $\{p_i[k_1, k_2], i = 1, 2, \ldots M\}$ are employed to spread a narrowband signal (watermark) over a much larger bandwidth signal (host frame) in such a way that the watermark energy is undetectable in any single frequency of the host image. This can be expressed as follows: $\omega_i^{(s)}[k_1, k_2] = \omega[k_1, k_2] \cdot p_i[k_1, k_2], i = 1, 2, \ldots, M$, where $\omega_i^{(s)}[k_1, k_2]$ is the spread version of the watermark to be embedded in the ith frame. Let $F_i[k_1, k_2] = \text{DCT}\{f_i[n_1, n_2]\}$ be the DCT transform of the ith frame $f_i[n_1, n_2]$. The spread watermark is embedded, in the DCT domain, in the middle-high frequency region S of $F_i[k_1, k_2]$.

The embedding is performed in the DCT domain according to the following rule:

$$F_i^{(\omega)}[k_1, k_2] = \begin{cases} F[k_1, k_2] + \alpha\omega_i^{(s)}[k_1, k_2], & (k_1, k_2) \in S \\ F[k_1, k_2], & (k_1, k_2) \notin S \end{cases} \tag{21.1}$$

where $F_i^{(\omega)}[k_1, k_2]$ represents the DCT of the ith watermarked frame $f_i[n_1, n_2]$, and α is a scaling factor that determines the watermark strength. Parameter α must be chosen in such a way as to compromise between not degrading the picture on one side and being detectable by the tracing algorithm on another.

The ith watermarked frame is then obtained by performing the IDCT transform $f_i^{(\omega)}[n_1, n_2] = \text{IDCT}\{F_i^{(\omega)}[k_1, k_2]\}$. Finally, the whole sequence is coded and then transmitted through a noisy channel. If $\{\hat{f}_i^{(\omega)}[n_1, n_2], i = 1, 2, \ldots M\}$ is the received video sequence then the DCT transform in the receiver gives $\hat{F}_i^{(\omega)}[k_1, k_2] = \text{DCT}\{\hat{f}_i^{(\omega)}[n_1, n_2]\}$.

The middle–high frequency region of embedding S is selected. Then, the corresponding portion of $\hat{F}_i^{(\omega)}[k_1, k_2]$ is multiplied by the watermark $\omega[k_1, k_2]$, which is known at the receiving side, giving an estimation $\hat{\omega}_i^{(s)}[k_1, k_2]$ of the spread version of the watermark embedded in the ith frame, $\hat{\omega}_i^{(s)}[k_1, k_2] = \hat{F}_i^{(\omega)}[k_1, k_2] \cdot \omega[k_1, k_2]$. The dispreading operation, for the generic ith frame gives $\hat{\omega}_i[k_1, k_2] = \hat{\omega}_i^{(s)}[k_1, k_2] \cdot p_i[k_1, k_2]$. Finally, the watermark is estimated by averaging the dispread watermark over the M transmitted frames

$$\hat{\omega}[k_1, k_2] = \frac{1}{M} \sum_{i=1}^{M} \hat{\omega}_i[k_1, k_2] \tag{21.2}$$

The QoS is evaluated by comparing the extracted watermark with the original one. Formally this can be represented as

$$\mathrm{MSE}_i = \frac{1}{K_1 K_2} \sum_{K_1=1}^{K_1} \sum_{k_2=1}^{K_2} (\omega_i[k_1, k_2] - \hat{\omega}_i[k_1, k_2])^2 \tag{21.3}$$

and

$$\mathrm{MSE} = \frac{1}{M} \sum_{i=1}^{M} \mathrm{MSE}_i \tag{21.4}$$

In experiments presented in References [26, 28], the dimensions of the video sequences employed have been properly chosen in order to simulate a multimedia service in a UMTS scenario. Therefore, QCIF (144 × 176) video sequences, which well match the limited dimensions of a mobile terminal's display, have been employed. Sample results are shown in Figures 21.3 and 21.4. As shown in the figures, the quality degradation of the watermark embedded into the host video has the same behavior as the one affecting the video.

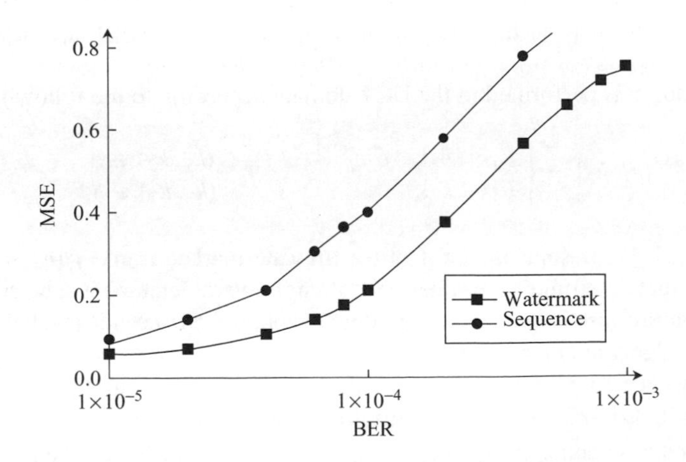

Figure 21.3 Watermark MSE and video sequence MSE (normalized to 1) vs the BER for the sequence 'Akiyo' MPEG-2 coded at 600 kb/s.

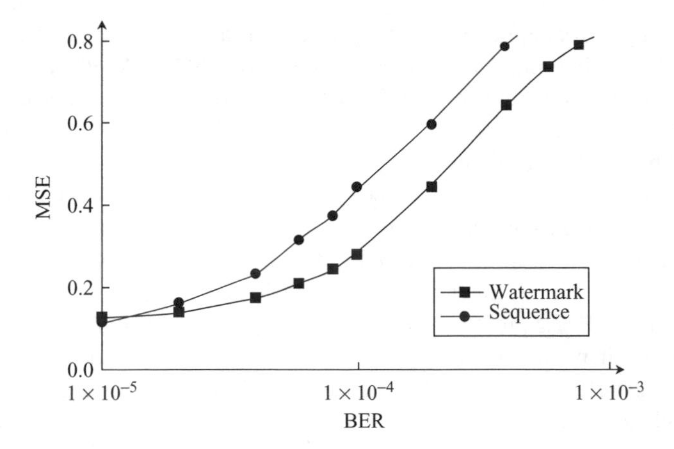

Figure 21.4 Watermark MSE and video sequence MSE (normalized to 1) vs the BER for the sequence 'Akiyo' MPEG-2 coded at 200 kb/s.

21.2 QoS PROVISIONING IN WLAN

As an example system, the 802.11 WLAN consists of *basic service sets* (BSS), each of which is composed of wireless stations (STA). The WLAN can be configured as an *ad hoc* network (an independent BSS) or an infrastructure network (composed of an access point and the associated STAs).

The channel access for the STAs in a BSS is under the control of a coordination function. The 802.11 MAC protocol provides two coordination functions: *distributed coordination function* (DCF) and *point coordination function* (PCF). The DFC is a contention-based access scheme using *carrier sense multiple access with collision avoidance* (CSMA/CA).

Priority levels for access to the channel are provided through the use of *interframe spaces* such as short interframe space (SIFS) and distributed interframe space (DIFS). The backoff procedure is used for collision avoidance, where each STA waits for a *backoff time* (a random time interval in units of slot times) before each frame transmission. The PCF provides contention-free frame transmission in an infrastructure network using the *point coordinator* (PC), operating at the access point (AP), to determine which STA currently gets the channel access. The DCF and the PCF can coexist by alternating the *contention period* (CP), during which the DCF is performed, and the *contention-free period* (CFP), during which the PCF is performed. A CFP and a CP are together referred to as a *repetition interval* or a *superframe*. Different aspects of 802.11 WLAN are discussed in References [29–40]. The performance analysis of the DCF was studied in References [29, 30, 35]. The performance of DCF degrades in high traffic loads due to serious collisions. The CSMA/CA is not suitable for data traffic at higher channel speeds [29] due to the large waste on the backoff time. The influence of various sizes of the backoff time on the channel throughput and the optimal setting of the backoff time was studied in Cali *et al.* [31]. A common technique is to adjust the backoff time according to the traffic priority. The PCF based on a polling scheme is suitable for time-bounded real-time traffic. The simple polling schedules for voice traffic and video traffic are presented in Crow *et al.* [35]. Some complex polling schedules are proposed in References [32–34, 36, 39].

To expand support for applications with QoS requirements, the IEEE 802.11E task group [37] is proceeding to build the QoS enhancements of the 802.11 MAC. The techniques for providing prioritized and parameterized QoS data deliveries have been discussed in the task group. An enhanced DCF was proposed, where each traffic flow is assigned with a different backoff time whose value decreases with increasing traffic priority, to achieve the prioritized QoS data delivery. To guarantee the bounded delay requirements in the parameterized QoS data delivery, the hybrid coordination function (HCF) was proposed [42]. In the HCF, an STA can be guaranteed to issue the frame transmission even during the CP using the contention-free burst (CFB). The CFB can be considered a temporary CFP during which the transmissions of STAs are coordinated by the polling scheme as the PCF. Moreover, a multipolling mechanism (called contention-free multipoll, CF-multipoll) was proposed [41] to reduce the polling overhead that the traditional PCF suffers from. In the multipoll, the PC can poll more than one STA simultaneously using a single polling frame.

In this section, we consider how to efficiently serve real-time traffic in the IEEE 802.11 WLAN by using the multipolling mechanism. The multipolling mechanism can increase the channel utilization and is robust in mobile and error-prone environments. The mechanism can be used in the PCF and the HCF. Moreover, a polling schedule is provided to guarantee the bounded delay requirements of real-time flows.

21.2.1 Contention-based multipolling

In the PCF or the HCF, each STA in the polling list takes a polling frame when polled. This polling scheme is called *SinglePoll.* The number of polling frames for a polling list can be reduced if a multipolling mechanism is used. Here, we discuss an efficient multipolling mechanism that has the advantages of high channel utilization and low implementation overhead. This multipolling mechanism, referred to as *contention period multipoll (CP-Multipoll)*, incorporates the DCF access scheme into the polling scheme.

In the DCF, any contending STA for the channel will select a backoff time in units of slot times and execute the backoff procedure as follows: if the channel is sensed idle for a DIFS period, an STA starts the transmission immediately. Otherwise, an STA should defer until the channel becomes idle for a DIFS period and then the backoff time is decreased. If the channel is idle for a slot time, an STA decreases the backoff time by one or else freezes the backoff time. When the backoff time becomes zero, an STA begins the frame transmission. If a collision occurs, the STA duplicates the backoff time used in the last transmission and executes the backoff procedure again. In the DCF, the virtual carrier sensing using the *Network Allocation Vector* (NAV) is performed at the MAC sublayer. The information on the duration of a frame exchange sequence for one STA is included in the frame header (duration field) and is announced to other STAs. Other contending STAs will wait for the completion of the current frame exchange by updating their NAVs according to the announced duration information. For a communication the RTS (*request to send*) and CTS (*clear to send*) frames are exchanged before the data frame transmission. Other STAs defer their channel access by setting their NAVs according to the duration field in the RTS, the RTS, or the data frame. The exchange of RTS/CTS frames can also avoid the hidden terminal problem [35].

The basic idea of CP-Multipoll is to transform the polling order into the contending order which indicates the order of winning the channel contention. Different backoff time values are assigned to the flows in the polling group and the corresponding STAs execute the backoff procedures after receiving the CP-Multipoll frame. The contending order of these STAs is the same as the ascending order of the assigned backoff time values. Therefore, to maintain the polling order in a polling group, we can assign the backoff time value incrementally according to the expected polling order.

The CP-Multipoll has the following advantages: a polled STA can hold the channel access flexibly depending on the size of local buffered data, so it becomes easy to deal with the data burst; if a polled STA makes no response to the CP-Multipoll, other STAs in the same polling group will detect the channel idle right away and advance the starting of channel contention. Therefore, the CP-Multipoll can decrease the waste of channel space and can afford any polling error.

21.2.2 Polling efficiency

We consider the environment with overlapping BSSs. In the overlapping BSS, we distinguish the STAs associated with one BSS from the STAs associated with neighboring BSSs by using the terms 'internal STAs' and 'external STAs', respectively. Moreover, a STA that will cause an internal collision is called a nonbehaved STA; otherwise, the STA is called a behaved one.

The following terminology is used in the performance analysis:

- *frame_num*, maximum number of data frames allowed to be transmitted in a TXOP (transmission opportunity);
- l_{type}, number of bits in a 'type' frame;
- t_{type}, transmission time on the channel for a 'type' frame;
- *single-poll*, the single polling frame in the single poll (i.e. CF-Poll frame);
- n, poll size (i.e. the number of poll records in a multipolling frame);
- n-poll, the multipolling frame with n poll records;
- ERR_{type}, probability that a 'type' frame is dropped due to bit errors;
- α, probability that an STA has no data to send when polled;
- β, probability that an STA becomes a nonbehaved STA;
- h, number of PCs operating in the overlapping space;
- *InitBT*, initial backoff time;
- E, polling efficiency.

The following assumptions will be used in the analysis:

(1) The PC (point coordinator) always performs the initial backoff before sending any polling frame.

(2) The data frame is transmitted without any acknowledgment. When an STA is polled, the STA either sends a half of *frame_num* data frames on average or sends a null data frame if there is no data to send.

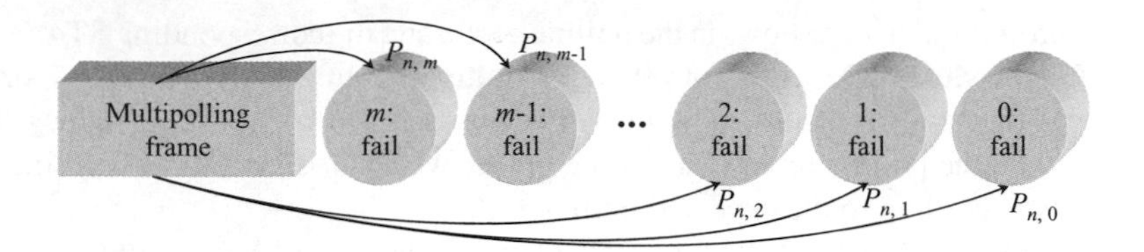

Figure 21.5 The state diagram of the multipolling mechanism. (Reproduced by permission of IEEE [43].)

(3) There is an equal probability BER for a bit error to occur due to the channel noise (interference, fading or multipath). Hence, ERR_{type} can be expressed as $1-(1-BER)^{l_{\text{type}}}$.

(4) An STA becomes a nonbehaved one if the STA fails to receive the CTS frame from the PC. That is, $\beta = ERR_{\text{CTS}}$.

If *AvgD* and *AvgT* denote the total number of bits in the data frames successfully sent from the polled STAs and the average complete time in time units for a poll, respectively, then the polling efficiency is defined as $E = AvgD/AvgT$. This represents the average uplink data rate during a poll.

In the single poll, a polled STA contributes data frames if the STA successfully receives a CF-Poll frame and has pending data frames to be successfully transmitted. The polled STA may suffer frame error due to interference from external STAs. Therefore,

$$\begin{aligned} AvgD = {} & (1-\alpha)\cdot frame\_num/2 \cdot l_{\text{data\_frame}} \\ & (1-ERR_{\text{data\_frame}})\cdot(1-ERR_{\text{single\_poll}}) \end{aligned} \tag{21.5}$$

The polled STA will give a response to the PC after an SIFS period for a successful CF-Poll. If a polled STA does not respond to the PC after a PIFS period for a failed CF-Poll, the PC takes over the channel control and may send the next CF-Poll frame. In the HCF, the RTS/CTS frames should be exchanged before the data transmission to prevent the interference from other STAs. Since the HCF is being substituted for the PCF in the 802.11E, we consider the single poll of the HCF in our analysis. Therefore,

$$\begin{aligned} AvgT = {} & InitBT + (t_{\text{single-poll}} + PIFS)\cdot ERR_{\text{single-poll}} + [t_{\text{single-poll}} + SIFS \\ & + t_{\text{RTS}} + t_{\text{CTS}} + 2SIFS + (1-\alpha) frame\_num/2 \cdot (t_{\text{data\_frame}} + SIFS) \\ & + \alpha(t_{\text{null\_frame}} + SIFS)]\cdot(1-ERR_{\text{single-poll}}) \end{aligned} \tag{21.6}$$

Next, we use a state diagram to represent the situation after sending a multipolling frame with poll size n. In Figure 21.5, the state $m : fail(0 \le m \le n)$ represents the event that there are m STAs in the polling group which failed to receive the multipolling frame. Let $P_{n,m}$ denote the probability of the state $m : fail$. Let $D_{n,m}$ and $T_{n,m}$ denote the amount of uplink data frames in bits and the total time duration in time units under the state $m : fail$, respectively. Generally, we have the following values:

$$P_{n,m} = \binom{n}{m}\cdot(ERR_{n-\text{poll}})^m \cdot (1-ERR_{n-\text{poll}})^{n-m}$$

$$AvgD = \sum_{m=0}^{n} P_{n,m} \cdot D_{n,m} \tag{21.7}$$

$$AvgT = InitBT + \sum_{m=0}^{n} P_{n,m} \cdot T_{n,m}$$

For the CF-Multipoll, the external STAs in the overlapping BSS will have their TXOPs overlap the ones allocated to the internal STAs. We assume that the interference from external STAs has been included in the parameter BER. Hence,

$$D_{n,m} = (n - m) \cdot (1 - \alpha) \cdot frame\_num/2 \cdot l_{\text{data\_frame}} \cdot (1 - ERR_{\text{data\_frame}})$$

Also, each successive TXOP starts an SIFS period after the predecessor's TXOP limit expires in the CF-Multipoll. Note that the time spent by a polled STA is fixed regardless of the number of pending data frames. Hence,

$$T_{n,m} = t_{n-\text{poll}} + n \cdot SIFS + n \cdot frame\_num \cdot (t_{\text{data\_frame}} + SIFS) \tag{21.8}$$

In the CP-Multipoll, there are two processing phases: one is for the normal multipoll and the other is for the error recovery. In the normal phase, there are $(n - m)$ STAs successfully receiving the multipolling frame under the state m : *fail*. Among these STAs, $(1 - \beta)(n - m)$ STAs are behaved STAs and $\beta(n - m)$ STAs are nonbehaved ones. Each nonbehaved STA is assumed to destroy one data frame transmitted by a behaved one. In the recovery phase, m STAs failed to receive the multipolling frame and $(n - m)$ nonbehaved STAs will be served individually using the CP-Multipoll with poll size one. The analysis of this phase is similar to the one in the single poll. Hence, $D_{n,m}$ has the following value:

$$\begin{aligned} D_{n,m} &= D_{n,m}^{\text{normal}} + D_{n,m}^{\text{recovery}} \\ D_{n,m}^{\text{normal}} &= [(1 - \beta) \cdot (n - m) \cdot frame\_num/2 - \beta(n - m)] \\ &\quad (1 - \alpha) \cdot l_{\text{data\_frame}} \cdot (1 - ERR_{\text{data\_frame}}) \\ D_{n,m}^{\text{recovery}} &= [m + \beta(n - m)] \cdot (1 - \alpha) \cdot frame\_num/2 \\ &\quad l_{\text{data\_frame}} \cdot (1 - ERR_{\text{data\_frame}}) \cdot (1 - ERR_{1-\text{poll}}) \end{aligned} \tag{21.9}$$

The length of the time of the normal phase is dominated by the transmission time of those $(1 - \beta)(n - m)$ behaved STAs. The total backoff time consumed in the normal phase is $h \times n + 1$, regardless of the value m. Hence,

$$\begin{aligned} T_{n,m} &= T_{n,m}^{\text{normal}} + T_{n,m}^{\text{recovery}} \\ T_{n,m}^{\text{normal}} &= t_{n-\text{poll}} + (h \cdot n + 1) \cdot Slot + (1 - \beta) \cdot (n - m) \\ &\quad (t_{\text{RTS}} + t_{\text{CTS}} + 2\,SIFS + (1 - \alpha) \cdot frame\_num/2 \cdot (t_{\text{data\_frame}} + SIFS) \\ &\quad + \alpha \cdot t_{\text{null\_frame}}) T_{n,m}^{\text{recovery}} = [m + \beta(n - m)] \cdot [(t_{1-\text{poll}} \\ &\quad + 2\,Slot) \cdot ERR_{1-\text{poll}} + (t_{1-\text{poll}} + 2\,Slot + t_{\text{RTS}} + t_{\text{CTS}} + 2\,SIFS + (1 - \alpha) \\ &\quad frame\_num/2 \cdot (t_{\text{data\_frame}} + SIFS) + \alpha \cdot t_{\text{null\_frame}}) \\ &\quad (1 - ERR_{1-\text{poll}})]. \end{aligned} \tag{21.10}$$

The superpoll can poll a group of STAs together as the CF-Multipoll and the CP-Multipoll. In the superpoll, the polled STA will attach the poll records of those polled ones whose polling orders are after it to its current transmitted data frame. This scheme can be considered

Table 21.1 System and poll-related parameters. (Reproduced by permission of IEEE [43])

Parameter	Value
Channel rate	11 Mbs
PHYheader	192 b
MAC header	272 b
Slot	20 μs
SIFS	10 μs
DIFS	50 μs
PIFS	30 μs
Superframe	25 μs
CFP_Max_Duration	20 μs
frame_num	3
ldata_frame(octets)	200(default), 400, 600, 800
lnull_frame(octets)	34
lsingle_poll(octets)	34
ln_poll(octets)	$16 + 8n$
n	1–20
BER	10^{-3}, 10^{-4}, 10^{-5}(default), 10^{-6}
α	0.2 (default), 0.4, 0.6, 0.8
h, the number of PCs operating in the overlapping space	1, 2 (default), 3, 4
InitBT(μs)	90(default), 110, 130, 150

as one with replicated poll records. To keep the correct polling order, each polled STA in the polling group should monitor the channel and check whether the previous STA has finished sending a data frame. If a polled STA fails to receive its poll record, the next following STA in the polling group will wait for a timeout before its own frame transmission. However, the situation where some STAs cannot listen to other STAs' channel activities is not considered. This may cause inconsistent setting of timers among polled STAs and may cause internal collisions.

21.2.2.1 Performance example

The parameter settings related to system and the polling schemes are listed in Table 21.1 [43]. The performance improvement of CP-Multipoll in percentage over other schemes is shown in Figure 21.6.

21.3 DYNAMIC SCHEDULING ON RLC/MAC LAYER

In wireless networks, the fading characteristics of the wireless physical channel may introduce location-dependent channel errors. A scheduling algorithm, referred to as *channel-condition independent packet fair queueing* (CIF-Q), is presented in Ng *et al.* [44] to solve the problem of location-dependent channel errors in wireless networks by suspending

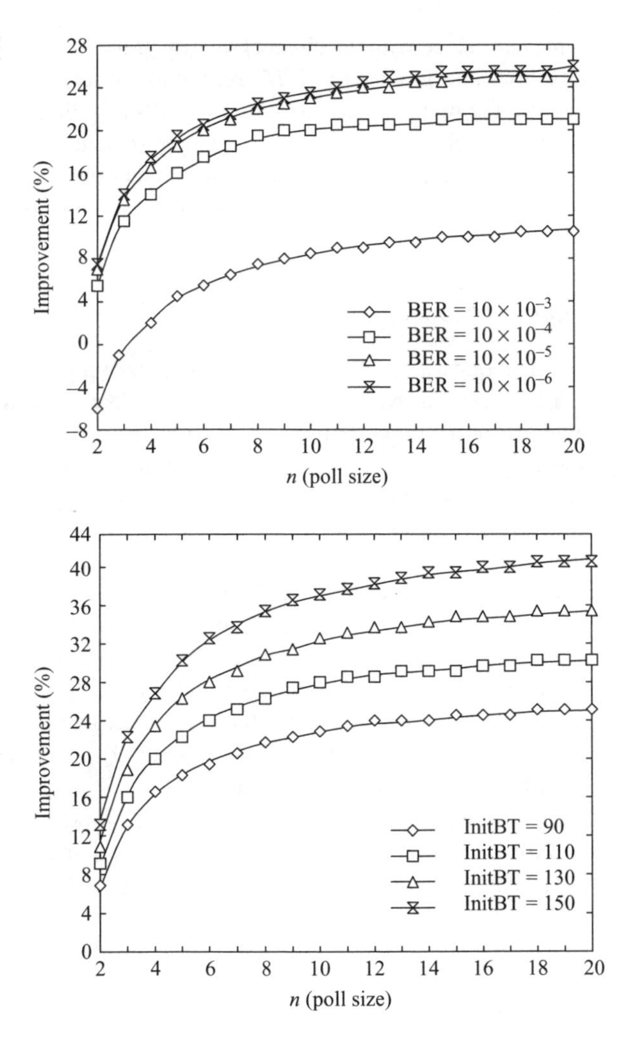

Figure 21.6 Comparison between single poll and CP-Multipoll.

transmission service of a connection of a mobile station (MS) when the MS is in a high BER region. *To compensate for the service loss of the MS in a high BER region, the CIF-Q scheduling algorithm increases the service priority after the MS returns to a low BER region to fulfill its QoS specification.* However, this suspension may cause long delays for the connection of the MS, because the duration that an MS resides within a high interference region may be unpredictably long. In many services, such as video transmissions, minor errors are acceptable to the receiver-side applications, but a long delay is not.

In Yars *et al.* [45], a scheduling scheme at the RLC/ MAC layer is proposed, referred to as a dynamic scheduling mechanism for mobile communication (DSMC). The scheme is discussed in the context of GPRS applications. DSMC is a dynamic scheduling architecture that can conform to a variety of QoS requirements in the networks without changing the coding rate on RLC layer. The DSMC scheduling scheme, which is based on the self-clocked fair queuing (SCFQ) algorithm [46], can dynamically adjust the service rate (weight) for a particular connection in accordance with the channel quality. It will use the low service rate when an MS is within a high interference (*high-IF*) region to reduce bandwidth waste due to retransmissions, and use the high service rate when the MS is within a low interference (*low-IF*) region to fulfill the QoS requirements (e.g. delay bound and loss ratio) of the MS. The high and low service rates are both determined at connection setup time. Thus, the complexity of scheduling algorithm in wireless networks can be reduced.

21.3.1 DSMC functional blocks

An MS issues a connection request to the CAC controller by specifying its QoS requirements $(D_{e,k}, R_p, R_m)$, its priority level e, as well as the number of blocks m to be transmitted. $D_{e,k}$ is the delay bound required for the requested session k with a priority of e on air interface, U_m. R_p and R_m, respectively, denote the peak and the minimum data rates requested by the session k. In other words, the MS requests that m blocks of the session k should be scheduled in a queue of priority level e, and transmitted at a minimum rate higher than R_m or a peak rate not higher than R_p before the delay bound $D_{e,k}$ is reached. The minimum data rate R_m is used as the low service rate when the MS is within a high interference region. On the other side, the peak data rate R_p is not taken for granted as the high service rate when the MS is within a low interference region. Instead, the high service rate is determined by a simple calculation performed by the service rate (SR) calculator inside the radio resource manager. The SR calculator determines the admissible peak data rate (high service rate) R'_p that can be supported by the BSS according to the delay bound $D_{e,k}$, and the minimum data rate R_m under a hypothetical interference model of the MS.

As in many examples throughout the book, the interference model is based on a two-state Markov chain with transition probabilities α and β. These parameters α and β can be collected from a user behavior profile and can be updated dynamically. After receiving the R'_p from the SR calculator, the CAC controller can optionally accept, reject or renegotiate with the MS. If the connection request is accepted, the CAC controller stores the parameters (R'_p, R_m, m, e) to a parameter database. The rate selector can then select and send the current data rate R to the CCU (channel codec unit) associated with the MS according to the interference level measured by the interference monitor.

The *DSMC scheduling architecture* can be situated in a BSS of the GPRS networks and applied to either uplink or downlink transmission. The system will adopt a certain number of priority levels. As an example, in GPRS, block transmission is classified into five scheduling priority levels, including four data priority levels and a signal priority level, which is the highest priority level [47]. The DSMC scheduling architecture follows the specification but includes a new data priority level for the retransmission blocks. The DSMC scheduling architecture [45] consists of five scheduling servers for data block transmission, one server for each priority level. Each server is responsible for scheduling data blocks transmitting through a single packet data channel (PDCH). The highest priority level, $P1$, is for the retransmission blocks and the lowest priority level, $P5$, is for the best-effort data blocks. Both $P1$ and $P5$ schedule data blocks in a first-come-first-served order. However, the scheduling servers of priority levels $P2$–$P4$ adopt the SCFQ scheduling algorithm in scheduling data blocks of QoS specific connections. In other words, multiple queues may exist in each priority of $P2$–$P4$. The queues with the same priority will be served in accordance with the SCFQ scheduling algorithm, to be explained in below.

The block requests of a particular priority can not be served until all the blocks of the higher priority have been served. The transmission of a block is nonpreemptive.

The *SCFQ scheduling algorithm* is basically a packet-based general processor sharing scheme [48, 49] without the complex virtual clock tracking mechanism. The elimination of the virtual clock tracking mechanism makes SCFQ easier to implement on a high-speed network. In SCFQ, an arrival block request is tagged with a service finish time (FT) before it is placed in a queue. The service FT tag of a block request is computed from the service time and the service starting time of the block as

$$\mathrm{FT} = \frac{\text{block length}}{\text{service rate}} + \max(\text{FT of the tail block, FT of the serving block})$$

The service starting time of the block can be the FT of the tail block of the queue if the queue is nonempty, or it is the FT of the serving block. The block requests among the heads of the queues will be picked up to be served one by one in accordance with the increasing order of FT tags, and in a round-robin fashion if more than two heading blocks have the same FT tags.

21.3.2 Calculating the high service rate

In order to determine the service rate, we need to first obtain the total delay of a block. The total delay depends on the service discipline and the input traffic. Owing to the bursty nature of multimedia traffic streams, a leaky-bucket regulator, as discussed in Chapter 8, is assumed at the network interface of the sender site to regulate the block request flow of each session.

In the DSMC scheduling scheme, a block of a connection will experience queueing delay, block transmission delay, and retransmission delay. A block at the head of a queue (henceforth referred to as a 'heading block') may experience a priority delay, which, in turn, consists of an *interpriority delay* caused by all block transmissions in the higher priority levels and an *intrapriority delay* resulted from the SCFQ scheduling.

Each preceding block of the newly arrived block will experience a priority delay and a block transmission delay when the preceding block becomes a heading block. In addition, the newly arrived block will experience an intrapriority delay when it becomes the heading block of the queue itself. Hence, the queueing delay of a block is, thus, the summation that the priority delay and the block transmission delay experienced by all preceding blocks, plus the intrapriority delay experienced by the block. In analytical modeling the following notation will be used:

- sj, session j
- L, block length (b);
- $P_{\mathrm{hi}}^{\mathrm{be}}$, block error probability when the block is transmitting within a high interference region;
- $P_{\mathrm{li}}^{\mathrm{be}}$, block error probability when the block is transmitting within a low-interference region;
- P_{hi}, stationary probability that the MS is within a high-interference region;
- P_{li}, stationary probability that the MS is within a low-interference region;

- $(r_{e,k})^{\rm hi}$, low service rate of session k of priority e;
- $(r_{e,k})^{\rm li}$, high service rate of session k of priority e;
- $(HD_{e,k})^{\rm hi}$, heading-block delay of session k of priority e within a high-interference region;
- $(HD_{e,k})^{\rm li}$, heading-block delay of session k of priority e within a low-interference region;
- $\overline{b}_{e,k}$, mean queue length of the session k of priority e;
- $\overline{w}_{e,k}$, mean queueing delay with the session k of priority e;
- dp, data priorities.

The *interpriority delay* is the time that the selected heading block waits for the blocks from higher priority levels to be served. In the following, we refer to the priority level under discussion as the priority e. Let $A_{p,s}[t_1, t_2]$ denote the amount of blocks arrived to a session with a priority p during a time interval (t_1, t_2) and $W_{p,s}[t_1, t_2]$, is the number of blocks served for a session with a priority p during a time interval (t_1, t_2). As shown in Figure 21.7, after the $(i-1)$th selected heading block has been served, the ith heading block selected by the server of the priority e may encounter an interpriority delay, which is the time period (t_1', t_2). Each session j with a priority g higher than e must be served up within the time period (t_1', t_2). Thus, the amount of traffic served during the time period (t_1', t_2), $W_{g,g}[t_1', t_2]$, is equal to the amount of arrival traffic to the session during the time period (t_1, t_2), $A_{g,j}[t_1, t_2] = A_{g,g}[t_1, t_1' + \Delta t_e]$, as in Equation (21.11). Here, t_1' is the epoch of the final time of serving the $(i-1)$th heading block selected from the priority e and Δt_e the interpriority delay encountered by each heading block selected from the priority e

$$W_{g,j}[t_1', t_2] = A_{g,j}[t_1, t_2], \forall t_2 > t_1' > t_1, t_1 \geq 0 \tag{21.11}$$

Let H represent the set of all priority levels higher than e. Summing up Equation (21.11) for all sessions of the priorities in H, we have the following inequality:

$$\sum_{g\in H}\left[\sum_{sj\in g} W_{g,j}[t_1', t_2]\right] = \sum_{g\in H}\left[\sum_{sj\in g} A_{g,j}[t_1, t_2]\right] \forall t_2 > t_1' > t_1,\ t_1 \geq 0 \tag{21.12}$$

where sj stands for session j. The traffic $A_{g,j}[t, t+\tau]$ during a time period $(t, t+\tau)$ has an upper bound since the DSMC scheduling architecture uses a leaky bucket to regulate the flow of each session [8]. Let $A^*_{g,j}(\tau)$ denote the upper bound of $A_{g,j}[t, t+\tau]$. From Chapter 8 and also Cruz [50], we have the following inequality according to the *leaky-bucket*

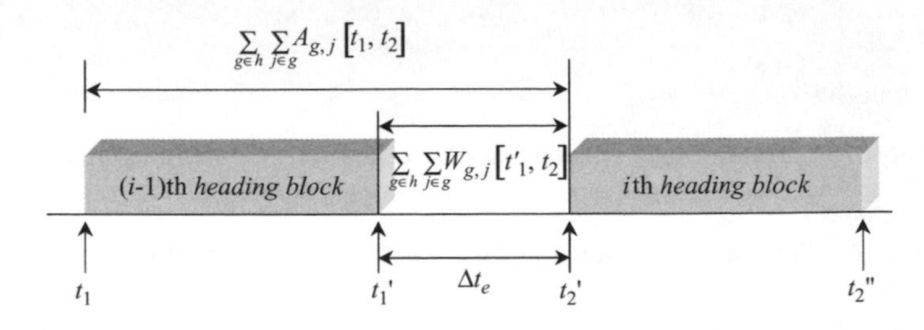

Figure 21.7 Dealy caused by higher priority session to the particular priority. (Reproduced by permission of IEEE [45].)

constrained envelope function:

$$A_{g,j}[t, t+\tau] \le A^*_{g,j}(\tau) = \sigma_{g,g} + \rho_{g,j} \cdot t \quad \forall t \ge 0, \quad \forall \tau \ge 0 \tag{21.13}$$

where $\sigma_{g,g}$ is the leaky-bucket size and $\rho_{g,j}$ is the token arrival rate for the session j of the priority g. Therefore, from the above leaky-bucket constraint, we can derive the following inequality:

$$\sum_{g\in H}\left[\sum_{sj\in g} A_{g,j}[t_1, t_2]\right] \le \sum_{g\in H}\left[\sum_{sj\in g} A^*_{g,j}(t_2 - t_1)\right] \quad \forall t_2 > t_1' > t_1,\ t_1 \ge 0 \tag{21.14}$$

Since the retransmission of data blocks has the highest priority and there is only one retransmission queue, we can denote the arrival traffic to the retransmission queue during the period (t_1, t_2) as $A_{1,1}[t_1, t_2]$. The retransmission traffic is the aggregated traffic of the retransmission of all sessions. By observing the system over a substantially long period of time, we can find that $A_{1,1}[t_1, t_2]$ also conforms to the leaky-bucket constraint, and we rewrite the above inequality Equation (21.1) as

$$A_{1,1}[t_1, t_2] + \sum_{g\in h, g\ne 1}\left[\sum_{sj\in g} A_{g,j}[t_1, t_2]\right] \le A^*_{1,1}(t_2 - t_1) + \sum_{g\in h, g\ne 1}\left[\sum_{sj\in g} A^*_{g,j}(t_2 - t_1)\right]$$
$$\forall t_2 > t_1' > t_1,\ t_1 \ge 0 \tag{21.15}$$

By using notation specified earlier we can derive the equation for $A^*_{1,1}(t_2 - t_1)$ as

$$A_{1,1}[t_1, t_2] \le A^*_{1,1}(t_2 - t_1) = L \times \sum_{s\in\{dp\}}\left\{\left\lceil \frac{\sum_{sj\in s} \sigma_{s,j} + \rho_{s,j}\cdot(t_2 - t_1)}{L}\right\rceil \times P^{\text{be}}_{\text{hi}} \times P_{\text{hi}}\right.$$
$$\left. + \left\lceil \frac{\sum_{sj\in s} \sigma_{s,j} + \rho_{s,j}\cdot(t_2 - t_1)}{L}\right\rceil \times P^{\text{be}}_{\text{li}} \times P_{\text{li}}\right\} \tag{21.15a}$$

By combining Equation (21.12) and inequality Equation (21.15), we have

$$\sum_{g\in h}\left[\sum_{sj\in g} W_{g,j}[t_1', t_2]\right] \le A^*_{1,1}(t_2 - t_1) + \sum_{g\in h, g\ne 1}\left[\sum_{sj\in g} A^*_{g,j}(t_2 - t_1)\right]$$
$$\forall t_2 > t_1' > t_1, t_1 \ge 0$$

Finally, if C_i is the link capacity of the radio band (channel) i, by applying the inequality Equation (21.13) we have

$$C_i \times \Delta t_e \le A^*_{1,1}(\Delta t_e + t_1' - t_1) + \sum_{g\in h, g\ne 1}\left[\sum_{sj\in g} \sigma_{i,j} + \rho_{i,j} \times (\Delta t_e + t_1' - t_1)\right]$$
$$\forall t_2 > t_1' > t_1, t_1 \ge 0 \tag{21.16}$$

From the above inequality, we calculate the maximum value of Δt_e.

The *intrapriority delay* is delay contributed from the SCFQ scheduling delay. It is the delay during which a heading block of a session waits for the heading blocks of some other sessions with the same priority to be served. Following the results presented by

Golestani [51], we can derive the maximum intrapriority delay of the heading block of a session k with priority e as $(|\kappa_e| - 1) \cdot (L/C_i)$, where C_i represents the capacity (b/s) of link (band) i, κ_e represents the set of backlog sessions for priority e, and $|\kappa_e|$ represents the number of backlog sessions in the priority e.

21.3.3 Heading-block delay

A heading block will encounter an intrapriority delay, an interpriority delay, and a block transmission delay of its own. The heading-block delay, denoted HD, for an MS depends on the transmission rates and can be derived as

$$(HD_{e,k})^{\mathrm{li}} = |\kappa_e| \cdot \Delta t_e + (|\kappa_e| - 1) \cdot \frac{L}{C_i} + \frac{L}{(r_{e,k})^{\mathrm{li}}} \tag{21.17}$$

$$(HD_{e,k})^{\mathrm{hi}} = |\kappa_e| \cdot \Delta t_e + (|\kappa_e| - 1) \cdot \frac{L}{C_i} + \frac{L}{(r_{e,k})^{\mathrm{hi}}} \tag{21.18}$$

where $(r_{e,k})^{\mathrm{li}}$ and $(r_{e,k})^{\mathrm{hi}}$ are the service rate (b/s) of session k of priority e, i.e. the high and low service rate, within low-IF region and high-IF region, respectively. The item on the further right-hand side in each of the above equations represents the transmission delay of the heading block of the session k with the priority e. Since Δt_e is likely to be short under the control of the leaky-bucket regulator, HD is relatively small compared with the mean duration that an MS will stay in an interference region. Therefore, it was assumed that the interference condition will not change during a heading-block delay.

The *queueing delay* of a data block is the time during which the block waits until the block is selected for transmission. Clearly, a data block newly arrived in a queue cannot be served by the corresponding SCFQ server until all the proceeding data blocks in the same queue have been served. Therefore, the queueing delay of a data block includes the time that the block waits for it to become a heading block itself, plus the interpriority and intrapriority delays that the block encounters when it becomes a heading block.

If $\overline{b}_{e,k}$ represent the mean queue length of a session k with a priority e then from Little's result, from Chapter 6, the mean queue length encountered by a newly arrived data block is equal to the block arrival rate multiplied by the mean heading-block delay time. Parameter $\overline{b}_{e,k}$ is given as

$$\overline{b}_{e,k} = \lceil \frac{\rho_{e,k}}{L} \rceil \cdot \left[(HD_{e,k})^{\mathrm{li}} \cdot P_{\mathrm{li}} + (HD_{e,k})^{\mathrm{hi}} \cdot P_{\mathrm{hi}} \right] \tag{21.19}$$

Now, the mean queueing delay $\overline{w}_{e,k}$ for the session k with a priority e can be calculated as

$$\overline{w}_{e,k} = \lceil \frac{\rho_{e,k}}{L} \rceil \cdot \left[(HD_{e,k})^{\mathrm{li}} \cdot P_{\mathrm{li}} + (HD_{e,k})^{\mathrm{hi}} \cdot F_{\mathrm{hi}} \right]^2 + |\kappa_e| \cdot \Delta t_e + (|\kappa_e| - 1) \frac{L}{C_i} \tag{21.20}$$

21.3.4 Interference model

We assume that the general interference model is an interrupted poisson process with transition probabilities of α and β. Hence, the duration that an MS is within the low- region or the high-interferece region can be represented, respectively, by an exponential distribution $1 - \mathrm{e}^{-\alpha \cdot t}$, denoted $L(t)$, or $1 - \mathrm{e}^{-\beta \cdot t}$, denoted $H(t)$.

The interference state in which a block is served is determined by the starting state and the waiting time of the block. Therefore, we use an alternating renewal process to calculate the probability of the interference state in which a block is served. In this alternating renewal process, the block waiting time can be divided into several renewal intervals. A renewal interval consists of two exponential distributions, $L(t)$ and $H(t)$. Let $F(t)$ be the convolution sum of $L(t)$ and $H(t)$. We use the notation $P_{\mathrm{ls}}^{\mathrm{hi}}(W)$ to represent the probability that a block with a waiting time W arrives when an MS is within a low-interference (IF) region and is served when the MS is within a high-IF region. Following the same convention, the notations of probabilities $P_{\mathrm{ls}}^{\mathrm{li}}(W)$, $F_{\mathrm{hs}}^{\mathrm{hi}}(W)$, and $F_{\mathrm{hs}}^{\mathrm{li}}(W)$ should be self-explanatory. These probabilities can be derived as [52]:

$$
\begin{aligned}
P_{\mathrm{ls}}^{\mathrm{li}}(W) &= [1 - L(W)] + \int_0^w [1 - L(W - y)]\mathrm{d}\left[\sum_{n=1}^{\infty} F_n(y)\right] \\
P_{\mathrm{ls}}^{\mathrm{hi}}(W) &= 1 - P_{\mathrm{ls}}^{\mathrm{li}}(W) P_{\mathrm{hs}}^{\mathrm{hi}}(W) = [1 - H(W)] + \int_0^w [1 - H(W - y)]\mathrm{d}\left[\sum_{n=1}^{\infty} F_n(y)\right] \\
P_{\mathrm{hs}}^{\mathrm{li}}(W) &= 1 - P_{\mathrm{hs}}^{\mathrm{hi}}(W)
\end{aligned} \tag{21.21}
$$

21.3.5 Normal delay of a newly arrived block

Both the MS-terminated downlink data block and the MS-originated uplink block transmission requests may arrive at an SCFQ queue when the MS is within either a low-IF region with a probability P_{li} or a high-IF region with a probability P_{hi}. A block may be served when the MS is within a high-IF region or a low-IF region. So, the normal delay (without retransmissions) of a newly arrived data block, is equal to $(\overline{b}_{e,k} + 1) \cdot HD_{e,k}$, where $HD_{e,k}$ can be $(HD_{e,k})^{\mathrm{li}}$ or $(HD_{e,k})^{\mathrm{hi}}$. By using the above Equation (21.21), we can obtain the normal delay ($ND_{e,k}$) of a block with a mean waiting time of $\bar{w}_{e,k}$ as

$$
\begin{aligned}
ND_{e,k} = {} & P_{\mathrm{li}} \cdot [(\overline{b}_{e,k} + 1) \cdot (HD_{e,k})^{\mathrm{li}} \cdot P_{\mathrm{ls}}^{\mathrm{li}}(\bar{w}_{e,k}) \\
& + (\overline{b}_{e,k} + 1) \cdot (HD_{e,k})^{\mathrm{hi}} \cdot F_{\mathrm{ls}}^{\mathrm{hi}}(\bar{w}_{e,k})] + P_{\mathrm{hi}} \cdot [(\overline{b}_{e,k} + 1) \cdot (HD_{e,k})^{\mathrm{hi}} \cdot P_{\mathrm{hs}}^{\mathrm{hi}}(\bar{w}_{e,k}) \\
& + (\overline{b}_{e,k} + 1) \cdot (HD_{e,k})^{\mathrm{li}} \cdot P_{\mathrm{hs}}^{\mathrm{li}}(\bar{w}_{e,k})]
\end{aligned} \tag{21.22}
$$

The *retransmission delay of* each retransmission consists of two delays, the waiting time T_{a} of a selective ARQ request, and the retransmission time T_{r}. The transmission of a data block and the retransmission of this block may occur in either interference condition. In other words, a retransmission cycle may start and end in either interference region. Let A denote the mean retransmission delay starting from a low-IF region, whereas B denotes the mean retransmission delay starting from a high-IF region. Hence, we can describe the mean retransmission delays A and B by two cross recursive equations, as shown below. As a consequence, the retransmission delay ($RD_{e,k}$) of a block can be obtained as

$$
\begin{aligned}
A &= P_{\mathrm{li}}^{\mathrm{be}}\{(T_{\mathrm{a}} + T_{\mathrm{r}}) + A \cdot P_{\mathrm{ls}}^{\mathrm{li}}(T_{\mathrm{a}} + T_{\mathrm{r}}) + B \cdot P_{\mathrm{ls}}^{\mathrm{hi}}(T_{\mathrm{a}} + T_{\mathrm{r}})\} \\
B &= P_{\mathrm{hi}}^{\mathrm{be}}\{(T_{\mathrm{a}} + T_{\mathrm{r}}) + B \cdot P_{\mathrm{hs}}^{\mathrm{hi}}(T_{\mathrm{a}} + T_{\mathrm{r}}) + A \cdot P_{\mathrm{HS}}^{\text{low-if}}(T_{\mathrm{a}} + T_{\mathrm{r}})\} \\
RD_{e,k} &= P_{\mathrm{li}} \cdot \left\{A \cdot P_{\mathrm{ls}}^{\mathrm{li}}\left(\bar{w}_{e,k} + \frac{L}{(r_{e,k})^{\mathrm{li}}}\right) + B \cdot P_{\mathrm{ls}}^{\mathrm{hi}}\left(\bar{w}_{e,k} + \frac{L}{(r_{e,k})^{\mathrm{hi}}}\right)\right\} \\
&\quad + P_{\mathrm{hi}} \cdot \left\{B \cdot P_{\mathrm{hs}}^{\mathrm{hi}}\left(\bar{w}_{e,k} + \frac{L}{(r_{e,k})^{\mathrm{hi}}}\right) + A \cdot P_{\mathrm{hs}}^{\mathrm{li}}\left(\bar{w}_{e,k} + \frac{L}{(r_{e,k})^{\mathrm{li}}}\right)\right\}
\end{aligned} \tag{21.23}
$$

21.3.6 High service rate of a session

By summing up the normal delay, Equation (21.22), and the retransmission delay, Equation (21.23), of a newly arrived block, the SR calculator can calculate the total delay ($TD_{e,k}$) for a newly arrived block in a session k with a priority e. For simplicity it is assumed that the low service rate of a session when the MS is within a high-IF region is a predefined value. The value of this low service rate can be chosen by a user application in accordance with the required characteristics of the media stream used in the application. For example, the low service rate can be assigned as the minimum tolerable decoding rate of a Motion Pictures Expert Group (MPEG) video stream.

If we assume that a session k with a priority e has a QoS specification for m block transmissions with air-interface delay bound $D_{e,k}$, then the SR calculator can calculate the high service rate of the session k with a priority e under the constraint of the inequality $TD_{e,k} \leq D_{e,k}/m$.

21.3.6.1 Performance example

The three algorithms are compared by simulation with the same parameters as in Yang *et al.* [45]. The results are shown in Figure 21.8.

21.4 QoS IN OFDMA-BASED BROADBAND WIRELESS ACCESS SYSTEMS

Vector orthogonal frequency division multiplexing (VOFDM) is considered as a base for BWA systems by the Broadband Wireless Internet Forum (BWIF) [1]. The BWA system is used with existing wireless LAN technologies such as IEEE802.11 (a, b) and IEEE 802.16 Group aims to unify the BWA solutions [56]. 802.16 Group issued standards in the 10–66 GHz bands and IEEE802.16a Group was formed to develop standards to operate in the 2–11 GHz bands in which channel impairments, multipath fading and path loss become more significant with the increase in the number of subscribers.

System performance at high transmission rates depends on the ability of BWA system to provide efficient and flexible resource allocation. Recent studies [55,57] on resource allocation demonstrate that significant performance gains can be obtained if frequency hopping and adaptive modulation are used in subcarrier allocation, assuming knowledge of the channel gain in the transmitter.

The resource allocation problem has been considered in many studies. Almost all of them define the problem as a real-time resource allocation problem in which QoS requirements are fixed by the application. QoS requirement is defined as achieving a specified data transmission rate and BER of each user in each transmission. In this regard, the problem differs from the water-feeling schemes wherein the aim is to achieve Shannon capacity under the power constraint [57].

Therefore, in this section we consider the problem where K users are involved in the OFDMA system to share N subcarriers. Each user allocates nonoverlapping set of subcarriers S_k where the number of subcarriers per user is $J(k)$. In the following, $X_k(l)$ represents the lth subcarrier of the FFT block belonging to the kth user. $X_k(l)$ is obtained by coding the assigned c bits with the corresponding modulation scheme. In the downlink the $X_k(l)$ are multiplexed to form the OFDM symbol of length $(N + L)$ with the appended guard

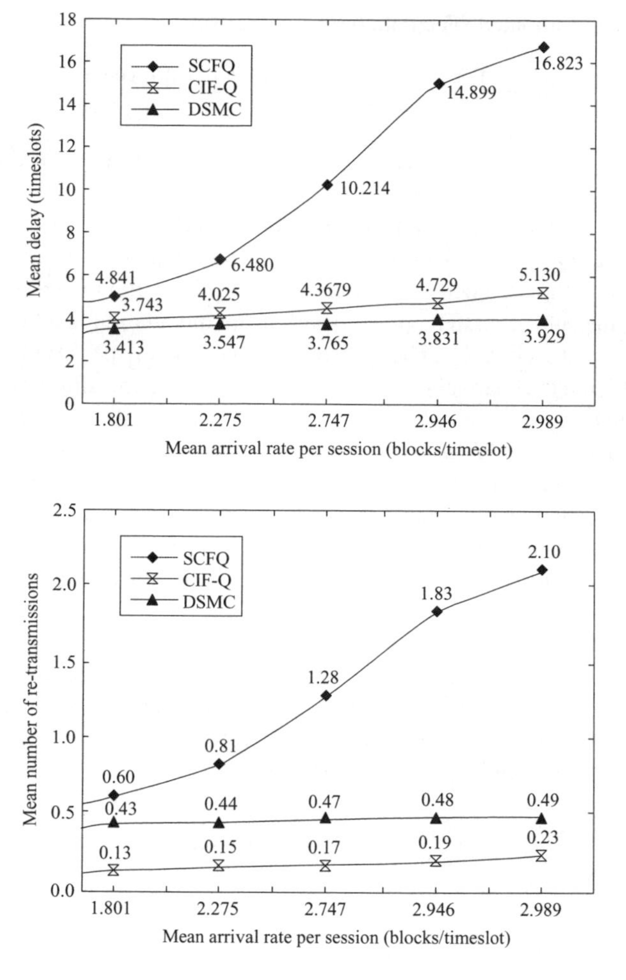

Figure 21.8 System performance. (Reproduced by permission of IEEE [45].)

prefix L in order to eliminate ISI. At the uplink, the equivalent overall OFDM symbol (with a synchronization error) has the form

$$x(l) = \sum_{k=0}^{K-1} \sum_{n=0}^{j(k)-1} X_k(n) e^{j} (2\pi/N)[I_k(n)]l \tag{21.24}$$

where $n = -L, \ldots, N-1$, and $I_k(n)$ denotes the subcarrier assigned to the kth user. Resource allocation mechanism associates the set of subcarriers to the users with different bits loaded into them. The received signal from the jth user can be represented as

$$y_j(l) = x(l) \otimes h_j(l) + w(l) \tag{21.25}$$

where $h_j(t)$ is the baseband impulse response of the channel between BS and the jth user. $y_j(l)$ is the received signal $y_j(t)$ sampled at rate $1/T$. The first L samples are discarded and the N-point FFT is computed. The data of the jth user is

$$Y_j(n) = \begin{cases} X_j(n)H_j(i_j(n)) + W(n), & \text{if } i_j(n) \in S_j \\ 0, & \text{otherwise} \end{cases} \tag{21.26}$$

where $H_j(n) := \sum_i h_j(i) \exp[j2(\pi/N)ni]$ is the frequency response of the channel of the kth user.

The allocation module of the transmitter assigns subcarriers to each user according to some QoS criteria. QoS metrics in the system are rate and BER. Each user's bit stream is transmitted using the assigned subcarriers and adaptively modulated for the number of bits assigned to the subcarrier. The power level of the modulation is adjusted to meet QoS for given fading of the channel. The transmission power for the AWGN channel can be predicted. In addition the channel gain of subcarrier n to the corresponding user k should be known. The channel gain of the subcarrier is defined as $\alpha_{k,n} = H_k(n) * PL_k$, where PL is the path loss, defined by $PL_k = PL(d_o) + 10\alpha \log_{10}(d_k/d_o) + X_\sigma$. Parameter d_o is the reference distance, d_k is the distance between transmitter and receiver, α is the path loss component and X_σ is a Gaussian random variable for shadowing with a standard deviation σ [58, 59].

With the known channel information, the objective of resource allocation problem can be defined as maximizing the throughput subject to a given total power constraint regarding the user's QoS requirements.

If $\gamma_{k,n}$ is the indicator of allocating the nth subcarrier to the kth user, the transmission power allocated to the nth subcarrier of kth user can be expressed as $P_{k,n} = f_k(c_{k,n}, BER_k)/\alpha_{k,n}^2$, where $f_k(c_k, n)$ is the required received power with unity channel gain for reliable reception of $c_{k,n}$ bits per symbol. Therefore, the resource allocation problem with an imposed power constraint can be formulated as

$$\begin{aligned} &\max_{C_{k,n},\gamma_{k,n}} R_k = \sum_{n=1}^{N} c_{k,n}\gamma_{k,n} \quad \text{for all } k \\ &\text{subject to } P_T = \sum_{k=1}^{K}\sum_{n=1}^{N} \frac{f_k(c_{k,n}, BER_k)}{\alpha_{k,n}^2}\gamma_{k,n} \le P_{max} \end{aligned} \tag{21.27}$$

The limit on the total transmission power is expressed as $P_{\max}$ for all $n \in \{1, \ldots, N\}$, $k \in \{1, \ldots, K\}$ and $c_{k,n} \in \{1, \ldots, M\}$. If there is no power constraint, Equation (21.27) is changed in order to minimize P_T subject to allocating R_k bits for all k. In other words problem is to find the values of the $\gamma_{k,n}$ and the corresponding $c_{k,n}$ while minimizing P_T.

The *optimal solution* in a multiuser environment with multiple modulation techniques is complicated since it needs to pick the subcarriers in balance. The problem can be classified according to each set of bits assigned to a subcarrier. For a user k, $f_k(c_{k,n}) \in \{f_k(1, BER_k), \ldots, f_k(M, BER_k)\}$and M times $[K * N]$ power matrices $\{P^c\}$ can be constructed for each c. For a constant c, $\{f(c)\}$ can be computed and the transmission power requirement can be found. The dimension of the indicator function is incremented and represented now by $\gamma_{k,n,c}$ and defined as

$$\gamma_{k,n,c} = \begin{cases} 1, & \textit{if } c_{k,n} = c \\ 0, & \text{otherwise} \end{cases} \tag{21.28}$$

The above problem can be solved with integer programming (IP). We refer to the IP approach as the optimal solution to the resource allocation problem. There are $K \cdot N \cdot M$ indicator variables and M power matrices where the entries of each matrix for a given c can be found from

$$P_{k,n}^{c} = \frac{f_k(c, BER_k)}{\alpha_{k,n}^{2}} \tag{21.29}$$

Using Equation (21.29) as an input, the cost function now can be written as

$$P_T = \sum_{k=1}^{K} \sum_{n=1}^{N} \sum_{c=1}^{M} P_{k,n}^{c} \gamma_{k,n,c} \tag{21.30}$$

and the description of the IP problem is

$$\min_{\gamma_{k,n,c}} P_T, \text{ for } \gamma_{k,n,c} \in \{0, 1\} \tag{21.31}$$

subject to

$$R_k = \sum_{n=1}^{N} \sum_{c=1}^{M} c \cdot \gamma_{k,n,c}, \text{ for all } k, \text{ and } 0 \le \sum_{k=1}^{K} \sum_{c=1}^{M} \gamma_{k,n,c}, \le 1, \text{ for all } n$$

Although the optimal solution gives the exact results, from an implementation point of view, it is too complex. This leads to searching suboptimal solutions that are fast and close to the optimal solution. *Suboptimal solutions*, in most attempts to simplify the resource allocation, decompose the problem into two procedures, a subcarrier allocation with fixed modulation, and bit loading. Subcarrier allocation with fixed modulation deals with one P^c matrix with fixed c and then, by using bit loading scheme, the number of bits is incremented.

Subcarrier allocation is based on the fact that $f_k(x, y)$ is a convex function [57]. We can start with $P_{k,n}^{1}$ and we can define new $\bar{R}_k$ with $\sum_{k=1}^{K} \bar{R}_k \le N$, which can be obtained by decrementing R_k properly. Then the solution to this problem can be solved with Linear Programming or Hungarian problem.

Linear programming

For simplicity, we briefly restate the problem description,

$$P_T = \min \sum_{k=1}^{K} \sum_{n=1}^{N} P_{k,n}^{1} \rho_{k,n} \quad \rho_{k,n} \in [0, 1] \tag{21.32}$$

subject to

$$\sum_{n=1}^{N} \rho_{k,n} = \overline{R}_k \forall k \in \{1, \ldots, K\} \quad \text{and} \quad \sum_{k=1}^{K} \rho_{k,n} = 1 \quad \forall n \in \{1, \ldots, N\}$$

After linear programming, the $[K \cdot N]$ allocation matrix has entries ranging between 0 and 1. The entries are converted to integers by selecting the highest $\bar{R}_k$ nonzero values from N columns for each k and assigning them to the kth user.

Hungarian algorithm

The problem described above can also be solved by an assignment method such as the Hungarian algorithm [58]. The Hungarian algorithm works with square matrices. Entries of the square matrix can be formed by adding $\bar{R}_k$ times the row of each k. The problem formulation is as

$$P_T = \min \sum_{k=1}^{N} \sum_{n=1}^{N} P_{k,n}^1 \rho_{k,n} \quad \rho_{k,n} \in \{0, 1\} \tag{21.33}$$

and the constraints become

$$\sum_{n=1}^{N} \rho_{k,n} = 1 \quad \forall n \in \{1, \ldots, N\} \quad \text{and} \quad \sum_{k=1}^{N} \rho_{k,n} = 1 \quad \forall k \in \{1, \ldots, N\}$$

Although the Hungarian method has computation complexity $O(n^4)$ in the allocation problem with fixed modulation, it may serve as a base for adaptive modulation.

The *Bit loading algorithm* (BLA) is used after the subcarriers are assigned to users that have at least $\bar{R}_k$ bits assigned. The bit loading procedure is as simple as incrementing bits of the assigned subcarriers of the users until $P_T \leq P_{\max}$. If$\Delta P_{k,n}(c)$ is the additional power needed to increment one bit of the nth subcarrier of kth user $\Delta P_{k,n}(c_{k,n}) = [f(c_{k,n} + 1) - f(c_{k,n})]/\alpha_{k,n}^2$, then the bit loading algorithm assigns one bit at a time with a greedy approach to the subcarrier as $\{\arg \min_{k,n} \Delta P_{k,n}(c_{k,n})\}$.

BL algorithm

(1) *For all n, Set* $c_{k,n} = 0$, $\Delta P_{k,n}(c_{k,n})$, *and* $P_T = 0$;

(2) *Select* $\bar{n} = arg\ min_n \Delta P_{k,n}(0)$;

(3) *Set* $c_{k,\bar{n}} = c_{k,\bar{n}} + 1$ *and* $P_T = P_T + \Delta P_{k,n}(c_{k,n})$;

(4) *Set* $\Delta P_{k,n}(c_{k,\bar{n}})$;

(5) *Check* $P_T \leq P_{max}$ *and* R_k *for* $\forall k$, *if not satisfied GOTO STEP 2.*

(6) *Finish.*

The Hungarian approach and LP approach with bit loading appear as two different suboptimal solutions to the resource allocation with adaptive modulation. In the sequel they will be referred to as GreedyHungarian (GH) and GreedyLP (GLP).

21.4.1 Iterative solution

The GreedyLP and GreedyHungarian methods both first determine the subcarriers and then increment the number of bits on them according to the rate requirements of users. This may not be a good schedule in some cases, like a user with only one good subcarrier and low rate requirement. The best solution for that user is allocating its good carrier with high number of bits. However, if GreedyLP or GreedyHungarian is used, the user may have allocated more than one subcarrier with lower number of bits and, in some cases, its good subcarrier is never selected. Consider another scenario where a user does not have any good subcarrier (i.e. it may have a bad channel or be at the edge of the cell). In this case, rather than pushing more bits and allocating fewer subcarriers, as in GreedyLP and GreedyHungarian, the opposite strategy is preferred since fewer bits in higher number of subcarriers give a better result. Another difficulty arises in providing fairness. Since GreedyLP and GreedyHungarian are based on a greedy approach, the user in the worst condition usually suffers. In any event, these are complex schemes and simpler schemes are needed to finish the allocation within the coherence time. To cope with these challenges, in the following a simple, efficient and fair subcarrier allocation scheme is introduced with iterative improvement [53].

The scheme is composed of two modules, referred to as *scheduling* and *improvement modules*. In the scheduling section, bits and subcarriers are distributed to the users and passed to the improvement module where the allocation is improved iteratively by bit swapping and subcarrier swapping algorithms.

The *fair scheduling algorithm* starts the allocation procedure with the highest level of modulation scheme. In this way, it tries to find the best subcarrier of a user to allocate the highest number of bits. In Koutsopoulos and Tassiulas [55] the strategy is described by an analogy: 'The best strategy to fill a case with stone, pebble and sand is as follows. First filling the case with the stones and then filling the gap left from the stones with pebbles and in the same way, filling the gap left from pebbles with sand. Since filling in opposite direction may leave the stones or pebbles outside". With this strategy more bits can be allocated and the scheme becomes immune to uneven QoS requirements. The fair scheduling algorithm (FSA) runs greedy release algorithm (GRA) if there are nonallocated subcarriers after the lowest modulation turn and the rate requirement is not satisfied. GRA decrements one bit of a subcarrier to gain power reduction, which is used to assign higher number of bits to the users on the whole. FSA is described as follows.

FS algorithm

(1) *Set* $c = M$, *Select a* k, *and* $P_T = 0$;

(2) *Find* $\bar{n} = \arg\min_n P_{k,n}^c$;

(3) *Set* $R_k = R_k - c$ *and* $\rho_{k,\bar{n}} = 1$, *update* P_T, *shift to the next* k;

(4) *If* $P_T > P_{\max}$, *step out and set* $c = c - 1$, *GOTO STEP 2.*

(5) *If* $\forall k$, $R_k < c$, *set* $c = c - 1$, *GOTO STEP 2.*

(6) *If* $\{c == 1\}$, $\sum_{k=1}^{K} \sum_{n=1}^{N} \rho_{k,n} < N$, $P_T > P_{\max}$, *run "greedy release" and GOTO STEP 2.*

(7) *Finish.*

The *greedy releasing algorithm* tends to fill the unallocated subcarriers. It releases one of the bits of the most expensive subcarrier to gain power reduction in order to drive the process. GRA works in the opposite direction to BLA. GRA is described as follows.

GR algorithm

(1) *Find* $\{\bar{k}, \bar{n}, \bar{c}_{\bar{k},\bar{n}}\} = \arg\max_{k,n,c} P_{k,n}^c \rho_{k,n} \forall c$;

(2) *Set* $\bar{c}_{k,n} = \bar{c}_{k,n} - 1$, $P_T = P_T - \Delta P_{\bar{k},\bar{n}}(c_{\bar{k},\bar{n}})$;

(3) *Set* $c = c_{\bar{k},\bar{n}} - 1$;

(4) *Finish.*

The *horizontal swapping algorithm* (HSA) aims to smooth the bit distribution of a user. When the subcarriers are distributed, the bit weight per subcarrier can be adjusted to reduce power. One bit of a subcarrier may be shifted to the other subcarrier of the same user if there is a power reduction gain. Therefore, variation of the power allocation per subcarrier is reduced and a smoother transmission is performed. HSA is described as follows.

HS algorithm

(1) *Set* $P_C = \infty$;

  (1a) *Find* $\{\bar{k}, \bar{n}, \bar{c}_{\bar{k},\bar{n}}\} = arg\ max_{k,n,c}(P^c_{k,n}\rho_{k,n}) < P_C \forall c$;

(2) *Define* $n \in S_k$, *where* $\{\rho_{k,n} == 1\}$ *for* $\forall n_i$;

(3) *Set* $\Delta_{\dot{n}} = max_n \Delta P_{\bar{k},\bar{n}}(c_{\bar{k},\bar{n}} - 1) - \Delta P_{\bar{k},\dot{n}}(c_{\bar{k},\dot{n}}), \dot{n} \in S_k$;

(4) *Set* $P_C = P^{\bar{c}}_{\bar{k},\bar{n}}$;

  (4a) *If* $\Delta_{\dot{n}} > 0$, *set* $P_T = P_T - \Delta_{\dot{n}}$

  (4b) *Set* $c_{\bar{k},\bar{n}} = c_{\bar{k},\bar{n}} - 1, c_{\bar{k},\dot{n}} = c_{\bar{k},\dot{n}} + 1$ *GOTO Step (1a)*

(5) *If* $\{P_C = min_{k,n,c}(P^c_{k,n}\rho_{k,n})\}$, *finish.*

The *vertical swapping algorithm* (VSA) does vertical swapping for every pair of users. In each iteration, users try to swap their subcarriers such that the power allocation is reduced. There are different types of vertical swapping. For instance, in triple swapping, user i gives its subcarrier to user j and in the same way user j to user k and user k to user i. In Koutsopoulos ad Tassiulas [55], pairwise swapping is modified to cope with the adaptive modulation case. In this case, there is more than one class where each class is defined with its modulation (i.e number of bits loaded to a subcarrier) and swapping is only within the class. Each pair of user swap their subcarriers that belong to the same class if there is a power reduction. In this way, adjustment of subcarrier is done across users, to try to approximate the optimal solution. VSA is described as follows.

VS algorithm

(1) $\forall$ *pair of user* $\{i, j\}$;

  (1a) *Find* $\partial P_{i,j}(n) = P^{\dot{c}}_{i,n} - P^{\dot{c}}_{j,n}$ *and* $\Delta^{\hat{n}} P_{i,j} = max \partial P_{i,j}(n), \forall n \in S_i$;

  (1b) *Find* $\partial Pj, i(n) = P^{\dot{c}}_{j,n} - P^{\dot{c}}_{i,n} \Delta^{\breve{n}} P_{j,i} = max\ \partial P_{j,i}(n), \forall n \in S_j$;

  (1c) *Set* $\Omega^{\hat{n},\breve{n}} P_{i,j} = \Delta^{\hat{n}} P_{i,j} + \Delta^{\hat{n}} P_{j,i}$;

  (1d) *Add* $\Omega^{\hat{n},\breve{n}} P_{i,j}$ *to the* $\{\Lambda\}$ *list*;

(2) *Select* $\Omega = max_{(i,\hat{n}),(j,\breve{n})} \Lambda$;

(3) *If* $\Omega > 0$, *Switch subcarriers and* $P_T = P_T - \Omega$ *GOTO STEP (1a)*;

(4) *If* $\Omega \leq 0$, *finish.*

21.4.2 Resource allocation to maximize capacity

Suppose there is no fixed requirements per symbol and the aim is to maximize capacity. It has been shown in Viswanath *et al.* [60] that, for point-to-point links, a fair allocation strategy maximizes total capacity and the throughput of each user in the long run, when the user's channel statistics are the same. This idea underlying the proposed fair scheduling algorithm is exploiting the multiuser diversity gain.

With a slight modification, the fair scheduling algorithm for point-to-point communication was extended to an algorithm for point-to-multipoint communication [53]. Suppose the user time varying data rate requirement $R_k(t)$ is sent by the user to the base station as feedback of the channel condition. We treat symbol time as the time slot, so t is discrete, representing the number of symbols. We keep track of average throughput $t_{k,n}$ of each user for a subcarrier in a past window of length t_c. The scheduling algorithm will schedule a subcarrier $\bar{n}$ to a user $\bar{k}$ according to the criterion

$$\{\bar{k}, \bar{n}\} = \arg\max_{k,n}(r_{k,n}/t_{k,n})$$

where $t_{k,n}$ can be updated using an exponentially weighted low-pass filter described in Viswanath *et al.* [60]. Here, we are confronted with determining the $r_{k,n}$ values. We can set $r_{k,n}$ to R_k/N, where N is the number of carriers. With this setting, the peaks of the channel for a given subcarrier can be tracked. The algorithm schedules a user to a subcarrier when the channel quality in that subcarrier is high relative to its average condition in that subcarrier over the time scale t_c. When we consider all subcarriers the fairness criterion matches with the point-to-point case as

$$\bar{k} = \max_k R_k/T_k, \text{ where } T_k = \sum\nolimits_{n=1}^{N} t_{k,n}$$

The theoretical analysis of fairness property of the above relation for point-to-point communication is derived in Viswanath *et al.* [60]. Those derivations can be apply for point-to-multipoint communication.

21.4.2.1 Performance example

The required transmission power for c bits/subcarrier at a given BER with unity channel gain is [57]:

$$f(c, BER) = \frac{N_0}{3}\left[Q^{-1}\left(\frac{BER}{4}\right)\right]^2 (2^c - 1)$$

where $Q^{-1}(x)$ is the inverse function of$Q(x) = \frac{1}{\sqrt{2\pi}} \int\limits_x^{\infty} e{-}t^2/2\mathrm{d}t$

Figure 21.9 shows the average data rates per subcarrier vs total power constraint when there are four users. Each user has a rate requirement of 192 b/symbol (maximum rate) and BER requirement of 10^{-4}. The performance of the iterative approach is close to that of the optimal and difference between suboptimal and iterative approaches decreases as the total transmit power increases.

21.5 PREDICTIVE FLOW CONTROL AND QoS

Even if the dimensioning of network resources has been done correctly and the admission control mechanism is good, the network may go into periods of congestion due to the transient oscillations in the network traffic. For this reason it is necessary to develop a mechanism to quickly reduce the congestion or pre-empt it, so as to cause the least possible

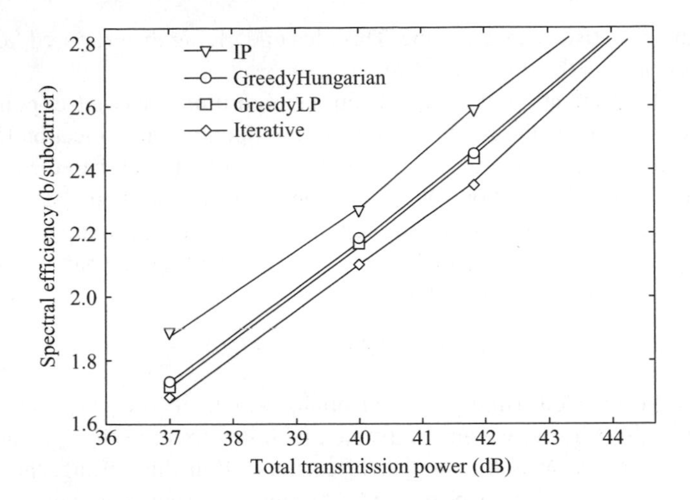

Figure 21.9 Spectral efficiency vs total transmission power.

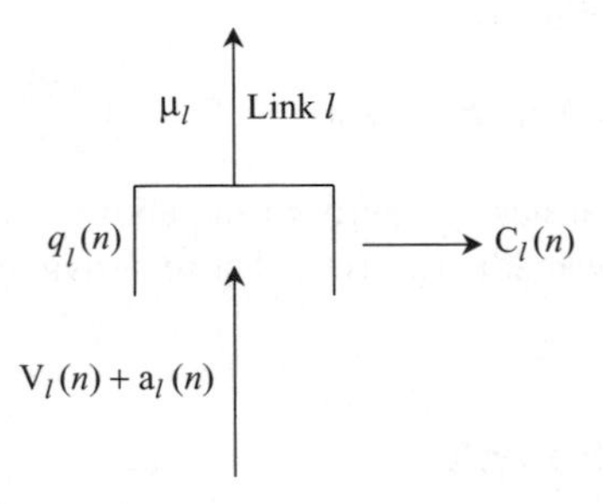

Figure 21.10 Individual link-level model.

degradation of QoS to the underlying applications [61–81]. 4G networks will carry a mixture of *real-time* (RT) traffic, like video or voice traffic, and *nonreal-time* (NRT) traffic, like data. One approach to controlling the NRT traffic is to be able to predict the RT traffic (at the link of interest) at some time in the future, then, based on this prediction, control the NRT traffic.

In Figure 21.10, $V_l(n)$ and $\alpha_l(n)$ correspond to the aggregate RT traffic and NRT traffic, respectively, arriving at a link of interest (link l having capacity μ_l), at time n. One can then estimate $C_l(n)$, the available link capacity for NRT traffic, at some time in the future. This information would then be used at the *network-level* to distribute the available link capacities to the NRT flows.

On the network level the available link capacities for the NRT flows is then distributed to maximize throughput (or more generally some utility function), based on appropriate fairness requirements. An example network is shown in Figure 21.11, where flows traverse links with available capacities for the NRT flows calculated at the individual link level. In Chapter 7, the network-level problem has been investigated in the case when the available link capacity for NRT flows at each node is a constant. The problem remains open in the case when the available capacity is time-varying.

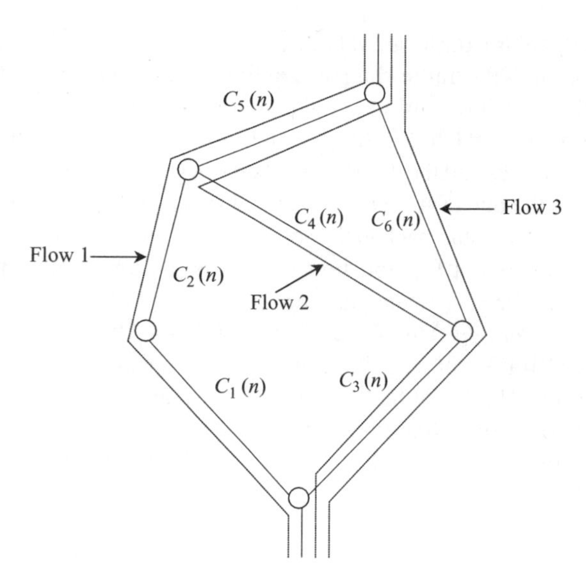

Figure 21.11 Network-level model.

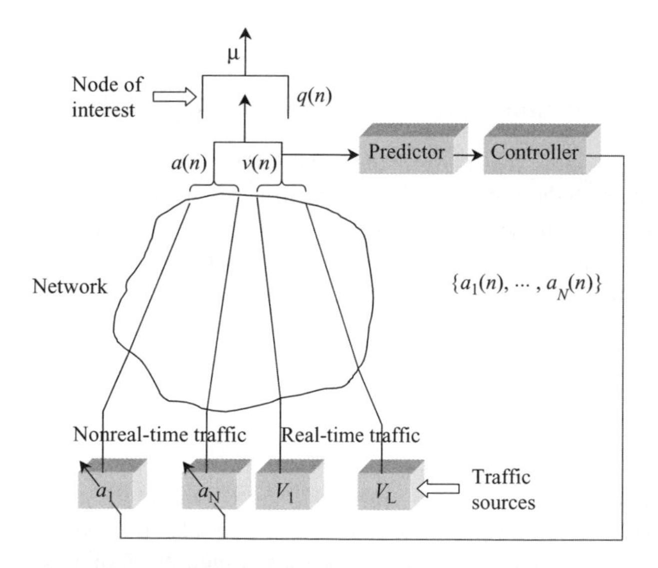

Figure 21.12 System diagram of predictive flow control.

21.5.1 Predictive flow control model

In this section we focus on the individual link-level problem, a single multiplexing point in the network which consists of a link and an associated buffer that serve both RT and NRT traffic. The multiplexing point in the network could be an output port of a router/switch or a multiplexer. The system diagram is shown in Figure 21.12.

In the following $V(n)$ will represent the aggregate amount of RT traffic that arrives at the queue of interest at time n. $V_{\max} := \sup_{n>0}\{V(n)\}$ will be assumed finite and $V(n)$ stationary in the mean, i.e. $\bar{V} := \mathrm{E}\{V(n)\}$. The goal is to control the NRT traffic based on predicting the aggregate RT traffic arrival rate at the queue. $a_i(n)$ will refer to the available link capacity for the ith NRT traffic computed at time n based on the predicted value of the RT traffic rate. This explicit rate information is sent back to the ith NRT traffic source.

If N is the number of NRT traffic sources and $n_i, \imath = 1, \ldots, N$ is the round-trip delay between the ith NRT source and the destination, then $a(n) = \Sigma_{i=1}^{N} a_i(n - n_i)$ is the aggregate NRT traffic arrival to the queue at time n. A control message is propagated from the queue of interest to the destination and back to the source. $\hat{V}_i(n)$ will represent the predicted value of $V(n)$ based on the history of V before time $n - n_i$. We assume that the predictor is linear. A simple example can be represented as $\hat{V}_i(n + n_i) = \Sigma_k h_k V_i(n - k)$. If $V(z)$ is the Z-transform of $V(n) - \bar{V}$ and $\hat{V}_i(z)$ the Z-transform of $\hat{V}_i(n) - \bar{V}$ then $\hat{V}_i(z) = z^{-n_i} H_i(z) V(z)$ where $H_i(z)$, for NRT traffic source i, is a causal, stable, linear, time-invariant system [74]. It should be noted that $H_i(z)$ will be the same for all sources with the same RTT n_i. For example, if there is only one NRT flow with round-trip delay 5, a possible predictor (in time domain) could be $\hat{V}_1(n + 5) = (1/2)V(n) + (1/2)V(n - 1)$. In this case, we will have $H_1(z) = (1/2)(1 + z^{-1})$.

The queue-length, $q(n)$, at time n at the queue of interest will be determined by $a(n)$, $V(n)$, and the service rate (link rate) of the queue μ. We assume that the queue process begins at time $n = 0$ and $q(0) = 0$. For stability, we also require that $\bar{V} < \mu$. The feedback control scheme is as follows. We predict the aggregate RT traffic rate and use the predicted value to compute $a_i(n)$, $1 \leq n \leq N$.

21.5.1.1 Predictive flow control algorithm for single NRT traffic model

To start with, we assume that there is only one NRT traffic source $a_1(n)$ (or a group of NRT traffic loops with the same round-trip delay n_1). So, by definition, $a(n) = a_1(n - n_1)$. Ideally, what we want to achieve is $a(n) + V(n) = \mu$ at all time n. However, there are two difficulties in achieving this. First, since we do not know $V(n)$ in advance, we need to estimate its value through prediction, resulting in a certain possibility of error. Second, $V(n)$ could be greater than μ but since $a(n)$ cannot be negative, the sum $a(n) + V(n)$ cannot be made equal to μ. Having in mind the possibility of prediction error and the possibility that $V(n) > \mu$, the NRT traffic $a_1(n)$ will be controlled as

$$a_1(n) = [p\mu - \hat{V}_1(n + n_1)]^+ \tag{21.34}$$

where p is the percentage of output link capacity that we would like to utilize $[p > (\bar{V}/\mu)]$ and $[x]^+ = x$, if $x \geq 0$, and $[x]^+ = 0$, otherwise. Consider now the situation of perfect prediction, and let $V(n)$ have exceeded μ for some time. During the period that $V(\cdot)$ has exceeded μ, the above equation correctly sets $a_1(\cdot)$ to zero, but even after $V(\cdot)$ is no longer larger than μ, there could still be a substantial backlog in the queue, during which time $a_1(\cdot)$ should be set to zero. However, according to Equation (21.34), the moment $V(n)$ is less than μ, the NRT source is allowed to transmit, thus potentially causing unnecessary congestion at the queue. So, what we will attempt to do is to keep the queue length at the node of interest small, while maintaining a certain level of throughput given by

$$\lim_{n\to\infty} \frac{\sum_{j=1}^{n}[a(j) + V(j)]}{n} = p\mu \tag{21.35}$$

For this reason we define the control algorithm [81] as follows.

Control algorithm ($N = 1$ case)

(1) *Define a virtual queueing process $q_1(n)$ and set $q_1(0) = 0$.*

(2) $q_1(n) = [q_1(n-1) + \hat{V}_1(n) - p\mu]^+$. *For $n \leq 0$, we let $V(n) = 0$.*

(3) $a_1(n) = [p\mu - \hat{V}_1(n + n_1) - q_1(n + n_1 - 1)]^+$. *For $n \leq 0$, we let $a_1(n) = 0$.*

For N NRT traffic sources the algorithm is modified as follows.

Control algorithm ($N > 1$ case)

(1) *Set $q_i(0) = 0,\ 1 \leq i \leq N$.*

(2) $q_i(n) = [q_i(n-1) + \hat{V}_i(n) - p\mu]^+$. *For $n \leq 0$, we let $V(n) = 0$.*

(3) $a_i(n) = [p\mu - \hat{V}_i(n + n_i) - q_i(n + n_i - 1)]^+/N$.

(4) *For $n \leq 0$, we let $a_i(n) = 0$.*

21.5.1.2 Performance example

Three different alternatives for predictor are used [81]. In A1, the fixed low-pass filter is chosen as $H_{\mathrm{LPF}}(z) = (1/4)(1 + z^{-1} + z^{-2} + z^{-3})$. The MMSE predictor is designed as follows. First, a low-pass filter A1 is applied to the high-priority RT traffic. Next, a standard minimizing mean square error linear predictor $H_{\mathrm{MMSE}}(z)$ with the form $\sum_{m=0}^{M_i} B_m^{(i)} z^{-n_i-m}$ is calculated based on the low-frequency part of the RT traffic. The final MMSE predictor is $z^{-n_i} H_i(z) = H_{\mathrm{LPF}}(z) H_{\mathrm{MMSE}}(z)$. Note that $H_{\mathrm{MMSE}}(z)$ will require explicit knowledge of the round-trip delays per-flow information n_i. In A3 that information is approximated by an average value $n_0$for all flows. The prediction error is defined as $\Delta(j) = V(j) - \hat{V}_1(j)$.

The queue length at the node of interest is bounded by [79]

$$q(n) \leq q_0(n) + \sup_{0 \leq n_0 \leq n} \sum_{j=n_0+1}^{n} \Delta(j) - \inf_{0 \leq n_0 \leq n} \sum_{j=n_0+1}^{n} \Delta(j)$$

Under the definitions and predictive flow control algorithm defined above, if $V_{\max} < \infty$ and $[(p\mu - \bar{V})/(\mu - \bar{V})] \leq H_1(1) \leq 1$, we have [81] $q(n) - q_0(n) \leq 2C_1$, where C_1 is a constant that does not depend on n.

For $l = n - n_0$ we define the accumulated error as

$$X_{n,l} \triangleq \sum_{j=n-l+1}^{n} \Delta(j) = \sum_{j=n_0+1}^{n} \Delta(j)$$

where $\mathrm{E}\{X_{n,l}\} = 0$ (because the predictor is unbiased) and

$$\mathrm{Var}\{X_{n,l}\} = \mathrm{Var}\left\{ \sum_{j=n-l+1}^{n} \Delta(j) \right\}$$
$$= \sum_{j1=n-l+1}^{n} \sum_{j2=n-l+1}^{n} C_\Delta(j1 - j2) = \sum_{j1=-l+1}^{0} \sum_{j2=-l+1}^{0} C_\Delta(j1 - j2) \quad (21.36)$$

For the results shown below, $V(n)$ is a generated Gaussian process which is multitimescale-correlated with $C_v(k) = 479.599 \times 0.999^{|k|} + 161.787 \times 0.99^{|k|} +$

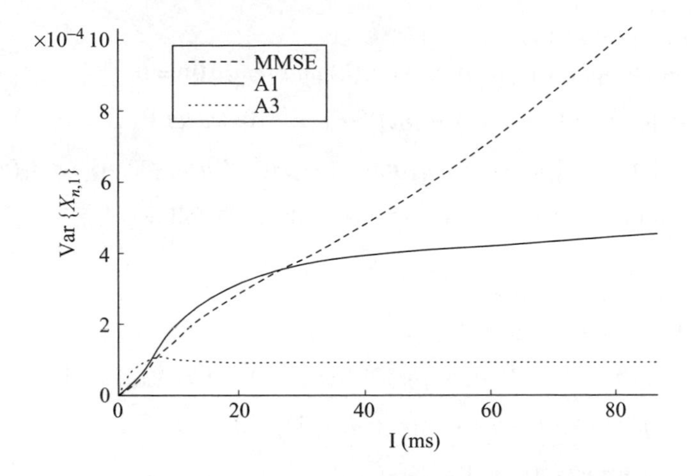

Figure 21.13 Var$\{X_{n,l}\}$ with different predictors. (Reproduced by permission of IEEE [81].)

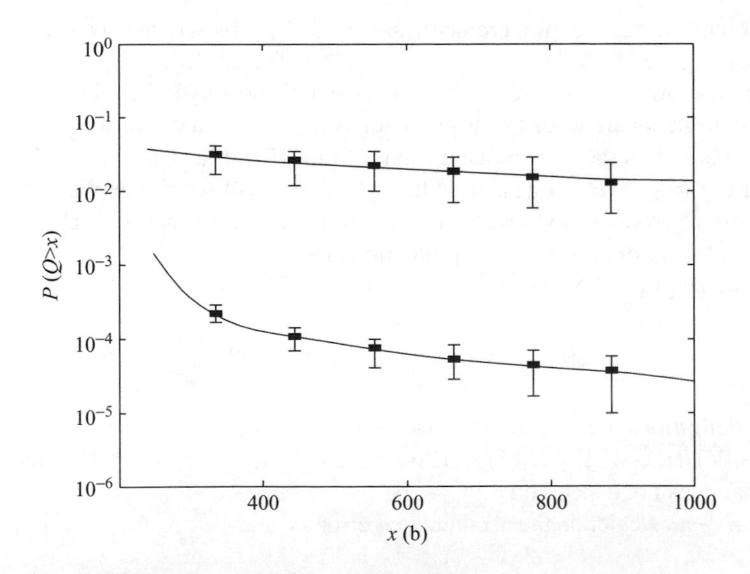

Figure 21.14 Tail (cumulative queue length) probabilities with different control algorithms. (Reproduced by permission of IEEE [81].)

$498.033 \times 0.9^{|k|}$ and $\bar{V} = 100$ kb/s. This type of source has often been used to represent the multiple time-scale correlation in network traffic [71, 80]. The link capacity is 200 kb/s, and the utilization is set to 98%. The time unit is 1 ms and the unit of the queue length is 1 bit [81]. Var$\{X_{n,l}\}$is shown in Figure 21.13. We can see that, for A1 and A3 predictors, Var$\{X_{n,l}\}$ converges to some constant when l is large enough. The asymptotic variance of A3 is also smaller than that of Al. In Figure 21.14, we compare the above control algorithm with the control algorithm that uses Equation (21.34). As we mentioned before, when $\hat{V}_1(n) \leq p\mu$ for all n, the control algorithms reduce to the same linear equation. However, when this condition is not true, the two algorithms are quite different. We can see this difference in performance in Figure 21.14. In this simulation, there is only one NRT source with round-trip delay of 5 ms. The RT source and the link capacity are still the same as in Figure 21.13 To see the difference when the condition $\hat{V}_1(n) \leq p\mu$ is violated, we use the same A3 predictor in both control algorithms. In the figure, the control algorithm is marked C1, and the one that uses Equation (21.34) is marked C2. The utilization is set to $p = 98\%$. Note that when using control algorithm C2, given p, the utilization is not equal to p. In this simulation, we set p such that the measured utilization for C2 is 98%, which is the same as in C1. From this figure, using the same predictor, we can see that the improved control algorithm outperforms the one that uses Equation (21.34).

REFERENCES

[1] R. Dugad, K. Ratakonda and N. Ahuja, A new wavelet-based scheme for watermarking images, in *Proc. IEEE Int. Conf. Image Processing*, Chicago, IL, 4–7 October 1998, pp. 419–423.

[2] D. Kundur and D. Hatzinakos, Digital watermarking using multiresolution wavelet decomposition, in *Proc. IEEE Int. Conf. Acoustics, Speech, Signal Processing*, vol. 5, 1998, pp. 2969–2972.

[3] H. Inoue, A. Miyazaki and T. Katsura, An image watermarking method based on the wavelet transform, in *Proc. IEEE Int. Conf. Image Processing*, Kobe, 25–28 October, 1999, pp. 296–300.

[4] H.J.M. Wang, P.C. Su and C.C.J. Kuo, Wavelet-based blind watermark retrieval technique, in *Proc. SPIE, Conf. Multimedia System Applications*, vol. 3528, Boston, MA, November 1998.

[5] P. Campisi, A. Neri and M. Visconti, A wavelet based method for high frequency subbands watermark embedding, in *Proc. SPIE Multimedia System Applications III*, Boston, MA, November 2000.

[6] M.M. Yeung and F. Mintzer, An invisible watermarking technique for image verification, in *Proc. IEEE Int. Conf. Image Processing*, Santa Barbara, CA, 1997, pp. 680–683.

[7] D. Kundur and D. Hatzinakos, Toward a telltale watermarking tech- nique for tamperproofing, in *Proc. IEEE Int. Conf. Image Processing*, Chicago, IL, 4–7, October 1998, pp. 409–413.

[8] R.H. Wolfgang and E.J. Delp, Fragile watermarking using the *VW2D* watermark, in *Proc. SPIE, Security Watermarking Multimedia Contents*, vol. 3657, San Jose, CA, January 1999.

[9] P.D.F. Correira, S.M.M. Faria, and P.A.A. *Assunqão*, Matching MPEG-1/2 coded video to mobile applications, in *Proc. Fourth Int. IEEE Symp. Wireless Personal Multimedia Communications*, Aalborg, 9–12, September 2001.

[10] Information technology – coding of moving pictures and associated audio for digital storage media at up to about 15 *Mb/s* – Part 2: Video. ISO, ISO/IECII I72–2, 1993.

[11] F. Yong Li, N. Stol, T.T. Pham and S. Andresen, A priority-oriented QoS management framework for multimedia services in IJMTS, in *Proc. Fourth Int IEEE Symp. Wireless Personal Multimedia Communication*, Aalborg, 9–12, September 2001.

[12] L. Hanzo, P.J. Cherriman and J. Streit, *Wireless Video Mommunication. Second to Third Generation Systems and Beyond*, IEEE Series on Digital and Mobile Communication. IEEE: New York, 2001.

[13] Hanjalic, G.C. Langelaar, P.M.B. van Roosmalen, J. Biemond and R.L. Lagendijk, *Image and Video Databases. Restoration, Watermarking and Retrieval*. Elsevier: New York, 2000.

[14] I.J. Cox, M.L. Miller and J.A. Bloom, *Digital Watermarking*. Morgan Kaufmann: San Francisco, CA, 2002.

[15] N. Nikolaidis and I. Pitas, Robust image watermarking in the spatial domain, *Signal Process.*, vol. 66, no. 3, 1998, pp. 385–403.

[16] M. Bami, F. Bartolini, V. Cappellini and A. Piva, A DCT-domain system for robust image watermarking, *Signal Process.*, vol. 66, no. 3, 1998, pp. 357–372.

[17] I. Cox, J. Kilian, F. Leighton and T. Shamoon, Secure spread spectrum watermarking for multimedia, *IEEE Trans. Image Processing*, vol. 6, 1997, pp. 1673–1687.

[18] H. Hartung and B. Girod, Watermarking of uncompressed and com- pressed video, *Signal Process.*, vol. 66, 1998, pp. 283–301.

[19] M.D. Swanson, B. Zhu and A.H. Tewfik, Transparent robust image watermarking transform, in *Proc. IEEE Int. Conf. Image Processing*, Lausanne, 16–19, September 1996, pp. 211–214.

[20] R.M. Wolfgang and E.J. Delp, A watermark for digital images, in *Proc. IEEE Int. Conf. Image Processing*, Lausanne, 16–19, September 1996, pp. 219–222.

[21] T. Ebrahimi, MPEG-4 video verification model: A video encoding/de coding algorithm based on content representation, perceived quality: Toward comprehen- sive metrics, *Proc. SPIE*, vol. 4299, 2001.

[22] R. Gold, Optimal binary sequences for spread spectrum multiplexing, *IEEE Trans. Inform. Theory*, vol. IT-13, 1967, pp. 619–621.

[23] K.T. Tan, M. Ghanbari and D.E. Pearson, An objective measurement tool for MPEG video quality, *Signal Process.*, vol. 70, no. 3, 1998, pp. 279–294.

[24] S. Winkler, Visual fidelity and perceived quality: toward comprehensive metrics, *Proc. SPIE*, vol. 4299, 2001.

[25] P. Campisi, M. Carli, G. Giunta and A. Neri, Tracing watermarking for multimedia communication quality assessment, in *Proc. IEEE Int. Conf. Communications*, New York, 28 April to 2 May 2002.

[26] P. Campisi, G. Giunta and A. Neri, Object based quality of service assessment for MPEG-4 videos using tracing watermarking, in *Proc. IEEE Int. Conf. Image Processing*, Rochester, NY, 22–25 September, 2002.

[27] H. Liu, *Signal Processing Applications in CDMA Communications*. Artech House: Norwell, MA, 2000.

[28] P. Campisi, M. Carli, G. Giunta and A. Neri, Blind quality assessment system for multimedia communications using tracing watermarking, *IEEE Trans. Signal Processing*, vol. 51, no. 4, April 2003, pp. 996–1002.

[29] G. Anastasi and L. Lenzini, QoS provided by the IEEE 802.11 wireless LAN to advanced data applications: a simulation analysis, *Wireless networks*, vol. 6, no. 2, 2000, pp. 99–108.

[30] G. Bianchi, Performance analysis of the IEEE 802.11 distributed coordination function, *IEEE J. Select. Areas Commun*, vol. 18, no. 3, 2000, pp. 535–587.

[31] F. Cali, M. Conti and E. Gregori, Dynamic tuning of the IEEE 802.11 protocol to achieve a theoretical throughput limit, *IEEE/ACM Trans. Networking*, vol. 8, no. 6, 2000, pp. 785–799.

[32] C. Coutras, S. Gupta and N.B. Shroff, Scheduling of real-time traffic in IEEE 802.11 wireless lAN, *Wireless Networks*, vol. 6, no. 6, 2000, pp. 457–466.

[33] S. Choi and K.G. Shin, A cellular wireless local area network with QoS guarantee for heterogeneous traffic, *ACM Mobile Networks Applic.*, vol. 3, no. 1, 1998, pp. 89–100.

[34] S. Choi and K.G. Shin, A unified wireless LAN architecture for real-time and non-real-time communication services, *IEEE/ACM Trans. Networking*, vol. 8, no. 1, 2000, pp. 44–59.

[35] B.P. Crow, I. Widjaja, J.G. Kim and P.T. Sakai, IEEE 802.11 wireless local area networks, *IEEE Commun. Mag.*, vol. 35, no. 9, 1997, pp. 116–126.

[36] A. Ganz and A. Phonphoem, Robust superpoll with chaining protocol for IEEE 802.11 wireless LANs in support of multimedia applications, *Wireless Networks*, vol. 7, no. 1, 2001, pp. 65–73.

[37] IEEE 802.11 Task Group, http://grouper.ieee.org/groups/802/11/, 2000.

[38] R.C. Meier, An integrated 802.11 QoS model with point-controlled contention arbitration, IEEE Document 802.11-00/448, November 2000.

[39] O. Sharon and E. Altman, An efficient polling MAC for wireless LANs, *IEEE/ACM Trans. Networking*, vol. 9, no. 4, 2001, pp. 439–451.

[40] S.T. Sheu and T.F. Sheu, A bandwidth allocation/sharing/extension protocol for multimedia over IEEE 802.11 ad hoc wireless LANs, *IEEE J. Select. Areas Commun.*, vol. 19, no. 10, 2001, pp. 2065–2080.

[41] M. Fischer, QoS baseline proposal for the IEEE 802.11E, IEEE Document 802.11-00/360, November 2000.

[42] M. Fischer, EHCF normative text, IEEE Document 802.11-01/110, March 2001.

[43] S.-C. Lo, G. Lee and W. –T. Chen, An efficient multipolling mechanism for IEEE 802.11 wireless LANs, *IEEE Trans. Comput.*, vol. 52, no. 6, 2003, pp. 764–778.

[44] T.S.E. Ng, I. Stoica and H. Zhang, Packet fair queueing algorithms for wireless networks with location-dependent errors, in *Proc. INFOCOM' 98*, vol. 3, 1998, pp. 1103–1111.

[45] J. -S. Yang, C. -C. Tseng and R. -G. Cheng, Dynamic scheduling framework on an RLC/MAC layer for general packet radio service, *IEEE Trans. Wireless Commun.*, vol. 2, no. 5, 2003, pp. 1008–1016.

[46] S.J. Golestani, A self-clocked fair queuing scheme for broadband applications, in *Proc. IEEE INFOCOM'94*, April 1994, pp. 636–646.

[47] *GSM 03.60 General Packet Radio Service (GPRS); Service Description;Stage 2*, ETSI Standard v. 7.0.0, 1999.

[48] W. Stallings, *High-Speed Networks: TCP/IP and ATM Design Principles*. Prentice-Hall: Englewood Cliffs, NJ, 1998, pp. 325–330.

[49] A. Demers, S.Keshav and S. Shenkar, Analysis and simulation of a fair queuing algorithm, in *Proc. SIGCOMM'89*, Austin, TX, September 1989, pp. 1–12.

[50] R.L. Cruz, A calculus for network delay–Part I: Network elements in isolation, *IEEE Trans. Inform. Theory*, vol. 37, 1991, pp. 114–131.

[51] S.J. Golestani, Network delay analysis of a class of fair queuing algorithms, *IEEE J. Select. Areas Commun.*, vol. 13, 1995, pp. 1057–1070.

[52] S. Ross, *Stochastic Processes*, 2nd edn. Wiley: New York, 1980, pp. 114–115.

[53] M.Ergen, S.Coleri and P.Varaiya , QoS aware adaptive resource allocation techniques for fair scheduling in OFDMA based broadband wireless access systems, *IEEE Trans. Broadcasting*, vol. 49, no. 4, 2003, pp. 362–370.

[54] I. Koffman and V. Roman, Broadband wireless access solutions based on OFDM access in IEEE 802.16, *IEEE Commun. Mag.*, April 2002.

[55] I. Koutsopoulos and L. Tassiulas, Channel state-adaptive techniques for throughput enhancement in wireless broadband networks, in *INFOCOM 2001*, vol. 2, 2001, pp. 757–766.

[56] *IEEE Standard for Local and Metropolitan Area Networks Part 16: Air Interface*, IEEE Std 802.16-2001.

[57] C.Y. Wong, R.S. Cheng, K.B. Letaief and R.D. Murch, Multiuser OFDM with adaptive subcarrier, bit, and power allocation, *IEEE J. Select. Areas Commun.*, vol. 17, no. 10, 1999, pp. 1747–1758.

[58] S.G. Glisic, *Adaptive WCDMA, Theory and Practice*. Wiley: Chichester, 2003.

[59] S.G. Glisic, *Advanced Wireless Communications, 4G Technology*., Wiley: Chichester, 2004.

[60] P. Viswanath, D.N.C. Tse and R. Laroia, Opportunistic beamforming using dumb antennas, *IEEE Trans. Inform. Theory*, vol. 48, no. 6, 2002, pp. 1277–1294.

[61] J. Bolot and A. Shankar, Dynamic behavior of rate-based flow control mechanisms, *ACM Comput. Commun. Rev.*, vol. 20, no. 2, 1992, pp. 35–49.

[62] M. Hluchyj and N. Yin, On closed-loop rate control for ATM networks, in *Proc. IEEE INFOCOM*, 1994, pp. 99–108.

[63] E. Altman, T. Basar and R. Srikant, Congestion control as a stochastic control problem with action delays, *Automatica*, vol. 35, no. 12, 1999, pp. 1937–1950.

[64] L. Benmohamed and S. Meerkov, Feedback control of congestion in store-and-forward networks: the case of single congestion node, *IEEE/ACM Trans. Networking*, vol. 1, 1993, pp. 693–798.

[65] S.Q. Li, S. Chong and C.Hwang, Link capacity allocation and network control by filtered input rate in high speed networks, *IEEE/ACM Trans. Networking*, vol. 3, 1995, pp. 10–15.

[66] D. Bertsekas and R. Gallager, *Data Networks*. Prentice-Hall: Englewood Cliffs, NJ, 1992.

[67] S.H. Low and D.E. Lapsley, Optimization flow control, I: basic algorithm and convergence, *IEEE/ACM Trans. Networking*, vol. 7, 1999, pp. 861–874.

[68] F.P. Kelly, A. Maulloo and D. Tan, Rate control for communication networks: Shadow prices, proportional fairness and stability, *J. Opns. Res. Soc.*, 1998, pp. 237–252.

[69] H. Yaiche, R. R. Mazumdar and C. Rosenberg, A game theoretic framework for bandwidth allocation and pricing in broadband networks, *IEEE/ACM Trans. Networking*, vol. 8, 2000, pp. 667–678.

[70] O. Ait-Hellal, E. Altman and T. Basar, Rate based flow control with bandwidth information, *Eur. Trans. Telecommun.*, vol. 8, no. 1, 1997, pp. 55–65.

[71] H.S. Kim and N.B. Shroff, Loss probability calculations at a finite buffer multiplexer, *IEEE/ACM Trans. Networking*, vol. 9, 2001, pp. 765–768.

[72] H.S. Kim and N.B. Shroff, On the asymptotic relationship between the overflow probability and the loss ratio, *Adv. Appl. Prob.*, vol. 33, no. 4, 2001, pp. 810–835.

[73] A.P. Zwart, A fluid queue with a finite buffer and subexponential input, *Adv. Appl. Probabil.*, vol. 32, 2000, pp. 221–243.

[74] A.V. Oppenheim, A.S. Willsky and S.H. Nawab, *Signals and Systems*. Prentice-Hall: Englewood Cliffs, NJ, 1997.

[75] P.W. Glynn and W. Whitt, Logarithmic asymptotics for steady-state tail probabilities in a single-server queue, *J. Appl. Prob.*, 1994, pp. 131–155.

[76] R.M. Loynes, The stability of a queue with nonindependent inter-arrival and service times, in *Proc. Cambridge Phil. Soc.*, vol. 58, pp. 497–520, 1962.

[77] J. Choe and N.B. Shroff, Use of the supremum distribution of Gaussian processes in queueing analysis with long-range dependence and selfsimilarity, *Stochast. Models*, vol. 16, no. 2, 2000.

[78] N.B. Likhanov and R. Mazumdar, Cell loss asymptotics in buffers fed by heterogeneous longtailed sources, in *Proc. IEEE INFOCOM*, 2000, pp. 173–180.

[79] D. Qiu and N.B. Shroff, Study of predictive flow control, Purdue University, Technical Report Available at: http://min.ecn.purdue.edu/~dongyu/paper/techrep.ps, May 2001.

[80] J. Choe and N.B. Shroff, A central limit theorem based approach for analyzing queue behavior in high-speed networks, *IEEE/ACM Trans. Networking*, vol. 6, 1998, pp. 659–671.

[81] D. Qiu and N.B. Shroff, A predictive flow control scheme for efficient network utilization and QoS, *IEEE/ACM Trans. Networking*, vol. 12, no. 1, 2004, pp. 161–172.

Index

Advanced Wireless Networks: 4G Technologies. Savo G. Glisic
© 2006 John Wiley & Sons, Ltd.